ROCKFALL
CHARACTERIZATION AND CONTROL

A. Keith Turner
Robert L. Schuster
Editors

TRANSPORTATION RESEARCH BOARD
OF THE NATIONAL ACADEMIES

Washington, D.C.
2012
www.TRB.org

Transportation Research Board
Miscellaneous Publication

Transportation Research Board publications are available by ordering individual publications directly from the TRB Business Office, through the Internet at www.TRB.org, or by annual subscription through organizational or individual affiliation with TRB. Affiliates and library subscribers are eligible for substantial discounts. For further information, contact the Transportation Research Board Business Office, 500 Fifth Street, NW, Washington, DC 20001 (telephone 202-334-3213; fax 202-334-2519; or e-mail TRBsales@nas.edu).

Printed in the United States of America.

Library of Congress Cataloging-in-Publication Data

Rockfall : characterization and control / A. Keith Turner, Robert L. Schuster, editors.
pages cm. — (Transportation Research Board miscellaneous publication)
Includes bibliographical references and index.
ISBN 978-0-309-22306-5 (hardcover) —
ISBN 978-0-309-22312-6 (softcover) 1. Landslide hazard analysis—United States. 2. Landslides—United States. I. Turner, A. Keith, 1941- editor of compilation. II. Schuster, Robert L., editor of compilation.
QE599.U5R63 2012
551.3'07—dc23

2012045373

ISBN 978-0-309-22306-5 (hardcover)
ISBN 978-0-309-22312-6 (softcover)

COVER PHOTOGRAPHS

Front main: An extremely fast-moving rockslide blocked a section of Interstate 40 in North Carolina in October 2009 and closed the roadway for nearly 6 months. (Photograph courtesy of the North Carolina Department of Transportation.)

Front insert: Remedial slope stabilization work under way after a rockfall event that killed three people on Interstate 90, east of Seattle, Washington, in September 2005. (Photograph courtesy of the Washington State Department of Transportation.)

Back upper: Opened in October 1993, the tunnel through the Pen-Y-Clip headland on the coast of North Wales, United Kingdom, was constructed through microdiorite and under substantial scree deposits. It incorporates a variety of rockfall protection measures, including check fences, gabion walls, barrier fences, scree retaining walls, parapet walls, rock trap walls, boulder attenuation areas, and portal canopies. (Photograph by David Giles, University of Portsmouth, United Kingdom.)

Back lower: A rockfall estimated at 3 to 4.5 meters (10 to 15 feet) high, 2 meters (6 feet) wide, and 1.2 meters (4 feet) deep occurred just inside the Yosemite National Park boundary on December 30, 2010. The rockfall caused no injuries or property damage but closed Highway 140 for 24 hours. (Photograph by Fletcher Ogg, National Park Service.)

THE NATIONAL ACADEMIES

Advisers to the Nation on Science, Engineering, and Medicine

The **National Academy of Sciences** is a private, nonprofit, self-perpetuating society of distinguished scholars engaged in scientific and engineering research, dedicated to the furtherance of science and technology and to their use for the general welfare. On the authority of the charter granted to it by the Congress in 1863, the Academy has a mandate that requires it to advise the federal government on scientific and technical matters. Dr. Ralph J. Cicerone is president of the National Academy of Sciences.

The **National Academy of Engineering** was established in 1964, under the charter of the National Academy of Sciences, as a parallel organization of outstanding engineers. It is autonomous in its administration and in the selection of its members, sharing with the National Academy of Sciences the responsibility for advising the federal government. The National Academy of Engineering also sponsors engineering programs aimed at meeting national needs, encourages education and research, and recognizes the superior achievements of engineers. Dr. Charles M. Vest is president of the National Academy of Engineering.

The **Institute of Medicine** was established in 1970 by the National Academy of Sciences to secure the services of eminent members of appropriate professions in the examination of policy matters pertaining to the health of the public. The Institute acts under the responsibility given to the National Academy of Sciences by its congressional charter to be an adviser to the federal government and, on its own initiative, to identify issues of medical care, research, and education. Dr. Harvey V. Fineberg is president of the Institute of Medicine.

The **National Research Council** was organized by the National Academy of Sciences in 1916 to associate the broad community of science and technology with the Academy's purposes of furthering knowledge and advising the federal government. Functioning in accordance with general policies determined by the Academy, the Council has become the principal operating agency of both the National Academy of Sciences and the National Academy of Engineering in providing services to the government, the public, and the scientific and engineering communities. The Council is administered jointly by both Academies and the Institute of Medicine. Dr. Ralph J. Cicerone and Dr. Charles M. Vest are chair and vice chair, respectively, of the National Research Council.

The **Transportation Research Board** is one of six major divisions of the National Research Council. The mission of the Transportation Research Board is to provide leadership in transportation innovation and progress through research and information exchange, conducted within a setting that is objective, interdisciplinary, and multimodal. The Board's varied activities annually engage about 7,000 engineers, scientists, and other transportation researchers and practitioners from the public and private sectors and academia, all of whom contribute their expertise in the public interest. The program is supported by state transportation departments, federal agencies including the component administrations of the U.S. Department of Transportation, and other organizations and individuals interested in the development of transportation. **www.TRB.org**

www.national-academies.org

This report is dedicated to the memory of
three internationally recognized pioneers
in rockfall research:

Arthur M. Ritchie (United States),
Douglas R. Piteau (Canada),
and
Raymund M. Spang (Germany).

Their contributions form the foundation of our current understanding
of rockfall events and continue to influence rockfall research.
Details of their contributions are found throughout the chapters of this report.

This report is also dedicated to the memory of

Steve M. Lowell,

lead author of Chapter 18, "Rockfall Management Programs,"
who profoundly influenced U.S. practice
on the topic of rockfall over the past several decades.

PREFACE

In 1958 the Highway Research Board, the predecessor of the Transportation Research Board (TRB), produced Special Report 29, *Landslides and Engineering Practice*, the first in a series of comprehensive reports that synthesize information related to landslides and rockfall. In 1972 a task force was organized with members drawn from several of the standing committees within the Geology and Properties of Earth Materials Section to revise Special Report 29. It took 6 years to prepare and publish TRB Special Report 176, *Landslides: Analysis and Control*. By 1989, members of the TRB standing committees within the Soil Mechanics and the Geology and Properties of Earth Materials sections clearly identified the need for a revised and expanded report to address the latest advances in the methods for investigation and mitigation of landslides. Thus, a study committee was established in 1990 under the chairmanship of A. Keith Turner of the Colorado School of Mines, and TRB Special Report 247, *Landslides: Investigation and Mitigation*, was published in 1996.

These reports have been widely used in the geotechnical engineering community as comprehensive, practical sources of information on all aspects of landslide investigations and measures that may be used to mitigate the hazards posed by landslides. Among the most widely distributed of TRB publications, these reports have enjoyed wide international appeal and have been translated into several languages. The scope and content of this series of reports illustrate the remarkable growth in knowledge of and experience with landslide phenomena over the past half-century, as well as the expanded capabilities of new technologies that may be used to assist in their investigation, characterization, and mitigation.

TRB Special Report 247 featured an expanded treatment of several topics, including risk analysis, rock strength properties, and the stability of rock slopes, that are important in evaluating rockfall. In the decade following the 1996 publication of Special Report 247, however, several factors contributed to a greatly expanded interest in rockfall on the part of the transportation community. Thus, in May 2005 TRB organized a task force (see box on page viii), once again under the chairmanship of A. Keith Turner, to develop an entirely new TRB report that would address all aspects of rockfall while complementing TRB Special Report 247.

Rockfall: Characterization and Control provides a comprehensive review of rockfall research and of methods for investigating, characterizing, predicting, and mitigating rockfall events. Considerable information is based on unpublished sources and the experiences of individual task force members. No accessible comprehensive source of rockfall information existed, so the authors of individual chapters have compiled information from many and diverse locations.

Task Force on Rockfall Characterization and Control

A. Keith Turner, Colorado School of Mines (emeritus), Golden, *Chair*
Richard D. Andrew, Yeh and Associates, Inc., Denver, Colorado
Thomas C. Badger, Washington State Department of Transportation, Olympia
John D. Duffy, California Department of Transportation, San Luis Obispo
Jerry D. Higgins, Colorado School of Mines, Golden
Lawrence A. Pierson, Landslide Technology, Portland, Oregon
Robert L. Schuster, U.S. Geological Survey (emeritus), Golden, Colorado (ex officio)
Michael P. Vierling, New York State Thruway Authority, Albany

Transportation Research Board Staff

G. P. Jayaprakash, Senior Program Officer—Soils, Geology, and Foundations Engineer
Michael A. DeCarmine, Program Officer

The report contains 18 chapters prepared by internationally recognized experts. Approximately 30% of the report (five chapters) describes the fundamentals of rockfall phenomena, with special emphasis on the methods used to assess rockfall hazard and risk. An additional 40% of the report (five chapters) discusses the scientific and technical aspects of rockfall investigation, including rockfall site characterization, instrumentation and new investigation technologies, rockfall kinematic principles, rockfall prediction and modeling options, and field-test operations. The final eight chapters (about 30% of the report) describe rockfall mitigation and management options, with special reference to rockfall protection provided along transportation routes. A DVD contains original color versions of illustrations, video clips of individual rockfall field tests, video records of selected rock slope mitigation activities, and a reproduction of the original 16-mm movie on rockfall field tests conducted by Arthur Ritchie in 1963 for the Washington State Highway Commission (now the Washington State Department of Transportation).

The rockfall task force first met in July 2005 to begin the process of developing this report. Individual task force members undertook the preparation of specific chapters. Responsibilities included developing chapter outlines, identifying expertise outside the task force to provide specific material or assist in writing chapters, writing material, coordinating the efforts of multiple authors, and reviewing chapter drafts before submittal for review by the entire task force membership. In addition to holding meetings in Washington, D.C., during the TRB annual meetings throughout the period from 2006 to 2010, the task force held additional formal meetings in 2006 and 2007. These discussions greatly improved the quality of the individual chapters.

Although the material contained in each chapter is solely attributed to the author(s) of the chapter, each chapter underwent a rigorous peer review process by task force members and by Robert L. Schuster, who conducted an independent review of the whole report. A. Keith Turner, Robert Schuster, and G. P. Jayaprakash provided a technical and editorial review of all the draft materials. The final product incorporates numerous revisions made throughout the iterative review process.

The International System of Units (SI) is used to define all measurements in this report; an equivalent value specified in U.S. customary units is also provided. A table of conversion factors for SI and inch-pound (U.S. customary) units of measurement is included for reference.

ACKNOWLEDGMENTS

Coordination and technical editing of the efforts of multiple authors are not easy tasks. TRB and the future users of this report worldwide are indebted to the members of the task force for their efforts.

Sincere appreciation is expressed to all those who contributed information in the form of data and photographs, ideas, and advice, without which this volume could not have been completed. To list all contributors would be impossible; to list the most important would be unfair to the others.

This report was produced by busy individuals who volunteered their time and talents to its creation. The task force members were aided by several individuals who agreed to write, review, and supervise the development of particular chapters or portions of chapters. Brief biographies of all members of the task force and the chapter authors are provided at the end of the report. The editors wish to thank this dedicated group for their encouragement, unfailing goodwill, and patience during the lengthy editorial process.

Special thanks go to TRB staff members G. P. Jayaprakash, Senior Program Officer—Soils, Geology, and Foundations Engineer; Mark R. Norman, Director, Technical Activities; and Michael DeCarmine, Program Officer. In the TRB Publications Office, Senior Editor (retired) Norman Solomon, Cay Butler, and Naomi Kassabian edited the text; Senior Editor Janet M. McNaughton managed the editorial production; Paul R. Burley compiled the index; Mary McLaughlin proofread the pages; Jennifer J. Weeks prepared the manuscript for editing; and Juanita Green managed the production and printing, under the supervision of Javy Awan, Director of Publications.

—*A. Keith Turner and Robert L. Schuster*
Editors

CONTENTS

METRIC CONVERSION TABLE

The narratives comprising the chapters of this publication initially report all measurements using the International System of Units (SI), a modernized metric system of measurement, and then provide an equivalent value (in parentheses) using the customary units (inches/pounds) employed in the United States. Illustrations and tables, however, retain the units of their source materials. For the convenience of the reader, the following table summarizes the appropriate conversion factors employed within the chapter narratives; these may be used to convert values shown in figures and tables.

To convert the unit in the first column to the unit in the second column, multiply by the factor in the third column.					
Customary Unit	*SI Unit*	*Factor*	*SI Unit*	*Customary Unit*	*Factor*
LENGTH			**LENGTH**		
inches	millimeters	25.4	millimeters	inches	0.039
inches	centimeters	2.54	centimeters	inches	0.394
feet	meters	0.305	meters	feet	3.281
yards	meters	0.914	meters	yards	1.094
miles	kilometers	1.609	kilometers	miles	0.6214
AREA			**AREA**		
square inches	square millimeters	645.1	square millimeters	square inches	0.00155
square feet	square meters	0.093	square meters	square feet	10.764
square yards	square meters	0.836	square meters	square yards	1.196
acres	hectares	0.405	hectares	acres	2.471
square miles	square kilometers	2.59	square kilometers	square miles	0.386
VOLUME			**VOLUME**		
gallons (U.S.)	liters	3.785	liters	gallons (U.S.)	0.264
cubic inches	cubic centimeters	16.39	cubic centimeters	cubic inches	0.061
cubic feet	cubic meters	0.028	cubic meters	cubic feet	35.314
cubic yards	cubic meters	0.765	cubic meters	cubic yards	1.308
cubic miles	cubic kilometers	4.168	cubic kilometers	cubic miles	0.2399
MASS AND WEIGHT[a]			**MASS AND WEIGHT**[a]		
ounces	grams	28.35	grams	ounces	0.035
pounds	kilograms	0.454	kilograms	pounds	2.205
short tons[b]	megagrams[c]	0.907	megagrams[c]	short tons[b]	1.102
long tons[b]	megagrams[c]	1.016	megagrams[c]	long tons[b]	0.984
FORCE AND PRESSURE OR STRESS			**FORCE AND PRESSURE OR STRESS**		
poundforce	newtons	4.45	newtons	poundforce	0.225
poundforce per square inch	kilopascals	6.89	kilopascals	poundforce per square inch	0.145
ENERGY[d]			**ENERGY**[d]		
foot-pounds	joules	1.356	joules	foot-pounds	0.738
foot-tons	kilojoules	2.712	kilojoules	foot-tons	0.369
TEMPERATURE			**TEMPERATURE**		
To convert Fahrenheit temperature (°F) to Celsius temperature (°C), use the following formula: °C = (°F − 32)/1.8.			To convert Celsius temperature (°C) to Fahrenheit temperature (°F), use the following formula: °F = (°C × 1.8) + 32.		
ACCELERATION DUE TO GRAVITY			**ACCELERATION DUE TO GRAVITY**		
The standard average acceleration due to gravity (*g*) is 32.174 ft/s^2. This is used in converting mass to weight and in computing rockfall trajectories.			The standard average acceleration due to gravity (*g*) is 9.80665 m/s^2. This is used in converting mass to weight and in computing rockfall trajectories.		

[a]*Mass* is a measurement of the amount of matter an object contains, while *weight* is the measurement of the pull of gravity on an object. Thus, weight is a force and is not an intrinsic attribute of the object. However, in commercial and everyday use, the term weight is usually used to mean mass, and the verb "to weigh" means "to determine the mass of" or "to have a mass of." Used in this sense, weight is expressed in the non-SI unit of measure known as the kilogram-force. The equivalent unit of force in customary units of measurement is the poundforce or, simply, the pound. In strict scientific contexts, forces are measured within the SI system as newtons. The conversion factors shown in this table between ounces and grams, and between kilograms and pounds, reflect the everyday commercial use of these terms.

[b]A short ton (used in the United States) is 2,000 pounds; a long ton (used in the United Kingdom) is 2,240 pounds. This book uses short tons when referring to customary units.

[c]The megagram is the official SI designation of 1,000 kilograms. This is often referred to as a *metric ton* in the United States or as a *tonne* within the SI system, although it is not an official SI unit definition.

[d]*Energy* is defined as the ability of a physical system to influence (or "do work") on other physical systems. Thus, energy is always defined in terms of the amount of applied force acting over a given displacement distance. In the SI system, energy is defined in joules. One joule is equal to the energy expended (or work done) in applying a force of 1 newton through a distance of 1 meter. In customary units, energy is defined in units of foot-pound force (or, simply, foot-pounds), which is the energy transferred by applying a 1-pound force through a displacement of 1 foot. Because typical rockfall impacts involve large energies, they are reported in terms of kilojoules (1 kilojoule = 1,000 joules) or as foot-tons (1 foot-ton = 2,000 foot-pounds).

PART 1

RECOGNITION OF ROCKFALL HAZARD

Chapter 1

A. Keith Turner and
G. P. Jayaprakash

INTRODUCTION

Rockfall is a natural phenomenon on steep natural or constructed slopes that are exposed to weathering and erosion. Fallen rock fragments accumulate as talus or scree deposits below rock cliffs (Figure 1-1). Progressive degradation of rock cliffs or rock cuts results in rockfall that causes minor to devastating damage to transportation and other civil and mining facilities. The title of this report, *Rockfall Characterization and Control*, was selected to reflect the state of knowledge of rockfall processes, the procedures available for rockfall investigation, and the regulatory and economic climate under which rockfall investigations and corrective actions are typically undertaken.

In many ways, this report may be considered a companion volume to *Special Report 247: Landslides: Investigation and Mitigation* (Turner and Schuster 1996). In the intervening years, there have been many advances in the way rockfalls are investigated and in the options available for rockfall hazard mitigation. Rockfall has developed into a topic with considerable economic importance, especially for transportation agencies. This economic interest, coupled with the remarkable expansion in the ability to evaluate rockfall phenomena and new technologies for designing and constructing rockfall mitigation procedures, has led to the need for a much more in-depth discussion of rockfall issues than in the earlier publication.

Engineering assessments of excavations and natural slopes in rock masses generally concentrate on identifying and evaluating mechanisms that may lead to failure of significant volumes of rock. Because the subject of rock slope stability has been comprehensively addressed in recent years, this volume emphasizes new developments in the analysis of rockfall problems and the design and execution of measures to contain them.

1. ROCKFALL AND TRANSPORTATION

Failures involving relatively small volumes of rock often pose significant hazards, particularly in areas where steep terrain is near developed areas or transportation corridors. Spang (1987) reported that rockfalls, rather than deep-seated slides, were the major cause of interruption to the then West German Federal Railway's 28,000-km (17,500-mi) track network. Martin (1988) stated that rockfalls, small rockslides, and raveling are the most chronic slope stability problems along transportation routes in mountainous areas of North America; tens of millions of dollars are spent annually on maintenance and remedial measures to protect against such hazards.

Transportation facilities are typically blocked for some time, often several days, when very large rock blocks fall from natural or excavated rock faces (Figure 1-2). Erosion readily removes the relatively fine-grained matrix of glacial, fluvial, and colluvial deposits and releases boulders and rock blocks contained within these deposits. Rockfall

FIGURE 1-1 Relatively fresh rockfall from Kaibab Formation in Grand Canyon below Lees Ferry, Arizona. Rockfall is reported to have occurred in spring 1970; photograph was taken in summer 2004. (PHOTOGRAPH COURTESY OF R. LEIGHTY.)

from these sources sometimes involves very large boulders (Figure 1-3). In some cases, the impacts of large rockfall blocks have caused severe and extensive damage to transportation infrastructure, which has led to considerable repair costs and extended traffic delays and detours while the repairs are under way (Figure 1-4).

While large rockfall blocks pose a significant hazard and often require considerable expense to remove them, even fairly small rockfall fragments are a significant hazard when they come to rest on a highway pavement or railroad track (Figure 1-5). The presence of quite small rock fragments often results in severe damage to vehicles and injuries to their occupants. Small rock fragments often are not readily observed by motorists, and avoidance maneuvers cause accidents and crashes (Figure 1-6).

Although a relatively rare occurrence, falling rock blocks sometimes directly hit a road vehicle or a train. Direct impacts of large, or even small, rocks on vehicles often cause fatalities (Figure 1-7). In a few cases, the vibrations caused by a passing train appear to have triggered rockfall and caused unstable rock materials to hit a moving train (Figure 1-8).

1.1 Rockfall Concerns of Railroads

Railroads are especially vulnerable to rockfall; while only 2% of train-related accidents are related to ground-movement hazards, including rockfall, they represent 12% of the direct costs of all rail accidents (Lemay 2007). Even relatively small rock fragments, and especially those with a maximum dimension in the range of 0.3 to 1.0 m (1 to 3 ft), are likely to lift equipment off the rails and cause a derailment (Abbott et al. 1998). Two rockfall events illustrate this hazard. On January 4, 2007, a westbound 105-car Canadian National freight train hit a small pile of rockfall debris on the track that follows the Thompson River near Lytton, British Columbia. It happened around 1:00 a.m.; it was dark and a previous train had cleared the same track an hour earlier. The lead locomotive, with the crew, left the rails and almost stopped, before falling more than 60 m (200 ft) to the edge of the river below (Figure 1-9). The crew received only minor injuries and were rescued the next morning. The second locomotive also derailed, but luckily did not fall (Figure 1-10). However, there were minor fuel leaks that required environmental cleanup. A similar incident occurred with a Burlington Northern Santa Fe Railway train in the Wind River Canyon south of Thermopolis, Wyoming, on May 12, 2010 (Figure 1-11). In this case, the accident occurred at midday. The train was traveling at about 50 km/h (30 mph) when it hit a boulder about the size of a pickup truck. Two locomotives and one car fell down the 15-m (50-ft) embankment into the Wind River. The crew were unhurt and managed to escape from the badly damaged and partially submerged locomotives. Limited access to the accident site hampered rescue efforts, and a large quantity of fuel spilled into the river and affected water supplies to Thermopolis and other users downstream.

FIGURE 1-2
In 1977, large rockfall blocks closed Colorado State Highway 133 near Redstone, Colorado, for several days, pending their breakup by blasting and removal of material.
(PHOTOGRAPH COURTESY OF THE COLORADO DEPARTMENT OF TRANSPORTATION.)

FIGURE 1-3 *(center)*
Large boulder, 5.5 m (18 ft) long, 3.7 m (12 ft) high, and 3.7 m (12 ft) wide, that landed on North Cascades Highway, Chelan County, Washington State, on June 5, 2010. It was so large that two loaders were required to push it off the road. No one was injured and the road was reopened later that evening.
(PHOTOGRAPH COURTESY OF THE WASHINGTON STATE DEPARTMENT OF TRANSPORTATION.)

FIGURE 1-4 *(bottom left)*
On Monday, March 8, 2010, impacts of large rockfall blocks caused severe damage to bridge deck located just west of Hanging Lake Tunnel on Interstate 70 in Glenwood Canyon, Colorado. Interstate was closed from early Monday morning until Thursday afternoon while blocks were broken up by blasting and removed and while unstable residual rock block on slope was removed. Traffic lanes were restricted for extended period before bridge deck was repaired.
(PHOTOGRAPH COURTESY OF THE COLORADO DEPARTMENT OF TRANSPORTATION.)

FIGURE 1-5 *(above)*
Even relatively small rockfall fragments pose significant hazard to motorists. This block landed on Colorado State Highway 7 in St. Vrain Canyon, west of Longmont, Colorado.
(PHOTOGRAPH COURTESY OF THE COLORADO DEPARTMENT OF TRANSPORTATION.)

FIGURE 1-6
Eastbound tractor-trailer encountered rockfall material on Interstate 70 in Georgetown, Colorado, in April 2004. Truck jackknifed and crashed through median barrier into westbound lanes.
(PHOTOGRAPH COURTESY OF THE COLORADO DEPARTMENT OF TRANSPORTATION.)

FIGURE 1-7 *(center)*
In May 2006, large rockfall blocks fell onto A2 Motorway in Switzerland near northern entrance to Gotthard Tunnel. Rocks hit car, killing its two occupants. Closure of motorway caused major travel disruptions for several days.
(PHOTOGRAPH BY S. TISCHLER. © KEYSTONE, ZURICH, SWITZERLAND. USED WITH PERMISSION.)

FIGURE 1-8 *(bottom left)*
Burlington Northern Santa Fe Railway train was hit by rockfall debris near Libby, Montana, on June 25, 2010. Sixteen cars derailed. Line was reopened one day later after emergency-response crews removed more than 5,400 m^3 (7,000 yd^3) of debris from track.
(PHOTOGRAPH COURTESY OF THE U.S. GEOLOGICAL SURVEY.)

FIGURE 1-9 *(above)*
Lead locomotive of Canadian National train plunged more than 60 m (200 ft) to edge of river below when train encountered small rockfall debris pile on tracks near Lytton, British Columbia, in January 2007 (http://www.arizonarails.com/cn-derailment.html).

1.2 Initiation of Rockfall Hazard Evaluation Programs

FIGURE 1-10
Small rockfall debris pile on tracks near Lytton, British Columbia, caused train derailment in January 2007 (http://www.arizonarails.com/cn-derailment.html).

FIGURE 1-11
One Burlington Northern Santa Fe locomotive lies partly in Wind River while another lies on its side halfway down embankment after derailment on May 12, 2010, just south of Thermopolis, Wyoming.
(PHOTOGRAPH COURTESY OF THE WYOMING HIGHWAY PATROL.)

In September 1974, an empty eastbound Canadian Pacific Railway unit coal train encountered about 15 m^3 (20 yd^3) of rock on the track located along the Thompson River near Spences Bridge, about 170 km (106 mi) northeast of Vancouver, British Columbia. As in the 2007 accident that occurred nearby and is illustrated in Figures 1-9 and 1-10, the two lead locomotives and several cars derailed and plunged over a steep rock face 30 m (100 ft) high. The results of the 1974 accident were much more serious than the 2007 accident; both locomotive operators were killed and the derailed locomotives and cars blocked the Trans-Canada Highway located below the track.

The severity of this accident led Canadian Pacific officials to anticipate a government inquiry. Thus, they decided not only to assess the accident site but also to institute a program to evaluate the stability of rock cuts on the entire Canadian Pacific rail system across Canada. This program for inspection and classification of potential rock slope instability was the first of its kind in the transportation field. The program inspected and evaluated all rock cuts in western Canada in 1974 and in the rest of Canada in 1975. Initially, the inspections were done by track motor car and by walking through the cuts. In later years, helicopter inspections were added to examine high rock cliffs adjacent to the track. These inspections often identified tension cracks that were not observable from track level (Brawner 1998).

The initial rating system was purposely kept simple; the emphasis was to rate slope stability according to the perceived potential urgency of stabilization (Brawner and Wyllie 1975; Brawner 1998). Five levels of priority ratings were assigned (Table 1-1) based on relatively rapid field observations that recorded location, alignment, climate, train frequency, slope dimensions, structural geology, history of past rockfall, evidence of seepage, work space available, stabilization recommendations, and site photography (Brawner 1998).

At several locations, the inspection and classification program identified the potential for rockfall to be so high that urgent stabilization was recommended. At two locations, the potential for rockfall was considered to be so serious that train traffic was halted until the unstable area was stabilized. On the basis of these inspections, Canadian Pacific allocated substantial funds to undertake stabilization programs at 10 rock slopes with the most critical stability ratings. These stabilization programs included rock scaling; increasing ditch dimensions; installing dowels, rock bolts, or cable mesh; applying shotcrete; constructing buttresses; and installing warning fences (Brawner 1998).

Approximately 6 months after the 1974 accident, the Canadian Transport Board held a hearing in Vancouver. The accident report issued by the Board expressed concern about the accident but commended Canadian Pacific on its quick and

Table 1-1
Priority Ratings Used by Canadian Pacific Railway Comparative Rock Slope Stability Classification System Developed in 1974–1975

Priority Category	Description
A	Moderate probability of failure of sufficient volume to result in derailment if failure undetected.
B	Some probability of failure of sufficient volume to result in derailment if failure undetected.
C	Moderate probability of failure of small volumes that might reach the track.
D	Moderate probability of localized rocks or rockfalls occurring during extreme climatic conditions, causing very heavy rainfall or runoff, extreme freeze–thaw cycles, and so forth.
E	Slight possibility of localized failures under extreme climatic conditions, generally shallow cuts.

Source: Brawner 1998.

professional response. The major recommendations of the report were a requirement to inspect rock cuts annually and to continue the stabilization program (Brawner 1998).

1.3 Development of Rockfall Hazard Rating Systems for Highways

As highway transportation volumes continued to increase, the importance of paying greater attention to rock slope stability along transportation routes became much clearer. These requirements and the desire to establish a comparable standard of practice across the United States led to the adoption of a modified version of the Canadian Pacific program for inspection and classification of potential rock slope instability to reflect the needs of highway agencies. These modifications formed the basic concepts of the Rockfall Hazard Rating System (RHRS), a process for managing rock slopes along transportation systems (Pierson 1992).

Development of the RHRS began in 1984 in the state of Oregon; it was eventually completed and implemented as part of a Federal Highway Administration (FHWA) national pooled-fund research project, led by the Oregon Department of Transportation, and completed in 1989 (Pierson et al. 1989). Recognizing the value of a proactive rockfall mitigation program based on a system that rates the relative hazard posed by individual rockfall sections, FHWA sponsored preparation of *Rockfall Hazards Mitigation Methods*, a manual developed in practical terms that were understandable and usable by highway engineers and staff at the foreman and field superintendent levels (Brawner 1993). Seminars to review the manual were presented in many states (Pierson and Van Vickle 1993). The comparative data provided by RHRS serve as a decision tool in setting priorities for mitigation work at the more critical locations (Pierson et al. 1989; Pierson 1992; Pierson et al. 2005). The RHRS has been widely implemented both nationally and internationally. It was designed to be straightforward and easy to understand and implement. Because development funding was provided by many states from various parts of the country, the RHRS was created to be adaptable and easily customized for local conditions and uses. Chapter 3 provides a detailed explanation of the RHRS.

1.4 Current and Anticipated Extensions to Rockfall Evaluation Methods

Currently, 28 transportation agencies in North America have implemented a rockfall management system based on the RHRS. The Washington Department of Transportation has further modified the RHRS and created an "Unstable Slope Management System," which is used to prioritize mitigation work on all forms of geological and construction-related slope instabilities, including problematic rockfalls, embankments, and landslides (Lowell and Morin 2000). Chapter 4 of this book provides extended discussions of the various RHRS modifications and the other types of rating systems.

These hazard rating systems are qualitative and compare the likelihood, or subjective probability, that a hazardous condition exists at any given site. They are not based on a numerical analysis of the risk of an accident or a specific rockfall event occurring, as can be evaluated by the probabilistic procedures proposed by Roberds (2005). In recent years, several organizations have begun to adapt the concepts of risk management to assessing landslide phenomena (Australian Geomechanics Society 2000; Fell et al. 2005). Chapter 5 describes how these risk management concepts are being applied to rockfall.

2. INTENDED AUDIENCE

Although rockfall events along transportation facilities are stressed, most of the discussions and examples in this report apply equally well to all cases of rockfall characterization and control. The factors of geology, topography, and climate that interact to cause rockfalls are the same regardless of the use to which humans put a given piece of land. Thus, the methods for evaluating rockfall hazards remain the same, regardless of associated human activities, while the methods for preventing or correcting rockfall hazards, within economic limits, remain largely independent of nearby land uses. Therefore, it is hoped that, despite the narrow range of much of its exemplary material, this volume will be useful to any geologist or engineer whose practice includes evaluating rockfall hazards. The text has been written and organized to appeal to a diverse audience, including the following:

- Transportation engineers responsible for rockfall investigations throughout the world,
- Students in geoscience and geotechnical fields with an interest in rockfalls, and
- Researchers needing a definitive source for rockfall investigation and mitigation procedures.

This diverse audience is very similar to the audience identified for the earlier TRB landslide volume (Turner and Schuster 1996). Each of the three audience groups has different information needs and expectations; this report attempts to address them while maintaining a balance and some brevity in the presentation. For example, the report includes comprehensive, practical discussions of how rockfall hazards and risks are assessed followed by extensive reviews of appropriate methods of rockfall analysis and investigation, with special emphasis on the conduct of field investigations and the utility of new technologies, such as terrestrial light detection and ranging (LiDAR) and radar, for assessing and monitoring rock slopes. These topics are important to both practicing engineers and students of rockfall phenomena.

In addition, this volume addresses the need of many readers for a reasonably complete single source of much of the information on mitigation activities. Thus, the report provides comprehensive discussions of the criteria used to select rockfall mitigation options and the characteristics of each mitigation method. It concludes with an assessment of the roles of rockfall maintenance and management programs.

Many students and researchers desire comprehensive references to the literature and discussions of case studies, state-of-the-art techniques, and research directions. Accordingly, considerable effort was expended in identifying suitable literature citations and in providing some discussion of recent developments. References to specialized and hard-to-obtain sources were avoided as much as possible; most of the cited references are readily available through university and special libraries.

3. DEFINITIONS AND RESTRICTIONS

The meaning of the term *rockfall* has been the subject of considerable debate as interest in rockfall events has evolved and intensified and rockfall investigations and assessments have been encouraged by transportation agencies. There has even been some disagreement as to whether one or two words should be used to describe these natural phenomena. Multiple definitions of rockfall have been proposed; several have suggested that the term rockfall be restricted to events involving relatively small volumes of rock material and only a few individual rock blocks.

3.1 One Word or Two?

There is some uncertainty within the English language as to whether rockfall phenomena should be referred to as one word (rockfall) or two words (rock fall). The slope movement classification system described by Cruden and Varnes (1996) in *Special Report 247* (Turner and Schuster 1996) used two nouns to describe any landslide: the first describes the material (i.e., rock) and the second describes the type of movement (i.e., fall). Cruden and Varnes described a fall as "the detachment of soil or rock from a steep slope along a surface on which little or no shear displacement takes place. The material then descends mainly through the air by falling, bouncing, or rolling. Movement is very rapid to extremely rapid." In support of this classification system, *Special Report 247* uniformly used two words (rock fall) when referring to rockfall events.

However, most dictionaries recommend that a single word be used. Thus, this book uses the single-word spelling unless a cited reference uses two words.

3.2 Rockfall Definitions

Ladd (1935) provided an extended and very insightful assessment of slope instability issues facing American railroads in the 1930s and included the following rather comprehensive definition of rockfall:

> Rock-falls are precipitated loosened rock groups, or individual boulders, initially from faces of cuts, or from nearby or remote cliffs and rock outcrops which present vertical or steeply sloped faces. They are distinguished from landslides by being distinctly extreme surface-phenomena; solid rock; usually very small in volume; and consisting, generally, of individual rather than massed units. The repeated fall of rock fragments and shale particles (dribble) from cliffs or approximately vertical faces must be included in this class of superficial earth-movement. Such material, accumulated, forms what is known as "talus."

Ladd (1935) further subdivided rockfall into four types:

1. "Dribble" of fine material;
2. Persistent fall of coarse material (often combined with fine), leading, in nature, to talus accumulations;
3. Falls of loosened rock from jointed or blast-shattered faces of cuts; and
4. Occasional fall of single boulders, or a group of them, often of huge size, and sometimes from a great height (and long horizontal distance), loosened from rock outcrops or cliff faces by undermining as a result of weathering, rain, and seepage wash and by frost-heaving in joint planes.

In his groundbreaking report, Ritchie (1963) somewhat surprisingly did not define rockfall or discriminate among rockfall types. He emphasized the influence of slope angle on the type of falling motion. When the rock slope was steeper than 76° (or 0.25:1), Ritchie observed that even the minimal forward motion of rock blocks at their source is often sufficient for free fall. With lesser slope angles, the falling rock hits the slope and bounces, and when the slope angle is less than 45°, rolling motions dominate (Ritchie 1963).

Spang (1987) and Spang and Rautenstrauch (1988) reviewed the definition of the term rockfall and noted that there is no general agreement as to the volume that characterizes a rockfall event. However, the phenomenon is generally accepted as having the following characteristics:

1. The event involves a single block or group of blocks that become detached from the rock face.
2. Each falling block behaves more or less independently of other blocks.
3. There is temporary loss of ground contact and high acceleration during the descent.
4. The blocks attain significant kinetic energy during their descent.

3.3 Restrictions of Rockfall Size

Spang (1987) suggested that the term rockfall be restricted to events that have a maximum kinetic energy of 500 kJ (185 ft-ton). This value is equivalent to a 5-tonne (5.51-ton) block dropping vertically from a height of 10 m (32.8 ft). Spang proposed this size limitation because he believed potential rockfalls with greater kinetic energy would require active stabilization as it was impractical to contain them with protective structures such as rock catchment fences. In comparison with Spang's suggested upper limit of 500 kJ (185 ft-ton) for kinetic energies produced by rockfall, rockfall fences designed at about that time in Hong Kong (Chan et al. 1986) were stated to be able to withstand an impact energy of only 100 kJ (37 ft-ton); locations liable to experience greater energies were required to be stabilized in situ.

Rather than adopt arbitrary values of kinetic energy, it is considered preferable to maintain the definition of rockfall as one involving significant velocities and some measure of free flight but without an upper limit on the volume or kinetic energy of individual blocks. This convention is adopted throughout this book. Movement may be by free falling, bouncing, rolling, or sliding or by a combination of those four factors. Rockfall may involve more than one rock but typically does not include very large volumes of rock such as rock avalanches or landslides that may include rock. The boundary between a slope movement that may be called a rockslide versus a rockfall is a continuum, and many of the rock mass properties and geological processes that contribute to rockfall that involve one or a few rock blocks may also contribute to the instability of a large portion of a slope. Hsü (1975) applied the term rockfall to even extremely large and rapidly moving rock failures, such as those that triggered the 1881 Elm landslide in Switzerland.

Most analytical studies of the behavior of falling rock blocks consider only the interactions between an individual rock block and the ground surface during successive impacts or as it rolls or bounces along the surface. Collisions and interactions among multiple falling blocks are not evaluated. Such interactions among multiple rock blocks occur during many actual rockfall events; thus, most analytical models represent a significant simplification over reality. However, experiments and analyses described by Okura, Kitahara, and Sammori (2000) and Okura, Kitahara, Sammori, and Kawanami (2000) involved multiple blocks and so provide some limited analytical data. As the volumes of rock involved in a single event become larger, the potential for interactions among rock fragments of all sizes becomes ever more significant until, in extremely large and rapidly moving rockfalls (called *stürzstroms*), these interactions significantly increase the maximum distance traveled by the rockfall material (Hsü 1975).

4. HISTORIC INFORMATION ON ROCKFALLS

Because rockfalls generally are very local events, historic records are relatively scarce, especially compared with descriptions of much larger landslides, rockslides, and other natural disasters. Nevertheless, it is clear that rockfall has caused the destruction of buildings, fatalities, and injuries to many individuals and has posed a danger to travelers on roads and railroads for many years. The following sections describe a serious rockfall event in Quebec City in 1889 for which ample records exist, followed by a review of contemporary reports of rockfall hazards to transportation facilities in the United States in the 1930s. This section concludes with two recent noteworthy rockfall events—the collapse of the "Old Man of the Mountain" natural rock profile in New Hampshire in 2003 and the relatively severe rockfall events triggered by a rather moderate earthquake in Italy in 2009. These two examples provide an assessment of modern rockfall evaluation procedures and the economic and social consequences of rockfall events.

4.1 Quebec City Rockfall of September 19, 1889

Founded in 1608 by Samuel de Champlain, Quebec City, Canada, is one of the oldest cities in North America. Old Quebec (Vieux-Québec) was situated mostly behind fortified city walls on high ground overlooking the St. Lawrence River, but some houses and port facilities were located on a narrow strip of land below a rock cliff (Figure 1-12). On September 19, 1889, a section of the cliff at Cap Diamant fell 90 m (300 ft) onto the houses below. The failure occurred after a day of very heavy rain. The homes of 28 families on Champlain Street were crushed, leaving about 100 people buried beneath 24 m (80 ft) of rock debris. More than 40 people were killed by the rockfall, a number of them children. Rescuers were quickly on the scene, but in some cases it

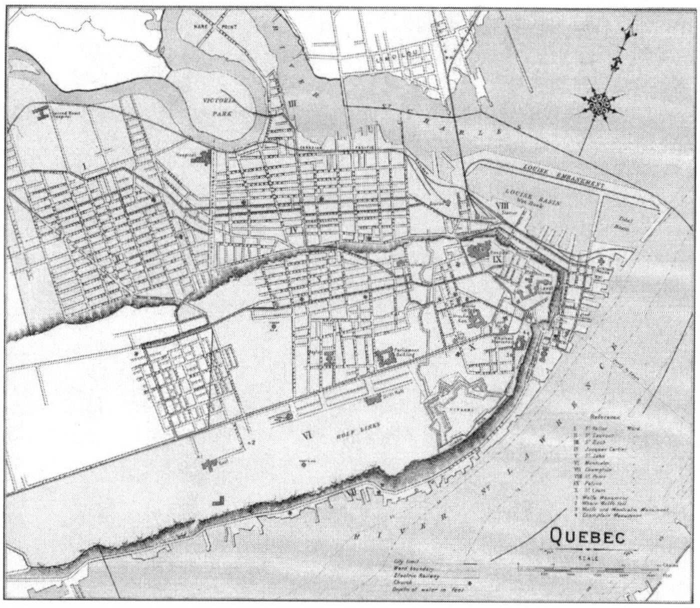

FIGURE 1-12 Map of Quebec City in 1906 shows fortified city walls and cliff that includes Cap Diamant, source of 1889 rockfall that destroyed houses along Champlain Street on St. Lawrence River bank (http://en.wikipedia.org/wiki/File:Quebec_City_Map_1906.jpg).

took days to remove the rocks, rescue the survivors, and retrieve the bodies of those who died.

Sadly, this rockfall event was not the first in Quebec City. Rockfalls have been part of Quebec City history since the 1700s, with most of the damage being confined to the Quartier Champlain at the foot of Cap Diamant. A similar rockfall event on May 17, 1841, again after heavy rain, destroyed eight buildings on Champlain Street and killed 32 people. In 1852 and 1864, similar, but smaller, rockfall events caused additional loss of life and property destruction. In 1881, the Canadian federal government expended considerable funds to purchase and remove several houses located immediately below the cliff and to build a retaining wall designed to protect houses across the street from falling rock. Potentially unstable portions of the rock face were removed and the largest fissures in the rock face were filled with cement. Before the 1889 rockfall, the houses on the opposite side of Champlain Street, farther away from the cliff, were considered to be safe. After the 1889 rockfall, the ruins of the houses and rock debris were left in place to help support the cliff (Figure 1-13).

The September 20, 1889, edition of the *Quebec Daily Mercury* newspaper (*Quebec Daily Mercury* 1889) contains additional details, including an eyewitness account from a surviving resident of Champlain Street:

> Portions of the rock fell at intervals of about five minutes, first in small quantities, then more profusely. After the second slide occurred, I said it was time for us to move, and managed to get away.

It also provides details about the size and character of the 1889 rockfall:

> The great avalanche of rock runs for a space of about one hundred yards (says the City Engineer) commencing at the east end of the last Kiosk of Dufferin Terrace and running in a direct line westward as far as the Officers Quarters on the Citadel. The face of the now

FIGURE 1-13
View looking east along Champlain Street in Quebec City taken several months after 1889 rockfall. Damaged buildings were removed or left as rubble at base of cliff, and street was reopened. Part of 1881 retaining wall is visible. Structure at top of cliff is last kiosk on Dufferin Terrace pedestrian promenade on cliff top.
(PHOTOGRAPH BY WILLIAM NOTMAN & SON, VIEW-2337. © McCORD MUSEUM OF CANADIAN HISTORY, MONTREAL.)

standing cliff forms a smooth perpendicular wall along the whole line, with one or two cracks that indicate the unsolid nature of the whole mass, which may soon follow, now that its base and outer supports are gone or exposed to the wearing influence of the elements.

At 7:20 p.m. an enormous mass was detached from the cliff and hurled with the noise of thunder, slowly at first, but rapidly increasing in momentum through its slide of a couple of hundred feet, over the retaining wall, pushing half a dozen houses out of its way and crushing most of them beneath its weight as though they had been so many paper boxes. Some of those who saw the slide were standing at the doors of their houses and were too much paralysed to move out of its way. Others ran to a place of safety. . . . The debris of rock filled up the narrow street to the depth of some thirty five feet, and cut off all communication between the portions of the city north and south of it. On the wharves behind where the houses had stood, scattered timbers and the earth and rock that partially covered them were mingled with huge piles of coal that left little passage way between them and the river. (*Quebec Daily Mercury* 1889)

4.2 Rockfall Assessment Methods in the 1930s

Ladd (1935) provided a detailed assessment of slope instability issues, including rockfall, faced by American railroads and, to a lesser extent, highway engineers in the 1930s. His rather comprehensive definition of rockfall was presented above in Section 3.2 (Ladd 1935). Slightly earlier, he wrote a two-part series of reports for *Public Roads* magazine that summarized his experience with landslides along highways in West Virginia, Ohio, and southwestern Pennsylvania (Ladd 1927, 1928). Together, these reports present a rather clear picture of the slope stability problems and the state of practice in remediating them in the United States in the late 1920s and early 1930s.

Ladd's reports deal with all classes of landslides, but he devoted some attention to rockfall. In his 1935 report, he described three recent highway rockfall incidents:

In Ohio, two enormous boulders have fallen upon highways, along the Ohio River. They fell from the bluffs and did serious damage to concrete pavements. In West Virginia a boulder fell upon a concrete road pavement, broke it, and threw it out of line for a distance of two hundred, feet. In New York State, during 1934, three people were killed by a rock-fall on the Storm King Highway.

Ladd (1935) also provided some personal experiences:

In a suburb of Pittsburgh, Pennsylvania, a few years ago, I saw a truck crushed by a boulder that fell from the face of a cut, on one side of the street. . . . I was once on an auto trip in West Virginia, for the purpose of examining landslides. An Engineer followed a few hundred feet behind in another car. A small boulder fell from a face high above us, plunged through the top of his car, struck his head a glancing blow and fractured his skull. . . . I once saved the lives of two men, and nearly lost my own, by a boulder fall. I had aural warning—a slight grinding noise in the rock face above us.

Ladd also identified several train derailments caused by rockfalls that damaged the track or when a train encountered even relatively small volumes of rock debris on the track. These experiences are essentially identical to the recent rockfall incidents resulting in train accidents discussed earlier in this chapter. Ladd concluded by stating that he included these rockfall discussions because he believed it was imperative to impress very strongly upon the minds of engineers the following facts:

- Rockfalls are frequent occurrences,
- They occasionally do great damage and cause disastrous train wrecks, and
- Most rockfalls are preventable.

Ladd (1935) concluded his discussion of rockfalls with the following statement, which seems just as relevant today as when it was written:

Prevention of railroad accidents such as have been enumerated above lies in surveys of all dangerous ground in which boulders may become unsupported, or from which they may become detached; in recognition of such rocks or rock masses; and in safe disposal of them. Such work cannot be left to section-men or untrained workers. But who, now, among railroad men, is properly trained for such responsibility? I have often had this thought, on trips of inspection connected with landslide

FIGURE 1-14
"Old Man of the Mountain" on April 26, 2003, 7 days before its collapse. Late-spring snowfall occurred night before (http://en.wikipedia.org/wiki/File:Old_Man_4-26-03.jpg). (PHOTOGRAPH BY J. JOSEPH.)

problems, when accompanied by competent Engineers. As we climbed bluffs and roamed hill and mountain slopes they have blindly passed boulders, without notice or comment, which have fairly screamed at me—"I'm ready to leap. Watch out!" It is proof, again, that a little practical knowledge of field-geology can be of great service to the Engineer.

4.3 Collapse of the "Old Man of the Mountain"

A series of five granite cliff ledges on Cannon Mountain at Franconia Notch, New Hampshire, when viewed from the north, naturally formed a remarkable jagged profile of a man's face (Figure 1-14). Popularly named "Old Man of the Mountain," the rock formation was 370 m (1,200 ft) above Profile Lake and measured 12 m (40 ft) tall and 7.6 m (25 ft) wide. The first recorded observation of the "Old Man" was in 1805. The profile became New Hampshire's state emblem in 1945. It was put on the state's license plate, on state highway route signs, on a U.S. postage stamp issued in 1955, and on the reverse of New Hampshire's statehood quarter coin in 2000 (Figure 1-15). Special viewing areas along Interstate 93 in Franconia Notch State Park allowed visitors to see the profile.

The profile collapsed between midnight and 2:00 a.m. on May 3, 2003. This natural event ended a nearly 200-year relationship between the people of New England and the "Old Man" that included remarkable efforts to understand its formation, its stability, and how it could be secured and preserved. Fowler (2005) provided a detailed history of these stabilization methods and the best technical assessment of its collapse. Fowler especially noted the financial constraints that limited geotechnical investigations and stabilization efforts.

For more than 100 years before the mid-1970s, geotechnical activities were undertaken by private individuals relying on limited, mostly personal, resources or by fiscally limited governmental initiatives. In the mid-1970s, this situation changed when greater but indirect funding became available for preparation of the draft environmental impact statement for Interstate 93 and its various alternatives that were then being proposed to pass through Franconia Notch. However, no substantial funding was ever provided to study the profile's stability or long-term security. Fowler suggested that this lack of funding resulted largely from fears by those responsible for appropriating such funding that field implementation of preservative schemes might endanger the delicate security of the important landmark. As a result, no formal or

(a)

(b)

FIGURE 1-15
Two examples of popularity of "Old Man of the Mountain" profile: (*a*) U.S. stamp issued in 1955 and (*b*) reverse of New Hampshire's Statehood Quarter coin issued in 2000.

systematically rigorous geomechanical study of this complex rock mass was ever conducted.

Fowler (2005) described the results of a 10-week reconnaissance of the profile undertaken in 1976 by geologists and engineers of the New Hampshire Department of Transportation who were also experienced rock climbers. This study was undertaken to measure and document, for the first time, the actual dimensions and structural relationships among the blocks composing the profile; to make a systematic estimate of their composite stability; to estimate ways dynamic stress, such as that from blasting during construction of Interstate 93, might affect the rock mass; and to make recommendations for the security of the profile during the proposed construction below. As shown in Figure 1-16, the reconnaissance revealed the profile to be composed of five uniquely shaped blocks, formed by fortuitous weathering and selective breakage along five discrete sets of structural features (joints, fractures, and a shear zone). These blocks formed, respectively, the forehead, eyebrow, nose, upper lip, and chin of the profile. One set of joints formed the subhorizontal planes that cut through the rock mass, dividing it into the profile features that were formed by selective fractures on their easterly edges.

Earlier geotechnical stabilization efforts on the profile concentrated on the stability of the partially separated, obviously unstable, forehead block and several smaller blocks on its surface. In the 1920s, a crack toward the rear of the forehead was spanned with chains; in 1957 the State Legislature passed a $25,000 appropriation for a more elaborate weatherproofing, including fast-drying cement, plastic covering, steel rods with turnbuckles, and a concrete gutter to divert runoff from above. Each summer a team from the New Hampshire State Park Division and the Department of Transportation provided annual maintenance for these measures. However, no attempts had been made to stabilize the lower blocks, which were critical to the stability and security of the profile. Some initial analyses undertaken in 1975 by faculty and students at Dartmouth College suggested for the first time that the state of static stress in the rock mass as a unit, and in these lower blocks in particular, might be delicate and that detailed dimensional and spatial information was needed for more complete analysis. The 1976 studies by the New Hampshire Department of Transportation yielded the estimated structural–mechanical relationships for these lower blocks, as shown in Figure 1-17.

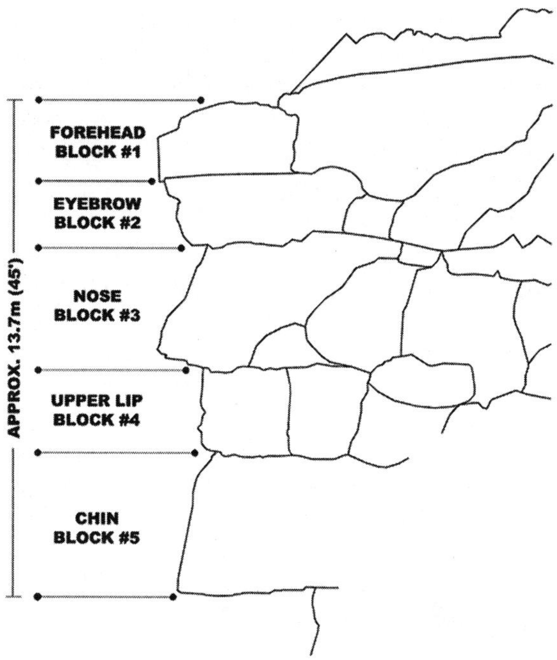

FIGURE 1-16 Diagram of component block combinations forming "Old Man of the Mountain" profile (Fowler 2005).

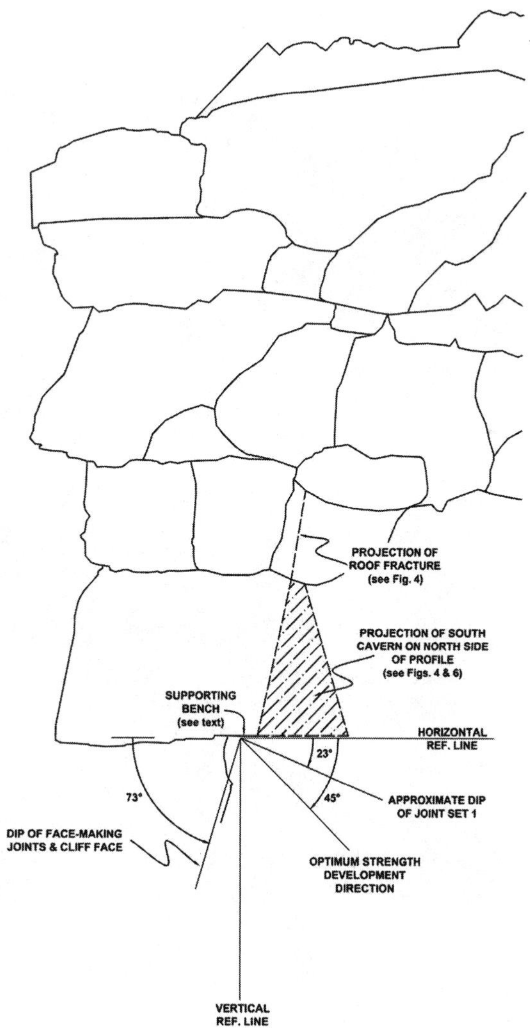

FIGURE 1-17 Estimated structural–mechanical relationships immediately beneath chin (Fowler 2005).

Figure 1-17 shows that the blocks forming the chin are supported by only a very small ledge and are retained in place, along with those forming the upper lip, by a cantilevering action of the nose, and higher, blocks. Thus, the entire profile was in a precarious state of balance, and stability depended on the strength and durability of the very small supporting ledge. Vibrations from the blasting required to construct Interstate 93 in 1985 and 1986 were carefully monitored and no instability of the profile was observed.

However, Fowler (2005) noted that later maintenance visits revealed many incipient changes in the condition of the surfaces and fractures in the rock mass near the base of the chin, including kaolinization and decomposition that were steadily deteriorating the granite and vigorous freeze–thaw cycling that was quarrying off small blocks from nearby parts of the cliff. These processes ultimately caused the ledge to fail in 2003, as shown by careful analysis of before-and-after photographs (Figure 1-18). The blocks marked A through D in both photographs show clearly that the chin and upper lip blocks fell away, probably when the supporting ledge failed, breaking along joints that separated them from blocks A and B. The failure line drawn on Figure 1-18*a* shows the approximate back line of these blocks and portions of the rest of the blocks that collapsed. All portions of the original composite rock mass collapsed except for the rear portion of the forehead slab.

Dismay over the collapse was so great that people left flowers at the base of the cliffs in tribute. In 2004, the state legislature considered a proposal to change New Hampshire's state flag to include the profile, an idea that was eventually shelved. A nonprofit group is building a memorial at ground level near the site.

(a) (b)

FIGURE 1-18 "Old Man of the Mountain" profile (*a*) before and (*b*) after collapse, showing extent of toppling failure. Note forehead turnbuckles in both views (Fowler 2005).

4.4 Rockfall Triggered by Earthquake

Rockfall is increasingly reviewed as an important consequence of earthquakes, particularly those that affect mountainous regions. Post-disaster reports of recent earthquakes in mountainous regions commonly include discussions of rockfall events. This section provides an example of rockfall events triggered by a 2009 earthquake in Italy.

On April 6, 2009, an earthquake with a magnitude of 6.3 on the moment-magnitude scale and a magnitude of 5.8 on the older Richter or local-magnitude scale affected the central Italian region of Abruzzo. Much of the damage occurred in the capital city of L'Aquila, a city with a population of approximately 73,000; thus, this earthquake is referred to as the L'Aquila Earthquake. Unfortunately, about 300 people were killed, making it the deadliest earthquake to strike Italy since 1980. More than 1,500 people were injured and about 17,000 people were evacuated to emergency shelters. The total cost of damage to buildings is estimated to be 2 to 3 billion euros (Salamon et al. 2010). Though not of very high magnitude compared with some worldwide earthquakes, the resulting consequences are significant for a European country. Post-disaster reconnaissance teams from several countries visited the region immediately after the earthquake and published extensive reports on the resulting damage (Blumetti et al. 2009; Stewart 2009; Salamon et al. 2010).

Collapsed and damaged structures in L'Aquila caused most of the fatalities. Older masonry buildings and relatively modern reinforced concrete structures were severely damaged. However, numerous rockfalls occurred throughout the region, especially from limestone slopes. A most impressive rockfall affected the village of Fossa, built on talus debris from a ridge of Mesozoic limestone that forms Cavalletto Mountain on the eastern flank of the Aterno Valley. Fossa was directly damaged by a rockfall that brought huge blocks into the town (Figure 1-19).

Near Fossa, fallen rockfall debris made roads impassable to vehicles, and at one location several cars were damaged by falling blocks (Figures 1-20 and 1-21). Blocks continued falling from the slope for a long period, often independently of any aftershocks, posing a significant residual risk for the area. Residual risk of rockfalls caused the temporary closure of several important roads (Blumetti et al. 2009). Other important rockfalls were located within the Gran Sasso mountain range and along the northeastern slope of Mt. Bazzano.

Other notable rockfalls of large blocks, up to 3 m^3 (106 ft^3), occurred at vulnerable locations with steep cliffs, especially at two tourist destinations where cave attractions are found. Stewart (2009) reported on a relatively detailed investigation and mapping of a rockfall event at the Caves of Stiffe (Grotte di Stiffe). In this case, a large block was liberated from the slope high above this popular tourist area and hit a concession building (Figure 1-22). The last impact mark of the block before hitting the building was about 15 m (49 ft) from the building (Figure 1-23). Between this impact point and the building, the block apparently cleared a small tree. These observations indicate that the block was moving at a considerable velocity. Additional impact

FIGURE 1-19 *(top)* Panoramic view of rockfall above Fossa, Italy, triggered by L'Aquila earthquake on April 6, 2009. Recent (light-colored) rockfall scar is clearly visible (Blumetti et al. 2009).

FIGURE 1-20 *(bottom)* Rockfall debris blocked roads near Fossa, Italy, after L'Aquila earthquake (Salamon et al. 2010).

marks were traced up the hillside, permitting a further analysis of the trajectory of this block (Stewart 2009).

5. OVERVIEW OF REPORT

The 18 chapters in this report are organized into Parts 1 through 4. This arrangement was adopted to group chapters according to related rockfall characterization and control topics. It is hoped that this grouping will assist readers in identifying those chapters most likely to address their immediate needs. Although considerable effort was expended to eliminate repetition of material in different chapters, some reiteration was necessary to provide continuity of thought and to allow adequate explanation of specific topics. Such repetition was judged more acceptable than excessive referral within the text to other sections and chapters.

Part 1, Recognition of Rockfall Hazard, includes this introductory chapter plus four chapters that cover topics related to the identification and evaluation of rockfall hazards. Chapter 2 provides a brief review of failure types and mechanisms. It supplies the reader with a single unified source of rockfall information by summarizing and referencing more extensive published sources. Chapter 3 introduces the development and principles of the RHRS as developed originally in Oregon and supported by FHWA since 1990 (Pierson et al. 1989; Pierson 1992; Pierson and Van Vickle 1993; Pierson et al. 2005). This approach has been widely adopted by many jurisdictions. Within North America, more than 25 transportation agencies currently use the RHRS concept. Several have modified the original system to better reflect their topographic, climate,

FIGURE 1-21 *(top left)*
Cars were hit by rockfall debris at Fossa, Italy, when rockfall was triggered by L'Aquila earthquake (Salamon et al. 2010).

FIGURE 1-22 *(center left)*
Rockfall triggered by L'Aquila earthquake on April 6, 2009, caused large block to hit concession building at Caves of Stiffe (Grotte di Stiffe) (Stewart 2009).

FIGURE 1-23 *(bottom left)*
Last impact mark made by block before it hit building is located about 15 m (49 ft) from building, and block trajectory apparently was above small tree. These observations indicate that block achieved high velocity (Stewart 2009).

or geology characteristics as well as to expand upon its evaluation of mitigation alternatives and possible economic impacts of rockfall events. Chapter 4 provides details of eight systems that represent significant adjustments to the original RHRS system. Chapter 5 introduces some additional alternative approaches to quantitative assessment of the risks posed by rockfall hazards. These approaches represent a number of state-of-the-art directions to the evaluation and quantification of rockfall and to the selection of rockfall mitigation measures.

Part 2, Fundamentals of Rockfall Analysis and Investigation, contains five chapters that collectively review, in some detail, the entire rockfall investigation process. This section begins with Chapter 6, which describes the rockfall characterization process, including organization of the investigation process and the importance of obtaining the appropriate descriptive data during field investigation. Subsequent chapters focus on various aspects of an ideal rockfall investigation process. Chapter 7 discusses the range of new technologies for measuring and monitoring rockfalls, including photogrammetric, LiDAR, radar, and Global Positioning System surveying systems. Chapter 8 presents the fundamentals of Newtonian mechanics and how they are used to describe the motions of falling, bouncing, and rolling rock blocks. Chapter 9 discusses and compares approaches for quantitative modeling and prediction of rockfall. It describes the various modeling approaches, ranging from empirical understanding and models based on field observation used to establish rockfall hazard zonations to simple computer-based approaches incorporating the above empirical relationships. It then compares the numerous two-dimensional simulation models that use basic physics and mechanics concepts along a two-dimensional profile to provide energy and bounce height data that can be used in designing mitigation measures. It concludes with descriptions of the recent development of three-dimensional models that allow a full spatial evaluation of rockfall events. Chapter 10 provides a detailed description of conditions affecting the conduct of rockfall field tests.

Part 3, Rockfall Mitigation, consists of six chapters that begin with an overview of rockfall mitigation options (Chapter 11), including engineered versus nonengineered solutions, and discussions of important criteria affecting the selection of a mitigation option, such as security and reliability, constructability, durability, suitability, aesthetics, cost-effectiveness, and environmental considerations. Five additional chapters describe the major mitigation options in turn. Considerable effort has been expended to explain and contrast the most appropriate methods.

Part 4, Rockfall Maintenance and Management Programs, consists of two concluding chapters. Chapter 17 discusses how various maintenance and monitoring activities significantly reduce the risk associated with rockfall and improve public safety by reducing the exposure of road users to hazard(s) during critical periods of slope instability. Chapter 18, the final chapter, discusses the importance of rockfall management programs based on the Washington State experience.

REFERENCES

Abbott, B., I. Bruce, T. Keegan, F. Oboni, and W. Savigny. 1998. Application of a New Methodology for the Management of Rockfall Risk Along a Railway. In *Proc., 8th Congress International Association of Engineering Geology, A Global View from the Pacific Rim*, Vol. 2, Balkema, Vancouver, British Columbia, Canada, pp. 1201–1208.

Australian Geomechanics Society. 2000. Landslide Risk and Management Concepts and Guidelines. *Australian Geomechanics*, Vol. 35, pp. 49–92.

Blumetti, A. M., V. Comerci, P. DiManna, L. Guerrieri, and E. Vittori. 2009. *Geological Effects Induced by the L'Aquila Earthquake (6 April 2009, MI = 5.8) on the Natural Environment. Preliminary Report*. Dipartimento Difesa del Suolo, Servizio Geologico d'Italia, Rome. http://www.apat.gov.it/site/_files/Inqua/2009_abruzzo_earthquake_report.pdf.

Brawner, C. O. 1993. *Rockfall Hazard Mitigation Methods*. Publication No. FHWA SA-93-085, Federal Highway Administration, U.S. Department of Transportation, Washington, D.C.

Brawner, C. O. 1998. Case Examples of Rock Stability on Transportation Projects. In *Proc., 49th Highway Geology Symposium*, Prescott, Ariz., pp. 26–38.

Brawner, C. O., and D. C. Wyllie. 1975. Rock Slope Stability on Railway Projects. In *Proceedings of the American Railway Engineering Association Regional Meeting*, Vancouver, British Columbia, Canada.

Chan, Y. C., C. F. Chan, and S. W. C. Au. 1986. Design of Boulder Fence in Hong Kong. In *Proc., Conference on Rock Engineering and Excavation in an Urban Environment*, Feb. 23–26, 1986, Hong Kong, Institution of Mining and Metallurgy, London, pp. 87–96.

Cruden, D. M., and D. J. Varnes. 1996. Landslide Types and Processes. In *Special Report 247: Landslides: Investigation and Mitigation* (A. K. Turner and R. L. Schuster, eds.), Transportation Research Board, National Research Council, Washington, D.C., pp. 36–75.

Fell, R., K. K. S. Ho, S. Lacasse, and E. Leroi. 2005 A Framework for Landslide Risk Assessment and Management. In *Landslide Risk Management: Proceedings of the International Conference on Landslide Risk Management* (O. Hungr, R. Fell, R. Couture, and E. Eberhardt, eds.), Vancouver, Canada, May 31–June 3, Taylor & Francis, London, pp. 3–26.

Fowler, B. K. 2005. Stability and Collapse, Old Man of the Mountains, Franconia Notch, New Hampshire. *Environmental and Engineering Geoscience*, Vol. 11, No. 1, pp. 17–27.

Hsü, K. J. 1975. Catastrophic Debris Streams (Stürzstroms) Generated by Rockfalls. *Geological Society of America Bulletin*, Vol. 86, pp. 129–140.

Ladd, G. E. 1927. Landslides and Their Relation to Highways, Part 1. *Public Roads*, Vol. 8, pp. 21–31, 35.

Ladd, G. E. 1928. Landslides and Their Relation to Highways, Part 2. *Public Roads*, Vol. 9, pp. 153–163.

Ladd, G. E. 1935. Landslides, Subsidences and Rockfalls as Problems for the Railroad Engineer. In *Proc., 36th Annual Convention of the American Railway Engineering Association*, Vol. 36, pp. 1091–1162.

Lemay, P. 2007. Listening for Rockfalls: Microseismic Technology May Help Railway Better Detect Rockfalls at a Distance. *Transportation Development Centre, Transport Canada, R & D Updates Bulletin*, Vol. 17, No. 1. http://www.tc.gc.ca/eng/innovation/tdc-publication-updates-v17n1-827.htm.

Lowell, S., and P. Morin. 2000. Unstable Slope Management: Washington State. *TR News*, No. 207, Transportation Research Board, National Research Council, Washington, D.C., pp. 11–15.

Martin, D. C. 1988. Rockfall Control: An Update. *Bulletin of the Association of Engineering Geologists*, Vol. XXV, No. 1, pp. 137–139.

Okura, Y., H. Kitahara, and T. Sammori. 2000. Fluidization in Dry Landslides. *Engineering Geology*, Vol. 56, No. 3, pp. 347–360.

Okura, Y., H. Kitahara, T. Sammori, and A. Kawanami. 2000. The Effects of Rockfall Volume on Runout Distance. *Engineering Geology*, Vol. 58, No. 2, pp. 109–124.

Pierson, L. A. 1992. Rockfall Hazard Rating System. In *Transportation Research Record 1343*, Transportation Research Board, National Research Council, Washington, D.C., pp. 6–13.

Pierson, L. A., and R. Van Vickle. 1993. *Rockfall Hazard Rating System Participant's Manual*. Prepared for the National Highway Institute, U.S. Department of Transportation, Publication No. FHWA SA-93-057, Federal Highway Administration, U.S. Department of Transportation, Washington, D.C.

Pierson, L. A., D. L. Beckstrand, and B. A. Black. 2005. *Rockfall Hazard Classification and Mitigation System*. Publication No. FHWA/MT-05-011/8174. Federal Highway Administration, U.S. Department of Transportation, Washington, D.C.

Pierson, L. A., S. Davis, and R. Van Vickle. 1989. *The Rockfall Hazard Rating System Implementation Manual*. FHWA Publication No. FHWA-OR-EG-90-01, Federal Highway Administration, U.S. Department of Transportation, Washington, D.C.

Quebec Daily Mercury. 1889. A Dreadful Catastrophe! (Microfilm), Canadian Microfilming Co., Montreal. http://www.collectionscanada.gc.ca/sos/002028-2200-e.html.

Ritchie, A. M. 1963. Evaluation of Rockfall and Its Control. In *Highway Research Record 17, Stability of Rock Slopes*, Highway Research Board, National Research Council, Washington, D.C., pp. 13–28.

Roberds, W. 2005. Estimating Temporal and Spatial Variability and Vulnerability. In *Landslide Risk Management: Proceedings of the International Conference on Landslide Risk Management* (O. Hungr, R. Fell, R. Couture, and E. Eberhardt, eds.), Vancouver, Canada, May 31–June 3, Taylor & Francis, London, pp. 129–158.

Salamon, A., R. Amit, G. Baer, Y. Hamiel, and A. Mushkin. 2010. *The MW6.3, 2009, L'Aquila Earthquake, Central Italy*. Report GSI/13/2010, Geological Survey of Israel.

Spang, R. M. 1987. Protection Against Rockfall—Stepchild in the Design of Rock Slopes. In *Proc., 6th International Conference on Rock Mechanics*, Montreal, Quebec, Canada, Balkema, Rotterdam, Netherlands, pp. 551–557.

Spang, R. M., and R. W. Rautenstrauch. 1988. Empirical and Mathematical Approaches to Rockfall Protection and Their Practical Application. In *Proc., 5th International Symposium on Landslides*, Lausanne, Switzerland, pp. 1237–1243.

Stewart, J. P. (ed.). 2009. *Preliminary Report on the Seismological and Geotechnical Aspects of the April 6 2009 L'Aquila Earthquake in Central Italy (Version 2.0)*. Report of the National Science Foundation–Sponsored GeoEngineering Extreme Events Reconnaissance Team, Report No. GEER-016, GEER Association, University of California, Berkeley.

Turner, A. K., and R. L. Schuster (eds.). 1996. *Special Report 247: Landslides: Investigation and Mitigation*, Transportation Research Board, National Research Council, Washington, D.C.

Chapter 2

Jerry D. Higgins and
Richard D. Andrew

ROCKFALL TYPES AND CAUSES

1. INTRODUCTION

Rockfall is a gravitationally driven geomorphic process that occurs on steep natural or constructed slopes. While the usage and classification of the term rockfall is quite varied, the term is generally thought to include the detachment or failure of a rock block or mass and its subsequent trajectory of downslope movement. Geomorphic classifications, such as those provided by Selby (1993), treat rockfalls as a type of rock slope failure that initiates on a very steep slope and whose trajectory path is dominated by free fall on characteristically very steep slopes. Toppling and sliding mechanisms of initiation are classified separate from rockfalls. The term *rockfall* is recognized as a collection of processes that occur over the full spectrum of material sizes, ranging from large rock masses to gravel- and boulder-sized particles. Selby differentiated falls as rock mass falls, slab and block falls, and, the most common type, rock particle falls.

A more engineering-based definition that relates to mitigation of rockfall hazards was proposed by McCauley et al. (1985); it described rockfall as the movement of rock of any size from a cliff or slope that is steep enough for the mass to move downslope. Movement may involve any combination of free falling, bouncing, rolling, and sliding. McCauley et al. noted that rockfall may involve more than one rock, but they excluded slope failures involving large volumes of rock, such as rock avalanches and landslides that may include rocks.

Extremely large rockfalls on very steep slopes may generate catastrophic debris streams, known as *stürzstroms*, which are described by Hsü (1975). As discussed in Chapter 9, Hungr and Evans (1988) classified rockfalls of relatively small volumes of material as "fragmental rockfalls." They described these fragmental rockfall events as being characterized by more or less independent movements of the individual fragments, whereas extremely large-volume events involved different movement mechanisms with particle interactions that affected the sliding or mass flow of large volumes of coherent or broken rock.

In the context of this book, and for the practice of rockfall hazard characterization and mitigation, the authors favor a more encompassing and slightly modified version of the definition of rockfall presented by Cruden and Varnes (1996), which includes a variety of initiation mechanisms and trajectory paths. They described rockfall as a very rapid to extremely rapid slope movement in which bedrock material is detached from a steep slope and descends the slope by falling, bouncing, or rolling. Detachment of the rock debris may be initiated by falling, if a slope is undercut, but it may also be the result of sliding or toppling movements.

In North American practice, cobbles and boulders that erode from oversteepened slopes composed of large clast-supported (block-in-matrix) deposits

FIGURE 2-1
Small-volume rockfall from low slope in block-in-matrix materials. Note rockfall deposits in ditch (Yeh and Associates, Inc. 2002).

are also often classified as rockfall. Rockfall volumes considered here fall within the limits defined as fragmental rockfalls by Hungr and Evans (1988). This broader encompassing definition recognizes a continuum of the varied initiation mechanisms, a range of event sizes or volumes, and the complex trajectories that are involved with rockfall. These factors define the extent of rockfall hazard and risk to life, property, and infrastructure. Figures 2-1 through 2-3 illustrate the considerable range of rockfall size and volume described by this modified version of the definition proposed by Cruden and Varnes.

FIGURE 2-2
Source zone where 1,530-m^3 (2,000-yd^3) block of quartzite detached and fell from cliff in November 2004 along Interstate 70 in Glenwood Canyon, western Colorado.
(PHOTOGRAPH BY J. D. HIGGINS, COLORADO SCHOOL OF MINES.)

FIGURE 2-3
Block that detached from source zone in Figure 2-2 broke into numerous blocks at base of cliff, and they tumbled onto Interstate 70, blocking traffic. Their removal required blasting.
(PHOTOGRAPH COURTESY OF T. ORTIZ, COLORADO DEPARTMENT OF TRANSPORTATION.)

It has become common practice for engineering geologists and geotechnical engineers to evaluate the rockfall hazard on slopes and to assess the risk to transportation and other civil and mining facilities as a basis for decisions on mitigation. The first steps in developing an effective mitigation plan are as follows:

1. Identify the type of rockfall that may occur in various geological settings.
2. Determine the cause(s) that may initiate the rockfall.

To aid practitioners in developing these skills, this chapter describes the basic rockfall types and the typical causes of rockfall.

2. PRIMARY ELEMENTS CONTROLLING ROCKFALL

Rockfall results from the processes of detachment followed by relatively rapid downslope movement. Thus, material and slope characteristics are the primary intrinsic elements that control the occurrence of rockfall.

2.1 Material Characteristics

It is rare for a rock mass to be free of zones or planes of weakness, referred to as *discontinuities*. If discontinuities are not present, very high steep slopes can form (or be constructed) that are highly stable. The steep rock slopes towering over Yosemite Valley, California, illustrate this point (Figure 2-4). More typically, rock masses are anisotropic and heterogeneous, and they are separated into distinct blocks by numerous discontinuities. Discontinuities are naturally occurring structural defects within the rock mass, such as joints, bedding planes, foliation, faults, and similar geological structures, but they can also be induced by aggressive or ill-conceived construction practices, particularly the result of overblasting during excavation work.

Joints are a common type of rock discontinuity; they are well-defined cracks where little or no displacement has occurred. They often occur as regular sets of planar fractures with near-parallel strike and regular spacing. Multiple joint sets divide rocks into blocks or wedges.

Bedding is a prominent discontinuity that plays a role similar to joint structure. Bedding planes occur in stratified layers of rock and signify a change in depositional environment or source material. They constitute interruptions in the continuity of the rock fabric, and consequently the strength along them is usually lower than that of the intact rock on either side, except in very weak rock.

Foliation is the banding or lamination within some metamorphic rocks formed by a segregation of certain minerals and is an important influence on rock strength properties when some bands are formed of low-strength minerals or those with platey shapes such as micas, chlorite, and talc. Foliations typically produce planes of weakness in the rock mass that become more evident when the rock mass is loaded parallel to foliation.

FIGURE 2-4 Many rock slopes in Yosemite Valley in northern California contain relatively few discontinuities and are very steep, high slopes. (PHOTOGRAPH BY J. D. HIGGINS, COLORADO SCHOOL OF MINES.)

Faults are fractures where displacement (from a few centimeters to many kilometers) has occurred due to external tectonic compression, tension, or torsion forces. Faults may form a narrow to wide zone of crushed and sheared rock or fault gouge. The fault zone may act as an aquitard if it is clayey or fine grained; if coarse grained, it may facilitate groundwater flow.

A rock mass may be broken by several joint orientations or a combination of the discontinuity types listed above. They form weak zones in a rock mass along which blocks may detach from the slope. Important properties of these discontinuities include orientation, spacing, persistence, roughness, weathering, aperture, filling, and seepage. Discontinuities and their properties are described in detail in Chapter 6 and thus are introduced only briefly here.

Discontinuities are the primary elements that affect the stability of a rock mass. Rock blocks may separate from the rock mass if bounded by discontinuities that dip out of the slope at an angle steeper than the effective friction angle (resistance to sliding) along the discontinuity. Detachment along discontinuities may also be initiated by water pressure, ice, wind, vibration, root growth, or other triggering mechanisms. Even though a rock mass may include numerous discontinuities, their orientations frequently play

FIGURE 2-5
Walls of Vermont granite quarry are stable to heights in excess of 100 m (330 ft) because of a lack of discontinuities and a rock mass that is very strong. Observable horizontal features are created during mining of blocks; they are not preexisting discontinuities.
(PHOTOGRAPH COURTESY OF T. C. BADGER.)

an important role in determining the stability of a slope by forming sliding planes or planes of release. Figure 2-5 shows quarrying operations in a granite mass containing very few discontinuities. The quarry walls are stable because there are virtually no discontinuities and the rock is very strong.

Discontinuities control many processes that influence the detachment of rock blocks and failures of rock masses. They are referred to as *structurally controlled rockfalls*. Piteau and Peckover (1978) suggested that the following primary elements influence structurally controlled rockfall:

1. Slope characteristics (orientation, slope angle, height),
2. Relation of the discontinuity system to possible shear surfaces,
3. Discontinuity shear strength, and
4. Driving forces such as water pressure and freeze–thaw within discontinuities.

Additional elements to be considered are the persistence of the discontinuity, which refers to its areal extent, and the spacing between discontinuities, which defines the size of the blocks. A rock mass cut by highly persistent, widely spaced discontinuities that are adversely oriented to the slope face may generate large-volume rockslides or rockfalls. Alternatively, rock masses that have closely spaced discontinuities with very low persistence are more prone to produce small-volume rockfall events.

In contrast, environmentally controlled rockfalls are controlled not by discontinuities but by erosional and weathering processes. Geological materials more likely to form this type of rockfall include interbedded mudstone–sandstone sequences, highly fractured rock masses, and "block-in-matrix" deposits, such as talus, colluvium, and glacial outwash and till. Environmentally controlled rockfalls primarily result from the following:

1. Slope characteristics, and
2. Resistance to environmental elements such as erosion and weathering of the matrix material, interbedded materials, or, in the case of talus, the interlocking of loose blocks.

2.2 Slope Characteristics

For a rock to detach and move, the slope from which it initiates must be sufficiently steep to expose the controlling discontinuities or the weak erodible material that bounds the block or mass. For planar failures to occur in a structurally controlled rock mass, the slope face must be oriented nearly parallel (generally within ±20° to ±30°) to the single discontinuity that forms the failure surface. For wedge-shaped failures, created by the intersection of at least two planes, failure can occur when the orientation of the line of intersection of the two discontinuities dips out of the slope steeply enough to overcome the shear strength along the intersection. Once detachment of the block has occurred, the slope below must be steep enough to allow the dislodged block to overcome any conditions that may inhibit downslope movement, such as slope irregularities, vegetation (trees and brush), and other obstructions.

3. TYPES OF ROCK SLOPE FAILURES

Numerous descriptions of rock slope failure types have been published (Brawner 1993; Goodman and Kieffer 2000; Hoek and Bray 1981; Norrish and Wyllie 1996; Selby 1993; Wyllie and Mah 2004). Table 2-1 summarizes the types of failures

Table 2-1
Types of Rock Slope Failure That May Contribute to Rockfall

Failure Mode	Failure Description	Susceptible Material
Simple structurally controlled		
Planar sliding	Sliding of one or multiple rock blocks on a single discontinuity surface.	Rocks with discontinuities daylighting in and striking nearly parallel to the slope face. All rock types may slide on joint sets; sedimentary and volcanic rocks may slide on weak beds or bedding planes; metamorphic rocks may slide along foliation.
Wedge sliding	Sliding of one or multiple blocks along the line of intersection of two nonparallel planes.	Rock masses with well-defined orthogonal discontinuity sets, foliation, bedding planes, or weak beds in which the intersection of two discontinuities plunges into (daylights) a slope face. May occur in any igneous, sedimentary, and metamorphic rocks.
Toppling	Rotation of column(s) or block(s) of rock about a fixed base and out of the slope.	Hard rocks with regular, parallel discontinuities dipping steeply into the slope, with or without cross joints. Common in foliated metamorphic rocks and steeply dipping layered sedimentary rocks, block-jointed granites, steep faces of columnar basalts, and thinly bedded slate.
Circular sliding	Failure occurs along a surface that approaches a circular shape and is not a zone of particular weakness, merely the line of least resistance.	Highly fractured or low intact-strength material; highly weathered rock or highly fractured rocks with so many discontinuities that it is effectively homogeneous; weak sedimentary rocks with bedding that does not dip out of slope; clay-rich rocks softened by loosening and wetting; smectite-rich clay shales and altered tuffs, rocks adjacent to major fault zones, and coal-bearing formations.
Complex structurally controlled		
Buckling and kink-band slumping	Compressional failure of columns or slabs parallel to the rock slope face.	Thinly bedded, weak sedimentary rocks inclined steeply and parallel to the slope surface; shale–sandstone and shale–chert sequences, coal measures, and foliated metamorphic rocks.
Block torsion	Failure block spins about a hinge point where sliding has been impeded.	Blocky rock that contains a series of discontinuities and obstructions.
Rock slump	Backward rotation of single or multiple blocks as they fail in series, resembling a classic soil slump.	Materials with vertical and horizontal zones of weakness, allowing for sliding on the basal rupture surface and between blocks or along bedding at the same time. Commonly, sedimentary units or stratified units.

(continued on next page)

Table 2-1 *(continued)*
Types of Rock Slope Failure That May Contribute to Rockfall

Failure Mode	Failure Description	Susceptible Material
Sheet failure	Failure in tension of thin sheets along "sheet joints" that result in rockfall or rockslide.	Steeply dipping, preexisting sheet joints typical in granitic rocks and sandstones; new sheet joints in weathered rocks, friable massive sandstones, and pyroclastic sediments on steep slopes.
Bilinear wedge failure	In a system with two blocks formed by multiple discontinuities, the lower block is forced out by the weight of the upper block.	Blocky material that contains multiple discontinuities.
Key block failure	Removing the "key" that supports or holds rocks above and behind the key block generally leads to rapid raveling.	Blocky hard rock masses.
Secondary toppling	Undercutting the toe of the rock slope or toppling caused by large-scale slope movement.	Hard rocks with regular, parallel joints dipping into slope face, with or without cross joints; foliated metamorphic rocks and steeply dipping layered sedimentary rocks; also block-jointed granites, steep faces of columnar basalts.
Environmentally controlled		
Differential erosion	Weak material underlying or intermixed with more competent rock erodes and the more competent material subsequently fails.	Layered materials with variable strength or rate of weathering characteristics. Multilayered volcanic rocks; basalt with weak interlayers; sedimentary rocks; interbedded shales and sandstones; conglomerates and volcanic mudflows; and unconsolidated or weakly consolidated talus, alluvium, glacial till, etc.
Boulder fall	Rounded cobbles and boulders are released from their matrix by erosion and fall due to gravity.	Hard rocks held by a weaker matrix that is subject to weathering and erosion, unconsolidated or weakly consolidated natural slope deposits, such as talus, alluvium, glacial till, colluvium, conglomerates–breccias, and debris-flow deposits.
Raveling	Progressive loosening and erosion of blocks from the surface of an outcrop.	Very highly fractured hard rocks, such as columnar jointed basalts or brittle sandstones, overlying soft shale, poorly cemented conglomerates and breccia, thinly bedded sedimentary and volcanic rocks.

Source: Adapted from Goodman and Kieffer 2000.

that are most applicable to rockfall. In this chapter, rockfall failures are classified as simple structurally controlled, complex structurally controlled, and environmentally controlled failures.

Careful and thoughtful consideration is important when one is observing rock slopes to identify rockfall failure types, as described below. The discontinuities that control more frequent, smaller-volume rockfalls from a slope may also pose the potential for much larger-scale slope instability. In such situations, rockfall mitigation alone is unlikely to reduce the hazard significantly, and stability analyses and mitigation options for the entire rock slope may be needed.

3.1 Simple Structurally Controlled Failures

Many textbooks that introduce basic concepts of rock slope engineering present the first four failure types shown in Table 2-1. These structurally controlled rockfall failures include planar, wedge, and circular sliding and toppling (Figure 2-6). They are common, easy to recognize, and relatively simple to identify with stereographic techniques. All are controlled by discontinuity properties, except in the case of circular sliding. Circular sliding typically occurs when a shear surface (*a*) develops in highly fractured rock with no identifiable discontinuity pattern, (*b*) develops across discontinuities in very weak materials, or (*c*) partially follows discontinuities in moderately weak materials. For this reason, some practitioners instead consider circular failure to be a nonstructurally controlled failure. It is included here because it is one of the simple failure modes and is partially controlled by the presence of discontinuities in some instances.

The underlying principles governing kinematic and equilibrium rock slope stability analysis methods are not described in this text. Both are covered in detail by Hoek and Bray (1981), Norrish and Wyllie (1996), and Wyllie and Mah (2004). However, these principles are frequently employed during analyses of various types of planar failure facilitated by the use of stereographic projections (*stereonets*), as discussed in Sections 3.1.1, 3.1.2, and 3.1.3. Norrish and Wyllie (1996) summarized the four structural conditions necessary for planar failure as follows:

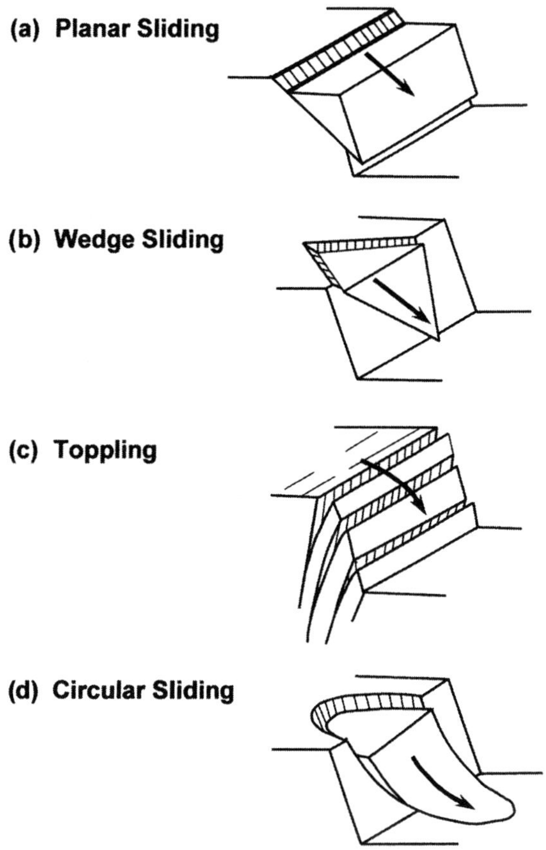

FIGURE 2-6 Simple structurally controlled failure types: (*a*) planar sliding, (*b*) wedge sliding, (*c*) toppling, and (*d*) circular sliding (adapted from Hoek and Bray 1981).

1. The dip direction of the planar discontinuity must be within 20° of the dip direction of the slope face.
2. The dip of the planar discontinuity must be less than the dip of the slope face and thus the discontinuity must "daylight" in the slope face.
3. The dip of the planar discontinuity must be greater than the angle of friction of the surface.
4. The lateral extent of the potential failure mass must be defined either by lateral release surfaces that do not contribute to the stability of the mass or by the presence of a convex slope shape that is intersected by the planar discontinuity.

Evaluation of the orientations of planar discontinuities is facilitated by the use of stereographic projections that allow three-dimensional orientations to be represented and analyzed in two dimensions. The principles and applications of stereographic analysis for rock slope stability are described in detail by Goodman (1976) and Hoek and Bray (1981) and are summarized in considerable detail by Norrish and Wyllie (1996). The following sections contain illustrations that use stereographic

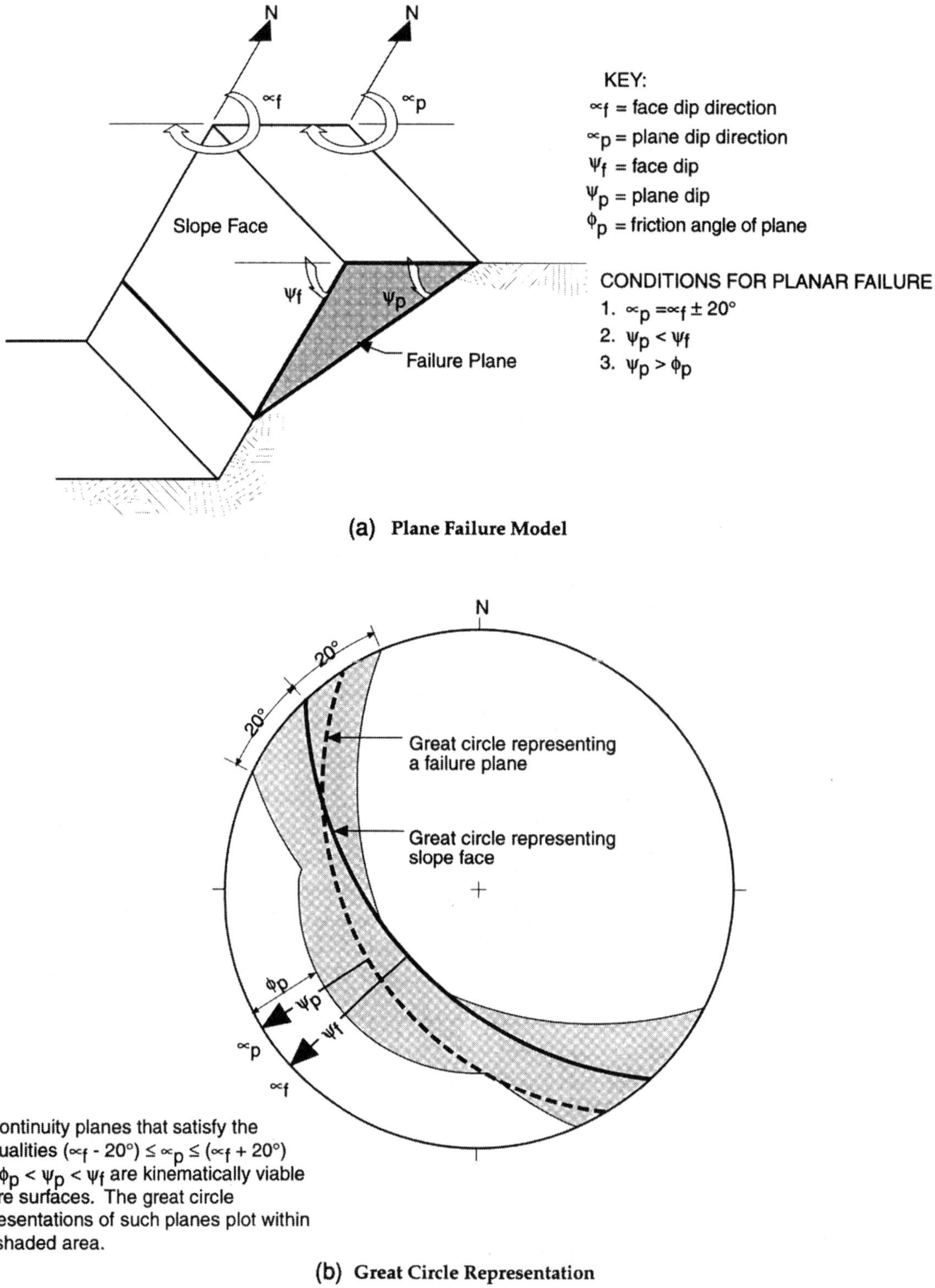

FIGURE 2-7
Kinematic analysis for planar failure: (*a*) plane failure model and (*b*) great circle representation (Norrish and Wyllie 1996).

analyses that were originally presented by Norrish and Wyllie. Appendix A in this volume provides stereonets and briefly defines their use for kinematic analysis.

3.1.1 Planar Sliding

Planar sliding is illustrated in Figure 2-6*a*. This failure mode may define the overall stability of a slope or smaller rockfall instabilities for one or several blocks, depending on the persistence and orientation of discontinuities. Sliding occurs along a discrete discontinuity or in a layer of weak material that daylights in the slope face; the failure surface generally strikes parallel or nearly parallel to the slope face. Norrish and Wyllie (1996) suggested that the strike of the failure plane must be within ±20° of the strike of the slope face for a planar failure to initiate. Additional discontinuities oriented more perpendicular to the slope face may provide release surfaces that allow planar failure of smaller rock masses or individual rock blocks. Alternatively, the planar failure may be located on the nose of a rock outcrop so that the curving slope forms the lateral boundaries of the failure mass. A tension crack may develop as a plane of release at the head of a slide mass or between individual rock blocks along existing discontinuities. Figure 2-7 illustrates this definition and the kinematic conditions for planar failure.

If discontinuities are well-defined, planar sliding will occur in many types of low- and high-strength rocks. Planar sliding is especially common in layered sedimentary and volcanic rocks. The dip of the failure plane may be steep to shallow, depending on the shear strength of the basal discontinuity. Figure 2-8 illustrates a small planar failure along the weathered and altered contact between a steeply dipping pegmatite dike and biotite gneiss. Figure 2-9 shows a sedimentary rock outcrop in Tennessee where planar sliding sheds small rockfall volumes onto a roadway.

Planar failures often occur as large rockslides; however, they may generate individual rockfalls as well. The Ferguson rockslide, near Yosemite National Park, California, is a recent example of this case (Figure 2-10). In the spring of 2006, the rockslide moved slowly downslope primarily as a translational slide (planar failure). At the intersection of the slide plane with the slope face, the movement caused oversteepening of the slide toe that resulted in several months of rockfall. Approximately 800,000 m^3 (1,046,360 yd^3) of rockfall accumulated in a large talus slope that covered California State Highway 140 and extended 10 m (33 ft) into the Merced River (Harp et al. 2008).

FIGURE 2-8
Individual blocks and small planar slides of pegmatite failed along weathered and altered contact between pegmatite dike and biotite gneiss.
(PHOTOGRAPH BY J. D. HIGGINS, COLORADO SCHOOL OF MINES.)

FIGURE 2-9
Potential planar failure producing rockfall in sedimentary rock.
(PHOTOGRAPH COURTESY OF THE TENNESSEE DEPARTMENT OF TRANSPORTATION.)

FIGURE 2-10
Oblique aerial photograph of Ferguson rockslide. Rockfall source zone is at toe of landslide and rockfall deposit (talus) covers roadway.
(PHOTOGRAPH COURTESY OF E. L. HARP, U.S. GEOLOGICAL SURVEY.)

3.1.2 Wedge Sliding

Wedge sliding (Figure 2-6*b*) occurs on slopes with discontinuities striking obliquely to the slope face. The intersection of two discontinuities forms a V-shaped wedge with the line of intersection of the two discontinuities plunging toward, and intersecting, the slope face. Sliding occurs along the plunge line and is resisted by the shear strength mobilized along both discontinuity surfaces. The kinematic conditions for wedge failure are illustrated in Figure 2-11.

Rock masses with well-defined orthogonal joint sets, bedding, or foliation are commonly prone to wedge failures. Shale, thin-bedded siltstones, claystones, limestones, and slate are rock units that exhibit conditions that are most conducive to wedge failures (Norrish and Wyllie 1996). The size of wedge failures can range from a single or a few blocks to large rockslides. Figures 2-12 through 2-14 show wedge failures in metamorphic and sedimentary rock slopes. The large wedge in Figure 2-12 formed by retrogressive failure in the first decade after the cut slope was constructed. The wedge formed by a combination of frequent failures of one or several blocks interspersed with several larger wedge failures. Figures 2-13 and 2-14 illustrate potential wedge failures of single or several rock blocks from sedimentary rock formations.

3.1.3 Toppling

Toppling failure on rock slopes is a common source of rockfalls. Toppling results from the rotation of columns of rock about a fixed base (Figure 2-6*c*). The center of gravity of the block must fall outside the dimension of the base for toppling to occur. Generally, toppling occurs where sets of closely spaced discontinuities strike approximately parallel to a slope face and dip into the slope at an angle greater than about 50° (Brawner 1993). If multiple columns are toppling, there may be slippage between them. The kinematic conditions controlling toppling failure are illustrated in Figure 2-15.

Goodman and Kieffer (2000) suggested that toppling failures can be deep, large, and potentially rapid. The period of rapid movement is typically preceded by slowly accelerating displacements as the rock weight is gradually redistributed to the toe of the column. As toppling movement continues, the rock mass may break up on the slope face and raveling of loosened blocks may occur. Topples can produce steep talus fronts of highly broken and very loose rock blocks that can threaten facilities located below with rockfall (Goodman and Kieffer 2000).

Toppling mechanisms that most commonly result in rockfall have been subdivided into block, flexural, and a combination of the two, as illustrated in Figure 2-16 (Hoek and Bray 1981; Norrish and Wyllie 1996; Wyllie and Mah 2004). Recognition of these types is important because there are different stability analysis techniques for flexural and block toppling. *Block toppling* (Figure 2-16*a*)

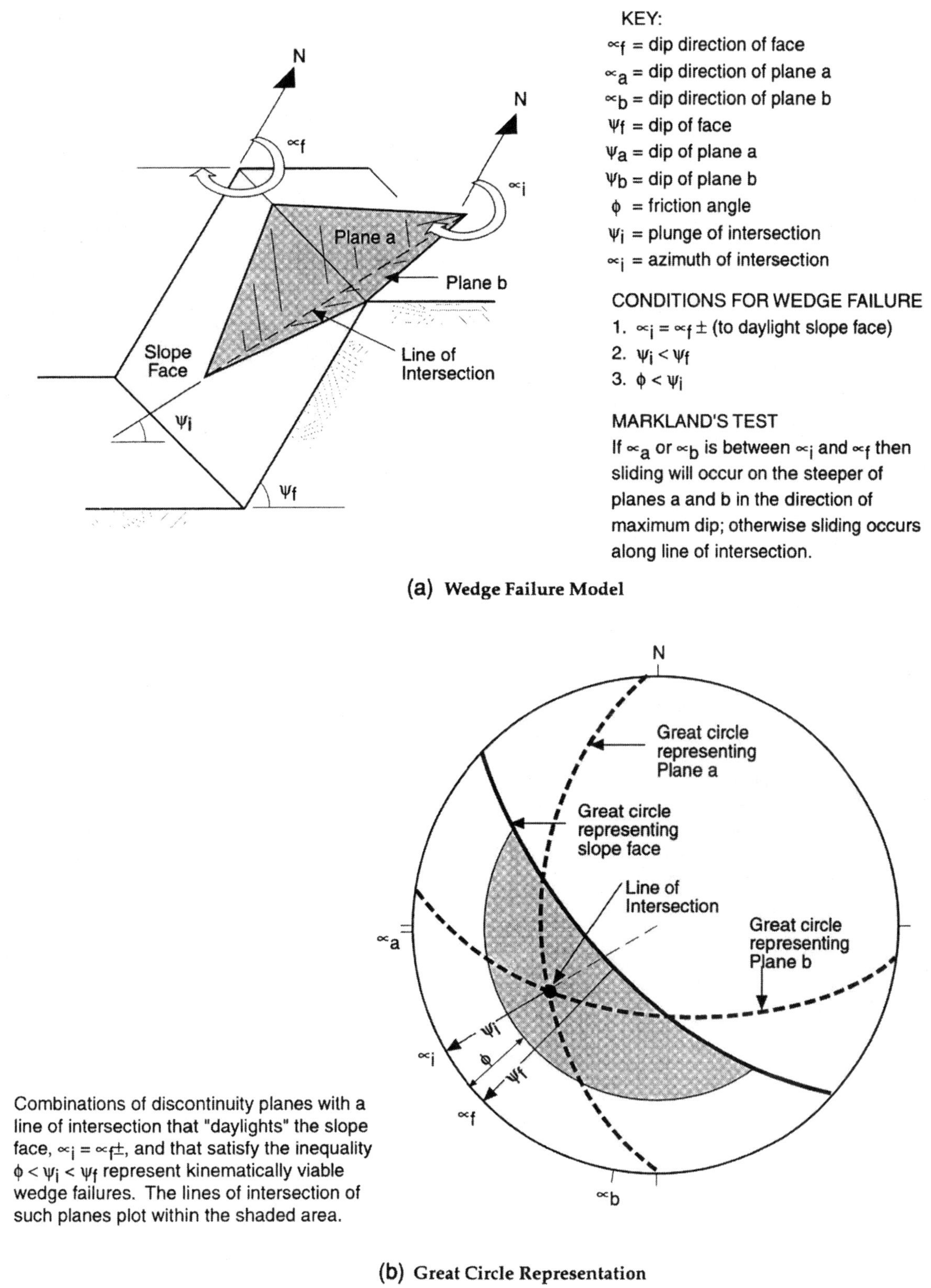

FIGURE 2-11
Kinematic analysis for wedge failure: (*a*) wedge failure model and (*b*) great circle representation (Norrish and Wyllie 1996).

FIGURE 2-12
Wedge failure and rockfall in strongly jointed metamorphic rocks along U.S. Highway 2 in Montana.
(PHOTOGRAPH COURTESY OF L. PIERSON, LANDSLIDE TECHNOLOGY.)

FIGURE 2-13
Cut slope showing potential for rockfall generated by wedge failure along Tennessee roadway.
(PHOTOGRAPH COURTESY OF THE TENNESSEE DEPARTMENT OF TRANSPORTATION.)

FIGURE 2-14 *(below)*
Rock slope in highly fractured sedimentary rocks with potential for rockfalls initiated by wedge sliding of one or several rock blocks.
(PHOTOGRAPH COURTESY OF THE TENNESSEE DEPARTMENT OF TRANSPORTATION.)

occurs in hard-rock masses with widely spaced, steeply dipping joints forming columns dissected by orthogonal joints. Typical examples are found in sandstones (Figure 2-17) and columnar-jointed basalts. *Flexural toppling* (Figure 2-16*b*) commonly occurs on steep slopes composed of high-strength rock with continuous columns separated by discontinuities steeply dipping into the slope face. The columns break in flexure as they bend forward. Flexural toppling is common in thinly bedded slate and shale that lack well-developed jointing perpendicular to bedding (Wyllie and Mah 2004) and in columnar-jointed rocks as illustrated in Figure 2-18. *Block-flexure toppling* (Figures 2-16*c* and 2-19) is characterized by flexure along relatively long, discontinuous columns. The toppling is the result of displacements along many cross joints (Wyllie and Mah 2004).

3.1.4 Circular Sliding

Circular sliding failures are characterized by sliding along an arcuate-shaped failure surface and typically occur in soils, highly weathered rock, weak rock types, and highly fractured rock masses (Figures 2-6*d* and 2-20). These have been called *soil-type slumping* by Goodman and Kieffer (2000). The circular failure surface develops along a path of least resistance in weak materials rather than along material discontinuities. Weak rock materials, such as shale, mudstone, and nonwelded

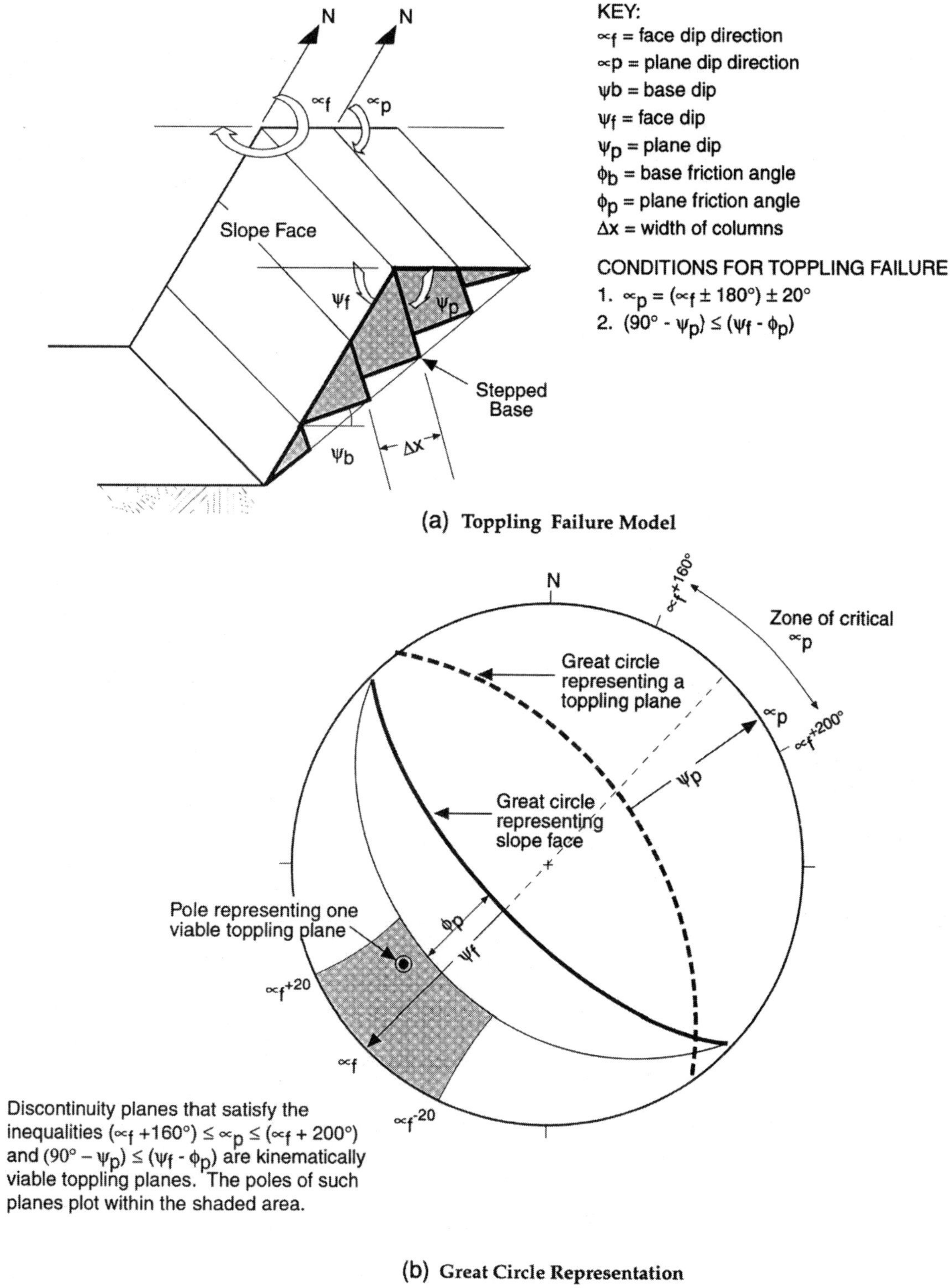

FIGURE 2-15
Kinematic analysis for toppling failure: (*a*) toppling failure model and (*b*) great circle representation (Norrish and Wyllie 1996).

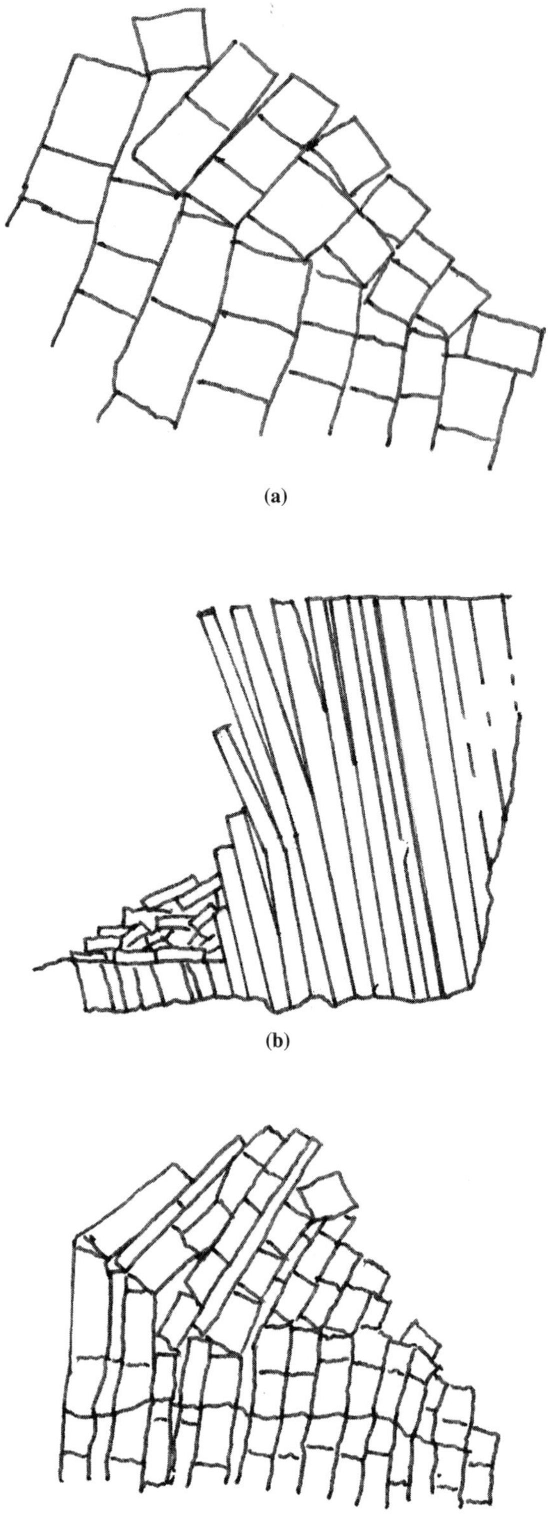

FIGURE 2-16
Toppling modes: (*a*) block toppling, (*b*) flexural toppling, and (*c*) block-flexure toppling (adapted from Goodman and Bray 1976; Hoek and Bray 1981; and Wyllie and Mah 2004).

FIGURE 2-17
Sedimentary rocks dipping steeply away from slope face may fail by toppling.
(PHOTOGRAPH COURTESY OF THE TENNESSEE DEPARTMENT OF TRANSPORTATION.)

FIGURE 2-18
Flexure toppling in columnar-jointed andesite.
(PHOTOGRAPH COURTESY OF T. C. BADGER, WASHINGTON STATE DEPARTMENT OF TRANSPORTATION.)

tuff, may contain discontinuities (bedding planes or thin interbedded materials), but if these discontinuities do not daylight in the slope face, the failure surface will cut through them seeking the path of least resistance. In weathered rock, the failure surface may partially follow discontinuities. Circular failure surfaces also develop in

FIGURE 2-19
Block-flexure toppling in columnar-jointed andesite.
(PHOTOGRAPH COURTESY OF T. C. BADGER, WASHINGTON STATE DEPARTMENT OF TRANSPORTATION.)

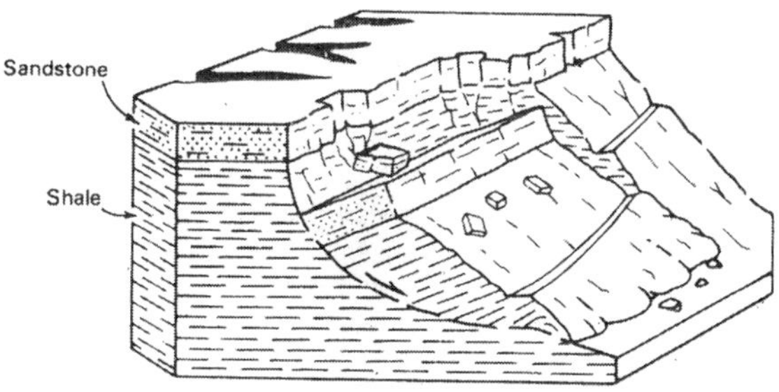

FIGURE 2-20
Circular sliding in low-strength bedrock (Cruden and Varnes 1996).

highly fractured rocks that exhibit closely spaced and randomly oriented fracturing. These conditions are commonly found within fault zones and altered volcanics.

Circular failures range in scale from a few meters in length and width to large failures incorporating a significant portion of a slope. Rockfall may be generated from a slow-moving landslide failure, whether circular or planar, simply by the movement of the landslide dislodging individual blocks from the material mass (Figure 2-21). In this case, the rockfall is symptomatic of the larger slope instability.

FIGURE 2-21
Rockfall source zone at intersection of toe of slow-moving circular failure and steep slope. Slope face has become oversteepened and rockfall has raveled from slope onto Interstate 70 below.
(PHOTOGRAPH BY J. D. HIGGINS, COLORADO SCHOOL OF MINES.)

3.2 Complex Structurally Controlled Failures

Many structurally controlled rockfalls originate on slopes exhibiting more complex patterns of failure than those described above. In general, the slope and discontinuity geometries make it much more difficult to analyze the stability of rock blocks in these cases, compared with the simple structurally controlled failures discussed in Section 3.1. The following sections provide brief descriptions and illustrations of the more common types of complex structurally controlled rockfall failures.

3.2.1 Buckling and Kink-Band Slumping

Buckling and *kink-band slumping* (Figure 2-22) result from compression failures of thin columns or slabs of rock exposed on a steep rock face. A *buckling failure* (Figure 2-22*a*) occurs when one discontinuity boundary of the column or slab is oriented nearly parallel to the slope face (Goodman and Kieffer 2000). The load within a rock layer is sufficient to buckle the thin layer(s) at locations where weak zones or existing flexures occur in the rock. High water pressure in the slope may contribute to the failure (Brawner 1993). The movement may be rapid, and the rock slab above the buckled zone may suddenly slide down the slope. *Kink-band slumping*

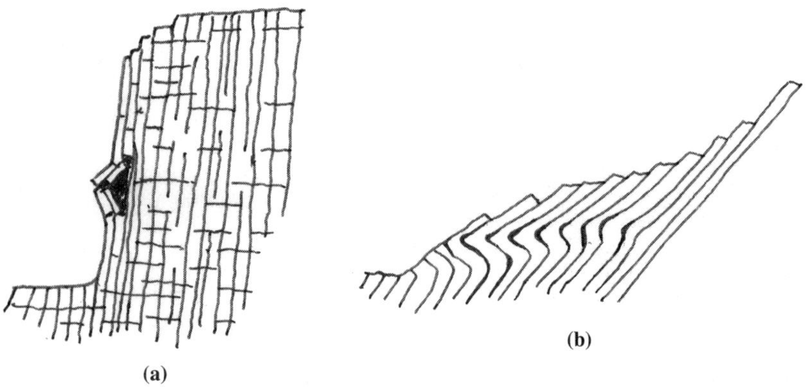

FIGURE 2-22
Complex structurally controlled failures: (*a*) buckling failure and (*b*) kink-band slumping (Goodman and Kieffer 2000).

(Figure 2-22*b*) was identified by Kieffer (1998) and described by Goodman and Kieffer (2000) as a failure mode intermediate between rock buckling and slumping. This failure mode may break up the rock mass near the slope surface and make it more susceptible to rockfall.

3.2.2 Block Torsion

Goodman and Kieffer (2000) described the *block-torsion failure* mode as a planar or wedge failure where an individual rock block has a local sliding restriction that causes the block to rotate around a hinge point (Figure 2-23). As sliding stops, the block moves with a rotary motion along the impeded contact surface. Blocks may also rotate about a corner. Analyses of these types of failures have been described by Wittke (1990) and Mauldon and Goodman (1996).

3.2.3 Rock Slump

Another type of circular failure, called *rock slumping* by Goodman and Kieffer (2000), involves backward rotation of one or more rock blocks, as illustrated in Figure 2-24. Goodman and Kieffer described this mode of failure as the "analog to toppling" as there is progressive destruction of the rock slope from multiple blocks being rotated and moved; however, the layers dip toward the free face of the slope rather than into the slope as in toppling failure. The characteristic slow movement and resulting breakup of rock at the surface of the slope may dislodge rockfall or produce a highly broken and dilated rock mass that is then more subject to raveling processes. In this case, rockfall may be a symptom of active movement and instability of a larger portion of the slope.

3.2.4 Sheet Failure

Sheet failure describes a slope movement process that involves tensional failure of thin layers of rock on very steep slopes that then slide or fall along sheet joints oriented nearly parallel to the slope. *Sheet joints* (also known as *dilation joints*, *stress-relief fractures*, *relaxation joints*, or *exfoliation joints*) may have developed over a considerable

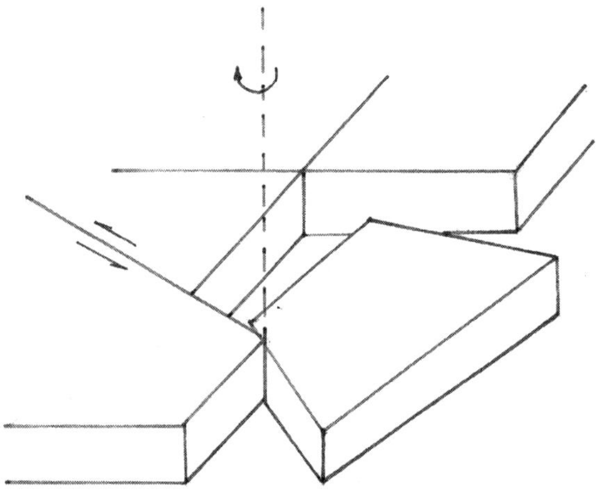

FIGURE 2-23
Block-torsion failure (Goodman and Kieffer 2000).

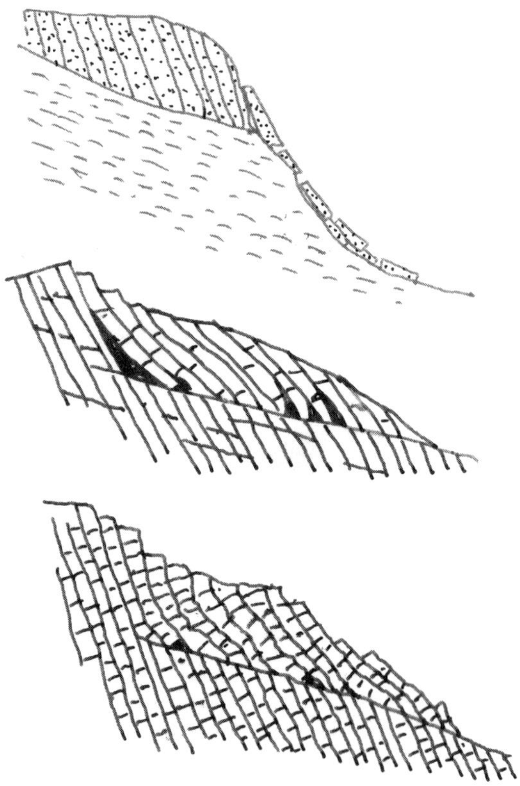

FIGURE 2-24
Three examples of rock-slumping failures (Goodman and Kieffer 2000).

period of time and thus are a preexisting condition of the rock mass; alternatively, they may form shortly after excavation of a rock cut (Waltham 2002). Preexisting sheet joints are common in massive sandstones and granitic rocks that have been exposed by tectonism or deep erosion. A classic site in the United States for failures along sheet joints on natural slopes is in Yosemite National Park, where, in several locations, the characteristic arch-shaped features are temporarily supporting large unstable slabs that are "hanging" on the slope (Figure 2-25). Waltham (2002) described the rockfall hazards due to sheet failures in Norway's fiords.

When rock excavations are made into weak unjointed rock, new sheet joints tend to form parallel to the surface because of the removal of confining pressure (Figure 2-26). Subsequent, relatively shallow, rock excavations into the slope surface may intersect these sheet joints and create hanging slabs. If the dip of the sheet joint exceeds the friction angle corresponding to the shear strength mobilized along the sheet joint, then the thin rock sheets are held in place only by the tensile strength at the top of the sheets. These failures are typically sudden, and the scale of the failure depends on the height of the slope (Goodman and Kieffer 2000).

FIGURE 2-25
Sheet failure in steep granite walls of Yosemite Valley, California.
(PHOTOGRAPH BY J. D. HIGGINS, COLORADO SCHOOL OF MINES.)

FIGURE 2-26
Sheet joint in massive weak sandstone developed after recent cut excavation.
(PHOTOGRAPH COURTESY OF THE WASHINGTON STATE DEPARTMENT OF TRANSPORTATION.)

3.2.5 Bilinear Wedge Failure

Bilinear wedge failure involves several discontinuities that create two blocks where interblock movement can occur along a steeply dipping discontinuity between the blocks (Figure 2-27). The lower block is passive and is wedged, or forced, out of the slope face by the driving force of the upper (active) block. This failure mechanism most commonly results in rockfall rather than producing larger slope instability. Norrish and Wyllie (1996) discussed the stability analysis of a bilinear wedge.

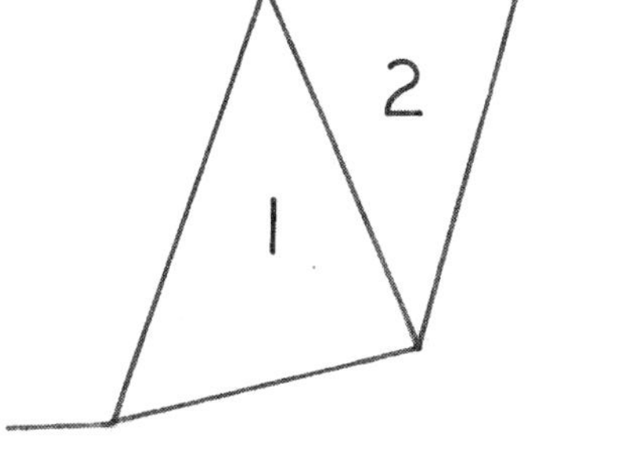

FIGURE 2-27
Bilinear wedge failure where Block 2 pushes Block 1 out of slope (Norrish and Wyllie 1996).

3.2.6 Key-Block Failure

One or more rock blocks bounded by discontinuities in a rock slope may create a "key" that supports or holds rock blocks in the slope above it. Removing the key block(s) may allow numerous rocks to fall or ravel from the slope. Block 1 in Figure 2-28 is

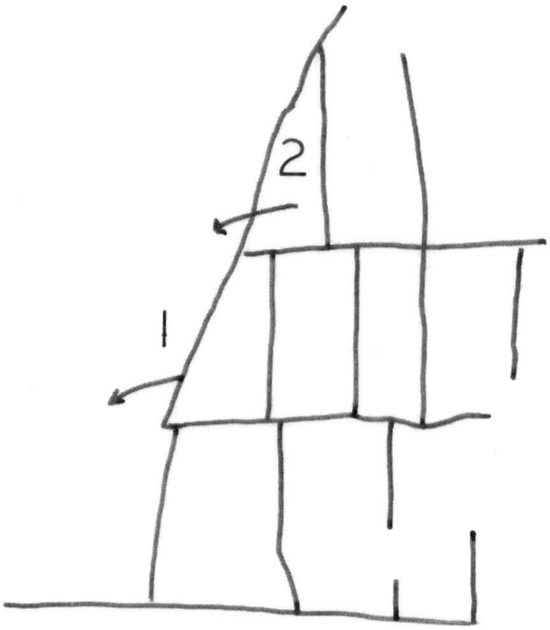

FIGURE 2-28
Key block failure where Block 1 slides out of slope, removing support for Block 2, which then falls from slope (Yeh and Associates, Inc. 2002).

FIGURE 2-29
(*a*) Key block (arrow) supports several thousand cubic meters of dilated gneissic mass resting on adverse-oriented planar discontinuity that dips out of slope at 55°. (*b*) Key block (KB) is situated below adverse-oriented planar discontinuity (D); it is about 1.5 m thick and is defined by two discontinuities (arrows).
(PHOTOGRAPHS COURTESY OF THE WASHINGTON STATE DEPARTMENT OF TRANSPORTATION.)

the key block. The extent of raveling or failure depends on the volume of rock involved as illustrated in Figure 2-29. Key blocks may be prime candidates for rock bolting and should be identified before rock scaling. Goodman and Shi (1985) described key-block theory and its application to stability analysis.

Chapter 1 describes the failure in 2003 of the rock cliff formation in Franconia Notch, New Hampshire, which formed a remarkable profile of a man's face, popularly called the "Old Man of the Mountain" (Fowler 2005). A series of five granite cliff ledges, measuring 12 m (40 ft) tall and 7.6 m (25 ft) wide, were supported by a very narrow ledge just below the "chin" of the profile. When this ledge failed, due to normal weathering phenomena, the entire sequence of blocks forming the profile toppled (Figures 1-17 and 1-18). In this case, the ledge and chin block supported the entire sequence of blocks and thus formed the key block.

3.2.7 Secondary Toppling

Secondary toppling includes failures caused by undercutting of the toe of a rock slope or toppling that results from large-scale slope failures. Four secondary toppling modes that may be associated with rockfall have been identified by investigators (Goodman and Bray 1976; Hoek and Bray 1981; Wyllie and

Mah 2004; Wyllie and Munn 1978). Figure 2-30*a* illustrates *slide-head toppling,* where movement in the lower slope results in opening of discontinuities near the head of the slope, allowing toppling to occur. Figure 2-30*b* illustrates *slide-toe toppling*. In this case, toppling failure is induced at the toe of the slope as a result of unstable material loading steeply dipping beds of high-strength rock at the toe. This condition may be induced by removing material at the toe or placing material on the slope crest.

Figures 2-30*c* and 2-30*d* illustrate toppling induced by differential erosion, a process typical of many areas in the southwestern United States. When weak material, such as shale, mudstone, or clay, underlies tall blocks of stronger rock, such as sandstone or volcanic rocks, removal of the underlying material by weathering and erosion allows the upper material to topple or the toe to "kick outward" (slump). This process removes lateral support for the next column, which can lean outward. Topples of massive units, as shown in Figure 2-31, may easily topple beyond the width of typical roadside catchment ditches. Chapter 7 includes a case study that describes the monitoring of the 1935 failure of "Threatening Rock," a sandstone column resting on shale at its toe located in Chaco Culture National Historical Park, New Mexico. This monitoring provides excellent documentation of the progressive nature of secondary toppling failures.

3.3 Environmentally Controlled Failures

Environmental processes, such as differential erosion and weathering of bedrock or soil deposits and ice-jacking in rock fractures, are common causes of rockfall on steep slopes. Rockfalls generated by these processes may range in scale from small rocks up to large boulders several meters in diameter. The following sections describe the more common environmentally controlled failures.

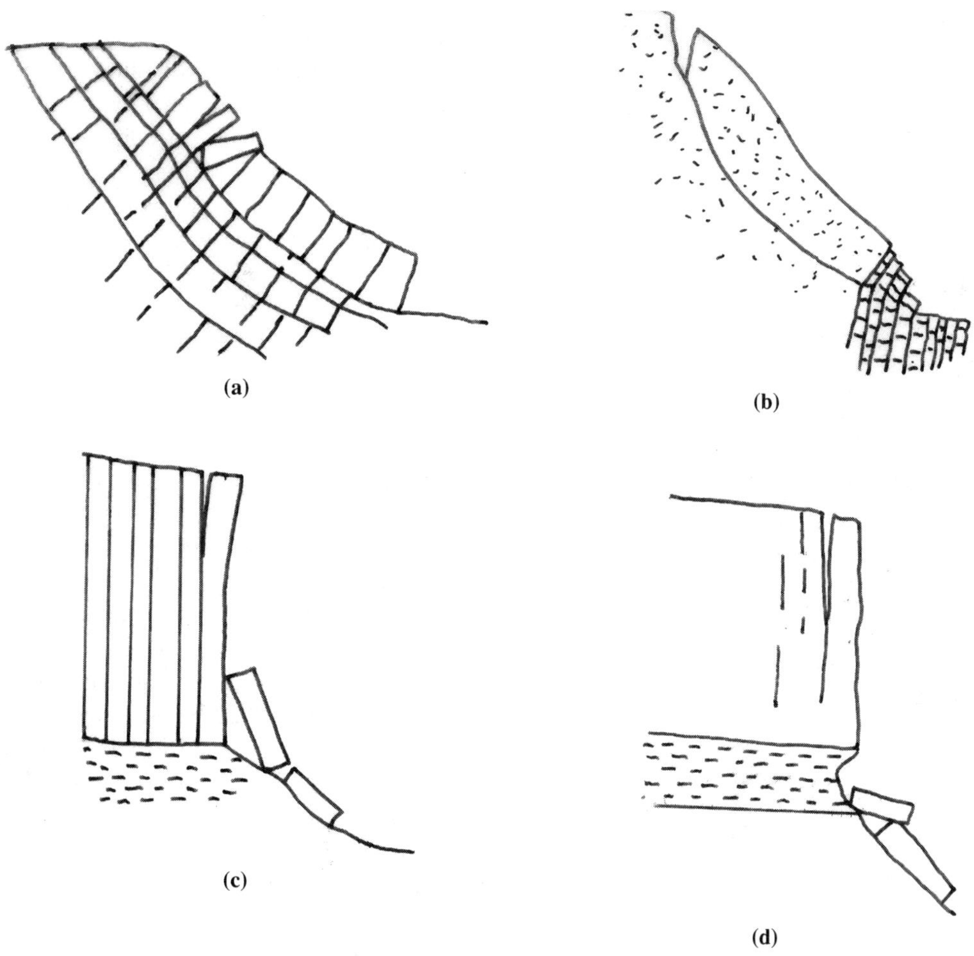

FIGURE 2-30 Secondary toppling modes: (*a*) slide-head toppling; (*b*) slide-toe toppling; (*c*) toppling and slumping of jointed rock; and (*d*) toppling of columnar-jointed, but strong, overlying rock (Goodman and Bray 1976; Hoek & Bray 1981; Wyllie and Mah 2004; Wyllie & Munn 1978).

FIGURE 2-31
Secondary toppling of large sandstone blocks in DeBeque Canyon, Colorado. Underlying shale layer has differentially weathered and undercut tall, outward-leaning sandstone blocks.
(PHOTOGRAPH BY J. D. HIGGINS, COLORADO SCHOOL OF MINES.)

FIGURE 2-32
Differential erosion of interbedded shale, siltstone, and sandstone creates very large sandstone rockfall blocks.
(PHOTOGRAPH BY J. D. HIGGINS, COLORADO SCHOOL OF MINES.)

3.3.1 Differential Erosion and Weathering

Rockfall can be initiated by degradation of the slope due to surface erosion from concentrated runoff or internal erosion (piping), especially in loosely cemented silty and sandy materials, weathered granitic rocks, and saprolites (Goodman and Kieffer 2000). Removal of material by erosion may leave large blocks unsupported; these blocks then become detached from the slope. Rockfall is also common where differential erosion and weathering degrade slopes composed of multilayered rock units of variable resistance to weathering and erosion processes. As the less-resistant layers degrade at the slope face, the unsupported resistant rock units detach by rotating along discontinuities or sliding from the slope on the weak underlying layers. Steep slopes composed of sandstones that are interbedded with shale or mudstones are routinely prone to this type of rockfall. The shale or mudstone layers typically weather and erode more rapidly than the sandstones, which results in unsupported sandstone ledges (Figure 2-32).

Volcanic rocks, such as basalt and andesite flows, are often interlayered or underlain with pyroclastic deposits and buried soil horizons or were deposited on other rock types. Layers formed by volcanic flows tend to be highly jointed and permeable, which allows surface water to infiltrate through them to the less-permeable underlying materials. Accumulation of ground water can weaken the underlying materials, causing blocks at the face to slide out (Goodman 1993). The interbedded or underlying materials are often less resistant to weathering and erosion processes and thus typically erode and weather more rapidly than the volcanic flows, which can lead to unsupported blocks forming at the face of the slope.

Over time, the relatively competent volcanic rock becomes unstable and topples, or the toes of these columns become unsupported and slide out of the slope (Figure 2-33). Ash layers and relatively loose and nonlithified buried soils may be eroded by concentrated groundwater flow onto the slope surface. In the Columbia River basalts of eastern Washington and Oregon, the first author has observed many examples of rock ledges formed from differential erosion within flows or between different flows. Differences in jointing resulting from the

FIGURE 2-33
Cliff experiencing toppling failures of volcanic flow rock caused by differential erosion of underlying tuffaceous unit.
(PHOTOGRAPH COURTESY OF THE WASHINGTON STATE DEPARTMENT OF TRANSPORTATION.)

different cooling patterns of thick flows versus thin flows often cause variable resistance to weathering. Clinker or highly vesicular zones tend to weather more rapidly and thus cause overhanging ledges.

3.3.2 Boulder Fall

Boulder fall results from erosion and weathering of unconsolidated deposits composed of cobbles and boulders embedded in a finer-grained matrix or from coarse talus deposits. Water and wind erosion, raveling of particles, freeze–thaw cycles, and wet–dry cycles tend to erode the matrix and expose and loosen the larger clasts (Figures 2-34 through 2-36). Boulder fall typically originates in deposits of colluvium, talus, glacial outwash and till, stream terraces, debris flows, and alluvial fans. Cobbles and boulders within some of these deposits can be subrounded to rounded and therefore may quickly develop momentum and travel farther than angular blocks originating in bedrock. Rounded boulders are also more likely than angular blocks to roll through a rockfall catchment ditch. Occasionally, extremely large boulders are released from glacial or colluvial deposits (Figure 2-35).

FIGURE 2-34
Boulder fall from colluvial soils along U.S. Highway 550 in southwestern Colorado.
(PHOTOGRAPH BY J. D. HIGGINS, COLORADO SCHOOL OF MINES.)

3.3.3 Raveling

Probably the most common type of rockfall failure is *raveling*, also defined as *rock particle fall* by Selby (1993). This mass-wasting process results from slow, continuous weathering and erosion that gradually loosen rock blocks from a slope. Once detached, blocks may then rapidly move downslope because of gravity (Figures 2-37 and 2-38). The degradation processes that constantly act on slope exposures include the following:

- Freezing and thawing of water (ice-jacking) in cracks, joints, faults, and shear zones;
- Slow chemical and physical weathering of cementing materials;
- Wetting and drying of clay cement; and
- Physical loosening of rock blocks caused by tree roots and animals.

Raveling is a common process within highly fractured strong rocks such as basalts, mica-rich metamorphic rocks, and brittle sandstones interbedded with soft shale or mudstones. It also occurs in weakly cemented conglomerates and breccias (Brawner 1993; Goodman and Kieffer 2000).

FIGURE 2-35
Glacial and colluvial deposits sometimes release very large boulders, such as this 5.5-m (18-ft) long, 3.7-m (12-ft) high, and 3.7-m (12-ft) wide boulder that landed on North Cascades Highway in Washington on June 5, 2010.
(PHOTOGRAPH COURTESY OF THE WASHINGTON STATE DEPARTMENT OF TRANSPORTATION.)

FIGURE 2-36
In Silverado Canyon, Orange County, California, talus deposits derived from outcrops of meta-sandstone upslope are mobilized during rainstorms and tumble down steep slope as rockfall.
(PHOTOGRAPH BY J. D. HIGGINS, COLORADO SCHOOL OF MINES.)

FIGURE 2-37
Raveling of thinly bedded and highly fractured sedimentary rock.
(PHOTOGRAPH BY J. D. HIGGINS, COLORADO SCHOOL OF MINES.)

FIGURE 2-38
Raveling of road cut in highly weathered sedimentary rock.
(PHOTOGRAPH BY J. D. HIGGINS, COLORADO SCHOOL OF MINES.)

4. ROCKFALL CAUSES

The initiation of rockfall results from a combination of internal and external influences. Internal influences relate to the rock mass properties such as rock type, discontinuity properties, and groundwater. External influences, or triggering factors, are conditions that change the forces acting on a rock (Pantelidis 2009). These forces may include the influence of rainfall, snowmelt, seepage, channeled water runoff, weathering, erosion, freeze–thaw and heating–cooling cycles, tree roots, wind, disturbance by animals, and earthquakes. Additional events may result from human activities such as construction practices, blasting, vibration from equipment and trains, and stress relief due to excavation (Hoek 2007). These external, or triggering, influences are much more likely to induce failures when the internal conditions favor rock mass instability.

McCauley et al. (1985) reported on a study of 308 rockfall incidents along California highways that identified 14 causes of rockfalls (Table 2-2). Among the possible causes, those related to the presence of water were responsible for most of the failures. Rainfall, freeze–thaw cycles, snowmelt, channeled runoff, and springs and seeps totaled 67% of the listed causes. Wyllie and Norrish (1996) reported that rockfall records for a 19-year period on a major railroad in western Canada showed that about 70% of the events occurred during the fall, winter, and spring when weather conditions include heavy rainfall, prolonged periods of freezing temperatures, and daily freeze–thaw cycles in the fall and spring. Peckover (1975) reported that rockfalls were most frequent on railway lines in the Fraser Canyon, British Columbia, Canada, between October and March, the wettest and coldest time of the year for the area. Guzzetti et al. (2003) examined 153 slope failure events in Yosemite National Park and attributed triggering factors to each event (Table 2-3). Most slope failures in the inventory were rockslides (55%) and rockfalls (30%). Eight triggering factors were identified, but water-related factors (rainfall, rainfall and snow, and freeze–thaw) contributed to 73% of the failures. Earthquakes triggered rockfalls in 14% of these slope failures.

Hungr et al. (1999) studied the magnitude and frequency of rockfalls and rockslides in the same general area as Peckover (1975), mentioned above. The study found that about 50 rockfall and slide events occurred per year with volumes of less than 1 m^3 (35.3 ft^3). Volumes of 10,000 m^3 (353,000 ft^3) occurred much less often, every 10 to 50 years. These studies suggest that there are many factors to be considered as possible causes of rockfall. Clearly, the presence of water within discontinuities and the geological properties of rock masses are the most significant factors affecting rockfall failures. In addition, smaller events are much more likely to occur than larger ones. The following sections discuss rockfall causes in more detail.

Table 2-2
Causes of 308 Rockfalls on Highways in California

Cause of Rockfall	Percentage of Total[a]
Rain	30
Freeze–thaw	21
Fractured rock	12
Wind	12
Snowmelt	8
Channeled runoff	7
Adverse planar fracture	5
Burrowing animals	2
Differential erosion	1
Tree roots	0.6
Springs or seeps	0.6
Wild animals	0.3
Truck vibrations	0.3
Soil decomposition	0.3

[a]May not sum due to rounding.
Source: McCauley et al. 1985.

Table 2-3
Triggering Factors of Slope Failures in Yosemite National Park

Triggering Factor	Number	Percent
Rainfall	78	51.0
Rainfall and snow	15	9.8
Freeze–thaw	18	11.8
Earthquakes	21	13.7
Blasting and construction	12	7.8
Lightning, wind storms, spring runoff	9	5.9

Source: Guzzetti et al. 2003.

4.1 Internal Influences

Rock mass properties control the susceptibility of a slope to produce rockfalls. These properties include intact rock strength, discontinuities, and groundwater. Lithology and discontinuities control the geometry of failure surfaces and are key factors for determining the shear strength of a rock mass. Groundwater tends to increase driving forces on the potentially unstable blocks while reducing the mobilized shear strength along discontinuities; thus, the presence of groundwater promotes instability.

4.1.1 Lithology

Lithology refers to the physical characteristics of intact rock. The properties of intact rock do not usually play a critical role in initiating a rockfall event, because the shear strength of intact rock is much greater than the rock mass shear strength, which is largely controlled by the shear strength developed along discontinuities. However, in the cases of low-strength rock (such as shale, claystone, mudstone, or any highly weathered rock), lithology may contribute to rockfall failure by, for example, the bearing capacity failure of weak rock underlying strong rock. Lithology descriptions include rock type, mineral assemblages, fabric, and texture. In general, the strength and deformation characteristics of intact rock are determined by its fabric, texture, and mineralogy (Deere and Miller 1966).

Relative rock strengths based on laboratory tests of different lithologies at room temperatures and low confining pressures are shown in Figure 2-39 (Davis 1984). Intact rock may be composed of minerals that are unstable or metastable under certain conditions. For example, swelling clays that cement clasts in a conglomerate, when exposed in the walls of a recently constructed rock cut, may undergo a volume change due to increases or decreases in moisture and reduced confining pressure, causing the clasts to ravel out of the slope face. This action can also undermine more-competent overlying material and lead to rockfall.

Strongest	Quartzite
↓	Granite
	Quartz-Cemented Sandstone
	Basalt
	Limestone
	Calcite-Cemented Sandstone
	Schist
	Marble
	Shale/Mudstone
	Anhydrite
Weakest	Salt

FIGURE 2-39
General ranking of rock types according to strength (Davis 1984).

Minerals that compose a rock mass can cause special engineering problems, including rapid weathering, swelling, chemical attack of neighboring rocks, and very low friction properties on discontinuity surfaces (Goodman 1989). Table 2-4 provides a partial list of potentially problematic minerals.

4.1.2 Discontinuities

An in-depth discussion of important discontinuity attributes is included in Chapter 6. As noted in Section 2 of this chapter, rock mass discontinuities control the stability of rock slopes in most cases, and their existence within a rock mass is usually a prerequisite for rockfall. External events or circumstances often trigger the actual failure by changing the forces on the rock, but the type of failure is characteristically a function of the discontinuities that are present for structurally controlled rockfall. Rockfall susceptibility is thus influenced by prior geological events that determine the number of discontinuities and their orientations and properties.

Badger (2002) studied the effect of anticlinal folding on rock slope stability. He found that folding causes both bedding parallel shear stresses and tensional stresses radial to the fold axis. A discontinuity set is formed coincident with bedding and another is formed perpendicular to bedding and strikes parallel to the fold axis. Slopes that strike parallel to the fold axis may experience planar and toppling failures along these discontinuities. Coe and Harp (2007) studied rockfall susceptibility of folded strata in the Sevier fold–thrust belt in American Fork Canyon, Utah. They found the greatest rockfall susceptibility to be in fold hinge zones, and this susceptibility increased as the curvature became tighter. The rockfall susceptibility on fold limbs increased as the steepness of the dip of the beds increased. Coe and Harp demonstrated that rockfall susceptibility is sufficiently influenced by folding in these situations; they recommended that investigators include fold positions and

Table 2-4
Partial List of Minerals That May Cause Special Problems in Rock Cuts

POTENTIAL INSTABILITY	MINERALS
Soluble minerals	Calcite, dolomite, gypsum, anhydrite, salts, and zeolite
Unstable minerals	Marcasite and pyrrhotite
Potentially unstable minerals	Nontronite (iron-rich montmorillonite), nepheline, leucite, micas rich in iron
Weathering of minerals releases sulfuric acid	Pyrite, pyrrhotite, and other sulfides
Minerals with low friction coefficients	Clays (especially montmorillonites), talc, chlorite, serpentine, micas, graphite, and molybdenite
Potentially swelling minerals	Montmorillonites, anhydrite, vermiculite
Minerals that react or interfere with portland cement	Opal, volcanic glass, some cherts, gypsum, zeolite, and mica

SOURCE: Goodman 1989.

geometries as important variables in rockfall hazard assessments in similar locations.

4.1.3 Groundwater

Groundwater has a considerable impact on the stability of rock slopes (Hoek and Bray 1981). An understanding of the hydrogeology of a site, including seepage and pore-pressure distribution, is vital to a rock slope stability assessment. A wide range of hydrostatic pressures can be generated in a slope due to fluctuating groundwater levels, which can be caused by external influences such as seasonal rainfall, runoff, snowmelt or ice on the slope face, and seepage from surface-water bodies. Water-saturated discontinuities apply hydrostatic forces to rock blocks. Thus, measurement of water pressure may be essential during site investigations for establishing the stability of rock slopes. Observations of seepage rates at the slope face may not be adequate, as these rates may be lower than the evaporation rate, making the slope appear to be dry, while there may still be significant water pressure within the rock mass. In terms of the slope stability in these materials, the presence of a small volume of water at high pressure trapped within the rock mass is more important than a large volume of water discharging from a free-draining aquifer (Hoek and Bray 1981).

Figure 2-40 provides an illustration of the influence of changes in water pressure within rock mass discontinuities on slope stability. It is based on a slope stability modeling effort undertaken by Bergamin et al. (2004) and discussed in Chapter 9. Bergamin et al. analyzed the stability of a partially overhanging slope along a highway at the entrance to the Maderaner valley in Switzerland. The unstable rock slope extends about 30 m (98 ft) along the

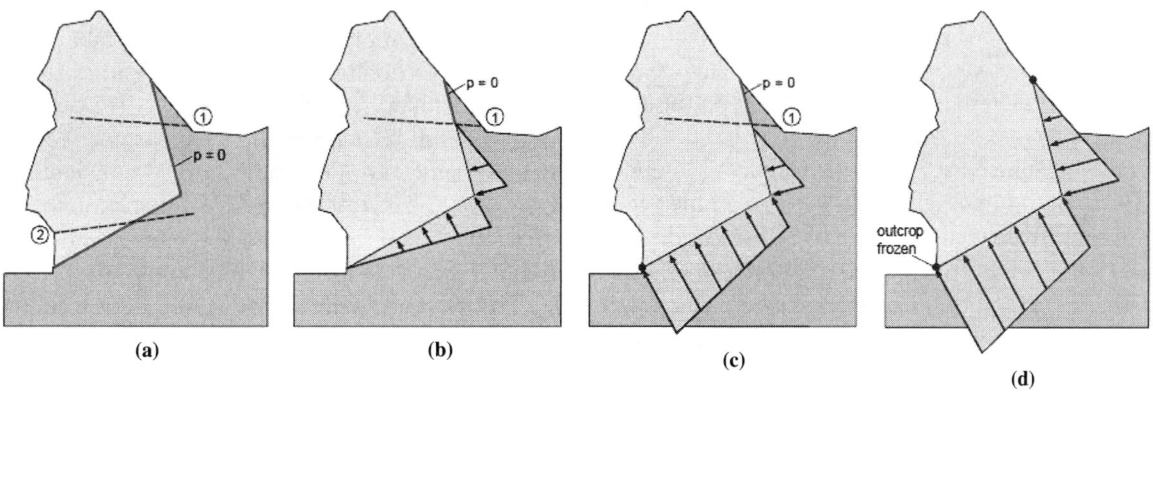

FIGURE 2-40
Four examples for assumed water-pressure distributions: (*a*) completely drained by boreholes from top (1) and from road (2), (*b*) partially drained by boreholes from top (1) with linear pressure reduction along slip surface, (*c*) partially drained as in Image *b* but with completely frozen outcrop line, and (*d*) undrained with completely frozen outcrop line (Bergamin et al. 2004).

road and rises 8 to 22 m (26 to 72 ft) above it. Behind it is a small topographic low point where water can collect. The gneissic rock mass is cut by a complex system of discontinuities, including three different cleavage planes, two sets of joints, a fault, and various shear surfaces.

Figure 2-40 shows four assumed alternative drainage conditions that were considered during these stability analyses. The modeling evaluated stability along an assumed basal failure plane that is exposed along the highway under the overhanging structure. Figure 2-40*a* shows a completely drained condition; two sets of boreholes are used to drain water from the discontinuities within the rock, while additional drainage of the basal failure plane at the toe of the slope along the highway occurs. In this case, there is no water pressure acting to reduce normal pressures along the potential basal failure plane, so maximum levels of shear strength from friction can be mobilized along this plane.

Figure 2-40*b* shows a partially drained condition; the upper portion of the rock mass is drained by boreholes from the top and from the road, but the lower portion has developed water pressures in nearly vertical discontinuities at the back of the rock mass and along the potential basal failure plane, as shown by the pressure diagram. Seepage is allowed to continue at the toe of the potential sliding block. The water pressure developed in the discontinuity at the back of the block (on the right) exerts a force that adds to the gravitational force attempting to drive the block down the potential basal failure plane. The water pressure along the potential basal failure plane also acts against the normal force of the block and further reduces the shear strength from friction that can be mobilized along this basal failure plane.

Figure 2-40*c* represents a slightly worse partially drained case because it is now assumed that seepage along the potential basal failure plane has been totally blocked by freezing, which may occur in cold climates. By removing the toe drainage, water pressures along the potential basal failure plane have increased. This action further reduces the normal force acting between the block and potential basal failure plane over the conditions in Figure 2-40*b*, thus reducing the frictional resistance to sliding that can be mobilized. However, the driving force from water in the discontinuities at the back of the rock mass has not increased over the case in Figure 2-40*b*, because the borehole drains continue to function.

Figure 2-40*d* illustrates the assumed worst-case groundwater conditions. The borehole drainage has been eliminated, while the drainage at the toe remains blocked by freezing. The driving force from the water pressure in the back bounding discontinuity is increased over the other scenarios due to blockage of drainage at the slope face and the lack of borehole drains in the upper slope. Increased water pressure along the potential basal failure plane further reduces the frictional resistance to sliding along the basal failure plane.

The natural drainage conditions considered in this example can be realistic in climates where freeze–thaw processes are common. Although completely blocked drainage of discontinuities by ice may last for short periods, it may not take long to recharge the system. Soft rocks, such as mudstones and shale, are subject to considerable reductions in cohesion and friction with changes in moisture content. Clay discontinuity fillings can also be influenced by moisture changes.

4.2 External Triggering Mechanisms

External rockfall mechanisms are considered as triggering mechanisms because typically they cannot cause a failure independently of internal conditions. The internal conditions provide a potential for rockfall that can be significantly increased by one or more external factors. Climatic conditions, weathering, and human activities are among the more common external triggers.

4.2.1 Climatic Conditions

Climatic conditions, such as rain, snow, and freeze–thaw, are common factors that trigger the instability of rock slopes. Introductory textbooks typically discuss the relationship of rock slope instability to climatologic conditions (Cruden and Varnes 1996; Wyllie and Norrish 1996). Terzaghi (1962) attributed rockfall "chiefly to the weakening effect of frost wedging and important seasonal temperature changes." Selby (1993) noted that, in mountainous areas, maximum rockfall frequency typically occurs during spring snowmelt and wet autumn seasons.

The natural processes of precipitation, runoff, wind, freeze–thaw, devegetation, and chemical weathering drive the climatic process of erosion.

Erosion causes rockfall by the removal of supporting materials from the rock mass and may be involved in transporting rockfall debris. Differential erosion of fine-grained matrix materials can lead to rockfall in glacial till or talus slopes. Erosion of exposed surfaces progressively subjects more of the rock mass to water and frost penetration.

Rockfalls commonly occur during severe rainstorms when fine-grained supporting materials are washed away, leaving larger rocks unsupported. Channels may form as a result of insufficient drainage on cut-slope benches or in the natural ground above slopes, and channeled runoff can cause rockfall by undercutting the sides of the channel and removing material from the channel bed. Precipitation is also often linked to the recharge of groundwater within a slope. Dramatic spikes in pore pressures are not uncommon in rock slopes during and shortly after precipitation events, especially when high-permeability fracture flow dominates.

Snowmelt, by the same mechanism as rainfall and channeled runoff, may cause rockfall by removing smaller particles that support larger rocks and by elevating pore pressures within a rock mass. However, a blanket of snow may help to insulate underlying material and moderate temperature fluctuations during a portion of the year.

Wind may exacerbate rockfall potential by plucking small particles that support larger particles. Wind may be a significant trigger in dry climates, particularly on steep slopes and on the oversteepened soil zones that typically occur at the top of rock cuts. Wind throw of trees on a rock slope can cause trees to blow over or can cause leveraging movements in the root zone, thus loosening the material, which may trigger rockfalls. Wind tends to be funneled through canyons, which, in mountainous terrain, are often the only available routes for highways. Steep cuts in these locations may be susceptible to wind-related rockfall, because those with south-facing slopes can become very dry during the summer, which enables wind to erode fine-grained particles more easily.

Freeze–thaw cycles are frequently cited as a major trigger of rockfall. Water from melting snow enters rock discontinuities and, if the temperature drops below freezing, the water freezes and expands. Since water undergoes an approximately 9% volume increase when it freezes, the expansion can create large pressures in a confined space, resulting in widening of cracks and joints and movement of blocks away from the slope face.

The critical times for freeze–thaw action are usually late fall and early winter and late winter and early spring (Peckover 1975; Wyllie and Norrish 1996). This correlation has been found to be important in predicting rockfall severity in Colorado, where the highest frequency of rockfall occurs during months of maximum freeze–thaw activity (Yeh and Associates Inc. 2002). Figure 6-26 in Chapter 6 summarizes the Colorado data. The data also show that the period of May through July has the second highest rockfall frequency, primarily related to intense rainfall during thunderstorms.

Freeze–thaw cycles vary with location. McCauley et al. (1985) noted that in some areas the temperature may not vary enough or the time span is insufficient to complete the freeze–thaw cycle. Douglas (1980) and Douglas et al. (1991) found that peak rockfall yields from basalt ledges in Northern Ireland occurred from February through March and November through December, during the time of maximum freeze–thaw cycles. Matsuoka and Sakai (1999) studied the cirque wall retreat in the Japanese Alps. The cirque walls remained frozen all winter and experienced a seasonal thawing rather than numerous freeze–thaw cycles. Matsuoka and Sakai found that maximum seasonal rockfall was a result of seasonal thawing to a depth of about 1 m (3 ft). A study of a west-facing rock slope in the Swiss Alps, reported by Stoffel et al. (2005), found that 88% of rockfalls occur between October and the end of May during freeze–thaw cycles. However, the highest activity is in April and May, when ice lenses formed from meltwater slowly disappear in the joints and fissures of rocks.

Krautblatter and Moser (2009) conducted a detailed study of rockfall in the German Alps, where they collected rockfall in nets at the base of steep slopes between 1999 and 2003. They found that rockfall increased up to 218 times during wet freeze–thaw cycles compared with dry-weather cycles. They also identified what they described as secondary rockfall events, which are defined as fine-grained rockfall material that originates from accumulated loose rock material on small ledges or outcrops within the rock wall and is released by fluvial processes during heavy runoff. Krautblatter and Moser attribute these events to numerous injuries in the study area over the last century.

4.2.2 Weathering

Weathering is caused by atmospheric conditions and ultimately reduces the strength of rock by degrading minerals to secondary products, such as feldspar being converted to kaolinite, or by breaking up rock into smaller particles. In this way, weathering can decrease the stability of an existing slope and lead to rockfall. Weaker materials are more sensitive to weathering processes. Weathering in layered sequences, such as shale interbedded with more competent rock types, can lead to undercutting of the stronger beds and ultimately turn the more competent layers into a source of rockfall.

Weathering of rock takes the form of both *disintegration* (physical-mechanical weathering) and *decomposition* (chemical weathering). Disintegration is the result of conditions such as wetting–drying, freezing–thawing, and heating. Disintegration is most prevalent in sedimentary rocks such as sandstone and shale, particularly if they contain swelling clay, and in metamorphic rock with a high mica content. Montmorillonite, illite, and kaolinite are examples of minerals that have swell potential and can occur in the intact rock or, less commonly, as infilling in discontinuities. When these materials are freshly exposed to the atmosphere, or moisture is increased or decreased, they swell or shrink. Swelling exerts high pressures that can cause differential movement in the rock mass and slaking. Shrinking can cause cracking, resulting in slaking of the rock mass. Decomposition refers to changes in rocks produced by chemical actions as a result of oxidation, hydration, and carbonation.

In North America, there is a substantial gradient to the depth of weathering from north to south (Brawner 1993), increasing from temperate climates to tropical climates. Recently glaciated areas tend to have shallow depths of weathering in bedrock. Oxidation due to fluctuating water conditions can result in expansion of minerals. Groundwater can also cause dissolution of various rocks and minerals, such as limestone, dolomite, gypsum, calcite, and rock salt.

Weathering and the associated processes that transport weathered materials from a rock slope cause slopes to deteriorate over time. Nicholson (2004) described the development of a three-stage approach to the assessment of rock slope conditions, called *rock slope deterioration assessment,* which is based on evaluations of 210 rock slopes in quarries and along highways in the United Kingdom. Nicholson (2004) defined deterioration as:

> shallow, progressive, physical and chemical alteration of rock material and its subsequent detachment and removal or redistribution by transporting agents. It is usually small in scale relative to more deep-seated failures but it is difficult to specify precisely the size of event involved.

Figure 2-41 summarizes the nine indicators of the deterioration process that Nicholson suggested may be readily observed on a rock slope face. They include confirmation of weathering processes that produce large-scale evidence, such as rebound fracturing, production of perched blocks, scaling and flaking of rock surfaces, differential erosion of weaker strata, and widening of fractures. They also include features that may require detailed examination of the rock mass, such as mineral alteration, granular disintegration, and blast-induced fractures.

Nicholson described deterioration characteristics according to seven "primary rock mass types" (Figure 2-42). They reflect important engineering properties such as strength (weak versus strong), variations in composition, and structure (whether the rocks contain layers, or are fissile, or are blocky). Nicholson also recognized that some conditions, such as intensely fractured zones, soluble rocks, and interbedded sequences of rocks with highly variable resistances to weathering and erosion, posed additional deterioration concerns. They are defined as *subsidiary rock mass types* (Figure 2-42).

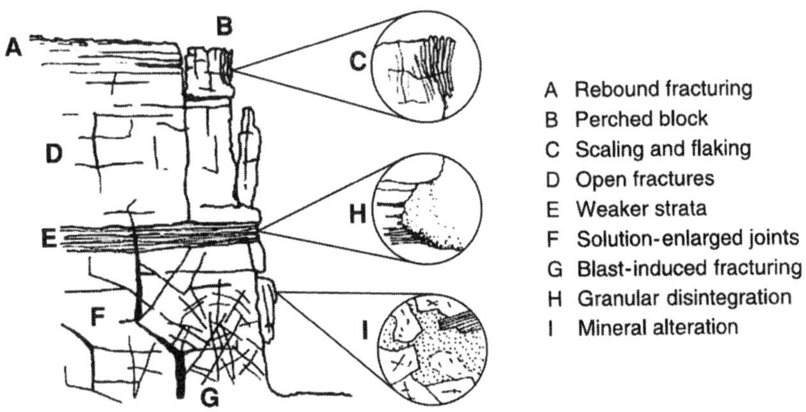

FIGURE 2-41 Indicators of deterioration process (adapted from Nicholson 2004).

Deterioration transport mechanisms		Description and deterioration characteristics
PRIMARY ROCK MASS TYPES		
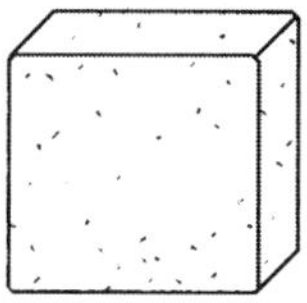	**WEAK MASSIVE** Grain raveling, grainfall, wash erosion, contour scaling (minor: *stonefall*)	Weak rock masses with no dominant structure or very thickly bedded. Slopes prone to erosion by surface runoff and to penetrative material weathering.
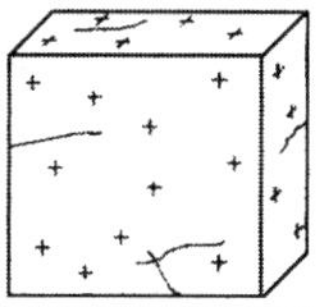	**STRONG MASSIVE** Wash erosion, contour scaling, stonefall, blockfall (minor: *stone raveling, flaking, slabfall, rockfall*)	Strong rock masses with no dominant structure or a very wide fracture spacing of 1 to 3 m (Deere 1963). Slopes prone to solution, wash erosion, scaling, and fragmentation from root wedging. Rebound fracturing in deep excavations.
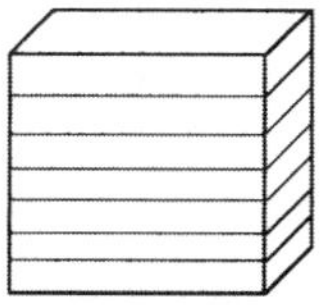	**LAYERED** Stonefall, stone raveling, wash erosion, blockfall (minor: *grain raveling, flaking, solution, contour scaling, rockfall*)	Repeated lithological or structural layering (any angle). Can be very stable, but prone to erosion, flaking, and collapse along chutes where folded or vertically layered.
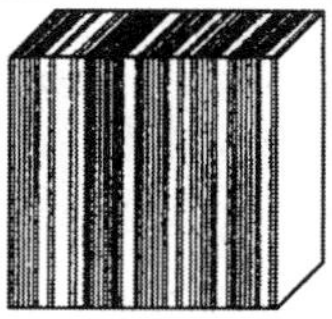	**FISSILE LAYERED** Flaking, grain raveling, wash erosion (minor: *stone raveling, stonefall, rockfall*)	Very thinly layered rock due to thin bedding, schistosity, cleavage, or lamination. Rapid breakdown, debris production, and slope regression common.
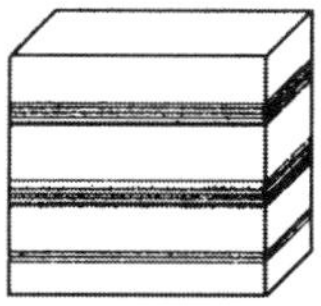	**COMPOSITE LAYERED** Stonefall, stone raveling (minor: *block raveling, flaking, wash erosion, contour scaling, rockfall*)	Interlayered strata having contrasting material properties. Differential weathering leads to undermining, producing stonefall and occasional overhang collapse. Groundwater flow may concentrate in weaker layers.
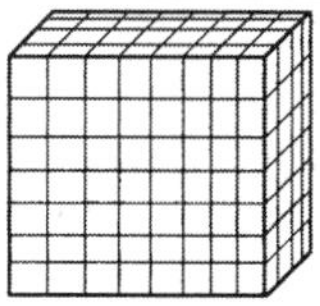	**REGULAR BLOCKY** Stone raveling, stonefall, blockfall, rockfall (minor: *grain raveling, flaking, wash erosion, contour scaling*)	Orthogonal blocky structure (or rubbly in chalk and oolitic limestone) due to intense and regular intersection of three or more sets of fractures. Deterioration relates to rock mass properties. Blocky rock masses experience high-frequency falls.
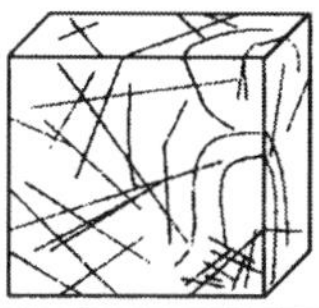	**IRREGULAR BLOCKY** Stone raveling, stonefall, rockfall (minor: *grain raveling, flaking, wash erosion, blockfall*)	Irregularly shaped and variably sized blocks due to nonpatterned intersection of fractures. Deterioration involves large blocks or large volumes of material. Rarely, small-scale debris flows in weathered regolith.

SUBSIDIARY ROCK MASS TYPES

INTENSELY FRACTURED ZONES	**SOLUBLE ROCK MASSES**	**COMPOSITE STRUCTURE**
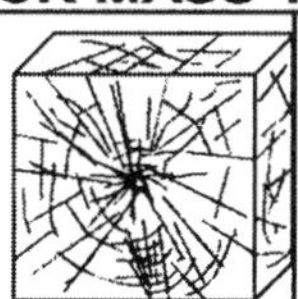	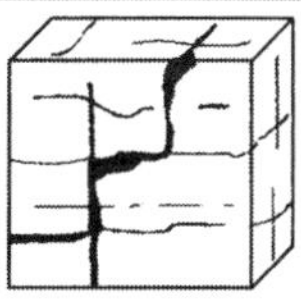	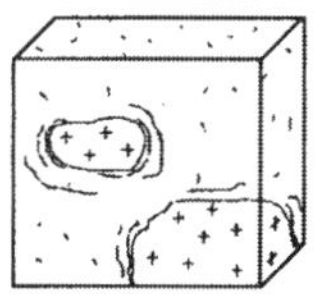
Intense, localized fracturing and shattering, producing loose structure. Vulnerable to root growth. Intense raveling, rockfall, and rare debris flows.	Dissolution of rock leading to fracture enlargement, microsolution, and karstification. Collapse of solution cavities leads to rockfall and blockfall.	Contrasting rock mass structures in close proximity, in a nonuniform distribution. Vulnerable to differential weathering and overhang collapse.

FIGURE 2-42
Rock mass deterioration characteristics (adapted from Nicholson 2004).

Nicholson identified deterioration transport mechanisms as critically important for determining appropriate slope treatment and maintenance measures. These mechanisms are classified primarily by frequency of occurrence and then according to constituent material size and relative velocity of movement (Figure 2-43). Frequency of occurrence is defined by three classes: *semicontinuous* (progressive erosion or mass movement from a given location), *sporadic* (erosion or mass movement at cyclic or other recurring intervals), and *isolated* (infrequent erosion or mass movement events). Semicontinuous erosion and mass movement processes include raveling, flaking, wash erosion, solution, karstification, and flexural toppling. Sporadic processes include various types of falls (subdivided according to material size) and toppling. Isolated processes include small-scale, weathering-related rockfalls, debris flows, and rockslides.

4.2.3 Anthropogenic Causes

A wide range of human activities may result in rockfall. The most common activities are blasting and excavation practices, stress relief, vibration, water-table drawdown, and deforestation. Rockfall hazards resulting from these factors potentially can be better controlled than those arising from natural conditions.

Damage of the rock face from blasting involves the fragmentation of rock primarily by gas energy, and by shock energy to a lesser extent, generated during detonation of the explosive (Konya and Walter 1991). Controlled blasting techniques and sequentially timed blasts are an important way to limit damage to the excavated slope (Hoek and Bray 1981). Section 6 of Chapter 13 provides additional details about blasting applications to rock slopes. The amount of resulting damage induced depends on the stress and strain induced in the rock and the gas pressure generated by the burning of the explosive. Along North American highways, many rockfall problems occur on slopes that were excavated through the use of poor blasting techniques that preceded the advent of modern controlled blasting procedures (Brawner 1993). Blast damage can extend possibly 5 to 10 m (16 to 33 ft) behind the face (Wyllie and Mah 2004), which can cause many problems, including overbreak, extensive shattering of the rock mass, development of tension cracks, and pathways for groundwater recharge. Gases from blasts can migrate along preexisting discontinuities or new cracks well behind the rock face, enlarging them. This process can permit excessive water infiltration, allow excess pore-water pressure, and increase the frost susceptibility and weathering potential of the slope face (Brawner 1993).

Excavations that expose adversely oriented discontinuities, or that are overly steep in low-strength rock masses, are frequent sources of rockfall. Another frequent cause of rockfall is excavating and oversteepening the toe of a slope, such as during routine maintenance of ditches. Surcharging the top of the slope (i.e., stockpiled material) may also lead to slope failure.

Vibrations from blasting and other sources can induce rockfall in highly jointed rock masses. Wyllie and Mah (2004) discussed blasting design and vibration control for rock slopes. Some sources of vibrations other than blasting include trains, heavy trucks, construction equipment, and pile driving. They can be analyzed by monitoring peak particle velocities to estimate their impact on the rock slope. The vibration frequency plays a role in how large a block can be mobilized, as higher frequencies affect a smaller portion of a discontinuity surface at a given time. In addition, the decay of blast energy with distance reduces far-field effects. In other words, blasting vibrations are more likely to result in smaller rockfall events nearer the blast site.

Reservoirs and lakes present a unique situation in which confining pressure on a rock slope can be attributed to water. In addition, this surface water recharges the groundwater by filling the discontinuities and pore spaces in the surrounding rock. When the surface water level drops, the groundwater pressure may result in discharges back into the reservoir or lake. If the water pressure in the rock mass discontinuities does not equilibrate as quickly as the confinement pressure declines, elevated groundwater pressures can occur within the rock mass and rockfall or slope failures can result. Contributions to the water table from leaking irrigation ditches, agricultural watering, underground utilities, and septic systems can increase pore-water pressure in rock masses and increase rockfall susceptibility.

Vegetation typically helps reduce erosion and may even stabilize slopes, although the effects are difficult to quantify because of the complex interactions of mechanics and hydrology. Detailed studies have shown that vegetation provides a

Semicontinuous Erosion and Mass Movement Processes

Raveling is the frequent and semi-continuous fall of material. Three size divisions are recognized: grain (<20 mm), stone (20 to 200 mm) and block (>200 mm). *Grain raveling* occurs in weak, granular, and soil-like materials and can be transitional with wash erosion. *Stone raveling* occurs in moderately strong, intensely fractured rock masses. *Block raveling* occurs in strong, fractured rock masses such as layered sandstone or blocky limestone.

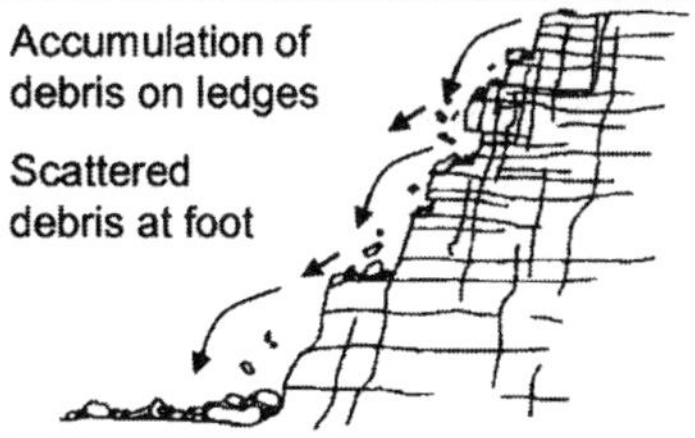

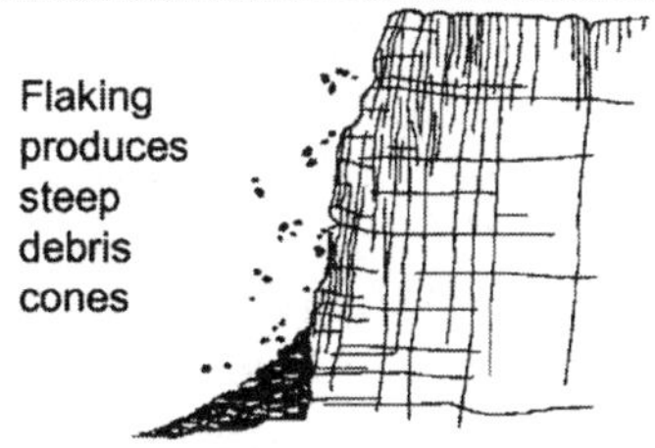

Flaking is a form of raveling involving the frequent and semi-continuous fall of material with a distinctive platy form, and occurs predominantly in fissile rocks such as shales and slates. Flaking can rapidly produce steep, extensive debris cones at the slope foot.

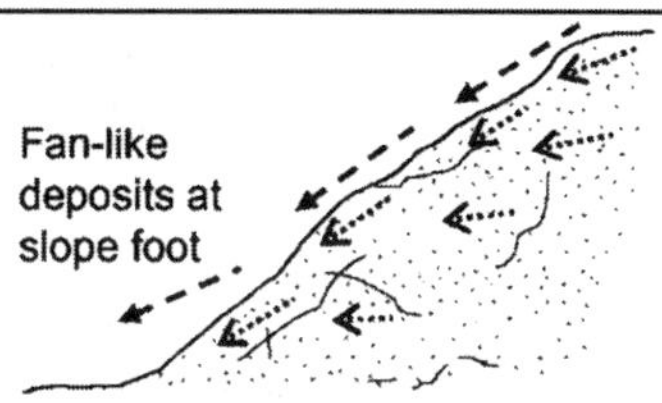

Wash erosion involves the detachment and transport of fine material entrained in surface water runoff and occurs in weak, soil-like rock masses.

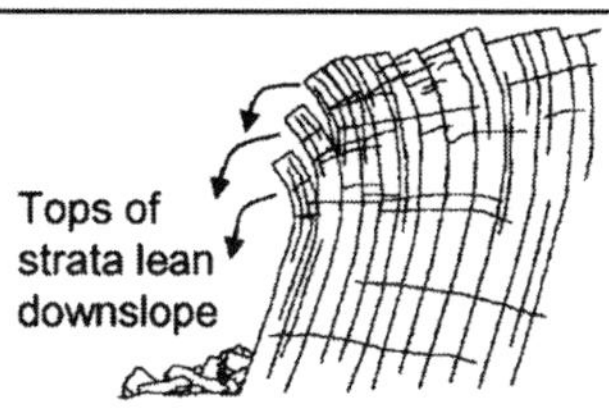

Flexural toppling is a slow, progressive deformation and sliding of layered strata due to gravity upon removal of lateral constraint. Occurs in steeply dipping, fissile, thinly bedded, and composite rock masses.

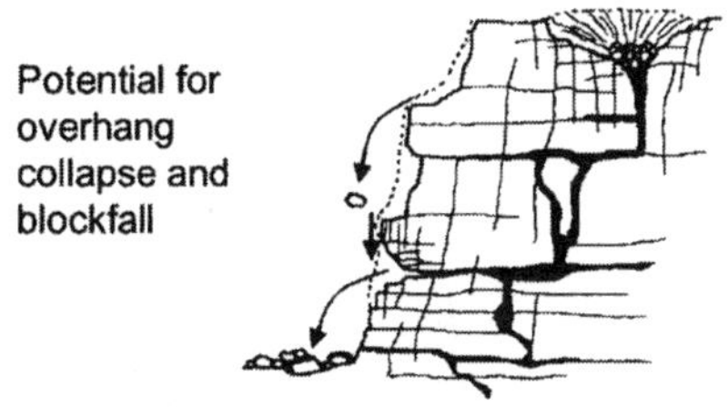

Solution and karstification involve the dissolution of soluble mineral grains and cementing material in aggressive, acid solutions, including rainwater. Solution can develop karst forms in some soluble rocks.

Sporadic Processes

Fall describes the occasional release of individual rock fragments. Three subtypes are recognized: Grainfall (weak massive rock masses with a granular texture); stonefall (stronger rock masses) and blockfall (strong, blocky rock masses).

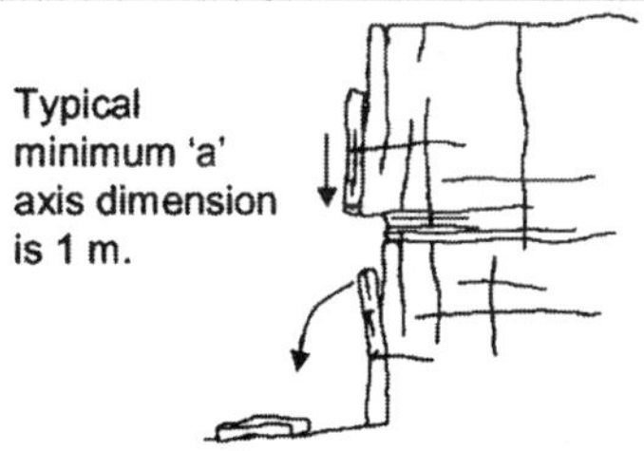

Slabfall and toppling are also forms of fall involving isolated and infrequent freefall of large, tabular slabs and rotation of large prismatic blocks. These occur in strong rock masses with discontinuities parallel to a steep slope plane.

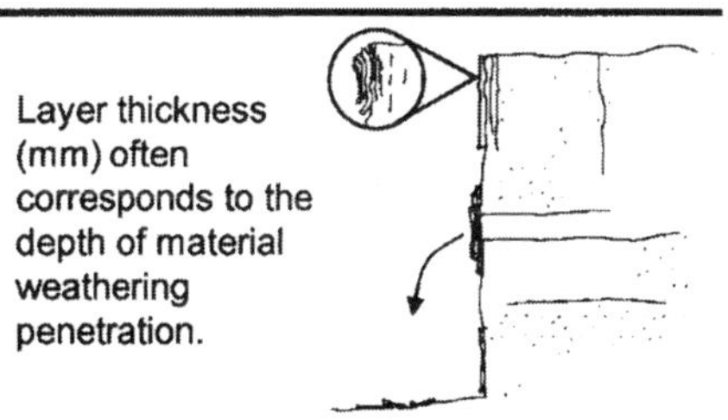

Contour scaling is a special form of fall involving the occasional exfoliation of thin layers of rock material formed parallel to the slope surface. Contour scaling occurs most commonly in moderately strong, massive rocks.

Isolated Processes

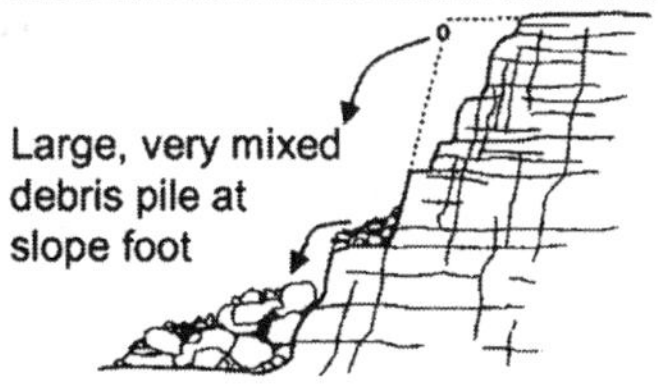

Rockfall is used here as a specific term to describe the fall of many blocks of varying sizes in a single identifiable event and involves slide, roll, bounce, and freefall (including overhang collapse). In the context of deterioration, rockfalls are infrequent, shallow, small in scale, and involve the movement of weathered debris.

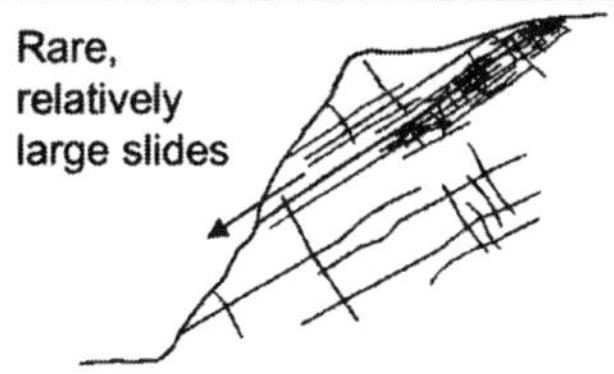

Rockslides are rapid translational movements of intact rock along a distinct, planar discontinuity. These are not strictly deterioration transport mechanisms but are included for completeness and because small rockslides occur that are fundamentally produced and triggered by weathering processes.

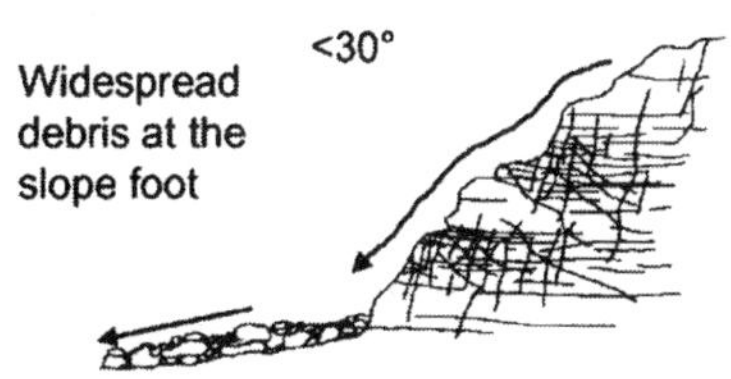

Debris flow is the rapid transport of a mixture of coarse and fine particles in a partially saturated, grain-supported flow and may involve initial sliding. Deterioration-related debris flows are rare, shallow, small in scale, and involve the movement of weathered debris.

FIGURE 2-43
Deterioration transport mechanisms (adapted from Nicholson 2004).

net improvement in slope stability, primarily due to root reinforcement in soil slopes (Gray 1978; Gray and Leiser 1982; Gray and Megahan 1981; Sotir and Gray 1989). Experience has shown that removal of the root support by devegetation, deforestation, careless clearing, and logging practices can contribute to rockfall. Wildfires (which may not be anthropogenic) can contribute to rockfall with falling and burning trees dislodging precarious blocks in the short term and with increased surface runoff and erosion in the longer term. In the case of wildfires, high burn intensity may remove vegetation that at one time tended to catch rockfall before it could reach a road or structure. A spike in rockfall occurrence has been widely observed during and immediately after wildfires, such as the example discussed in Chapter 17.

Roots of large trees growing in discontinuities in rock slopes can increase joint aperture over time by wedging blocks of rock apart. The increase in joint aperture decreases the frictional resistance along discontinuity surfaces and increases the susceptibility of the rock mass to greater frost penetration and water infiltration. However, forested slopes can reduce the potential for rockfalls to reach the base of a slope. Perret et al. (2006) found that an uneven-aged, multilayered stand of forest increases the benefits of a forested slope.

Burrowing animals have been known to trigger rockfalls by simply loosening the soil on steep slopes and undermining blocks. Loosening of the soil also facilitates erosion by wind and water. The movement of humans and animals walking on a slope can also initiate rockfalls.

4.2.4 Earthquakes

Strong ground motion from earthquakes can produce long-period vibrations with long durations that often trigger rockfall. Keefer (1984) identified five types of earthquake-induced landslides that occur on rock slopes; more than 90% of these failures are rockfalls and rockslides. Keefer (1993) developed a method for assessing the seismic stability of rock slopes that is useful for transportation agencies, since it can be applied on a regional scale with data from maps, reports, aerial photographs, and reconnaissance-level field work. Harp and Noble (1993) modified the rock mass quality designation (Q-rating system), originally established for tunnel design, for use in rating the susceptibility of rock slopes to seismically induced failure. This method was further developed by Harp and Jibson (2002). The method may be applied on a regional scale suitable for transportation studies and is discussed in Chapter 6 as an approach for evaluating and comparing potential rockfall source zones.

Thresholds of shaking for rock slope failures are discussed by Jibson and Harp (1995) and Harp and Wilson (1995). The California Department of Transportation considers earthquakes of magnitude greater than 5 to be capable of triggering rockfall. A study in Israel on rockfall hazards along a railway found that local magnitude greater than 6 could trigger rockfall (Katz et al. 2011).

4.2.5 Stress Relief

Stress relief displacements can change with time when the loads or pressures on the rock change (e.g., due to the flow of water), the geometry of the loaded or excavated region changes (e.g., by further excavation), the deformability properties of the rock change (e.g., by weathering or hydration), or the rock responds slowly to changes in stress or strain (Goodman 1989). In Figure 2-44, immediate strain is followed by primary creep in which strain occurs at a decreasing rate with time. In some rocks, the primary creep curve approaches a steady rate of creep, termed *secondary creep*. In specimens stressed near peak strength, secondary creep may turn upward into tertiary creep, lead-

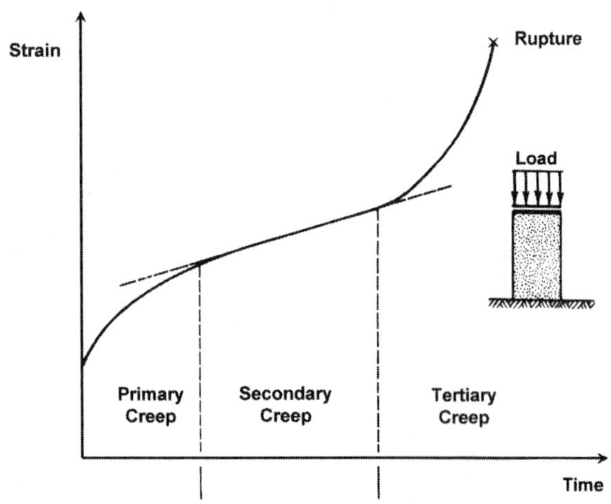

FIGURE 2-44
General form of creep curve for rock (adapted from Goodman 1989).

ing to a strain rate increase that results in failure (creep rupture) (Goodman 1989).

When a slope is first excavated or exposed, there is a period of initial response as a result of elastic rebound, relaxation, and dilation of the rock mass due to changes in stress induced by the excavation (Zavodni 2000). In contrast to the rapid excavation, and the relatively fast movements that take place in open pit mines, mountain slopes may creep over periods of hundreds of years. Long-term creep, such as a toppling failure, may occur when there is no defined failure surface or when the change in slope geometry is very slow—for example, due to stress relief after glacial retreat or erosion of the slope toe by a river.

The cumulative strain of such slopes can be considerable (Wyllie and Mah 2004). Therefore, when a new free face is created on a rock, there will be movement toward a new state of equilibrium. The movement will be in two phases: an immediate elastic rebound that is largely volumetric and a time-dependent further relaxation. This movement will occur first in the actual new surface and will consist of expanding, moving, and rotation of grains. As the outer layer relaxes, the next layer has less restraint and it will relax somewhat but not as much as the first layer. As relaxation progresses, unless the grains are completely decoupled, a strain gradient is established from the outside into a region where the original strain condition is approached (Emery 1966). This strain may result in strength loss to the rock mass and instability in the form of rockfall (Brawner 1993).

REFERENCES

Badger, T. C. 2002. Fracturing Within Anticlines and Its Kinematic Control on Slope Stability. *Environmental and Engineering Geoscience*, Vol. VIII, No. 1, pp. 19–33.

Bergamin, S., J. Kolberg, and P. Fritz. 2004. Stability Analysis of a Rock Slope on the Bristen Road Using AutoBlock. *Proc., ISRM International Symposium, 3rd ARMS*, Mill Press, Rotterdam, Netherlands, pp. 127–132.

Brawner, C. O. 1993. *Rockfall Hazard Mitigation Methods*. FHWA Publication No. FHWA SA-93-085. Federal Highway Administration, U.S. Department of Transportation, Washington, D.C.

Coe, J. A., and E. L. Harp. 2007. Influence of Tectonic Folding on Rockfall Susceptibility, American Fork Canyon, Utah, USA. *Natural Hazards & Earth Systems Science*, Vol. 7, pp. 1–14.

Cruden, D. M., and D. J. Varnes. 1996. Landslide Types and Processes. In *Special Report 247: Landslides: Investigation and Mitigation* (A. K. Turner and R. L. Schuster, eds.), Transportation Research Board, National Research Council, Washington, D.C., pp. 36–75.

Davis, G. H. 1984. *Structural Geology of Rocks and Regions*. John Wiley & Sons, New York.

Deere, D. U. 1963. Technical Description of Rock Cores for Engineering Purposes. *Felsmechanik und Ingenieurgeologie*, Vol. 1, No. 1, pp. 16–22.

Deere, D. U., and R. P. Miller. 1966. *Engineering Classification and Index Properties of Intact Rock*. Report AFWL-TR-65-116. Air Force Weapons Laboratory, Kirkland Air Force Base, N. Mex.

Douglas, G. R. 1980. Magnitude Frequency Study of Rockfall in County Antrim, Northern Ireland. *Earth Surface Processes*, Vol. 5, No. 2, pp. 123–129.

Douglas, G. R., W. B. Whalley, and J. P. McGreevy. 1991. Rock Properties as Controls on Free-face Debris Fall Activity. *Permafrost and Periglacial Processes*, Vol. 2, No. 4, pp. 311–319.

Emery, C. L. 1966. The Strain in Rocks in Relation to Highway Design. In *Highway Research Record 135*, Highway Research Board, National Research Council, Washington, D.C., pp. 1–9.

Fowler, B. K. 2005. Stability and Collapse, Old Man of the Mountains, Franconia Notch, New Hampshire. *Environmental and Engineering Geoscience*, Vol. 11, No. 1, pp. 17–27.

Goodman, R. E. 1976. *Methods of Geological Engineering in Discontinuous Rocks*. West Publishing Co., St. Paul, Minn.

Goodman, R. E. 1989. *Introduction to Rock Mechanics*, 2nd ed. John Wiley & Sons, New York.

Goodman, R. E. 1993. *Engineering Geology: Rock in Engineering Construction*. John Wiley & Sons, New York.

Goodman, R. E., and J. W. Bray. 1976. Toppling of Rock Slopes. *Proc., Specialty Conference on Rock Engineering for Foundations and Slopes*, Aug. 15–18, 1976, Boulder, Colo., American Society of Civil Engineers, New York, pp. 201–231.

Goodman, R. E., and S. D. Kieffer. 2000. Behavior of Rock in Slopes. *Journal of Geotechnical and Geoenvironmental Engineering*, Vol. 126, No. 8, pp. 675–684.

Goodman, R. E., and G. H. Shi. 1985. *Block Theory and Its Application to Rock Engineering*. Prentice Hall, Englewood Cliffs, N.J.

Gray, D. H. 1978. Role of Woody Vegetation in Reinforcing Soils and Stabilizing Slopes. *Symposium on Soil Reinforcing and Stabilising Techniques in Engineering Practice*, Oct. 16–19, 1978, Sydney, Australia, Transport and Road Research Laboratory, Crowthorne, Berkshire, United Kingdom, pp. 253–306.

Gray, D. H., and A. T. Leiser. 1982. *Biotechnical Slope Protection and Erosion Control*. Van Nostrand Reinhold, New York.

Gray, D. H., and W. F. Megahan. 1981. *Forest Vegetation Removal and Slope Stability in the Idaho Batholith*. Forest Service Research Paper INT-271. U.S. Department of Agriculture, pp. 11–12.

Guzzetti, F., P. Reichenbach, and G. F. Wieczorek. 2003. Rockfall Hazard and Risk Assessment in the Yosemite Valley, California, USA. *Natural Hazards & Earth Systems Science*, Vol. 3, No. 6, pp. 491–503.

Harp, E. L., and R. W. Jibson. 2002. Anomalous Concentrations of Seismically Triggered Rock Falls in Pacoima Canyon: Are They Caused by Highly Susceptible Slopes or Local Amplification of Seismic Shaking. *Bulletin of the Seismological Society of America*, Vol. 92, No. 8, pp. 3180–3189.

Harp, E. L., and M. A. Noble. 1993. An Engineering Rock Classification to Evaluate Seismic Rock Fall Susceptibility and Its Application to the Wasatch Front. *Bulletin of the Association of Engineering Geologists*, Vol. 30, No. 3, pp. 293–319.

Harp, E. L., and R. C. Wilson. 1995. Shaking Intensity Thresholds for Rock Falls and Slides: Evidence from the 1987 Whittier Narrows and Superstition Hills Earthquake Strong Motion Records. *Bulletin of the Seismological Society of America*, Vol. 85, No. 6, pp. 1739–1757.

Harp, E. L., M. E. Reid, J. W. Godt, J. V. DeGraff, and A. J. Gallegos. 2008. Ferguson Rock Slide Buries California State Highway near Yosemite National Park. *Landslides*, Vol. 5, pp. 331–337.

Hoek, E. 2007. *Practical Rock Engineering*. Rocscience Inc., Toronto, Ontario, Canada. http://www.rocscience.com/education/hoeks_corner.

Hoek, E., and J. W. Bray. 1981. *Rock Slope Engineering*. Institution of Mining and Metallurgy, London.

Hsü, K. J. 1975. Catastrophic Debris Streams (Stürzstroms) Generated by Rock-Fall. *Geological Society of America Bulletin*, Vol. 86, pp. 129–140.

Hungr, O., and S. G. Evans. 1988. Engineering Evaluation of Fragmental Rockfall Hazards. *Proc.*, *5th International Symposium on Landslides*, July 10–15, 1988, Lausanne, Switzerland (C. Bonnard, ed.), Vol. 1, A. A. Balkema, Leiden, Netherlands, pp. 685–690.

Hungr, O., S. G. Evans, and J. Hazzard. 1999. Magnitude and Frequency of Rock Falls and Rock Slides Along the Main Transportation Corridors of Southwestern British Columbia. *Canadian Geotechnical Journal*, Vol. 36, pp. 224–238.

Jibson, R. W., and E. L. Harp. 1995. *Inventory of Landslides Triggered by the 1994 Northridge, California Earthquake*. U.S. Geological Survey Open-File Report 95-213. U.S. Geological Survey, Reston, Va.

Katz, O., P. Reichenbach, and F. Guzzetti. 2011. Rock Fall Hazard Along the Railway Corridor to Jerusalem, Israel, in the Soreq and Refaim Valleys. *Natural Hazards*, Vol. 56, pp. 649–665.

Keefer, D. K. 1984. Landslides Caused by Earthquakes. *Geological Society of America Bulletin*, Vol. 95, No. 4, pp. 406–421.

Keefer, D. K. 1993. The Susceptibility of Rock Slopes to Earthquake-Induced Failure. *Technical Note*, *Bulletin of Association of Engineering Geologists*, Vol. 30, No. 3, pp. 353–361.

Kieffer, S. D. 1998. *Rock Slumping—A Compound Failure Mode of Jointed Hard Rock Slopes*. Ph.D. dissertation. Department of Civil and Environmental Engineering, University of California, Berkeley.

Konya, C. J., and E. J. Walter. 1991. *Rock Blasting and Overbreak Control*. FHWA Report FHWA-HI-92-001. Federal Highway Administration, U.S. Department of Transportation, Washington, D.C.

Krautblatter, M., and M. Moser. 2009. A Nonlinear Model Coupling Rockfall and Rainfall Intensity Based on a Four Year Measurement in a High Alpine Rock Wall (Reintal, German Alps). *Natural Hazards & Earth Systems Science*, Vol. 9, pp. 1425–1432.

Matsuoka, N., and H. Sakai. 1999. Rockfall Activity from an Alpine Cliff During Thawing Periods. *Geomorphology*, Vol. 28, Nos. 3–4, pp. 309–328.

Mauldon, M., and R. E. Goodman. 1996. Vector Analysis of Keyblock Rotations. *Journal of Geotechnical Engineering*, Vol. 122, No. 12, pp. 976–987.

McCauley, M. L., B. W. Works, and S. A. Naramore. 1985. *Rockfall Mitigation*. FHWA Report FHWA/CA/TL-85/12. Federal Highway Administration, U.S. Department of Transportation, Washington, D.C.

Nicholson, D. T. 2004. Hazard Assessment for Progressive, Weathering-Related Breakdown of Excavated Rockslopes. *Quarterly Journal of*

Engineering Geology and Hydrogeology, Vol. 37, pp. 327–346.

Norrish, N. I., and D. C. Wyllie. 1996. Rock Slope Stability Analysis. In *Special Report 247: Landslides: Investigation and Mitigation* (A. K. Turner and R. L. Schuster, eds.), Transportation Research Board, National Research Council, Washington, D.C., pp. 391–428.

Pantelidis, L. 2009. Rock Slope Stability Assessment Through Rock Mass Classification Systems. *International Journal of Rock Mechanics & Mining Sciences*, Vol. 46, pp. 315–325.

Peckover, F. L. 1975. Treatment of Rock Falls on Railway Lines. *American Railway Engineering Association Bulletin 653*, pp. 471–503.

Perret, S., M. Baumgartner, and H. Kienholz. 2006. Inventory and Analysis of Tree Injuries in a Rockfall-Damaged Forest Stand. *European Journal of Forest Research*, Vol. 125, pp. 101–110.

Piteau, D. R., and F. L. Peckover. 1978. Engineering of Rock Slopes. In *Special Report 176: Landslides, Analysis and Control* (R. L. Schuster and R. J. Krizek, eds.), Transportation Research Board, National Research Council, Washington, D.C., pp. 192–228.

Selby, M. J. 1993. *Hillslope Materials and Processes*. Oxford University Press, New York.

Sotir, R. B., and D. H. Gray. 1989. Fill Slope Repair Using Soil Bioengineering Systems. *Proc., XX International Erosion Conference*, Vancouver, British Columbia, Canada, pp. 413–429.

Stoffel, M., I. Lievre, M. Monbaron, and S. Perret. 2005. Seasonal Timing of Rockfall Activity on a Forested Slope at Taschgufer (Swiss Alps)—A Dendrochronological Approach. *Zeitschrift für Geomorphologie N.F.*, Vol. 49, No. 1, pp. 89–106.

Terzaghi, K. 1962. Stability of Steep Slopes on Hard Unweathered Rock. *Geotechnique*, Vol. 12, No. 4, pp. 251–270.

Waltham, T. 2002. Unloading Joints and Rockfalls in Norway's Fiordlands. *Geology Today*, Vol. 18, No. 6, pp. 220–225.

Wittke, W. 1990. *Rock Mechanics: Theory and Applications with Case Histories*. Springer-Verlag, Berlin.

Wyllie, D. C., and C. W. Mah. 2004. *Rock Slope Engineering*, 4th ed. Spon Press, New York.

Wyllie, D. C., and F. J. Munn. 1978. Use of Movement Monitoring to Minimize Production Losses due to Pit Slope Failure. *Proc., 1st International Symposium on Stability in Coal Mining*, Miller Freeman Publications, Vancouver, British Columbia, Canada, pp. 75–94.

Wyllie, D. C., and N. I. Norrish. 1996. Stabilization of Rock Slopes. In *Special Report 247: Landslides: Investigation and Mitigation* (A. K. Turner and R. L. Schuster, eds.), Transportation Research Board, National Research Council, Washington, D.C., pp. 474–504.

Yeh and Associates, Inc. 2002. *Rockfall Manual of Practice, Vol. 1: Methods*. Consulting Report to the Colorado Department of Transportation, Denver.

Zavodni, Z. M. 2000. Time-Dependent Movements of Open Pit Slopes. In *Slope Stability in Surface Mining* (W. A. Hustrulid, M. K. McCarter, and D. J. A. Van Zyl, eds.), Society of Mining, Metallurgy and Exploration, Littleton, Colo., pp. 81–87.

Chapter 3

LAWRENCE A. PIERSON

ROCKFALL HAZARD RATING SYSTEMS

1. INTRODUCTION

In recognition of the seriousness of rockfall problems in mountainous areas and the difficulty of carrying out detailed investigations and analyses along extensive sections of rights-of-way, highway and railway transportation agencies have developed rockfall hazard rating systems. These systems, which are in wide use, are based on visual inspections and simple calculations. Rockfall hazard rating systems serve as tools that, together with commonsense engineering judgment, give reasonable assessments of the relative hazards due to rockfalls from cut slopes (Hoek 2000). The hazard rating systems are typically adopted to provide a relative ranking of potentially hazardous slopes, especially for agencies with a large stock of slopes, to set priorities and direct resources. The main implementation goal is to obtain consistent and comparable information for a series of locations so that those responsible for operating the transportation system can plan for safety improvements, optimize the benefits from investing limited safety budgets, and reduce maintenance and other operational costs. Rockfall hazard rating systems are receiving increasing attention by many transportation agencies in several countries (Wong 2005).

Transportation agencies need specific response policies and guidelines for mitigation actions for all natural and man-made hazards, such as rockfall. These documents define the planning process for developing rockfall mitigation projects. Some agencies rely solely on a reactionary response in the belief that such a response and lack of knowledge concerning possible hazards reduce their accountability and liability exposure. However, the public reaction to such a position can be negative. In addition, as the understanding of the causes of rockfall and the science of rockfall control advance, an "unawareness defense" likely provides diminishing protection during rockfall-related litigation.

A proactive approach to rockfall management and mitigation requires a consistent, long-term program under the direction of a team experienced in both the engineering design and the construction aspects of mitigation measures. Although the analytical and computing tools available today are more powerful than in the past, experience is still the best gauge of how effective a specific mitigation measure will be for a particular rockfall problem. Effective rockfall mitigation designs consider the operational parameters of the facility, the constructability of the measures, and the life-cycle costs of construction versus the costs associated with continued maintenance and monitoring of the problem area.

2. TERMINOLOGY

The hazard rating systems discussed here are qualitative and compare the likelihood or subjective probability that a hazardous condition exists at any given site. They are not based on a numerical analysis of the risk of an accident or the occurrence of a specific rockfall event, which can be evaluated by the probabilistic procedures proposed by Roberds (2005). Many terms have been used in discussions of hazard, risk, vulnerability, and so forth. Their usage has not been consistent among various authors, and the variation may lead to confusion even among rockfall practitioners. For clarity in this chapter and in the book, the following definitions, adapted from Fell et al. (2005), are used:

- *Consequence*. The outcome or result of a hazard being realized.
- *Danger (threat)*. A natural phenomenon that may lead to damage, described in terms of its geometric, mechanical, and other characteristics. The danger from rockfall may be an existing one (such as an actively rotating column) or a high and unacceptable potential for rockfall.
- *Hazard*. The likelihood that a particular danger (threat) exists or may occur.
- *Likelihood*. A qualitative description of probability or frequency of a given outcome based on an established set of criteria, data, assumptions, and information.
- *Risk*. A quantitative value derived from an analysis of the numerical values of probability, vulnerability, and consequences.
- *Vulnerability*. The degree of loss within the area affected by a hazard.

The distinction between risk and hazard should be clearly understood. In simple terms, *hazard* may be defined as a measure of the likelihood of rockfall occurrence, while rockfalls may have *consequences*, defined above as the outcome or result of a rockfall event. In conformance with this logic, *risk* is defined as a measure of likelihood and consequence of rockfall occurrence. Both hazard and consequence (the "losses") are characterized to arrive at an approximation of risk. For example, a rock cut with a high likelihood of producing rockfall (and thus high hazard) but low consequence (e.g., little traffic) might be identified as a cut of low to moderate risk. Figure 3-1 illustrates the relationship between hazard, consequence, and risk. Basically, risk is the product of consequence and hazard.

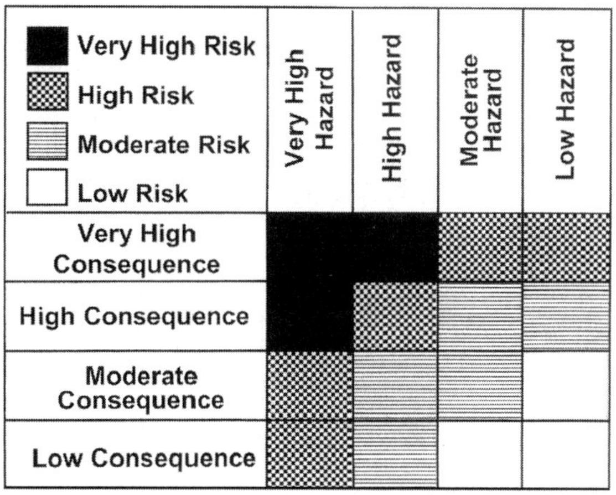

FIGURE 3-1 Relationship between hazard, consequence, and risk.

Fell et al. (2005) recommended that additional qualitative descriptors adapted from the Australian Geomechanics Society (2000) be used to clarify the level of likelihood and consequence. They are defined in Tables 3-1 and 3-2. The descriptors correspond to ranges within a continuous span of possible outcomes. Rockfall hazard rating systems follow a similar method in categorizing and then scoring the hazards associated with specific criteria related to the likelihood of a rockfall event and the potential consequences for the transportation system.

Once implemented, the rockfall hazard rating system provides qualitative site scores that serve as a powerful tool in managing rock slopes that produce rockfall. The qualitative site scores provide a cost-effective comparison of the relative hazard posed by various sites. The results, together with experienced judgment, allow sites to be rated and

Table 3-1
Qualitative Measures of Likelihood

Descriptor	Description
Almost certain	The event is expected to occur.
Likely	The event will probably occur under adverse conditions.
Possible	The event could occur under adverse conditions.
Unlikely	The event could occur under very adverse circumstances.
Rare	The event is conceivable but only under exceptional circumstances.
Not credible	The event is inconceivable.

Source: Australian Geomechanics Society 2000.

Table 3-2
Qualitative Measures of Consequences

Descriptor	Description
Catastrophic	Complete destruction or severe damage to a facility that will require major repair and stabilization. Injuries or fatalities are likely.
Major	Extensive damage to a facility requiring significant maintenance effort and stabilization work. Injuries or fatalities are possible.
Medium	Moderate damage to a facility requiring routine and unscheduled maintenance. Injuries are possible.
Minor	Limited damage to a facility. Activity is primarily related to routine maintenance. Injuries are unlikely.
Insignificant	Little damage. Injuries would be rare or inconceivable.

Source: Australian Geomechanics Society 2000.

prioritized so that mitigation planning can be accomplished without the need for a more elaborate and expensive quantitative analysis.

3. BENEFITS OF IMPLEMENTING A ROCKFALL RATING SYSTEM

A rockfall mitigation program is often justified for transportation systems that extend through mountainous areas where rockfalls occur and the cost of rockfall maintenance is high (Wyllie 1987; Wyllie and Norrish 1996). Because the majority of transportation systems have been in service for decades and the number of vehicles traveling on them has dramatically increased, the problems related to rockfall have become more severe and common. Rockfall activity from older slopes is often more frequent than rockfall activity from younger or fresher slopes, and it tends to increase over time unless some form of intervention is undertaken. The increase is due to weathering (loss of strength) along joint surfaces, other environmental influences such as rainfall and freeze–thaw cycles, and the effects of trees and root growth. Older slopes also have inherent problems related to the way they were constructed. Overly aggressive production blasting and ripping methods have been used to complete the excavation, but these techniques often have left the cut slopes more damaged than modern construction practices using controlled blasting. An example is shown in Figure 3-2. In addition, rock catchment areas were not often included at the base of these slopes, or they were not large enough to restrict fallen rocks from entering the roadway.

FIGURE 3-2 Example of how blasting methods can affect stability of rock cut. Cut is in strong rock, and photograph presents comparison between normal production blasting (right side) and blasting that uses presplitting as controlled blasting technique (left side) (Hoek 2000).

The implementation of a rock slope stabilization program can be daunting for an agency responsible for a large transportation system with hundreds and sometimes thousands of rock slopes with histories of rockfall. Nevertheless, such programs have been initiated by many agencies. Even though they may require a commitment of many years, the value they provide is measured in terms of improved safety for users, lower operational costs, and reduced litigation exposure. Development of a program begins with collection of information on the number of rockfall sites, their locations, and which sites deserve priority attention. If this information is not obtained, the program may provide little real value or may be terminated before any measurable progress is made.

To complete a rockfall hazard assessment, rockfall sections and the hazards associated with each section are identified and evaluated. The factors related to rockfall activity, as well as those that address the vulnerability of the transportation facility and its users, are part of the assessment. Vulnerability serves as the link between rockfall activity and the recommended level of mitigation. For example, a highway or railroad segment with a high likelihood of rockfall occurrence but a low consequence from the event may warrant only a minor, low-cost mitigation measure, or perhaps only continued monitoring. In contrast, a more complex and costly mitigation measure may be justified for a site with a lower likelihood of rockfall occurrence but a much higher potential for serious consequences.

The benefits of implementing and using a rockfall rating system are discussed below. They include the following:

- Creation of a rockfall inventory,
- Education of the public,
- Reduction of liability exposure for the transportation agency,
- Provision of tools for strategic planning, and
- Usefulness for other applications.

3.1 Creation of a Rockfall Inventory

Transportation agencies are typically responsible for many miles of roadway. Within the roadway systems, agencies are often faced with many potential rockfall sites. Without good systemwide information indicating where rockfalls are troublesome or could be hazardous, investments in rockfall mitigation are likely to be reactionary. In such cases, sites where an accident has already occurred or where unacceptable maintenance requirements are reported tend to be the targets of efforts and expenditures.

For strategic planning, an agency should first understand the scope of its rockfall problem, including where rockfalls occur and the relative hazards they represent. In most cases, local roadway maintenance personnel have detailed, firsthand information concerning rockfall activity, but that knowledge may not be documented or available for planning. Furthermore, personnel turnover due to promotion or retirement may result in loss of the information. During the implementation of a rockfall hazard rating system, geographic and historical rockfall activity information will be gathered. With this information and the knowledge that comes from evaluating sites relative to each other, a transportation agency can obligate its limited resources judiciously and better address public safety.

3.2 Education of the Public

The public is clearly aware that there are many hazards associated with driving and that most are the result of human error or poor judgment behind the wheel. In this respect, accidents involving rockfall are different because they are unrelated to factors under a driver's control. Few highway accidents draw more public attention or adverse reaction than one caused by a rockfall. Rockfalls are widely perceived as a problem that should have been known about and addressed.

To temper this reaction, a transportation agency must demonstrate good faith and show not only awareness of the potential for rockfalls but also that prudent steps are being taken to reduce rockfall hazards. An excellent way to do this is to implement a rockfall rating system to define the problem and then to use the system for project development planning.

3.3 Reduction of Liability Exposure for the Transportation Agency

Many types of roadside and highway hazards can lead to litigation if an accident occurs. Demonstrating that a rational program is being followed to reduce these hazards on a systemwide basis can lead to a reduction in exposure to litigation. Courts have indicated that it is unreasonable to expect any transportation agency to have enough funds at its disposal to deal immediately with all safety-related issues. The courts also recognize that design standards and safety requirements have changed over time, as have traffic volumes, which can significantly increase the risk of occurrence of many safety problems. The main issues of concern are the following:

1. Is the agency taking prudent steps to reasonably reduce rockfall hazards?
2. Are the steps to reduce the hazard systematic and consistently applied?
3. Does the methodology include a recognized approach for identifying and developing rockfall mitigation projects as funding becomes available?

A "yes" answer to each of these questions has been demonstrated to increase the likelihood of cases being terminated or settled out of court. In cases that proceeded to trial, a properly implemented and followed rockfall hazard rating system has proved beneficial to the agency's defense.

3.4 Provision of Tools for Strategic Planning

A strategic plan determines where an organization is going and how it will get there. The plan guides decisions to attain a desired outcome on the basis of good information. In the case of transportation-related rockfall issues, the goal of strategic planning is to correct or reduce the systemwide rockfall problems faced by a transportation agency. To accomplish this, specific rockfall management tools are needed. A rockfall hazard rating system can provide those tools, including a means of identifying rockfall sites, documenting rockfall history and potential, assigning hazard scores so that sites can be categorized and differentiated from each other, and prioritizing rockfall mitigation construction projects on the basis of relative hazard and cost.

3.5 Usefulness for Other Applications

The primary purpose of a rockfall rating system is to help a transportation agency manage rockfall hazards in the context of a large and diverse system. However, rockfall hazard rating systems have also been applied to evaluate individual roadway sections and other projects with rockfall problems, including quarry and housing developments; to make decisions on new highway alignments and construction projects; and to mitigate hazards faced by energy-generating facilities. These applications suggest that rockfall hazard rating systems can be customized to meet specific project needs regardless of project size. The rockfall hazard rating system selected by an agency should be customizable to local conditions, organizational requirements, and, in some cases, specific project criteria.

4. DEVELOPMENT OF ROCKFALL HAZARD RATING SYSTEMS

Identification of an increasing rockfall maintenance burden (primarily rising costs) and the necessary response to rockfall-related accidents form the basis for reactive rockfall mitigation programs. Although these items are important components of a rockfall hazard evaluation, they are not in themselves the tools needed to guide a proactive rockfall mitigation program to identify and stabilize problematic slopes. Such a program is based on a system that rates the relative hazard posed by individual rockfall sections so that comparisons can be made among all sections. The comparative data serve as a decision tool in setting priorities for mitigation work at the more critical locations.

Brawner and Wyllie (1975) and Wyllie (1987) undertook the initial development of rockfall hazard rating systems. These early efforts were principally tailored for use along railway lines, but they served as the basis for the Rockfall Hazard Rating System (RHRS), a process for managing rock slopes along transportation systems and adjacent to other developed properties (Pierson et al. 1989; Pierson and Van Vickle 1993).

Development of the RHRS began in the state of Oregon in 1984 and was eventually completed and implemented as part of a Federal Highway Administration national pooled-fund research project by the Oregon Department of Transportation in 1989. It has since been widely implemented nationally and internationally. The RHRS was designed to be straightforward and easy to implement. Because development funding was provided by many states from various parts of the country, the RHRS was created to be customized easily for local conditions and uses.

To date, 28 transportation agencies in North America have implemented a rockfall management system based on the RHRS. The Washington State Department of Transportation has further modified the RHRS and created an Unstable Slope Management System, which is used to prioritize mitigation work on all forms of geological and construction-related slope instabilities, including problematic rockfalls, embankments, and landslides (Lowell and Morin 2000). Two other agencies have created their own systems. Further discussion of the various RHRS modifications and the other types of rating systems in use is provided in the next chapter.

5. DESCRIPTION OF THE RHRS

Because of the overall similarities and wide use of the qualitative rating systems, the RHRS is described as an example of a rockfall and rock slope management tool. The system includes the following six main implementation and maintenance steps:

- A uniform method of slope inventory,
- A preliminary rating of all slopes (optional),

- The detailed rating of all hazardous slopes,
- A preliminary design and cost estimate for the most serious sections,
- Rockfall mitigation project identification and development, and
- A routine review of rockfall sites and a regular update of database information.

The rockfall areas identified during the creation of the inventory are rated and then ranked by scoring categories that represent the significant elements contributing to the overall hazard. If a large number of sites are to be evaluated, a preliminary rating during the inventory process may be appropriate. The preliminary rating separates the slopes into categories of more manageable size on the basis of their rockfall history and potential. This step would be followed by the more time-consuming and costly detailed rating of higher-hazard sites. That is, the slopes deemed to exhibit a high priority for mitigation would be focused on first. The preliminary and detailed ratings are described in Sections 5.3.1 and 5.3.2, respectively.

Once sites have been rated and the rockfall inventory and rating information entered into the database, preliminary designs and cost estimates are made for a reasonable number of rockfall sites to help with planning and project development decisions. With this information, a strategic master plan for annual stabilization work can be developed with the goal of improving safety and reducing operating costs. The plan needs to be flexible and should consider the economic impacts of decisions on the basis of systemwide benefits. It may seem appropriate that more highly rated sites receive attention first, but these sites may require more complex mitigation measures and demand a greater percentage of available resources. The project development process should take this into account. The final step is to perform routine reviews of sites to keep the information current and maintain the value of the database.

Keeping accurate records and photographs of slope conditions, rockfalls, and completed stabilization work is important. This information documents the location and condition of hazardous areas and can be used to determine the long-term effectiveness of the program in reducing the incidence or effect of rockfalls. The records can be most conveniently handled with database programs that readily allow updating and retrieval.

5.1 Rock Slope Inventory

The first step is to perform an inventory of potential rockfall sites. Typically, systemwide rockfall problems are so extensive that any one person is unlikely to know the extent and location of all problem areas. The inventory is most efficient when it is combined with an initial slope-screening assessment, which is described in Section 5.2. Location data based on the agency's identification system should be obtained for the start and the end of the rockfall section. In addition, it is highly desirable to have Global Positioning System (GPS) coordinates so that systemwide rockfall location maps can be created with geographic information system (GIS) technology. The maps can be helpful in understanding the extent of the agency's rockfall problem and can serve as planning tools.

Because much rockfall activity is not obvious during the on-site inventory process, it is important to obtain and document the experiences and observations of persons (typically local maintenance personnel) who routinely respond to rockfall incidents at each site. Their experience represents firsthand information that is not usually documented or available except through an interview process. The information is helpful in establishing the starting and ending points of each rockfall section and in understanding historical rockfall activity and behavior.

For the purpose of the inventory, the information is tied to a defined segment of the transportation system. Such sections are called *rockfall sections*, and each section should be defined as any uninterrupted slope where the level and occurring mode of rockfall are the same. The distinctions provided by this definition allow roadway sections where rockfall problems appear continuous to be subdivided into discrete elements. This is important, since the hazard posed by combined sites would be an average value, and the appropriate mitigation methods would likely be too diverse or extensive to be affordable. The rockfall section approach protects the investment in implementing the rockfall hazard rating for planning purposes because it considers the hazard posed by each site, along with the type and cost of mitigation. Decisions on grouping several rockfall sections together are better made later in the project development process.

A rockfall section should not include areas without rockfall potential. Locations where drainages or crossroads intercept the roadway, as well as low points in the overall slope topography, are used to define the starting and ending points of each rockfall section. Although rockfall problems can persist for considerable distances, this criterion allows seemingly continuous sections to be subdivided. Even within apparently continuous sections, the frequency and amount of rockfall affecting the roadway can vary sufficiently so that these sections should be further subdivided. Rockfall on the roadway can be greater where fallout areas are narrower or the ditch shape less effective. Launch features on the slope can significantly increase the likelihood of rockfall reaching the roadway.

The level of rockfall activity may be much greater where the causes of the rockfall change or the source of the rockfall is concentrated. This can result from several factors, including the relationship between the slope orientation and structural discontinuities, changes in rock type, resistance to differential erosion, and other factors as discussed in Chapter 2. As the mode and activity level change, the severity of the rockfall hazard changes, and the type of remediation will likely vary as well. All these factors are considered in defining the number and extent of rockfall sections.

Identifying these differences and establishing rockfall sections are best accomplished when both a maintenance person and an experienced geologist or geotechnical engineer perform the inventory. The inventory is the first indication of the magnitude of the rockfall problem and establishes the basis for further work. The desired result is a flexible database of problem sites.

The rockfall history is obtained and documented during the creation of the slope inventory. Rockfall location, frequency, time of year, size or quantity, rock type, source areas on the slope, where rockfalls come to rest, accident history, frequency of ditch cleaning or road patrols, annual cost of rockfall-related maintenance, and an opinion on why rockfall occurs are all part of this documentation. This historical perspective gives an excellent indication of future slope performance and rockfall behavior.

5.2 Initial Screening Assessment of All Slopes

High-quality video or detailed photographic roadway logs, if available, can be used to perform an initial screening to identify potential rockfall sites. In many cases, the video logs include GPS coordinates and roadway linear-reference information, so the location information can be well documented. Obviously, agencies do not have potential rockfall sites throughout all parts of their system. The prescreening can eliminate many sections of highway from field inspection and thus save many hours of driving and implementation costs.

5.3 Rating All Hazardous Slopes

Rockfall ratings are best developed in two stages: establishment of a preliminary rating followed by a detailed rating assessment. The preliminary rating groups sites on the basis of their rockfall history (activity) and potential for future rockfalls, and especially the likelihood that rockfall material may reach the road. The detailed rating numerically differentiates the sites and provides a relative rating by scoring several hazard categories according to established criteria. The preliminary rating can be accomplished quickly, while the detailed rating requires more time and resources. Depending on the number of sites to be evaluated, creation of the slope inventory and the preliminary ratings can be performed with a single effort. If the number of sites is small, the preliminary and detailed ratings may be combined.

5.3.1 Preliminary Rating

The purpose of the preliminary rating is twofold:

1. It groups sites inspected during the slope inventory into three broad, more manageably sized hazard categories as described below.
2. It provides a quick initial rating that allows the application of the detailed rating at the more critical sites.

The preliminary rating improves efficiency and helps prevent resources (time and budget) from being directed toward sites with too low a potential of producing a hazardous rockfall. The rating is subjective and requires valid judgments on the part of experienced, insightful personnel. The preliminary rating defines the rockfall hazard of any slope as high, medium, or low according to three classes labeled A, B, or C on the basis of the criteria shown in Table 3-3. Figures 3-3, 3-4,

Table 3-3
Preliminary Rating Criteria

	Class		
Criterion	A	B	C
Historical rockfall activity	High	Moderate	Low
Estimated potential for rock on roadway	High	Moderate	Low

Source: Pierson 1992.

FIGURE 3-3
Typical A-rated, near-vertical rock cut without adequate catch ditch near Beaver, Utah (Pack et al. 2006).

and 3-5 illustrate typical A, B, and C slopes, respectively.

In rating the estimated potential for rockfall on the roadway, the following factors are considered:

- Estimated size of material,
- Estimated quantity of material per event,
- Amount available, and
- Ditch effectiveness.

In rating the historical rockfall activity, the following factors are considered:

- Frequency of rockfall on highway,
- Quantity of material,
- Size of material, and
- Frequency of clean out.

Because the RHRS supports a proactive approach to rockfall management, the estimated-potential-for-rock-on-roadway criterion is often the controlling element of the preliminary rating. For example, if a slope contains a large block that displays evidence of active displacement and limited fallout area is available, it would receive an A rating, regardless of past rockfall activity. The historical-rockfall-activity criterion is used where clarification is needed. For example, assume that, after inspection of a site, classification as an A or a B slope on the basis of the potential for rockfall reaching the roadway is unclear. If the site history is that rockfall activity remains high, the slope is rated A.

A rating of C is applied if a rock is unlikely to fall at a site, or, if one should fall, it is unlikely to reach the roadway. In other words, the risk of a hazardous situation occurring is nonexistent to low. Once a C rating is determined, the site information is typically not recorded or entered in the rockfall database.

FIGURE 3-4
Typical B-rated slope near Colton, Utah (Pack et al. 2006).

FIGURE 3-5
Typical C-rated slope near Emery, Utah (Pack et al. 2006).

The preliminary rating evaluation is a critical step in the RHRS process, especially where large numbers of slopes are involved. The need for further efforts is determined at this step. Initially only the A-rated sections are advanced for further evaluation with the detailed rating. This will economize the effort while ensuring that it is directed toward the most critical areas. The sections rated B are evaluated as time and funding allow. The sections rated C will receive no further rockfall mitigation attention and are often not retained in the statewide database.

The A and B preliminary rating categories provide a mechanism for sites to be reassigned or downgraded once rockfall mitigation measures have been applied. For example, after hazard-mitigation measures have been completed, an A site may become a lower-priority B or even a C site, and it will be reclassified in the database. The detailed rating scores described in the next section do not significantly change on the basis of mitigation efforts and thus do not provide a means for reassessing the level of hazard following mitigation efforts. Ideally, a complete detailed rating assessment is conducted at the site before an updated hazard score is entered into the database.

5.3.2 Detailed Rating

The purpose of the *detailed rating* is to differentiate the hazard posed by different slopes numerically so that sites can be scored, ranked, and prioritized for possible mitigation work. Once rated, the sites can be sorted and prioritized on the basis of their scores. The lists are then used to help make decisions on where safety projects are best initiated.

The detailed rating, shown in Table 3-4, includes nine categories by which slopes are evaluated and scored. The individual category scores are then totaled. Higher scores indicate greater rockfall hazard. The nine categories represent selected significant elements within a rockfall section that contribute to the overall hazard. The four columns

Table 3-4
Scoring Sheet for Detailed Ratings by RHRS

	Rating Criteria and Score			
Category	Points 3	Points 9	Points 27	Points 81
1. Slope height (ft)	25	50	75	100
2. Ditch effectiveness	Good catchment	Moderate catchment	Limited catchment	No catchment
3. Average vehicle risk	25% of the time	50% of the time	75% of the time	100% of the time
4. Percentage of decision sight distance	Adequate sight distance, 100% of low design value	Moderate sight distance, 80% of low design value	Limited sight distance, 60% of low design value	Very limited sight distance, 40% of low design value
5. Roadway width including paved shoulders (ft)	44	36	28	20
6. Geological character				
Case 1: structural condition	Discontinuous joints, favorable orientation	Discontinuous joints, random orientation	Discontinuous joints, adverse orientation	Continuous joints, adverse orientation
Case 1: rock friction	Rough, irregular	Undulating	Planar	Clay infilling or slickensided
Case 2: structural condition	Few differential erosion features	Occasional differential erosion features	Many differential erosion features	Major differential erosion features
Case 2: difference in erosion rates	Small difference	Moderate difference	Large difference	Extreme difference
7. Block size or volume of rockfall event				
Block size (ft)	1	2	3	4
Volume of rockfall event (yd^3)	3	6	9	12
8. Climate and presence of water on slope	Low to moderate precipitation, no freezing periods, no water on slope	Moderate precipitation or short freezing periods or intermittent water on slope	High precipitation or long freezing periods or continual water on slope	High precipitation and long freezing periods or continual water on slope and long freezing periods
9. Rockfall history	Few falls	Occasional falls	Many falls	Constant falls

Source: Pierson 1992.

to the right of each category describe logical breaks (benchmark criteria) within the continuum of conditions that are found in nature or have been constructed. The associated risk increases from left to right. Some of the categories require a subjective evaluation, whereas others can be directly measured and then scored.

As the risk increases from left to right, the set scores above each column are increased. Note that the set scores increase exponentially from 3 to 81 points. An exponential scoring system is used because it provides a rapid increase in score that distinguishes the more hazardous sites. The set scores are merely representatives of a continuum of possible points from 1 to 100. In rating a slope, use of the full range of possible points instead of only the set points above each column allows the rater greater flexibility in evaluating the relative impact of conditions that can be extremely variable. Initially, novice raters feel more comfortable using the set points, since less judgment is required. However, continuing to rely on the set points is not an optimal use of the system and is not common practice.

A brief description of each of the categories from the RHRS is included below. The system was developed on the basis of customary units; however, conversion to metric specifications is easily made. The following categories are considered:

Category 1. Slope height. Category 1 represents the vertical height of the slope. Measurement is to the highest point from which rockfall is expected. If rocks come from the natural slope above the cut, regardless of right-of-way limits depending on agency policy, the cut height plus the additional slope height (vertical distance) is used.

Category 2. Ditch effectiveness. The effectiveness of a ditch is measured by its ability to prevent falling rock from reaching the traveled way. In estimating ditch effectiveness, factors to consider are (*a*) slope height and angle; (*b*) ditch width, depth, and shape; (*c*) anticipated block size or volume of rockfall per event; and (*d*) the effect of slope irregularities (launching features) on falling rocks. A launching feature can negate the benefits expected from a fallout area. Valuable information on ditch performance can be obtained from maintenance personnel and from the ditch design criteria discussed in Chapter 14.

Category 3. Average vehicle risk. Average vehicle risk (AVR) represents the percentage of time that a vehicle will be present in the rockfall section, which is obtained from the following relationship:

$$\text{AVR}(\%) = \frac{\begin{array}{c}\text{average daily traffic (cars/day)}\\ \times \text{ slope length (mi)}\end{array}}{\begin{array}{c}\text{posted speed limit (mph)}\\ \times\ 24(\text{h/day})\end{array}} \times 100$$

A rating of 100% means that on average a vehicle can be expected to be within the rockfall section 100% of the time.

Category 4. Decision sight distance. Sight distance is the shortest distance along a roadway that an object is continuously visible to the driver. The sight distance can change appreciably throughout a rockfall section. Horizontal and vertical curves, together with obstructions such as rock outcrops and roadside vegetation, can severely limit the available sight distance. Decision sight distance is used to determine the length of roadway (in feet) needed to make a complex or instantaneous decision. It is critical when obstacles on the road are unexpected or difficult to perceive and unplanned vehicle maneuvers are required to avoid them.

The relationship between decision sight distance and the posted speed limit used in the inventory system has been modified from *A Policy on Geometric Design of Highways and Streets* of the American Association of State Highway and Transportation Officials (AASHTO 1984).

Category 5. Roadway width. The available maneuvering room to avoid a rockfall is measured perpendicularly to the highway centerline from one edge of the pavement to the other and includes the shoulders if they are paved. When the roadway width is not constant, the minimum width throughout the section is used.

Category 6. Geological case characteristics. Generally, there are two conditions, or cases, that can lead to rockfall occurrence. The case that best fits the slope is used in the evaluation. Each case considers two components. If both cases are present, both are scored, but only the worst case (combined highest case score) is used in the rating. Case 1 considers structural conditions and rock friction; Case 2 considers erosional structural conditions and differential erosion rate. Thus Case 1 is more often applied to slopes composed of igneous and metamorphic rocks, while Case 2 is more useful in dealing with slopes composed of sedimentary rocks.

Case 1 includes slopes in which joints, bedding planes, or other discontinuities are the dominant structural features (Part 1). It also includes an evaluation of the resistance to movement provided by friction and the interlocking of rock blocks along the discontinuities (Part 2). The major elements that affect the structural condition evaluation are the presence of joints, bedding planes, or fractures with orientations and lengths that promote planar, circular, block, wedge, or toppling failures. Continuous joints are those greater than 10 ft (3 m) in length. Rock friction along a discontinuity is governed by the characteristics of the rock material as well as by the surface roughness and properties of any infilling (see Chapter 6, Section 3.3).

Case 2 considers differential erosion features or oversteepened slopes (Part 1) and the relative rate of erosion (Part 2) as the dominant conditions that control rockfall. Case 2 can be applied to oversteepened colluvium/talus slopes, where the oversteepening may have been caused by construction or maintenance, and to slopes made up of more than one rock type or a combination of soil with embedded rocks that become unsupported and exposed because of variable resistance to weathering or erosion. Differential erosion within a slope due to layers of variable erosion resistance is a frequent cause of rockfall. The evaluation score reflects how quickly erosion is occurring; the size of the rocks, blocks, or units being exposed; the frequency of rockfall events; and the amount of material released during a rockfall event.

Category 7. Block size or volume of rockfall per event. The measurement and score for this category are based on the type of rockfall event more likely to occur. Ideally, the scores also take into account the tendency for rockfalls to break up as they move down the slope. If both types of events occur, the one providing the higher score is used.

Category 8. Climate and presence of water on slope. Water and freeze–thaw cycles both contribute to the weathering and movement of rock materials. If water is known to flow continually or intermittently on the slope, the slope is rated accordingly. Areas receiving less than 25 in./year (635 mm/year) are low-precipitation areas. Areas receiving more than 50 in./year (1,270 mm/year) are considered high-precipitation areas. This category and these values are adjusted to reflect local conditions. The category may not be rated at all if the conditions vary only slightly across the agency's area.

Category 9. Rockfall history. Historical information is an important gauge of the potential for future rockfalls. As the database of rockfall occurrences is developed, more accurate conclusions for rockfall potential can be reached. Rockfall history is an excellent indicator of what future slope performance can be expected, and it serves as part of a valuable overall evaluation, or check, of the total score a site receives.

5.4 Preliminary Design and Cost Estimates

To reduce systemwide rockfall potential and demonstrate that the agency is committed to addressing its rockfall mitigation responsibilities, some projects need to be part of the regular construction and maintenance programs. In deciding which rockfall sites to develop as a project, preliminary information on mitigation designs and cost estimates is needed. The information should be included in the RHRS database. In the planning of highway construction projects, including rockfall mitigation projects, definitions of the desired project goals and limitations, such as design criteria, project limits, the construction budget, and right-of-way needs, are important. Substitution of a different, more appropriate rockfall mitigation design at a later stage can frustrate the process and in some cases make it unworkable.

5.4.1 Rockfall Remediation Designs

During the detailed rating, the raters should gather enough site-specific information to allow them to ascertain which rockfall remediation measures are appropriate for the site and the approximate quantities required. More than one design element will likely be needed. To make informed decisions on whether to take a correction or hazard reduction approach, management will need to understand the costs and benefits associated with each. Correcting a hazard may require significant work elements, such as realigning the roadway or flattening a slope, and therefore tend to be more expensive. Hazard reduction measures can vary from limited duration improvements such as slope scaling to more positive interventions such as installing slope screening, rock bolts, or shotcrete.

Frequently, a combination of several techniques will work best. At this early stage, for the purpose of the rockfall rating system, the goal is to provide an appropriate method to deal with the rockfall problem. The preliminary concepts can later be refined by more detailed investigation and analysis, but they define appropriate techniques that properly address rockfall conditions.

The value of skilled personnel in making these field determinations cannot be overemphasized. The effectiveness of a rockfall mitigation design is best judged by someone with experience. There are too many gaps in the understanding of the arrangement and mechanical properties of rock masses to rely solely on an analytical approach. Several techniques, which are described in Chapters 11 through 16, are routinely used to mitigate rockfall hazard. The choice of techniques depends on several factors, including the size or volume of anticipated rockfall, access to the rockfall source, maintenance limitations, the construction budget, seasonal and environmental limitations, and the desired result. Although most of the techniques are familiar, inexperience can result in the wrong application. If the field designs are developed by inexperienced staff, a review of the conceptual design by an experienced person is desirable before the information is entered into the database or used to make project development decisions.

5.4.2 Preliminary Cost Estimates

The cost estimate is an important element of the rockfall database and will be considered as final project development priorities are established. For comparison purposes, the rockfall mitigation design cost calculated at this step is preliminary and represents only the cost of the rockfall remedial measures. A project may eventually include pavement widening, guardrail installation, structural pavement overlay, and so forth. These additional items, as well as mobilization, traffic control, engineering, and contingency items, are not included in the cost estimate, which simplifies and expedites the estimating process. The construction costs of the various design elements can vary a great deal nationally because of the availability of experienced contractors and the competitive environment between contractors.

5.5 Rockfall Mitigation Project Identification and Development

There are many ways to use the information contained in the database to make rational and defensible decisions on which sites to advance to final design and construction. Four suggested methods are described below.

5.5.1 Projects Can Be Advanced on the Basis of Their Scores

This method is an obvious use of the system and one adopted by many agencies. Since the most hazardous slopes are those with the highest scores, obligating funds to the highest-rated slopes as the funds become available is reasonable.

The main drawback is that frequently the highest-rated slopes are among the most costly to repair. Another equally important drawback to this method is that the highest-rated slopes may be on low-volume facilities, and thus realize little system benefit. The agency faces a decision as to whether to proceed from the top of the list down or to attempt to improve as many high-priority sites as possible with the funds that are available.

5.5.2 Projects Can Be Advanced on the Basis of Their Estimated Construction Costs Relative to Their RHRS Scores

This is, in effect, a variation of a benefit–cost methodology. The preliminary cost estimate for the top-rated slopes is divided by the RHRS score, and a list is generated with the lowest dollar-to-RHRS-point ratios at the top. Projects developed from the list will provide the greatest systemwide hazard reduction with a fixed investment amount. The 100 highest-rated slopes are a reasonable starting point in implementing this approach.

If this approach is used, some of the highest-rated slopes may be left unattended because of their higher cost. If that is a concern, development of projects from both lists may be a solution. With either method, significant progress is made toward reducing the risk associated with rockfall.

5.5.3 Projects Can Be Developed on the Basis of Having Similar Remedial Designs

Rockfall sections containing similar construction elements, as identified during the preliminary design and cost estimate process, can be grouped into a single project. It is common for a contractor to specialize in two or three mitigation techniques. Finding several qualified contractors versed in constructing a broad array of rockfall mitigation measures is difficult. Limiting the complexity of the contract alleviates this problem and improves competition.

An example of this approach would be to take a length of highway and combine all of the sections that are earmarked for slope screening into one project. A larger quantity of slope screening will be contracted. This may result in a more straightforward, easily managed contract with larger bid quantities, which typically translates into lower unit-bid prices.

5.5.4 Projects Can Be Developed on the Basis of Proximity of Rockfall Sites Along a Section of Roadway

Because the rockfall sites have been identified and remedial measures properly determined, a larger contract can be let. This method may result in lower unit-bid prices and an overall savings for mobilization.

An example of this method would be to group all A-rated slopes along a 20-mi (32-km) stretch of roadway into one project. In using the approach, it may be useful to consider including the B-rated slopes within the section. The method will remove an entire section of roadway from the RHRS database and will clearly demonstrate that progress is being made.

5.6 Routine Review of Rockfall Sites and Regular Updating of the Rockfall Database

The final but ongoing step in the rockfall rating system process is to review the rated slopes in the database, with an emphasis on the higher-rated or A slopes. The review should be conducted regularly to protect an agency's original investment in creating the database and the value of the information as both a project development tool and a reliable source of data in the event of litigation. The review can be completed quickly by comparing past site photographs with current conditions. Updated information from maintenance is also considered.

5.6.1 Database Review

Conditions in nature can change unexpectedly. A severe winter or heavier-than-normal precipitation in the spring or fall can hasten changes that affect the potential for rockfall events. Human actions can also cause changes in slope conditions. At times, the potential for rockfall can decrease.

Not all changes that affect a slope's rating are obvious. At times, the changes can be difficult to discern. The rater should refer to photographs taken during previous inspections to help identify changes. Review of recent maintenance information to determine whether there has been any significant work at the site that could affect the slope's rating is also beneficial. The review and update process is ongoing. For the sake of economy, much of the work can be accomplished throughout the year in the course of travel to and from other project sites.

5.6.2 Database Update

If any slopes in the database have been modified by construction or maintenance activities, they are reviewed and either removed from the RHRS system or rerated. Any newly constructed slopes are rated and, if necessary, added to the database. Eventually, all slopes in the database, including the B-rated slopes, are evaluated with the detailed rating, and preliminary designs and cost estimates are prepared.

Existing preliminary designs for any rockfall section whose rating has changed because of site modifications are reviewed. Changes in site conditions can make the old conceptual designs inappropriate. Cost estimates are updated periodically as the unit-bid prices change.

Once all slopes have been rated, an agency may redefine what constitutes an A or B slope by using a range of scores established by the agency rather than the subjective evaluation criteria applied during the preliminary rating.

6. IMPORTANCE OF PARTICIPATION BY MAINTENANCE PERSONNEL

Local maintenance personnel retain most of the rockfall activity and behavior information throughout an area, but too often the information is not documented, which makes it unavailable to designers or construction personnel. The rockfall rating process provides an excellent opportunity to document the information and to initiate working relationships and rockfall reporting strategies.

Maintenance staffs are charged with performing road patrols and removing rockfall from the roads, cleaning out roadside ditches and fallout areas, repairing rockfall damage, and responding to rockfall accidents. The experience and understanding of the rockfall problems generated by these activities should be tapped as part of the rockfall hazard rating process and in considering various design concepts. At times, designs aimed at controlling rockfall can complicate other maintenance activities. Involvement of maintenance personnel in the rating and design process can eliminate some of these conflicts. The use of mitigation measures where rockfalls create the greatest maintenance problems can help preserve maintenance operating budgets for other services and roadway improvements.

6.1 Maintenance Survey

Where rockfall ratings are to be applied over a large geographic area containing thousands of miles of roads, the involvement of many local maintenance personnel is required. Difficulties can arise in efficiently scheduling joint field meetings. In one implementation example, the maintenance involvement for the Montana Department of Transportation was accomplished by using a web-based questionnaire (Pierson et al. 2005). This was done in conjunction with a prescreening process, as described in Section 5.2. On the basis of the prescreening results, each of the 123 local maintenance sections received a list of sites in its area to review. The site location data were already entered on each questionnaire. Sites were added, deleted, or divided as necessary, rockfall activity was detailed, and the preliminary historical rating (A, B, or C) for each site was provided. The process provided local maintenance personnel the opportunity to schedule an on-site meeting to discuss specific problem areas or issues.

6.2 Reporting of Rockfall Occurrences

A procedure should be initiated for reporting rockfall events and recording site-specific maintenance costs. The information allows recent increases in rockfall activity to be considered during rating updates and project development decisions. Typically, the procedures established for reporting site-specific rockfall activity and maintenance costs are limited. In addition, maintenance personnel tend to be more focused on addressing roadway problems and improving safety and less interested in complex and time-consuming reporting requirements that do not provide immediate tangible benefits. A training program that explains how the rockfall information will be used, anticipated long-term safety benefits, and anticipated reductions in rockfall maintenance requirements will improve compliance with reporting procedures.

A streamlined reporting strategy based on a checklist and brief fill-in-the-blank system for use in the field can improve efficiency, reporting cooperation, and reliability of the data. Typically, information is given concerning the location where rocks are found on the road, date, weather conditions, accidents, general frequency or recurrence interval of events, size of rockfalls, and approximate maintenance hours or budget expended. Also reported are any maintenance fixes used to address rockfalls: ditch improvements, placement of barriers, rock scaling, and so forth.

7. ROCKFALL RATING SYSTEM DATABASE

Implementing a rockfall rating system will generate a large amount of valuable data that must be effectively managed and made readily accessible to realize the goals of implementation. The information will be used in selecting rockfall mitigation projects, determining how maintenance resources are being expended, establishing appropriate responses in case of a rockfall accident, and demonstrating the agency's commitment to addressing rockfall hazards. The database would contain the information needed for selecting appropriate mitigation measures as discussed in Chapter 11. To use this information properly, a system of data entry, archival, retrieval, and editing is needed.

7.1 Importance of the Database

The rockfall rating system database is the repository for rock slope and rockfall management information. The information will be applied in reaching specific goals, such as reducing overall rockfall hazard; managing the cost of rockfall maintenance; limiting litigation exposure; and reducing the effects of rockfall on the agency's infrastructure, whether it includes highways, railway lines, ports, property developments, or power-generating facilities.

The fully functional database must allow for the data to be queried on the basis of multiple criteria, such as conceptual mitigation cost, overall RHRS rating, or specific category ratings. For example, an agency's long-term goal may be to mitigate risk at sites that have a relatively high rating and relatively low conceptual mitigation costs. The database should be constructed to allow the calculation of a quantifiable measure, such as a cost-to-RHRS-score ratio. This flexibility will help the agency meet its long-term strategic plan.

7.2 Development of the Database

The goals for the database should include ease of data management, protection from misuse or vandalism, and the ability to produce useful reports for a wide range of queries. Each agency or company will need to decide whether the system should be part of a mainframe system placed on a server or contained on a single PC or laptop. The decision would be based on the need for security and accessibility by the appropriate staff, including data entry, geotechnical, design, planning, management, and information system personnel. A properly constructed database will allow the querying of the data for rockfall sites to be evaluated on the basis of multiple criteria, for the site data to be linked to photographic records, and for various GIS mapping functions.

Mainframe systems typically have more rigorous system and data format requirements. Creating the mainframe database structure to accept the rating system information can add dramatically to the cost of implementation, but most agencies maintain department asset and roadside hazard data within their mainframe system. This functionality adds a step for transferring or entering field data that are normally gathered on a laptop or handheld computer during field operations.

The database can be constructed by using a variety of software and computing platforms. It can be developed by implementing simple architecture in a spreadsheet program, by using moderately complex architecture and error-checking functions of database programs within standard office software suites, or by integrating the data collected into the agencywide mainframe computing systems. Computing platforms can be based on office data entry from paper field forms with a desktop computer. Alternatively, a laptop in the field could be used, or a version of the database for data collection using handheld computers could be developed.

7.3 Management of the Database

The responsibility for managing the database generally rests with the geotechnical staff that will populate the database and perform data updates. Updates at regular intervals are desirable, whenever slope or rockfall activity changes are observed or reported, or whenever specific rockfall mitigation maintenance actions occur. Additional edits should be made whenever a major rockfall event or accident occurs or new rock cuts are constructed.

8. LIMITATIONS OF ROCKFALL HAZARD RATING SYSTEMS

Rockfall rating systems are intended to help agencies manage their rock slope assets. The systems in use today provide a rational means of accomplishing this, but they are qualitative by design and include subjective categories that require judgment in their application. Although it has been demonstrated that well-trained practitioners can consistently apply the rating criteria, the latitude provided in scoring, with a broad range of possible scores and the subjective categories, increases the likelihood that different raters will obtain different scores. The differences are manageable and generally not significant, and the anticipated variation does not diminish the value of the rating system as a rock slope management tool.

Despite an apparent significant numerical difference between a site with a score of 550 and one rated 700, the difference actually only provides a method for planning and prioritizing projects and is not to be perceived as a predictor of where the next rockfall or rockfall accident will occur. Far too many variables enter into the occurrence of a rockfall incident for a simple rating system to provide that predictive ability.

By definition, even a B-rated slope has the potential to produce a rockfall onto the roadway. If such an event leads to an accident, the agency will be expected to react to it, no matter where the affected section appears on the rockfall priority list. However, an agency should resist the tendency to overreact at these times. Sites where an accident has occurred should be reevaluated with the detailed rating to determine whether the rockfall incident has increased or decreased the rockfall potential. Ideally, the level of investment at the site will be responsive but also consistent with the redefined rockfall hazard relative to other sites where the hazard potential may be much greater.

REFERENCES

ABBREVIATION

AASHTO American Association of State Highway and Transportation Officials

AASHTO. 1984. *A Policy on Geometric Design of Highways and Streets*. Washington, D.C.

Australian Geomechanics Society. 2000. Landslide Risk Management Concepts and Guidelines. *Australian Geomechanics*, Vol. 35, No. 1, pp. 49–92.

Brawner, C. O., and D. Wyllie. 1975. Rock Slope Stability on Railway Projects. In *Proceedings of the American Railway Engineering Association Regional Meeting*, Vancouver, British Columbia, Canada.

Fell, R., K. K. S. Ho, S. Lacasse, and E. Leroi. 2005. A Framework for Landslide Risk Assessment and Management. In *Landslide Risk Management: Proceedings of the International Conference on Landslide Risk Management* (O. Hungr, R. Fell, R. Couture, and E. Eberhardt, eds.), Vancouver, Canada, May 31–June 3, Taylor & Francis, London, pp. 3–26.

Hoek, E. 2000. *Practical Rock Engineering*. Course notes. www.rocscience.com/hoek/PracticalRockEngineering.asp.

Lowell, S., and P. Morin. 2000. Unstable Slope Management: Washington State. *TR News*, No. 207, March–April, pp. 11–15.

Pack, R. T., K. Boie, S. Mather, and J. Farrell. 2006. *UDOT Rockfall Hazard Rating System: Final Report and User's Manual*. Report UT-06.07. Utah Department of Transportation.

Pierson, L. A. 1992. Rockfall Hazard Rating System. In *Transportation Research Record 1343*, Transportation Research Board, National Research Council, Washington, D.C., pp. 6–13.

Pierson, L. A., and R. Van Vickle. 1993. *Rockfall Hazard Rating System Participant's Manual*. Publication FHWA SA-93-057. Federal Highway Administration, U.S. Department of Transportation, Washington, D.C.

Pierson, L. A., D. L. Beckstrand, and B. A. Black. 2005. *Rockfall Hazard Classification and Mitigation System*. Publication FHWA/MT-05-011/8174. Federal Highway Administration, U.S. Department of Transportation, Washington, D.C.

Pierson, L. A., S. Davis, and R. Van Vickle. 1989. *The Rockfall Hazard Rating System Implementation Manual*. Publication FHWA-OR-EG-90-01. Federal Highway Administration, U.S. Department of Transportation, Washington, D.C.

Roberds, W. 2005. Estimating Temporal and Spatial Variability and Vulnerability. In *Landslide Risk Management: Proceedings of the International Conference on Landslide Risk Management* (O. Hungr, R. Fell, R. Couture, and E. Eberhardt, eds.), Vancouver, Canada, May 31–June 3, Taylor & Francis, London, pp. 129–158.

Wong, H. N. 2005. Landslide Risk Assessment for Individual Facilities. In *Landslide Risk Management: Proceedings of the International Conference on Landslide Risk Management* (O. Hungr, R. Fell, R. Couture, and E. Eberhardt, eds.), Vancouver, Canada, May 31–June 3, Taylor & Francis, London, pp. 237–298.

Wyllie, D. 1987. Rock Slope Inventory System. *Proceedings of the Federal Highway Administration Rockfall Mitigation Seminar*, Federal Highway Administration Region 10, Portland, Ore.

Wyllie, D. C., and N. I. Norrish. 1996. Stabilization of Rock Slopes. In *Special Report 247: Landslides: Investigation and Mitigation* (A. K. Turner and R. L. Schuster, eds.), Transportation Research Board, National Research Council, Washington, D.C., pp. 474–504.

Chapter 4

Lawrence A. Pierson
and A. Keith Turner

IMPLEMENTATION OF ROCK SLOPE MANAGEMENT SYSTEMS

1. INTRODUCTION

Characterization of rockfall hazards requires consideration of several attributes of each rockfall site, including rock slope geometry, vehicular traffic volumes, roadway geometry, geological and climate conditions, presence of water on the slope, and rockfall history. In the context of rockfall hazard management, a clear procedure for characterizing geology and its relationship to the roadway is especially important, because these relate so strongly to the rockfall potential.

The Rockfall Hazard Rating System (RHRS) is described in detail in Chapter 3. The RHRS is a tool to systematically inventory and rank hazardous rock slopes adjacent to facilities and various types of infrastructure. It was developed at the Oregon Department of Transportation (DOT) on the basis of an initial prototype proposed by Wyllie (1987). Subsequently, FHWA supported the adoption of the RHRS as a proactive method for managing rock slopes along highways (Pierson et al. 1990; Pierson and Van Vickle 1993).

As described in Chapter 3, the RHRS uses a two-phase slope-categorization process (Pierson and Van Vickle 1993). The first phase is a preliminary rating, in which slopes are assigned a rating of A, B, or C based on the estimated potential for rock to reach the roadway and on historical rockfall activity. The most hazardous, A-rated slopes are then characterized with a detailed rating that considers nine categories, as described in Chapter 3. Table 4-1 illustrates the relationships between the RHRS detailed rating categories and the four primary characterization attributes.

Since the original development and implementation of the RHRS (Pierson et al. 1990; Pierson and Van Vickle 1993), numerous transportation agencies in the United States and Canada have evaluated the RHRS concepts to determine if the system could meet their rockfall management needs. Most of these agencies have used the RHRS without modification, whereas others have made only minor adjustments. In some cases, a few agencies have made more substantive modifications. A recent survey of North American transportation agencies provides important insights into the role and utility of the RHRS concepts and approach.

2. SURVEY OF ROCK SLOPE MANAGEMENT SYSTEMS IN 2008

A survey was conducted in 2008 by one of the chapter authors (Pierson) to ascertain the current state of the practice in the use of rock slope management systems within the United States and Canada. The survey yielded responses from 50 agencies (8 from Canada and 42 from the United States); of these, 42 identified the presence of rockfall sites within their jurisdictions that represented some level of safety concern or a burden

Table 4-1
RHRS Categories Related to Characterization Attributes

RHRS Category	Characterization Attribute
1. Slope height	Rock slope geometry
2. Ditch effectiveness	Rock slope geometry
3. Average vehicle risk	Traffic and road geometry
4. Decision sight distance	Traffic and road geometry
5. Roadway width including paved shoulders	Traffic and road geometry
6. Geological character	Geological conditions
7. Block size or volume of rockfall per event	Geological conditions
8. Climate and presence of water on slope	Climate, water, and rockfall history
9. Rockfall history	Climate, water, and rockfall history

on maintenance activities, or both. These results indicated a much wider distribution of perceived rockfall problems than anticipated. Naturally, the hazards posed by these rockfall sites varied, and even though hazards were present, not all agencies considered these a significant concern. The survey results provided the following generalized rankings of the rockfall hazards by these 42 jurisdictions:

- *High hazard.* 15 jurisdictions (36%).
- *Medium hazard.* 17 jurisdictions (40%).
- *Low hazard.* 10 jurisdictions (24%).

The definitions of high, medium, or low hazard were not rigorous, and thus these designations are subjective, as characterized by the respondents. Nevertheless, it is useful to compare these interpreted hazard levels with several survey topics, including what condition prompts a rock slope assessment (Table 4-2), the annual level of maintenance expenditures for rockfall mitigation (Table 4-3), and the personnel who identify the slopes to be evaluated (Table 4-4).

The responses in Table 4-2 show logical trends. Those jurisdictions that consider rockfall hazards to be of relatively low importance report rockfall mitigation efforts on a case-by-case basis as the dominant response, and those jurisdictions with high or medium rockfall hazard rankings show that a much wider variety of conditions, including accidents, rockfall events, and systematic rock slope rating procedures, result in rockfall mitigation efforts. Jurisdictions with low rockfall hazard rankings have not instituted systematic regional rock slope rating procedures since none report this level as a condition that has prompted a rockfall mitigation response.

Table 4-3 shows a logical correlation between the perceived level of rockfall hazard and annual maintenance expenditures; jurisdictions defined as having low rockfall hazard rankings report relatively minor annual expenditures, whereas jurisdictions defined as having high rockfall hazard rankings spend much larger annual amounts with several reporting annual expenditures exceeding $1 million. Those jurisdictions with a medium rockfall hazard ranking show a wide range of annual expenditures; some report expenditures exceeding $1 million, whereas the largest number report annual expenditures in the range of $500,000 to $1 million. Several other jurisdictions report modest annual expenditures, which likely reflect competing demands for limited resources faced by these agencies.

Table 4-2
Conditions Prompting Response for Rockfall Mitigation

	Condition Prompting Response					
Hazard Rank	Accident or Rockfall Event	Case-by-Case Basis	Undertake Rock Slope Rating or Ranking Process	Other Reasons	No Condition Identified	Total
High	4	9	7	3	0	15
Medium	5	8	9	2	0	18
Low	0	9	0	1	1	11
All hazard ranks	9	26	16	6	1	44

Table 4-3
Annual Maintenance Expenditures Reported by Rockfall Hazard Ranking

	Annual Maintenance Expenditures ($ thousands)						
Hazard Rank	<$100	$100–$250	$250–$500	$500–$1,000	>$1,000	No Value Identified	Total
High	0	3	3	3	6	0	15
Medium	3	4	2	5	3	1	18
Low	8	2	0	0	0	1	11
All hazard ranks	11	9	5	8	9	2	44

Table 4-4 contains responses from the majority of the jurisdictions with high and medium rockfall hazard rankings (Tables 4-2 and 4-3). Since the jurisdictions in which rockfall hazards are low indicated that they have not instituted systematic procedures for identifying rock slopes that may pose rockfall hazards, they did not respond to this question. Table 4-4 clearly shows a strong preference for geological or geotechnical personnel to be involved in this process, but the responses also show that several jurisdictions have developed a shared responsibility among different sectors of the agency in order to ensure that the appropriate rock slopes are identified for evaluation. In contrast, a subsequent question concerning the responsibility for developing and managing a rockfall database revealed that all jurisdictions assigned this effort to their geotechnical or geological personnel, although one indicated a shared responsibility with maintenance personnel.

The survey also attempted to ascertain how these rock slope management systems were being used to assist in rockfall mitigation efforts. Responses from the selected jurisdictions with high and medium rockfall hazard rankings revealed that only 39% had any formal rockfall mitigation program in place that would reduce the rockfall hazards in any proactive fashion. When these jurisdictions were asked if they relied on the information contained within their rock slope management system to identify projects for funding and construction, the responses revealed that 23% always used the management system, 31% often used it, 31% sometimes used it, and 15% never used it.

In spite of the considerable range in the level of application of these rock slope management systems, about 86% of the jurisdictions responded that they have dedicated specific staff to implementing and maintaining the system. However, the level of staff committed is typically quite low; the average commitment is slightly less than 1.0 full-time-equivalent (FTE) position, and 62% of the responses show less than 0.5 FTE. Although one jurisdiction reports assigning 7 FTE to this task, several report 0.1 FTE or less.

Table 4-4
Personnel Responsible for Identifying Slopes Requiring Rockfall Evaluation

	Responsible Personnel[a]			
Hazard Rank	Maintenance	Geotechnical or Geological	Others	Number of Responses
High	8	10	3	13
Medium	4	13	2	13
Low	—	—	—	—
Total of high and medium hazard ranks	12	23	5	26
Proportion of total response (%)[a]	43	82	18	

Note: — = no responses at this hazard rank.
[a]Several jurisdictions reported that responsibility for identifying rock slopes is shared among multiple classes of personnel; hence the numbers in the "Responsible Personnel" columns total more than the number of responses, and the percentages total more than 100%.

3. DEVELOPMENT OF ROCK SLOPE MANAGEMENT SYSTEMS

Several transportation agencies evaluated the RHRS approach in the early 1990s, aided and encouraged by training programs supported by FHWA (Pierson and Van Vickle 1993). Bateman (2003) reported that a survey conducted by the Tennessee DOT approximately a decade after the introduction of the RHRS revealed that at least 18 transportation agencies had adopted some form of rock slope management system by that time. The more recent and comprehensive survey conducted in 2008 revealed that 67% of jurisdictions, or 25 agencies, with rockfall concerns have a rock slope management plan in place or plan to implement such a system in the near future.

Fourteen agencies (33%) stated that they did not use a rock slope management system nor did they have any plans at that time to implement such a system. This response included all 10 agencies with low rockfall concerns and four with medium rockfall concerns. These responses probably indicate that an agency that perceives rockfall problems to be of low or moderate concern either is more likely to remain reactive to rockfalls or is not willing, or perhaps is unable, to dedicate the agency resources required to implement a more proactive approach.

Of the 25 agencies with a rock slope management system in place, 22 (88%) said their system was based on the RHRS. Of the three agencies that intend to implement a management system, all indicated that they plan to base theirs on the RHRS. However, most agencies have made some modifications to the published RHRS methodology, but this reaction is not unexpected. The system was designed for use by a variety of organizational structures and to be modified as needed for different applications or to conform to local conditions such as the topography, type of geology, and weather effects. Many of the more recent changes have involved methods for collecting and storing site-location data. With the advent of the Global Positioning System (GPS) and the use of laptop computers and devices such as hand-held computers and digital cameras, many agencies have automated portions of the data collection process and consider these to be changes to the original RHRS.

Several jurisdictions adopted the RHRS in its original form, whereas others modified the system to a greater or lesser degree so that it suited their local conditions and user requirements (Bateman 2003). States that adopted the RHRS without significant modification include California, Kentucky, Nevada, New Hampshire, North Carolina, Pennsylvania, Virginia, and Wyoming (Bateman 2003). Since 2003, Montana has completed implementation of the RHRS without significant modifications.

Following the initial introduction of the RHRS in the early 1990s, two jurisdictions—the Ontario Ministry of Transportation and the New York State DOT—decided to approach the slope management process with systems that differed from the concepts contained within the RHRS (Franklin and Senior 1997b; Senior 1999; New York State DOT 1996).

Subsequently, other jurisdictions independently assessed the available methodologies and then developed systems that combined components from these early systems and added other modifications to address local conditions or individual procedural preferences. Colorado, Ohio, Missouri, Tennessee, and Utah followed this approach. Washington State has developed a program aimed at addressing all types of slope instabilities including landslides, debris flows, and fill failures that also incorporates economic considerations based on relative repair and user costs. A similar approach has been implemented as a pilot study by Oregon (Oregon DOT 2001). Alaska is following this trend as part of their geotechnical asset management program.

4. NEW YORK STATE ROCK SLOPE RATING PROCEDURE

In 1988, after having a fatality caused by rockfall, the New York State DOT evaluated the rockfall hazards along state highways with a modified version of the early rating system suggested by Wyllie in a presentation at an FHWA-sponsored seminar (Wyllie 1987) and produced a list of about 1,700 rock slopes. Although this effort produced a ranking of perceived rock slope hazards within the state, no formal statewide implementation policy was established to schedule and fund mitigation efforts. Subsequently, a revised working draft rating system was developed in 1993 (New York State DOT 1993). This draft had three changes to the initial system (New York State DOT 1996):

- Isolation of three components of a possible rockfall–vehicle accident as independent factors,
- More objective attention to the question of how much risk is associated with a falling rock that

hits a vehicle as well as with a vehicle that hits a fallen rock, and

- Consideration of not only the hazard posed by an existing rock slope but also the level of hazard remaining after remediation.

Although never fully implemented statewide, the system has continued to evolve, and an updated report was issued in 2007 (New York State DOT 2007).

The system design is based on the assumption that the rockfall hazard at any particular rock slope site poses a relative risk to the public resulting from the interaction of geological, cross-sectional, and traffic-related factors at that site. Each factor is assumed to be independent of the others. Thus the factors can be combined (multiplied) to create a number, defined as the *total relative risk* (TRR), which is representative of the risk that a rockfall will cause a vehicular accident at each rock slope location on the statewide inventory. Three factors are used to assess the comparative risk that an accident will be caused by a rockfall (Hadjin 2002):

1. Geological properties of the rock slope (*geological factor*, GF);
2. Ditch configurations and slope offset from the pavement edge (*section factor*, SF); and
3. Traffic volume and stopping sight distance on the highway approaching the site (*human exposure factor*, HEF).

Each of the three factors is quantified on the basis of site observations and measurements. The TRR value for each rockfall site is determined by multiplying the GF, SF, and HEF values:

$$TRR = GF \times SF \times HEF$$

4.1 Geological Factor

The GF uses six parameters to represent the hazard posed by rockfall based on the specific geological and physical characteristics of the slope (Figure 4-1). Five of these parameters are similar to those used in the RHRS: geology, block size, rock friction, water and climate, and rockfall history. The sixth parameter, backslope conditions, measures the hazard contributed from the natural slopes located above the roadway cut slope. When these slopes are steeper than 25°, the presence of boulders on the slope influences this score. As with the RHRS, two forms of evaluation are used for the geology parameter: one for crystalline rocks, in which instabilities are assumed to be related to the presence, frequency, and orientation of fractures, and a second for sedimentary rocks, in which the rockfall hazard is assumed to be related to the characteristic layering. Each site evaluation is based on the score produced by the most appropriate form.

Similar to the RHRS, each parameter within the GF is scored by using five steps weighted according to an exponential (3^x) scale, resulting in values

SCORE			1 point	3 points	9 points	27 points	81 points
or	1a	Geology (crystalline)	Massive, no fractures dipping out of slope	Discontinuous fractures, random orientations	Fractures forming wedges	Discontinuous fractures dipping out of slope	Continuous fractures dipping out of slope
	1b	Geology (sedimentary)	Horizontal to slightly dipping	Raveling, occasional small blocks	Small overhangs or columns, numerous small blocks	Overhangs, some large unstable blocks, high columns	Bedding or joints dipping out of slope
and	2	Block Size	≤ 0.15 m	0.15 – 0.3 m	0.3 – 0.6 m	0.6 – 1.5 m	1.5 m +
	3	Rock Friction	Rough, irregular	Undulating	Planar	Smooth, slickensided	Clay, gouge filled
	4	Water / Climate	Dry	Some seepage	Moderate seepage	High seepage, brush vegetation	High seepage with long backslope and high vegetation
	5	Rockfall History	No falls (0-4 / yr)	Occasional minor falls (5-14 / yr)	Occasional falls (15-24 / yr)	Regular falls (25-34 / yr)	Major falls / slides (35 + falls / yr)
	6	Backslope Conditions	Flat / gentle 0-15 deg.	Moderate 15-25 deg.	Steep 25-35 deg.	Very steep > 35 deg. Or 25-35 deg. with boulders	Very steep slope > 35 deg. with backslope boulders

FIGURE 4-1 GF scoring categories in New York State Rock Slope Rating Procedure (adapted from New York State DOT 2007).

ranging from 1 point for the best case to 81 points for the worst case. The final numerical value for the GF consists of the sum of the points assigned to each of the six parameters, but the sum is then divided by 10 in order to reduce that numerical value.

4.2 Section Factor

The SF represents the risk that fallen rock would actually reach the pavement by comparing actual ditch geometry and rock slope offset with the widely accepted Ritchie ditch criteria (Ritchie 1963). The underlying assumption is that ditches with a geometry that meets the Ritchie criteria will prevent all but 15% of the falling rocks from reaching the traffic lanes (New York State DOT 1996, 2007). The evaluation of SF involves the measurement of slope angle, slope height, actual ditch depth below pavement level, and the actual ditch width from the pavement edge to the beginning of the cut face. The slope angle and slope height values are used with an empirical chart developed by Ritchie to determine ideal ditch width and depth dimensions. Slope height and ditch dimensions can be measured either in U.S. customary units (feet) or in the International System of Units (SI) (meters). By using the appropriately scaled empirical chart, the Ritchie ideal dimensions will be reported in the same measurement units. The SF numerical value is then computed as follows:

$$SF = (DR + WR)/(DA + WA)$$

where

DR = ditch depth, obtained from empirical Ritchie graph;
WR = ditch width, obtained from empirical Ritchie graph;
DA = actual ditch depth, measured in the field; and
WA = actual offset distance, from toe of rock slope to pavement edge, or shoulder edge where one exists, with a minimum value of 1 m (3 ft).

Values for this calculation range from a best-case value of 1 to a worst-case value of about 11. When the sum of the existing ditch values (DA + WA) is greater than or equal to the sum of the Ritchie values (DR + WR), the SF is set to 1 and the TRR score for the site is defined as 1. This value is based on an assumption that should a rockfall occur, the ditch will trap the fallen material and it will not reach the roadway.

4.3 Human Exposure Factor

The HEF represents the statistical probability of an accident caused by a collision between rockfall and passing vehicles. If rock debris falls and reaches the roadway, a vehicle is threatened with impact by two separate mechanisms:

1. The falling rock hits a vehicle or lands so close to an approaching vehicle that the vehicle runs into the rock, or
2. The vehicle hits a previously fallen rock that has come to rest on the roadway.

The rock in the first situation is considered to be an *active condition* because it is falling as the vehicle approaches or passes under the point of impact. The second situation is termed a *passive condition* because the rock has landed before the vehicle approaches and is then hit by the vehicle. The HEF is accordingly broken into two parts referred to as the active factor (F_a) and the passive factor (F_p).

F_a includes two possibilities: a moving vehicle is hit by falling rock, and the driver sees the rock falling but is unable to stop the vehicle in time to avoid a collision. Documentation for the New York State Rock Slope Rating Procedure (New York State DOT 1996, 2007) demonstrates that for these two active condition cases, the probability of a vehicle's being hit by a falling rock or running into one that is actively falling can be expressed as follows:

For U.S. customary units:

$$F_a = AADT[(L + SSD)/(V \times 126{,}720)]$$

For SI:

$$F_a = AADT[(L + SSD)/(V \times 24{,}000)]$$

where

F_a = active condition factor,
AADT = annual average daily traffic (two-way for two-lane undivided highways or one-way for divided highways),
L = length of rockfall zone [ft (m)],
SSD = stopping sight distance [ft (m)], and
V = travel speed [mph (km/h)].

The two conversion factors account for the measurement units—feet and miles per hour or meters and kilometers per hour, respectively.

F_p characterizes the relative risk of a vehicle striking a rock that has previously fallen and is within the traveled way. If the highway section has adequate SSD, a driver may perceive the problem and react to avoid hitting the rock, and no accident will occur. As the SSD shortens, a collision with the fallen rock becomes more likely. The governing factors in this situation are the available decision sight distance (DSD) and the required SSD. Both can be determined by field observation using methods established by the American Association of State Highway and Transportation Officials and by reference to highway design manuals. The documentation for the New York State rock slope rating procedure (New York State DOT 1996) demonstrates that the relative risk of a vehicle's hitting a fallen rock lying on the road can be expressed as

$$F_p = \log_{10}(\text{AADT}) \times \log_{10}(L)[a/(\text{SSD} - a)]$$

where a is the maximum of either (SSD − DSD) or zero and the other variables are as defined earlier.

HEF computation thus requires the following three values for each rock slope being evaluated:

- Average travel speed: established by the regional staff,
- AADT: readily available from transportation agency databases, and
- DSD: must be measured in the field for each site.

Recorded plan information often does not suffice because in many instances vertical sight distance will be the controlling factor.

The HEF value is then defined as the sum of the active and passive risk values divided by 3:

$$\text{HEF} = (F_a + F_p)/3$$

In the foregoing equation, the sum is divided by 3 solely to reduce its numerical value.

4.4 Appropriate Mitigation Treatment

The TRR values, computed by the procedures just described, are used for comparing the rockfall hazards and consequences that exist at a series of rock slopes, but they are of limited value for evaluating possible benefits of undertaking specific mitigation treatments at these sites. For that purpose, the developers of the New York State Rock Slope Rating Procedure proposed the use of the *risk reduction concept*, defined as the benefit provided by one of several possible treatments applicable to a given rock slope. The concept introduces two additional values:

- *Residual risk* is the much smaller TRR value that can be expected after a slope has received a mitigation treatment, and
- *Risk reduction* can then be defined as the difference between the TRR value of the existing unmitigated slope and the estimated residual risk value for the same slope after mitigation.

In brief, the concept can be defined as follows:

risk reduction = TRR − residual risk

The New York State Rock Slope Rating Procedure (New York State DOT 1996) specifies that an optimum residual risk target value can be achieved by

- Presplitting the rock slope at a properly designed angle, and
- Providing a catchment zone at the slope base in accordance with the Ritchie ditch criteria.

These mitigation approaches are but two of a much wider suite of mitigation techniques described elsewhere in this report, and they are impractical in many cases. However, when their use is possible, they provide a beneficial amount of risk reduction. Presplitting potentially reduces values for all three factors (GF, SF, and HEF). The GF parameters of geology, block size, rockfall, and backslope above the cut are reduced when a new smooth rock face is created. The SF value is reduced to 1 when the new ditch dimensions meet their adopted Ritchie criteria. If the site distance is improved by widening the fallout area along a section of road with a horizontal curve, the HEF value will be reduced accordingly.

This optimum residual risk value should not be treated as a goal that must be achieved, but it serves as an estimate of what can be accomplished. Other remedial treatments—rock scaling, rock bolting, use of a rock catchment fence, and others—would tend to produce less risk reduction solely on the basis of the applied rating criteria. Computing risk reduction values for a series of alternative treatments and then comparing these reductions with the estimated cost of the alternatives provides a form of cost–benefit analysis, which can help decision makers select a potentially more cost-effective mitigation treatment.

5. ONTARIO ROCKFALL HAZARD RATING SYSTEM

The Ontario road system is one of the most extensive and heavily traveled in Canada. Although Ontario is located in a region of comparatively low topographic relief, the hazards resulting from rockfall became evident in 1992 when a fatality due to rockfall occurred (Franklin and Senior 1997b).

5.1 Ontario Perspective

In northern Ontario, highway construction has required excavations in the igneous and metamorphic rocks of the Canadian Shield (Figure 4-2). In southern Ontario, routes crossing the Paleozoic sedimentary rocks of the Niagara escarpment and other smaller escarpments may also require rock cuts. Still, the topographic relief in northern Ontario is fairly subdued (Senior 1999), and the rock cuts in this region rarely exceed 15 m (50 ft) high. Although rock cuts along the eastern shore of Lake Superior are generally higher, they seldom exceed heights of 30 m (100 ft). In southern Ontario, the Niagara escarpment typically rises 84 m (275 ft) above its base and forms the largest topographic barrier within the province. However, several major valleys provide relatively straightforward transportation corridors through this barrier, and the escarpment is partially or completely buried by glacial moraine deposits at several locations. Thus, rock cuts in southern Ontario rarely exceed heights of 15 m (50 ft), and most are much lower (Figure 4-3).

Many of the rock cuts along Ontario highways were built before the introduction of controlled blasting methods, and roadside ditches were designed to carry water and retain snow, not falling rock. As a result, rock cuts in the northern region consist mainly of blocky, often overblasted and heavily fractured rock faces. Although most are predominantly stable, they are still prone to raveling and ice jacking, which continually deposit small amounts of rock into roadside ditches and onto shoulder areas. As these failures progress, larger rockfall volumes may occur and the rockfall debris can spill out onto the pavement. In the southern portion of the province, differential weathering of sedimentary rocks, often composed of interbedded sequences of more resistant formations (limestone, dolomite, or sandstone) and more easily weathered shale or heavily fractured or thinly bedded rocks, yields roadside rock cuts that degrade and ultimately deposit rock fragments seasonally onto the highway.

FIGURE 4-2
Typical rock cut along northern Ontario highway crossing Canadian Shield showing section of Provincial Highway 69 near Parry Sound, Ontario, in May 2006.
(PHOTOGRAPH COURTESY OF THE ONTARIO MINISTRY OF TRANSPORTATION.)

FIGURE 4-3
Typical rock cut in southern Ontario exposing Paleozoic sedimentary rocks; rock cut located along divided highway (MacDonald–Cartier Freeway, Highway 401) at Kingston, Ontario.
(PHOTOGRAPH COURTESY OF THE ONTARIO MINISTRY OF TRANSPORTATION.)

In 1992, the Ontario Ministry of Transportation (MTO) initiated a review of its rockfall hazard policy and construction standards following a fatal accident caused by a rock slope failure. The review began with a preliminary assessment of rock slopes along provincial highways using the RHRS. Over 12,000 lane-km (7,500 lane-mi) were evaluated within the MTO's Northern Region, and the preliminary assessment identified roughly 200 Class A rock slopes, the highest hazard category (Senior 1999).

MTO subsequently engaged a consultant to undertake detailed hazard scores following the RHRS procedures and to provide preliminary treatment recommendations and costs. Initially no substantial modifications to the existing RHRS were anticipated. However, the initial detailed RHRS ratings showed that the system presented several limitations based on specific Ontario conditions, which became apparent with the use of the proposed exponential (3^x) scoring system. The results were not sensitive enough within the smaller range of conditions encountered within the province, which tended toward the middle to lower end of the ranges covered by the rating criteria.

It was found that only six of the 10 RHRS parameters made a meaningful contribution to the rating results. For example, the RHRS score changed by less than 1% as the heights of otherwise identical rock cuts varied from 5 m (16 ft) to 15 m (50 ft), the typical range of heights of most highway cuts in Ontario. This finding could imply that the careful estimation of slope height was unnecessary; yet a rock cut that is 15 m (50 ft) high is significantly more hazardous and more expensive to treat than cuts of lesser height. Because of similar observations with other categories, it was decided that the Ontario rating system would abandon the use of the exponential weighting approach. This decision amounted to a significant departure from a major element of the RHRS that was deemed valuable for agencies in which a wider range of natural conditions existed. Once this decision was made, further modifications were considered and adopted.

5.2 Ontario Rockfall Hazard Rating System

The Ontario Rockfall Hazard Rating (RHRON) system, initially developed in the early 1990s, thus involved major revisions to the RHRS parameters to improve the evaluation of rockfall conditions in Ontario (Franklin and Senior 1997a, 1997b; Senior 1999). The RHRON process was recently reviewed, based on user feedback, and an updated field procedures manual has been released (Franklin et al. 2012). The RHRON system is organized around four factors:

F1: magnitude ("How much rock is potentially unstable?");
F2: instability ("How long before the next fall?");
F3: reach ("What are the chances of this fall reaching or blocking the highway?"); and
F4: consequences ("What damage will it do when it gets there?").

These factors address some basic questions and are evaluated by the criteria shown in Table 4-5. Figure 4-4 illustrates schematically the relationships between the four RHRON factors and how these factors are related to index measurements and site descriptions that define the assessment criteria in Table 4-5. Within RHRON, parameters and factors are rated on a scale from 0 (good) to 9 (bad). The basic RHRON score is computed as

$$\text{RHRON} = (F1 + F2 + F3 + F4)/4$$

An important perceived benefit of the RHRON system is its inclusion of parameters that are related to the types of failure mechanisms common to rock cuts in Ontario. Field observations have identified the types and frequencies of potential failure mechanisms as follows: raveling (25%), toppling (23%), overhang (15%), ice jacking (11%), block roll (11%), two-dimensional sliding (8%), three-dimensional sliding (2%), and multiple causes (5%) (Franklin and Senior 1997a). While the RHRS includes raveling within its Case 2 geological character category, raveling is not explicitly recognized as a separate failure mechanism. MTO deemed it important to recognize raveling explicitly because raveling represented a major cause of rockfall in Ontario. Further analysis revealed other parameters that appeared to be fundamental to the assessment of raveling as well as ice jacking and similar modes of rock face instability:

- A measure of intensity of jointing, which could be obtained from block size observations or by evaluating rock cores with rock quality designation

Table 4-5
Four Factors Used by RHRON System

Factor	Question	Assessment Criteria
Magnitude	How much rock is likely to become unstable? How much potential energy would be released?	Volume and mass of rock; height, velocity, and energy of fall
Instability	How unstable is the material? How likely and how frequent are rockfall events?	Instability mechanisms; face looseness and irregularity; rock quality; shear strength; joint roughness, filling, persistence, and orientation; rock durability and strength; water pressure; seismic accelerations; severity of freeze–thaw process
Reach	What are the chances that this rock will reach the pavement? How much of the pavement will be blocked?	Kinetic energy; height of fall; size and shape of debris; presence of launching features; ditch width and depth; history and evidence of previous rockfalls
Consequences	How serious are the consequences of the blockage?	Traffic density, rockfall visibility, speed limit and stopping distance, pavement width

Source: Senior 1999.

(RQD) methods [RQD is a standard method of estimating the intensity of jointing by examining the characteristics of drilled rock cores. See Section 3.4.1 in Chapter 6 or refer to work by Deere and Deere (1989).];

- A suitable description of face looseness obtained by quantifying the characteristics of joint apertures; and
- Documentation of the height of the water table at or immediately behind the rock face as a percentage of face height.

Within the RHRON system, most of the original RHRS parameters have been redefined to some extent. For example, on the basis of index measurements and descriptions, joint shear strength has been linked to peak friction angle, while difference in erosion rates (durability) has been expressed in terms of a slake–durability index to allow estimation of erosion rates (Franklin 1983; Shakoor and Rodgers 1992). The required laboratory testing becomes a frame of reference for several parameters, but because this testing adds another level of complexity in completion of the ratings, it is only performed when necessary. New parameters used by the RHRON system include the total quantity of unstable rock, face looseness, crest angle (ratio of slope height to width of ditch plus shoulder), and "overspill," an estimate of how much of the road will become blocked by the rockfall. Several of these parameters require a subjective evaluation.

As with the RHRS, the RHRON procedure includes both a preliminary and a detailed rating:

- The preliminary screening is used to identify Class A sites, which require a detailed inspection by a rock hazard specialist; and
- A subsequent detailed rating of the A-rated sites includes identification of potential instability mechanisms, development of a detailed RHRON system rating, preparation of preliminary mitigation recommendations along with preliminary cost estimates, and photographic or video documentation of the hazards.

5.3 RHRON Preliminary Screening Process

The RHRON preliminary screening process is the first step in creating an inventory of potentially hazardous rockfall locations within a geographical area or highway maintenance district. The process begins with a compilation of the rock cuts that require evaluation on one or more RHRON Highway Rockfall Hazards Log sheets (Figure 4-5). These identify the each rock cut location and summarize its basic RHRON scores and rockfall hazard class (A, B, or C). Each rock cut is subjected to a preliminary assessment that is recorded on

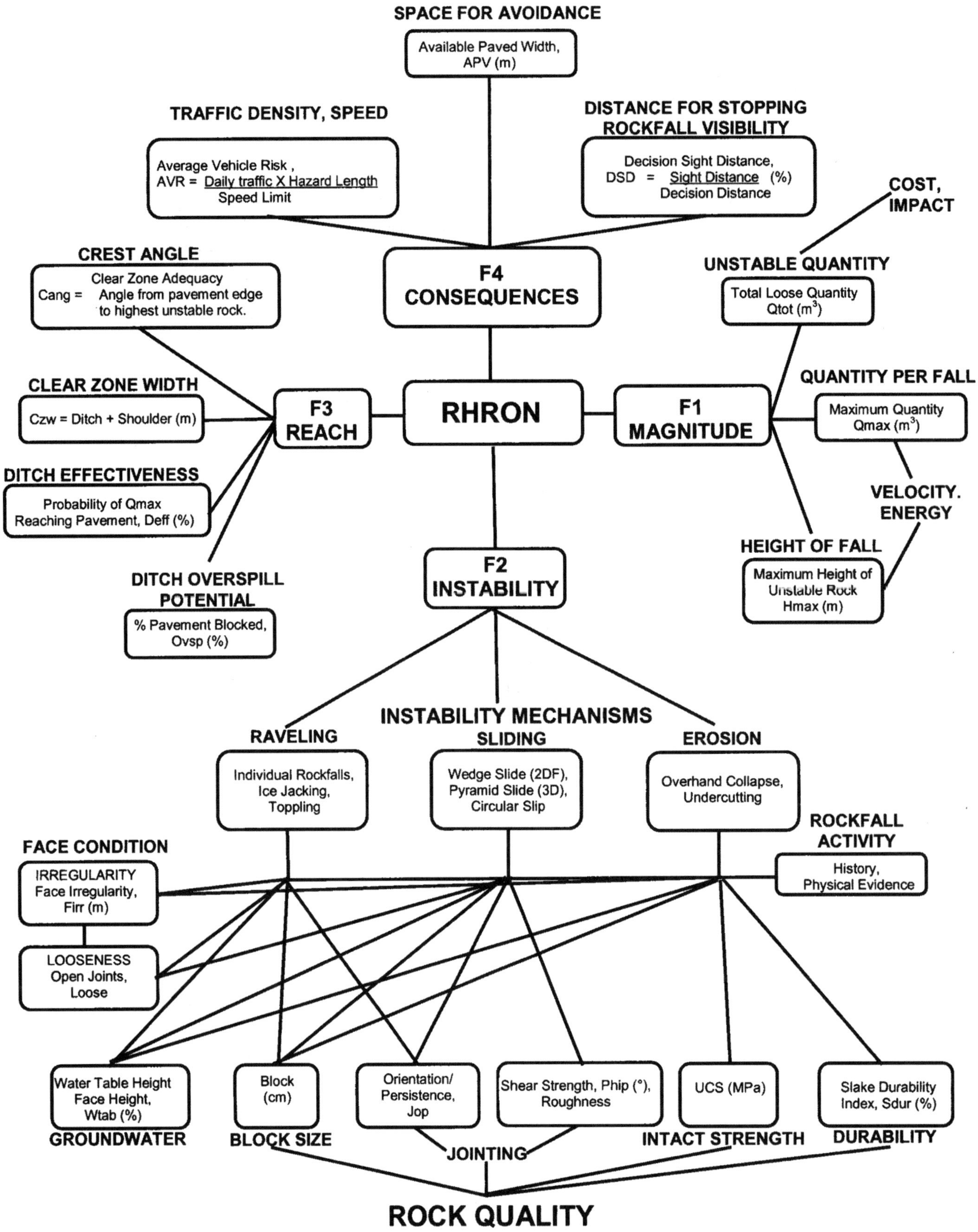

FIGURE 4-4
Schematic illustration of relationships among four factors and parameters used by RHRON (Franklin et al. 2012).

Page ____ of ____

				LHRS[1]		Alternative GPS Location Details			Mechanism[2]		Factor[3]					Hazard Class	
Hwy #	Site #	Direction of Lane(s)	Side NSEW	Base Point	Offset	Zone N/A	Northing Latitude	Easting Longitude	SROIBWPT	Cang	1	2	3	4	Basic RHRON	A:B:C	Notes and Remarks

NOTES:

[1] **LHRS is Ontario's Linear Highway Reference System that uses offsets (in km) from basepoints such as road junctions to identify a roadway location.**

[2] **Mechanism defines rockfall failure: Stable, Raveling, Overhang, Ice-jacking, Block (single), Wedge (2D), Pyramid (3D), Toppling.**

[3] **Factor refers to the four RHRON Factors (Figure 4-4): F1 = Magnitude, F2 = Instability, F3 = Reach, F4 = Consequences.**

FIGURE 4-5
RHRON Highway Rockfall Hazards Log Sheet (Franklin et al. 2012). (Note: This form and the others reproduced in this chapter are available on the DVD that accompanies this book.)

an RHRON Preliminary Screening Data Sheet (Figure 4-6).

The preliminary assessment involves the determination of the *crest angle*, which is defined as "the upward angle from the edge of pavement to the highest potentially unstable block" (Franklin et al. 2012). Conservatively, this is usually taken as the crest of the cut slope. If a rock face is vertical, the crest angle is defined by the height of the rock face and the combined width of the ditch and shoulder. The crest angle is the basis for assessing the lateral extent, or *reach*, of falling rock fragments, and thus is shown as the parameter used to assess RHRON Factor F3 in Figure 4-6. During the preliminary assessment, each of the four RHRON factors is evaluated by a single parameter. These parameters have been selected because they are readily observed or estimated, which makes the preliminary screening process rapid and relatively consistent from individual to individual while also providing a useful rockfall hazard assessment. These simple evaluation steps are in contrast to the much more complex evaluations required in the RHRON detailed rating process described in Section 5.4.

The basic RHRON score is the average of the four factor scores, and so will range between 0 and 9. Perhaps because of its limited numerical range, Franklin et al. (2012) specify it should be recorded to two decimal places. Franklin et al. (2012) also specify how a rock cut is classed into categories A, B, or C on the basis of the values of the crest angle and basic RHRON score. Those locations that belong to Class A are then subjected to a detailed rating assessment.

5.4 RHRON Detailed Rating Process

The RHRON detailed rating process is supported by a two-sided Field Data Sheet (Figure 4-7). Detailed RHRON assessments are based on the same concepts as the simplified RHRON rating scheme; they provide equal weight to the four factors of magnitude (F1), instability (F2), reach (F3), and consequences (F4). However, each of these factors is now assessed by using a more comprehensive set of 20 parameters (Franklin et al. 2012).

Side 1 of the RHRON Field Data Sheet (Figure 4-7*a*) allows segmentation of the site into as many as eight *hazard segments*, which are uninterrupted lengths along which the potential instability mode and degree of hazard are accepted as sufficiently constant to allow the hazard condition to be recorded as a single number. For each hazard segment, the slope height and dimensions are recorded along with other characteristics that tend to be variable, such as instability mechanisms, remedial treatments, and costs. The most hazardous segment is then given a complete detailed RHRON rating. The detailed RHRON rating process also gives equal and separate attention to the three main instability mechanisms: raveling, sliding, and erosion. The bottom portion of Figure 4-4 shows the complex interrelationships between several parameters and these instability mechanisms.

During the detailed RHRON rating process, each of the four factors is evaluated on the basis of several individual parameters. Side 2 of the RHRON Field Data Sheet (Figure 4-7*b*) allows the assessment of the most hazardous segment identified on Side 1 of the form by rating the parameters identified on the upper part of the form. The lower part of the form provides definitions of appropriate numerical ratings (ranging from 0 to 9) for those parameters that are not explicitly quantifiable. Once these ratings are recorded, the four RHRON factors (F1 to F4) are calculated using a series of defined parameter formulas. This process is explained in detail in the RHRON field procedures manual (Franklin et al. 2012).

5.5 Selecting Mitigation Method

The lower portion of Side 1 of the detailed rating Field Data Sheet (Figure 4-7*a*) provides a checklist to guide the selection of mitigation methods that may be proposed as a result of the RHRON evaluation. A full range of treatment alternatives is listed, including measures for hazard reduction by trimming and regrading, stabilization and reinforcement, catchment and control systems, and warning and monitoring systems. Trimming and regrading methods will reduce the magnitude factor (F1), while stabilization methods are designed to reduce the instability factor (F2). Catchment and control methods are intended to reduce the reach factor (F3) and warning and monitoring systems are used to reduce the results of a rockfall, and thus affect the consequences factor (F4).

Ministry of Transportation

Ontario Rockfall Hazard Rating System (RHRON) - PRELIMINARY SCREENING DATA SHEET

Highway #:	AADT:
Site #:	GPS Unit:
Direction of Lanes:	GPS Zone: (NAD 83)
NSEW:	GPS Coordinates: N
LHRS Base point:	E
LHRS Offset Distance:	Waypoint #:

ROCKFALL WARNING SIGNS: Yes / No — Digital Image Time:

	Yes / No	
Stable		Measured crest angle (B) =
Ravelling		Measured crest angle (A) =
Overhang		Slope Distance S_A (if measured) =
Ice-Jacking		Czw =
Block (single)		Roadway width (between white lines) X =
2d **W**edge		H.I. =
Pyramid		Cang =
Toppling		Basic RHRON required (Cang>27) Yes / No

Check (✓) all that apply

COMMENTS:

UTILITIES:

RHRON FACTORS:

F1=

F2=

F3=

F4= **F4(Avg)=**

RHRON = (F1+F2+F3+F4) / 4 =

Rated by:

HAZARD CLASS (A, B or C): **Date:**

	F1 - MAGNITUDE	F2 - INSTABILITY	F3 - REACH	F4 - CONSEQUENCES	
RATING	**Qtot (m^3)**	**Frequency of falls**	**Crest Angle - Cang (degrees)**	**Traffic Density (AADT)**	**Visibility of rock on highway**
0	1	> 100 yr (slight chance of a fall)	20	0-750	> 250 m
1	2	50 yr	27	750-1500	233
2	3	10 yr	33	1500-3000	217
3	5	5 yr	40	3000-6000	200
4	8	Annually	47	6000-11000	183
5	13	6 Monthly	53	11000-20000	167
6	21	Monthly	60	20000-35000	150
7	36	Bi Weekly	67	35000-65000	133
8	60	Weekly	73	65000-110000	117
9	> 100	Daily - Highly unstable, collapse imminent	80 (high face, narrow shoulder)	> 110,000	< 100 m (very limited visibility)

FIGURE 4-6
RHRON Preliminary Screening Data Sheet (Franklin et al. 2012).

SEGMENT	1	2	3	4	5	6	7	8	SITE#: Hwy:		
Start segment m									Direction: Side:		
End segment m									LHRS: Offset:		
Length of segment m									ΣLhaz m:		
Digital image, # & times									GPS Waypoint number:		
ROCKFALL TYPE AND QUANTITY **S**table, **A**dequate ditch, **R**aveling, **O**verhang, **I**ce-jacking, **B**lock [single], 2D **W**edge, **P**yramid, **T**oppling											
Mechanism									V2 Qmax Maximum fall m^3:		
Quantity Qtot m3									V3 ΣQtot Total falls m^3:		
SITE GEOMETRY AND GPS COORDINATES									UTILITIES		
Offside angle (B)									Type	Side	Offset
Nearside angle (A)											
Slope distance (S)									POSITIVE SITE ID?		
Paved width (X)									Yes / No		
Height m									V12 Height (worst seg):		
Czw m									V14 Czw (worst seg):		
PROPOSED METHOD(S) AND UNIT COSTS OF REMEDIATION									POCKMARKS (circle quantity)		
(preliminary assessment for prioritization only) Quantities per segment									Many	Some	None

TRIMMING AND REGRADING		(To reduce F1 magnitude factor by ___ % = REMX1)									Total	
Hoe Ram	crew-hr.											
Backhoe Exc	crew-hr.											
Trim blast	m drilled											
Control blast	cu. m.											
STABILIZATION AND REINFORCEMENT		(To reduce F2 instability factor by ___ % = REMX2)										
Rockbolts	each											
HT anchors	m drilled											
Drainholes	m drilled											
Shotcrete	sq. m.											
Mesh	sq. m.											
Straps	lin. m.											
CATCH, CONTROL & AVOIDANCE		(To reduce F3 reach factor by ___ % = REMX3)										
Clean ditch	lin. m											
New ditch	lin. m.											
J. barrier	lin. m											
Catch fence	lin. m.											
Draped mesh	sq. m											
SIGNPOSTING, MONITORING ALARMS etc.		(To reduce F4 consequences factor by ___ % = REMX4)									ΣREMX = 90%	
Fallen Rock	each											
WARNING POSTED		Yes	No	Paved Shoulder		Ditch Depth						
Continge = cost contingency multiplier											1.5	
Inspector's notes:												
Recommendations by:		Date:										

(a)

FIGURE 4-7
RHRON detailed rating Field Data Sheet: (*a*) Side 1 (Franklin et al. 2012).

SITE IDENTIFICATION		Hwy: Side:	SITE #:
Location: km N S E W of LHRS Basepoint:			
PARAMETER		VALUE V	RATING R
Identification	Detail		good bad
P1 Hist	History/evidence of falls	Rating only	0 1 2 3 4 5 6 7 8 9
P2 Qmax	Largest potential fall	m^3	0 1 2 3 4 5 6 7 8 9
P3 Qtot	Total of potential fall/slides	m^3	0 1 2 3 4 5 6 7 8 9
P4 Firr	Face irregularity	m	0 1 2 3 4 5 6 7 8 9
P5 Loose	Face looseness (open joints)	Rating only	0 1 2 3 4 5 6 7 8 9
P6 Jop	Joint orientation-persistence	Rating only	0 1 2 3 4 5 6 7 8 9
P7 UCS	Uniax. compressive strength	MPa	0 1 2 3 4 5 6 7 8 9
P8 Phip	Shear strength	Peak friction angle degrees	0 1 2 3 4 5 6 7 8 9
P9 Block	Block size	cm	0 1 2 3 4 5 6 7 8 9
P10 Sdur	Slake-durability index (I_{d2})	%	0 1 2 3 4 5 6 7 8 9
P11 Wtab	Water table (% height of face)	%	0 1 2 3 4 5 6 7 8 9
P12 Height	Slope height (to highest hazard)	m	0 1 2 3 4 5 6 7 8 9
P13 Cang	Crest angle = $\tan^{-1}$(V12/V14)	degrees	0 1 2 3 4 5 6 7 8 9
Traffic Information	AADT		
	Posted speed	Psl km/h	
	Sight distance	m	
P15 Deff	Ditch effectiveness	Estimate %	0 1 2 3 4 5 6 7 8 9
P16 Ovsp	Ditch overspill potential	Estimate %	0 1 2 3 4 5 6 7 8 9
Rating by:		Date:	

Estimates for P1, P4, P5, P6, P7, P8, P15 and P16

P1, P5, P6, P16		P15 Deff		P4 Firr
Term	Rating			
Ext Low	1	Ext Good	11	0.28
Very Low	2	Very Good	22	0.56
Low	3	Good	33	0.83
Mod Low	4	Mod Good	44	1.11
Moderate	5	Moderate	56	1.39
Mod High	6	Mod Poor	67	1.67
High	7	Poor	78	1.94
Very High	8	Very Poor	89	2.22
Ext High	9	Ext Poor	100	2.50

P7 UCS (Mpa)	Rating	Base Phi	P8 Phip
19	4	30	48
25	3.5	32	51
34	3	35	53
62	2	40	59
111	1	45	64
200	0	50	70

(b)

FIGURE 4-7 *(continued)*
RHRON detailed rating Field Data Sheet: (*b*) Side 2 (Franklin et al. 2012).

6. OREGON UNSTABLE SLOPE RATING SYSTEM

The Oregon Unstable Slope Rating System is a method for rating landslide and rockfall hazards along the Oregon state highway system that follows the asset management principles promoted by FHWA. Previously, only rock slopes along Oregon roads were inventoried and rated with the RHRS. In the new Oregon unstable slope rating system, all types of unstable slopes are inventoried with a single system, with rock slopes continuing to have RHRS ratings (Oregon DOT 2001; Pack et al. 2006).

The RHRS focuses on the potential for failure and the specific site conditions that contribute to a possible hazardous condition, but it does not address the consequences of failure related to the function of the overall highway system. Oregon highways span numerous physiographic and geographic boundaries, each having its own particular stability issue, such as high rock cuts in mountainous areas and large embankments on steep slopes in areas of high annual precipitation. It was decided that the evaluation of the broader range of failure types and conditions would benefit from a more generalized rating approach that could normalize the differences between physiographic provinces.

6.1 Assessing Hazard

The Oregon Unstable Slope Rating System is an adaptation of the Washington State Unstable Slope Management System (Lowell and Morin 2000). It uses an additive scoring system with five sets of parameters, as shown in Figure 4-8:

1. Failure type or hazard,
2. Roadway impact,

1	Failure type/ hazard	Very small or insignificant failures that do not affect the roadway (not scored)	Low hazard: slower slides with low potential for causing a road hazard (9 points)	Medium hazard: slides that have not moved suddenly in the past, but have the potential to cause a road hazard. (27 points)				High hazard: rapid slides that have created a road hazard in the past. Includes debris flows and rockfall. (81-100 points based on sight distance)		
			Low hazards receive 0 points	Medium hazards receive a maximum of 54 points				High hazards can receive full point range		
2	Roadway impact (pick one)	Landslides:	All low hazard slides above (0 points)	Would only affect shoulder during major failure (3 points)	Two-way traffic would remain after major failure (9 points)	One-way traffic would remain after major failure (27 pts)	Total closure in the event of a major failure: 0- to 3-mile detour (54 pts)	Total closure in the event of a major failure: 3- to 10-mile detour (70 pts)	Total closure in the event of a major failure: 10- to 60-mile detour (88 pts)	Total closure in the event of a major failure: >60-mile detour (100 pts)
		OR								
		Rockfall:	Rocks are completely contained in ditch (3 points)	Rocks fall into shoulder only (9 points)	Rocks enter roadway (27 points)	No ditch, all falling rocks enter roadway (81 points)		Rocks occasionally fill part or all of a lane (100 points)		
3	Annual maintenance frequency	0–5 Failures per year: sliding scale 0–100 points								
4	Average daily traffic	0–40,000 cars per day: sliding scale 1–100 points								
5	Accident history	No accidents (3 points)		Vehicle or property damage (9 points)		Injury (27 points)		Fatality (100 points)		

FIGURE 4-8 Summary sheet for Oregon Unstable Slope Rating System (adapted from Oregon DOT 2001 and Pack et al. 2006).

3. Annual maintenance frequency,
4. Average daily traffic (ADT), and
5. Accident history.

As summarized in Figure 4-8, each of the five parameters is evaluated with a maximum score of 100. These scores are then summed to produce a hazard score, with a maximum value of 500. The first factor (failure type or hazard) is based on the speed of the slope movement. This factor appears only partially applicable to rockfalls, since most rockfalls occur rapidly and so belong within the high hazard category. However, roadway impact (Factor 2) will typically provide compensating lower scores.

Factor 2 is divided into two sections: landslides and rockfall. Rockfall conditions are scored according to ditch effectiveness (summarized in Figure 4-8). The score for annual maintenance frequency (Factor 3) is developed by using two formulas. If maintenance occurs at the site less than once per year, the score is 50 times the annual frequency. When maintenance occurs more than once each year, the frequency is multiplied by 12.5 and then added to 37.5 but may not exceed a maximum score of 100.

ADT (Factor 4) is based on a sliding scale from zero to 100 points, as traffic increases from zero to 40,000 cars per day on the roadway. The use of 40,000 as a benchmark is arbitrary. Accident history (Factor 5) is based on historical data for accidents caused by slope failures. An accident involving a fatality receives a score of 100 because of the liability associated with such severe consequences.

6.2 Assessing Consequences

The Oregon system multiplies the hazard score by two additional factors that are intended to provide a measure of consequences concerning costs and user impacts. A final site ranking is produced that is considered during development of the Oregon State Transportation Improvement Program. These additional factors are

1. Maintenance benefit–cost (MBC) factor, and
2. Highway classification factor (HCF).

The MBC factor is related to the 20-year maintenance and failure repair costs for the site and varies between 0.5 and 1.5. As a factor, this step has a dramatic effect on the hazard score, since it can produce results ranging from a 50% reduction to a 50% increase in the hazard score. The HCF weights the final score depending on the significance of the route on the basis of highway classification, which includes district, regional, statewide, or Interstate designations. Each category has a weighted factor that reflects the importance of that classification, with the lowest classification (district roads) using a factor of 1.0 and the progressively higher categories of roadways using larger factors.

As described in Section 5.5 of Chapter 3, the original RHRS proposed dividing the estimated remediation cost by the hazard score and thus generating a list of projects ranked according to the greatest hazard reduction for a given investment. The new Oregon Unstable Slope Rating System assesses the hazard and consequences in a more equal fashion, but these assessments are separable for various analyses conducted under the asset management framework. It is anticipated that more effective decisions concerning landslide and rockfall project priority ranking will result once the new system is implemented and the overall financial needs of a management program are estimated and a long-term funding strategy is developed.

7. UTAH ROCKFALL HAZARD RATING SYSTEM

The Utah DOT began developing a statewide rockfall hazard inventory in 2001 by applying the RHRS preliminary rating procedures described in Section 5.3 of Chapter 3. This survey identified a total of 1,099 sites, composed of 479 A-rated sites, 569 B-rated sites, and 51 C-rated sites (Pack and Boie 2002). Following RHRS practice, Category A sites were perceived as the most hazardous and Category C sites the least. Examples of Utah sites that belong to these A, B, or C categories are shown in Figures 3-3, 3-4, and 3-5.

The Utah DOT subsequently undertook detailed rockfall investigations at sites categorized by the earlier evaluations as either A, A-, or B+. Approximately 500 sites were subjected to more detailed evaluations during 2002 and 2003. Since a definitive system for conducting detailed investigations had not yet been adopted, this second study evaluated each site by using three different rock slope management systems: the RHRS (described in Chapter 3), the Oregon Unstable Slope Management System (described in Section 6 of this chapter), and the New York State Rock Slope Rating Procedure (described in Section 4). At each site, the data collection procedures included all parameters required by each of the three systems. This method allowed a comparison of the results of each system relative to the Utah context (Pack et al. 2006).

From these efforts, a detailed hazard rating system was developed that met the needs of the Utah DOT. The system incorporates the RHRS preliminary rating procedures to create a statewide inventory of rockfall hazards and a modified version of the New York State Rock Slope Rating Procedure to develop detailed rock slope ratings.

The Utah evaluators concluded that the New York State Rock Slope Rating Procedure was limited by the subjective nature of the GF compared with the more objectively determined SF and HEF. Two parameters—the known rockfall history of a site and the observed block size—appeared to have a pivotal role and yet were equally weighted with four other more subjectively determined parameters. Upon review, Utah elected to double the weight of the rockfall history and frequency and block-size parameters over the other four factors. Hence, in the Utah system, the GF is calculated by

$$\begin{aligned} GF = &(\text{geology}) + 2\times(\text{block size}) \\ &+ (\text{rock friction}) + (\text{water}) \\ &+ 2\times(\text{history}) + (\text{backslope}) \end{aligned}$$

One additional concern related to terminology used in the New York State parameter definitions. The New York State system differentiates between sedimentary geological conditions and crystalline geological conditions when the geology parameter is rated within the GF (Figure 4-1). The purpose was to define physical characteristics associated with the rock type, but performance of these rocks varies within these lithologic classifications. For example, igneous basalt flows (volcanic) often experience more sedimentary-like toppling failures, whereas rockfalls from massive sandstones are similar to those from igneous crystalline rocks. The physical characteristics of metamorphic rocks vary from slate (sedimentary-like) to gneiss (crystalline). To avoid confusion, Utah adopted descriptive terms such as *layered* to replace the term *sedimentary* used in the New York State system. The goal of these terms is to better describe a rock type in terms of its mode of failure.

The New York State Rock Slope Rating Procedure proposed the use of the risk reduction concept to evaluate the cost-effectiveness of different mitigation schemes (Section 4.4). The Utah DOT project undertook a similar assessment at eight Utah rockfall locations by using the new Utah system (Pack et al. 2006). The GF, SF, and HEF scores for each of these locations were carefully reevaluated, and appropriate mitigation methods were selected for each location from a list of methods that included slope reshaping, ditch excavation, rock bolting, roadside barriers, controlled blasting, mechanical scaling, and control options such as fences, netting, and draping of rock faces with mesh.

The recommended mitigation methods were specific to each site, and with the assumption that they had been constructed, the influences on the GF, SF, and HEF were evaluated. These analyses provided useful examples of how the new Utah system could be applied to rank the mitigation of hazardous rockfall sites by priority while also illustrating the sensitivity of the parameters used by the system (Pack et al. 2006).

8. OHIO ROCKFALL HAZARD RATING MATRIX

In 2001 the Ohio DOT Office of Research and Development contracted with Kent State University to develop the Rockfall Hazard Rating Matrix for Ohio (Woodard 2004; Shakoor 2005). The study evaluated 108 rock slopes. Data characterizing each site were analyzed statistically to establish relationships between field observations and rockfall hazard ratings. These relationships were then used to design the Rockfall Hazard Rating Matrix (Figure 4-9), which contained four categories: geological parameters, indicating the potential for rockfall occurrence and size; geometric parameter, evaluating the potential for falling rock material to land on the roadway; traffic parameters, estimating the hazard posed to vehicles, and a rockfall history parameter. Figure 4-10 is the scoring sheet used for evaluating a rock slope in the field according to the criteria shown in the Ohio matrix.

8.1 Ohio Perspective

The geology in Ohio is characterized by the presence of gently dipping, resistant, competent strata (siltstone, sandstone, limestone) alternating with

EVALUATION PARAMETERS			RATING SCORES FOR DIFFERENT CATEGORIES OF EVALUATION CRITERIA			
			3 Points/(1)	9 Points/(2)	27 Points/(3)	81 Points/(4)
GEOLOGIC PARAMETERS						
Geologic Character	Differential Weathering	Slake Durability Index	90-100%	75-90%	50-75%	<50%
		Max. Amount of Undercutting	0-1 ft	1-2 ft	2-4 ft	>4 ft
	Discontinuity Role	Discontinuity Extent/Orient.	Discontinuous joints, favorable orientation	Discontinuous joints, random orientation	Discontinuous joints, adverse orientation	Continuous joints, adverse orientation
		Discontinuity Surface Features	Very rough JRC=20	Rough JRC=15	Undulating JRC=10	Smooth JRC=5
Block Size/Volume of Rock Fall			1 ft/ 3 yd^3	2 ft/ 6 yd^3	3 ft/ 9 yd^3	4 ft/ 12 yd^3
Hydrologic Conditions			No water seeps on slope	A few water seeps on slope	Many water seeps on slope	Numerous water seeps on slope
GEOMETRIC PARAMETERS						
Ritchie Score			<1	1-1.5	1.5-2.5	>2.5
TRAFFIC PARAMETERS						
ADT x Slope Length / 24 hrs / Posted Speed Limit 100%			25% of time (very low)	50% of time (low)	75% of time (medium)	100% of time (high)
% Decision Sight Distance			Adequate sight distance, ≥100%	Moderate sight distance, 75%	Limited sight distance, 50%	Very limited sight distance, <50%
Pavement Width			50 feet	40 feet	30 feet	20 feet
ROCKFALL HISTORY			No falls	A few falls	Many falls	Numerous falls

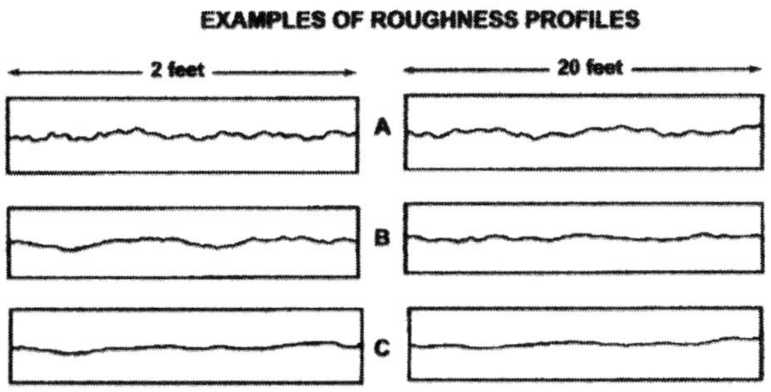

Joint Roughness Coefficient

A. Rough undulating – tension joints, rough sheeting, rough bedding — JRC = 20

B. Smooth undulating – smooth sheeting, non-planar foliation, undulating bedding. — JRC = 10

C. Smooth nearly planar – planar shear joints, planar foliation, planar bedding — JRC = 5

FIGURE 4-9
Ohio Rockfall Hazard Rating Matrix (Woodard 2004).

GEOLOGIC PARAMETERS					
Differential Erosion					
SDI	________	(a)	Greater Value (c or f)*	________	(g)
Maximum Amount of Undercutting	________	(b)			
Total (a + b)	________	(c)*	Block size	________	(h)
Discontinuities Role			Hydrologic	________	(i)
Discontinuity Extent/Orientation	________	(d)			
Discontinuity Surface Features	________	(e)	Total (g+h+i)/4	________	(j)
Total (d + e)	________	(f)*			
GEOMETRIC PARAMETER					
Ritchie's Score	________	(n)			
TRAFFIC PARAMETERS					
AVR	________	(o)			
% DSD	________	(p)			
Pavement Width	________	(q)			
Total (o+ p+q)/3	________	(r)			
ROCK FALL HISTORY					
History	________	(s)			
OVERALL SCORE					
Lines (j+n+r+s)	________	(t)			

FIGURE 4-10
Scoring sheet for Ohio Rockfall Hazard Rating Matrix (Woodard 2004) (SDI = slake–durability index, AVR = average vehicle risk, and DSD = decision sight distance).

erodible, less-competent strata (claystone, mudstone, shale). Differential weathering of rock slopes comprising such strata causes undercutting of the competent layers (Figure 4-11). Subsequent plane and wedge failures in the competent strata produce rockfalls. The frequency and size of these events depend on joint spacing within the competent unit and the extent to which it has been undercut.

Rock slopes composed of closely jointed rock strata do not show the effects of differential weathering and undercutting as much, but they still experience rockfalls similar to those on a raveling slope

(Figure 4-12). These rockfalls tend to occur more frequently and are typically composed of smaller blocks. Because they tend to involve smaller volumes that can be retained in the roadside catchment ditch, they generally pose less danger to the traveling public. In contrast, the undercutting-induced failures are quite hazardous because of their instantaneous occurrence, high speed, and occasionally large volumes of rock (Figure 4-13).

8.2 Ohio Rockfall Matrix Parameters

The Ohio Rockfall Hazard Rating Matrix contains four parameter categories (Figure 4-9) that correspond to eight of the nine categories contained in the standard RHRS (Table 4-1). Only the RHRS slope height category is not explicitly recorded or used by the Ohio approach. This variation may be due to the limited range in heights of Ohio rock cuts. However, the Ohio matrix reorders the sequence of RHRS categories in order to create its four parameter categories.

8.2.1 Geological Parameters

The Ohio matrix uses three parameters to define the geological conditions: geological character, block size or volume of rockfall, and hydrologic conditions. These parameters correspond to the RHRS Categories 6, 7, and 8. As discussed in Section 8.1, the occurrence and severity of rockfalls in Ohio are controlled by the presence of differential weathering and by the characteristics of discontinuities within the rock strata. Thus, the first geological parameter, geological character (Figure 4-9), is subdivided to allow evaluation of these two conditions.

The first condition is differential weathering, which plays a critical role in determining the potential for rockfall on slopes composed of interbedded competent and less-resistant strata (Figure 4-11). The statistical analysis by Woodward (2004) showed that the slake–durability index was useful in determining the rate of erosion of weak rock strata; this conclusion was supported by earlier research (Franklin 1983; Shakoor 1995; Shakoor and Brock 1987; Shakoor and Rodgers 1992; Shakoor and Weber 1988). The value of the slake–durability index has also been reported by Santi (2006). However, in many locations where the weak rock strata are inaccessible, it is difficult to obtain samples for testing. Accordingly, the Ohio matrix provides a second observational option to be used in lieu of slake durability testing, based on the maximum amount of undercutting, which can be measured directly or visually estimated.

The second condition, discontinuity role, is for slopes (rock types) in which jointing, bedding planes, or other discontinuities are the dominant structural features of the rock mass. The role of discontinuities in producing rockfall is evaluated by defining their orientation and continuity as well as their shear strength. The orientation and continuity definitions are identical to those used by the original RHRS (Pierson 1991). The shear strength of discontinuities is evaluated by the joint roughness coefficient (JRC) proposed by Barton (1974)

FIGURE 4-11 *(top)* Differential weathering of rock slope in Ohio (Woodard 2004).

FIGURE 4-12 *(bottom)* Differential weathering and undercutting of competent strata in closely jointed Ohio rock strata; only small rockfall fragments are produced (Woodard 2004).

FIGURE 4-13
Large rockfall blocks produced where widely spaced joints and thick competent strata are exposed (Woodard 2004).

and included at the bottom of the Ohio Rockfall Hazard Rating Matrix (Figure 4-9).

The remaining two geological parameters, the block size or volume of rockfall and the hydrologic conditions, are equivalent to RHRS Categories 7 and 8 (Table 4-1). The block size or volume of rockfall is determined by using the original RHRS terminology; the hydrologic conditions are qualitatively evaluated in terms of the number of water seeps observed, which appeared to be most appropriate for Ohio conditions on the basis of field observations and experience (Woodard 2004).

The geological parameters thus typically include four observational scores, each ranging from 3 to 81; these are summed and then divided by 4 to produce a geological parameter score ranging between 3 and 81 (as shown by the scoring sheet, Figure 4-10).

8.2.2 Geometric Parameter

The Ohio geometric parameter corresponds to the ditch effectiveness category (Category 2) in the original RHRS. It evaluates the potential for fallen rock material to land and remain on the roadway by following the procedures developed by the New York State DOT to evaluate the SF, as described in Section 4.2. In the Ohio system the New York SF is renamed the *Ritchie score*. Observations at many sites have revealed that this value ranges from <1 when the existing ditch exceeds the Ritchie recommended dimensions to values of around 3.5 for the most inadequate ditches. Thus these Ritchie scores are converted to geometric parameter ratings in the range 3 to 81 either by following the recommended assignments shown in the rating matrix (Figure 4-9) or by using a best-fit exponential equation and graph (Figure 4-14).

Davis and Shakoor (2005) reported the results of additional evaluations of the effectiveness of Ohio catchment ditches. On the basis of evaluations of 100 ditches, they concluded that although the Ritchie criteria provide numerical values for ditch dimensions, the criteria have some limitations and should be used with caution. Their studies used four evaluation methods and found that 48% of the evaluated ditches provided adequate protection, 28% provided marginal protection, and 24% provided inadequate protection.

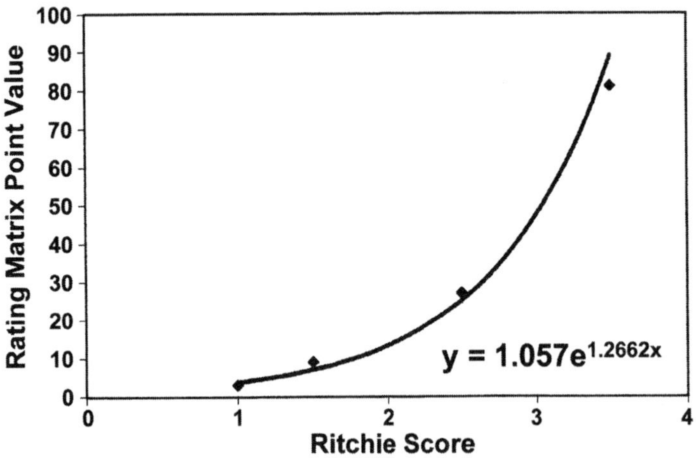

FIGURE 4-14
Relationship between Ritchie score and Ohio Rockfall Hazard Rating Matrix geometric parameter score (Woodard 2004).

8.2.3 Traffic Parameters

The Ohio traffic parameters include three components that correspond to RHRS Categories 3, 4, and 5. As with the original RHRS, these components are included to provide an estimate of the hazard posed by rockfall to roadway traffic. The parameters are evaluated according to the original RHRS terminology and methods, with only slight modifications to the definitions of the

Table 4-6
Classification of Evaluated Ohio Rock Slopes

Rockfall Hazard Potential	Overall Score	Number of Slopes[a]
Low	<50	26
Medium	50–100	51
High	>100	31

[a]Total number of slopes evaluated = 108.
Source: Woodard 2004.

RHRS classifications for decision sight distance and pavement width.

8.2.4 Rockfall History Parameter

When the Ohio Rockfall Hazard Rating Matrix was developed, rockfall history information for Ohio locations was not available. However, this information was added to the matrix as a fourth parameter because it was considered to be an important aspect of rockfall hazard evaluation and thus these data were expected to be collected and made available in the future (Woodard 2004).

8.3 Compilation of Overall Rating Scores

An overall hazard rating score is computed as the sum of the four parameter scores (geological parameters, geometric parameter, traffic parameters, and rockfall history). Since each parameter score varies exponentially from 3 to 81, the overall score theoretically ranges from 12 to 324. However, since rockfall history information was unavailable during the study that developed the Ohio Rockfall Hazard Rating Matrix, the study scores were based on only the first three parameters and the theoretical range of the overall scores ranged from 9 to 243.

The actual scores for the 108 sites evaluated during the development of the matrix ranged from a low of 22.5 to a high of 156.6, with an average value of 80.3 (Woodard 2004). This finding appears to realistically reflect the range of rock slope conditions experienced in Ohio. Subsequently, it was decided to classify Ohio rock slopes as summarized in Table 4-6. Figures 4-15 through 4-17 show examples of typical Ohio rock slopes with low, moderate, and high potentials for rockfall hazards, respectively.

8.4 Summary of Ohio System

Ohio conditions are typical of many midcontinent regions underlain by relatively undeformed

FIGURE 4-15
Example of a low-hazard-potential Ohio site. The rock slope produces small to moderately sized blocks that are retained in a large ditch (Woodard 2004).

FIGURE 4-16
Example of a moderate-hazard-potential Ohio site. The rock slope produces mostly small to moderately sized blocks; however there is a possibility for some larger blocks. The ditch appears adequate to retain anticipated rockfall volumes (Woodard 2004).

FIGURE 4-17
Ohio rock slope site with high hazard potential that produces large blocks and has very narrow, inadequate catchment area (Woodard 2004).

sedimentary strata and having only moderate topographic relief. In many locations, a sequence of strata with varying resistance to erosion makes evaluation of differential erosion and weathering rates of weaker strata important. The response to these conditions was to accept the basic RHRS premise but to adapt its methods to relate to Ohio geological and topographic conditions:

- The available documentation for the Ohio approach (Woodard 2004; Shakoor 2005) does not describe a two-stage assessment approach as recommended by the RHRS (Pierson 1991). The preliminary assessment stage is not discussed, and the proposed Ohio procedure and the Ohio Rockfall Hazard Rating Matrix (Figure 4-9) correspond to the more detailed assessment phase of the RHRS.
- The research project to develop the rating matrix employed statistical evaluations of field observations to define the critical parameters in the Ohio rating matrix.
- The resulting Ohio matrix reorders the sequence of rockfall categories in the original RHRS to form four evaluation parameters: geological, geometric, traffic, and rockfall history.
- The fairly consistent and usually moderate heights of rock exposures make rock slope height (RHRS Category 1, Table 4-1) relatively unimportant in comparing slope hazards; consequently, this category was not included in the Ohio matrix.
- The Ohio project adopted some ideas from existing RHRS modifications proposed by others. In particular, it accepted the New York State method for quantifying ditch effectiveness (see Section 4-2).
- The importance of utilizing some laboratory tests to evaluate the durability of weak and easily eroded strata had been suggested by earlier research in Ohio (Shakoor 1995; Shakoor and Brock 1987; Shakoor and Rodgers 1992; Shakoor and Weber 1988) and by the Ontario studies (Franklin and Senior 1997a); thus the use of the slake–durability index is a recommended part of the Ohio rating system where practical.

9. MISSOURI ROCKFALL HAZARD RATING SYSTEM

Around 2000, the Missouri University of Science and Technology (formerly the University of Missouri, Rolla), in collaboration with the Missouri DOT, began to design a system for rating rock cuts and priority-ranking mitigation efforts within the state of Missouri. The research team was familiar with the rockfall hazard rating concepts and procedures incorporated within existing systems (Maerz 2000), including the original RHRS and modifications adopted by New York State (Section 4) and Ontario (Section 5). In addition, they had considerable experience with the use of mobile video-logging systems (Maerz and McKenna 1999). The Missouri Rockfall Hazard Rating System (MORFH RS) was developed to be responsive to Missouri conditions and to utilize mobile video-logging capabilities (Maerz et al. 2004, 2005; Youssef et al. 2003, 2007).

Missouri conditions that influenced the MORFH RS design included the following:

- Most Missouri rock cuts have relatively low heights in predominantly flat-lying carbonate bedrock; consequently the stability of rock cuts is typically affected by highly weathered rocks and special problems resulting from karst features including filled sinkholes; and
- Missouri has constructed many rock cuts with benches, which in some instances have been shown to increase the tendency for rockfall to reach the roadway.

Thus the MORFH RS design and documentation include some useful suggestions for assessing these two conditions. Sections 9.3 and 9.4 provide additional information on the MORFH RS approaches to these assessments.

9.1 System Limitations

The MORFH RS design is somewhat constrained by its use of nonstandard terminology. As an example, MORFH RS defines risk as "a predictor of the likelihood of failure," which is the same as the standard definition of a hazard (see Chapter 3, Section 2). In addition, MORFH RS uses consequence factors as "a predictor of the consequence of failure," when according to more widely accepted terminology, these actually evaluate vulnerability (see Chapter 3, Section 2) rather than consequence. The standard definition of risk, which is "a quantitative value derived from an analysis of the numerical values of probability, vulnerability, and consequences" (see Chapter 3, Section 2), is actually fairly similar to the MORFH RS definition of consequence. The confusion resulting from these definitions makes it difficult for MORFH RS to be intuitively followed. Therefore, the following discussion will use *hazard* and *risk* based on their more standard and widely accepted meanings; thus, *hazard* will refer to the MORFH RS risk factors and *risk* will refer to the MORFH RS consequence factors.

The MORFH RS design uses 23 factors, including nine hazard factors, 10 risk factors, three adjustment factors, and one internally calculated value (Maerz et al. 2005). Two hazard parameters, slope angle and block size, are also included in the risk category. Most parameters are rated by using four categories to classify the observations, and the scores for these categories generally follow a 0, 4, 8, 12 scale; however, this scale is not always used. Because of this discrepancy, the person doing the evaluation is faced with scoring variations that complicate the assessment task. For example, the evaluation of the effectiveness of mitigation efforts is divided between the hazard and risk categories such that the catchment ditch effectiveness is measured by parameters defining ditch dimensions, which are in the risk category, whereas face irregularity and face looseness parameters, which account for blasting effectiveness and scaling effectiveness, respectively, are in the hazard category. A detailed description of these relationships and the scoring methods is provided by Maerz et al. (2005).

Table 4-7
Missouri Preliminary Rating Process

Factor	Condition Triggering Detailed Assessment[a]
Weathering and karst	1. Highly weathered rating on video image or 2. Any indication of karst (voids, filled sinks) or 3. Any significant differential erosion (cutback voids, overhangs) or
Face irregularity and looseness	1. Highly irregular face or moderately irregular face high on cut or 2. Highly loose face or moderately loose face high on cut or
Fallen rock in ditch or on cut	Significant amount of loose rock visible in ditch or
Ditch effectiveness	Very low ditch effectiveness (too small, too narrow) or
Adversely oriented discontinuities	Indication of adversely oriented discontinuities or
Bench or benches	Presence of benches

[a]No detailed assessment is triggered if there is slope height <10 ft or less than width of ditch plus shoulder.
SOURCE: Maerz et al. 2005.

9.2 Missouri Preliminary Rock Slope Rating Procedure

As with the original RHRS and several other modified rating procedures, MORFH RS includes a preliminary rating process to determine whether a detailed investigation of a rock slope is warranted. Table 4-7 provides a summary of the preliminary rating process used to determine if the more detailed rating is warranted.

MORFH RS was one of the earliest systems to propose collecting many of the physical measurements from standard photo-logging images or by mounting a video camera on a vehicle dashboard and driving past a potential rockfall site (Maerz and McKenna 1999; Maerz 2000). A special computer program, called RockSee, has been developed to streamline the process of obtaining measurements such as slope height and ditch width (Youssef et al. 2003, 2007). Maerz et al. (2004) also reported on the efficiencies achieved by combining the video data analysis procedures with digital field data acquisition by using a hand-held Pocket PC that supports a geographic information system (GIS) database interface.

9.3 Assessment of Karst Conditions

Karst features, typically in the form of infilled sinkholes, are very common in Missouri. In a highway rock cut, the most important characteristic of these features is the type of fill material, which can range from cemented to easily eroded materials. If they are filled by well-cemented materials, these cuts are treated as normal cuts without adjustments. If the sinkholes are filled by more easily eroded materials, a karst adjustment is used (Table 4-8).

9.4 Assessment of Rock Cuts with Benches

In the past, many rock cuts have been constructed as benched slopes because benches were believed to be beneficial and were intended to contain falling rocks or reduce the velocity and horizontal displacement of a falling rock. However, over

Table 4-8
Missouri System Rating Procedure for Karst Features

Karst Effect	Class Number	Consequence Rating
Noncarbonate rocks (igneous, sandstone)	0	0
Carbonate rocks that could possibly have karst features but are not evident on rock cut face	1	3
Karst features that appear on rock cut face; width, 50 ft; filled by boulders and cobbles or undercut with weak materials	2	6
Karst features that appear on rock cut face; width, 100 ft; filled by boulders and cobbles with weak materials	3	9
Karst features that appear on rock cut face; width, 150 ft; filled by boulders and cobbles with weak materials	4	12

Source: Maerz et al. 2005.

time, benches that are typically difficult to access and maintain may fill with debris and become launching zones that redirect falling rocks toward the road. Therefore, when rock cuts with benches are assessed, the benches are only considered to be beneficial if

- They are properly maintained and cleaned of accumulated debris;
- Rock faces above are in good condition, with little or no weathering, adversely oriented discontinuities, or loose materials (i.e., they produce little rockfall);
- Their surfaces are horizontal or have a reverse back slope to increase the likelihood that a falling rock will be retained on the bench;
- They contain soft materials that will absorb the energy of the falling rock;
- They feature trees or other obstacles that will help retain falling rock; and
- They are wide enough to retain nearly all falling rock.

To determine whether a bench has a good or a bad effect, an assessment is made with six parameters: three parameters to evaluate the conditions on the slopes above the bench and three parameters to assess the geometry and conditions of the bench (Table 4-9). Each parameter is scored on a scale of 0 to 4; if the sum of the scores exceeds 12, the bench is considered bad, and the MORFH RS rating is adjusted.

10. TENNESSEE ROCKFALL MANAGEMENT SYSTEM

The Tennessee DOT experiences recurring rock cut failures, especially in the eastern and central portions of the state, where the geological and topographic conditions combine to produce a large number of potentially unstable rock slopes of various lithologies, discontinuity characteristics, degrees of weathering, and topographic profiles.

Table 4-9
Missouri System Assessment Procedure for Benches

	Presence of Bench? Yes or No[a]			
Characteristic	Score	4	2	0
Faces above bench	Weathering	High	Low	Fresh
	Face irregularity	High	Moderate	Smooth
	Face looseness	Large	Moderate	No
Bench characteristics	Width	Narrow <5 ft	Moderate 15 ft	Wide >20 ft
	Rock on bench	Large amount	Moderate amount	None
	Slope of bench	Toward road	Horizontal	Backward

Total Score ________ (if greater than 12, then bench is considered bad)

[a]If yes, look at bench and faces above bench.
Source: Maerz et al. 2005.

These potentially unstable slopes are adjacent to both major transportation routes and rural roads.

10.1 Tennessee Perspective

Along the eastern border of the state, within the Blue Ridge physiographic and geological province, the combination of topographic relief, resistant rock types, and plentiful rock discontinuities produces numerous planar and wedge failures. Most of the U.S. and state highway system in Tennessee was built before rockfall catchment standards were implemented, and many major road segments are narrow or sinuous with ineffective narrow ditches adjacent to tall and potentially hazardous rock cuts (Figure 4-18). To the west, in the Ridge and Valley and Cumberland Plateau regions, medium- to thick-bedded sedimentary rocks with inclined to nearly horizontal bedding produce large toppling failures after a period of differential weathering (Figure 4-19). In the Nashville area, the rock strata are relatively flat, but many rock cuts typically expose two or more lithologies, including more easily weathered mudstones, siltstones, clay shales, and argillaceous limestones (Figure 4-20), and commonly produce rockfalls caused by differential weathering and raveling.

10.2 Developing Tennessee RHRS

Initially, the Tennessee DOT developed a rockfall database with Access software to maintain the results of field surveys of rock slopes that used the

FIGURE 4-18 *(top right)*
Planar rockfall hazard, typical of Blue Ridge province of eastern Tennessee.
(PHOTOGRAPH COURTESY OF THE TENNESSEE DEPARTMENT OF TRANSPORTATION.)

FIGURE 4-19 *(center right)*
Rockfall hazard zone in Ridge and Valley Region north of Chattanooga, Tennessee, with combination of toppling and differential weathering.
(PHOTOGRAPH COURTESY OF THE TENNESSEE DEPARTMENT OF TRANSPORTATION.)

FIGURE 4-20 *(bottom right)*
Rockfall hazard zone in Cumberland Plateau west of Nashville, Tennessee, with combination of differential weathering and raveling.
(PHOTOGRAPH COURTESY OF THE TENNESSEE DEPARTMENT OF TRANSPORTATION.)

standard RHRS methodology (Bateman 2002). In 2000, a research project was initiated to develop a Tennessee Rockfall Management System that included preliminary and detailed rockfall hazard rating procedures suitable for Tennessee conditions, enhancements to the existing rockfall database, and data integration into a web-based GIS to allow for display, analysis, and priority-ranking of rockfall hazards. The ability to collect field data with mobile electronic devices was also a component of this development project.

During Phase I, completed in 2002, data from five counties demonstrated the desirability of modifying the RHRS to better reflect Tennessee rockfall conditions and specific Tennessee DOT requirements. The resulting Tennessee Rockfall Hazard Rating System (TRHRS) is used to manage road cuts that are potentially hazardous because of rockfall, and it forms the basis of the Tennessee Rockfall Management System (Bateman 2002). TRHRS retains many of the characteristics of the original RHRS, including a two-stage rating process with a preliminary rating procedure to classify rock slopes into A, B, and C categories followed by a more detailed rating procedure for the slopes with the highest hazard rating (A and possibly some B).

FIGURE 4-21
Comparison of total scores produced by RHRS and TRHRS procedures (adapted from Mauldon et al. 2007).

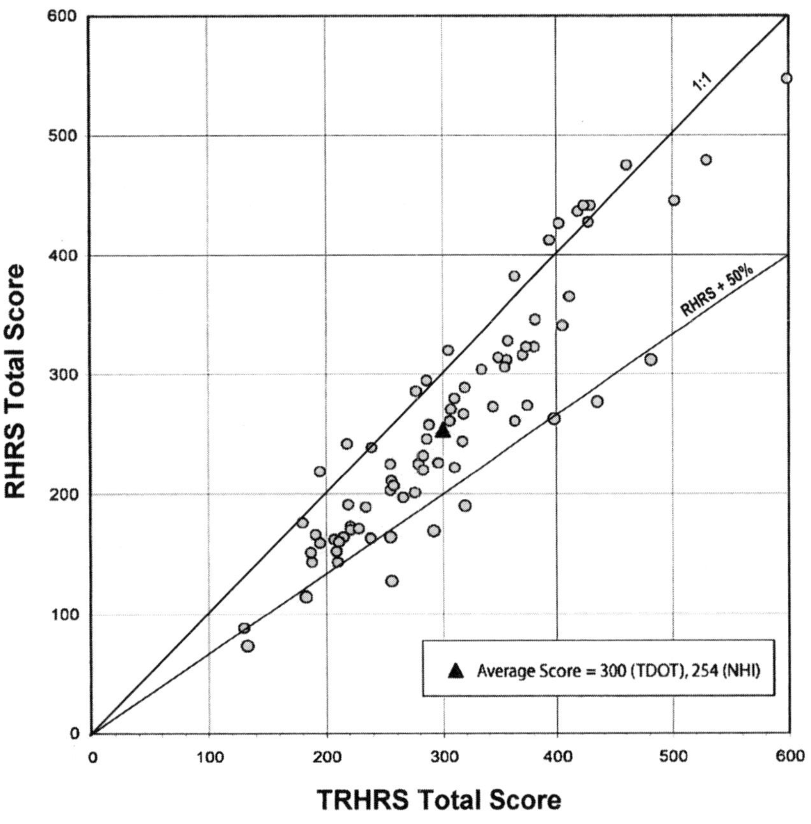

As discussed in more detail in Sections 10-3 and 10-4, only slight modifications were instituted to the preliminary rating procedures. However, the detailed rating procedures were changed to

- Better reflect the role of ditch width,
- Enhance the field evaluation of rockfall history since Tennessee maintenance records are frequently not easy to access, and
- Improve the geological characterization by explicitly recognizing potential rockfall modes, incorporating the estimated abundance of each rockfall mode, and combining the scores for all potential rockfall modes.

The various modifications to the detailed rating procedure within the TRHRS naturally affected the final rating scores. To evaluate the significance, if any, of these changes in rating scores, a comparison was made between scores developed by the TRHRS and those from the original RHRS. A series of A-rated slopes received two detailed ratings, one based on the new TRHRS system and a second based on the original RHRS. The TRHRS scores ranged from about 130 to 600, with a mean score of 300, whereas the RHRS scores averaged about 15% lower (Figure 4-21). A perfect correlation between the two systems would have all data falling on the 1:1 line in Figure 4-21; although this result did not occur, the RHRS + 50% line bounds nearly all the TRHRS scores.

It was concluded that the TRHRS produced higher overall scores compared with those produced by the original RHRS, but no major change in the relative ranking between rockfall sites was apparent. Since the new TRHRS rating system included enhancements that better reflected Tennessee conditions, it was concluded that the TRHRS approach was appropriate for Tennessee.

Phase II of the project was undertaken during 2002 to 2005. Rockfall ratings were completed for the remaining 73 counties located in eastern and central Tennessee, where the most serious rockfall hazards are located. The TRHRS procedures and concepts were further evaluated with this larger data compilation, and highlights and initial conclusions from this data analysis process were presented in a comprehensive final report (Mauldon et al. 2007). This report includes full documentation of all procedures, instructions on using the supporting software to perform electronic field data acquisition or to access the rockfall database, and a field training manual for identifying rockfall hazards. Additional information concerning the Tennessee methods and concepts is presented by Bateman (2003), Bellamy et al. (2003), Rose et al. (2001, 2003), Rose (2005), and Vandewater et al. (2005).

10.3 TRHRS Preliminary Rating

A hazardous road cut or rock slope has the potential for rockfall events that can deposit debris onto the roadway. In Tennessee, the process of identifying potentially hazardous slopes begins with a virtual drive-through using the Tennessee Roadway Information Management System (TRIMS). TRIMS is an integrated roadway management tool that incorporates video logging of all state routes with photographs captured at increments of one-hundredth of a mile (Figure 4-22). Potentially hazardous road cuts are identified during the virtual drive-through, and the corresponding locations are recorded by identifying the route and mileage log. Other relevant roadway information items, such as ADT and speed limit, are also recorded so that they can be used during subsequent detailed rating evaluations.

The potentially hazardous locations identified during the TRIMS review are visited, evaluated, and assigned preliminary ratings according to the RHRS guidelines (Pierson and Van Vickle 1993). Minor adjustments are included in the TRHRS instructions. Because records of historical rockfall activity are incomplete for many Tennessee locations, the potential for fallen rock materials to reach the roadway is evaluated by using the following four criteria:

1. Impact marks on the road;
2. Ditch effectiveness, including width and shape of catchment;
3. Estimated size and amount of material per event; and
4. Presence of launching features.

FIGURE 4-22 User interface of Tennessee Roadway Information Management System. (PHOTOGRAPH COURTESY OF THE TENNESSEE DEPARTMENT OF TRANSPORTATION.)

If the potential for rocks to reach the roadway is moderate or greater, a limited decision sight distance is hazardous, because drivers may be unable to avoid fallen rocks on the road. The TRHRS recommends that the road cut be assigned a preliminary A-rating in this situation. TRHRS guidelines suggest that historical rockfall activity, where known, should be evaluated by using the following criteria:

1. Frequency and presence of rockfall on the roadway,
2. Frequency of removal of rock debris from catchment or roadway,
3. Amount of material in the catchment (particularly in the absence of maintenance reports), and
4. Number of impact marks in the road surface (particularly in the absence of maintenance reports).

10.4 TRHRS Detailed Rating

As with RHRS systems in general, the purpose of the detailed rating is to numerically differentiate the potential hazard at a series of identified road cuts. The numerical rating scores allow for the sorting and priority-ranking of maintenance or remediation measures. The TRHRS recommends detailed ratings only for road cuts that have received an A-rating by the preliminary rating process.

The initial TRHRS system permitted the use of personal digital assistants, such as Palm Pilots, to record relevant detailed rating data in the field. Later implementation and expansion of the system included data-gathering forms that could be used with a laptop computer, tablet PC, or Windows-based mobile GPS receiver. These electronic user interfaces support data entry and direct importation of information into the Tennessee rockfall database (Bellamy et al. 2003). Data collection with mobile electronic devices has several advantages, including

- Low-cost hardware and software,
- Emulation of Windows-based database applications,
- Easy synchronization with desktop computers and spatial databases on computer servers,
- Elimination of paper forms,
- Elimination of office data entry,
- Error checking performed in real time in the field, and
- Integration with GIS through GPS receivers.

Procedures that employ mobile electronic devices are the primary method of data collection for the TRHRS detailed ratings (Bellamy et al. 2003; Mauldon et al. 2007, Appendix C). However, a paper form is also maintained as a backup (Figure 4-23).

Most categories and scoring techniques for the TRHRS detailed ratings are identical to those in the original RHRS. However, the detailed rating system for the TRHRS has only eight categories:

1. Slope height,
2. Average vehicle risk (AVR),
3. Percentage of decision sight distance,
4. Roadway width,
5. Ditch effectiveness,
6. Rockfall history,
7. Presence of water on slope, and
8. Geological characteristics.

Each category is assigned a score that increases with degree of hazard, and the scores of all categories are summed to yield an overall score. An exponential (3^x) scoring system assigns values ranging from 3 to 81 points for most categories, but some categories use equations for determining scores within a continuum of 1 to 100 points. Categories modified by TRHRS use somewhat modified scales to maintain consistency with the original RHRS.

Three categories were extensively modified in TRHRS: Category 5, ditch effectiveness; Category 7, presence of water on the slope; and Category 8, geological characteristics. Minor modifications were made to Category 6, rockfall history. Detailed descriptions of these modifications, and the justifications for instituting them, are provided by Mauldon et al. (2007, Appendix A).

The TRHRS characterizes the geology category by evaluating all potential failure modes at a road cut, scores each failure mode, and sums these scores. In contrast, the original RHRS included the scores from the two categories that best define the geological character at the site that leads to rockfall. The RHRS distinguishes between two geological characters, structurally controlled and differential erosion-controlled rockfall. It requires an evaluation of whichever character is dominant or provides the highest score, or both (i.e., the one that poses the greatest potential for rockfall). The TRHRS methodology takes this evaluation a step further and assesses specific failure modes within each of these two broader categories. These include planar, wedge, and toppling failures within the structurally controlled category and differential weathering and raveling failures within the weathering-controlled category.

The evaluations are undertaken by assessing appropriate characteristics. Characteristics pertinent to all failure modes are the relative abundance of the failure as a percentage of the total rock cut surface area and block size, which is the longest dimension of the rockfall blocks. Characteristics unique to planar and wedge failure are steepness of failure planes and the microfriction and macrofriction profiles of the failure planes. The amount of relief is a characteristic unique to differential weathering, and block shape is unique to raveling (Vandewater et al. 2005). Table 4-10 summarizes the relatively complex rating scores for these various failure modes. Scores for the different failure modes are summed. However, the maximum score is limited to 300 points to keep the contribution from this category to the overall rockfall hazard score roughly similar to that of the original RHRS (Mauldon et al. 2007). The Tennessee DOT also added a rockfall closure impact rating and an economic assessment that are used to select sites for mitigation (see Section 12).

TDOT RHRS FIELD SHEET v1.1

I. TRIMS/Preliminary Data

Date ______

File No. ______

County No.	______	Rater	______
Route No.	______	Speed Limit	______
Beg. L.M.	______	District	______
Ref C/L	______	ADT	______
County	______	Latitude	______
Region	______	Longitude	______

SCORING

1. Slope Height ______
2. AVR ______
3. % DSD ______
4. Road Width ______
5. Ditch Effectiveness ______
6. Rockfall History ______
7. Water ______
8. Geologic Character ______

TOTAL SCORE ______

II. Site and Roadway Geometry

1. Slope Height (ft)

estimated ______

alpha (a) ______ beta (b) ______

width (x) ______ ft instrument height (H.I.) ______ ft

$$\text{Slope Height} = \frac{\sin a * \sin b * X}{\sin(a-b)} + \text{H.I.}$$

2. Average Vehicle Risk (AVR)

$$\text{AVR} = \frac{\text{ADT (cars/day)} * \text{(Rock Slope Length/5280)}}{\text{((24hpd)} * \text{Speed Limit (mph))}}$$

Slope Length ______ ft Speed Limit ______ ft **AVR =** ______ %

3. % Decision Site Distance (% DSD)

Choose one: adequate, moderate, limited, very limited
3 9 27 81

OR

Calculate: ______ / ______ X 100 = ______ %
(observed DSD) / (AASHTO DSD)

4. Road Width (ft)

______

5. Ditch Effectiveness

Design Catchment Width (feet)

Slope Height (ft)	Recommended width for vertical slope	Recommended width for non-vertical slope
0 – 40	18	18
40 – 50	18	24
50 – 60	24	30
60 – 70	28	34
70 – 80	32	38
80 – 100	36	42
100 – 125	36	42
125 – 175	40	48
> 175	52	60

Effective catchment width (ft) ______ **Launching Features? (yes or no)** ______
6:1 catchment shape? (yes or no) ______

Percent of Design Catchment Width from Table	>90%	70%-90%	50%-70%	<50%
Score with 6:1 or greater catchment slope	3	9	27	81
Score w/ Poor Catchment **OR** Launch Features	9	27	81	81
Score w/ Poor Catchment **AND** Launch Features	27	81	81	81

6. Rockfall History

Benchmark	Frequency	Field Judgment	Score
Few	1 or less per year	No impact marks in the road, no rocks in the road, few rocks in ditch	3
Several	2 per year	No impact marks in the road, no rocks in the road, many rocks in the ditch	9
Many	3 – 4 per year	Few impact marks or few rocks in the road	27
Constant	5 or more per year	Many impact marks and/or many rocks in the road	81

7. Presence of Water on Slope

(choose one) none 3, seeping 9, flowing 27, gushing 81 ______

NOTES:

III. Geologic Characteristics (circle all that apply; modes are additive)

	Planar				Wedge			
Abundance	<10%	10-20%	20-30%	>30%	<10%	10-20%	20-30%	>30%
score	3	9	27	81	3	9	27	81
Block size	<1ft	1-3ft	3- 6ft	>6ft	<1ft	1-3ft	3- 6ft	>6ft
score	3	9	27	81	3	9	27	81
Steepness (degrees)	0-20	20-40	40-60	>60	0-20	20-40	40-60	>60
score	2	5	14	41	2	5	14	41
Friction (micro/macro)	rough/undulating	smooth/undulating	rough/planar	smooth/planar	rough/undulating	smooth/undulating	rough/planar	smooth/planar
score	2	5	14	41	2	5	14	41

8. Geology Score = ______

	Topple/B. Release				Differential Weathering			
Abundance	<10%	10-20%	20-30%	>30%	<10%	10-20%	20-30%	>30%
score	5	14	41	122	3	9	27	81
Block size	<1ft	1-3ft	3- 6ft	>6ft	<1ft	1-3ft	3- 6ft	>6ft
score	5	14	41	122	3	9	27	81
Relief					<1ft	1-3ft	3- 6ft	>6ft
score					3	9	27	81

Raveling				
Abundance	<10%	10-20%	20-30%	>30%
score	3	9	27	81
Block size	<1ft	1-2ft	2- 3ft	>3ft
score	3	9	27	81
Block Shape	tabular	blocky	round	
score	3	9	27	

FIGURE 4-23
TRHRS detailed rating form, Version 1.1.
(FORM COURTESY OF THE TENNESSEE DEPARTMENT OF TRANSPORTATION.)

Table 4-10
Scoring Method for TRHRS Failure Modes for Geology Category

	Ranking			
Criteria and Score	Low	Medium Low	Medium High	High
Planar Rockfall Mode				
Abundance (%)	<10	10–20	20–30	>30
Score	3	9	27	81
Block size [ft (m)]	<1 (<0.3)	1–3 (0.3–0.9)	3–6 (0.9–1.8)	>6 (>1.8)
Score	3	9	27	81
Steepness (degrees)	0–20	20–40	40–60	>60
Score	2	5	14	41
Friction (micro/macro)	Rough/undulating	Smooth/undulating	Rough/planar	Smooth/planar
Score	2	5	14	41
Wedge Rockfall Mode				
Abundance (%)	<10	10–20	20–30	>30
Score	3	9	27	81
Block size [ft (m)]	<1 (<0.3)	1–3 (0.3–0.9)	3–6 (0.9–1.8)	>6 (>1.8)
Score	3	9	27	81
Steepness (degrees)	0–20	20–40	40–60	>60
Score	2	5	14	41
Friction (micro/macro)	Rough/undulating	Smooth/undulating	Rough/planar	Smooth/planar
Score	2	5	14	41
Topple Rockfall Mode				
Abundance (%)	<10	10–20	20–30	>30
Score	5	14	41	122
Block size [ft (m)]	<1 (<0.3)	1–3 ft (0.3–0.9)	3–6 (0.9–1.8)	>6 (>1.8)
Score	5	14	41	122
Differential Weathering Rockfall Mode				
Abundance (%)	<10	10–20	20–30	>30
Score	3	9	27	81
Block size [ft (m)]	<1 (<0.3)	1–3 ft (0.3–0.9)	3–6 (0.9–1.8)	>6 (>1.8)
Score	3	9	27	81
Relief [ft (m)]	<1 (<0.3)	1–3 (0.3–0.9)	3–6 (0.9–1.8)	>6 (>1.8)
Score	3	9	27	81
Raveling Rockfall Mode				
Abundance (%)	<10	10–20	20–30	>30
Score	3	9	27	81
Block size [ft (m)]	<1 (<0.3)	1–3 ft (0.3–0.9)	3–6 ft (0.9–1.8)	>6 (>1.8)
Score	3	9	27	81
Shape	Tabular	Blocky	Round	
Score	3	9	27	

Note: TRHRS scores were selected to maintain the geology category score close to the original score in the RHRS. For the planar and wedge failure modes, the steepness and friction criteria were assigned scores of about half the value of the RHRS scores of 3, 9, 27, and 81. Because topple failure mode has only two criteria, these were assigned scores of about 1.5 times the value of the RHRS scores.
Source: Adapted from Vandewater et al. 2005.

11. COLORADO ROCKFALL HAZARD RATING SYSTEM

Colorado experiences considerable and varied rockfall hazards in the central and western parts of the state. As population and development have increased within the mountains, so has the severity of the rockfall hazard. Consequently, the Colorado Department of Transportation has extensive experience with developing and using rockfall hazard rating systems and since the late 1980s has employed a series of increasingly quantitative evaluation systems that evolved over at least three cycles (Stover 1992; Andrew 1994; Santi et al. 2009). The latest version of the Colorado Rockfall Hazard Rating System (CRHRS-III) is presented.

11.1 Colorado Perspective

Eastern Colorado is part of the Great Plains physiographic province and is largely free of rockfall hazards, but central and western Colorado lie within the Southern Rocky Mountains and Colorado Plateau physiographic provinces, which are characterized by rugged mountains, plateaus, and deep river canyons that are subject to numerous rockfall events. Figure 4-24 shows the number of vehicle accidents related to rockfall along Interstate 70, which crosses Colorado east to west, and demonstrates the relative severity of rockfall hazards in these three physiographic provinces.

The range of rockfall hazard reflects the different climate, topographic, and geological conditions within the Southern Rocky Mountains and Colorado Plateau. Igneous and metamorphic crystalline rocks dominate in the Rocky Mountains, but sedimentary rocks produce considerable rockfall hazards on the eastern and western margins of the mountains. Interbedded sedimentary rocks form high cliffs in the Colorado Plateau and present considerable rockfall hazards. Additional rockfall hazards are produced by the erosion and raveling of "block-in-matrix" materials composed of glacial and debris-flow deposits and colluvium, which are found throughout the region. Because of this geological variability, Colorado rockfall assessment schemes have increasingly developed fairly precise descriptions and scoring techniques for a wide range of conditions.

As in other mountainous regions, Colorado rockfall hazards are materially affected by the extreme local topographic relief, narrow canyons, and high elevations, which provide recurring freeze–thaw cycles and extensive weathering of

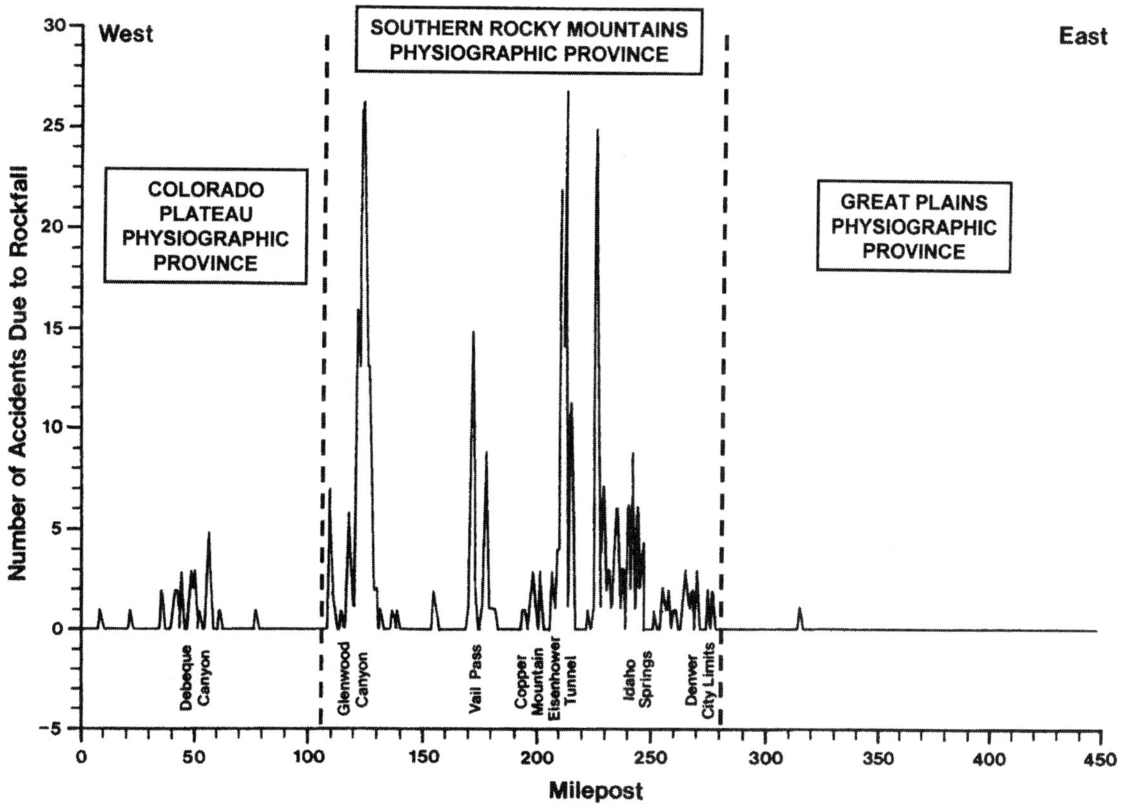

FIGURE 4-24 Number of vehicle accidents attributed to rockfall along Interstate 70 in Colorado from 1976 to 1989; greatest number is in Southern Rocky Mountains; those in Colorado Plateau are found within river canyons (adapted from Stover 1992).

exposed rock masses. The severity of freeze–thaw cycles is strongly influenced by slope orientation, with south-facing slopes experiencing more such cycles than north-facing slopes. As a consequence, the latest Colorado rockfall assessment procedure addresses additional factors related to rock mass discontinuities, climate and exposure conditions, and slope geometry characteristics.

11.2 Latest Version of Colorado Rockfall Hazard Rating System

The latest version of the CRHRS (CRHRS-III) was developed at the Colorado School of Mines (Russell 2006). It includes a number of significant changes to earlier versions. The descriptions for several rating categories—including launching features, ditch catchment, precipitation–seepage–exposure, fractures and orientation, erosional features, differences in erosion, and observed history—have been changed to make them less subjective. Furthermore, CRHRS-III addresses the entire slope above the roadway, which is important where a natural slope extends above a constructed rock cut and rockfall may initiate from either area.

CRHRS-III includes 18 rating categories that contribute to a hazard score and three rating categories that contribute to a risk score. The hazard score and the risk score are not combined into a final total score but are kept separate to distinguish between hazard and risk for each site. Figure 4-25 shows the rockfall hazard field worksheet for this latest version of CRHRS.

Russell (2006) and Santi et al. (2009) provide detailed definitions for each of the parameters and procedures used by CRHRS-III. Initially, the procedures were evaluated by assessing a total of 200 slopes within the Colorado Front Range (Russell 2006). Subsequently, during these tests, statistical analyses based on 355 slopes demonstrated that for rock cuts composed of crystalline (igneous or metamorphic) rock, the total hazard score could be fairly accurately predicted by using an equation that incorporates only four rating categories (Russell 2006; Santi et al. 2009).

12. ECONOMIC IMPACTS AND MITIGATION COSTS

The original RHRS and the modified systems described in this chapter are designed to rank sites according to their rockfall hazard potential in order to provide a tool to help decision makers move forward in a logical manner. Still, most of these systems do not take into account the costs of mitigation at these sites or the disruption of traffic caused by a rockfall event. Some rock slopes may only shed small rocks, which are a hazard to travelers but are easily removed. These events are typically not as critical from a transportation system perspective as larger potential failures that could cause significant road closures.

On July 1, 1997, heavy rainfall contributed to a large rockfall that closed all four lanes of Interstate 40 in western North Carolina for more than 4 months. The rockfall event trapped two cars and injured five people. As a result of the road closure, tragically, seven people were subsequently killed on the detour route in truck-related accidents that were attributed to the poor level of service provided by the highway used as the detour facility (DiMillo 1998). Indirect costs resulting from the closure were much greater than the $3 million cost of the removal of the rockfall material or the $1 million spent on additional stabilization measures. For example, the tourist industry claimed that it lost $40 million to $50 million in business revenue during the shutdown period. A review of the remaining 32 km (20 mi) of this section of Interstate estimated that $40 million would be required to correct the remaining potential rockfall problems (DiMillo 1998).

In Colorado, rockfall events in Glenwood Canyon closed Interstate 70 for 2 days in 2004 and for 4 days in 2010. In each case, the shortest detour added 320 km (200 mi) to the route between Denver and western Colorado. The additional travel time and distance resulted in considerable economic impacts to local economies and the trucking industry.

The original RHRS and the various derivative systems are intended to provide an understanding of the scope of the rockfall problem systemwide and to identify the more hazardous sites, but they do not address facility degradation issues, such as those raised by these Interstate rockfall incidents.

A more comprehensive rockfall management system, such as the one implemented by the Tennessee DOT (Bateman 2004), is needed to address facility degradation issues. The Tennessee Rockfall

Colorado Rockfall Hazard Rating Field Worksheet

Site Information				
ROUTE NO	SEGMENT ID NUMBER	ENGINEERING REGION	MAINTENANCE SECTION	DATE
COUNTY	BEGIN MILE POST	END MILE POST	R / L OF CENTERLINE	RATER

Actual Values		Remarks
Slope Height ______	Ditch Depth ______	Major rockslide potential:
Slope Angle ______	Sight Distance ______	Dominating rockfall mode: Block-in-matrix / Sedimentary rock / Crystalline rock
Ditch Slope ______	Speed Limit ______	Dominating sight distance: Horizontal / Vertical
Ditch Width	ADT	Mitigation effectiveness: A B C D F

Rating Cut Slope / Total Slope (Circle one)

			3 Points	9 Points	27 Points	81 Points
Slope		Slope Height	25 to 50 ft	50 to 75 ft	75 to 100 ft	>100 ft
Slope		Rockfall Frequency	> 2 years	1 to 2 years	Yearly, seasonal	Year round / severe events
Slope		Average Slope Angle Score	0 to 2	2 to 4	4 to 8	> 8
Slope		Launching Features	None (smooth slope)	Minor (< 2 ft. surface variation)	Many (2 to 6 ft. surface variation)	Major (> 6 ft. surface variation)
Slope		Ditch Catchment	95% to 100% / Class 1	65% to 94% / Class 2	30% to 64% / Class 3	< 30% / Class 4 / ≥ Major launching features
Climate		Annual Precipitation	< 10 inches	10 to 20 inches	20 to 35 inches	> 35 inches
Climate		Annual Freeze Thaw Cycles	1 to 5	6 to 10	11 to 15	16 or more
Climate		Seepage / Water	Dry	Damp / wet	Dripping	Running water
Climate		Slope Aspect	N	E, W, NE, NW	SE, SW	S
Geology	Sed. Rock	Degree of Under-Cutting	0 to 1 ft	1 to 2 ft	2 to 4 ft	> 4 ft
Geology	Sed. Rock	Jar Slake	6	5	3 to 4	1 to 2
Geology	Sed. Rock	Degree of Interbedding	1 to 2 weak interbeds, < 6 in.	1 to 2 weak interbeds, > 6 in.	> 2 weak interbeds, < 6 in.	> 2 weak interbeds, > 6 in.
Geology	Crys. Rock	Rock Character	Homogenous / massive	Small faults / strong veins	Schist / shear zones < 6 in.	Weak pegmatite / micas / shear zones > 6 in.
Geology	Crys. Rock	Degree of Overhang	0 to 1 ft	1 to 2 ft	2 to 4 ft	> 4 ft
Geology	Crys. Rock	Weathering Grade	Fresh	Surface staining	Slightly altered / softened	Core stones
Geology	Discontinuities	Block Size / Volume	<1 ft / <1 cy	1 to 2 ft / 1 to 3 cy	2 to 5 ft / 3 to 10 cy	>5 ft / >10 cy
Geology	Discontinuities	Number of Sets	1	1 plus random	2	> 2
Geology	Discontinuities	Persistence, Orientation	< 10 ft and dips into slope	> 10 ft and dips into slope	< 10 ft and daylights out of slope	> 10 ft and daylights out of slope
Geology	Discontinuities	Aperture	Closed	0.1 to 1 mm	1 to 5 mm	> 5 mm
Geology	Discontinuities	Weathering Condition	Fresh	Surface staining	Granular infilling	Clay infilling
Geology	Discontinuities	Friction	Rough	Undulating	Planar	Slickensided
Geology	Block in Mat.	Block Size (x3)	< 1 ft	1 to 2 ft	2 to 5 ft	> 5 ft
Geology	Block in Mat.	Block Shape (x3)	Tabular	Blocky	Blocky to angular	Rounded and smooth
Geology	Block in Mat.	Vegetation (x3)	Fully vegetated	Patchy vegetation	Isolated plants	None
Total Hazard Score:						
Traffic		Sight Distance	> 80 %	60 % to 80 %	40 % to 60 %	< 40 %
Traffic		Avg. Vehicle Risk	0 to 24%	25 to 49%	50 to 74%	75% or more
Traffic		No. of Accidents	0 to 2	3 to 5	6 to 8	9 and over
Total Risk Score:						

Additional Rater's Comments:

FIGURE 4-25
Field worksheet for current version of CRHRS-III (Russell 2006).

Management System includes six components, one of which is the TRHRS:

1. A database for storage of rockfall information,
2. GIS capabilities for viewing and analyzing rockfall data in a spatial manner,
3. Computerized field data-gathering methods,
4. An RHRS to rank sites in order of hazard,
5. Economic tools for assessing the impact of a rockfall incidents, and
6. Tools for estimating mitigation costs.

Component 3 includes automated field data collection, and the data collected are incorporated into the larger GIS environment designed for spatial evaluation of rockfall sites (Component 2). The economic tools (Components 5 and 6) not only include methods for assessing the economic impacts of rockfall road closures but also identify *hazard reducers*. The hazard reducers include various techniques that maintenance personnel can incorporate into their workflow to provide some mitigation of rockfall sites until larger remediation projects can be undertaken (Bateman 2004). The Tennessee DOT system was developed after a review of the Unstable Slope Management System developed by the Washington State DOT (Lowell and Morin 2000).

Supplementary tools were added to the TRHRS, described in Section 10, to assess the impact of rockfall incidents at the most highly ranked sites. These supplementary tools were the Rockfall Closure Impact (RCI) procedure and the mitigation cost assessment procedure (Bateman 2004). When combined with TRHRS, these tools provide methods not only for assessing the hazards associated with rockfall, such as a car striking or being struck by a falling rock, but also for assessing and comparing the impact of rockfall road closures on the traveling public and the cost of mitigation. With this information, planners can make informed decisions concerning the expenditure of limited highway construction funds.

12.1 Tennessee Rockfall Closure Impact Assessment

The Tennessee RCI assessment procedure ranks sites by using the exponential scoring system employed in the original RHRS (Pierson 1991). The RCI evaluates five criteria: ADT, detour length, impedance, impedance duration, and facility degradation (Figure 4-26). The total RCI score is an indicator of the significance of a specific roadway closure due to a rockfall. Higher RCI scores mean that a rockfall incident will have a greater impact on traffic and the surrounding communities. The ADT criterion rates traffic counts to reflect the rural nature of some critical Tennessee roadways.

Criteria	Score = 3	Score = 9	Score = 27	Score = 81
ADT (Average Daily Traffic)	Little Traffic ADT>300	Some Traffic ADT 300-1000	Moderate Traffic ADT 1000-3000	Major Traffic ADT >3000
Impedance	Shoulder	1 lane	2 lanes	>=3 lanes or Total
Impedance Duration	Hours	1 Day	Days	Weeks
Detour Length	Very Short <1 mile or lane still open	Short 1-2 miles	Medium 3-4 miles	Long or None >4 miles
Facility Degradation RF-DF= DOF	0	1	2	3

DOF=Degree of Facility (RF=Degree of Road Facility; DF= Degree of Detour Facility)

0 = Local Roads / 1 Lane Road
1 = 2 Lane, no shoulder
2 = 2 lane, adequate shoulder
3 = 3 lane
4 = 4 lane
5 = 4 lane, divided highway, 5 lane highway
6 = Interstate

FIGURE 4-26 Tennessee RCI rating form (Bateman 2004).

The RCI score was developed to be used with the TRHRS score to choose sites for mitigation logically. Cost and the RCI and TRHRS scores are used to select sites for mitigation under a program that has a minimum annual budget of $2 million. The impedance criterion provides an estimate of the amount of road that will be closed, and impedance duration assesses the length of time that a roadway may be affected. Detour length reflects the additional travel required to detour a closed segment of road [with a detour of more than 6.5 km (4 mi) receiving the highest score]. Facility degradation is a measure of the change in service level from the original road to the detour facility.

12.2 Tennessee Mitigation Cost Assessment Procedure

A preliminary estimate of mitigation costs completes the planning and assessment process. This estimate is accomplished by using two complementary approaches. A cost notes form (Figure 4-27) serves several functions: it gathers information on possible mitigation alternatives, acts as the basis for mitigation cost estimates performed in the office, and provides a list of tasks that can be sent out to district maintenance offices. It provides quick guidance for different mitigation strategies that might be used and provides a rough estimate of the benefits that are expected to result from a particular repair. Bateman (2004) suggests that, with experience, the cost notes form can be quickly completed on the basis of the detailed hazard rating provided by the TRHRS and a single visit to the site. The Tennessee Rockfall Management System relies on digital field acquisition of information with mobile electronic devices. Accordingly, a standard electronic worksheet equivalent to the cost notes form was developed to facilitate fast and accurate data gathering and storage in the field (Figure 4-28).

Minor Repair Options	Amount of Option			
Hand Scale Slope?	>10% of Slope Face	10-30% of Slope Face	30-60% of Slope Face	>60% of Slope Face
Machine Scale Slope?	>10% of Slope Face	10-30% of Slope Face	30-60% of Slope Face	>60% of Slope Face
Remove Vegetation?	>10% of Slope Face	10-30% of Slope Face	30-60% of Slope Face	>60% of Slope Face
Any Ditch Effectiveness?				
Clean Out Ditch?	Few Rocks <1 truckload	Some Rocks 1-2 truckloads	Many Rocks 2-3 truckloads	Major cleanup >3 truckloads
Select Trim Blasting?	>10% of Slope Face	10-30% of Slope Face	30-60% of Slope Face	>60% of Slope Face
Major Repair Options	**Amount of Option**			
Recut Slope to State Standards?	>10 % of Slope Length	10-30% of Slope Length	30-60% of Slope Length	>60% of Slope Length
Extra ditch width needed?				
Any structures taken?	Type and number			
Rockfall Fence?	>10 % of Slope Length	10-30% of Slope Length	30-60% of Slope Length	>60% of Slope Length
Draped Wire Mesh?	>10% of Slope Face	10-30% of Slope Face	30-60% of Slope Face	>60% of Slope Face
Retaining Wall?	>10 % of Slope Length	10-30% of Slope Length	30-60% of Slope Length	>60% of Slope Length
Field Estimation of Best Repair				
Field Estimation of Secondary Repair				
Comments				

FIGURE 4-27 Tennessee cost notes form (Bateman 2004).

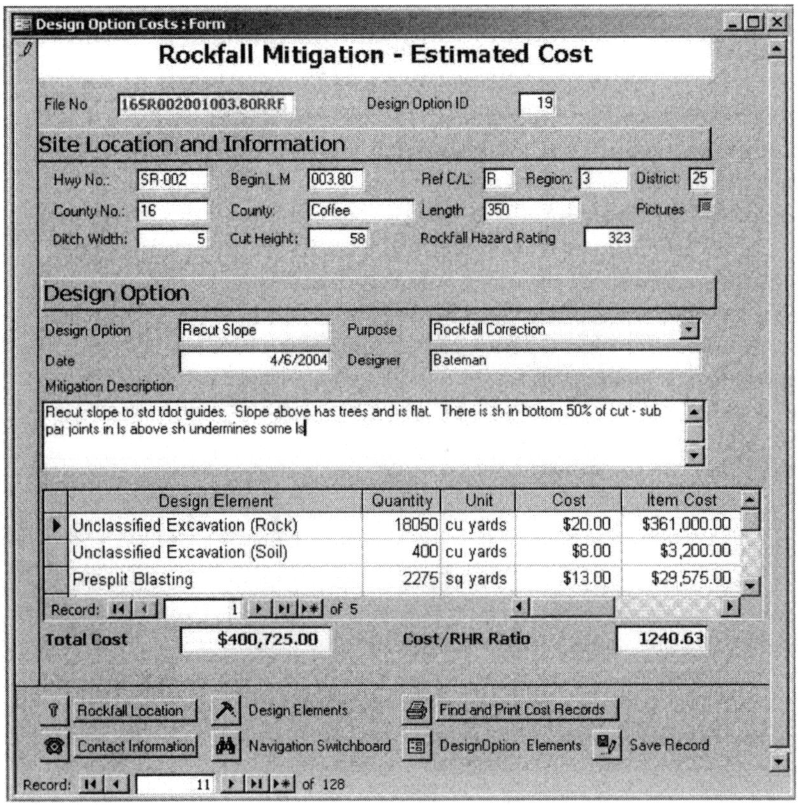

Design Option Costs : Form

Rockfall Mitigation - Estimated Cost

File No 16SR002001003.80RRF Design Option ID 19

Site Location and Information

Hwy No.: SR-002 Begin L.M 003.80 Ref C/L: R Region: 3 District: 25

County No.: 16 County: Coffee Length 350 Pictures

Ditch Width: 5 Cut Height: 58 Rockfall Hazard Rating 323

Design Option

Design Option Recut Slope Purpose Rockfall Correction

Date 4/6/2004 Designer Bateman

Mitigation Description

Recut slope to std tdot guides. Slope above has trees and is flat. There is sh in bottom 50% of cut - sub par joints in ls above sh undermines some ls

Design Element	Quantity	Unit	Cost	Item Cost
Unclassified Excavation (Rock)	18050	cu yards	$20.00	$361,000.00
Unclassified Excavation (Soil)	400	cu yards	$8.00	$3,200.00
Presplit Blasting	2275	sq yards	$13.00	$29,575.00

Record: 1 of 5

Total Cost $400,725.00 Cost/RHR Ratio 1240.63

Rockfall Location Design Elements Find and Print Cost Records

Contact Information Navigation Switchboard DesignOption Elements Save Record

Record: 11 of 128

FIGURE 4-28 Tennessee DOT rockfall database interface for estimating cost of rockfall mitigation. Data gathered on cost notes form (Figure 4-27) are combined with estimated costs to evaluate sites with high TRHRS or RCI scores (Bateman 2004).

This worksheet allows the user to enter the individual items, prices, and quantities needed for a particular mitigation strategy.

13. DISCUSSION

Rockfall is a common highway safety hazard in mountainous terrain. Loss of life and property, serious injury, disruption of traffic, and high maintenance costs are major problems associated with rockfall in many states. Costly litigation against states due to rockfall-caused accidents defers large sums from construction budgets. DiMillo (1998) noted that experiences with the RHRS approach have been very favorable. Courts have ruled that it is unreasonable to expect a transportation agency to have at its disposal all the funds necessary to immediately deal with all the rockfall safety-related issues. However, having a proactive system in place provides useful information to designers for project development and helps management make rational decisions on how to prioritize projects so as to best allocate scarce repair funds. The response by transportation agency management has been one of relief and acceptance (DiMillo 1998). Managers believe that public safety is now being served appropriately and that greater legal protection is afforded to the agency by having the RHRS in place in their state.

The RHRS and derivative systems are designed to rank sites in order of hazard to the public. Several of the modified RHRS systems have separated scores that evaluate the hazard from those that evaluate the consequence, or risk, associated with rockfall. However, even these modified systems tend to focus only on the immediate effects of rockfall. They do not take into account the costs of mitigation for rated sites or the impact of a rockfall event on the traffic flow. Consequently, a complete proactive rockfall management system requires additional components to address these issues.

Only a few such systems have been reported. The Unstable Slope Management System developed by the Washington State Department of Transportation takes some of these factors into account and focuses on the impact of both landslides and rockfall on a highway facility (Lowell and Morin 2000). More recently, the Oregon Unstable Slope Rating System (Section 6) is an adaptation of the Washington State system that follows the asset management principles promoted by FHWA. All types of unstable slope are inventoried with a single system, with rock slopes continuing to have RHRS ratings (Oregon DOT 2001; Pack et al. 2006). As discussed in Section 12, Tennessee is actively developing additional components to the TRHRS, described in Section 10, that explicitly address rockfall closure impacts and provide preliminary cost estimates for mitigation of rockfall hazards (Bateman 2004).

REFERENCES

ABBREVIATION

DOT Department of Transportation

Andrew, R. D. 1994. *The Colorado Rockfall Hazard Rating System*. Report CTI-CDOT-2-94. Colorado Department of Transportation, Denver.

Barton, N. R. 1974. *A Review of the Shear Strength of Filled Discontinuities in Rock*. Publication 105. Norwegian Geotechnical Institute, Oslo.

Bateman, V. 2002. *Managing Rockfall Hazard: The Tennessee Rockfall Hazard Database*. Master's thesis. University of Idaho, Moscow, Idaho.

Bateman, V. 2003. Development of a Database to Manage Rockfall Hazard: The Tennessee Rockfall Hazard Database. Presented at 82nd Annual Meeting of the Transportation Research Board, Washington, D.C.

Bateman, V. 2004. Rockfall Closure Impact and Economic Assessment for Rockfall Sites in Tennessee. *Proc., 55th Annual Highway Geology Symposium*, Kansas City, Missouri, pp. 357–372.

Bellamy, D., V. Bateman, E. C. Drumm, W. M. Dunne, C. Vandewater, M. Mauldon, and B. Rose. 2003. Electronic Data Collection for Rockfall Analysis. In *Transportation Research Record: Journal of the Transportation Research Board, No. 1821*, Transportation Research Board of the National Academies, Washington, D.C., pp. 97–103.

Davis, J. W., and A. Shakoor. 2005. Evaluation of the Effectiveness of Catchment Ditches Along Ohio Roadways. In *Transportation Research Record: Journal of the Transportation Research Board, No. 1913*, Transportation Research Board of the National Academies, Washington, D.C., pp. 197–204.

Deere, D. U., and D. W. Deere. 1989. *Rock Quality Designation (RQD) After Twenty Years*. Report GL-89-1. U.S. Army Corps of Engineers, Washington, D.C.

DiMillo, A. 1998. Rockfall Hazard Mitigation. In *A Quarter Century of Geotechnical Research*, Report FHWA-RD-98-139, Section 4.92, Federal Highway Administration, U.S. Department of Transportation. http://www.tfhrc.gov/structur/gtr/century/chap04/chpt04.htm.

Franklin, J. A. 1983. *Evaluation of Shales for Construction Projects: An Ontario Shale Rating System*. Report RR229. Ministry of Transportation and Communications, Ontario, Canada.

Franklin, J. A., and S. A. Senior. 1997a. Rockfall Hazards: Strategies for Detection, Assessment and Remediation. In *Proceedings, International Symposium on Engineering Geology and the Environment, Organized by the Greek National Group of IAEG*, Athens, Greece (P. G. Marinos, G. C. Koukis, G. C. Tsiambaos, and G. C. Stournaras, eds.), Balkema, Rotterdam, Netherlands, Vol. 1, pp. 657–663.

Franklin, J. A., and S. A. Senior. 1997b. The Ontario Rockfall Hazard Rating System. In *Proceedings, International Symposium on Engineering Geology and the Environment, Organized by the Greek National Group of IAEG*, Athens, Greece (P. G. Marinos, G. C. Koukis, G. C. Tsiambaos, and G. C. Stournaras, eds.), Balkema, Rotterdam, Netherlands, Vol. 1, pp. 647–656.

Franklin, J. A., D. F. Wood, S. A. Senior, and J. A. Blair. 2012. *RHRON: Ontario Rockfall Hazard Rating System: Field Procedures Manual*. Materials Engineering and Research Office Report MERO-043. Ontario Ministry of Transportation, Downsview, Ontario, Canada. https://www.raqsb.mto.gov.on.ca/login/raqs.nsf/English/Graphic/ViewConstructionDocs?OpenForm.

Hadjin, D., 2002. New York State Department of Transportation Rock Slope Rating Procedure and Rockfall Assessment. In *Transportation Research Record: Journal of the Transportation Research Board, No. 1786*, Transportation Research Board of the National Academies, Washington, D.C., pp. 60–68.

Lowell, S., and P. Morin. 2000. Unstable Slope Management, Washington State. *TR News*, No. 207, pp. 11–15.

Maerz, N. H. 2000. Highway Rock Cut Stability Assessment in Rock Masses Not Conducive to Stability Calculations. *Proc., 51st Annual Highway Geology Symposium*, Seattle, Washington, pp. 249–259.

Maerz, N. H., and S. McKenna. 1999. Surveyor: Mobile Highway Inventory and Measurement System. In *Transportation Research Record: Journal of the Transportation Research Board, No. 1690*, Transportation Research Board, National Research Council, Washington, D.C., pp. 135–142.

Maerz, N. H., A. Youssef, and R. Lauer. 2004. MORFH RS: A Rockcut Rating System for Missouri Highways. *Proc., 55th Annual Highway Geology Symposium*, Kansas City, Missouri, pp. 406–424.

Maerz, N. H., A. Youssef, and T. W. Fennessey. 2005. New Risk-Consequence Rockfall Hazard Rating System for Missouri Highways Using Digital Image Analysis. *Environmental and Engineering Geoscience*, Vol. 11, No. 3, pp. 229–249.

Mauldon, M., E. Drumm, W. M. Dunne, V. Bateman, B. Rose, and M. Kim. 2007. *Rockfall Management System for Tennessee: Final Report*. Division of Materials and Tests, Tennessee Department of Transportation, Nashville. http://www.tdot.state.tn.us/longrange/reports/res1189.pdf. Accessed December 2009.

New York State DOT. 1993. *Rock Slope Hazard Rating Procedure*. Working Draft. Soil Mechanics Bureau, Albany.

New York State DOT. 1996. *Rock Slope Rating Procedure*. Geotechnical Engineering Manual 15. Geotechnical Engineering Bureau, Albany.

New York State DOT. 2007. *Rock Slope Rating Procedure*. Geotechnical Engineering Manual 15, Revision #1. Geotechnical Engineering Bureau, Albany. https://www.dot.ny.gov/divisions/engineering/technical-services/technical-services-repository/GEM-15b.pdf.

Oregon DOT. 2001. *ODOT Landslide and Rockfall Pilot Study Final Report*. Geo-Hydro Section, Salem.

Pack, R. T., and K. Boie. 2002. *Utah Rockfall Hazards Inventor—Phase I*. Research Division Report

UT-03.01. Utah Department of Transportation, Salt Lake City.

Pack, R. T., K. Boie, S. Mather, and J. Farrell. 2006. *UDOT Rockfall Hazard Rating System: Final Report and User's Manual*. Research Division Report UT-06.07. Utah Department of Transportation, Salt Lake City.

Pierson, L. A. 1991. The Rockfall Hazard Rating System. In *Proc., National Symposium on Highway and Railroad Slope Maintenance* (D. F. Hambley, ed.), Symposium Series No. 6, Association of Engineering Geologists, Denver, Colo., pp. 1–22.

Pierson, L. A., and R. Van Vickle. 1993. *Rockfall Hazard Rating System Participant's Manual*. FHWA Publication SA-93-057. Federal Highway Administration, U.S. Department of Transportation.

Pierson, L. A., S. A. Davis, and R. Van Vickle. 1990. *Rockfall Hazard Rating System: Implementation Manual*. FHWA Report OR-EG-90-01. Federal Highway Administration, U.S. Department of Transportation.

Ritchie, A. M. 1963. Evaluation of Rockfall and Its Control. In *Highway Research Record 17*, Highway Research Board, National Research Council, Washington, D.C., pp. 13–28.

Rose, B. T. 2005. *Tennessee Rockfall Management System*. Ph.D. dissertation. Virginia Polytechnic Institute and State University, Blacksburg, 100 pp.

Rose, B., M. Mauldon, V. Bateman, D. Bellamy, E. Drumm, C. Vandewater, and W. Dunne. 2001. Data Collection, Visualization, and Distribution for Rockfall Management Systems. *CE World: ASCE's First Virtual World Congress for Civil Engineering*. http://www.ceworld.org.

Rose, B., M. Mauldon, V. Bateman, E. Drumm, and W. Dunne. 2003. Rockfall Management System and Spatial Analysis of Rock Cuts. In *Proc., Soil and Rock America 2003: 12th Panamerican Conference on Soil Mechanics and Geotechnical Engineering and 39th U.S. Rock Mechanics Symposium*, June 22–26, Massachusetts Institute of Technology, Cambridge, pp. 193–198.

Russell, C. P. 2006. *Modification and Statistical Analysis of the Colorado Rockfall Hazard Rating System*. MS Thesis T-6253. Department of Geology and Geological Engineering, Colorado School of Mines, Golden.

Santi, P. M. 2006. Field Methods for Characterizing Weak Rock for Engineering. *Environmental and Engineering Geoscience*, Vol. 12, No. 1, pp. 1–11.

Santi, P. M., C. P. Russell, J. D. Higgins, and J. I. Spriet. 2009. Modification and Statistical Analysis of the Colorado Rockfall Hazard Rating System. *Engineering Geology*, Vol. 104, pp. 55–65.

Senior, S. A. 1999. Rockfall Hazard Remediation Along Ontario Highways. *Proc., 50th Annual Highway Geology Symposium*, Roanoke, Virginia, Virginia Department of Transportation, Richmond, pp. 276–286.

Shakoor, A. 1995. Slope Stability Considerations in Differentially Weathered Mudrocks. In *Clay and Shale Slope Instability*, Reviews in Engineering Geology No. 10 (W. C. Haneberg and S. A. Anderson, eds.), Geological Society of America, Boulder, Colo., pp. 131–138.

Shakoor, A. 2005. *Development of a Rockfall Hazard Rating Matrix for the State of Ohio*. Executive Summary Report. Ohio Department of Transportation, Columbus.

Shakoor, A., and D. Brock. 1987. Relationship Between Fissility, Composition, and Engineering Properties of Selected Shales from Northeast Ohio. *Bulletin of the Association of Engineering Geologists*, Vol. 24, No. 3, pp. 363–379.

Shakoor, A., and J. P. Rodgers. 1992. Predicting the Rate of Shale Undercutting Along Highway Cuts. *Bulletin of the Association of Engineering Geologists*, Vol. 24, No. 1, pp. 61–75.

Shakoor, A., and M. W. Weber. 1988. Role of Shale Undercutting in Promoting Rockfalls and Wedge Failures Along Interstate 77. *Bulletin of the Association of Engineering Geologists*, Vol. 25, No. 2, pp. 219–234.

Stover, B. K. 1992. *Highway Rockfall Research Report*. Special Publication 37. Colorado Geological Survey, Denver.

Vandewater, C. J., W. M. Dunne, M. Mauldon, E. C. Drumm, and V. Bateman. 2005. Classifying and Assessing the Geologic Contribution to Rockfall Hazard. *Environmental and Engineering Geoscience*, Vol. 11, No. 2, pp. 141–154.

Woodard, M. J. 2004. *Development of a Rockfall Hazard Rating Matrix for the State of Ohio*. PhD dissertation. Kent State University, Kent, Ohio.

Wyllie, D. C. 1987. Rock Slope Inventory. *Proc., Federal Highway Administration Rockfall Mitigation Seminar*, Federal Highway Administration Region 10, Portland, Oreg.

Youssef, A., N. H. Maerz, and M. A. Fritz. 2003. A Risk-Consequence Rockfall Hazard Rating System for Missouri Highways. *Proc., 54th Annual Highway Geology Symposium*, Burlington, Vermont, pp. 175–195.

Youssef, A., N. H. Maerz, and Q. Xiang. 2007. RockSee: Video Image Measurements of Physical Features to Aid in Highway Rock Cut Characterization. *Computers and Geosciences*, Vol. 33, pp. 437–444.

Chapter 5

A. KEITH TURNER

ROCKFALL RISK ASSESSMENT AND RISK MANAGEMENT

1. INTRODUCTION

Beginning in the late 1970s, highway and railway agencies began to develop rockfall hazard rating systems based on visual inspections and simple calculations. The intent of these systems was to provide a rating of slope hazard according to the perceived potential urgency of stabilization (Brawner and Wyllie 1975; Brawner 1998). As highway transportation volumes continued to increase, the importance of paying greater attention to rock slope stability along highways led the Oregon Department of Transportation, beginning in 1984, to develop the Rockfall Hazard Rating System (RHRS), which reflected the needs of highway agencies (Pierson et al. 1989). Chapter 3 describes the concepts and procedures of the RHRS, and Chapter 4 describes the subsequent modifications and adjustments to the original RHRS procedures by several North American transportation agencies as they implemented rockfall hazard rating procedures into management and programming systems.

The RHRS and derivative hazard rating systems are used by many organizations today. They provide a relative ranking of existing, potentially hazardous slopes in order to set priorities and direct resources, especially for agencies with a large stock of slopes. The main implementation goal is to obtain consistent and comparable information for a series of locations so that those responsible for operating the transportation system can plan for safety improvements, optimize the benefits from investing limited safety budgets, and reduce maintenance and other operational costs. Such applications are receiving increased attention by many transportation agencies in several countries (Wong 2005).

Although useful for priority-ranking funding and mitigation efforts, the systematic assessment provided by the RHRS and similar rating systems does not provide a means of comparison of these hazards with other common hazards to which the traveling public is subject, nor does it fully provide all the information necessary for a complete asset management approach to rock slopes. In addition, such systems do not provide a measure of actual risk reduction due to mitigation efforts, which could be used in cost–benefit analyses. Because relative risk is characterized differently by the various rating systems, these measures cannot be directly compared because of their nonquantitative nature.

Formal risk-based approaches are increasingly being proposed as a solution for setting priorities for rockfall mitigation, but currently the practical application is often difficult to achieve in real-world situations because of the investment requirements, especially for those agencies faced with many problematic sites. Still, it is likely that these systems will play a larger role in the decision and management processes in the future. This chapter provides a summary of the basic concepts of risk analysis, risk assessment, and risk management and reviews applications of these techniques to rockfall events.

2. TERMINOLOGY

Many terms have been used in the discussion of hazard, risk, vulnerability, and other aspects of rockfall characterization and control. Their usage has not been consistent among various authors, and this variation may lead to confusion, even among rockfall practitioners. For clarity, the following definitions, adapted from Fell et al. (2005) and introduced in Chapter 3, are repeated here:

- *Consequence*. The outcome or result of a hazard being realized.
- *Danger (threat)*. A natural phenomenon that may lead to damage, described in terms of its geometric, mechanical, and other characteristics. The danger from rockfall may be an existing one (such as an actively rotating column) or a high and unacceptable potential for rockfall.
- *Hazard*. The likelihood that a particular danger (threat) exists or may occur.
- *Likelihood*. A qualitative description of probability or frequency of a given outcome based on an established set of criteria, data, assumptions, and information.
- *Risk*. A quantitative value derived from an analysis of the numerical values of probability, vulnerability, and consequences.
- *Vulnerability*. The degree of loss within the area affected by a hazard.

The distinction between risk and hazard must be clearly understood. In simple terms, hazard may be defined as a measure of the likelihood of rockfall occurrence, whereas rockfalls may have *consequences*, defined earlier as the outcome or result of a rockfall event.

Vulnerability is the reverse of durability and is a critical aspect when risk is evaluated. Lee and Jones (2004) define vulnerability as "the potential to suffer harm, loss, or detriment from a human perspective." They introduce the following general relationship:

hazard event × vulnerability = adverse consequences

This relationship can be expressed by the following equation:

$$H \times (E \times V) = C \qquad (1)$$

where

H = specific hazard event;
E = total value of all threatened items valued by humans, known as elements at risk and including people, buildings, infrastructure, economic activity, and other elements;
V = vulnerability, or proportion of E reduced by hazard event; and
C = adverse consequence of hazard event.

Since *risk* is concerned with the likelihood and scale of adverse consequences, risk is the product of hazard and vulnerability (Figure 5-1). Thus the risk of rockfall will be defined as a measure of the likelihood of and the adverse consequences resulting from a rockfall occurrence. Both hazard and adverse consequences (losses) must be characterized in order to arrive at an approximation of risk. For example, a rock cut with a high likelihood of producing rockfall (and thus high hazard) but low consequence (e.g., very little traffic) might be identified as a low- to moderate-risk rock cut.

Conditions change over time and both the elements at risk and their vulnerability typically change also. Slight changes in the timing of a hazard event such as a rockfall can result in very different adverse consequences, depending on exactly what the situation is at the moment of the event. A single initiating hazard event may produce a sequence of consequences. For example, earthquake impacts are frequently classed as *primary* (those resulting from ground shaking), *secondary* (those resulting from subsequent fires), and *tertiary* (including delayed economic disruptions). Even in simple cases, such as rockfall, the ramifications can spread spatially and temporally because of the chain of events. Thus, many experts suggest that vulnerability consists of two distinct aspects (Roberds 2005), namely,

- *Spatial vulnerability*. The level of potential damage, loss, or disruption will vary spatially

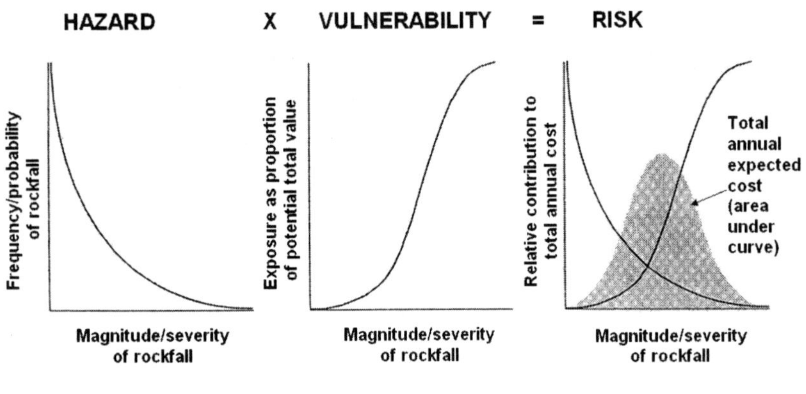

FIGURE 5-1 Diagrammatic representation demonstrating rockfall risk as product of rockfall hazard and rockfall vulnerability (adapted from Lee and Jones 2004).

depending on the area influenced by the hazard (in the case of rockfall, this level is reflected by the runout zone) and by the distribution and characteristics of the elements at risk.
- *Temporal vulnerability*. Mobile elements at risk, such as people walking below a cliff or traffic moving through a rock cut, will suffer adverse consequences only if they are affected by the hazard event; obviously this vulnerability can vary considerably during a 24-h day or seasonally.

In the case of temporal vulnerability, it is not the vulnerability that changes in the short term but the value of the elements at risk. An individual is always vulnerable to injury or even death if hit by rockfall, but the individual's risk is entirely dependent on whether the individual is present when the rockfall occurs. This variation is defined as *exposure*. Thus Equation 1 becomes modified to

$$H \times (E \times V \times E_x) = C \qquad (2)$$

where H, E, V, and C are as defined previously, and E_x is the exposure—the proportion of each category of elements at risk present at the time of the hazard event or influenced by it.

Adverse consequences are also frequently influenced by the sequence of human decisions or actions, usually taken completely independently of the hazard event. Thus, an element of chance often plays a major role in the determination of adverse consequences, producing complex *event sequences* (Figure 5-2), which are often exceptionally difficult to enumerate and evaluate quantitatively, even by teams of experts armed with extensive data.

In summary, both the magnitude and frequency characteristics for a hazard such as rockfall and the variations of vulnerability and exposure over space and time contribute to risk (Figures 5-1 and 5-2). Basic statements of risk can be developed provided that the probabilities of specific magnitudes of events are known. For landsliding in general, and also rockfall, the following relationship is relevant (Carrara 1983; Varnes 1984; Wu et al. 1996; Lee and Jones 2004):

$$R_s = P(H_i) \times (E \times V \times E_x) \qquad (3)$$

where E, V, and E_x are as defined previously; R_s is the specific risk, or the expected degree of loss due to a particular magnitude of rockfall (H_i) occurrence within a specific area over a given period of time; and $P(H_i)$ is the hazard, or the probability that a particular magnitude of rockfall (H_i) will occur within a specific area and time frame.

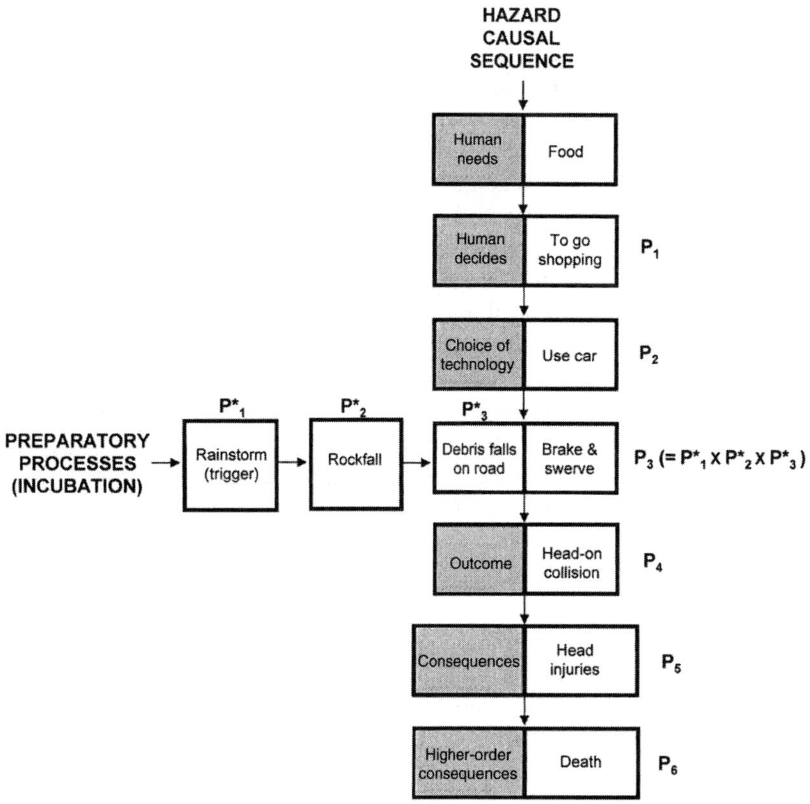

Probability of death = $P_1 \times P_2 \times P_3 (= P^*_1 \times P^*_2 \times P^*_3) \times P_4 \times P_5 \times P_6$

FIGURE 5-2 Rockfall event sequence (horizontal sequence of boxes) interacts with accident sequence (vertical sequence of boxes) to result in a highway accident leading to a fatality (adapted from Lee and Jones 2004).

3. RISK MANAGEMENT PROCESS

There is no single well-established terminology in risk management. To further complicate matters, *risk* means different things to various people and professions, and thus risk management terminology is often misinterpreted and misused. The Working Group on Landslides of the International Union of Geological Sciences (IUGS) formed the Committee on Risk Assessment, which developed and published a specific terminology to be used internationally for landslide risk management (IUGS 1997). This terminology, which includes the definitions provided in Section 2 and in Chapter 3, was designed to be consistent with various national standards for risk management.

3.1 Sources

The topic of landslide risk assessment was introduced by Einstein (1988) at the Fifth International Symposium on Landslides. Fell (1994) provided definitions for risk and hazard suitable for landslide risk assessment. Using case studies of landslide-prone areas in eastern Australia, Fell combined elements of probability, vulnerability, and risk to determine a total risk value of a landslide event and compared that risk with acceptable risks. Fell suggested that risk could be evaluated by one of the following methods:

- Making a risk comparison, or comparison of the risk calculated for the project with socially acceptable risks;
- Determination of the cost-effectiveness of risk reduction, or the cost to save lives and reduce the risk to a socially acceptable level; or
- Use of a cost–benefit analysis of the annual cost of risk reduction measures with the annual reduction in risk based on assigned values.

Fell suggested that the decisions involved are best made not by engineers but rather by owners, politicians, or others because these decisions typically involve consideration of legal, political, social, and financial issues. Fell also noted that considerable evidence suggests that the population at large will accept a higher level of voluntary risk than involuntary, imposed risk, but in general there is often a lack of understanding of risk by the public, and in some cases by practitioners. This lack can complicate the decision process, especially when an uninformed and unrealistic public voices zero tolerance for any risk whatsoever.

Wu et al. (1996) provided detailed explanations of the fundamental principles used to describe uncertainty, how probability theory is used to estimate landslide hazards, the relationship between hazard and risk, and applications for landslide investigations. Two international conferences, held in 1997 and 2005 (Cruden and Fell 1997; Hungr et al. 2005), are important sources for detailed discussions of the fundamental concepts and applications of risk analysis for landslides in general and, by extension, for rockfall.

The 1997 conference proceedings includes a state-of-the-art review of quantitative risk assessment produced by the risk analysis committee of the IUGS Working Group on Landslides (IUGS 1997), which still provides pertinent information. The 1997 conference proceedings also contains key papers by Einstein (1997), Fell and Hartford (1997), and Morgenstern (1997), as well as additional important contributions by Baynes (1997), Lee (1997), Moon (1997), and Roberds et al. (1997). The 2005 conference proceedings includes state-of-the-art papers by Fell et al. (2005), Nadim et al. (2005), Roberds (2005), and Wong (2005) and also papers addressing critical issues by Düzgün and Lacasse (2005), Moon et al. (2005), Stewart and Buys (2005), and Wilson et al. (2005).

Chapter 3 of this volume includes a discussion of hazard and risk concepts, in which Figure 3-1 illustrates the relationship between hazard, consequence, and risk. Fell et al. (2005) recommended that additional qualitative descriptors adapted from the Australian Geomechanics Society (AGS 2000) be used to clarify the level of likelihood and consequence. These descriptors are defined in Tables 3-1 and 3-2. Australian researchers have been active in evaluating risk from landslides and rockfall; the AGS issued guidelines for landslide risk management in 2000 (AGS 2000), and these were subsequently revised and reissued in 2002 and again in 2007 (AGS 2002, 2007a, 2007b, 2007c). The Road and Traffic Authority for the State of New South Wales developed a guide to risk analysis of slopes (Stewart et al. 2002). Case studies have been provided by Baynes (1997), Baynes et al. (2002), Flentje et al. (2007), Michael-Leiba et al. (2000), Moon (1997), Moon et al. (2005), Stewart and Buys (2005), and Wilson et al. (2005).

Lee and Jones (2004) provided an extended discussion of landslide risk assessment and management topics with specific reference to the state of the practice and state of the art in the United Kingdom. Glade et al. (2005) offered a broader international perspective of the entire field of landslide hazard and risk with an extensive series of papers organized to discuss conceptual models for the approach to landslide risk, evaluation of risk, management of landslide risk, and applications of landslide risk assessment. Although these authors discuss applications of risk analysis to all types of landslides, most of the principles and analysis techniques are relevant to rockfall assessments.

3.2 Defining the Risk Management Process

Numerous published sources (Lee and Jones 2004; Fell et al. 2005; Glade et al. 2005) concur that the entire risk management process is best understood as a hierarchical process that incorporates stages of risk analysis and risk assessment and that these stages in turn involve sequences of clearly defined tasks (Figure 5-3). Although these sources

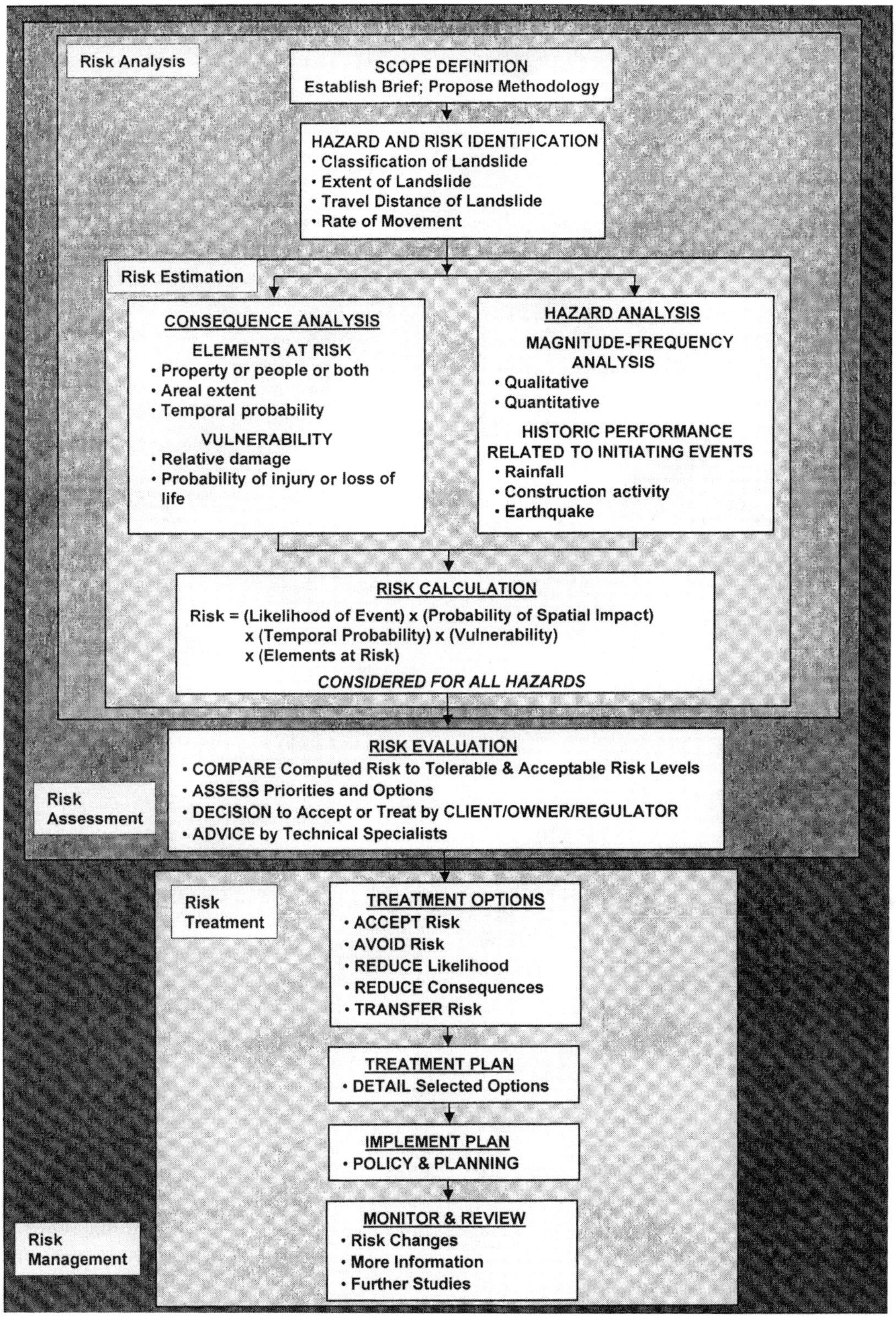

FIGURE 5-3
Flowchart for landslide risk management process (adapted from Glade et al. 2005).

offer some slightly different definitions of some of the component tasks, all refer to the definitive guidelines for landslide risk management issued by the AGS in 2000 (AGS 2000). The following discussion represents a consensus of the conclusions reached in these sources.

3.2.1 Risk Analysis

As illustrated in Figure 5.3, *risk analysis* is the first stage of the risk management process. It contains five main tasks. The first task is *scope definition*, a critical first step that ensures that the risk analysis addresses the relevant issues, satisfies the needs of those concerned, and avoids subsequent misunderstandings. This task answers such questions as "What sort of information is required?" "What type of methodology should be employed?" "What resources are required?" "What is the assessment to be used for?"

The second task is *hazard and risk identification*. In the case of rockfall, this task requires identification of the basic site conditions, rockfall source areas, and what elements are exposed to the rockfall events.

The risk analysis then moves into a *risk estimation phase*, which consists of three tasks: consequence analysis, hazard analysis, and risk calculation. The consequence analysis and hazard analysis tasks are typically undertaken in parallel. *Consequence analysis* involves

- Identification and quantification of the elements at risk (including both property and persons),
- Assessment of the spatial and temporal probabilities for each element at risk [$P(S{:}H)$ and $P(T{:}S)$], and
- Assessment of the vulnerability of the elements at risk, including, as appropriate, both property damage [$V(PROP{:}T)$] and loss of life or injury [$V(D{:}T)$].

Hazard analysis identifies and characterizes the potential rockfalls together with their frequency of occurrence to provide the probability of a rockfall event (*PH*). As shown in Figure 5-3, this task typically includes an assessment of the triggering mechanisms, review of historical rockfall records, and rockfall magnitude and frequency analysis by qualitative, semiquantitative, or quantitative methods. The risk calculation task combines information obtained in the previous tasks by using the following equation:

$$R = P(H) \times P(S{:}H) \times P(T{:}S) \times V \times E \tag{4}$$

where

R = annualized risk (which may be thought of as annual probability of fatality or any other unacceptable outcome);
$P(H)$ = annual probability of hazardous event (rockfall);
$P(S{:}H)$ = probability of spatial impact (probability that rock reaches lane on which vehicles travel given occurrence of a rockfall);
$P(T{:}S)$ = temporal probability that consequence will occur (i.e., probability that vehicle will be present on road below site when rock falls);
V = vulnerability of element at risk given presence of element at risk within zone of impact because of hazardous event (rockfall) [vulnerability will reflect property damage ($V(PROP{:}T)$) or injury or loss of life ($V(D{:}T)$), as appropriate for each case]; and
E = element at risk (i.e., passenger or valued facility).

3.2.2 Risk Assessment

Risk assessment takes the results from the risk analysis stage and assesses them against societal norms for various classes of risk tolerance criteria to determine if the estimated risks are low enough to be acceptable or if remedial mitigation measures are required. In either case, these results are presented to the owner or regulator to decide on any further course of action. The decision as to what, if any, further mitigation is appropriate is usually the responsibility of the owner or regulator; this responsibility typically includes consultation with the affected public or stakeholders. Although nontechnical clients may seek guidance from a risk analyst, it is important from a legal viewpoint that the owner or regulator make the final decision. Cruden (1997) presented a detailed analysis of the inherent variability of risk estimates, which he suggested may be greater than the variability of either the hazard or the vulnerability of the elements at risk. If proven to be the case, this increased variability in risk estimates makes the subsequent risk management tasks significantly more difficult. A clear recognition of the difference between acceptable and tolerable risk is important.

Acceptable risk describes the likelihood of an event whose probability of occurrence is small and whose consequences are so slight that individuals or societal groups are willing to be subjected to the possibility that the event might occur. The concept of acceptable risk evolved partly from the realization that absolute safety is generally an unachievable goal and that even very low exposures to certain toxic substances may confer some level of risk. One example of acceptable risk is the probability that an individual will be struck by lightning. However, most people accept some mitigation measures in response to this risk, such as lightning rods on buildings or temporary suspension of sporting events during thunderstorms, that further reduce the risk of personal injury. Acceptable risk of rockfall events includes situations in which the potential for rockfall is so low or damage is expected to be so slight that no further remediation is required unless some very low-cost measures (in terms of time and effort) are available to further reduce the risk.

Tolerable risk includes those risks that society does not regard as negligible or something that should be ignored but which society is willing to live with so as to secure certain benefits. Tolerable risk includes various situations that society is confident are being properly controlled by the owner or by regulators; the situations are under constant review and steps to further reduce them will be undertaken if these are economically and technically feasible. Thus, tolerable risk falls within a range that society finds acceptable, even if some individuals are likely to be affected. Examples of responses to ensure that the public accepts tolerable risks include transportation safety regulations, food and product testing and certifications, and financial oversight reporting. Levels of tolerable risk are typically higher if the risk is undertaken voluntarily by the individual (e.g., the potential for injury when rock climbing or sky diving is accepted by individuals undertaking these activities) than if the risk is involuntary (e.g., travel on a roadway). Typical limits of tolerable risk related to loss of life are in the range of 1×10^{-4} to 1×10^{-6} per annum for existing and newly constructed engineered slopes (Fell et al. 2005). Additional discussions of these risk criteria concepts are provided in Sections 5-5 and 5-6.

3.2.3 Risk Management

Risk management uses the results from the risk assessment stage to evaluate risk mitigation options. Chapter 18 provides a more extensive discussion on the role and benefits of instituting a rockfall management program.

In Figure 5-3 a series of tasks is listed as part of a risk treatment action, which requires a consideration of mitigation options:

- Accept the risk,
- Avoid the risk,
- Reduce the likelihood of rockfall,
- Reduce the consequences, or
- Transfer the risk.

Acceptance of the risk is typically only feasible if the estimated risk level falls below the tolerable risk criteria. The remaining mitigation options are discussed at length in Chapters 11 to 16. Mitigation measures often used to reduce the likelihood of rockfall events include various rock slope scaling and stabilization measures and other mitigation measures such as rockfall barriers, catchment fences, or drapery systems that reduce the likelihood that a rockfall will reach a roadway or other critical area. The consequences of rockfall are reduced by implementing monitoring and warning systems or by developing evacuation plans for residents at risk or diversion plans for transportation facilities blocked by rockfall. Transferral of risk implies the use of insurance, development of a risk mitigation plan, or the implementation of regulatory controls (such as the closure of a highly exposed road during times of heightened rockfall activity).

4. RISK ANALYSIS METHODS

Rockfall hazard analysis and risk analysis are not easy to achieve in practice. Rockfalls originate from cliffs of various heights and characteristic lithologies and structures and involve a considerable range of rock volumes. Further complications arise from the numerous variables that influence the trajectory of rockfalls. Topographic and geometric considerations at various locations require rockfall risk assessments to consider point-like, linear, or areal distributions of rockfall (Figure 5-4).

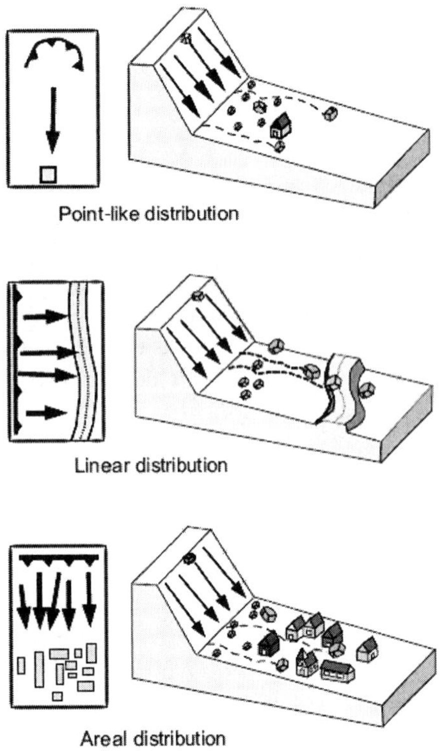

FIGURE 5-4 Different categories of rockfall distributions that affect rockfall hazard and risk (Crosta and Agliardi 2003).

Among the most destructive of mass movements, rockfalls often pose a severe threat to individuals, property, and infrastructure (Bunce et al. 1997; Guzzetti et al. 2004; Fell et al. 2005). For these reasons, rockfall hazard and risk assessments are of considerable interest to those responsible for ensuring public safety and safeguarding property and infrastructure where rockfall hazards are present (Fell 1994; Fell and Hartford 1997; Crosta and Agliardi 2003; Guzzetti and Reichenbach 2010). Lee and Jones (2004) proposed a three-way categorization of risk assessment: qualitative, semiquantitative, and quantitative. This classification appears particularly relevant to the evaluation of rockfall risk and is adopted in the subsequent discussion.

FIGURE 5-5 Typical risk matrix (adapted from Lee and Jones 2004).

Likelihood of Event	Adverse Consequences			
	Severe	Moderate	Mild	Negligible
High	high	high	medium, low	near zero
Medium	high	medium	low	near zero
Low	high, medium	medium, low	low	near zero
Negligible	high, medium, low	medium, low	low	near zero

4.1 Qualitative Risk Assessment

Qualitative risk assessments express the likelihood of rockfall and the potential for adverse consequences in qualitative terms. A common approach is to construct a risk matrix (Figure 5-5) that qualitatively scores the adverse consequences in a series of columns and the likelihood (or probability) of occurrence in a series of rows. Often, four categories are used for the columns and rows. This method yields a matrix with 16 cells; if five categories are used, they produce a rather complex matrix of 25 cells. Typically the consequences are labeled *severe*, *moderate*, *mild*, and *negligible*, while the likelihood categories are labeled *high*, *medium*, *low*, and *negligible*. The combinations of consequence and likelihood categories then allow the cells to be labeled with qualitative risk levels such as *high*, *medium*, or *low*, or as intermediate levels, as shown in Figure 5-5.

Sometimes a numerical scheme is used to define the consequence and likelihood categories and, in some cases, the values of the cells are computed by multiplication of the relevant category values. However, although these risk matrix procedures provide a comparison of different relative risk conditions, they represent a gross simplification of reality. They represent a procedure based on professional intuition and judgment, a process long defended as appropriate in many cases, particularly by the engineering community.

Qualitative risk assessments are more easily interpreted and used if the consequence and likelihood categories are defined appropriately with clearly articulated measures of their significance. Examples of definitions used for the consequence and likelihood categories proposed and adopted by the AGS (2000) are presented in Chapter 3 (Tables 3.1 and 3.2). Fell et al. (2005) concluded that qualitative risk assessments represent the state of the practice for many land use planning and zoning applications. Raetzo et al. (2002) and Lateltin et al. (2005) described this application in Switzerland; a summary of the Swiss methods is provided in Section 6.1. Fell et al. (2005) also concluded that the principal value of qualitative risk assessments is related to their support of systematic studies of landslide phenomena and uniform classification of hazards and risks that are understandable by those responsible for risk management.

4.2 Semiquantitative Risk Assessment

Semiquantitative risk assessments combine qualitative and quantitative measures of the likelihood and consequences. Usually the probabilities of frequency of occurrence are defined quantitatively, and levels of consequence remain less well defined. In some cases, the probabilities of frequency of occurrence are known; in other cases they are assumed. Moon et al. (2005) described a semiquantitative approach for evaluating the frequency of occurrence by constructing and using *landslide size–frequency models* (see Section 5.4.2).

Semiquantitative risk assessments often employ statistical and probability theory (see Section 5.1) and various graphical representational methods, such as event trees, fault trees, and influence diagrams (see Sections 5.2 and 5.3). As an example, the Road and Traffic Authority of New South Wales, Australia, developed a semiquantitative methodology for a variety of landslide and rockfall conditions (Stewart et al. 2002). Several semiquantitative case history examples are described in Section 6.

4.3 Quantitative Risk Assessment

Quantitative risk assessments provide quantitative measures of both the likelihood and consequences. They generally do not provide a single answer because there are often probability distributions for both the magnitude and the frequency of the hazard event as well as a range of possible losses or consequences. The fully quantitative approaches to risk estimation are much more demanding than either the qualitative or the semiquantitative approach. Considerable computational effort and adequate data sources are necessary to complete a quantitative risk assessment. The numerical results often conceal the uncertainties that result from the assumptions made. Also, as discussed by Cruden (1997), the inherent variability of risk estimates may be greater than the variability of either the hazard or the vulnerability of the elements at risk. For these reasons, quantitative assessment of risk due to rockfall has been applied relatively infrequently, usually in situations where societal or economic sensitivities require a particularly rigorous rockfall risk analysis. Section 6 includes several case histories of fully quantitative risk assessments of rockfall, including examples originally reported by Bunce (1994), Bunce et al. (1997), and Roberds (2001, 2005).

5. TOOLS AND TECHNIQUES USED FOR RISK ASSESSMENT

Risk assessments typically use one or more tools and techniques based on statistical and probability theory and established graphical representational methods. They also frequently use widely accepted definitions of societal and individual risk and established criteria for tolerable and acceptable risk levels. Recommendations for mitigation are often based on the application of the as-low-as-reasonably-practicable, or ALARP, principle. The following sections describe these tools and techniques; many of these form part of the case histories presented in Section 6.

5.1 Probability Analysis

Semiquantitative and quantitative risk analysis methods (see Section 4) rely on calculation of the probability of occurrence of rockfall events and the probabilities of the ensuing consequences. These calculations use the concepts of marginal, conditional, and joint probabilities and the application of the binomial probability model. These concepts are presented in most statistics textbooks (Benjamin and Cornell 1970; Davis 2002; Navidi 2006).

5.1.1 Marginal, Conditional, and Joint Probabilities

The probability of an outcome from a single event is referred to as the *marginal probability*. For example, the marginal probability of obtaining an even number (2, 4, or 6) from the single roll of a die is $3/6$, or .5. Sometimes, prior knowledge of information about an outcome changes the probability of the outcome. The probability of one event's occurring, given that another event has already occurred, is called a *conditional probability*. For example, if one has rolled a die and obtained an odd number, what is the probability that the number is 5? The conditional probability of obtaining a 5, given that the answer is an odd value (i.e., it is 1, 3, or 5), is $1/3$, or .33. The *joint probability* of obtaining a 5 on a single roll of a die is equal to the marginal probability of achieving any odd number multiplied by the marginal probability of obtaining 5, given that the answer is an odd number, or $1/2 \times 1/3 = 1/6$. In this case, this joint probability equals the probability that any single number will occur.

However, the computation of joint probabilities from the multiplication of a series of marginal and conditional probabilities is very useful in assessing the sequences of events and consequences from a rockfall site. Consider the following simple case, the probability that a rockfall fragment, triggered by rainfall, will fall onto a road:

- First, the probability of a triggering event, such as a precipitation event, needs to be established. Suppose that it rains above a given threshold 90 days in a year. Then the annual marginal probability of a rainfall is 90/365 = .25. Thus, the probability of no rainfall, and therefore no possibility of rockfall from this cause, is .75.
- Next, the conditional probability that a rainfall event actually triggers a rockfall must be assessed; this conditional probability is set as .33.
- Then, the probability of a falling rock reaching the road is needed, set as .2.

Then the probabilities are calculated as follows:

- Probability that rockfall will reach the road = $.25 \times .33 \times .2 = .0165$,
- Probability that rock will fall short of the road = $.25 \times .33 \times .8 = .066$,
- Probability that no rockfall will result from rainfall = $.25 \times .67 = .1675$, and
- Probability of no rainfall and therefore no possibility of rockfall = .75.

Because these four possible outcomes cover all possible outcomes, their probabilities must sum to 1.0 ($.0165 + .066 + .1675 + .75 = 1.0$). The calculation of these sequences of marginal and conditional probabilities is easily visualized by using event trees and fault trees, as described in Section 5.2.

5.1.2 Binomial Probability Model

When undertaking quantitative risk analysis for rockfall, numerous authors (Bunce 1994; Bunce et al. 1997, 1998; Hungr and Beckie 1998; Roberds 2001, 2005) use the binomial theorem to calculate the probability that a vehicle will be in the path of a falling rock and thus capable of being hit by the rock. This procedure is usually supported with reference to statistical sources (Benjamin and Cornell 1970; Navidi 2006), but without any further elaboration. Because this procedure is fundamental to undertaking risk analysis of rockfall, a brief explanation of the relevant statistical concepts is provided here.

A series of experiments with only two possible outcomes—one labeled a success and the other a failure—are referred to as *Bernoulli trials*. If the probability of success is denoted by p, then the probability of failure is $1 - p$. For rockfall studies, if the probability of a falling rock's hitting a vehicle is p, then the probability of the rock's missing a vehicle is $1 - p$. If a number of rockfall events occur, each can be considered an independent Bernoulli trial, and the number of cases of rocks hitting a vehicle is a random variable, which is said to have a binomial distribution. The binomial distribution requires the following four conditions:

1. There is a fixed number of trials involving a random process,
2. The outcome of each trial is classified into either success or failure,
3. The probability of success is the same for each trial, and
4. The trials are independent, meaning that the outcome from one trial does not influence the outcome of any other trial.

The occurrence of a falling rock hitting a vehicle meets these four criteria.

A nine-step procedure defines the underlying rationale for applying the *binomial theorem*, which is the formal definition of the binomial distribution, to the analysis of accidents caused by falling rocks. This procedure has been adapted from discussions by Davis (2002). The style used to define the various parameters corresponds to those used by Bunce et al. (1997). The nine steps are as follows:

1. The spatial probability of a stationary vehicle being hit by a falling rock is $P(S{:}H)$.
2. The spatial probability of a stationary vehicle not being hit by a falling rock is thus $1 - P(S{:}H)$.
3. The spatial probability that N_R falling rocks do not hit a stationary vehicle is $(1 - P(S{:}H))^{N_R}$.
4. The spatial probability that a final falling rock will hit a vehicle after $N_R - 1$ falling rocks do not hit a stationary vehicle is $(1 - P(S{:}H))^{(N_R-1)} \times P(S{:}H)$.
5. The spatial probability that one falling rock will hit a vehicle in a series of N_R falling rocks is $N_R \times (1 - P(S{:}H))^{(N_R-1)} \times P(S{:}H)$.

6. The spatial probability that N_a falling rocks will hit vehicles after a series of $(N_R - N_a)$ misses is $(1 - P(S{:}H))^{(N_R - N_a)} \times P(S{:}H)^{N_a}$, but
7. The N_a rockfall hits on vehicles by a sequence of N_R rockfalls can occur in many different patterns, specifically $N_R!/[(N_R - N_a)! \times N_a!]$ ways, where $N_R!$ defines the N_R factorial, $N_a!$ defines the N_a factorial, and so forth. Thus the spatial probability of N_a vehicles being hit by N_R rockfalls is

$$P(S,\ N_a) = \frac{N_R!}{(N_R - N_a)! \times N_a!}(1 - P(S{:}H))^{(N_R - N_a)} \times P(S{:}H)^{N_a}$$

   which is the binomial theorem that defines the binomial probability distribution.
8. The spatial probability that no vehicles will be hit ($N_a = 0$) by a series of N_R falling rocks is obtained from the binomial theorem, which, in such a case, simplifies to $P(S, 0) = (1 - P(S{:}H))^{N_R}$.
9. Then the spatial probability that one or more vehicles will be hit by a series of N_R falling rocks is determined as $P(S) = 1 - P(S, 0) = 1 - (1 - P(S{:}H))^{N_R}$.

An impact by a falling rock on a stationary vehicle will occur only if the vehicle and path of the falling rock coincide in time and space. The annual temporal probability, $P(T{:}S)$, of a stationary vehicle being hit is equal to the part of the year the vehicle occupies the space affected by the rockfall. Then the annual probability that an accident will involve one or more vehicles is $P(A) = P(T{:}S) \times P(S)$.

The same analysis procedure permits calculation of annual probabilities for additional typical accidents resulting from rockfall: the impact of a falling rock on a moving vehicle and the collision of a moving vehicle with a previously fallen rock. However, the methods of calculating the spatial and temporal probabilities, $P(S)$ and $P(T{:}S)$, are different for each type of accident. Bunce et al. (1997) provided details of these calculations. Hungr and Beckie (1998) and Bunce et al. (1998) further discussed these calculation procedures and other options for determining probabilities for these events.

5.2 Event Trees and Fault Trees

Two related graphical approaches—event trees and fault trees—are widely used tools in semiquantitative and quantitative risk analyses.

5.2.1 Event Tree Analysis

Event tree analysis provides an inductive approach for bringing together and considering several different parameters and for analyzing initiating events that result in a variety of outcomes. An event tree is a graphical representation of the underlying logic model. The tree is constructed using forward logic, beginning at an initiating event and then identifying and quantifying a sequence of possible outcomes. The tree consists of branches and nodes (Figure 5-6). All possible states of the system are considered at each node and each state or branch of the event tree, and each of these is characterized by a probability value that defines its likelihood of occurrence. Events beginning at a node are mutually exclusive and the sum of their respective probabilities must be equal to 1. Thus, at Node A in Figure 5-6, $P(1A) + P(1B) + P(1C) = 1$.

Event trees are typically constructed so that each node has only two outcomes, but in some cases three, or occasionally even more, outcomes are defined, such as at Nodes A and C in Figure 5-6. The probabilities of the outcomes being assessed are defined by computing the conditional probabilities of all the branches that are included in the path from initiating event to outcome. Thus, with reference to Figure 5-6,

$$P(\text{Outcome}1) = P(1A)$$
$$P(\text{Outcome}2) = P(2A) \times P(1B)$$
$$P(\text{Outcome}6) = P(3A) \times P(3C)$$

Applications of event trees to the semiquantitative analysis of rockfall are provided by Hoek (2000, Chap. 9) and Peila and Guardini (2008). The Peila and Guardini (2008) case history is described in Section 6.3.

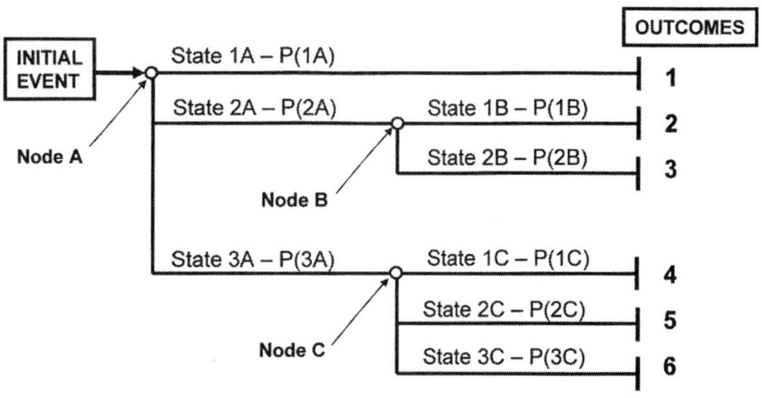

FIGURE 5-6 General scheme of event tree (adapted from Peila and Guardini 2008).

5.2.2 Fault Tree Analysis

Fault tree analysis techniques were first developed in the early 1960s. They have been used in a wide range of engineering disciplines as one of the primary methods for performing reliability and safety analyses. Fault tree analysis breaks down an undesirable event into its contributing factors and investigates combinations of events and conditions leading to the undesirable event. Fault tree analysis uses a deductive approach to reach a valid conclusion. The fault tree is constructed on the basis of an accident or undesirable event that is to be avoided, and then a backward logic is used to identify the immediately preceding causes. Each of the immediate causes is examined in turn until the analyst has identified the basic causes of each event. In fact, fault tree analysis is essentially the reverse of the inductive approach used by event tree analysis. Event tree analysis and fault tree analysis are, however, closely linked. Fault trees are often used to quantify system events that are part of event tree sequences.

Fault trees graphically represent the underlying logic model that describes the interactions of failures and other events within a system. Basic events at the bottom of the fault tree are linked via logic symbols (known as *gates*) to one or more initiating events. These events represent identified hazards or system failure modes for which predicted reliability or availability data are required. The use of fault trees is common in many industrial and manufacturing applications, where information is readily available to describe all outcomes, both successes and failures, within a system. This broad base of information is usually not available for rockfall applications. Thus, the use of fault tree analysis for evaluating rockfall risk is relatively uncommon. However, Figure 5-7 provides an example of a fault tree developed to assess rockfall events along Interstate 70 near Georgetown, Colorado.

5.3 Influence Diagrams

An *influence diagram*, or *relevance diagram*, is a compact graphical and mathematical representation of

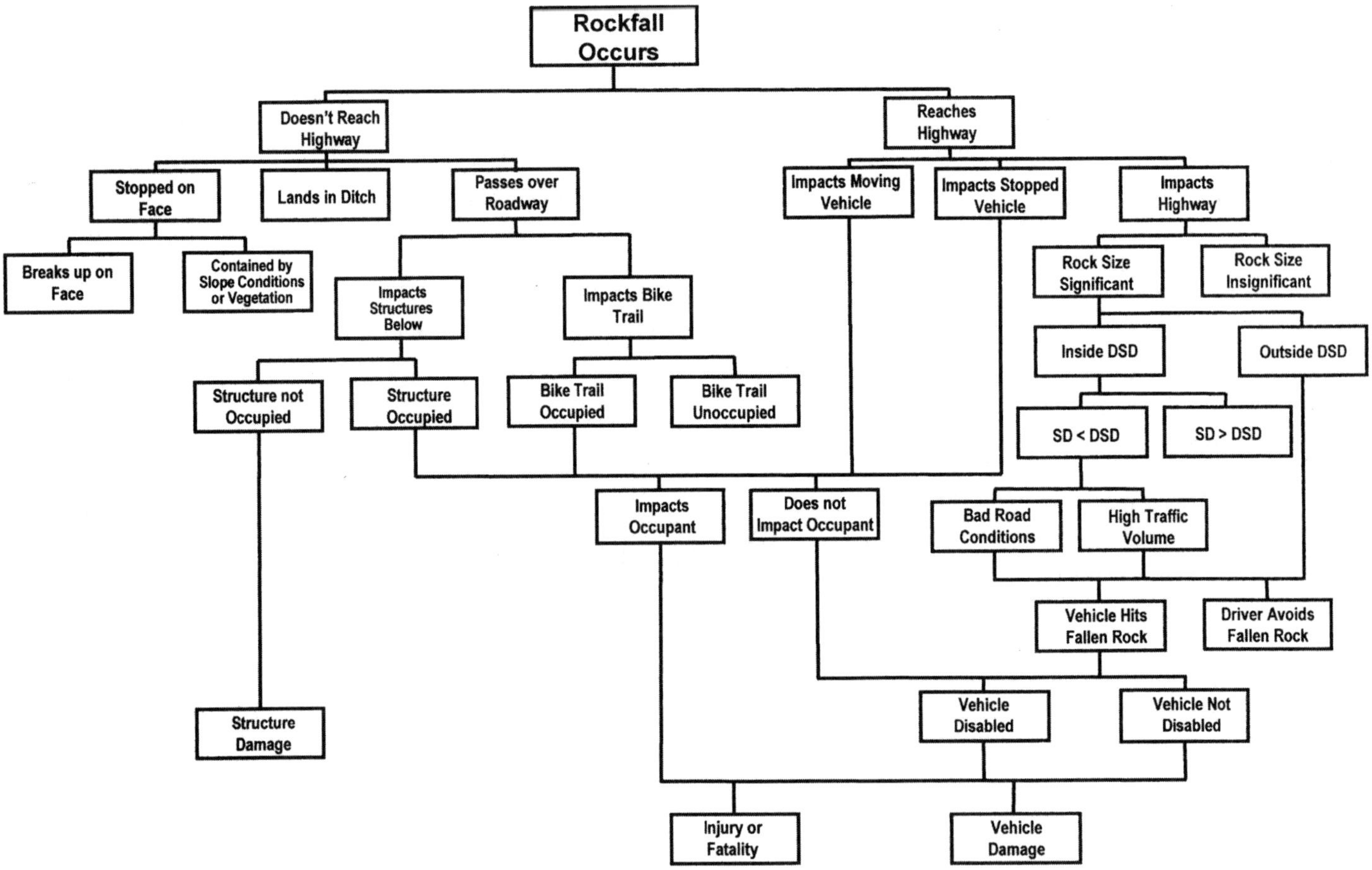

FIGURE 5-7
Fault tree analysis of rockfalls along Interstate 70, Georgetown, Colorado (SD = sight distance, DSD = decision sight distance).
(DIAGRAM COURTESY OF YEH AND ASSOCIATES, DENVER, COLORADO.)

a decision situation. Influence diagrams were first developed in the mid-1970s and are now widely adopted by the decision analysis community as an alternative to event or fault trees, which may become excessively complex with very large numbers of branches.

A typical influence diagram uses three types of nodes and three types of arcs (or arrows) connecting the nodes. In this way it can represent the probabilistic relationships between, for example, rockfall events, sequential outcomes, and their consequences. The node types are as follows:

- Decision node corresponding to each decision to be made is drawn as a rectangle,
- Uncertainty node corresponding to each uncertainty to be modeled is drawn as an oval, and
- Value node is drawn as an octagon.

The three types of arcs are

- Functional arcs ending at a value node,
- Conditional arcs ending at an uncertainty node, and
- Informational arcs ending at a decision node.

Figure 5-8 is a simple example of an influence diagram in which someone is planning a vacation and is assessing the potential for weather disruptions. There is one decision node (Vacation Activity), two uncertainty nodes (Weather Condition, Weather Forecast), and one value node (Satisfaction). There are two functional arcs (ending in Satisfaction), one conditional arc (ending in Weather Forecast), and one informational arc (ending in Vacation Activity). The functional arcs ending in Satisfaction indicate that Satisfaction is a utility function of Weather Condition and Vacation Activity. In other words, satisfaction with the vacation can be quantified if the weather and the choice of activity are known.

Figure 5-9 is an influence diagram employed during a qualitative risk assessment of slope instability on coastal cliffs in Cornwall, southwest England (Westgate et al. 2003).

5.4 Estimating Frequency of Occurrence

As described in Section 3.2.1 and illustrated in Figure 5-3, risk analysis includes a risk estimation phase, which includes a hazard analysis task. A key factor in developing a hazard analysis is the

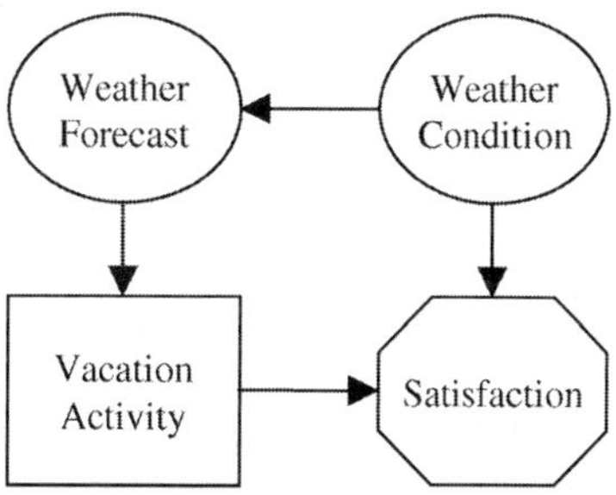

FIGURE 5-8 Simple influence diagram (http://en.wikipedia.org/wiki/File:Figure22.png).

estimation of the frequency of occurrence of a rockfall event. In some cases, the probabilities of frequency of occurrence are known; in other cases they are assumed. The IUGS Working Group on Landslides (IUGS 1997) identified the following alternative methods for calculating frequency of occurrence:

1. Use of historical data records for the area of interest, or from other areas with similar characteristics;
2. Empirical methods based on correlations reflecting known instability conditions;
3. Geomorphological evidence of slope retreat or estimate based on expert judgment;
4. Established relationships with frequency and intensity of triggering events, such as rainfall;
5. Direct assessment based on expert judgment, often with reference to a conceptual model, and employing event tree or fault tree assessment methods;
6. Modeling of the primary variable, such as piezometric pressures, coupled with some knowledge of rock mass geometry and shear strength;
7. Application of probabilistic methods; or
8. A combination of the foregoing methods.

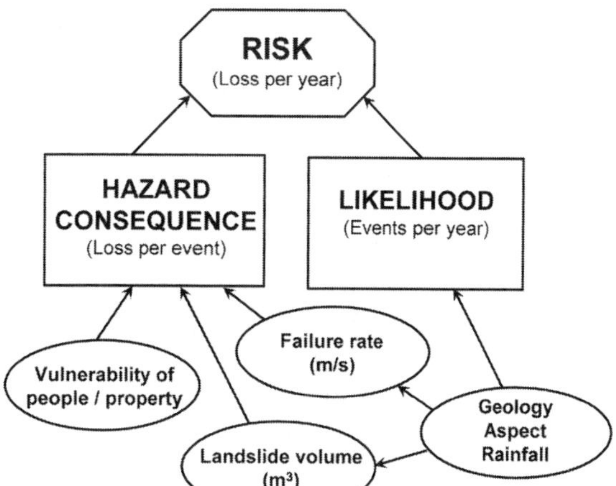

FIGURE 5-9 Influence diagram used during qualitative risk assessment of Cornish coastal cliffs (adapted from Westgate et al. 2003).

5.4.1 Use of Historical Data

Historical archives, including newspapers, journals, church archives, postcards, and personal records, reveal at least a minimum level of activity in an area. These records tend to report only larger, and thus noteworthy, events, so the frequency of smaller events is underrepresented. In spite of these limitations, historical data often provide a valuable insight into event frequency; consequently, a number of comprehensive efforts have been undertaken to create such inventories for several classes of natural hazards. For example, Guzzetti et al. (1999) reported on the AVI (Aree Vulnerate Italiane) Project, in Italy, which created an inventory of historical floods and landslides for the period 1918 to 1990 by interviewing 150 experts, reviewing 350,000 newspaper issues and almost 1,500 published and unpublished technical reports, as well as utilizing numerous other sources.

An inventory of historical rockfalls reported in Yosemite National Park has been developed by the U.S. Geological Survey (Wieczorek et al. 1992; Wieczorek and Snyder 2004). The latest report (Wieczorek and Snyder 2004) references an accompanying Excel worksheet (http://pubs.usgs.gov/of/2003/of03-491) that contains data on the location, date, type of slope movement, trigger, size or volume, and types of damage, along with narrative descriptions and references, for 519 rockfalls that occurred between 1857 and early 2004. About 330 of these events occurred within Yosemite Valley and most others were in the near proximity. This information was collected by reviewing published and unpublished historical accounts and field studies of recent rockfalls. Wieczorek and Snyder concluded that many more than 500 rockfalls undoubtedly occurred during this historical period, but many went unnoticed or unreported because of their small size or the lack of impact on trails, roads, structures, or utilities.

Historical records typically reveal a relationship between frequency of occurrence and size or volume of rockfall. Small-volume rockfalls occur frequently, whereas larger rockfalls are infrequent events. Hungr et al. (1999) suggested that the size–frequency distributions of rockfalls from homogeneous areas could be defined by a power law similar to that used to define earthquake magnitudes. Dussauge-Peisser et al. (2002) used this approach to develop simple hazard models for three cliff sites based on analysis of three long-duration historical rockfall inventories. They suggested that the cumulative distribution of rockfall volumes could be adequately defined by using a power law with the following form:

$$f(V) = \alpha V^{-b} \qquad (5)$$

where $f(V)$ is the cumulative frequency of rockfalls with a volume greater than V, and α and b are constants.

The three data sets used by Dussauge-Peisser et al. (2002) were the following:

1. A subset of the Yosemite Valley data set from Wieczorek et al. (1992). Dussauge-Peisser et al. considered 101 events that occurred in the 78 years between 1915 and 1992 and involved volumes in the range of 1 m^3 to 6 × 10^5 m^3 (35 ft^3 to 2.1 × 10^7 ft^3). Yosemite Valley has about 100 km (60 mi) of granite cliffs, reaching heights of 1,000 m (3,300 ft).
2. Rockfall records for the period 1935 to 1995, which identified 87 rockfall events for a 120-km (75-mi) length of cliffs 50 to 400 m (165 to 1,320 ft) high developed in limestone and marl near Grenoble, France.
3. Rockfall records of all rockfalls larger than 1 m^3 (35 ft^3) that threaten a minor road, which identified 59 rockfall events in the period 1954 to 1976 for a cliff 7 km (4.4 mi) long developed in mica schist overlain by Carboniferous sedimentary rock in the Val d'Arly Gorge, in the French Alps.

When plotted as log-log graphs, the power law can be seen as a straight line (Figure 5-10). Table 5-1 summarizes the three rockfall hazard model characteristics developed by Dussauge-Peisser et al. (2002) using these three data sets. The authors made the following conclusions:

- Above a site-specific minimum volume, the cumulative frequency of occurrence of rockfall sizes (volumes) could be defined by a simple power law, regardless of the period of observation.
- Underreporting of small-volume rockfall events in the historical records is evident in the lack of fit of the data to the power-law line for all three cases. Thus the power law corresponds to reported values only for volumes over about 50 m^3

(1,765 ft³) for the Yosemite and Grenoble locations and 20 m³ (706 ft³) for the Val d'Arly Gorge. This "roll-over" pattern has been reported by numerous other investigators.

- The b constant (the slope of the fitted line in Figure 5-10) is independent of geological setting and is fairly constant at about 0.45 (their examples range between 0.41 and 0.46).
- The α constant varies with the site but represents the cumulative frequency of 1-m³ (~1-yd³) rockfall events. However, this characteristic may differ significantly for the true value because of the historical underreporting of small-volume events, as noted earlier.
- Thus, for risk estimation purposes, Dussauge-Peisser et al. (2002) proposed using the $f(_{100})$ frequency value—the annual cumulative frequency of rockfall events larger than 100 m³ (~100 yd³). As shown in Table 5-1, when the $f(_{100})$ frequency value was normalized by dividing it by the surface area of each cliff subject to rockfall at each site, the estimated frequency $f(_{100})/10\ km^2$ ranged from 0.26/10 km² for Grenoble to 19.45/10 km² for the Val d'Arly Gorge.

The approach used by Dussauge-Peisser et al. (2002) provides one example of how historical data can be statistically analyzed to quantify the overall frequency of rockfalls in a given area. Hantz et al. (2003) proposed a combined historical, geomechanical, and probabilistic approach to estimating rockfall frequencies and volumes, and therefore rockfall hazards. They combined information provided by historical rockfall inventories with geomechanical evaluations of potential rockfall failure modes to produce probabilities of rockfall failures and their volumes. They demonstrated the approach with an example from near Grenoble, France.

5.4.2 Size–Frequency Models

Moon et al. (2005) described a different semiquantitative approach to evaluating the frequency of occurrence by constructing and using landslide

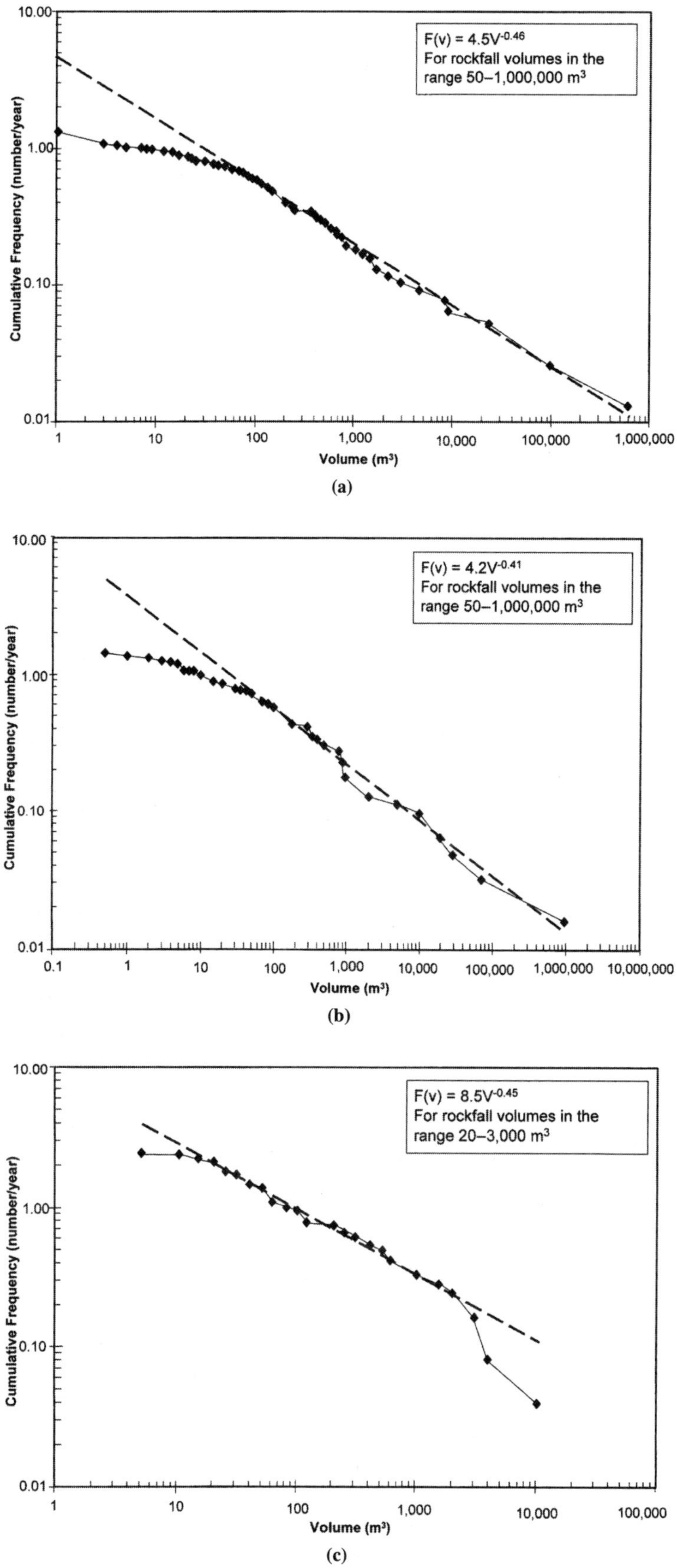

FIGURE 5-10
Power law rockfall frequency–volume models for rockfall hazard analysis: (*a*) Yosemite Valley, California; (*b*) Grenoble, France; and (*c*) Val d'Arly Gorge, France (Dussauge-Peisser et al. 2002).

Table 5-1
Characteristics of Rockfall Frequency–Volume Models for Grenoble, France; Yosemite Valley, California; and Val d'Arly Gorge, France

	Grenoble, France	Yosemite Valley, California	Val d'Arly Gorge, France
Geology	Calcareous cliffs	Granite cliffs	Metamorphic and sedimentary rocks
Number of rockfalls	87	101	59
Approximate cliff surface (km^2)	24	30	2.2
Time period (years)	60	78	22
Range of power law fit (m^3)	50–10^6	50–10^6	30–3,000
Exponent b	0.41 ± 0.11	0.46 ± 0.11	0.45 ± 0.15
Exponent α	4.2	18	8.5
f_{100}	0.62	2.16	1.07
f_{100}/10 km^2	0.26	0.72	19.45

NOTE: f_{100} = cumulative annual frequency of events with volumes >100 m^3.
SOURCE: Dussauge-Peisser et al. 2002.

size–frequency models. They proposed that the presentation of graphical size–frequency models has the advantage of showing how observations, interpretations, and judgments are interrelated; allows patterns to be recognized and understood; and allows easy comparison of models for different situations. Figure 5-11 is an example of Moon et al.'s recommended graphical presentation of a landslide size–frequency model, showing their original explanatory notes.

The x-axis of Figure 5-11 shows landslide volume in order-of-magnitude categories, although smaller divisions (such as a half-order of magnitude) can be used if they are relevant for particular projects. The y-axis shows the landslide frequency, or the average number of landslides per year in each of the landslide volume categories. This feature differs from the previous discussion of the method proposed by Dussauge-Peisser et al. (2002), which plotted cumulative frequency.

The underlying structure of Figure 5-11 is a histogram on a log-log scale. Showing the underlying log-log structure is useful when a graph is developed but does not need to be shown on the final drawing. For example, the hatched column in Figure 5-11 indicates that on average there are judged to be 20 landslides per year with volumes in the range 0.1 to 1 m^3 (3.5 to 35 ft^3). This finding typically means that there will be fewer than 20 landslides in most years. Different periods (such as design life) can be used on the y-axis if required. It is often useful to show the average annual number of landslides in each volume category above the graph. The average annual volume of landslides in each volume category can be calculated by multiplying the number of landslides by the average volume on a log scale (e.g., the log average of the 0.1- to 1-m^3 category is 0.3 m^3). The area of the histogram allows calculation of the average annual total number and volume of landslides.

Although Figure 5-11 is a histogram with discrete volume categories, typically the model is also shown as a curve connecting the midpoints of each histogram class. This method allows several models to be presented and compared on a single graph. The average volume of landslides per year (or landslide process rate) can then be thought of as the area under the curve, but it is actually calculated from the histogram, as described previously.

The size–frequency model has several other important characteristics, potential limitations, and possible calibration factors, which are referenced by the explanatory notes on Figure 5-11:

- For any particular slope, the maximum credible volume and the shape of the landslide size–frequency curve depend on the slope geometry (e.g., height and orientation), slope geology, and failure mechanisms.
- Landslide size–frequency models may include several failure mechanisms.
- When size–frequency models are developed, there is often good evidence of small frequent

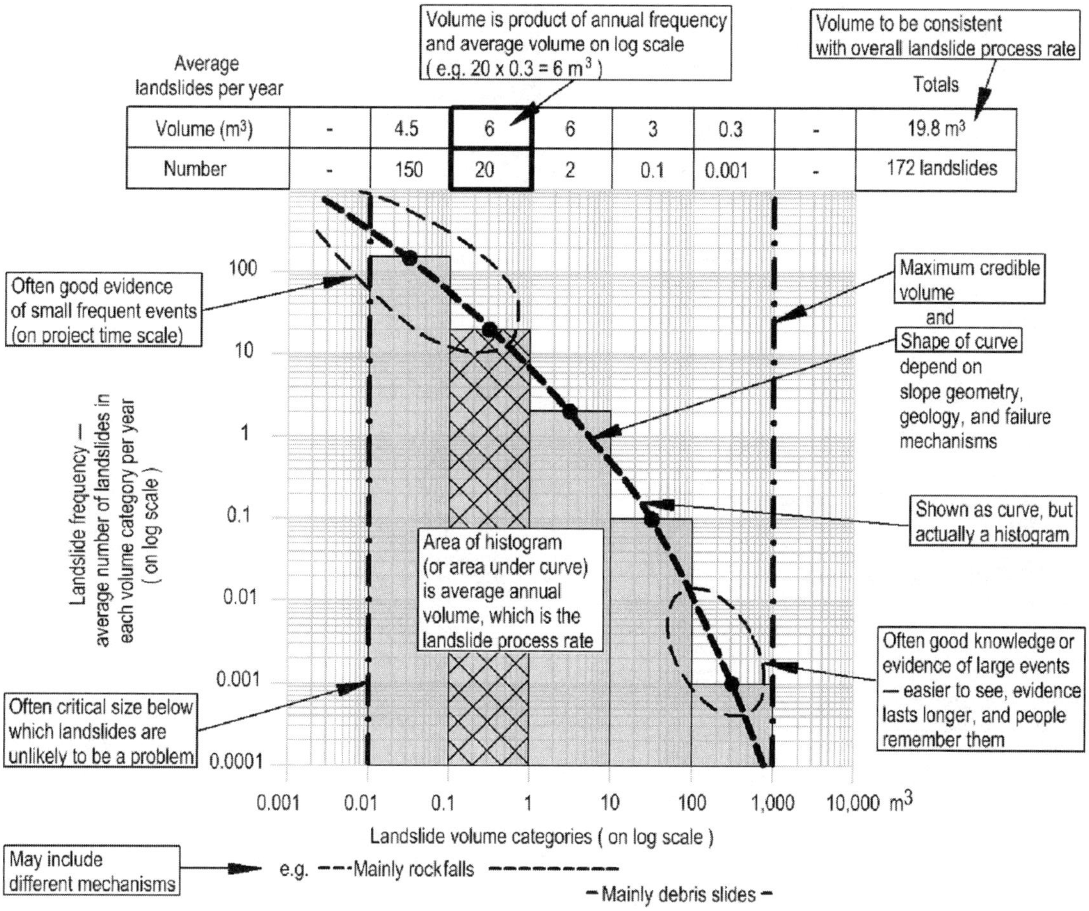

FIGURE 5-11 Graphical presentation of landslide size–frequency model with explanatory notes (Moon et al. 2005).

events (sometimes on a project time scale), which is helpful in developing and calibrating the model.

- There is also sometimes good knowledge or evidence of some of the larger landslide events that affect the slope of concern (or similar slopes in the area). Depending on the time scale, people remember the larger events and the evidence of larger events is easier to see and lasts longer.
- The average annual volume (or area under the size–frequency curve) needs to be consistent with the overall landslide process rate (which may vary with time) as represented by slope retreat rate models or other slope evolution models.

Size–frequency models will vary with slope-specific conditions and will reflect those conditions. Figure 5-12 shows a typical condition found along many roads. The left cut face may suffer rockfalls due to raveling and toppling, as well as possible small wedge failures or failures from differential weathering of rock strata. The right cut face is lower and less steeply inclined but has the same geological materials. It is subject to rockfalls from small rock topples, which may be quite common because of the inclination of the strata. The inclined strata are likely to produce moderate rockfalls and small rock slides, and the possibility of a much larger rock slide failure cannot be ruled out. The size–frequency models related to these two contrasting cut faces are shown in Figure 5-12 (*inset*).

Moon et al. (1996, 2005) used size–frequency models to study rockfall hazards caused by toppling failures along a 48-year-old, very steep, 30-m (100-ft) high railway cut in granite at Bethungra, New South Wales, Australia (Figure 5-13). Rockfalls, mostly caused by toppling of individual blocks or columns defined by persistent near-vertical joints, had damaged trains and caused at least one derailment. As part of the design of slope stabilization work, landslide size–frequency models were developed based on a probabilistic toppling failure model calibrated against the history of rockfalls (Figure 5-14).

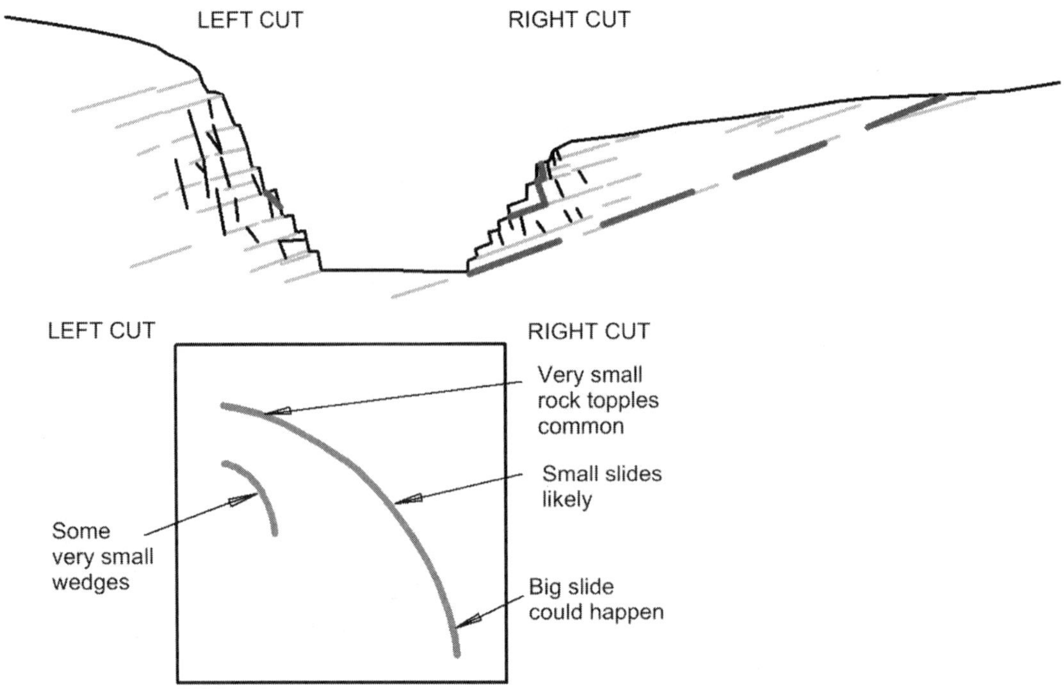

FIGURE 5-12
Effects of site-specific conditions on size–frequency models (presentation by R. A. Wilson at International Conference on Landslide Risk Management, Vancouver, Canada, May 31–June 3, 2005).

FIGURE 5-13
Railway cut at Bethungra, New South Wales, Australia (Moon et al. 1996).

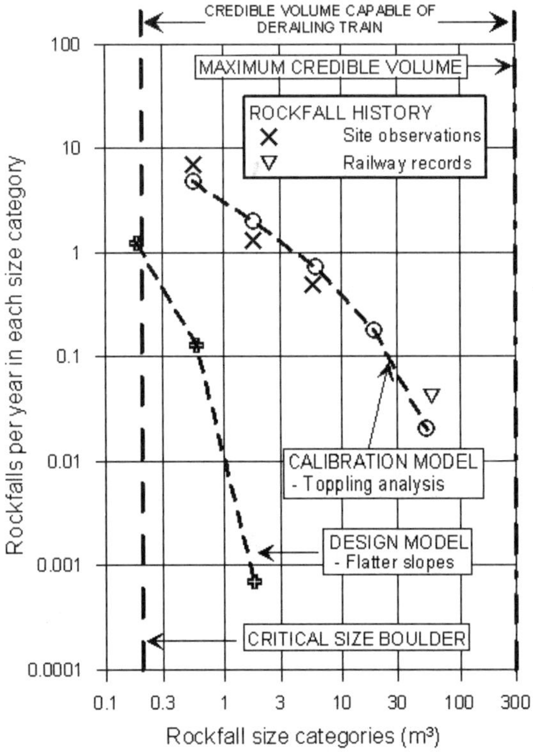

FIGURE 5-14
Rockfall size–frequency relationship for railway cut at Bethungra, New South Wales, Australia (Moon et al. 2005).

Moon et al. (2005) summarized this application of size–frequency models with the following points:

- Half-order magnitude volume categories were used;
- The critical project element was the railway track;
- The critical-size rockfall was judged to be 0.2 m^3 (7 ft^3) because anything smaller on the track was unlikely to derail a train traveling at the low speed limit in place at the site;
- The maximum credible rockfall volume, based on slope height, potential column widths, and stability analysis, was judged to be 300 m^3 (1,060 ft^3);
- There was a good record of rockfalls of up to 10 m^3 (353 ft^3) for the previous 2 years and knowledge of two past failures in the 30- to 100-m^3 (1,060- to 3,500-ft^3) range;
- Long-term railway inspections confirmed that there had been no decrease in rockfalls with time; and
- It was possible to assess the average annual rockfall volume, and thus the rate of cliff erosion, because the number of wagon loads of rocks that had been removed was known.

With these criteria, a successful stabilization plan to redesign the slopes was completed. An additional example of the application of size–frequency models is provided by the case history in Section 6.5.

5.5 Quantifying Risk

The chief aim of quantitative risk assessment methods is to reduce risk to a mathematical value, preferably a single mathematical value. If this quantification can be made, it will facilitate better communication, improve decision making, and make it possible to compare rockfall risk with other risks to determine their relative significance. However, a calculated risk value is the product of probability (likelihood) of occurrence multiplied by a consequence value. Thus, a highly probable but low-damage event, such as a small rockfall, and a very infrequent (low probability) but much larger, catastrophic rockfall or rock slide event may produce identical risk values. The different scale of potential losses related to these two situations produces vastly different perceived threats; thus, society typically demands recognition of this difference by decision makers and expects an appropriate priority ranking of responses.

Because of the inherent uncertainties in the various assumptions regarding the estimation of frequency of occurrence and the valuation of consequences, producing a single risk value for a rockfall event is often inappropriate. As noted by Cruden (1997), the variability in the computed risk value estimate is likely to exceed those of the component factors. So rather than a single value from a risk assessment, it is typically useful to perform multiple repeat analyses with a range of rockfall frequencies and outcomes from a worst-case estimate to some lower-bound estimate. Techniques for undertaking and managing such multiple risk scenarios are discussed in Section 5.7.

Discussions of quantitative estimation of risk typically use two concepts: individual risk and societal risk. Neither concept is entirely intuitive; therefore, each is briefly described in the following sections. Lee and Jones (2004) devoted an entire chapter (Chapter 6) to discussing all aspects of risk quantification, with special reference to landslides and rockfall.

5.5.1 Individual Risk

The term *individual risk* is generally used to define the frequency with which individuals, such as road users, are expected to suffer harm from some hazard, such as rockfall (Lee and Jones 2004). It is a somewhat abstract concept and is not to be confused with the risk faced by a specific individual, which is best termed *personal risk* (Lee and Jones 2004). Individual risk has traditionally been taken to be the risk of death and is expressed as the probability per year. In the transportation area, it is sometimes expressed in terms of per journey or per passenger mile. In any case, it is the risk expected to be experienced by a single individual in a given time period and typically reflects the amount of time the individual is exposed to the hazard and the severity of the hazard.

Discussions in the literature concerning individual risk are often contradictory. Some calculate individual risk in terms that reflect those individuals who are most at risk, typically those who are directly exposed to the hazard because of their location, their work or recreation activities, or the duration of time they remain vulnerable.

This is the preferred method of calculation (Lee and Jones 2004). However, others calculate individual risk as a value related to an average or shared value representative of an entire group of potentially affected individuals. This value is only valid if the risk is uniformly distributed over the entire affected population; in the case of rockfall hazards, such an average value may conceal the high risk for a few individuals when it is averaged over a large number of people at relatively low risk.

5.5.2 Societal Risk

The term *societal risk* defines the frequency and number of people suffering a given level of harm from the occurrence of specified hazards (Lee and Jones 2004). As with individual risk, societal risk is usually concerned with the risk of death and is expressed as the probability per year.

The data on frequencies and number of fatalities are usually presented as so-called *F-N* diagrams. These plot the frequency (*F*) of *N* or more fatalities per year versus the number of fatalities (*N*), both on logarithmic scales (Figure 5-15). There are particularly severe risks associated with rockfalls and landslides in Hong Kong. Many natural slopes are covered with large boulders, especially on colluvial slopes or below rock cliffs, and boulder falls are common after intense rainstorms. Landslides are particularly common on man-made slopes or excavated slopes, both of which are frequently very steep. Figure 5-15 illustrates two *F-N* curves, one for boulder falls and one for landslides on man-made slopes, based on Hong Kong data shown in Table 5-2. In 69 years (1926 to 1995) three boulder-fall events caused one, three, and five deaths, respectively. In a 78-year period (1917 to 1995) 117 landslides on man-made slopes caused one or more fatalities, with the worst case of 73 deaths occurring in 1917.

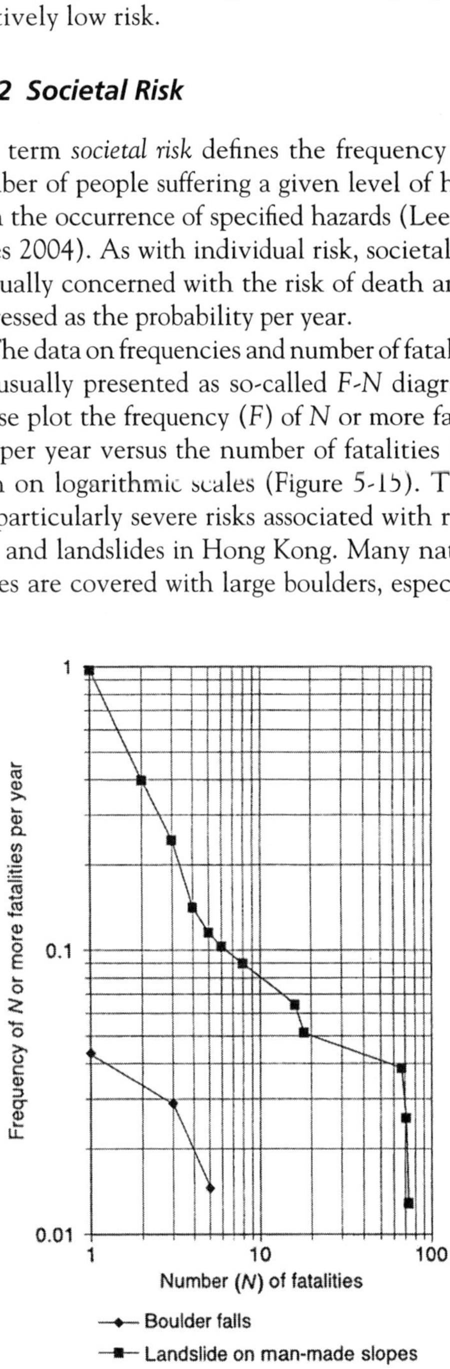

FIGURE 5-15
F-N curves for boulder falls and landslides on man-made slopes in Hong Kong (Lee and Jones 2004).

5.6 ALARP Principle

The *as-low-as-reasonably-practicable*, or *ALARP, principle* requires that risk mitigation measures be implemented until the reduction in risk is disproportionate to the mitigation costs and the risks have been reduced to a level that is as low as reasonably practicable. This philosophy is becoming increasingly used within an industry where civil safety is the main concern, such as the analysis of dam safety, and in hazardous industries where the risks are seen to be involuntary (i.e., the people at risk do not have direct control over their exposure).

The ALARP principle involves defining what is reasonable, which will be unique to each organization. An integral part of any test of reasonableness typically includes a balance between the investment required to achieve a certain level of risk reduction and the benefits gained by undertaking the mitigation works. This balance is carefully examined when the benefit is in terms of reduced risk of fatalities. Unfortunately there are no hard and fast rules that can be applied. However, it is vital that the reasoning and underlying justifications be documented. The ALARP assessment process is based on detailed knowledge of the factors contributing to the identified risk, the various means available to mitigate the risks, and the constraints pertaining to their implementation.

As discussed in Section 3.2.2, there is an important difference between acceptable and intolerable risks. Acceptable risks are generally defined as those that everyone likely to be affected by a rockfall event is prepared to accept, and thus no further remediation is required. Tolerable risks fall within a range that society finds acceptable, even if some individuals are likely to be affected. Higher levels

Table 5-2
***F-N* Data for Historic Boulder Falls and Landslide Incidents in Hong Kong**

	Boulder Falls 1926–1995 (69 years)		Landslides on Man-Made Slopes 1917–1995 (78 years)	
Number of Fatalities (*N*)	Number of Events with *N* or More Fatalities	Frequency (*F*) of *N* or More Fatalities	Number of Events with *N* or More Fatalities	Frequency (*F*) of *N* or More Fatalities
1	3	0.0435	117	0.974
2			31	0.397
3	2	0.0290	19	0.244
4			11	0.141
5	1	0.0145	9	0.115
6			8	0.103
8			7	0.0897
16			5	0.0641
18			4	0.0513
67			3	0.0385
71			2	0.0256
73			1	0.0128

Source: Lee and Jones 2004.

of risk are judged to be intolerable. The upper limits of tolerable risk are typically higher if the risk is undertaken voluntarily by the individual (e.g., rock climbing, sky diving) than if the risk is involuntary. Typical limits of tolerable risk related to loss of life typically range between 1×10^{-4} and 1×10^{-6} per year for existing and newly constructed engineered slopes (Fell et al. 2005). The ALARP principle is applied to evaluate alternative mitigation works and can bring intolerable risks into the tolerable range.

Figure 5-16 illustrates these concepts using an *F-N* diagram. Lines of constant risk fall along diagonals. The tolerable-risk region is labeled *ALARP*, as this is the region where the ALARP principle is most readily applied. The Australian National Committee on Large Dams advocates a truncation of tolerable risk to reflect the practical limitations of what risk reduction measures can achieve for facilities with very high potential loss of life. Similar *F-N* plots have been used to evaluate rockfall risk and mitigation alternatives along the Alpine Way and other park roads by the New South Wales National Parks Service.

A typical ALARP evaluation involves multiple risk assessments to evaluate a range of scenarios, with methods discussed in Section 5.7. Once several alternative mitigation measures have been identified that reduce risk below the intolerable threshold, the implementation of the ALARP principle typically employs benefit–cost ratio analyses to evaluate these alternatives and to help decide whether further risk treatment work is justified (see Section 5.8).

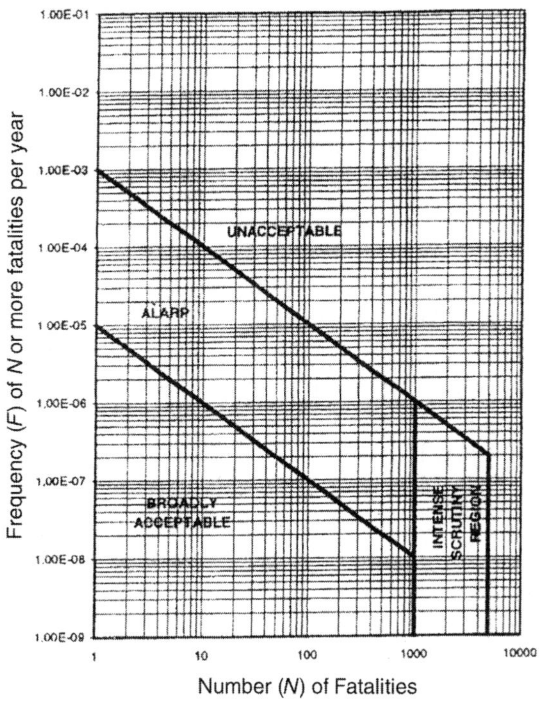

FIGURE 5-16
Societal risk tolerance criteria and primary region of applicability of ALARP principle as defined by Geotechnical Engineering Office, Hong Kong (Geotechnical Engineering Office 1998; Fell et al. 2005).

5.7 Managing Multiple Risk Analysis Scenarios

Many quantitative risk analyses use single-point estimates for frequencies of hazard occurrence and for the consequences. Occasionally, three scenarios are developed to test the outcomes for a perceived worst case, best case, and most-likely case. Such an approach has several limitations:

- It considers only a few discrete outcomes, ignoring hundreds or thousands of others;
- It gives equal weight to each outcome; that is, no attempt is made to assess the likelihood of each outcome; and
- Interdependence between inputs, impact of different inputs relative to the outcome, and other nuances are ignored, oversimplifying the model and reducing its accuracy.

Monte Carlo simulation offers a better way to undertake multiple quantitative risk analysis scenarios. In Monte Carlo simulation, uncertain model inputs are represented by using ranges of possible values known as probability distributions. Commonly employed probability distributions include the following:

- *Normal, or bell curve*. The user simply defines the mean or expected value and a standard deviation to describe the variation about the mean. This symmetrical distribution allows values near the mean to be most likely to occur.
- *Lognormal*. Values are positively skewed, not symmetrical as with a normal distribution. This distribution is used to represent values that do not go below zero but have unlimited positive potential.
- *Uniform*. All values have an equal chance of occurring, and the user simply defines the minimum and maximum.
- *Triangular*. The user defines the minimum, most likely, and maximum values. Values around the most likely value are more likely to occur.
- *Discrete*. The user defines specific values that may occur and the likelihood of each.

The Monte Carlo simulation method uses random or pseudorandom numbers to sample from the input probability distributions, and the results from each such iteration are recorded. After hundreds or thousands of such iterations, the results form a probability distribution of possible outcomes. In this way, Monte Carlo simulation provides a much more comprehensive view of what may happen, including the likelihood of any particular outcome.

A number of commercial software programs are available for generating random Monte Carlo samples from different types of input probability distributions and performing risk assessment scenarios. These include the program @RISK (Palisade Corporation 2010), which is an add-on to Microsoft Excel, and the program Crystal Ball (Oracle Corporation 2010).

The Latin hypercube sampling technique gives results comparable with those of the Monte Carlo technique but with fewer samples (Hoek 2000, Chap. 8). The method is based on stratified sampling with random selection within each stratum. Hoek (2000, Chap. 8) states that typically an analysis using 1,000 samples obtained by the Latin hypercube technique will produce results comparable to those of an analysis using 5,000 samples obtained with the Monte Carlo method. Both techniques are incorporated into commercial computer programs such as @RISK. Both the Monte Carlo and the Latin hypercube techniques require either that the distribution of all the input variables be known or that they be assumed. When no information on the distribution is available, it is usual to assume a normal or a truncated normal distribution (Hoek 2000, Chap. 8).

5.8 Strategic Risk Management

The final step in risk management for rockfall hazards to road users (Figure 5-3) uses the results of the risk assessment involving the ALARP principle for multiple scenarios to evaluate a suite of rockfall hazard mitigation options. If the mitigation alternatives are ranked according to their level of risk reduction, a graphical comparison (*risk profile*) is produced for the alternative risk reduction measures.

Subsequently, incremental benefit–cost ratios for the various risk reduction options are determined by relating the estimated costs to the potential level of risk reduction achieved by each measure. Usually, these analyses are displayed on a single graph with two lines plotted, one showing the cumulative cost of the sequential implementation of the mitigation options, and the other showing their benefit in terms of reduction of risk.

This process can be expanded to an entire portfolio of rockfall locations, and in fact it is even more useful if it is expanded beyond a single site. The approach provides considerable management flexibility and allows funds to be applied incrementally to various locations within a road system. For example, funds might be used to reduce the highest risks first and the remaining risks as funds become available. This worst-case-first process was originally proposed by the RHRS documentation (Pierson et al. 1989; Pierson 1992) but is not consistent with modern asset management principles. The strategic risk management process allows all factors affecting particular roads or networks to be factored into the decision-making process and so provides considerable flexibility in the assessment of management options.

Figures 5-17 and 5-18 demonstrate the results of such a strategic risk assessment process, which was applied to a series of large dams in the state of Victoria, Australia (Davidson et al. 2002). They illustrate how the first few actions are able to reduce most of the risk and subsequent actions achieve far less risk reduction benefit.

A sliding scale is typically developed to assess whether further investment is warranted. For example, risk reduction measures are implemented if the benefit–cost ratio is greater than 1.0 when the costs of the mitigation measures are relatively low, such as less than $50,000, but are implemented only if the benefit–cost ratio is greater then 2.0 when their costs exceed $200,000. These criteria often are related to significant environmental, strategic, or economic factors within or across regions.

6. CASE HISTORIES

The foregoing discussion suggests that formal probability and risk approaches typically involve too much computation, or recomputation, to be

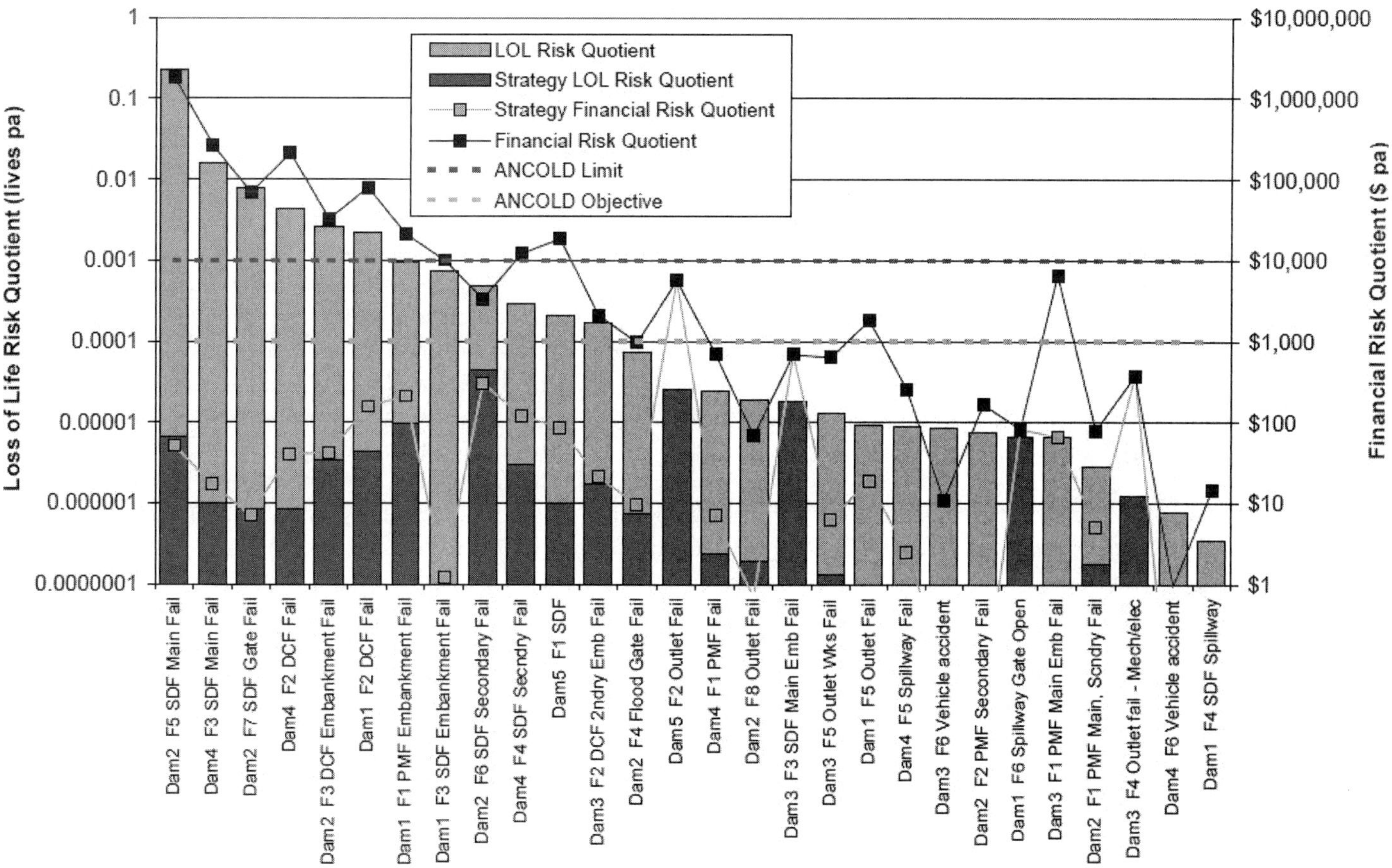

FIGURE 5-17
Risk profile for 13 dams in state of Victoria, Australia, following proposed risk-reduction actions (adapted from Davidson et al. 2002) [LOL = loss of life, ANCOLD = Australian National Committee on Large Dams, SDF = spillway design flood, DCF = damage cost factor (ratio of mean damage repair cost to cost of entire dam structure), PMF = probable maximum flood, pa = per annum].

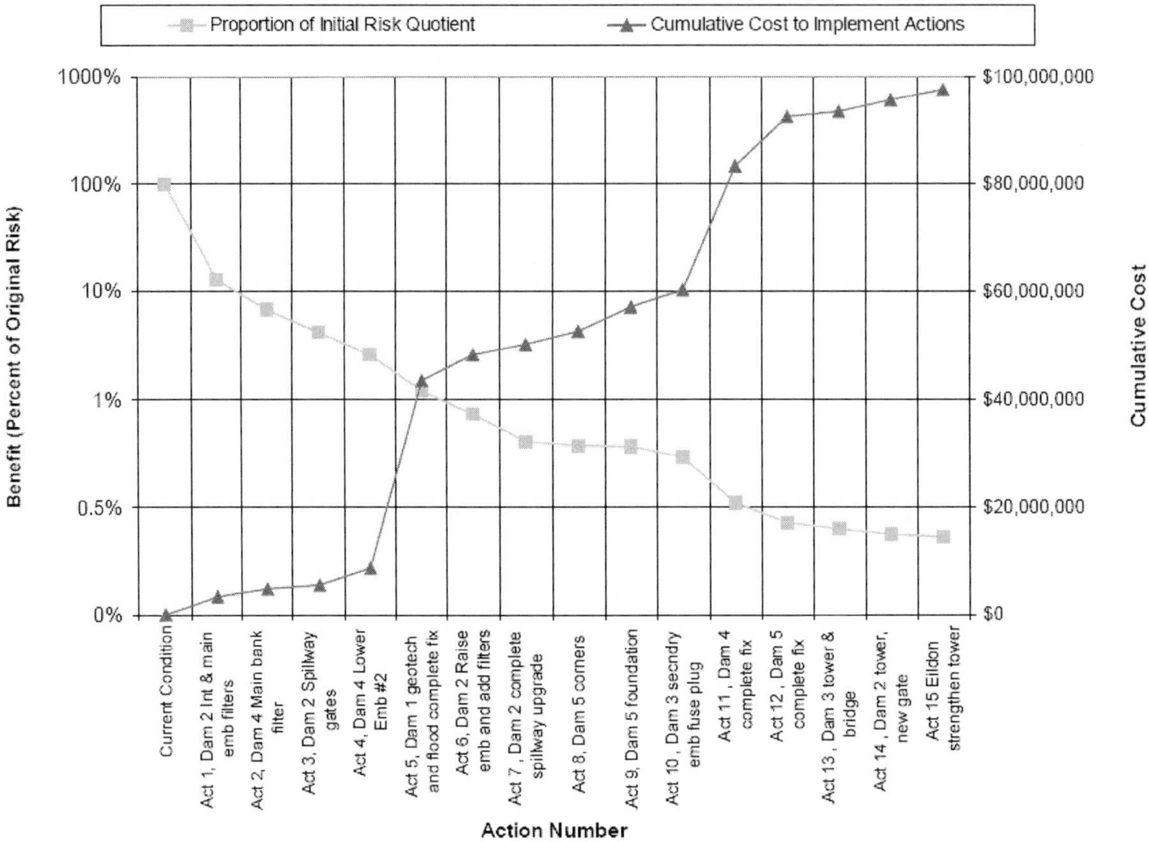

FIGURE 5-18 Risk-reduction benefit versus cumulative cost for proposed mitigation options at 13 dams in State of Victoria, Australia (adapted from Davidson et al. 2002).

practical in many situations. Thus, alternative, more-simplified quantitative approaches seem necessary. The case histories reviewed here provide some insights into how various degrees of quantification can be successfully employed to support rockfall risk assessment and risk management.

6.1 Landslide Risk Management in Switzerland

At least 6% of Swiss territory is affected by landslide phenomena (Lateltin et al. 2005). Swiss landslide events belong to four categories:

- Reactivation of ancient landslides,
- Rockfalls,
- Large and rapidly moving rock avalanches, and
- Debris flows and very shallow landslides.

Reactivation of ancient landslides generally involves relatively deep-seated, slow-moving displacements that result in property damage but rarely pose any significant risk to individuals. In contrast, the high velocities of rockfalls, rock avalanches, and debris flows make them dangerous and annually cause fatalities and traffic disruptions (Lateltin et al. 2005). The Swiss are increasingly vulnerable to landslide hazards because of rapidly increasing population densities and numbers of economic assets placed in at-risk locations. Changing environmental conditions, such as global warming and loss of protective forests on lower valley slopes, may be aggravating the threat posed by landslides in the Alps.

Since 1991, new federal regulations have required the 26 Swiss cantons to establish hazard maps and zoning for mass movements so as to restrict development on hazard-prone land. Although the cantons are responsible for initiating hazard mapping, the federal authorities subsidize up to 70% of the cost of this activity. These laws promote the protection of the environment, particularly human life and high-value property, from natural hazards with a minimum of structural countermeasures. Therefore, nonstructural preventive measures are considered especially desirable. The four main elements of the natural risk management strategy in Switzerland are (Figure 5-19)

- Hazard assessment,
- Definition of protection requirements,
- Plan for mitigation measures, and
- Emergency planning.

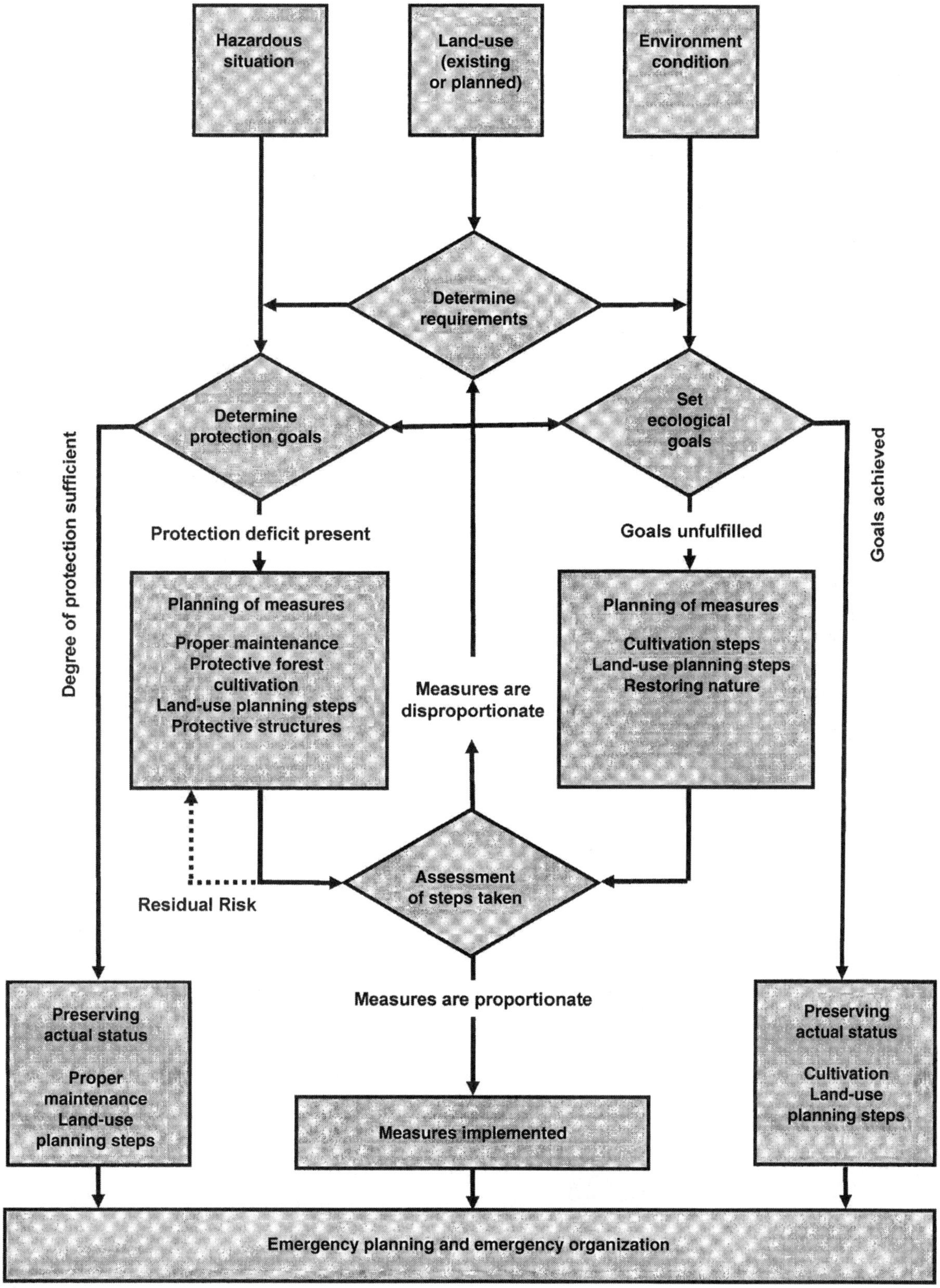

FIGURE 5-19
Strategy for landslide risk management in Switzerland (Lateltin et al. 2005).

Raetzo et al. (2002) and Lateltin et al. (2005) reported that about 50% of Swiss territory was at that time covered by maps showing landslide phenomena at different scales and that these mapping efforts were continuing. These landslide maps form the basis of a qualitative landslide risk assessment procedure that is applied to land use zoning. The Swiss assessment procedure consists of three main steps (Raetzo et al. 2002):

Step 1. Hazard identification.
Step 2. Hazard assessment.
Step 3. Risk management and land use planning.

6.1.1 Step 1: Hazard Identification

Hazard identification involves the classification of Swiss landslides, the creation of landslide phenomena maps, and the development of official registers of events.

Swiss landslides are classified according to their estimated depth. Landslides that are less than 2 m (6 ft) deep are termed *shallow*. Those with depths of 2 to 10 m (6 to 30 ft) are termed *intermediate*, whereas those with depths greater than 10 m (30 ft) are termed *deep*. Landslides with long-term mean velocities of less than 2 cm/year (1 in./year) are classified as *substabilized*, those with movements of 2 to 10 cm/year (1 to 3 in./year) are classified as *slow*, and those moving at rates greater than 10 cm/year (3 in./year) are classified as *active*. Rockfalls are characterized by their speed, the size of their rock fragments, and their volumes. Rock avalanches, although rare occurrences, are always considered high-risk events because of their huge volumes (frequently more than 1 million m^3, or 1.33 million yd^3) and high velocities (frequently more than 40 m/s, or 132 ft/s). Debris flows and very shallow landslides are frequent in Switzerland. Although typically of moderate volume (usually less than 20,000 m^3, or 26,600 yd^3), these phenomena are very dangerous because of their relatively high velocities (1 to 10 m/s, or 3 to 33 ft/s).

Swiss federal regulations require the cantons to establish registers documenting slope instability events and maps denoting areas of landslide phenomena. The landslide maps use a uniform legend to define landslide classes and features, permitting easy comparison of maps from different parts of the country. Similar recommendations provide for the production of uniform registers describing slope instability events, which will, in the future, assist in the historical assessment of event frequencies and volumes.

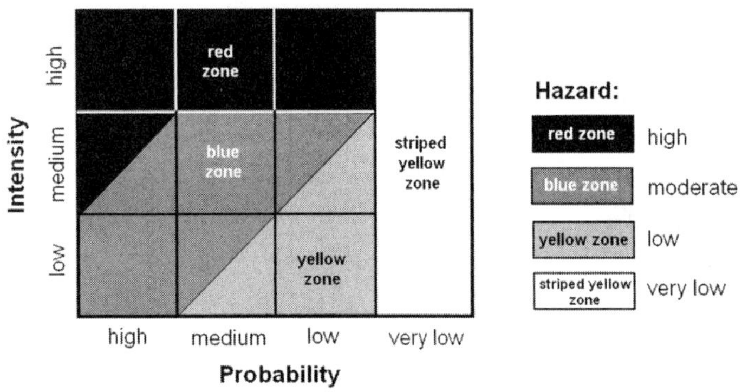

FIGURE 5-20 Swiss hazard levels as function of probability and intensity (Lateltin et al. 2005).

6.1.2 Step 2: Hazard Assessment

Qualitative landslide *hazard assessment* implies the estimation of the intensity of a landslide event (reactivation of ancient landslide, rockfall, rock avalanche, or debris flow) and its probability of occurrence over time. For simplicity, the intensity and probability of occurrence are defined with three levels: high, medium, and low. The intensity and probability levels form a hazard matrix (Figure 5-20), which is then used to define a series of hazard categories that are shown on landslide hazard maps with a series of standard colors, including red for high hazard, blue for moderate hazard, yellow for low hazard, yellow with white hatching for residual risk, and uncolored or white for no hazard. Table 5-3 provides the definitions or characteristics of the hazard levels associated with each zone.

Table 5-4 summarizes the intensity criteria for rockfalls, landslides, and debris flows. For rockfalls, the impact energy (the sum of translation and rotation energy) on any exposed structure defines the intensity levels. Limits are set at 30 kJ and 300 kJ (11 ft-tons and 110 ft-tons). The 300-kJ limit corresponds to the impact energy that can be resisted by a reinforced concrete wall as long as the structure is properly built. The 30-kJ limit corresponds to the maximum energy that typical timber barriers can resist (Raetzo et al. 2002). Rock avalanches always produce high-intensity impacts, greater than 300 kJ (110 ft-tons) in the runout zone; therefore, they are always defined as high-intensity events.

Table 5-3
Swiss Federal Recommendations for Landslide Hazard Map Degrees of Hazard, Represented by Corresponding Colors

Color Code	Definitions
Red: high hazard	The red zone designates a prohibition domain (an area where development is prohibited). • People are at risk of injury both inside and outside buildings. • A rapid destruction of buildings is possible. Or • Events with a lower intensity, but a higher probability of occurrence. In this case: – People are mainly at risk outside buildings, or – Buildings can no longer house people.
Blue: moderate hazard	The blue zone designates a prescriptive domain (area where severe damage can be reduced by means of appropriate protective measures; use of restrictive regulations). • People are at risk of injury outside buildings. • Risk is considerably lower inside buildings. • Damage to buildings should be expected, but not a rapid destruction as long as the construction type has been adapted to the present conditions.
Yellow: low hazard	The yellow zone designates an alerting domain (an area where people are notified of the possible hazard). • People are at slight risk of injury. • Slight damage to buildings is possible.
Yellow, white hatching: residual risk	The yellow zone with white hatching is mainly an alerting domain (highlights the existence of residual risk). The low probability of a high intensity event can be designated by this zone.
White: no apparent hazard	No danger or negligible danger (according to currently available information)

Source: Adapted from Raetzo et al. 2002 and Lateltin et al. 2005.

For landslides, a low-intensity event has an annual mean velocity of less than 2 cm/year (1 in./year), a medium-intensity event has a velocity ranging from 2 cm/year to approximately 10 cm/year (3 in./year), and the high-intensity class includes very rapid, shallow slides, with velocities greater than 10 cm/day (3 in./day). The high-intensity class is also assigned to shear zones or zones with clear differential movements or to locations where horizontal displacements greater than 1 m (3 ft) per event are likely to occur. For debris flows, the intensity depends on the thickness of the debris flow. The limits of the three intensity classes are set at 0.5 m and 2 m (1.5 ft and 6.5 ft).

The probability of landslide occurrence is related to three return periods; 1 to 30 years (high

Table 5-4
Swiss Criteria for Landslide Intensity Assessment

Landslide Class	Low Intensity	Medium Intensity	High Intensity
Rockfall (kJ)	$E < 30$	$30 < E < 300$	$E > 300$
Rock avalanche (kJ)			$E > 300$
Landslide (cm/year)	$v \leq 2$	$v > 2$	Large differential movements; or $v > 0.1$ m/day for shallow landslides; or displacement > 1 m per event
Debris flows (m)	$t < 0.5$	$0.5 < t < 2$	$t > 2$

Note: E = impact energy; v = long-term mean speed; t = thickness of debris flow deposit.
Source: Adapted from Raetzo et al. 2002.

probability), 30 to 100 years (medium probability), and more than 300 years (low probability). The probability of occurrence within a given period of time (such as the life expectancy of a structure) is related to the return period according to the following formula (Raetzo et al. 2002):

$$P = 1 - (1 - 1/T)^n \qquad (6)$$

where

P = probability of occurrence within a given period of interest,
T = return period of landslide event (years), and
n = period of interest (years).

With this formula; the probability of a landslide occurring within a 30-year period is

P = .64 if the landslide has a return period of 30 years,
P = .26 if the landslide has a return period of 100 years, and
P = .10 if the landslide has a return period of 300 years.

In principle, the probability scale does not exclude very rare events, nor does the intensity scale exclude high-magnitude events. Potential hazards with a very low probability of occurrence, such as hazards with a return period of more than 300 years, are usually shown on the hazard maps by yellow-white hatching. Complete safety cannot be ensured in such cases, and a residual risk remains. Assessment of the impact of a rare rock avalanche, which exceeds the design event, provides information about residual risk.

6.1.3 Step 3: Risk Management and Land Use Planning

The Swiss landslide hazard maps form the basis for land use planning. Raetzo et al. (2002) suggested that their primary utility is in the following situations:

- Elaboration and improvement of cantonal master plans and communal local plans for land use;
- Planning, construction, and modification of existing buildings and infrastructure elements;
- Granting planning approval for construction and infrastructure installations, as well as providing a technical basis for laws related to land use; and
- Granting of subsidies for building and development (such as road and rail networks or residences), as well as for slope stabilization and protection measures.

6.2 Physically Based Method for Rockfall Hazard Assessment

Crosta and Agliardi (2003) reexamined procedures used for rockfall hazard assessment to identify the best way to compute, represent, and rank different hazard levels. They concluded that since elements at risk may be located throughout a rockfall trajectory, estimates of intensity and probability of rockfall occurrence are needed along the entire rockfall trajectory. Rockfall probabilities and intensities vary along this path as a result of complex interactions involving the mechanical and geometrical characteristics of both the slope and the falling blocks. Thus, a physically sound modeling of rockfall motion along a slope offers the most feasible approach to rockfall hazard assessment.

Crosta and Agliardi (2003) used a previously developed three-dimensional (3-D) rockfall simulation model called STONE (Guzzetti et al. 2002; Agliardi and Crosta 2003) to provide estimates of rockfall probabilities and intensities for selected study areas in the Lombardy region of northern Italy. Chapter 9 provides details of the STONE simulation program. STONE produces maps showing three parameters that can be used to evaluate the probabilities and intensity of rockfall events:

- The count of rockfall trajectories (c) at each location is a proxy for the probability of being struck by a falling or rolling boulder at that location; the greater the concentration of computed trajectories, the higher the expected frequency of rockfall occurrence; and
- The computed maximum bounce height (h) and the maximum velocity (v) provide information on rockfall kinematics, and therefore the intensity of the expected rockfall.

Figure 5-21 shows two example maps produced by STONE for one study site near Lecco, Italy. The STONE program computes the three parameters (c, h, and v) for rockfall block trajectories passing through each raster cell of the model.

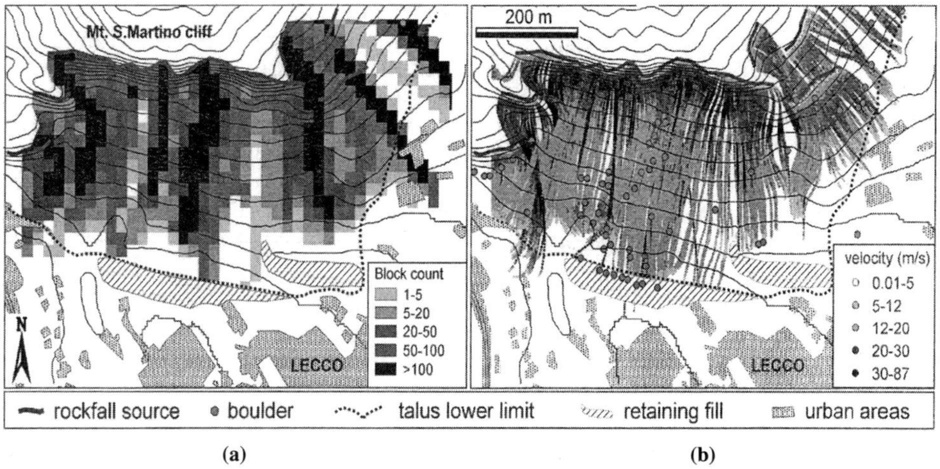

FIGURE 5-21
Maps of Monte San Martino cliff area north of Lecco urban area in Italy showing results of rockfall simulations by STONE model (Crosta and Agliardi 2003): (*a*) count of rockfall trajectories passing through series of raster map cells with 20-m (66-ft) ground resolution; (*b*) maximum estimated rockfall velocities passing through series of raster map cells with 5-m (16.5-ft) ground resolution.

Figure 5-21*a* shows the trajectory counts for cells with a 20- × 20-m (66- × 66-ft) ground resolution, and Figure 5-21*b* shows the estimated maximum velocities within a series of 5- × 5-m (16.5- × 16.5-ft) cells in the same area.

Maximum velocity is readily converted to an estimate of the maximum kinetic energy (k):

$$k = 0.5mv^2 \quad (7)$$

where m is the acceleration due to gravity. The three components (c, h, and k) are conveniently represented in a 3-D space by a 3-D matrix that defines the rockfall hazard index (RHI), which portrays hazard as a function of these parameters (Figure 5-22).

Any cell positioned within the RHI cube is defined with a three-digit index, RHI(ckh), if the values of the controlling parameters (c, k, and h) are ranked into n classes. RHI(000) corresponds to the origin of the 3-D hazard space and represents a condition of no rockfalls. Thus, any RHI value including a zero digit is unrealistic, since it indicates no rockfall occurrence, and the lowest-level hazard RHI digit value is 1. Rockfall hazard increases at a maximum rate along the diagonal

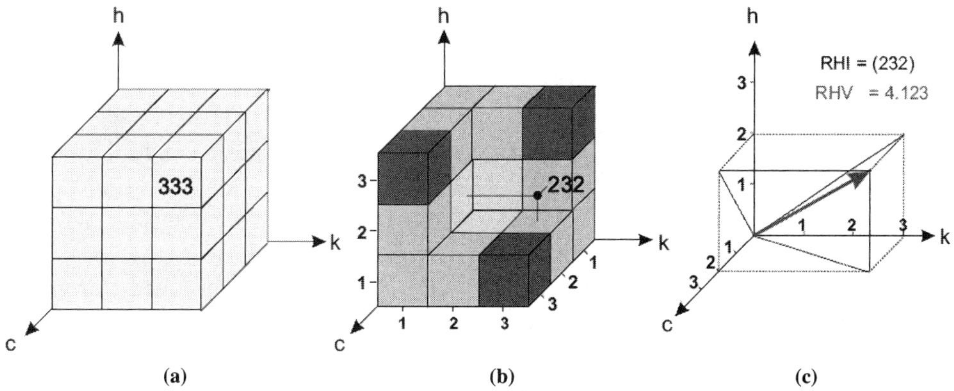

FIGURE 5-22
Three-dimensional matrix with axes *c*, *k*, and *h* used to define rockfall hazard (Crosta and Agliardi 2003): (*a*) general definition of RHI, (*b*) partial RHI cube illustrating ranked RHI values, and (*c*) rockfall hazard vector (RHV) concept.

line that trisects the hazard space. From a purely geometrical point of view, all the points lying on planes perpendicular to the diagonal line should exhibit the same hazard level. Since the input parameters are reclassified into discrete classes, the hazard index is also discrete.

RHI values lying off the diagonal represent intermediate hazard conditions. Many different real situations may be located on a single *constant hazard plane*, each involving different occurrence probabilities and different amounts of kinetic energy and thus requiring different mitigation approaches. The use of a positional index allows the contribution of each variable to the hazard to be assessed, thus defining why a particular location on the slope is characterized by a given hazard level.

Nevertheless, the positional nature of the RHI index hampers the direct ranking of hazard. For example, in Figure 5-22*b* it is difficult to decide if RHI(113) represents a more severe hazard than RHI(311) or RHI(121). Thus, Crosta and Agliardi (2003) translated the RHI index into a single sequential value by invoking the concept of rockfall hazard vector (RHV), as shown in Figure 5-22*c*. The magnitude of the RHV is defined as

$$|RHV| = \sqrt{c^2 + k^2 + h^2} \quad (8)$$

Because *c*, *k*, and *h* are discrete unitary values, the RHV magnitude is discrete. Nevertheless, the magnitude of the RHV vector allows the hazard level to be ranked into classes and to produce an objective and clear rockfall hazard map (Crosta and Agliardi 2003).

To employ the RHI and RHV concepts, the physical values of *c*, *k*, and *h*, each with its different physical meaning and magnitude, are converted to a small number of classes. Crosta and Agliardi (2003) used three classes, which simplified the classification and ranking of the computed RHI values, as this produced 27 RHI values. They suggested that this level of subdivision produced more readily interpreted rockfall hazard maps but conceded that other classification schemes might be better in other studies.

Crosta and Agliardi (2003) also conceded that the definition of class limits for each parameter is an intrinsically uncertain task that is usually performed on a subjective basis. They selected class limits by evaluating the potential destructiveness of simulated rockfalls. For example, increasing kinetic energy implies a higher capability of a falling rock to damage structures or barrier systems, whereas higher maximum bounce height results in a higher probability for overtopping barriers. Using such considerations, Crosta and Agliardi (2003) decided to reclassify the parameters *c*, *k*, and *h* as shown in Table 5-5. The classification process involved two normalization procedures for the count of rockfall trajectories (*c*) to convert values produced by the STONE program simulations, which were influenced by the number of simulation trials, into classes that defined the local concentrations of rockfalls into chutes. Two methods were required; one for regional-scale models and one for local-scale models, because of differing assumptions concerning rockfall source areas.

Crosta and Agliardi (2003) tested the application of the RHI-RHV method at regional and local scales in two study areas located in the Lombardy region of northern Italy. They overcame the tendency for the RHI-RHV values to produce noisy and fragmented map displays by smoothing the results using a local filtering operation that replaced the existing value of a cell with the mean and maximum values in the local neighboring cells. This method improved the visual appearance of the rockfall hazard maps, but Crosta and Agliardi caution that the overzealous use of these smoothing operations may lead to unrealistic hazard scenarios; depending on the chosen smoothing process, they may be too conservative or too optimistic.

Table 5-5
Parameter Reclassification Scheme Used in RHI-RHV Procedure for Studies in Lombardy Region of Northern Italy

	c (normalized)[a]			
Class	Regional Scale	Local Scale	*k* (kJ)	*h* (m)
1	<0.2	<0.01	≤700	≤4
2	0.2–1.0	0.01–0.1	700–2,500	4–10
3	>1.0	>0.1	>2,500	>10

[a]The values of *c* are normalized according to different approaches, depending on the modeling scale.
Source: Crosta and Agliardi 2003.

Figures 5-23 and 5-24 illustrate an example rockfall hazard assessment for a road running along the lower portions of the Monte San Martino and Corno Medale slopes near the town of Lecco. Figure 5-23 provides a 3-D representation of the RHV magnitude values sampled from a smoothed hazard map along the road, and Figure 5-24 plots the computed RHV values along the road profile. The RHV values were sampled every 5 m (16.5 ft) and contain values smoothed by the mean and maximum values in the local neighboring cells on the rockfall hazard map; they clearly define the most hazardous sectors along the road. Crosta and Agliardi (2003) suggested that these methods of analysis may be useful in designing mitigation works or for optimizing maintenance efforts.

Guzzetti et al. (2004) reported additional applications of the RHI-RHV method to rockfall hazard and risk assessment along a transportation corridor in the Nera valley of central Italy. They identified several additional strategies for combining mapped estimates of the three parameters used in the RHI-RHV method:

- Combine the two maps providing information on the intensity of the expected rockfall (i.e., the map of the maximum computed bounce height and the map of the maximum computed velocity), and then combine the result with the map of the count of rockfall trajectories;
- Analyze the three maps jointly, considering the three variables defining landslide hazard (rockfall count, velocity, and bounce height) to be of equal importance; and
- Use more complex options, such as multivariate statistical techniques (e.g., cluster analysis) to combine the three maps.

However, they then elected to use a version of the RHI-RHV method previously defined by Crosta and Agliardi (2003). They also faced decisions concerning appropriate values to use in reclassifying the estimated values for the three RHI parameters: rockfall count (c), rockfall maximum bounce height (h), and rockfall maximum velocity (v). They decided to use a five-level classification scheme rather than the three-level scheme proposed by Crosta and Agliardi. Table 5-6 provides their reclassification criteria for the evaluation sites in the Nera Valley of central Italy.

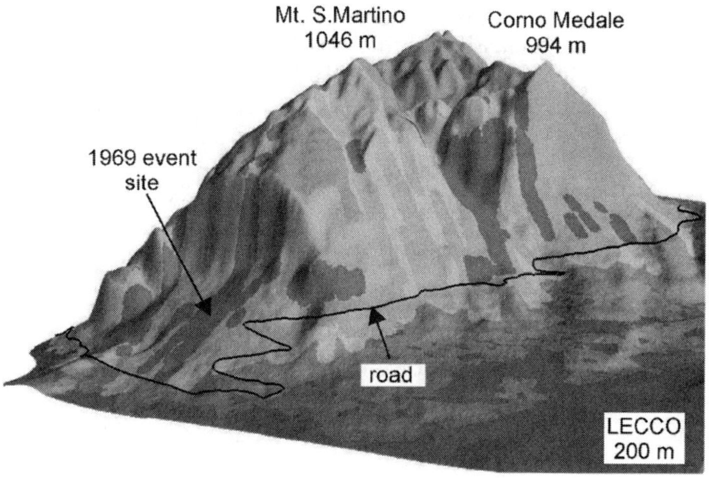

FIGURE 5-23 *(above)* Three-dimensional representation of rockfall hazard map based on smoothed RHV values for road along Monte San Martino and Corno Medale slopes near town of Lecco, Italy (Crosta and Agliardi 2003).

6.3 Semiquantitative Assessment of Risk Reduction by Rockfall Protection Systems

Peila and Guardini (2008) illustrated the use of event trees to produce a semiquantitative assessment of the potential risk reduction benefit provided by the installation of a rockfall fence along an Italian mountain road. The situation used as an example of their procedures is a local mountainside road that has a rock cut 400 m (1,312 ft) long and 200 m (656 ft) high, from which rockfall endangers the road.

Review of a historical rockfall catalog covering a period of 30 years revealed that one rockfall event per year is likely to occur at this site, with each event involving between 10 and 30 blocks. Then, with the approach proposed by Jaboyedoff et al. (2005), a two-dimensional (2-D) rockfall simulation program, with a simulation run involving 5,000 events, computed that the number of

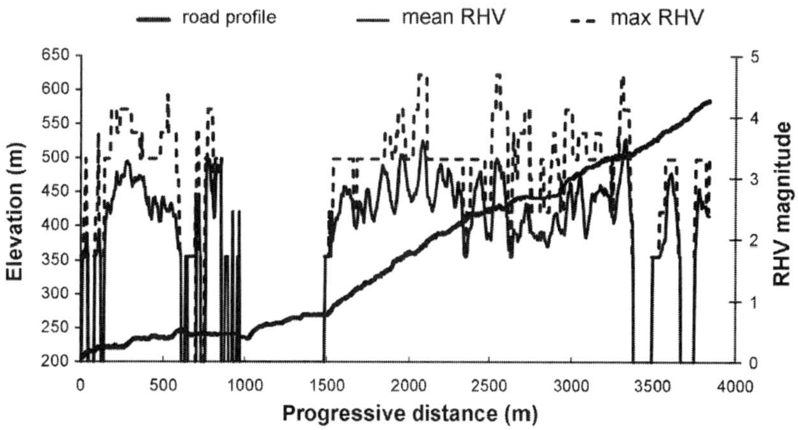

FIGURE 5-24 *(below)* Plot of computed RHV values along road profile smoothed by both mean and maximum values in local neighboring cells (Crosta and Agliardi 2003).

Table 5-6
Reclassification of Rockfall Counts, Rockfall Bounce Heights, and Rockfall Velocities Used to Compute RHI Parameters in Nera Valley, Central Italy

RHI Parameter	Value	Class	Notes
Rockfall count (c) (number)	1–10	1	Very low frequency
	11–100	2	Low frequency
	101–250	3	Medium frequency
	251–500	4	High frequency
	>500	5	Very high frequency
Rockfall maximum bounce height (h) (m)	<1	1	Boulder is rolling.
	1–5	2	Height of most of the elastic rockfall fences is 5 m.
	5–10	3	Rockfalls are difficult to stop by retaining structures.
	10–30	4	
	>30	5	Rockfalls cannot be stopped by retaining structures.
Rockfall maximum velocity (v) (km/h)[a]	1.5–25	1	1.5 m/s is the lowest limit of rockfall velocity.
	25–40	2	Below this threshold a boulder comes to rest.
	40–70	3	Blocks of 5 m^3 in this velocity range can be stopped by fences capable of absorbing up to 3,000 kJ.
	70–115	4	Blocks of 2 m^3 in this velocity range can be stopped by fences capable of absorbing up to 2,500 kJ.
	>115	5	Rockfalls with these velocities cannot be stopped by fences.

[a]This can be converted into kinetic energy (k) by the formula $k = 0.5\ mv^2$, where m = acceleration due to gravity.
SOURCE: Adapted from Guzzetti et al. 2004.

rocks reaching the road per year (N_r) would be 2.5 and that approximately half these blocks traveled across the road and farther down the slope. Peila and Guardini (2008) used an event tree to evaluate the sequence of possible events leading to a fatal accident (Figure 5-25).

The calculation of the probabilities associated with the event tree (Figure 5-25) begins with the assumption that 2.5 rock blocks reach the road along this rock cut each year and involves 13 steps:

1. Estimate the probability that one or more vehicles traveling through the cut will be hit by a falling rock. This event is a function of two simultaneous events: spatial and temporal correspondence between the rock and the vehicle (for the accident to occur, the block and the vehicle have to fill the same space at the same moment). Thus, the probabilities for both spatial and temporal correspondence must be computed.
2. Estimate the probability of spatial correspondence. If rockfall is assumed to be uniformly distributed along the length of the cut (400 m, or 1,312 ft), the probability of spatial correspondence between a falling rock and a moving vehicle is the relationship between the length of the vehicle and the length of the cut. With a vehicle length of 5.4 m (18 ft) this probability is $5.4/400 = 1.35 \times 10^{-2}$. By applying the binomial theorem and knowing the number of rocks that reach the road each year, the probability can be computed that one or more vehicles will be in a position to be hit by a falling rock during a year. The annual generic probability for a single vehicle being in a position where it would be hit by a falling rock while traveling through the cut (assuming 2.5 rocks reach the road each year) is 3.34×10^{-2}.
3. Estimate the probability of temporal correspondence. Peila and Guardini (2008) assumed that the annual probability of temporal correspondence between a falling rock and a moving vehicle was equal to the part of a year occupied by a single pass of a vehicle through the rock cut. The length (L_p) of the rock cut is 400 m (0.4 km) and the vehicle speed (V_v) is 80 km/h (50 mph), so this annual probability is $(0.4/80)/8{,}760 = 5.71 \times 10^{-7}$.

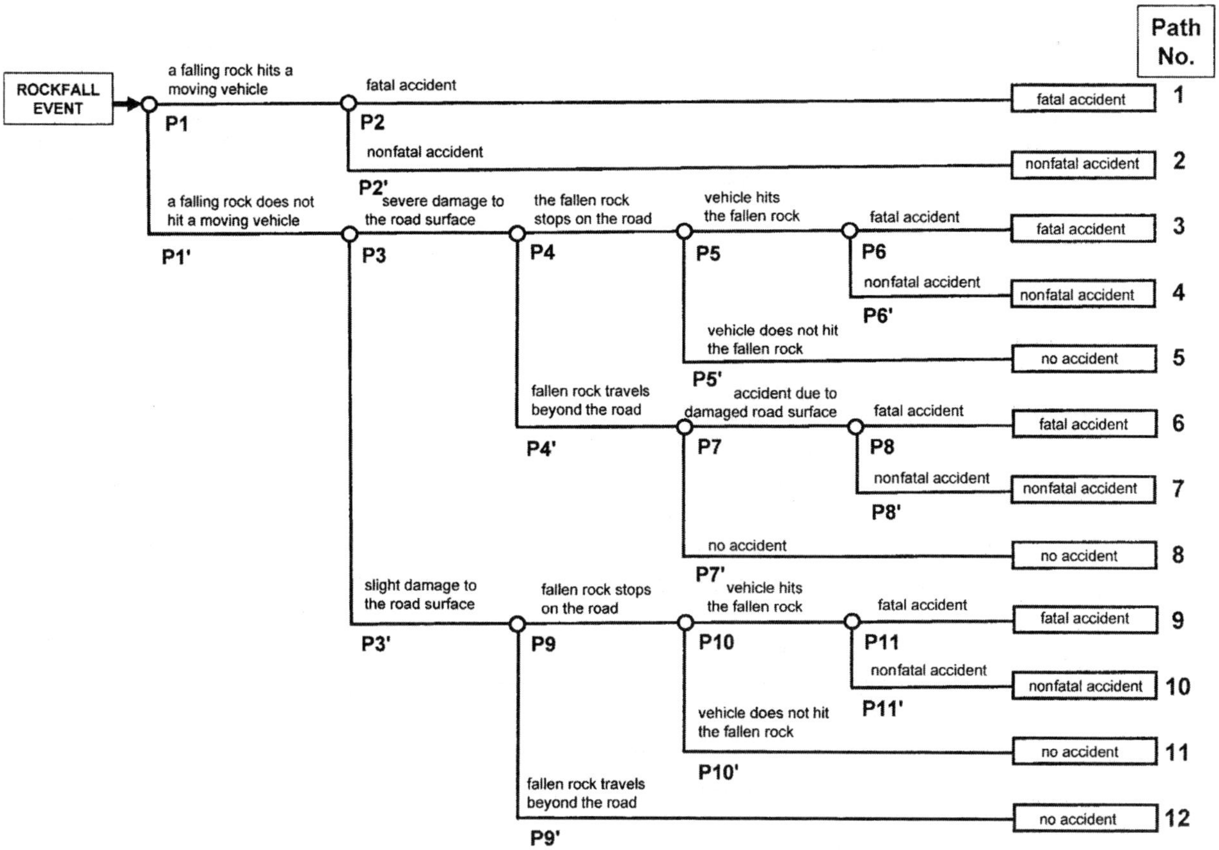

FIGURE 5-25
Event tree used to analyze probability of fatal accident on Italian mountain road (Peila and Guardini 2008).

4. Estimate the annual probability of an individual vehicle being hit by a falling rock. This probability is the product of the annual probabilities of spatial and temporal correspondence, or $3.34 \times 10^{-2} \times 5.71 \times 10^{-7} = 1.91 \times 10^{-8}$.
5. Estimate the annual probability ($P1$) of a moving vehicle being hit by a falling rock. This probability ($P1$, Figure 5-25) can be calculated as the product of this generic probability for one vehicle and the number of vehicles traveling on the road per year. Thus, if 250 vehicles use this road per day, $P1 = 1.91 \times 10^{-8} \times 250 \times 365 = 1.74 \times 10^{-3}$.
6. Estimate the annual probability ($P1'$) of falling rocks missing all vehicles and hitting the roadway surface. This probability is the one's complement of the $P1$ value, or $1 - 1.74 \times 10^{-3} = .99826$.
7. Estimate the annual probability of a fatality once an accident has occurred (moving vehicle hit by falling rock). Using the same assumptions proposed earlier by Bunce et al. (1997), Peila and Guardini (2008) assumed the probability of a fatality, given that this type of accident has occurred, to be .2 ($P2$ in Figure 5-25). Thus the outcomes for Paths 1 and 2 in Figure 5-25 are
   - Probability of fatal accident (Path 1) = $1.74 \times 10^{-3} \times .2 = 3.48 \times 10^{-4}$, and
   - Probability of nonfatal accident (Path 2) = $1.74 \times 10^{-3} \times .8 = 1.39 \times 10^{-3}$.
8. Estimate the annual probability of damage to the road surface from a falling rock. When a falling rock does not hit a vehicle, its impact with the pavement may damage the road surface so as to cause an accident when a vehicle encounters the damage. This probability ($P3$ in Figure 5-25) requires knowledge of the size distribution of the falling rocks. Peila and Guardini (2008) obtained rock block size information from a historical rockfall catalog covering a 30-year period, which showed that 30% of the falling rocks are large enough to leave significant pavement damage upon impact; thus $P3$ was set to .3.

9. Estimate the annual probability of a rock remaining in the roadway. Peila and Guardini (2008) analyzed the rockfall trajectories produced by a 2-D rockfall simulation program to determine the probability that a fallen rock would remain on the road (*P*4 in Figure 5-25). They assumed a fifty-fifty chance for the rock remaining on the road; thus, *P*4 = *P*4′ = .5 for this case example.
10. Estimate the annual probability of an accident in which a vehicle hits a rock lying in the roadway. Peila and Guardini (2008) used the procedures and assumptions proposed earlier by Bunce et al. (1997) to estimate the annual probability that a moving vehicle would hit a previously fallen rock lying in the road (*P*5 in Figure 5-25). This method uses traffic speeds (80 km/h, or 50 mph) and associated decision-sight distance values (270 m, or 886 ft) to determine the probability that a vehicle would be unable to stop before it hit a fallen rock. For this case example, $P5 = 3.35 \times 10^{-2}$.
11. Estimate the annual probability of a fatality following an accident in which a vehicle hits a fallen rock. Using the same assumptions proposed earlier by Bunce et al. (1997), Peila and Guardini (2008) assumed that the probability of a fatality given that this type of accident has occurred would be .1 (*P*8 in Figure 5-25). This value is lower than the probability assigned for a moving vehicle being hit by a falling rock (Case *P*1), because it takes into account the possibility that the reactions of the driver may slow the vehicle before impact, reducing the severity of the accident.
12. Estimate the annual probability of an accident caused by a severely damaged road surface. Peila and Guardini (2008) used Italian road accident statistics to estimate that the probability of a road accident resulting from a severely damaged road surface (*P*7) would be 1.11×10^{-2} and the probability of a fatality resulting from such an accident (*P*8) would be 4.9×10^{-2}.
13. Estimate the annual probability of an accident caused by minor damage to the road surface. This situation, represented in the lower branches of the event tree (Paths 9 to 12), corresponds to events analyzed in previous steps. Thus, Peila and Guardini (2008) assumed that the probabilities that the rock stops on the road (*P*9) or not (*P*9′) correspond to *P*4 and *P*4′, and the probabilities of subsequent events are the same as in the earlier situations (*P*10 = *P*5 and *P*11 = *P*6). It could be argued that these probabilities should perhaps be adjusted, since these events are likely to involve smaller rock blocks and thus the probabilities of an accident, or a fatality resulting from an accident, are less.

The event tree approach shown in Figure 5-25 results in 12 distinct outcomes, each defined by a different path through the event tree. The probability of occurrence of each outcome is calculated by the conditional probabilities of the events forming the path. These are the product of the probability for each single event that constitutes the path. For example, the probability of the outcome for Path 1, which results in a fatal accident, is given by the product of the annual probability that a falling rock hits a vehicle (*P*1) and the probability that this accident results in a fatality (*P*2):

$$P(\text{Path1}) = P1 \times P2 = 1.74 \cdot 10^{-3} \times .2 = 3.48 \cdot 10^{-4}$$

The probability values associated with all 12 paths are calculated with the same procedure, and they are summarized in Table 5-7.

However, several paths have the same outcome; in fact, there are only three outcomes: (*a*) fatal accident, (*b*) nonfatal accident, and (*c*) no accident. The annual probability of each of these three outcomes can be obtained by summing the values of identical outcomes. Table 5-8 summarizes the outcome situation for the case example before any rockfall mitigation measures are installed.

Peila and Guardini (2008) used an event tree to compare the reduction in risk offered by various mitigation measures. Each mitigation measure reduces the risk by preventing some of the falling rocks from reaching the road. This reduction in the number of rocks reaching the road (N_r) thus reduces the probability of spatial correspondence between falling rocks and vehicles on the road (calculated in Step 2) and in turn changes the annual probability (*P*1) of a moving vehicle being hit by a falling rock, or the annual probability (*P*1′) of a falling rock hitting and damaging the roadway. The type of rockfall mitigation measure, its location on the slope, and the slope characteristics will influence the degree to which the number of rocks reaching the road is reduced. For example,

Table 5-7
Annual Probabilities for Outcomes for 12 Paths Analyzed Using Event Tree Approach

Path Number	Outcome	Annual Probability Before Mitigation	Annual Probability After Mitigation
1	Fatal accident	3.48×10^{-4}	7.08×10^{-6}
2	Nonfatal accident	1.39×10^{-3}	2.83×10^{-5}
3	Fatal accident	5.01×10^{-4}	1.59×10^{-5}
4	Nonfatal accident	4.51×10^{-3}	1.43×10^{-4}
5	No accident	.145	.150
6	Fatal accident	8.14×10^{-5}	8.16×10^{-5}
7	Nonfatal accident	1.58×10^{-3}	1.58×10^{-3}
8	No accident	.148	.148
9	Fatal accident	1.17×10^{-3}	3.71×10^{-5}
10	Nonfatal accident	1.05×10^{-2}	3.34×10^{-4}
11	No accident	.338	.350
12	No accident	.349	.350

Note: Probabilities before installation of rockfall mitigation and after a rockfall catchment fence was installed allow for comparison of effectiveness of mitigation.
Source: Peila and Guardini 2008.

at some locations a rockfall catchment fence can be overtopped by a significant percentage of falling rocks, and at other locations the kinetic energy of falling blocks will cause significant damage to the installation or rocks may not be stopped before they reach the road.

Rockfall simulation procedures, described in Chapter 9, allow the evaluation of alternative rockfall mitigation installations. Peila and Guardini (2008) used a 2-D rockfall simulation program, with a simulation run involving 5,000 events, to evaluate a number of potentially feasible rockfall catchment fence locations. At one location, the simulation indicated that the fence would stop 98% of the rocks from reaching the road. Thus, the original value of 2.5 rocks per year reaching the road was reduced to $2.5 \times 0.02 = 0.05$ rocks per year, and the entire event tree was recalculated. The new, considerably reduced, annual probabilities for each of the 12 outcomes are shown in the rightmost column of Table 5-7, where they can be easily compared with the original annual probabilities for each outcome before installation of the mitigation. Table 5-9 summarizes the outcome situation for the case example after the proposed rockfall catchment fence was installed.

Comparison of the annual risks for fatal and nonfatal accidents to road users at this location, as summarized in Tables 5-8 and 5-9, shows that the installation of a rockfall fence at the proposed

Table 5-8
Annual Probabilities Prior to Installation of Rockfall Mitigation Measures

Outcome	Component in Figure 5-25	Annual Probability
Probability that a moving vehicle is hit by a falling rock	$P1$	1.74×10^{-3}
Probability that a moving vehicle hits a fallen rock	$P5$	3.35×10^{-2}
P(fatal accident)	P(Path 1) + P(Path 3) + P(Path 6) + P(Path 9)	2.10×10^{-3}
P(nonfatal accident)	P(Path 2) + P(Path 4) + P(Path 7) + P(Path 10)	1.80×10^{-2}
P(no accident)	P(Path 5) + P(Path 8) + P(Path 11) + P(Path 12)	.980

Source: Peila and Guardini 2008.

Table 5-9
Annual Probabilities After Installation of Proposed Rockfall Catchment Fence

Outcome	Component in Figure 5-25	Annual Probability
Probability that a moving vehicle is hit by a falling rock	*P*1	3.57×10^{-5}
Probability that a moving vehicle hits a fallen rock	*P*5	1.06×10^{-3}
P(fatal accident)	*P*(Path 1) + *P*(Path 3) + *P*(Path 6) + *P*(Path 9)	1.42×10^{-4}
P(nonfatal accident)	*P*(Path 2) + *P*(Path 4) + *P*(Path 7) + *P*(Path 10)	2.09×10^{-3}
P(no accident)	*P*(Path 5) + *P*(Path 8) + *P*(Path 11) + *P*(Path 12)	.998

Source: Peila and Guardini 2008.

location is likely to substantially reduce the incidence of rockfall-induced accidents. Figure 5-26 provides a visual comparison of the probabilities of fatal and nonfatal accidents before and after installation of the proposed rockfall fence.

6.4 Semiquantitative Risk Analysis Procedure

The Road and Traffic Authority (RTA) of New South Wales, Australia, has developed a semiquantitative procedure for evaluating risk from all classes of natural slopes (Stewart et al. 2002). Baynes et al. (2002) provided additional information concerning the accuracy and precision of estimates produced by this procedure (see Section 6.4.5). The RTA procedure was developed on the basis of several guiding principles (Stewart et al. 2002); these define a procedure that

- Is applicable to a large number of sites, covering a great range of conditions;
- Produces consistent results when used by many practitioners;
- Is based on visual inspection of the slope, supplemented by other information when available;
- Is reasonably economical to apply, preferably involving short on-site visits of no more than a few hours; and
- Concentrates on risks involving road users but also is applicable to risks involving development of adjacent land.

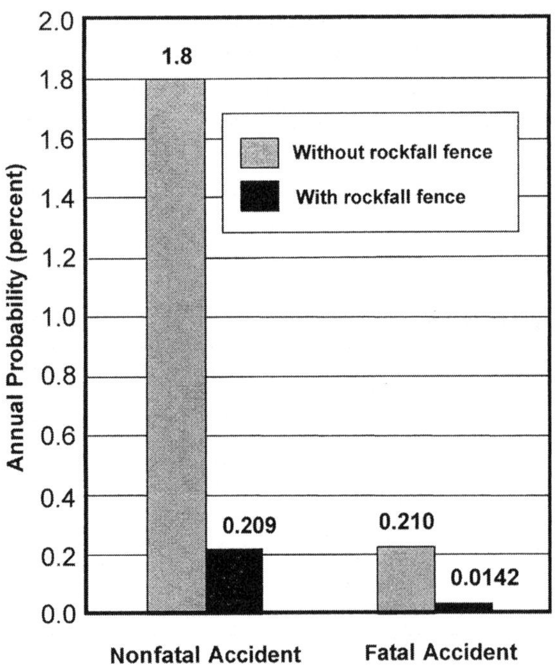

FIGURE 5-26 Annual accident probability with and without proposed rockfall catchment fence (Peila and Guardini 2008).

Therefore, the RTA procedure focuses on hazards that most commonly affect road users, such as rockfalls, but it is applicable to a wide variety of slope stability hazards.

The RTA procedure follows the general recommendations for landslide risk management promoted by AGS (2002), as described in Section 3-2 and Figure 5-3. Figure 5-27 provides a graphical summary of the entire RTA procedure. The procedure begins with identification of the hazards and assessment of their dynamics; these actions are shown as two boxes in the top center portion of Figure 5-27.

Hazard identification begins with site description, which is supplemented by information defining the history of slope instability, often by reference to incidence reports. The site description process emphasizes the use of field sketches that contain all relevant information (Figure 5-28). Stewart et al. (2002) considered sketches to be superior to photographs because they selectively record the features the practitioner thinks are important. This capture of the practitioner's reasoning makes

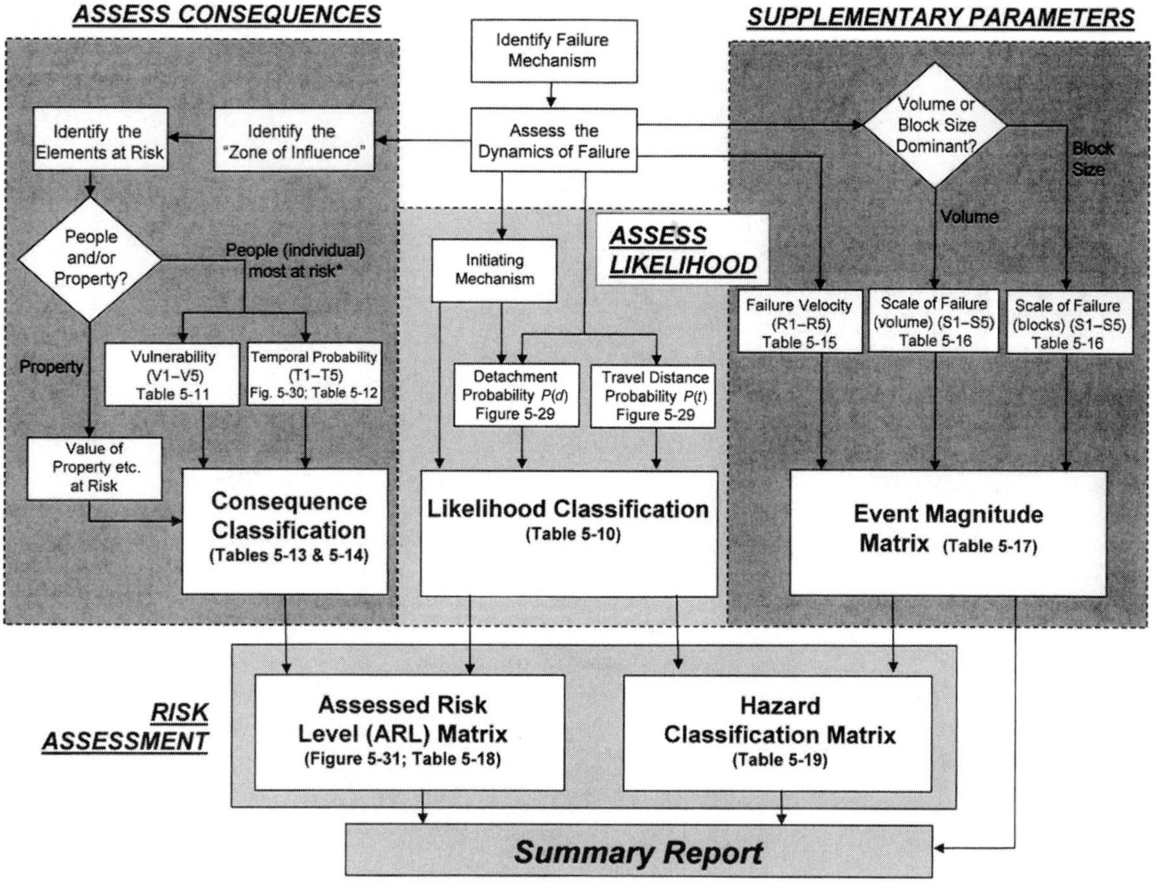

FIGURE 5-27 Flowchart for RTA of New South Wales slope risk analysis guide (adapted from Stewart et al. 2002). [*Where loss of life is identified as consequence, total (societal) risk, based on probability of different numbers of people being present, shall also be considered in determining any responses to assessed mechanism and its possible consequences.]

desk auditing an effective tool in maintaining consistency of the analyses. The sketches may be supplemented by photographs documenting observational details.

The RTA procedure continues with an assessment of the likelihood of failure (central portion of Figure 5-27), an assessment of the consequences (left portion of Figure 5-27), and the use of supplementary parameters to evaluate the magnitude of failure (right portion of Figure 5-27). These assessments use a series of defined ratings to classify the likelihood and magnitude of actual or potential failure mechanisms that have been identified for a slope and to evaluate their consequences. These classifications are used to conduct a risk assessment, shown in the lower portion of Figure 5-27. Thus, the RTA procedure is a semiquantitative risk analysis process. A summary report provides documentation of the conclusions reached by applying this procedure at each location.

The definitions associated with the ratings reflect a specified probability range and guide practitioners to the appropriate rating. Each rating category identifies a change in likelihood or consequences that differs by an order of magnitude. The RTA procedure ratings are defined as objectively as possible. They avoid terms that imply bias or those terms where the common technical definition differs widely from popular usage (e.g., *possible* is often assigned a probability range of .1 to .3 by risk applications, but in popular usage is often inferred to mean any outcome that is not absolutely impossible).

6.4.1 Assessment of Likelihood of Failure

The *likelihood of failure* refers to the probability that a failure will occur and extend over a given zone of influence. The likelihood assessment begins with an evaluation of the initiating mechanism, which may include earthquakes and weather-related or construction-related triggering events. Because historical data concerning the frequency of events are often lacking, the RTA procedure uses a series of six likelihood classes (L1 to L6), which are defined in Table 5-10. The classes are defined in terms of

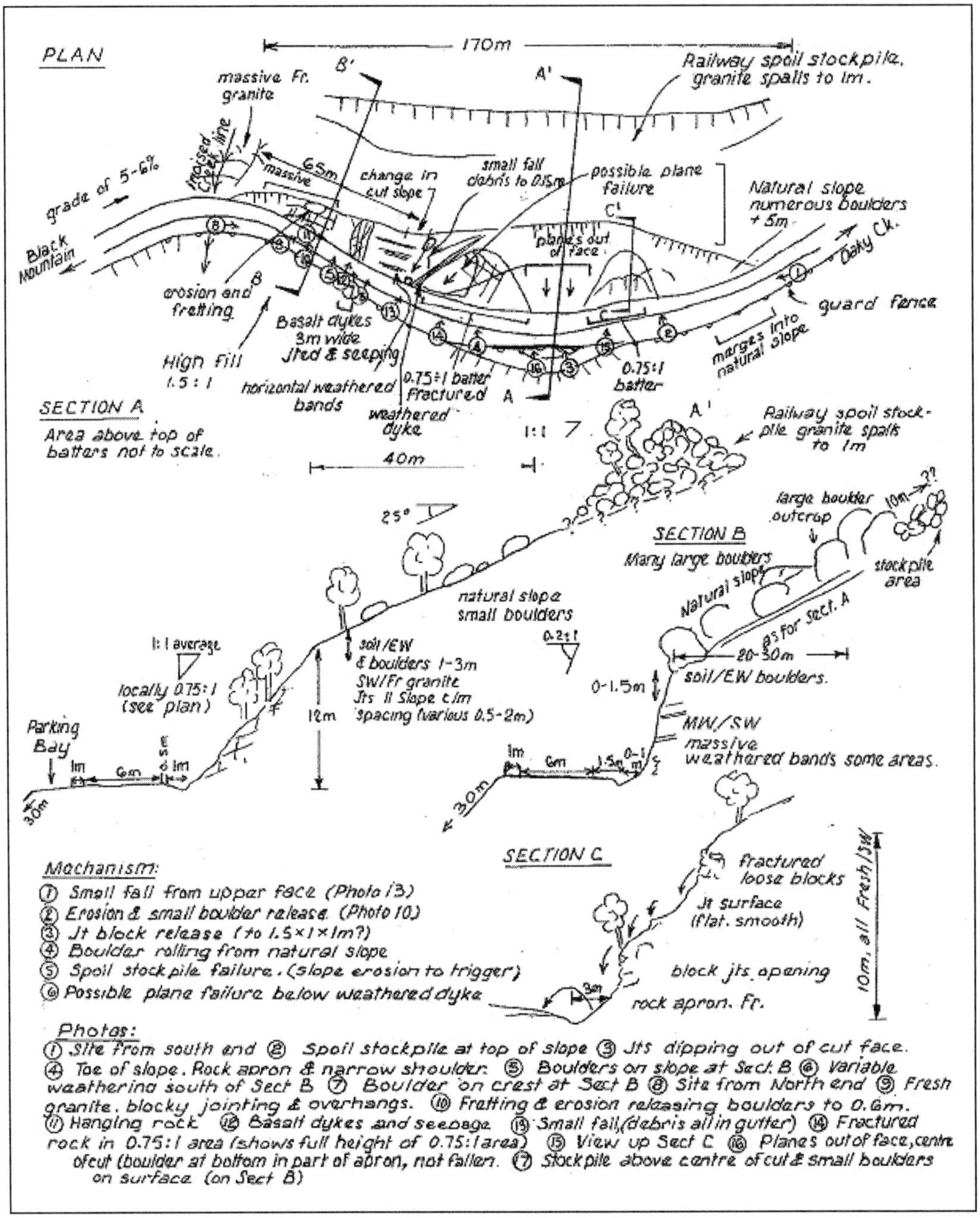

FIGURE 5-28
Example of site sketch (Stewart et al. 2002).

likelihood of occurrence within relatively short periods, such as before the next regular inspection cycle (e.g., 5 years), and within much longer periods, such as the life expectancy of a transportation infrastructure component (e.g., 100 years). An estimated annual probability is indicated for each class to provide guidance for the assignment of subjective probability.

It is often desirable to independently assess the probabilities of detachment and travel distance, as shown in Figure 5-29. The *detachment probability*, $P(d)$, defines the probability that rockfall blocks will be detached from their original locations and begin to move downslope. The *travel distance probability*, $P(t)$, defines the probability that, once dislodged, the debris will travel as far as the element

Table 5-10
Likelihood Probability Scale

	Likelihood of Landslide Event Occurring		
Likelihood Class	By Next Inspection (5 years)	Within Facility Lifetime (100 years)	Annual Probability (P)
L1	Very likely	Definitely within 3 years	.9
L2	Probable	Likely within 30 years	.1
L3	Unlikely	Likely within 100 years	.01
L4	Very unlikely	Maybe in 100 years	.001
L5	No occurrence	Not expected	.0001
L6	No occurrence	Very unlikely	.00001

Source: Stewart et al. 2002.

at risk; in the case of rockfall along transportation facilities, this is the probability that blocks will land and remain on the traveled way. These two conditional probabilities are then combined (multiplied together) to obtain the likelihood ratings (L1 to L6) shown in Table 5-10. The guidance document for the RTA procedure recommends that both probabilities be reported.

6.4.2 Assessment of Consequences

The *assessment of consequences* identifies the effects of failure. Consequences are expressed separately for risk to life and for economic loss.

The consequences for risk to life use separate ratings for the vulnerability (V1 to V5) of individuals, as defined in Table 5-11, and for the temporal probability (T1 to T5) of an individual (Table 5-12). Tables 5-11 and 5-12 define the ratings of vulnerabilities and temporal probabilities for both individuals and road users, and also show associated estimated annual probabilities. Figure 5-30 graphically shows the relationship between temporal probabilities and temporal probability ratings (T1 to T5) for road users. Consequence ratings for risk to life (C1 to C5) are determined by using a matrix defined by the ratings of vulnerabilities and temporal probabilities (Table 5-13). In contrast, the economic consequences ratings (also C1 to C5) are allocated directly on the basis of several criteria, including road closure and repair costs (Table 5-14). Table 5-14 shows that these consequence ratings correspond to economic loss levels that range over several orders of magnitude.

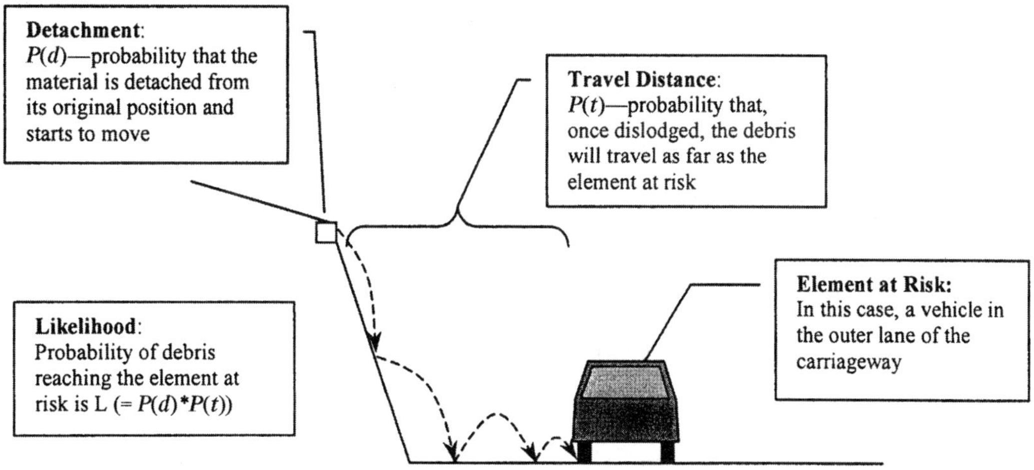

FIGURE 5-29
Defining detachment and travel distance probabilities for RTA of New South Wales slope risk analysis guide (Stewart et al. 2002).

Table 5-11
Vulnerability Probability Scale

Vulnerability Class	Element at Risk: People	Element at Risk: Road Users[a]	Annual Probability (*P*)
V1	Persons in open unable to evade rockfall; building collapse	Vehicle impacts rock > 1 m high at highway speed	>.5
V2	Person in open able to evade debris; only partial building collapse	Vehicle impacts rock 0.5 to 1 m high at highway speed or 1 m block at urban speed	.1–.5
V3	Building penetrated, no collapse	Vehicle impacts rock ~ 0.2 m high at highway speed or 0.5 to 1 m block at low speed	.01–.1
V4	Building struck and damaged but not penetrated	Vehicle impacts rock ~ 0.2 m high at highway speed or 0.5 to 1 m block at low speed	.001–.01
V5	Building struck but only minor damage	Vehicle impacts rock ~ 0.2 m high at urban speed or smaller block at highway speed	<.001

[a]Vehicle speed categories are defined in three classes: *highway* (100–110 km/h or 65–70 mph), *urban* (60–80 km/h or 35–50 mph), and *low* (20–30 km/h or 15–20 mph).
SOURCE: Stewart et al. 2002.

Table 5-12
Temporal Probability Scale

Temporal Class	Element at Risk: Presence of People in Buildings	Element at Risk: Traffic Level	Annual Probability (*P*)
T1	Usually present	Heavy urban	>.5
T2	Often present	Major urban arterial roads and heavily used rural roads	.1–.5
T3	Sometimes present	Urban arterials and major rural roads	<.1
T4	Unlikely to be present	Suburban roads and minor rural arterials	<.01
T5	Very unlikely to be present	Low-volume roads	<.001

SOURCE: Stewart et al. 2002.

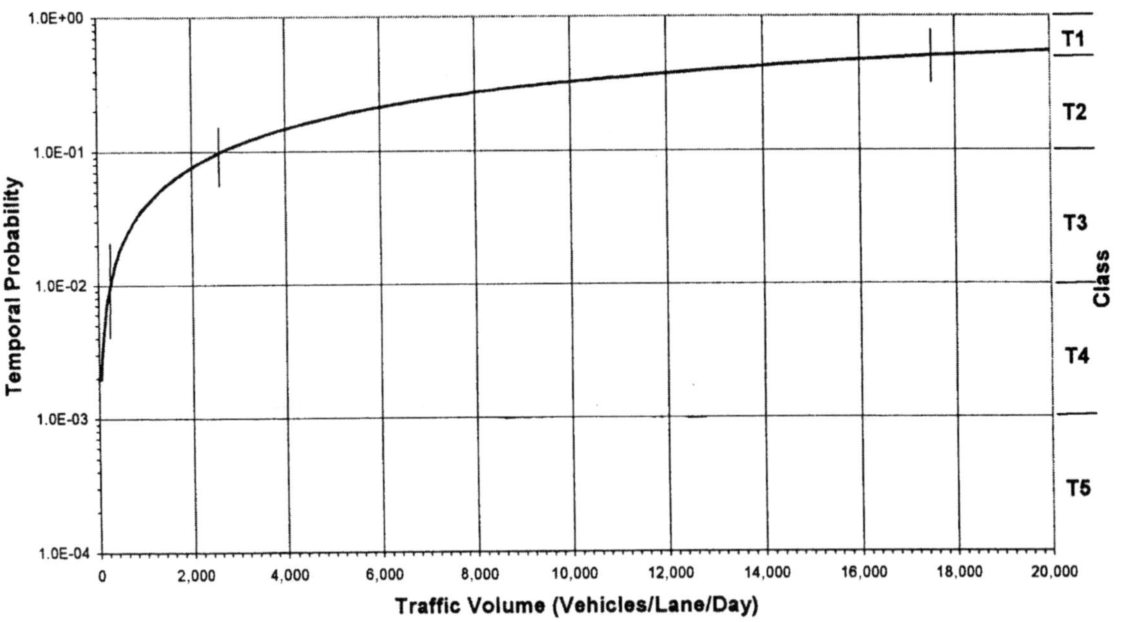

FIGURE 5-30 Temporal probability rating for road users from RTA of New South Wales slope risk analysis guide (Stewart et al. 2002).

Table 5-13
Consequence Matrix for Risk to Life

	Vulnerability Class (Table 5-11)	Low ⟵⟶ High Temporal Class Defining Probability of Individual Being Present at Time of Failure (Table 5-12) T5	T4	T3	T2	T1
High	V1	C4	C3	C2	C1	C1
↕	V2	C4	C3	C2	C1	C1
	V3	C5	C4	C3	C2	C2
	V4	C5	C5	C4	C3	C3
Low	V5	C5	C5	C5	C4	C4

Source: Stewart et al. 2002.

6.4.3 Supplemental Parameters

The RTA procedure was developed to assess all types of landslides and thus includes a process for evaluating the *magnitude of a failure event*, which is defined in terms of the velocity and size characteristics of the failure. Shown on the right-hand side of Figure 5-27, this evaluation uses supplementary parameters to define the velocity and scale, or size, of the failure. Table 5-15 defines the failure velocity ratings (R1 to R5). Because rockfalls are rapidly moving events, they belong to Class R1, and this rating procedure does not allow for differentiation of rockfalls.

The ratings for scale of failure (Table 5-16) are allocated either for the volume of failed material or for the size of the largest blocks. The volume scale is used for landslides involving mass movements of relatively large volumes of soil, rock, or both. The block-size scale is used for events such as rockfalls, which result in the release of single blocks of rock or small groups of such blocks. The ratings for velocity of failure (Table 5-15) and scale of failure (Table 5-16) are combined in the event

Table 5-14
Consequence Scale for Property Damage

Consequence Class	Property Damage and Consequential Effects	Loss Level ($ millions)
C1	Total closure of major road for extended period Major infrastructure damage Very high disruption and repair costs	>10
C2	Partial closure of major road or total closure of important road for extended period Substantial infrastructure damage Large disruption or high repair costs, or both	2–5
C3	Partial or total closure of important road for a short period Moderate infrastructure damage Moderate disruption or repair costs	0.5–2
C4	Partial or total closure of minor road for short period Minor infrastructure damage Minor disruption or repair costs	0.1–0.5
C5	Partial or total closure of local road for short period Negligible infrastructure damage Little or no disruption or repair costs	<0.1

Source: Stewart et al. 2002.

Table 5-15
Failure Velocity Scale

Failure Velocity Class	Description	Velocity (mm/s)	Typical Velocity
R1	Extremely rapid		
		5×10^3	5 m/s
	Very rapid		
		5×10^1	3 m/min
R2	Rapid		
		5×10^{-1}	1.8 m/h
R3	Moderate		
		5×10^{-3}	13 m/month
R4	Slow		
		5×10^{-5}	1.6 m/year
R5	Very slow		
		5×10^{-7}	16 mm/year
	Extremely slow		

Source: Stewart et al. 2002.

magnitude classification matrix (Table 5-17), which defines event magnitudes M1 to M5.

6.4.4 Assessment of Risk

Risk is defined as the product of likelihood of hazard occurrence and the consequences arising from the failure event. Figure 5-31 illustrates the conditional probabilities that characterize the chain of events that must occur for a rockfall to result in a fatality from an interaction between a vehicle and rockfall debris; these are defined by the ratings for likelihood (L), temporal probability (T), and vulnerability (V). The ratings of likelihood (Table 5-10) and consequences (Tables 5-13 and

Table 5-16
Scale of Failure Ratings

Scale of Failure Class	Likelihood of Landslide Event Occurring: Volume of Failure (m^3)	Likelihood of Landslide Event Occurring: Dimensions of Single Blocks
S1	>20,000	Minimum dimension > 1 m
S2	>2,000 and <20,000	Maximum dimension > 1 m
S3	>200 and <2,000	Maximum dimension < 1 m
S4	>20 and <200	Maximum dimension < 0.5 m
S5	<20	Maximum dimension < 0.2 m

Source: Stewart et al. 2002.

Table 5-17
Event Magnitude Classification Matrix

Velocity of Failure Class (Table 5-15)		Scale of Failure Class (Table 5-16): S5	S4	S3	S2	S1
Fast	R1	M3	M2	M2	M1	M1
↑	R2	M4	M3	M2	M2	M1
	R3	M4	M4	M3	M2	M2
↓	R4	M5	M4	M4	M3	M2
Slow	R5	M5	M5	M4	M4	M3

Source: Stewart et al. 2002.

5-14) are combined in a matrix (Table 5-18) to define the five classes of assessed risk level (ARL). In addition, the ratings for likelihood (Table 5-10) and event magnitude (Table 5-17) are combined in the hazard classification matrix (Table 5-19) to produce a five-category definition of hazard severity (H1 to H5) that assists with risk evaluation and mitigation decisions. The rating that represents the highest risk associated with the site and the associated hazard rating are defined in the summary report, along with details of the analysis procedure.

6.4.5 Evaluating RTA Procedure

Results of risk analyses developed by applying initial versions of the RTA procedure raised some concerns as to whether the procedure was an effective approach for risk analysis and for setting mitigation priorities (Baynes et al. 2002). Following development of a new version (Version 3.1) of the RTA procedure, Baynes et al. (2002) evaluated the accuracy and precision of landslide risk analyses developed by applying this new procedure at about 600 sites. Their study was designed to evaluate the ARL ratings in terms of

- *Precision*, defined as the repeatability of results developed by several individuals at a variety of sites;
- *Accuracy*, defined by the ability to correctly differentiate high-risk and low-risk sites; and
- *Applicability*, an assessment of the usefulness of the RTA procedure in setting mitigation priorities.

This evaluation is a relatively rare quality-assurance assessment of risk analysis processes. About 25 individuals from several geotechnical firms in New South Wales received 2 days of

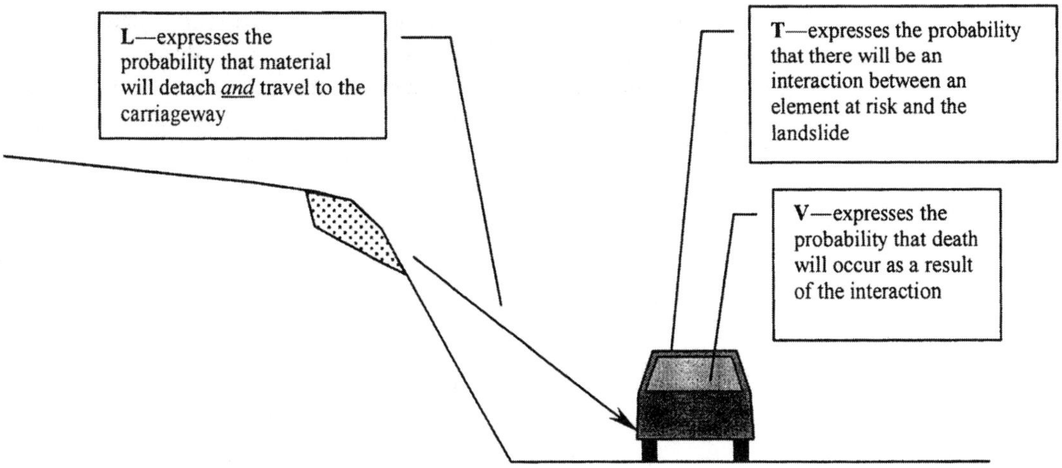

FIGURE 5-31
Parameters used to determine assessed risk level (ARL) for slope failures that deposit debris on road [from RTA of New South Wales slope risk analysis guide (Stewart et al. 2002)].

relatively rigorous training in the use of the RTA procedure. After a number of sites had been assessed by these individuals, a series of "field audits" were conducted by two "experts" who had developed the RTA procedure. These field audits consisted of additional independent assessments of each site by these two individuals, comparison of these new assessments with each other to assess their consistency, and comparisons with the assessments produced by the recently trained geotechnical consultants to assess the presence of bias.

These comparisons suggested that several of the recently trained geotechnical consultants were overestimating risk. They also revealed that at about 20% of the sites, the two expert system developers produced assessments with significantly different ratings. The two experts then revisited several of the field sites, where they conferred, reviewed, discussed, and revised their previous individual assessments and produced a consensus assessment for each site. Similar consultations and discussions with the recently trained geotechnical consultants during a half-day workshop resulted in clarification and slight modification of some aspects of the RTA procedure and revision of the previous assessments. These consensus-building actions resulted in a significant reduction of bias and increased consistency in the evaluation results (Baynes et al. 2002). These experiences led Baynes et al. (2002) to draw the following conclusions:

- Risk assessments based on the RTA procedure that are produced by untrained assessors are unreliable and unacceptable.
- Training in the RTA procedure requires at least a 2-day course plus a follow-up half-day workshop to review results after a few sites have been evaluated.

Table 5-18
ARL Matrix

Likelihood Class (Table 5-10)	Consequence Class (Tables 5-13 and 5-14) C5	C4	C3	C2	C1
L1	ARL3	ARL2	ARL1	ARL1	ARL1
L2	ARL4	ARL3	ARL2	ARL1	ARL1
L3	ARL5	ARL4	ARL3	ARL2	ARL1
L4	ARL5	ARL5	ARL4	ARL3	ARL2
L5	ARL5	ARL5	ARL5	ARL4	ARL3
L6	ARL5	ARL5	ARL5	ARL5	ARL4

Source: Stewart et al. 2002.

Table 5-19
Hazard Classification Matrix

Likelihood Class (Table 5-10)	Event Magnitude Class (Table 5-17) M5	M4	M3	M2	M1
L1	H3	H2	H2	H1	H1
L2	H4	H3	H2	H2	H1
L3	H4	H4	H3	H2	H2
L4	H5	H4	H4	H3	H2
L5	H5	H5	H4	H4	H3
L6	H5	H5	H5	H4	H4

Source: Stewart et al. 2002.

FIGURE 5-32 Location map for Lawrence Hargrave Drive project, Australia (Wilson et al. 2005).

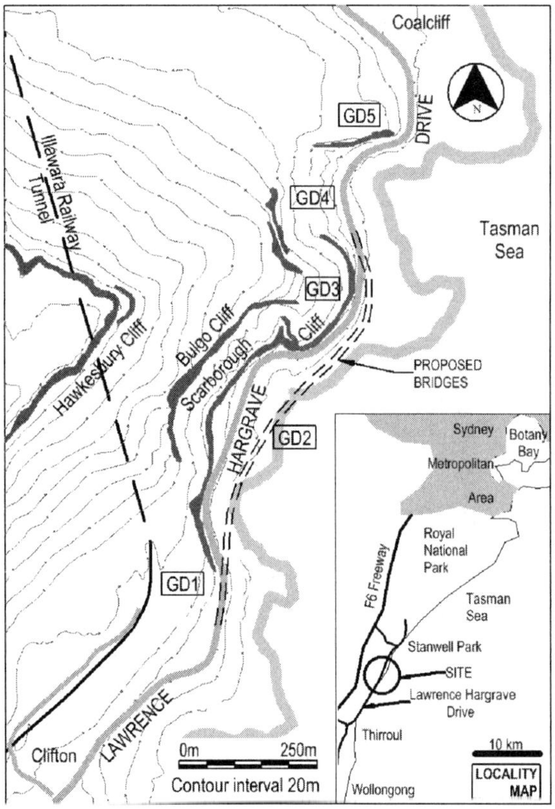

- Precision of evaluations results from consensus assessments conducted by pairs of trained assessors visiting a site and initially preparing independent evaluations and then discussing and revising these to produce a consensus rating. Independent evaluations produced by experienced assessors differed by less than one ARL class at about 80% of the sites (but significant differences occurred at the remaining 20% of the sites). In contrast, consensus evaluations produced by teams of experienced assessors always differed by less than one ARL class.
- Accuracy of risk evaluations produced by trained assessors using the RTA procedure is acceptable. Trained assessors produce risk evaluations that are within one ARL class of the correct value, independently estimated by a review of historical records, at about 80% of the sites.
- Applicability of the RTA procedure seems ensured, since it portrays risks with acceptable accuracy. The estimated ARL levels provide a slightly conservative impression of the relative levels of risk; thus, the RTA procedure is suited to establishing priorities for maintenance and mitigation programs.

6.5 Quantitative Risk Assessment for Lawrence Hargrave Drive Project, New South Wales, Australia

Lawrence Hargrave Drive is a coastal road that extends between Stanwell Park and Thirroul on the South Coast of New South Wales, about 30 km south of the southern outskirts of Sydney, Australia, (Figure 5-32). Constructed in 1878, the road has experienced frequent slope failures including rockfalls, debris flows, and embankment collapses (Hendrickx et al. 2005). In 2003 a section of the road was closed for safety reasons. This section of Lawrence Hargrave Drive is located about 20 to 45 m (65 to 147 ft) above the base of a coastal escarpment consisting of steep cliffs that rise to a height of about 300 m (1,000 ft) above the road (Figures 5-33 and 5-34). Aggressive marine erosion at the base of the cliffs is undercutting the coastal platform, causing this section of Lawrence Hargrave Drive to experience a history of severe embankment instability, rockfall, and debris flow problems (Figure 5-35).

Following the road closure, quantitative risk analysis procedures were employed to develop an engineering solution to reduce the risk to acceptable levels (by reducing the annual probability of death from about 1×10^{-2} to about 1×10^{-5}). The quantitative risk assessment was based on evaluating landslide size–frequency models, as described in Section 5.4.2, in combination with aspects of the semiquantitative procedure for evaluating risk from all classes of natural slopes developed by the RTA of

FIGURE 5-33 Aerial photograph looking south along coast showing Lawrence Hargrave Drive near base of cliffs in 1967 (Moon et al. 2005).

FIGURE 5-34
December 2003 view of cliffs composed of Scarborough and Bulgo sandstone above Lawrence Hargrave Drive along Sections GD1 and GD2 shown in Figure 5-32 (Hendrickx et al. 2005).

New South Wales, Australia (Stewart et al. 2002), as described in Section 6.4. An Excel workbook was used to review the slope risks associated with the existing road and to assess the effectiveness of the proposed engineering works (Moon et al. 2005).

Assessment of slope hazards was completed with the aid of geological mapping, interpretation of aerial photographs, archived government reports, historical photographic and newspaper collections, and a landslide inventory based on a geographic information system (Wilson et al. 2005). Historical photographs provided important evidence on erosion rates and the size, nature, and frequency of landslide events (Hendrickx et al. 2005).

FIGURE 5-35
Rockfall catchment ditch along Lawrence Hargrave Drive in Section GD2 (Figure 5-32) showing rockfall debris collected between July 2003 and March 2004 (Hendrickx et al. 2005).

Table 5-20
Geotechnical Domains Along Lawrence Hargrave Drive

Geotechnical Domain Identifier	Geographic Description
GD1	Southern headland
GD2	Southern amphitheatre (bay)
GD3	Middle headland
GD4	Northern amphitheatre (bay)
GD5	Northern headland

Source: Wilson et al. 2005.

The closed section of the road crosses three headlands separated by two bays. These features were used to separate the site into five geotechnical domains (Figure 5-32 and Table 5-20). Hendrickx et al. (2005) described the geology of this location in considerable detail. The bedrock geology comprises a thick sequence of sandstone and claystone units of Permian to Triassic age that generally dip between 0° and 5° obliquely inland. The sandstone units within the succession form prominent subvertical cliffs, while the claystones form intervening slopes (Figures 5-33, 5-34, and 5-36). The entire sedimentary sequence was subdivided into seven slope units (Table 5-21). Figure 5-36 illustrates the relationships between the bedrock geology and these slope units.

Hendrickx et al. (2005) defined the slope instability in the following simple terms. Marine erosion is removing the Bulli coal seam, which leads to toppling failures of the overlying Coalcliff sandstone. This action in turn leads to oversteepening of the overlying Wombarra claystone, which fails by landsliding. This process is repeated up the slope to the Hawkesbury sandstone forming the cliffs at the crest of the escarpment (Figure 5-36). The processes are further driven by differential erosion and undercutting of the claystone intervals within the major sandstone units, and these lead to rockfalls from the sandstone units. Excavation to form the road has both removed material from the Wombarra claystone and resulted in accelerated regression of the Wombarra claystone and subsequently the Otford and Scarborough sandstones.

A landslide process rate model was developed by bringing together knowledge and judgments about erosion rates for the differing materials and landforms on the escarpment. Process rate curves were developed for slope units and integrated into

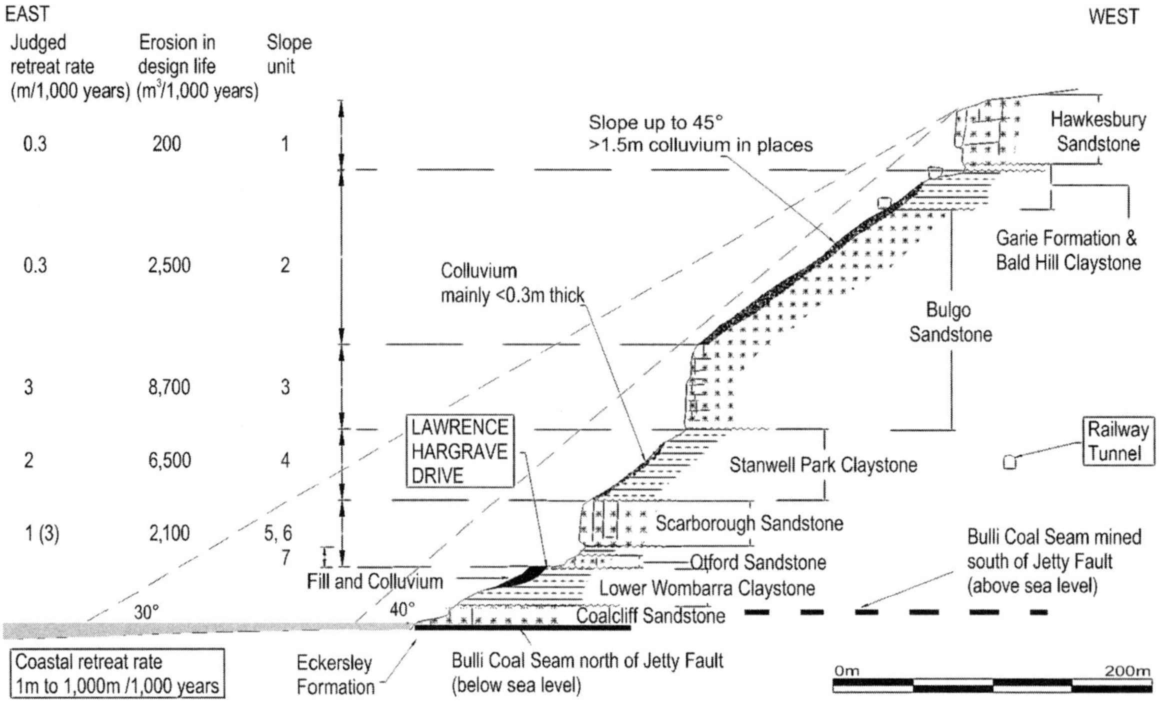

FIGURE 5-36 Geological cross section and slope units defined for the Lawrence Hargrave Drive project (Moon et al. 2005).

the quantitative risk assessment utilizing landslide size–frequency models, as described in Section 5.4.2. Figure 5-37 shows the size–frequency curves for each of the slope units defined in Table 5-21 and Figure 5-36. Figure 5-37 also shows several of the assumptions adopted to create the size–frequency models, including typical slope retreat rates, typical height of the slopes, and the percentage of erosion due to slope failures.

The quantitative risk assessment process is described in detail by Wilson et al. (2005). The

Table 5-21
Slope Units Located Above Lawrence Hargrave Drive

Slope Unit Identifier	Description (Figure 5-36)
1	Hawkesbury Cliff
2	Bulgo Slope
3	Bulgo Cliff
4	Stanwell Park Slope
5	Typical Scarborough Cliff
6	Slabbing Scarborough Cliff[a]
7	Otford Cliff

[a]Caused by persistent near-vertical joints oriented parallel to rock face.
Source: Wilson et al. 2005.

adopted assessment procedure followed the semi-quantitative procedure for evaluating risk from all classes of natural slopes of the RTA of New South Wales, Australia (Stewart et al. 2002), as described in Section 6.4. The ARL rating for the road before remediation was judged to be ARL1, the highest risk category, as defined in Table 5-18. The objective of the evaluation project was to determine mitigation measures that would reduce the risk to life to ARL3 or better. The study did not address the risk to property using the ARL criteria. To define practical differences between the differing ARL levels, specific ARL criteria were adopted for this project (Table 5-22). As described in Section 6.4, the development of ARL ratings requires establishing the vulnerability of road users. Two levels of vulnerability were considered: the vulnerability of people in a vehicle that is directly hit by a rockfall and that of people in a car that hits rockfall debris in the road. Table 5-23 summarizes these adopted vulnerability values. The risk assessment considered four cases:

- Case A represented the road conditions before 2002, when no mitigation measures had been installed;
- Case B represented road conditions just before its closure; a rockfall catchment fence had been

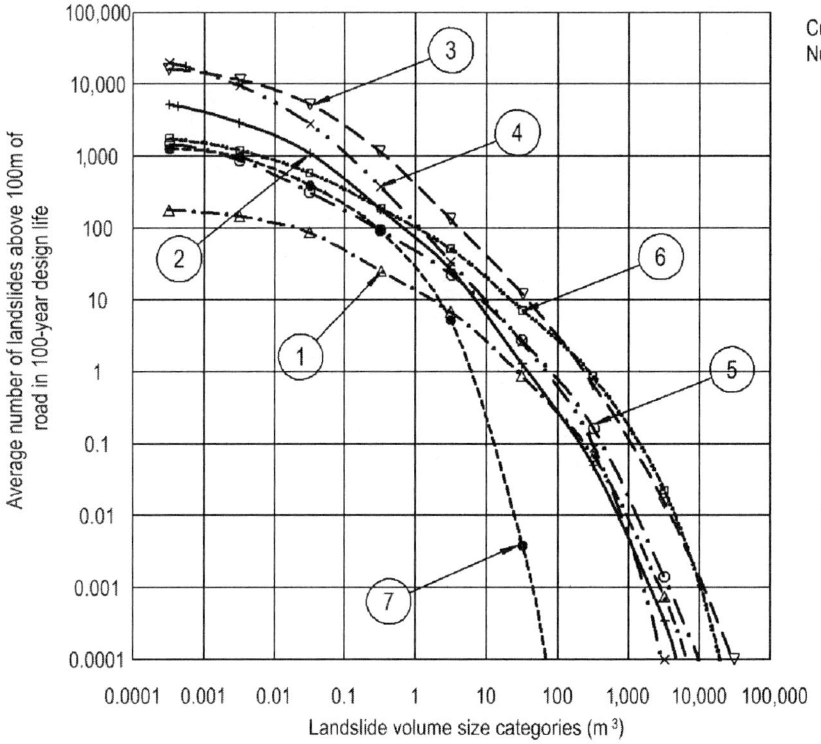

Curve Number	Slope unit	Adopted in model: Typical slope retreat rate ** (m/1,000 years)	Typical height (m)	% erosion by landslide
1	Hawkesbury Cliff	0.3	30	95
2	Bulgo Slope	0.3	110	70
3	Bulgo Cliff	3	60	90
4	Stanwell Park Slope	2	40	60
5	Typical Scarborough Cliff	1	30	90
6	Slabbing Scarborough Cliff *	3	30	95
7	Otford Cliff	1	8	80

* Persistent joints parallel to face
** Adopted rates varied (e.g., lower where slopes locally flatter)

FIGURE 5-37
Adopted size–frequency models for seven slope units above Lawrence Hargrave Drive (Wilson et al. 2005).

installed and rockfall catchment ditches had been placed in Geotechnical Domains GD2 and GD4;

- Case C represented conditions after installation of proposed bridge diversions in Geotechnical Domains GD1, GD2, and GD3, but without any other mitigation works; and
- Case D represented conditions following installation of a full suite of mitigation works.

Wilson et al. (2005) provided an extensive discussion of these four cases. Cases A and B represent historical conditions and thus provided a method for calibrating the risk assessment process. Cases C and D allow for consideration of the effectiveness of the proposed mitigation options. Figure 5-38 provides a graphical summary of the risk assessment process.

Table 5-22
Assessed Risk Levels Used in Lawrence Hargrave Drive Project

Assessed Risk Level	Annual Probability of Death
ARL3	3×10^{-5} to 3×10^{-6}
ARL4	3×10^{-6} to 3×10^{-7}
ARL5	$<3 \times 10^{-7}$

Source: Wilson et al. 2005.

6.6 Quantitative Assessment of Rockfall Risk from Coastal Cliffs, England

Coggan et al. (2001) undertook rockfall hazard appraisal and qualitative and quantitative risk

Table 5-23
Adopted Vulnerabilities for Rockfall Events, Lawrence Hargrave Drive Project

Magnitude of Rockfall Crossing Road (volume in m³)	Rockfalls from Scarborough Cliff (P): Rockfall Hits Car	Car Hits Rockfall
0.03	.05	.0006
0.3	.1	.002
3	.3	.03
30	.7	.03
300	1	.03
3,000	1	.03

Source: Wilson et al. 2005.

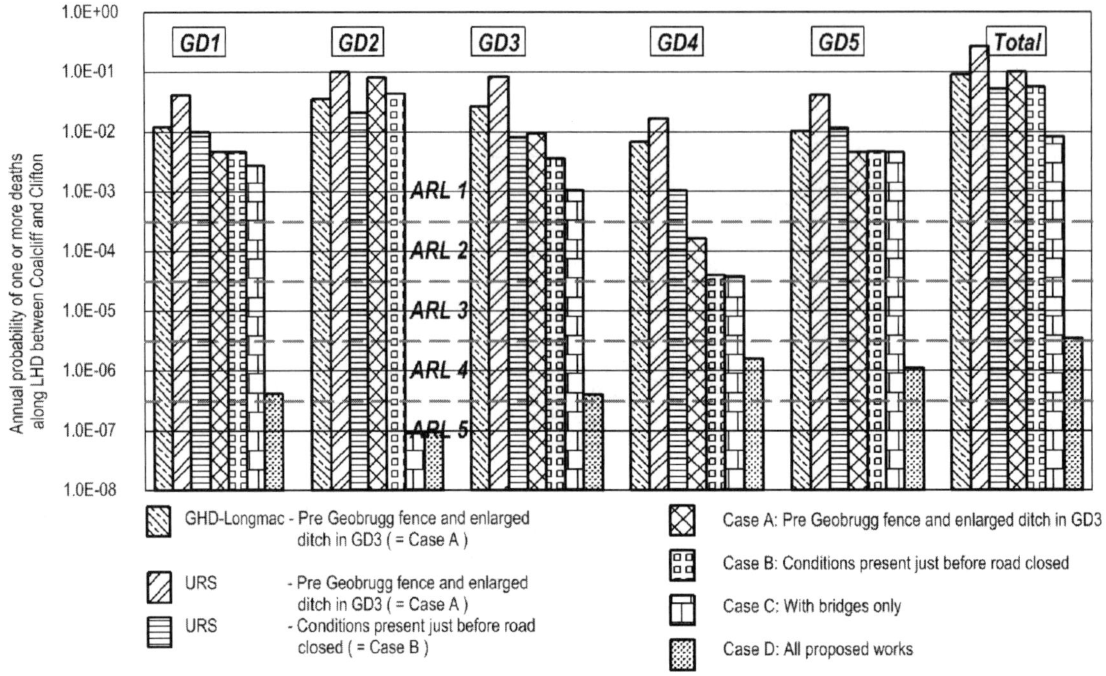

FIGURE 5-38 Summary of results of quantitative risk assessment calculations, Lawrence Hargrave Drive project (Wilson et al. 2005).

assessment techniques along a section of Cornish coast in southwestern England. This coast has beaches backed by cliffs composed predominantly of sandstone and mudstone; these areas are important tourist destinations, so cliff instability poses a significant threat to the public on the adjacent beaches. A series of wet winters and associated storm damage appeared to have increased the incidents of cliff instability, and some sections of beach had been closed to public access because of the perceived dangers of cliff collapse.

The study methods were developed on the basis of the approach developed and used to evaluate rockfall risk along British Columbia highways (Bunce et al. 1997). Existing shoreline management plans provided qualitative hazard assessments of rockfall activity and cliff instability. A qualitative risk assessment methodology (Table 5-24) developed by Boggett et al. (2000) was applied to rank the cliff instability along a section of the coast. This initial assessment was based on the existing hazard assessment documents and limited field observations, and it identified a series of zones with moderate to high risk for rockfall.

Although useful, this methodology still did not quantify the risk posed by rockfall, and the zone classifications were somewhat arbitrary. Thus a quantitative risk assessment methodology was applied to assess the rockfall risk for a typical high-hazard cliff zone. Figure 5-39 is an influence diagram that summarizes the key factors believed to affect the likelihood of rockfall resulting in injury or loss of life to a beach user. Existing estimates of the coastal recession rate were combined with field estimates of block size to give an indication of the potential annual total number of rockfalls. These estimates were then adjusted to reflect the probability of rockfall occurrence during occupancy of the beach; this step excluded storm periods, high tides, and most of the winter months. It was decided that 10% of the total rockfall might occur during beach occupancy periods.

To assess the potential for harmful impact, an impact danger zone was defined that was 2 m (6 ft) high and extended 5 m (15 ft) from the cliff base. This zone was subdivided into a number of cells with an equivalent volume to a person's head. The likelihood of a cell being occupied by an individual at the same time as a traversing block was calculated on the basis of estimates of beach user activity and time spent in the impact danger zone. These estimates were based on beach popularity, or use, and nature of the recreational pursuit. Different assumptions are required, for example,

Table 5-24
Ranking Scores Used for Qualitative Risk Assessment of Coastal Rockfall

SCORE NUMBER	HAZARD (H)	PROBABILITY (P)[a]	RISK VALUE (R)	VULNERABILITY (V)
1	Small failure/erosion	Unlikely	Hardstanding areas not in use	Little or no effect
2	Moderate failure and occasional small falling blocks	Possible	Unoccupied building or public right-of-way (beach)	Minor damage or injury
3	Substantial failure and occasional large falling blocks	Likely	Road, footpath	Major damage or injury
4	Deep failure > 30 m and large rockfall		Major buildings	Loss of life
5	Major failure		Residential area	

NOTE: Hazard and risk values are scored from 1 to 5. Probability is scored from 1 to 3. Vulnerability is scored from 1 to 4. The hazard number (H_N) = $H \times P$. Thus it ranges between 1 (1×1) and 15 (5×3). The risk number (R_N) = $H \times P \times R \times V$. Thus it ranges between 1 ($1 \times 1 \times 1 \times 1$) and 300 ($5 \times 3 \times 5 \times 4$). Boggett et al. (2000) used the H_N and R_N values to assign sections of coastal cliffs under evaluation to one of five risk categories. The same method was used subsequently by Coggan et al. (2001).
[a]Chance of occurrence in 10 years.
SOURCE: Adapted from Boggett et al. 2000.

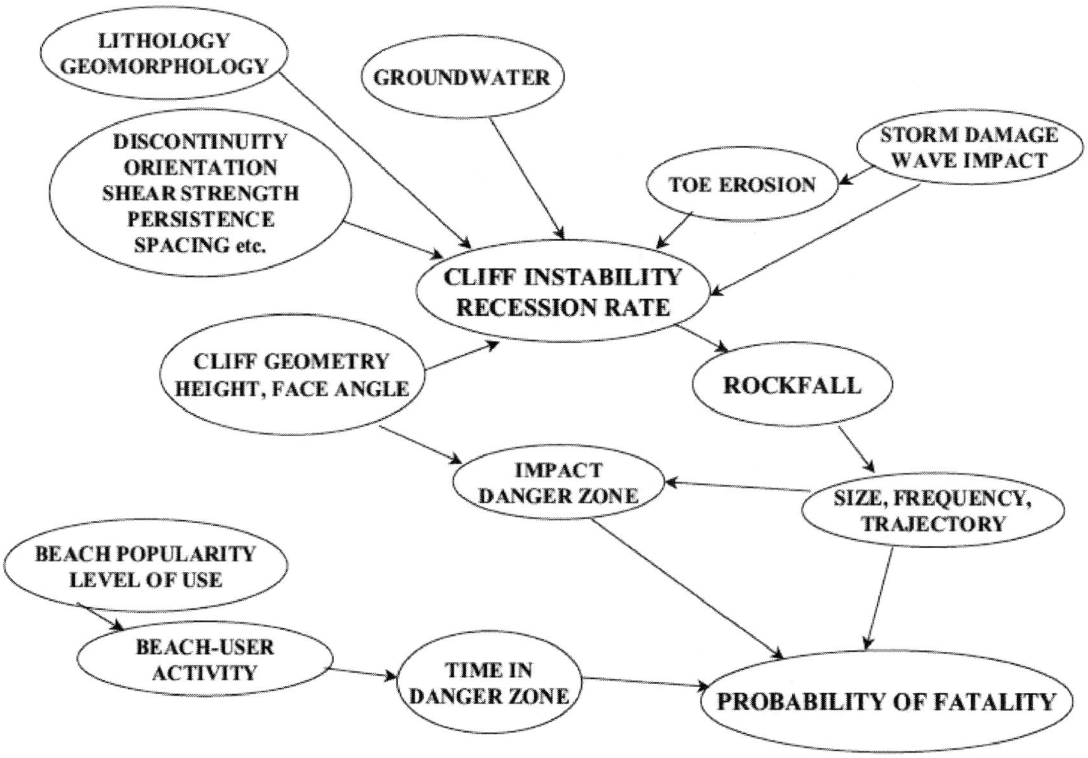

FIGURE 5-39
Influence diagram showing key factors affecting likelihood of rockfall from Cornish coastal cliffs and potential of rockfall to result in harm or loss of life to beach users (Coggan et al. 2001).

for an occasional beach user (tourist sunbathing for a few hours or an individual walking at the base of the cliff) compared with a professional fossil hunter, who may spend extended periods at the base of a cliff. To simplify the calculations, vulnerability of the individual was assumed to be unity, in that a direct hit from a rockfall would always result in loss of life, which is clearly a worst-case scenario.

From the foregoing assumptions, the annual probability of loss of life from rockfall for an occasional beach user at the site investigated was determined to be of the order of 5×10^{-6}. The annual probability of a single fatality within the whole population of beach users was determined to be of the order of 1×10^{-2}. Coggan et al. (2001) considered this value to be probably a conservative (high) estimate. Different assumptions would substantially change this value. For example, the percentage of rock volume falling within critical periods may be substantially less than 10%, and the postulated rockfall may cause injury but not death in many cases.

To assess the implications of this risk and put it into perspective, Coggan et al. compared it with published acceptance criteria. The 1×10^{-2} value for a single fatality lies at the limit of tolerable industrial risks and is relatively high for voluntary activities. Identification of a possible risk posed by rockfall to beach users raises issues as to where the responsible authority lies and whether the risk can be considered voluntary or imposed. In either case, the question remains, "Does the landowner, maritime authority, or responsible body have a duty of care to provide safe recreational facilities?" Risk management in these circumstances typically involves provision of signs informing the general public of risks, periodic inspections, and periodic scaling or other remedial measures taken to reduce the risk to more acceptable levels.

FIGURE 5-40 Argillite Cut on British Columbia Highway 99, Canada (Hoek 2000).

6.7 Rockfall at Argillite Cut, Highway 99, British Columbia, Canada

In 1982 road construction along British Columbia Highway 99, which follows the eastern margin of Howe Sound just north of Vancouver, resulted in temporary road closures and vehicle queues for periods of a half-hour or more. One queue extended through the Argillite Cut (Figure 5-40). A rock fell from the cut onto a waiting vehicle, killing a woman and disabling her father. The father successfully sued the provincial Ministry of Transportation and Highways (MOTH) for damages. In 1989, the Supreme Court of Canada found that the Ministry could readily foresee the risk that harm might befall users of the highway if it were not reasonably maintained and that maintenance could be found to extend to the prevention of injury from falling rock.

6.7.1 Initial Scoping Evaluation of Relevant Hazards

Subsequently, a quantitative risk analysis methodology was used to assess the probability of loss of life due to rockfall at the Argillite Cut under the 1982 traffic-delay patterns. The estimated risk of rockfall was compared with accepted societal risks (Bunce 1994; Bunce et al. 1997, 1998; Hungr and Beckie 1998).

The quantitative risk analysis process began with a scoping evaluation of rockfall hazards that are significant to highway users in locations and circumstances similar to those encountered along Highway 99 in 1982. Five hazards were identified:

1. A moving vehicle may be struck by a falling rock,
2. A stationary vehicle may be struck by a falling rock (the actual situation in 1982),
3. A moving vehicle may hit a previously fallen rock or rock debris that is obstructing or blocking the highway,

4. Rockfall may block a highway causing costly traffic delays and diversions, or
5. Rockfall may damage the highway infrastructure.

Although additional hazards exist for pedestrians, cyclists, or maintenance workers, these were not assessed. In fact, the study quantified the risk for only the first three hazard situations just itemized; roadway damage and blockage by rockfall were not evaluated.

6.7.2 RHRS Assessment of Argillite Cut

The frequency and magnitude of rockfall events had to be determined. In 1993 the British Columbia MOTH adopted the RHRS (Pierson et al. 1990; Pierson 1992) as the preferred method of comparing and ranking the severity of hazards of highway rock slopes. Because of the legal cases following the 1982 rockfall event, the Argillite Cut was among those sites that were assessed with the RHRS procedures. The Argillite Cut was divided into six zones for RHRS assessment, and these received scores ranging from 245 to 402 (Bunce et al. 1997). Block dimensions were 0.3 to 1.0 m (1 to 3 ft), but the major factors affecting the level of perceived hazard were the considerable height of the cut, with most zones exceeding 30 m (100 ft); relatively ineffective ditches; and poor sight distances, which were approximately 40% of the required decision sight distance for the posted speed limit (Bunce et al. 1997). The RHRS score for the entire cut was 493, higher than that for any individual zone. This score was due primarily to the extended length of the cut (480 m, or 1,575 ft) and the relatively large scores for each of the zones (Bunce et al. 1997). More recently, Highway 99 has been extensively realigned, and the deficiencies of the Argillite Cut have been removed.

6.7.3 Developing Estimates of Rockfall Frequency

Although the RHRS assessment provided valuable information concerning the degree of rockfall hazard in the Argillite Cut, more precise information concerning the frequency of rockfall events was needed. Examination of rockfall records maintained by MOTH and by Capilano Highway Services Limited (CHSL), which assumed the maintenance contract for the highway in 1988, provided information for the period 1955 to 1992. However, all such records have limitations. The CHSL records after 1988 were more complete, but it was learned that only rockfalls involving rocks with diameters larger than about 0.15 m (6 in.) were recorded, and some rockfalls had been removed by private citizens or police without being recorded. The records identified nine rockfall events during the approximately 4-year period (1988 to 1992); thus the minimum rockfall frequency is 2.2 per year.

A second approach was used to confirm this rockfall frequency estimate. Damage to a section of asphalt pavement in a section of the Argillite Cut, pavement that had been placed about 4.75 years earlier, was reviewed to identify the number, size, and distribution of rockfall impacts. Bunce et al. (1997) provided details of this effort. In summary, 84 impact marks were attributed to rockfall. Of these, 24 were outside the traffic lanes and thus were presumably not a major hazard to highway users. Several rockfalls apparently involved multiple fragments and produced multiple impact marks, which reduced the number of rockfall incidents that posed a hazard to motorists from about 60 to about 35. When these estimates were adjusted for the length of the Argillite Cut, a best estimate of rockfall frequency along the cut was 9.1 rockfall events per year, with an upper-bound (worst-case) estimate of 15.4 rockfall events per year.

However, these higher frequencies include rockfall events of all sizes. Further examination of the sizes of the rockfall impact marks on the pavement revealed an exponential distribution of sizes. This distribution was used to estimate the number and frequency of rockfall events that could harm vehicles. As a consequence, the estimated frequency of rockfalls involving fragments large enough to significantly damage vehicles was between 2.0 and 2.4 events per year for the entire Argillite Cut. These values are consistent with the 2.2 rockfall events per year obtained from the historical records.

6.7.4 Risk Calculations

Bunce et al. (1997) provided detailed explanations of their risk calculations for three situations: (*a*) a stationary vehicle is hit by a falling rock, (*b*) a moving vehicle is hit by a falling rock, or (*c*) a moving vehicle hits a previously fallen rock lying in the roadway. Each of these risks was estimated by using simplifying assumptions about the traffic and the rockfalls. The traffic was assumed to be uniformly distributed in time and space and to be independent of rockfalls. All vehicles were assumed to have the same mean length. In a similar fashion, rockfalls were assumed to be uniformly distributed in time and space along the Argillite Cut and to be independent of the traffic. These assumptions allowed the probability of a collision between a rock and a vehicle to be derived from the binomial theorem. Table 5-25 shows the risk estimates produced by Bunce et al. (1997) using these assumptions.

For example, the situation in which a line of vehicles is stopped through the cut for half an hour, each vehicle is assumed to be 5.4 m (17.7 ft) long, and the vehicles stop with a 1.5-m (5-ft) space between them may be calculated as follows:

- The temporal probability that a vehicle occupies the trajectory of a rockfall when a rock falls is the number of half-hour periods in a year, or $0.5/8{,}760 = 5.7 \times 10^{-5}$.
- The proportion of the road length occupied by vehicles is $5.4/(5.4 + 1.5) = 0.78$, which corresponds to the probability that a rock will hit a vehicle once a fall occurs.
- The spatial probability that a vehicle occupies a section of road where a rockfall occurs is computed by applying the binomial theorem; for 2.2 rockfalls per year, this value is .96.
- Thus the spatial and temporal probabilities are combined (multiplied) to produce an estimate of the annual probability that a vehicle stopped in the cut for half an hour will be hit by a rockfall; this value is $.96 \times 5.7 \times 10^{-5} = 5.5 \times 10^{-5}$, and the probability corresponds to about a 1 in 19,000 chance of occurrence.

The probability that a rock will hit an individual vehicle in a line of cars stopped in the cut corresponds to the length of a single vehicle compared with the length of the cut (which is 476 m,

Table 5-25
Probability of Accident at Argillite Cut Assuming Different Rockfall Frequencies

	Rockfall Frequency Estimate		
Type of Accident	Reported Rockfalls (2.2 rockfalls/year)	Best Estimate from Rock Impact Marks (9.1 rockfalls/year)	Upper Bound Estimate from Rock Impact Marks (15.4 rockfalls/year)
Stationary Vehicle Struck by Falling Rock			
Annual probability of impact with one or more vehicles during single 0.5-h stoppage	5.5×10^{-5}	5.7×10^{-5}	5.7×10^{-5}
Annual probability of impact with individual vehicle during single 0.5-h stoppage	1.4×10^{-6}	5.6×10^{-6}	9.1×10^{-6}
Moving Vehicle Struck by Falling Rock			
Annual probability of impact	2.9×10^{-2}	1.2×10^{-1}	1.9×10^{-1}
Probability of impact during single trip through Argillite Cut	1.7×10^{-8}	6.7×10^{-8}	1.1×10^{-7}
Moving Vehicle Hits Previously Fallen Rock			
Annual probability of impact	5.6×10^{-1}	5.6×10^{-1}	5.6×10^{-1}
Probability of impact during single trip through Argillite Cut	3.2×10^{-7}	3.2×10^{-7}	3.2×10^{-7}

Source: Bunce et al. 1997.

or 1,562 ft). This probability is thus 5.4/476 = .011. Then, by using the binomial theorem and 2.2 rockfalls per year, the spatial probability of this individual vehicle being struck by a rockfall is .025 (compared with the probability of .96 for the entire line of vehicles computed earlier). Thus, the annual probability of impact to an individual vehicle during a single half-hour stoppage is $.025 \times 5.7 \times 10^{-5} = 1.4 \times 10^{-6}$.

The computations for rock impacts with moving vehicles used similar logic. However, the computations of the annual probability of an impact with a moving vehicle require estimates of the proportion of the highway that is instantaneously occupied by a vehicle. This calculation required values for the average daily traffic volumes and assumptions that the traffic is uniformly distributed throughout the 24-h day and that vehicles travel at the posted speed. The probability of a rockfall impact during a single trip through the cut can be approximated by dividing this annual probability by the total number of trips per year.

Estimating the probabilities of an impact by a moving vehicle on a previously fallen rock involved some additional assumptions concerning (*a*) the ability of a driver to see the obstruction and avoid it and (*b*) the proportion of fallen rocks large enough to cause an accident. Bunce et al. (1997) assumed that rocks falling at a distance greater than the decision sight distance in front of the vehicle posed no accident hazard and that rocks falling immediately in front of the vehicle were certain to cause an accident. Thus, they assumed that, on average, an accident will occur if a rock falls within half the decision sight distance of the vehicle. This assumption allowed computation of the annual probability of an impact with a fallen rock and then the probability of an impact with a fallen rock during a single trip through the cut.

6.7.5 Estimating Risk of Fatalities

Bunce et al. (1997) used additional assumptions to determine the probability of fatalities due to rockfall, which are presented in Table 5-26. Because it appeared that any rock that caused damage to the pavement had sufficient energy to penetrate the external skin of a vehicle and harm the occupants, they used the probabilities derived from the best estimates of rockfall frequency from analysis of the rock impact marks (9.1 rockfalls/year), which are shown in Table 5-25. These probability values are repeated in Table 5-26.

It was also necessary to estimate the probability of a fatality given that the vehicle suffered an impact (Column 4 of Table 5-26). For stationary vehicles, this probability was estimated to be .125 on the basis of the assumptions that about 0.25

Table 5-26
Probability of Fatality at Argillite Cut

Type of Accident	Impact Probability (from Table 5-25): Best Estimate Annual Probability	Impact Probability (from Table 5-25): Best Estimate Single-Trip Probability	Assumed Probability of Fatality Once Accident Occurs	Annual Probability of Fatality	Probability of Fatality per Single Trip	Probability of Fatality per 500 Trips
Vehicle stationary for 0.5 h, hit by falling rock	5.7×10^{-5}	5.6×10^{-5}	.125	7.1×10^{-6}	7.0×10^{-6}	—
Moving vehicle hit by falling rock	1.2×10^{-1}	6.7×10^{-8}	.2	2.4×10^{-2}	1.3×10^{-8}	6.7×10^{-6}
Moving vehicle hits previously fallen rock	5.6×10^{-1}	3.2×10^{-7}	.1	5.6×10^{-2}	3.2×10^{-8}	1.6×10^{-5}
Sum of moving vehicle hit by falling rock and hitting fallen rock	—	—	—	7.9×10^{-2}	4.5×10^{-8}	2.3×10^{-5}

NOTE: — = value not computed.
SOURCE: Adapted from Bunce et al. 1997.

of the length of a vehicle is occupied by passengers and that no more than one-half the rockfall impacts result in a fatality. For the case of a moving vehicle hit by a falling rock, a probability of .2 (or 1 in 5) was assumed. This higher value was selected to take into account the potential detrimental effects of a rock impact on the performance of a vehicle, such as the driver losing control and crashing. For the case of a moving vehicle encountering a fallen rock, a probability of .1 (1 in 10) was assumed, thereby allowing for situations in which the driver was able to reduce the impact severity and thus reduce the potential of a fatality.

6.7.6 Summary

Bunce et al. (1997) noted that their estimated probability of one or more fatalities is nearly twice the value actually experienced (one fatality in 26 years) at the Argillite Cut. However, they also reported that when the entire length of Highway 99 is considered, their estimates fall below the actual recorded events.

Bunce et al. (1997) compared their estimates of fatalities with commonly accepted societal standards for the annual probability of death of an individual; they concluded that acceptable values of annual probability for individual fatalities are in the range of 1×10^{-4} to 1×10^{-5}. Such values placed the estimated probabilities of fatalities involving rockfall and moving vehicles in the Argillite Cut at the limits of acceptable risk or even in the marginally acceptable range.

Although the analysis suffered from uncertainties concerning the rockfall history at the Argillite Cut and from unknown relationships between rockfall impacts and injuries and fatalities, the procedure provided a useful application of quantitative risk analysis of rockfall and demonstrated its place within a rockfall management process.

FIGURE 5-41 Main road along west side of Lake Garda, Italy, heading toward Riva; narrow two-lane road serves important tourist destinations by following base of steep rock slopes up to 1,000 m (3,300 ft) high (http://www.panoramio.com/photo/3561360). (PHOTOGRAPH COURTESY OF YORY, PANORAMIO.)

6.8 Establishing Risk-Based Safety Criteria for Lake Garda Road, Northern Italy

La Gardesana is a narrow two-lane roadway that serves an important tourist area in northern Italy. Extending along the western shore of Lake Garda, it lies at the base of steep rock slopes up to 1,000 m (3,300 ft) high (Figure 5-41). For many years, numerous rockfalls have occurred on this road, occasionally resulting in fatalities. Rock protection sheds have been built over some of the roadway, and various rockfall mitigation measures have been installed above some of the remaining 2.4 km (1.5 mi) exposed to rockfall hazards. Despite these efforts, a rockfall event in 2001 resulted in a fatality, and the road was closed for safety concerns. The provincial authorities sought advice concerning appropriate mitigation actions so that the road could be safely reopened as quickly as possible, thereby minimizing economic disruption of the tourist industry.

The response to this situation was based on a quantitative baseline risk assessment procedure that has been reported by Roberds (2001, 2005). Roberds (2005) included detailed computations for this vulnerability assessment as a 10-page appendix.

The quantitative baseline risk assessment procedure first evaluated the existing rockfall conditions and assumed unrestricted traffic flow. This assessment was performed independently for each of 13 exposed roadway sections. The results were then combined to establish the risk for the entire roadway. For each section, the hazard (i.e., frequency that various magnitudes of debris would land and then remain in the road) was based on estimates of

1. The average frequency of various sizes of detachments for each of three zones above the road (these were based on geology, stabilization

effectiveness, and historical evidence specifically for that road section), and
2. The likelihood that various debris volumes would land and then remain in the roadway for each detachment volume and height (these were based on topography, barrier effectiveness, and historical evidence specifically for that road section).

Figure 5-42 provides a diagrammatic overview of these hazard scenarios. First the sizes of potential detached blocks (M_D) from the source areas were estimated. Because it was observed that many blocks became smaller as they traveled down the slopes, the sizes of the blocks landing in the roadway (M_L) were also estimated. It was also observed that some blocks did not reach the road because of intermediate ledges and some rockfall protection barriers. This outcome was assigned a probability $P1$. A second probability ($P2$) defined the likelihood that a block would land in the roadway. It was also observed that at some locations, some blocks traveled across the road and then continued farther down the slope. Thus a third probability ($P3$) defined the likelihood that a block would remain in the roadway. Because the sizes of the blocks remaining in the roadway were different from the sizes of all blocks landing in the roadway, their sizes were defined by a third size parameter (M_R).

The quantitative baseline risk assessment procedure then addressed the vulnerability (i.e., number of casualties for each magnitude of debris that lands or remains in the roadway) for each road section on the basis of

1. The likelihood of each of six types of accidents caused by debris of a specific magnitude that lands or remains in the roadway (these evaluations were based on roadway and traffic conditions such as sight and braking distance and frequency of different types of vehicles), and
2. The expected value of the number of casualties associated with each type of accident (these values were determined by evaluating roadway and traffic conditions, including the average number of occupants per vehicle, vehicle protection, and impact velocity for different types of vehicles).

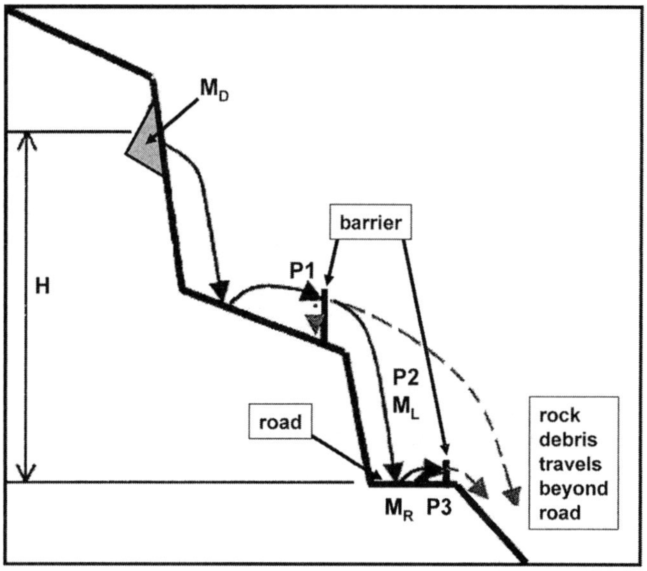

FIGURE 5-42 Parameters used to establish rockfall hazard scenarios (Golder Associates 2002).

The vulnerability was thus computed by using the accident scenarios illustrated in Figure 5-43 and summarized in Table 5-27. These scenarios assessed the outcomes for a number of alternative accident situations: (*a*) a rockfall directly hitting a vehicle, (*b*) a vehicle stopping before hitting a rock lying in the roadway, (*c*) a vehicle directly hitting rockfall debris remaining in the roadway, or (*d*) vehicle accidents due to several situations,

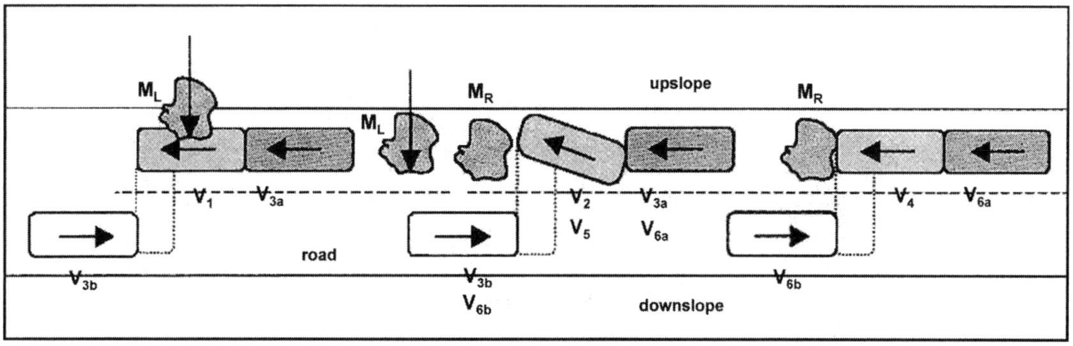

FIGURE 5-43
Parameters used to establish accident scenarios (Roberds 2005).

Table 5-27
Definition of Accident Scenarios

Accident Scenario Identifier[a]	Description of Accident
V1	Falling rock directly hits vehicle.
V2	Vehicle stops before hitting rockfall debris in roadway.
V3a	Second following vehicle rear-ends either Vehicle V1 or Vehicle V2.
V3b	Oncoming vehicle hits either Vehicle V1 or Vehicle V5.
V4	Vehicle hits rockfall debris lying in roadway.
V5	Vehicle avoids rockfall debris in roadway but stops in oncoming lane.
V6a	Second following vehicle rear-ends Vehicle V4.
V6b	Oncoming vehicle hits Vehicle V4.

[a]Identifiers and accident scenarios are shown in Figure 5-43.
Source: Roberds 2005.

including subsequent vehicles hitting stopped or damaged vehicles, vehicles crossing into opposing lanes, or accidents resulting from distractions due to rockfall.

For each of the accident scenarios, vulnerability regarding casualties was assessed in terms of

1. The likelihood that falling debris of a specific magnitude would hit a vehicle (a bus, an automobile, a motorcycle, or a truck) or cause an accident (because of driver distraction or because a vehicle runs into a previous accident) and in either case, the likely number of casualties; or
2. The likelihood that debris of a specific magnitude would remain in the roadway, either being run into by a vehicle or causing an accident (because of driver avoidance or because a vehicle runs into a previous accident) and in either case, the likely number of casualties.

These assessments were based on estimated vehicle and road characteristics for each section of the road (e.g., reaction times, braking coefficients, sight distances at various times, vehicle speed, number of vehicles per hour for various times, number of lanes, and number of vehicle occupants and their protection).

The quantitative baseline risk (i.e., the expected number of casualties per year) for each specific road section was determined by mathematically combining the hazards and the vulnerability values for that road section, as shown in Figure 5-44. The risk for the entire roadway (i.e., the expected number of casualties per year) was then determined by summing the risks for each of the 13 road sections.

Consistent with historical data, the baseline risk assessment indicated that the risk associated with reopening the road without additional mitigation was too high compared with typical public safety criteria, such as those used in Hong Kong (Geotechnical Engineering Office 1998). Therefore, various potential mitigation activities were identified and evaluated, including traffic restrictions and installation of additional debris retention and slope stabilization measures. Nine combinations of activities were evaluated by

- Assessing the expected changes in the various hazard or vulnerability factors if those activities were carried out, and then
- Determining the residual risk after those changes were implemented.

Figure 5-45 illustrates the estimated risk levels following implementation of each mitigation option along with the original baseline risk estimate for the unmitigated conditions (status quo). It was found that the risk could be reduced to an acceptable level by one specific combination of activities, which could be implemented relatively quickly. These activities consisted primarily of (*a*) traffic restrictions and control and (*b*) slope or roadway inspections, especially during and immediately after inclement weather.

6.9 Quantitative Assessment of Residual Risk in Rockfall Protected Area, Principality of Andorra

The principality of Andorra is a small country of 468 km^2 (182 mi^2) located in the heart of

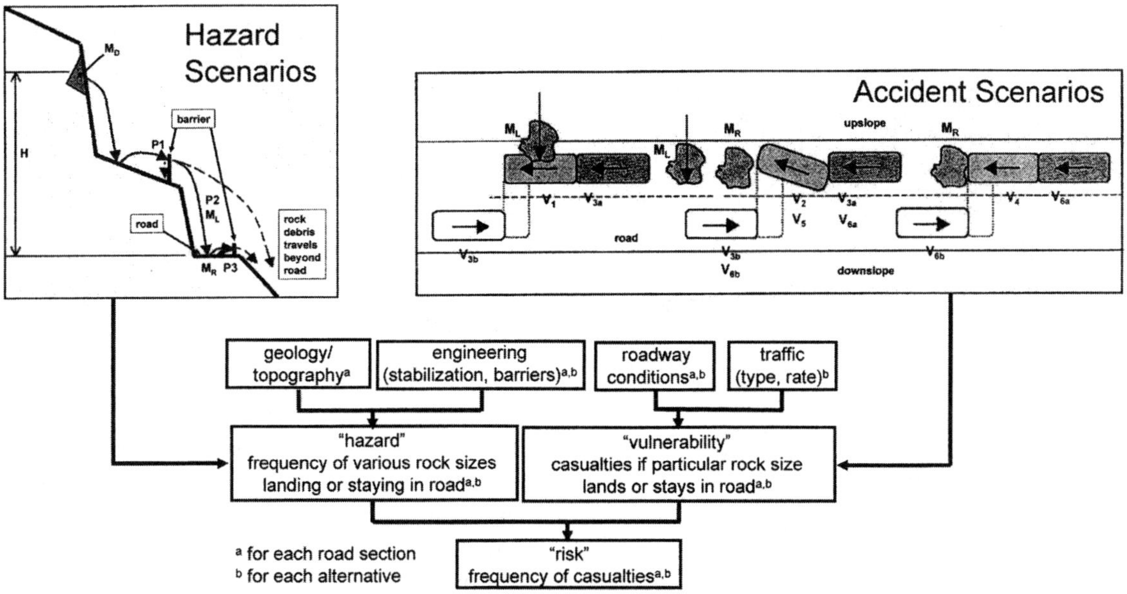

FIGURE 5-44 Quantitative baseline assessment of risk to road users from rockfall determined by combining results of hazard scenarios with those of accident scenarios (Golder Associates 2002).

the Pyrenees between France and Spain. It is a mountainous country and is drained by two main rivers, the Valira del Nord and Valira d'Orient, which merge at the capital Andorra la Vella to form the Gran Valira River, which flows southward to the Spanish border. Glaciation during the Pleistocene period affected the Valira valleys and steepened their side slopes. At Andorra la Vella, the northern side of the valley is formed by the Carroi massif, composed of granodiorite rocks, with the slope of the Solà d'Andorra located below it. This slope experiences frequent rockfall activity due to a combination of triggering mechanisms such as stress relief, freeze–thaw weathering, wedging by tree roots, and elevated water pressures in fractures after rainfall. The occurrence of rockfall events triggered by earthquakes also cannot be disregarded. Repeated rockfalls have produced fan-shaped talus cones at the foot of the rock cliffs (Figure 5-46).

Since the 1970s, the Andorran Principality has experienced rapid urban growth, especially along the Solà d'Andorra, where the capital and its neighborhoods of La Margineda and Santa Coloma are located. Some new developments have extended into areas exposed to rockfalls. In December 1983 and January 1994, falling rocks damaged several buildings, but the most severe event occurred on January 21, 1997, when a 25-m^3 (800-ft^3) block hit a building and two cottages in Santa Coloma, causing considerable damage and injury to one person. Following this event, local authorities and the Ministry of Public Works decided to implement several risk mitigation measures, including land use zoning and restrictions on development, construction of protective rockfall fences, and establishment of a surveillance plan (Copons et al. 2005).

Corominas et al. (2005) used quantitative risk assessment methods to evaluate the rockfall risk at the Solà d'Andorra slope before and after the implementation of these risk mitigation works, in particular the construction of two protective

FIGURE 5-45 Mitigated risk levels for nine evaluated mitigation options compared with risk without mitigation (status quo) (Golder Associates 2002).

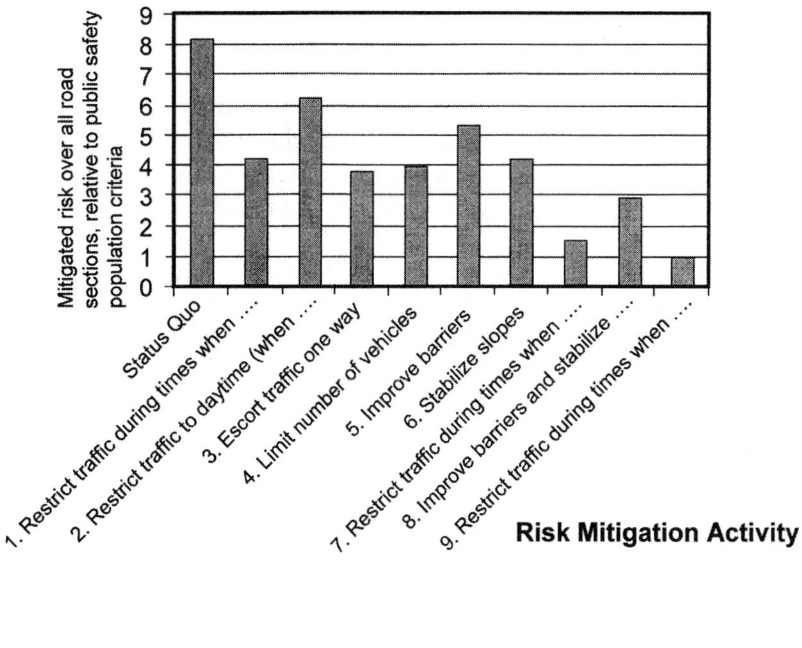

FIGURE 5-46 General view of Solà d'Andorra slope including steep slopes of Carroi massif, source of many rockfalls, and at-risk communities located on talus cones below (Corominas et al. 2005).

fences. The risk levels were calculated following a staged process that

- Identified the potential rockfall release areas,
- Determined the volume distribution of the falling rocks,
- Estimated the frequency of the rockfall events, and
- Evaluated the trajectories of falling blocks using a 3-D simulation model that provided both the expected travel distances and the kinetic energy of the blocks.

The level of risk to both persons and property in the developed area located at the foot of the rock cliff was calculated by taking into account the nature of the exposed elements and their vulnerability (Figure 5-47).

The most dangerous section of the Solà d'Andorra has now been protected by the construction of two lines of rockfall catchment fences. Corominas et al. calculated that placement of these fences has reduced individual risk for the most exposed person inside the buildings from the intolerable level of 3.8 $\times 10^{-4}$ fatalities per year before installation of the fences to an acceptable level of 9.1 $\times 10^{-7}$ fatalities per year after mitigation. Corominas et al. also determined that the fences reduced the societal risk from 1.5 $\times 10^{-2}$ fatalities per year to the marginally acceptable level of 1.2 $\times 10^{-5}$ fatalities per year. They also noted that the continued viability of these fences requires continuous monitoring and maintenance.

FIGURE 5-47 Geometrical assumptions used to define spatial probabilities of rockfall impacts on (*a*) buildings and (*b*) individuals within buildings (Corominas et al. 2005).

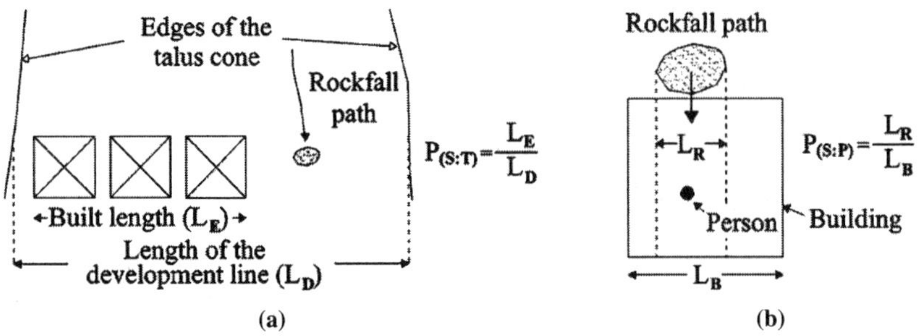

ACKNOWLEDGMENTS

The author thanks Alan Moon, Fred Baynes, and Bill Roberds for several interactions and discussions about risk assessment and risk management procedures. Their guidance and willingness to share published sources have made it possible to develop this chapter.

REFERENCES

ABBREVIATIONS

AGS	Australian Geomechanics Society
IUGS	International Union of Geological Sciences

Agliardi, F., and G. Crosta. 2003. High Resolution Three-Dimensional Numerical Modelling of Rockfalls. *International Journal of Rock Mechanics and Mining Sciences*, Vol. 40, pp. 455–471.

AGS. 2000. Landslide Risk Management Concepts and Guidelines. *Australian Geomechanics*, Vol. 35, No. 1, pp. 49–92.

AGS. 2002. Landslide Risk Management Concepts and Guidelines. *Australian Geomechanics*, Vol. 37, No. 2, pp. 1–44.

AGS. 2007a. Australian GeoGuides for Slope Management and Maintenance. *Australian Geomechanics*, Vol. 42, No. 1, pp. 159–182.

AGS. 2007b. Guideline for Landslide Susceptibility, Hazard and Risk Zoning for Land Use Planning. *Australian Geomechanics*, Vol. 42, No. 1, pp. 13–36.

AGS. 2007c. Practice Note Guidelines for Landslide Risk Management. *Australian Geomechanics*, Vol. 42, No. 1, pp. 63–114.

Baynes, F. 1997. Problems Associated with Geological Characterisation for Quantitative Landslide Risk Assessment. In *Landslide Risk Assessment: Proceedings of the International Workshop on Landslide Risk Assessment*, Honolulu, Hawaii, Feb. 19–21 (D. Cruden and R. Fell, eds.), Balkema, Rotterdam, Netherlands, pp. 153–164.

Baynes, F. J., I. K. Lee, and I. E. Stewart. 2002. A Study of the Accuracy and Precision of Some Landslide Risk Analyses. *Australian Geomechanics*, Vol. 37, No. 2, pp. 149–155.

Benjamin, J. R., and C. A. Cornell. 1970. *Probability, Statistics and Decision for Civil Engineers*. McGraw-Hill, New York.

Boggett, A. D., N. J. Mapplebeck, and R. J. Cullen. 2000. South Shore Cliffs, Whitehaven—Geomorphological Survey and Emergency Cliff Stabilization Works. *Quarterly Journal of Engineering Geology and Hydrogeology*, Vol. 33, pp. 213–226.

Brawner, C. O. 1998. Case Examples of Rock Stability on Transportation Projects. In *Proc., Forty-Ninth Highway Geology Symposium*, Prescott, Ariz., pp. 26–38. http://www.highwaygeologysymposium.org/default.asp.

Brawner, C. O., and D. C. Wyllie. 1975. Rock Slope Stability on Railway Projects. In *Proceedings of the American Railway Engineering Association Regional Meeting*, Vancouver, Canada.

Bunce, C. M. 1994. *Risk Analysis for Rock Fall on Highways*. MSc thesis. University of Alberta, Edmonton, Canada.

Bunce, C. M., D. M. Cruden, and N. R. Morgenstern. 1997. Assessment of the Hazard from Rock Fall on a Highway. *Canadian Geotechnical Journal*, Vol. 34, pp. 344–356.

Bunce, C. M., D. M. Cruden, and N. R. Morgenstern. 1998. Assessment of the Hazard from Rock Fall on a Highway: Reply. *Canadian Geotechnical Journal*, Vol. 35, p. 410.

Carrara, A. 1983. Multivariate Models of Landslide Hazard Evaluation. *Mathematical Geology*, Vol. 15, pp. 403–427.

Coggan, J. S., R. J. Pine, and D. Stead. 2001. A Proposed Methodology for Rockfall Risk Assessment Along Coastlines. *Geoscience in South-West England, Proceedings of the Ussher Society*, Vol. 10, No. 2, pp. 190–194.

Copons, R., J. M. Vilaplana, J. Corominas, J. Altimir, and J. Amigó. 2005. Rockfall Risk Management in High Density Urban Areas: The Andorran Experience. In *Landslide Hazard and Risk* (T. Glade, M. Anderson, and M. J. Crozier, eds.), Wiley, New York, pp. 675–698.

Corominas, J., R. Copons, J. Moya, J. M. Vilaplana, J. Altimir, and J. Amigó. 2005. Quantitative Assessment of the Residual Risk in a Rockfall Protected Area. *Landslides*, Vol. 2, pp. 343–357.

Crosta, G. B., and F. Agliardi. 2003. A Methodology for Physically Based Rockfall Hazard Assessment. *Natural Hazards and Earth System Sciences*, Vol. 3, pp. 407–422.

Cruden, D. M. 1997. Estimating the Risks from Landslides Using Historical Data. In *Landslide Risk Assessment: Proceedings of the International Workshop on Landslide Risk Assessment*, Honolulu, Hawaii, Feb. 19–21 (D. Cruden and R. Fell, eds.), Balkema, Rotterdam, Netherlands, pp. 177–184.

Cruden, D., and R. Fell (eds.). 1997. *Landslide Risk Assessment: Proceedings of the International Workshop on Landslide Risk Assessment*, Honolulu, Hawaii, Feb. 19–21, Balkema, Rotterdam, Netherlands.

Davidson, R., S. McGrath, A. Bowden, and A. Reynolds. 2002. Strategic Risk Assessment for Thirteen Victorian Dams. In *Proc., 2002 ANCOLD Conference on Dams: Future Directions in Demanding Environments*, Adelaide, South Australia (T. M. Daniell and A. M. Parsons, eds.), Australian National Committee on Large Dams Incorporated (ANCOLD Inc.).

Davis, J. C. 2002. *Statistics and Data Analysis in Geology*, 3rd ed. John Wiley and Sons, New York, 638 pp.

Dussauge-Peisser, C., A. Helmstetter, J-R. Grasso, D. Hantz, P. Desvarreux, M. Jeannin, and A. Giraud. 2002. Probabilistic Approach to Rock Fall Hazard Assessment: Potential of Historical Data Analysis. *Natural Hazards and Earth System Sciences*, Vol. 2, pp. 15–26.

Düzgün, H. S. B., and S. Lacasse. 2005. Vulnerability and Acceptable Risk in Integrated Risk Assessment Framework. In *Landslide Risk Management: Proceedings of the International Conference on Landslide Risk Management* (O. Hungr, R. Fell, R. Couture, and E. Eberhardt, eds.), Vancouver, Canada, May 31–June 3, Taylor & Francis, London, pp. 505–516.

Einstein, H. H. 1988. Special Lecture: Landslide Risk Assessment Procedure. In *Proc., Fifth International Symposium on Landslides*, Lausanne, Switzerland (C. Bonnard, ed.), Balkema, Rotterdam, Netherlands, Vol. 2, pp. 1075–1090.

Einstein, H. H. 1997. Landslide Risk: Systematic Approaches to Assessment and Management. In *Landslide Risk Assessment: Proceedings of the International Workshop on Landslide Risk Assessment*, Honolulu, Hawaii, Feb. 19–21 (D. Cruden and R. Fell, eds.), Balkema, Rotterdam, Netherlands, pp. 25–50.

Fell, R. 1994. Landslide Risk Assessment and Acceptable Risk. *Canadian Geotechnical Journal*, Vol. 31, No. 2, pp. 261–272.

Fell, R., and D. Hartford. 1997. Landslide Risk Management. In *Landslide Risk Assessment: Proceedings of the International Workshop on Landslide Risk Assessment*, Honolulu, Hawaii, Feb. 19–21 (D. Cruden and R. Fell, eds.), Balkema, Rotterdam, Netherlands, pp. 51–110.

Fell, R., K. K. S. Ho, S. Lacasse, and E. Leroi. 2005. A Framework for Landslide Risk Assessment and Management. In *Landslide Risk Management: Proceedings of the International Conference on Landslide Risk Management* (O. Hungr, R. Fell, R. Couture, and E. Eberhardt, eds.), Vancouver, Canada, May 31–June 3, Taylor & Francis, London, pp. 3–26.

Flentje, P., D. Stirling, and R. Chowdhury. 2007. Landslide Susceptibility and Hazard Derived from a Landslide Inventory Using Data Mining—An Australian Case Study. In *Proceedings of the First North American Landslide Conference*, Vail, Colo. (V. R. Schaefer, R. L. Schuster, and A. K. Turner, eds.), Association of Engineering Geologists Special Publication 23, pp. 284–293.

Geotechnical Engineering Office. 1998. *Landslides and Boulder Falls from Natural Terrain: Interim Risk Guidelines*. GEO Report No. 75. Government of Hong Kong, Special Administrative Region.

Glade, T., M. Anderson, and M. J. Crozier (eds.). 2005. *Landslide Hazard and Risk*. J. Wiley and Sons, Ltd., Chichester, UK.

Golder Associates. 2002. Brochure.

Guzzetti, F., and P. Reichenbach. 2010. Rockfalls and Their Hazard. In *Tree Rings and Natural Hazards* (M. Stoffel, M. Bollschweiler, D. R. Butler, and B. H. Luckman, eds.), Advances in Global Change Research, Vol. 41, Springer, pp. 129–137.

Guzzetti, F., M. Cardinali, and P. Reichenbach. 1999. The AVI Project: A Bibliographical and Archive Inventory of Landslides and Floods in Italy. *Environmental Management*, Vol. 18, No. 4, pp. 623–633.

Guzzetti, F., G. Crosta, R. Detti, and F. Agliardi. 2002. STONE: A Computer Program for the Three-Dimensional Simulation of Rock-falls. *Computers and Geosciences*, Vol. 28, pp. 1079–1093.

Guzzetti, F., P. Reichenbach, and S. Ghigi. 2004. Rockfall Hazard and Risk Assessment Along a Transportation Corridor in the Nera Valley, Central Italy. *Environmental Management*, Vol. 34, No. 2, pp. 191–208.

Hantz, D., C. Dussauge-Peisser, M. Jeannin, and J.-M. Vengeon. 2003. Rock Fall Hazard Assessment: From Qualitative to Quantitative Failure Probability. In *Proc., International Conference on Fast Slope Movements Prediction and Prevention for Risk Mitigation*, Naples, Italy, Associazione Geotechnica Italiana, Rome, pp. 263–267.

Hendrickx, M., R. Wilson, A. Moon, I. Stewart, and P. Flentje. 2005. Slope Hazard Assessment on a Coast Road in New South Wales, Australia. In *Landslide Risk Management: Proceedings of the International Conference on Landslide Risk Management* (O. Hungr, R. Fell, R. Couture, and E. Eberhardt, eds.), Vancouver, Canada, May 31–June 3, Taylor & Francis, London. CD-ROM.

Hoek, E. 2000. *Practical Rock Engineering* (course notes). http://www.rocscience.com/hoek/PracticalRockEngineering.asp.

Hungr, O., and R. D. Beckie. 1998. Assessment of the Hazard from Rock Fall on a Highway: Discussion. *Canadian Geotechnical Journal*, Vol. 35, p. 409.

Hungr, O., S. G. Evans, and J. Hazzard. 1999. Magnitude and Frequency of Rock Falls Along the Main Transportation Corridors of Southwestern British Columbia. *Canadian Geotechnical Journal*, Vol. 36, pp. 224–238.

Hungr, O., R. Fell, R. Couture, and E. Eberthardt (eds.). 2005. *Landslide Risk Management: Proceedings of the International Conference on Landslide Risk Management*, Vancouver, Canada, May 31–June 3, Taylor & Francis, London.

IUGS. 1997. Quantitative Risk Assessment for Slopes and Landslides: The State of the Art. In *Landslide Risk Assessment: Proceedings of the International Workshop on Landslide Risk Assessment*, Honolulu, Hawaii, Feb. 19–21 (D. Cruden and R. Fell, eds.), Balkema, Rotterdam, Netherlands, pp. 3–12.

Jaboyedoff, M., J. P. Dudt, and V. Labiouse. 2005. An Attempt to Refine Rockfall Hazard Zoning Based on the Kinetic Energy, Frequency and Fragmentation Degree. *Natural Hazards and Earth System Sciences*, Vol. 5, pp. 621–632.

Lateltin, O., C. Haemmig, H. Raetzo, and C. Bonnard. 2005. Landslide Risk Management in Switzerland. *Landslides*, Vol. 2, pp. 313–320.

Lee, E. M. 1997. Landslide Risk Management: Key Issues from a British Perspective. In *Landslide Risk Assessment: Proceedings of the International Workshop on Landslide Risk Assessment*, Honolulu, Hawaii, Feb. 19–21 (D. Cruden and R. Fell, eds.), Balkema, Rotterdam, Netherlands, pp. 227–238.

Lee, E. M., and D. K. C. Jones. 2004. *Landslide Risk Assessment*. Thomas Telford, London.

Michael-Leiba, M., F. Baynes, and G. Scott. 2000. Quantitative Landslide Risk Assessment of Cairns, Australia. In *Landslides: Research, Theory, and Practice* (E. N. Bromhead, N. Dixon, and M.-L. Ibsen, eds.), Thomas Telford, London, pp. 1059–1064.

Moon, A. T. 1997. Predicting Low Probability Rapid Landslides at Roxburgh Gorge, New Zealand. In *Landslide Risk Assessment: Proceedings of the International Workshop on Landslide Risk Assessment*, Honolulu, Hawaii, Feb. 19–21 (D. Cruden and R. Fell, eds.), Balkema, Rotterdam, Netherlands, pp. 227–284.

Moon, A. T., M. D. Robertson, and W. N. Davies. 1996. Quantifying Rockfall Risk Using a Probabilistic Toppling Failure Model. In *Proc., Seventh International Symposium on Landslides*, Trondheim, Norway (K. Senneset, ed.), Balkema, Rotterdam, Netherlands, Vol. 2, pp. 1311–1316.

Moon, A. T., R. A. Wilson, and P. N. Flentje. 2005. Developing and Using Landslide Frequency Models. In *Landslide Risk Management: Proceedings of the International Conference on Landslide Risk Management* (O. Hungr, R. Fell, R. Couture, and E. Eberhardt, eds.), Vancouver, Canada, May 31–June 3, Taylor & Francis, London, pp. 681–690.

Morgenstern, N. R. 1997. Toward Landslide Risk Assessment in Practice. In *Landslide Risk Assessment: Proceedings of the International Workshop on Landslide Risk Assessment*, Honolulu, Hawaii, Feb. 19–21 (D. Cruden and R. Fell, eds.), Balkema, Rotterdam, Netherlands, pp. 15–24.

Nadim, F., H. Einstein, and W. Roberds. 2005. Probabilistic Stability Analysis for Individual Slopes in Soil and Rock. In *Landslide Risk Management: Proceedings of the International Conference on Landslide Risk Management* (O. Hungr, R. Fell, R. Couture, and E. Eberhardt, eds.), Vancouver, Canada, May 31–June 3, Taylor & Francis, London, pp. 63–98.

Navidi, W. C. 2006. *Statistics for Engineers and Scientists*. McGraw-Hill, New York.

Oracle Corporation. 2010. Oracle Crystal Ball Enterprise Performance Management: Oracle Data Sheet. http://www.oracle.com/us/products/middleware/bus-int/crystalball/oracle-crystal-ball-epm-ds-068276.pdf. Accessed December 2010.

Palisade Corporation. 2010. *@RISK: Risk Analysis Software Using Monte Carlo Simulation for Excel*. http://www.palisade.com/risk. Accessed December 2010.

Peila, D., and C. Guardini. 2008. Use of the Event Tree to Assess the Risk Reduction Obtained from Rockfall Protection Devices. *Natural Hazards and Earth System Sciences*, Vol. 8, pp. 1441–1450.

Pierson, L. A. 1992. Rockfall Hazard Rating System. In *Transportation Research Record 1343*, Transportation Research Board, National Research Council, Washington, D.C., pp. 6–13.

Pierson, L. A., S. Davis, and R. VanVickle. 1989. *The Rockfall Hazard Rating System Implementation Manual*. Report FHWA-OR-EG-90-01. Federal Highway Administration, U.S. Department of Transportation, 119 pp.

Raetzo, H., O. Lateltin, D. Bollinger, and J. P. Tripet. 2002. Hazard Assessment in Switzerland: Codes of Practice for Mass Movements. *Bulletin of Engineering Geology and the Environment*, Vol. 61, pp. 263–268.

Roberds, W. 2001. Quantitative Landslide Risk Assessment and Management. In *International Conference on Landslides: Causes, Impacts and Countermeasures*, June, Davos, Switzerland (M. Kühne, H. H. Einstein, E. Krauter, H. Kapperich, and R. Pöttler, eds.), VGE, Essen, Germany, pp. 585–595.

Roberds, W. 2005. Estimating Temporal and Spatial Variability and Vulnerability. In *Landslide Risk Management: Proceedings of the International Conference on Landslide Risk Management* (O. Hungr, R. Fell, R. Couture, and E. Eberhardt, eds.), Vancouver, Canada, May 31–June 3, Taylor & Francis, London, pp. 129–157.

Roberds, W. J., K. Ho, and K. W. Leung. 1997. An Integrated Methodology for Risk Assessment and Risk Management for Development Below Potential Natural Terrain Landslides. In *Landslide Risk Assessment: Proceedings of the International Workshop on Landslide Risk Assessment*, Honolulu, Hawaii, Feb. 19–21 (D. Cruden and R. Fell, eds.), Balkema, Rotterdam, Netherlands, pp. 333–346.

Stewart, I. E., and H. G. Buys. 2005. Managing Slope Risk for a Large Highway Network. In *Landslide Risk Management: Proceedings of the International Conference on Landslide Risk Management* (O. Hungr, R. Fell, R. Couture, and E. Eberhardt, eds.), Vancouver, Canada, May 31–June 3, Taylor & Francis, London, pp. 599–607.

Stewart, I. E., F. J. Baynes, and I. K. Lee. 2002. The RTA Guide to Slope Risk Analysis Version 3.1. *Australian Geomechanics*, Vol. 37, No. 2, pp. 115–147.

Varnes, D. J. 1984. *Landslide Hazard Zonation: A Review of Principles and Practice*. Engineering Geology Commission on Landslides and Other Mass Movements on Slopes, UNESCO, Paris, Natural Hazards Series, No. 3.

Westgate, B. M., J. S. Coggan, and R. J. Pine. 2003. Development of a Risk-based Approach to

Coastal Slope Instability Assessment in Cornwall. *Geoscience in South-West England, Proceedings of the Ussher Society*, Vol. 10, No. 4, pp. 390–396.

Wieczorek, G. F., and J. B. Snyder. 2004. *Historical Rock Falls in Yosemite National Park, California*. U.S. Geological Survey Open-File Report 03-491.

Wieczorek, G. F., J. B. Snyder, C. S. Alger, and K. A. Isaacson. 1992. *Rock Falls in Yosemite Valley, California*. U.S. Geological Survey Open-File Report 92-387.

Wilson, R. A., A. T. Moon, M. Hendrickx, and I. E. Stewart. 2005. Application of Quantitative Risk Assessment to the Lawrence Hargrave Drive Project, New South Wales, Australia. In *Landslide Risk Management: Proceedings of the International Conference on Landslide Risk Management* (O. Hungr, R. Fell, R. Couture, and E. Eberhardt, eds.), Vancouver, Canada, May 31–June 3, Taylor & Francis, London, pp. 589–598.

Wong, H. N. 2005. Landslide Risk Assessment for Individual Facilities. In *Landslide Risk Management: Proceedings of the International Conference on Landslide Risk Management* (O. Hungr, R. Fell, R. Couture, and E. Eberhardt, eds.), Vancouver, Canada, May 31–June 3, Taylor & Francis, London, pp. 237–298.

Wu, T. H., W. H. Tang, and H. H. Einstein. 1996. Landslide Hazard and Risk Assessment. In *Special Report 247: Landslides: Investigation and Mitigation* (A. K. Turner and R. L. Schuster, eds.), Transportation Research Board, National Research Council, Washington, D.C., Chap. 6, pp. 106–118.

Part 2

FUNDAMENTALS OF ROCKFALL ANALYSIS AND INVESTIGATION

Chapter 6

Jerry D. Higgins and
Richard D. Andrew

SITE CHARACTERIZATION

1. INTRODUCTION

Site characterization for rockfall analysis and mitigation relies heavily on accurate identification of hazard areas and on the definition of slope conditions by a well-designed engineering geological investigation. The characterization may include assessment of factors such as terrain characteristics, geological and geotechnical properties of earth materials, rock mass structure, groundwater conditions, and past and present geological processes. Investigation findings are used to construct a four-dimensional conceptual model, in which three spatial dimensions are combined with time as the fourth dimension to analyze the interactions of slope properties and processes. The conceptual model forms the basis for assessment of the hazard potential and analysis of rockfall behavior, and these assessments provide the basis for designing mitigation procedures.

An adequate conceptual geological model for a slope of limited area and uniform gradient may be developed from an investigation consisting of a relatively quick and simple reconnaissance study and prediction of rockfall behavior from back-analysis of past events or by empirical methods. At the other extreme, a slope of large area, irregular profile, and frequent rockfall events that presents high risk to a transportation facility and the public often requires a much more extensive investigation that includes detailed fieldwork and mapping and, possibly, drilling to obtain subsurface information and samples behind the slope face. This information is then used to evaluate the rockfall potential at the site, which may include computer modeling to determine rockfall behavior along the slope.

Essential components of a rockfall site characterization include identification of likely rockfall hazard areas along roadways and, within those hazard areas, identification of rockfall source zones, determination of the relative susceptibility of geological materials in source zones to rockfall initiation, and identification and description of rockfall travel paths and runout zones (Figure 6-1) (Mazzoccola and Sciesa 2000; Dorren 2003; Guzzetti et al. 2003; Coe et al. 2005). For convenience of discussion, these zones are defined as follows:

- The rockfall *source zone* is the area of potential rockfall initiation, which may be composed of bedrock outcrops, cut rock slopes, or coarse surficial deposits.
- The rockfall *travel path* is the likely path rockfall will take down a slope from the source zone.
- The rockfall *runout zone* is the area of deceleration and stopping for rockfall, located on or at the base of the slope.

Ideally, these zones will be separate, but on many slopes boundaries overlap. For example, rockfall may be initiated not only from the near vertical cliff in Figure 6-1 but also along the travel

FIGURE 6-1
Recent rockfall from Dakota Sandstone cliff in Animas River valley, Colorado, illustrates relationships between rockfall source zone, travel path, and runout zone.
(PHOTOGRAPH BY J. D. HIGGINS, COLORADO SCHOOL OF MINES.)

path where boulders may be dislodged by impacts from falling rock or by erosion or similar processes. Also, some rocks may stop along the travel path and not travel all the way to the base of the slope. In the case of a steep rock cut, the cut face may be the source zone and the ditch and highway may be the travel path and runout zone.

Typically, the potential for rocks to dislodge from rock outcrops or surficial materials within a rockfall source zone is evaluated. Also, observations pertaining to the general geology of the site are required, and rock mass properties—such as lithology, detailed geological structure, and rock discontinuities—are accurately described and quantified. Information on groundwater conditions, topography, and past instabilities can provide insight into the characteristics and likely performance of the rock mass. Rockfall failure types can be predicted by using kinematic (stereonet) analysis once discontinuity orientations and properties are recorded. Rock mass characterization systems may be used qualitatively to assess rockfall susceptibility of potential source zones. The topography and slope material properties are defined along travel paths and runout zones. Predictions of rockfall behavior as rocks fall, bounce, or roll from the source zone along the rockfall travel path and stop in the runout zone are based on these observations. Inadequate data collection reduces the accuracy of the conceptual model, which affects the accuracy of the analysis and ultimately the selection of appropriate mitigation measures.

The purpose of this chapter is to indicate the data that are typically obtained during a rockfall site investigation and to describe appropriate methods to collect these data, both of which greatly assist in developing an accurate conceptual geological model.

2. ROCKFALL BEHAVIOR

The susceptibility of geological units to the initiation of rockfall depends on a number of material properties and environmental conditions of the slope. Chapter 2 discussed them in some detail. Susceptibility is affected by the properties of discontinuities within a rock or soil mass, such as joint spacing, persistence, and orientation; weathering and alteration conditions; and type of infilling materials. Environmental conditions that contribute to rockfall initiation are climate, amount and form of precipitation, freeze–thaw cycles, presence and type of vegetation, and steepness of the slope.

Ritchie (1963) noted that the behavior of a rock bounding down a slope is influenced by slope geometry, slope material properties, rock geometry, and rock material properties. Table 6-1 lists several important parameters for each of these factors. Rockfall events originating from the same

Table 6-1
Parameters Determining Behavior of Rockfall

Factor	Parameter
Slope geometry	Slope inclination
	Slope length
	Surface roughness
	Lateral variability
Slope material properties	Slope coefficients
	Rock coefficients
Rock geometry	Rock size
	Rock shape
Rock material properties	Rock durability
	Rock mass

SOURCE: Pfeiffer et al. 1995.

source location may behave differently as a result of the interaction of these factors. The geometry of the slope, especially the slope inclination and length of the topographic profile, will affect the path a rock takes down the slope and the velocity (and energy) that the bounding and rolling rock can attain. Chapter 8 describes the fundamental physical laws and concepts that control the mechanical behavior and energies of falling rocks. Abrupt changes in the slope profile may cause rocks to be launched out from the slope; this factor will determine the appropriate location of barriers on the slope and whether ditches near the base of a slope will be effective rockfall catchment areas. The surface roughness of the slope affects the amount of bouncing and rolling; rough surfaces cause more bouncing but also may slow down the bounding rock. *Surface roughness* can be defined as the small irregularities of a slope surface within a relatively uniform segment of the slope profile. Surface roughness alters the angle at which a bounding rock strikes the slope surface and largely determines the character of the bounce. Slope variations perpendicular to the slope profile are important; channels or gullies often direct the rockfall path. Slope material properties affect the interaction of the bounding rock and slope surface. Soft soils tend to dampen impact and slow the rock, while hard, smooth rock surfaces have little damping effect. Rock size and density affect the velocities that can be achieved, while the shape of the rock block affects bounce heights and the distance the rock will roll. Finally, the material properties of the bounding rock determine whether it is likely to break up as it travels down the slope.

3. INVESTIGATION AND DATA COLLECTION

A rockfall investigation, at a minimum, answers six basic questions:

1. Is there a rockfall problem and where is it located?
2. What is the nature of past rockfall events at these sites?
3. Where are the source area(s), travel path(s), and runout zone(s) and what are the geological properties of each?
4. What is the likely motion of the rock traveling down the slope (rolling, bouncing, or sliding)?
5. What size rocks typically reach the base of the slope (or point of interest) versus their size in the source zone?
6. How far do the rocks roll past the base of the slope (or point of interest)?

Typically, the characteristics of individual rockfall locations define the level of detail at which each question is answered. For most locations, these questions can be at least partially answered by assessing past rockfall events on existing slopes and by using backanalysis to predict the behavior of future events. But backanalysis may not answer all questions, especially if the slope will be modified; thus, the investigator may wish to use computer modeling as a predictive tool. Fieldwork and mapping are required to develop a conceptual model of geological conditions and processes with sufficient detail to support computer modeling. A number of methods can be used, either separately or in combination, to assist in identifying rockfall susceptibility and to characterize the rockfall hazard zone. These methods are summarized in Table 6-2. The sequence of investigation tasks required to obtain adequate data for constructing a rockfall conceptual model is similar to that for other types of slope stability studies. Rockfall investigations normally include the following tasks:

1. Review of past rockfall events, including literature review;
2. Aerial photographic analysis or field-reconnaissance mapping;
3. Rock mass descriptions;
4. Rock mass characterization; and
5. Slope profile and materials.

Smaller and relatively uniform slopes may be adequately characterized by completing only Tasks 1 and 2; however, all five tasks are often required to provide sufficient data to apply backcalculations of past events accurately or to calibrate and apply a rockfall computer model for prediction and design purposes. These tasks are explained in detail in the following sections.

3.1 Review of Past Rockfall Events, Including Literature Review

An assessment of transportation agency records and other information sources describing past

Table 6-2
Methods Used to Identify and Describe Rockfall Susceptibility and Character of Various Rockfall Zones

Component of Rockfall Hazard	Methods
Identification of susceptible rockfall sites along existing transportation routes	Literature and records: Transportation agency maintenance records, rockfall accident records, eyewitness or local resident accounts, legal depositions Fieldwork: Helicopter reconnaissance, aerial photographic mapping, field observations and mapping
Identification of rockfall source zones, travel paths, and runout zones	Fieldwork: Helicopter reconnaissance, aerial photographic mapping, field observations and mapping
Characterization of susceptibility to rockfall initiation in source zone	Literature and records: Some information possible from transportation agency maintenance records, rockfall accident records, legal depositions, eyewitness or local resident accounts Fieldwork: Most information from aerial photographic mapping, helicopter reconnaissance, field observations and mapping, rock mass classifications, geotechnical reports and maps
Description of rockfall travel paths and runout zones and rockfall behavior	Literature and records: Some information possible from compilation of historic rockfall data Fieldwork: Most information from aerial photographic mapping, helicopter reconnaissance, field observations, surveying, and geological mapping

Source: Adapted from Coe et al. 2005.

rockfall events is often a useful starting point for rockfall studies along existing transportation corridors. Some of these records may provide valuable information on rockfall locations, magnitude, and frequency, and even on rockfall behavior, which can be useful for predicting future events and can provide information that is important for rockfall analysis and selection of mitigation procedures.

Transportation agencies often maintain records for rock slopes along their routes that include accident and maintenance reports, aerial and oblique photographs, and evaluations and numerical scores developed by applying rockfall hazard rating systems (RHRSs), as discussed in Chapter 3. For existing roadways, accident and maintenance reports normally provide locations of rockfalls and may indicate whether single or multiple rocks were involved, specify the size of the rock blocks, and possibly provide some indication of runout distance (Figures 6-2 through 6-4). A lack of rockfall maintenance or accident records may suggest that rockfall is infrequent or that rocks are detaching but not traveling far enough down the slope to reach the facility and pose a significant hazard. Dated photographs and RHRS scores are useful for assessing changes in rock slope conditions over time. They often provide some sense of rate and degree of slope deterioration, and this information may help predict future rockfall frequency. They may also identify the likely source zone and whether there is a single source or multiple sources.

Published geological reports and maps can provide a general sense of lithology, structure, and seismic activity. Project-specific engineering geological and geotechnical reports from the site or from nearby similar geological environments can

FIGURE 6-2 Single large rock on road, Topanga Canyon, California. (PHOTOGRAPH COURTESY OF J. DUFFY, CALIFORNIA DEPARTMENT OF TRANSPORTATION.)

provide information on rock mass properties, performance of slopes, and performance of mitigation measures that have been used in the past.

Other sources of information such as legal depositions, precipitation and temperature records, eyewitness accounts, and verbal reports from local residents often provide insightful information on rockfall behavior. For example, records might reveal that a free-falling or airborne rock hit a vehicle or that a vehicle hit a rock after the rock came to rest on the roadway. Either situation provides some evidence of bounce heights, block velocities, and trajectories as well as the frequency of occurrence. Rockfalls may be most prominent during midmorning as the sun heats up the slope, during high winds or rain, or during daily freeze–thaw cycles. Most historic events may be tied to typical or atypical seasons or events (i.e., spring thaw, El Niño cycles, hurricanes, strong seismic motion). These assessments of past rockfalls and associated literature sources assist in answering several of the important questions listed previously and provide a basis for a backanalysis of past rockfall behavior.

3.2 Aerial Photographic Analysis and Reconnaissance Mapping

Analysis of aerial and oblique photographs in combination with site reconnaissance is useful in

FIGURE 6-3
Group of rocks in roadway, Cow Cliffs, California. Rocks deposited on road and shoulder indicate size range and runout distance for rockfall.
(PHOTOGRAPH COURTESY OF J. DUFFY, CALIFORNIA DEPARTMENT OF TRANSPORTATION.)

FIGURE 6-4
Numerous rocks along shoreline below ocean cliff document frequent rockfall events in area of intense erosion. A range of rock sizes is an important consideration in designing mitigation systems.
(PHOTOGRAPH COURTESY OF J. DUFFY, CALIFORNIA DEPARTMENT OF TRANSPORTATION.)

establishing whether a rockfall hazard exists on a slope. Frequent rockfall will leave field evidence such as impact marks and broken trees and brush due to recent events (Figure 6-5), or individual rocks or accumulations of rocks will be scattered

FIGURE 6-5
Recent rockfall left broken limbs mixed with rock along rockfall path.
(PHOTOGRAPH BY J. D. HIGGINS, COLORADO SCHOOL OF MINES.)

throughout the area (Figure 6-6). In contrast, if no rocks are lying on the lower slopes (as in Figure 6-7), rockfall is likely infrequent, or rocks are detaching but not moving far from the base of the slope. The investigator should observe the pattern of rockfall deposits to determine whether single or multiple rocks fall in individual events. Very large groups of rocks may indicate a rockslide and a global stability problem rather than a rockfall hazard.

Engineering geological mapping on relatively large-scale aerial photographs (1:12,000 or larger) in combination with site reconnaissance is useful for identifying the location of potential rockfall source zones, travel paths, and runout zones. However, aerial photography is less useful if vegetation masks the ground surface. Ideally, these three zones are identified because they all require additional data collection. For slopes that have drainage channels or chutes forming well-defined rockfall travel paths, the mapping can provide a means of correlating source zones with associated travel paths and runout zones. Figure 6-8 provides an example of a reconnaissance engineering geological map based on an aerial photograph. The map shows locations of cut slopes, bedrock outcrops (potential source zones), travel paths (chutes), and other rockfall-related data. Depending on the complexity of the site, it may be important to map geological detail in addition to that shown in Figure 6-8. However, at a minimum, this reconnaissance map identifies various zones on the slope where specific observations or data collection may be required.

The review of past rockfall events and associated literature and preparation of a reconnaissance engineering geological map identify the general properties, rockfall processes, and locations of source zones, travel paths, and runout zones. The level of the potential hazard and risk dictates the detail of the data collection and investigation methods necessary for selecting appropriate mitigation for a site. For a cursory review of the site, the tasks described above may be adequate. However, for a full analysis and design of a rockfall mitigation system for a large, complex slope,

(a)

(b)

FIGURE 6-6
Accumulation of rocks (*a*) in chute below rock outcrop and (*b*) in runout zone near base of slope provides evidence of frequent rockfall activity.
(PHOTOGRAPHS COURTESY OF R. BARTINGALE, COLORADO SCHOOL OF MINES.)

FIGURE 6-7
Rocks are absent on lower slopes of Madonna Mountain near San Luis Obispo, California, suggesting low potential for rockfall.
(PHOTOGRAPH COURTESY OF J. DUFFY, CALIFORNIA DEPARTMENT OF TRANSPORTATION.)

a record of rock mass structure in the source zone and details of the rockfall travel paths and runout profiles may be necessary. The following sections discuss the engineering geological observations that may be required in comprehensive studies, including descriptions and classification of rock mass properties in the source zones that affect the frequency, size, and shape of rockfall and the slope topography and materials in the rockfall path and runout zones that affect the travel distance and behavior.

3.3 Rock Mass Descriptions

Describing the rock types exposed in the source zone is important for site characterization because rock types can be directly correlated with various types of rockfall behavior. Geological structure affects properties such as rock mass strength, size and shape of rock blocks, and stability of blocks in an outcrop exposed on a steep slope. The structural features that are most important to rockfall site characterization are commonly called *discontinuities* (or rock defects). These features include joints, bedding planes, foliation, cleavage, schistosity, faults, shear zones, clay seams, and similar geological structures. They form weak zones in a rock mass along which blocks may separate and develop into a rockfall event. Important properties of these discontinuities include orientation, spacing, persistence, roughness, wall condition, aperture, filling, and seepage. Discontinuities often form in distinct orientations; those having a consistent orientation belong to a discontinuity set. The number of discontinuity sets often controls the resulting block size and shape. Lithology and discontinuity characteristics are discussed below.

3.3.1 Lithology

Identifying rock type is important because it provides a direct correlation to rock properties and rockfall behavior. Typically, rock type can be correlated with a range of intact rock strength (Table 6-3), mineralogy (which can be related to susceptibility to weathering), and joint patterns and number of joints. For example, sedimentary units are particularly susceptible to planar failure of blocks along bedding planes. Often sedimentary stratigraphic sequences consist of alternating low- and high-intact-strength units, which can result in differential weathering of the low-strength layers and undercutting of the high-strength layers, creating a large potential for rockfall. Metamorphic and igneous rocks generally exhibit moderate to high intact strength, and rockfall failures occur along joints and foliation boundaries.

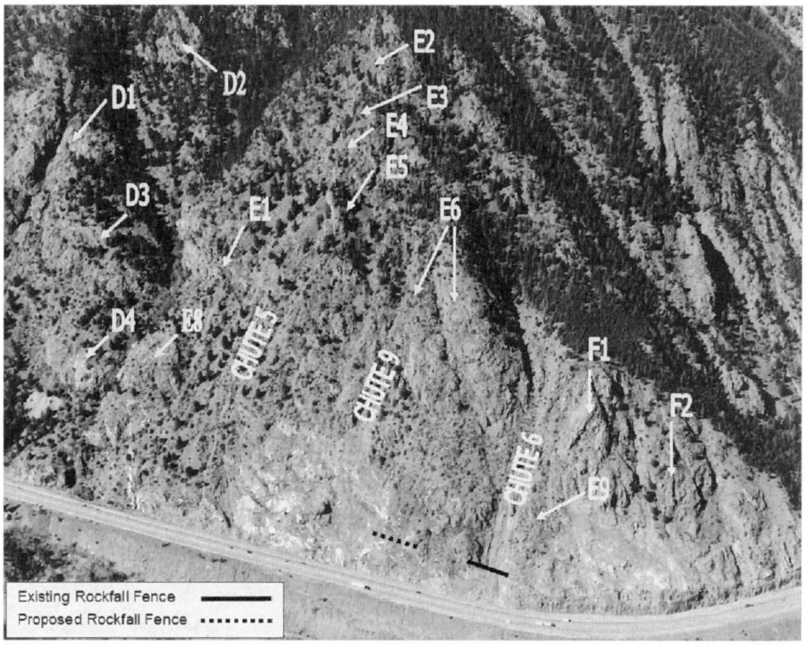

FIGURE 6-8
Rockfall chutes and source areas with fence locations on slope above Interstate 70 near Georgetown, Colorado.
Source zones are designated by D4, E6, and so forth.
(PHOTOGRAPH COURTESY OF YEH AND ASSOCIATES, INC., DENVER, COLORADO.)

3.3.2 Orientation

In rock engineering work, the *orientation*, or *attitude*, in space of a discontinuity is typically described by the dip and dip direction. *Dip* is the maximum angle of a plane (discontinuity) with a horizontal plane, as illustrated in Figure 6-9 by the angle ψ. *Dip direction* is the azimuth of the dip direction (α), in the horizontal, measured clockwise from true north (Figure 6-9). For example, with the notation suggested by the International Society for Rock Mechanics (ISRM) Working Group (1981), 090°/76° is the typical notation for a discontinuity that dips 76° to the east. Dip direction ranges from 0° to 360° and dip value can range from 0° to 90°. This notation form is convenient, in that it can be plotted directly on a stereonet for analysis and many software programs accept input in this format. It is important to note on data sheets that the measurements are dip direction and dip, as they can be easily confused with the notation of strike and dip that geologists commonly use, which for the same plane would be 0°/76°.

Table 6-3
Typical Soil and Rock (Intact) Strength Values

Type	Material	Unit Weight (saturated/dry) lb/ft³	Unit Weight (saturated/dry) kN/m³	Friction Angle (°)	Cohesion lb/ft²	Cohesion kPa
Cohesionless						
Sand	Loose sand, uniform grain size	118/90	19/14	28–34[a]		
	Dense sand, uniform grain size	130/109	21/17	32–40[a]		
	Loose sand, mixed grain size	124/99	20/16	34–40[a]		
	Dense sand, mixed grain size	135/116	21/18	38–46[a]		
Gravel	Gravel, uniform grain size	140/130	22/20	34–37[a]		
	Sand and gravel, mixed grain size	120/110	19/17	48–45[a]		
Blasted or broken	Basalt	140/110	22/17	40–50[a]		
	Chalk	80/62	13/10	30–40[a]		
	Granite	125/110	20/17	45–50[a]		
	Limestone	120/100	19/16	35–40[a]		
	Sandstone	110/80	17/13	35–45[a]		
	Shale	125/100	20/16	30–35[a]		
Cohesive						
Clay	Soft bentonite	80/30	13/6	7–13	200–400	10–20
	Very soft organic clay	90/40	14/6	12–16	200–600	10–30
	Soft, slightly organic clay	100/60	16/10	22–27	400–1,000	20–50
	Soft glacial clay	110/76	17/12	27–32	600–1,500	30–70
	Stiff glacial clay	120/105	20/17	30–32	1,500–3,000	70–150
	Glacial till, mixed grain size	145/130	23/20	32–35	3,000–5,000	150–250
Rock	Hard igneous rock: granite, basalt, porphyry	160–190[b]	25 to 30	35–45	720,000–1,150,000	35,000–55,000
	Metamorphic rock: quartzite, gneiss, slate	160–180[b]	25 to 28	30–40	400,000–800,000	20,000–40,000
	Hard sedimentary rock: limestone, dolomite, sandstone	150–180[b]	23 to 28	35–45	200,000–600,000	10,000–30,000
	Soft sedimentary rock: sandstone, coal, chalk, shale	110–150[b]	17 to 23	25–35	20,000–400,000	1,000–20,000

[a]Higher friction angles in cohesionless materials occur at low confining or normal stresses.
[b]For intact rock, the unit weight of the material does not vary significantly between saturated and dry states with the exception of materials such as porous sandstones.
SOURCE: Hoek and Bray 1981.

3.3.3 Spacing

The true spacing between discontinuities is the perpendicular distance between adjacent features belonging to a single set (Figure 6-10); thus, different discontinuity sets will have distinct spacing characteristics. Spacing measurements are normally made during a scanline survey along the rock outcrop surface to evaluate measurements made by using rock cores or by borehole televiewer. Either method of measurement will likely produce an "apparent" spacing measurement, because the measured distance will not be oriented perpendicular to the discontinuities and must be converted to "true" spacing. This conversion can be accomplished by applying Equation 1, which contains parameters defined in Figures 6-9 and 6-10.

$$S_2 = d_2 \sin\alpha_2 \qquad (1)$$

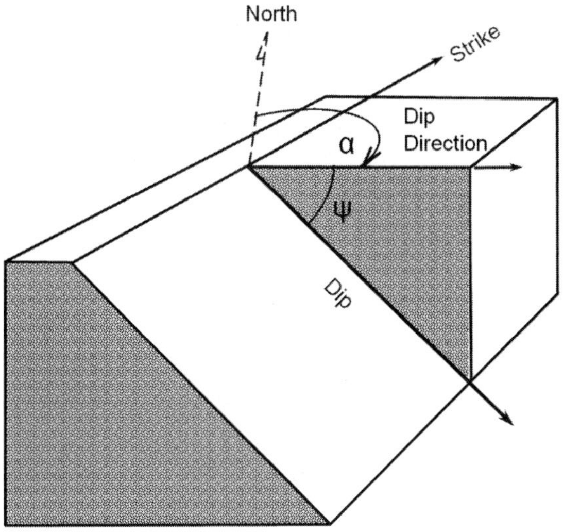

FIGURE 6-9
Diagram illustrating dip and dip direction (adapted from Hoek and Bray 1981).

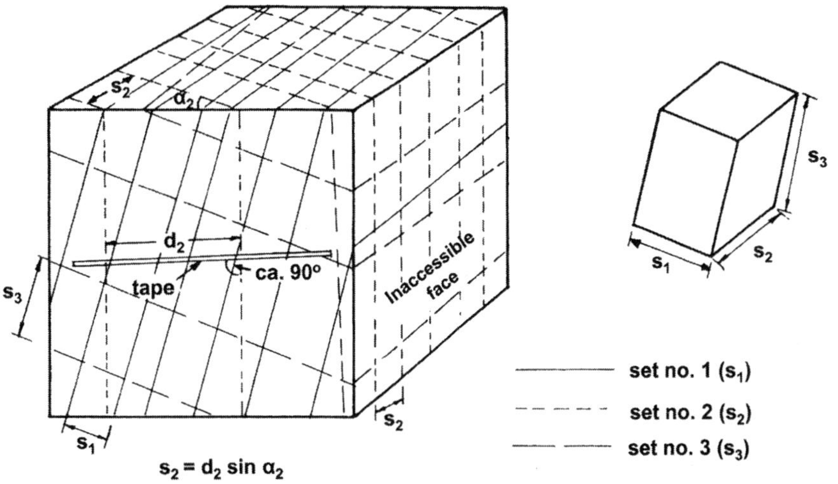

FIGURE 6-10
Measurement of discontinuity spacing on rock exposure (adapted from ISRM Working Group 1981).

where S_2 is true spacing between discontinuities forming Set 2 and d_2 is apparent spacing.

To quantify the spacing characteristics further, rockfall investigations typically report minimum, average, and maximum spacing values for each discontinuity set. Spacing may be used to define the range of block sizes that make up the rock mass and the likely range in size of rockfall originating from the rock mass. The closeness of spacing affects rock mass strength and seepage characteristics and may be indicative of modes of failure such as toppling, sliding of blocks, and raveling. Closely spaced discontinuities generally weaken a rock mass and increase seepage potential, because individual discontinuities may connect to form a continuous zone of weakness and a preferential flow path. Also, closely spaced discontinuities provide more surface area than widely spaced ones and thus promote weathering processes that attack intact rock and reduce shear strength over time. The ISRM Working Group (1978, 1981) has adopted standard terminology to describe the spacing of discontinuities for rock mass mapping (Table 6-4).

3.3.4 Persistence

Persistence is a measure of the areal extent or area of a discontinuity and contributes to defining block size (along with spacing), length of sliding surfaces, and hydraulic conductivity of rock masses. Persistence is crudely estimated in the field by observation of discontinuity trace lengths on the surface of exposures and notation of discontinuity termination points. Barton (1987) suggested that discontinuities that terminate beyond the rock exposure (*x*) should be differentiated from those that terminate in solid rock (*r*) or against other discontinuities (*d*) as illustrated in Figure 6-11. He noted that a high count of *x* (systematic set of discontinuities) is more persistent than a subsystematic set (dominant count in *d*) and that nonsystematic discontinuities will have low to very low persistence and thus high counts in *r*. The ISRM Working Group (1978, 1981) suggested a classification of persistence based on measurement of modal trace length (Table 6-5). This classification of persistence requires extensive rock exposures and so is not reasonable when only small rock exposures are available.

3.3.5 Roughness

Opposing discontinuity surfaces influence rock mass behavior. The *roughness* of these surfaces is

Table 6-4
Suggested Terminology Describing Spacing of Discontinuities

Description	Spacing (mm)
Extremely close	<20
Very close	20–60
Close	60–200
Moderate	200–600
Wide	600–2,000
Very wide	2,000–6,000
Extremely wide	>6,000

Source: ISRM Working Group 1978, 1981.

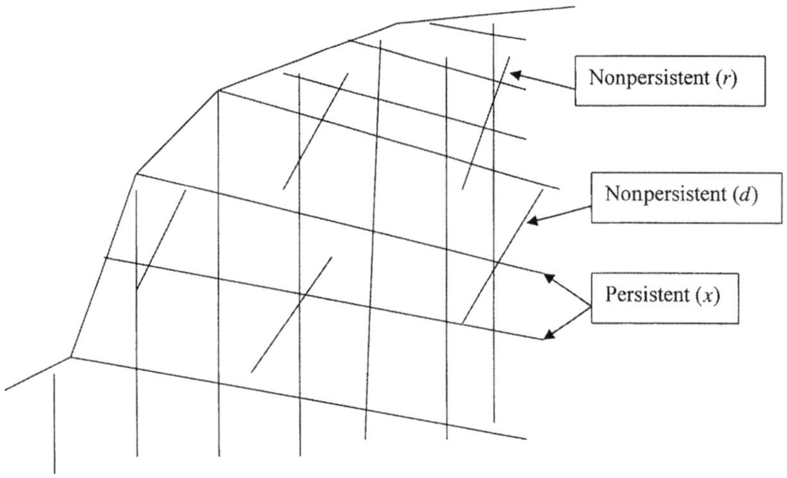

FIGURE 6-11
Diagram illustrating persistence of discontinuity sets on rock exposure: *r* indicates termination on solid rock, *d* indicates termination against other discontinuities, and *x* indicates termination unknown beyond boundaries of exposure.
(DIAGRAM BY J. D. HIGGINS.)

an important component of shear strength along discontinuities that have no infilling material, and the surfaces are mostly in contact with each other and have not been displaced. The purpose in describing the roughness of discontinuity walls is generally to aid in estimating shear strength between blocks in the source zone. Smooth surfaces exhibit lower shear strength than rough surfaces. Roughness is characterized by two scales of discontinuity surface characteristics: waviness and unevenness. *Waviness* is defined as large-scale undulations that are too large to be sheared off during displacement. Waviness modifies the apparent angle of dip along a discontinuity and is important for large rock slides (global stability) but is not so significant a control for the sliding of individual or several medium-to-small surficial blocks from a rock exposure. *Unevenness* is small-scale roughness that is usually damaged during shear displacement unless stress levels are low or wall strength is high. Unevenness along discontinuity surfaces controls the shear strength of blocks most often associated with rockfall. During rockfall site characterization, roughness is normally classified on the basis of observation and is used in empirical classification schemes to characterize the relative potential for rockfall to originate from a given rock formation.

Table 6-5
Field Classification of Persistence of Discontinuities

DESCRIPTION	MODAL TRACE (m)
Very low	<1
Low	1–3
Medium	3–10
High	10–20
Very high	>20

SOURCE: ISRM Working Group 1978, 1981.

A quantitative procedure for measuring joint roughness in the field has been presented in reports by the ISRM Working Group (1978, 1981). Descriptive terms of roughness based on two scales of observation have been suggested in these reports and are shown in Table 6-6. They are *small-scale roughness* (several centimeters) and *intermediate-scale roughness* (several meters). Intermediate-scale roughness includes stepped, undulating, and planar features. Small-scale roughness, often superimposed on intermediate-scale roughness, includes rough, smooth, and slickensided surfaces. Typical roughness profiles are illustrated in Figure 6-12. The directions of striations and slickensides should be observed, as they may cause directional variations in shear strength.

Shear strength of unfilled discontinuities may be estimated on the basis of roughness (ISRM Working Group 1981). Shear strength may exist at a maximum (peak) value, a minimum (residual) value, or some intermediate value, depending on previous shear displacement. The shear strength is defined by Equation 2:

$$\tau = \sigma_n' \tan(\varphi + i) \qquad (2)$$

where

τ = shear strength (peak or residual),
φ = friction angle (peak or residual),
σ_n' = effective normal stress, and
i = waviness (if present).

Table 6-6
Classification of Roughness of Discontinuities

CLASS	ROUGHNESS CATEGORY
I	Rough (or irregular), stepped
II	Smooth, stepped
III	Slickensided, stepped
IV	Rough, undulating
V	Smooth, undulating
VI	Slickensided, undulating
VII	Rough, planar
VIII	Smooth, planar
IX	Slickensided, planar

NOTE: Class I has the highest shear strength and Class IX has the lowest.
SOURCE: ISRM Working Group 1978, 1981.

Values of peak shear strength depend on the effective normal stress and roughness. Peak friction angles range from 30° to 70°. Joints with vertical or very steep step features (Profiles I, II, and III in Figure 6-12), or with less than 100% persistence, will have cohesion, which increases the shear strength. The residual friction angle depends on the weathering character of discontinuity walls and the rock type. The residual friction angle for unweathered discontinuity walls ranges from 25° to 35°, while highly weathered walls have a friction angle of about 15° (ISRM Working Group 1981). Residual friction angles can be estimated by using Schmidt hammer rebound values obtained from weathered rock wall (Barton and Choubey 1977). Values of peak friction angle can be estimated by using the joint roughness profiles and joint compressive strength (Barton and Choubey 1977; ISRM Working Group 1981).

The waviness angle (i) is most applicable in analysis of large rock slides (global stability) rather than in analysis of common rockfall problems. Most rockfall blocks sliding from a steep slope would be smaller than the scale of the waviness angle, but small- and intermediate-scale roughness may control the stability of individual rocks.

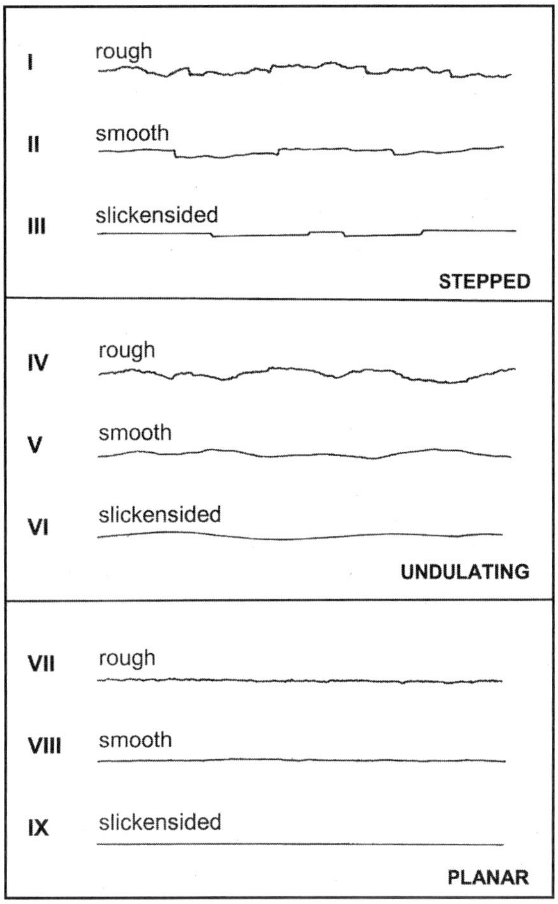

FIGURE 6-12 Typical roughness profiles and nomenclature suggested by ISRM Working Group (1978, 1981). Length of each profile is in range of 1 to 10 m (3 to 30 ft) with equal vertical and horizontal scales.

3.3.6 Wall Condition and Strength

The *condition*, or compressive strength, of the discontinuity walls is an important factor affecting shear strength and deformability of a rock mass if the walls are in rock-to-rock contact and are affected by the infilled material in the discontinuities. Weathering tends to attack rock where it is exposed to surface conditions and along discontinuities where water and air can penetrate the rock mass. Also, hydrothermal processes may alter the rock walls adjacent to discontinuities. In addition, weathering and alteration may reduce the rock mass shear strength by reducing the strength of the intact rock in the interior of the blocks. However, the walls of the discontinuities are the most vulnerable to weathering and alteration, while the interior of the block may be unchanged.

A description of the degree of weathering or alteration of the rock mass is essential for empirical classification of rockfall potential. The ISRM Working Group (1978, 1981) developed a visual classification of weathering grades that classifies rock mass according to the degree of disintegration and decomposition and is suitable for application to rock mass classification schemes. The grade of weathering (or alteration) of the rock mass is described according to the definitions provided in Table 6-7. In contrast, the weathering or alteration grade of the rock material forming discontinuity walls is described according to the classification system defined in Table 6-8. This description may be especially important when applied to discontinuities with unfavorable orientations relative to the slope face.

The strength of the rock forming discontinuity walls may influence the shear strength of rough surfaces. High stresses compared with wall strength may cause asperities to be sheared off and result in a reduction of the joint roughness component of the friction angle. Compressive strength of discontinuity walls is often determined by simple manual index tests conducted during field inspections (Table 6-9). More quantitative index tests to estimate compressive strengths, such as point load and Schmidt hammer, may be performed on discontinuity wall materials in situ or by using core or lump

Table 6-7
Weathering and Alteration Grades of Rock Mass

Term	Description	Grade
Fresh	No visible sign of rock material weathering; perhaps slight discoloration on major discontinuity surfaces.	I
Slightly weathered	Discoloration indicates weathering of rock material and discontinuity surfaces. All the rock material may be discolored by weathering and may be somewhat weaker externally than in its fresh condition.	II
Moderately weathered	Less than half the rock material is decomposed or disintegrated to a soil. Fresh or discolored rock is present either as a continuous framework or as corestones.	III
Highly weathered	More than half the rock material is decomposed or disintegrated to a soil. Fresh or discolored rock is present either as a discontinuous framework or as corestones.	IV
Completely weathered	All rock material is decomposed or disintegrated to soil. The original mass structure is still largely intact.	V
Residual soil	All rock material is converted to soil. Mass structure and material fabric are destroyed. There is a large change in volume, but the soil has not been significantly transported.	VI

Source: ISRM Working Group 1978, 1981.

samples. These procedures are described in the reports of the ISRM Working Group (1978, 1981).

3.3.7 Aperture

Aperture is defined as the perpendicular distance separating opposing walls of an open discontinuity in which the space is filled with air or water, as illustrated in Figure 6-13, *a* and *b*. Aperture does not apply to filled discontinuities (Figure 6-13*c*) unless the infilled material has been removed by erosion (ISRM Working Group 1978, 1981). Aperture potentially can have an important effect on shear strength and hydraulic conductivity of the rock mass.

An aperture may range from "closed" to rather large. Large apertures may result from processes such as shear displacements along discontinuities with significant roughness or waviness, solution weathering of carbonates or salts, tensile openings, or erosion of weak filling materials (ISRM Working Group 1978, 1981; Wyllie and Mah 2004). Generally, apertures are more likely to be largest near the ground surface, where weathering has most modified the discontinuities, and smaller, perhaps less than a fraction of a millimeter, at

Table 6-8
Weathering and Alteration Grades of Rock Material Comprising Discontinuity Walls

Term	Description
Fresh	No visible sign of weathering of rock material.
Discolored	Color of original fresh rock material is changed. Degree of change from original color should be indicated. If color change is confined to particular mineral constituents, it should be mentioned.
Decomposed	Rock is weathered to the condition of a soil in which original material fabric is still intact, but some or all the mineral grains are decomposed.
Disintegrated	Rock is weathered to the condition of a soil in which original fabric is intact. Rock is friable, but the mineral grains are not decomposed.

Note: Stages of weathering may be subdivided with qualifying terms—for example, slightly discolored, moderately discolored, or highly discolored.
Source: ISRM Working Group 1978, 1981.

Table 6-9
Manual Index Tests for Discontinuity Walls

Grade	Description	Field Identification	Approximate Range of Uniaxial Compressive Strength (MPa)
S1	Very soft clay	Easily penetrated several inches by fist	<0.025
S2	Soft clay	Easily penetrated several inches by thumb	0.025–0.05
S3	Firm clay	Penetrated several inches by thumb with moderate effort	0.05–0.10
S4	Stiff clay	Readily indented by thumb but penetrated only with great effort	0.10–0.25
S5	Very stiff clay	Readily indented by thumbnail	0.25–0.50
S6	Hard clay	Indented with difficulty by thumbnail	>0.50
R0	Extremely weak rock	Indented by thumbnail	0.25–1.0
R1	Very weak rock	Crumbles under firm blows with point of geological hammer, can be peeled by a pocket knife	1.0–5.0
R2	Weak rock	Can be peeled by a pocket knife with difficulty, shallow indentations made by a firm blow with point of geological hammer	5.0–25
R3	Medium strong rock	Cannot be scraped or peeled with a pocket knife, specimen can be fractured with single firm blow of geological hammer	25–50
R4	Strong rock	Specimen requires more than one blow of geological hammer to fracture it	50–100
R5	Very strong rock	Specimen requires many blows of geological hammer to fracture it	100–250
R6	Extremely strong rock	Specimen can be chipped only with geological hammer	>250

Source: ISRM Working Group 1978, 1981.

depth, where surface weathering has had less effect and stresses tend to be greater than at the surface.

Apertures may be measured at rock outcrops, by using drill core, provided recovery is excellent, or in clean-walled boreholes with borehole cameras and televiewers (Wyllie and Mah 2004). The fact that apertures near the surface often differ significantly from those at depth makes it difficult for engineering geologists to project outcrop measurements and observations to the subsurface. However, in the case of rockfall investigations, the near-surface rocks are often the focus of interest, unless deep cuts are planned.

The reports of the ISRM Working Group (1978, 1981) suggest recording the modal apertures for each discontinuity set and, for those discontinuities with noticeably wider apertures, recording the aperture, location, and orientation. The suggested field description of apertures is presented in Table 6-10.

3.3.8 Filling

Filling (Figure 6-13c) results when the width between the rock walls of a discontinuity is filled with mineral materials (ISRM Working Group 1978, 1981). The width and physical properties of the filling are likely to affect rock mass shear strength, deformability, and permeability as well as to influence the ease with which surface blocks can be loosened from rock slopes and become rockfall. Filling materials vary considerably in physical properties; they commonly consist of materials such as calcite, chlorite, silt, clay, fault gouge, and breccia. The most important factors associated with filling material that affect rockfall potential, and thus that should be described in detail, include mineralogy, grain-size distribution and particle size, overconsolidation ratio, water content and permeability, previous shear

FIGURE 6-13
Illustrations showing (*a*) closed, (*b*) open, and (*c*) filled discontinuities and definitions of aperture and width (ISRM Working Group 1978, 1981).

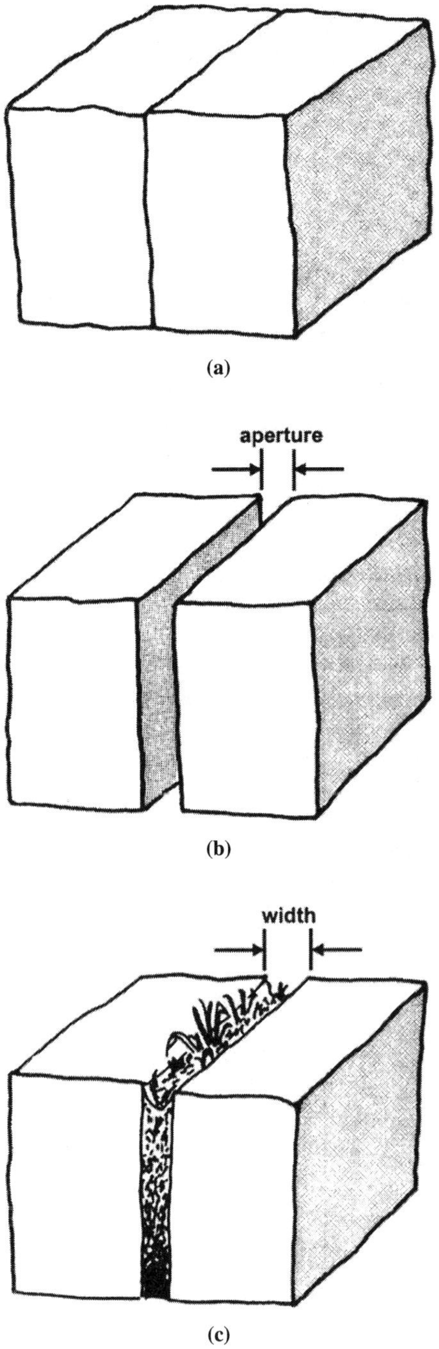

displacement, wall roughness, width, and fracturing or crushing of wall rock. Filling that results from weathering along the discontinuity may be composed entirely of decomposed or disintegrated rock, as defined in Table 6-11.

If the filling material potentially forms a sliding surface, samples may be required for shear-strength analysis (ISRM Working Group 1978, 1981; Wyllie and Mah 2004). Erodibility of the material may be of interest, since removal of infilling potentially causes a significant aperture.

Table 6-10
Aperture Classification

Aperture Width	Description	Classification
<0.1 mm	Very tight	Closed features
0.1–0.25 mm	Tight	
0.25–0.5 mm	Partly open	
0.5–2.5 mm	Open	Gapped features
2.5–10 mm	Moderately wide	
>10 mm	Wide	
1–10 cm	Very wide	Open features
10–100 cm	Extremely wide	
>1 m	Cavernous	

Source: Adapted from ISRM Working Group 1978, 1981.

3.3.9 Seepage

Seepage of water in a rock mass normally occurs along discontinuities, except in some sedimentary rocks, where a significant proportion of flow may occur through pores. Seepage from discontinuities provides indirect information on apertures and on the position of the water table or multiple water tables. Seepage also indicates a possible seasonal reduction in stability by hydraulic pressure in discontinuities, by softening of infilling material, or by ice wedging in climates with significant freeze–thaw cycles (Barton 1987; ISRM Working Group 1978, 1981; Wyllie and Mah 2004). Seepage from

Table 6-11
Discontinuity-Filling Material Weathering Grades

Class	Definition
Decomposed	Rock is weathered to the condition of soil, original material fabric is intact, some or all of the mineral grains are decomposed.
Disintegrated	Rock is weathered to the condition of soil, original material fabric is intact, material is friable, mineral grains are not decomposed.

Source: Adapted from ISRM Working Group 1978, 1981.

Table 6-12
Seepage Ratings for Unfilled and Filled Discontinuities

Rating	Description
Unfilled Discontinuities	
I	Discontinuity is very tight and dry; water flow does not appear possible.
II	Discontinuity is dry with no evidence of water flow.
III	Discontinuity is dry but shows evidence of water flow—rust staining, etc.
IV	Discontinuity is damp but no free water is present.
V	Discontinuity shows seepage, occasional drops of water, but no continuous flow.
VI	Discontinuity shows a continuous flow of water (estimate L/min and describe pressure as low, medium, or high).
Filled Discontinuities	
I	Filling materials are heavily consolidated and dry, significant flow appears unlikely due to very low permeability.
II	Filling materials are damp, but no free water is present.
III	Filling materials are wet, occasional drops of water.
IV	Filling materials show signs of erosion, continuous flow of water (estimate volume/min).
V	Filling materials are washed out locally, considerable water flow along outwash channels (estimate volume/min and describe pressure as low, medium, or high).
VI	Filling materials are washed out completely, very high water pressures experienced, especially on first exposure (estimate volume/min and describe pressure as low, medium, or high).

SOURCE: ISRM Working Group 1978, 1981.

individual or sets of discontinuities, whether filled or unfilled, can be described in the field according to the descriptive rating system shown in Table 6-12.

The rockfall site investigation may not be conducted during the wet season; therefore, it is important not only to examine rock exposures for evidence of seasonal seepage but also to investigate roadway maintenance records, rock slope rating systems, and regional experience in similar geological settings. This information may help predict conditions at the site that cannot be directly observed in the field during the dry season.

Because hydraulic pressure in discontinuities can be critical to the stability of surface blocks or the entire rock slope, it is often important to assess the potential effectiveness of surface and horizontal drains for keeping discontinuities from becoming saturated. Orientation, spacing, and aperture of relevant discontinuities will influence drain effectiveness. The influence of ice blockage of discontinuities should be considered in areas where significant moisture is available and temperatures fluctuate around the freezing point.

3.3.10 Number of Discontinuity Sets

Two classes of discontinuities may be important to rock slope stability: (*a*) *systematic joint sets*, which are persistent features with individual joints parallel or subparallel in plan and cross section, and (*b*) *nonsystematic discontinuities* that display random patterns in plan and cross section. They may include features such as fractures, shear zones, and faults.

The number of discontinuity sets that intersect a rock mass influences rockfall stability, and, in combination with orientation and spacing, determines the size and shape of rocks that can detach from the slope. As the number of discontinuity sets increases and block size decreases, there is a greater opportunity for blocks to rotate, translate, and detach from a rock face. If insufficient sets exist (fewer than three), the probability of rockfall instability may be low, because intact rock breakage would be required for blocks to be released from a rock face (Figure 6-14). Also, major persistent and random discontinuities, such as faults and shear zones, may significantly increase the potential for rockfall failure in locations where

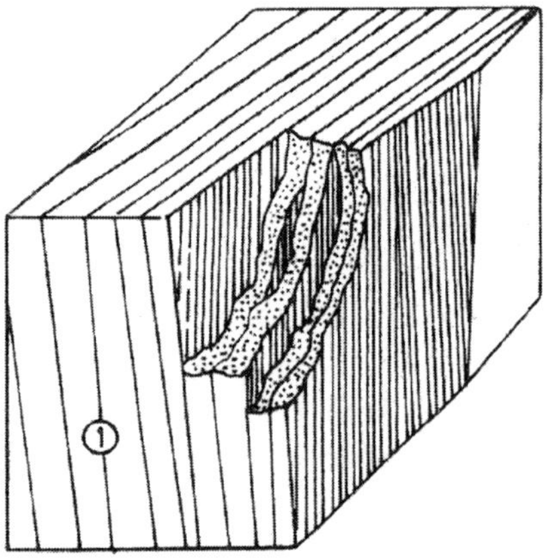

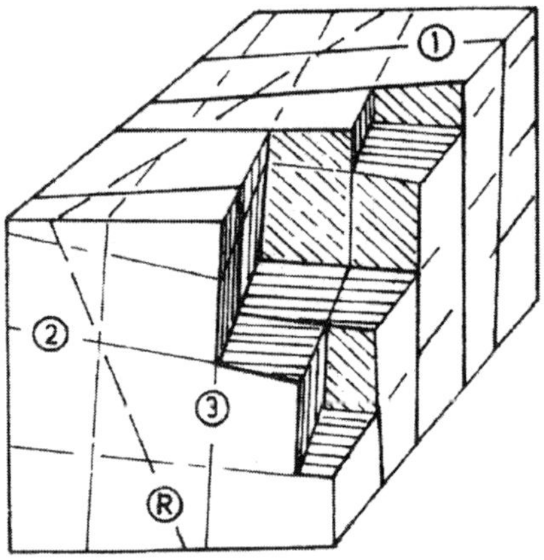

FIGURE 6-14 Effect of number of joint sets on rock block size and shape: (*a*) one joint set and (*b*) three joint sets (ISRM Working Group 1978, 1981).

Table 6-13
Description of Joint Sets

CLASS	DESCRIPTION OF JOINT SET
I	Massive, occasional random joints
II	One joint set
III	One joint set plus random
IV	Two joint sets
V	Two joint sets plus random
VI	Three joint sets
VII	Three joint sets plus random
VIII	Four or more joint sets
IX	Crushed rock, earthlike

SOURCE: ISRM Working Group 1978, 1981.

they happen to intersect the slope face. These discontinuities, often in combination with joint sets, may set up rockfall failure modes that are not consistently present throughout the rock mass.

The recognition of individual random discontinuities and discontinuity sets is normally accomplished by field survey and analysis of orientation data through scanline surveys described in Section 3.3.12. It is important to record all discontinuity sets present. The report by the ISRM Working Group recommends that systematic joint sets be distinguished from nonsystematic discontinuities when the number of sets or individual discontinuities is recorded.

The number of sets is determined by contouring discontinuity pole plots on stereonets to determine likely orientations and the range of orientations for each set. Random, nonsystematic, and persistent discontinuities will be obvious from the pole plots and will be important only if they are located so that they intersect the slope of interest. This detail of analysis may not be necessary for small slopes in simple geological terrains, as the potential sources of rockfall, numbers of discontinuity sets, and sizes of blocks may be obvious from field inspection. The report of the ISRM Working Group suggests that the number of joint sets occurring within the rock slope of interest can be described by field inspection, as shown in Table 6-13.

3.3.11 Block Size

Discontinuity spacing, persistence, and number of sets determine the *block size*. Blocks can range from very large to very small, perhaps from >8 to <0.0002 m^3 (282 to 0.007 ft^3) (Wyllie and Mah 2004). The block size and interblock shear strength determine the mechanical behavior of a rock mass, which is important for both global rock slope stability and failure of individual blocks from the slope surface. In general, the smaller the block size, the lower is the rock mass strength. Also, large numbers of fractures exposed at the surface allow moisture to penetrate the rock mass more easily and promote rockfall due to processes such as weathering and ice wedging. Block size will also affect blasting efficiency.

The discontinuity orientation and number of sets determine the shape of the blocks, which may resemble cubes, rhombohedrons, tetrahedrons, sheets, or irregular shapes, as illustrated in Figures 6-14 and 6-15. Block shape affects the bouncing and rolling characteristics of rockfall. The reports of the ISRM Working Group (1978, 1981) suggest standard terminology for field descriptions of rock masses to give an impression of block size and shape (Table 6-14).

Block size is described by selecting several typical blocks and measuring each dimension, which allows for estimates of modal, maximum, and minimum sizes. If only one or two joint sets are present, estimating a typical block size is unrealistic, because size may vary considerably. An alternative method of describing block size is use of the volumetric joint count method (ISRM Working Group 1978, 1981). A volume joint count (J_v) consists of a summation of the number of joints per meter for each joint set. The joints of each set should be counted along a line perpendicular to the strike of the set along a sampling length of 5 to 10 m (16 to 32 ft).

The procedure for calculating J_v is illustrated in Figure 6-16. The rock mass has three discontinuity sets, and thus the joints belonging to each set are counted along three sampling lines; two lines (Lines 1 and 2) are 10 m (32 ft) long, and one line (Line 3) is 7 m (23 ft) long. As shown in Figure 6-16, the joint counts were Line 1 = 6, Line 2 = 3, and Line 3 = 4. Then the calculation for J_v is performed as follows:

$$J_v = 6/10 + 3/10 + 4/7$$
$$= 0.6 + 0.3 + 0.6 = 1.5 \text{ joints/m}^3$$

Although the units are not truly "joints per cubic meter," the method gives a relative assessment of the size of blocks formed. On the basis of the ISRM Working Group (1978, 1981) descriptive terms shown in Table 6-15, the example rock mass with J_v = 1.5 is composed of large blocks.

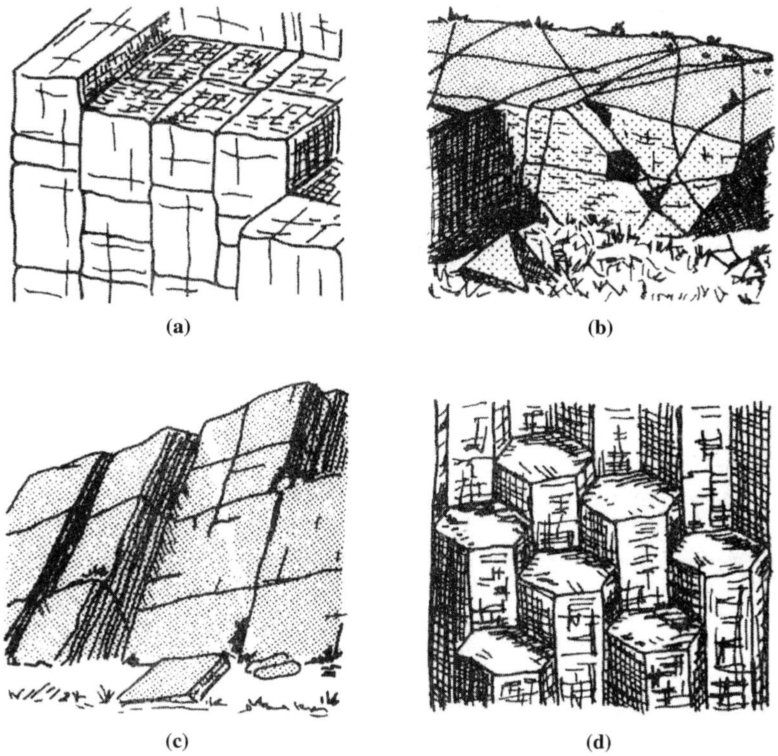

FIGURE 6-15 Block shapes discussed in Table 6-14: (*a*) blocky, (*b*) irregular, (*c*) tabular, and (*d*) columnar (ISRM Working Group 1978, 1981).

3.3.12 Measurement of Geological Structure: Scanline and Window Mapping

Normally, rockfall hazard characterizations are conducted for two scenarios: natural (weathered) slopes and new slopes constructed by rock excavation. Rock mass structure data collected from existing rock exposures are likely to be representative of conditions that affect rockfalls originating from a natural slope. Joint conditions may be markedly different at shallow depths behind the slope; therefore, preconstruction measurements on a natural slope may not represent characteristics of rock buried in the slope that will be exposed after excavation. The deeply buried rock may not be as weathered, and discontinuities may exhibit different properties at depth because they have been under stress and less exposed to weathering. To reduce uncertainties for deep excavation projects, it may be important to conduct subsurface drilling or rock trenching.

Beyond these uncertainties, the sampling and measurement of the rock mass structure properties present a basic sampling problem (Brady and Brown 1985). There is always a question of the proportion of the rock mass that must be sampled

Table 6-14
Rock Mass Description

Descriptive Term	Definition
Massive	Few joints or very wide spacing
Blocky	Approximately equidimensional
Tabular	One dimension considerably smaller than the other two
Columnar	One dimension considerably larger than the other two
Irregular	Wide variation in block size and shape
Crushed	Heavily jointed to "sugar cube"

Source: ISRM Working Group 1978, 1981.

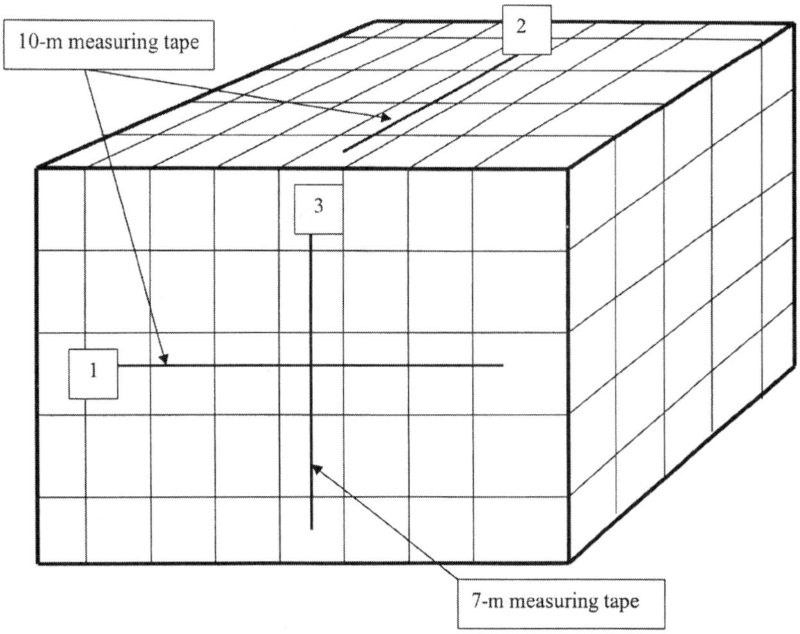

FIGURE 6-16 Schematic diagram illustrating measurement of volumetric joint count (J_v) on rock exposure, with measurements made perpendicular to strike of joint set: Line 1 intersects six joints, Line 2 intersects three joints, and Line 3 intersects four joints. (DIAGRAM BY J. D. HIGGINS.)

to obtain sufficient results and the degree of confidence that can be placed on mean values. Statistical techniques have been developed for guidance (Priest and Hudson 1981; Brady and Brown 1985). However, practical considerations in the field (such as lack of rock exposures in the case of deep excavation projects) force the engineering geologist to use judgment in interpreting results. Generally, measurement techniques that systematically examine all important structural features in a mapped zone are helpful in reducing some of the uncertainty. The methods typically used to map the rock mass structure include *scanline* (line) and *window mapping*.

Wyllie and Mah (2004) recommend use of a tape between 50 and 100 m (164 and 328 ft) long in conducting a scanline survey. The tape is stretched across the bedrock exposure (Figure 6-17). The distance along the scanline is noted for each discontinuity intersection, and the discontinuity properties are described as discussed in previous sections. Each scanline should include general information, such as date of survey, surveyor name, scanline location (accurate position of each end of the tape), rock face orientation, and rock type (Brady and Brown 1985). Two orthogonal scanlines should be measured to characterize each geological zone to avoid bias in the survey, because discontinuities parallel to a scanline will not be measured.

Window mapping consists of mapping all discontinuities in a "window," typically approximately 10 m^2 (108 ft^2) in area, on a rock face (Wyllie and Mah 2004). Properties of structural features are described as noted above for scanline mapping. Preferably, two mutually perpendicular rock exposures should be mapped to avoid bias with respect to discontinuities that are nearly parallel to the rock face, and the window should be large enough to intersect a large sample of discontinuities.

Table 6-15
Descriptive Terms for Block Size

DESCRIPTION	J_v (joints/m^3)
Very large blocks	<1.0
Large blocks	1–3
Medium-size blocks	3–10
Small blocks	10–30
Very small blocks	>30

NOTE: Values of J_v > 60 represent crushed rock, typical of a clay-free crushed zone.
SOURCE: ISRM Working Group 1978, 1981.

Structural zones are established by grouping areas of similar lithology and geological structure. A set of orthogonal scanlines, or window maps, should be used in each zone to describe the rock mass properties. Subsequent analysis of the structural features may allow for grouping of these zones according to failure potential, types of failure, and slope orientation.

3.3.13 Subsurface Investigation

Line and window mapping can be supplemented with a subsurface investigation, particularly where there are few or limited surface outcrops. There are several advantages in gathering subsurface information when potential rockfall characteristics are assessed, including collecting samples for laboratory testing, verifying surface mapping, recording information not available at the surface such as variation of properties with depth or groundwater conditions, and determining rock structure properties including the orientation and dip angle of any discontinuities.

Water can greatly increase the potential for rockfall by reducing the shear strength along a failure surface, decreasing cohesion in infilling materials, and increasing the forces that tend to induce pressure within tension cracks as well as by producing forces that result from volume changes during freeze–thaw cycles. If adverse groundwater conditions are suspected as a result of surface observations, it may be important to monitor groundwater by means of borehole instrumentation. Slotted polyvinyl chloride standpipes can be inserted into a boring and water levels can be read periodically to determine the seasonal fluctuations of the phreatic surface. More complex instruments, such as piezometers, can be permanently installed in boreholes in areas with complex hydrologic conditions. A packer test can be done to determine the hydrologic permeability or transmissivity characteristics. It can also be used to measure the water pressure in confined zones of the subsurface. Selecting appropriate mitigation techniques often depends on accurately quantifying the surface and groundwater flow at a site.

FIGURE 6-17 Orthogonal scanlines on nearly vertical rock face. (PHOTOGRAPH COURTESY OF A. K. TURNER, COLORADO SCHOOL OF MINES.)

3.4 Rock Mass Characterization

When possible, it is desirable to investigate instabilities in rockfall source zones with analytic tools and engineering-mechanics-based approaches. However, obtaining the rock mass strength data required for these analyses is usually difficult because of the large number of samples and extensive testing required to define the mechanical properties of the materials adequately and then characterize the stability of a slope. Therefore, several easily quantified rock mass properties are typically used to predict rock mass behavior. Often, in the case of rockfall along existing transportation routes, maintenance and accident records readily identify where instabilities exist and rockfalls occur. The required analysis focuses on identifying likely source zones, the rock mass characteristics of these zones, and the relative potential for each zone to generate rockfall. This approach can be used to identify areas that need attention and to assess the effectiveness of potential mitigation measures.

Empirical rock mass classification systems are often used either to estimate the strength and deformation properties of the rock mass or to evaluate the general engineering quality of the rock mass. These estimates can then be used to compare potential rockfall source zones. Several rock mass classification schemes have been developed; most were originally applied to underground excavations and stabilization requirements. Some have been modified to include rock slope stability, and at least one has been modified to predict rockfall potential. One, or several, of these classifications is commonly used to provide general numeric indices for the assessment of rockfall source zones and to quantify the relative potential for rocks to be dislodged and become rockfall.

The following sections briefly describe three common rock mass classifications: rock quality designation (RQD), rock mass rating (RMR), and Q-rating system. Typically, these rock mass classification schemes are used to estimate stabilization requirements in a relatively homogeneous rock mass where no single discontinuity controls the stability. Rock mass classification cannot replace detailed design procedures. Applying rock mass classifications to rock engineering problems beyond their original designated purpose should be done with caution.

3.4.1 Rock Quality Designation

The modified core logging procedure, RQD (Deere et al. 1967; Deere 1968; Deere and Deere 1989), serves as an index of the general engineering quality of a rock mass. Highly weathered or altered, soft, fractured, sheared, and jointed rock is not counted in the core recovery; therefore, RQD is a measurement of "good rock" versus "low-strength (low-quality) rock" recovered from an interval of a borehole (Deere and Deere 1989).

RQD is measured by examining core recovered from wire-line core drilling. It is defined as the percentage of hard and sound rock core longer than 100 mm (4 in.) divided by the length of the core run. The lengths of core pieces should be measured

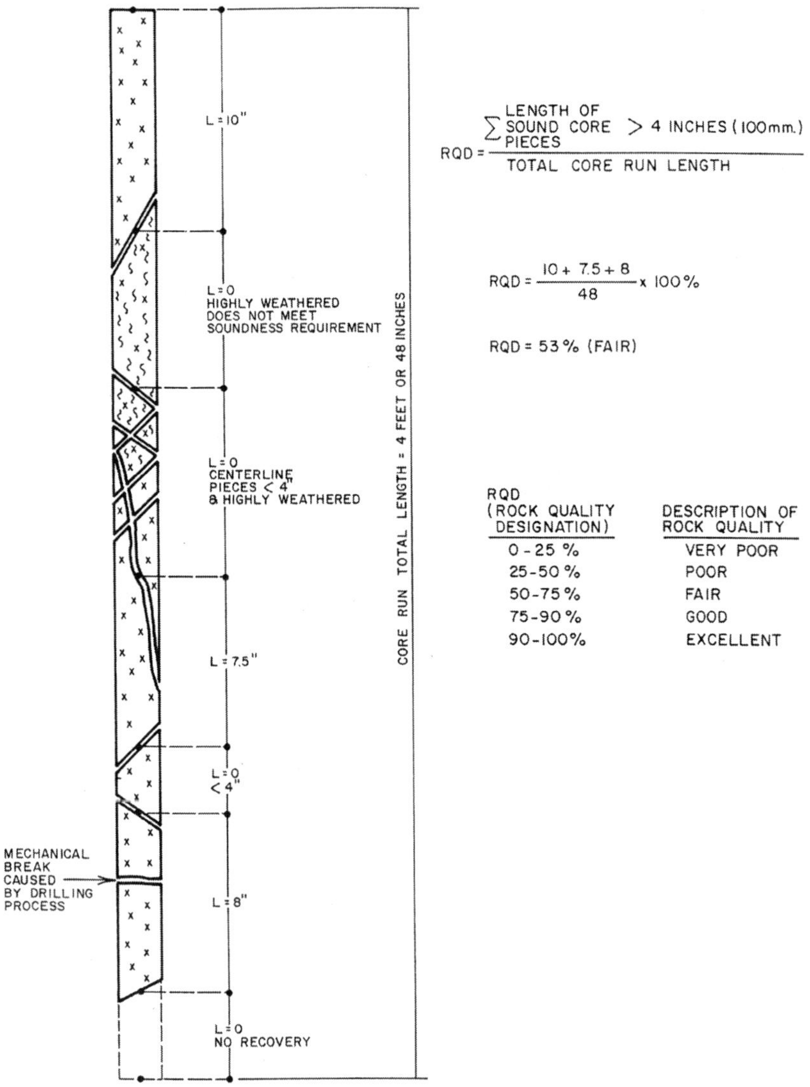

FIGURE 6-18 RQD logging procedure (adapted from Deere and Deere 1989).

along their centerlines. If the core is broken from handling or the drilling process (mechanical breaks), the broken pieces should be fitted together and counted as one piece. The procedure for measuring RQD is shown in Figure 6-18.

RQD was originally developed for use on a minimum NX-size core (54.7-mm or 2.155-in. diameter) and drilled with a double-tube core barrel. However, experience has demonstrated that consistent RQD values can be obtained from smaller and larger cores, ranging from an NQ-size core (47.6-mm or 1.875-in. diameter) up to cores as large as 150-mm (6-in.) diameter (Deere and Deere 1989). The requirement of "hard and sound" rock eliminates weathered or altered core from being evaluated. Several ways to deal with the hard and sound issue are discussed in detail by Deere and Deere (1989).

If drill core is not available, Barton et al. (1974) demonstrated that the order of magnitude of RQD can be estimated from outcrops by determining the number of discontinuities per unit of volume of rock on the basis of the J_v index described in Section 3.3.11. RQD is then calculated by using the Barton et al. (1974) relationship presented in Equation 3:

$$RQD = 115 - 3.3 \cdot J_v \quad (3)$$

RQD may also be established from measurements on rock outcrops by applying the following method:

- Lay out several lines, each 1.5 to 3 m (5 to 10 ft) long and defined by a measuring tape and having several different orientations, on a rock outcrop. These lines are used to replicate drill core runs.
- Determine the total length of hard and sound rock that exists in segments greater than 100 mm (4 in.) long along the line.
- Calculate the RQD by dividing the resulting sum of these lengths of hard and sound rock by the length of the line.

Using several orientations of lines in this method reduces bias that may occur in selecting a single orientation. Both of these methods offer an advantage over RQD determined from core in that all discontinuity orientations are considered (U.S. Bureau of Reclamation 1989).

Although RQD was originally used as a standalone classification, more recently it has often been used as a component of more comprehensive rock mass classification schemes such as the RMR and the Q-rating system.

3.4.2 Rock Mass Rating

RMR, developed by Bieniawski (1976, 1989), is based on six parameters: (*a*) uniaxial compressive strength of rock material, (*b*) RQD, (*c*) spacing of discontinuities, (*d*) condition of discontinuities, (*e*) groundwater conditions, and (*f*) orientation of discontinuities. In addition, ratings can be adjusted for discontinuity orientation and applied to underground excavations, foundations, or slopes. These parameters are rated by using descriptive tables, and the values are summed to determine the RMR. The sum of the values can be used to classify the rock mass ranging from "very good rock" to "very poor rock," and the results can be used to predict stand-up time for mines and tunnels, which was the original purpose of the RMR system.

RMR with adjustments for discontinuity orientation relative to the orientation of the slope face can be useful for rockfall site investigations. To apply RMR to a rockfall investigation, the rock mass is divided into a number of structural regions and each region is classified separately. The structural regions are usually bounded by major structural features, such as faults or a change in rock type, and exhibit similar lithology and structure. Significant changes of discontinuity properties within a single rock type, such as discontinuity spacing or condition, require subdivision into several smaller structural regions. When each of the structural regions is rated and scores are compared, the general rock mass quality of different zones provides a semiquantitative index of the relative potential for rockfall originating in those zones.

3.4.3 Modified Q-Rating System

The original *Q-rating system* (Barton et al. 1974) is used as an engineering classification for the preliminary design of underground excavation support and cost estimates, much like the RMR. Because the Q-rating system depends on most of the discontinuity properties that control rock slope stability, Harp and Noble (1993) selected this system to be modified for applications to rockfall. They proposed use of this system for describing rock quality in potential source zones and for predicting rockfall susceptibility. Because it may not be feasible to access some slopes directly, this method can be used to evaluate the quality of the bedrock outcrops from a distance, often on the basis of oblique or aerial photographs.

The six factors used to calculate Q are RQD, joint set number (J_n), joint roughness number (J_r), joint alteration number (J_a), joint water reduction (J_w), and aperture factor (AF). AF replaces the stress reduction factor in the original Q-rating system. Each factor has an associated rating for different conditions. On the basis of these factors, the value for Q is calculated as shown by Equation 4.

$$Q = \left[\frac{\text{RQD}}{J_n}\right] * \left[\frac{J_r}{J_a}\right] * \left[\frac{J_w}{\text{AF}}\right] \qquad (4)$$

The first input parameter is RQD, defined in Section 3.4.1. Because cores are seldom available for rockfall site investigations, RQD may be determined from outcrops according to methods described in Section 3.4.1 (and by using Equation 3). J_n is based on the number of strongly developed discontinuity sets in the rock mass (Table 6-16). Sporadic discontinuities of no particular orientation are considered random. J_r is based on the departure of the joint from being planar. This classification uses roughness descriptions presented in Section 3.3.5. J_a is either a measure of the weathering on discontinuity surfaces or a measure of the type of weathering material that fills a discontinuity. J_w is a measure of the outflow of water from discontinuities and is inversely proportional to the water pressure within the discontinuity network. AF is related to the amount of openness of the discontinuities. Values for all these factors may be selected from Table 6-16.

The first quotient in the equation represents the structure of the rock mass and is a measure of block or particle size. The second quotient expresses the shear strength of the block mass and represents the roughness and frictional characteristics of the joint walls or filling materials. The third quotient is a measure of the active stress conditions and the joint aperture of the rock mass. The lower the value of Q, the more susceptible the outcrop is to initiating rockfall. High values of Q reflect a rock mass that has few or tight discontinuities and is generally less susceptible to rockfall. Therefore, this system can be applied over small to large portions of rock slopes by first establishing structural regions and then determining Q for each region. A comparison of Q-ratings establishes the most likely rockfall source zones. Harp and Jibson (2002) suggested a relative classification of rockfall susceptibility to seismic failure that uses the modified Q-rating system (Table 6-17). The relationship of the Q-ratings to the susceptibility categories may vary for aseismic areas; however, it is the authors' experience that Q-ratings of less than 1 to 2 fall in the high-susceptibility ranking.

3.5 Rock Mass Strength Properties

When stability analysis is appropriate for rockfall mitigation, rock mass strength parameters are required. Once the lithology and geological structures have been identified and classified in

Table 6-16
Modified *Q*-Rating System for Determination of Rockfall Susceptibility

Rating Category	Rating	Notes
1. Rock quality designation (RQD)		Where RQD is measured as 10 or less (including 0) a nominal value of 10 is used to evaluate Q
Very poor	0–25	
Poor	25–50	
Fair	50–75	RQD intervals of 5; 100, 95, 90, etc., are sufficient
Good	75–90	
Excellent	90–100	
2. Joint set number (J_n)		For intersections ($3 \times J_n$)
Massive, no or few joints	0.5–1.0	
One joint set	2	
One joint set plus random	3	
Two joint sets	4	
Two joint sets plus random	6	
Three joint sets	9	
Three joint sets plus random	12	
Four joint sets, random, heavily jointed, "sugar cube", etc.	15	
Crushed rock, earthlike	20	
3. Joint roughness number (J_r)		
(*a*) Rock wall contact and rock wall contact before 10-cm shear		
Discontinuous joints	4	
Rough or irregular, undulating	3	
Smooth, undulating	2	
Slickensided, undulating	1.5	
Rough or irregular, planar	1.5	
Smooth, planar	1	
Slickensided, planar	0.5	
(*b*) No rock wall contact when sheared		
Zone containing clay minerals thick enough to prevent rock wall contact	1	
Sandy, gravelly, or crushed zone thick enough to prevent rock wall contact	1	
4. Joint alteration number (rock wall contact) (J_a)		
Tightly healed, hard, nonsoftening, impermeable filling, such as quartz or epidote	0.75	
Unaltered joint walls, surface only	1 ($\varphi_r = 20–35$)	Values of φ_r are intended as an approximate guide to mineralogic properties of the alteration products, if present.
Slightly altered joint walls, nonsoftening mineral coatings, sandy particles, clay-free disintegrated rock, etc.	2 ($\varphi_r = 25–30$)	
Silty or sandy coatings, small clay fraction (nonsoftening)	3 ($\varphi_r = 20–25$)	
Softening or low-friction clay mineral coatings (discontinuous coatings, 1–2 mm or less)	4 ($\varphi_r = 8–16$)	

(continued)

Table 6-16 *(continued)*
Modified *Q*-Rating System for Determination of Rockfall Susceptibility

Rating Category	Rating	Notes
5. Joint water reduction factor (J_w)		
Dry	1.0	
Seeps present (≤5 gal/min)	0.7	
Well-defined seeps (>5 to ≤10 gal/min)	0.3	
Well-established groundwater flow (>10 gal/min)	0.1	
6. Joint aperture (AF)		If perched or loose rocks are common, increase by 1.
All joints tight	1	
Most joints tight, a few open as much as 2 cm (<1 in.)	2.5	If pervasive joints dip out of slope, increase by 1.
Most joints tight, a few loose, open as much as 5 cm (2 in.)	5	
Significantly (20%) open, as much as 10 cm (4 in.)	7.5	
Greatly (60%) open, as much as 20 cm (8 in.)	10	
Gaping open, many joints open >20 cm (8 in.)	15	

NOTE: $Q = (RQD/J_n) \times (J_r/J_a) \times (J_w/AF)$. Values of φ_r are approximate.
SOURCE: Adapted from Harp and Noble 1993.

the rockfall source zone, the rock mass strength properties can be estimated from a combination of engineering judgment, laboratory testing, and backcalculation.

The most effective method for evaluating the strength of the on-site rock is through a comprehensive laboratory testing program. Laboratory testing can be used to determine both intact rock strength properties and discontinuity strengths. Intact rock strengths are determined from Brazilian disk tension tests, unconfined compressive strength tests, and triaxial tests. Discontinuities are tested by direct shear tests. Density, corrosivity, and sulfates can also be tested; the results can then be used to evaluate the design life of the mitigation system.

Once intact rock strengths and discontinuity strengths are determined, the source area can be evaluated for its rock mass strength. The rock mass strength is related to the potential for generating rockfall. Empirical relationships have been studied over the years to estimate rock mass strengths. Researchers (Hoek and Brown 1988, 1997; Hoek et al. 2002) have developed an empirical failure criterion for fractured rock masses.

Table 6-17
Descriptive Ranking of Rockfall Susceptibility to Seismic Failure Based on Modified *Q*-Values

Range of Modified Q-Values	Descriptive Ranking of Rockfall Susceptibility
<0.1	Very high
0.1–1	High
1–10	Moderate
>10	Low

SOURCE: Harp and Jibson 2002.

The Hoek–Brown failure criterion (Hoek and Brown 1988, 1997; Hoek et al. 2002) is based on the properties of intact rock and introduces the character of discontinuities in a rock mass to reduce these properties. A Windows-based computer program called RockLab has been developed to solve and plot the equations (Rocscience 2009). The method assumes isotropic rock and rock mass behavior and, therefore, is applicable to rock masses only where closely spaced discontinuities with similar surface characteristics are likely to cause isotropic behavior. The scale used is such that the structure being analyzed is large relative to the block size. If one discontinuity set is significantly weaker than the others, the Hoek–Brown failure criterion method is not used. In that case, the stability of the structure is analyzed by considering failure mechanisms involving sliding or rotation of blocks and wedges defined by a structural feature.

3.6 Slope Profile and Materials

Ideally, the slope profile and materials along the travel path of a rock moving from the source zone to the farthest extent of the runout zone are characterized so that rockfall may be accurately modeled by empirical or computer methods and protection methods may be considered for mitigation. It is important to recognize the complexity of a rockfall-prone slope at the beginning of an investigation, so that, if computer modeling is anticipated, the site investigation is designed to collect the data required to conduct the modeling. Software packages for rockfall modeling commonly use specific measurements and observations to account for the geometric and material parameters discussed in previous sections. Chapter 9 provides additional details on modeling and evaluating rockfall behavior.

A key part of characterizing a rockfall site is construction of topographic profiles connecting the source zones to the runout zones (Figure 6-19) and descriptions and measurements of surface properties along the profiles. For projects that include a large area and many rockfall paths, a large-scale engineering geological map with a topographic base that includes the three zones provides a convenient representation. The topographic profiles or maps are constructed on the basis of conventional surveying, Global Positioning System, or existing topographic data. Large natural slopes often contain many gullies (or chutes) that can direct rockfall down the slope; thus, it is important to map and correlate them with source and runout zones (Figure 6-8). It is essential that the topographic profile and map be of sufficient detail (scale) to define not only the overall slope profile but also potential launch points (Figure 6-19). Currently, several two-dimensional models are widely used for rockfall analysis, and therefore slope profiles of likely rockfall paths are adequate.

FIGURE 6-19 Slope profile identifying launch points and source areas (adapted from Brawner 1994).

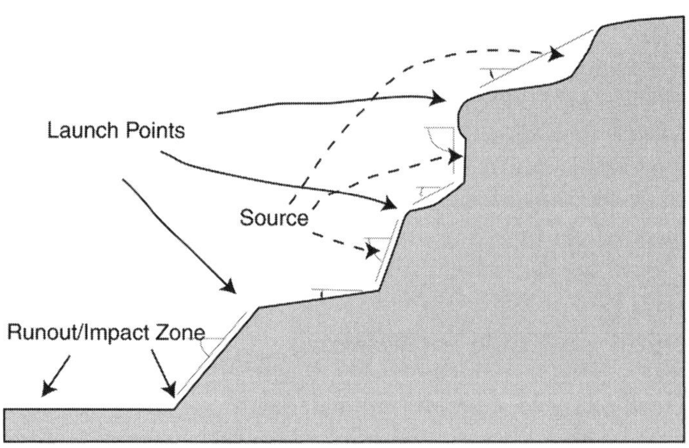

In addition, three-dimensional models have been developed that evaluate not only the bounce height and energy of a rockfall but also include the dispersion characteristics of the event. These models require a detailed topographic base to determine rockfall paths.

During profile construction and mapping, existing impact locations on structures or trees, or in soft ground, are noted in the field (Figure 6-20). This information can be used to approximate bounce heights and launch points. Although impact marks on the slope or trees do not necessarily represent maximum bounce heights, they do provide some indication about whether rocks are bouncing or rolling and some basic insight into bounce heights.

It is essential that surface materials along the rockfall profiles be described in enough detail to provide property characteristics required by the computer simulation model. Descriptions of unconsolidated surficial materials include information on grain-size distribution and some indication of the relative "hardness" of the surface, while bedrock descriptions include information on lithology, relative hardness and strength of surface, and percent exposure. Tables 6-18 and 6-19 may be helpful for describing soils and bedrock exposed at the surface. Some example descriptions follow.

- Coarse boulder talus, 0.5 to 1 m (1.6 to 3.3 ft) in diameter, composed of hard, angular interlocking clasts of granitic gneiss, very strong and hard surface.
- Mixture of talus and colluvium ranging from silt to cobble in size, loose and moderately soft surface.
- Bedrock slope composed of very strong basalt, smooth dip slope, 100% rock exposure at the surface and no rockfall or rubble on the slope.

As discussed in Section 2, the behavior of a rock traveling down a slope will be affected by the coarse materials accumulated along the rockfall profile. The surface roughness produced by these coarse materials, and by other irregularities of similar scale, controls the rockfall trajectory and the degree of damping of the rock block motion. Surface roughness is easily estimated in the field by measuring the greatest repeatable perpendicular variation of the slope surface from the average slope angle over a distance of 2 m (6 ft) (Figure 6-21). Typically, vegetation on the slope

(a) (b)

FIGURE 6-20
(*a*) Rockfall-damaged tree on slope above Interstate 70 at Georgetown, Colorado. Bark is scarred and tree trunk is broken off 5.5 m (18 ft) above ground surface. (*b*) Aspen tree damaged by rockfall, including scars on bark. Tree trunk is broken off 2.3 m (7.5 ft) above ground surface.
(PHOTOGRAPH COURTESY OF R. BARTINGALE, COLORADO SCHOOL OF MINES.)

Table 6-18
Soil Hardness Classification for Constructing Slope Profiles

Soil Hardness	Depth of Penetration of Pick End of Rock Hammer from Firm Blow to Soil Surface
Soft	Entire head of the hammer is buried. Only hammer neck is visible.
Moderately soft to moderately firm	Entire pick is buried. Blunt end of hammer is visible.
Firm	50% to 75% of pick penetrates soil surface.
Very firm	Less than 50% of pick penetrates soil surface.

Source: Bartingale 2009.

is also described, because it can reduce rockfall velocity and even help stabilize the materials that form the slope.

Accurate description of the runout zone, the zone of deceleration and deposition of rockfall, is important for understanding the behavior of past rockfall and for calibrating computer models. Information required to describe the runout zone includes size, density, shape, and distribution of rocks that appear likely to have been deposited from previous rockfalls. Coe et al. (2005) proposed a method for obtaining consistent rockfall runout zone characteristics. A tape is extended longitudinally through the runout zone. The profile is separated into cells based on similar slope and surface materials, and all rocks 0.3 m (1 ft) in diameter or larger that are observed lying on

Table 6-19
Rock Strength Classification for Constructing Slope Profiles

CLASS	STRENGTH	FIELD TEST	APPROXIMATE RANGE OF UNIAXIAL COMPRESSION STRENGTH (kg/cm^2)
I	Extremely strong	Many blows with geological hammer required to break intact specimen.	>2,000
II	Very strong	Handheld specimen breaks with hammer end of pick under more than one blow.	2,000–1,000
III	Strong	Cannot be scraped or peeled with knife, handheld specimen can be broken with single moderate blow with pick.	1,000–500
IV	Moderately strong	Can just be scraped or peeled with knife, indentations 1–3 mm show in specimen with moderate blow with pick.	500–125
V	Moderately weak to weak	Material crumbles under moderate blow with sharp end of pick and can be peeled with a knife but is too hard to hand-trim for triaxial test specimen.	125–12

NOTE: 1 kg/cm^2 = 1.024 $tons/ft^2$.
SOURCE: Adapted from Naval Facilities Engineering Command 1982.

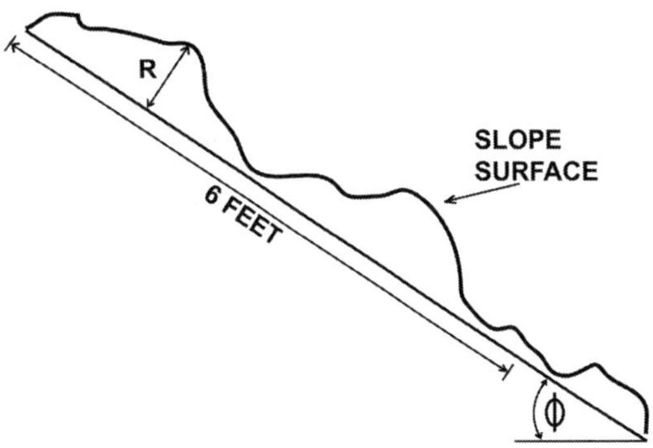

FIGURE 6-21
Estimation of surface roughness (*R*) established as largest repeatable perpendicular variation of slope surface from average slope angle (ϕ) over distance of 2 m (6 ft) (Bartingale 2009).

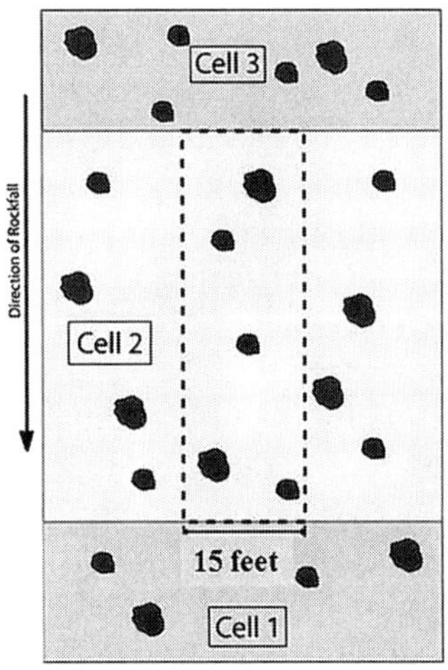

FIGURE 6-22
Rock distribution for each cell is measured by counting all rocks in section 4.6 m (15 ft) wide (in direction perpendicular to rockfall path). Rocks at least 0.3 m (1 ft) in diameter and located completely on top of soil are considered to be from rockfall processes (adapted from Bartingale 2009 and Coe et al. 2005).

the surface within a 4.6-m (15-ft) width along the profile are counted. The diameter of each rock is estimated (Figure 6-22). Any rocks that are partially buried in the soil matrix are assumed to have been deposited by processes other than rockfall. This method provides details of runout distance for different sizes of rocks from past rockfalls.

The size, shape, and makeup of rocks are important. Rock shapes can be generalized as spherical, tabular, and disklike to facilitate analysis (Figure 6-23). These ideal shapes behave somewhat differently in motion, especially at lower velocities during acceleration or deceleration. Spherical rocks will accelerate at a faster rate and decelerate at a slower rate than angular rocks. The size of rock blocks is important for energy calculations and depends primarily on the nature of the source material and its tendency to disintegrate. Disintegration, or fragmentation, of rock blocks further dissipates energy, and the smaller blocks so created have travel characteristics considerably different from those of the original larger rock blocks. Chapter 8 provides detailed discussions of the fundamental physical laws and concepts that control the mechanical behavior and energies of falling, bouncing, and rolling rock blocks.

4. CASE HISTORY: ROCKFALL SITE CHARACTERIZATION FOR GEORGETOWN INCLINE

Yeh and Associates, Inc. (2005) performed a rockfall mitigation feasibility study for a 3.54-km (2.2-mi) section of Interstate 70 west of Denver, Colorado. The work included a site characterization study, construction of a conceptual model, and computer modeling of rockfall to determine possible mitigation scenarios. The approach to site characterization and development of the conceptual model is summarized below as a demonstration of the various steps required to characterize a relatively complex, and high-risk, rockfall location. In general, the study followed the suggested methods outlined earlier in this chapter.

The site is located along a section of Interstate 70 between Georgetown and Silver Plume, Colorado, approximately 72 km (45 mi) west of Denver in the Front Range of the Colorado Rocky Mountains. This section of Interstate 70 is locally referred to as the Georgetown Incline because the roadway climbs approximately 150 m (500 ft) from Georgetown to Silver Plume with a continuous steep grade ranging from 5% to 8%, very steep for Interstate design standards (Figure 6-24). The highway alignment is cut into steep mountainous slopes that exceed 1V:1H in many places. The natural back slopes rise more than 525 m (1,700 ft) above the current highway alignment along most of the project. Rock cuts and disturbed areas just above the highway, created to accommodate the multilane roadway, generally range from 5 to 45 m (20 to 150 ft) high and have slope angles ranging from vertical to 60° (Figure 6-25).

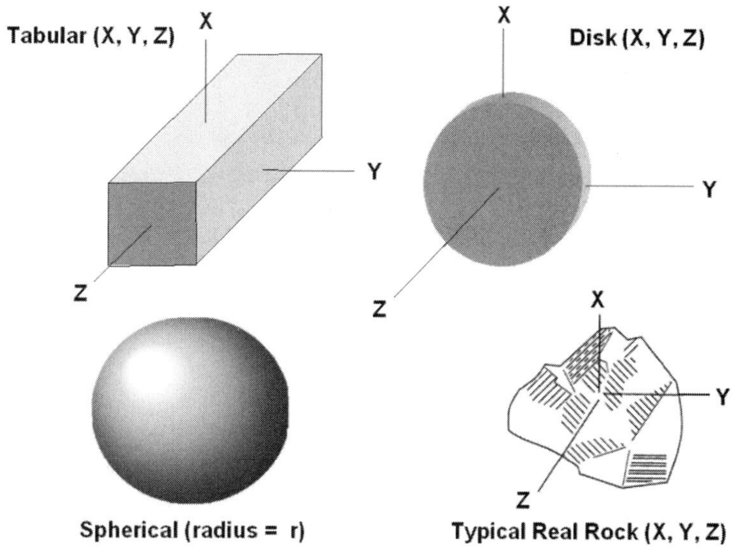

FIGURE 6-23 Common rockfall shapes (tabular, disk, and spherical) compared with typical real rock (adapted from Brawner 1994).

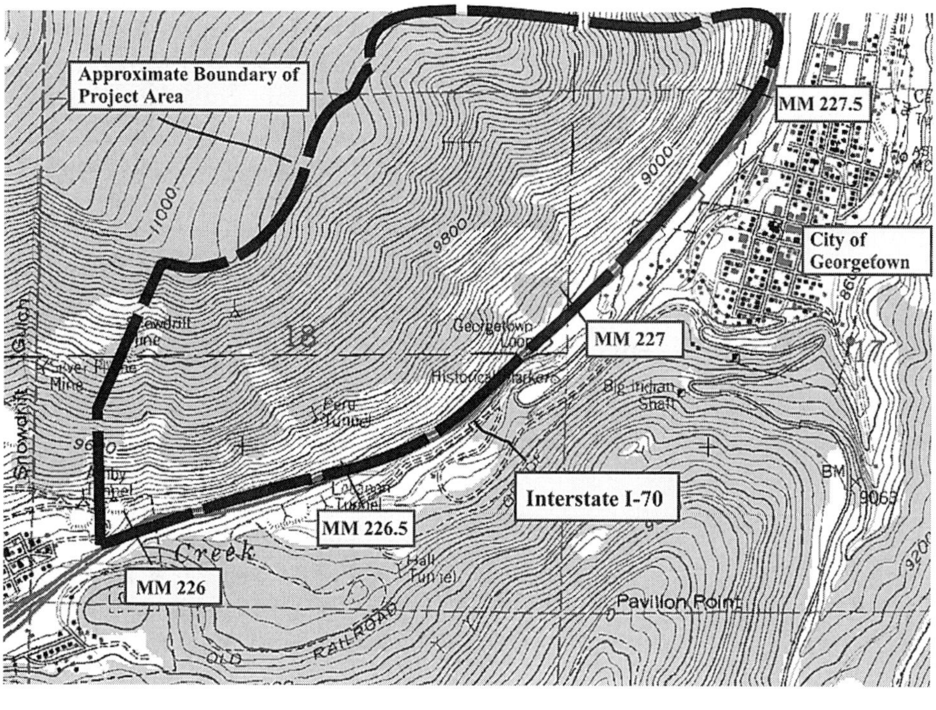

FIGURE 6-24 Topographic map of Georgetown Incline, showing approximate boundary of rockfall area (Arndt et al. 2003).

FIGURE 6-25 Oblique aerial photograph showing overview of Georgetown Incline. (PHOTOGRAPH COURTESY OF R. BECK, COLORADO DEPARTMENT OF TRANSPORTATION.)

The bedrock geology consists of Idaho Springs granite gneiss that includes pegmatite intrusions and Silver Plume granite that has been intruded into the granite gneiss. The surficial materials covering the steep back slopes above the roadway consist of colluvium, talus, and isolated mine tailings; these deposits include materials ranging from silt to boulders. Numerous bedrock outcrops form vertical cliff faces, creating potential source areas for rockfall. In addition, the site is located at an elevation between 2,590 and more than 3,350 m (between 8,500 and more than 11,000 ft) and thus experiences climate conditions that are conducive to rockfall, including periods of intense precipitation and extended freeze–thaw events. The combination of climate, steep slopes, fractured bedrock outcrops, and relatively loose surficial materials that include many large fragments creates an extensive rockfall hazard area that has experienced numerous rockfall events. Interstate 70 is a major transportation corridor through the Rocky Mountains and thus experiences heavy traffic volumes. Consequently, high potential for rockfall poses considerable risk to the traveling public.

4.1 Review of Past Rockfall Events and Literature

Rockfall events have led to accidents and fatalities along the Georgetown Incline. Approximately 100 rockfall-related accidents have occurred in the past 24 years, and they have resulted in 17 injuries and three fatalities. Rock-filled ditches along the westbound lanes and numerous rock blocks located on the embankments along the eastbound lanes provide physical evidence that there is a much greater incidence of rockfall activity than is indicated by reported accidents (Yeh and Associates 2005).

Two fatalities in 1999 and two accidents that resulted in injury in 2001 were caused by rockfall activity at the Georgetown Incline. On May 16, 1999, a woman was killed after boulders struck her car at Mile Marker 226.40. Some of the boulders in this event were reportedly the size of bathtubs. On December 9, 1999, a man was killed when a rock came through the window of the shuttle van in which he was riding. This accident occurred at Mile Marker 227.35. A rockfall event that occurred on May 6, 2001, caused two accidents when cars encountered rock debris in the road. The 2001 event occurred after 4 days of mixed rain and snow precipitation. Rockfall debris from this event was observed in the rockfall ditch of the westbound lane, adjacent to the slope; on the shoulder of the eastbound lane, meaning that the rocks had crossed the entire multilane roadway; and on a bicycle path located below the highway embankment, indicating that some rocks traveled even farther from the base of the slope. Rockfall debris that affects the eastbound lanes, or the adjacent bicycle path, must have experienced high energies and extreme bounce heights.

There are typically many causes and triggers for rockfall in mountainous areas such as the Georgetown Incline; however, climate and precipitation combined with rock and soil conditions often are prominent factors. The effect of climate and precipitation was examined with respect to triggering rockfall events. Figure 6-26 plots reported rockfall accidents, average snowfall, average maximum and minimum temperatures, average precipitation, and average maximum wind velocity for each month. Although rockfall-related accidents may occur anytime on the Georgetown Incline and may be triggered by several factors, the months experiencing the highest frequency of rockfall-related accidents appear to correlate to elevated levels of precipitation, snowfall, and snowmelt. For March and April, snowfall peaks and average maximum and minimum temperatures fluctuate above and below freezing (freeze–thaw cycles). The warm temperatures, generally during the daytime, melt snow and water enters shallow

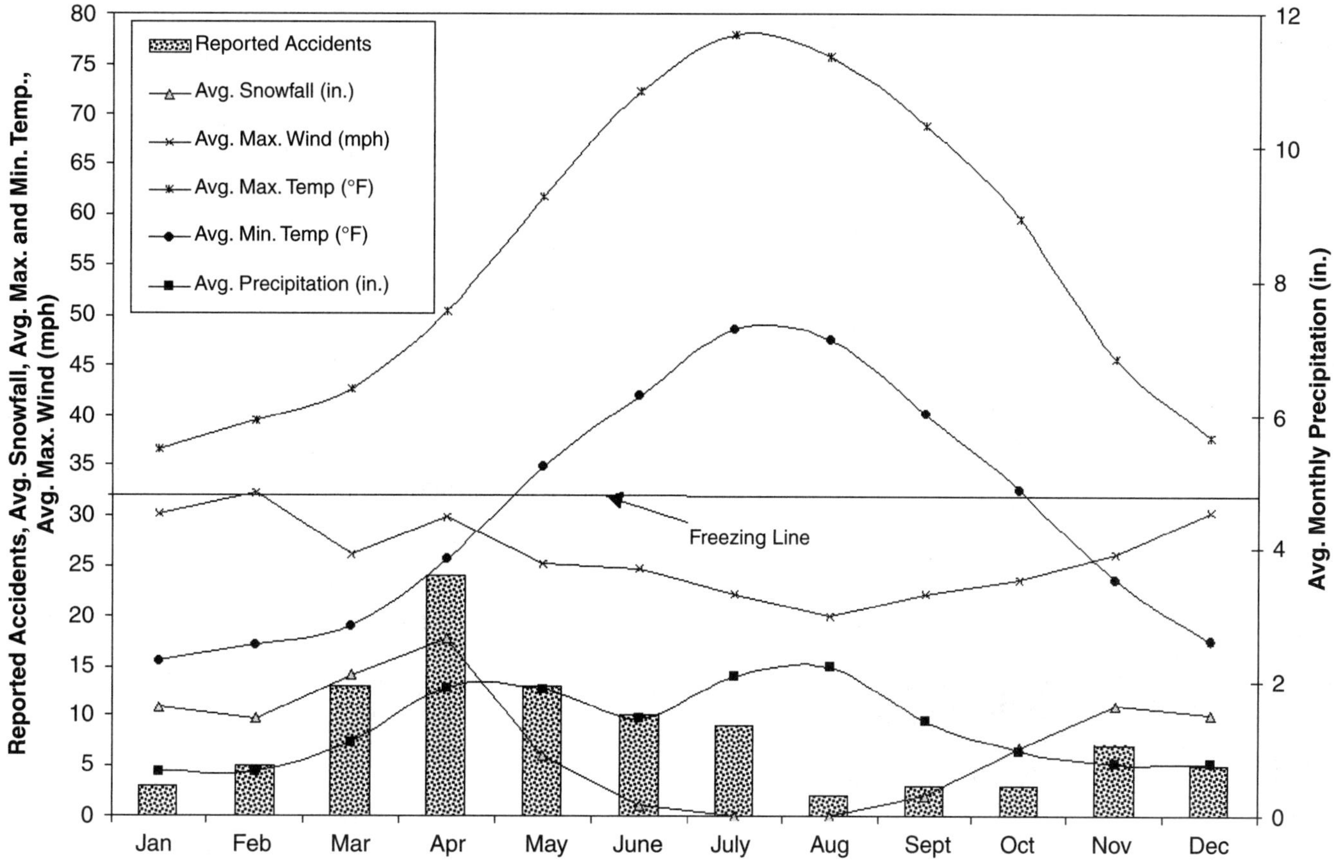

FIGURE 6-26 Relationship of reported accidents, average snowfall, wind velocity, average maximum and minimum temperatures, and average precipitation for each month on Georgetown Incline. Weather measurements are for 1948 to 1979 and 1995 to 2000. Colorado DOT accident information is from 1976 to 1999. Data are from Western Regional Climate Center and W. Wilson, Colorado DOT (Arndt et al. 2003).

rock fractures. Temperatures drop at night, or for several days during the passage of cold fronts, and the water freezes in shallow fractures. Rocks may be dislodged by ice wedging, or ice forming at the surface may block drainage and elevate water pressures within fractures, promoting rockfall. Also, warm periods during March and April can cause rapid melting of snow, resulting in surface runoff that may erode fractured bedrock blocks or dislodge boulders from unconsolidated deposits. May, June, and July also experience elevated numbers of rockfall-related accidents, which appear to correlate with elevated precipitation levels. Much of this precipitation comes in the form of intense rainfall, or several days of rain, that can trigger rockfall and even debris flow, because of erosion of fractured bedrock and unconsolidated material (Yeh and Associates 2005). Wind does not appear to be a major trigger of rockfall at this site.

The extensive rockfall history at the Georgetown Incline defined a general pattern for a conceptual model. It also provided insight for how best to proceed to characterize the site. Accident reports provided the locations at the roadway of the rockfall events, the sizes of rocks involved, the locations of rocks and size distribution in the runout zone, and the extent of the runout zone. These factors are a useful part of the conceptual model on which to base analysis.

Accident reports, maintenance experience, and several unpublished Colorado Department of Transportation (DOT) studies suggested that existing rockfall catchment ditches on the north side of Interstate 70 (next to the westbound lanes) were usually adequate to stop rockfall originating on the cut slopes immediately above the roadway (Figure 6-27*a*). Rockfall that traveled beyond the catchment ditch typically originated high on the steep natural slopes (above the cut slopes) in very steep and jointed bedrock outcrops (Figure 6-27*b*) and traveled at high velocities (and often achieved high bounce heights) down

(a)

(b)

(c)

topographic chutes and well-established rockfall travel paths to the roadway. Coarse colluvium and talus deposits (Figure 6-27*c*) tend to occur in less steep areas and chutes and produce less rockfall that reaches Interstate 70 because the lower slope angle prevents the high launch velocities required to keep the rock blocks moving down the slope (Arndt et al. 2003; T. Ortiz, personal communication, Aug. 5, 2009).

4.2 Rockfall Hazard Rating System

An initial evaluation of the rockfall hazard for the rock cuts along the roadway was previously conducted by using the Colorado Rockfall Hazard Rating System (CRHRS). The CRHRS, described by Andrew (1994), is an adaptation of the RHRS developed by Pierson in 1989 for the Oregon Department of Transportation (Pierson et al. 1989; Pierson 1992). Chapter 3 discusses, in some detail, the concepts guiding the development of the RHRS; Chapter 4 discusses adaptations of the RHRS concepts, including CRHRS, by several transportation agencies. The version of the CRHRS used in the Georgetown Incline study was updated by the Colorado DOT (1997) and by Santi et al. (2009).

For many rockfall projects, CRHRS scores are useful tools for evaluating the most hazardous locations, and, if scoring is available for multiple years, they may provide a sense of rate and degree of slope deterioration with time (see Section 3.1). CRHRS scores were available for the cut slopes on the Georgetown Incline from 1992, 1998, and 2003, but there were no ratings on the natural slopes above the cut slopes. As mentioned above, rockfall incident reports and previous Colorado DOT studies indicated that nearly all the rockfall that traveled past the catchment ditch located at the shoulder of the roadway originated on the natural slopes above the cut slope. The Colorado DOT experience and modeling indicated that the ditches were effective in stopping most rockfall originating on the cut slopes (T. Ortiz, personal

FIGURE 6-27
Rockfall source zones on slopes above Georgetown Incline: (*a*) cut slopes immediately above Interstate 70, (*b*) fractured granite gneiss, and (c) colluvial deposits collecting and eroding in steep drainages.
(PHOTOGRAPHS COURTESY OF B. ARNDT, YEH AND ASSOCIATES, INC., DENVER, COLORADO.)

communication, Aug. 5, 2009). For these reasons the CRHRS did not play an important role in this study, and other means were used to assess rockfall hazard zones.

4.3 Engineering Geological Mapping

Since it had been determined that most rockfall to reach Interstate 70 traveled down the slope in well-established travel paths and topographic chutes, engineering geological mapping was conducted to identify the numerous potential source zones and rockfall travel paths. The steep, rugged terrain limited access to the slope for mapping; thus, aerial photographic mapping was used with limited field checking. Figure 6-28 shows a portion of the map that was prepared for the Georgetown Incline site. Because vegetation did not mask the bedrock and colluvial deposits on the slope, the engineering geologist could easily identify the important landforms and materials. The original map showed locations of cut slopes, bedrock outcrops, mile markers, and rockfall travel paths (pathways and chutes in Figure 6-28). As work progressed, additional information was added to the map, such as Q-ratings of bedrock outcrops and the location of profiles for rockfall modeling, to complete the conceptual model. Topographic profiles for the identified rockfall pathways were prepared by using topographic maps. The fieldwork and mapping allowed the engineering geologist to determine the condition of the bedrock in potential source zones and whether rockfall originating in the bedrock had a reasonable travel path to the road or stopped on the slope. Also, field descriptions of material properties along the rockfall chutes were obtained.

Figure 6-28 shows the location of numerous bedrock outcrops and rockfall travel paths (chutes and pathways). The chutes identified by Yeh and Associates (2005) are topographic gullies that act as travel paths for rockfall. The pathways are broader zones on the slope representing rockfall travel paths connecting source zones with Interstate 70. Some of the bedrock outcrops are adjacent to chutes or pathways and some are isolated so that rocks falling from the outcrops do not have a clear travel path to the roadway. Accident and maintenance reports show that few rocks from these source zones reached the highway. Therefore, they were considered to pose a low risk to Interstate 70.

Once the source zones and associated travel paths were identified, the relative hazard on Interstate 70 at each chute or pathway location was evaluated on the basis of the rockfall susceptibility of the bedrock outcrops above and the history of events.

4.4 Rock Mass Rating

To determine the relative rockfall susceptibility of bedrock outcrops high on the slope, the modified Q-rating system (Harp and Noble 1993) was applied to rate the rock mass. Because of limited access to the steep slopes and safety considerations with respect to climbing on the slope above heavy traffic, most Q-ratings were accomplished from across the valley by using binoculars and telescopes or from aerial photographic analysis. Many of the bedrock outcrops were rated two or three times from different angles, and the ratings were averaged to reduce error from the remote measurement method. (The Harp and Noble method was developed by using a similar remote measurement method.) Some outcrops were rated only once either because the site was observed from a close distance or because it was not possible to view it from alternative angles.

Table 6-20 shows a sample of modified Q-ratings for outcrops, associated chutes, and rockfall accidents at the base of the chutes. The table was used to identify and rank the rockfall hazard for each chute. The chute numbers and outcrop designations (e.g., A3) listed in Table 6-20 correspond to those on the engineering geological map, a portion of which is shown in Figure 6-28. For example, the site topography directs rockfall originating in Source Area A4 and some of Source Area A3 to enter Chute 1, which provides a pathway to Interstate 70 for falling rock. One rockfall-related accident was recorded on the roadway at the base of Chute 1. This accident provided evidence that rockfall can be transported all the way to the road.

Q-ratings for the two bedrock outcrop source zones are listed in Table 6-20. The two A3 ratings are 0.44 and 0.56, and the A4 rating is 1.78; all were considered to rank as having high rockfall susceptibility. Therefore, Chute 1 was considered to be a likely source of rockfall and to pose a hazard to Interstate 70. Some chutes were ranked more

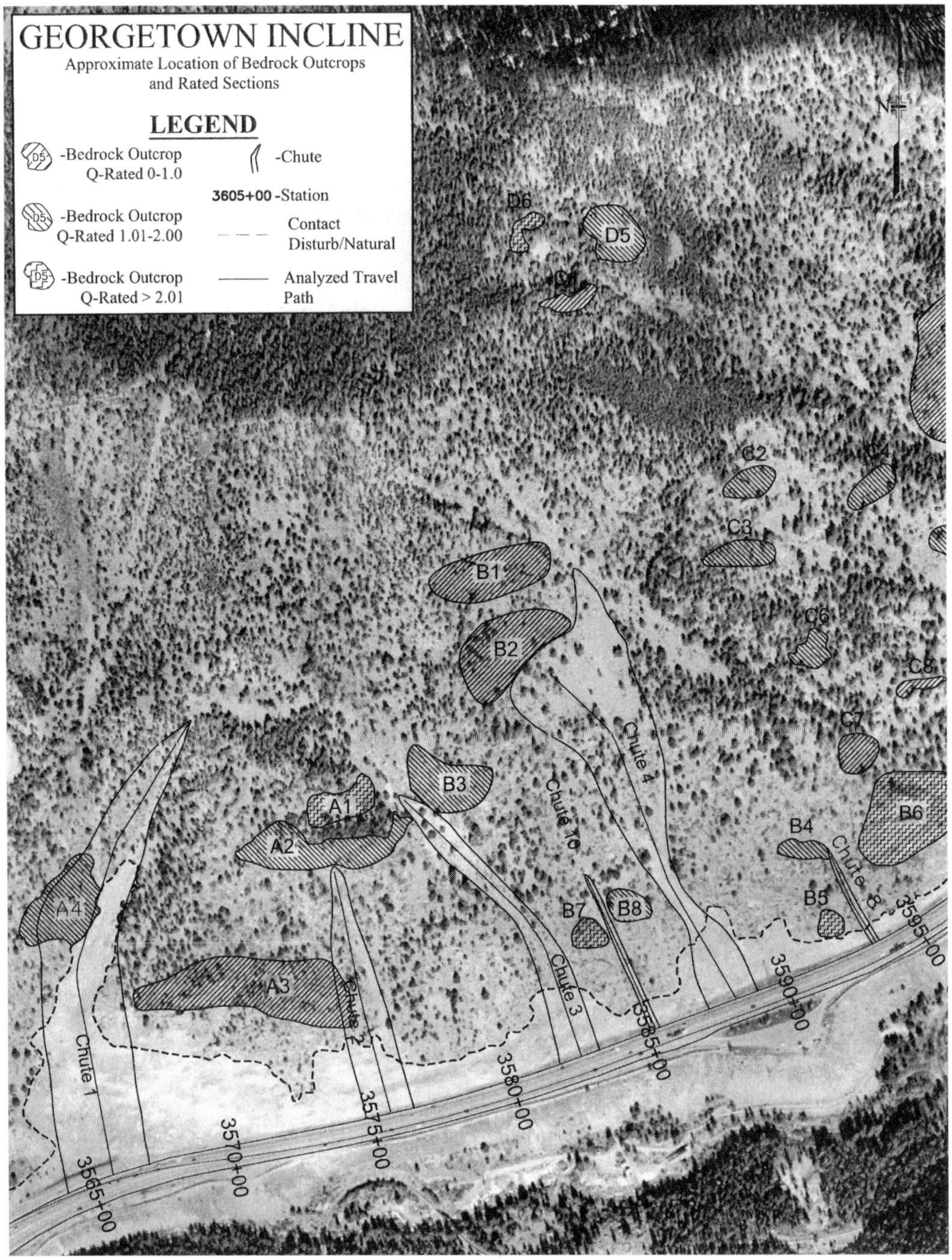

FIGURE 6-28 Engineering geological map of Georgetown Incline showing outcrops, rockfall pathways, and *Q*-ratings. Colorado DOT survey stations are spaced every 500 ft along Interstate 70 (Yeh and Associates, Inc., 2005).

hazardous or less hazardous because of the *Q*-rating scores of the source zones and, possibly, the number of accidents. The number of accidents was high for some chutes that connected to source zones with *Q*-ratings of slightly greater than 1 (e.g., Chute 2). In these cases, there were more moderate *Q*-ratings; however, factors other than the rock mass condition in the source zone contributed to the actual hazard at the roadway. Such factors included the size of the source zone that contributed rockfall to the chute and the number of rocks that made it through the chute to the roadway.

In general, only outcrops rating less than 1 and associated with well-formed travel paths that allow rockfall to reach the roadway were considered for the conceptual model. Field and mapping observations identified several major outcrops with *Q*-ratings of less than 1 where

Table 6-20
Modified *Q*-Ratings of Outcrops, Associated Chutes, and Accidents for Georgetown Incline

Chute	Mile Range	Outcrop	Multiple Q-Ratings Recorded[a]	Average Rating	Number of Accidents
1	226.0–226.06	A3, A4	0.44, 0.56, 1.78	0.93	1
2	226.17–226.29	A1, A2	4.00, 1.56, 1.33, 3.00, 2.00	2.38	14
3	226.36–226.39	A1, A2, B3, B7, B8	4.00, 1.56, 1.33, 3.00, 2.00, 1.00, 1.11, 3.50, 1.73	2.14	9
4	226.46–226.57	B1, B2, B3, B4	0.52, 0.69, 0.69, 0.64, 1.00, 1.11, 0.92, 2.33, 1.88	1.09	9
5	227.16–227.22	E1, E2, E3, E4, E5, E8	2.22, 1.11, 0.56, 0.56, 1.67, 5.00	1.85	4
6	227.37–227.44	E6, E7, E9, F1, F4	4.00, 1.35, 2.50, 0.50, 0.27, 0.50, 0.59	1.39	11
8	226.60–226.64	B4, B6, C7	0.92, 2.33, 1.88, 2.31, 2.00, 0.98, 0.93	1.62	1
9	227.25–227.32	E1, E2, E3, E5, E6	2.22, 1.11, 0.56, 1.67, 4.00	1.91	4
10	226.42–226.44	B3, B7, B8	1.00, 1.11, 3.50, 1.73	1.84	0

NOTE: Chute 8 was already meshed. Chute 4 was fenced and meshed November 2000. Chute 6 was fenced and meshed January 2001. Accidents were reported from 1976 to 1999. Outcrops that do not have a path to highway are not included.
[a]The multiple Q-ratings were obtained by several individuals observing each chute from multiple observation points located across the valley from the rock slope. At least three different observation points were used for each chute, but the number of observation points was determined by the availability of suitable sight lines from accessible locations. The average of these ratings was used to evaluate the rock mass conditions.
SOURCE: Arndt et al. 2003.

rockfall appeared to be stopped on the slope because no travel path connected the source zone and the roadway. The Colorado DOT did not consider them to be at high risk requiring mitigation.

4.5 Conceptual Model

The various steps of the characterization study, summarized above, contributed to defining a conceptual model of rockfall hazard along the Georgetown Incline section of Interstate 70; this conceptual model is illustrated in Figure 6-28. The characterization study found that the greatest risk of rockfall to Interstate 70 was along well-established rockfall travel paths that connected the roadway to steep fractured bedrock outcrops. The Q-ratings for these outcrops were high or very high susceptibility. With the addition of topographic profiles, material properties along the rockfall chutes, and details of rock size and characteristics of past events, adequate data were available for computer modeling of rockfall. The modeling could be compared with data collected from past events for calibration and used to predict likely events as a basis for design of mitigation.

REFERENCES

ABBREVIATIONS

DOT	Department of Transportation
ISRM	International Society for Rock Mechanics

Andrew, R. 1994. *Colorado Rockfall Hazard Rating System*. Report No. CTI-CDOT-2-94. Colorado Department of Transportation, Denver.

Arndt, B. P., R. Andrew, and K. Herring. 2003. Rockfall Potential Along I-70 at the Georgetown Incline. In *Engineering Geology in Colorado: Contributions, Trends, and Case Histories* (D. D. Boyer, P. M. Santi, and W. P. Rogers, eds.), Special Publication No. 15, Association of Engineering Geologists, Denver, Colo., pp. 1–18.

Bartingale, R. 2009. *Calibration of the Colorado Rockfall Simulation Program (CRSP) Version 5.0*. MS thesis. Colorado School of Mines, Golden.

Barton, N. 1987. Discontinuities. In *Ground Engineer's Reference Book* (F. G. Bell, ed.), Butterworths, London, pp. 5/1–5/15.

Barton, N., and V. Choubey. 1977. The Shear Strength of Rock Joints in Theory and Practice. *Rock Mechanics*, Vol. 10, pp. 1–54.

Barton, N., R. Lien, and J. Lunde. 1974. Engineering Classification of Rock Masses for the Design of Tunnel Support. *Rock Mechanics*, Vol. 6, pp. 189–236.

Bieniawski, Z. T. 1976. Rock Mass Classification in Rock Engineering. In *Proceedings of the Symposium on Exploration for Rock Engineering* (Z. T. Bieniawski, ed.), Balkema, Leiden, Netherlands, pp. 97–106.

Bieniawski, Z. T. 1989. *Engineering Rock Mass Classifications*. John Wiley and Sons, New York.

Brady, B. H. G., and E. T. Brown. 1985. *Rock Mechanics for Underground Mining*. George Allen and Unwin Ltd., London.

Brawner, C. O. 1994. *Rockfall Hazard Mitigation Methods Participant Workbook*. National Highway Institute Course No. 13219, Publication No. FHWA SA-93-085. Federal Highway Administration, U.S. Department of Transportation, Washington, D.C.

Coe, J. A., E. L. Harp, A. C. Tarr, and J. A. Michael. 2005. *Rock-Fall Hazard Assessment of Little Mill Campground, American Fork Canyon, Uinta National Forest, Utah*. Open-File Report 2005-1229. U.S. Geological Survey, Reston, Va.

Colorado DOT. 1997. *Rockfall Rating Field Worksheet*. Staff and Materials Branch, Denver.

Deere, D. U. Geologic Considerations. 1968. In *Rock Mechanics in Engineering Practice* (K. Stagg and O. Zienkiewicz, eds.), Wiley, New York, pp. 1–20.

Deere, D. U., and D. W. Deere. 1989. *Rock Quality Designation (RQD) After Twenty Years*. Report GL-89-1. U.S. Army Corps of Engineers, Washington, D.C.

Deere, D. U., A. J. Hendron, F. D. Patton, and E. J. Cording. 1967. Design of Surface and Near Surface Construction in Rock. In *Failure and Breakage of Rock, Proceedings of the 8th Symposium on Rock Mechanics* (C. Fairhurst, ed.), Society for Mining Engineers and American Institute of Mining, Metallurgical, and Petroleum Engineers, New York, pp. 237–302.

Dorren, L. K. A. 2003. A Review of Rockfall Mechanics and Modeling Approaches. *Progress in Physical Geography*, Vol. 27, pp. 69–87.

Guzzetti, F., G. Crosta, R. Detti, and F. Agliardi. 2003. STONE: A Computer Program for the Three-Dimensional Simulation of Rock Falls. *Computers and Geosciences*, Vol. 28, No. 9, pp. 1079–1093.

Harp, E. L., and R. W. Jibson. 2002. Anomalous Concentrations of Seismically Triggered Rock Falls in Pacoima Canyon: Are They Caused by Highly Susceptible Slopes or Local Amplification of Seismic Shaking? *Bulletin of the Seismological Society of America*, Vol. 92, No. 8, pp. 3180–3189.

Harp, E. L., and M. A. Noble. 1993. An Engineering Rock Classification to Evaluate Seismic Rockfall Susceptibility and Its Application to the Wasatch Front. *Bulletin of the Association of Engineering Geologists*, Vol. 30, No. 3, pp. 293–319.

Hoek, E., and J. W. Bray. 1981. *Rock Slope Engineering*. Institution of Mining and Metallurgy, London.

Hoek, E., and E. T. Brown. 1988. The Hoek-Brown Failure Criterion—A 1988 Update. In *Rock Engineering for Underground Excavations, Proc., 15th Canadian Rock Mechanics Symposium* (J. C. Curran, ed.), Department Civil Engineering, University of Toronto, Toronto, Ontario, Canada, pp. 31–38.

Hoek, E., and E. T. Brown. 1997. Practical Estimates of Rock Mass Strength. *International Journal of Rock Mechanics and Mining Science and Geomechanics Abstracts*, Vol. 34, No. 8, pp. 1165–1186.

Hoek, E., C. Carranza-Torres, and B. Corkum. 2002. Hoek–Brown Failure Criterion. *Proc., North American Rock Mechanics Symposium–Tunneling Association of Canada Conference*, Vol. 1, Toronto, Ontario, Canada, pp. 261–273.

ISRM Working Group. 1978. Suggested Methods for the Quantitative Description of Discontinuities in Rock Masses. *International Journal of Rock Mechanics and Mining Science and Geomechanics Abstracts*, Vol. 15, pp. 319–368.

ISRM Working Group. 1981. Suggested Methods for the Quantitative Description of Discontinuities in Rock Masses. In *Rock Characterization Testing and Monitoring: ISRM Suggested Methods* (E. T. Brown, ed.), Pergamon Press, New York, pp. 3–52.

Mazzoccola, D., and E. Sciesa. 2000. Implementation and Comparison of Different Methods for Rockfall Hazard Assessment in the Italian Alps. In *Landslides in Research, Theory and Practice, Proceedings of the 4th Congress of the International Association of Engineering Geologists* (E. Bromhead, N. Dixon, and M. L. Ibsen, eds.), New Delhi, India, pp. 221–228.

Naval Facilities Engineering Command. 1982. *Soil Mechanics Design Manual*. NAVFAC DM-7.1. Department of the Navy, Alexandria, Va.

Pfeiffer, T. J., J. D. Higgins, R. D. Andrew, R. J. Schultz, and R. B. Beck. 1995. *Colorado Rockfall Simulation Program, Version 3.0a User's Manual*. Colorado Department of Transportation, Denver.

Pierson, L. A. 1992. Rockfall Hazard Rating System. In *Transportation Research Record 1343*, Transportation Research Board, National Research Council, Washington, D.C., pp. 6–13.

Pierson, L. A., S. Davis, and R. Van Vickle. 1989. *The Rockfall Hazard Rating System Implementation Manual*. FHWA Publication No. FHWA-OR-EG-90-01. Federal Highway Administration, U.S. Department of Transportation, Washington, D.C.

Priest, S. D., and J. A. Hudson. 1981. Estimation of Discontinuity Spacing and Trace Length Using Scanline Surveys. *International Journal of Rock Mechanics and Mining Science*, Vol. 18, pp. 183–197.

Ritchie, A. M. 1963. Evaluation of Rockfall and Its Control. In *Highway Research Record 17*, Highway Research Board, National Research Council, Washington, D.C., pp. 13–28.

Rocscience. 2009. *RockLab v. 1.0: Rock Mass Strength Analysis Using the Generalized Hoek–Brown Failure Criterion*. Toronto, Ontario, Canada. www.rocscience.com/products/Rocklab.asp. Accessed May 27, 2009.

Santi, P. M., C. P. Russell, J. D. Higgins, and J. I. Spriet. 2009. Modification and Statistical Analysis of the Colorado Rockfall Hazard Rating System. *Engineering Geology*, Vol. 104, No. 5, pp. 55–65.

U.S. Bureau of Reclamation. 1989. *Engineering Geology Field Manual*. U.S. Department of the Interior, Denver, Colo.

Wyllie, D. C., and C. W. Mah. 2004. *Rock Slope Engineering: Civil and Mining*, 4th ed. Spon Press, New York.

Yeh and Associates, Inc. 2005. *I-70 Georgetown Incline Rockfall Mitigation Feasibility Study*. Report to Colorado Department of Transportation, Region 1.

Chapter 7

Richard D. Andrew,
Ben Arndt, and
A. Keith Turner

INSTRUMENTATION AND MONITORING TECHNOLOGY

1. INTRODUCTION

The field of instrumentation is a specialty within the larger realm of engineering sciences. The technology continually improves and a significant project value is associated with implementing a successful instrumentation program. Also, substantial capital investment is involved in having a protocol to install, read, and maintain instruments. To a considerable extent, implementing a successful instrumentation program depends on having the appropriate applied knowledge and experience. Such a program provides critical insight into the design of mitigation systems. Yet, if incorrectly administered, instrumentation can raise more questions and uncertainty than originally existed.

Changes to a rock mass—by weathering, erosion, or construction activities—result in a redistribution of stresses and strain. This readjustment is reflected in a range of displacements, deflections, and loads that can be detected and measured. Instrumentation techniques provide measurements that can be used to investigate the mechanical properties of the mass, the interaction between the mass and any associated artificial structures, and the effectiveness of any proposed remedial measures.

Several questions nearly always arise when a rockfall event occurs on a road or railway. One of the primary questions is whether to close the facility. Another question is when the facility will be reopened and under what conditions it will remain open. The first is likely to be answered conservatively since considerable judgment is required to assess the potential for near-future rockfall from an area of recent rockfall, especially in adverse conditions. In assessing the condition of the rock mass, an instrumentation and monitoring plan can provide valuable insight and a basis for reasonable recommendations to decision makers.

Broadly, there are two potential rockfall-producing scenarios: first, the whole slope is unstable and second, local displacement within the slope causes individual blocks to be dislodged from the face. In the first case, the individual rockfall events often are a sign of a larger and possibly hazardous situation involving some or even the entire affected slope. In such a case, instrumentation may assist in providing monitoring and warning information. In the second case, the stability of a single rock, or a cluster of rocks, in the vicinity of a recent rockfall is in question. In this case, a number of monitoring techniques may be used to provide precise assessments of further ground deformation that may not be apparent from surficial observation.

The applications of instrumentation to rockfall are diverse and unique to a particular site. Much of the time, situations exist in which the engineer or geologist tries to evaluate the time at which some external influence will cause a rockfall. Instrumentation can be used to measure, record, and extrapolate past data to infer the likelihood and timing of a future event. The shortcoming in this approach is that the correct extrapolation is most likely not linear. As is the case with many other geological and natural processes, changes may occur in discrete, rapid events with long periods of time between events. Alternatively, deformation may develop at an ongoing or progressively increasing rate over a long period.

Instrumentation can be targeted to measure some value or influence, and these measurements are compared with a predicted value to establish a degree of stability or potential for impending failure. The disadvantage to this approach is that although obtaining measurements is usually feasible, predicting a critical or threshold value is judgmental.

Monitoring is the process of actively observing and measuring conditions at a rockfall site, and a typical rock slope monitoring program is undertaken to address one or more of the following objectives:

- To maintain safe operation of the transportation facility and to protect the traveling public;
- To provide advance notice of any instability, thus allowing for the introduction of appropriate hazard mitigation strategies;
- To provide for the protection of personnel and equipment during construction operations in the vicinity and for the modification of any excavation actions to minimize the impact of the instability; and
- To provide geotechnical information in order to analyze slope failure mechanisms, design appropriate remedial measures, or conduct a redesign of the rock slope.

Monitoring can involve observing changes to surface exposures or measurements of movements or stresses within the rock mass. Climatic influences, such as temperature, precipitation, and groundwater conditions, are also commonly monitored. The techniques available to measure the various components of rock deformation belong to two general categories: *surface measurements* and *ground displacement measurements*. Surface measurement techniques range from simple visual observations and photographic recording to more sophisticated approaches that include optical and electronic surveying, use of differential Global Positioning System (GPS) techniques, terrestrial photogrammetric approaches, and the newest technologies—laser scanning (LiDAR) and radar scanning using differential interferometry. Ground displacement measurement techniques include the application of mechanical and electronic instruments, such as extensometers, strain gauges, and crack gauges.

1.1 Instrumentation and Monitoring Technologies

Instruments for geotechnical applications have evolved greatly during the past 20 years. By the late 1980s, geotechnical instrumentation had evolved to the point that a major textbook was devoted to the subject (Dunnicliff 1988), and the Transportation Research Board published a *Transportation Research Record* volume containing reports of new applications and methods (Green and Mikkelsen 1988; Hinckley et al. 1988). Patton (1983) described the use of instrumentation in analyzing the stability of rock slopes, and Hanna (1985) provided an extensive review of geotechnical field instrumentation during this period. The last two Special Reports of the Transportation Research Board on landslides include chapters devoted to landslide instrumentation practices (Wilson and Mikkelsen 1978; Mikkelsen 1996). Kliche (1999) provided one of the most recent syntheses of the field of rock slope stability, including instrumentation and monitoring topics.

Numerous authors have reported on the full impact of new instruments and computer automation in data collection and data reduction, mostly in connection with landslide investigations (e.g., Dunnicliff and Davidson 1991; Tatchell 1991; Gillon et al. 1992; Mikkelsen 1992; Proffitt et al. 1992). Kane and Beck (1994) and O'Connor and Wade (1994) described the use of time-domain reflectometry (TDR) in identifying the locations of buried failure surfaces. Dowding and Huang (1994), McKenna et al. (1994), and McKenna (1995) described experiences with large-scale instrumentation and automation of data collection for monitoring slope stability at the huge open-pit mines of Syncrude, Canada, Ltd., in northern Alberta. A monitoring and instrumentation field laboratory has been established by the Alberta Geological Survey to support the long-term monitoring and continued characterization of the structure and instability of Turtle Mountain in southern Alberta, the site of a catastrophic rock avalanche that buried the town of Frank in 1903 (Froese and Moreno 2006). Although the instabilities at Turtle Mountain are of a much larger scale than typical rockfalls, the long-term nature of the monitoring efforts, and the relatively stable funding base, have allowed experimentation with a wide range of instrumentation and monitoring technologies. Turtle Mountain thus provides a source of experience and documentation of the relative merits of various instrumentation and monitoring procedures and technologies. A brief summary of the Turtle Mountain Monitoring Project and Field Laboratory is provided in Section 10.3. Most published descriptions of instrumentation relate to landslides and large-scale earth and rock movements and may be useful in some circumstances to rockfall but also provide valuable background information on instrumentation procedures.

1.2 Potential for Adopting New Instrumentation Technologies

During the past decade, geotechnical monitoring techniques and instrumentation have been significantly influenced by the general proliferation of miniature electronics that has revolutionized many aspects of data acquisition, data processing, and the communication of data from the field to central offices. The widespread availability of powerful and inexpensive personal computing devices, coupled with the dramatic expansion of web-based communication and collaboration capabilities, has resulted in expectation of the speed and convenience of information without delay.

New technologies for rock slope characterization and observation are being used at an increasing number of rockfall locations. These methods include a variety of terrestrial photogrammetry, LiDAR, and interferometric radar applications. The new technologies offer significant economic and personnel safety benefits for field data acquisition and provide more extensive and detailed geometric information on rock mass characteristics, thereby allowing for better analysis and design of mitigation measures.

2. ESTABLISHING AN INSTRUMENTATION PROGRAM

Instrumentation to monitor a rock mass or unstable rock outcrop typically occurs either as part of a corridor-wide evaluation program along sections of transportation facilities with a history of rockfall or as an emergency response to a specific rockfall event. During emergency rockfall events, additional funding may be available to instrument sites and provide insights in planning and design that may not have been available under normal funding scenarios. Overall, whether the instrumentation program is funded from programmed or emergency funds, the plan typically addresses the following items:

- Determination of types of measurements required,
- Selection of the specific types of instruments best suited to make the required measurements,
- Definition of the location and depth of instrumentation and number of instruments,
- Development of the data acquisition techniques,
- Maintenance of the instrumentation system, and
- Long-term management and presentation of the acquired data.

Instrumentation involves not just purchasing and installing the hardware but also collecting the data, maintaining the instruments, and reducing the data. Each step requires a commitment of resources and is equally important. It is necessary to fully understand and account for the long-term reading and maintenance commitments of installing instrumentation. Unfortunately, the act of installing instrumentation and data collection systems does not improve the stability of the rocks. High-quality instrumentation may appear to be excessively expensive. However, labor costs for installing, reading, maintaining, and reducing data generally far exceed the initial cost of the hardware. Data loggers and communications systems can, even though they impose additional initial costs, typically reduce the amount of labor for reading the instruments. The most economical choice will be site specific and will depend on such factors as distance to the site and access requirements, reading frequency, size of area to be monitored, types of sensors, and other factors.

Ideally, instruments collect data accurately, reliably, efficiently, and in a timely manner; otherwise, corrective treatment may be started too late, resulting in even greater economic losses than would have been experienced without it. Instrumentation, if properly and regularly maintained, can provide significantly more and higher-quality monitoring data than what is generally achievable with observational methods. A variety of instrumentation is available to accurately detect and monitor movement.

3. ROCK SLOPE DEFORMATION MEASUREMENT

Instrumentation to detect ground deformation is the first level that can be implemented in a localized area. Instrumentation is typically employed in emergencies to evaluate pending rockfall events suggested by visual identification of unstable rock blocks or masses or adversely oriented and open discontinuities. Small movements of the rock mass before, or even during, incipient failure are usually not visually evident. Thus, the value of visual information alone is limited. Ground deformation instrumentation can provide valuable information on incipient as well as fully developed movement. In this respect, use of instrumentation is not intended to replace field observations and other investigative procedures. Instead, it augments other data by providing supplementary information and by warning of impending major movements. Typical situations for which various instruments have been used are the following:

- Determination of lateral and vertical movements within a rock mass,
- Determination of the rate of movement so as to provide a warning of impending danger,
- Monitoring the activity of marginally stable natural or cut slopes and identification of effects of construction activity or precipitation,
- Provision of remote digital readout or information to a remote alarm system, and
- Monitoring and evaluation of the effectiveness of various control measures.

Many of the systems are readily available from instrumentation manufacturers but may require a specialty contractor or specially trained personnel to install. These instrumentation systems can also be set up for longer-duration monitoring within transportation corridors with a history of rockfall, rock slopes, or both, that have been rated with the potential to generate rockfall.

3.1 Crack Meters

Crack meters are among the most basic instruments used to monitor rock slope stability. These instruments typically consist of potentiometric sensors or vibrating wire sensors. Either sensor system can be connected to a data logger.

Crack meters are designed to monitor movements across discontinuities or cracks that have specific ranges of opening, with individual models capable of measuring different widths between 0.001 and 1.0 m (0.04 and 40 in.). Apertures or cracks with anticipated opening movements greater than 1 m (40 in.) typically require other extensometer instruments, which are described later. Figure 7-1 shows a crack meter installed across a rock joint.

A common type of crack meter used for rock slopes consists of a calibrated rod. As the rod is elongated or compressed, the movement in an internal spring is sensed by a vibrating wire element. The movement in the wire is transmitted to an external readout box and downloaded or uploaded via cellular or satellite services to be processed. The data are then plotted to determine time-related movements.

3.2 Tiltmeters

Tiltmeters measure the inclination (tilt or rotation) of discrete parts of rock cuts and structures

FIGURE 7-1 Example of installed crack meter.
(PHOTOGRAPH BY B. ARNDT.)

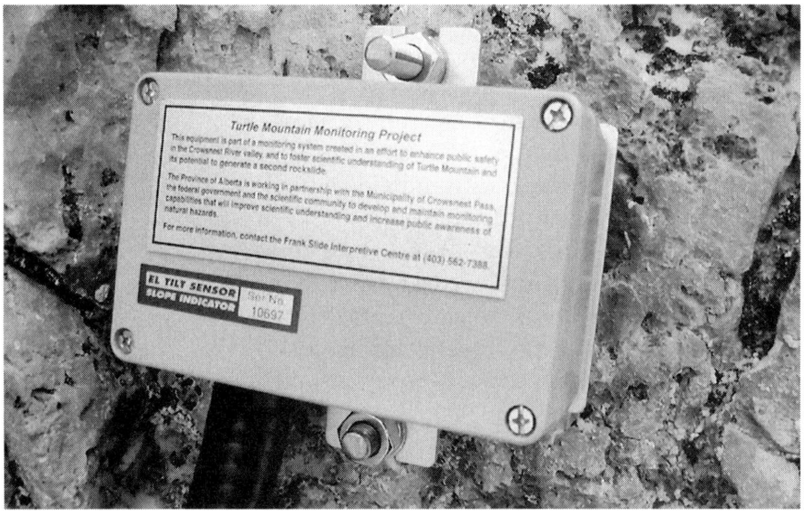

FIGURE 7-2
Tiltmeter installed on Turtle Mountain, Alberta, Canada (Moreno and Froese 2006).
(PHOTOGRAPH COURTESY OF THE ALBERTA GEOLOGICAL SURVEY.)

from the established datum. They are most commonly used to monitor tilting of bridge abutments, retaining walls, and marginally stable slopes or rock outcrops. When mounted to a rock face, tiltmeters can make very accurate orientation measurements (Figure 7-2). Changes in orientation over time infer movement or distress. Tiltmeters are particularly useful for assessing toppling failures, where tilt is a major component of movement.

There are several types of tiltmeter instruments, ranging from a simple plumb line or a machinist's level, both requiring individual observation and data recording, to more sophisticated automated equipment with accelerometers or electrolytic-level transducers that allow for data to be recorded remotely. Modern electronic tiltmeters offer accurate, high-resolution measurements that can be read remotely and are readily automated. The units are compact and are not too expensive. Two technologies are used. Many units use an arrangement of electrodes to sense the exact position of a bubble in the electrolytic solution to a high degree of precision; they typically can detect a change in tilt as small as 1 arc-second. These units are quite insensitive to temperature and can be fully compensated by using built-in thermal electronics. A newer technology uses microelectromechanical systems (MEMS) electronics. These MEMS tiltmeters provide medium-resolution measurements over a wider measurement range of ±10°. Biaxial models have a second sensor oriented at 90° to the first, allowing tilting in any direction to be determined.

Beam sensors are a type of tiltmeter in which the tilt sensor is mounted on a rigid beam with a defined length, typically 1 to 2 m (3 to 6 ft). Each end of the beam is anchored to the structure or rock face being assessed (Figure 7-3). The rigid beam increases the sensitivity of the tilt measurements. However, beam sensors only measure tilt along the axis of the beam; they are thus single-axis tiltmeters. The rigid beam allows changes in tilt to be converted to actual movements (settlement, heave, convergence, or lateral displacement). Also, as shown in Figure 7-3, beam sensors can be linked end to end to monitor differential movements and provide absolute displacement and settlement profiles along the axis of the beams.

In many cases access restrictions favor the permanent installation of tiltmeters, as shown by the Turtle Mountain installation in Figure 7-2. Fixed-in-place tiltmeters provide more accurate measurements because once the tiltmeter is mounted, it is normally adjusted so that it reads close to zero (no tilt). The initial reading is used as a baseline (Figure 7-4). Changes in inclination are found by comparing subsequent readings with the initial reading. However, in some cases a single portable tiltmeter is used to monitor tilting movements at a number of locations. Portable tiltmeters typically

FIGURE 7-3
Beam sensor tiltmeter (Durham Geo Slope Indicator).

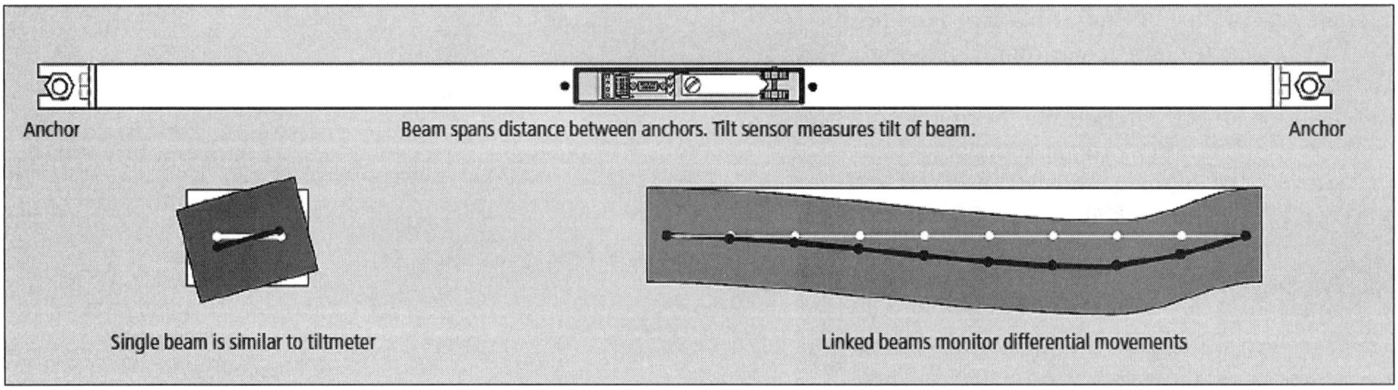

use a servo-accelerometer sensor similar to that used by inclinometer probes. A relatively inexpensive ceramic or metal tilt base plate is permanently mounted at each measurement site. Mechanical connections ensure that the portable tiltmeter frame is precisely located on the base plate (Figure 7-5). Using these portable instruments introduces some inaccuracy due to the use of the base plate and the limitations of the mechanical connections. Portable tiltmeters typically provide a measurement resolution of 8 arc-seconds with a system repeatability of ±40 arc-seconds. However, portable tiltmeters allow a single tiltmeter to be used to monitor any number of sites, although readings can only be made manually at each site.

FIGURE 7-4 Initializing tiltmeter on Mount Pinatubo, Philippines.
(PHOTOGRAPH COURTESY OF THE U.S. GEOLOGICAL SURVEY.)

3.3 Extensometers

An *extensometer* measures the opening (or closing) of a crack either in one, two, or three dimensions; changes in the opening dimensions infer movement. A wide variety of extensometers are commercially available; alternatively, simple systems can be easily created. Three types of extensometers are described in the following sections: (*a*) very simple displacement-monitoring systems, often termed *poor-man extensometers*, (*b*) *tape or wire extensometers* designed to measure distances between defined monitoring points located on the ground surface, and (*c*) *borehole (or rod) extensometers* that are used to monitor deformations within a rock mass (Dutro and Dickinson 1974).

Some publications refer to borehole extensometers as *strainmeters*, but the term strainmeter is properly used to refer to a different class of instruments used by geophysicists to measure deformation within rock masses by evaluating very small changes in the diameter of a borehole (Agnew 1986). These strainmeters are typically installed at considerable depths below the ground surface and are used in scientific studies of solid earth tides and changes in stress regimes following earthquakes; thus, the most extensive network of strainmeters is installed in Japan. However, strainmeters have also been used to a very limited extent to evaluate deformations in deep underground structures and tunnels. Gladwin et al. (2000) reported on the use of strainmeters to monitor rock mass deformations over an active long-wall coal mine in Australia. The use of strainmeters in most rock slope stability investigations, or for monitoring rockfall hazards, appears to offer no economic or technical advantages over other instrumentation options.

3.3.1 Poor-Man Extensometers

Extensometers can be as simple as two pins secured across an open crack with the distance between the pins manually and regularly measured. In many applications, the rate of movement or remote location of an unstable rock mass may not allow for the installation of conventional extensometers. A relatively simple method is lapping two polyvinyl chloride pipes over the area of displacement and staking the ends of the pipes to the opposing rock masses. Figure 7-6 shows a typical installation. The pipes are then marked and measured at regular

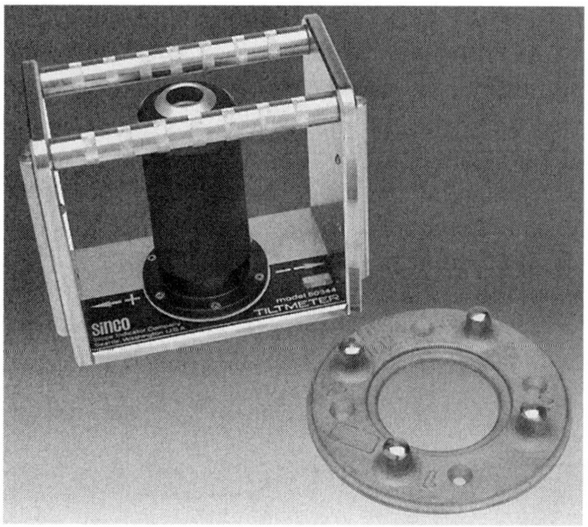

FIGURE 7-5 Portable tiltmeter and base plate (Durham Geo Slope Indicator).

FIGURE 7-6
Typical poor-man extensometer installation.
(PHOTOGRAPH BY B. ARNDT.)

FIGURE 7-7
Obtaining distance measurement with poor-man extensometer.
(PHOTOGRAPH BY B. ARNDT.)

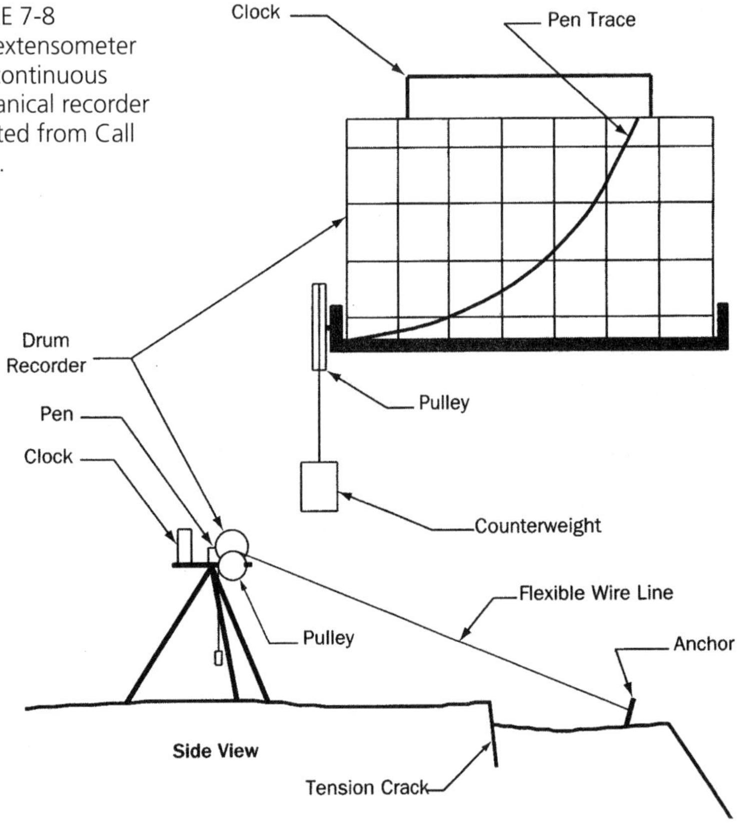

FIGURE 7-8
Wire extensometer with continuous mechanical recorder (adapted from Call 1982).

intervals to determine the amount of movement (Figure 7-7). Although this is not a precise measurement of movement, it can provide a basis for determining the potential hazard of an unstable rock mass in an emergency situation.

3.3.2 Tape or Wire Extensometers

The change in distance between two fixed points that are too widely spaced for a crack meter can be measured with high repeatability by using various types of surface extensometers and convergence and crack gauges. The most common type is the tape or wire extensometer. These devices have been used for many years to monitor instabilities around open-pit mines (Call 1982; Kliche 1999). Figure 7-8 illustrates the principles of these systems, which consist of a steel tape in combination with a dial indicator and a spring system that holds the tape at a constant tension.

Figure 7-8 shows an early system that involved a mechanical recording system. Modern systems have replaced the mechanical recording with various electronic sensors, such as vibrating wire sensors, which allow for automated digital recording and telemetry of the results to remote data collection centers. Commercially available tape–extensometer systems (Figure 7-9) use a readily portable tape reel that is only 0.6 m (2 ft) long and weighs only 2 kg (4.5 lb). Measurements are made between pairs of permanently mounted eyebolts. The operator stretches the tape between two reference points, hooking the free end of the tape to one point and the instrument body to the other. The operator then tensions the tape by turning a

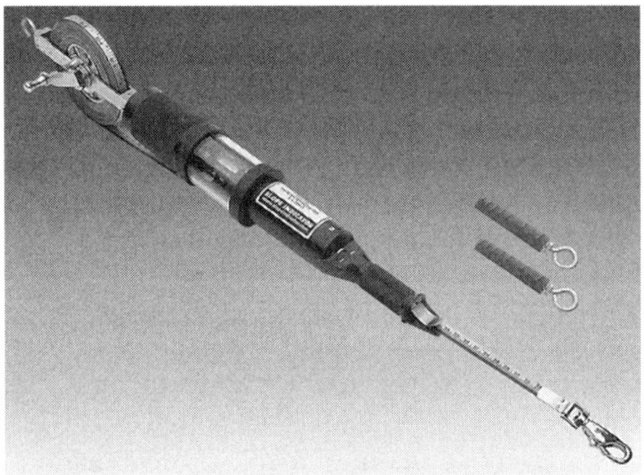

FIGURE 7-9
Digital tape extensometer with eyebolts used to establish reference points (Durham Geo Slope Indicator).

knurled collar until two index marks are aligned and notes the reading from the tape and the digital display. The sum of these readings is the distance between the two reference points. Tapes are typically provided in 20- to 30-m (66- to 100-ft) lengths. The precision of these tape extensometer installations depends on the sensitivity of the recording sensors and the distance between the points, but for a 10-m (30-ft) span, detection of changes as small as ±0.13 mm (±0.005 in.) can be expected under good conditions. Wind and precipitation can also cause unstable readings.

The Turtle Mountain monitoring efforts constructed a number of wire extensometers with head assemblies containing a weight and a transducer measuring the movement of the weight located on the upslope side of the crack and anchors located on the downslope side of the crack (Figure 7-10). Measurements are made hourly and the information is temporarily stored in a data logger before being telemetered back to a central data collection point.

3.3.3 Borehole Extensometers

Borehole extensometers are typically used to monitor rock mass deformations around tunnels or other underground openings or displacements of retaining structures. The main components of a borehole extensometer are anchors, rods inside protective pipe, and a reference head (Figure 7-11). The anchors are installed at appropriate locations within the borehole. Individual rods connect each anchor to the reference head at the surface. A 76-mm (3-in.) borehole can accommodate six different anchor points and their associated rods. A protective plastic pipe prevents bonding between rods and grout within the borehole.

Readings are obtained at the reference head by measuring the distance between the end of each rod and a reference surface. Any change in this distance indicates that movement has occurred. Typically the deepest downhole anchor is used as a stable reference point. The resulting data allow for

FIGURE 7-10
Typical extensometer installation across major fissure on Turtle Mountain, Alberta, Canada (extensometer wire has been digitally enhanced).
(PHOTOGRAPH COURTESY OF THE ALBERTA GEOLOGICAL SURVEY.)

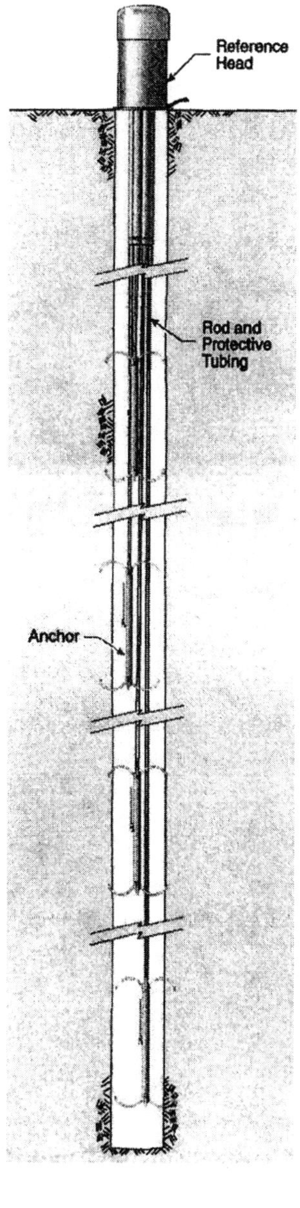

FIGURE 7-11
Typical borehole extensometer installation (Durham Geo Slope Indicator).

determination of the locations, rates, and accelerations of movements, and thus the calculation of strain. Mechanical reference heads are used when there is easy access to the extensometer reference head. A micrometer is used to obtain measurements. Electric reference heads are used when access to the reference head is difficult or when continuous monitoring is required. In this case, measurements are obtained with displacement sensors connected to a data logger.

Rods are made of fiberglass or stainless steel. Fiberglass rod extensometers are factory assembled and shipped to the site, ready to install. The flexibility of these extensometers also makes them easier to install in confined areas such as tunnels. Stainless steel rod extensometers must be assembled on site. However, their stiffer rods can be used for deeper anchor depths. In general, rods in tension can be longer than rods in compression, and steel rods can be longer than fiberglass rods. Typical maximum lengths of fiberglass rods under tension are 20 to 30 m (65 to 100 ft), while steel rods under tension may extend for lengths of 30 to 50 m (100 to 164 ft). These maximum lengths may not be achievable if the borehole is subjected to bending or shearing or in nonvertical holes. In such cases, friction between rods and the protective pipe becomes a limiting factor.

4. CONVENTIONAL SURVEYING TECHNOLOGIES

Conventional surveying techniques that are commonly used in rock slope applications employ either a total station or, more recently, GPS. The *total station* is an electronic theodolite that is integrated with an electronic distance meter (EDM) to read distances from the instrument to a particular point. GPS operates through satellites that send out signals to the receiver. The measurements for both systems are taken on a series of established points, such as bolts cemented to various rock bodies or survey targets previously installed within suspected areas of instability on a slope. These points are resurveyed according to an established schedule. The survey data are then analyzed to define the locations and amount of movement within portions of the slope being monitored. A three-dimensional (3-D) tomogram (Figure 7-12) provides a visual summary of such displacement data. Areas of movement are depicted by contours (usually with color codes) within a three-axis coordinate system.

4.1 Surveys with Total Stations and Electronic Distance Meters

Modern surveying techniques use total stations, electronic theodolites, or both, that are integrated with EDMs. Some models include internal electronic data storage to record measured distances and horizontal and vertical angles, and other models are equipped to record these measurements on an external data collector, typically a hand-held computer. Some total stations also have a GPS interface that combines the advantages of these two technologies. Figure 7-13 shows examples of recent models of total-station instruments that provide these capabilities. The use of total stations requires a clear line of sight between the instrument location, or locations, and the monitoring points. However, with knowledge of the exact coordinates of the total-station instrument, use of a total station provides high-precision measurements of distances and angles. This capability allows for the accurate determination of elevation differences so that movements in the vertical axis can be determined as accurately as those in the horizontal plane.

The best-quality total stations are capable of measuring angles to 0.5 arc-second. Inexpensive "construction-grade" total stations can generally measure angles to 5 or 10 arc-seconds. Distances are determined by sending and receiving a modulated microwave or infrared carrier signal, generated by a small solid-state emitter within the instrument's optical path. The signal is reflected by a prism or by the object under survey, and the distance is computed by determining the integer number of wavelengths to the target for each of multiple frequencies.

Measurements can be made with or without specific reflecting points. The use of prisms or other reflector devices allows more accurate distance measurements and assists in the obtaining of repeatable sightings on the same reference point. *Reflectorless techniques* require sighting the instrument on reference points that are identifiable and reasonably light in color. When prism reflectors are used, a typical total station can measure distances with an accuracy of about 1 mm (0.04 in.)

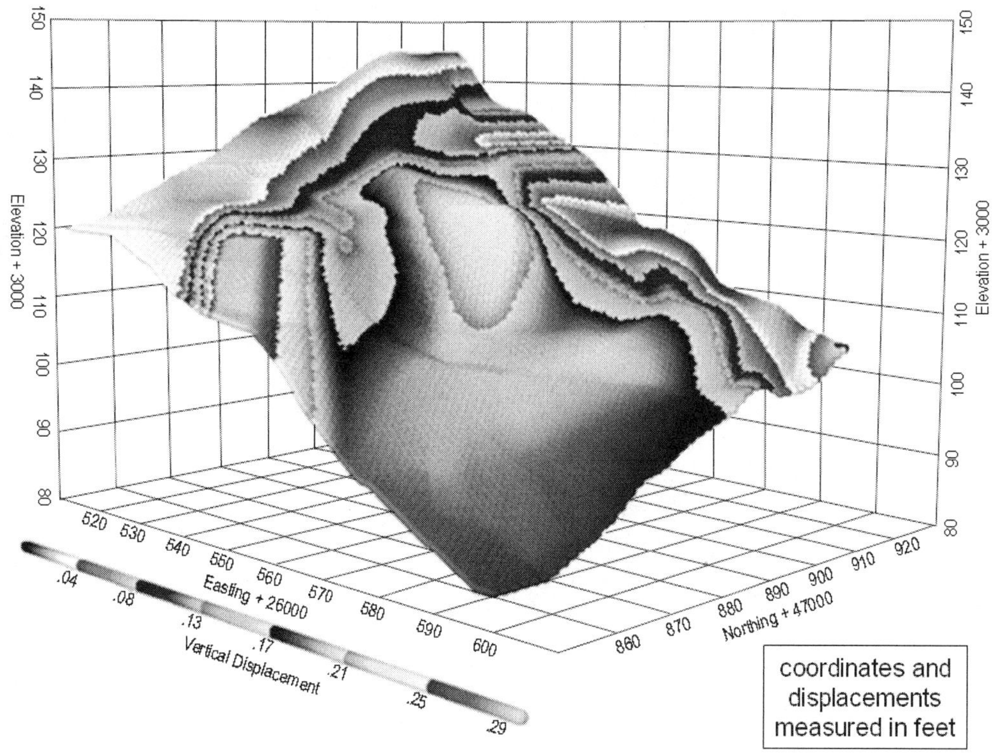

FIGURE 7-12 Areas of displacement on unstable slope visually summarized with 3-D tomogram. (PHOTOGRAPH BY R. ANDREW.)

+ 1.5 ppm of distance over distances ranging up to about 1,500 m (5,000 ft) (Leica 2009). However, with reflectorless total-station operation, the accuracy of distances measured is typically 2 to 3 mm (0.08 to 0.12 in.) and the maximum range is reduced.

Installing prisms on rock faces can be problematic and may require specialized training in the use of rope-based climbing techniques for their placement. Figure 7-14 shows the drilling and placement of a reflector prism on a rock cliff. Figure 7-15 is a close-up view of a reflector prism mounted on an overturning rock.

Robotic total stations allow the operator to control the instrument from a distance via remote control. This procedure eliminates the need for an assistant; the operator holds the reflector and controls the total station from the observed point. Robotic stations can also be programmed to automatically access and determine the locations of permanently located prism reflectors, or other defined target locations, according to a predetermined schedule. This operation allows for the automated collection of monitoring observations; however, the rather expensive total-station instruments usually require protection from the weather, vandals, or theft. Therefore, automated unattended data collection sequences are rarely feasible at most rockfall sites.

In addition to slope stability monitoring applications, total stations have also been used for a variety of mapping and geological evaluations of exposed rock faces. Ilsley (1994) used an earlier model of a total station, which was referred to as a

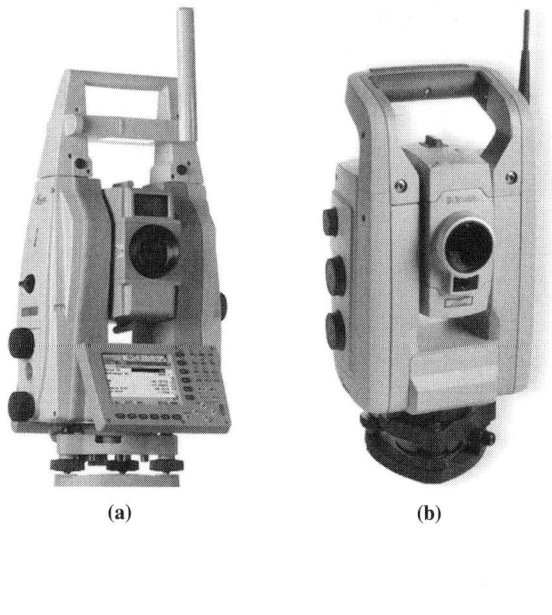

FIGURE 7-13 Two recent models of total-station instruments: (*a*) Leica TS30 and (*b*) Trimble S3. (PHOTOGRAPHS COURTESY OF LEICA GEOSYSTEMS AG AND TRIMBLE NAVIGATION LIMITED.)

FIGURE 7-14
Placement of reflector prism.
(PHOTOGRAPH BY J. WHITE FOR THE COLORADO GEOLOGICAL SURVEY.)

FIGURE 7-15
Reflector prism mounted on unstable rock.
(PHOTOGRAPH COURTESY OF THE COLORADO GEOLOGICAL SURVEY.)

laser-transit, to undertake detailed engineering geological mapping of rock slopes. Feng et al. (2001, 2003) discussed methods for determining fracture orientations and fracture roughness by scanning rock faces with reflectorless total stations. Their methods used total stations to scan sections of the rock face to define the geometry of the fractures and the surface roughness, but they also used digital photography to record their observations.

4.2 Surveys with GPS

In 1978, the U.S. Department of Defense began to deploy the satellite-based GPS, designed to meet the positioning needs of the military and national security on a 24-h-per-day basis globally and under all weather conditions. The system was named the Navigation Satellite Timing and Ranging (NAVSTAR) GPS and became fully operational in April 1995, although a full constellation of 24 satellites had been established by January 1994. Originally conceived as a military navigation system, the potential benefits of GPS for civilian applications rapidly became apparent. In 1996, U.S. President Clinton issued a policy directive declaring GPS to be a "dual-use" system and establishing an Interagency GPS Executive Board to manage it as a national asset. Subsequently, in May 2000, the "selective availability" (SA) feature of the original GPS was discontinued. The SA added intentional, time-varying errors of up to 100 m (330 ft) to the publicly available navigation signals and was intended to deny an enemy the use of civilian GPS receivers for precision weapon guidance. Its discontinuance allowed all users to receive a nondegraded signal and reflected the improved methods of applying GPS measurements to a variety of civilian uses. Further modernization of the NAVSTAR GPS has been authorized that will lead to even more precise global positions by 2013 (Shaw et al. 2000).

The standard generic term for all similar satellite navigation systems is *global navigation satellite system* (GNSS). Currently, the NAVSTAR GPS is the only fully operational GNSS. Russia is restoring its Global'naya Navigatsionnaya Sputnikovaya Sistema, or Global Navigation Satellite System (GLONASS), to full operation. The European Union is developing the Galileo GNSS. Currently in its initial deployment phase, it is scheduled to be operational in 2014. China has indicated that it will expand its regional Beidou navigation system into a GNSS called Compass. Other countries have also proposed establishing systems. New ground-based receivers are being produced that can use signals from several such systems (Langley 2000). The increased number of satellites visible at a given moment from a given location will increase the accuracy and speed of determining a global position.

4.2.1 Basic Principles of Operation

Each NAVSTAR GPS satellite continually transmits at two frequencies: L1 (1575.42 MHz) and L2 (1227.6 MHz); each is modulated with two types of codes and a navigation message that provides

- The time the message was sent,
- Precise orbital information (the ephemeris), and
- The general system health and rough orbits of all GPS satellites (the almanac).

The codes are a civilian code and an encrypted military code (Langley 1991). Currently the L1 carrier is modulated with both types of codes, whereas L2 is modulated with only the military code. Modernized GPS will transmit a second civilian code on L2 and a third civilian code on a new frequency, termed L5 (1176.45 MHz) (Shaw et al. 2000).

A ground-based GPS receiver calculates its global position by precisely timing the signals sent by the GPS satellites. The receiver measures the transit time of each message and computes the distance to each satellite. Geometric trilateration combines these distances with knowledge of the precise location of each satellite to obtain the global position of the receiver. Two types of observations are of interest to users (Langley 1993, 1999).

The first type of observation is termed *pseudo-range* because it is a measure of the true range, or distance, between the satellite and the receiver that also contains some small corrective terms to account for distance estimation errors caused by receiver and satellite clock discrepancies, the impacts of the ionosphere and troposphere on signal propagation, and multipath reflections of the signal from nearby reflective surfaces (Weill 1997). Given the geometric positions of the satellites as a function of time, four pseudo-ranges are in principle sufficient to compute the location of the receiver and its clock correction. However, additional satellite observations are desirable, especially if the minimum number of satellites (four) are not in an ideal geometric configuration, thus providing a less-than-optimal solution, or if multipath reflections exist (Hoffman-Wellenhof et al. 2001; Leick 2003). Government policies (U.S. Department of Defense 2008; U.S. National Technical Information Service 2008) define a Standard Positioning Service (SPS) based on the civilian code observations and a Precise Positioning Service (PPS) based on the military code observations. Both SPS and PPS address "classical satellite" navigation methods in which a single ground-based receiver observes several satellites to determine its geocentric position.

The second type of observation involves an assessment of the *carrier phase*. Since the L1 and L2 frequencies are composed of oscillating waves, the range from satellite to receiver can be calculated by summing the number of full and fractional cycles and multiplying by the wavelength. This distance will be much more accurate than that determined by the pseudo-range observations because the wavelength, and therefore the resolution, of the carrier phase [7.48 in. (19 cm) in the case of L1] is quite small. However, all carrier cycles are identical, and the GPS receiver cannot differentiate one cycle from another (Langley 1993). The receiver can only measure a fraction of a cycle very accurately, typically to less than 2 mm (0.08 in.), whereas the total number of complete cycles remains unknown. This feature is commonly referred to as the *initial cycle ambiguity*. Fortunately, the receiver is able to keep track of the phase changes over the entire period of time since it was turned on. Thus, provided no signal loss occurs, the initial cycle ambiguity remains unchanged. The initial cycle ambiguity can be resolved by using so-called relative-positioning techniques, but these require the use of at least two GPS receivers simultaneously tracking the same satellites. The much greater precision provided by carrier-phase observations, and therefore the use of various relative-positioning methods and multiple GPS receivers, forms the basis of essentially all surveying applications of GPS, including the use of GPS for rock exposure monitoring and rockfall hazard studies.

4.2.2 Resolving GPS Errors and Biases

Both GPS pseudo-range and carrier-phase measurements are affected by several random errors and biases, or systematic errors (Hoffman-Wellenhof et al. 2001; Leick 2003). These errors and biases may be classified as follows:

- *Satellite errors*. Ephemeris (orbital) errors, satellite clock errors, geometrical effects of the configuration of the satellite constellation.
- *Receiver errors*. Receiver clock errors, antenna limitations, system noise.
- *Signal propagation errors*. Ionospheric delay, tropospheric delay, multipath effects.

Several of these errors or biases can be eliminated or reduced by combining GPS observations. Two GPS receivers that are tracking the same satellite will contain essentially the same satellite-related and atmospheric-related errors and biases, especially if they are in relatively close proximity. Thus differences between carrier-phase measurements will yield precise estimates of their relative positions, and therefore the distance separating them. This method is referred to as the *between-receiver single difference*. Similarly, two measurements taken by a single receiver tracking two satellites contain the same receiver errors, known as the *between-satellite single difference*.

When two receivers track two satellites simultaneously, it is possible to thus combine the effects of the previous two methods, thereby removing, or substantially reducing, satellite-related, atmospheric-related, and receiver-related errors and biases. This *double difference* method also provides a method for overcoming the initial cycle ambiguity concerns and so is the basis for all high-precision carrier-phase-based GPS positioning methods. Many high-performance GPS receivers used for precise surveying are able to analyze both the L1 and L2 frequencies. This ability allows for further improvements in measurement precision and also additional error checks, especially on the effects of ionospheric delay (Leick 2003).

4.2.3 Defining Precise Locations Using GPS Relative-Positioning Methods

Essentially all surveying applications of GPS use relative-positioning, also called *differential positioning*, methods and carrier-phase observations. These methods require the use of a minimum of two GPS receivers simultaneously tracking the same satellites, although some methods allow for the use of additional receivers and some use available GPS data and correction services.

In the most fundamental arrangement, one of the two receivers becomes a reference, or base, station, which remains stationary over a position with precisely known coordinates. The second receiver, known as the *rover receiver*, is placed at one or more locations with unknown coordinates. Depending on the type of GPS operation, this rover receiver may be stationary or moving (Leick 2003). The methods used are typically defined as

- Static GPS surveying,
- Fast (rapid) static GPS surveying,
- Stop-and-go GPS surveying,
- Real-time kinematic (RTK) GPS, and
- Real-time differential GPS.

The first three methods all rely on postprocessing the data collected by the two receivers; thus they do not provide location information results until after the survey is completed and the data are analyzed.

Static GPS surveying provides the most precise location determinations. It uses carrier-phase measurements of at least four satellites and employs two or more static receivers. Observation times range from 20 min to several hours, depending on the distance (baseline) separating the two receivers, the number of visible satellites, and the constellation geometry. The method typically produces accuracies comparable with geodetic survey standards and thus is used primarily for establishing primary survey control points.

Fast (rapid) static surveys use similar methods, but typically only the base station remains stationary. Multiple rover receivers are sometimes used. They move frequently, occupying their stations for only relatively short periods, usually only 2 to 10 min at each location. At the beginning of a survey, all receivers are included in an initialization procedure to determine the parameters for resolving the initial cycle ambiguity (Leick 2003). The survey is usually restricted to locations extending up to about 15 km (9 mi) from the base station. Stop-and-go GPS surveying extends the same principles to even more rapidly roving surveys, typically using backpack portable survey units and very brief (30-s) periods of observation at each point. With either method, the rover receiver units are not switched off during moves and remain in contact with all the visible satellites throughout the process. If this connection does not occur, the initialization process has to be repeated. A further check on accuracy is made by reoccupying the first station and the end of the survey. With these precautions, locations are typically determined with a precision of 1 cm (0.5 in.) by these surveys when carrier-phase measurements are used.

A variation of the stop-and-go survey involves continuously moving rover GPS units and thus is termed *kinematic GPS*. The precision of locations obtained by kinematic GPS is naturally lower than that with the methods in which a station is occupied for even a brief period.

In many cases, it is desirable to obtain real-time location coordinates in the field as the survey progresses. To do this, the rover receivers must be able to communicate with the base station and exchange information so that the relative positions between the fixed (known) base station and the roving receiver can be immediately computed. Radio communication links between the base station and the roving station provide the means of transferring the required information. This combination of roving receivers and fixed base station with radio communication links is RTK GPS (Figure 7-16). It employs carrier-phase measurements and similar precautions and procedures as the stop-and-go or kinematic surveys, and locations are routinely defined with a precision of 2 to 5 cm (1 to 2 in.). Even greater precision is possible if stations are occupied for brief periods or if multiple base stations are used (Raquet and Lachapelle 2001). Relative positions within a survey, such as distances between two objects, can be determined to even greater accuracies.

FIGURE 7-16
Example of RTK GPS survey showing base station on right and rover receiver being used to locate point.
(PHOTOGRAPH COURTESY OF BOSTON UNIVERSITY.)

A second method, termed *real-time differential GPS*, undertakes similar survey techniques but over somewhat larger distances. Real-time differential GPS typically relies on pseudo-range distance measurements and thus has lower absolute global location accuracy (precision) that typically ranges from 0.5 to 5 m (20 to 200 in.). However, relative distance measurements obtained by this method are typically considerably more accurate, with values in the range of 2 to 5 cm (1 to 2 in.) often reported.

4.2.4 Applications of GPS to Rock Slope Monitoring

GPS surveying methods support rock slope monitoring programs by providing the following:

1. An economical and rapid, yet reasonably precise, method of determining geographic coordinates for rockfall hazard or rockfall-monitoring locations for data entries within regional (statewide) rockfall databases. These coordinates allow for spatial queries and displays of these data within geographic information systems (GIS). Chapter 4 describes several such applications currently employed by transportation agencies.
2. Precise survey control for other rock slope monitoring and instrumentation techniques, including the various remote-sensing monitoring methods described in Sections 5 and 6.
3. Direct identification and monitoring of even minor movements of rock masses, particularly in cases where routine access is difficult or dangerous.

Each of these applications typically involves different GPS surveying methods and practices. In many cases, consultation with GPS experts will ensure that the most efficient methods are employed and that the surveys provide measurements with the appropriate levels of precision and accuracy.

For example, most GIS applications, and thus rockfall databases, require only moderately precise geographical coordinates, which are often obtained relatively simply with low-cost portable GPS receivers. However, if the locations are to be later used in preparing more precise construction or design documents, or are to be used in computer-assisted drafting (CAD) systems for

FIGURE 7-17 Differential GPS survey station on Turtle Mountain, Alberta, Canada. Application uses highly accurate static GPS method and long observation times (Moreno and Froese 2008b).
(PHOTOGRAPH COURTESY OF THE ALBERTA GEOLOGICAL SURVEY.)

more detailed analyses, then much more precise GPS surveying methods are appropriate.

These higher-precision locations also are used by the survey control applications supporting other monitoring and instrumentation programs. These applications may also be adversely affected if limitations of GPS methods prevent the establishment of precise locations. The accuracy of GPS surveys is adversely affected by two factors commonly encountered in rockfall-monitoring locations:

- Inadequate view of the satellites and
- Multipath reflections of the GPS signals.

GPS will not provide accurate locations unless at least four satellites are in direct view of the receiver or receivers. If a large portion of the sky is blocked, such as at locations within deep canyons or immediately below high cliffs, the number of satellites in direct view is likely to be insufficient to provide a good trilateration solution. Even if the number of satellites is four or more, the geometry of the satellite constellation is likely to offer a weak trilateration solution. GPS receivers are also adversely affected by heavy tree cover. The potential for multipath reflections of the satellite signals increases in locations near reflective surfaces such as larger rock faces, buildings, and water bodies.

However, GPS survey methods also have some unique advantages that make them potentially very useful in some direct-monitoring applications. Relative-positioning GPS methods retain high accuracies in distance measurement even over long distances. For example, under ideal circumstances, and with careful attention to best procedures, GPS can establish true distance measurements with accuracies of 5 mm (0.2 in.) over long distances. In addition, GPS surveys do not require a direct line of sight between the points being measured; all that is required is for each point to have a clear and simultaneous view of a given satellite constellation. In contrast, total-station survey methods may provide a more accurate distance measurement, with typical accuracies of 2 to 3 mm (0.08 to 0.12 in.), but only over much shorter distances, up to less than 300 m (1,000 ft), unless reflectors are used. Total stations also require a direct, unobstructed line of sight along the distance being measured. Accordingly, relative-positioning GPS methods, referred to as *differential* GPS (without always requiring real-time solutions), have been used for direct monitoring of large unstable rock masses. Moreno and Froese (2008b) reported on the use of highly accurate static GPS survey methods to monitor the stability at Turtle Mountain, Alberta, Canada, as illustrated by Figure 7-17. Section 10.3 provides additional discussion of the instrumentation options installed at Turtle Mountain.

5. TERRESTRIAL PHOTOGRAMMETRY

Terrestrial photogrammetry has proved to be a good way of obtaining remotely sensed information from exposed rock faces. The introduction of digital cameras, combined with new image-processing techniques, has allowed digital terrestrial photogrammetry to become very attractive compared with previous analog analytical photogrammetric techniques. With relatively inexpensive equipment and easy-to-use software operating on personal computers, it is possible to develop accurate 3-D models of rock slopes, tunnels, and structures based on photogrammetric principles (Beraldin et al. 2000; Gaich et al. 2004, 2006; Haneberg 2008).

Photogrammetry is a measurement technology in which the 3-D coordinates of points defining the geometry of objects are determined by measurements made in two or more photographic images taken from different positions (Figure 7-18). Three-dimensional coordinates are computed for each point in the image; thus higher-resolution images provide a more detailed model of objects such as a rock face.

5.1 Accuracy of Terrestrial Photogrammetry

Much research has been done to determine the accuracy and precision of airborne and terrestrial photogrammetric techniques (Wickens and Barton 1971; Franklin et al. 1988; Harrison 1993; Krosley and Shaffner 2003; Maas and Hampel 2006). These studies reveal that the accuracy of 3-D models developed by photogrammetric techniques depends on several factors, including camera type, focal length of the camera lens, length of the baseline, range to the objects being evaluated, and the size and geometry of these objects.

Maas and Hampel (2006) found that typical and well-calibrated modern photogrammetric systems, under controlled conditions, may achieve accuracies on the order of 1:10,000 in terms of distance to the object. This accuracy means, for example, that at a distance of 100 m (328 ft), the measured position of a single point on the object will be within 1 cm (0.4 in.) of its true position. According to Poropat (2006), robust, accurate image-matching algorithms and appropriate calibration enable modern high-quality digital cameras to achieve a precision of 1 to 10 cm (0.4 to 4.0 in.) at distances of 100 m (328 ft) with lenses of moderate focal length. However, although accuracy varies linearly with the focal length or pixel size, it also varies nonlinearly with other parameters such as the length of the camera baseline. When rock faces are evaluated, the ability to accurately determine the attitudes, or orientations, of planar surfaces that represent fractures, joints, or faults influencing the stability of the rock mass depends to a considerable degree on their orientations relative to the camera position (Krosley and Shaffner 2003).

5.2 Applications to Rock Slope Evaluation

Haneberg et al. (2006) described the use of close-range terrestrial photogrammetry in the evaluation of a series of rock slopes along Interstate 90 in the Cascade Mountains of Washington State. Figure 7-19 is a 3-D photogrammetric representation of one of these rock slopes. This outcrop model consists of 425,523 *x*,*y*,*z* points located in the local coordinate system established for the highway mitigation project. The model represents approximately 740 m^2 (8,000 ft^2) of outcrop surface, so the average point spacing is about 4 cm (1.6 in.). The estimated root mean square error in the locations of these points is ±2 cm (±0.75 in.). Figures 7-20 and 7-21 illustrate the 3-D geometrical framework of this outcrop model as a point

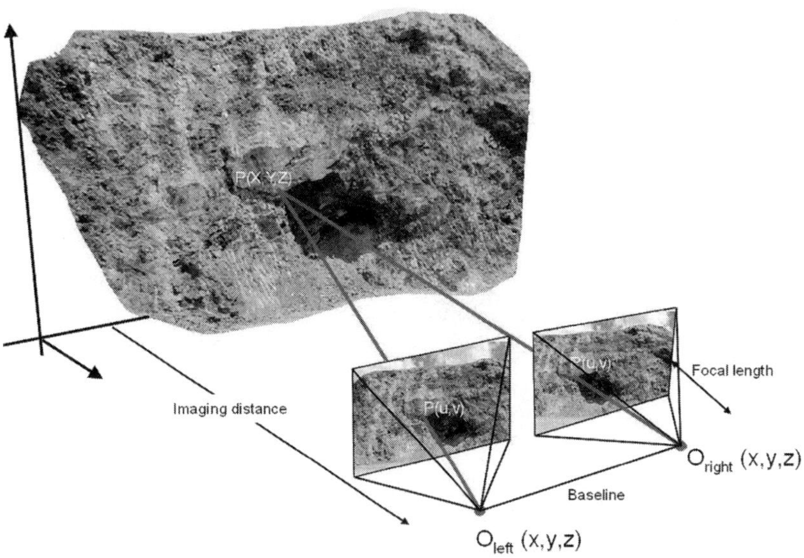

FIGURE 7-18
Terrestrial photogrammetry, based on analysis of stereoscopic image pair. Two corresponding image points *P*(*u*,*v*) in left and right images relate to single 3-D object point *P*(*x*,*y*,*z*). Modern photogrammetric computational techniques do not require length of baseline to be known (Gaich et al. 2006).

FIGURE 7-19
Typical 3-D rock outcrop model produced by digital terrestrial photogrammetry methods. Model consists of 3.8 million pixels draped over 425,523 *x*,*y*,*z* points (Haneberg et al. 2006).

FIGURE 7-20 Point cloud representation of geometrical framework of rock outcrop model shown in Figure 7-19. To improve visibility, only every fifth point has been plotted (Haneberg et al. 2006).

cloud and as a triangulated mesh, respectively. To improve the visibility of these models, only every fifth point was plotted in Figure 7-20 or used to create the triangular mesh in Figure 7-21. The mesh contains 23,360 triangular facets, which were used in subsequent calculations.

Haneberg et al. (2006) fully described the subsequent structural analysis of this outcrop model to evaluate probable rockfall failure mechanisms and thus the selection of remedial measures to improve the stability of the rock face. Because these methods are essentially identical to those described in Section 6 with geometrical data obtained by terrestrial LiDAR, further illustrations from the study by Haneberg et al. (2006) are not included here.

A considerable number of applications of digital terrestrial photogrammetry, using similar techniques, have been reported (Aerden and San Roman 2002; Gaich et al. 2004; Poropat 2006; Haneberg and Medley 2007; Pötsch et al. 2007; Kemeny and Turner 2008). Roberts and Poropat (2000) and Grobler et al. (2003) reported on applications of digital terrestrial photogrammetry to the analysis of rock faces in open-pit mines. Haneberg (2007) described a procedure for evaluating the surface roughness of exposed fracture surfaces using 3-D point cloud information derived from either photogrammetric or LiDAR sources. Such information is important in evaluating the potential for sliding failures of rock masses along such fractures.

The photogrammetric approach provides an accurate, highly detailed visual representation of the rock outcrop within a 3-D context (Figure 7-19). This model allows analysis of fractures that appear only as linear features on the rock outcrop face, even when they do not produce any exposed planar surfaces whose orientations can be evaluated (Reid and Harrison 2000; Post et al. 2001; Kemeny and Post 2003; Kemeny, Mofya, et al. 2003).

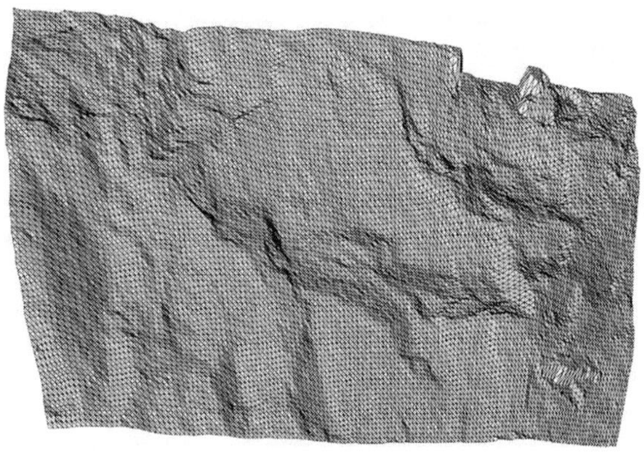

FIGURE 7-21 Triangulated mesh representation of rock outcrop model shown in Figure 7-19. Mesh consists of 23,360 facets (Haneberg et al. 2006).

5.3 Advantages and Limitations of Terrestrial Photogrammetry

The biggest advantage of terrestrial photogrammetry for evaluating rock slopes is that stereoscopic measurements can be performed on images acquired with readily available and affordable photographic equipment. A very-high-resolution professional digital camera typically costs about $25,000, but a typical 10-megapixel high-quality digital camera costs only $1,000 and yet is capable of acquiring images that can be used for rock slope evaluations (Haneberg et al. 2006; Haneberg 2008). Several sources offer the appropriate photogrammetric software that can analyze the digital images, create the 3-D models, and assist the user in performing the required evaluations of geological structures (ADAM Technology 2009; 3G Software and Measurement 2009; Terrasource Software 2009). Some sources offer turnkey systems consisting of a digital camera and software to support the photogrammetric analysis. Terrestrial photogrammetry is therefore a convenient and relatively inexpensive application.

Terrestrial photogrammetry also provides photographic documentation of the rock slope being investigated. The generated 3-D stereoscopic model corrects geometric distortions in the photographs; this feature is called *orthorectification* and allows the rock slope image to be projected onto a vertical reference plane, which can then be used as a base for engineering design drawings.

Several comparisons have been conducted between terrestrial photogrammetry and other data collection systems, chiefly the terrestrial LiDAR techniques that are discussed in the following section (Kemeny, Monte, et al. 2003; Kadobayashi et al. 2004; Kemeny and Turner 2008). Tonon and Kottenstette (2006) reported on a special workshop that compared LiDAR and photogrammetric techniques for evaluating rock slopes. The workshop included field tests of several systems. Zhang et al. (2004) reported on the applications of terrestrial photogrammetry with data collected using total stations.

6. TERRESTRIAL LiDAR AND RADAR TECHNOLOGIES

Terrestrial LiDAR and terrestrial radar technologies support quantitative assessment of rock slopes and reduce or eliminate sampling bias and safety concerns associated with conventional rock slope characterization methods (Kemeny and Turner 2008; Sturzenegger and Stead 2009). By using their own energy source, these technologies can operate at night and are not affected by sunlight or other sources of illumination. However, LiDAR beams are affected by atmospheric clarity conditions; their maximum range is reduced in smoky or hazy conditions, and they may be unable to produce satisfactory results under rainy, snowy, or foggy conditions. In contrast, radar operates at wavelengths that are unaffected by clouds or moisture conditions. While radar can provide very precise ranging measurements, it has a much lower spatial resolution compared with LiDAR. Digital photographic images are typically collected to supplement the LiDAR or radar data sets, and many rock slope assessments are based on combinations of digital image analysis, terrestrial photogrammetry (described in Section 5), and 3-D LiDAR or radar models (Kemeny, Mofya, et al. 2003; Kemeny, Monte et al. 2003; Kemeny, Norton, et al. 2006; Tonon and Kottenstette 2006; Turner et al. 2006; Haneberg 2007; Martin et al. 2007; Ferrero et al. 2009; Sturzenegger and Stead 2009).

LiDAR systems, described in Sections 6.1 and 6.2, use the speed of light and very precise timing devices to calculate the distance between a laser emitter–receiver device and an object reflecting the beam (Slob and Hack 2004). Applications of mobile terrestrial LiDAR devices (Section 6.2) have recently become feasible as LiDAR capabilities have evolved. Terrestrial radar systems (Section 6.3) have only relatively recently been used to monitor rock slopes. Modifications to typical airborne or satellite-based radar systems produced terrestrial interferometric radar systems, which were first used to monitor displacements of unstable slopes at open-pit mines in Australia (Reeves et al. 2000).

6.1 Terrestrial LiDAR

Terrestrial LiDAR devices perform a progressive scan of a desired scene by sending out a sequence of narrowly focused laser beams with gradually changing orientations. Thus, provided the unit is mounted on a stable platform during the scanning operation, a series of closely spaced but slightly offset scan lines are illuminated by the laser beam and a dense point cloud is generated that can be processed to yield 3-D [x, y, z] definitions of the features being scanned. Point clouds containing thousands to millions of 3-D data points produce images of near-photographic quality, as shown by the example in Figure 7-22, which is composed of about 1 million points.

There are two primary types of terrestrial LiDAR devices: time-of-flight scanners and phase-shift scanners. Figure 7-23 illustrates recent models of both types of scanners produced by several manufacturers.

6.1.1 Time-of-Flight Scanners

Time-of-flight laser scanners emit a pulsed laser beam that is reflected by the scanned object. A

FIGURE 7-22 Typical point cloud representation of highway rock cut. Location is along Mount Lemmon Highway near Tucson, Arizona. Visible rock cut is about 10 m (33 ft) high (Kemeny and Turner 2008).

FIGURE 7-23 Examples of recent models of terrestrial LiDAR scanners produced by several manufacturers. Units are time-of-flight scanners unless "phase shift" is noted (illustrations from manufacturer websites).

sensor measures the time of flight for the optical pulse to travel to and from the reflected surface. The distance the pulse traveled is then calculated by using the following equation:

$$\text{distance} = (\text{speed of light} \times \text{time of flight})/2$$

This technique allows measurements with an accuracy on the order of 1 cm (0.5 in.) or better over distances as great as 1 km (3,300 ft) under ideal conditions. Some time-of-flight scanners have the ability to measure several arrival times for a single emitted pulse, allowing for the identification and separation of objects at differing distances. For example, when a slope with vegetation is scanned, the *first arrival* would indicate the distance to the vegetation in front of the slope, and the *last arrival* would indicate the distance to the ground surface.

6.1.2 Phase-Shift Scanners

Phase-shift scanners use a continuous, sinusoidally modulated laser beam that is reflected by an object. The phase shift is determined by comparing the emitted and reflected signals. The time of flight is then determined from the following equation:

$$\text{time of flight} = \frac{\text{phase shift}}{2\pi \times \text{modulation frequency}}$$

These calculated time-of-flight values are then substituted into the equation given in Section 6.1.1 to determine the distance. Multiple modulation frequencies are often used to increase the accuracy of the time-of-flight determination. Although the maximum range of phase-shift scanners is less than the maximum range of time-of-flight scanners, the phase-shift scanners are capable of measuring distances with much greater accuracy—within 5 mm (0.25 in.)—and they can sample at much higher rates, up to 100 times those of time-of-flight laser scanners.

6.1.3 Scanner Characteristics

Multiple manufacturers located in several countries produce terrestrial LiDAR devices. Not all units are suitable for assessing rock slopes in the field; there are several trade-offs between accuracy, maximum distance, and precision (Jacobs 2005a). Innovations and new models continue to be introduced at a rapid pace, making it difficult to gather accurate information on the capabilities of these devices. Comprehensive product surveys provide guidance at regular intervals; two recent examples were published by Point of Beginning (POB 2010) and GIM-International (GIM-International 2009). They show that at least a dozen manufacturers are producing a large number of terrestrial LiDAR units. Table 7-1 summarizes the basic characteristics of two typical time-of-flight scanners and a typical phase-shift scanner, based partly on information abstracted from the POB 2010 product survey.

Terrestrial LiDAR units have remained relatively expensive since their inception; a scanner costs roughly 10 times as much as a total-station survey instrument. In spite of this cost difference, several state transportation agencies have purchased terrestrial LiDAR units, often for the purpose of assessing bridge structures, rock slopes, and landslides. Terrestrial LiDAR units purchased by private surveying firms have yielded productivity increases that more than offset their purchase costs (Jacobs 2005b). Slob et al. (2005) and Kemeny and Turner (2008) reported similar economies for rock slope studies.

6.1.4 Conducting LiDAR Field Surveys

Figure 7-24 shows a typical field setup for a terrestrial LiDAR survey of a rock cut along a road.

Table 7-1
Basic Characteristics of Three Terrestrial LiDAR Scanners

Characteristic	Scanner Manufacturer and Model: Optech, Inc.[a] ILRIS 3-D	Riegl USA, Inc.[b] LMS Z420i	Leica Geosystems AG[c] HDS6100
Scanner type	Time of flight	Time of flight	Phase shift
Laser wavelength and eye safety class	1,535 nm (near infrared), Class 1	Near infrared, Class 1	650–690 nm, Class 3R
Accuracy	8 mm at 100 m	10 mm at 50 m	0.1 mm at 50 m
Precision repeatability	7 mm (single shot) at 50 m	8 mm (single shot) at 50 m	≤2 mm at 25 m, 90% reflectivity
	4 mm (averaged) at 50 m	4 mm (averaged) at 50 m	≤3 mm at 50 m, 90% reflectivity
Minimum range	3 m	2 m	0.4 m
Maximum range	1,100 m at 80% reflectivity 400 m at 10% reflectivity	1,000 m at 80% reflectivity 350 m at 10% reflectivity	79 m
Horizontal field of view	40°	360°	360°
Vertical field of view	40°	80°	310°
Scan-angle step size	0.00115°	0.0025°	0.0018°
Spot size	17 mm at 50 m 29 mm at 100 m	16 mm at 50 m 25 mm at 100 m	14 mm at 50 m
Data sample rate	2,500 points/s	8,000 to 12,000 points/s	500,000 points/s
Additional functionality	Integrated digital camera (20-megapixel CMOS sensor)	Calibrated, oriented, 12-megapixel digital camera	Integrated 5- to 6-megapixel digital camera
Operational control	Handheld PDA for wireless control	Standard laptop PC via TCP–IP	Handheld PDA for wireless control, or tablet or notebook PC

Note: CMOS = complementary metal–oxide–semiconductor; PDA = personal digital assistant; TCP/IP = transmission control protocol–Internet protocol.
[a]Optech, Inc., Vaughan, Ontario, Canada, www.optech.ca.
[b]Riegl USA, Inc., Orlando, Florida, www.rieglusa.com.
[c]Leica Geosystems AG, Heerbrugg, Switzerland, www.leica-geosystems.com.
Source: Manufacturers' literature (from Slob 2010).

FIGURE 7-24
Typical field setup of terrestrial LiDAR unit scanning rock slope along road in Catalan Province, Spain (unit shown is Optech 3-D Ilris scanner). Umbrella keeps scanner cool and makes digital display easier to read (Slob et al. 2005).

The field LiDAR surveying procedure involves decisions concerning (*a*) the placement of the scanner, (*b*) the use of multiple overlapping scans, (*c*) suitable scanning parameters to control the point density and resolution of the rock face, (*d*) collection of digital images of the rock face, and (*e*) collection of supplemental information. Additional important decisions concerning the establishment of survey control points or the use of other measures for subsequent georeferencing of the point cloud are discussed in Section 6.1.5. Kemeny and Turner (2008) provided considerable information on these topics, which have also been addressed by several other reports of case studies (Ferrero et al. 2009; Lato, Hutchinson, Diederichs, and Harrap 2009).

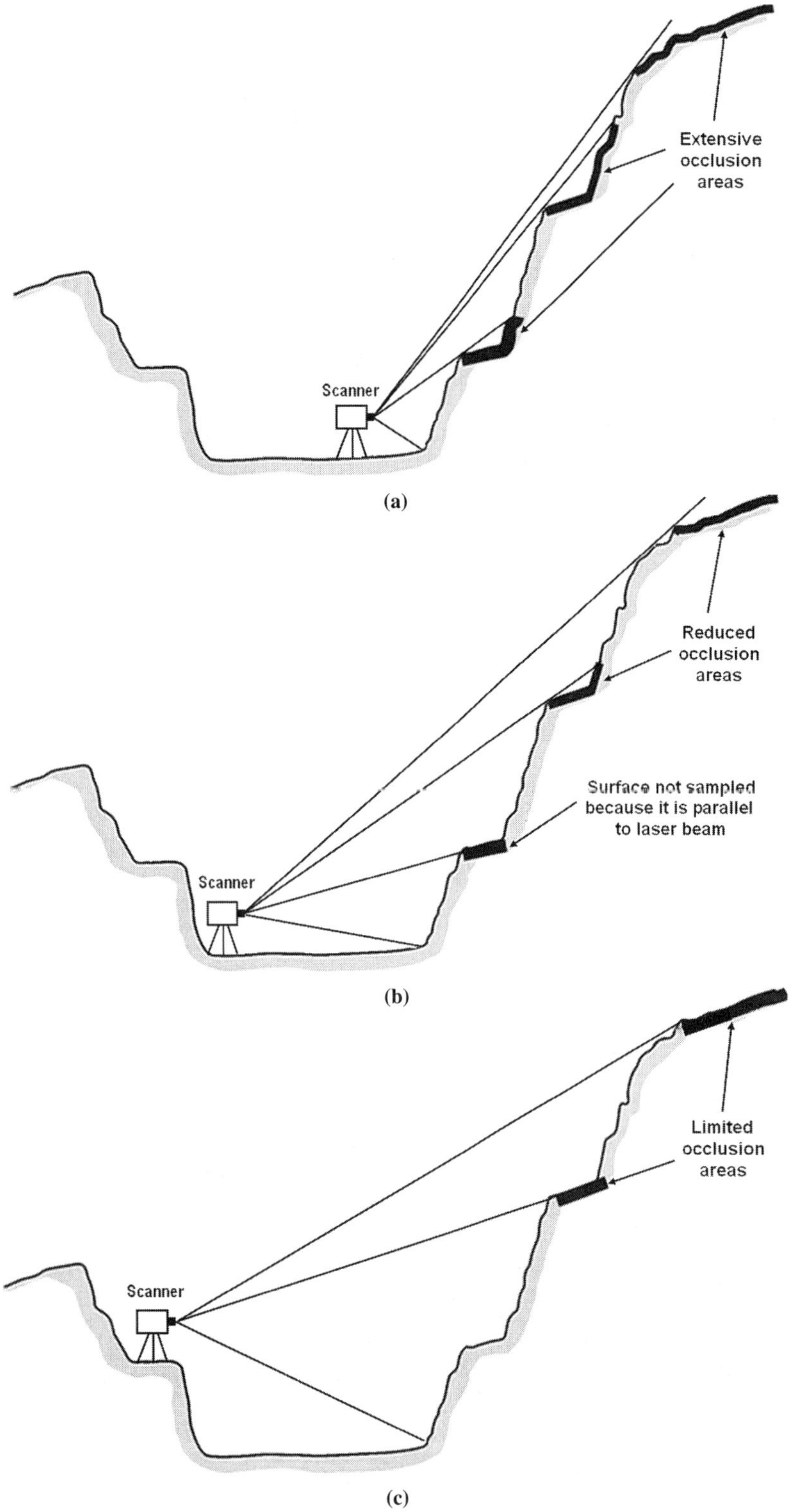

FIGURE 7-25
Laser scanner position may result in vertical sampling bias caused by occlusion (shadowing of surfaces): (*a*) low scanner position close to slope produces considerable occlusion; (*b*) scanner moved away from slope reduces occlusion; (*c*) raised scanner position further reduces occlusion but this position may not always be achievable along many transportation routes (adapted from Slob 2010).

6.1.4.1 Optimal Scanner Placement

Scanner placement is a critical decision. Since LiDAR relies on line-of-sight distance measurement, the distance of the unit to the rock face and the orientation of the rock face discontinuity surfaces influence the resolution, precision, and accuracy of the point cloud data. Whenever the LiDAR beam illuminates the rock face at a low angle, both the point density and the strength of the reflected beam are reduced, which results in poor resolution of these rock surface features and loss of spatial accuracy and precision. Some portions of the slope may be invisible to the LiDAR (Figures 7-25 and 7-26). A general rule of thumb is that the distance from the scanner to the rock face should be at least as great as the height of the rock face. This guideline suggests that whenever a rock face is higher than about 30 m (100 ft), the optimal scanner location is farther away than the right-of-way width of a typical two-lane road; this location is likely to present access and placement problems, depending on topographic conditions and land ownership. The laser beam expands as it travels from the LiDAR unit. Recent LiDAR units have a beam divergence of only 0.25 mrad, or 25 mm per 100 m (approximately 1 in. per 300 ft), but even this small divergence means that the distance between the scanner and the rock face influences the resolution of the features identified by the point cloud data (Lichti and Jamtsho 2006).

6.1.4.2 Multiple Scans

LiDAR scans from different locations overcome the occlusion problems in the evaluation of rock cuts with discontinuity surfaces oriented as shown in Figure 7-26. Multiple scans are typically used at many rock cuts along transportation routes to maintain the desired angle of at least 50° between the laser beam and the rock face (Kemeny and Turner 2008). Figure 7-27 diagrams the guiding principle governing the selection of multiple scans; at least a 20% overlap is desired between adjacent scans. Section 6.1.5 describes procedures for combining multiple scans to form a single data set.

6.1.4.3 Scanning Parameters

The vertical and horizontal extent of the area of interest on the rock face that is to be scanned is usually specified before a LiDAR survey starts.

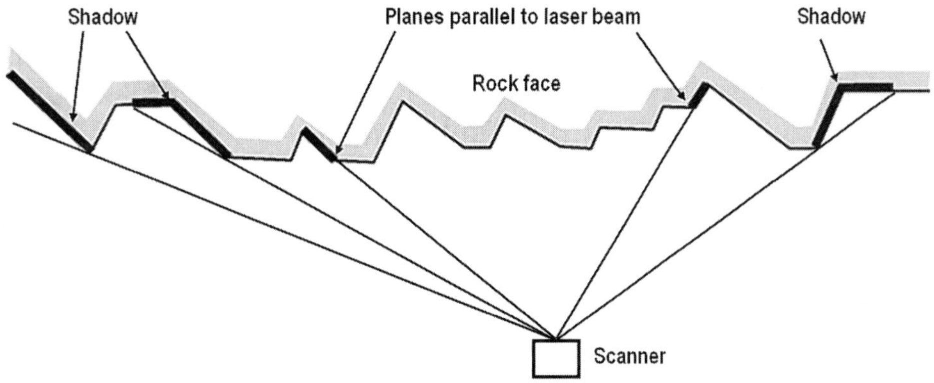

FIGURE 7-26
Laser scanning from single location may produce horizontal sampling bias caused by effects of shadowing and presence of semiparallel discontinuities (adapted from Slob 2010).

Limiting the area being scanned reduces the scanning duration and minimizes the volume of data collected. Terrestrial LiDAR scanners also allow the user to specify the angular offset or separation between adjacent scanning profiles, as well as the point-sampling rate. Combined with the average distance, these two parameters establish the nominal point spacing for a rock surface oriented normal to the laser beam. Reported rock mass characterization studies have typically used point cloud data with points spaced between 1.5 and 2 cm (0.5 and 0.75 in.) (Kemeny and Turner 2008; Ferrero et al. 2009; Sturzenegger and Stead 2009; Lim et al. 2010). Kemeny and Turner (2008) suggested that LiDAR surveys of larger rock exposures that can only be evaluated from great distances, such as parts of Glenwood Canyon in Colorado, may yield useful information with points spaced at 5 cm (2 in.). Similar point densities have been collected with mobile LiDAR surveys and used successfully in subsequent analyses of rock cut stability (Lato, Hutchinson, Diederichs, et al. 2009).

6.1.4.4 Digital Images

Current terrestrial LiDAR scanners automatically capture high-resolution digital images with a built-in digital camera either installed inside the scanner or mounted externally. The camera, which typically records images containing at least 12 megapixels, is calibrated and aligned, or *bore-sighted*, relative to the laser; thus individual points in the point cloud are provided with red-green-blue (RGB) digital values to produce a colored point cloud (Kemeny and Turner 2008). Once the point cloud is transformed into a 3-D surface model (see Section 6.1.8.1) the much-higher-resolution original digital images can be draped onto the 3-D surface using texture-mapping techniques.

6.1.4.5 Supplemental Information

At the field location, additional information is typically recorded concerning the weather conditions, basic dimensions of the rock face, locations of important features such as survey control points, and geological descriptions of the rock

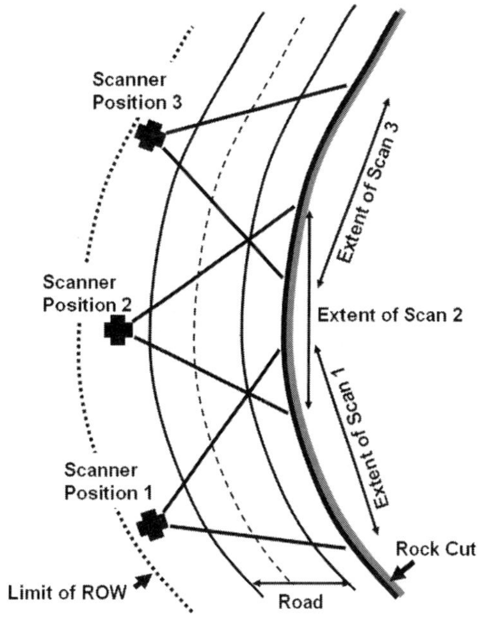

FIGURE 7-27
Typical arrangement of multiple LiDAR surveys of curving rock cut (ROW = right-of-way).

face. Representative discontinuity information often is collected with traditional methods. Such information is useful during subsequent processing of the LiDAR data.

6.1.5 Managing and Georeferencing Point Cloud Data

Terrestrial LiDAR point cloud data sets are voluminous and often unlike any data sets familiar to geologists, geotechnical engineers, or design engineers. Thus the subsequent use, storage, and management of point cloud data sets typically require new data-handling software and training of personnel. At this time, an ASCII-formatted, comma-delimited 3-D coordinate file (often referred to as an *xyz file*) is the only generic format available for data transfer among various postprocessing functions. Because these files are extremely large and are slow to process, each LiDAR scanner manufacturer has its own specialized and proprietary binary file format (Kemeny and Turner 2008).

A terrestrial LiDAR unit initially produces 3-D data referenced relative to the scanner's orientation and position. The scanner's position is defined as the origin [0, 0, 0], the default *y*-axis is aligned to the mean instrument direction, and the scanner axis is the vertical axis. If these data are to be used by existing spatial CAD or GIS databases, or if multiple scans are to be merged to form a single point cloud, then the data have to be referenced to a regional or a local coordinate grid. Terrestrial LiDAR system manufacturers provide software to allow real-time or subsequent georeferencing of the point cloud data, typically by using reflectors at locations determined by a total station or other survey methods. The distribution and number of such control points and the accurate determination of their positions in the field and in the point cloud are especially critical for accurate volume and engineering design calculations (Lichti et al. 2002, 2005; Feng and Röshoff 2004; Fröhlich and Mettenleiter 2004; Sturzenegger et al. 2007).

However, when LiDAR data are used to evaluate rockfall or characterize rock mass conditions, the point cloud data can be georeferenced with somewhat less precision. After all, geologists have traditionally made structural observations with hand-held compasses that have accuracies of about 1° at best, and slope stability calculations rely on the relative orientation of joints and bedding planes to the slope orientation and geometry (Kemeny and Turner 2008; Ferrero et al. 2009; Sturzenegger and Stead 2009). Similar requirements exist for quarry or tunnel stability evaluations, which depend on analyses of block size and block stability.

For the evaluation of discontinuity orientations, it is only necessary to level and reorient the data relative to the true north. If the LiDAR scanner unit, mounted on a tripod, is leveled perfectly horizontal and the bearing of the mean scanner direction is determined, the data set can be simply rotated. The accuracy of this process depends entirely on the precision with which the bearing of the scanner can be measured and the accuracy with which the scanner can be leveled. When the LiDAR scanner has to be inclined to capture the top of a steep and high rock face, the point cloud data have to undergo a more complex transformation. Schuhmacher and Böhn (2005) defined three alternative methods for georeferencing terrestrial LiDAR point clouds and provided accuracy comparisons when these were used in urban settings. Lichti et al. (2005) provided additional information on error budgets for direct georeferencing and described common artifacts within LiDAR point clouds. Additional georeferencing methods that are appropriate for rock slopes have been described by Mohamed and Wilkinson (2009) and Olsen et al. (2009).

Slob et al. (2005) performed several successful geological structure analyses by placing two or more small, 60- by 60-cm (2- by 2-ft) white plywood boards with known orientations within the scanning field (Figure 7-28). Postprocessing of the scanned data and evaluation of the points on the reference boards allowed transformation of the data with accuracies comparable with those of traditional geological field surveys.

Individual LiDAR point clouds are combined to produce a single merged point cloud that accurately represents a lengthy rock face exposure, such as along a road or extensive shore cliff (Olsen et al. 2009). Typically, adjacent scanned regions that overlap by at least 20% are matched by use of the iterative closest point (ICP) algorithm introduced by Besl and McKay (1992) and since improved by many others. Applications of the ICP process to LiDAR data are widely reported (Schuhmacher and

Böhm 2005; Kemeny and Turner 2008; Olsen et al. 2009). The merged point clouds typically contain much smaller occluded regions and more accurate representations of various discontinuity surfaces.

Kemeny and Turner (2008) reported that merging individual point clouds into a single very large point cloud introduces data management and processing problems. For example, at the optimal point spacing of 1.5 to 2 cm (0.5 to 0.75 in.) described in Section 6.1.4, the LiDAR scan of a section of rock face 30 m (100 ft) high by 40 m (130 ft) long will contain 3 million points. A series of six to seven such scans when merged into a single point cloud will contain around 20 million points. This merged point cloud will be very difficult to visualize and rotate within most point cloud–processing software. Thus, although merged point clouds provide better data for analysis of rock mass structural discontinuities, the creation of extensively merged point clouds often does not allow for efficient analyses. The conversion of the point cloud data to 3-D surface representations, as described in Section 6.1.8, often reduces the total number of points and therefore the size of the files. The ICP algorithm can then be used to merge these 3-D surface representations to cover large regions.

6.1.6 Applications of Terrestrial LiDAR

Terrestrial LiDAR techniques are increasingly used to quantitatively evaluate the 3-D characteristics of natural and engineered rock slopes because they offer several advantages, including

- The ability to sample extended areas beyond the base of an outcrop, thereby providing more representative statistical samples;
- The ability to survey inaccessible steep and high rock faces;
- Risk reduction for workers, because the survey is conducted from a remote safe location protected from traffic and rockfalls; and
- Creation of a permanent record of the rock slope condition at a specific time.

Nevertheless, terrestrial LiDAR has limitations. LiDAR uses wavelengths that are influenced by atmospheric conditions. The lack of calibrated targets introduces some uncertainty into the image registration process. The strength of the returning LiDAR signal is influenced by the roughness, wetness, and composition of the reflecting surface, its distance from the scanner, and its angle of incidence to the laser beam (Lichti et al. 2002; Lim et al. 2005; Buckley et al. 2008).

In spite of these limitations, several authors have successfully used LiDAR techniques for rock slope characterization (Slob et al. 2002, 2004; Feng and Röshoff 2004; Kemeny, Norton, et al. 2006; Kemeny, Turner, et al. 2006; Tonon and Kottenstette 2006; Sturzenegger et al. 2007; Kemeny and Turner 2008). Figure 7-29 defines two major applications of terrestrial LiDAR: the evaluation of rockfall and the evaluation of rock masses. Each application has several components, and these are typically completed by subsequent analytical applications. Rockfall evaluations are

FIGURE 7-28 Small rock cut in carboniferous slates with well-developed discontinuities along road in Catalan Province, Spain; two 60- by 60-cm (2- by 2-ft) white boards provide scale and orientation references to LiDAR data set (Slob et al. 2005).

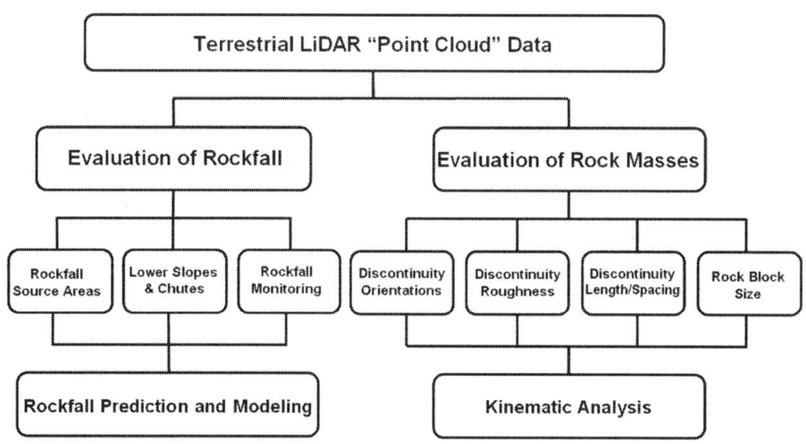

FIGURE 7-29 Applications of terrestrial LiDAR.

typically completed by a variety of rockfall prediction and modeling techniques that are described in Chapter 9. However, as described in Section 6.1.7, terrestrial LiDAR surveys provide fundamental quantitative data required by these prediction and modeling techniques. Similarly, the use of terrestrial LiDAR to evaluate rock mass characteristics, as described in Section 6.1.8, provides fundamental data required by the use of stereographic projections (*stereonets*) to analyze the stability of rock slopes, as described in Chapter 2.

6.1.7 Rockfall Evaluations with LiDAR

As shown by Figure 7-29, terrestrial LiDAR evaluates rockfall by characterizing rockfall source areas, evaluating rockfall chutes, and monitoring rockfall processes (Martin et al. 2007; Sturzenegger et al. 2007; Kemeny and Turner 2008).

6.1.7.1 Characterizing Rockfall Source Areas

Terrestrial LiDAR has quantitatively evaluated rockfall source areas in steep mountainous regions (Abellán et al. 2006; Rabatel et al. 2008) as well as in locations where rockfall poses hazards (Massey et al. 2006; Whitworth et al. 2006). Kemeny and Turner (2008), Lato, Hutchinson, Diederichs, and Harrap (2009), and Lan et al. (2010) provided examples of LiDAR evaluation of rockfall sources along transportation routes.

Cliff erosion studies along the northeastern English coast used a series of terrestrial LiDAR surveys over a 20-month period (Lim et al. 2005, 2010; Rosser, Dunning, et al. 2005; Rosser, Petley, et al. 2005). The LiDAR provided high-resolution measurements of cliff-face "erosional scars" resulting from rockfalls ranging from 1.25×10^{-4} m^3 (4.4×10^{-3} ft^3) to more than 2.5×10^3 m^3 (88.22×10^3 ft^3). The smallest volumes correspond to rock masses 5 cm × 5 cm × 5 cm (2 in. × 2 in. × 2 in.). An inventory of more than 100,000 discrete rockfalls provides insight into rockfall characteristics and mechanisms for a range of rock types and cliff conditions (Lim et al. 2010). Similar studies have been undertaken to evaluate the coastal cliffs near San Diego, California, which are composed of very different rock types (Olsen et al. 2009). The high-resolution LiDAR-derived data allowed these studies to gain new insights into the characteristic processes affecting rock slope failure (Rosser et al. 2007).

6.1.7.2 Evaluating Rockfall Chutes

Although the characteristics of the rockfall source area determine the size and frequency of rockfall events, the topography and roughness of lower slopes greatly influence the trajectories, bounce heights, and velocities of the falling rockfall blocks, and therefore are critical to the design of rockfall mitigation measures. At many locations, the shape of these lower slopes defines preferred rockfall paths, or chutes, over which most detached and falling rock blocks will travel. Until recently, rockfall modeling and prediction procedures used only a two-dimensional (2-D) longitudinal slope profile, but it has long been recognized that the 3-D character of these lower slopes greatly influences the extent and distribution of rockfall hazard zones (Lan et al. 2010). Terrestrial LiDAR provides an efficient method of collecting the detailed topographic information to define conditions on these lower slopes (Kemeny and Turner 2008).

Often a rock block does not travel all the way to the base of the slope but comes to rest partway down a chute. These rocks typically continue to slowly travel down the chute in a time-dependent fashion, being mobilized during periods of rainfall, for example. Since rock blocks larger than about 0.5 m (2 ft) can be readily identified in LiDAR images and their positions accurately determined, LiDAR can be used to understand this behavior by taking periodic repeated scans and using change-detection algorithms (Kemeny and Turner 2008; Rabatel et al. 2008).

6.1.7.3 Monitoring Rockfall Processes

Repeated LiDAR surveys over time allow for identification of changes in the rock slope, including movements or erosion of the rock face as well as movements of detached rock blocks (Rosser, Petley, et al. 2005; Rosser et al. 2007; Rabatel et al. 2008; Olsen et al. 2009). Kemeny and Turner (2008) described monitoring activities using LiDAR on rock slopes above transportation facilities. LiDAR data are also capable of providing many of the parameters used by the rockfall hazard rating systems to evaluate rock slopes (Kemeny and Turner 2008; Lato, Hutchinson, Diederichs, and Harrap 2009).

6.1.8 Evaluating Rock Mass Characteristics with LiDAR

As shown by Figure 7-29, terrestrial LiDAR characterizes rock masses by evaluating discontinuity orientations, roughness, length, and spacing, as well as by evaluating the size of potential rockfall blocks (Sturzenegger et al. 2007; Kemeny and Turner 2008; Sturzenegger and Stead 2009). The LiDAR data provide quantitative values that are used subsequently by stereographic projections (stereonets) to analyze the stability of rock slopes and the potential for rockfall, as described in Chapter 2.

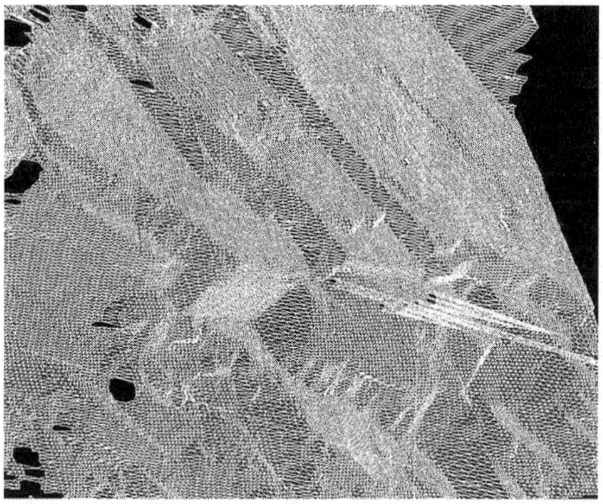

FIGURE 7-30 Detail of triangulated mesh representing small portion of rock surface (Turner et al. 2006).

6.1.8.1 Discontinuity Orientations

Conversion of point cloud data into a 3-D surface model of the rock face permits the identification of rock mass discontinuities (Lemy and Hadjigeorgiou 2004; Ferrero et al. 2009; Sturzenegger and Stead 2009). Some techniques produce an interpolated surface consisting of regular triangular facets, whereas alternative parametric techniques quickly construct smooth, manifold surfaces from noisy point cloud data and can repair incomplete meshes. Detailed discussions of alternative 3-D surface reconstruction algorithms are provided in several reports (Slob et al. 2002, 2004, 2005; Kemeny, Mofya, et al. 2003; Kemeny, Monte, et al. 2003; Kemeny, Norton, et al. 2006; Kemeny, Turner, et al. 2006; Turner et al. 2006; Buckley et al. 2008; Kemeny and Turner 2008; Ferrero et al. 2009; Lato, Hutchinson, Diederichs, and Harrap 2009).

Figure 7-30 shows a section of a triangulated mesh representing a 3-D rock surface that contains several large planar discontinuities. The original point cloud was collected from a single scanning location; thus some areas were shadowed or poorly exposed to the LiDAR unit and some "holes" occur in the mesh. Also, a few extraneous points are visible; they are readily identified and removed from the data set. Although it is possible to manually examine this mesh to identify areas having a relatively uniform orientation, the use of efficient automated methods is much more attractive (Slob et al. 2005; Ferrero et al. 2009). Figure 7-31 shows a typical result.

The automated extraction of 3-D discontinuity surfaces is only possible when these discontinuities are clearly exposed on the rock face. When joints or bedding planes are oriented more or less perpendicular to the rock face, they typically are not exposed as surfaces but are visible as lines or cracks in the rock face (Kemeny and Post 2003). Figure 7-32 illustrates two sections of a rock cut along a road in the mountains of southwestern Colorado. The rock joints are clearly visible at one exposure (Figure 7-32*a*), whereas around the corner the rock face, oriented almost at right angles to the first exposure, exhibits a very different character (Figure 7-32*b*). At this second exposure, the rock discontinuities are merely dark lines on the cliff face and rarely show any surface offsets. Turner et al. (2006) described how structural information on these discontinuities was extracted from these two rock faces with a combination of 3-D model analysis and digital image analysis.

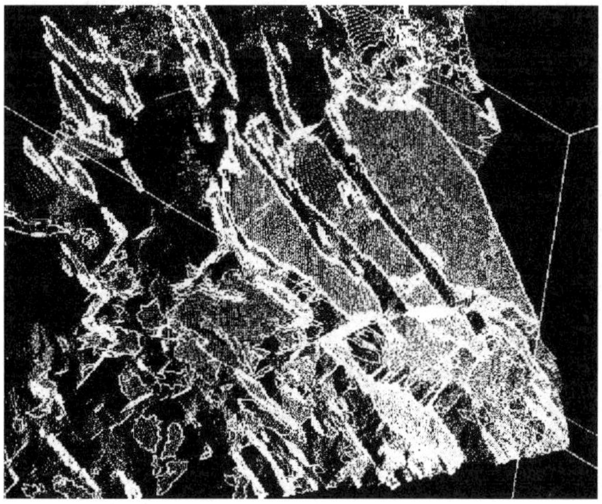

FIGURE 7-31 Interactive identification of joint set or bedding plane (Turner et al. 2006).

(a)

(b)

FIGURE 7-32 Two adjacent rock cuts in Colorado showing distinctive exposure characteristics that require different analysis approaches: (*a*) rock cut with several joints and foliations forming prominent exposed surfaces [height of view about 8 m (25 ft)]; (*b*) rock face around corner from that in Image *a* is oriented approximately perpendicular to that exposure; prominent large exposed surfaces visible in Image *a* now appear only as dark lines, and rock face has few prominent offsets (Turner et al. 2006).

Combined analysis of digital images and 3-D surface models is especially valuable when highly fractured rock masses or those that show little surface relief are examined (Kemeny and Post 2003; Kemeny, Mofya, et al. 2003; Kemeny, Monte, et al. 2003; Kemeny, Norton, et al. 2006; Haneberg et al. 2006; Baker et al. 2008; Haneberg 2008; Kemeny and Turner 2008; Sturzenegger and Stead 2009). It also offers considerable potential for evaluating the presence and origin of seepage on rock cuts (Haneberg et al. 2006; Sturzenegger et al. 2007) and for additional rock mass characterizations, including assessment of blasting and weathering effects on rock slopes (Sturzenegger and Stead 2009).

Once several discontinuity surfaces have been identified, further analysis permits their classification into sets, which are defined by the common orientation of several discontinuities. Typically rock mass discontinuities are not randomly oriented but exhibit a few preferred orientations that reflect previous stress regimes and include bedding planes in sedimentary rocks. They often control the potential instability and failure mechanisms of a rock slope and thus influence the degree of rockfall hazard. Several authors have described how discontinuities defined by LiDAR data are further classified by using stereonet analysis to produce geostructural evaluations (Slob et al. 2004, 2005; Turner et al. 2006; Kemeny and Turner 2008; Ferrero et al. 2009; Sturzenegger and Stead 2009).

6.1.8.2 Discontinuity Roughness

Detailed analysis of point cloud data permits quantification of discontinuity roughness (Feng et al. 2003; Fardin et al. 2004; Haneberg 2007, 2008; Baker et al. 2008; Poropat 2008; Sturzenegger and Stead 2009). The surface roughness of discontinuities influences the failure behavior of fractured rock masses. Small-scale roughness on fractures, joints, or bedding planes determines the shear strength and sliding friction along the discontinuity, whereas larger-scale roughness, sometimes referred to as *waviness*, influences the direction of shear and the dilation of the rock mass during failure (Hobbs 1993; Grasselli et al. 2002).

6.1.8.3 Discontinuity Length and Spacing

Fracture length and spacing can be determined from either point clouds or digital images. Only an "apparent" spacing is determined from a 2-D image; 3-D evaluations allow for the computation of the "true" spacing—the distance perpendicular to the average orientation of the discontinuity set (Kemeny and Turner 2008). Discontinuity length and spacing are typically interrelated. Persistent discontinuities, those that are long in relation to their spacing, typically allow for accurate spacing measurement, whereas determination of the spacing of nonpersistent discontinuities is more difficult. Sturzenegger and Stead (2009) demonstrated that terrestrial LiDAR and terrestrial photogrammetry data provide superior measurements for persistent discontinuities compared with traditional field observations. Observations are obtained in areas inaccessible to ground-based observers and provide the correct lengths of extended discontinuities. Automatic trace delineation involves image-processing algorithms; the use of edge-detection algorithms for rock fractures is described by several authors (Hadjigeorgiou et al. 2003; Kemeny and Post 2003; Sturzenegger et al. 2007; Kemeny and Turner 2008; Sturzenegger and Stead 2009). Kemeny and Turner (2008) pointed out that automated fracture delineation by edge-detection algorithms does not readily differentiate between discontinuities belonging to different sets.

6.1.8.4 Size of Rock Blocks

The orientation and spacing of a series of discontinuity sets control the distribution of sizes of individual blocks within a rock mass. Thus, inspection of a point cloud allows the direct identification of individual blocks in a rockfall source zone and the computation of their shapes, dimensions, and volumes (Abellán et al. 2006; Whitworth et al. 2006; Kemeny and Turner 2008). It is also possible to develop a statistical size distribution of potential rockfall blocks by probabilistic evaluations of the identified discontinuity sets.

6.2 Mobile Terrestrial LiDAR

Traditional terrestrial LiDAR units, described in Section 6.1, are statically mounted on tripods. Aerial LiDAR employs similar scanning devices mounted on helicopters and airplanes, which are additionally equipped with a GPS and an inertial navigation system (INS), also known as an *inertial measurement unit* (IMU), to accurately determine location and orientation of the scanning unit during the flight. Recent innovations, especially very high rates of data collection, have allowed some of the previously static terrestrial LiDAR sensors to be deployed on vehicles to create mobile terrestrial applications (Glennie 2007a).

The use of mobile terrestrial LiDAR to identify and examine potential rockfall areas along transportation corridors offers significant advantages over traditionally employed static terrestrial LiDAR techniques at many locations. Mobile terrestrial LiDAR units provide capabilities to dynamically collect data from within the right-of-way while avoiding interference with normal traffic operations within the corridor.

Mobile terrestrial LiDAR is an evolving technology. The earliest reported application was in 2003, when a 400-km (250-mi) section of Afghanistan Highway 1 was scanned with a mobile LiDAR unit originally designed for helicopter deployment. The unit was placed on a truck when a suitable helicopter could not be acquired (Newby and Mrstik 2005). On the basis of the success of this Afghanistan application, and subsequent interest expressed by potential customers for applications in North America, several manufacturers began to develop complete mobile systems (Redstall 2006a, 2006b; Alshawa et al. 2007; Hunter et al. 2008; Morgan 2009). Initial applications produced inventories of roadside infrastructure elements, especially in urban or congested areas with high traffic volumes, but some applications were in open-pit mines (Lato and Hutchinson 2008).

Several manufacturers are now offering mobile terrestrial LiDAR units (Table 7-2). A typical mobile terrestrial LiDAR system consists of the following components:

- A GPS receiver combined with a high-accuracy navigation-grade IMU to provide continuous monitoring of platform location and orientation,
- An array of LiDAR scanners to provide a 360° field of view,
- Multiple digital video or digital frame-grabber cameras,
- A state-of-the-art timing electronics unit that provides internal synchronization and calibration of the sensor subsystems, and
- A computer control and data-storage system.

Figure 7-33 illustrates the operational linkages between components of a typical mobile LiDAR system. The GPS receiver, the IMU, the LiDAR scanners, and the video recording units are typically contained on a flexible mounting assembly (Figure 7-34) that allows the mobile LiDAR to be deployed on a variety of vehicles (Figures 7-35 to 7-37). At least one manufacturer offers the mobile LiDAR instrument pod mounted on the end of a hydraulic lift that can raise or lower the unit to between 2 and 4 m (6.5 and 13 ft) above the ground (Figure 7-38). Usually the systems controlling the

Table 7-2
Characteristics of Mobile Terrestrial LiDAR Systems

	Scanner Manufacturer			
Characteristic	Optech, Inc.[a]	Riegl USA, Inc.[b]	3-D Laser Mapping Ltd.[c]	Terrapoint Division of Ambercore Software Inc.[d]
System name	Lynx Mobile Mapper	VMX-250 Mobile Mapper	StreetMapper 360	Titan Mobile Mapper
Cameras	Maximum two cameras (2- or 5-megapixel)	Maximum eight cameras (4-megapixel typical)	Maximum four cameras (each 4 megapixels)	One camera (1.3-megapixels)
Number of sensors	One to four LiDAR sensors	Two Riegl VQ-250 LiDAR sensors	Two Riegl VQ-250 LiDAR sensors	Four Riegl VQ-210 LiDAR sensors
Measurement rate	100,000 shots/s per sensor[e]	300,000 shots/s per sensor[e]	300,000 shots/s per sensor[e]	30,000 shots/s per sensor[e]
Absolute accuracy	±5 cm depending on vehicle speed and quality of GPS	1–5 cm depending on vehicle speed	1–5 cm depending on vehicle speed	±5 cm depending on vehicle speed and quality of GPS
Point density, spatial resolution (points/m^2)	≤1,000 at 100 km/h, depending on range	100–2,000, depending on range and vehicle speed	100–2,000, depending on range and vehicle speed	100–800, depending on range and vehicle speed
Range precision	±7 mm (1 sigma)	±5 mm (1 sigma)	±5 mm (1 sigma)	Not specified
Maximum range	100–200 m (with 20% reflectivity target)	300 m with 90% reflectivity; 100 m with 10% reflectivity	300 m with 90% reflectivity; 100 m with 10% reflectivity	100–200 m depending on reflectivity of target

[a]Optech, Inc., Vaughan, Ontario, Canada, www.optech.ca.
[b]Riegl USA, Inc., Orlando, Florida, www.rieglusa.com.
[c]3-D Laser Mapping Ltd., Nottingham, United Kingdom, www.3dlasermapping.com.
[d]Terrapoint USA, Inc., The Woodlands, Texas, www.ambercore.com/titan.php.
[e]Laser units send out a rapid series of light pulses, or "shots." Each shot results in locating a single "point" location in three dimensions (x, y, z). Thus, the units collect data on point locations with scanning rates quoted in points/s; however, the operation of the units is usually reported in terms of shots/s.
Source: Manufacturers' literature.

mobile LiDAR sensors—the timing electronics, the computer and operator interface, and the data-storage component—are located inside the vehicle. The units are powered from the vehicle's electrical system, with battery backup provided for convenience. Units are thus usually deployed on small trucks or vans (Figures 7-35 to 7-37) but have been deployed on "hi-rail" or "speeder" trucks to survey railroad rights-of-way (Figures 7-38 and 7-39) or on boats to survey coastal areas.

6.2.1 Positional Accuracy of Mobile Terrestrial LiDAR

Because the mobile scanner collects data dynamically, the final positional accuracy of the points within the point cloud generated by the scanner or scanners depends on both scanner precision and accurate knowledge of the instantaneous position and orientation of the scanner. The basic global position is determined by RTK GPS methods, as discussed in Section 4.2.3. Carrier-phase GPS measurements provide sub-5-cm (better than 2-in.) positioning. Nevertheless, the generation of an accurate trajectory is often a fairly difficult problem, especially in urban areas, because of the unreliability of obtaining a strong GPS positioning solution when the view of some satellites is occluded.

Because of the vibration and movement of the vehicle, complex accelerations of millimeters or centimeters per second (up to several inches per second) are superimposed on the overall vehicle path. As a result, a simple interpolation between GPS positions (collected once per second) is inadequate for properly positioning each LiDAR data point. A navigation-grade IMU, consisting of coupled gyroscopes and accelerometers, is used to solve this problem. The IMU also provides path reconstruction during short intervals when the GPS signals are unavailable because of obstructed views of the sky; thus mobile LiDAR systems are capable of scanning short tunnels and underpasses

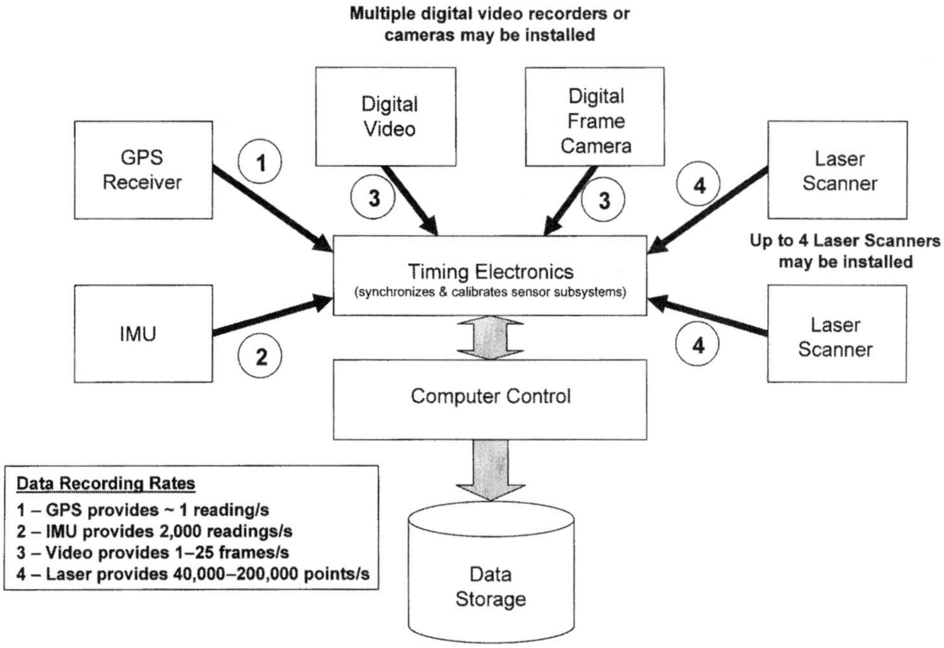

FIGURE 7-33 Operational linkages between components of typical mobile LiDAR system.

or canyons without loss of precision. Figure 7-33 illustrates the relative roles of GPS and IMU in establishing position and orientation.

Mobile LiDAR systems produce a 3-D point cloud reported in a geographic coordinate system (Figure 7-40). Although Figure 7-40 appears to be a grayscale photograph, it is actually a view of tens of millions of 3-D points surveyed across the face of the bounding rock cut and highway surface. Recorded intensity values allow visible discrimination of painted lines on the road surface because of their highly reflective properties. This point cloud has several important characteristics:

- Points are limited to features visible from the perspective of the scanner because it is a line-of-sight technology;
- Point density decreases as distance from the scanner increases; mobile LiDAR systems have maximum ranges of 100 to 300 m (330 to 1,000 ft), varying with different system designs and the level of reflectivity of the objects being scanned (Table 7-2);
- Points are collected at a high rate, with effective scanning rates for various systems ranging from 30,000 to 300,000 points/s (Table 7-2); and
- The strength of the return is measured and reported as an intensity value, which allows genuine visual discrimination of different materials.

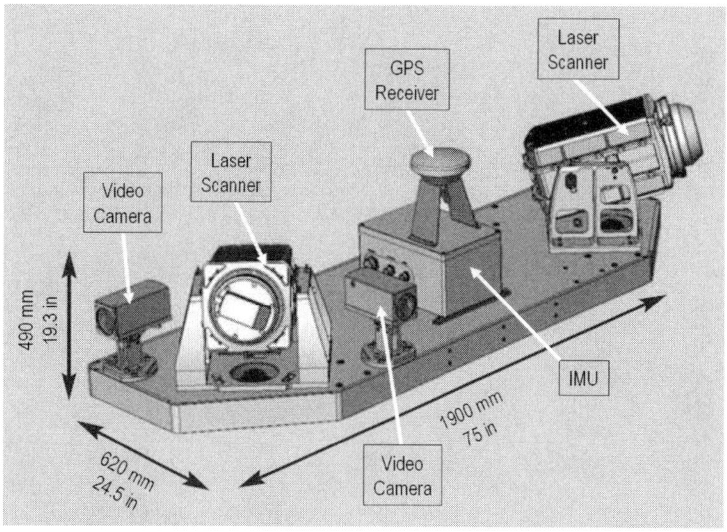

FIGURE 7-34 Typical integrated mobile LiDAR sensor system on flexible mounting assembly (Optech 2008).

FIGURE 7-35 Optech Lynx mobile LiDAR system (Optech 2008).

FIGURE 7-36
StreetMapper mobile LiDAR system with two laser scanners installed (Hunter et al. 2008).

FIGURE 7-37
Riegl VMX-250 mobile LiDAR system (Riegl USA 2010).

FIGURE 7-38
Terrapoint TITAN mobile LiDAR system mounted on hydraulic boom in bed of hi-rail truck for deployment along Algoma Central Railway in northern Ontario, Canada (Lato, Hutchinson, Diederichs, et al. 2009).

FIGURE 7-39
Optech Lynx mobile LiDAR system mounted on railway speeder (Morgan 2009).

These globally referenced LiDAR data inherently have two associated accuracy values, commonly known as local and global accuracy. *Local accuracy* is primarily a factor of sensor error, which is the observed error of the LiDAR sensor itself. Table 7-2 gives some reported values, which are approximately 1 cm (0.4 in.) at a typical 100-m (330-ft) distance between the LiDAR sensor and the objects being scanned. *Global accuracy* depends on the GPS–IMU solution; it is a measure of how well the determined locations of 3-D points match their true geographic locations. With RTK GPS methods, global accuracy has a base error of 0.5 cm (0.2 in.) and this accuracy decreases as the distance from the differential GPS base station increases, at

FIGURE 7-40
LiDAR data collected by mobile LiDAR system along Provincial Highway 15 near Ottawa, Ontario, Canada, in June 2007. Rock mass present on right-hand side of point cloud is roughly 5 m (15 ft) high (Lato, Hutchinson, Diederichs, et al. 2009).

a rate of about 0.5 cm per 1 km (or 0.32 in. per mile); this rate is usually defined as 0.5 cm + 5 ppm (0.2 in. + 5 ppm) (Glennie 2007b). A thorough overview of accuracy considerations for mobile terrestrial LiDAR was conducted and published by Barber et al. (2008). Additional evaluations on the accuracy of mobile terrestrial LiDAR are provided by Lichti et al. (2005), Glennie et al. (2006), and Glennie (2007b).

Global accuracy does not affect the spatial resolution of the data or most analyses of rock exposures to determine their potential for rockfall. Global accuracy does affect the ability to accurately relate the identified features to their true locations in space and so to other GIS data sets. It is essential for the geotechnical use of these data that the data collection errors be minimized. If the data are to be used for orientation-based measurements, any significant positional errors will propagate through subsequent analyses.

6.2.2 Using Mobile Terrestrial LiDAR to Evaluate Rock Exposures

As noted previously, the applications of mobile terrestrial LiDAR have largely focused on inventorying roadside infrastructure elements, especially in urban or congested areas with high traffic volumes (Redstall 2006a, 2006b; Glennie 2007a; Hunter et al. 2008; Optech 2008; Riegl USA 2010), but some applications have been in open-pit mines (Lato and Hutchinson 2008).

Lato, Hutchinson, Diederichs, et al. (2009) provided a comprehensive description of the application of mobile terrestrial LiDAR to the monitoring of rockfall hazards along transportation corridors. They reported on several applications along both highways and a railway in Ontario, Canada. They also provided valuable comparisons between mobile terrestrial LiDAR and both traditional static-terrestrial and airborne LiDAR systems; these comparisons are summarized in Section 6.2.3.

The largest application of mobile LiDAR to rockfall monitoring, reported by Lato, Hutchinson, Diederichs, et al. (2009), is an evaluation of a 20-km (12.5-mi) section of the Algoma Central Railway, which is located in northern Ontario and connects Sault Ste. Marie to Hearst, Ontario, about 475 km (300 mi) to the north. The route parallels the Algoma River, whose course is controlled by a major northeast-southwest-trending fault. Consequently the railway is bordered by rocky slopes that rise up to 900 m (almost 3,000 ft) above track level along much of the route.

Rockfall hazards are common along the Algoma Central, which was constructed over 100 years ago with relatively unsophisticated blasting and excavation techniques. The railway operator had previously undertaken an extensive evaluation of potential rockfall hazards by using typical field investigative methods. Thus in 2007–2008, it was decided to test a combination of terrestrial and helicopter-based mobile LiDAR surveys by comparing the results obtained with these systems with the existing rockfall hazard database resulting from earlier field investigations.

For the terrestrial mobile LiDAR data collection, a Terrapoint TITAN system was used mounted on a hi-rail truck that could travel along the rails (Figure 7-38). The data were collected during a 1.5-h transit of the 20-km (12.5-mi) section. Thus, data collection did not impede or disrupt scheduled traffic along the railway.

The TITAN system collects data at a rate of 30,000 points/s; this rate is, of course, independent of truck speed. To produce data of sufficient density to extract discrete geological features, sections of track known to be of high rockfall hazard potential on the basis of earlier field investigations were assessed at a slower speed than that used for the nonhazardous locations. The most significant operational issue encountered during the planning and collecting of the data was the quality of the received GPS signals. Because of the location of the railway within a canyon, sky visibility was often restricted. The GPS data were supplemented by data produced by the navigation-grade IMU that is part of the TITAN system. The global accuracy of the TITAN data collected along the test section of the Algoma Central was ±15 cm (6 in.) and the local accuracy of the data for any given section was approximately ±3 cm (1.2 in.). Following collection, data were delivered in a time-stamped, globally referenced x, y, z intensity point cloud format. The file size was 15 gigabytes (Lato, Hutchinson, Diederichs, et al. 2009).

Lato, Hutchinson, Diederichs, et al. (2009) fully discussed the subsequent processing workflow used to identify and measure the orientations of structural features observed in the LiDAR data. The adopted procedure allowed an individual rockfall zone to be structurally evaluated in roughly 15 to 20 min. The density of the TITAN data in regions of interest along the railway was

FIGURE 7-41 Typical point cloud representation of rock cut along Algoma Central Railway in northern Ontario that was collected by mobile terrestrial LiDAR system (Lato, Hutchinson, Diederichs, et al. 2009).

roughly 500 points/m^2 (47 points/ft^2); thus the minimum size of a detectable discontinuity was roughly 20 cm × 20 cm (8 in. × 8 in.). Figure 7-41 shows a typical point cloud representation of a rock cut along the Algoma Central Railway that was collected by the TITAN mobile terrestrial LiDAR system.

6.2.3 Comparing Mobile Terrestrial LiDAR Data with Traditional LiDAR Data Sources

In the past, LiDAR data have been collected from static terrestrial platforms or from mobile aerial vehicles. Static terrestrial LiDAR systems collect data with densities of about 10,000 points/m^2 (900 points/ft^2), whereas airborne systems collect much lower point densities, usually about 10 points/m^2 (0.9 point/ft^2). Mobile terrestrial LiDAR systems collect data at densities that vary depending on the velocity of the moving platform and the speed of the data collection rate (Table 7-2). Typical resulting data densities range between 50 and 500 points/m^2 (4.5 and 45 points/ft^2) (Lato, Hutchinson, Diederichs, et al. 2009).

Figure 7-42 shows a direct comparison of data collected with mobile and static terrestrial LiDAR systems. In this case, the mobile terrestrial LiDAR system scanned the rock face with a data density of about 400 points/m^2 (36 points/ft^2), which corresponds to a point spacing of 5 cm (2 in.). In contrast, the static terrestrial LiDAR system scanned the rock face with a data density of 10,000 points/m^2 (900 points/ft^2), which corresponds to a point spacing of 1 cm (0.4 in.). Figure 7-42, *a* and *b*, clearly illustrates these two point densities. When profiles are selected from these two scans (Figure 7-42, *c* and *d*), it is clear that the mobile scan has a greater overall error variation between adjacent points when compared with the statically scanned data. This greater variability results from uncertainty in the position and orientation of the mobile scanner as compared with its statically located counterpart rather than any basic differences in the inherent measurement accuracy of the two LiDAR scanners.

The two profiles shown in Figure 7-42, *c* and *d*, also demonstrate one additional potential advantage of collecting rock surface scans with a mobile LiDAR system rather than a static one. The profile collected by the mobile system (Figure 7-42*c*) has a much greater continuity compared with the profile in Figure 7-42*d*, which has several large data gaps where the rock surface was not visible from the static scanner location. Of course, multiple scans from a variety of static locations combined with data from these separate scans will decrease the potential for occlusion when static terrestrial LiDAR systems are used. However, this method requires more time and effort in the field, whereas the mobile LiDAR systems, with their constantly changing viewing positions, can produce the same result much more rapidly and easily.

Figure 7-42 provides a clear demonstration of the relative strengths and limitations of the mobile and static terrestrial LiDAR systems for evaluating rockfall zones. Although the lower density of the scans produced by the mobile systems makes it more challenging to visualize the rock surfaces and limits the minimum size of the features that can be identified, it appears that mobile LiDAR systems have some advantages and provide data that are sufficient to perform accurate geomechanical structural analyses.

6.3 Terrestrial Interferometric Radar

The term *radar* stands for "radio detection and ranging." Radar operates in the radio and microwave portions of the electromagnetic spectrum at wavelengths ranging from a few millimeters to about a meter. The radar spectrum is itself divided into a number of bands designated by letters of the alphabet, which were originally adopted for security reasons during World War II but are still used as a shorthand reference. Radar is an active

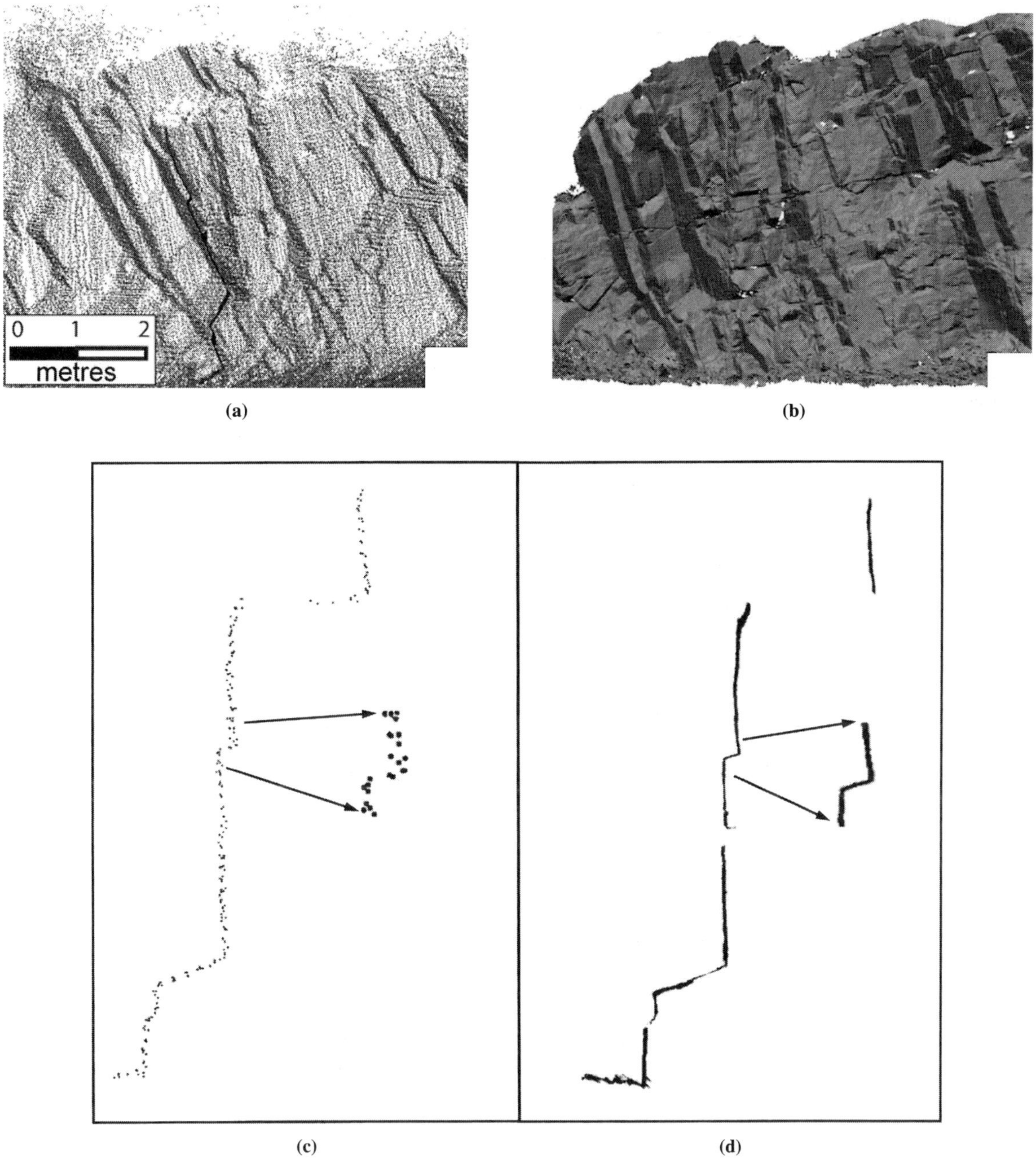

FIGURE 7-42 Direct comparison of data collected with mobile and static terrestrial LiDAR systems: (*a*) mobile terrestrial LiDAR data collected at outcrop of well-fractured granite near Ottawa, Ontario, Canada; (*b*) same outcrop scanned with static terrestrial LiDAR system; (*c*) profile of mobile terrestrial LiDAR data shown in Image *a;* (*d*) same profile represented by static terrestrial LiDAR data shown in Image *b* (Lato, Hutchinson, Diederichs, et al. 2009).

remote-sensing system that provides its own energy to illuminate targets of interest and records the return signal reflected from a target.

Resolution of a radar image improves as the imaging wavelength becomes shorter with respect to antenna length. Longer wavelengths are less sensitive to atmospheric attenuation and thus produce better images, but their use implies the need for very large antennas. In the 1950s, however, it was discovered that a long antenna could be synthesized by using a transported small antenna. This process was named *synthetic aperture radar* (SAR). With SAR technology, imaging radar systems could be deployed on mobile (airborne and orbital) platforms.

A second important advancement involved the adoption of radar interferometry; the use of radar signals traveling along two different paths allows very accurate range information to be derived from the coherent interference between the two signals. The resulting *interferograms* record minute differences on a subwavelength scale for corresponding points on the image pair. Very accurate topographic maps are produced when radar interferometry techniques are applied to SAR images collected from airborne or orbital platforms.

Radar interferometry can be developed by using radar images recorded from two or more spatially separated locations or from radar images recorded from a single static location at different times. This latter technique is termed "differential interferometry" and is used for velocity mapping and change detection because it can identify very small changes in position over time. One of the earliest and most spectacular applications of radar interferometry for change detection was accomplished at the site of the Landers earthquake of 1992 in the Mojave Desert of California (Massonnet 1997).

The application of radar techniques to detect rockfall hazards requires a much shorter repeat time and greater spatial resolution than what is provided by airborne and satellite-based radar systems. Therefore, ground-based interferometric radar systems have been developed and used to monitor displacements of unstable slopes at both landslide-prone areas and open-pit mines. Ground-based interferometric radar techniques include at least three approaches, summarized in the following sections.

6.3.1 Static Differential Interferometric Radar

Malassingne et al. (2001) conducted field experiments in 1997 to test the potential of static differential interferometric radar for monitoring active volcanoes. Because corner reflectors provide a much stronger backscatter signal than natural surfaces, they were used in these tests, and absolute distances were measured to an accuracy of about 1 mm (0.04 in.) at a range of 1,500 m (4,920 ft). Differential measurements at that range typically could detect changes as small as 0.1 mm (0.004 in.). Over a period of 4 days, changes within a 40- by 40-m (131- by 131-ft) active rockfall area were compared with those from a relatively stable bedrock slope.

6.3.2 Interferometric SAR

Reeves et al. (2000) provided some of the earliest descriptions of ground-based radar systems that use differential interferometric SAR (InSAR) to detect changes in rock surfaces in open-pit mines. The application was encouraged by earlier research (Carnec et al. 1996). Initial tests conducted at two Australian coal mines demonstrated that after compensation for atmospheric effects, submillimeter accuracy could be obtained over distances of 70 to 150 m (230 to 492 ft). The potential accuracies revealed by these tests were better than those achieved by typical reflectorless EDM systems, and so further development of ground-based radar systems that could detect small changes in a rock surface continued with support from the mining industry.

Tarchi et al. (2003) reported successfully monitoring landslide displacements in Italy with ground-based SAR interferometry. Their radar system, called *linear SAR*, consists of rail-mounted radar antennas that produced a SAR response by being moved along the rail for a distance of 2.4 m (7.9 ft) in 6-mm (0.24-in.) steps for a total of 401 azimuth points. Acquisition time was about 14 min to scan a region about 800 m (2,624 ft) by 350 m (1,148 ft).

This system, in conjunction with other sensors, serves as an early-warning system for the safety of residential areas below the 30 million m^3 (39 million yd^3) Ruinon rockslide in the Italian Alps. The radar installation was placed at a stable location facing the rockslide area at an average distance of 500 m (1,640 ft). Slope areas were evaluated with a resolution of about 2 m (6.6 ft), and displacements as small as 0.5 mm (0.02 in.) were identified. Repeated scans over a period of several days yielded about 400 SAR images. Pairs of images were used to generate differential interferograms. One benchmark demonstrated a nearly constant 11.7-mm/h (0.46 in./h) rate of displacement over a 48-h period. This finding agreed very closely with the results obtained by traditional field surveys over the same time interval.

6.3.3 Scanning Radar Systems Using Differential Interferometry

Two approaches to the application of differential interferometry for monitoring rock faces, and therefore rockfall hazards, can be distinguished on the basis of the number, characteristics, and movement of the radar antennas. Both approaches progressively scan the rock face and produce images defining areas where change is occurring, making the interpretations much quicker and easier. One approach uses a single scanning antenna that progressively scans the rock face with a 2-D pattern (Figure 7-43); the second approach uses dual receiving antennas and a one-dimensional scanning pattern (Figure 7-44). Both approaches have advantages and disadvantages.

The first approach uses a single pencil-beam real-aperture antenna to scan the rock face in two dimensions and compares two radar images obtained at different times from a fixed position (Figure 7-43). At each scanned location within the overall image, the currently measured phase is compared with the phase recorded by the previous scan or to the time-averaged phase measured by several previous scans. These phase differences between scans are related to movements of the corresponding scanned locations of the rock face in the interval between scans. However, a correction must be estimated to account for changing weather or atmospheric conditions that can affect the transmission of the radar signals. This approach requires a high-precision, 2-D scanning system and exceptionally phase-stable radar. Both conditions add to the expense of the system. Nevertheless, this approach forms the basis of the commercial slope stability radar (SSR) monitoring system developed in Australia (University of Queensland 2002). The SSR system is described further in Section 6.3.4.

The second approach uses two spatially separated fan-beam radar antennas to illuminate the entire vertical face over a narrow horizontal distance. The face is scanned repeatedly in a horizontal, one-dimensional sweep (Figure 7-44). The phase difference that occurs as a result of the small path-length difference between the target and each antenna is recorded for each scan position. Since the radar is stationary, the interferometric phase difference between scans can be easily determined and is very sensitive to movements of the rock surface. However, because the path lengths for a given pair of simultaneous scans are nearly identical, atmospheric effects are similar for both channels and can be ignored. Furthermore, this approach allows the use of SAR techniques and does not have the stringent long-term phase-stability requirements of the pencil-beam, 2-D scan design, although two radar systems are required. Because scanning is in a single dimension, the scanning system is simpler than for a 2-D system. Brigham Young University employed this approach to develop a low-cost interferometric SAR system (called YINSAR) that has been field tested to detect and measure displacements on mine slopes (McHugh et al. 2006).

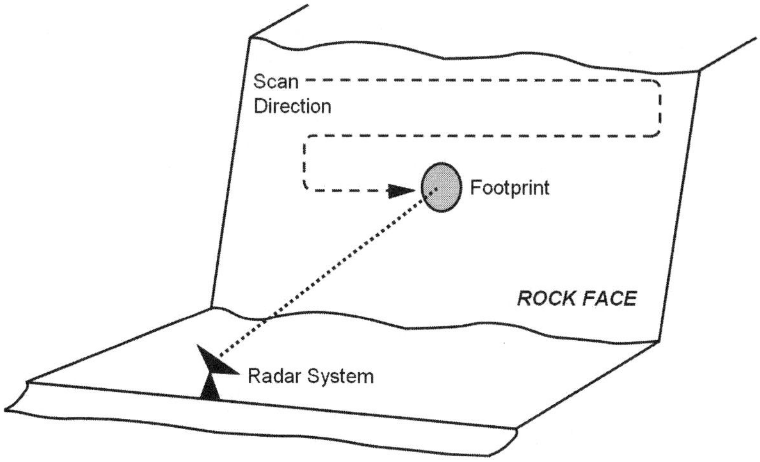

FIGURE 7-43 Two-dimensional scanning radar configuration.

6.3.4 Slope Stability Radar

A scanning radar system using differential interferometry has been designed by researchers at the University of Queensland, Australia, specifically for monitoring mine slopes (University of Queensland 2002). The system, SSR, uses real-aperture radar with a narrow 2° beam to scan a desired target area, such as a rock face, mine pit wall, or unstable slope, following a 2-D scanning pattern such as that shown in Figure 7-43. Scanning at a rate of 10°/s over a range of ±60° vertically and 340° horizontally, the system continuously monitors the slope face for deformation. Typical scan repeat time is 15 min. The phase of the returning signal is recorded for each pixel in the resulting image, and differential interferometry

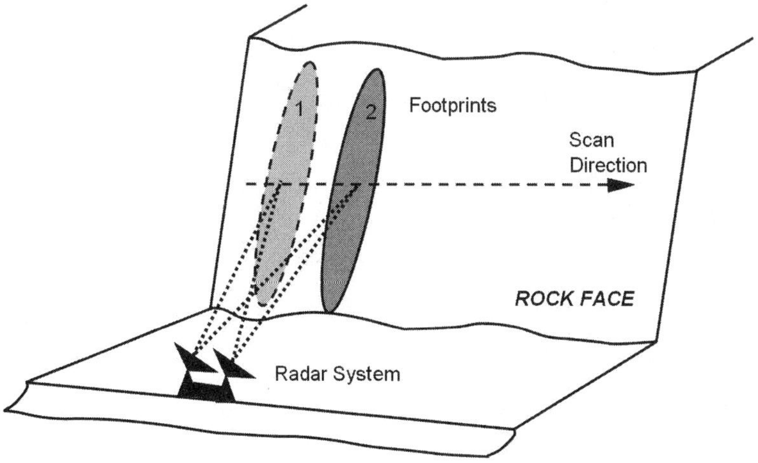

FIGURE 7-44 One-dimensional fan-beam scanning radar configuration.

(a) (b)

FIGURE 7-45
SSR monitoring of exposed mine or quarry rock face revealing areas of accumulating talus.
(IMAGES BY J. CATSOULIS, GROUNDPROBE PTY LTD.)

between sequential images identifies areas undergoing rapid change. The system can operate at a range up to 450 m (1,475 ft) from the target slope (Reeves et al. 2000). Line-of-sight displacement can be measured to ±0.2 mm (±0.008 in.) without the use of reflectors installed on the slope. Figure 7-45 shows results of an SSR scan of a relatively stable exposed mine or quarry rock face in which areas of talus accumulation below the face are clearly identified.

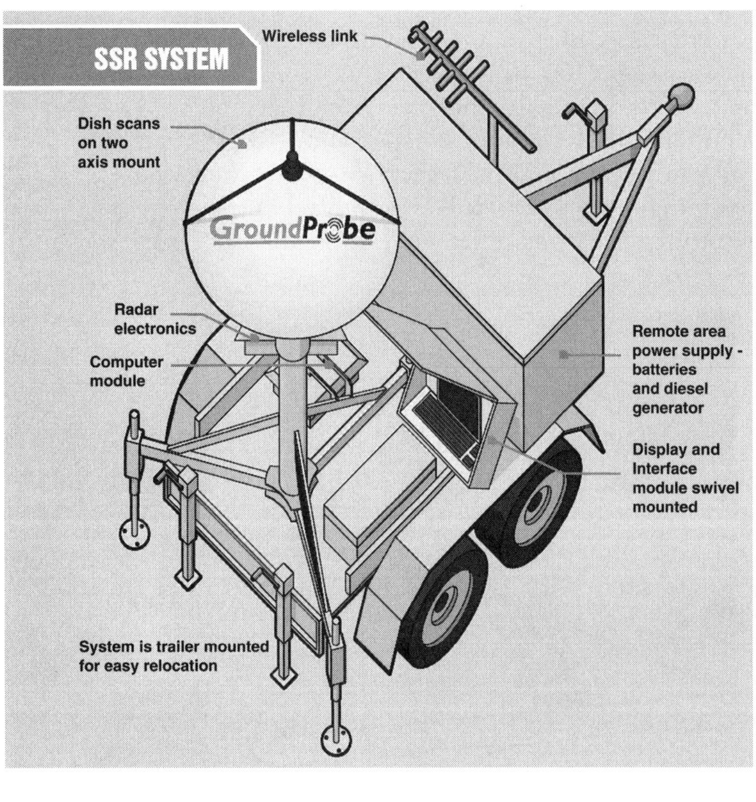

FIGURE 7-46
SSR functional components (Harries and Holmstrom 2007).

Commercial truck- and trailer-mounted SSR units are available for lease and purchase from GroundProbe (GroundProbe 2009), and comparable radar systems are also commercially available for lease or purchase. The unit features a scanning parabolic dish antenna 0.92-m (2.2-ft) in diameter, a computer for controlling the system operation and data collection, remote-area power supply, warning siren and lights, charge-coupled device (CCD) camera, communication links, and Internet compatibility (Figure 7-46). Currently SSR systems are used to monitor exposed faces in open-pit mines, where total displacements and rates of change of the slope configuration are valuable factors in assessing the risk to operations on the mine floor below the slopes and for determining when or if operations should be discontinued to protect miners and equipment from rockfall. More than 60 SSR units are reported to be currently in use worldwide to monitor mine faces (GroundProbe 2009).

Because the cost of leasing an SSR unit is very high, the use of these units by transportation agencies has been quite limited and restricted to evaluating highly hazardous situations along critically important routes. In 2006, the California Department of Transportation employed one SSR unit to monitor a reactivated rockfall zone along California State Highway Route 140 in the Merced River Canyon just west of Yosemite National Park (see Box 17-3 in Chapter 17). In this case the SSR unit was placed across the canyon from the rockslide area. It scanned the entire slide every 8 min from a distance of about 180 m (600 ft). Figure 7-47 is a reproduction of the computer display produced by the proprietary SSR interpretation software. The rectangle on the rockfall zone image and on the interferogram surrounds the area of greatest movement. Figure 7-47 also shows a number of other graphs and monitoring displays that are available to aid the data evaluation process. In this case, threshold movement rates were selected to warn of critical movements, and the SSR system was connected to traffic signals that controlled detour traffic. The signals were switched to red and traffic stopped whenever critical movement rates were exceeded.

In the fall of 2009, the Washington State Department of Transportation employed a ground-based radar unit to monitor a massive landslide that buried Washington State Highway 410 west of the

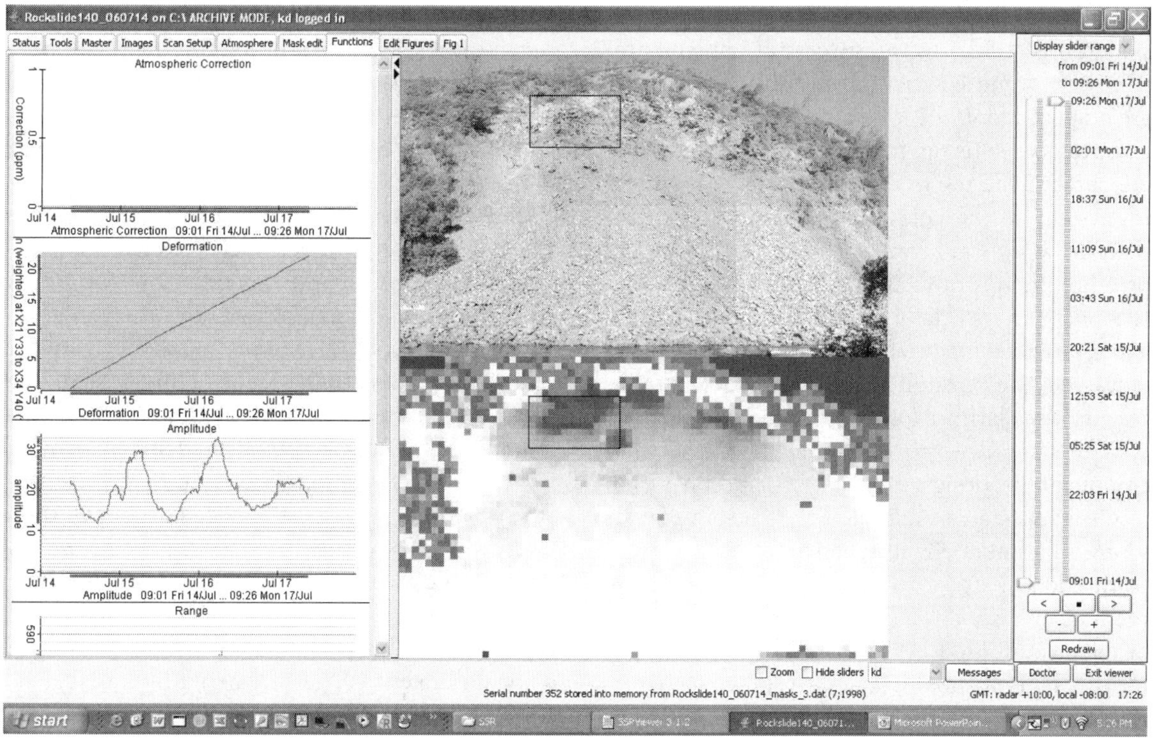

FIGURE 7-47 Computer display of SSR monitoring of rock slide area along California State Highway Route 140 in Merced River Canyon just west of Yosemite National Park.
(IMAGE COURTESY OF J. DUFFY, CALIFORNIA DEPARTMENT OF TRANSPORTATION.)

community of Naches on October 11, 2009 (Figure 7-48) (T. Badger, Washington State Department of Transportation, personal communication, June 25, 2012). Because of the large areal extent of the landslide and the limited availability of suitable monitoring sites, a long-distance unit was sited about 1,500 m (5,000 ft) across the valley. Although this radar-monitoring application was not for a rockfall, it demonstrated that transportation agencies might find radar monitoring to be justified in high-risk and economically important situations. In this case, the proximity of nearby detour traffic and the safety of the geotechnical drilling crews working on the landslide led to the decision to employ a radar system.

7. AUTOMATED ROCKFALL-MONITORING SYSTEMS

Rockfall poses a safety hazard to traffic on all types of transportation systems, especially in mountainous regions where rockfall often occurs at locations remote from maintenance centers. Consequently, there is a clear advantage to having automated monitoring systems that can reliably identify rockfall events along both railways and highways. In response to this perceived need, railways have relied for decades on rockfall warning systems that monitor the integrity of the rails forming the track or slide-detector fences containing triggering wires that, if broken, identify a potential impediment on the track. Chapter 17 provides an overview of the role of various rockfall warning systems within the broader context of maintenance, monitoring, and response activities for mitigating rockfall hazards.

FIGURE 7-48 Aerial view of landslide that blocked Washington State Highway 410 west of Naches, Washington, on October 11, 2009. Ground radar was employed to monitor landslide movements.
(PHOTOGRAPH COURTESY OF THE WASHINGTON STATE DEPARTMENT OF TRANSPORTATION.)

In this section the design and operational considerations related to the successful implementation of automated rockfall-monitoring systems are emphasized. Recent advances in instrumentation, remote sensing, and communications have encouraged the use of remote real-time monitoring systems to warn of hazardous conditions along roads as well as railways and to protect buildings or other important facilities. Systems using older technologies, especially those relying on triggering wires, provided many false alarms. Newer technologies offer the possibility of greatly reduced numbers of false alarms. However, some proposed new systems have not proved suitable because their complexity in design, installation, and operation results in excessive equipment and personnel costs for installation and maintenance.

In addition to providing reliable warnings of rockfall events, any acceptable solution to the problem presented by automated monitoring of rockfall hazards should have the following characteristics (Brackett 2002):

- Is economical to construct;
- Is inexpensive to repair and reconstruct;
- Reduces the time required to physically remove accumulated debris;
- Avoids the costs associated with removal of older, preexisting monitoring systems; and
- Introduces safer conditions for maintenance personnel.

The following sections cover the traditional methods of remote detection of rockfall events along railways, some fairly recent experiments by the railway industry with new technologies for remote rockfall detection, and some new technologies recently employed to remotely monitor damage to rockfall barrier fences and similar infrastructure elements installed in rockfall hazard zones along highways or other structures.

7.1 Detection of Rockfall Events Along Railways

Although only 2% of train-related accidents are related to ground movement hazards, including rockfall, they represent 12% of the direct costs of all rail accidents (Lemay 2007). Thus, railways have, over an extended period, attempted to detect rockfall events by using either slide-detector fences (SDFs) (see Section 7.1.1) or by monitoring the continuity of the rails forming the track (see Section 7.1.2). Events identified by either method send messages to the trackside interlocked signals controlling the affected block of track, causing them to exhibit a danger signal.

Yet these systems are far from ideal remote detection systems. The fence systems are especially high-maintenance installations, and they typically suffer from a false alarm rate as high as 70% (Lemay 2007). Railways are able to monitor and control their traffic by using block signaling systems (electrically isolated sections of track) in a manner that is not easily replicated on public highways. However, the signals do not, by themselves, stop any train; they must be acted on by the train crew, who must witness such a signal before entering the block where a rockfall event was detected. Crews on a train that has already traveled past a signal when a break occurs receive no warning of a rockfall event. When triggered by a perceived rockfall event, the signals will prevent train movements entirely or result in speed restrictions until the site can be visited and the severity of the damage, if any, determined. This verification may take some time because of the remoteness of many of the rockfall sites and the poor weather conditions that contribute to rockfalls.

Rail traffic is relatively vulnerable to derailment by even relatively small volumes of rockfall debris on the track. Abbott et al. (1998) described a number of rockfall characterization factors that were considered during development of the Canadian National Railway's Rockfall Hazard and Risk Assessment (CNRHRA) system. These factors also influence how an automated rockfall-monitoring system protects a railway route. The factors include the following:

1. *The relative probability that the largest rock fragments will have a maximum dimension in the range of 0.3 to 1.0 m (1 to 3 ft)*. Fragments within this size range are most likely to lift equipment off the rails and cause a derailment. Locomotives typically have sufficient clearance to pass over smaller fragments, and larger fragments are often pushed aside, provided there is adequate lateral deflection space.
2. *The concentration of the rockfall debris*. Concentrated debris is more likely to cause a derailment than scattered debris fragments.

3. *The shape of the rock fragments*. Slab-shaped fragments are more likely to wedge under a locomotive and derail it.
4. *The estimated impact energy of the rockfall fragments*. Initially, any rockfall event originating from a height greater than 8 m (26 ft) above the rail was assumed to be a high-impact-energy event and was considered to have a greater likelihood of completely breaking a rail, thus triggering the central traffic control (CTC) circuit and sending a warning to the block signal system.

However, the ability to simply define a potential rockfall event as high impact did not account for situations in which a rockfall event damaged the track sufficiently to make it impassable, for example, by pushing the track out of alignment or by damaging the railhead without breaking the rail completely. In these circumstances, the CTC circuit would not be affected and no warning would result. The original concept also did not consider the relative strength of the rail compared with the size and strength of potential rockfall fragments; it is desirable to be able to assess whether the fragments have enough energy to break a rail or whether the impacts would further break up the fragments. Given these imponderables, the CNRHRA did not employ any impact energy assessments in its ratings.

According to Brackett (2002), an acceptable automated rockfall-monitoring system for railways should

- Detect objects that are hazardous to train movements with the same or a greater degree of accuracy as existing SDFs without interfering with rail operations,
- Minimize both the probability that rocks exceeding the specified size will enter the detection zone undetected and the occurrence of false (nonhazardous) activations,
- Reduce the time spent by maintenance personnel in hazardous areas repairing the system and establish a situation in which all repairs can be done at grade level,
- Provide a reduced per-foot installation cost so that more areas can be equipped with the system,
- Not require extensive site preparation for installation, and
- Continue to function after a minor event and accommodate remote resetting to minimize train delays.

This final functional requirement is especially important to railway operators. It allows the first train entering a block where detection has occurred to proceed at a reduced speed while being prepared to stop for an obstruction. If the alarm is caused by a nonobstructive event, the train crew radios this information to the traffic control center and the system is reset so that subsequent trains are able proceed at normal track speed.

7.1.1 Slide-Detector Fences

An SDF consists of a number of horizontal wires strung about 20 to 25 cm (8 to 10 in.) apart on a series of vertical poles typically spaced about 5 to 20 m (15 to 65 ft) apart. The poles are placed parallel to the railway track on the side that is susceptible to rockfalls. SDFs are primarily deployed along the tracks where near-vertical cliffs pose severe rockfall hazards (Figure 7-49).

SDFs are not suitable for areas in which slopes composed of talus or relatively easily eroded materials exist near their natural angle of repose, because these slopes typically have numerous boulders that are often released down the slope by a freeze–thaw cycle or rain. These slopes may also be inhabited by animals that can often break the

FIGURE 7-49 Electrical SDF located along Union Pacific tracks in Byers Canyon near Hot Sulphur Springs in western Colorado (http://users.sisna.com/jimbobnay/GrandCounty/Images/ByersCan1.jpg). (PHOTOGRAPH BY J. R. GRIFFIN 2007.)

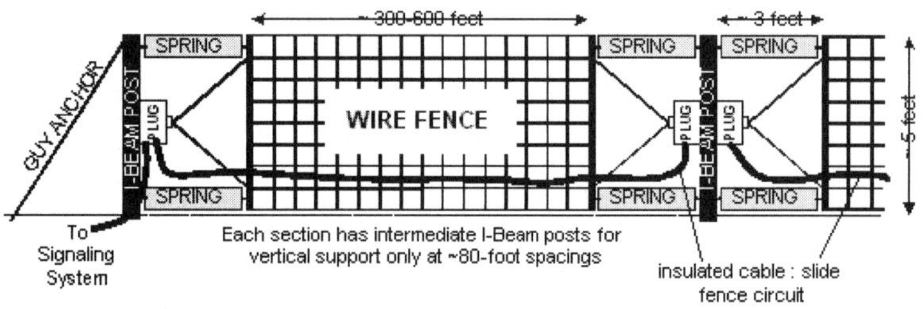

FIGURE 7-50 Typical mechanical SDF installation (http://en.wikipedia.org/wiki/Railway_slide_fence).

SDF wires. Together, the actions of the animals and the dislodgement of boulders result in large numbers of false alarms that make the SDF unreliable and expensive at these locations.

When falling objects such as rocks come into contact with the fence, they break the wires and trigger an alarm that is transmitted to the block signal system, which issues a visible warning to train crews to restrict train speeds through the slide zone. These restrictions remain in effect until the fence is repaired. As a result, for falling materials to be detected, they must first damage the SDF, which must then be repaired before it can detect another falling object. Brackett (2002) defined the generic problems with slide-fence technology as follows:

- Fences need frequent repairs because of damage as a result of their operation;
- Fences must also be taken down and reconstructed whenever rock scaling is required in the area; furthermore, fences are prone to false activations that require repair;
- Because of the location of slide fencing, maintenance personal must work in potentially hazardous conditions when repairs are necessary;
- During the time a fence is inoperative, trains are required to travel at restricted speed, resulting in delays to rail traffic; and
- Slide-fence construction is labor intensive, and anchoring posts in rock is expensive.

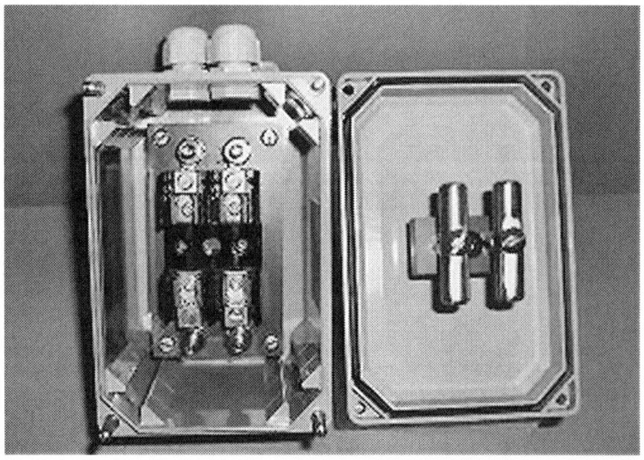

FIGURE 7-51 Details of typical pull-apart sensors used in SDFs [S & C Distribution Company (http://www.sandcco.com)].

There are two types of SDFs: mechanical and electrical. Typical mechanical SDFs consist of a series of tensioned wires strung about 25 cm (10 in.) apart on poles. When a rockfall breaks one or more of the wires that are maintained under tension, a spring-loaded contact box is pulled apart and a warning is issued. An alternative mechanical SDF design consists of a series of fence sections, as shown in Figure 7-50. Each fence section is tensioned in place by strong springs. Electromechanical "pull-apart" plug sensors at the ends of each fence section provide a warning that is monitored by signaling equipment. With this type of fence, the wires do not have to break, and so it is easier to maintain and reset. Figure 7-51 shows a modern design for the pull-apart sensors. Contained in a rugged durable plastic housing, the stainless steel pull-apart pins typically require a fairly strong impulse of 54 joules (40 ft-lb) to become disengaged. This feature prevents minor impacts on the fence from causing false alarms.

Electrical SDF designs use a series of parallel conductive wires strung about 20 cm (8 in.) apart on poles, as shown in Figure 7-49. The wires go back and forth within a given section of the fence, thereby creating a series of fence sections with independent electrical circuits. In normal operation, the flow of current in the fence wires causes a relay connected to that circuit to be energized, indicating that each fence circuit is continuous. When a rockslide occurs, the rock breaks one or more of the wires, interrupting the current flow. This occurrence causes the relay to become deenergized, and a warning message is issued indicating that a rockslide has occurred.

7.1.2 Rail Continuity Detection Systems

Railways routinely monitor the continuity of the rails forming the track by creating electrically isolated sections of track called *blocks*, passing an electrical current through the rails within each block, and then detecting any loss of electrical continuity in this circuit. This system supports the CTC process that is implemented on most railway routes with significant train activity. If either of the rails breaks, the loss of electrical continuity changes the trackside block signals to their most restrictive indication. However, a break may result from several causes, including service stresses on trains and equipment, thermal contraction on the coldest of winter days, or roadbed subsidence. As noted previously, rockfall events may not completely break the rails but may still cause severe damage that makes the track impassible, or rockfall debris may be sufficient to cause derailment. Thus, rail continuity detection systems, although in common use for other purposes, are not especially reliable for automated rockfall monitoring.

However, railways are also concerned about erosion or subsidence of their roadbeds caused by flooding or slope failure. Only some of these occurrences are detected by monitoring the continuity of the rails. With continuously welded rails and concrete ties, the track may remain connected even when considerable volumes of the subgrade subside or are removed by erosion. Thus, railway operators are evaluating a number of other approaches and devices to detect these hazards although they do not appear to have any application to rockfall detection.

7.2 Electronic Field Disturbance Sensing

In Canada, the Transportation Development Centre of Transport Canada has established the Railway Ground Hazard Research Program, a collaborative effort among industry, government, and academic researchers to advance the mapping, analysis, forecasting, and mitigation of natural hazards that affect railway safety, including rockfalls. Two innovative approaches for remotely monitoring rockfall are under field evaluation.

The first approach has been under experimental development and testing since 1990. The purpose of the *electromagnetic field disturbance* (EMFD) rockfall-detection project is to design and develop an electronic system that detects rocks falling onto railway tracks while being compatible with railway operations and capable of augmenting or replacing existing SDFs (Brackett 2002). The EMFD sensing system works by transmitting a radio frequency (RF) signal down a cable that is configured in a loop between the transmitter and receiver. The cable is placed between the rails (Figure 7-52). The speed of transmission of the waves is affected by the presence of materials close to the cable. When the EMFD system receives an RF signal modified by its environment, the signal is filtered to remove unwanted interference, amplified, and compared with an internally generated constant-reference frequency. A rock landing near the cable will slow down the wave and result in a phase shift. A rapid change in the phase shift indicates a possible obstruction on the track (Brackett 2002). The EMFD system shows promise. Field tests are under way. The sensing cable is relatively low in cost, easily repaired, rapidly installed, and compatible with normal railway operations, including snow removal.

FIGURE 7-52 EMFD installation undergoing field tests on Canadian National Railway at Lasha, near Lytton, British Columbia. EMFD cable is visible between rails; system controls are housed in small equipment cabinet located to left of signal mast (Brackett 2002).

The second approach uses *geophones*, microseismic sensors routinely used in geophysical exploration, to detect and locate rockfall impacts on or near the track (Lemay 2007). In a test conducted in a railroad yard near Montréal, geophones attached to ties successfully detected the seismic energy produced by rocks dropped by researchers as well as by passing trains, and even by people walking (Figure 7-53). The durability of the hardware in an active railway environment was also evaluated. Installation of an improved prototype alongside an existing detection system on a Canadian National Railway line in the Rocky Mountains is answering such questions as where to place the sensors (on the rail, on the ballast, or on the ties), how to filter the signals to allow the system to distinguish between rockfalls that present a danger and those that do not, and how the system will communicate with a central control system (Lemay 2007).

7.3 Remote Monitoring of Damage to Rockfall Barrier Fences

Chapter 15 contains detailed descriptions of rockfall barrier fences and their components. Fences

FIGURE 7-53 Geophone attached to railroad tie to evaluate potential for locating rockfall impacts (Lemay 2007).

are typically constructed with a variety of energy-absorbing components, so that they are able to withstand impacts from falling rock blocks up to a predefined *service energy level* (SEL), the level of impact loading that a fence can withstand within accepted maintenance requirements, and a *maximum energy level* (MEL), the maximum level of impact load that a fence can contain. Following an MEL impact, the fence typically needs major refurbishment.

Thus, fence performance is evaluated by comparing the relationships between impact loading, maintenance, and efficiency. *Impact loading* is simply the amount of load transferred to the fence from a rockfall impact. *Maintenance* refers to necessary repairs from an impact. *Efficiency* represents the effectiveness of the system to distribute load while minimizing damage. As discussed in Chapter 15, the impact location strongly influences efficiency. The most efficient impact location for a fence is in the center of a fence panel, where the panel can fully flex. The flexing increases the time provided to dissipate energy and arrest the motion of the rock and thus reduces the loads imparted to fence components and generally increases the overall ability of the fence to absorb greater impact energies with less maintenance. When impacts occur at or near more rigid fence components, such as posts, flexibility and energy dissipation decrease. Consequently, the requirement for maintenance of the fence system following the impact event increases.

Because fences typically experience numerous rockfall impacts, the energy-absorbing components included within their design are progressively affected over time, and their ability to absorb major impacts becomes degraded. Routine visitation and inspection of fences at some locations may be difficult and expensive, because they are not readily accessible. In such locations, it is often desirable to have some form of remotely observable automated monitoring system that can evaluate the state of the energy-absorbing fence components.

The most commonly applied technique for remotely recording impacts on rockfall barrier fences, thereby assessing the potential for damage, is to install sensors on the fence support ropes, usually adjacent to and spanning one or more of the friction brakes on these support ropes (Chapter 15 describes the use of friction brakes). An example of these sensors is the Impact Sentinel system that was functionally tested in Switzerland in 2006 (Swiss Federal Research Institute 2006) and is distributed by Geobrugg (Geobrugg 2010).

The Impact Sentinel system consists of a network of easily mounted and maintenance-free wireless sensors (Figure 7-54) and a compact receiver with data logger and communication modules (Figure 7-55). A single data logger supports up to 40 sensors within 300 m (1,000 ft) of its location. The logger analyzes signals from the sensors and creates alarm signals that can be forwarded as text messages over cellular wireless networks to appropriate individuals. The sensors work on two levels:

1. *Shock sensors* that record and transmit small rockfall events, which typically do not require maintenance work on the protection system; and
2. *Pull ropes* installed to span the friction brake rings, thus recording their elongation during a severe rockfall (Figure 7-56); the critical threshold value can be individually defined for each friction brake assembly.

The sensors operate independently from batteries or external electrical power. An integrated solar cell supplies a cache condenser that permits continuous sensor function for more than 90 h of darkness. A piezogenerator or a lithium battery provides the additional power required to transmit data following impact events. Wireless communication between the sensors and the data logger makes wiring unnecessary within a monitored site.

FIGURE 7-54
Impact Sentinel sensor.
(PHOTOGRAPH COURTESY OF GEOBRUGG.)

8. FIELD DATA ACQUISITION AND MANAGEMENT

Instrumentation and monitoring programs are often initiated at rockfall sites as part of the emergency response activities established following a rockfall event. Unfortunately, many of these installations are abandoned relatively quickly, usually because funding for the emergency response program is depleted. Routine operating budgets within many transportation agencies are insufficient to support the continued monitoring on a long-term basis, even though the rockfall hazards and slope instability remain.

However, the value of long-term rockfall monitoring is recognized at selected rockfall locations known to pose high hazard levels to transportation facilities. At these locations, relatively sophisticated data acquisition and management systems are increasingly employed to economically and reliably collect instrumentation data and transmit them from remote field locations to central locations where the data are evaluated. As noted in this section, railways have used automated rockfall warning systems over a considerable period of time to a much greater degree than have any other organizations. This use is largely due to the potential for rockfall to cause severe track damage, the likelihood of derailment by even modest volumes of rockfall debris, and the characteristics of railway signaling systems, which can be used to transmit warnings of rockfall events.

FIGURE 7-55
Data logger for Impact Sentinel sensor network.
(PHOTOGRAPH COURTESY OF GEOBRUGG.)

Establishment of a field data collection and communication system depends on the type of instrumentation installed, desired frequencies of obtaining readings, accessibility to and within the site, safety of individuals visiting the site, and whether the data are monitoring imminent rockfall events, thus requiring immediate review, or are part of a long-term characterization of a site, thus allowing for some delay in evaluation. Of course, an overriding concern is the cost of the

FIGURE 7-56
Installation of Impact Sentinel sensor across friction brake. Circle shows location of sensor, and arrows show extent of pull-rope triggering element.
(PHOTOGRAPH COURTESY OF GEOBRUGG.)

FIGURE 7-57 Data logger for EMFD installation undergoing field tests on Canadian National Railway at Lasha, near Lytton, British Columbia (Brackett 2002).

data collection and communication equipment. Although cost and reliability considerations previously favored the use of irregularly scheduled and relatively infrequent manual data collection procedures by individuals visiting the site, the rapid evolution of inexpensive, powerful, and reliable electronic devices supporting data collection, storage, and communication tasks has made automation of field data acquisition and management entirely feasible for most installations.

A typical automated field data acquisition monitoring system installed at a rockfall location includes a group of interacting and interrelated sensors connected to a data logger that stores and subsequently transmits the information to an office location where the data are evaluated. Once set up, the field installation continues to operate for extended periods without human intervention. Section 7.3 describes the use of data loggers in connection with sensors used to warn of rockfall impacts on rockfall barrier fences; Figure 7-55 shows a typical data logger for such an application. Figure 7-57 shows the installation of the data logger for the experimental rockfall-detection system being tested by the Canadian National Railway, which is described in Section 7.2 and illustrated in Figure 7-52.

Typically a single data logger controls the operation of the entire array of sensors installed at a field location. The unattended and remote nature of many data-logger applications results in modern data loggers that are extremely rugged and reliable devices that operate with very low power consumption, which allows them to operate for extended periods from solar-recharged battery sources. They also typically feature a time-stamping mechanism that accurately associates a time with each recorded data value.

Data loggers range from simple single-channel devices connected to a single sensor and costing only a few dollars to fully programmable multichannel devices that cost several hundred dollars. Multiplexers contained within multichannel data loggers allow the unit to sequentially monitor a large number of sensors. Technological advances are improving the capabilities of data loggers, even as their prices continue to drop. The older concept of a stand-alone data logger is changing to one of a device that not only collects data but also has access to wireless communications for automatic data reporting and remote control of the monitoring system. Modern data loggers serve web pages with current readings, e-mail their warning alarms to reporting locations, and transmit their daily results into databases or directly to the users. This data exchange has been improved with a wider range of communications options and standardization of transmission protocols and data formats.

Figure 7-58 schematically shows the communications links used by a relatively large and long-term field monitoring installation, the Turtle Mountain Monitoring Project and Field Laboratory, located near Frank, Alberta, Canada (Bjorgan and Froese 2008). This installation, described in more detail in Section 10.3, uses a wide variety of instrumentation to monitor the stability of Turtle Mountain, the location of a catastrophic rock avalanche that buried the town of Frank in 1903 (Froese and Moreno 2006). Figure 7-58 shows how the data collected from multiple GPS stations are collected by a 2.4-GHz WiFi system connected to a single data logger on the mountain that then transmits the data about 3.7 km (2.25 mi) by a 5-GHz radio link to a second data logger located in a building in the nearby town of Blairmore. A dedicated total station located at the base of the mountain takes laser readings from reflector prisms mounted on the mountain. The second data logger, in Blairmore, collects all the GPS and total-station data and transmits the information by Internet connections to a processing server computer located about 640 km (400 mi) away in Vancouver, where the data are analyzed.

Figure 7-59 is a schematic map of the wireless communication links used at Turtle Mountain. The Provincial Building in Blairmore connects the local network to the Internet by a router. The seven wireless links shown in Figure 7-59 connect the two main communications hubs, at the Blairmore Provincial Building and at the Frank Slide Interpretive Centre (FSIC), with a secondary data collection site at the pump house where the total-station data originate and with the multiple sensor networks on Turtle Mountain. Figure 7-60 shows the installation of the radio antennas for the link between the FSIC and the top of Turtle Mountain.

Although the communications network at Turtle Mountain is much larger and more complex than systems used at most rockfall locations, remote rockfall-prone sites along transportation corridors that require frequent monitoring readings may justify a communications network based on similar technologies. A variety of communications alternatives is possible depending on range, data rate, and cost. Options include various radio links, traditional telephone lines, dedicated data lines (including fiber-optic options), or the use of various local area networks (LANs), wireless LAN (WLAN), or mobile communications, including the use of Worldwide Interoperability for Microwave Access (WiMax), a broadband telecommunications protocol that provides fixed and fully mobile Internet access. Depending on the option selected, communications permit data to be received on a set schedule, or the system can be queried for data, or the field installations can send a signal when prescribed threshold values are recorded. This last capability provides a warning similar to that of the older slide-fence systems used on rail lines throughout the world.

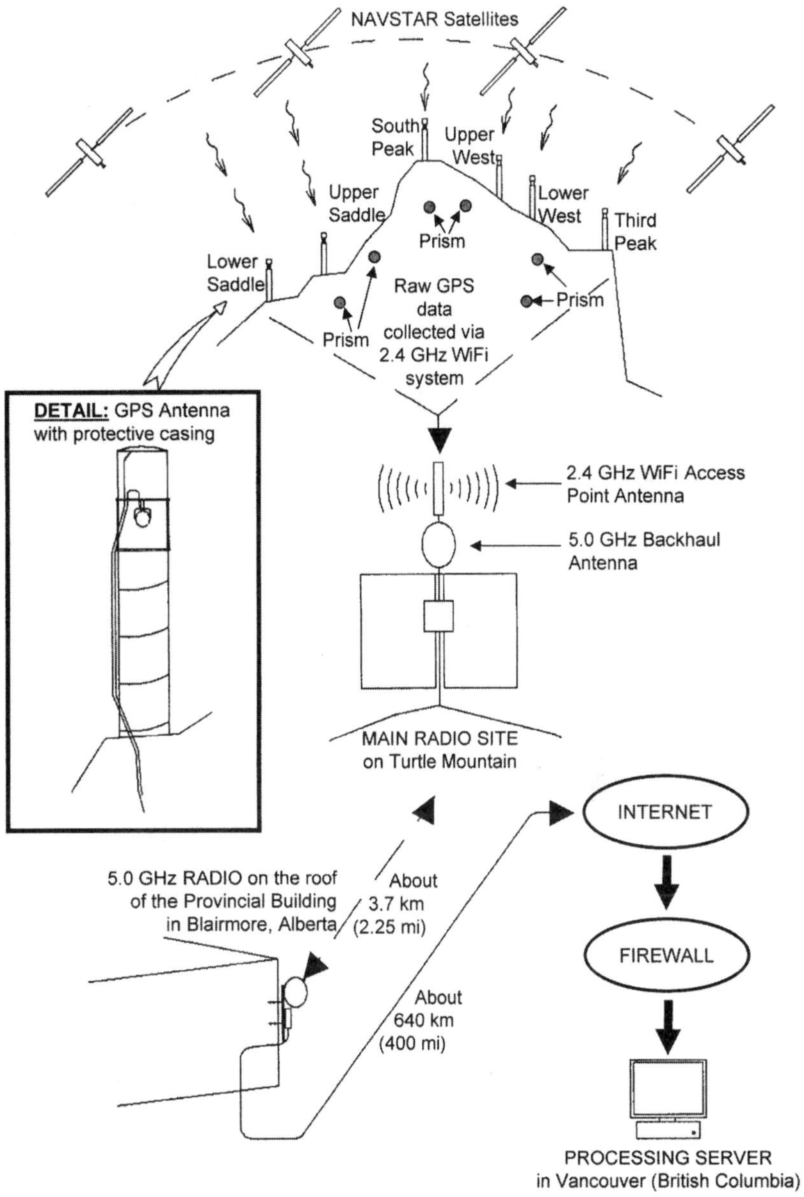

FIGURE 7-58 Schematic diagram of data collection and management system for GPS and laser-ranging data at Turtle Mountain Monitoring Project and Field Laboratory, Frank, Alberta, Canada (adapted from Bjorgan and Froese 2008).

9. MONITORING OF ROCK-ROLLING FIELD EXPERIMENTS AND BARRIER FENCE TESTS

High-speed digital video-recording technologies are applied to the capture and evaluation of images of rock-rolling field experiments and tests of rockfall barriers. These technologies continue to evolve rapidly as advances lead to a reduction in data file sizes and provide increasingly user-friendly data acquisition and analysis systems.

Sequences of images from digital video cameras provide a tremendous amount of geometrical

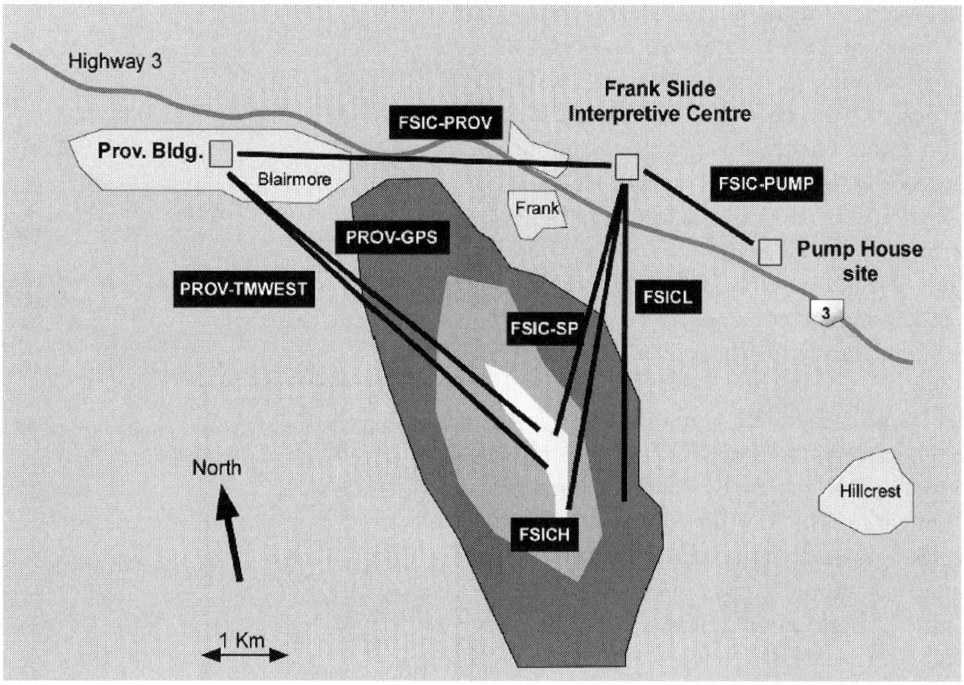

FIGURE 7-59
Schematic map of main communication links used at Turtle Mountain (Moreno and Froese 2009b): FSIC-PROV = point-to-point link connecting FSIC and Provincial Building (PROV); FSIC-PUMP = point-to-point link connecting FSIC to pump house; PROV-GPS = point-to-point link connecting PROV to new radio tower; PROV-TMWEST = point-to-multipoint link connecting PROV with two sites on Turtle Mountain; FSICL = wireless-access network connecting FSIC to seismic-monitoring stations and GPS station at bottom of Turtle Mountain; FSIC-SP = point-to-point link connecting FSIC to South Peak webcam at top of Turtle Mountain; FSICH = wireless-access network connecting FSIC to recently installed GPS stations at top of Turtle Mountain.
(MAP COURTESY OF THE ALBERTA GEOLOGICAL SURVEY.)

information about the scene that is being represented (Forsyth and Ponce 2002; Hartley and Zisserman 2004). Video photogrammetry, or videogrammetry, permits the extraction of accurate 3-D geometrical information from these image sequences (Balch 2002). When the 3-D coordinates of points on an object are observed and measured in two or more video images taken from different angles, projective geometry principles are used to establish the epipolar geometry—the geometry of stereo vision—that relates two images of the same object (Figure 7-61). The two images can be obtained from two cameras that simultaneously view a fixed object or from successive images captured by a single camera viewing a moving object (Criminisi 2001; Heuel 2004). Epipolar geometry requires either that the locations of the cameras be known or that fixed points in the field of view be known (Xu and Zhang 1996; Zhang 1998). A system of linear equations is solved by using the coordinates of at least eight noncoplanar points. After the equations are solved, the coordinates of any point (or object) within the field of view can be obtained as long as the point remains visible in the sequence of images.

9.1 High-Speed Photography and Photonics

High-speed photography is an established aspect of many scientific investigations. A series of images is collected that record an event. The establishment and growth of digital cameras and electronic image recording have led to the need to broaden the definition of high-speed photography to include the term *photonics*. Ray (2002) defined

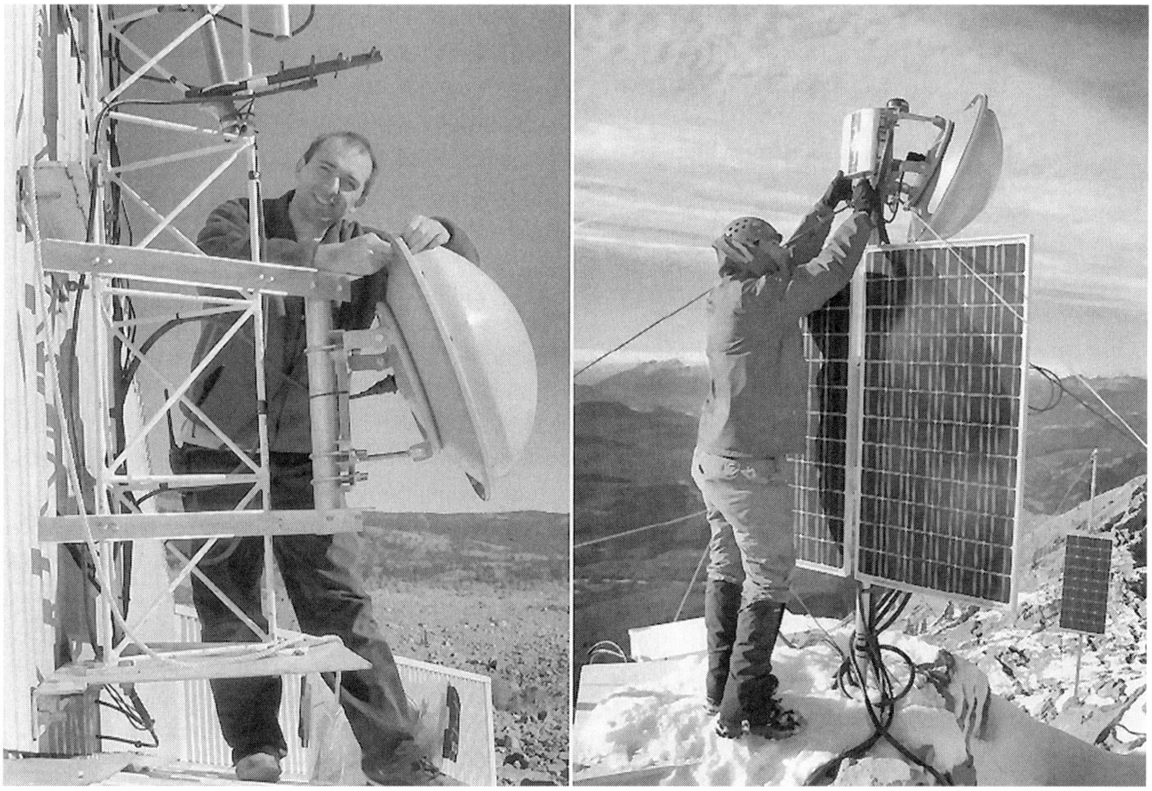

(a) (b)

FIGURE 7-60 Installing radio-link antennas: (*a*) FSIC and (*b*) South Peak of Turtle Mountain (Moreno and Froese 2009b). (PHOTOGRAPHS COURTESY OF THE ALBERTA GEOLOGICAL SURVEY.)

photonics as "the technology of generating and harnessing light and other forms of radiant energy whose quantum unit is the photon." The topic is now routinely referred to as *high-speed photography and photonics*, and numerous international conferences and reference sources use these terms (Ray 2002).

A definition of the high-speed aspect depends to some degree on the needs of the experimenter; typical temporal resolution requirements are one to three orders of magnitude smaller than the duration of the event being studied. Paisley (1993) proposed the following generic definition of high-speed photography: "recording optical or electro-optical information fast enough for an event to be evaluated with a temporal resolution which satisfies the experimenter."

In 1948 the American Society of Motion Picture and Television Engineers proposed that exposures of 1 ms or less and framing rates of 128 frames/s (fps) or more constituted high-speed photography. This criterion contrasted with the standard 24 fps used by motion pictures and the 25- to 30-fps standard of the television industry. However, many industrial and scientific applications today routinely require nanosecond exposures and 100 million fps, so the selection of appropriate methods and equipment depends largely on the application requirements.

For many applications, the temporal resolution, defined by the frame rate measured in frames per second, is not the only important criterion; spatial resolution is equally important. Fuller (1994) proposed a definition of high-speed photography

FIGURE 7-61 Definition of epipolar geometry: two cameras take pictures of same scene from different points of view. Epipolar geometry describes relationship between two resulting views (Source: http://en.wikipedia.org/wiki/File:Aufnahme_mit_zwei_Kameras.svg). (IMAGE BY A. NORDMANN.)

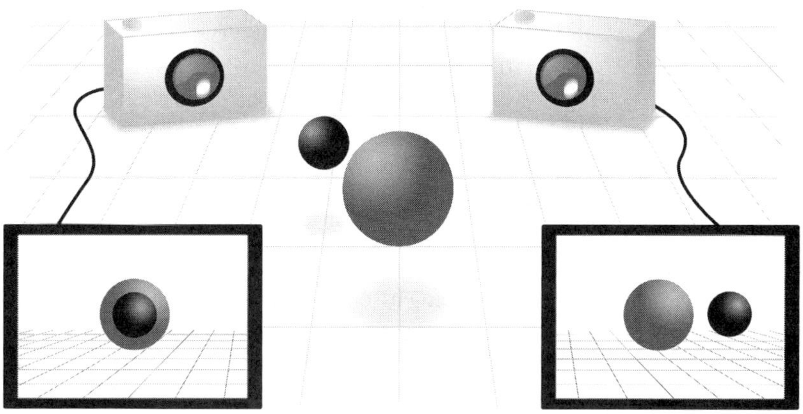

and photonics that adequately covers all aspects: "recording optical or electro-optical information with adequately short exposures and fast enough framing rates for an event to be evaluated with a temporal and dimensional resolution which satisfies the experimenter."

9.2 Digital Cameras and Video Recordings

All digital cameras, both standard cameras and video cameras, use a 2-D array of miniature sensors that transform the photons of the incident energy to electrons (Smith 2002). Typically, the incident energy is from the visible portion of the electromagnetic spectrum, but digital sensors exist that evaluate the infrared (or thermal) portions of the spectrum, whereas other sensors, including LiDAR and radar sensors (see Section 6), use other portions of the spectrum. All of these sensors use similar hardware design; the energy recorded by 2-D sensor arrays is processed to produce a digital image composed of a large number of individual picture elements called *pixels*.

9.2.1 Characteristics of Digital Cameras

In basic terms, the number of pixels forming an image defines the spatial resolution of the image; thus the larger the number of pixels a digital camera system produces, the more detail it can capture. The term *megapixel* is a measure of 1 million pixels, but for digital cameras this term actually defines the number of image sensor elements in the camera. For example, a camera with an array of 2,048 × 1,536 sensor elements is commonly said to have 3.1 megapixels (2,048 × 1,536 = 3,145,728). Although this megapixel count is often used to define the capability of digital cameras, it only partially determines image resolution, especially in the case of color cameras.

In color digital cameras, the sensor array is covered with a color-filter mosaic so that each sensor element records the intensity of a single primary color (red, green, or blue). The camera interpolates color information from neighboring sensor elements to create the final color image. Thus, at each sensor location, two of the three color channels are being interpolated and the resolution quality of the color image may be degraded, depending on the allocation of the primary colors within the color-filter mosaic; the most commonly used arrangement directly samples green with twice as many elements as either red or blue. In contrast, a black-and-white digital camera does not use a color-filter mosaic; each sensor element records the light intensity without the need for any interpolation. The megapixel count for a black-and-white (noncolor) digital camera provides a good measure of the image resolution.

The quality of the image produced by a digital camera is influenced by the camera design and its method of operation. The principal controlling influences are the following:

- *Lens quality and design*. Lenses with superior optical properties improve the quality of recorded images. The focal length of the lens, the distance between the lens and the sensor array, controls the scale of the resulting image and the extent of the field of view, which defines the area included within the image. Increasing the focal length will increase the scale and magnification of the image while simultaneously reducing the field of view. C-mount lenses typically have a focal length of 17.54 mm (0.69 in.).
- *Lens aperture*. The size of the opening in the lens at the front of a camera (the aperture) is adjusted by the camera user to reflect current lighting conditions. Although it controls the amount of light reaching the sensor surface, the aperture also influences the depth of field, which is the range of distance over which the image appears in focus. A smaller aperture produces a larger depth of field, producing an image that is in focus over a larger range of distance, whereas a larger aperture reduces the depth of field.
- *Shutter speed*. A second fundamental control chosen by the camera user, the shutter speed, controls how long light is allowed to pass through the aperture. A faster (shorter) shutter speed will freeze the motion of objects moving within the scene; a slower (longer) shutter speed will result in blurred images of moving objects.

Together, the shutter speed and lens aperture control the intensity and duration, and therefore the total amount, of incident energy provided to each sensor element for creating an image. Too much energy will result in light or washed-out images, whereas not enough energy will result in dark images.

Video cameras have an additional fundamental design parameter: the *frame rate*, or the number of frames recorded each second. Standard video cameras record at 25 to 30 fps. High-speed video (HSV) cameras record at much higher frame rates (Smith 2002). Depending on the selected camera model and manufacturer, HSV cameras can be obtained that record at speeds ranging upward of 1,000 fps with resolutions of 1,280 × 800 pixels, which is equivalent to one of the current high-definition TV display standards. Such characteristics enable the reproduction of fast-moving action and graphics without blurring of the image, but, as discussed subsequently, storage of large numbers of these high-resolution images requires large-capacity digital storage devices. HSV cameras with much higher recording rates are manufactured, but these rates are typically unnecessary for analyzing rockfall events, and at the higher speeds, their ability to monitor more than a few seconds of activity is restricted by digital storage capacities. HSV cameras used for recording rock-rolling field tests or tests of barrier fences generally operate at rates between 250 fps and 1,000 fps. However, the data storage capacities required for the images become very large as the frame rate increases, so that 250 fps is considered adequate for most rockfall-recording scenarios.

9.2.2 Video Display and Recording Standards

HSV cameras record images containing several image resolutions; most models allow the user to select from a variety of resolutions ranging between 512 × 512 pixels and 1,920 × 1,080 pixels. The user chooses the resolution on the basis of the application characteristics; lower resolutions provide less detail but typically allow higher frame rates. The lower resolutions also allow longer recording periods before the maximum image storage capacity within the camera is reached.

However, analysis of the recorded images is performed on standard computer display devices, so the recorded images are typically converted to standard video files by use of one of the normal standards used by DVDs or streaming video. These applications may have different, and often lower, resolutions than those produced by the HSV cameras. Thus, the selection of the image resolution used by the HSV camera typically depends on the anticipated methods and display devices used during the analysis of the recorded images as well as the characteristics of the objects being recorded.

Three display devices capable of showing HSV recordings are widely used: computer monitors, TV monitors, and video projectors. These devices offer a similar range of display resolutions that reflect common industry or broadcasting standards and the fact that much video imagery is exchanged between broadcast (TV) sources and computer networks. They also expect to receive images in one of several standard file formats, including

- *Computer sources*. JPEG, TIFF, or BMP files for still images and AVI, MPEG, or MOV files for video images; and
- *Broadcast sources*. CD or DVD products containing files according to the standards of the National Television System Committee (NTSC) in North America, phase-alternating line (PAL) in Europe, or sequential color with memory (SECAM) in France.

All these formats involve image compression techniques that make the files much smaller than those of initial raw digital images recorded by the HSV cameras, and repeated reprocessing of the data through a series of these formats may result in image degradation. Table 7-3 summarizes the typical resolutions used by these display devices and therefore the resolutions used during the subsequent analysis of the recorded rockfall images.

Typical HSV cameras initially record digital images as raw data files. For technical reasons, these raw data files are directly stored, at very high transfer rates, on internal temporary memory within the camera (Smith 2002). These raw data files are binary files that store the pixel values without data compression; thus, a single 1,280- × 800-pixel black-and-white image frame requires 1 megabyte of storage and a series of images recorded at 250 fps for a total of 10 s (2,500 frames) produces 2.5 gigabytes of data. Color images are three times as large as the black-and-white images. These files are not viewable on any standard display devices without the use of specialized software provided by the camera manufacturer.

Current HSV cameras and associated software typically allow the user to record and download images as raw data files that can be manipulated later, either as working raw data sets or as any of

Table 7-3
Typical Display Resolutions for Digital Display Devices

Display Code	Display Name	Aspect Ratio	Resolution: Width × Height (pixels)	Description
Computer Monitor				
SVGA	Super Video Graphics Array	4:3	800 × 600	Not used in new monitors but exists in basic video projectors
XGA	eXtended Graphics Array	4:3	1,024 × 768	Typical PC resolution; entry-level plasma TV monitors
SXGA	Super eXtended Graphics Array	5:4	1,280 × 1,024	Most commonly used general-purpose PC monitors
WXGA	Widescreen eXtended Graphics Array	8:5	1,280 × 800	PC resolution compatible with basic HDTV; also used by 720p video projectors
High-Definition TV Display				
HD-720i	High-definition interlaced	16:9	1,280 × 720	Standard HDTV resolution; compatible with WXGA monitors and 720p video monitors; both interlaced and progressive scan options exist
HD-720p	High-definition progressive scan	16:9	1,280 × 720	
HD-1080i	Full high-definition interlaced	16:9	1,920 × 1,080	Highest HDTV resolution; compatible with highest-quality 1080p video monitors; both interlaced and progressive scan options exist
HD-1080p	Full high-definition progressive scan	16:9	1,920 × 1,080	
Video Projector				
480p	Low-cost and portable projectors	16:9	854 × 480	Low-cost basic projectors
576p	Widely used older projectors	16:9	1,024 × 576	Good for viewing simple graphics
720p	Typical recent projectors; 1080p models becoming price competitive	16:9	1,280 × 720	Good for standard HDTV and complex graphics
1080p	Highest-quality commonly available projectors	16:9	1,920 × 1,080	Compatible with commonly available highest-resolution HDTV

a multitude of standard video files—such as AVI, MOV, MPEG, or MPEG4—that are accepted by various motion analysis software products. Two widely used motion analysis software products are Motion Studio (Integrated Design Tools 2010) and Imaging Studio, which is supplied with HSV cameras produced by AOS Technologies. Both of these software programs save raw data files produced by HSV cameras or convert them into uncompressed AVI files, compressed AVI files, MPEG files, or other industry standard video file formats. Alternatively, depending on the software product, the raw data files are converted to TIFF files containing event-sequence times linked to a series of individual 3- to 5-megabyte images. These TIFF images can then be viewed as video files or be converted to other standard video file formats. Section 9.4.2 describes the subsequent evaluation process with motion analysis software.

The raw data files produced by HSV cameras for a 3- to 5-s event recorded at frame rates of 250 fps typically range between 1 and 1.5 gigabytes for black-and-white images and 3 and 3.5 gigabytes for color images. In contrast, conversion of the same black-and-white imagery to uncompressed AVI files typically results in a file of around 900 megabytes, whereas compressed AVI files are even smaller, typically 100 megabytes or less. Various other standard video file formats generally are similar in size to compressed AVI files. Many current media and video players are unable to accept files larger than a few gigabytes, so file conversion and compression are often necessary.

9.3 Recording Rock-Rolling Field Tests with Digital HSV Systems

Chapter 10 describes the procedures for conducting rock-rolling field test experiments, with Sections 7, 8, and 9 of Chapter 10 providing information to guide the on-site actions that are fundamental to successful monitoring of rock-rolling events, including the establishment of suitable spatial reference systems, preparation of field data records, and image-collection arrangements. This section complements the material presented in Chapter 10 by expanding on the technical aspects of using HSV camera systems to successfully capture rock-rolling field tests.

Videogrammetric evaluation of a digital video image of a falling rock allows for the accurate determination of the 3-D coordinates of the position of the rock when the frame was captured (Balch 2002). Analysis of a sequence of digital video images obtained from two or more video cameras determines the 3-D trajectory of a rock, from its initial position to the impact zone. The location of the required fixed control points on the slope and the positions of the cameras are usually established by appropriate survey methods (see Section 4).

Bounce height is an important rockfall parameter and is most reliably and safely determined by observation of video recordings of field rock-rolling events. Markers must be set out on the slope and compared with a video to see where rocks landed. As described in Chapter 10, an HSV camera oriented perpendicular to the rockfall path with previously installed poles of known dimensions and locations visible in the field of view allows videogrammetric methods to accurately determine the distance (or height) of a falling rock above the ground surface or slope face.

The HSV camera (Figure 7-62) typically used for recording rock-rolling events has an internal memory capable of receiving the sequence of images recording an entire individual rock-rolling event in real time. These memory units are capable of accepting digital images with extremely high data transfer rates but typically have sufficient capacity to hold sequences of relatively modest duration (Smith 2002). The cameras typically have multiple options for lens mounts, shutter speeds, aperture settings, image resolutions, frame rates, and length of recording sequence. Most rockfall applications use a frame rate of at least 250 fps. Higher recording rates are available, but the data manipulation becomes cumbersome for higher frame rates. Many commonly available cameras can easily record 1,000 fps but create data files that are in excess of 5 gigabytes for a few seconds of recording.

HSV cameras typically record continuously and overwrite to their internal memory until a preset time interval is reached, when recording stops and images are allowed to be retained. These retained images are then downloaded to an external computer or hard drive before the camera is turned off or before a further recording sequence is initiated. If this download is not done, many of the current HSV cameras will not retain the images.

Using HSV cameras in a field application to monitor rock-rolling events presents significant

FIGURE 7-62
Typical HSV camera (AOS Technologies).

challenges that are not typically encountered when these cameras are used in an indoor laboratory setting. Considerations when an HSV camera system is set up to record a rock-rolling test include those discussed in the following subsections.

9.3.1 Determining Appropriate Number of Required Cameras

For rock-rolling exercises, barrier testing, or both, at least two HSV cameras are typically used (see Section 9 of Chapter 10). Two viewpoints set at least 90° apart provide useful information delineating rockfall trajectories and support detailed backanalysis of tested barrier systems. In addition, if one camera does not operate, a backup camera allows the testing process to continue. Recent rockfall attenuator tests used three camera locations to obtain video (Arndt et al. 2009).

9.3.2 Selection of HSV Camera Locations

Selection of appropriate camera locations and provision of stable mounting systems are also important. Front and side views are typically used because these combinations provide the best information on the trajectories and rotation characteristics of the rock blocks. Fixed field mounts are preferred (Figure 7-63) but are often not possible. Many field tests use cameras mounted on tripods (Figure 7-64). Tripods are preferred if multiple setups are required during the field experiments.

9.3.3 Establishment of External Measurement References

As described in Section 7 of Chapter 10, external measurement references are located beyond the anticipated trajectories of the rockfall blocks and within the field of view in the image, as close to the event as possible and extending in an

FIGURE 7-63
Location of post (*white arrow*) that provides permanent camera mount for rock-rolling field test of rockfall attenuator systems in Colorado. Concrete test rocks were rolled down concrete pan on left side of photograph.
(PHOTOGRAPH BY B. ARNDT FOR THE COLORADO DEPARTMENT OF TRANSPORTATION.)

FIGURE 7-64
Camera mounted on tripod at temporary location during standby interval in rockfall barrier fence test. USB cords are linked to laptop computer beyond left edge of photograph.
(PHOTOGRAPH BY B. ARNDT FOR THE COLORADO DEPARTMENT OF TRANSPORTATION.)

x-*y* plane. These references provide scaling information used during the postprocessing of the HSV camera data. In many instances, one or several long rods showing marks at regular intervals are sufficient (Figure 7-65).

9.3.4 Power Sources for Camera and External Computers

Providing electrical power to the HSV cameras and external computers can be difficult under field conditions. Generators are frequently required to ensure a continuous power supply. Battery power

FIGURE 7-65
Two external measurement rods within frame of video image obtained during rockfall barrier fence test.
(PHOTOGRAPH BY B. ARNDT FOR THE COLORADO DEPARTMENT OF TRANSPORTATION.)

sources are suitable only for relatively short time periods and are often unreliable, especially in cold or inclement weather. Batteries typically discharge during relatively extended standby periods between individual rockfall tests, making the HSV camera system unavailable to compile data when rock-rolling tests are resumed. Dependable and continuous power sources are imperative to successfully capture rockfall test data.

9.3.5 Selection of Color or Black-and-White HSV Camera

Several HSV camera systems provide only black-and-white images. Other models offer color images, or even a selection of either black-and-white or color recording. As noted previously, purely black-and-white HSV cameras may provide images with greater resolution than color images, and black-and-white image files are typically smaller than color files. In addition, black-and-white images offer, in many instances, greater contrast when rocks are evaluated in motion, especially under dim or poor lighting conditions or if the falling or rolling rock does not contrast well with the background. However, color imaging is commonly used. When color images are recorded, HSV camera settings typically have to be adjusted more often to provide appropriate contrast in the images. Even with these adjustments, analysis of changes or movements of the rock blocks often proves difficult.

9.3.6 Selection of Lenses and Lens-Mount Combinations for Individual HSV Camera

Before initiation of rock-rolling tests, the selection of lenses for each of the multiple HSV cameras is evaluated and reviewed to ensure that the best quality of imagery is obtained. In many instances, the availability on site of a wide range of lenses, with focal lengths ranging from 28 to 100 mm (1.1 to 3.9 in.), allows the various cameras to be equipped to provide the best field of view and resolution. Typically, a variety of at least three different lenses for each camera allows field adjustments to be made quickly.

9.3.7 Ensuring Familiarity with HSV Camera Operations

Training or hiring of qualified and experienced personnel ensures a successful program when HSV cameras are used for capturing motion data. As subsequently described in Section 9.4, most of the current software and camera combinations are becoming more user friendly but still require fairly extensive training in their use. A clear understanding of the capabilities and limitations of HSV cameras is important before rockfall field tests are monitored with this equipment. It is especially critical to have continuity of personnel during the testing sequence, so the data are collected in a consistent and logical manner. A detailed written set of procedures that are carefully and consistently followed also aids in ensuring a successful result.

9.3.8 Data Storage During and After Testing

Technological improvements to external hard drives have made the storage of large image files much more feasible. Multiple storage devices provide redundancy if a storage device suffers a hardware failure. Typically the raw data files produced by HSV cameras are stored in spite of their large size, but conversion of these raw data files to other standard video file formats is also performed, as these files are much smaller and are more readily evaluated by visual inspection or processed by motion analysis software. The time required to accomplish these data storage and conversion procedures influences the availability of the HSV system to monitor further tests.

9.3.9 On-Site Data Review

As soon as a rock-rolling event has been recorded, the video images are reviewed to ascertain that the event has been successfully recorded. Typically, this review is undertaken by using the replay capability of the HSV camera or by using support software provided by the camera manufacturer installed on an external computer. This visual review process not only serves to check on the adequacy of the record of the just completed rockfall test but also allows the camera operator to make any needed adjustments to the camera settings so that subsequent tests are successfully recorded. After this review, the images are typically transferred from the internal memory storage of the HSV camera to multiple external storage devices and saved in one or more standard data formats. The imagery is now available for further detailed analysis, as described in Section 9.4, and the HSV camera system is ready to record the next rockfall test event.

9.4 Analyzing Digital Video Imagery

Detailed quantitative evaluations of the motions of falling rock blocks are obtained by employing videogrammetric techniques to provide detailed analyses of the recorded video imagery sequences. These analyses evaluate translational and rotational velocities of the blocks, allowing computation of their kinetic energies. Frame-by-frame measurements provide additional useful design information, such as the distances between impacts, bounce heights, rates of acceleration or deceleration, ratios of rotational to translational kinetic energy, and spatial changes to all these values as the blocks travel down the slope. Assessments of impacts of blocks with rockfall barrier systems evaluate their performance characteristics and provide invaluable design information (Arndt et al. 2009).

There are two main approaches to the analysis of digital video sequences: (*a*) direct visual analysis of the imagery combined with manual measurement and recording of pertinent dimensions and locations and (*b*) automated image analysis supported by motion analysis software applications. When evaluating imagery documenting field tests of rockfall attenuator systems, Arndt et al. (2009) discovered that in cases where the rocks simply rolled down the inclined slope of a concrete ramp without bouncing more than 0.3 m (1 ft) in height, it was possible to accurately estimate average translational and rotational velocities by direct visual assessment of the images with manual measurements scaled to calibration marks painted on the ramp. However, if the boulder bounced more than about 0.3 m (1 ft) above the ramp, it was difficult to estimate the distance traveled between two points by using visual means, and it was then necessary to use automated image analysis methods based on motion analysis software.

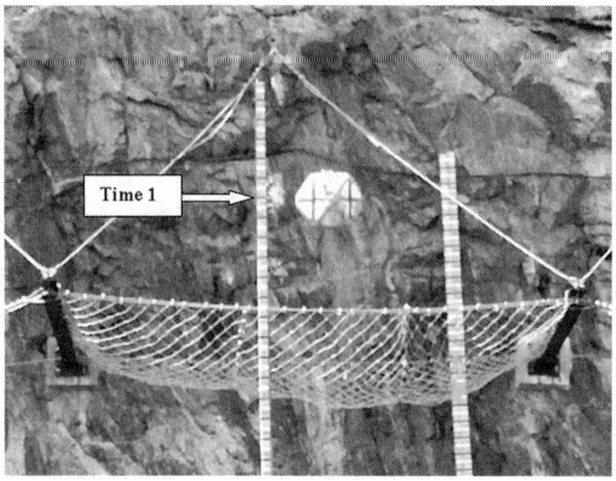

(a)

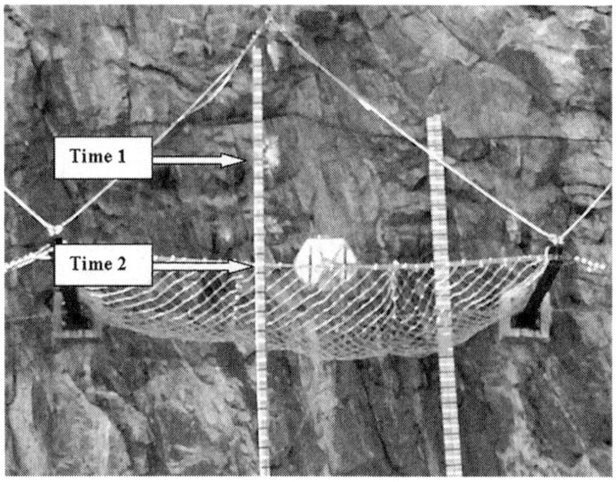

(b)

FIGURE 7-66 Visual methods using time to assess velocity. Time separating Images *a* and *b* is 0.113 s. White arrows in Image *b* depict vertical displacement of rock during this time interval. Measurements on the two images determined rock fell approximately 1.82 m (5.97 ft), producing estimated velocity of 16 m/s (53 ft/s) prior to impact with panel. (PHOTOGRAPHS BY B. ARNDT FOR THE COLORADO DEPARTMENT OF TRANSPORTATION.)

9.4.1 Visual Methods to Assess Rock Velocity

When rock block motions are relatively simple and do not involve complex rotations, it is generally possible to quite accurately determine the translational or rotational velocities of the rock block by visually analyzing the sequence of frames, along with their recorded capture times. By comparing the distance moved by the block with the time index shown on the HSV frames, it is possible determine the distance–time relationship and thus to compute velocity. Figure 7-66 illustrates two intervals during a test involving a rock dropped vertically into a barrier system. The measurement device in the image is marked in regular increments.

9.4.2 Using Motion Analysis Software Technology

Over the past two decades, motion analysis software has become an important diagnostic tool for noninvasive evaluations of motion and performance in industrial, scientific, and clinical studies. Aided by rapid technological advances in digital image collection and storage, motion analysis software has become the perfect complement to HSV cameras for monitoring because it provides quantitative information on any movement-based activity. Applications of motion analysis software include

- Animation used by the entertainment industry in producing films and video games;
- Human motion analysis for medicine, sports, and video surveillance;
- Monitoring of industrial and manufacturing processes, especially to detect inefficiencies or malfunctions in assembly lines or production machinery; and
- Scientific applications using object and feature detection techniques to count and track particles in motion ranging from bacteria and viruses in biomedical studies to flight of projectiles and collisions.

Given this wide range of applications, the capabilities of available motion analysis software also span a considerable range.

A number of low-cost systems are designed typically as sports-training tools or to provide support for health care activities. In these roles, they allow athletes to evaluate their golf or tennis swings or physical therapists to evaluate the movements of patients during walking or running exercises. These systems are typically restricted to evaluating video recordings made at relatively low frame rates, usually 60 fps. Although some systems accept video recordings with higher frame rates, the motion analysis capabilities are typically based on the 60-fps frame rate, which is all that is required by the sports training and health care applications. These low-cost motion analysis software systems do not provide capabilities suitable for detailed analysis of rockfall-monitoring data.

More sophisticated, and expensive, motion analysis software systems support relatively automated 3-D tracking and evaluation of object motion by utilizing epipolar solutions. These systems analyze video recordings collected with high frame rates, in any of several standard formats, and produce quantitative reports and statistical summaries that are readily incorporated in subsequent evaluations. Thus, these systems support the required postprocessing of video imagery from rockfall field tests to determine rotational and translational velocities of falling rock blocks.

Arndt et al. (2009) used one of these motion analysis software systems, ProAnalyst, developed by Xcitex, Inc. (Xcitex, Inc. 2010), to evaluate field tests of rockfall attenuator systems (Figure 7-67).

Arndt et al. (2009) described their experience with the ProAnalyst motion analysis software. They concluded that this technology holds considerable promise but that its use requires some experience and skill. Superior evaluations were achieved by an experienced professional user of the ProAnalyst software. When other less-experienced individuals used the software, their calculated velocities differed by about 30%. This finding indicates that the accuracy of the results depends on the assumptions used to develop the ProAnalyst model.

The ProAnalyst motion analysis software evaluates an instantaneous velocity at any given position along the trajectory, whereas visual analysis determines an average velocity between two known points. Thus, these two methods provide distinctly different estimates of kinetic energies. Arndt et al. (2009) suggested that the ProAnalyst values are potentially more accurate, although considerable uncertainty and random variability remain in the data. They also noted that the ProAnalyst motion analysis methods worked best when the block rotation was relatively simple and in the plane of the image. Deviations and wobbling of the rock toward or away from the camera as the block moved down the concrete ramp reduced computational accuracy.

FIGURE 7-67 Automated rockfall trajectory analysis using ProAnalyst motion software. Black line indicates translational velocity and white line rotational velocity, as modeled in program for rock rolling down slope (Arndt et al. 2009).

10. CASE STUDIES

Three case studies provide insights into several aspects of the use of instrumentation and monitoring technology to evaluate rockfall conditions.

The collapse of Threatening Rock in Chaco Canyon, New Mexico, in 1941 provides a historical perspective. Documentation concerning the use of an improvised extensometer system shows the value of even rather simple and inexpensive devices in providing quantification of the movements of unstable rock masses. The retrospective geological investigations by Schumm and Chorley (1964) demonstrated the ability of such studies to provide the scientific context for rockfall hazards. Their extrapolation of the available measurement data provides an estimate of the long-term development of the Threatening Rock column and shows how some unstable rock masses may actually exist for several thousands of years. Such observations have considerable importance to those attempting to establish realistic probabilities of failure and thus to estimate risk from rockfall events.

FIGURE 7-68 Threatening Rock, Chaco Canyon, New Mexico, circa 1935, before toppling failure (Hamilton 1939).

In contrast, the second case history, relating recent rockfall-monitoring activities along Interstate 70 near Idaho Springs, Colorado, provides an example of the typical value gained from information on the movements in fractured and potentially unstable rock masses that is collected by commonly used instrumentation. The third case history, a description of the Turtle Mountain Monitoring Project and Field Laboratory in southwestern Alberta, Canada, provides information on a readily accessible source of information on the design, operation, and long-term maintenance of a wide range of field instrumentation networks and associated data collection and communications systems. The size of Turtle Mountain is much larger than essentially all rockfall situations. However, the concern about a catastrophic rockslide event at Turtle Mountain has led to the establishment of a well-funded installation with a considerable public information capability. Experience with a wide range of instruments and communications systems in a location characterized by rugged terrain and subject to relatively severe climate conditions provides much useful information for those considering installing instrumentation and data communications for monitoring a hazardous rockfall location.

10.1 Collapse of Threatening Rock, Chaco Canyon, New Mexico

An early documented case history (Hamilton 1939) described movement of a detached block of sandstone that threatened historical structures located in Chaco Culture National Historical Park in New Mexico (Figure 7-68). The site is located in Chaco Canyon, which is an ancient ancestral pueblo administered by the National Park Service. In January 1941, a large monolithic slab of Cliff House sandstone, part of the Mesa Verde group of sedimentary rocks, toppled onto one of the ancient ruins of Pueblo Bonito (Figures 7-69 and 7-70). The monolithic block, commonly referred to as Threatening Rock, was estimated to be 30 m (100 ft) high, 45 m (150 ft) wide, and 9 m (30 ft) thick (Figure 7-68). The base of the sandstone block was bearing on the Menefee formation, which is characterized by weak, friable shale material.

In 1935 the park superintendent set up an improvised extensometer system consisting of two steel bars set in concrete to record movement of the detaching sandstone block. One steel bar was set on the cliff face and the other steel bar was set on the toppling block (Schumm and Chorley 1964). A series of relatively accurate measurements was made between the two bars between 1936 and the final collapse in January 1941; measurements were made almost every month following March 1937. As shown in Figure 7-71, these measurements indicated that the block moved outward 550 mm (22 in.) during the 5 years before failure; the measurements during the final month before the ultimate toppling failure, on January 22, 1941, indicated an accelerating additional outward movement of 250 mm (10 in.). These data suggest an exponential rate of movement before the toppling event. Total measurements suggested that the detached block moved at least 1,750 mm (70 in.) from the initial split from the cliff face. Schumm and Chorley (1964) extrapolated the measurement data to estimate how long it might have taken to achieve this total displacement. They obtained an estimate of about 2,500 years, which suggested that the initial split from the cliff face occurred around 550 BC.

It is interesting to note that the pueblo inhabitants were well aware of the toppling nature of the block and had constructed earth and masonry terraces at its base to reduce the erosion of the friable shales (Schumm and Chorley 1964). The construction of the ancestral pueblo is estimated to have begun in 850 AD, and it was abandoned in 1145 AD.

Because of the engineering interest in this collapse at the time, the 1935 *National Park Service Region III Quarterly* summed up the engineering issues facing the protection of Pueblo Bonito before the failure of the rock (Hamilton 1939) as follows:

> Most engineers who are familiar with the rock believe the best solution is to make use of the discoveries of modern science, and to use steel, air-hammers, treading machines, cement grout, the principle of the lever and the wheel, and many other skills, and processes unknown to the local pueblo inhabitants. Briefly, it is tentatively planned to drill holes in the back of the rock and face of the cliff near the top, exactly opposite, and to grout steel rods into these holes. The outer ends of these rods would be threaded and these connected by a turnbuckle. As this is turned the rods would be drawn together.

FIGURE 7-69
Remnants of Threatening Rock in 2010.
(PHOTOGRAPH BY B. ARNDT.)

> The question of how many rods to use, what size to make them, and how deeply to place them in the rock, has not been settled. Some experiments on pulling rods from sandstone will have to be made so that the engineers can be sure we will use rods enough, long enough and placed deep enough.

Unfortunately, in 1935 the science and engineering principles for rockfall and rock slope engineering had not been developed enough to preserve the pueblo features. However, this early example demonstrates the intuitive use of the principles of extensometer use by the National Park personnel.

FIGURE 7-70
Aerial view of Pueblo Bonito showing remnants of fallen Threatening Rock (*upper right*).
(PHOTOGRAPH BY B. ARNDT.)

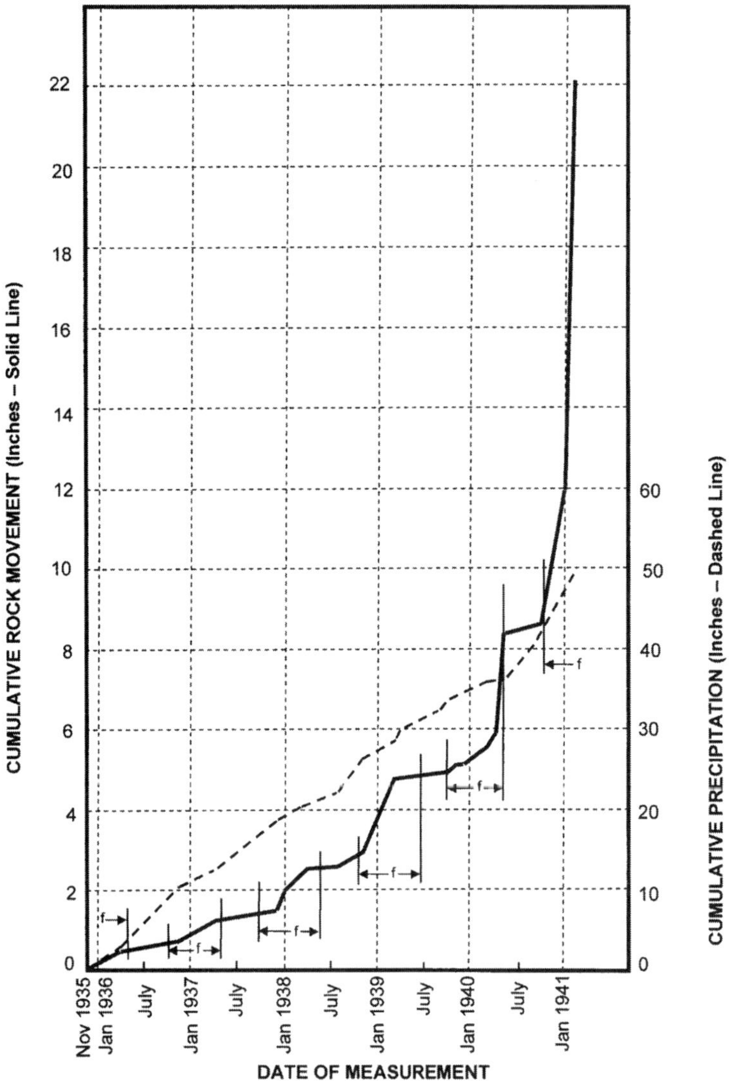

FIGURE 7-71
Recorded movements of Threatening Rock from November 1935 to collapse on January 22, 1941: cumulative movement (*solid line*); cumulative precipitation (*dashed line*). Freeze–thaw periods (*f*) are limited by dates of first and last killing frost for each winter season (adapted from Schumm and Chorley 1964).

FIGURE 7-72
Installing crack meters along joint plane above Interstate 70 near Idaho Springs, Colorado.
(PHOTOGRAPH BY B. ARNDT.)

10.2 Rockfall Monitoring Along Interstate 70, Idaho Springs, Colorado

Crack meters were used for monitoring movement along a major Interstate corridor when multiple rockfall events occurred during the night of August 13, 2005, just west of Idaho Springs, Colorado, along westbound Interstate 70 near Mile Marker 238. Three rockfall events that occurred during a 9-h period generated approximately 535 m^3 (700 yd^3) of debris materials. As a result of the rockfall events, westbound Interstate 70, a major east–west transportation route through Colorado, was closed for approximately 4 days. Alternative westbound routes were used; however, adverse impacts occurred to transportation, tourism, and the local communities.

The rockfall occurred along well-defined foliation layers in metamorphic schist with well-defined pegmatite intrusions. The failure occurred along a planar surface that dipped from 55° to 65° unfavorably into Interstate 70. Five vibrating wire crack meters were installed spanning this failure surface to monitor short-term rock slope movements (Figures 7-72 to 7-74). The crack meters were connected to a data logger that could be remotely accessed via cellular phone service. Collected data indicated movements of the unstable rock mass of 0.25 to 1.27 mm (0.01 to 0.05 in.) during approximately a 2-month period (September–October 2005), but movement ceased and no significant additional change was observed during the winter months and following spring (Arndt and Ortiz 2006). A posttensioned ground-anchor buttress was subsequently constructed to support the potentially unstable rock mass perched on the existing failure plane. The crack meters were used to assess the stability of the rock mass before and during construction and for approximately a year after completion of the buttress.

10.3 Turtle Mountain Monitoring Project and Field Laboratory, Alberta, Canada

On April 29, 1903, a considerable section of Turtle Mountain, in southwestern Alberta, failed catastrophically, producing a rock avalanche of about 30 million m^3 (39.25 million yd^3) that buried a portion of the town of Frank and killed more than 70 persons (Figure 7-75). Subsequent studies revealed that a remaining portion of Turtle Mountain, referred to as South Peak, has similar geological conditions and is surrounded by a series of deep fissures (Froese et al. 2009). Starting in 1933, monitoring of movements across major fractures began at 18 locations; these measurements were repeated at irregular intervals for more than 70 years. These data suggested, but could not conclusively prove, slow movements of the South Peak rock mass. In the early 1980s renewed funding established a series of mechanical crack gauges and additional remote-sensing studies, but by 1994 these crack gauges were mostly inoperative because of exposure and weathering of the rock surfaces.

However, these monitoring efforts, although not yielding entirely conclusive data, continued to raise concerns about the hazardous nature of South Peak of Turtle Mountain. Should South Peak fail catastrophically, as in 1903, estimates suggest that it would produce a rock avalanche of approximately 5 million m^3 (6.5 million yd^3) (Mei et al. 2008). Such a slide is projected to cover Alberta Highway 3, the Canadian Pacific Railway mainline, substantial areas of residential and commercial properties, and a sports and recreation complex currently under construction (Moreno and Froese 2009a). To provide early warning for such an event, in 2003 the government of Alberta funded $1.1 million to establish the Turtle Mountain Monitoring Project, which was responsible for installing a near-real-time warning system at South Peak. This system of more than 80 sensors, connected to data loggers and using radio communications, is designed to send out warnings based on predetermined thresholds for each sensor as defined by a system of four alert levels (green, yellow, orange, and red). A description of the threshold determination rationale has been provided by Froese et al. (2005) and Moreno and Froese (2009a).

In 2005, once the initial instrumentation system was completed, Alberta Geological Survey assumed responsibility for the long-term operation, maintenance, and upgrading of the Turtle Mountain Monitoring Project installation and the facilitation of research on the mountain with the monitoring system. The name was changed to Turtle Mountain Field Laboratory (TMFL) to reflect a broader mandate with three objectives:

- To provide an early warning for a failure of South Peak,
- To investigate the mechanism for a failure of South Peak, and
- To provide an educational resource for the public and researchers regarding the conditions on the mountain.

FIGURE 7-73 Crack meter installed across potential failure surface; monitoring cable extends downward. (PHOTOGRAPH BY B. ARNDT.)

FIGURE 7-74 More detailed view of crack meter installed across potential failure surface. (PHOTOGRAPH BY B. ARNDT.)

Descriptions of the TMFL project, access to all the data, and descriptions of the instrumentation installed there are accessible via the Internet (http://www.ags.gov.ab.ca/geohazards/turtle_mountain/turtle_mountain.html).

Recently a web–GIS application was created for the TMFL to enhance public awareness of the hazards and monitoring of the mountain, its effect on

FIGURE 7-75 Southwest view of Turtle Mountain showing locations of South and North Peaks in relation to historic Frank slide of 1903 (dashed lines delineate maximum slide extent) (adapted from Moreno and Froese 2008a).

the surrounding communities, and efforts by the Alberta government to mitigate damage similar to that caused by the 1903 event (Chao et al. 2009). It provides a user-friendly method for the public to display and manipulate maps and other available data for the Turtle Mountain area. In addition to the Internet access, the Turtle Mountain site has an interpretive visitor center—the FSIC—that provides public outreach functions.

Although the Turtle Mountain instability is much larger than that of a typical rockfall situation, the TMFL provides a uniquely available long-term, funded public source on the installation and operation of an entire suite of instrumentation and monitoring technologies. For this reason, a brief summary of the design of the TMFL installation and of some of the experience gained by those responsible for the TMFL is provided here.

More detailed summaries on the TMFL have been provided by Froese and Moreno (2007) and Moreno and Froese (2006, 2007). Bjorgan and Froese (2008) provided details of differential GPS and total-station distance-monitoring activities. Mei et al. (2008) described experiments that have identified very small ground deformations at Turtle Mountain by using satellite-based radar (InSAR) technology. Recent reports described the roles and responsibilities of those affected by the emergency-warning functions of the TMFL (Moreno and Froese 2009a) and the details of the data management and communications system for TMFL, including troubleshooting of the instruments and the communications (Moreno and Froese 2009b). A considerable volume of additional information is readily accessible from the Alberta Geological Survey, including annual data reports for 2005 through 2008 (Moreno and Froese 2006, 2008a, 2008b, 2009c).

FIGURE 7-76 Combination theodolite prism and differential GPS antenna mounted on concrete pillar near South Peak of Turtle Mountain; solar panel supplying power is on right (Moreno and Froese 2006).

The monitoring system on Turtle Mountain includes a number of different types of instruments that provide a near-real-time data stream. Four considerations affected the selection, design, layout, and implementation of the instrumentation on Turtle Mountain (Moreno and Froese 2006):

1. *Robustness*. The site is located at a remote mountain location, where the instruments are subjected to harsh climatic conditions and potential vandalism. Thus, the instruments need to be capable of stand-alone operations and be fairly robust in design. Attempts were made to either hide the installations or provide appropriate signage to educate the public regarding the importance of the system. A typical sign is attached to the tiltmeter illustrated in Figure 7-2. Vandalism has not been a problem, but extreme weather has affected some instruments.
2. *Redundancy*. It was considered desirable to have different types of sensors providing information over the same crack network to provide confidence in the observations. Large cracks are monitored by a combination of crack meters, tiltmeters, and extensometers. This arrangement has proved to be effective in validating and interpreting results. At several locations, reflector prisms, used for distance measurements with the total station, and digital GPS units are colocated on the same concrete pillar (Figure 7-76). This arrangement permits the distance measurements obtained by the two methods to serve as quality assurance checks on the two monitoring systems.
3. *Sensor type*. Instruments can be classified as mechanical, optical, or electrical, and each responds differently to environmental conditions. Although mechanical devices are more reliable than electrical instruments in high-humidity locations, electrical instruments are more reliable than mechanical instruments, such as crack meters, during snow precipitation. Thus, it was considered desirable to have instruments from different groups monitoring at most locations, so as to ensure continuous monitoring under various anticipated environmental conditions.
4. *Range of motion*. A variety of instruments with different ranges of motion and sensitivity allows detection of subtle movement trends as well as large-scale deformations. Thus, crack meters detect the smallest movements, whereas

somewhat larger movements are detected by extensometers. Once movements become even larger, tiltmeters, differential GPS, and distance measurements by total stations become relevant.

The TMFL monitoring network includes a considerable variety of instruments. Table 7-4 summarizes descriptions of the instrumentation; further details are readily obtained by accessing the previously cited TMFL literature sources. The instruments belong to three categories (Moreno and Froese 2006):

1. *Primary sensors*. These are tiltmeters, wire extensometers, and crack meters, instruments that provide a reliable and year-round data stream measuring easily interpretable parameters of the rock mass, such as changes in length or rotation.
2. *Secondary sensors*. These measure quantifiable parameters but may be subject to more variation due to environmental effects. They include a network of reflector prisms for distance measurements from a robotic total station located at the base of the mountain and a network of differential GPS receivers. Several reflector prisms were installed on the east face of Turtle Mountain, a location that is not easily accessible (Figure 7-77). Once installed, these reflectors have permitted accurate monitoring of movements with considerable efficiency and safety within an area of concern on Turtle Mountain that cannot be readily monitored by other methods.
3. *Tertiary sensors*. These provide additional data that are useful in the interpretation of results from the primary and secondary sensors, such as a meteorological station, an outflow weir at the base of the mountain providing an assessment of the pore-pressure buildup following precipitation, and a microseismic network consisting of two sensors within a borehole and six near-surface passive microseismic stations. The microseismic network was only partially successful, since the seismicity on Turtle Mountain is minor; only a single significant localized event was detected. By 2008 the microseismic network no longer was operational, but all data from 2003–2007 have been archived.

Examples of several of the instruments installed at TMFL have been used in previous sections of this chapter. Figure 7-2 shows a tiltmeter installed on Turtle Mountain, while Figure 7-10 illustrates one of the wire extensometer installations and Figure 7-17 shows a typical differential GPS station on the mountain. Section 8 provides some details of the communications systems used by the TMFL (including Figures 7-58, 7-59, and 7-60).

As is the case at many rockfall locations, Turtle Mountain requires the instruments to operate unattended for extended periods. It is subject to harsh winter weather conditions. The electrical power required by these instruments is relatively modest and is supplied by batteries recharged by solar panels (Figure 7-78). Solar panels are installed adjacent to many instruments and are thus visible in several illustrations of typical TMFL installations, such as Figure 7-10 (extensometer), Figure 7-17 (differential GPS), and Figure 7-76 (dual reflector prism and GPS).

Surface wire extensometers were custom designed for this project. They were installed across major fractures, as shown in Figure 7-10, and consist of a head assembly anchored into bedrock on one side of the fissure and an anchor end that is situated on the opposite side. The head assembly contains a weight that is connected to the anchor end by a steel cable that passes over a rotary potentiometer (Figures 7-79 and 7-80). Widening or closing of the fissure causes the suspended weight to rise or fall; this motion results in rotation of the potentiometer and these values are converted to displacement measurements. Extensometers were located so that the head assembly was on the upslope side of the fissure and the anchor was on the downslope side. Extensometer readings are made hourly. A signal cable from the head assembly of each extensometer runs through a protective conduit to a data logger (Figure 7-81), which also collects tiltmeter data. The data logger transmits the stored data by radio link to the FSIC.

The extensometers have operated quite reliably and provide measurements with a resolution of about 2 mm (0.08 in.). However, wind and snow loadings on the wires, particularly on longer spans, cause some noise in the data. Observations to date have shown that the fissures are gradually widening, with most of the movement occurring at the onset of cold weather. Freezing of water within cracks in the fissured rock mass is the apparent cause of these movements.

Table 7-4
Instrumentation Installed and Monitored at TMFL

Instrumentation		
Type	Number	Comments
Primary Sensors		
Tiltmeter	10	Placed along highly fractured summit ridge near South Peak; resolution is about 10 arc-seconds; hourly readings sent to data logger, then transmitted to FSIC.
Crack meter	22	Installed across large cracks in accessible areas on west side of South Peak; sets of three crack meters used at six locations, with the crack meters oriented at different angles across cracks, to allow determination of direction and magnitude of movement vector. Readings taken every 15 s and averaged once an hour before being transmitted to FSIC. Snow-and-ice buildup caused problems; improved protective roofs installed in 2007. Lightning strikes caused some damage and resulted in installation of surge protectors.
Wire extensometer	5	Four units installed in 2004; fifth unit installed in 2006. All custom-built and located across major fissures. They share the data-logger system with tiltmeters. Resolution is about 2 mm (0.08 in.). Observations show movements related to onset of cold weather, suggesting that freezing of water in fissures is cause.
Secondary Sensors		
Reflector prism read by total station	19	Four prisms are colocated on concrete differential GPS stations, keeping them above the snow. Other 15 prisms are located on east face of South Peak, lower on the mountain. A computer-controlled total station located in a building at the base of the mountain, about 3 km (1.9 mi) distant, measures distance to each prism once per hour. Accuracy is about 3 mm (0.12 in.).
Differential GPS station	11	Six differential GPS stations installed in 2006 along South Peak crest were supplemented by five additional stations installed in 2008 (Moreno and Froese 2009c). Raw GPS data collected at central point on the mountain, then transmitted and stored for postprocessing (Bjorgan and Froese 2008). Horizontal accuracy is 1 to 2 mm (0.04 to 0.08 in.).
Tertiary Sensors		
Weather station	1	Located on ridge crest near South Peak, weather station provides measurements of air temperature, humidity, rainfall, wind speed and direction, barometric pressure, solar radiation, and near-surface rock temperature. Data monitored by data logger and transmitted to data repository.
Thermistor	7	String of thermistors within borehole complements weather station data. They show that significant rock temperature fluctuations are limited to depths of less than 8 m (26 ft).
Outflow weir	1	V-notch weir and pressure transducer installed at old Frank mine entrance at base of mountain measures groundwater response to precipitation events.
Microseismic network	6	Network consisted of two sensors placed within borehole and six near-surface passive microseismic stations. Since seismicity on Turtle Mountain is minor, network detected only one significant localized event in 2006. By 2008 the network became inoperative, but all data from 2003–2007 have been archived.

FIGURE 7-77
Installing reflector prism on east face of Turtle Mountain in July 2007 (Moreno and Froese 2008b).

FIGURE 7-78
Example of solar-power installations used on Turtle Mountain. Dual solar panels and two battery boxes provide twice the normal generating and storage capacity (Moreno and Froese 2006).

FIGURE 7-79
Head assembly of extensometer EX-5 (extensometer wire digitally enhanced) (Moreno and Froese 2008a).

FIGURE 7-80
Head assembly of typical surface extensometer on South Peak of Turtle Mountain. Housing has been removed to illustrate suspended weight (Moreno and Froese 2006).

FIGURE 7-81 Central data logger for South Peak extensometer and tiltmeter networks. Instrument cables are visible feeding into data logger (Moreno and Froese 2006).

Twenty vibrating wire crack meters were installed in eight arrays that cross major fissures on the west side of Turtle Mountain, below South Peak. Five of these arrays have crack meters installed in triplets to determine a true movement vector. The crack meter instrumentation is the one system on Turtle Mountain that is most severely affected by snow loading and lightning strikes. These factors have constantly compromised the quality of the data, making this system the least reliable of the primary systems. Crack meters are also limited by the relatively small range of displacement they can accommodate.

Lightning protection (Figure 7-82) was installed following a lightning strike on July 26, 2004, that destroyed six of the installed instruments. Each lightning protection module covers four crack meters and includes a three-stage transient protection board and an enclosure with cable glands and grounding lug that allows grounding to be achieved with a short rock bolt. The transient protection board contains gas-discharge tubes, fast-response transient suppressors, and an isolation-fusing resistor.

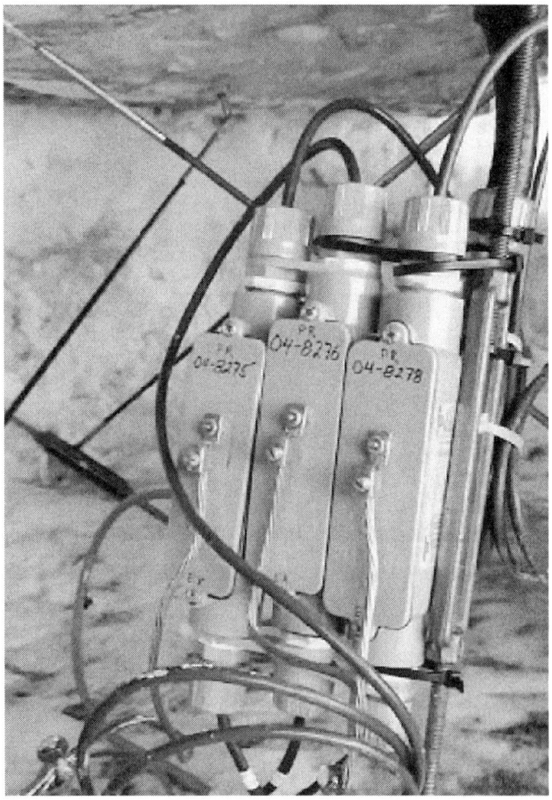

FIGURE 7-82 Electrical surge suppressors installed to protect crack meter arrays from lightning strikes on South Peak of Turtle Mountain. Grounding is accomplished through connection to small rock bolt (Moreno and Froese 2006).

REFERENCES

ABBREVIATION

POB Point of Beginning

3G Software and Measurement. 2009. *Contact Free Acquisition and Assessment of Rock and Terrain Surfaces by Metric 3-D Images*. http://www.3gsm.at. Accessed December 2009.

Abbott, B., I. Bruce, T. Keegan, F. Oboni, and W. Savigny. 1998. Application of a New Methodology for the Management of Rockfall Risk Along a Railway. In *Proceedings of the 8th Congress of the International Association of Engineering Geology: A Global View from the Pacific Rim*, Vancouver, British Columbia, Canada, A. A. Balkema, Rotterdam, Netherlands, Vol. 2, pp. 1201–1208.

Abellán, A., J. M. Vilaplana, and J. Martínez. 2006. Application of a Long-range Terrestrial Laser Scanner to a Detailed Rockfall Study at Vall de Núria (Eastern Pyrenees, Spain). *Engineering Geology*, Vol. 88, pp. 136–148.

ADAM Technology. 2009. *3-DM Analyst: Mining Analysis Software*. 12 pp. http://www.adamtech.com.au/3dm/3-DMAnalystMining.pdf. Accessed December 2009.

Aerden, D. G. A., and J. S. San Roman. 2002. STEREOFIT: A Program for Fitting Directional Data to Multiple Planes and Some Applications in Structural Geology. *Computers and Geosciences*, Vol. 28, pp. 81–85.

Agnew, D. C. 1986. Strainmeters and Tiltmeters. *Reviews of Geophysics*, Vol. 24, pp. 579–624.

Alshawa, M., E. Smigiel, P. Grussenmeyer, and T. Landes. 2007. Integration of a Terrestrial LiDAR on a Mobile Mapping Platform: First Experiences. In *Proc., 5th International Symposium on Mobile Mapping Technology*, Padua, Italy.

Arndt, B., T. Ortiz, and A. K. Turner. 2009. *Circular E-C141: Colorado's Full-Scale Testing of Rockfall Attenuator Systems*. Transportation Research Board of the National Academies, Washington, D.C.

Arndt, B. P., and L. T. Ortiz. 2006. Closure of Interstate 70 Near Idaho Springs, Colorado, Due to Multiple Rockslope Failures, and Subsequent Rockslope Evaluation, Analysis, Mitigation, and Instrumentation. In *Proc., 41st U.S. Symposium on Rock Mechanics*, Golden, Colorado, June 17–21, ARMA/USRMS 06-1076.

Baker, B. R., K. Gessner, E.-J. Holden, and A. P. Squelch. 2008. Automatic Detection of Anisotropic Features on Rock Surfaces. *Geosphere*, Vol. 4, No. 2, pp. 418–428.

Balch, K. 2002. High Speed Videography. In *High Speed Photography and Photonics*, SPIE Press Monograph PM120, International Society for Optical Engineering, Bellingham, Wash., Chap. 6, pp. 99–123.

Barber, D., J. Mills, and S. Smith-Voysey. 2008. Geometric Validation of a Ground-Based Mobile Laser Scanning System. *ISPRS Journal of Photogrammetry and Remote Sensing*, Vol. 63, No. 1, pp. 128–141.

Beraldin, J-A., F. Blais, P. Boulanger, L. Cournoyer, J. Domey, F. El-Hakim, G. Godin, M. Rioux, and J. Taylor. 2000. Real World Monitoring Through High Resolution Digital 3-D Imaging of Objects and Structures. *ISPRS Journal of Photogrammetry and Remote Sensing*, Vol. 55, pp. 230–250.

Besl, P., and N. McKay. 1992. A Method for Registration of 3-D Shapes. *IEEE Transactions on Pattern Analysis and Machine Intelligence*, Vol. 14, No. 2, pp. 239–256.

Bjorgan, G., and C. R. Froese. 2008. Combined GPS and EDM Monitoring on Turtle Mountain, Alberta. In *Conference Proceedings, GeoEdmonton'08, 61st Canadian Geotechnical Conference and 9th Joint CGS/IAH-CNC Groundwater Conference*, Sept. 21–24, Canadian Geotechnical Society, Edmonton, Alberta, pp. 1236–1242.

Brackett, P. 2002. *Railway Rockfall Electromagnetic Field Disturbance Sensing System Development and Test Results*. Technical Report TP 13928E. Transportation Development Centre, Transport Canada, 17 pp. (Available as PDF document from Communications Services, Transportation Development Centre, Transport Canada, tdccdt@tc.gc.ca.)

Buckley, S. J., J. A. Howell, H. D. Enge, and T. H. Kurz. 2008. Terrestrial Laser Scanning in Geology: Data Acquisition, Processing and Accuracy Considerations. *Journal of the Geological Society of London*, Vol. 165, pp. 625–638.

Call, R. D. 1982. Monitoring Pit Slope Behavior. In *Stability in Surface Mining* (C. O. Brawner, ed.), Society of Mining Engineers of the American Institute of Mining, Metallurgical, and Petroleum Engineers, New York, Vol. 3, pp. 229–248.

Carnec, C., D. Massonnet, and C. King. 1996. Two Examples of the Use of SAR Interferometry on Displacement Fields of Small Spatial Extent. *Geophysical Research Letters*, Vol. 23, No. 24, pp. 3579–3582.

Chao, D. K., S. Chowdhury, F. Moreno, and C. R. Froese. 2009. *Implementation of a Web-GIS Application for the Turtle Mountain Monitoring Project with ArcGIS Server 9.2*. ERCB/AGS Open File Report 2009-05. Alberta Energy Resources Conservation Board, Calgary, Alberta, Canada.

Criminisi, A. 2001. *Accurate Visual Metrology from Single and Multiple Uncalibrated Images*. Distinguished Dissertation Series XIII. Springer-Verlag, Berlin.

Dowding, C. H., and E. C. Huang. 1994. Early Detection of Rock Movement with Time Domain Reflectometry. *Journal of Geotechnical Engineering*, ASCE, Vol. 120, No. 8, pp. 1413–1427.

Dunnicliff, C. J. 1988. *Geotechnical Instrumentation for Monitoring Field Performance*. John Wiley and Sons, Inc., New York.

Dunnicliff, C. J., and D. Davidson. 1991. Lessons Learned from Automating Instrumentation at Some U.S. Dams. *Geotechnical News*, Vol. 9, No. 2, pp. 46–51.

Dutro, H. B., and R. O. Dickinson. 1974. Slope Instrumentation Using Multiple-Position Borehole Extensometers. In *Transportation Research Record 482*, Transportation Research Board, National Research Council, Washington, D.C., pp. 9–17.

Fardin, N., Q. Feng, and O. Stephansson. 2004. Application of a New In Situ 3-D Laser Scanner to

Study the Scale Effect on the Rock Joint Surface Roughness. *International Journal of Rock Mechanics and Mining Sciences*, Vol. 41, pp. 329–335.

Feng, Q., and K. Röshoff. 2004. In Situ Mapping and Documentation of Rock Faces Using a Full-Coverage 3-D Laser Scanning Technique. *International Journal of Rock Mechanics and Mining Sciences*, Vol. 41, No. 3, pp. 139–144.

Feng, Q., N. Fardin, L. Jing, and O. Stephansson. 2003. A New Method for In Situ Non-contact Roughness Measurement of Large Rock Fracture Surfaces. *Rock Mechanics and Rock Engineering*, Vol. 36, pp. 3–25.

Feng, Q., P. Sjögren, O. Stephansson, and L. Jing. 2001. Measuring Fracture Orientation at Exposed Rock Faces by Using a Non-Reflector Total Station. *Engineering Geology*, Vol. 59, pp. 133–146.

Ferrero, A. M., G. Forlani, R. Roncella, and H. I. Voyat. 2009. Advanced Geostructural Survey Methods Applied to Rock Mass Characterization. *Rock Mechanics and Rock Engineering*, Vol. 42, pp. 631–665.

Forsyth, D. A., and J. Ponce. 2002. *Computer Vision: A Modern Approach*. Prentice Hall, Upper Saddle River, N.J.

Franklin, J. A., N. H. Maerz, and C. P. Bennett. 1988. Rock Mass Characterisation Using Photoanalysis. *International Journal of Mining and Geological Engineering*, Vol. 6, pp. 97–112.

Froese, C. R., and F. Moreno. 2006. Turtle Mountain Field Laboratory (TMFL), Part 1: Concept and Research Activities. In *Proc., 59th Canadian Geotechnical Conference*, Canadian Geotechnical Society, Richmond, British Columbia, pp. 544–551.

Froese, C. R., and F. Moreno. 2007. Turtle Mountain Field Laboratory (TMFL), Part 1: Overview and Activities. In *Conference Proceedings, 1st North American Landslide Conference*, Vail, Colo., AEG Special Publication No. 23, Association of Environmental and Engineering Geologists, pp. 971–980.

Froese, C. R., F. Moreno, M. Jaboyedoff, and D. M. Cruden. 2009. 25 Years of Movement Monitoring on South Peak, Turtle Mountain: Understanding the Hazard. *Canadian Geotechnical Journal*, Vol. 46, pp. 256–269.

Froese, C. R., C. Murray, D. S. Cavers, W. S. Anderson, A. K. Bidwell, R. S. Read, D. M. Cruden, and C. W. Langenberg. 2005. Development and Implementation of a Warning System for the South Peak of Turtle Mountain. In *Landslide Risk Management: Proceedings of the International Conference on Landslide Risk Management* (O. Hungr, R. Fell, R. Couture, and E. Eberhardt, eds.), Vancouver, Canada, May 31–June 3, Taylor & Francis, London, pp. 705–712.

Fröhlich, C., and M. Mettenleiter. 2004. Terrestrial Laser Scanning—New Perspectives in 3-D Surveying. *International Archives of Photogrammetry, Remote Sensing and Spatial Information Sciences*, Vol. 36, Part 8/W2, Oct. 2004.

Fuller, P. W. W. 1994. Aspects of High Speed Photography. *Journal of Photographic Science*, Vol. 42, pp. 42–43.

Gaich, A., M. Pötsch, A. Fasching, and W. Schubert. 2004. Contact-Free Measurement of Rock Mass Structures Using the Joint MetriX3-D System. *International Journal of Rock Mechanics and Mining Science*, Vol. 41, No. 3.

Gaich, A., M. Pötsch, and W. Schubert. 2006. Basics and Application of 3-D Imaging Systems with Conventional and High-Resolution Cameras. In *Laser and Photogrammetric Methods for Rock Face Characterization* (F. Tonon and J. T. Kottenstette, eds.), American Rock Mechanics Association, Alexandria, Va., pp. 33–48.

Geobrugg. 2010. *Remote Monitoring of Geobrugg Rockfall Protection Systems*. Geobrugg Technical Brochure. Romanshorn, Switzerland, 6 pp. http://www.impact-sentinel.com/Impact_Sentinel_Broschure_en.pdf.

Gillon, M. D., P. P. Foster, G. T. Proffitt, and A. P. Smits. 1992. Monitoring of the Cromwell Gorge Landslides. In *Proc., Sixth International Symposium on Landslides*, Christchurch, New Zealand, A. A. Balkema, Rotterdam, Netherlands, Vol. 2, pp. 1135–1140.

GIM-International. 2009. Terrestrial Laser Scanners. *GIM-International*, Vol. 23, Issue 8, Aug., pp. 20–23. (Extended list available for download at http://www.gim-international.com/files/productsurvey_v_pdfdocument_33.pdf.)

Gladwin, M. T., R. L. Gwyther, and M. W. Mee. 2000. Remote Monitoring of Rock Mass Deformation during Mining. In *Proc., Massmin 2000 Conference*, Brisbane, Australia, Australasian Institute of Mining and Metallurgy, pp. 209–216.

Glennie, C. 2007a. Reign of Point Clouds: A Kinematic Terrestrial LiDAR Scanning System. *Inside GNSS*, Fall 2007, pp. 21–31.

Glennie, C. 2007b. Rigorous 3-D Error Analysis of Kinematic Scanning LiDAR Systems. *Journal of Applied Geodesy*, Vol. 1, pp. 147–157.

Glennie, C., K. Kusevic, and P. Mrstik. 2006. Performance Analysis of a Kinematic Terrestrial LiDAR Scanning System. Presented at MAPPS/ASPRS Fall Conference, Nov. 6–10, San Antonio, Tex.

Grasselli, G., J. Wirth, and P. Egger. 2002. Quantitative Three-Dimensional Description of a Rough Surface and Parameter Evolution with Shearing. *International Journal of Rock Mechanics and Mining Sciences*, Vol. 36, No. 6, pp. 789–800.

Green, G. E., and P. E. Mikkelsen. 1988. Deformation Measurements with Inclinometers. In *Transportation Research Record 1169*, Transportation Research Board, National Research Council, Washington, D.C., pp. 1–15.

Grobler, H. P., G. Poropat, and A. R. Guest. 2003. Photogrammetry for Structural Mapping in Mining. In *ISRM 2002 Technology Roadmap for Rock Mechanics*, South African Institute of Mining and Metallurgy, Marshalltown, pp. 431–436.

GroundProbe. 2009. http://www.groundprobe.com. Accessed Dec. 22, 2009.

Hadjigeorgiou, J., F. Lemy, P. Côté, and X. Maldague. 2003. An Evaluation of Image Analysis Algorithms for Constructing Discontinuity Trace Maps. *Rock Mechanics and Rock Engineering*, Vol. 36, No. 2, pp. 163–179.

Hamilton, J. 1939. Threatening Rock. *National Park Service Region III Quarterly*, Vol. 1, No. 1. http://www.nps.gov/history/history/online_books/region_111/vol1-1e.htm.

Haneberg, W. C. 2007. Directional Roughness Profiles from Three-Dimensional Photogrammetric or Laser Scanner Point Clouds. In *Rock Mechanics: Meeting Society's Challenges and Demands—Proc., 1st Canada–U.S. Rock Mechanics Symposium*, Vancouver, Canada (E. Eberhardt, D. Stead, and T. Morrison, eds.), Taylor & Francis, London, Vol. 1, pp. 101–106.

Haneberg, W. C. 2008. Using Close Range Terrestrial Digital Photogrammetry for 3-D Rock Slope Modeling and Discontinuity Mapping in the United States. *Bulletin of Engineering Geology and the Environment*, Vol. 67, pp. 457–469.

Haneberg, W. C., and E. W. Medley. 2007. Internal Structure of the San Andreas Fault Zone at the A. R. Wilson Quarry, Aromas, California, as Inferred from 3-D Digital Outcrop Modeling. *Abstracts with Programs* (Geological Society of America), Vol. 39, p. 454.

Haneberg, W. C., N. I. Norrish, and D. P. Findley. 2006. Digital Outcrop Characterization for 3-D Structural Mapping and Rock Slope Design Along Interstate 90 Near Snoqualmie Pass, Washington. In *Proc., 57th Highway Geology Symposium*, Breckenridge, Colo., Sept. 26–29, pp. 146–160.

Hanna, T. H. 1985. *Field Instrumentation in Geotechnical Engineering*. Trans Tech Publications, Clausthal-Zellerfeld, Germany.

Harries, N., and M. Holmstrom. 2007. The Use of Slope Stability Radar in Monitoring Slopes and Managing Slope Instability Hazards. *Queensland Roads*, Edition No. 4, Sept., pp. 39–44.

Harrison, J. P. 1993. *Improved Analysis of Rock Mass Geometry Using Mathematical and Photogrammetric Methods*. PhD thesis. Imperial College, London.

Hartley, R. I., and A. Zisserman. 2004. *Multiple View Geometry in Computer Vision*, 2nd ed. Cambridge University Press.

Heuel, S. 2004. *Uncertain Projective Geometry: Statistical Reasoning for Polyhedral Object Reconstruction*. Lecture Notes in Computer Science 3008. Springer-Verlag, Berlin.

Hinckley, A. L., B. D. Tanner, and E. C. Campbell. 1988. Automation of Monitoring of Geotechnical Instrumentation. In *Transportation Research Record 1119*, Transportation Research Board, National Research Council, Washington, D.C., pp. 26–29.

Hobbs, B. E. 1993. The Significance of Structural Geology in Rock Mechanics. In *Comprehensive Rock Engineering 1: Fundamentals* (E. T. Brown, ed.), Pergamon Press, Oxford, U.K., pp. 25–62.

Hoffman-Wellenhof, B., H. Lichtenegger, and J. Collins. 2001. *Global Positioning System: Theory and Practice*, 5th ed. Springer-Verlag, Berlin.

Hunter, G., C. Cox, and J. Kremer. 2008. Development of a Commercial Laser Scanning Mobile Mapping System: StreetMapper. Presented at Second International Workshop: The Future of Remote Sensing, Oct. 17–18, Antwerp, Belgium.

Ilsley, R. C. 1994. Engineering Geological Mapping of Rock Slopes Using a Laser-Transit. In *Proceedings of the 7th International Congress of the International Association of Engineering Geology* (R. Oliveira, L. F. Rodrigues, A. G. Coelho, and A. P. Cunha, eds.), Sept. 5–9, Lisbon, Portugal, Vol. 2, pp. 1117–1123.

Integrated Design Tools. 2010. *Motion Studio Cross-platform User Manual*. Integrated Design Tools, Inc., Tallahassee, Fla., 229 pp. http://www.idtvision.com/documents/mstudio_man_en.pdf. Accessed August 2010.

Jacobs, G. 2005a. Understanding Laser Scanning Terminology. *Professional Surveyor*, Vol. 25, No. 2, pp. 26–31.

Jacobs, G. 2005b. Laser Scanning: Big Advantages, Even for Small Projects. *Professional Surveyor*, Vol. 25, No. 10, pp. 30–35.

Kadobayashi, R., N. Kochi, H. Otani, and R. Furukawa. 2004. Comparison and Evaluation of Laser Scanning and Photogrammetry and Their Combined Use for Digital Recording of Cultural Heritage. *International Archives of the Photogrammetry, Remote Sensing and Spatial Information Sciences*, Vol. 35, Part B5.

Kane, W. E., and T. J. Beck. 1994. Development of a Time Domain Reflectometry System to Monitor Landslide Activity. In *Proc., 45th Highway Geology Symposium*, Portland, Ore., Aug. 17–19, Geology Department, Portland State University, pp. 163–173.

Kemeny, J., and R. Post. 2003. Estimating Three-Dimensional Rock Discontinuity Orientation from Digital Images of Fracture Traces. *Computers and Geosciences*, Vol. 29, No. 1, pp. 65–77.

Kemeny, J., and A. K. Turner. 2008. *Ground-Based LiDAR: Rock Slope Mapping and Assessment*. Publication FHWA-CFL/TD-08-006. Central Federal Lands Highway Division, Lakewood, Colo., Sept., 103 pp. http://www.cflhd.gov.

Kemeny, J., E. Mofya, and J. Handy. 2003. The Use of Digital Imaging and Laser Scanning Technologies for Field Rock Fracture Characterization. In *Proceedings of Soil and Rock America 2003, 12th Pan American Conference on Soil Mechanics and Geotechnical Engineering and the 39th US Rock Mechanics Symposium* (J. Culligan, H. Einstein, and A. White, eds.), Massachusetts Institute of Technology, Cambridge, pp. 117–122.

Kemeny, J., J. Monte, J. Handy, and S. Thiam. 2003. The Use of Digital Imaging and Laser Scanning Technologies in Rock Engineering. Presented at International Symposium on the Fusion Technology of Geosystem Engineering, Rock Engineering and Geophysical Exploration, Seoul, Korea, Nov. 18–19.

Kemeny, J., B. Norton, and K. Turner. 2006. Rock Slope Stability Analysis Using Ground-Based LiDAR and Digital Image Processing. *Felsbau—Rock and Soil Engineering*, No. 3/06, pp. 8–15.

Kemeny, J., K. Turner, and B. Norton. 2006. LiDAR for Rock Mass Characterization: Hardware, Software, Accuracy and Best-Practices. In *Report on a Workshop held June 17–18, 2006 in Golden, Colorado*, American Rock Mechanics Association (ARMA), pp. 49–61. http://www.usbr.gov/pmts/geology/goldenrocks.pdf.

Kliche, C. A. 1999. *Rock Slope Stability*. Society for Mining, Metallurgy and Exploration Inc. (SME), Littleton, Colo.

Krosley, L., and P. Shaffner. 2003. Applications and Accuracy of Photogrammetry for Geological and Geotechnical Data Collecting. *Abstracts, 2003 Annual Meeting*, Association of Engineering Geologists.

Lan, H., C. D. Martin, C. Zhou, and C. H. Lim. 2010. Rockfall Hazard Analysis Using LiDAR and Spatial Modeling. *Geomorphology*, Issues 1–2, May 15, pp. 213–233.

Langley, R. B. 1991. The GPS Receiver: An Introduction. *GPS World*, Vol. 2, No. 1, pp. 50–53.

Langley, R. B. 1993. The GPS Observables. *GPS World*, Vol. 4, No. 4, pp. 52–59.

Langley, R. B. 1999. Dilution of Precision. *GPS World*, Vol. 10, No. 5, pp. 52–59.

Langley, R. B. 2000. Smaller and Smaller: The Evolution of the GPS Receiver. *GPS World*, Vol. 11, No. 4, pp. 54–58.

Lato, M., and J. Hutchinson. 2008. Mining for Lidar Data. *Professional Surveyor*, Vol. 28, No. 2, pp. 10–15.

Lato, M., D. J. Hutchinson, M. S. Diederichs, D. Ball, and R. Harrap. 2009. Engineering Monitoring of Rockfall Hazards Along Transportation Corridors: Using Mobile Terrestrial LiDAR. *Natural Hazards and Earth Systems Science*, Vol. 9, pp. 935–946.

Lato, M., D. J. Hutchinson, M. S. Diederichs, and R. Harrap. 2009. Optimization of LiDAR Scanning and Processing for Automated Structural Evaluation of Discontinuities in Rock Masses. *International Journal of Rock Mechanics and Mining Sciences*, Vol. 46, pp. 194–199.

Leica. 2009. *Flexline TS02/06/09 Datasheet*. http://www.leica-geosystems.us/en/Construction-Surveying-TPS-Leica-FlexLine-TS020609_72053.htm. Accessed Nov. 2009.

Leick, A. 2003. *GPS Satellite Surveying*, 3rd ed. John Wiley and Sons, New York.

Lemay, P. 2007. Listening for Rockfalls: Microseismic Technology May Help Railway Better Detect Rockfalls at a Distance. *R and D Updates Bulletin* (Transportation Development Centre, Transport Canada), Vol. 17, No. 1, July, 1 p. http://www.tc.gc.ca/eng/innovation/tdc-publication-updates-v17n1-827.htm.

Lemy, F., and J. Hadjigeorgiou. 2004. A Field Application of Laser Scanning Technology to Quantify Rock Fracture Orientation. In *Proc., ISRM Regional Symposium EUROCK 2004 and 53rd Geomechanics Colloquy*, pp. 435–438.

Lichti, D. D., and S. Jamtsho. 2006. Angular Resolution of Terrestrial Laser Scanners. *The Photogrammetric Record*, Vol. 21, No. 114, pp. 141–160.

Lichti, D. D., S. J. Gordon, and M. P. Stewart. 2002. Ground-based Laser Scanners: Operations, Systems and Applications. *Geomatica*, Vol. 56, No. 1, pp. 21–33.

Lichti, D. D., S. J. Gordon, and T. Tipdecho. 2005. Error Models and Propagation in Directly Georeferenced Terrestrial Scanner Networks. *Journal of Surveying Engineering*, Vol. 131, No. 4, pp. 135–142.

Lim, M., D. N. Petley, N. J. Rosser, R. J. Allison, A. J. Long, and D. Pybus. 2005. Combined Digital Photogrammetry and Time-of-Flight Laser Scanning for Monitoring Cliff Evolution. *The Photogrammetric Record*, Vol. 20, pp. 109–129.

Lim, M., N. J. Rosser, R. J. Allison, and D. N. Petley. 2010. Erosional Processes in the Hard Rock Coastal Cliffs at Staithes, North Yorkshire. *Geomorphology*, Vol. 114, pp. 12–21.

Maas, H.-G., and U. Hampel. 2006. Photogrammetric Techniques in Civil Engineering Material Testing and Structure Monitoring. *Photogrammetric Engineering and Remote Sensing*, Vol. 72, No. 1, pp. 39–45.

Malassingne, C., F. Lemaître, P. Briole, and O. Pascal. 2001. Potential of Ground Based Radar for the Monitoring of Deformation of Volcanoes. *Geophysical Research Letters*, Vol. 28, No. 5, pp. 851–854.

Martin, C. D., D. D. Tannant, and H. Lan. 2007. Comparison of Terrestrial-based, High Resolution, LiDAR and Digital Photogrammetry Surveys of a Rock Slope. In *Rock Mechanics: Meeting Society's Challenges and Demands—Proc., 1st Canada–U.S. Rock Mechanics Symposium*, Vancouver, Canada (E. Eberhardt, D. Stead, and T. Morrison, eds.), Taylor & Francis, London, Vol. 1, pp. 37–44.

Massey, C. I., I. F. Hodgson, and D. N. Petley. 2006. A Rockfall Simulation Study for Housing Development in Gibraltar. In *Proc., 10th International Congress, International Association for Engineering Geology and the Environment (IAEG)*, Nottingham, UK, Sept. 6–10, Geological Society of London, Paper 377 (on CD-ROM).

Massonnet, D. 1997. Satellite Radar Interferometry. *Scientific American*, Vol. 276. pp. 32–39.

McHugh, E. L., J. Dwyer, D. G. Long, and C. Sabine. 2006. *Applications of Ground-based RADAR to Mine Slope Monitoring*. NIOSH Publication No. 2006-116. Centers for Disease Control and Prevention, U.S. Department of Health and Human Services, 32 pp. http://www.cdc.gov/niosh/mining/pubs/pubreference/outputid1721.htm. Accessed Dec. 2009.

McKenna, G. 1995. Advances in Slope-Monitoring Instrumentation at Syncrude Canada Ltd.: A Followup. *Geotechnical News*, Vol. 13, No. 1, pp. 42–44.

McKenna, G., G. Livingstone, and T. Lord. 1994. Advances in Slope-Monitoring Instrumentation at Syncrude Canada Ltd. *Geotechnical News*, Vol. 12, No. 3, pp. 66–69.

Mei, S., V. Poncos, and C. Froese. 2008. *Mapping Millimetre-Scale Ground Deformation over the Frank Slide and South Peak of Turtle Mountain, Alberta, Using Spaceborne InSAR Technology*. ERCB/AGS Earth Sciences Report 2007-09. Alberta Energy Resources Conservation Board, Calgary, Alberta, Canada.

Mikkelsen, P. E. 1992. Instrument Innovations for Monitoring Natural Hazards. *Geotechnical News*, Vol. 10, No. 1, pp. 46–49.

Mikkelsen, P. E. 1996. Field Instrumentation. In *Special Report 247: Landslides: Investigation and Mitigation* (A. K. Turner and R. L. Schuster, eds.), Transportation Research Board, National Research Council, Washington, D.C., pp. 278–316.

Mohamed, A., and B. Wilkinson. 2009. Direct Georeferencing of Stationary LiDAR. *Remote Sensing*, Vol. 1, pp. 1321–1337.

Moreno, F., and C. Froese. 2006. *Turtle Mountain Field Laboratory Monitoring and Research Summary Report, 2005*. EUB/AGS Earth Sciences Report 2006-07. Alberta Energy and Utilities Board.

Moreno, F., and C. R. Froese. 2007. Turtle Mountain Field Laboratory (TMFL), Part II: Review of Trends—2005 to 2006. In *Conference Proceedings, 1st North American Landslide Conference*, Vail, Colorado, AEG Special Publication No. 23 (CD-ROM), Association of Environmental and Engineering Geologists, pp. 981–990.

Moreno, F., and C. R. Froese. 2008a. *Turtle Mountain Field Laboratory: 2006 Data and Activity Summary*. ERCB/AGS Open File Report 2008-01. Alberta Energy Resources Conservation Board, Calgary, Alberta, Canada.

Moreno, F., and C. R. Froese. 2008b. *Turtle Mountain Field Laboratory: 2007 Data and Activity Summary*. ERCB/AGS Open File Report 2008-07. Alberta Energy Resources Conservation Board, Calgary, Alberta, Canada.

Moreno, F., and C. R. Froese. 2009a. *ERCB/AGS Roles and Responsibilities Manual for the Turtle*

Mountain Monitoring Project, Alberta. ERCB/AGS Open File Report 2009-06. Alberta Energy Resources Conservation Board, Calgary, Alberta, Canada.

Moreno, F., and C. R. Froese. 2009b. *ERCB/AGS System Manual for the Turtle Mountain Monitoring Project, Alberta.* ERCB/AGS Open File Report 2009-14. Alberta Energy Resources Conservation Board, Calgary, Alberta, Canada.

Moreno, F., and C. R. Froese. 2009c. *Turtle Mountain Field Laboratory, Alberta (NTS 82G): 2008 Data and Activity Summary.* ERCB/AGS Open File Report 2009-15. Alberta Energy Resources Conservation Board, Calgary, Alberta, Canada.

Morgan, D. 2009. Using Mobile Lidar to Survey Railway Infrastructure, Lynx Mobile Mapper. In *Proc., International Federation of Surveyors (FIG) Commissions 5, 6 and SSGA Workshop: Innovative Technologies for an Efficient Geospatial Management of Earth Resources* (A. G. Novitskaya, ed.), 23–30 July, Lake Baikal, Listvyanka, Russian Federation, Siberian State Academy of Geodesy (SSGA), Novosibirsk, Russia, pp. 33–40.

Newby, S., and P. Mrstik. 2005. LiDAR on the Level in Afghanistan. *GPS World,* Vol. 16, No. 7, pp. 16–22.

O'Connor, K. M., and L. V. Wade. 1994. Applications of Time Domain Reflectometry in the Mining Industry. In *Proc., Symposium and Workshop on Time Domain Reflectometry in Environmental, Infrastructure, and Mining Applications,* Evanston, Ill., Special Publication 19-94, Bureau of Mines, U.S. Department of the Interior, Sept., pp. 494–506.

Olsen, M. J., E. Johnstone, N. Driscoll, S. A. Ashford, and F. Kuester. 2009. Terrestrial Laser Scanning of Extended Cliff Sections in Dynamic Environments: Parameter Analysis. *Journal of Surveying Engineering,* Vol. 135, No. 4, pp. 161–169.

Optech, Inc. 2008. *Lynx Mobile Mapper.* Toronto, Ontario, Canada.

Paisley, D. W. 1993. What Constitutes High Speed Photography? *OE Reports* (International Society for Optical Engineering), Bellingham, Wash., Nov. 13, 1993.

Patton, F. D. 1983. The Role of Instrumentation in the Analysis of the Stability of Rock Slopes. In *Proc., International Symposium on Field Measurements in Geomechanics,* Zurich, Sept. 5–8, pp. 719–748.

POB. 2010. *Point of Beginning 2010 Laser Scanner Survey.* BNP Media, Troy, Mich. http://laser.jadaproductions.net.

Poropat, G. V. 2006. Remote 3-D Mapping of Rock Mass Structure. In *Laser and Photogrammetric Methods for Rock Face Characterization* (F. Tonon and J. T. Kottenstette, eds.), American Rock Mechanics Association, Alexandria, Va., pp. 63–75.

Poropat, G. V. 2008. Remote Characterization of Surface Roughness of Rock Discontinuities. In *Proc., 1st Southern Hemisphere International Rock Mechanics Symposium,* Sept. 16–19, Perth, Australia (Y. Potvin, J. Carter, A. Dyskin, and R. Jeffrey, eds.), Vol. 2, pp. 447–458.

Post, R., J. Kemeny, and R. Murphy. 2001. Image Processing for the Automatic Extraction of Rock Joint Orientation Data from Digital Images. In *Proceedings of the 38th US Rock Mechanics Symposium,* Washington, D.C., A.A. Balkema, Rotterdam, Netherlands, pp. 877–884.

Pötsch, M., W. Schubert, and A. Gaich. 2007. The Application of Metric 3-D Images for the Mechanical Analysis of Keyblocks. In *Rock Mechanics: Meeting Society's Challenges and Demands—Proc., 1st Canada–U.S. Rock Mechanics Symposium,* Vancouver, Canada (E. Eberhardt, D. Stead, and T. Morrison, eds.), Taylor & Francis, London, pp. 77–84.

Proffitt, G. T., G. J. Fairless, G. G. Grocott, and P. A. Manning. 1992. Computer Applications for Landslide Studies. In *Proc., Sixth International Symposium on Landslides,* Christchurch, New Zealand, A.A. Balkema, Rotterdam, Netherlands, Vol. 2, pp. 1157–1162.

Rabatel, A., P. Deline, S. Jaillet, and L. Ravanel. 2008. Rock Falls in High-Alpine Rock Walls Quantified by Terrestrial Lidar Measurements: A Case Study in the Mont Blanc Area. *Geophysical Research Letters,* Vol. 35, L10502, doi:10.1029/2008GL033424.

Raquet, J., and G. Lachapelle. 2001. Efficient Precision Positioning: RTK Positioning with Multiple Reference Stations. *GPS World,* Vol. 12, No. 4, pp. 48–53.

Ray, S. F. (ed.). 2002. *High Speed Photography and Photonics.* SPIE Press Monograph PM120. International Society for Optical Engineering, Bellingham, Wash.

Redstall, M. 2006a. Accurate Terrestrial Laser Scanning from a Moving Platform. *Geomatics World,* July/August, pp. 28–30.

Redstall, M. 2006b. Rapid 3-D Terrestrial Mapping from a Moving Vehicle. *Civil Engineering Surveyor,* Oct.

Reeves, B. A., G. F. Stickley, D. A. Noon, and I. D. Longstaff. 2000. Developments in Monitoring Mine Slope Stability Using Radar Interferometry.

In *Proc., International Geoscience and Remote Sensing Symposium, (IGARSS 2000)*, Honolulu, Hawaii, Vol. 5, pp. 2325–2327.

Reid, T. R., and J. P. Harrison. 2000. A Semi-automated Methodology for Discontinuity Trace Detection in Digital Images of Rock Mass Exposures. *International Journal of Rock Mechanics and Mineral Science*, Vol. 37, pp. 1073–1089.

Riegl, USA, Inc. 2010. *Preliminary Data Sheet, Riegl VMX-250 (08/04/2010)*. 6 pp. http://www.riegl.com/nc/products/mobile-scanning/produktdetail/product/scannersystem/6.

Roberts, G., and G. Poropat. 2000. Highwall Joint Mapping in 3-D at the Moura Mine Using SIROJOINT. In *Proc., Bowen Basin Symposium 2000 Coal and Mining in the New Millennium*, Geological Society of Australia, Rockhampton, Queensland, Oct.

Rosser, N. J., S. A. Dunning, M. Lim, and D. N. Petley. 2005. Terrestrial Laser Scanning for Quantitative Rockfall Hazard Assessment. In *Landslide Risk Management: Proceedings of the International Conference on Landslide Risk Management* (O. Hungr, R. Fell, R. Couture, and E. Eberhardt, eds.), Vancouver, Canada, May 31–June 3, Taylor & Francis, London. CD-ROM.

Rosser, N. J., M. Lim, D. N. Petley, S. Dunning, and R. J. Allison. 2007. Patterns of Precursory Rockfall Prior to Slope Failure. *Journal of Geophysical Research*, Vol. 112, F04014, doi:10.1029/2006JF000642.

Rosser, N. J., D. N. Petley, M. Lim, S. A. Dunning, and R. J. Allison. 2005. Terrestrial Laser Scanning for Monitoring the Process of Hard Rock Coastal Cliff Erosion. *Quarterly Journal of Engineering Geology and Hydrogeology*, Vol. 38, pp. 363–375.

Schuhmacher, S., and J. Böhn. 2005. Georeferencing of Terrestrial Laserscanning Data for Applications in Architectural Modeling. *International Archives of Photogrammetry, Remote Sensing and Spatial Information Sciences*, Vol. 36, Part 5/W17 (GITC bv, Lemmer, Netherlands).

Schumm, S. A., and R. J. Chorley. 1964. The Fall of Threatening Rock. *American Journal of Science*, Vol. 262, pp. 1041–1054.

Shaw, M., K. Sandhoo, and D. Turner. 2000. Modernization of the Global Positioning System. *GPS World*, Vol. 11, No. 9, pp. 36–44.

Slob, S. 2010. *Automated Rock Mass Characterisation Using 3-D Terrestrial Laser Scanning*. ITC Dissertation 172. University of Twente, Enschede, Netherlands.

Slob, S., and H. R. G. K. Hack. 2004. 3-D Terrestrial Laser Scanning as a New Field Measurement and Monitoring Technique. In *Engineering Geology for Infrastructure Planning in Europe: A European Perspective* (R. Hack, R. Azzam, and R. Charlier, eds.), Springer-Verlag, Berlin, pp. 179–190.

Slob, S., H. R. G. K. Hack, and K. Turner. 2002. An Approach to Automate Discontinuity Measurements of Rock Faces using Laser Scanning Techniques. In *Proc., ISRM EUROCK 2002* (C. Dinid da Gama and L. Ribeira e Sousa, eds.), Funchal, Portugal, 25–28 Nov. 2002, Sociedade Portuguesa de Geotecnia, Lisbon, pp. 87–94.

Slob, S., H. R. G. K. Hack, B. van Knapen, and J. Kemeny. 2004. Digital Outcrop Mapping and Determination of Rock Mass Properties Using 3-D Terrestrial Laser Scanning Techniques. In *Proceedings of EUROCK 2004 and 53rd Geomechanics Colloquium* (S. Schubert, ed.), Verlag Glückauf, Essen, Germany, pp. 439–443.

Slob, S., B. van Knapen, R. Hack, K. Turner, and J. Kemeny. 2005. Method for Automated Discontinuity Analysis of Rock Slopes with Three-Dimensional Laser Scanning. In *Transportation Research Record: Journal of the Transportation Research Board, No. 1913*, Transportation Research Board of the National Academies, Washington, D.C., pp. 187–194.

Smith, G. W. 2002. High Speed CCD Camera Technology. In *High Speed Photography and Photonics*, SPIE Press Monograph PM120, International Society for Optical Engineering, Bellingham, Wash., Chap. 5, pp. 81–98.

Sturzenegger, M., and D. Stead. 2009. Close-Range Terrestrial Digital Photogrammetry and Terrestrial Laser Scanning for Discontinuity Characterization on Rock Cuts. *Engineering Geology*, Vol. 106, Issues 3–4, pp. 163–182.

Sturzenegger, M., M. Yan, D. Stead, and D. Elmo. 2007. Applications and Limitations of Ground-based Laser Scanning in Rock Slope Characterization. In *Rock Mechanics: Meeting Society's Challenges and Demands—Proc., 1st Canada–U.S. Rock Mechanics Symposium*, Vancouver, Canada (E. Eberhardt, D. Stead, and T. Morrison, eds.), Taylor & Francis, London, Vol. 1, pp. 29–36.

Swiss Federal Research Institute. 2006. *Rockfall Test with Remote Monitoring System "Sentinel."* Birmensdorf, Switzerland, 2 pp. http://www.impact-sentinel.com/WSL_Zeugnis_E.pdf.

Tarchi, D., N. Casagli, S. Moretti, D. Leva, and A. Sieber. 2003. Monitoring Landslide Displacements by Using Ground-based Synthetic

Aperture Radar Interferometry: Application to the Ruinon Landslide in the Italian Alps. *Journal of Geophysical Research*, Vol. 108, No. B8, pp. ETG10.1–ETG10.14.

Tatchell, G. E. 1991. Automatic Data Acquisition Systems for Monitoring Dams and Landslides. In *Field Measurements in Geotechnics: Proceedings of the Third International Symposium on Field Measurements in Geomechanics* (G. Sørum, ed.), Oslo, Norway, Sept. 9–11, A. A. Balkema, Rotterdam, Netherlands, Vol. 1, pp. 349–360.

Terrasource Software. 2009. *Sirovision*. http://www.terrasourceonline.com/Sirovision/Sirovision.htm. Accessed Dec. 2009.

Tonon, F., and J. T. Kottenstette (eds.). 2006. *Laser and Photogrammetric Methods for Rock Face Characterization*. In *Report on a Workshop Held June 17–18, 2006 in Golden, Colorado*, American Rock Mechanics Association (ARMA). http://www.usbr.gov/pmts/geology/goldenrocks.pdf.

Turner, A. K., J. Kemeny, S. Slob, and R. Hack. 2006. Evaluation and Management of Unstable Rock Slopes by 3-D Laser Scanning. In *Proc., 10th International Congress, International Association for Engineering Geology and the Environment (IAEG)*, Nottingham, UK, Sept. 6–10, Geological Society of London, Paper 404 (CD-ROM).

University of Queensland. 2002. *Slope Stability Radar Goes Commercial*. News Release. University of Queensland, Australia, June 26. http://www.uq.edu.au/news/index.phtml?article=3279. Accessed Dec. 22, 2009.

U.S. Department of Defense. 2008. *Global Positioning System Standard Positioning Service Performance Standard*, 4th ed. Sept., 160 pp. http://www.navcen.uscg.gov/gps/geninfo/2008SPSPerformanceStandardFINAL.pdf.

U.S. National Technical Information Service. 2008. *Federal Radionavigation Plan*. DOT-VNTSC-RITA-0802/DoD-4650.5. Springfield, Va. http://www.navcen.uscg.gov/pubs/frp2008/2008_Federal_Radionavigation_Plan.pdf.

Weill, L. R. 1997. Conquering Multipath: The GPS Accuracy Battle. *GPS World*, Vol. 8, No. 4, pp. 59–66.

Whitworth, M. Z., D. Giles, I. Anderson, and M. Clewett. 2006. Terrestrial Laser Scanning for Applied Geoscience Studies in the Urban Environment. In *Proc., 10th International Congress, International Association for Engineering Geology and the Environment (IAEG)*, Nottingham, UK, Sept. 6–10, Geological Society of London, Paper 252 (CD-ROM).

Wickens, E. H., and N. R. Barton. 1971. Application of Photogrammetry to the Stability of the Excavated Rock Slopes. *Photogrammetric Record*, Vol. 7, pp. 46–54.

Wilson, S. D., and P. E. Mikkelsen. 1978. Field Instrumentation. In *Special Report 176: Landslides: Analysis and Control* (R. L. Schuster and R. J. Krizek, eds.), Transportation Research Board, National Research Council, Washington, D.C., pp. 112–138.

Xcitex, Inc. 2010. *ProAnalyst Brochure*. Cambridge, Mass., 6 pp. http://www.xcitex.com/downloads/ProAnalystBrochure.pdf. Accessed Aug. 10, 2010.

Xu, G., and Z. Zhang. 1996. *Epipolar Geometry in Stereo, Motion and Object Recognition: A Unified Approach*. Computational Imaging and Vision Series, Vol. 6, Springer-Verlag, Berlin.

Zhang, Z. 1998. Determining the Epipolar Geometry and Its Uncertainty: A Review. *International Journal of Computer Vision*, Vol. 27, No. 2, pp. 161–198.

Zhang, Z., S. Zhenga, and Z. Zhana. 2004. Digital Terrestrial Photogrammetry with Photo Total Station. *International Archives of the Photogrammetry, Remote Sensing and Spatial Information Sciences*, Vol. 35, Part B5.

Chapter 8

A. KEITH TURNER AND
JOHN D. DUFFY

EVALUATION OF ROCKFALL MECHANICS

1. INTRODUCTION

Rockfall occurs along both natural rock slopes and constructed rock cuts. A single rockfall usually involves a series of events (Figure 8-1):

1. Detachment of rock fragments from a rock face by toppling, sliding, erosion, or a combination of triggering factors;
2. Free, nearly vertical, fall of a single boulder or rock mass;
3. Impact at the base of the rock slope;
4. Bouncing, rolling, and sliding of the rock fragments along the surface of a talus slope, if present; and
5. Deceleration and stoppage of the rock fragments on a gently inclined valley bottom or (as shown in Figure 8-1) their catchment by a protecting structure such as a rockfall fence, net, gallery, or earth embankment barrier.

Several approaches are used in gathering information and analyzing rockfall trajectories. Quantitative analysis of block trajectories is based on the fundamental principles of Newtonian mechanics, which are summarized in Section 2 of this chapter. Sections 3 and 4 describe the application of these principles to the analysis of collisions of particles and the evaluation of bouncing and rolling processes. As discussed in Section 5, the calibration and verification of parameters and equations derived from Newtonian principles are an important step before acceptable and defendable predictions with regard to rockfall behavior and associated hazards that may be present at any given location can be presented. This step can be accomplished through backanalysis of natural rockfall events, but often these data are further analyzed and verified by field and laboratory tests. Chapter 10 describes the conduct of field tests to evaluate rockfall phenomena. Numerical models described in Chapter 9 are based on the application of the principles of Newtonian mechanics, and provided that they are accurately calibrated by backanalysis of historical rockfall events or by field tests as described in Chapter 10, they are a valuable tool in the quantitative evaluation of potential trajectories.

Four mechanisms of block movement are usually distinguished: free falling, bouncing, rolling, and sliding. The free-fall trajectory is a parabola (Ritchie 1963). The bouncing and rolling that occur after a falling block strikes a slope surface are undoubtedly the least understood and the most enigmatic portion of a trajectory. Bouncing is typically defined in a simplified way by using one or two coefficients of restitution (Wu 1985; Hungr and Evans 1988; Pfeiffer and Bowen 1989; Giani 1992; Azzoni and de Freitas 1995; Heidenreich 2004).

As discussed in Section 3.3, various investigators have defined coefficients of restitution in terms of velocities or energies or have invoked

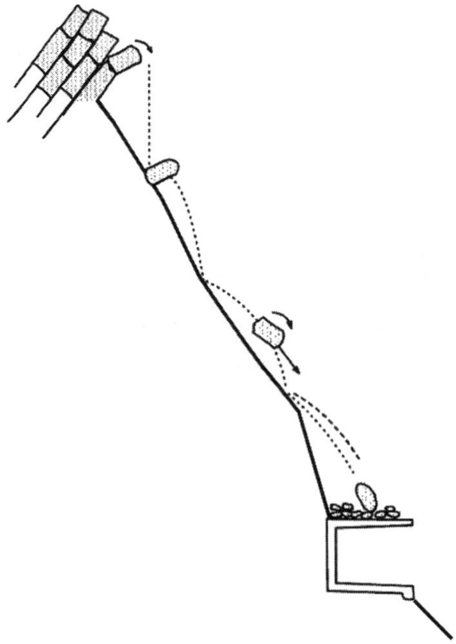

FIGURE 8-1
Typical rockfall trajectory (Heidenreich 2004).

impulse concepts. Each method yields a different value for the coefficients, and this, in turn, makes the application of published values for defining these coefficients within numerical models a complex and problematic process. Unrealistic predictions will result if the definitions of the coefficients used by the models do not match the definitions used in the published sources.

Collisions of particles are typically evaluated according to the principles of Newtonian mechanics by assuming a constant value for the coefficients of restitution regardless of the impact energy. For impacts of boulders on hard-rock slopes, this approach may be justified as long as fragmentation of the block is ignored. But for impacts of blocks onto granular materials found on the surface of most talus slopes below rockfall-prone areas, this assumption does not match reality (Heidenreich 2004; Labiouse and Heidenreich 2009). As discussed in Section 4.6, plastic deformation processes develop as the boulder penetrates the ground surface. Therefore, the energy losses, and thus the coefficients of restitution, depend on the block kinematics.

Rockfalls commonly accumulate to form talus slopes in mountainous areas. As a consequence, much geomorphic research has been devoted to the analysis of talus slopes, also known as *scree* slopes (Behre 1933; Chandler 1973; Kirkby and Statham 1975; Statham 1976, 1979; Statham and Francis 1986; Hétu and Gray 2000). Several of these studies provide valuable field-based observations that verify the results obtained by theoretical quantitative analysis of rockfall trajectories. Section 5.1 summarizes these aspects.

Prediction of the distance that rockfall material will travel beyond the base of talus slopes, termed the *runout distance*, for a range of return periods is important in assessing rockfall hazard and risk. This knowledge is mandated as part of the land use planning process in many countries and jurisdictions. The delineation of rockfall runout areas often involves the mapping of rockfall debris in the field or the use of aerial photography (Figure 8-2). As described in Chapter 9, numerous field observations have permitted the development of empirical relationships describing rockfall runout; numerical modeling based on these empirical relationships is used as part of rockfall hazard and risk assessments.

FIGURE 8-2
Delineation of rockfall runout area in Chilean Andes (Holm and Jakob 2009).

2. PRINCIPLES AND DEFINITIONS OF MASS, ENERGY, AND MOTION

The behavior of falling and striking rock blocks is governed by Newton's laws of motion, which are as follows:

1. Every body continues in its state of rest or of uniform motion in a straight line unless compelled by an external force to do otherwise.
2. The rate of change of momentum of a body is proportional to the applied force and takes place in the direction in which the force acts.
3. To every action there is an equal and opposite reaction.

These laws underpin all classical mechanics and form a common basis for the analytical studies of rockfall phenomena. For example, the following fundamental equations describing projectile motion (with motion oriented along the x-axis) are used to calculate the velocities of falling rock blocks:

$$x = x_0 + v_0 t \tag{1}$$

$$v = v_0 + at \tag{2}$$

$$x = x_0 + v_0 t + \frac{1}{2} a t^2 \tag{3}$$

$$v^2 = v_0^2 + 2a(x - x_0) \tag{4}$$

where

v_0 = initial velocity,
x_0 = initial position,
v = velocity at time t,
x = position at time t,
a = acceleration, and
t = time.

Equation 4 can be derived from Equations 2 and 3, as shown in Box 8-1.

2.1 Mass and Weight

The *mass* of an object is a fundamental property of the object; it is a fundamental numerical measure of inertia and the amount of matter in the object. In contrast, *weight* is a measurement of the gravitational force acting on an object. In modern scientific usage, weight and mass are fundamentally different quantities: mass is an intrinsic property defining the amount of matter contained in an object, whereas weight is a force that results from the action of gravity on an object. Weight thus measures how strongly the mass of the Earth attracts an object. Near the surface of the Earth, the acceleration due to gravity (g) is approximately constant; this means that an object's weight is roughly proportional to its mass (Marion and Thornton 1995).

However, the recognition of the difference between weight and mass is, historically, a relatively recent development, and in many everyday situations "weight" continues to be used when "mass" is meant. For example, it is often stated that an object weighs a certain number of kilograms, even though the kilogram is a unit of mass. Systems of units of weight (force) and mass have a tangled history, partly because the distinction was not properly understood when many of the units first came into use.

Box 8-1

Derivation of Equation 4: Defining Velocity in Terms of Acceleration and Displacement

Equation 4 has the form $v^2 = v_0^2 + 2a\,(x - x_0)$. It may be derived from Equations 2 and 3 as follows:

Step 1. Equation 2, with the form $v = v_0 + at$, should be squared, yielding

$$v^2 = v_0^2 + 2av_0t + a^2t^2$$

Step 2. Equation 3, with the form $x = x_0 + v_0t + (1/2)at^2$, may be used to provide an expression for t^2 by rearranging terms:

$$x - x_0 - v_0t = \frac{1}{2}at^2$$

Thus,

$$t^2 = 2\left(\frac{x - x_0 - v_0t}{a}\right)$$

Step 3. Substituting this expression for t^2 into the equation developed in Step 1 yields

$$v^2 = v_0^2 + 2av_0t + 2a^2\left(\frac{x - x_0 - v_0t}{a}\right)$$

Simplifying terms produces

$$v^2 = v_0^2 + 2av_0t + 2a(x - x_0) - 2av_0t$$

$$v^2 = v_0^2 + 2a(x - x_0)$$

2.1.1 International System of Units

In most modern scientific work, physical quantities are measured in the International System of Units (SI). The *kilogram* (kg) is the SI unit of mass. The associated SI unit of force (and hence weight in the mechanics sense) is the *newton* (N), with 1 kg weighing 9.8 N under standard conditions on Earth's surface. The newton can be expressed in SI base units as kg × m/s^2 (kilograms times meters per second squared). However, confusion is sometimes introduced when a non-SI unit of force, the *kilogram-force*, is used in analysis. The kilogram-force (usually defined by the symbol kg_F) is defined as the force exerted by a 1-kg mass in standard Earth gravity (in other words the weight of a 1-kg mass). It equals exactly 9.80665 N. The use of kilogram-force should be avoided.

2.1.2 Customary U.S. Units

In the United States, the *pound* (lb) is the unit of force (and therefore weight). The pound is widely used in commerce. The use of the pound as a weight constrains the mass unit to an inconveniently large unit called a *slug*, defined as the amount of mass that accelerates at 1 ft/s^2 when 1 lb of force is exerted on it. A slug is thus a mass that weighs about 32 lb.

Unfortunately, many analyses use the word "pound" to define both a force (i.e., a weight) and a mass. Such dual use of "pound" causes much confusion and should be avoided; however, when used in this fashion, the term *pound-force* (usually defined with the symbol lb_F) should be clearly distinguished from the term *pound-mass* (usually defined with the symbol lb_M). The slug is inconveniently large for many applications; thus, the use of the slug is discouraged, and the exclusive use of the SI system for scientific work is strongly encouraged.

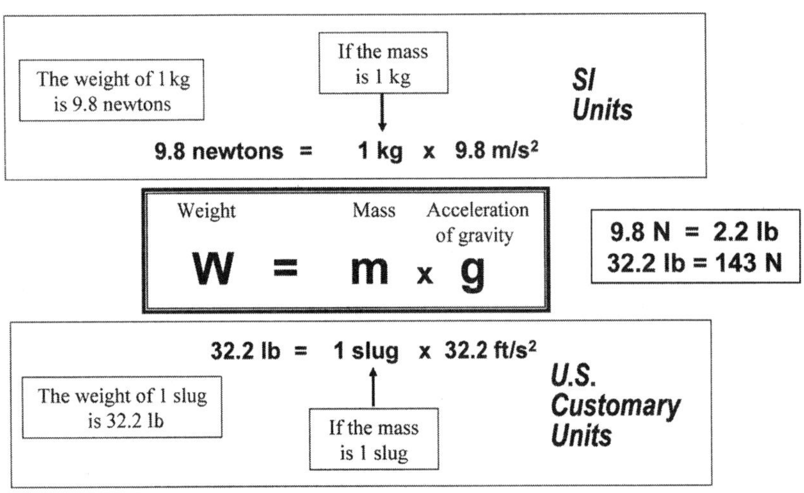

FIGURE 8-3
Weight and mass in SI and U.S. customary units.

2.1.3 Conversion Between Weight and Mass: Gravitational Acceleration

To convert between weight (force) and mass, Newton's second law is used:

$$F = ma \qquad (5)$$

where

F = force (weight) due to gravity,
m = mass of the object in question, and
a = acceleration due to gravity.

In this context the same equation is often written as $W = mg$, with W standing for weight and g for the acceleration due to gravity. Because the gravitational force (g) exerted on an object is proportional to the mass of the object, the strength of gravity can be defined in terms of force per unit mass, that is, *newtons per kilogram* (N/kg). However, the unit N/kg resolves to m/s^2 (meters per second per second), which is the SI unit of acceleration, and in practice gravitational strength, identified by g, is usually quoted as acceleration. On Earth, g is approximately 9.8 m/s^2. In customary U.S. units, as defined in Section 2.1.2, g = 32.2 ft/s^2. Figure 8-3 illustrates the relationships between weight and mass for both SI and U.S. customary units.

2.2 Velocity

Velocity is defined as the rate of change of position. *Speed* is the scalar absolute value of velocity. The rate of change of velocity is referred to as *acceleration*. There are two types of velocity: (*a*) *translational velocity*, v = distance/time, and (*b*) *angular velocity*, ω = radians/time. Both types must be quantified to determine the total kinetic energy of a bouncing and rolling rock.

Translational velocity (v) is the velocity of the center of mass of an object (such as a rock). It is a vector; both speed and direction are required to define it. In SI units, translational velocity is measured in meters per second: m/s or ms^{-1}. The average velocity of an object moving through a displacement Δx during a time interval Δt is described by the following formula:

$$\bar{v} = \frac{\Delta x}{\Delta t} \tag{6}$$

Angular velocity (ω) is the rotational velocity of a rock mass spinning around its center of mass. Angular velocity may be determined by measuring the time a rock takes to complete one revolution (360° or π radians). While this can be fairly easily accomplished in a two-dimensional analysis, with rotation occurring about a single axis, the observation and analysis become much more complex for three-dimensional situations. In three dimensions, several rotations can be under way simultaneously, each about a different principal axis. Such analyses have generally not been attempted in supporting the investigation of rockfall events.

2.3 Moment of Inertia

The *moment of inertia* is the rotational analog of mass; it plays much the same role in rotational dynamics as does mass in basic dynamics. The moment of inertia is used to define relationships between angular momentum and angular velocity, torque, angular acceleration, and several other quantities.

The moment of inertia of an object about a given axis quantifies the difficulty in changing the object's angular motion about that axis (Marion and Thornton 1995). Figure 8-4 illustrates this situation for two spinning disks (A and B). Both have the same mass, but Disk B has a larger radius than Disk A. Each disk has a uniform thickness and mass distribution and is rotating with the same number of revolutions per minute. Under these conditions, more effort will be required to accelerate (or decelerate) Disk B (change its angular velocity) than Disk A, because the mass of Disk B is distributed farther from its axis of rotation. A mass that is farther from the axis of rotation must, for a given angular velocity, move more quickly than a mass closer in; thus Disk B has a larger moment of inertia than does Disk A. This phenomenon is often observed in figure skating. A skater begins to spin with arms extended, and when the arms are pulled in close to the body, the rate of spin increases.

But rock blocks may not be as uniform in their mass distributions as the disks (or skaters) just described. The moment of inertia of a given rock block about a particular axis of rotation depends

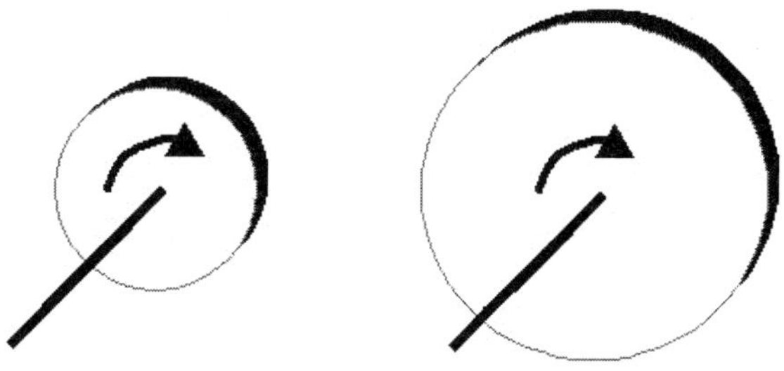

FIGURE 8-4 Moment of inertia for Disk A is less than moment of inertia for larger Disk B.

not only on the block's total mass but also on how the mass is distributed about the axis. In most field-based rock-rolling investigations, the rock blocks are considered to be homogeneous solids, so their center of mass is considered to be located at the intersection of the three principal axes used to define the block size and shape. The blocks are then assessed as appropriate idealized rectangular (cubical) or spherical bodies, as illustrated in Figure 8-5 and discussed below. As described in Chapter 10, the appropriate axes of rotation are determined by observing high-speed video recordings of field-based experimental trials (previously, high-speed movie films were used for this purpose). Field experiments show that, in many cases, rocks revolve around their longest axis while traveling approximately the first 45 m (150 ft). Then, as velocity increases, the rocks begin to revolve around their shortest axis like a bicycle wheel (Smith and Duffy 1990; Azzoni et al. 1995). It follows that, in this case, discoidal or tabular blocks basically behave like spheroidal blocks.

The symbol I is usually used to refer to the moment of inertia, and moments of inertia have units of dimension mass × length^2; thus the SI units are kg-m^2, while the customary U.S. units are slug-ft^2. On the basis of dimensional analysis alone, the moment of inertia of a nonpoint object must take the following form:

$$I = kmr^2 \tag{7}$$

where

I = moment of inertia,
k = dimensionless constant called the *inertia constant* that varies with the object under consideration,
m = mass of the object, and
r = radius of the object from the center of mass.

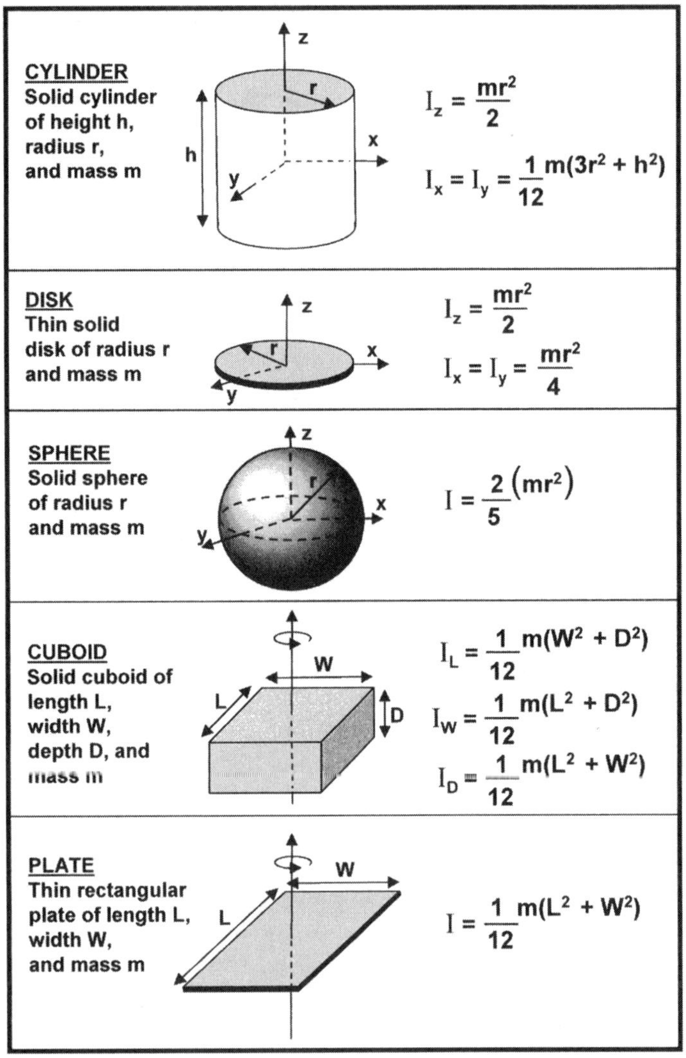

FIGURE 8-5
Moments of inertia for common solid objects.

The following are examples of k for common rock block shapes:

- $k = 1$, thin ring or thin-walled cylinder rotating around its center;
- $k = 2/5$, solid sphere rotating around its center; and
- $k = 1/2$, solid cylinder or disk rotating around its center.

Figure 8-5 provides additional details of the values of moment of inertia for commonly considered simple shapes.

2.4 Momentum

In classical mechanics, *momentum* is the product of the mass and the velocity of an object according to the following simple formula:

$$p = mv \tag{8}$$

where

p = momentum of the object,
m = mass of the object, and
v = velocity of the object.

Momentum is measured in SI units as kilogram-meters per second (kg-m/s) or as newtons per second (N/s).

2.4.1 Linear and Angular Momentum

The momentum of a nonrotating rigid body moving in a straight line is sometimes referred to as *linear momentum* to distinguish it from angular momentum, which is related to the rate of "spin" of rotating bodies. Linear momentum is a vector quantity, since it has a direction as well as a magnitude, whereas angular momentum is a pseudovector quantity, because it may undergo an additional sign reversal under an improper rotation.

2.4.2 Conservation of Momentum

Momentum is a conserved quantity, which means that the total momentum of any closed system (one not affected by external forces) cannot change. According to Newton's second law, the rate of change of momentum (acceleration) of a particle is proportional to the resultant force acting on the particle in the direction of that force.

2.5 Potential and Kinetic Energy Concepts

Energy can be defined as the capacity for doing work. Energy is measured in joules (J) in the SI system. It may exist in a variety of forms and may be transformed from one type of energy to another. However, these energy transformations are constrained by the fundamental conservation of energy principle, discussed in Section 2.6.

2.5.1 Potential Energy

Potential energy is stored energy held within the gravitational field. The lifting of a heavy object to a greater elevation requires the exertion of energy, which later will become kinetic energy when the object is dropped. The higher the object is lifted, the more potential energy it contains. Potential energy exists whenever an object with mass has a position within a force field—the universal situation on Earth is that all objects are in Earth's gravitational field. In this case, the potential energy of an object is given by the following:

$$PE = mgh \tag{9}$$

where PE is potential energy (J), h is height above Earth's surface (m), and other terms are as previously defined.

2.5.2 Kinetic Energy of Nonrotating Rigid Bodies

Kinetic energy is energy of motion. The word *kinetic* is derived from the Greek word meaning *to move*. The greater the mass of an object and the faster the object moves, the more kinetic energy it possesses. As an object falls, its potential energy is converted into kinetic energy.

In classical mechanics, the kinetic energy of a *point object* (a body so small that its size can be ignored) or a nonrotating rigid body is given by the equation describing its *translational kinetic energy:*

$$KE_T = \frac{1}{2}mv^2 \tag{10}$$

where KE_T is translational kinetic energy (J) and other terms are as previously defined. For example, the kinetic energy of an 80-kg mass traveling at 18 m/s (40 mph) is ½ × 80 × 18² = 12,960 J.

Note that the kinetic energy increases with the square of the velocity. This means, for example, that doubling an object's velocity will quadruple its kinetic energy. As a consequence, a rock (or any object) traveling twice as fast as another of the same mass requires four times as much distance to stop, if an equal stopping force is assumed.

The translational kinetic energy of an object is related to its momentum by Equation 11:

$$KE_T = \frac{p^2}{2m} \tag{11}$$

where all terms are as previously defined.

2.5.3 Kinetic Energy of Rotating Bodies

If a rigid body is rotating about any line through the center of mass, it has *rotational kinetic energy* (KE_R), which is simply the sum of the kinetic energies of its moving parts. Thus the rotational kinetic energy is equal to

$$KE_R = \int \frac{v^2 dm}{2} = \int \frac{(r\omega)^2 dm}{2} = \frac{\omega^2}{2}\int r^2 dm = \frac{\omega^2}{2}I = \frac{1}{2}I\omega^2 \tag{12}$$

where

KE_R = rotational kinetic energy (J),
ω = body's angular velocity,
r = distance of any small mass (dm) from that line, and
I = body's moment of inertia = $\int r^2$ dm.

In Equation 12, the moment of inertia must be computed about an axis through the center of mass, and the rotation measured by ω must be around that axis. More general equations exist for systems in which the object is subject to wobble due to its eccentric shape or for situations where the rotations do not occur about axes through the center of mass (Bedford and Fowler 1995). Such situations are generally not considered during quantitative analysis of rockfall events and thus are not discussed further.

2.5.4 Total Kinetic Energy

When a body is both moving laterally (translating) and rotating, its *total kinetic energy* is the sum of the body's center-of-mass translational kinetic energy and the rotational kinetic energy resulting from rotation around the center of mass:

$$KE = KE_T + KE_R \tag{13}$$

where KE is total kinetic energy and other terms are as previously defined.

Thus, the kinetic energy of a bouncing, rolling boulder is the sum of its translational kinetic energy (KE_T) and its rotational kinetic energy (KE_R). The total kinetic energy KE may be expressed mathematically by combining Equations 10 and 12:

$$KE = \frac{1}{2}mv^2 + \frac{1}{2}I\omega^2 \tag{14}$$

where all terms are as previously defined.

2.6 Conservation of Energy Principle

The conservation of energy principle constrains the actions that occur as one form of energy is transformed into another (Marion and Thornton 1995). The principle may be stated as "energy can neither be created nor destroyed." It has particular relevance to three aspects of rockfall analysis:

- The computation of the trajectories of free-falling rock blocks,
- The analysis of collisions and impacts of rock blocks, and
- The analysis of the mechanics of bouncing and rolling rocks.

The application of the conservation of energy principle to the computation of trajectories is discussed in the next paragraph; the applications of the conservation of energy principle to the mechanics of collisions and impacts of rock blocks and of bouncing and rolling rocks are discussed in Sections 3 and 4.

When a rock (or any object) experiences free fall, its initial potential energy is progressively converted into kinetic energy. If the effects of air resistance and aerodynamic lift are neglected (both are extremely minor influences on falling rocks), the vertical velocity can be computed as shown in Figure 8-6. This velocity is directly related to the height of the fall and is independent of the mass of the rock. The principle of energy conservation also means that any initial horizontal velocity imparted to the rock at its launch point will remain unchanged during the free fall. Thus, the trajectories of free-falling rocks will conform to a parabola.

FIGURE 8-6 Conservation of energy applied to free-falling rocks.

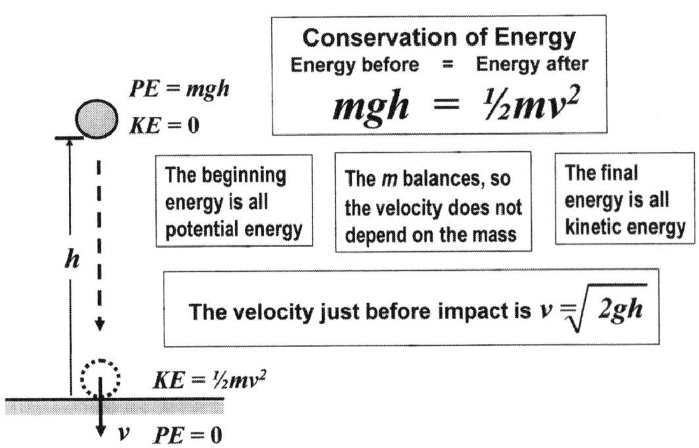

3. ANALYZING COLLISIONS AND IMPACTS

When two or more objects travel along intersecting paths, they will collide with or have an impact on each other. The magnitude of the velocity difference at impact is called the *closing speed*. Collisions or impacts exert strong forces for short periods of time that result in changes in the velocities of the objects. Collisions or impacts can be *elastic*, meaning that they conserve energy and momentum; *inelastic*, meaning that they conserve momentum but not energy; or *plastic*, meaning that they conserve momentum and the two objects stick together.

3.1 Analyzing Collisions

A *collision* is an isolated event in which two or more colliding bodies exert strong forces on each other that result in changes in velocity for each of the colliding objects. Since velocity is a vector, movement in one direction must be labeled "positive" and in the opposite direction "negative." The principle of conservation of momentum applies during a collision. Thus, the final velocities depend on the masses of the two objects as well as on their initial velocities and the elasticity of the collision. For an elastic collision, the conservation of momentum may be stated as

$$m_1v_1 + m_2v_2 = m_1v_{1f} + m_2v_{2f} \tag{15}$$

and the conservation of translational kinetic energy may be stated as

$$\frac{1}{2}m_1v_1^2 + \frac{1}{2}m_2v_2^2 = \frac{1}{2}m_1v_{1f}^2 + \frac{1}{2}m_2v_{2f}^2 \tag{16}$$

where

m_1 = mass of the first object,
m_2 = mass of the second object,
v_{1f} = scalar final velocity of the first object after impact,
v_{2f} = scalar final velocity of the second object after impact,
v_1 = scalar initial velocity of the first object before impact, and
v_2 = scalar initial velocity of the second object before impact.

After combination of Equations 15 and 16, the definition of the postcollision velocities of

Objects 1 and 2 with different masses m_1 and m_2 becomes

$$v_{1f} = \frac{v_1(m_1 - m_2) + 2m_2v_2}{m_1 + m_2} \quad (17)$$

and

$$v_{2f} = \frac{v_2(m_2 - m_1) + 2m_1v_1}{m_1 + m_2} \quad (18)$$

Figure 8-7 shows the consequences of collisions between two objects under several idealized circumstances.

To analyze a collision of two objects in two dimensions, the overall velocity of each object must be split into two perpendicular velocities: one tangent to the common normal surfaces of the colliding bodies at the point of contact and the other along the line of collision. Figure 8-8 illustrates the process for an elastic collision between Object 1 moving at velocity v in the x-direction and a stationary Object 2, with the resulting tangential velocity vector v_T and normal velocity vector v_N, as shown in Figure 8-8. After the collision, at time t_2, the objects will have moved along these vectors to the positions shown with the dashed circles.

The same process can be followed for an inelastic collision. If the collision is frictionless at the point of contact, there is no conversion of translational kinetic energy to rotational kinetic energy, and the two bodies will not spin. In an inelastic collision, some energy is dissipated (as heat, as vibrations of the objects causing noise, or as a permanent deformation of either or both objects), and the collision imparts a smaller force along the line of collision. The postcollision tangential velocity v_T remains unchanged, but the postcollision normal velocity v_N must be adjusted to account for the reduction in translational kinetic energy.

Figure 8-9 compares the normal and tangential pre- and postcollision velocities for two colliding spheres labeled M_i and M_j. Figure 8-9*a* illustrates

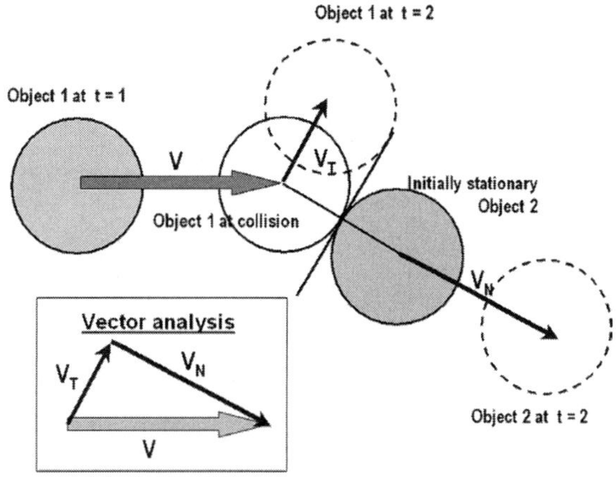

FIGURE 8-8
Two-dimensional analysis of elastic collision of two circular objects.

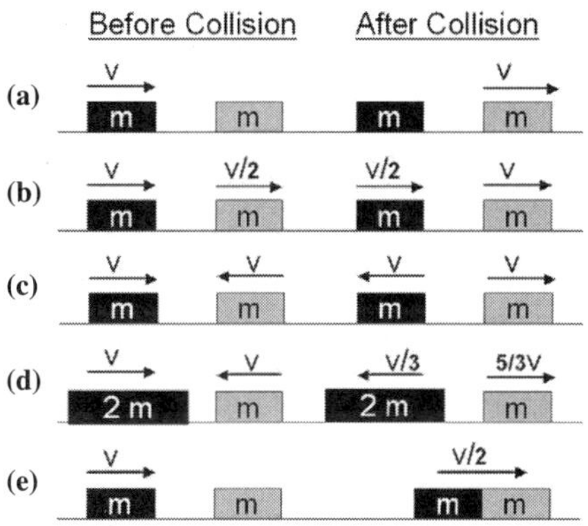

FIGURE 8-7
Idealized collisions in one dimension between two objects showing conservation of momentum principle: (*a*) elastic collision of equal masses; (*b*) elastic collision of equal masses within moving frame of reference; (*c*) elastic collision of equal masses moving with equal and opposite velocities; (*d*) elastic collision of unequal masses, with postcollision velocities computed by using Equations 17 and 18; and (*e*) plastic collision of equal masses.

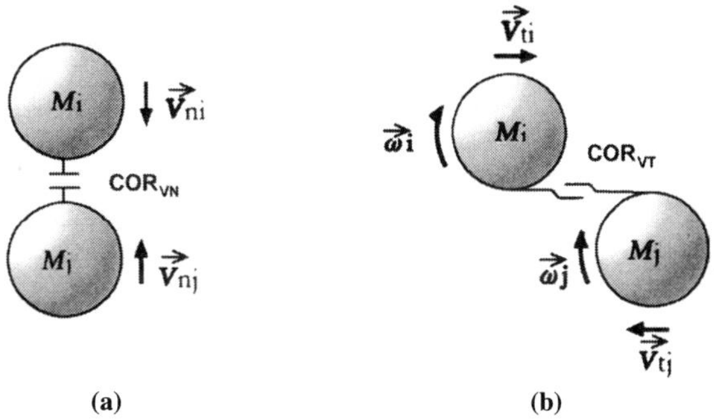

FIGURE 8-9
Rigid and frictional collision model for two particles (*i* and *j*): (*a*) velocity change in normal direction component transmits through normal coefficient of restitution (COR_{VN}); (*b*) velocity change in tangential direction component and angular velocity change transmit through tangential coefficient of restitution (COR_{VT}) (adapted from Okura, Kitahara, and Sammori 2000 and Okura, Kitahara, et al. 2000).

the normal component of the collision. The difference between the summation of the precollision velocity vectors V_{ni} and V_{nj} and the summation of the postcollision rebounding velocity vectors defines the *normal coefficient of restitution* (COR_{VN}), as discussed in Section 3.3.1. Figure 8-9*b* illustrates the tangential component of a collision that has a significant frictional surface component at the point of impact. If the collision is inelastic, additional calculations are required to adjust the tangential translational velocity vectors V_{ti} and V_{tj}. Several numerical rockfall analysis procedures use two parameters to account for the normal and tangential energy reductions in inelastic collisions. In addition, when there is a frictional surface component at the point of impact, a portion of the translational kinetic energy KE_T will be converted to rotational kinetic energy KE_R, and the two objects will spin. If the objects were already spinning, their rates of rotation will be modified. The pre- and postcollision rotations are related by the *tangential coefficient of restitution* (COR_{VT}), as discussed in Section 3.3.1.

By adopting the rigid and frictional collision model shown in Figure 8-9, Okura, Kitahara, and Sammori (2000) identified the responses of COR_{VT} to changes in both the *coefficient of kinetic (or sliding) friction* (defined as the ratio of the force required to induce sliding of an object and the weight of the object) and changes in the tangential and rotational velocities before and after the collision. These changes vary according to the frictional and strength properties of the colliding objects.

Figure 8-10*a* shows four conditions of COR_{VT} for impacts of a rock block on a surface. If the surface is very soft and easily deformed (Case 1 in Figure 8-10*a*), the impact will be essentially plastic, and the block will slide across the surface with little or no change in the tangential or rotational velocities. According to the definition used by Okura, Kitahara, and Sammori (2000), COR_{VT} will have a value of −1.0 under these conditions.

As the surface becomes harder and less deformable (Cases 2 and 3 in Figure 8-10*a*), the impact becomes progressively less inelastic, and the rotational velocities increase while the tangential velocities and the amount of sliding decrease. The value of COR_{VT} also increases through zero (Case 2) to positive values (Case 3). The theoretical

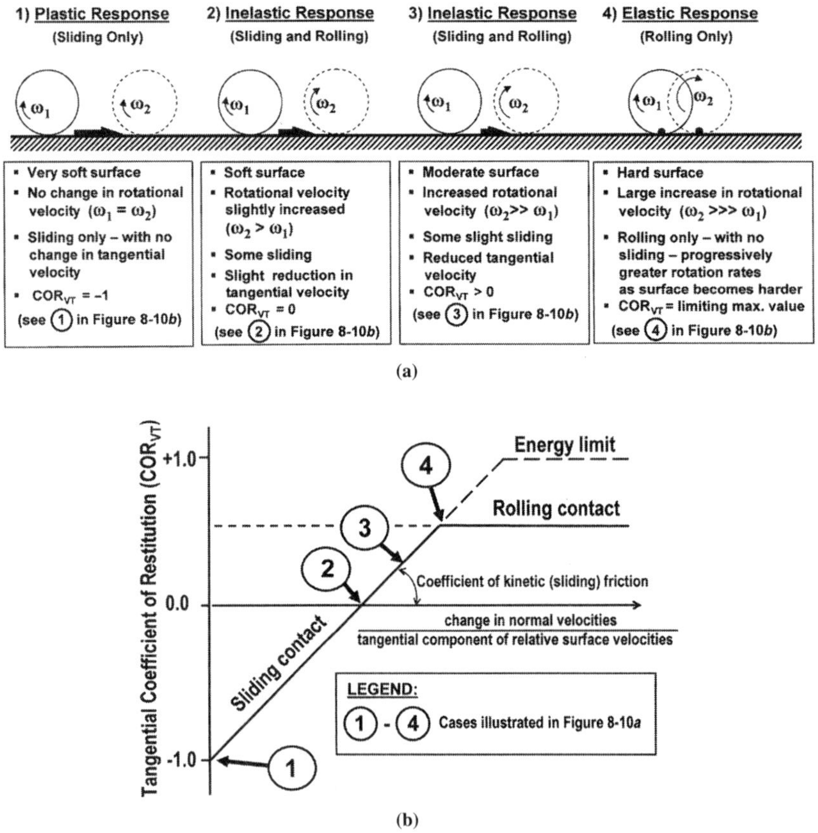

FIGURE 8-10 Changes in tangential coefficient of restitution reflect values of coefficient of kinetic (sliding) friction and surface characteristics that influence sliding and rolling responses to impacts: (*a*) four cases of rigid and frictional impacts of rock block on surface and (*b*) variation of tangential coefficient of restitution (COR_{VT}) (adapted from Okura, Kitahara, and Sammori 2000).

maximum value of COR_{VT} is +1.0, but this value is never reached in reality because impacts are never perfectly elastic. Instead, COR_{VT} reaches a stable value, and the block rolls along the surface without slippage at its point of contact (Case 4 in Figure 8-10*a*).

Okura, Kitahara, and Sammori (2000) summarized these relationships with a diagram reproduced as Figure 8-10*b*. The locations of the four cases in Figure 8-10*a* have been added to Figure 8-10*b*. The horizontal axis in Figure 8-10*b* defines the ratio between changes in the velocities normal to the impact (reflecting the elasticity or hardness of the surface) and changes in the tangential relative surface velocities (both rotational and tangential). These ratios reflect the deformability of the surface and its frictional properties, which influence the amount of slippage between the block and the surface.

These collision concepts are used to evaluate rolling and bouncing of a rock block as it continues to travel down a slope (Chau et al. 2002). Section 4 discusses these applications further. Chapter 9 describes the approaches of several commonly used rockfall simulation programs. Typically, additional conversion parameters are applied to account for the velocity transmissions at each impact. To date, little quantitative research has been conducted on the interactions and collisions among multiple rocks as they fall, bounce, or roll down a slope (Okura, Kitahara, and Sammori 2000; Okura, Kitahara, et al. 2000). Section 5.3 summarizes the results of the experiments reported by Okura, Kitahara, and Sammori (2000) and by Okura, Kitahara, et al. (2000), which validate these theoretical concepts and provide practical applications to rockfall mitigation design.

3.2 Analyzing Impacts

An *impact* occurs when a free-falling object, or a projectile, encounters a large immovable obstacle. An impact is thus a special case of a collision between two objects, and the same basic energy and momentum concepts apply. An important impact event for rockfall analysis is the impact of a falling rock block on a lower portion of the slope (Figure 8-11).

Impacts have been extensively studied (Goldsmith 1960; Stronge 2000), and the characteristics of bouncing balls have been the subject of considerable scientific investigation (Cross 1999, 2000). Figure 8-12 illustrates the path of a bouncing ball captured with a stroboscopic flash unit providing 25 images per second. Furthermore, because the velocities of bouncing objects are directly related to the height of the bounce, their velocities can be determined by measuring the heights of successive bounces; for example, the velocity just before impact is $v_N = \sqrt{2gh_N}$. Because it is frequently difficult to determine bounce heights accurately during experiments, or even from examination of images such as Figure 8-12, several experiments have used measurements of the time of flight between successive bounces and the relationship between velocity and time of flight:

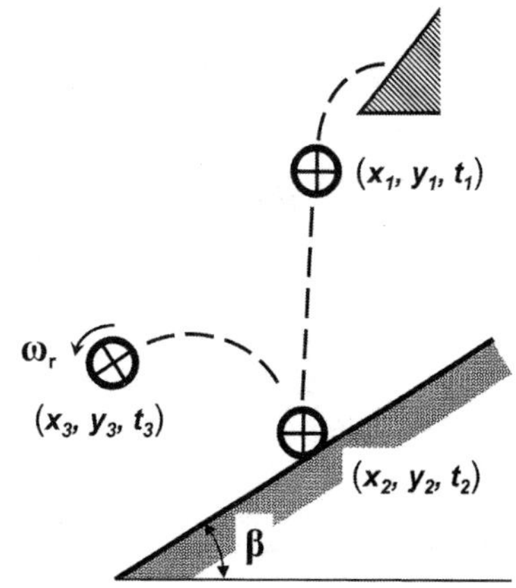

FIGURE 8-11
At least three sets of position and time observations are required to calculate coefficient of restitution of rock impact on lower slope. Lower slope angle is denoted by β and angular rotation by ω_r (Chau et al. 2002).

$$v_N = \frac{1}{2} g T_N \qquad (19)$$

where

v_N = velocity of bounce N,
g = acceleration due to gravity, and
T_N = time of flight of bounce N.

Figure 8-13 illustrates these relationships.

Time-of-flight measurements are readily obtained by using the microphone and sound card of a typical personal computer and processing the data with the sound-recorder program that comes with the Windows operating system (Aguiar and Laudares 2003). Current sound-recording techniques are capable of measuring time intervals with a resolution of 45 µs. This technology makes experimental procedures that originally required more expensive and customized equipment much easier and cheaper to conduct (Bernstein 1977; Smith et al. 1981; Stensgaard and Lægsgaard 2001).

This time-of-flight method of determining the coefficient of restitution has been used in several

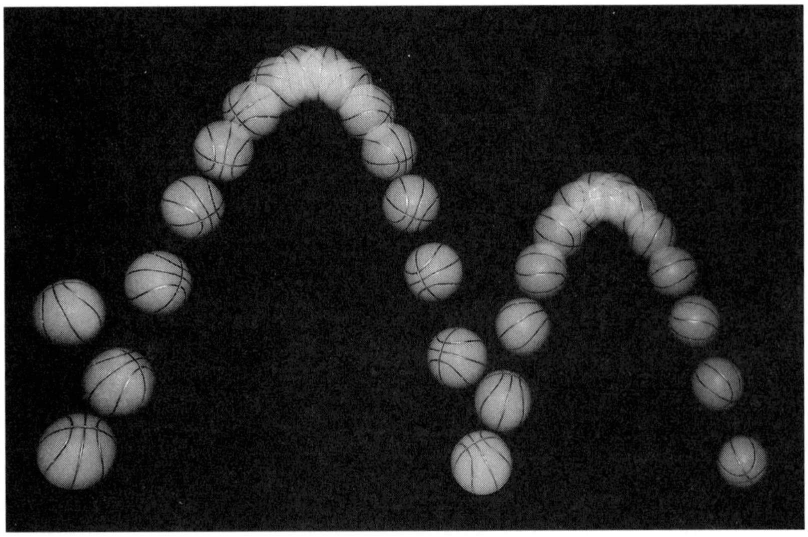

FIGURE 8-12 Bouncing ball captured with stroboscopic flash at 25 images per second (Michael Maggs, Wikipedia Commons).

experimental applications where the very accurate determination of coefficient values was desired. Imre et al. (2008) used this method to evaluate the coefficient of restitution of a Swiss limestone–dolomite–granite conglomerate.

3.3 Coefficient of Restitution

A certain amount of energy is dissipated when two (or more) rockfall blocks collide or a rockfall block strikes a slope surface. The height of bounce, the rebound direction, and the rotational rate induced to a block during impact depend on the impact conditions, which are specified by several parameters (Giani 1992; Chau et al. 2002; Heidenreich 2004; Labiouse and Heidenreich 2009). The *coefficient of restitution* is the most important of these parameters, since it estimates the loss of kinetic energy and thus represents the reduction in block velocity during collisions or impacts. An elastic collision has a coefficient of restitution of 1. An inelastic collision that transforms some of the initial kinetic energy into deformation of the material, heat, sound, and other forms of energy has a coefficient of restitution between 0 and 1. A plastic collision has a coefficient of restitution of 0. A collision between two lumps of clay that do not separate on colliding but stick together is an example of a plastic collision. Collisions of real-world objects are never fully elastic. For purposes of illustration, the coefficients of restitution of some common objects are presented in Table 8-1.

No consistent symbol is used to represent the coefficient of restitution; various investigators have used *e*, *R*, COR, or other symbols. For ease of discussion, COR has been selected as a standard for this chapter.

In general, there are two main energy dissipation modes (resistances) during and after an impact. The characteristics of the ground material determine the kinetic energy dissipation normal to the slope surface; sliding or rolling friction provides the resistance to motion parallel to the slope. Because of the different mechanisms involved in resisting motion normal and tangential to the slope, different parameters are used to define these two resistances. The *normal coefficient of restitution* (COR_N) measures the resistance normal to the slope, while the *tangential coefficient of restitution* (COR_T) measures the resistance to movement parallel to the slope. The analysis of collisions (Section 3.1 and Figure 8-9) introduced these concepts.

Figure 8-14 illustrates an impact of a rockfall block on a slope surface inclined at an angle β. The incident (before-impact) velocity (V_I) is resolved into normal and tangential velocity components (V_{IN} and V_{IT}), and the rebounding velocity (V_R) is resolved into normal and tangential velocity components (V_{RN} and V_{RT}). The angular velocities before and after impact are shown as ω_I and ω_R. The tangential velocity components (V_{IT} and V_{RT}) are shown as equal, and thus $COR_T = 1.0$, but the normal-rebound-velocity component (V_{RN}) is significantly smaller than the normal-incident-velocity component (V_{IN}), so that $COR_N < 1.0$. Thus, the rebound angle (θ_R) is smaller than the incident angle (θ_I).

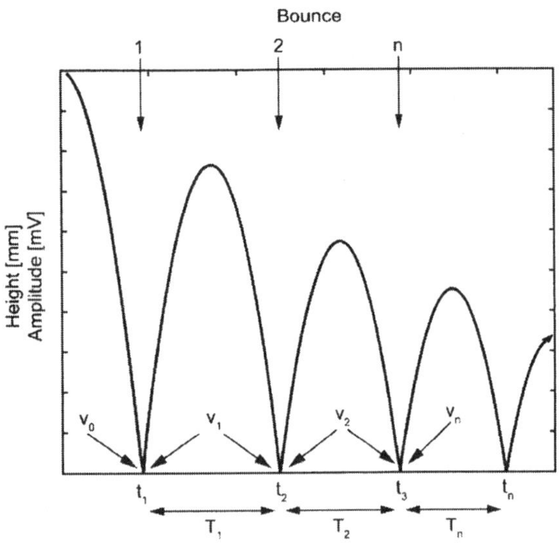

FIGURE 8-13 Schematic representation of ball bouncing on fixed surface, illustrating relationships between height, amplitude, velocity, and time of flight (Imre et al. 2008).

Table 8-1
Coefficient of Restitution for Some Common Objects

Object	Coefficient of Restitution
Tennis racquet	0.85
Tennis ball	0.53–0.58
Table tennis ball	0.94
Golf club face	0.83
Golf ball	0.78
Basketball	0.60
Baseball	0.55

Depending on the impact conditions, the impact is further differentiated according to the contact between block and ground. When the tangential velocity of the block in contact with the ground is 0, the impact is called a *sticking impact*, and the impact process is described as *pure instantaneous rolling*. When there is a nonzero tangential velocity component at the contact point, the impact is called a *sliding impact* or *slipping impact*. In this case, the block motion results in a combination of sliding and rolling.

Unfortunately, there is little standardization in the way the coefficient of restitution is defined; this reflects the current imperfect understanding of collision or impact mechanisms (Heidenreich 2004). The most common definitions of the coefficient of restitution are in terms of velocities; however, several investigations have defined the coefficient of restitution in terms of energies, while other less commonly used definitions, termed *impulse models*, have evaluated the compression and the restitution phases of the impact (Descoeudres and Zimmermann 1987; Bozzolo et al. 1988). There is no consensus as to which definition is more appropriate for rockfall prediction, but all of these definitions express the amount of velocity or energy, or both, dissipated during an impact. Thus, they are commonly called "coefficient of restitution" (Table 8-2), even though this terminology does not precisely conform to the concept proposed by Newton, who defined the coefficient of restitution as the ratio of velocities. Details of these alternative definitions are provided in the following sections.

3.3.1 Definitions Based on Velocity Ratios

A popular definition of the coefficient of restitution for two colliding objects is the ratio between the magnitudes of the rebounding and incoming velocities:

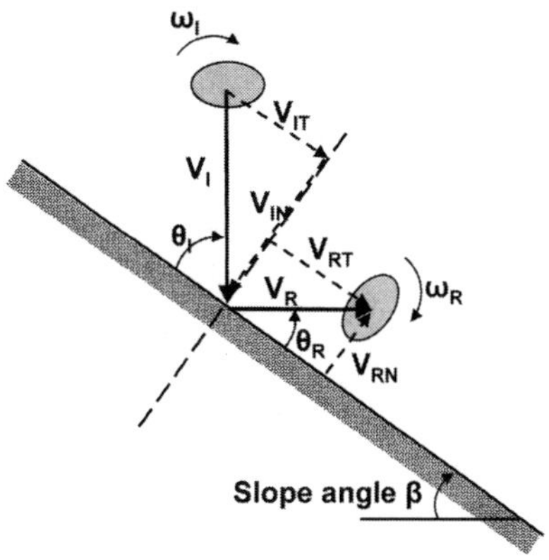

FIGURE 8-14 Idealized impact of rockfall block on slope surface allows for definition of translational and rotational velocities and their components before and after impact (adapted from Labiouse and Heidenreich 2009).

$$\mathrm{COR}_V = \frac{v_{1R} - v_{2R}}{v_{1I} - v_{2I}} \tag{20}$$

where

v_{1R} = scalar velocity of the first object after impact,
v_{2R} = scalar velocity of the second object after impact,
v_{1I} = scalar velocity of the first object before impact, and
v_{2I} = scalar velocity of the second object before impact.

For the case of an impact of a rock block onto a slope, the above equation simplifies to

$$\mathrm{COR}_V = \frac{v_R}{v_I} \tag{21}$$

where v_R is the scalar velocity of the block after impact and v_I is the scalar velocity of the block before impact.

Furthermore, because the velocity of a bouncing rock block just before impact is proportional to the height of its previous bounce, or its free-fall height (i.e., final $v_N = \sqrt{2gh_N}$), the coefficient of restitution can be determined by measuring the heights of successive bounces and using the following formula:

$$\mathrm{COR} = \sqrt{\frac{h_{N+1}}{h_N}} \tag{22}$$

where h_{N+1} is the height of the current bounce and h_N is the height of the last bounce (or the drop height for the first bounce).

Table 8-2
Alternative Definitions of the Coefficient of Restitution

Basis of Definition	Symbol Used in This Chapter	Definition
Generic	COR	Basic coefficient of restitution
	COR_N	Normal component of the coefficient of restitution
	COR_T	Tangential component of the coefficient of restitution
Velocities	COR_V	Coefficient of restitution defined by ratio of velocities
	COR_{VN}	Normal component of the coefficient of restitution defined by ratio of velocities
	COR_{VT}	Tangential component of the coefficient of restitution defined by ratio of velocities
Energies	COR_E	Coefficient of restitution defined by ratio of energies
	COR_{EN}	Normal component of the coefficient of restitution defined by ratio of energies
	COR_{ET}	Tangential component of the coefficient of restitution defined by ratio of energies
	COR_{TE}	Coefficient of restitution defined by ratio of total energies (translational and rotational)
Impulse	COR_I	Coefficient of restitution defined by ratio of the time-integrated normal components of the contact force during the compression and restitution phases of an impact

Accurate determination of the height of successive bounces during laboratory experiments is often difficult, but the duration of each bounce is related to the object's velocity, or its bounce height, and the precise time of each impact can be readily determined (Imre et al. 2008). Thus, the coefficient of restitution can be much more accurately obtained by using the following relationship:

$$\mathrm{COR} = \frac{T_{N+1}}{T_N} \qquad (23)$$

where T_{N+1} is the time of flight of the current bounce and T_N is the time of flight of the last bounce. Imre et al. (2008) used this method to evaluate the coefficient of restitution of a Swiss limestone–dolomite–granite conglomerate.

This basic velocity definition appears to correspond to the concept of Newton when impact dynamics were first considered and has since been used by many authors (Habib 1977; Japan Road Association 1983; Spang and Rautenstrauch 1988; Paronuzzi 1989; Azzoni and de Freitas 1995; Spang and Sonser 1995; Kamijo et al. 2000).

As shown in Figure 8-14 and discussed previously, differentiation of the incoming and rebounding velocity into normal and tangential components permits the evaluation of the rolling and bouncing motions of a rock block as it continues to travel down a talus slope. The coefficients of restitution COR_{VN} and COR_{VT} are defined as follows:

$$\mathrm{COR}_{VN} = \frac{v_{RN}}{v_{IN}} \qquad (24)$$

$$\mathrm{COR}_{VT} = \frac{v_{RT}}{v_{IT}} \qquad (25)$$

where

COR_{VN} = normal component of the coefficient of restitution defined by velocities,
COR_{VT} = tangential component of the coefficient of restitution defined by velocities,
v_{RN} = normal translational velocity after impact,
v_{IN} = normal translational velocity before impact,
v_{RT} = tangential translational velocity after impact, and
v_{IT} = tangential translational velocity before impact.

The subdivision into normal and tangential coefficients of restitution on the basis of velocity measurements has been used during field observations and the development of numerical simulation models by many authors (Piteau and Clayton 1977; Wu 1985; Urciuoli 1988; Pfeiffer and Bowen 1989; Fornaro et al. 1990; Kobayashi et al. 1990; Giani 1992; Evans and Hungr 1993; Budetta and Santo 1994; Ushiro et al. 2000).

These definitions refer to the normal and tangential components of the incident and the rebound velocities of the mass-center of the block and indicate the amount of translational velocity dissipated during impact. As described in Section 3.4, numerous attempts have been made to publish "recommended" values for the normal and tangential coefficients of restitution associated with typical ground materials forming slopes. However, these published values imply that the coefficients of restitution are independent of other factors, including the characteristics of the blocks (such as their mass and shape) or the kinematics of the impact (such as incidence angle, incident velocity, or block rotation before impact). These limitations led to further definitions of the coefficient of restitution.

3.3.2 Definitions Based on Energy Ratios

To overcome some of the limitations of the velocity-based definitions of the coefficient of restitution, some investigators have defined the coefficient of restitution for impacts according to the ratio of the translational energies before and after impact. This *energy-based coefficient of restitution* is thus

$$\mathrm{COR}_E = \frac{0.5mv_{N+1}^2}{0.5mv_N^2} = \frac{v_{N+1}^2}{v_N^2} \tag{26}$$

where

COR_E = energy-based coefficient of restitution,
m = mass of the object,
v_{N+1}^2 = square of scalar velocity of the object after impact, and
v_N^2 = square of scalar velocity of the object before impact.

The COR_E definition has been used by Bozzolo and Pamini (1986), Azzoni et al. (1995), and Chau, Wong, et al. (1999).

Obviously, Equation 26 is invalid when there is a change in the mass of the block, which will occur if the block breaks into multiple fragments. Urciuoli (1988) proposed a slight variation of this formulation, differentiating normal and tangential components for the energy-based coefficient of restitution:

$$\mathrm{COR}_{EN} = \frac{V_{RN}^2}{V_{IN}^2} \tag{27}$$

$$\mathrm{COR}_{ET} = \frac{V_{RT}^2}{V_{IT}^2} \tag{28}$$

where

COR_{EN} = normal component of the coefficient of restitution defined by velocities,
COR_{ET} = tangential component of the coefficient of restitution defined by velocities,
V_{RN}^2 = square of normal translational velocity after impact,
V_{IN}^2 = square of normal translational velocity before impact,
V_{RT}^2 = square of tangential translational velocity after impact, and
V_{IT}^2 = square of tangential translational velocity before impact.

As long as friction exists between the block and slope surface, some conversion of translational to rotational energy will result. Consequently, oblique impacts will result in rotations of the bouncing blocks, and this rotation will increase during a sequence of impacts. This effect should not be neglected. Thus, Bozzolo and Pamini (1986) and Azzoni et al. (1995) proposed an additional definition of the coefficient of restitution: the *total energy coefficient of restitution* (COR_{TE}). It accounts for the total energy of the block according to the following equation:

$$\mathrm{COR}_{TE} = \frac{0.5[m(V_{RN}^2 + V_{RT}^2) + I\omega_R^2]}{0.5[m(V_{IN}^2 + V_{IT}^2) + I\omega_I^2]} \tag{29}$$

where

COR_{TE} = total energy coefficient of restitution,
m = mass of block,
V_{RN}^2 = square of normal translational velocity of mass center after impact,
V_{IN}^2 = square of normal translational velocity of mass center before impact,
V_{RT}^2 = square of tangential translational velocity of mass center after impact,
V_{IT}^2 = square of tangential translational velocity of mass center before impact,

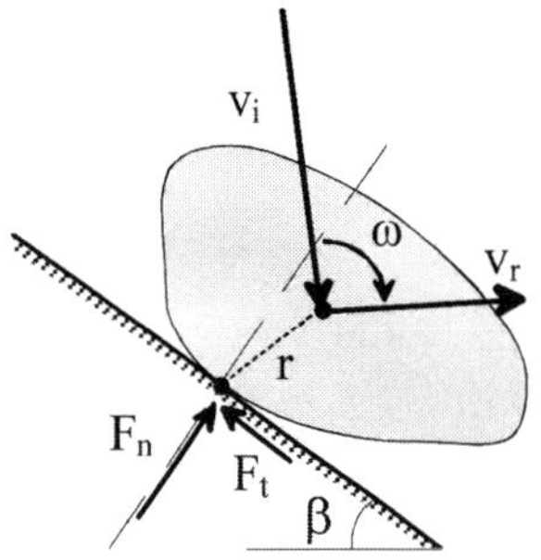

FIGURE 8-15 Forces acting on rock block during its impact with slope inclined at angle β and related velocity conditions (Heidenreich 2004).

I = moment of inertia of block,
ω_R^2 = square of angular velocity of block after impact, and
ω_I^2 = square of angular velocity of block before impact.

Azzoni and de Freitas (1995) and Azzoni et al. (1995) used COR_{TE} in their evaluation of parameters for rockfall analysis.

Few have investigated the rotational energy induced during each boulder impact (Japan Road Association 1983; Azzoni and de Freitas 1995; Chau et al. 2002). Calculation of the rotational velocities after an impact is problematic. For block impacts on hard-rock slopes, field observations tend to support the assumption that little or no sliding or slippage occurs between the contact surface of the block and the ground surface. This allows a relationship between translational and rotational energies to be established. However, impacts on granular slopes produce a combination of rolling and sliding, and the computations and analysis are much more complex (Heidenreich 2004; Labiouse and Heidenreich 2009). This aspect is discussed in more detail in Section 4.6.

Even though the coefficients of restitution based on energy or total energy concepts (COR_E or COR_{TE}) take into account the rotation of the block, they remain as material constant values that are related only to block and ground materials. They are not influenced by other parameters such as the incident velocity of the block or its impact configuration. Section 4 summarizes how several parameters typically interact to influence the motions of rolling and bouncing rock blocks.

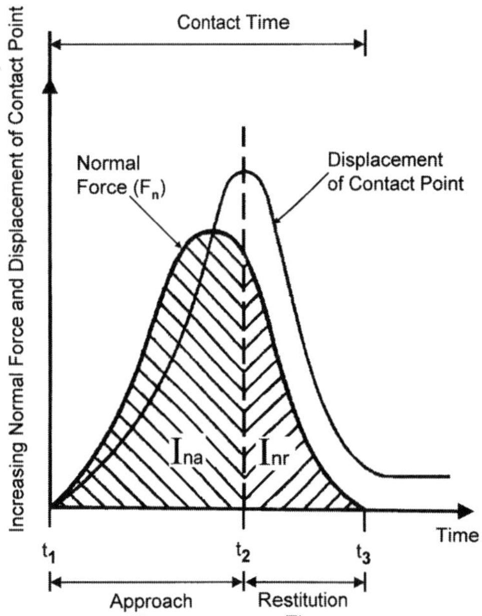

FIGURE 8-16 Temporal variations of normal contact force (F_n) and displacement of contact point divide contact time of impact into approach and restitution intervals. Areas under force curve define impulse components I_{na} and I_{nr} (adapted from Bozzolo et al. 1988).

3.3.3 Definitions Based on Impulse Concepts

Stronge (2000) proposed a further possible mathematical model to express the coefficient of restitution. He suggested that the ratio of work by the normal force during compression to that during restitution is a better candidate for a material constant than any of the coefficients of restitution previously defined by using ratios of energy or velocity. However, this definition is not yet used in the context of rockfall studies.

An alternative definition has been proposed and implemented by Descoeudres and Zimmermann (1987) and Bozzolo et al. (1988). They evaluated the temporal variation of the normal contact force (F_n in Figure 8-15) and the displacement of the contact point during the time a striking rock block is in contact with the ground. Figure 8-16 shows an idealized interpretation of these values. The maximum displacement occurs slightly later than the maximum contact force. The displacement curve in Figure 8-16 does not return to zero after the impact event but shows a residual permanent deformation. The separation of the peak force from the peak displacement and the amount of permanent deformation are mostly determined by the properties of the ground material.

The time of the peak displacement divides the contact time into two intervals, an initial approach time and a subsequent restitution time. The areas under the force curve corresponding to the approach and restitution intervals define the normal impulse components for these two intervals (I_{na} and I_{nr}). Thus a single *impulse coefficient of restitution* (COR_I) may be defined as

$$COR_I = \frac{I_{nr}}{I_{na}} = \frac{\int_{t_2}^{t_3} F_n(t)dt}{\int_{t_1}^{t_2} F_n(t)dt} \quad (30)$$

The tangential force F_t (Figure 8-15) is related to the normal force F_n by a coefficient of sliding friction (μ) that is described in Section 4.2. Thus, $F_t = \mu F_n$. The tangential component of the impulse I_{tr} is defined as

$$I_{tr} = \mu I_{nr} \quad (31)$$

When Equation 31 is satisfied, there is no instantaneous sliding and a sticking impact occurs, with considerable transfer of tangential translational kinetic energy to rotational energy, and the block spins more rapidly. Otherwise, sliding will occur. This is due primarily to plastic deformation and shearing of the slope material (Heidenreich 2004).

The impulse coefficient of restitution has been applied infrequently; the only published values are those of Descoeudres and Zimmermann (1987).

3.4 Published Values of Coefficients of Restitution

The majority of published sources provide values for the velocity-based coefficient of restitution for impacts that are based on the ratio of the actual incoming and rebounding velocities and thus define COR_V by using Equation 20, or they evaluate the normal and tangential component velocities to define COR_{VN} and COR_{VT} by using Equations 24 and 25. However, several sources provide values for the energy-based coefficient of restitution by evaluating actual incoming and rebounding velocities and thus define COR_E as specified by Equation 26. A few sources provide values for the total energy coefficient of restitution (COR_{TE}) as specified by Equation 29 or for the impulse coefficient of restitution (COR_I) as specified by Equation 30.

Table 8-3, originally compiled by Heidenreich (2004), summarizes the reported values provided by 17 sources and classifies each value according to the definition used to develop it. This classification is important because, as discussed in Section 3.5, the values of the coefficients of restitution developed according to the various definitions are not immediately comparable.

Reported values of the coefficient of restitution have been determined from field tests (Wu 1985; Evans and Hungr 1993; Robotham et al. 1995), by backanalysis (Descoeudres and Zimmermann 1987; Fornaro et al. 1990; Pfeiffer and Bowen 1989; Paronuzzi 1989; Evans and Hungr 1993; Budetta and Santo 1994; Azzoni and de Freitas 1995), or by theoretical estimation (Bozzolo and Pamini 1986). Despite these extensive efforts, knowledge of the coefficient of restitution remains fairly limited (Chau et al. 2002).

Heidenreich (2004) suggested that the most reliable values of the coefficients of restitution are those determined through experimental observations of in situ tests or by backanalysis of natural rockfalls. As described in Chapter 10, high-speed video recording of rockfall trajectories permits the evaluation of the translational and rotational velocities as a block approaches and rebounds from each relevant impact. Accurate coefficients of restitution are determined from such data. In the backanalysis of field tests, coefficients are evaluated by trial-and-error checking and comparison of numerical predictions to achieve the best possible match with field observations of runout distance and velocity, height, and distance of each bounce.

The reported values for the coefficient of restitution are frequently associated with certain ground characteristics, which are summarized in the "remarks" column of Table 8-3. The variability of the reported values, even for comparable ground materials, reveals the problematic nature of the coefficient of restitution and the relatively limited understanding of impact mechanics. As discussed in Section 4, the coefficient of restitution only partially defines a material constant, and other factors significantly influence bouncing and rolling motions.

In general, the published values indicate that slopes composed of hard materials, such as rock, have higher coefficients of restitution than slopes composed of debris, soil, or colluvium. A vegetation cover also seems to reduce the coefficient of restitution. Fornaro et al. (1990) analyzed the published values for the normal and tangential components of velocity-based coefficients of restitution (COR_{VN} and COR_{VT}) for slopes covered with three classes of materials: rock, medium to fine debris, and coarse detritus (Figure 8-17). Although the

Table 8-3
Published Values of the Coefficient of Restitution

Source	Value for COR_{VN}	Value for COR_{VT}	Value for COR_E	Value for COR_{TE}	Value for COR_I	Remarks
Habib 1977	0.75–0.80					Based on experience in Italy
	0.5–0.6					Based on experience in Norway
Piteau and Clayton 1977	0.8–0.9	0.65–0.75				Solid rock
	0.5–0.8	0.45–0.65				Detrital material mixed with large rock boulders
	0.4–0.5	0.35–0.45				Compact detrital material mixed with small boulders
	0.2–0.4	0.2–0.3				Grass-covered slopes
Wu 1985	0.2–0.8	0.5–0.75				Rock on rock or wood platform
Heierli et al. 1985	0.95		0.9			Rock
	0.55		0.3			Gravel layer (35 cm)
	0.45		0.2			Gravel layer (70 cm)
	0.45		0.2			Debris
Bozzolo and Pamini 1986				0.7		Rock at a slope angle of 44°
				0.55		Debris at a slope angle of 57°
Descoeudres and Zimmermann 1987					0.4	Vineyard slopes
					0.85	Rock slopes
Hoek 1987	0.53	0.99				Clean hard bedrock
	0.4	0.9				Asphalt roadway
	0.35	0.85				Bedrock outcrops with hard surface, large boulders
	0.32	0.82				Talus cover
	0.32	0.8				Talus cover with vegetation
	0.3	0.8				Soft soil, some vegetation
Urciuoli 1988[a]	0.05–0.35	0.5–1	0.02–1			Rock block striking limestone
	~0	0.24				Debris fan
Pfeiffer and Bowen 1989[b] (older versions of program CRSP)	0.37–0.42	0.87–0.92				Smooth, hard surface as paving, rock
	0.33–0.37	0.83–0.87				Bedrock or boulders with little soil or vegetation
	0.30–0.33	0.83–0.87				Talus with little vegetation
	0.30–0.33	0.80–0.83				Talus with some vegetation
	0.28–0.32	0.80–0.83				Soft soil slope with little vegetation
	0.28–0.32	0.78–0.82				Vegetated soil slope
Giani 1992; Barbieri et al. 1988	0.5	0.95				Bedrock
	0.35	0.85				Bedrock covered by large blocks
	0.3	0.7				Debris formed by uniform distributed elements
	0.25	0.55				Soil covered by vegetation
Evans and Hungr 1993	0.9	0.9				Small blocks on clean rock
	0.7	20.8[c]				Large blocks on talus
Azzoni et al. 1995				0.75–0.90		Rock (limestone)
				0.55–0.6		Fine angular debris and earth (compacted)
				0.35–0.45		Fine angular debris and earth (soft)
				0.45–0.50		Medium angular debris with angular rock fragments
				0.4–0.5		Medium angular debris with scattered trees

(continued)

Table 8-3 *(continued)*
Published Values of the Coefficient of Restitution

Source	Value for COR_{VN}	Value for COR_{VT}	Value for COR_E	Value for COR_{TE}	Value for COR_I	Remarks
				0.55–0.70		Coarse angular debris with angular rock fragments
				0.5–0.6		Earth with grass and some vegetation
				<0.20		Ditch with mud
				0.5–0.65		Flat surface of artificially compacted ground
				0.75		Road
Gerber 1995	0.17–0.43	0.45–0.88				Rock (limestone)
Kamijo et al. 2000	0.1–0.35					Vertical impact of a 220-kg rock on hard surface
	0.1					Vertical impact of an 800-kg rock on hard surface
Ushiro et al. 2000	0.1–0.4	0.71				Undefined materials
Jones et al. 2000[b] (values gathered by program calibration for CRSP 4.0)	0.6–1.0	0.9–1.0				Smooth hard surface and paving
	0.15–0.30	0.75–0.95				Bedrock and boulder fields
	0.12–0.2	0.65–0.95				Talus and firm soil slopes
	0.1–0.2	0.5–0.8				Soft soil slopes
Budetta and Santo 1994 (evaluated by program calibration)	0.2	0.53				Rock

[a]Urciuoli (1988) states that R_n, R_t, and R_E depend on the impact velocity.
[b]Pfeiffer and Bowen (1989) and Jones et al. (2000) use R_n and R_t as material constants but account for several other parameters by means of scaling factors.
[c]Values larger than 1 can be produced by backanalysis if the initial velocity component is very small or 0. The ratio defining R therefore yields large values not really expressing the restitution but rather an abrupt change of direction.
SOURCE: Heidenreich 2004.

published values exhibited a high degree of variability, especially in the values of COR_{VN}, a series of domains was assigned to each of the rough slope material descriptions. Several subsequent investigators have confirmed the basic validity of the domains shown in Figure 8-17 (Chau et al. 2002).

3.5 Significance of Alternative Definitions of the Coefficient of Restitution

Unfortunately, many reported values of the coefficient of restitution do not clearly define which definition, COR_V or COR_E, they represent. The information provided by Heidenreich (2004) and reproduced in Table 8-3 only partially addresses this issue—one that severely complicates the collection and use of historical data. Furthermore, some computer rockfall simulation programs, discussed in Chapter 9, use the COR_V definition as a parameter in the model, while other programs use the COR_E definition. A careful examination of the program documentation or of the computer code itself is often required to determine which definition is used.

This determination is critical because the two definitions, COR_V and COR_E, provide different

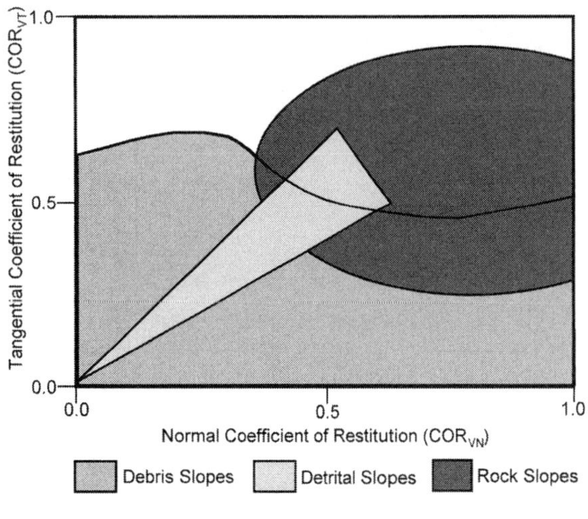

FIGURE 8-17 Domains for three slope materials defined by typical values for normal and tangential components of velocity-based coefficients of restitution (COR_{VN} and COR_{VT}) (Fornaro et al. 1990).

values for the coefficient of restitution under the same conditions. Consider the following example. An object falling 25 m will have an incoming terminal velocity of about 22 m/s. Suppose that after impact its rebound velocity is 18 m/s. Then, according to Equation 21, $COR_V = 18/22 = 0.82$; but according to Equation 26, $COR_E = 18^2/22^2 = 324/484 = 0.67$.

It is likely that some users of these programs have inadvertently used incorrect COR values. Obviously, the program will produce erroneous values if a value of 0.67 is used for COR when the program is designed to use a value of 0.82 to define the desired situation. The opposite mistake, that of entering a value of 0.82 when a value of 0.67 is the correct program value, is equally problematic.

Heidenreich (2004) emphasized that any coefficient of restitution taken from literature or obtained through in situ tests or backanalysis is likely to provide incorrect results when it is used at other sites. She therefore recommended against using these values for trajectory simulations without a thorough calibration and sensitivity analysis of the simulation model to evaluate the effects on the projected trajectories resulting from changes in the assigned values of the coefficients of restitution. Evans (1989) found that modifying the values for the tangential coefficient of restitution did not result in much variation of trajectories of rocks striking weathered bedrock slopes. However, Evans found that, for the same locations, the selection of appropriate values for the normal coefficient of restitution did significantly affect the overall trajectory outcomes. These conclusions are similar to those of Fornaro et al. (1990) and the domains shown in Figure 8-17.

Table 8-4
Parameters Influencing the Bouncing Phenomenon

Slope Characteristic	Block Characteristic	Kinematics
Strength	Strength	Velocity (translational and rotational)
Stiffness	Stiffness	Collision angle
Roughness	Weight	Configuration of block at impact
Inclination	Size	
	Shape	

Source: Labiouse and Heidenreich 2009.

4. ANALYZING RESISTANCE TO BOUNCING AND ROLLING

The analysis of the bouncing and rolling motions of rock blocks as they travel down a slope after a free fall is complex. It involves assessments of the impact mechanics, the shape and size of the blocks, the composition and characteristics of surface materials, the size and character of surface irregularities, and the presence of vegetation.

Although a rock block may not experience much rotational motion during its initial free-fall trajectory, it often begins to spin rapidly after the initial impact. Each bounce is completed by an inelastic impact with the ground surface; this reduces the total kinetic energy of the block but transfers a portion of its translational kinetic energy to rotational kinetic energy. As the rock continues to bounce and roll down the slope, the components of kinetic energy parallel to the slope and the rotational energy are attenuated by friction along the slope and collisions with features perpendicular to the slope (Jones et al. 2000).

4.1 Influences on Bouncing and Rolling

Observations of natural rockfall events, as well as extensive field experimental rock-rolling tests, demonstrate that the characteristics of motion after impact are affected by a combination of slope characteristics, several parameters related to the block, and the kinematics during impact (Azimi et al. 1982; Falcetta 1985; Bozzolo and Pamini 1986; Hungr and Evans 1988; Giani 1992; Azzoni and de Freitas 1995; Dorren, Berger, and Putters 2006; Dorren, Berger, et al. 2006; Labiouse and Heidenreich 2009). Table 8-4 summarizes these relationships.

Numerous systematic laboratory experimental investigations have been carried out on small- and medium-scale models to improve understanding of the bouncing phenomena and to determine the most important parameters influencing block impacts. In some cases, a whole trajectory is simulated (Statham 1979; Ujihira et al. 1993; Murata and Shibuya 1997; Chau et al. 2002; Heidenreich 2004; Labiouse and Heidenreich 2009), while other experiments concentrate on a single impact (Kawahara and Muro 1999; Kamijo et al. 2000; Ushiro et al. 2000; Wong et al. 2000). Most of these investigations involved impact tests on hard ground materials such as rock or concrete. Some laboratory experiments have used plaster to represent both the rock blocks and the hard ground materials, mainly to address at least partially the "scaling effects" of strength parameters when experiments with reduced-scale models are conducted (Chau et al. 2002). Surprisingly few investigations have been made into impacts with soft ground materials (Heidenreich 2004; Kawahara and Muro 2006; Labiouse and Heidenreich 2009). Further discussion of the role of laboratory experiments is provided in Section 5.3, while Chapter 10 discusses the conduct of field rock-rolling experiments.

Table 8-5 summarizes the main conclusions obtained from these studies concerning the influence of parameters related to block, ground, and kinematic conditions on coefficients of restitution. In some cases, the influence is not clear or is even contradictory.

Table 8-5
Parameters Influencing the Coefficients of Restitution and Their Corresponding Trends as Stated in the Literature

Parameter	COR_N	COR_T	COR_{TE}	Ground Material	Source
Related to Rock Block					
Increasing block mass	Decreases	No information	No information	Concrete, rock	Ushiro et al. 2000
Increasing angularity (shape changing from sphere to angular block)	Decreases	Increases (slightly)	Decreases	Granite	Wong et al. 2000
	No change when $\theta = 30°$ to $60°$; increases when $\theta = 15°$ to $30°$	Decreases (slightly)	Decreases (slightly)	Plaster	Chau, Wu, et al. 1999
Related to Ground Material					
Increasing Young's modulus	Increases	Increases	No information	Natural slopes	Pfeiffer and Bowen 1989
	Increases	Increases	No information	Natural slopes	Fornaro et al. 1990
	Increases	Increases	No information	Plaster, rock, soil	Chau et al. 2002; Wu 1985
Increasing dry density	Increases when density > optimum density; no change when density < optimum density	Increases when density > optimum density; no change when density < optimum density	Increases when density > optimum density; no change when density < optimum density	Soil, plaster	Chau, Wong, et al. 1999
Related to Kinematics					
Increasing impact angle (θ) and thus a reduction in the slope angle (β) (this is caused by $\theta + \beta = 90°$)	Decreases	Increases (slightly)	No information	Wood, rock	Wu 1985
	Decreases (slightly)	No information	No information	Granite, shotcrete	Wong et al. 2000
	Decreases	No information	No information	Plaster, soil	Wong et al. 2000
	Decreases (slightly)	No change	Decreases	Granite, shotcrete	Chau et al. 2002
	Decreases	No change	Decreases	Plaster, soil	Chau et al. 2002
Increasing drop height (H); this leads to increasing impact velocity	Decreases	Decreases	Decreases	Limestone	Urciuoli 1988
	Decreases (slightly)	No change	No information	Concrete, rock	Ushiro et al. 2000

Note: Sources may define normal and tangential coefficient of restitution (COR_N and COR_T) by velocity or energy ratios. COR_{TE} is the total energy coefficient of restitution. (See Section 3.3 for definitions.)
Source: Adapted from Labiouse and Heidenreich 2009.

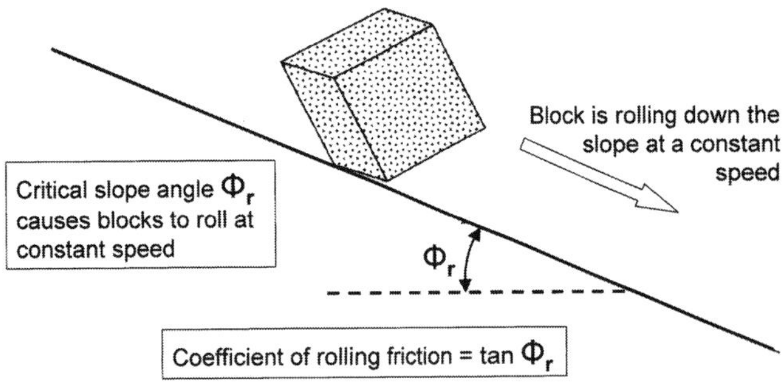

FIGURE 8-18
Coefficient of rolling friction (adapted from Okura, Kitahara, et al. 2000).

Table 8-6
Value of the Rolling Friction Coefficient for Various Geological Units on a Specific Slope

Slope Surface		Rolling Friction Coefficient
I.	Rock limestone	0.40
II.	Hard debris, earth and angular rock fragments with some boulders, compacted	0.55
III.	Soft debris, angular rock fragments and earth with some boulders, loose	0.75
IV.	Medium debris, medium debris covering smooth rock, compacted	0.60
V.	Soft debris, earth with rock fragments, loose	0.60
VI.	Mud in the ditch	0.85
VII.	Yard, artificial fill	0.60

Source: Azzoni et al. 1995.

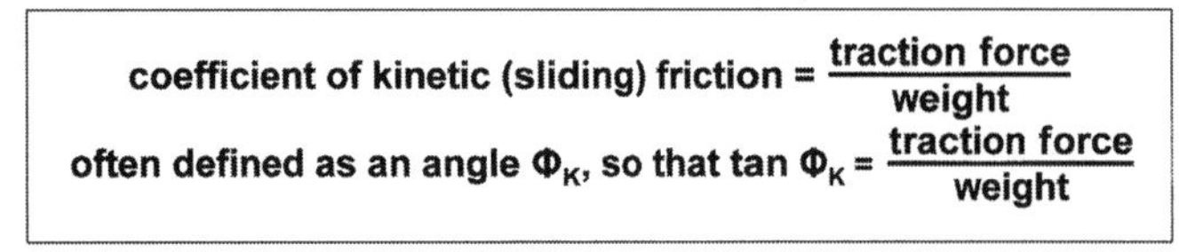

FIGURE 8-19
Coefficient of kinetic (sliding) friction (adapted from Okura, Kitahara, et al. 2000).

4.2 Coefficients Used to Define Bouncing and Rolling

Azzoni et al. (1995) evaluated the mechanics of bouncing and rolling blocks through experimental tests and by backanalysis of natural and artificially induced rockfalls. They concluded that the assessment of bouncing and rolling rockfall motions required two coefficients, the total energy coefficient of restitution (COR_{TE}) and a rolling friction coefficient.

The *rolling friction coefficient* is a parameter used to simulate the loss of energy occurring during the rolling motion of a falling block (Statham 1979; Bozzolo and Pamini 1986; Azzoni and de Freitas 1995; Okura, Kitahara, et al. 2000). The rolling friction coefficient is defined in Figure 8-18 as the tangent of the angle at which a block can be considered to move with a steady velocity on a slope, a so-called critical slope angle. On a steeper slope, the block accelerates, while on flatter slopes it decelerates and finally stops. Thus, the rolling friction coefficient accounts for all the processes of retardation affecting the rolling of a block down the slope, including minor slope irregularities discussed further in Section 4.5, and vegetation, as discussed further in Section 4.7. The rolling friction coefficient varies with both the type and shape of the materials and the ratio between the size of the block and the size of the debris (Azzoni et al. 1995). Table 8-6 gives values of the rolling friction coefficient for various geological units on a specific slope.

Okura, Kitahara, and Sammori (2000) and Okura, Kitahara, et al. (2000) conducted a series of experiments to evaluate the role of interactions (collisions) among multiple blocks in determining the maximum runout distance for rockfalls and rockslides. Their experiments are described in Section 5.3. They included consideration of a third coefficient, the coefficient of kinetic (or sliding) friction (Figure 8-19). Okura, Kitahara, et al. (2000) conducted experiments with granite blocks on a slope paved with granite slabs, and thus Figure 8-19 shows these materials. However, the concept of a kinetic friction coefficient is the same for all materials.

Because rolling is generally much easier than sliding, the coefficient of rolling friction is typically less than the coefficient of kinetic friction, and thus it is a prominent parameter required in most simulations of rolling and bouncing rocks. However, the coefficient of kinetic friction controls the final stages of rockfall runout and plays an important role in the rates of energy dispersion by bouncing rocks on soil-covered slopes. Most impacts on soil-covered slopes involve relatively low incident angles; thus, considerable shearing and slippage of the near-surface soil layers occur. This reduces the apparent normal and tangential coefficients of restitution.

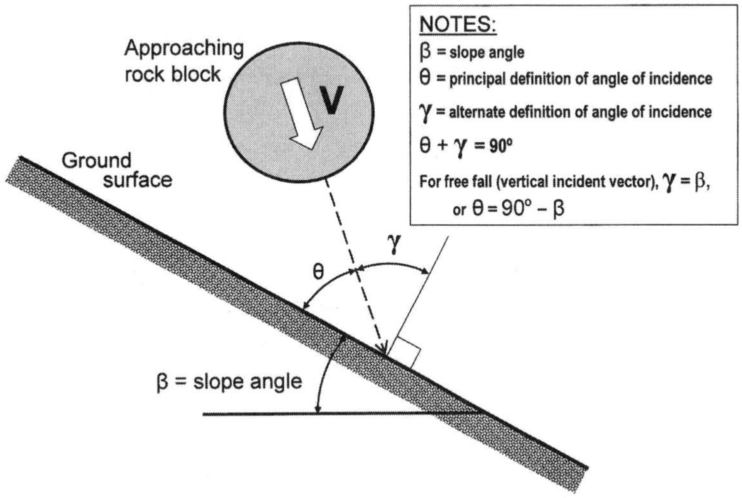

FIGURE 8-20 Standard definitions of impact angle geometry.

4.3 Influence of Impact Angle

Even though it is an important factor affecting postimpact block motion, most authors ignore the rotational velocity or energy of the block before and after impact. As long as friction exists between the block and the slope surface, partial locking is likely to occur. Consequently, rotation will be induced during an oblique impact. Relatively few studies have investigated the rotational rate induced after a series of block impacts and how the characteristics of impacts of bouncing rocks on slopes change as the angle of impact changes.

There is some lack of consistency in defining the angle of incidence of a rock just before impact, so Figure 8-20 provides the standard definitions used in this chapter. The symbol β is commonly used to indicate the slope inclination angle. The symbol θ is used to indicate the angle between the trajectory and the slope. This is the most commonly used definition of the impact angle, but some investigators use the angle between the trajectory and the normal to the slope, shown with the symbol γ in Figure 8-20. This terminology is consistent with the definitions used in Figure 8-14 and Section 3.3. Many published reports describe laboratory tests, and in some cases field tests, that involve dropping an object (a rock or a ball) onto an inclined planar surface. In these cases the angle γ is equal to the slope angle, and the angle θ = 90° – β.

4.3.1 Experiments by Wu (1985)

Wu (1985) undertook an intensive study to determine the influence of varying impact angles on the values of the velocity-defined normal and tangential coefficients of restitution (COR_{VN} and COR_{VT}). Two sets of data were collected. In the first experimental series, rocks were bounced on an inclined wooden platform that was adjusted to angles of 30°, 40°, 45°, and 60°. Thirteen rocks of various shapes and sizes were dropped at each angle. Figure 8-21 shows this experimental setup. The second experimental series dropped rocks from heights of approximately 10 m (40 ft) onto each of three actual rock cut slopes with angles of 28°, 40°, and 75°. A movie camera was set up to record the complete movement of each test drop; the values of H, L, and t in Figure 8-21 were determined by direct observation, while the movie images permitted the determination of the distances h or h_I in Figure 8-21. Figure 8-22 summarizes the values of the normal and tangential coefficients of restitution.

The mean values of the multiple observations and the values of their standard deviation are

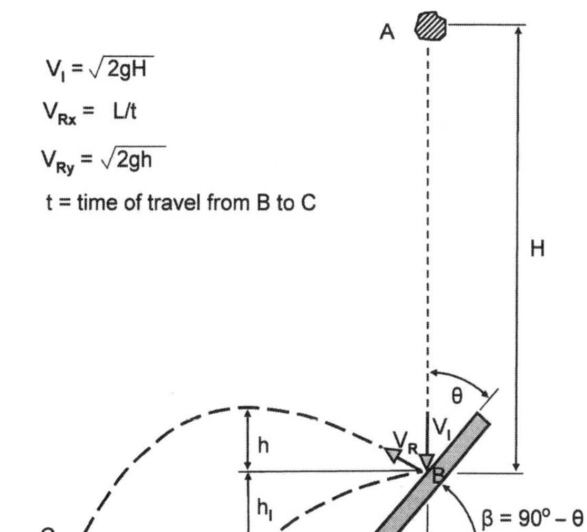

FIGURE 8-21 Experimental setup with inclined wooden platform (Wu 1985).

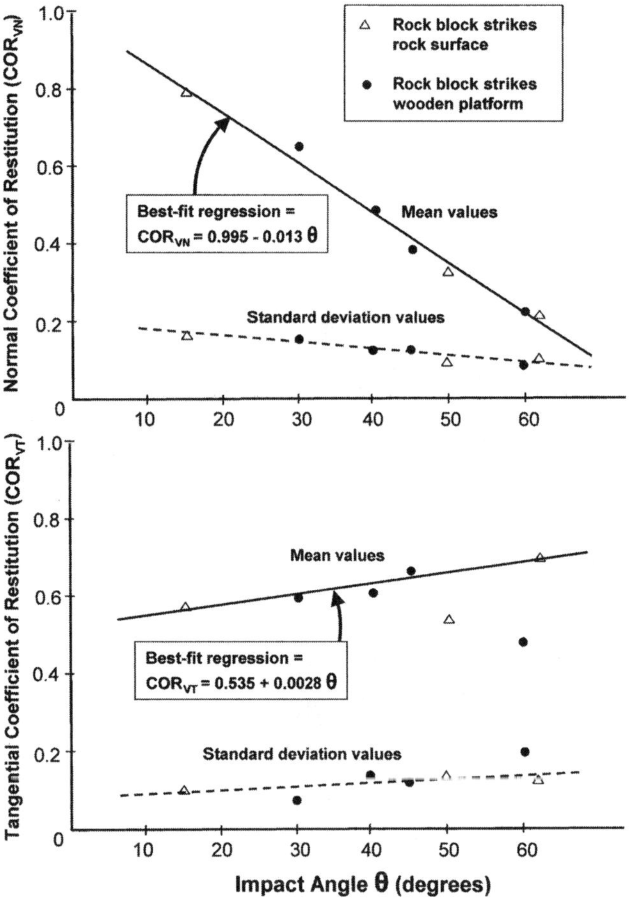

FIGURE 8-22
Observed relationships between normal and tangential coefficients of restitution (COR_{VN} and COR_{VT}) and changes in impact angle (Wu 1985).

shown for each impact angle (θ). The data clearly show a decrease in the normal coefficient of restitution (COR_{VN}) as the impact angle increases (in other words, as the slope angle decreases). The relationship for the tangential coefficient of restitution (COR_{VT}) is less clear. Wu suggested that it increased very slightly as the angle of incidence (θ) increased. However, the observations at the higher angles of incidence are erratic. Wu's conclusion appears to be questionable and has not been supported by subsequent investigations, such as those by Chau et al. (2002) and Labiouse and Heidenreich (2009), which are described subsequently. Figure 8-22 shows the correct best-fit regression equations for the relationships identified by Wu; these equations are incorrectly reported by Wu (1985).

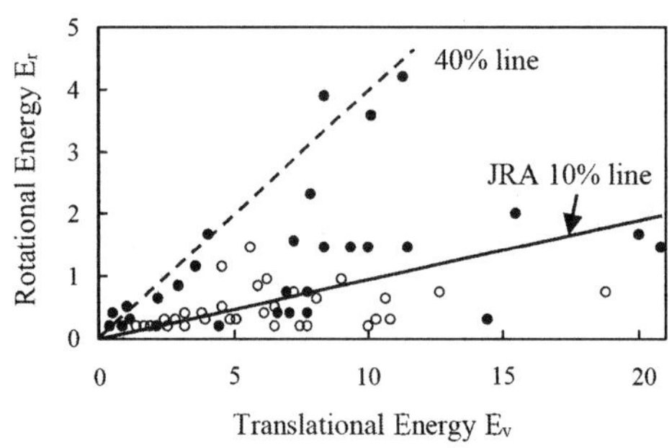

FIGURE 8-23
Observed relationship between rotational and translational kinetic energies of striking rocks. Solid circles indicate field data obtained from 52° rock slope at Takamatsu, Japan, while open circles indicate data from three other locations with smaller slope inclinations [data of the Japan Road Association (JRA) presented by Chau et al. 2002].

4.3.2 Evaluations by Japan Road Association

Some knowledge concerning the rotational energy of bouncing and rolling blocks was gained by field impact tests in Japan (Japan Road Association 1983). Although this report is in Japanese, it has been cited by several investigators, including Wong et al. (2000), Chau et al. (2002), and Heidenreich (2004).

The Japan Road Association (1983) performed 60 in situ rockfall tests with round to angular boulders traveling down slopes with inclinations between 32° and 52° at four locations in Japan. The investigations of these trajectories showed that the average ratio between the rotational kinetic energy (E_r) and the translational kinetic energy (E_v) was about 0.1 (10%). Figure 8-23 shows a plot of E_r versus E_v for all the Japanese Road Association field tests. The solid circles indicate the field data obtained from a 52° rock slope at Takamatsu, while the open circles indicate data from three other locations with smaller slope inclinations. Because the tests involved free-falling blocks onto the rock slopes, the impact angles were complements of the slope angle. Thus the impact angle for the steep Takamatsu slope (52°) was 38°. The impact angles for the slopes with lower slope inclinations were larger than 38°.

Figure 8-23 includes lines identifying ratios of rotational to translational energy of 10% and 40%. The Japanese Road Association reported that about one-half the data points were below the 10% line and that essentially all data were below the 40% line. The tests with the most oblique impacts (the Takamatsu slope, shown as solid circles) generally have higher ratios, which corresponds to more rapid rotations, than do the tests with less oblique impacts.

The 40% line can be interpreted as the upper bound for the case of spherical boulder impact.

This relates to the definition of moment of inertia for a sphere, as described in Section 2.3 and Figure 8-5. As shown in Figure 8-23, this upper limit agrees with the Japanese field data, although the shapes of the blocks used in the field tests were not perfectly spherical.

4.3.3 Experiments by Chau et al. (2002)

The research group led by Professor K. T. Chau in Hong Kong conducted probably the most detailed series of small-scale impact tests to evaluate the influence of several parameters on the coefficient of restitution (Chau et al. 1998; Chau, Wong, et al. 1999; Chau, Wu, et al. 1999; Chau et al. 2002; Wong et al. 2000).

Although the tests were not entirely able to satisfy the scaling requirements, they used various ground materials (plaster, compacted soil, shotcrete, and granite) to investigate the effect of the shape of the boulder, the impact angle, the impact velocity and energy, and the mechanical properties of the slope material on the coefficients of restitution COR_{VN}, COR_{VT}, COR_V, and COR_E. The blocks were dropped onto a horizontal or inclined slope composed of various ground materials, and the rotational motion induced to the blocks during impact was investigated. The free fall, impact, and subsequent trajectory of each block were recorded by a high-speed video camera.

In general, COR_{VN} increases slightly as the slope angle (β) increases, and thus the impact angle (θ) decreases (Figure 8-24). The trend corresponds to that reported by Wu (1985) for in situ field tests, but the amount of change is much less. This is verifiable by comparing Figure 8-24 with Figure 8-22, although the directions of the two *x*-axis values are reversed, since one investigator used the impact angle (θ) and the other used the slope angle (β). Figure 8-25 shows no change in COR_{VT} values as the slope angle changes. However, clear trends appear in Figures 8-26 and 8-27 for the coefficients of restitution defined in terms of velocity (COR_V) and energy (COR_E), respectively.

Chau et al. (2002) evaluated the data provided by the Japan Road Association (1983) and used the same analysis approach (described

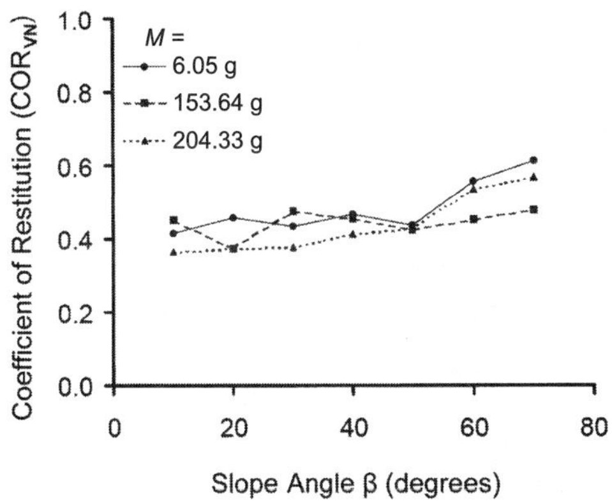

FIGURE 8-24
Changes in normal coefficient of restitution (COR_{VN}) with changes in slope angle β [where impact angle (θ) is 90° – β] for various spherical masses (*M*, measured in grams). Blocks and slope surface are both made of plaster (adapted from Chau et al. 2002).

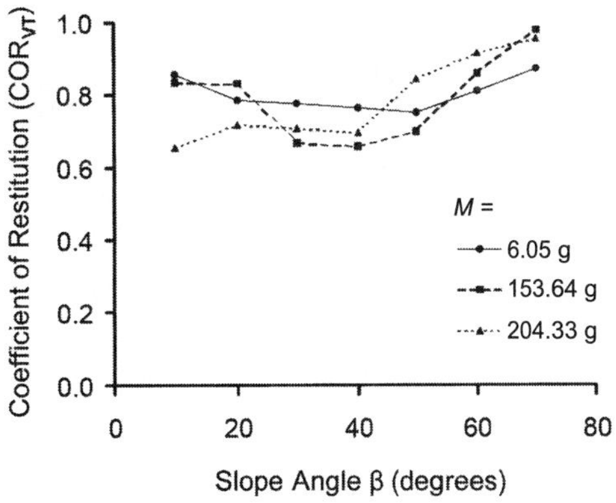

FIGURE 8-25
Changes in tangential coefficient of restitution (COR_{VT}) with changes in slope angle β [where impact angle (θ) is 90° – β] for various spherical masses (*M*, measured in grams). Blocks and slope surface are both made of plaster (adapted from Chau et al. 2002).

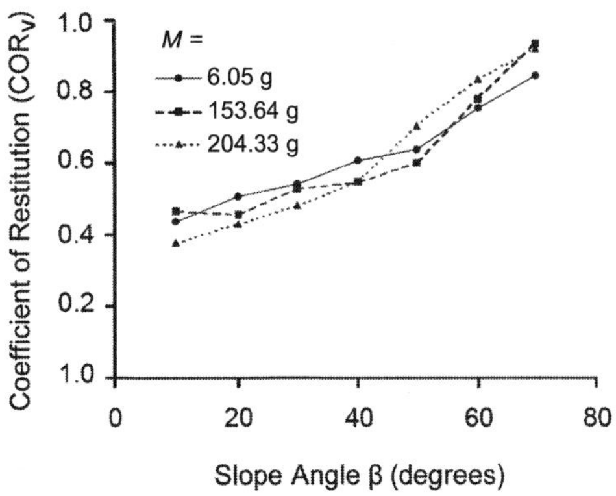

FIGURE 8-26
Changes in velocity-defined coefficient of restitution (COR_V) with changes in slope angle β [where impact angle (θ) is 90° – β] for various spherical masses (*M*, measured in grams). Blocks and slope surface are both made of plaster (adapted from Chau et al. 2002).

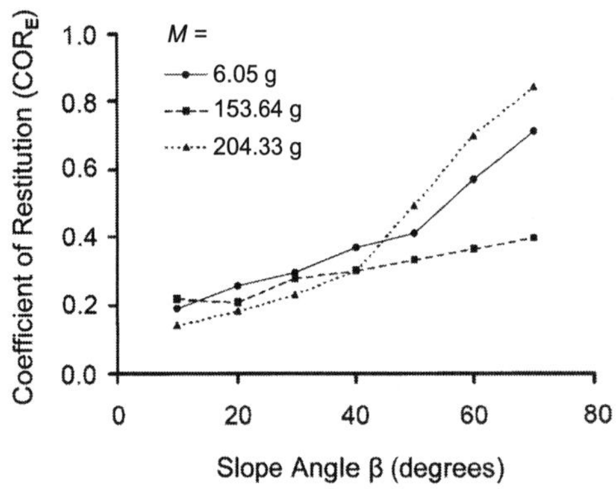

FIGURE 8-27
Changes in energy-defined coefficient of restitution (COR_E) with changes in slope angle β [where impact angle (θ) is 90° – β] for various spherical masses (*M*, measured in grams). Blocks and slope surface are both made of plaster (adapted from Chau et al. 2002).

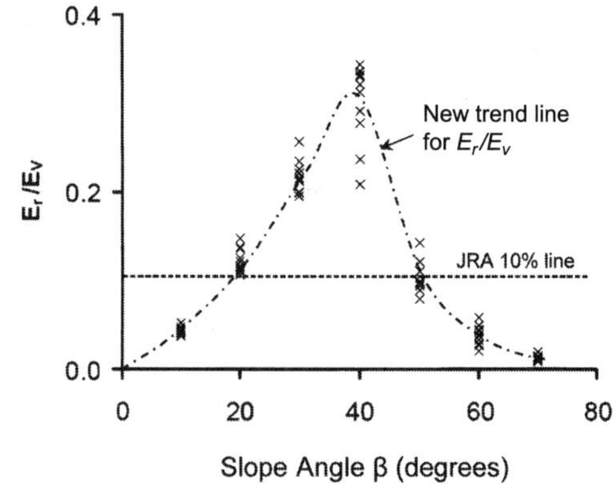

FIGURE 8-29
Data from Figure 8-28 showing relationship between ratio of rotational and translational energies and changes in slope angle (β) (adapted from Chau et al. 2002). (Note that JRA 10% line appears as horizontal line in this figure.)

in Section 4.3.2) to assess their own results (Figure 8-28). When their data are plotted to show the relationship between the ratio of rotational and translational energies and changes in slope angle (Figure 8-29), it is clear that the rotational rate acquired after each block impact depends strongly on the angle of impact. Chau et al. (2002) concluded that the values given earlier by the Japan Road Association (1983) are valid only for slope angles at about 20° or 50° (Figure 8-29).

4.3.4 Analyses by Kuninaka and Hayakawa (2004)

Kuninaka and Hayakawa (2004) undertook theoretical numerical modeling of the oblique impact of a disk with a layered surface (which they called a *wall*). They used a finite element representation of both the disk and the wall (Figure 8-30). Both were considered to be elastic objects; the disk was represented by 800 mass points and the wall by 4,000 mass points. The mass points were placed randomly within the disk and the wall, and neighboring mass points were connected by undeformed nonlinear springs to form a Delauney triangular network.

During an impact, their model distributed a portion of the initial translational energy into vibrations within the disk and the wall. They found it necessary to introduce surface roughness to make the disk rotate after an impact. This was achieved by introducing a degree of randomness in the positions of the mass points defining the surfaces of the disk and wall, which made the surfaces "rough." They discovered an increase in the normal coefficient of restitution as the impact angle γ increased (as the angle of impact became more oblique), with a peak value occurring at around 80°.

Kuninaka and Hayakawa (2004) also evaluated how the coefficient of sliding friction changed as the impact angle became increasingly oblique. They found that the coefficient of sliding friction reached its highest value when the impact angle γ was about 70°. The theoretical modeling procedures used by Kuninaka and Hayakawa illustrate

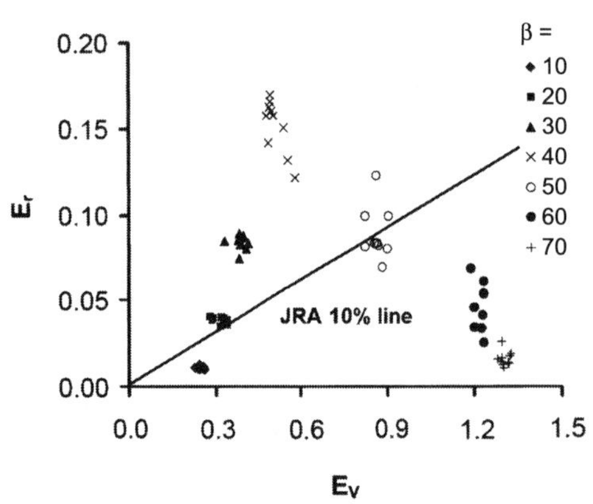

FIGURE 8-28
Data from experiments with different slope angles plotted according to their rotational kinetic energy (E_r) and translational kinetic energy (E_v) values. β = slope angle (adapted from Chau et al. 2002).

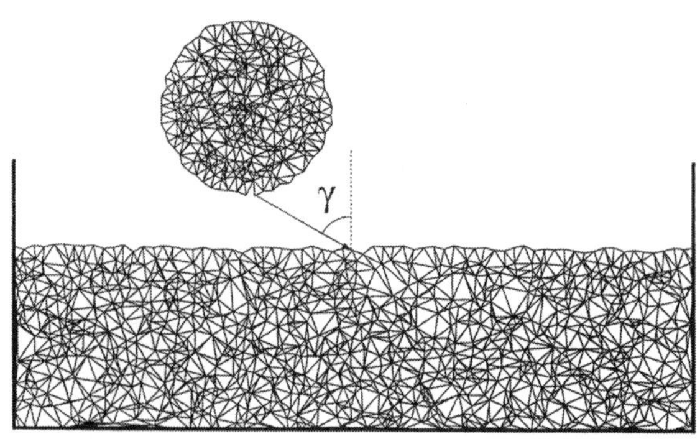

FIGURE 8-30
Oblique impact of elastic disk and wall modeled with finite element lattice (Kuninaka and Hayakawa 2004).

the potential for further studies of impact and collision mechanics. However, their results were directed toward much higher-velocity impacts where different physical principles apply and so cannot be directly applied to rockfall evaluations.

4.4 Influence of Shape and Size of Rocks

Rock shape contributes to the randomness of rockfall behavior. It also influences the apportionment of translational and rotational energy through the moment of inertia (Jones et al. 2000). The size and shape of rock blocks are also important influences on the runout distance. The shape and size of the block influence the length and height of block trajectories, the velocity of rolling, and the point of conversion from bouncing to rolling (Petje et al. 2005). Chapter 10 discusses the procedures for collecting shape and size data for rock blocks used in rock-rolling field tests. As discussed in Section 4.5, rock size and shape influence the degree to which surface roughness affects rockfall behavior.

Larger rocks have greater momentum and are less likely to lodge among irregularities; thus, on the basis of momentum and kinetic energy principles, they are expected to travel farther down a slope than smaller rocks. Ritchie (1963) observed this to be the case in most instances. However, field observations have shown that in some cases larger blocks stop more quickly than smaller rocks. Petje et al. (2005) suggested that this may be explained by the following:

- Larger blocks sink more deeply into the ground.
- Interactions between large blocks are often significant, and the blocks may stop sooner because of these kinetic energy dissipations.
- Larger blocks can disintegrate into smaller but still sizable blocks, resulting in the release of considerable kinetic energy and consequentially the cessation of movement.

Typically, natural rocks have irregular shapes. For many numerical analyses, the shapes are idealized by simple regular shapes. A commonly adopted shape is an ellipsoid (Figure 8-31). Many laboratory experiments have used regularly shaped objects; most often spheres are used. Several rockfall simulation programs (Chapter 9) allow for the selection of one block shape from several standard rock shapes, including spherical, discoidal, and tabular (Figure 8-32). Each standard regular shape has a defined moment of inertia (Section 2.3 and Figure 8-5).

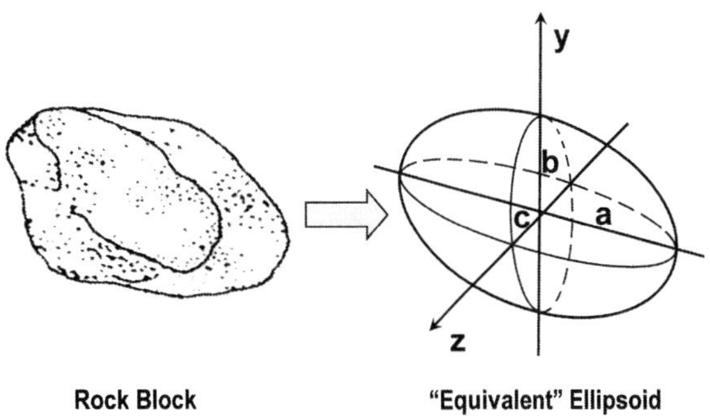

FIGURE 8-31 Shape of irregular rock is often represented by equivalent ellipsoid.

Observations of natural and artificially induced rockfalls have indicated that spheroidal blocks usually move faster than discoidal or tabular ones, and the velocity achieved by bigger blocks is larger than that of smaller ones (on the same slope) since the former are less affected by slope roughness. Azzoni et al. (1995) conducted tests on slopes where rolling movement was prevalent, and rockfall motions were analyzed to assess the effects of block shape and dimension on this type of movement. The block shape was quantified by a shape coefficient, defined as the ratio between the volume of the block and the volume of the minimum sphere included within the block. Both the shape and the dimensions of the block affected the maximum and the mean rolling velocity (Azzoni et al. 1995).

These results confirmed the results obtained by laboratory tests carried out in the 1970s at the Istituto Sperimentale Modelli e Strutture (ISMES)

FIGURE 8-32 Common idealized shapes (tabular, disk, and spherical) used in numerical analysis of rockfall, compared with typical rock. Principal axes are defined as shown. Sphere has no principal axis; all directions are equal, and controlling dimension is radius.

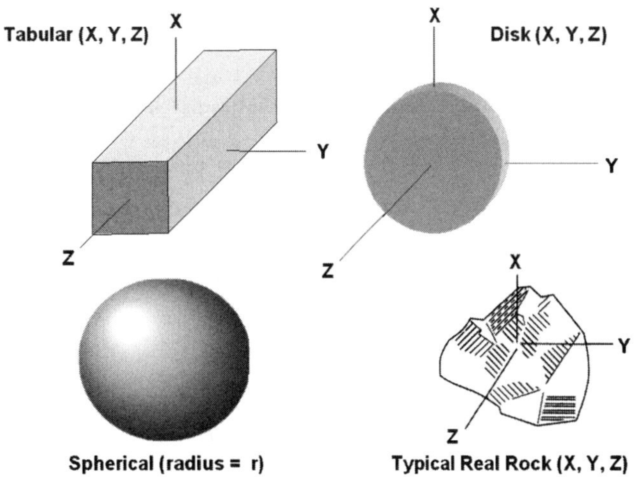

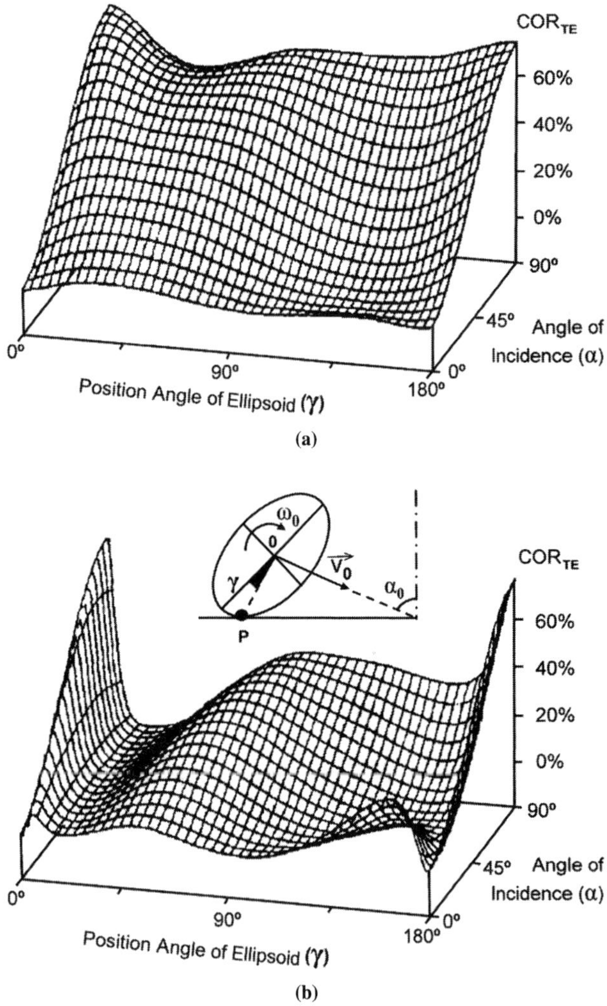

FIGURE 8-33
Axonometric views representing coefficient of restitution (COR$_{TE}$) plotted as function of position angle γ of ellipsoid relative to slope (0° ≤ γ ≤ 180°) and of its angle of incidence α (0° ≤ α ≤ 90°): (*a*) values obtained for slightly discoidal ellipsoid block with semiaxis ratio *b*/*a* = 0.76 and semiaxis ratio *c*/*a* = 0.40 and (*b*) values obtained for much thinner and more elongated ellipsoid block with semiaxis ratio *b*/*a* = 0.40 and semiaxis ratio *c*/*a* = 0.20. Inset defines α and γ (adapted from Bozzolo and Pamini 1986).

SpA located in Bergamo, Italy (Camponuovo 1977). The earlier experiments assessed the relationship between shape and volume of blocks and their free-falling velocities by throwing fragments of different shapes and dimensions down an inclined table covered by finer debris. The table was inclined at different angles. These tests, as well as subsequent tests by Statham (1979), were relatively simple and empirical, yet they remain a milestone in the study of rockfall mechanics because they clearly show the fundamental relationship between the shape and dimension of the blocks and their velocity.

Bozzolo and Pamini (1986) reported on the observations recorded for a series of detailed rockfall field tests conducted at several locations near Lugano, Switzerland, and on their development of a numerical model to quantify the observed behavior. They examined rocks bouncing and rolling on slopes formed with hard-rock surfaces and slopes covered with fine gravel debris. They analyzed the bouncing and rolling motions on the basis of the total energy definition for the coefficient of restitution (COR$_{TE}$) defined in Section 3.3. Figure 8-33 reproduces their assessment of the variation of the value of COR$_{TE}$ as an ellipsoidal block strikes with different orientations (defined by a position angle γ relative to the slope) and its angle of incidence, which they symbolize by α. The definition of α corresponds to 90° – θ, where θ is the standard definition of the impact angle. Figure 8-33*a* shows the results for a slightly discoid ellipsoid block with a semiaxis ratio *b*/*a* = 0.76 and a semiaxis ratio *c*/*a* = 0.40 (see Figure 8-31 for definitions). Figure 8-33*b* shows the results for a much thinner and more elongated ellipsoid block with a semiaxis ratio *b*/*a* = 0.40 and a semiaxis ratio *c*/*a* = 0.20. The two diagrams clearly show how the variability in the coefficient of restitution increases as the block becomes less spherical.

4.5 Influence of Surface Roughness

Numerous field observations, laboratory experiments, and computer simulations have demonstrated that slope irregularities materially influence the trajectory of a falling or bouncing rock block after an impact with the ground surface (Azzoni et al. 1992; Azzoni et al. 1995; Chau et al. 2002; Crosta and Agliardi 2004; Dorren, Berger, et al. 2004; Dorren, Maier, et al. 2004; Dorren, Heuvelink, et al. 2004; Dorren, Berger, and Putters 2006; Dorren, Berger et al. 2006). Jones et al. (2000) attributed most of the variability observed among rockfall events originating from a single-source location to irregularities in the slope surface. Several investigators have reported that surface irregularities have a greater influence on the motions of bouncing and rolling blocks than do variations in the values of the coefficients of restitution.

Surface irregularities serve as obstacles to a rolling rock (Figure 8-34) and thus increase the coefficient of rolling friction that was defined in Section 4.2 and Figure 8-18. A slope surface composed of mostly large fragments provides hollows and occasional steep adverse surfaces that impede the lateral downslope progress of the block (Giani 1992). This means that slope surfaces covered with larger fragments will tend to stop rolling blocks more quickly, as shown in Figure 10-16 (in Chapter 10), or that steeper slope angles are necessary to allow rock blocks to continue to roll down the slope.

Surface irregularities influence the motion of a bouncing rock because they frequently produce an increased impact angle, and as discussed in Section 4.3, observations and laboratory experiments show that larger impact angles result in a larger dissipation of kinetic energy (Bozzolo and Pamini 1986; Crosta and Agliardi 2004; Bourrier et al. 2009). Rough slopes typically increase the height and number of bounces a rock block undergoes as it travels down the slope but also tend to reduce the velocities and impact energies of the rocks (FHWA 1993).

The distribution and dimensions of surface irregularities are referred to as *surface roughness*. Characterization of surface roughness relates the size and spacing of surface irregularities to the size of the falling rock block. Field observations on talus (or scree) slopes have shown that smaller rocks tend to remain near the top of the slope, while progressively larger blocks are found downslope (Statham 1979; Briolli 1977; Evans and Hungr 1993). This demonstrates that larger blocks tend to continue to roll or bounce down a surface made up of smaller blocks presenting minor surface variations relative to the large block, but falling blocks, regardless of size, tend to be stopped as they progress over a surface made up of fragments of a size similar to or larger than the moving block (see Figure 10-16).

Most efforts to characterize surface roughness define it as the perpendicular variation from an average plunge line of the slope surface within a distance equal to the radius of a typical falling block. This procedure is used by several rockfall simulation programs, such as the Colorado Rockfall Simulation Program (Pfeiffer and Bowen 1989; Pfeiffer and Higgins 1990; Jones et al. 2000). Section 3.2.3 of Chapter 10 and Figure 10-17 define this process, which has been discussed by several sources (Pfeiffer and Bowen 1989; Pfeiffer and Higgins 1990; Jones et al. 2000).

Dorren (2010) proposed a new approach for characterizing slope surface roughness in support of a three-dimensional simulation model of rockfall into forested slopes called Rockyfor3D. Section 6.7 of Chapter 9 includes further discussion of the Rockyfor3D program. Only the process for characterizing surface roughness is presented here. Homogeneous zones of surface characteristics are determined in the field and mapped as polygons within the model study area. The surface roughness characteristics of each polygon zone are described by three size–probability classes that correspond respectively to the maximum obstacle height (MOH), in meters, of a representative obstacle that a falling block will encounter during impact in 70%, 20%, and 10% of the cases (Figure 8-35). The size–probability class values for each polygon are recorded as three raster maps, one for each class. During each rebound calculation, the size of the material encountered by the striking block within a defined raster cell is randomly chosen from the three size–probability classes according to their probability of occurrence (thus, the 70% class is selected most frequently, the 20% class occasionally, and the 10% class infrequently).

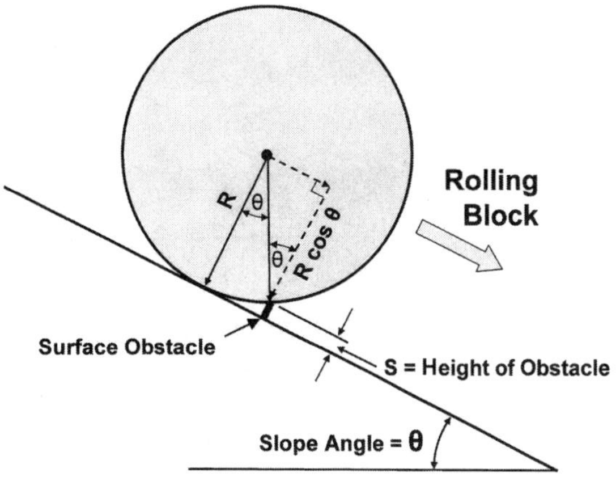

FIGURE 8-34 Conceptual diagram of spherical block of radius *R* rolling down slope at constant velocity encountering surface obstacle that increases coefficient of rolling friction (adapted from Okura, Kitahara, and Sammori 2000).

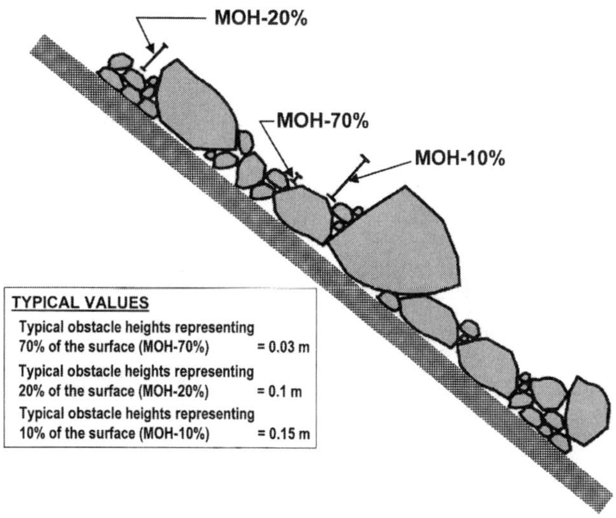

FIGURE 8-35 Surface roughness for Rockyfor3D model is defined by three size–probability classes that correspond respectively to height of representative obstacle in meters (MOH) that falling block will encounter during impact and rebound in 70%, 20%, and 10% of cases. MOH measurements are made looking down slope (adapted from Dorren 2010).

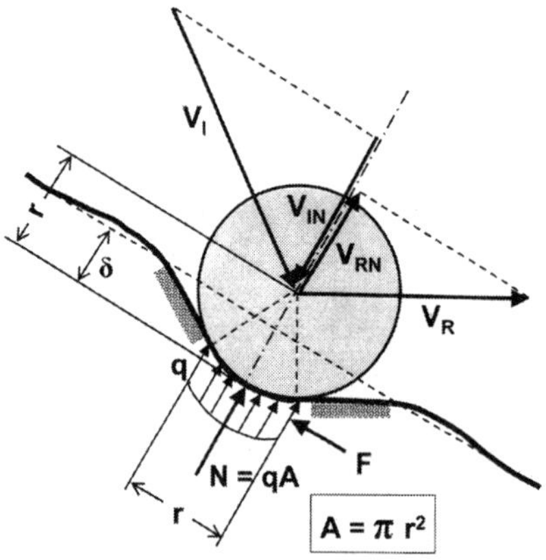

FIGURE 8-36 Block striking soft ground material (adapted from Ushiro et al. 2000).

Higgins et al. (2003) suggested that, for analyzing rockfall motions across relatively smooth surfaces, such as paved surfaces, the surface roughness is related to the irregularity of the block shape, and in such cases the surface roughness is adequately represented by a value of 25% to 50% of the rock radius.

4.6 Block Impacts on Soft Ground

Ushiro et al. (2000) evaluated impacts on both hard and soft ground material. Their computations of translational and rotational rebound velocities take into account the incident block rotation. For sticking impacts, where no slip occurs at the contact point, the incident tangential and rotational velocity components define the tangential rebound velocity. When a combination of rolling and sliding occurs at the contact point, the incident rotation of the block has no influence on the tangential rebound velocity. The rotation after impact is calculated from the initial rotation and the normal velocity component before and after impact.

For impacts on harder surfaces, the tangential and rotational rebound velocities depend more on the contact friction. They are reduced from incident values by a predefined factor when the contact friction values are above a predefined limit so that no slippage occurs. They are determined from the normal coefficient of restitution (COR_{VN}) and the contact friction coefficient when lower contact friction values exist so that a combination of sliding and rolling occurs.

For impacts on softer surfaces, Ushiro et al. (2000) assumed that the block penetrates into the ground material during impact (Figure 8-36). The penetration depth is determined by the normal force acting between block and ground, which depends on both the area of the contact surface and the ground stiffness. The normal velocity component after impact therefore depends on the normal contact force (N), the block mass (M), the area of the contact surface (A), the ground stiffness (k), the slope angle (β), and the penetration depth (δ).

Heidenreich (2004) and Labiouse and Heidenreich (2009) described the results of two series of laboratory tests undertaken to evaluate the mechanics of rockfall impacts on sandy slopes. Initially a series of small-scale tests was conducted. They used an inclinable box that contained a variety of soil materials and two block release mechanisms that allowed for vertical or inclined impacts. The impacts were recorded with a digital high-speed camera. Spheres with a diameter of 7.5 cm (3 in.) and masses of 130, 380, and 750 g were dropped to produce an impact velocity of about 4.4 m/s (14.4 ft/s), which corresponds to a free fall of about 1 m (3 ft). A total of 210 experiments evaluated a series of material properties (angle of internal friction, degree of compaction, and inclination angle), block mass, and impact angle and impact direction (inclined impact onto a horizontal surface, vertical impact onto an inclined surface, and inclined impact onto an inclined surface), as shown in Figure 8-37. The tests demonstrated that the coefficients of restitution were influenced by the degree of compaction of the slope material and the impact angle between the slope surface and the incoming trajectory. It was also observed that more massive blocks had less rebound (and thus a smaller coefficient of restitution). These conclusions substantially replicated the results of previous investigations. However, they suffered from a lack of similitude with actual rockfall because of scaling effects, and thus a second series of half-scale tests was undertaken.

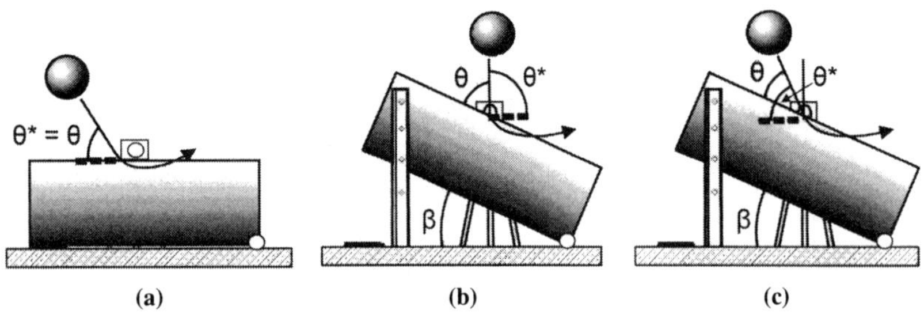

FIGURE 8-37 Experimental setup for initial small-scale tests: (*a*) inclined impact onto horizontal soil; (*b*) vertical impact onto inclined soil; and (*c*) inclined impact onto inclined soil (Labiouse and Heidenreich 2009).

The half-scale tests were conducted in a shaft 5 m (16.4 ft) in diameter and 8 m (26 ft) deep.

Two shapes of steel objects were tested: spheres and cylinders with a spherical base (Figure 8-38). They were of different sizes and filled with concrete to provide their mass (Table 8-7). They were dropped from various heights up to 10 m (32.8 ft) onto a 1-m-thick bed of uniform sand that could be inclined from the horizontal to as much as 30°. The tests evaluated three coefficients of restitution (COR_{VN}, COR_{VT}, and COR_{TE}).

Considerable variation in these values was found (Table 8-8). Labiouse and Heidenreich (2009) provided a detailed parametric study of these variations, and Heidenreich (2004) provided even greater detail, including evaluations of the effects of block size and shape. For slopes inclined more than 10°, cylindrical blocks exhibited higher coefficients of restitution than did the spheres. This was attributed to the larger radius of the spherical base of the cylinders compared with the spheres of similar mass (Figure 8-38). Thus, the cylinder penetrates less into the ground, which promotes its tangential propagation during impact and its normal rebound velocity. Furthermore, because of the lower penetration, less energy is dissipated during impact by plastic ground deformation. The cylinders also had a higher rotation rate than the spheres. This is due to the cylinder having a larger eccentricity between the vertical centerline and the first contact point (Figure 8-38), as well as the lower lateral rolling resistance due to a smaller penetration.

Labiouse and Heidenreich (2009) identified three main mechanisms that controlled the impact mechanics on slopes composed of soft materials:

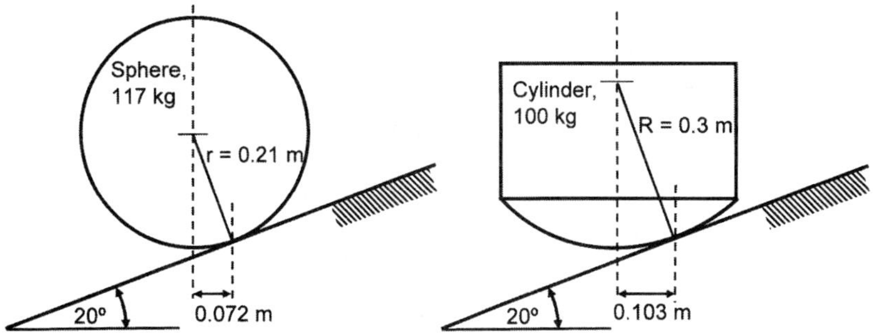

FIGURE 8-38 Eccentricity influences observed in half-scale tests; 100-kg cylinder has greater eccentricity than 117-kg sphere (Labiouse and Heidenreich 2009).

- Penetration into the slope is related to the bearing capacity of the ground material.
- Sliding on or within the uppermost layers is dependent on the mass and shear resistance of the ground material in front of the block.
- Rotation of the block, which is a function of the slope inclination, the impact direction, and the impact velocity, is influenced by any crater developed during impact.

Their observations demonstrated interdependence among these mechanisms. The following are examples:

- Vertical impacts on horizontal ground produce no rotation or lateral translation.
- Low-velocity impacts on a stable slope do not produce a distinct crater; however, the block experiences lateral translation and rotation. The ground behaves elastically for the most part, and the block rolls on the slope surface.
- High-velocity impacts on the same slope cause the block to penetrate into the soft ground, creating a distinct crater and causing local or

Table 8-7
Characteristics of the Blocks Used in the Half-Scale Experiments

Block Shape	Block Mass (kg)	Unit Weight (kN/m^3)	Diameter (m)	Radius (of Sphere or of Base of Cylinder) (m)
Cylinder with spherical base	100	25.8	0.42	0.30
	500	25.6	0.72	0.52
	1,000	26.2	0.90	0.60
Sphere	117	30.1	0.42	0.21
	152	7.5	0.73	0.365
	553	27.1	0.73	0.365

Source: Labiouse and Heidenreich 2009.

Table 8-8
Computed Coefficients of Restitution from Half-Scale Experiments

Coefficient of Restitution	Range of Values	Average Value from All Tests
Velocity-based normal coefficient of restitution (COR_{VN}) (see Equation 24)	0.003–0.257	0.04
Velocity-based tangential coefficient of restitution (COR_{VT}) (see Equation 25)	0.07–1.00	0.48
Total energy coefficient of restitution (COR_{TE}) (see Equation 29)	0.003–0.336	0.03

Source: Labiouse and Heidenreich 2009.

even global failure of the ground. Because the crater rim hinders rolling, normal translation is enhanced, while lateral translation and rotation are inhibited.

- If the slope inclination is close to the internal friction angle of the ground material, the ground fails superficially at impact. This causes a distinct lateral movement of the block, and penetration and rotation are comparatively less important.

In summary, these experiments showed that block motion during impact is governed by mechanisms of penetration, sliding, and rotation, acting antagonistically in part. Under different impact conditions, one or another of these mechanisms is favored, and this governs the resulting block motion after impact.

4.7 Rockfall and Forests

Many communities located within valleys of the Alps have relied on forests to provide protection against rockfalls and snow avalanches (Brang et al. 2001). Local regulations dating back to the 13th century reveal that mountain residents have historically used the concept of "protective forests" to stabilize slopes and limit damage from extreme events (Motta and Haudemand 2000). Substantial economic and social changes in mountain areas over the past few decades have drastically modified forest use, while construction in previously nonurban areas, the growing importance of road systems, the need to maintain road and rail viability the year round, new industries, and winter tourism have greatly increased the requirements for protection. These trends have resulted in extensive research into forest management practices (silviculture) in Austria, France, Germany, Italy, and Switzerland, and a large number of publications have been produced. Relevant recent publications include those of Motta and Haudemand (2000); Ottitsch and Weiss (2000); Weiss (2000); Brang et al. (2001); Dorren, Berger, et al. (2004); Dorren, Maier, et al. (2004); Dorren, Heuvelink, and Berger (2004); Le Hir et al. (2004); Brauner et al. (2005); Dorren et al. (2005); and Bigot et al. (2009).

Protection forests are legally defined in most Alpine countries in Europe. However, the term is used for forests with different functions (Ottitsch and Weiss 2000). Generally, a protection forest has either a site-protection function or a direct protective function. The *site-protection function* refers to a forest stand that protects its site against excessive soil erosion or debris flows (Dorren, Berger, et al. 2004; Dorren, Maier, et al. 2004; Dorren, Heuvelink, and Berger 2004). The *direct protective function* of a forest implies that the forest directly protects people, buildings, and infrastructure against the impact of natural hazards such as snow avalanches and rockfall (Brang et al. 2001). Regional authorities have thus concluded that forests with a direct protective function are a high priority for silvicultural intervention and cannot be abandoned (Motta 1998).

In the Swiss Alps, rockfall and snow avalanches are the most common hazards; 31% of the mountain forest plots in the National Forest Inventory show evidence of rockfall, and 37% of the plots show evidence of snow avalanche events (Stoffel et al. 2006). While the large volumes and high energies associated with snow avalanches often limit the protective effect of forest stands, the small masses that are generally involved in rockfall events (usually less than 5 m^3 or 175 ft^3) allow mountain forests to significantly reduce the energies of falling rocks (Hétu and Gray 2000). Both living and dead trees can stop falling rocks, while tree trunks and limbs lying on the ground act as barriers to rocks moving downslope (Schönenberger et al. 2005). Stokes et al. (2005) reported that experience of European foresters suggests that forests composed mostly of conifers provide better protection against snow avalanches, while deciduous forests provide better protection against rockfall. These differences are attributable to differences in the mechanical behavior of the wood structures and root systems of the various species.

4.7.1 Field Evaluation of Interactions of Rock Blocks with Forest Stands

Before about 2000, little quantitative information existed concerning how individual trees or groups of trees (a forest stand) reacted to rockfall impacts, and thus little was known of the degree to which a forest dissipated rockfall energies or influenced rockfall trajectories. The lack of empirical data made an assessment of the level of protection against rockfall provided by a forest at any specific site uncertain (Dorren et al. 2005).

The only mechanical strength information relating to the material resistance of timber came from laboratory experiments involving wooden test specimens. An initial series of field test experiments indicated that the laboratory experiments severely underestimated the energy dissipation capabilities of groups of trees in the field (Berger et al. 2002). These results led to the establishment of a multiyear collaborative research program (the ROCKFOR project) at the Cemagref Research Institute at Grenoble, France, to coordinate the efforts of researchers in France, Austria, Spain, and Switzerland.

Initial field observations indicated that individual trees dissipate rockfall energies by three mechanisms, depending on the height and severity of the impact (Figure 8-39). The earlier laboratory tests mostly reflected only the first mechanism, the shearing of the tree trunk, and thus underestimated the energy dissipation capacity of a tree. Field tests indicated that the tree trunk is most frequently sheared when the impact is a "direct hit" and located fairly close to the ground. Indirect impacts or impacts at lower or higher levels above the tree's base result in different outcomes. The tree is generally uprooted or its root system is sheared if the impact is close to the tree's base, while shaking and bending of the entire tree occur if the impact is relatively high in the tree. When impacts occur at relatively high velocities at elevated heights above the base of the tree, the tree trunk or limbs often "whip" and break (Berger et al. 2002; Le Hir et al. 2004). Stokes et al. (2005) provided considerable information on the mechanical resistance of various tree species to rockfall impacts. They tested the mechanical resistance of trees by winching them to failure, defined as shearing of the stem or uprooting of the tree.

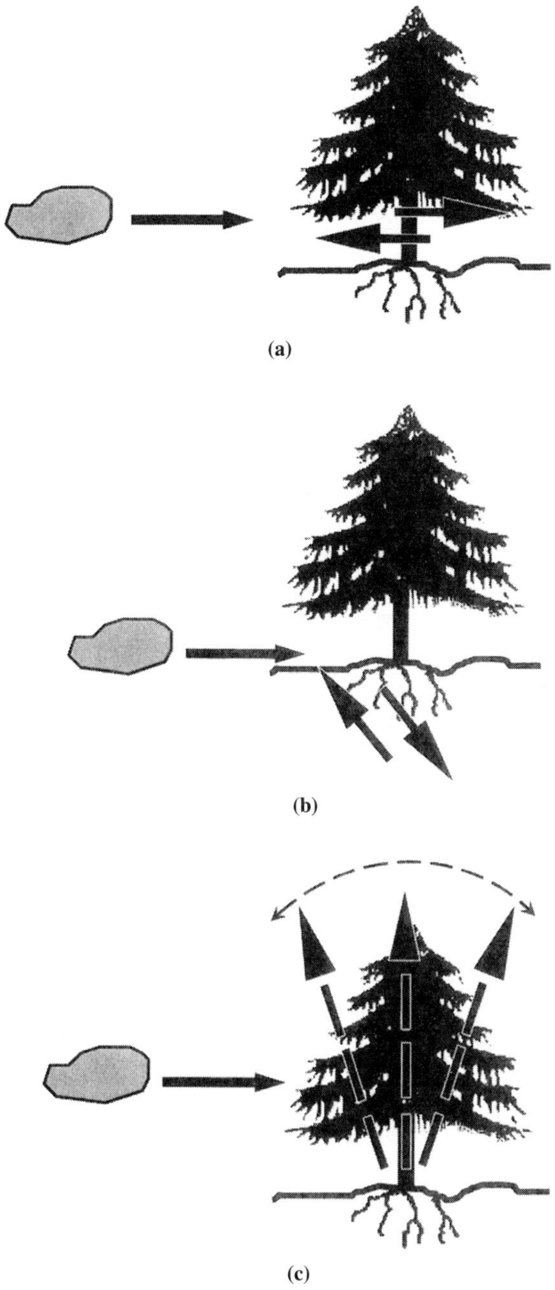

FIGURE 8-39 Three mechanisms for dissipating rock block energies after impact with individual tree: (*a*) shearing of trunk; (*b*) shearing of roots or uprooting, or both; and (*c*) shaking and bending (adapted from Berger et al. 2002).

Dorren et al. (2005) and Dorren and Berger (2006) examined impacts resulting from a series of field trials and classified them as frontal impact, lateral impact, or scratch (the falling rock glances off the tree) (Figure 8-40). Table 8-9 summarizes the observed frequency of each class of impact and its efficacy in stopping rocks at their field site. When rocks were not stopped by impact, each

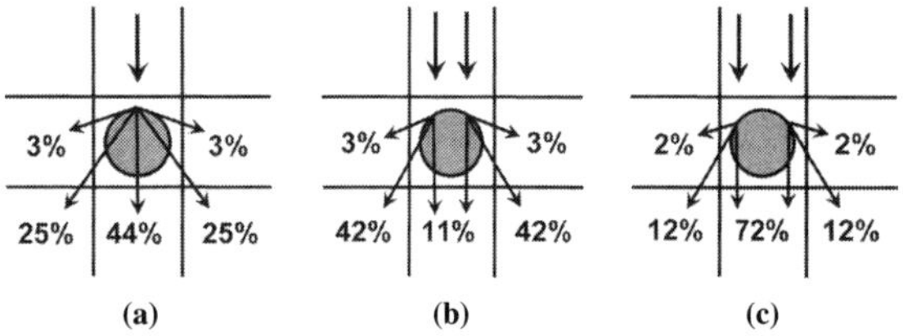

FIGURE 8-40
Deviation percentages for three impact classes: (*a*) frontal, (*b*) lateral, and (*c*) scratch (Dorren et al. 2005).

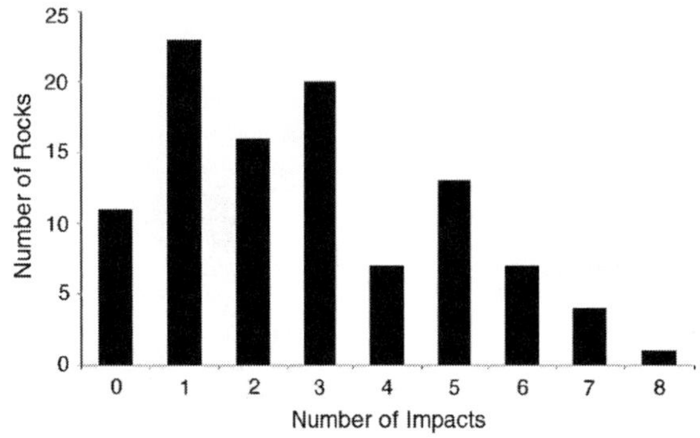

FIGURE 8-41
Histogram showing distribution of number of tree impacts by 102 rocks traveling down forested slope (Dorren et al. 2005).

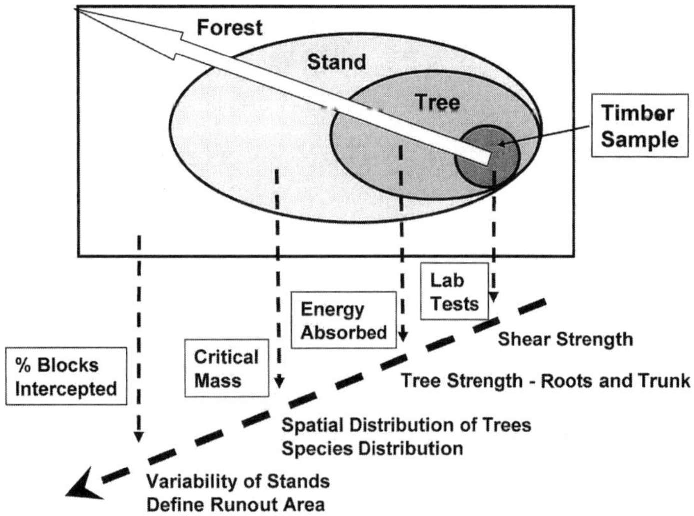

FIGURE 8-42
Changes in response to rockfall impacts and energy dissipation processes between laboratory and forest scales (adapted from Berger et al. 2002).

class of impact resulted in a different degree of energy dissipation and caused different deviations in the rock trajectory. Figure 8-40 shows the probabilities of trajectory directions after each type of impact. For example, a frontal impact will continue without any deflection in 44% of the cases, while this occurs in only 11% of the cases after a lateral impact and in 72% of the cases after a scratch. The cases without significant deflection included all the rocks that (*a*) stopped behind a tree, (*b*) broke or uprooted a tree and continued in the same direction, and (*c*) hit a tree and subsequently bounced slightly to the side or turned behind the tree and continued in more or less the same direction.

Dorren et al. (2005) also reported on the frequency of tree impacts as each rock traveled down the forested slope (Figure 8-41). On average, each rock had 2.8 impacts against trees (286 impacts/102 rocks); one rock struck eight different trees, while 11 rocks (11% of the total) struck no tree at all.

These initial rock-rolling field tests also indicated that the processes controlling the response to rockfall impacts became increasingly diverse as the focus changed from testing a timber sample in a laboratory to an impact on a tree, from a tree to a forest stand, and from a stand to an entire forested slope (Figure 8-42). The dissipation of kinetic energy exhibited by the falling rock blocks increased as the scale changed and the reactions became more diverse and the interactions more complex. In the laboratory, a test of a timber sample gives information only on the shear strength of a particular type of wood of defined cross section. An individual tree, when struck by a rock, resists by several mechanisms, as shown previously in Figure 8-39, so the effective energy dissipation is much greater. At the scale of the forest stand, the effective energy dissipation is further

Table 8-9
Occurrence Frequency of the Three Impact Classes and Their Efficacy in Stopping Rocks, Based on 286 Observed and Analyzed Tree Impacts

Type of Impact (See Figure 8-40)	Number of Impacts	Percentage of Total Impacts	Number of Rocks Stopped After Impact	Percentage of Impacts That Stopped Rocks
Frontal	183	63.8	25	14
Lateral	57	20.0	12	21
Scratch	46	16.2	0	0

Source: Dorren and Berger 2006.

increased because individual trees not only absorb energy from impacts but also interact with each other, depending on their spatial distribution and the mixture of tree species within the stand. Combinations of spatial and species distributions are typically evaluated by using a critical mass measure. A forested slope generally has several forest stands, so at this scale their combined influences define the percentage of rock blocks that are intercepted on the slope as well as the shape and size of the rockfall runout zone.

Field tests of rockfalls on forested slopes helped advance the procedures for conducting rock-rolling field tests; Chapter 10 summarizes these procedures. Observations from such tests provided empirical data, which were then used to develop quantitative values defining the strength and energy dissipation characteristics of individual trees (Figure 8-43). Results are reported as the maximum kinetic energy dissipation that can be achieved by a tree with a given diameter at breast height (DBH); this height is variously defined as 1.3 to 1.4 m, or about 4.5 ft above the base of the tree. Most observations focused on two common tree species in the Alps: the European silver fir (*Abies alba*), representing conifers, and the European beech (*Fagus sylvatica*), representing deciduous species. Dorren and Berger (2006) reported values for several species and related them to these two reference species. They also reported that the impact height influences these values for trees with a DBH less than 15 cm (0.5 ft), but impact height has little influence on larger trees.

In summary, the various field research observations of rockfall trajectories on forested slopes reported by Dorren, Berger, et al. (2004); Dorren, Maier, et al. (2004); Dorren, Heuvelink, and Berger (2004); Perret et al. (2004); and Stokes et al. (2005) found that forested slopes with average slope gradients of 33° to 40° result in rockfall velocities of 15 to 25 m/s (50 to 80 ft/s) and rebound heights between 1 and 2 m (3 and 6 ft).

4.7.2 Modeling Rockfall in Forests

The lack of extensive empirical data on rockfall processes within mountain forests led to the development of dynamic models that provide valuable data for evaluating the protective effects of various forest stands against rockfall and for improving forest management practices (Berger and Dorren 2006a, 2006b, 2007; Bigot et al. 2009; Bourrier et al. 2009). Several models and versions of models have been developed and described, but two are most commonly applied. Chapter 9 provides additional details on these models.

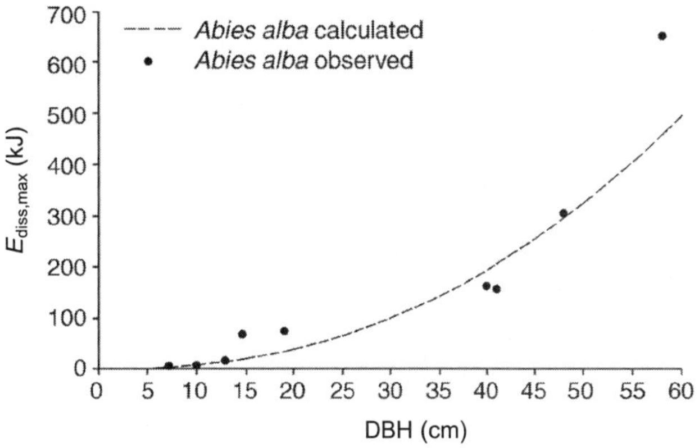

FIGURE 8-43 Relationship between maximum kinetic energy that can be dissipated by European silver fir (*Abies alba*) and its DBH, based on nine rockfall impacts causing trunk breakage (Dorren and Berger 2006).

The first model is called Rockfor$^{\cdot NET}$. It is a rapid-assessment tool, freely and publicly available on the Internet (http://www.ecorisq.org/en/rockfornet.php). It is based on findings from field rockfall experiments on forested and nonforested slopes (Dorren et al. 2005) and was described in some detail by Berger and Dorren (2006a, 2006b, 2007). Rockfor$^{\cdot NET}$ quantifies the protective capacity of a forest stand against rockfall by using easily acquired data to define the characteristics of the rockfall and slope. The forest is defined as a sequence of open rockfall nets, or "tree curtains," each consisting of a row of trees spaced at slightly less than the average rockfall block diameter (Figure 8-44).

More recently a combined deterministic–probabilistic three-dimensional rockfall simulation

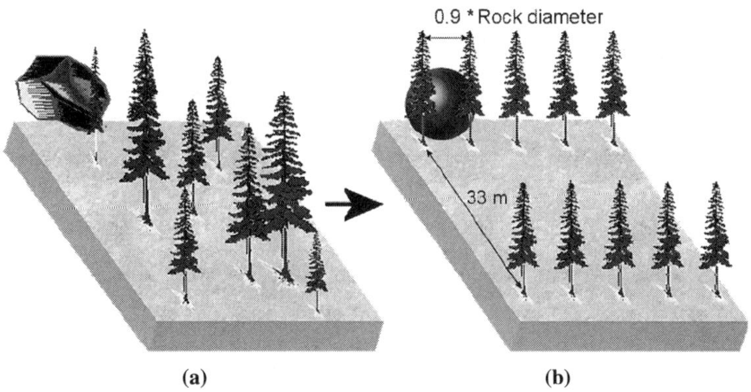

FIGURE 8-44 Defining forest in Rockfor$^{\cdot NET}$ model: (*a*) reality and (*b*) virtual representation (Berger and Dorren 2006b).

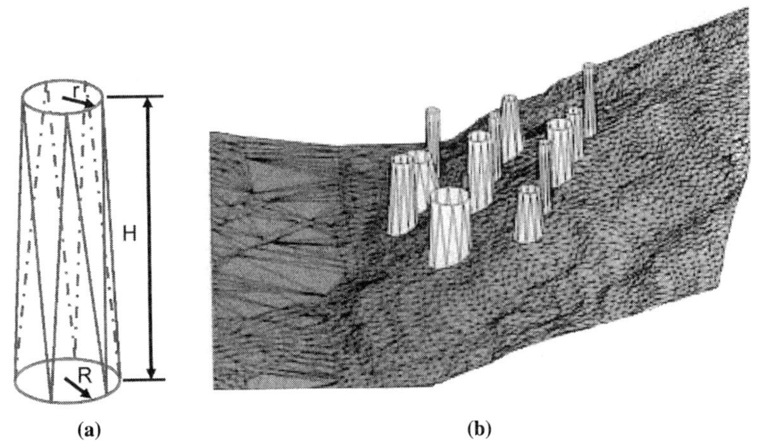

FIGURE 8-45
More sophisticated tree-modeling procedures: (*a*) triangulated model of tree trunk and (*b*) spatial distribution of trees on slope (Le Hir et al. 2004).

model called Rockyfor3D has been developed that simulates trajectories of falling, bouncing, and rolling rocks (Dorren, Berger, and Putters 2006; Dorren, Berger, et al. 2006; Stoffel et al. 2006). Rockyfor3D permits a more thorough analysis of the rockfall hazard because slope irregularities are taken into account, and it provides spatially dispersed values. Furthermore, it explicitly simulates the impact of falling rocks against trees. The influence of the spatial distribution of trees is assessed by the mean tree-free distance (MTFD). The MTFD indicates the mean distance a falling rock can travel in a given forest stand along the fall line between two tree impacts. Perret et al. (2004) defined the MTFD as follows:

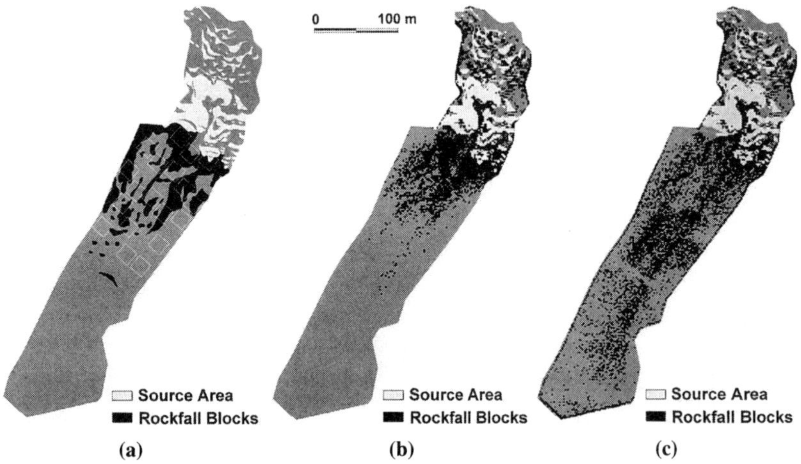

FIGURE 8-46
Evaluation of rockfall protection provided by forest in western Austria: (*a*) observed rockfall debris accumulations, (*b*) final locations of rockfall blocks according to simulation calibrated to current forest conditions, and (*c*) final locations of rockfall blocks according to simulation that has no forest (Dorren, Maier, et al. 2004).

$$\text{MTFD} = \frac{\text{evaluated area}}{[\text{number of trees}(\text{average DBH} + \text{average diameter of rock})]}$$

Le Hir et al. (2004) provided some insight into the initial concepts used in developing Rockyfor3D, including how to model a tree trunk to allow for its bending and shearing on impact and how to combine the spatial distribution of trees with slope geometry characteristics (Figure 8-45).

4.7.3 Evaluations of Rockfall Protection Provided by Forests

Numerous examples of evaluations of levels of rockfall protection provided by forests have been conducted by combining field tests and computer simulations (Dorren and Heuvelink 2004; Dorren, Berger, et al. 2004; Dorren, Maier, et al. 2004; Dorren, Heuvelink, and Berger 2004; Perret et al. 2004; Stoffel et al. 2006; Berger and Dorren 2007; Bigot et al. 2009; Bourrier et al. 2009; Rammer et al. 2010).

Dorren, Maier, et al. (2004) conducted field tests at a forested slope in western Austria and then used computer simulations with an early version of the Rockyfor3D model to evaluate the effectiveness of the protection provided by the forest (Figure 8-46). A map showing the limits of rockfall runout area was produced from field observations (Figure 8-46*a*). The computer simulations were then calibrated to reproduce closely the results of the mapped rockfall debris accumulations (Figure 8-46*b*). Subsequently the simulations were repeated with all forest influences removed (Figure 8-46*c*). The protection value of the forest was then assessed by comparing the distributions of rockfall blocks produced with and without the forest.

Perret et al. (2004) undertook a similar analysis for a forested slope in Liechtenstein with a two-dimensional profile-based rockfall simulation model developed at the University of Freiburg. The existing slope has only one fairly limited zone that is forested (Figure 8-47). Perret et al. calibrated their simulation model to match the observed or backcalculated energies and trajectory geometry of a single, almost cubical, limestone block with a volume of 3 m^3 (116 ft^3) that fell in April 2002. Their simulation achieved velocity and energy values, shown in Figure 8-47*b*, that relatively closely matched the backcalculated values. They then assumed that the entire slope was revegetated with a forest of mature trees. This forest was defined as "being of uniform composition, with trees having a DBH of 34 cm (~1 ft), with 350 trees per ha (140 trees/acre), and with MTFD of 15.5 m (~51 ft)." This was considered to represent the maximum feasible forest density and thus to offer the maximum rockfall protection. A simulation with this forest produced markedly smaller rockfall velocities and energies (Figure 8-47*c*).

Berger and Dorren (2007) conducted a series of rock-rolling field tests on a forested slope near Vaujany in the French Alps. The slope has a mean inclination of 38° and a total incline length of about 300 m (985 ft) and is completely forested. The mean stand density is 294 trees per hectare (117 trees per acre), and the mean DBH is about 37 cm (14.5 in.). The main tree species are silver fir (*Abies alba*, 57%), Norway spruce (*Picea abies*, 13%), European beech (*Fagus sylvatica*, 23%) and sycamore (*Acer pseudoplatanus*, 7%). A total of 57 rocks, averaging about 1 m (3 ft) in diameter, were released from a front-end loader at the top of the slope. Berger and Dorren used these experiments to validate the Rockfor·NET simulation program. Figure 8-48 summarizes their observations of cumulative percentages of rocks that were stopped with increasing distances down the slope below the release point. The percentage of rocks that were not stopped at each distance provides an estimate of the residual rockfall hazard.

The forest stopped all the test rocks before the base of the slope was reached at a distance of about 300 m (985 ft). However, a few rocks, about 5%, stopped within the first 20 m (65 ft). As the travel distance increased, the percentage of rocks that were stopped increased rapidly initially and then began to level off until about 80% were stopped at a distance of about 230 m (755 ft), where the slope was crossed by a forest road. Many of the rocks stopped on this level road surface. Those that continued farther down the slope apparently had lost some energy because there was an increase in the rate at which these rocks were stopped.

Berger and Dorren (2007) suggested that these observations could be estimated reasonably precisely by an empirical regression equation. They illustrated this phenomenon by a curved line on their published original version of Figure 8-48. However, the expression shown by Berger and

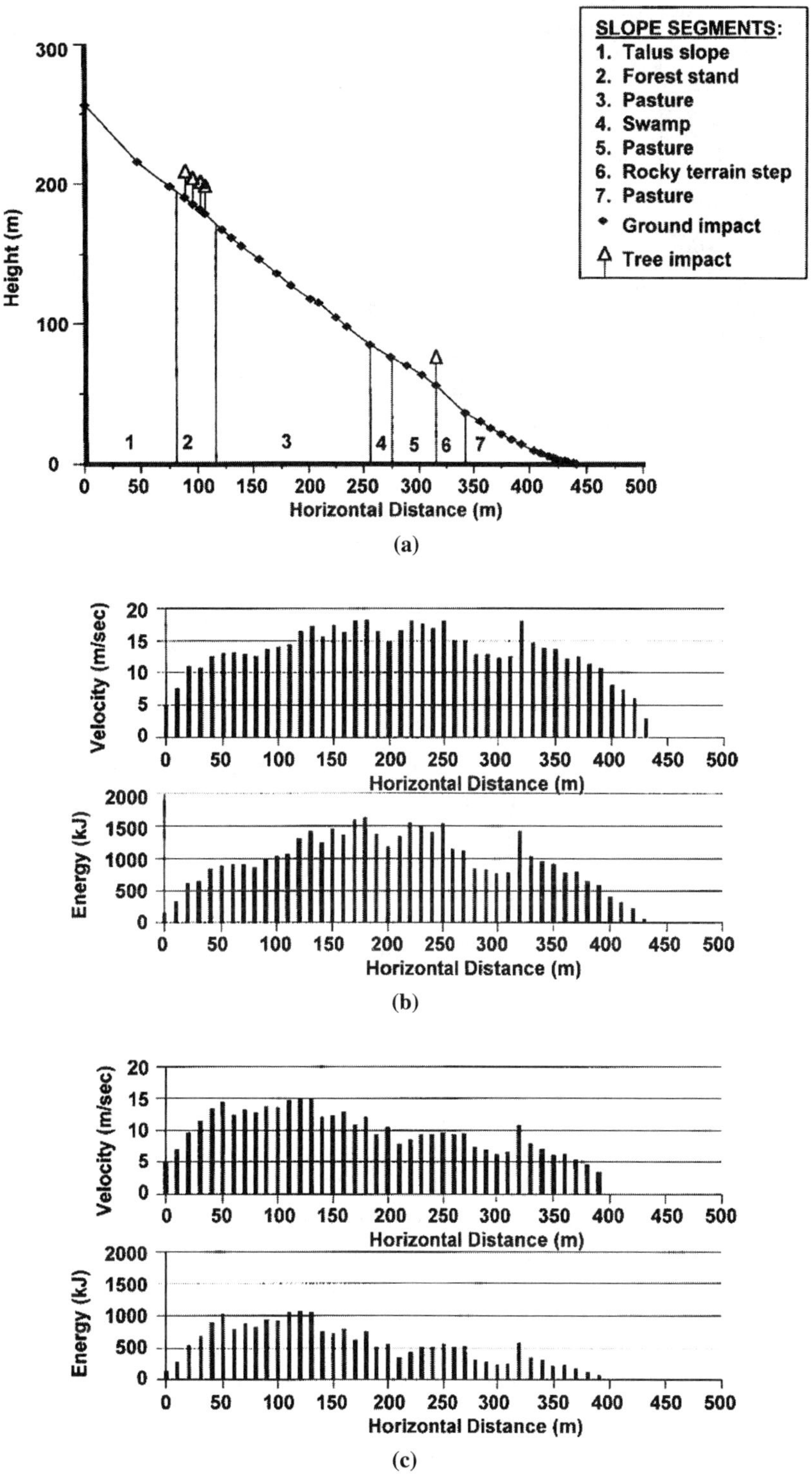

FIGURE 8-47
Assessment of rockfall protection that could be provided by reforestation of slope in Liechtenstein: (*a*) existing slope profile and conditions, (*b*) simulation model estimates of rockfall velocity and energy under existing conditions, and (*c*) simulation model estimates of rockfall velocity and energy after establishment of maximum reforestation (Perret et al. 2004).

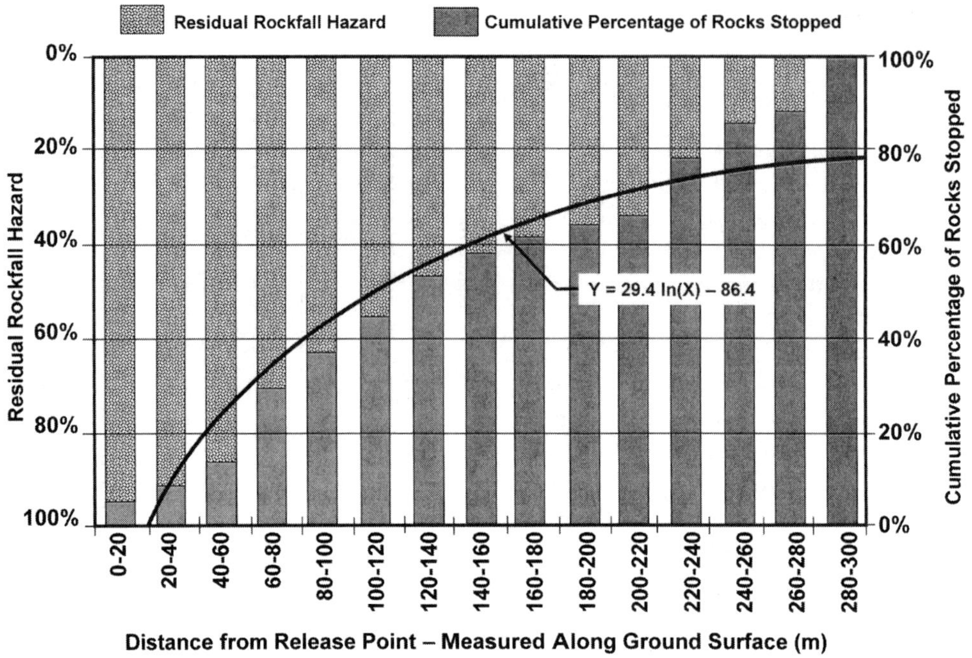

FIGURE 8-48 Cumulative percentages of rocks stopped, and thus residual rockfall hazard, with increasing distance from release point on a forested slope at Vaujany, France (Berger and Dorren 2007).

Dorren (2007) does not compute correct values, so, during the preparation of this chapter, the authors reanalyzed their data and derived the following equation:

$$Y = 29.4 \ln(X) - 86.4$$

where Y is the cumulative percentage of rocks that are stopped and $\ln(X)$ is the natural logarithm of the distance traveled down the slope (m).

Figure 8-48 shows this equation as a curved line that fairly closely fits the observed data within the range of 20 to 220 m (65 to 722 ft). It underestimates the observed percentages of rocks that stopped below the forest road and suggests that almost 20% of the rocks would continue beyond the base of the slope at a distance of 300 m (985 ft). It also fails to define the observed conditions at distances less than about 20 m (65 ft), and evaluation of the equation indicates that it predicts that no rocks will be stopped at a distance of 18.9 m (62 ft). Several investigators have suggested that forest stands with widths less than about 30 m (98 ft) provide little or no energy reduction for falling rocks passing through them. The tree curtains in the Rockfor^{NET} model of a forest are spaced at 33 m (108 ft), as illustrated in Figure 8-44.

Berger and Dorren (2007) suggested that this empirical relationship provides a reasonably accurate estimate of the energy-absorbing capability of any forest with slope and tree-density characteristics similar to those of the Vaujany site. The size of the rockfall blocks used in these tests was relatively modest [cubical blocks 1 m (3 ft) diameter on average]. Thus, this empirical relationship has several limitations, but it nevertheless provides a quick estimate for assessing possible forest rockfall hazard protection levels.

Rammer et al. (2010) presented the results of an iterative sequence of tests and improvements of a coupled rockfall and forest dynamics model. The forest dynamics component allows for the prediction of future forest characteristics as the trees mature. It allows for the development and analysis of alternative forest management and environmental scenarios that predict levels of rockfall hazard protection. The rockfall model component contained a recently modified rebound algorithm and included an algorithm to account for the sudden halt of falling rocks due to surface roughness. These capabilities greatly improved the correspondence between simulated and observed key rockfall variables such as runout distances, rebound heights, and bounce lengths. Rammer et al. used previous field test data from the Vaujany site described earlier and observations from actual rockfall events at Steg in Switzerland and Bad Isch in Austria.

Bigot et al. (2009) provided an instructive case history on the use of forests as a rockfall protection mechanism. A major rockfall event occurred in the French Alps in 1987, when a rock mass of 50 m^3 (1,765 ft^3) toppled and fractured into several individual blocks. Some blocks traveled through a forested slope and caused damage to the village of Saint Martin le Vinoux. Subsequently, rockfall fences were installed, but the forest was also considered to be a protection measure. In 1991, the forested slope was defined as protection forest, and an investigation to optimize the protective function of the existing forests was undertaken with a two-dimensional rockfall model. In 1992, forest mitigation efforts based on the recommendations from the investigation were undertaken; sections of the existing forests were cleared to create vegetation bands parallel to the contour lines separated by open spaces 20 m wide where additional trees were planted.

Bigot et al. (2009) reevaluated the site to assess whether this intervention was successful in improving the protective function of the forest. They evaluated the rockfall risk for the situation of 1987 (before the intervention), the current situation, and future situations as the forest evolves, by using model simulations with past, present, and future vegetation cover scenarios. They evaluated rockfall conditions with the simulation models Rockfor$^{\cdot NET}$ and Rockyfor3D. Both models correctly predicted that the forest was incapable of stopping rocks from the 1987 rockfall event. Furthermore, both models indicated an increase in the number of rocks reaching the base of the slope from 1987 onward. Rockyfor3D predicted that 11% of the rocks reached the base of the slope under 1987 conditions but predicted an increase to 19% in 2086. Rockfor$^{\cdot NET}$ predicted much larger values: 26% in 1987, increasing to 56% in 2086. Thus both models demonstrated that the 1992 forest mitigation methods did not improve the ability of the forest to reduce the rockfall hazard; the mitigation efforts had actually lessened the ability of the forest to reduce the rockfall hazard.

Bigot et al. concluded that a second attempt to increase the protective function of the forest was warranted with the aim of creating a dense coppice stand forest structure. A coppice stand refers to a traditional method of woodland management that takes advantage of the way in which many trees make new growth from the stump or roots when they are cut down. In a coppiced forest, young tree stems are repeatedly cut down to near ground level. In subsequent growth years, many new shoots will emerge, and after a number of years the coppiced tree, or stool, is ready to be harvested. Coppicing maintains trees at a juvenile stage; a regularly coppiced tree will never die of old age, and thus some coppice stools reach immense ages. Studies have shown that a forest composed of coppiced trees offers superior rockfall protection levels (Ciabocco et al. 2009).

5. VERIFICATION AND CALIBRATION OF PARAMETERS DESCRIBING ROCKFALL MECHANICS

Previous sections of this chapter describe the parameters and equations derived from Newtonian principles that control the motions of falling, bouncing, and rolling rock blocks. The calibration and verification of these parameters and equations are critically important in the development of acceptably accurate and defendable predictions concerning rockfall hazards.

Evaluation of rockfall hazards at any given location requires accurate determinations of the trajectories of falling rockfall blocks. These determinations require the best-possible knowledge of the mechanical principles that describe the motions of bouncing and rolling rocks and the energy losses during impact. However, the inherent heterogeneity present during real-world rockfall events, resulting from variations in the size, shape, and composition of blocks, the materials forming the slope, and the geometry of the slope surface, makes the selection of appropriate parameters extremely difficult. Numerical rockfall simulation models, described in Chapter 9, provide an effective, commonly used approach to the evaluation of rockfall hazards. Several approaches are used in verifying and calibrating the parameters used by these models:

- Backanalysis of natural rockfall events,
- Specifically designed field tests, and
- Laboratory experiments.

Labiouse and Heidenreich (2009) concluded that the complexity of the rebound process, particularly on sloping surfaces composed of relatively soft materials, makes it difficult to predict accurately the trajectories of bouncing and rolling rock blocks if they are described by only two unvarying coefficients of restitution and lumped-mass models. They suggested that predicting the motions of bouncing and rolling rocks traveling across surfaces covered with soft materials requires models that evaluate conditions at the contact surface between the block and the ground material. Furthermore, they recommended that predictions of rockfall trajectories always be checked against field observations before any delineation of areas at risk or the design of protection measures.

5.1 Backanalysis of Rockfall Events

After a natural rockfall event, detailed field identification and measurement of the tracks left by the blocks on the slope or on trees and observation of the distribution of previously fallen blocks provide valuable information that is used to analyze the kinematics of the rockfall by backanalysis. Several examples of these types of field observations have been published (Hungr and Evans 1988; Evans and Hungr 1993; Wieczorek et al. 1998; Wieczorek and Snyder 1999; Guzzetti et al. 2003; Jaboyedoff et al. 2005; Coe and Harp 2007). Backanalysis methods have been used for three distinct purposes:

- For determining the coefficients of restitution,
- For delineating rockfall hazard zones, and
- For evaluating geomorphological processes related to cliff retreat and talus (or scree) deposit development.

Backanalysis is a popular method for determining coefficients of restitution. Usually, the field data for the backanalysis provide a distance the rock traveled before coming to rest (this will be obvious), a mass for the rock (easy to measure), a source area (the original location of the rock may appear less weathered), and the spacing of a few impact points (marks or dents along the slope profile). Backanalysis of such events allows accurate estimation of the velocities and energies of the falling blocks as well as the rate of velocity or energy loss by each impact. Chapter 10 discusses the field procedures used to collect this information.

Backanalysis is also an important and commonly used method for delineating rockfall hazard zones. As discussed in Chapter 5 (Section 6.1), Switzerland has adopted regulations for delineating rockfall hazard zones that use backanalysis methods, at least in part (Raetzo et al. 2002). Several other Alpine countries have similar regulations. Hungr and Evans (1988) used backanalysis methods to evaluate their observations of rockfall hazards in southwestern British Columbia. This region has many glaciated valleys bordered by steep rock slopes, while the relatively level valley floors are populated and contain important transportation facilities. Talus deposits are common along the bases of the rocky slopes. Hungr and Evans defined areas of the valley floors that are covered with scattered large boulders, many of which have bounced or rolled well beyond the base of the talus slopes, as *rockfall shadow zones*. They used some energy conservation and dissipation concepts to provide an empirical evaluation of the rockfall hazard. Evans and Hungr (1993) expanded on these concepts. Similar backanalysis techniques have been used in evaluating the rockfall hazards in the Yosemite Valley (Wieczorek et al. 1998; Wieczorek and Snyder 1999; Guzzetti et al. 2003). Coe and Harp (2007) used backanalysis methods to evaluate rockfall susceptibility in the Wasatch Mountains of Utah. Jaboyedoff et al. (2005) described a number of refinements to basic backanalysis techniques that they believe provide improved rockfall hazard zonations.

Geologists and, more specifically, geomorphologists have long been interested in quantifying the processes that control the gradual erosion of mountains, and, of most interest to those concerned with rockfall, the processes that explain cliff retreat and the formation of talus (or scree) deposits (Behre 1933; Sharpe 1938; Dalrymple et al. 1968; Carson and Kirkby 1972; Chandler 1973; Statham and Francis 1986; Selby 1993). Initially, it was believed that the talus deposits exhibited an *angle of repose*, which was generally defined as the angle at which an avalanching, or sliding, mass of debris will come to rest (Carson and Kirkby 1972). This angle is perhaps better described as the *angle of residual shear* (Statham 1979). However, a number of field investigations (Chandler 1973; Kirkby and Statham 1975; Statham 1976) noted that most talus slopes have been at slightly less than this angle and that a number of other characteristics of talus deposits and slope geometry did not support the avalanching concept. A new model of the processes forming talus slopes that involved gradual accumulation of individual rockfall fragments was formulated by Kirkby and Statham (1975) and further defined by Statham (1976, 1979). The model development involved backanalysis evaluations of talus (scree) slopes on the Isle of Skye in western Scotland and model experiments. Statham (1979) used these investigations to introduce the concept of the *rolling friction coefficient* as a parameter to define the loss of energy that occurs during the rolling motion of a falling block along a talus slope (see Section 4.2).

Although backanalysis of natural events provides extremely useful and sometimes indispensable information allowing for verification and calibration of parameters that are then used in rockfall hazard evaluations at a given site, it has some limitations. Natural events frequently do not provide a sufficient number of observations to support statistically significant conclusions. The sizes and shapes of blocks in a single natural event often do not represent the full range of block characteristics that may be expected at the site. The observations result from natural events, not planned experiments; thus, they often do not encounter the full range of conditions at a site, including surface roughness, shape and strength of blocks, or variability of slope materials.

5.2 Field Tests

Full-scale field experiments, in which carefully selected rock blocks are released down a selected slope, provide critically important information concerning rockfall behavior. They supply not only valuable information on the kinematics of bouncing and rolling rocks, but also detailed information concerning the performance of proposed rockfall barrier systems. Consequently, many field experiments have been performed and evaluated worldwide (Ritchie 1963; Briolli 1977; Statham 1979; Japan Road Association 1983; Wu 1985; Bozzolo et al. 1988; Urciuoli 1988; Pfeiffer and Bowen 1989; Duffy et al. 1990; Fornaro et al. 1990; Kobayashi et al. 1990; Smith and Duffy 1990; Azzoni et al. 1992; Duffy 1992, 1996; Giani 1992; Evans and Hungr 1993; Kane and Duffy 1993; Azzoni and de Freitas 1995; Gerber 1995; Pierson et al. 2001; Giani et al. 2004; Bourrier et al. 2009; Giacomini et al. 2009).

Chapters 14, 15, and 16 discuss the testing and design of various types of rockfall protection systems. Chapter 14 (Section 2) provides extended discussion of rockfall catchment area design guidelines. Chapter 10 provides details on the methods used in conducting these field tests. Field tests are conducted by releasing selected rocks at the top of a slope that has been fully characterized with regard to material properties, geometry, and surface roughness. The block trajectories are recorded by high-speed video cameras, and the records are analyzed to determine block velocities and bounce heights. These analyses, described in Chapter 7, permit the computation of the coefficients of restitution and other pertinent parameters.

A large number of field rockfall tests were carried out at ISMES in Bergamo, Italy, between 1989 and 1995 on a variety of natural rock slopes. The tests supported the initial analyses of how rockfall trajectories are influenced by the characteristics of the block (size, shape, and mechanical properties) and those of the slope materials (nature, geometry). These analyses were reported by Azzoni et al. (1992) and Azzoni and de Freitas (1995). Additional field studies have evaluated both the coefficients of restitution and rolling friction (Statham 1979; Bozzolo and Pamini 1986; Giani 1992; Azzoni and de Freitas 1995; Okura, Kitahara, et al. 2000; Giani et al. 2004). These coefficients, together with the geometry of the slope and the falling blocks, are frequently identified as the most relevant parameters for defining the characteristics of rockfall motion.

Bourrier et al. (2009) used the results from full-scale field tests undertaken in Vaujany, France, to evaluate alternative methods of representing surface roughness in numerical simulation models. Figure 8-49 shows comparisons of the observed and simulated distributions of velocities, bounce heights, and translational kinetic energies of falling blocks. The field data values were calculated by backanalysis methods applied to digital video recordings taken at an observation point on the slope, while the simulation values were extracted from two simulation models and represent conditions at the same location. Model A used a three-class definition of slope roughness, as described in Section 4.5 and Figure 8-35, while Model B used a single value representing the average slope roughness. The results from the field observations and the simulations show good agreement. Bourrier et al. noted that the field observations were based on only 100 trials, while the simulations were based on at least 1,000 repetitions; thus, the simulations tended to produce more regular distributions and their maximum values were slightly larger.

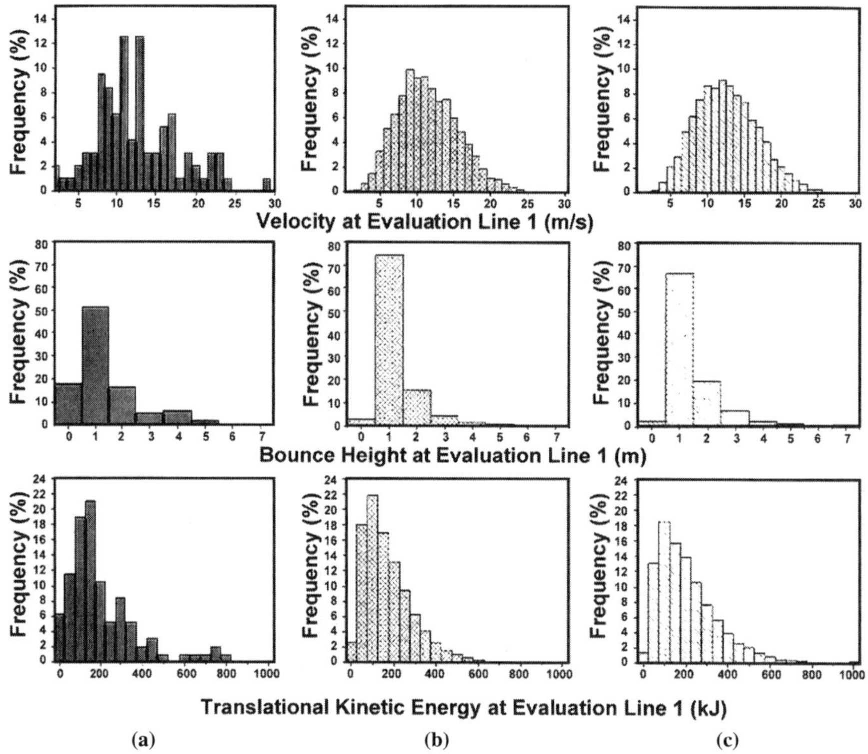

FIGURE 8-49 Distributions of velocities, bounce heights, and translational kinetic energies based on field trials and two simulation methods for fixed observation location on slope at Vaujany, France: (*a*) field observations, (*b*) Model A (three class values), and (*c*) Model B (average value) (Bourrier et al. 2009).

Field tests are important in the study of rockfalls because they alone allow the visualization of many different and complicated aspects of this highly random phenomenon. These tests permit a full-scale study of the rockfall mechanics, thus avoiding many of the limitations inherent in smaller-scale experiments. They allow the assessment of relevant physical parameters and so offer opportunities for the correct calibration of mathematical models (Azzoni and de Freitas 1995).

5.3 Laboratory Experiments

Field tests are expensive and often only produce relatively small numbers of observations. Back-analyses of natural rockfall events are also limited by small numbers of observations and frequently exhibit even less variation in block and slope characteristics than do field tests. Thus, both methods have limitations for statistical and parametric analysis, and, consequently, laboratory tests at various scales are often used to provide important supplemental information required for accurate parameter validation.

Laboratory rockfall studies use experiments with models of various scales. In some cases, a whole trajectory is simulated (Camponuovo 1977; Azimi et al. 1982; Statham 1979; Ujihira et al. 1993; Murata and Shibuya 1997), while other experiments concentrate on a single, usually the initial, impact (Kawahara and Muro 1999; Kamijo et al. 2000; Wong et al. 2000; Ushiro et al. 2000; Chau et al. 1998; Chau, Wong, et al. 1999; Chau, Wu, et al. 1999; Chau et al. 2002). Both classes of experiments are helpful in understanding rockfall mechanics and allow insight into the most important parameters. However, the quantitative interpretation of the results is not straightforward because of difficulties in matching the similitude requirements for all the parameters involved in the dynamic process. On the basis of small-scale experiments, Camponuovo (1977) emphasized that, while it is possible to determine and match the static characteristics of materials (such as deformability, strength, density, and static friction angle) even with small-scale models, it is not easy to match the dynamic characteristics (such as the coefficient of restitution, impact strength, and dynamic friction angle) with such models. According to Camponuovo, a small-scale model rock that meets mechanical similitude requirements typically provides a far lower value of the coefficient of restitution than the one observed with full-scale rebound tests.

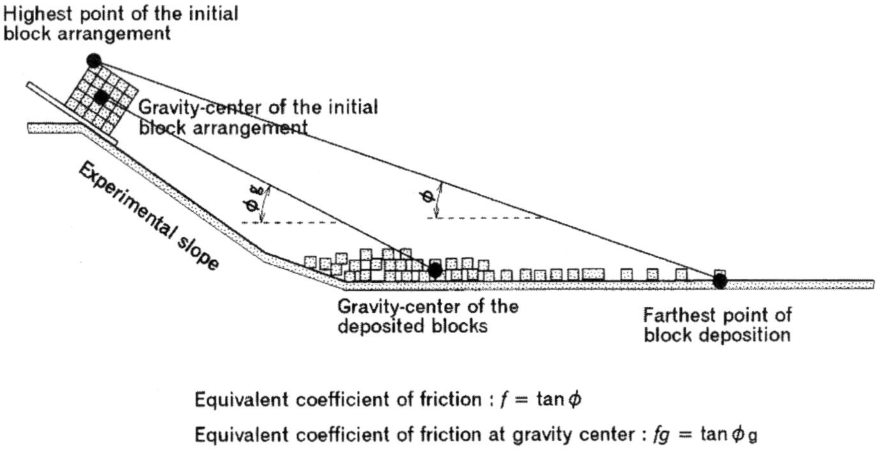

FIGURE 8-50 Experimental setup and resulting coefficient of friction concepts from tests with multiple blocks (Okura, Kitahara, et al. 2000).

Heidenreich (2004) and Labiouse and Heidenreich (2009) reported the results of an extensive series of relatively large-scale ("half-scale") laboratory tests that were carried out at the Rock Mechanics Laboratory of the Swiss Federal Institute of Technology, Lausanne. The investigations were designed to acquire better knowledge of the bouncing phenomenon of rock blocks on soft ground materials and to investigate the influence of various impact parameters. The tests used some test blocks previously used to evaluate the impacts of rocks on rock sheds (Labiouse et al. 1996). The division between relatively large-scale experiments conducted in a laboratory and in situ field tests depends to a considerable degree on the questions being investigated; several reported results combine laboratory and in situ field tests (Okura, Kitahara, and Sammori 2000; Okura, Kitahara, et al. 2000; Chau et al. 2002; Giani et al. 2004; Pichler et al. 2005).

Okura and colleagues undertook a series of tests to investigate the mechanical processes that influence rockfalls and rockslides that have multiple fragments. In this respect, their results are unique. They were especially concerned with the widely observed characteristic extensive runout distance exhibited by very large and very rapidly moving rockslides. They began with small-scale experiments in the laboratory (Okura, Kitahara, and Sammori 2000). They then undertook a series of large-scale experiments that are relevant to typical rockfall processes (Okura, Kitahara, et al. 2000). Figure 8-50 is a schematic representation of their experimental setup. They created an outdoor slope paved with granite slabs, down which they tipped various numbers of 10- and 20-cm (approximately 4- and 8-in.) cubical granite blocks.

The largest number of blocks tested was 1,000 of the smaller blocks that were initially stacked in a regular 10 × 10 × 10 cube arrangement. In each test the blocks on the front, top, and rear sides of the stack were color-coded with distinctive colors. Each experiment was recorded by high-speed video cameras. This permitted the tracking of each individual block, although the color codes were used to verify the accuracy of the tracking. The results showed that the blocks on the front face of the stack traveled the greatest distance down the slope; those on the top generally stopped in an average distance down the slope; and those on the rear face of the stack traveled the shortest distance. In other words, blocks did not overtake each other as they traveled down the slope. The sorting by distance was greatest with the largest number of blocks; this spreading effect was not influenced by the total mass of the blocks. The authors suggested that these results represent the exchange of energies as blocks strike each other. Blocks in the front of the mass tend to be accelerated by impacts from blocks behind them; those toward the rear are slowed down by the same impacts. Accordingly, rockfall events with many fragments should have a larger maximum runout distance than otherwise identical rockfall events with only a few fragments. Okura, Kitahara, et al. (2000) defined two coefficients of friction, one related to the center of gravity (or mass) of the distributed rock blocks and one related to the most distant rock block (Figure 8-50).

REFERENCES

ABBREVIATION

FHWA Federal Highway Administration

Aguiar, C. E., and F. Laudares. 2003. Listening to the Coefficient of Restitution and the Gravitational Acceleration of a Bouncing Ball. *American Journal of Physics*, Vol. 71, No. 5, pp. 499–501.

Azimi, C., P. Desvarreux, A. Giraud, and J. Martin-Cocher. 1982. Méthode de Calcul de la Dynamique des Chutes de Blocs—Application à l'Étude du Versant de la Montagne de La Pale (Vercors) [Calculation Method for Rockfall Dynamics—Application to the Study of the Side of La Pale Mountain (Vercors)]. *Bulletin de Liaison des Laboratoires des Ponts et Chaussées*, Vol. 122, pp. 93–102 (in French).

Azzoni, A., and M. H. de Freitas. 1995. Experimentally Gained Parameters, Decisive for Rock Fall Analysis. *Rock Mechanics and Rock Engineering*, Vol. 28, No. 2, pp. 111–124.

Azzoni, A., G. La Barbera, and A. Zaninetti. 1995. Analysis and Prediction of Rock Fall Using a Mathematical Model. *International Journal of Rock Mechanics and Mining Sciences and Geomechanics Abstracts*, Vol. 32, No. 7, pp. 709–724.

Azzoni, A., P. P. Rossi, E. Drigo, G. P. Giani, and A. Zaninetti. 1992. In Situ Observation of Rockfall Analysis Parameters. *Proc., 6th International Symposium on Landslides*, Christchurch, New Zealand, Balkema, Rotterdam, Netherlands, Vol. 1, pp. 307–314.

Barbieri, G., G. P. Giani, G. Uras, and A. Vernier. 1988. Modellizzazione della frana di crollo di Monte Oili in agro di Baunei (Nuoro) [Modeling of the Landslide Collapse in the Countryside of Monte Oili Baunei (Nuoro)]. Presented at plenary meeting of the Italian National Research Council, Rome, Dec. 19–20 (in Italian).

Bedford, A., and W. Fowler. 1995. *Dynamics: Engineering Mechanics*. Addison-Wesley, Reading, Mass.

Behre, C. H. 1933. Talus Behavior Above Timber in the Rocky Mountains. *Journal of Geology*, Vol. 41, pp. 622–635.

Berger, F., and L. K. A. Dorren. 2006a. Objective Comparison of Rockfall Models Using Real Size Experimental Data. In *Disaster Mitigation of Debris Flows, Slope Failures and Landslides* (H. Marui et al., eds.), Universal Academy Press, Inc., Tokyo, pp. 245–252.

Berger, F., and L. K. A. Dorren. 2006b. Rockfor^{NET}: A New Efficient Tool for Quantifying the Residual Rockfall Hazard of a Forested Slope. In *Disaster Mitigation of Debris Flows, Slope Failures and Landslides* (H. Marui et al., eds.), Universal Academy Press, Inc., Tokyo, pp. 229–235.

Berger, F., and L. K. A. Dorren. 2007. Principles of the Tool Rockfor^{NET} for Quantifying the Rockfall Hazard Below a Protection Forest. *Schweizerische Zeitschrift für Forstwesen*, Vol. 158, No. 6, pp. 157–165.

Berger, F., C. Quetel, and L. K. A. Dorren. 2002. Forest: A Natural Protection Means Against Rockfall, but with Which Efficiency? The Objectives and Methodology of the ROCKFOR Project. *Proc., Interpraevent 2002 in the Pacific Rim*, Matsumoto, Japan, pp. 815–826.

Bernstein, A. D. 1977. Listening to the Coefficient of Restitution. *American Journal of Physics*, Vol. 45, No. 1, pp. 41–44.

Bigot, C., L. K. A. Dorren, and F. Berger. 2009. Quantifying the Protective Function of a Forest Against Rockfall for Past, Present and Future Scenarios Using Two Modelling Approaches. *Natural Hazards*, Vol. 49, No. 1, pp. 99–111.

Bourrier, F., L. K. A. Dorren, F. Nicot, F. Berger, and F. Darve. 2009. Toward Objective Rockfall Trajectory Simulation Using a Stochastic Impact Model. *Geomorphology*, Vol. 110, pp. 68–79.

Bozzolo, D., and R. Pamini. 1986. Simulation of Rock Falls down a Valley Side. *Acta Mechanica*, Vol. 63, pp. 113–130.

Bozzolo, D., R. Pamini, and K. Hutter. 1988. Rockfall Analysis—A Mathematical Model and Its Test with Field Data. *Proc., 5th International Symposium on Landslides* (C. Bonnard, ed.), July 10–15, Lausanne, Switzerland, Balkema, Vol. 1, pp. 555–560.

Brang, P., W. Schönenberger, E. Ott, and R. H. Gardner. 2001. Forests as Protection from Natural Hazards. In *The Forests Handbook* (J. Evans, ed.), Blackwell Science Ltd., Oxford, United Kingdom, pp. 53–81.

Brauner, M., W. Weinmeister, A. Agner, S. Vospernik, and B. Hoesle. 2005. Forest Management Decision Support for Evaluating Forest Protection Effects Against Rockfall. *Forest Ecology and Management*, Vol. 207, Nos. 1–2, pp. 75–85.

Briolli, L. 1977. Relations Between Scree Slope Morphometry and Dynamics of Accumulation Processes. *Proc., Meeting on Rockfall Dynamics and Protective Works Effectiveness*, Istituto

Sperimentale Modelli e Strutture SpA, Bergamo, Italy, pp. 11–23.

Budetta, P., and A. Santo. 1994. Morphostructural Evolution and Related Kinematics of Rockfalls in Campania (Southern Italy): A Case Study. *Engineering Geology*, Vol. 36, pp. 197–210.

Camponuovo, G. F. 1977. ISMES Experience on the Model of St. Martino. *Proc., Meeting on Rockfall Dynamics and Protective Works Effectiveness*, Istituto Sperimentale Modelli e Strutture SpA, Bergamo, Italy, pp. 25–39.

Carson, M. A., and M. J. Kirkby. 1972. *Hillslope Form and Process*. Cambridge University Press.

Chandler, R. J. 1973. The Inclination of Talus, Arctic Talus Terraces, and Other Slopes Composed of Granular Materials. *Journal of Geology*, Vol. 81, pp. 1–44.

Chau, K. T., R. H. C. Wong, and C. F. Lee. 1998. Rockfall Problems in Hong Kong and Some New Experimental Results for Coefficients of Restitution. *International Journal of Rock Mechanics and Mining Sciences*, Vol. 35, Nos. 4–5, pp. 662–663.

Chau, K. T., R. H. C. Wong, J. Liu, J. J. Wu, and C. F. Lee. 1999. Shape Effects on the Coefficient of Restitution During Rockfall Impacts. *Proc., 9th International Congress on Rock Mechanics*, Paris, pp. 541–544.

Chau, K. T., R. H. C. Wong, and J. J. Wu. 2002. Coefficient of Restitution and Rotational Motions of Rockfall Impacts. *International Journal of Rock Mechanics and Mining Sciences*, Vol. 39, pp. 69–77.

Chau, K. T., J. J. Wu, R. H. Wong, and C. F. Lee. 1999. The Coefficient of Restitution for Boulders Falling onto Soil Slopes with Various Values of Dry Density and Water Content. *Proc., International Symposium on Slope Stability Engineering: Geotechnical and Geoenvironmental Aspects*, Nov. 8–11, Matsuyama, Shikoku, Japan, Vol. 2, pp. 1355–1360.

Ciabocco, G., L. Boccia, and M. Ripa. 2009. Energy Dissipation of Rockfalls by Coppice Structures. *Natural Hazards and Earth System Sciences*, Vol. 9, pp. 993–1001.

Coe, J. A., and E. L. Harp. 2007. Influence of Tectonic Folding on Rockfall Susceptibility, American Fork Canyon, Utah, USA. *Natural Hazards and Earth System Sciences*, Vol. 7, pp. 1–14.

Cross, R. 1999. The Bounce of a Ball. *American Journal of Physics*, Vol. 67, No. 3, pp. 222–227.

Cross, R. 2000. The Coefficient of Restitution for Collisions of Happy Balls, Unhappy Balls, and Tennis Balls. *American Journal of Physics*, Vol. 68, No. 11, pp. 1025–1031.

Crosta, G. B., and F. Agliardi. 2004. Parametric Evaluation of 3D Dispersion of Rockfall Trajectories. *Natural Hazards and Earth System Sciences*, Vol. 4, pp. 583–598.

Dalrymple, J. B., R. J. Blong, and A. J. Conacher. 1968. A Hypothetical Nine-Unit Land Surface Model. *Zeitschrift für Geomorphologie*, Vol. 12, pp. 60–76.

Descoeudres, F., and T. H. Zimmermann. 1987. Three-Dimensional Dynamics Calculation of Rock Fall. *Proc., 6th International Congress on Rock Mechanics*, Montreal, Canada (G. Herget and S. Vongpaisal, eds.), Balkema, Rotterdam, Netherlands, Vol. 1, pp. 337–342.

Dorren, L. K. A. 2010. *Rockyfor3D Revealed: Description of the Complete 3D Rockfall Model*. ecorisQ, Saint Hilaire du Touvet, France. http://www.ecorisq.org.

Dorren, L. K. A., and F. Berger. 2006. Stem Breakage of Trees and Energy Dissipation During Rockfall Impacts. *Tree Physiology*, Vol. 26, pp. 63–71.

Dorren, L. K. A., and G. B. M. Heuvelink. 2004. Effect of Support Size on the Accuracy of a Distributed Rockfall Model. *International Journal of Geographical Information Science*, Vol. 18, pp. 595–609.

Dorren, L. K. A., F. Berger, A. C. Imeson, B. Maier, and F. Rey. 2004. Integrity, Stability and Management of Protection Forests in the European Alps. *Forest Ecology and Management*, Vol. 195, Nos. 1–2, pp. 165–176.

Dorren, L. K. A., F. Berger, C. Le Hir, E. Mermin, and P. Tardif. 2005. Mechanisms, Effects and Management Implications of Rockfall in Forests. *Forest Ecology and Management*, Vol. 215, Nos. 1–3, pp. 183–195.

Dorren, L. K. A., F. Berger, E. Mermin, and P. Tardif. 2006. Results of Real Size Rockfall Experiments on Forested and Nonforested Slopes. In *Disaster Mitigation of Debris Flows, Slope Failures and Landslides* (H. Marui et al., eds.), Universal Academy Press, Inc., Tokyo, pp. 223–228.

Dorren, L. K. A., F. Berger, and U. S. Putters. 2006. Real Size Experiments and 3D Simulation of Rockfall on Forested and Nonforested Slopes. *Natural Hazards and Earth System Sciences*, Vol. 6, pp. 145–153.

Dorren, L. K. A., G. B. M. Heuvelink, and F. Berger. 2004. Effect of Support Size on the Accuracy of Spatial Models: Findings of Rockfall Simulations on Forested Slopes. *Proc., Joint Meeting,*

Environmetrics Society and the Spatial Accuracy Symposium, June 28–July 1, Portland, Maine.

Dorren, L. K. A., B. Maier, U. S. Putters, and A. C. Seijmonsbergen. 2004. Combining Field and Modelling Techniques to Assess Rockfall Dynamics on a Protection Forest Hillslope in the European Alps. *Geomorphology*, Vol. 57, No. 3, pp. 151–167.

Duffy, J. D. 1992. *Field Tests of Flexible Rockfall Barriers*. Brugg Technical Note.

Duffy, J. D. 1996. *Field Tests and Evaluation of Hi-Tech Low Energy Chain Link Rockfall Fence*. Report CA/05-96-01. California Department of Transportation, San Luis Obispo.

Duffy, J. D., D. D. Smith, and K. R. White. 1990. Field Tests and Evaluation of Rockfall Restraining Nets. *Proc., 41st Annual Highway Geology Symposium*, Albuquerque, N. Mex., pp. 125–141.

Evans, C. L. 1989. *The Design of Catch Bench Geometry in Surface Mines to Control Rockfall*. Master's thesis. University of Arizona, Tempe.

Evans, S. G., and O. Hungr. 1993. The Assessment of Rock Fall Hazard at the Base of Talus Slopes. *Canadian Geotechnical Journal*, Vol. 30, pp. 620–636.

Falcetta, J. L. 1985. Un Nouveau Modèle de Calcul de Trajectoires de Blocs Rocheux (A New Calculation Model of Rock Block Trajectories). *Revue Française de Géotechnique*, Vol. 30, pp. 11–17 (in French).

FHWA. 1993. *Rockfall Hazard Mitigation Methods*. Publication FHWA SA-93-085. U.S. Department of Transportation.

Fornaro, M., D. Peila, and M. Nebbia. 1990. Block Falls on Rock Slopes: Application of a Numerical Simulation Program to Some Real Cases. *Proc., 6th International Congress, International Association of Engineering Geology*, Amsterdam, Netherlands (D. G. Price, ed.), Balkema, Rotterdam, Netherlands, pp. 2173–2180.

Gerber, W. 1995. *Vitznau Auswertung von Feldversuchen (Vitznau Evaluation of Field Trials)*. Documentation, Forest FAN-Course Group on Natural Hazards with the Theme "Falling Rocks," Vitznau, Switzerland, Oct., Chapter 9 (in German).

Giacomini, A., O. Buzzi, B. Renard, and G. P. Giani. 2009. Experimental Studies on Fragmentation of Rock Falls on Impact with Rock Surfaces. *International Journal of Rock Mechanics and Mining Sciences*, Vol. 46, pp. 708–715.

Giani, G. P. 1992. *Rock Slope Stability Analysis*. Balkema, Rotterdam, Netherlands.

Giani, G. P., A. Giacomini, M. Migliazza, and A. Segalini. 2004. Experimental and Theoretical Studies to Improve Rock Fall Analysis and Protection Work Design. *Rock Mechanics and Rock Engineering*, Vol. 37, No. 5, pp. 369–389.

Goldsmith, W. 1960. *Impact: The Theory and Physical Behaviour of Colliding Solids*. Edward Arnold Publishers, London.

Guzzetti, F., P. Reichenbach, and G. F. Wieczorek. 2003. Rockfall Hazard and Risk Assessment in the Yosemite Valley, California, USA. *Natural Hazards and Earth System Sciences*, Vol. 3, pp. 491–503.

Habib, P. 1977. Note sur le Rebondissement des Blocs Rocheux (Note on the Rebound of Boulders). *Proc., Meeting on Rock Fall Dynamics and Protective Works Effectiveness*, Vol. 90, Istituto Sperimentale Modelli e Strutture SpA, Bergamo, Italy, pp. 123–125 (in French).

Heidenreich, B. 2004. *Small- and Half-Scale Experimental Studies of Rockfall Impacts on Sandy Slopes*. Doctoral thesis. École Polytechnique Fédérale de Lausanne, Lausanne, Switzerland.

Heierli, W., A. Merk, and A. Temperli. 1985. *Schutz Gegen Steinschlag/Protection Contre Chutes de Pierres (Protection Against Falling Rocks)*, 2nd ed. Research Report (Forschungsarbeit) 21/83. Federal Department of the Environment, Transport, Energy, and Communications, Federal Roads Office, Bern, Switzerland.

Hétu, B., and J. T. Gray. 2000. Effects of Environmental Change on Scree Slope Development Throughout the Postglacial Period in the Chic-Choc Mountains in the Northern Gaspé Peninsula, Québec. *Geomorphology*, Vol. 32, pp. 335–355.

Higgins, J. D., R. D. Andrew, and T. J. Pfeiffer. 2003. Computer Simulation for Mitigation of Colorado Rockfall. In *Engineering Geology in Colorado: Contributions, Trends, and Case Histories* (CD-ROM) (D. D. Boyer, P. M. Santi, and W. P. Rogers, eds.), Special Publication 15, Association of Engineering Geologists.

Hoek, E. 1987. *Rockfall: A Program in BASIC for the Analysis of Rockfall from Slopes*. Golder Associates and University of Toronto, Canada.

Holm, K., and M. Jakob. 2009. Long Rockfall Runout, Pascua Lama, Chile. *Canadian Geotechnical Journal*, Vol. 46, pp. 225–230.

Hungr, O., and S. G. Evans. 1988. Engineering Evaluation of Fragmental Rockfall Hazards. *Proc., 5th International Symposium on Landslides* (C. Bonnard, ed.), Vol. 1, Balkema, Rotterdam, Netherlands, pp. 685–690.

Imre, B., S. Räbsamen, and S. M. Springman. 2008. A Coefficient of Restitution of Rock Materials. *Computers and Geosciences*, Vol. 34, pp. 339–350.

Jaboyedoff, M., J. P. Dudt, and V. Labiouse. 2005. An Attempt to Refine Rockfall Hazard Zoning Based on the Kinetic Energy, Frequency and Fragmentation Degree. *Natural Hazards and Earth System Sciences*, Vol. 5, pp. 621–632.

Japan Road Association. 1983. *Rockfall Handbook*. Maruzen Publisher, Tokyo (in Japanese).

Jones, C. L., J. D. Higgins, and R. D. Andrew. 2000. *Colorado Rockfall Simulation Program Version 4.0*. Colorado Department of Transportation, Colorado Geological Survey, Denver.

Kamijo, A., S. Onda, H. Masuya, and Y. Tanaka. 2000. Fundamental Test on Restitution Coefficient and Frictional Coefficient of Rock Fall. *Proc., 5th Symposium on Impact Problems in Civil Engineering*, June 8–9, Japan Society of Civil Engineers, pp. 83–86.

Kane, W. F., and J. D. Duffy. 1993. *Brugg Low Energy Wire Rope Rockfall Net Field Tests*. Technical Research Report 93-01. Department of Civil Engineering, University of the Pacific, Stockton, Calif.

Kawahara, S., and T. Muro. 1999. Effect of Soil Slope Gradient on Motion of Rockfall. *Proc., International Symposium on Slope Stability Engineering*, Vol. 2, Matsuyama, Shikoku, Japan, pp. 1343–1348.

Kawahara, S., and T. Muro. 2006. Effects of Dry Density and Thickness of Sandy Soil on Impact Response due to Rockfall. *Journal of Terramechanics*, Vol. 43, pp. 329–340.

Kirkby, M. J., and I. Statham. 1975. Surface Slope Movement and Scree Formation. *Journal of Geology*, Vol. 83, pp. 349–362.

Kobayashi, Y., E. L. Harp, and T. Kagawa. 1990. Simulation of Rockfalls Triggered by Earthquakes. *Rock Mechanics and Rock Engineering*, Vol. 23, pp. 1–20.

Kuninaka, H., and H. Hayakawa. 2004. Anomalous Behavior of the Coefficient of Normal Restitution in Oblique Impact. *Physical Review Letters*, Vol. 93, No. 15, pp. 154301-1–154301-4.

Labiouse, V., and B. Heidenreich. 2009. Half-Scale Experimental Study of Rockfall Impacts on Sandy Slopes. *Natural Hazards and Earth System Sciences*, Vol. 9, pp. 1981–1993.

Labiouse, V., F. Descoeudres, and S. Montani. 1996. Experimental Study of Rock Sheds Impacted by Rock Blocks. *Structural Engineering International*, Vol. 6, No. 3, pp. 171–176.

Le Hir, C., F. Berger, L. K. A. Dorren, and C. Quetel. 2004. Forest: A Natural Means of Protection Against Rockfall, but How to Reach Sustainable Mitigation? *Proc., International Symposium Interpraevent 2004*, Riva–Trient, Italy, Vol. 2 (V), pp. 59–69.

Marion, J. B., and S. T. Thornton. 1995. *Classical Dynamics of Particles and Systems*, 4th ed. Saunders College Publishing, Fort Worth, Tex.

Motta, R. 1998. Protection Forests in the European Alps: Sustainability of Nonintervention and Minimal Tending Measures. *Proc., International Union of Forest Research Organizations Interdivisional Seoul Conference*, Oct. 12–17, Seoul, Korea, pp. 224–231.

Motta, R., and J. C. Haudemand. 2000. Protective Forests and Silvicultural Stability: An Example of Planning in the Aosta Valley. *Mountain Research and Development*, Vol. 20, No. 2, pp. 180–187.

Murata, S., and H. Shibuya. 1997. Measurement of Impact Loads on the Rockfall Prevention Walls and Speed of Falling Rocks Using a Middle Size Slope Model. *Proc., 2nd Asia–Pacific Conference on Shock and Impact Loads on Structures*, Melbourne, Australia, Nov. 25–27, pp. 383–393.

Okura, Y., H. Kitahara, and T. Sammori. 2000. Fluidization in Dry Landslides. *Engineering Geology*, Vol. 56, pp. 347–360.

Okura, Y., H. Kitahara, T. Sammori, and A. Kawanami. 2000. The Effects of Rockfall Volume on Runout Distance. *Engineering Geology*, Vol. 58, pp. 109–124.

Ottitsch, A., and G. Weiss. 2000. European Mountain Forest Policies: A Comparative Analysis. In *Forests in Sustainable Mountain Development: A State of Knowledge Report for 2000* (M. F. Price and N. Butt, eds.), International Union of Forest Research Organizations Research Series 5, CABI Publishing, Wallingford, United Kingdom, pp. 415–421.

Paronuzzi, P. 1989. Probabilistic Approach for Design Optimization of Rockfall Protective Barriers. *Quarterly Journal of Engineering Geology*, Vol. 22, pp. 175–183.

Perret, S., F. Dolf, and H. Kienholz. 2004. Rockfall into Forests: Analysis and Simulation of Rockfall Trajectories—Considerations with Respect to Mountainous Forests in Switzerland. *Landslides*, Vol. 2, pp. 123–130.

Petje, U., M. Ribčič, and M. Mikoš. 2005. Computer Simulation of Stone Falls and Rockfalls. *Acta Geographica Slovenica*, Vol. 45, No. 2, pp. 93–120.

Pfeiffer, T. J., and T. Bowen. 1989. Computer Simulation of Rock Falls. *Association of Engineering Geologists Bulletin*, Vol. 26, No. 1, pp. 135–146.

Pfeiffer, T. J., and J. D. Higgins. 1990. Rockfall Hazard Analysis Using the Colorado Rockfall Simulation Program. In *Transportation Research Record 1288*, Transportation Research Board, National Research Council, Washington, D.C., pp. 117–126.

Pichler, B., C. Hellmich, and H. A. Mang. 2005. Impact of Rocks onto Gravel: Design and Evaluation of Experiments. *International Journal of Impact Engineering*, Vol. 31, pp. 559–578.

Pierson, L. A., C. F. Gullixson, and R. G. Chassie. 2001. *Rockfall Area Design Guide*. Final Report SPR-3(032). Oregon Department of Transportation, Salem.

Piteau, D. R., and R. Clayton. 1977. Discussion of Paper "Computerized Design of Rock Slopes Using Interactive Graphics for the Input and Output of Geometrical Data by P. Cundall." *Proc., 16th Symposium on Rock Mechanics*, Minneapolis, Minn., American Rock Mechanics Association, pp. 62–63.

Raetzo, H., O. Lateltin, D. Bollinger, and J. P. Tripet. 2002. Hazard Assessment in Switzerland: Codes of Practice for Mass Movements. *Bulletin of Engineering Geology and the Environment*, Vol. 61, pp. 263–268.

Rammer, W., M. Brauner, L. K. A. Dorren, F. Berger, and M. J. Lexer. 2010. Evaluation of a 3-D Rockfall Module Within a Forest Patch Model. *Natural Hazards and Earth System Sciences*, Vol. 10, pp. 699–711.

Ritchie, A. M. 1963. Evaluation of Rockfall and Its Control. In *Highway Research Record 17*, Highway Research Board, National Research Council, Washington, D.C., pp. 13–28.

Robotham, M. E., H. Wang, and G. Walton. 1995. Assessment of Risk from Rockfall from Active and Abandoned Quarry Slopes. *Transactions of the Institute of Mining and Metallurgy Series A*, Vol. 104, pp. A25–A33.

Schönenberger, W., A. Noack, and P. Thee. 2005. Effect of Timber Removal from Windthrow Slopes on the Risk of Snow Avalanches and Rockfall. *Forest Ecology and Management*, Vol. 213, pp. 197–208.

Selby, M. J. 1993. *Hillslope Materials and Processes*, 2nd ed. Oxford University Press.

Sharpe, C. F. S. 1938. Landslides and Related Phenomena. Columbia University Press, N.Y.

Smith, D. D., and J. D. Duffy. 1990. *Field Tests and Evaluation of Rockfall Restraining Nets*. Final Report CA/TL-90/05. California Department of Transportation, San Luis Obispo.

Smith, P. A., C. D. Spencer, and D. E. Jones. 1981. Microcomputer Listens to the Coefficient of Restitution. *American Journal of Physics*, Vol. 49, No. 2, pp. 136–140.

Spang, R. M., and R. W. Rautenstrauch. 1988. Empirical and Mathematical Approaches to Rockfall Protection and Their Practical Applications. *Proc., 5th International Symposium on Landslides* (C. Bonnard, ed.), Vol. 2, Balkema, Rotterdam, Netherlands, pp. 1237–1243.

Spang, R. M., and H. Sonser. 1995. Optimized Rockfall Protection by ROCKFALL. *Proc., 8th International Congress of Rock Mechanics*, Vol. 3, Tokyo, pp. 1233–1242.

Statham, I. 1976. A Scree Slope Rockfall Model. *Earth Surface Processes*, Vol. 1, pp. 43–62.

Statham, I. 1979. A Simple Dynamic Model of Rockfall: Some Theoretical Principles and Field Experiments. *International Colloquium on Physical and Geomechanical Models*, Istituto Sperimentale Modelli e Strutture SpA, Bergamo, Italy, pp. 237–258.

Statham, I., and S. C. Francis. 1986. Influence of Scree Accumulation and Weathering on the Development of Steep Mountain Slopes. In *Hillslope Processes* (A. D. Abrahams, ed.), Allen and Unwin, Boston, Mass., pp. 245–267.

Stensgaard, I., and E. Lægsgaard. 2001. Listening to the Coefficient of Restitution—Revisited. *American Journal of Physics*, Vol. 69, No. 3, pp. 301–305.

Stoffel, M., A. Wehrli, R. Kühne, L. K. A. Dorren, S. Perret, and H. Kienholz. 2006. Assessing the Protective Effect of Mountain Forests Against Rockfall Using a 3-D Simulation Model. *Forest Ecology and Management*, Vol. 225, pp. 113–122.

Stokes, A., F. Salin, A. D. Kokutse, S. Berthier, H. Jeannin, S. Mochan, L. K. A. Dorren, N. Kokutse, M. A. Ghani, and T. Fourcaud. 2005. Mechanical Resistance of Different Tree Species to Rockfall in the French Alps. *Plant and Soil*, Vol. 278, Nos. 1–2, pp. 107–117.

Stronge, B. 2000. *Impact Mechanics*. Cambridge University Press, Cambridge, United Kingdom.

Ujihira, M., N. Takagai, and T. Iwasa. 1993. An Experimental Study on the Characteristics of the Impact Load of Falling Rock. *International Journal of Surface Mining and Reclamation*, Vol. 7, pp. 81–89.

Urciuoli, G. 1988. Experiments on the Fall of Blocks Along a Slope in the Dolomitic Limestone Formation of the Sorrento Peninsula. In *Seminar on Mapping and Monitoring of Landslides* (in Italian), Publication 164, pp. 35–54.

Ushiro, T., S. Shinohara, K. Tanida, and N. Yagi. 2000. A Study on the Motion of Rockfalls on Slopes. *Proc., 5th Symposium on Impact Problems in Civil Engineering*, Japan Society of Civil Engineers, pp. 91–96.

Weiss, G. 2000. Perception of Risks in Mountain Forests. In *Forests in Sustainable Mountain Development: A State of Knowledge Report for 2000* (M. F. Price and N. Butt, eds.), International Union of Forest Research Organizations Research Series 5, CABI Publishing, Wallingford, United Kingdom, pp. 359–366.

Wieczorek, G. F., and J. B. Snyder. 1999. *Rock Falls from Glacier Point Above Camp Curry, Yosemite National Park, California*. U.S. Geological Survey Open-File Report 99-385. http://pubs.usgs.gov/of/1999/ofr-99-0385/.

Wieczorek, G. F., M. M. Morrissey, G. Iovine, and J. Godt. 1998. *Rock-Fall Hazards in the Yosemite Valley*. U.S. Geological Survey Open-File Report 98-467. http://pubs.usgs.gov/of/1998/ofr-98-0467/.

Wong, R. H., K. W. Ho, and K. T. Chau. 2000. Shape and Mechanical Properties of Slope Material Effects on the Coefficient of Restitution on Rockfall Study. *Proc., 4th North American Rock Mechanics Symposium*, Seattle Wash., pp. 507–514.

Wu, S.-S. 1985. Rockfall Evaluation by Computer Simulation. In *Transportation Research Record 1031*, Transportation Research Board, National Research Council, Washington, D.C., pp. 1–5.

Chapter 9

A. KEITH TURNER AND
JOHN D. DUFFY

MODELING AND PREDICTION OF ROCKFALL

1. INTRODUCTION

Rockfalls are a concern in mountainous areas. They can threaten lives, settlements, equipment, facilities, and transportation corridors. They pose hazards along engineered slopes, even in regions of modest relief. Despite their typically limited volumes, rockfalls are characterized by high velocity and mobility, which makes them a potential cause of fatalities; hence, they pose a significant hazard and risk (Chapter 5). Rockfalls can be triggered by earthquakes (Kobayashi et al. 1990), by rainfall (Krautblatter and Moser 2009), by freeze–thaw cycles (Matsuoka and Sakai 1999), or by the progressive weathering of rock material, especially along discontinuities, under suitable climatic conditions (Nicholson 2004).

Chapters 1 and 2 have addressed the appropriate definition of rockfall. The convention adopted throughout this book is that rockfall involves significant velocities and some measure of free flight, but there is no universal upper limit on the volume or kinetic energy of rockfall. Movement may involve any combination of free-falling, bouncing, rolling, or sliding. Rockfall may involve more than one rock but typically does not include large volumes of rock such as rock avalanches or landslides that may include rock. There is a continuum between slope movements that may be called rockslides and rockfalls, and many of the rock mass properties and geological processes that contribute to a rockfall consisting of one or a few rock blocks may also contribute to the instability of a large portion of a slope. Rochet (1987) classified rockfall phenomena into four categories:

- *Single block falls*, which typically involve volumes ranging between 0.01 and 100 m^3 (0.35 and 3,500 ft^3);
- *Mass falls*, which typically involve volumes ranging between 100 and 100,000 m^3 (3,500 and 3.5 million ft^3);
- *Very large mass falls*, which typically involve volumes ranging between 100,000 and 10 million m^3 (3.5 million and 350 million ft^3); and
- *Mass displacements*, which typically involve volumes greater than 10 million m^3 (350 million ft^3).

Hungr and Evans (1988) defined rockfalls that involved relatively small volumes of material as *fragmental rockfalls*. They were characterized by more or less independent movements of the individual fragments, as opposed to the sliding or mass flow of coherent or broken rock typical of much larger rock mass failures. These large failures invariably disintegrate and are transformed into a rapid movement of unsorted fragments, termed a *stürzstrom* (Heim 1932), a *rock avalanche* (McConnell and Brock 1904), or a *rockfall avalanche* (Varnes 1978), where inclusion of the term *rockfall* implies an episode of free fall during the event.

Although fragmental rockfalls do not have a well-defined volume limit, this book emphasizes smaller-volume rockfall events. Transportation facilities often encounter locations where an almost continuously occurring series of small-volume rockfalls accumulate considerable volumes of debris. Thus, the procedures for predicting and modeling rockfall are closely related to those used in evaluating the likelihood that a slope will generate rockfall.

Rockfall prediction and modeling are often undertaken to assess the envelope of trajectories of rockfall blocks, their maximum runout distance, the distribution of kinematic parameters along a rockfall path, and the probability that a specified *design block* (the most likely to be encountered, or the largest anticipated block) will stop at specific distances from its source point. This information allows for the design of new rock slopes and remedial measures, including a variety of retaining fences and barriers, ditches and berms, or even rock sheds and tunnels. However, rockfall prediction and modeling, especially in combination with rock mass stability studies and regional geological and geomorphic studies, become important components of hazard assessments for land use planning purposes, especially in populated mountainous regions.

The following sections of this chapter summarize several aspects of rockfall prediction and modeling. Section 2 describes alternative methods for assessing rockfall source areas and various approaches used in modeling the stability of rock masses. Section 3 addresses the use of empirical evaluations and models based on field observations. These methods are particularly useful in providing rapid estimates of anticipated maximum rockfall runout distances and hence are often used in delineating the limits of rockfall hazard zones. Section 4 explores the expanding opportunities for developing rockfall hazard models based on geographic information system (GIS) principles. GIS methods are also used in defining the characteristics of slopes within rockfall-prone areas; thus, they are an important consideration in several of the numerical rockfall simulation models described in Sections 5 and 6.

The numerical simulation of rockfall trajectories is mostly based on the principles of Newtonian mechanics described in Chapter 8. Numerical simulation permits reasonably precise estimates of rockfall trajectories, velocities, and kinetic energies at any location on a slope that is subjected to rockfall events. These estimates are typically used in proposing and specifying rockfall mitigation or protection measures. A large number of computer models incorporating a variety of assumptions concerning the best way to describe the physical processes of falling, bouncing, rolling, and sliding have been developed over the past two or three decades. Their development has been encouraged by the wide availability of increasingly powerful, yet economical, personal computers. The predominant approach has been to conduct two-dimensional (2-D) simulations based on a "typical" profile to define the slope characteristics of a particular rockfall site. These methods are discussed in Section 5.

It has long been recognized, however, that these 2-D simulations only partially reflect the realities of a three-dimensional (3-D) slope. The much greater computational requirements of a 3-D analysis, combined with the difficulties of obtaining detailed topographic data to describe the slope geometry, have limited most applications to 2-D simulations. Recent computer advancements have greatly reduced the concern about the additional computational load, while technological advances in photogrammetry, light detection and ranging (LiDAR), and radar (Chapter 7, Sections 5 and 6) have largely resolved the data acquisition problem. Thus, a number of 3-D numerical rockfall simulation models have been developed in recent years, and additional models are under development. Section 6 describes these modeling approaches; most of them combine GIS terrain-modeling tools with sophisticated approaches to the evaluation of Newtonian mechanics principles in three dimensions.

This chapter is not intended to provide detailed instructions in the use of any computer program. Most programs have extensive user documentation. The purpose of the chapter is to provide an overview of the various rockfall prediction and modeling approaches, organized into the five major topics just described. The use of any of these numerical modeling methods requires experience and skill. Furthermore, the predictions produced by any simulation method are only to be considered reliable once they have been calibrated against field observations. The importance and methodology of the field calibration process are discussed in detail in Chapter 10.

2. ROCK MASS STABILITY MODELS

The analysis of a source of rockfall or the stability of an entire rock slope usually has two goals: (*a*) identification of kinematically admissible unstable blocks and (*b*) investigation of the factors influencing the stability of such blocks. The term *kinematically admissible* refers to locations where block movement is possible on the basis of kinematic considerations, without taking into account any forces. These considerations include the physical arrangement of the blocks relative to the slope face and the strength parameters between blocks. Investigations of potential failure mechanisms depend on some form of mathematical analysis to develop a coherent evaluation of the stability. Many potential rockfall sites are located where rock masses contain multiple sets of discontinuities, which produce a large number of potentially unstable blocks. The 3-D nature of the discontinuity orientations makes direct investigation of the size of the bounding surfaces, or the volume of potentially unstable blocks, extremely complex and difficult to quantify.

As discussed in Chapter 2, current practice for rock slope stability analysis invariably begins with an engineering geological investigation of the discontinuities, leading to kinematic and limit equilibrium stability assessments (John 1968; Londe et al. 1969). Norrish and Wyllie (1996) provided an extensive overview of these methods, as did Hoek and Bray (1981), Wittke (1990), Goodman (1993), Hudson and Harrison (1997), and Wyllie and Mah (2004). These methods were termed *conventional* for rock slope analysis by Coggan et al. (1998), who provided a summary of them, together with their advantages and limitations (Table 9-1).

Table 9-1
Conventional Methods of Rock Slope Analysis

Analysis Method	Critical Input Parameters	Advantages	Limitations
Stereographic and kinematic	Critical slope and discontinuity geometry; representative shear strength characteristics	Simple to use and show failure potential. Some methods allow analysis of critical key blocks. Can be used with statistical techniques to indicate probability of failure and associated volumes.	Suitable for preliminary design or for noncritical slopes using mainly joint orientations. Identification of critical joints requires engineering judgment. Must be used with representative joint–discontinuity strength data.
Limit equilibrium	Representative geometry, material and joint shear strength, material unit weights, groundwater and external loading–support conditions	Much software available for different failure modes (e.g., planar, circular, wedge, toppling). Mostly deterministic but some probabilistic analyses in 2-D and 3-D with multiple materials, reinforcement, and groundwater profiles. Suitable for sensitivity analysis of factor of safety to most inputs.	Factor of safety calculations must assume instability mechanisms and associated determinacy requirements. In situ stress, strains, and intact material failure not considered. Simple probabilistic analyses may not allow for sample or data covariance.
Rockfall simulation (discussed in Sections 5 and 6 of this chapter)	Representative slope geometry and surface condition; rock block sizes, shapes, unit weights, and coefficients of restitution	Practical tool for siting structures and catch fences. Can use probabilistic analysis. 2-D and 3-D codes available.	Limited experience in use relative to empirical design charts.

Source: Coggan et al. 1998.

Chapter 2 includes several examples of kinematic analysis of potential rock slope failures that use stereographic projections (*stereonets*) allowing 3-D orientations to be represented and analyzed in two dimensions. Appendix A in this volume includes stereonets and a brief summary of their use for kinematic analysis. Several commercial programs are available that support rock slope stability analyses by evaluating discontinuity orientations with stereographic procedures [Dips (RocScience 2011a), RockPack III (Rockware 2011)] or by applying key block theory [ROCK3D (Geo and Soft 2011)].

The use of probabilistic limit equilibrium tools to evaluate parameter uncertainty is increasing. This is a natural consequence of the increasing use of risk assessment in engineering practice. The availability of commercial software to support rock slope stability analyses by evaluating discontinuity orientations with stereographic procedures has promoted their combination with probabilistic algorithms to assess the influence of joint geometrical characteristics (e.g., discontinuity orientation, strength) on the factor of safety of rock slopes (Figure 9-1).

2.1 Numerical Modeling of Rock Slope Stability

Advances in computing power and the availability of inexpensive commercial numerical modeling codes have encouraged the extension of conventional methods of numerical modeling to assess rock slope stability by incorporating intact rock deformation (strain) during rock slope failure. Civil and mining engineers and academic researchers routinely use numerical modeling methods to evaluate the stability of rock slopes. Detailed discussion of such modeling applications is beyond the scope of this book. The interested reader is directed to the extensive technical literature on this topic, such as the following: Hoek and Bray (1981), Kovári and Fritz (1984), Wittke (1990), Goodman (1993), Hudson and Harrison (1997), Coggan et al. (1998), Goodman and Kieffer (2000), Eberhardt et al. (2002), Wyllie and Mah (2004), Eberhardt (2006), Pine et al. (2006), and Stead et al. (2006).

Many of these techniques address complexities relating to geometry, material anisotropy, nonlinear behavior, in situ stresses, and the presence of several coupled processes such as pore pressures and seismic loading. Coggan et al. (1998) subdivided numerical modeling into (*a*) continuum methods that include finite element and finite difference techniques and (*b*) discontinuum methods that include distinct element and discontinuous deformation analysis (DDA) techniques. Table 9-2, adapted from Coggan et al. (1998), shows the characteristics, advantages, and limitations of

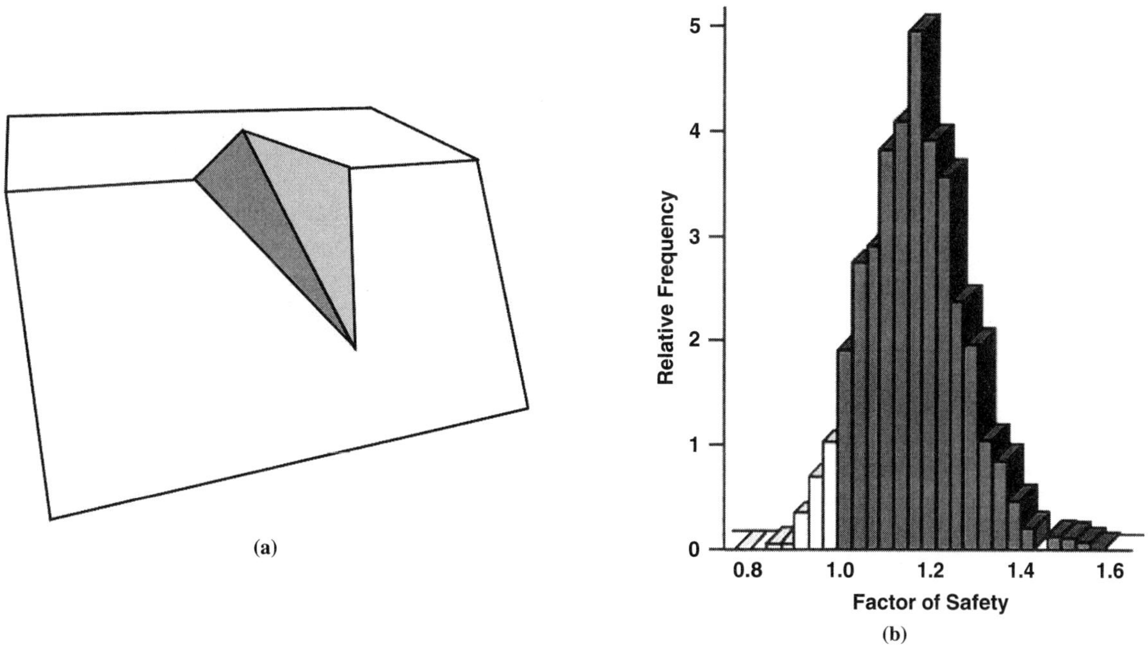

FIGURE 9-1 Example of probabilistic analysis of factor of safety of potentially unstable rock slope: (*a*) wedge failure; (*b*) histogram of distribution of factors of safety for all valid wedges generated by Monte Carlo sampling of input data (adapted from Stead et al. 2006).

Table 9-2
Numerical Methods of Rock Slope Analysis

Analysis Method	Critical Input Parameters	Advantages	Limitations
Continuum modeling: finite element, finite difference	Representative slope geometry, constitutive criteria (e.g., elastic, elastoplastic, creep), groundwater characteristics, shear strength of surfaces, in situ stress state	Allows for material deformation and failure, including complex behavior and mechanisms, in 2-D and 3-D with coupled modeling of groundwater. Can assess effects of critical parameter variations on instability mechanisms. Can incorporate creep deformation and dynamic analysis. Some programs allow use of embedded language so users can define their own functions and subroutines.	Users should be well trained and experienced, observe good modeling practice, and be aware of model and software limitations. Input data generally limited, and some required inputs are not routinely measured. Sensitivity analyses limited because of run-time constraints, but this is rapidly improving.
Discontinuum modeling: distinct element, DDA	Slope and discontinuity geometry, intact constitutive criteria (e.g., elastic, elastoplastic), discontinuity stiffness and shear strength, groundwater, and in situ stress conditions	Allows for block deformation and movement of blocks relative to each other. Can model complex behavior and mechanisms (combined material and discontinuity behavior, coupled with hydromechanical and dynamic analysis). Able to assess effects of parameter variations on instability. Some programs allow use of embedded language so users can define their own functions and subroutines.	As above, experienced users needed. General limitations similar to those listed above. Need to simulate representative discontinuity geometry (e.g., spacing, persistence). Limited data on joint properties available (e.g., joint stiffness).

Source: Adapted from Coggan et al. 1998.

numerical methods used in the analysis of rock and mixed rock and soil slopes.

Continuum methods include finite element and finite difference models. Both methods are in routine use in engineering landslide investigations and are most appropriate in the analysis of slopes involving weak rock or soils or rock masses where failure is controlled by the deformation of the intact material (in other words, where there is a continuum). They may also be used where failure occurs through a restricted number of discrete discontinuities such as a bedding plane or a fault (Stead et al. 2006). Kalkani and Piteau (1976) used the finite element method to analyze toppling of rock slopes at Hells Gate in British Columbia, Canada. Many finite difference applications have used the 2-D or 3-D versions of FLAC (Itasca International 2011b, 2011c). Coggan et al. (2000) successfully demonstrated the use of both 2-D and 3-D finite difference analyses in the backanalysis of highly kaolinized china clay slopes.

Discontinuum methods use what are commonly called "discrete element methods." They are most useful when the stability of the rock slope is controlled by movement of joint-bounded blocks and are now routinely used by civil and mining engineers to analyze rock slopes. Two principal methods are used: distinct element (Hart 1993; Coggan and Pine 1996) and DDA (Shi and Goodman 1989). DDA was used to analyze a major rockfall in Japan (Chen and Ohnishi 1999). The predominant distinct element code is the Universal Distinct Element Code (UDEC) developed by Itasca International (2011d). This code has been used to investigate a variety of rock slope failure mechanisms, ranging from simple planar mechanisms to complex, deep-seated toppling instability and buckling (Stead et al. 2006). The use of 3-D distinct element techniques has been more limited to date, but Adachi et al. (1991) used an early version of 3DEC (Itasca International 2011a) to analyze toppling slopes along a highway in Japan.

2.2 AutoBlock Program

Researchers at the Swiss Federal Institute of Technology in Zurich, Switzerland, developed a computer-aided design (CAD)-based program called AutoBlock to make the analysis of the stability of a rock slope by engineers much easier (Bergamin et al. 2004; Fritz and Bergamin 2004a, 2004b; Fritz and Hauser 2007). The standard numerical modeling programs described in Section 2.1 require considerable experience and training before they can be used productively. Many analyses have been restricted to 2-D cross-sectional models of rock slopes because 3-D applications depend on greater user experience to be modeled successfully, and when they are applied to assess the stability of rock slopes, the 3-D geometry of the slope surface has to be defined with precision. Topographic features are usually defined by digital terrain models, yet skill is needed in loading such topographic data into standard numerical modeling packages.

AutoBlock was developed to meet the requirements of engineers wishing to undertake rock slope stability analyses. The stability of discrete blocks is evaluated with respect to their relationships with the rock face geometry defined by a digital terrain model. The digital terrain model is intersected by identified sets of discontinuities, and the geometry and strength properties of each set are defined individually. In addition, the geometry of foundations or excavated openings may be combined with the terrain model to evaluate a future construction excavation. For each unstable block that may fail, on the basis of kinematic considerations, AutoBlock calculates the factor of safety against sliding by using the limit equilibrium method.

AutoBlock operates as an add-on to the popular program AutoCAD and exploits its possibilities and its power for 3-D visualizations. The user interface has menus and dialogue boxes that make AutoBlock easy to use for those who do not have a detailed knowledge of AutoCAD. It provides all the tools necessary for representing and visualizing the results and facilitates a stability analysis with any number of potentially unstable blocks. According to Fritz and Bergamin (2004a), AutoBlock has the following capabilities:

- The rock surface is defined by using a digital terrain model, which may be created to represent the surface with appropriate accuracy with data supplied from terrain contours, profiles, or grid measurements.
- Several potential sets of discontinuities are defined with an easy-to-use interface.
- Slippage (sliding) on one or two surfaces may be evaluated; further discontinuities are considered as tension cracks.
- Determination of all kinematically admissible unstable blocks is undertaken by successive combination of all given discontinuities and their intersections with the terrain model.
- Consideration is given to the important loading forces on the discontinuity surfaces (weight, anchor forces, water pressure).
- The failure mechanisms and factors of safety for all kinematically admissible potential unstable blocks with respect to sliding are determined.
- Extensive visualization possibilities provide a basis for the engineering interpretation of the results.

AutoBlock has been used to evaluate the stability of rock slopes and the potential for rockfall at several locations (Bergamin et al. 2004; Fritz and Bergamin 2004a; Fritz and Hauser 2007).

Bergamin et al. (2004) described the use of AutoBlock in analyzing the stability of a partially overhanging rock slope along a highway at the entrance to the Maderaner Valley in Switzerland

(a)

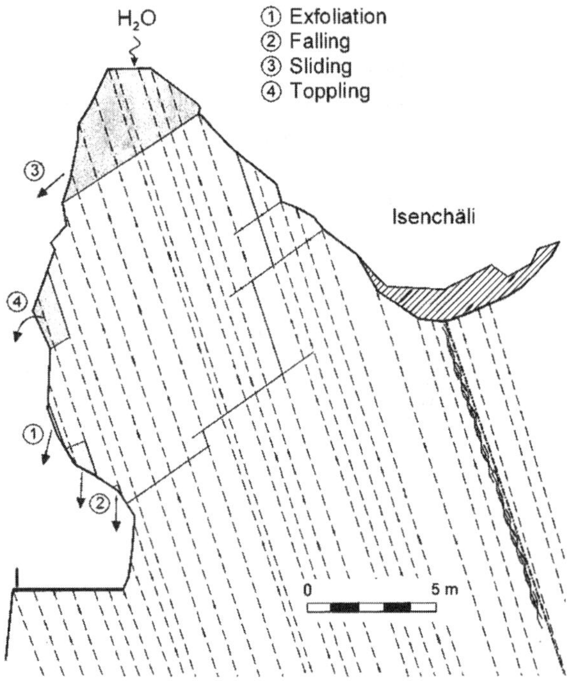

(b)

FIGURE 9-2 Bristen Road rock slope, Maderaner Valley, Switzerland: (*a*) view of rock slope with undercut; (*b*) compilation of hazard scenarios (Bergamin et al. 2004).

(Figure 9-2*a*). The unstable rock slope extends approximately 30 m (98 ft) along the road and rises 8 to 22 m (26 to 72 ft) above it. Behind it, there is a small topographic depression ("Isenchäli," Figure 9-2*b*) where surface water collects. The gneiss rock mass is cut by a complex system of discontinuities, including three cleavage planes, two sets of joints, a fault, and various shear surfaces (Figures 9-2*b* and 9-3).

This unusual overhanging section of rock experienced a variety of periodic rockfall events, including (*a*) small rock failures resulting from exfoliation and falls from the overhanging roof directly onto the road and (*b*) large toppling and sliding failures (Figure 9-2*b*). A detailed terrain model was constructed from several triangulated meshes to account for the overhanging areas. The high degree of detail and the correspondingly large amount of data (Figure 9-4*a*) made the stability analysis, which used the initial version of AutoBlock, cumbersome. Therefore, a less detailed mesh was constructed on the basis of a series of sections spaced 2.5 to 5 m (8 to 17 ft) apart (Figure 9-4*b*). Comparison of the two models revealed a difference in volume between the refined and the rough models of less than 5%. The overhanging zones of rock were modeled in AutoBlock by first defining them as independent solids and then "excavating" the roadway out of the original nonoverhanging terrain model.

Several stability analyses were undertaken with AutoBlock. Various combinations of discontinuity sets were evaluated, and only selected discontinuities were considered to be subject to sliding failures in each analysis. Blocks with volumes less than 0.5 m^3 (18 ft^3) were not evaluated, as they were assumed to be noncritical and within the range of resolution of the surface model. Depending on the choice of discontinuities, the stability of blocks that outcropped at the undercut or at the front face of the rock wall was evaluated. There was a

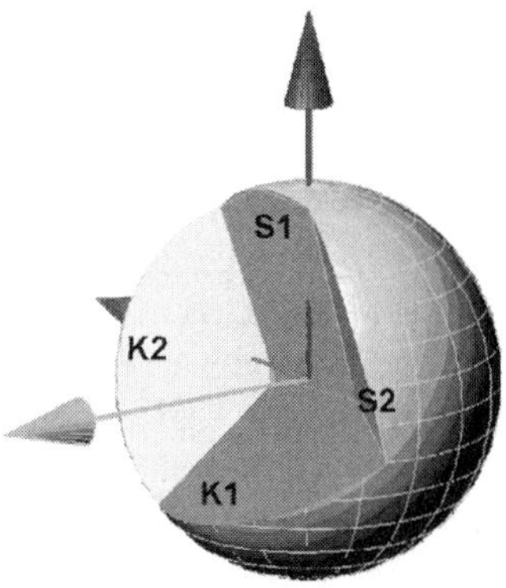

FIGURE 9-3 Representation of most important discontinuity sets (K1, K2, S1, and S2) by AutoBlock (Bergamin et al. 2004).

considerable difference in the maximum size of the potentially unstable blocks that were evaluated by the various analyses. Figure 9-5 illustrates one analysis in which blocks up to 1,600 m^3 (56,500 ft^3) could fail (slide out) from the front face. Bergamin et al. (2004) provided details of several analyses.

Bergamin et al. (2004) documented how the block instabilities were affected by various scenarios of water pressure along these failure planes. The scenarios ranged from fully drained to completely undrained and have been discussed in detail in Section 4.1.3 of Chapter 2 (see Figure 2-40).

Bergamin et al. (2004) reported that the AutoBlock analysis allowed for a recommended stabilization costing about 3 million Swiss francs (approximately \$3.42 million) that included drainage boreholes from the road and from the top (Isenchäli), along with a systematic arrangement of anchors to prevent a further exfoliation or toppling of small rock blocks. The AutoBlock analyses suggested that sliding of individual large blocks would not occur as long as the drainage was functioning. Thus, large anchorages to stabilize these large blocks were deemed unnecessary.

However, in the view of the client, this recommended solution carried too great a residual risk. The client preferred a solution that, even in the distant future, would require few inspections or maintenance. Therefore, in March 2003, the complete rock mass with a volume of 16,000 m^3 (21,000 yd^3) was removed by blasting at a cost of 9.6 million Swiss francs (approximately \$10.9 million) (Bergamin et al. 2004).

2.3 Matterock Methodology

In April and May 1991, two consecutive rockslides released a cumulative volume of approximately 30 million m^3 (1 billion ft^3) of debris from a cliff above the town of Randa in the Matter Valley of Switzerland. These Randa rockslides disrupted transportation links through the valley and destroyed the railway that provided access to the resort of Zermatt. Subsequent stability evaluations identified the important role that discontinuities within the rock mass had played in these rockslides (Sartori et al. 2003). Faced with continued threats of large-scale rock slope instabilities at this location, the Centre de Recherche sur l'Environnement Alpin, a research center supported by the Swiss canton of Valais and by Sion, its capital city, with research support from the universities at Lausanne and Neuchâtel, began to investigate better methods of detecting zones where discontinuities might produce rockfalls. The research led to the development of the *Matterock method*, which has been applied to the detection of several small- and large-scale rock instabilities.

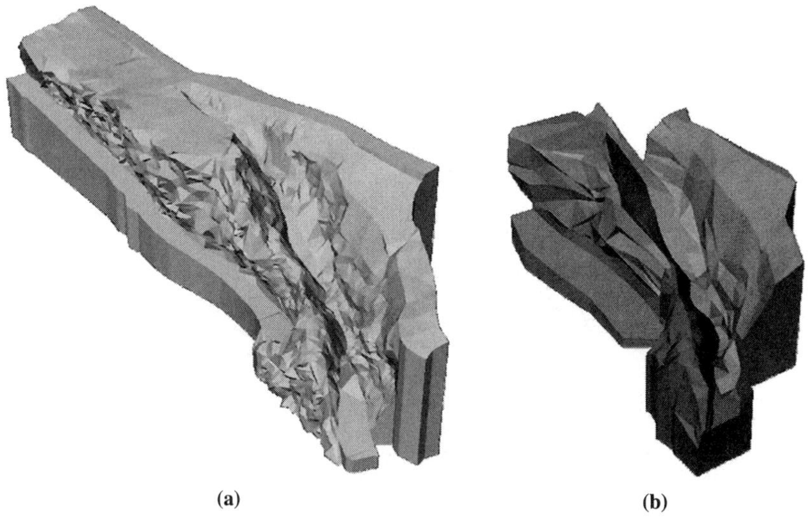

FIGURE 9-4 Digital terrain models for Bristen Road site: (*a*) refined model; (*b*) less detailed model (Bergamin et al. 2004).

The theory underlying the Matterock method has been presented by Rouiller and Marro (1997); Jaboyedoff et al. (1999); Jaboyedoff, Baillifard, Philippossian, et al. (2004); and Baillifard et al. (2001). The Matterock method consists of three steps:

1. Defining homogeneous structural domains,
2. Identifying and characterizing the discontinuity sets, and
3. Comparing the structural features with the topography by means of a digital elevation model (DEM).

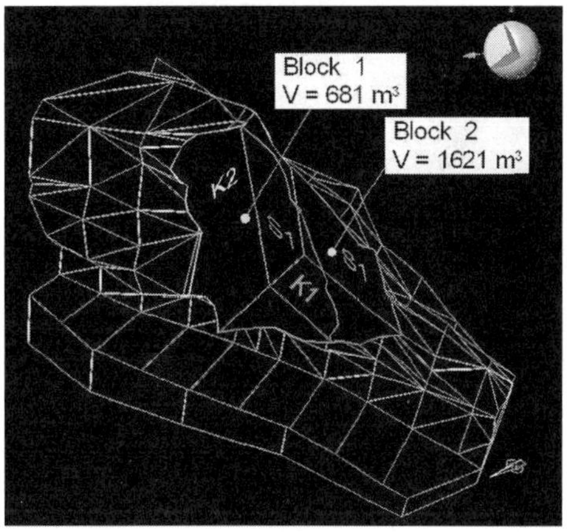

FIGURE 9-5 AutoBlock analysis showing bounding surfaces of potential sliding blocks formed by discontinuity sets K1, K2, and S1 (Bergamin et al. 2004).

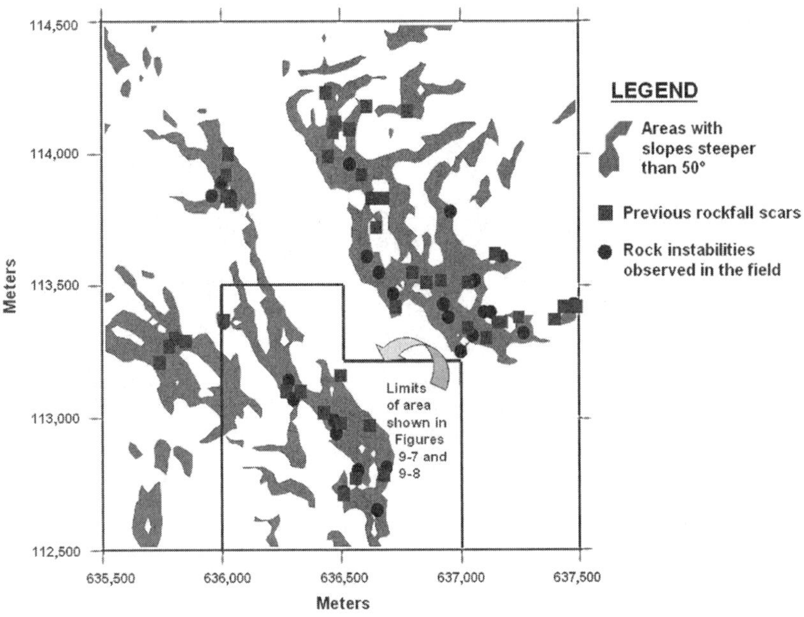

FIGURE 9-6
Map of area of Saas Valley, Valais Canton, southwestern Switzerland, shows majority of unstable rock slopes are located within areas of steepest topographic slopes. Gray shaded areas are slopes steeper than 50°; squares are previous rockfall scars; circles are unstable rock slopes observed in field; map coordinates are in meters defined according to 1903 Swiss Terrestrial Coordinate System (adapted from Jaboyedoff et al. 1999).

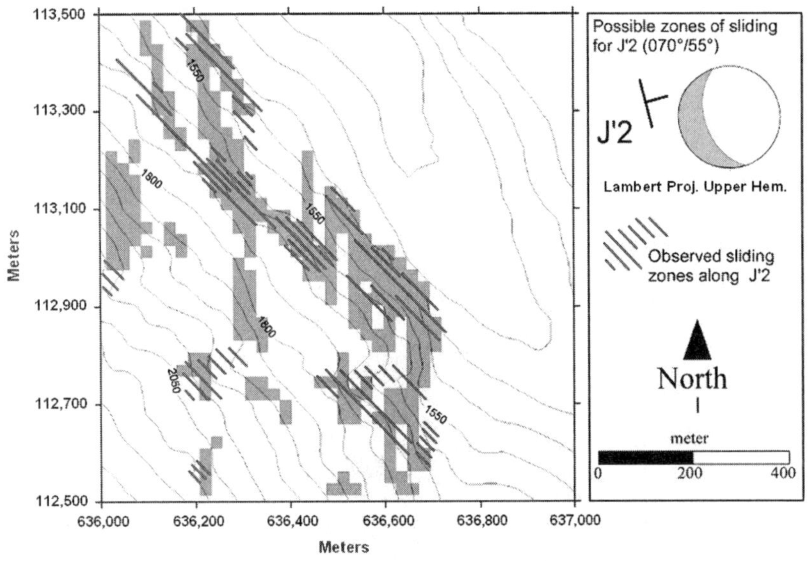

FIGURE 9-7
Shaded area represents topographic domain in Saas Valley, where discontinuity set J'2 (075°/55°) could allow small rockslides. Black lines define observed small rockslides on J'2 discontinuity; map coordinates are in meters defined according to 1903 Swiss Terrestrial Coordinate System (adapted from Jaboyedoff, Baillifard, Philippossian, et al. 2004).

Homogeneous structural domains are locations where geological structures, lithologies, and fracturing show consistent properties. Each homogeneous structural domain defines an area where potential rock instabilities are controlled by similar geometries; the occurrence of instability then depends on the relationship between the discontinuities and the orientation of topographic exposures. For example, various sliding or wedge failures will most likely be at locations where steep rock slopes are oriented to produce kinematically admissible unstable blocks.

The second step involves the identification and characterization of the discontinuity sets within each homogeneous structural domain. Special attention is paid to the orientation, spacing, trace length, undulation, infilling material, and aperture of discontinuities. Jaboyedoff et al. (1999) proposed the use of *window sampling* (Priest 1993) as the quickest way to evaluate the number, mean trace length, and mean spacing of the discontinuity sets. By assuming that, on average, the discontinuities are planar surfaces that are roughly circular in extent, it is possible to calculate the average surface area of each discontinuity set as well as the average volume of rock it will bound (Jaboyedoff et al. 1999). Once these average characteristics of a discontinuity set have been established, the probability of finding a discontinuity within a surface of any desired orientation or within a given volume may be calculated under the procedures defined by Priest (1993). The third step compares the orientations of the discontinuity sets with the topography defined by a DEM.

Jaboyedoff et al. (1999) and Jaboyedoff, Baillifard, Philippossian, et al. (2004) described the application of the Matterock method to an area in the Saas Valley of the canton of Valais in southeastern Switzerland. Figure 9-6 shows how the simple geomorphic criterion—the presence of steep slopes that can be readily determined from a DEM—often correlates closely to previous rockfalls or to observed potentially unstable locations. For a portion of the same Saas Valley study area, Figure 9-7 shows how the orientation of a given discontinuity set may be compared with the topographic conditions. In this case, the shaded areas represent locations where the topographic slopes could produce rock sliding along the discontinuity set labeled J'2, while the black lines indicate where actual rockfall activity was observed by sliding on discontinuities.

When accessibility problems, cost, or other reasons make direct observation of rock instabilities difficult, the geometrical characteristics of discontinuity sets and the procedures defined by Priest (1993) permit quantification of the likelihood that a given discontinuity set occurs within a surface exposure of any desired orientation or within a given volume. Figure 9-8 illustrates this approach for the same Saas Valley study area. The gray scale indicates the probable number of discontinuities of set J3 per unit of surface area. The outlined area contains approximately 0.26 J3 discontinuities, which corresponds to a 23% probability of finding at least one J3 discontinuity in this area. This probability can be used to quantify the probability of occurrence of this rockfall hazard. Of course the hazard scale must be calibrated to reflect the rock type and climatic conditions. A similar procedure can be used to evaluate wedge failures.

Jaboyedoff, Baillifard, Philippossian, et al. (2004) and Abbruzzese et al. (2009) described how these estimates produced by the Matterock method are incorporated into the Swiss hazard assessment and zonation procedures. Chapter 5 (Section 6.1) describes the Swiss landslide risk management procedures. In Switzerland, hazard assessment involves estimation of the intensity of an event and its probability of occurrence over time (Raetzo et al. 2002; Lateltin et al. 2005). The intensity and probability of occurrence form a hazard matrix (Figure 5-20), which then is used to define a series of hazard categories that are shown on landslide hazard maps with a series of standard colors. The Matterock method provides a way to assess the probability of occurrence of rockfall events. Abbruzzese et al. (2009) provided an extended discussion of how the intensity of events is combined with the probability-of-occurrence estimates to develop detailed hazard zonation maps.

The Matterock method is supported by a freely available computer program called Matterocking 2.0 (Jaboyedoff 2002). Matterocking 2.0 uses topographic data developed within commonly used GIS programs to model the terrain. It then supports evaluations of discontinuity sets and compares their orientations with the topographic surface to estimate the average number of discontinuities intersecting any topographic feature, such as a cliff or rock face. Locations where rock instabilities may occur by planar sliding or wedge failures are then easily identified. The software provides estimates of the number of potential failures (planar or wedge) and the densities of discontinuities within each cell of the DEM. Matterocking 2.0 is designed to perform calculations and not to generate graphic displays, although some visualization utilities permit a quick overview of the results. The Matterocking file formats allow the results to be evaluated by commonly used visualization programs (Jaboyedoff 2002). Matterocking software is downloadable from http://www.crealp.ch/fr/matterocking.html.

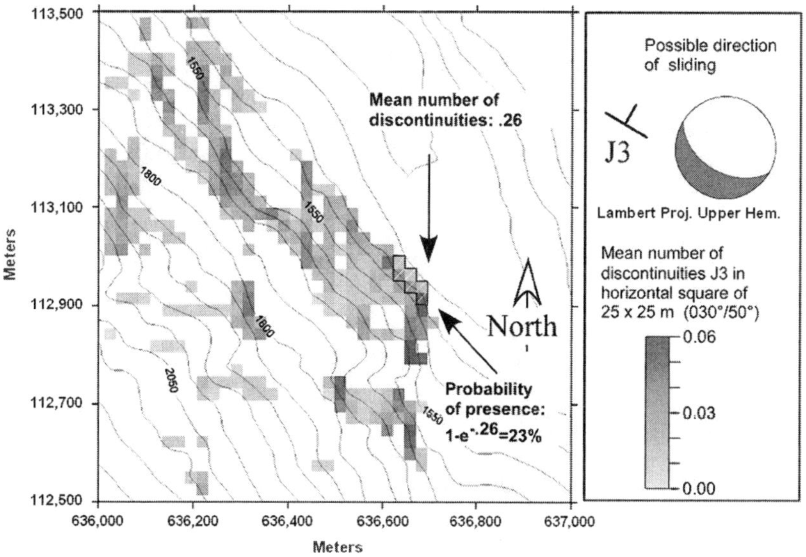

FIGURE 9-8 Gray scale displays estimated mean number of J3 discontinuities expected within areas of steep topography. These are locations of potential sliding failure. Number of discontinuities counted in outlined zone is 0.26; probability of finding at least one J3 discontinuity in this area is 23%; map coordinates are in meters defined according to 1903 Swiss Terrestrial Coordinate System (adapted from Jaboyedoff, Baillifard, Philippossian, et al. 2004).

3. EMPIRICAL EVALUATIONS BASED ON FIELD OBSERVATIONS

Empirical modeling approaches have been adopted by several authors to calculate the extent of rockfall runout and thus to define rockfall hazard zones (Lied 1977; Onofri and Candian 1979; Toppe 1987; Hungr and Evans 1988; Evans and Hungr 1993; Corominas 1996a, 1996b; Meißl 1998, 2001; Keylock and Domaas 1999). In evaluating the massive Elm, Switzerland, rockfall avalanche, Heim (1932) noted that the vertical and horizontal dimensions of such failures define a relatively consistent angle that can be used to estimate the maximum runout distance. Hungr and Evans (1988) defined a *rockfall shadow zone* as the area extending away from the base of a cliff to the most distant rock blocks that are identified as products of rockfall events originating on the cliff. Hungr and Evans noted that this zone typically extends beyond the limits of identifiable talus deposits onto some areas of older valley deposits, apparently because a few rock blocks achieve higher velocities and energies than most rock fragments and thus travel longer distances. Because the rockfall shadow zone is intended to include the most extreme rockfall events, its limits are appropriately defined by the most distant identifiable blocks. Hungr and Evans then defined a *minimum shadow angle* as the angle extending from the limits of the shadow zone to the apex of the talus deposit.

3.1 Alternative Definitions of the Rockfall Shadow Angle

Heim (1932) proposed one method for defining an angle relating the vertical and horizontal dimensions of rockfall. He called it the *fahrböschung* (subsequently termed the *travel angle*) and discovered that this angle is not a constant; it changes with the volume of the rock avalanche. Extremely large and rapidly moving rock avalanches, or *stürzstroms*, with volumes greater than 1 million m^3 (35 million ft^3) have smaller angles and thus greater proportional runout distances than do rockfalls with smaller volumes.

Hsü (1975) and numerous other investigators have confirmed Heim's conclusions. For rockfall applications, investigators have used four distinct methods for defining the angle, each based on different criteria to define both the vertical and the horizontal distances (Table 9-3). The four angles are defined as follows:

1. *Geometric angle* (β_g) is defined by the vertical difference between the rockfall source and the straight-line horizontal separation distance (L1 in Figure 9-9).
2. *Travel angle* (β_f) is defined by the vertical difference between the rockfall source and the horizontal projection of the actual traveled distance (L2 in Figure 9-9). This was defined as the fahrböschung by Heim (1932).
3. *Minimum shadow angle* (β_s) is defined by the vertical difference between the apex of the talus deposit and the straight-line horizontal separation distance (L1 in Figure 9-9).
4. *Average gradient angle* (β_n) is defined by the vertical difference between the apex of the talus deposit and the horizontal projection of the actual traveled distance (L2 in Figure 9-9).

The horizontal distance is always defined by the location of the rock block that has traveled farthest from the rockfall source area. Some investigators, however, use a straight-line distance to define this horizontal separation, while others use the horizontal projection to the actual traveled trajectory. This is always greater than the straight-line distance because blocks are influenced by minor declivities in the slope below a

Table 9-3
Definitions of Four Angles Used in Empirical Rockfall Models

Angle Name	Definition of Top Location	Definition of Horizontal Distance	Source
Geometric angle (β_g)	Rockfall source area	Straight-line horizontal projection (L1)	Heim 1932; Hsü 1975; Onofri and Candian 1979; Meißl 1998
Travel angle (β_f)	Rockfall source area	Horizontal projection of actual traveled distance (L2)	Heim 1932; Hsü 1975; Onofri and Candian 1979; Meißl 1998
Minimum shadow angle (β_s)	Apex of talus deposit	Straight-line horizontal projection (L1)	Lied 1977; Hungr and Evans 1988; Evans and Hungr 1993; Meißl 1998; Wieczorek et al. 1998, 2008
Average gradient angle (β_n)	Apex of talus deposit	Horizontal projection of actual traveled distance (L2)	Lied 1977; Hungr and Evans 1988; Evans and Hungr 1993; Meißl 1998; Wieczorek et al. 1998, 2008

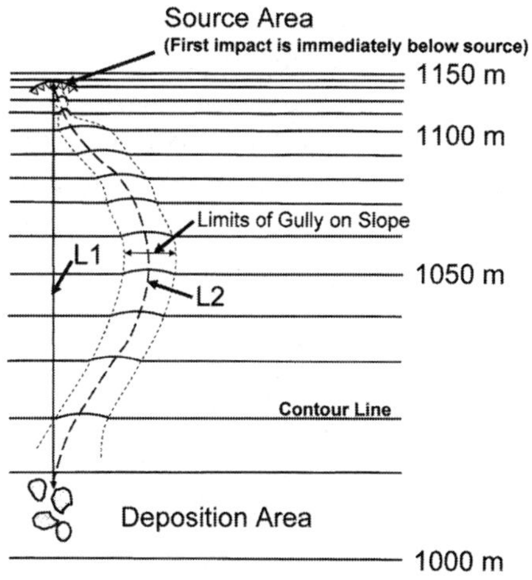

FIGURE 9-9
Map view of rockfall runout showing difference between straight-line horizontal distance (L1) and the horizontal projection of actual traveled distance (L2) (adapted from Petje et al. 2005).

rock cliff (Figure 9-9). In a similar fashion, some investigators have defined the vertical distance as the vertical separation between the rockfall source area and the final location of the farthest block (Figure 9-10). This distance typically includes some height of exposed rock face where the block experiences a free-fall trajectory before striking on or near the uppermost apex of a talus or scree deposit. Other investigators have considered the fact that a rock block loses most of its kinetic energy gained during the free fall on this first impact and thus have measured the vertical difference between the apex of the talus deposit and the final location of the farthest block (Figure 9-10).

Because the horizontal projection of the actual traveled distance (L2) is always greater than the straight-line horizontal distance (L1), the average gradient angle (β_n) will always be smaller than the minimum shadow angle (β_s), as shown in Figure 9-11. Similarly, the travel angle (β_f) will also be smaller than the geometric angle (β_g).

3.2 Establishing Appropriate Angle Values

Heim (1932) used both the geometric angle and the travel angle as characteristic quantities for determining the runout distance of the Elm rockfall avalanche. Many investigators assume that the tangent of the travel angle accurately represents the rockfall *friction coefficient*

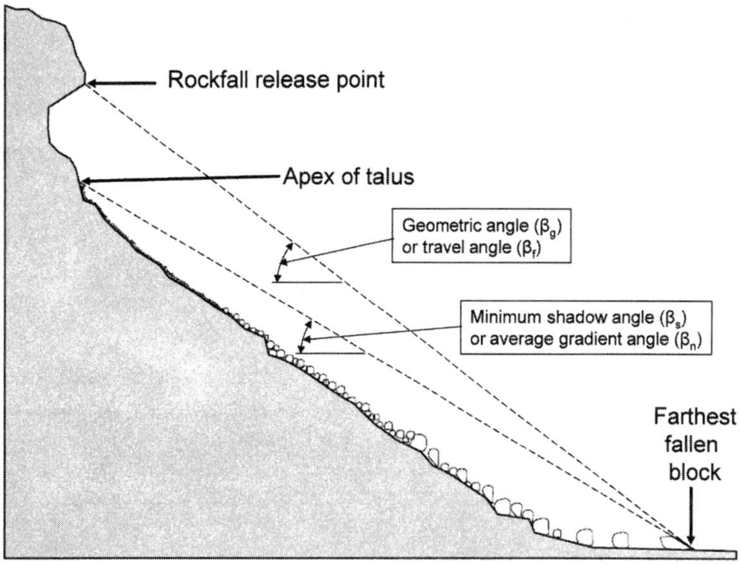

FIGURE 9-10
Vertical distance criteria used in empirical rockfall models. Choice of angle shown in each case depends on definition of horizontal distance traveled by rock block (distance L1 or L2 as shown in Figure 9-9) (adapted from Petje et al. 2005).

and thus predicts the rockfall runout distance. Nevertheless, such a relationship is only approximate because the friction coefficient also depends on the properties of the rockfall mass (including its volume, form, and material properties) and on the slope conditions (including geometry, roughness, presence of vegetation, and composition). Because the geometric angle can be more easily

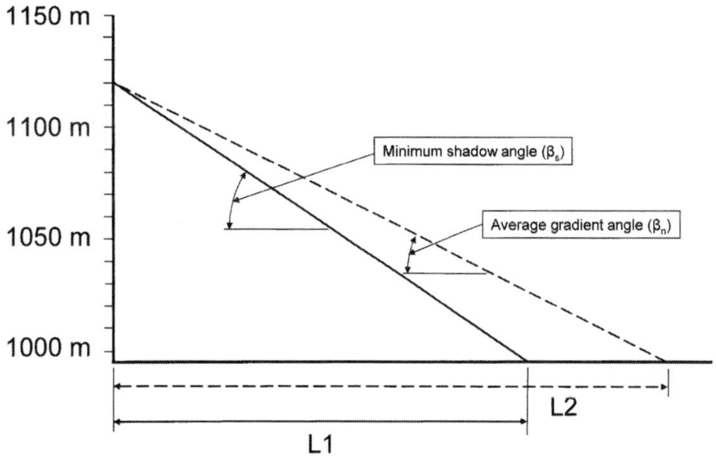

FIGURE 9-11
Comparison of average gradient angle (β_n) and minimum shadow angle (β_s) for rockfall situation shown in Figure 9-9 (adapted from Petje et al. 2005).

determined, it is often used instead of the travel angle, which typically requires extensive field investigation to determine the traveled path of each block. Hungr and Evans (1988) and Evans and Hungr (1993) noted two problems in determining the travel angle:

- The travel angle value is dependent on the total elevation difference from the source area to the final position of the farthest rock block. Yet a large part of kinetic energy stored in a free-falling block is lost on the first impact. Most blocks continue to roll down the slope, independent of the height of free fall.
- The precise position of the rockfall source location for any rock block in the deposition area is not always known.

Hungr and Evans (1988) thus introduced the concept of the minimum shadow angle (β_s), which was based on prior studies by Lied (1977). The vertical component of this angle is easily defined by the position of the apex of the talus slope rather than by the rockfall source area above the slope. Thus, this approach does not require knowledge of the precise location of each rockfall release. The horizontal component is defined by the straight-line horizontal distance to the farthest rockfall block, an easily determined value.

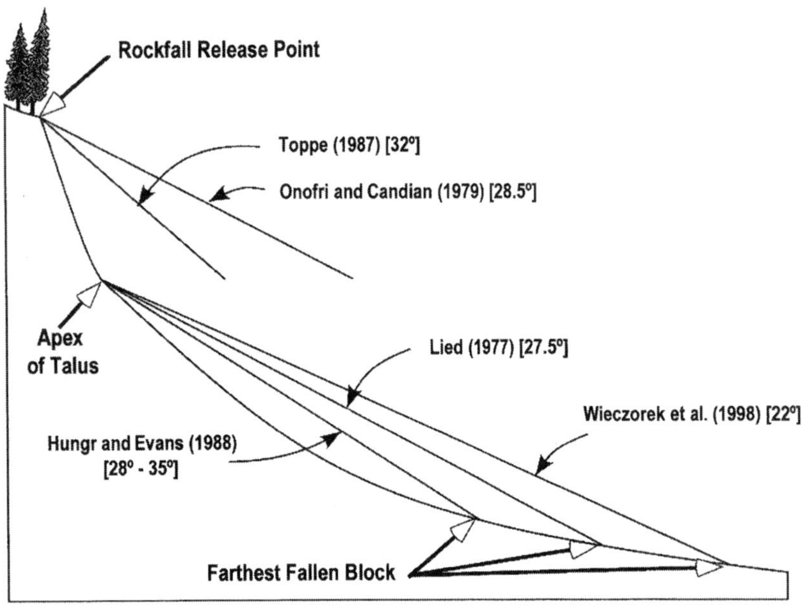

FIGURE 9-12
Reported ranges of angles used in empirical rockfall models (adapted from Jaboyedoff and Labiouse 2003).

The minimum shadow angle is readily defined but does not account for the actual traveled distance; thus, the average gradient angle has been proposed as an alternative. It connects the apex of the talus slope, assumed to be the point where a block first strikes the talus slope, with the horizontal projection of the actual traveled path along the slope to the point where the block stops. In this way, the real travel distance is taken into account, not the shorter straight-line distance as in the case of the minimum shadow angle. The average slope gradient closely matches the energy dissipation rate of a bouncing or rolling block, an important rockfall simulation parameter.

Several authors have determined that typical values of minimum shadow angles are between 22° and 30°. Lied (1977) concluded that the minimum shadow angle was between 28° and 30° for most situations; several authors support the use of a value of 25°. After field evaluations in British Columbia, Evans and Hungr (1993) concluded that the minimum shadow angle is at least 27.5°, regardless of rock face height, trajectory length, slope gradient, or whether the slope is vegetated. However, for smooth talus slopes, they suggested lower values of 23° or 24°. Meißl (1998) evaluated rockfall runout distances in several areas in the western Bavarian Alps. On the basis of field investigation and by using maps with a scale of 1:10,000, he discovered that geometric angles and travel angles differed by less than 1°. He observed travel angles between 29° and 47.5°, while geometric angles were between 29.5° and 48.5°, and both averaged about 38°. He also found that the minimum shadow angle and the average gradient angle had similar values; they differed by a maximum of 1° at locations with relatively low slope gradients. Figure 9-12 summarizes several reported values for angles based on either rockfall source locations (geometric and travel angles) or apex locations of talus slopes (minimum shadow angles).

These observations show that there is little difference between values of angles defined by using actual travel distances and those defined by using straight-line horizontal distances. Thus, little is gained by using the angles that are more difficult to obtain, and the easier-to-use geometric angle or minimum shadow angle appears adequate for empirical rockfall runout predictions. The choice of angle to be used appears to depend on the characteristics of the rockfall volume and the rockfall site. However, the minimum shadow angle, defined by using the location of the apex of the talus slope, does not require knowledge of the precise location of each rockfall release and so is the easiest to determine in the field at locations where talus is found below a rock cliff. Wieczorek et al. (1998, 1999, 2008) evaluated rockfall hazards in Yosemite Valley, California, and used a minimum shadow angle of 22° to define potential rockfall shadow zones on the valley floor (Figure 9-13).

3.3 Use of Empirical Models

Empirical models, based on the shadow angle approach, allow for the rapid estimation of the size of rockfall hazard zones at a regional scale (Lied 1977; Onofri and Candian 1979; Toppe 1987; Evans and Hungr 1993; Corominas 1996a, 1996b; Keylock and Domaas 1999; Wieczorek et al. 1998, 1999, 2008; Jaboyedoff and Labiouse 2003, 2011; Loye et al. 2008). Such models are readily applied to large areas to establish rockfall hazard zones on the basis of preliminary investigations of the potential rockfall propagation distances from known source areas. The evaluations begin with the identification of rockfall source areas; they are often undertaken by various methods based on the use of GIS techniques and DEM representations of the topography (Jaboyedoff and Labiouse 2003; Aksoy and Ercanoglu 2006; Loye et al. 2009). Section 4 discusses the applications of GIS technology to rockfall prediction and modeling. A rockfall modeling approach that combines the empirical shadow angle concept and DEM data, called CONEFALL, is described in Section 3.4.

The concept of producing simple estimates of rockfall runout distances based on the shadow angle approach has formed the basis of several rockfall hazard zonation studies (Parise 2002; Ayala-Carcedo et al. 2003; Hylland et al. 2003; Jaboyedoff and Labiouse 2003; Keylock and Domaas 1999; Wieczorek et al. 1999, 2008; Copons and Vilaplana 2008; Frattini et al. 2008; Copons et al. 2009). Rockfall shadow zones have been used to define rockfall hazards in Utah (Hylland et al. 2003) and in Yosemite Valley, California (Wieczorek et al. 1998, 1999, 2008). Holm and Jakob (2009) evaluated rockfall runout distances on an unvegetated, relatively smooth slope at high elevation in the Andes of Chile. They found that runout distances were considerably longer than estimated by published shadow angles; thus, they recommended the establishment of site-specific shadow angle values by direct observation rather than reliance on published average values. Detailed evaluations of rockfall data for Andorra allowed Copons and Vilaplana (2008) and Copons et al. (2009) to develop empirical regression equations that define maximum runout distances. Keylock and Domaas (1999) tested three empirical models on their ability to predict the maximum length of rockfall runout zones on the basis of simple topographic parameters.

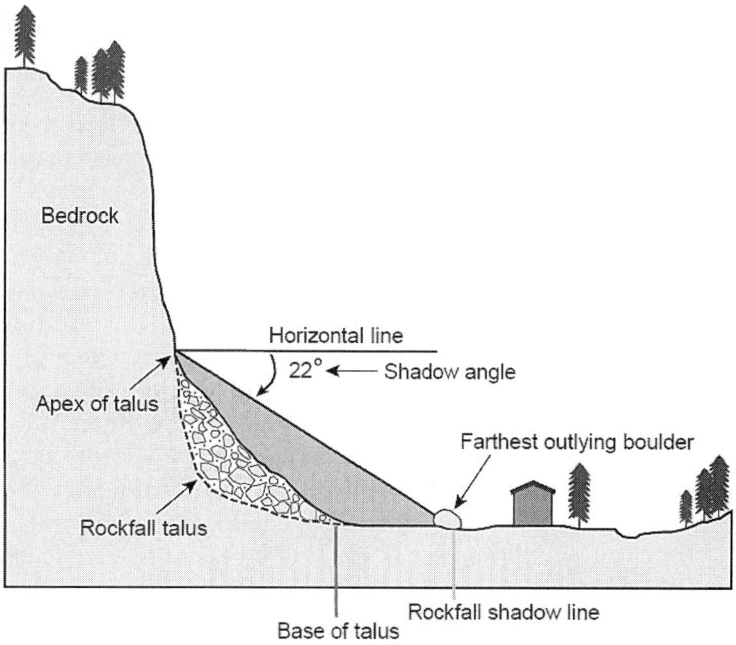

FIGURE 9-13 Schematic of rockfall conditions in Yosemite Valley, California, showing talus slope, base of talus, and farthest outlying boulder. Accepted minimum shadow angle of 22° determined location of rockfall shadow line (Wieczorek et al. 2008).

Many assessments combine empirical and GIS methods. For example, Ayala-Carcedo et al. (2003) used a combination of methods to produce maps at a large scale (1:5,000) showing maximum runout zones below a 5-km (3-mi) section of granite cliff near Madrid, Spain. Frattini et al. (2008) assessed rockfall susceptibility in the Val di Fassa in the Dolomites of the eastern Italian Alps, where rockfall hazards are severe. They compared estimated runout distances produced by empirical models using a 37° angle with distances produced by models that represented geomorphological factors within a version of the 3-D numerical modeling code STONE (described in Section 6.4). Guzzetti et al. (2003) evaluated rockfall runout distances in Yosemite Valley, California, by using the program STONE and compared the estimates with the rockfall shadow zone limits predicted by empirical models (Wieczorek et al. 1998). STONE provided runout estimates that were in good agreement with those produced by the empirical approach, although the runout distances determined by STONE were slightly smaller than those produced by the empirical approach on the basis of the 22° minimum shadow angle.

Loye et al. (2008) used empirical models to evaluate rockfall hazards in the canton of Vaud, Switzerland. Along the Rhône Valley, they discovered that these models overestimated the maximum runout distance at locations where the rockfall blocks would travel onto the Rhône alluvial plain. Rockfall blocks tended to lose speed and energy rapidly on reaching level sections of the valley floor. Thus, the empirical model required adjustment for these circumstances. Flat lands were identified within the GIS database whenever a surface larger than 2.5 km^2 (about 1 mi^2) had a slope gradient less than 11°. On the basis of analyses of aerial photographs, Loye et al. determined that, in this region, talus deposits never encroached more than 60 m (200 ft) onto the flat valley floors. Along the Rhône plain, where steep valley sides bound the flat valley floor without any intervening talus deposits, the maximum runout distance was set to 100 m (333 ft), partially to account for the anticipated greater block velocities expected at these locations. Figure 9-14 illustrates how Loye et al. (2008) adjusted their empirical model for this situation.

3.4 CONEFALL Model

CONEFALL (Jaboyedoff 2003b) was developed to combine empirical shadow angle concepts with available DEMs to provide rapid estimates of the limits of potential rockfall hazard zones. CONEFALL incorporates shadow angle theories defined by Heim (1932) with the DEM techniques described by Toppe (1987). Examples of CONEFALL applications have been provided by Jaboyedoff and Labiouse (2003), Jaboyedoff et al. (2005), and Loye et al. (2008).

CONEFALL computes the runout distance (L) of a rock block by determining the point of intersection with the lower-slope topography of a line connecting the rockfall source with a slope equal to $\tan \gamma = \Delta z / L$, where Δz is the difference in elevation (Figure 9-15). CONEFALL uses a fixed typical shadow angle (φ_p), as defined by Onofri and Candian (1979), to define the angle γ. The energy loss along a complex rockfall trajectory depends on several mechanisms, but the average rate of rockfall energy dissipation may be estimated by using a linear "friction" concept. This allows the shadow angle line to be considered as an *energy line* (Figure 9-15). Jaboyedoff and Labiouse (2011) showed that, at any distance X, the difference in height between the energy line and the topography (Δh in Figure 9-15) is related to the velocity of the rock block at that point [$V_{(x)}$] according to the following equation:

$$V_{(x)} = \sqrt{2g\Delta h}$$

where g is the acceleration due to gravity.

A rock block will travel down a slope as long as the slope elevation falls below the energy line. If the local slope is steeper than the limit φ_p, the

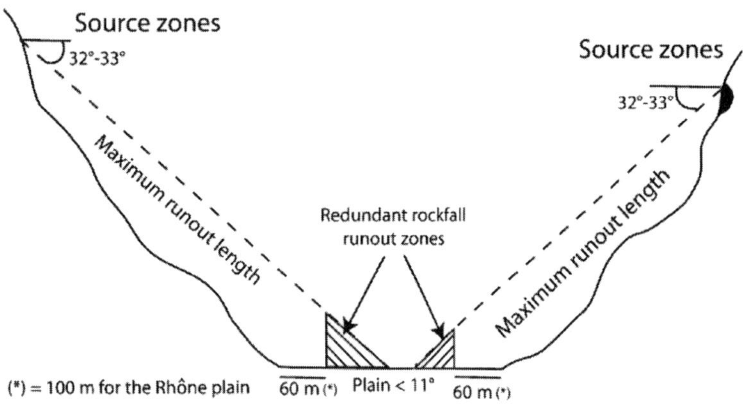

FIGURE 9-14 Cross section of Rhône Valley showing method of correcting empirical estimate of maximum runout distance over level valley floor (Loye et al. 2008).

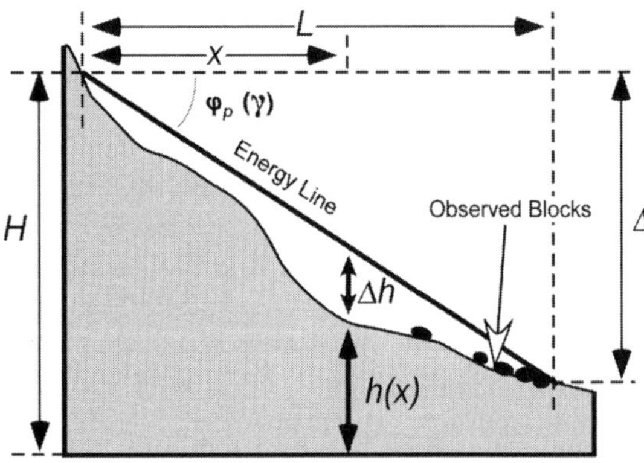

FIGURE 9-15
Variables used by CONEFALL calculations. Most distant block defines cone angle (energy line) and thus Δh, which is used to calculate block velocity (adapted from Jaboyedoff and Labiouse 2011).

block accelerates; if the slope is lower, the block decelerates. Blocks will travel from a source to the point of intersection with the topography of any line starting from the source point and making an angle φ_p with the horizontal. Therefore, a potential rockfall shadow zone for a single source point, defined by a grid point in the DEM, is located within a cone of slope φ_p with a summit placed at the source point (Figure 9-16). The cone is further defined in terms of its orientation and its lateral maximum dispersion, or aperture, by using two additional parameters (Figure 9-17).

CONEFALL applies this procedure to a series of rockfall source points defined within the DEM grid; they can be used to define an entire cliff face or its crest or base. Because, as defined above, the mean velocity of blocks, and thus their kinetic energy, is related to the difference in elevation Δh between the cone and the topography, CONEFALL can also produce estimates of block velocities and energies at each grid cell within the DEM. The principles and operation of CONEFALL have been provided by Jaboyedoff (2003b) and by Jaboyedoff and Labiouse (2003, 2011). CONEFALL software is freely available and is downloadable from http://www.quanterra.org/softs.HTM#soft01.

4. ROCKFALL HAZARD MODELS BASED ON GIS PRINCIPLES

Over the past two to three decades, methods of regional landslide hazard assessment have been increasingly based on data analyzed by GIS. These methods have been adapted to assess rockfall hazards. Soeters and van Westen (1996) provided a comprehensive review of GIS techniques and how they are used to assess slope instability and develop hazard zonation criteria. Fell et al. (2008a) provided a commentary that further defined the role and utility of GIS analysis procedures reported in "Guidelines for Landslide Susceptibility, Hazard and Risk Zoning for Land-Use Planning" (Fell et al. 2008b), which was developed by the Joint Technical Committee on Landslides and Engineered Slopes (JTC-1). Van Westen et al. (2008) supplemented the JTC-1 information by providing an extensive and up-to-date review of the role of spatial information and GIS techniques in landslide hazard assessment.

There are two distinct forms of GIS data structures: a *vector GIS* defines spatial relationships in terms of points, lines, and polygons that define homogeneous areas, while a *raster GIS* divides the mapped region into a large number of small cells (or rasters) and defines spatial relationships by assigning each cell an appropriate property value. Each approach has benefits and limitations. GIS technology has relevance to three rockfall hazard modeling applications:

1. The creation of thematic map products and the statistical evaluations of spatially distributed observations,
2. The modeling of spatially related phenomena, and
3. The use of GIS DEMs to represent the spatial characteristics of slopes in rockfall-prone areas.

In the first class of applications, GIS has been applied to a variety of regional hazard assessments at various scales. Many applications have evaluated multiple types of slope instabilities and have been oriented toward the production of maps delineating hazardous areas.

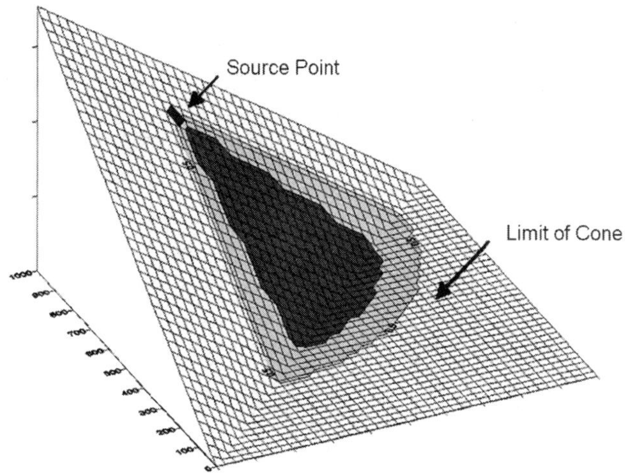

FIGURE 9-16 CONEFALL simulation of areas endangered by rockfalls starting from single source point. Cone angles of 35° (lighter, larger cone) and 40° (darker, smaller cone) are used (Jaboyedoff 2003b).

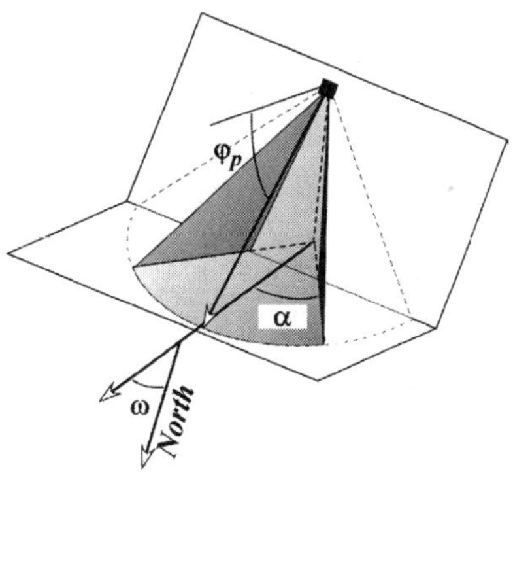

FIGURE 9-17 Basic CONEFALL parameter definitions: cone slope angle (φ_p), cone direction (ω), and aperture of cone ($\pm\alpha$) (Jaboyedoff 2003b).

In the second class of applications, GIS techniques are used to analyze pertinent factors that promote slope instability, such as unfavorably oriented discontinuities. This approach is followed by the Matterock method described in Section 2.3. Other applications use GIS terrain models combined with empirical rockfall models to identify the limits of rockfall shadow zones that reflect to some degree the anticipated trajectories of falling rockfall blocks. The program CONEFALL, described in Section 3.4, is an example of such an application.

The third application category is related to rockfall evaluations at a more detailed level, typically during a site-specific rockfall simulation analysis, when various 2-D and 3-D rockfall simulation models use GIS data to define the slope conditions and thus produce more representative trajectory estimates. These applications are described further in Sections 5 and 6 of this chapter. They use GIS concepts described in the following section and in Section 4.4.

4.1 GIS Raster Concepts

Topographic conditions are most frequently defined by a grid of elevation values and thus are represented as a raster GIS data set called a DEM, also known as a *digital terrain model*. Several analysis procedures related to DEM data are routinely used to predict rockfall trajectories; thus, they represent an important aspect of GIS-related rockfall prediction models.

4.1.1 Typical Resolutions of DEM Data

Applications of DEM analysis methods to rockfall evaluations have been encouraged by the development of GIS software and by the availability of free DEM data from government sources. Most of the DEM data sources have been at global or national scales and thus are at too coarse a resolution to support detailed site-specific studies. Typical national DEM data sets have cells specified by latitude and longitude. Thus, the exact cell dimensions vary with latitude, but typical cell resolutions are approximately 90 × 90 m (295 × 295 ft), 30 × 30 m (98 × 98 ft), or 10 × 10 m (33 × 33 ft), with the most widely accessible data having a cell size of 30 m. Such data are only suitable for regional rockfall hazard zonation studies. However, the evolution of airborne and terrestrial LiDAR scanning devices (see Chapter 7) has allowed for the economical production of DEM data of much higher resolution. Airborne LiDAR sources routinely produce DEMs with cells of 3 × 3 m (10 × 10 ft), while terrestrial LiDAR can produce DEMs of even higher resolution for specific rockfall zones.

4.1.2 Alternative Interpretations of DEM Data

Regardless of their resolution, DEM products typically are interpreted as providing "cell-centered" information. The terrain is represented by a series of equally spaced elevation values that are considered to be the average elevation of that grid cell located at the center of the cell (Figure 9-18*a*). An alternative interpretation, which is less commonly used, considers the series of elevation values to be located at the vertices of the grid cells (Figure 9-18*b*). Obviously, these two interpretations, if applied to the same data set, imply a half-cell-size offset in the definition of the raster grid boundaries (Figure 9-18*c*). This offset is usually only a concern during detailed site investigations.

Another consideration concerns the accuracy of typical DEM elevations. Most DEM data define elevations with integer values because the digital representation of integers requires less memory than does that of fractional values. However, the integer representation of elevation values, combined with the size of the cells, tends to make the DEM representation of the terrain conditions discrete rather than continuous.

4.1.3 Computing Derivatives of the DEM

The two first-order derivatives of the elevation (i.e., the rate of change of elevation) are slope and aspect. *Slope* (or *gradient*) defines the maximum rate of change in altitude, while the *aspect* (or *exposure*) defines the compass direction of this maximum rate of change. Slope may be recorded as a percentage or in degrees (or even radians); aspect is always recorded in degrees related to a compass bearing. Slope and aspect are sufficient for many rockfall analyses, but some applications require one or both of the second-order derivatives of elevation (i.e., the rate of change of slope and aspect). They are *profile convexity* (the change in slope) and *plan convexity* (the change in aspect). In

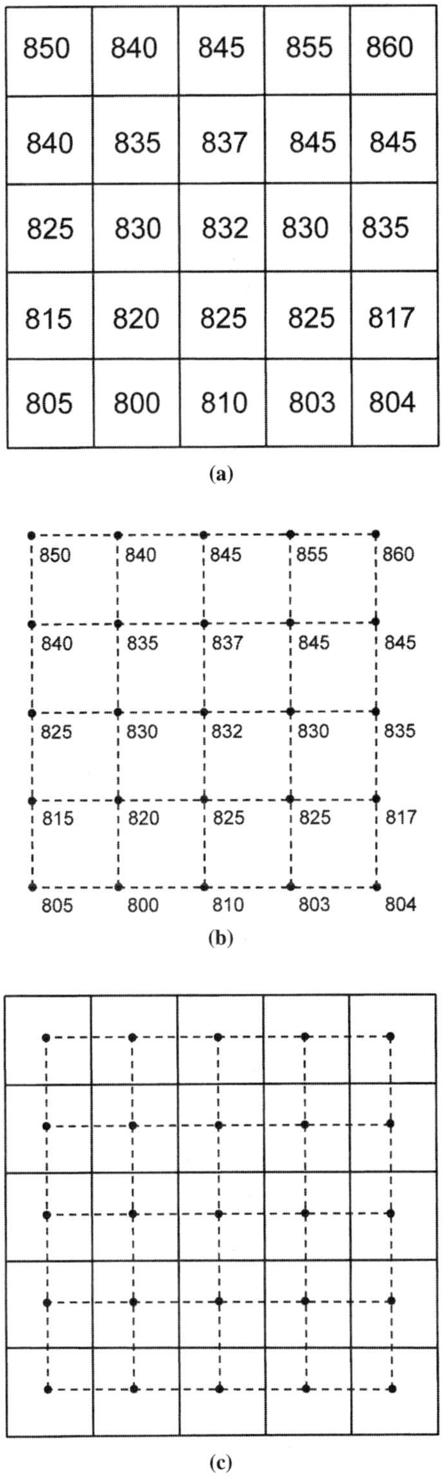

FIGURE 9-18
Two alternative interpretations of GIS DEM data: (*a*) elevations assumed to be cell centered, (*b*) elevations assigned to DEM cell vertices, and (*c*) alternative definitions that offset grids by one-half cell dimension. Only small sections of DEM are shown.

each case, the positive values representing a crest or spur are termed *convexities*, and the negative values representing valleys or hollows are termed *concavities*. They are generally recorded as degrees per unit of distance.

These derivative products are readily computed from DEM data. All GIS software products provide these capabilities. They are calculated for each cell by performing an analysis of a neighborhood of each cell and its eight surrounding neighbors. Figure 9-19 summarizes the most widely accepted and efficient computation procedure for slope and aspect. Burrough and McDonnell (1998) defined it as "a third-order finite difference estimator that uses all eight surrounding cell elevations." This procedure, based on a formulation originally proposed by Horn (1981), first calculates rates of change in the *x*- and *y*-directions, then combines these values to obtain the appropriate slope and aspect values that are stored in a new matrix in the positions corresponding to the central cell of each neighborhood (cell "e" in the example). This procedure is used by ArcGIS software with special adaptations for edges of the DEM data or for areas where some raster cells do not contain elevation data ("no data" cells) (ESRI 2011). Although they are not as frequently used, the two second-order convexity derivatives are readily obtained by reapplying the same slope and aspect calculations to the first-order (slope and aspect) derivatives.

Numerous sources provide additional evaluations of the potential errors and limitations of such calculations and their relationship to

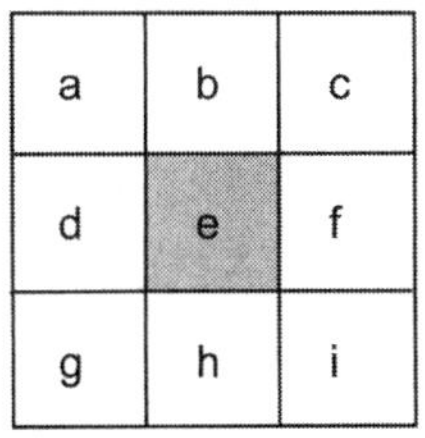

Calculation of Slope for Cell "e":

Rate of change in the x direction:
[dz/dx] = ((c + 2f + i) – (a + 2d + g)) / (8 * X-cell-size)

Rate of change in the y direction:
[dz/dy] = ((g + 2h + i) – (a + 2b + c)) / (8 * Y-cell-size)

Slope (in degrees) = $57.29578 * \arctan\sqrt{[dz/dx]^2 + [dz/dy]^2}$

Calculation of Aspect for Cell "e":

Rate of change in the x direction:
[dz/dx] = ((c + 2f + i) – (a + 2d + g)) / 8

Rate of change in the y direction:
[dz/dy] = ((g + 2h + i) – (a + 2b + c)) / 8

Aspect (in degrees) = 57.29578 * arctan([dz/dy] / –[dz/dx])

FIGURE 9-19
Derivation of slope and aspect values from DEM (adapted from ESRI 2011).

the resolution of DEM data sets (Burrough and McDonnell 1998; Zhou and Liu 2004). These slope and aspect computations belong to a class of neighborhood calculations that are further discussed in Section 4.4.

4.2 GIS Applications to Regional Hazard Assessment

Soeters and van Westen (1996) distinguished three approaches to regional hazard assessment:

1. *Heuristic methods* that associate weights with various instability factors on the basis of expert experience or by using multicriterion methods,
2. *Statistical methods* that compare the observed distribution of landslides (by means of an inventory) with the distribution of physical factors believed to cause landslides either directly or indirectly, and
3. *Physically based methods* that evaluate stability by using deterministic approaches based on physical laws.

4.2.1 Heuristic Methods

Heuristic methods use GIS techniques to produce hazard maps based on the association of various weighted instability factors, guided by expert experience. Leroi (1996) and Guzzetti et al. (1999) used GIS heuristic methods to evaluate landslide hazards; Meisina et al. (2001) and Baillifard et al. (2003) used these methods to assess rockfall hazards.

Guzzetti et al. (1999) combined heuristic methods with raster-based GIS techniques to evaluate the proficiency and limitations of multivariate statistical techniques and alternative methods of evaluating, at various scales, landslide hazards in the Umbria and Marche regions of central Italy. Their studies were hampered by the lack of availability of adequate data describing the many contributing factors and by the limitations of the GIS technology they were using. Nevertheless, the study provided important insights into the heuristic approach.

Meisina et al. (2001) evaluated a small area, about 12 km^2 (4.7 mi^2) in the Alps of northwestern Italy, where rockfall and slope instability pose severe hazards. They evaluated seven factors on the basis of a combination of field observation and GIS techniques. Field observations yielded maps showing lithology, geomorphology, structural geology (orientations of lineaments, schistosity, and joints), and land use. A DEM was used to develop slope and aspect maps and to evaluate the relationship between schistosity orientations and the shapes (topography) of the affected slopes. These factors were combined and then weighted to produce susceptibility maps for three classes of hazard: rockfall, landslide, and shallow soil slip failure.

Baillifard et al. (2003) used a multicriterion heuristic approach to assess rockfall hazards along a Swiss mountain road. Their study considered topographical, geomorphological, and geological parameters, which were selected on the basis of conditions observed at rockfall locations and the ease with which they could be mapped and integrated into a GIS. The locations of rockfall source areas were further refined by

1. Proximity to a fault,
2. Proximity to a talus slope,
3. Presence of a rock cliff,
4. Presence of a steep slope, and
5. Presence of a road.

These criteria were integrated into a GIS by using topographical, geomorphological, and geological vector and raster digital data. The model yielded a rating from 0 to 5, with 0 representing no hazard and 5 representing a severe hazard, and thus provided a relative hazard map (Figure 9-20). Areas with a high rating had to meet two additional criteria to be considered as locations of potential rockfall hazard: (*a*) the local structural pattern had to be unfavorable and (*b*) the area had to be susceptible to the effects of weathering and other triggering factors.

4.2.2 Statistical Methods

Carrara et al. (1995) and Chung et al. (1995) described early applications of GIS-based statistical models to landslide hazard evaluations. While they do not explicitly discuss rockfall, many of the described techniques are appropriate for rockfall applications. GIS-based statistical modeling of the factors controlling rockfall hazards has had only limited application to regional-scale rockfall hazard zonation (Duarte and Marquínez 2002; Marquínez et al. 2003; Chau et al. 2004; Frattini et al. 2008).

Duarte and Marquínez (2002) analyzed the spatial distribution of rockfall deposits at a regional scale for an area of 250 km² (about 100 mi²) in northern Spain. A cartographic database developed with GIS tools revealed several relationships between rockfall activity and environmental variables. This GIS database contained three basic thematic layers and a DEM (Figure 9-21).

The thematic layers were developed by a combination of field observations and interpretation of aerial photographs. Maps were developed at a scale of 1:25,000. The first thematic layer was bedrock geology, initially represented by 15 stratigraphic units. They were subsequently redefined into two major classes—siliciclastic rocks and calcareous rocks—within a derived layer called *lithology* (Figure 9-21). The second thematic layer represented the surficial deposits according to 30 deposit classes of glacial, fluvial, or colluvial origins. From this map, a derived layer showing the extent of active talus deposits was produced (Figure 9-21). The third thematic layer included slopes with exposed bedrock; they were termed *rocky slopes*.

The GIS database was completed by the addition of a DEM with a cell size of 25 × 25 m (82 × 82 ft), which was developed by analyzing contour maps at a scale of 1:25,000. The DEM was used to create slope and aspect derivative maps (see Section 4.1.3). It also provided topographic elevation information for each rocky slope location.

The rocky slopes were initially not classified as to whether they were locations of active or inactive rockfall. Subsequently, rockfall activity within each location was inferred by the existence of active talus deposits at the base of each rocky slope. The *flow direction* and *watershed* functions contained within the ArcInfo GIS software (ESRI 2011) were used to evaluate the DEM data to establish the direction of steepest descent (or flow direction) between the DEM cells (see Section 4.4.2). The cells defining the locations of active talus were then used with these flow direction estimates to define rockfall contribution areas, locations of potential rockfall sources to each talus deposit. By this process, the rocky slopes that were associated with active talus deposits were then reclassified as *active rockfall areas*; Duarte and Marquínez (2002) defined them as *rockfall basins*.

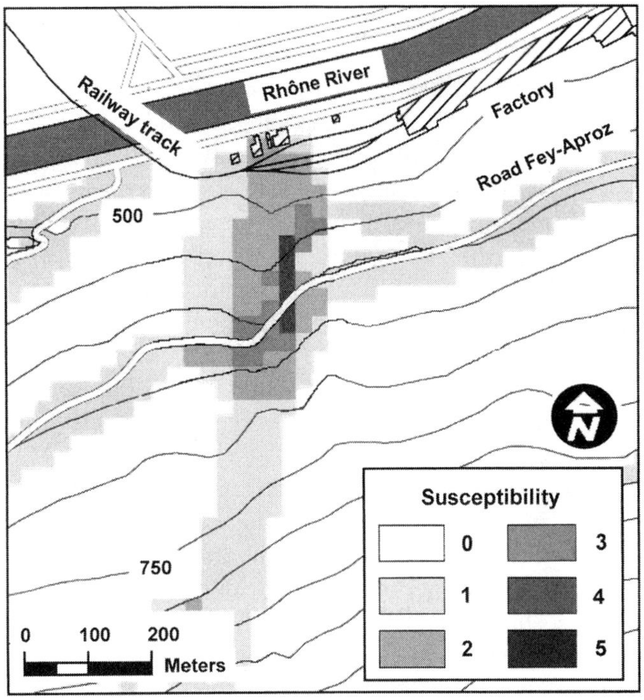

FIGURE 9-20
Relative hazard map produced by automatic assessment of rockfall susceptibility along road near Eterpas, Switzerland, on basis of five criteria: proximity to fault, proximity to talus slope, presence of rock cliff, presence of steep slope, and presence of road. Area has maximum susceptibility index value of 5 (Baillifard et al. 2003).

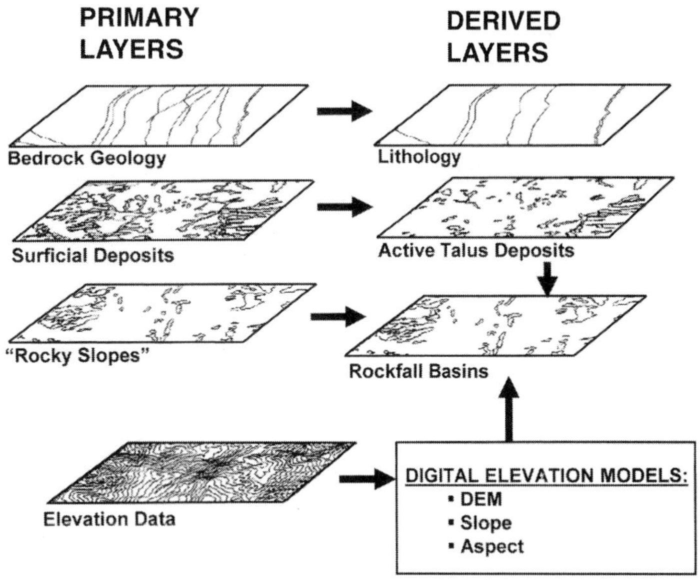

FIGURE 9-21
Thematic and derived layers forming GIS database used to analyze spatial distribution of rockfall deposits at regional scale for area in northern Spain (adapted from Duarte and Marquínez 2002).

Duarte and Marquínez concluded that this automated GIS method for defining active rockfall source areas offered considerable advantages over strictly manual mapping methods. To test the validity of this automatic assignation of rockfall basins, they manually constructed a rockfall basin map and compared the two maps. They reported that the results from each method showed considerable agreement; the main differences were associated with areas where relief features fell below the resolution level of the DEM. This result confirms the importance of using a DEM with as high a resolution as possible in evaluating rockfall potential.

Duarte and Marquínez (2002) then used the rockfall basins map to evaluate the characteristics of the lithology, elevation, and slope gradient and aspect conditions within the source areas. They attributed a difference in rockfall frequency between the siliciclastic rocks and calcareous rocks to differences in rock–joint orientations and fracture density. Elevation influenced rockfall frequency because of climatic variations with elevation in this area. The remaining variables, such as aspect and slope gradient variations, showed complex influences that Duarte and Marquínez found difficult to interpret.

Frattini et al. (2008) developed and tested an approach to rockfall hazard evaluation that combined a multivariate statistical model and a physically based 3-D simulation procedure. The multivariate statistical approach identified where rockfalls were most likely to take place, and the physically based 3-D simulation procedure was used to predict the runout distance of boulders and rock fragments and the anticipated velocities and bounce heights along their trajectories. A lack of reliable records describing historical rockfall events for the study area prevented the investigation from assessing the frequency or return period of rockfall events. The investigation focused on the Val di Fassa region, an area of about 300 km^2 (118 mi^2) located in the Dolomite Mountains of northeastern Italy—a region that is subjected to severe rockfall hazards and is renowned for cliffs more than 1,000 m (3,300 ft) high (Figure 9-22).

FIGURE 9-22 Typical high cliffs found in Val di Fassa; view of cliffs composed of carbonate rocks on south side of Stella Group, northern limit of Val di Fassa (http://www.flickr.com/photos/rossimarko81/1172482103/).

In the Val di Fassa, the high cliffs are mostly composed of thick carbonate sequences; the intermediate and lower slopes are formed on weaker marl and siltstone rocks or talus deposits. As a consequence, numerous rockfall sources seriously threaten the inhabited valley areas, which attract large numbers of tourists throughout the year because of the skiing and mountain hiking opportunities. This further complicates rockfall risk management efforts, as tourists engaged in skiing or hiking often reach remote locations that are normally uninhabited. The area exhibits the most serious rockfall hazard conditions.

Frattini et al. (2008) developed a detailed inventory of rockfall source areas and associated talus deposits for the entire Val di Fassa area on the basis of field surveys and interpretation of multitemporal aerial photographs. The maps were digitized and stored in a GIS database along with a detailed DEM that used a 10- × 10-m (33- × 33-ft) cell size. The DEM produced interpretive layers defining slope gradient, aspect, topographic curvature, and homogeneous subregional partitions. These data were used to develop a susceptibility model for rockfalls. GIS cells defining rockfall source area polygons were classified as active or inactive by using a 3-D rockfall simulation model (the HY-STONE program described in Section 6.4) to link each source grid cell to associated talus deposit polygons by calculating the trajectories of rockfall blocks.

Four scenarios, which were based on empirical model criteria (slopes greater than 37°) or geomorphological criteria and alternative rockfall initiation susceptibilities, were assessed with discriminant analysis to identify the environmental variables that were most important in the initiation of rockfalls. The process evaluated 23 factors and identified six as the most relevant. The factors defined conditions related to the lack of vegetation, unfavorable geological structures, and slope morphology.

4.2.3 Physically Based Methods

Physically based GIS techniques evaluate rockfall hazards deterministically by using simple approaches based on physical laws (Van Dijke and van Westen 1990; Terlien et al. 1995; Meißl 1998, 2001). They do not attempt to analyze the details of rockfall impacts, bounces, or rolling motions by using the Newtonian mechanics principles discussed in Chapter 8. Instead, they rely on the simpler physical concept of rockfall blocks tending to travel down the direction of steepest descent across the topography connecting a rockfall source to the final stopping point of each block. The velocity of the blocks is estimated by evaluating their kinetic energy represented by the steepness of the terrain, but this velocity is modified by frictional losses imposed by surface materials, terrain roughness, or the presence of vegetation.

Such concepts can be readily evaluated by using GIS models. Dorren (2003) defined these models as *distributed models*, since they are process-based but also take into account the spatial variability within a certain region or catchment. In a GIS-based distributed rockfall model, the terrain is usually represented by multiple rasters, which are derived from GIS data layers. Each raster represents a certain property of the terrain (for example, height above sea level, surface roughness, vegetation cover, and geology).

Van Dijke and van Westen (1990) developed a distributed model that represented an 80-km^2 (31-mi^2) region in western Austria. The model used a raster-based GIS with a cell size of 20 × 20 m (66 × 66 ft) that contained a DEM derived from topographic contour maps and maps representing engineering geology and vegetation conditions (Figure 9-23).

A neighborhood analysis used the D-8 method (described in Section 4.4.2) to calculate the fall direction of a rockfall block traversing each raster cell. Van Dijke and van Westen (1990) calculated the velocity on the basis of the energy

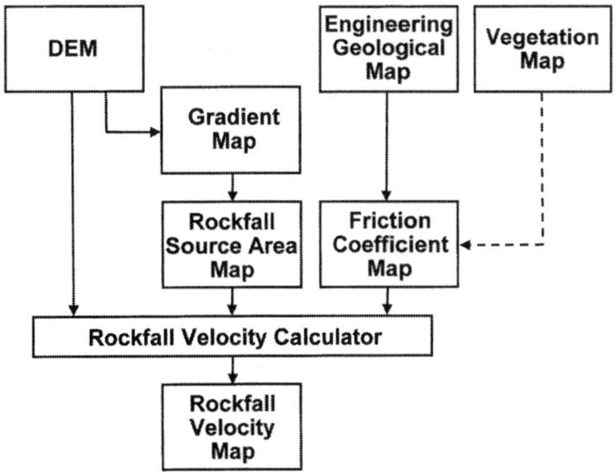

FIGURE 9-23 Schematic of GIS-based distributed rockfall model (adapted from Van Dijke and van Westen 1990).

conservation principle according to the following equation:

$$V_B = V_A + \sqrt{2g(\Delta h - \mu D)}$$

where

V_B = velocity of the block leaving the cell,
V_A = velocity of the block entering the cell,
g = acceleration due to gravity,
Δh = fall height (difference in height between cells),
μ = coefficient of friction, and
D = distance traveled over the slope surface.

The velocity was calculated for each raster cell forming a trajectory path, starting at a potential rockfall source cell. Potential rockfall source cells were defined by cells having a slope gradient greater than 60°. The value of the friction coefficient (μ) was assigned according to surficial conditions defined by the engineering geological and vegetation maps (Table 9-4). Van Dijke and van Westen tested the model by comparing its predicted rockfall hazard limits with those shown on detailed hazard maps of the area. They concluded that their model produced acceptably accurate distributions of rockfall hazard areas.

Meißl (1998, 2001) assessed rockfall hazards by using GIS-based distributed models that simulated the motions of rockfall blocks during both free fall and subsequent travel down talus slopes. The free-fall velocity was calculated with the following formula:

$$V = \sqrt{2gh}$$

Table 9-4
Friction Coefficients Used by the GIS-Based Distributed Rockfall Model

Material	Friction Coefficient
Tills[a]	0.35–0.5
Residual soils[a]	0.4–0.5
Fluvial materials[a]	0.4–0.5
Bare rock	0.4–0.9
Talus deposits	
Marl	0.35–0.45
Flysch	0.6–0.7
Sandstone	0.7–0.8
Dolomite	0.7–0.8
Limestone	0.8–0.9
Rockfall materials	0.9–1.0
Effect of vegetation	
Open forest[a]	1.0–2.0
Dense forest[a]	>2.0

Note: Most friction coefficients based on field observations.
[a]Coefficients based on estimated values.
Source: Van Dijke and van Westen 1990.

where V is free-fall velocity, h is height of free fall, and g is as previously defined. The energy lost during the first collision or bounce on the slope surface was accounted for by eliminating 75% of the velocity gained during the initial free fall, and the velocity of the block as it traveled downslope through a series of cells was computed with the equation proposed by Van Dijke and van Westen (1990) that was defined previously. Meißl (1998, 2001) designated rockfall source areas by delineating cliff faces on topographic maps or potential rockfall source areas shown on geomorphological and geological maps. Downslope trajectories were calculated by using a DEM with a 50- × 50-m (165- × 165-ft) cell size and the D-16 method defined in Section 4.4.2. The model predictions were compared with mapped rockfall patterns for an area in western Austria. On the basis of those comparisons, Meißl (1998) concluded that the simulated rockfall patterns for zones in mountainous regions with limited vegetation cover generally correspond to the mapped patterns, although he did not provide absolute model accuracies.

4.3 GIS Support for Modeling Spatially Related Phenomena

GIS techniques support the modeling of spatially related phenomena; thus, GIS methods are frequently used in analyzing pertinent factors that promote slope instability, such as the identification of locations dominated by unfavorably oriented discontinuities or the semiautomated definition of potential rockfall source areas.

Jaboyedoff, Baillifard, Philippossian, et al. (2004) described a GIS-based process for detecting and estimating the probability of the presence of fracture sets that may produce rockfalls or rockslides. They described a three-step process that is followed by the Matterock method described in Section 2.3. This probabilistic process has been used to generate rockfall hazard maps for the Saas Valley (Jaboyedoff et al. 1999; Jaboyedoff, Baillifard, Philippossian, et al. 2004), and maps produced by this process were compared with an actual rockfall event that occurred in 1994 near Fionnay, Switzerland (Jaboyedoff, Baillifard, Philippossian, et al. 2004). The comparison revealed that the process may be limited by the accuracy of the DEM data. DEM data with large cell sizes tend to smooth the terrain and thus prevent accurate relationship comparisons between the terrain and discontinuity orientations. They also adversely influence the correct prediction of the presence or absence of all but the longest discontinuities.

Jaboyedoff, Baillifard, Couture, et al. (2004) described another application of DEM analysis for identifying rockfall hazards. In cases where one or more inclined rock mass discontinuities form a potential sliding surface, such as illustrated on a small scale in Figure 9-24, the DEM data may be evaluated to identify the orientation of any

FIGURE 9-24
Example of inclined joint producing sloping local base level and thus potential rockfall by sliding failure.
(Photograph by M. Jaboyedoff.)

inclined surface that could isolate a spur or mass of rock and form a sliding failure plane. The "deepest" such plane along a major valley in mountainous regions can be defined by identifying a series of elevations along the channels of tributary streams and then using these points to define a planar surface, which Jaboyedoff, Baillifard, Couture, et al. called the *sloping local base level*. The method was evaluated by backanalysis of the massive 1991 rockslides near Randa, Switzerland (Sartori et al. 2003).

GIS terrain models may be combined with empirical rockfall models to identify the limits of rockfall shadow zones reflecting, in simple terms, the anticipated trajectories of falling rockfall blocks. The program CONEFALL, described in Section 3.4, is an example of such an application. As discussed in Section 4.2.3, Van Dijke and van Westen (1990) and Meißl (1998, 2001) used basic GIS techniques to assess rockfall hazards with models that use simple physical concepts.

Several rockfall assessments of Italian locations (Castelli and Scavia 2008; Nocilla et al. 2009; Longo and Oreste 2010) have used a GIS-based probabilistic modeling approach applying two programs, ISOMAP and ROTOMAP, developed by Geo and Soft International of Turin, Italy (Geo and Soft 2009). ISOMAP supports the development and display of DEM models. ROTOMAP uses the ISOMAP DEM data to define the terrain within a probabilistic 3-D "lumped-parameter" rockfall model (see Section 6.2) to simulate the trajectories of a large number of rockfall blocks and, through the analysis of the distribution of their average and maximum kinetic energies, to identify hazard zones and the optimal areas for the installation of protective systems. Figure 9-25 shows a typical display of results produced from a ROTOMAP rockfall hazard analysis of a rockfall area represented by an ISOMAP DEM. ISOMAP falls within the category of GIS applications described in this section, but ROTOMAP properly belongs to the category of 3-D rockfall simulation models described in Section 6.

4.4 Use of GIS DEM Data to Represent Slope Geometry

Whenever rockfall hazards are being evaluated at a site-specific level, GIS DEM data can be used to define slope conditions and produce more accurate estimates of rockfall block trajectories. The assessment of the mechanics of a series of rockfall block impacts on the basis of the principles described in Chapter 8 allows prediction of 3-D rockfall trajectories. However, such calculations require considerable information on the shapes of rockfall blocks in addition to the detailed description of surface irregularities provided by the DEM. Complex calculations must be conducted over multiple simulations, each representing a randomly selected initial block orientation, if acceptable representations of the variation in rockfall trajectories are to be achieved. Several of the 2-D and 3-D rockfall simulation models described in Sections 5 and 6 of this chapter conduct such extended calculations and rely on DEM data to provide information concerning the orientation and irregularity of the lower slopes across which rockfall blocks bounce and roll. Sections 5 and 6 of this chapter discuss the models in some detail; the GIS concepts used in assessing slope geometry are described here.

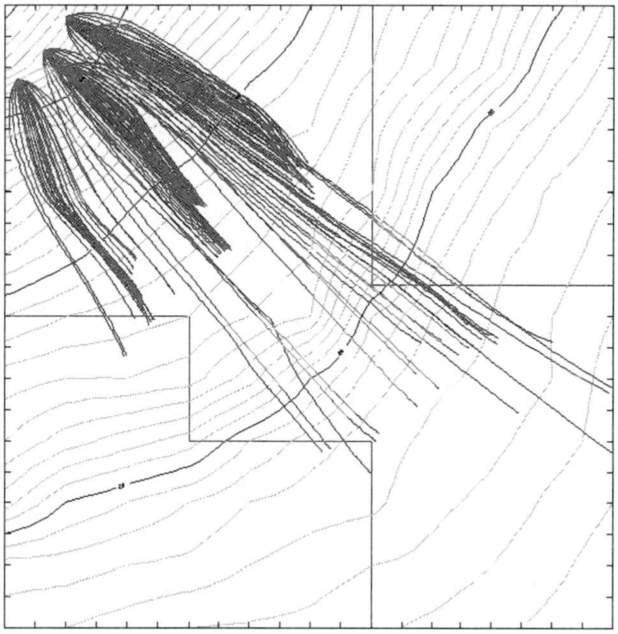

FIGURE 9-25 Rockfall trajectory paths from three source areas developed by ROTOMAP with DEM data produced by ISOMAP and represented by topographic contours (Geo and Soft 2009).

The basic derivation of slope and aspect information by evaluation of DEM data is defined in Section 4.1.3 and Figure 9-19. These calculations are a widely used GIS analytical tool and have obvious application to rockfall hazard analysis. They belong to a class of GIS techniques called *neighborhood functions* because they rely on the analysis of conditions in the area immediately surrounding a particular location (or raster cell).

A series of slightly more complex neighborhood functions is used by some rockfall models. These functions are discussed below.

4.4.1 Defining the GIS DEM Neighborhood

GIS DEM neighborhoods are defined by groups of cells near the location of interest. The most common method is to use a square window defined by an odd number of cells in each direction so that a central cell is readily defined. A 3 × 3 or 5 × 5 window (or *kernel* or *filter*) is most commonly used. The window is progressively passed over the entire DEM array, and the results are stored at the locations of the series of central cells in a new array with the same number of columns and rows (Figure 9-26). Obviously, special arrangements have to be made to compute appropriate values for cells around the margins of the DEM array where some of the cells that would normally be included within a window are missing. Similar arrangements have to be established if a window partly covers an area where cells do not have valid values, such as at the edges of an irregular map area where external cells are given a "no data" value. Most GIS systems automatically incorporate appropriate procedures for these circumstances.

4.4.2 Determining the Path of Steepest Descent

Many rockfall hazard investigations rely on the delineation of the paths of rockfall blocks across talus slopes. These paths are often referred to as *falltracks* and their delineation as *falltrack analysis*.

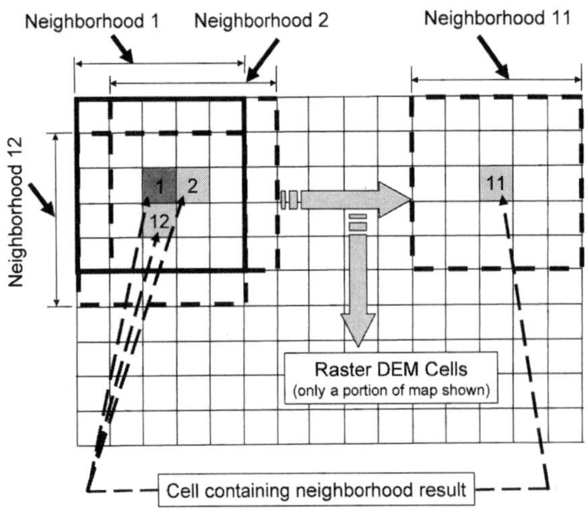

FIGURE 9-26 Neighborhood analysis process for raster DEM; 5 × 5 cell neighborhood is shown scanning across, then down, entire raster array. Only small portion of raster is shown. In this case, width of two cells around boundary must be calculated by using some special arrangements, because neighborhoods related to these cells extend beyond original data limits.

For many rockfall hazard applications, the complete detailed assessment of 3-D rockfall trajectories requires too great a computational effort and more information on rockfall block shapes than is readily available. A much simpler estimation of the falltrack distribution is all that is required.

Analysis of the direction of steepest descent across a sloping surface represented by a DEM with appropriate cell size resolution has been adopted by several investigators as an appropriate yet relatively simple approach to falltrack analysis (Dorren 2003; Petje et al. 2005). The direction of steepest descent relies on neighborhood analysis of DEM data to determine which of the cells surrounding the current cell has the lowest elevation. This cell is the location over which water would flow to discharge downslope; thus, steepest-descent analysis has been extensively investigated for watershed hydrology applications and forms the basis for the flow direction tool within the hydrology tools contained within the ArcGIS software (ESRI 2011).

The two most common methods for determining the direction of steepest descent are the D-8 method introduced by O'Callaghan and Mark (1984), which used a 3 × 3 neighborhood, and the D-16 method proposed by Meißl (1998), which used a 5 × 5 neighborhood (Figure 9-27). Both methods have been applied to rockfall path studies, on the basis of the concept that rock blocks, just like surface water, will typically follow the path of steepest descent down a slope (Van Dijke and van Westen 1990; Meißl 1998, 2001). Figure 9-27 shows the steepest-descent direction determined by the D-8 and D-16 methods for the same local DEM data. In each case, the selected cell having the lowest elevation and hence the steepest descent is shown with a darker shade. In implementations, the selected direction is recorded by an identification code stored in a new raster data set termed a *flow direction raster*. The basic implementations of these identification codes are shown in Figures 9-28*a* (for the D-8 method) and 9-28*b* (for the D-16 method).

The D-8 method evaluates eight possible directions (the directions along the columns and rows of the raster and the diagonal directions); thus, it tends to introduce systematic errors since it favors orientations with 45° separations. The consequence of such a systematic bias for a falltrack assessment is shown by Figure 9-29. Despite this bias, the D-8 method offers the simplest and

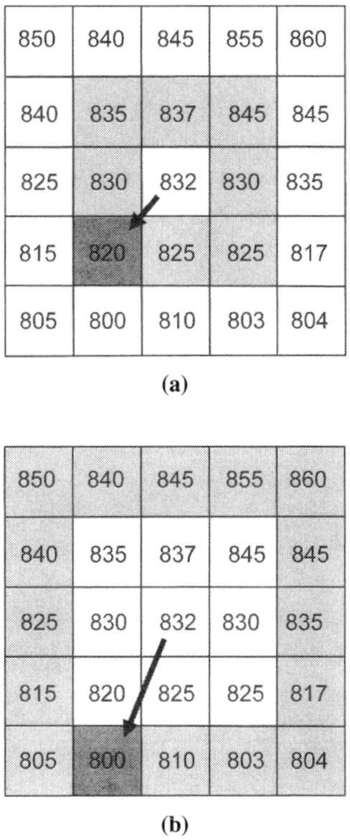

FIGURE 9-27
Two methods for determining direction of steepest descent by neighborhood analysis of DEM: (*a*) D-8 method uses 3 × 3 window to evaluate eight possible directions; (*b*) D-16 method uses 5 × 5 window to evaluate 16 possible directions (adapted from Dorren 2003).

therefore the most rapid calculation. Because the falltrack is calculated between adjacent cells, the velocity of the falling block is readily determined for every cell.

In contrast, the D-16 method reduces the directional bias of the D-8 method by using 16 orientations separated by 22.5°. However, its calculations are more complex because it evaluates positions beyond the eight neighboring cells and thus potentially ignores intervening ridges of higher elevations. For this reason, extra calculations must be undertaken to evaluate these internal conditions, to calculate block velocities, and to fill in gaps in the raster evaluations (Meißl 1998).

The basic D-8 and D-16 methods identify a single flow direction for each source cell; thus, a rock block in a given raster cell will leave by a single exit. DEM rasters typically contain local depressions (or *pits*), where individual cells or small neighborhoods have lower elevations than all surrounding cells.

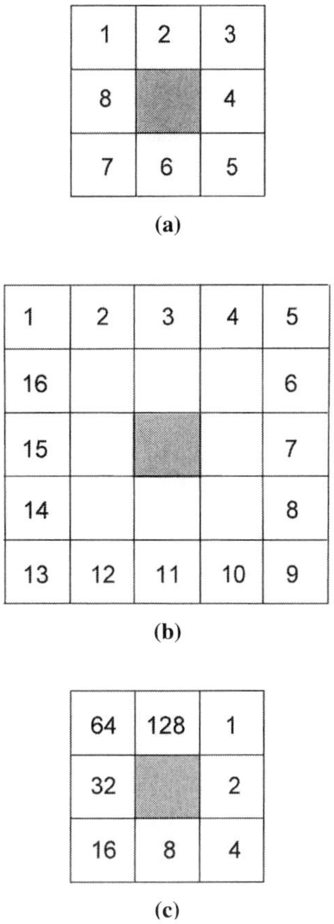

FIGURE 9-28
Flow-direction codes: (*a*) basic implementation for D-8 method, (*b*) basic implementation for D-16 method, and (*c*) more sophisticated implementation for D-8 method used by ArcGIS software.

These conditions naturally result from the regular grid spacing of the raster cells and become most noticeable on gently sloping areas. The D-8 and D-16 methods will fail to find a continuation of the downslope path in such circumstances, unless the rasters are evaluated to identify these pits and then

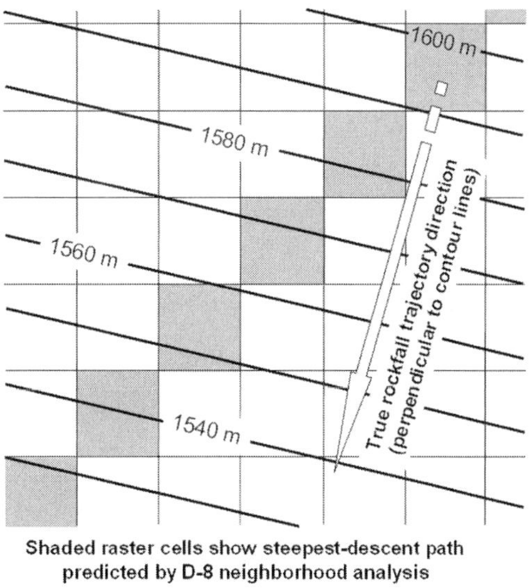

FIGURE 9-29
Arrow defines correct falltrack on slope depicted by contour lines. Raster-based falltrack determined with D-8 method, illustrated by gray cells, deviates from correct falltrack direction (Dorren 2003).

fill them by selectively raising DEM elevations of cells located in a pit (Jenson and Domingue 1988; Burrough and McDonnell 1998; ESRI 2011). The basic methods also have difficulty if several surrounding cells have the same elevations and thus offer identical discharge opportunities.

Fairfield and Leymarie (1991) provided a critical evaluation of the D-8 method and its limitations. They noted that the D-8 method would produce unrealistic parallel paths, or paths that followed the systematically preferred directions across smooth surfaces. As a result, the basic D-8 method will not develop appropriate falltrack dispersions on features such as talus cones. Partly for this reason, the CONEFALL program described in Section 3.4 (Jaboyedoff 2003b) defines the range of potential rockfall by using different methods. Fairfield and Leymarie (1991) proposed that path divergence could be achieved if the D-8 method was modified to include a slight randomization of the chosen directions. Quinn et al. (1991) simulated the lateral spreading of steepest-descent paths by using a multiple-flow-direction algorithm; Tarboton (1997) provided additional variations for this algorithm.

The ArcGIS software employs a slightly more sophisticated version of the D-8 method that allows for the identification and coding of multiple outflow directions on the basis of a procedure originally described by Greenlee (1987). Greenlee identified 256 possible surrounding conditions for a central pixel and its eight immediate neighbors. These conditions can be coded as a single data byte that permits values within the range 0 to 255. The value 0 is assigned to the condition where the central cell is entirely surrounded by higher-elevation neighbors (it is the bottom of a pit), and 255 is assigned to the condition where all the neighbors are at a lower elevation (the cell is the top of a peak). If the D-8 directional codes are assigned values as shown in Figure 9-28c, multiple-flow directions may be uniquely identified by adding together the relevant directional codes. For example, if the downslope direction is entirely toward the right, then the code of 7 (1 + 2 + 4) identifies the three exit cells located to the right of the central cell in Figure 9-28c.

4.4.3 Determining Rockfall Source Areas

Several early GIS-based rockfall hazard evaluations defined rockfall source areas in a simple manner by identifying locations where steep slopes coincided with the presence of rock outcrops (Van Dijke and van Westen 1990). These identifications were typically further refined by additional geomorphological studies or field observations (Van Dijke and van Westen 1990; Meißl 1998, 2001).

However, once the falltrack paths have been defined and their component cell-to-cell movements are stored in a flow direction raster, it is easy to trace backward through these paths from the cells defining the terminal locations of rockfall blocks to identify the rockfall source areas. This process is identical to that required in identifying the sources forming a stream network within a watershed. As a result, this analysis process is routinely provided in GIS; the ArcGIS software has a "watershed" function (ESRI 2011).

The backward tracing of simulated rockfall trajectory paths has been used by several investigators to identify rockfall source areas. Duarte and Marquínez (2002) used this automated GIS technique to define active rockfall source areas, which they called *rockfall basins*, and recommended its use. They believed it offered considerable advantages over strictly manual mapping methods. Frattini et al. (2008) used similar methods in assessing the rockfall hazards in the Val di Fassa area within the Dolomite Mountains of northeastern Italy.

4.4.4 Alternative Terrain Representations for Accurate Rockfall Trajectory Modeling

Agliardi and Crosta (2003) concluded that the "discrete description" of the topography provided by DEM data was insufficiently precise for the accurate 3-D simulation of rockfall trajectories by the program STONE (Guzzetti et al. 2002). They created a triangulated regular network by assuming that the DEM data were located at the corners of a grid and then defining a triangular network whose vertex coordinates were sampled from the DEM matrix (Figure 9-30). This allowed for the conversion of the discrete information provided by the DEM into a more continuous representation of the topography. Two reference coordinate systems (Figure 9-30) were used: a global Cartesian system (*X*, *Y*, *Z*) to define the rockfall block inertial properties and a second Cartesian system (x_p, y_p, z_p) that is local to each triangle. The axes x_p and y_p are parallel to the local triangle slope, while the axis z_p

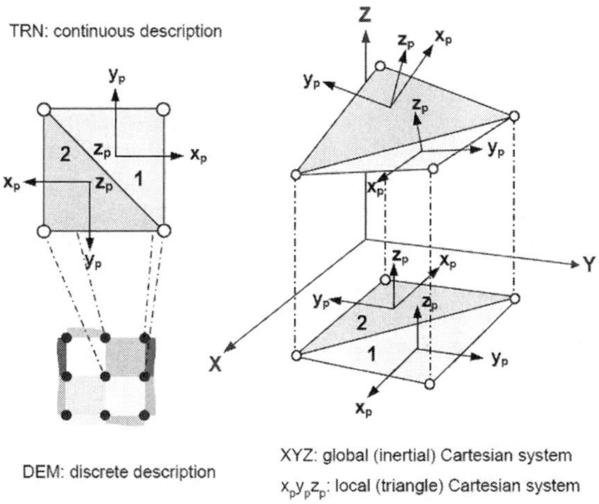

FIGURE 9-30
Development of triangulated regular network (TRN) from DEM to provide superior "continuous" representation of topography for 3-D rockfall trajectory modeling (Agliardi and Crosta 2003).

is perpendicular to the surface. STONE computes ballistic trajectories in the inertial system, which are then transformed to the local coordinates by means of rotation–translation matrices as required to determine the impact or rolling motions.

5. ROCKFALL SIMULATION MODELS INCORPORATING NEWTONIAN MECHANICS CONCEPTS ALONG A 2-D PROFILE

Interest in rockfall hazards rose during the 1970s because of the construction of new transportation facilities and increasing populations in areas at risk of rockfall and because of scientific developments in such fields as geomorphology and rock mechanics. The concurrent evolution of computer technology naturally led to the preparation and use of computer programs to simulate rockfall events and to provide quantitative verification of new concepts related to the kinematic behavior of falling rocks and field observations of rockfall phenomena.

Sections 2, 3, and 4 of this chapter have discussed rockfall prediction and modeling methods developed to evaluate the stability of rock masses and the empirical and GIS-based approaches that support regional rockfall hazard zonations. Practitioners charged with the assessment of rockfall hazards at specific locations and with the design of suitable rockfall mitigation measures at these sites required a different class of models that could gauge

- Potential rockfall trajectories;
- Bounce heights, velocities, and energies achieved by rock blocks as they fall; and
- Maximum runout distances to define rockfall hazard areas.

Capabilities of widely available computer resources, better understanding of rockfall behavior based on existing observational data from laboratory and field tests, and economic constraints related to the acquisition of detailed topographic descriptions of rockfall locations combined to encourage the development of 2-D rockfall simulation programs relying on a profile to define the site terrain and characteristics. Over the past 30 years, a number of 2-D rockfall simulation programs have been developed to predict rockfall trajectories and provide a basis for the analysis and specification of rockfall mitigation measures. Table 9-5 summarizes the characteristics of a series of 2-D rockfall simulation programs developed between 1977 and 2000.

This section discusses the concepts and characteristics of these programs. Their use has become an important and routine component in current rockfall hazard investigation and in the specification of rockfall mitigation methods. Valuable experience has been gained by observations of rockfalls during field and laboratory tests. The programs have significantly increased the knowledge and understanding of rockfall. If the programs are well calibrated, their simulation results, on evaluation by experienced professionals, produce information that is sufficiently accurate for identifying rockfall-prone areas and can form the basis for the design and specification of rockfall protective structures.

Since the advent of these 2-D programs, significant advances in computer technologies have occurred. In addition, development of new site investigation technologies, especially the rapid development of airborne and terrestrial LiDAR instruments (see Chapter 7), has made it economically feasible to obtain detailed representations of topographic conditions at rockfall locations. As a consequence, the development of 3-D rockfall simulation programs that overcome some of the inherent limitations of a 2-D analysis approach has

Table 9-5
Main Characteristics of 2-D Rockfall Simulation Programs

Initial Release[a]	Program Name	Kinematic Principle[b]	Simulation Method[c]	Definition of Coefficient of Restitution[d]	Source	Comments
1977	Computer Rockfall Model Simulation Program	Lumped mass	Partly probabilistic	COR_{VN}, COR_{VT}	Piteau and Clayton 1977; Piteau and Associates 1980	First documented 2-D simulation model.
1982	No name	Lumped mass	Probabilistic	COR_{VN}, COR_{VT}	Azimi et al. 1982	Differentiates between impacts on hard and soft ground.
1982	SASS and MASSI	Hybrid	Probabilistic	COR_{TE}	Bozzolo and Pamini 1982, 1986; Bozzolo et al. 1988	Evaluate translational and rotational energies with an energy-based definition of coefficient of restitution. Evaluate ellipsoidal block shapes.
1985	ROCKSIM	Lumped mass	Deterministic	COR_{VN}, COR_{VT}	Wu 1985	Developed to evaluate the influence of varying impact angles on the normal and tangential coefficients of restitution.
1985	No name	Rigid body	Deterministic	No information	Falcetta 1985	Uses a discrete element method.
1985	Rotolamento Salto Massi	Lumped mass	Deterministic	COR_{V}	Bassato et al. 1985	Deterministically evaluates block motions while free-falling, rolling, rebounding, and bouncing.
1987	Rockfall	Hybrid	Probabilistic	COR_{VN}, COR_{VT}	Spang 1987; Spang and Rautenstrauch 1988; Spang and Sönser 1995; Spang and Krauter 2001; Spang et al. 2008	First developed by Raymund Spang; this program is commercially distributed and supported by a German company. It has been widely used in Europe.
1988	CRSP Version 1	Hybrid	Probabilistic	COR_{VN}, COR_{VT}	Pfeiffer and Higgins 1988, 1990; Pfeiffer 1989; Pfeiffer and Bowen 1989; Pfeiffer et al. 1990	This is the initial version of a widely used 2-D rockfall simulation program.

1990	No name	Rigid body	Deterministic	COR_V	Kobayashi et al. 1990	Developed to evaluate rockfall hazards triggered by earthquakes.
1990	ROFMOD	Lumped mass	Probabilistic	COR_V	Zinggeler 1990; Zinggeler et al. 1991	2-D program that includes the influence of forests (trees) on rockfall trajectories.
1991	CADMA	Hybrid	Probabilistic	COR_{TE}	Azzoni et al. 1991, 1995; Azzoni and de Freitas 1995	Later modification of the Bozzolo and Pamini programs that uses the same energy-based definition of the coefficient of restitution.
1991	CRSP Version 2.1	Hybrid	Probabilistic	COR_{VN}, COR_{VT}	Pfeiffer et al. 1991	First revision of this popular program.
1993	CRSP Version 3	Hybrid	Probabilistic	COR_{VN}, COR_{VT}	Pfeiffer et al. 1993, 1995	Second revision of this popular program.
1998	RocFall	Lumped mass	Probabilistic	COR_{VN}, COR_{VT}	Stevens 1998	Originally developed as a thesis at the University of Toronto. This program has been continually revised. Current release is RocFall 7.1; it is distributed by RocScience.
1999	MOBYROCK	Lumped mass	Probabilistic	COR_V	Paronuzzi and Artini 1999	Used to evaluate block trajectories and impacts through backanalysis of rockfall events.
2000	CRSP Version 4.0	Hybrid	Probabilistic	COR_{VN}, COR_{VT}	Jones et al. 2000 Higgins et al. 2003	Extensively revised and recalibrated version of this popular program. The current version.

[a]Initial release dates are estimated on the basis of the earliest published descriptions of the program.
[b]Kinematic principles refer to methods used to analyze the rockfall block motions. *Rigid body* refers to motion analysis that takes into account the size and shape of the body; it allows for analysis of translational and rotational velocities and energies. *Lumped mass* refers to motion analysis that ignores the body shape and size; only translational velocities and energies are evaluated. *Hybrid* refers to a combination of lumped-mass and rigid-body methods; usually lumped-mass methods are used for free-fall portions of the rockfall trajectory while rigid-body methods are used to evaluate bouncing and rolling motions.
[c]Simulation methods are either deterministic or probabilistic. Under deterministic methods, rockfall trajectories are entirely defined by fixed properties and Newtonian mechanical principles. Under probabilistic methods, rockfall trajectories are defined by parameters that vary from simulation to simulation on the basis of predefined ranges of values and probability distributions; this introduces a range of uncertainty in rockfall trajectories.
[d]The following are methods of representing the coefficient of restitution:

- Separate definitions, based on velocity ratios, for the normal and tangential coefficients (defined as COR_{VN} and COR_{VT});
- A single definition, based on velocity ratios, for the coefficient of restitution (defined as COR_V); and
- A combined definition, based on changes of rotational and translational kinetic energies, called the total energy coefficient of restitution (COR_{TE}).

become more feasible. The 3-D programs are discussed in Section 6. They are appropriately termed "state of the art." They are not widely used in practice because they require significantly greater investments in gathering the required site characterization information and because their computations take longer and require enhanced computer resources.

Although research in rockfall simulation modeling has largely moved beyond the creation of new 2-D rockfall simulation models, new developments have been reported in the use of discrete element methods, which were briefly discussed in Section 2.1 of this chapter. These methods are routinely used in the civil and mining professions to analyze rock slopes. Several reports reference the early applications by Cundall (1971) and Cundall and Strack (1979) of discrete element methods in analyzing rockfall. There are two principal discrete element methods: distinct element, used by Cundall (1971) and Falcetta (1985) for rockfall evaluation, and DDA, which was used to analyze a major rockfall in Japan (Chen and Ohnishi 1999). The predominant distinct element code is UDEC, which was developed by Itasca International, Inc. (2011d).

Salciarini et al. (2009) used UDEC to evaluate rockfall from a natural rock slope near the town of Postignano in the central Italian Apennines. The UDEC model simulated individual falling blocks of various sizes and modeled their fragmentation as they moved down the slope. Comparison simulations were performed by the Colorado Rockfall Simulation Program (CRSP), which is discussed in Section 5.3. The two models produced realistic and similar predictions of runout distances; the UDEC model predicted shorter distances because of fragmentation of the blocks.

Further applications of discrete element methods to rockfall simulation modeling are likely to be undertaken. The availability of suitable computer codes, along with appropriate computer capacity, makes the use of these methods more feasible than in the past. The applications need not be restricted to 2-D simulations. Although the computations are even more extensive in 3-D, 3-D discrete element methods are being implemented in the newest version of CRSP, termed CRSP-3D, which is introduced in Section 6.8.

5.1 Alternative Simulation Approaches

Several investigators have classified 2-D rockfall simulation models according to the kinematic principles used to analyze rockfall block motions and whether the simulations are strictly deterministic or involve probabilistic approaches to incorporate uncertainty in the outcomes (Guzzetti et al. 2002; Dorren 2003; Heidenreich 2004). Although there is some variation in terminology, there is general agreement on the basic classification. A third important characteristic of these programs is how they define the coefficient of restitution. These three characteristics are summarized for each program listed in Table 9-5 and are discussed below.

5.1.1 Kinematic Principles

In terms of the kinematic principles used by the models, two main categories may be defined: those applying rigid-body analysis methods and those applying lumped-mass analysis methods. *Rigid-body methods* take into account the size and shape of the rock block; they use the Newtonian mechanical principles described in Chapter 8. These simulation models evaluate translational and rotational velocities of the rockfall blocks and the corresponding energies. In contrast, *lumped-mass simulation models* simplify the calculation process by ignoring the size and shape of the block. They model the block's mass as a concentrated single point (its center of mass) and evaluate rockfall motions by calculating the paths followed by these "infinitely small" objects. This simplification allows for more rapid calculations, but only translational velocities and energies are evaluated.

Some simulation programs take advantage of the fast and easy calculations of the lumped-mass methods to simulate block motions during the free-fall or rebound portions of rockfall trajectories but use rigid-body methods to evaluate bouncing, rolling, or sliding motions. Such models are classed as hybrid models in Table 9-5. Hybrid models generally use geometrical and mechanical characteristics of the slope and the block in modeling block impacts and computing rebound trajectories. Simulation models using rigid-body or hybrid methods generally are more capable of accurately reproducing rockfall trajectories that involve all phases of the rockfall phenomenon: free fall, bouncing, rolling, and sliding.

5.1.2 Deterministic or Probabilistic Models

Two-dimensional rockfall simulation programs are further differentiated according to whether they use probabilistic or deterministic approaches in the description and analysis of the block motions. *Deterministic programs* use approaches that are entirely defined by fixed properties and Newtonian mechanical principles; they typically use either mean input values or a worst-case assumption for the calculation of the rockfall trajectory. They produce definitive estimates for height of bounce, kinetic energy of the rockfall blocks, and runout distance. They typically evaluate individual trajectories for which the input parameters remain fixed. A small number of trajectories may be sequentially calculated by adjusting the input parameters to arrive at a worst-case trajectory with its associated characteristics.

In contrast, other rockfall simulation programs incorporate *probabilistic capabilities* to replicate the natural variations of some parameters, such as the shape and size of the block, the mechanical characteristics of the slope, the local slope angle at the point of impact, and the locations and characteristics of rockfall source areas. These variations, or uncertainties, are accounted for by varying the value of the parameter from simulation to simulation within a predefined range and by random sampling from probability distributions. This process produces a wider range of predicted rockfall trajectory outcomes.

Regardless of the type of program used, careful calibration of the model parameters is required to achieve simulation results providing realistic information that may be used for the identification of rockfall-prone zones or for the design of protective structures.

5.1.3 Coefficient of Restitution

To simulate rebound, most rockfall models represent the bouncing phenomenon in a simplified way by one or two overall coefficients, termed *coefficients of restitution*, that significantly affect the loss of energy or velocity during and after each impact. In some cases, energy losses resulting from blocks rolling or sliding as they fall are modeled by means of an additional friction coefficient. Other than slope shape and roughness, these coefficients constitute the most crucial modeling parameters. Chapter 8 discusses alternative methods of assessing the coefficients of restitution.

A definition that is based on the ratio of the velocities before and after an impact appears to correspond to the concept of Newton when impact dynamics were first considered and has since been used by many authors in developing 2-D rockfall simulation programs or in evaluating rockfall events (Spang and Rautenstrauch 1988; Azzoni and de Freitas 1995; Spang and Sönser 1995; Paronuzzi and Artini 1999). Many have concluded that a subdivision into two components, defined as the normal and tangential coefficients of restitution (COR_{VN} and COR_{VT}, respectively, in Table 9-5), but still based on velocity measurements, is the best approach. It has been used during field observations and the development of 2-D simulation models by many authors (Piteau and Clayton 1977; Wu 1985; Pfeiffer and Bowen 1989; Kobayashi et al. 1990; Pfeiffer et al. 1991, 1993, 1995; Stevens 1998; Jones et al. 2000).

When the block is in trajectory, conversion of some translational energy to rotational energy will occur because of friction between the block and slope surface. As a result, oblique impacts will produce rotation of the bouncing blocks, and this rotation will tend to increase during a sequence of impacts. Believing that this effect should not be neglected, Bozzolo and Pamini (1986) and Azzoni et al. (1995) proposed an additional definition of the coefficient of restitution (COR_{TE}) that accounts for the total energy of the block. Azzoni and de Freitas (1995) and Azzoni et al. (1995) used COR_{TE} in their evaluation of parameters influencing rockfall and in their CADMA model, which is discussed in Section 5.2.9.

Chapter 8 provides additional information on these alternative definitions of the coefficient of restitution, and Table 8-3 contains a collection of published values of the coefficient of restitution. Section 3.5 of Chapter 8 discusses the significance of ensuring that appropriate values of the coefficient of restitution are used when modeling is undertaken; in each case, the values must accurately reflect the assumptions employed by the model. Heidenreich (2004) emphasized that any coefficient of restitution taken from literature or obtained through in situ tests or backanalysis is likely to provide incorrect results when it is used at other sites. She therefore recommended against the use of published coefficient of restitution values without a calibration and sensitivity analysis of the simulation results to evaluate the model's response to changes in the values.

5.2 Review of 2-D Rockfall Simulation Models

Table 9-5 provides summary information for 16 rockfall simulation models that were developed, described, and used in support of rockfall investigations between 1977 and 2000. They include four versions of CRSP. Table 9-5 reports on the most commonly referenced models and is similar to several other published lists of rockfall simulation programs. However, the list is not exhaustive, and a large number of similar programs have been developed in support of rockfall research efforts at many institutions.

Most programs listed in Table 9-5 were research tools and thus had limited distribution, usually only within the community of active rockfall researchers. Many of the programs were developed with computer systems that are now obsolete, and some of the early programs operated on mainframe computers with user interfaces that were rudimentary compared with current standards. Thus, it should not be inferred that many of these programs will be used in further rockfall investigations. They are included in this discussion because they illustrate fundamental and important methods, issues, options, and limitations faced by those wishing to undertake rockfall simulation.

Only three of the programs have achieved widespread acceptance, received long-term technical support, or undergone progressive refinement over time:

1. *CRSP*. One of the most widely used 2-D rockfall simulation programs in the United States, it has been extensively used worldwide. The latest version of CRSP (Version 4.0) is distributed by the Colorado Geological Survey and may be ordered online at http://geosurveystore.state.co.us/p-676-colorado-rockfall-simulation-program-version-40.aspx.
2. *RocFall*. Originally developed as a master of science thesis at the University of Toronto, Ontario, Canada, it has been supported and continually revised and refined by RocScience, Inc., located in Toronto, which continues to handle the distribution of the program (RocScience 2011b). The latest version of RocFall is Version 4.0.
3. *Rockfall*. Originally developed as a research tool by Dr. Raymund Spang, a considerably enhanced and refined Rockfall program is currently supported and distributed by the consulting firm Dr. Spang Ingenieurgesellschaft für Bauwesen, Geologie und Umwelttechnik GmbH (Dr. Spang Engineering Company for Construction, Geology, and Environmental Technology, Ltd.), headquartered in Witten, Germany. The latest version of Rockfall is Version 7.1.

Most rockfall site investigations use one or several of these three simulation programs to provide a basis for the analysis and specification of rockfall mitigation measures. Accordingly, they are discussed in some detail in Sections 5.3, 5.4, and 5.5. The remainder of this section provides an overview of the other programs shown in Table 9-5.

5.2.1 Initial Rockfall Model (1977)

The first documented 2-D rockfall simulation program was described by Piteau and Clayton (1977) at a rockfall research meeting at the Istituto Sperimentale Modelli e Strutture SpA in Bergamo, Italy. Subsequently the use of the program was described in a reference manual published by the Federal Highway Administration (Piteau and Associates 1980).

As shown in Table 9-5, this program analyzed the block kinematics by using a lumped-mass approach, and thus the program did not evaluate rotational energies or the influences of block size or shape on the trajectory. However, the model did use partly probabilistic methods to assign values to several parameters whose values were not precisely known. Two modes of block motion were considered: bouncing and rolling. The bouncing mode computed standard parabolic free-fall trajectories, with energy dissipation at each impact accounted for by normal and tangential coefficients of restitution (COR_{VN} and COR_{VT}, respectively). In contrast, energy attenuation during the rolling mode was accounted for by assigning a friction coefficient that represented the resistance to rolling (or sliding) motions over the slope surface. This value was assigned on the basis of the steepness of the slope and its surface composition. Piteau and Associates (1980) noted that the mechanics of the bouncing mode were better understood and better represented by the program than those of the rolling mode.

In many respects, this program established the procedures subsequently used by the developers of most other 2-D simulation models. For example, the slope geometry and conditions were represented by dividing the slope into a series of segments or cells. The basic slope inclination geometry of each cell was defined as a straight line, but each slope segment was assigned values for the coefficients of restitution (COR_{VN} and COR_{VT}) and the friction coefficient on the basis of the materials forming the surface and the perceived surface roughness (Figure 9-31).

Piteau and Clayton (1977) provided recommended values for COR_{VN} and COR_{VT} for four classes of surface conditions (see Table 8-3). Probabilistic methods were used to "perturb" the average slope inclination value at each impact location, which allowed for uncertainties in slope roughness and rock block irregularities. Figure 9-32 shows results of investigations into the effects of varying values of the coefficients of restitution on the predicted rockfall trajectories. The program provided summaries of trajectory characteristics at selected evaluation locations on the slope; estimates of runout distances; and, by selectively adjusting slope geometry or assigned coefficient values for individual slope segments, evaluations of alternative rockfall mitigation strategies, such as benching and installation of barriers (Piteau and Associates 1980).

5.2.2 Modeling by Azimi et al. (1982)

Azimi et al. (1982) described the use of a 2-D rockfall simulation program that differentiates between impacts on soft- and hard-ground materials. It evaluates the depth of penetration into the ground by the block during impact. The block is simulated as a lumped mass. Impacts on hard ground assume the ground to be perfectly elastic. The block is modeled as being less than perfectly elastic, and the normal and tangential coefficients of restitution are used to establish the velocity reductions. The values of the normal coefficient of restitution depend on both ground material and block properties. Values of the tangential coefficient of restitution reflect the incident velocity, the slope angle, the impact duration, and a global friction coefficient. Additional block resistance is used to account for a possible fragmentation of the block. Impacts on soft ground are modeled by considering the block as a rigid body that penetrates into a perfectly plastic ground material. The rebound velocity depends on the block mass, the incident velocity, and a global friction coefficient. The impact is further evaluated to account for the depth of penetration into the ground, the creation of a crater, and the resistance to sliding and rolling motions of the block imposed by the crater.

5.2.3 SASS and MASSI Models

Bozzolo and Pamini (1982, 1986) described the SASS model, which was developed to simulate the 2-D motion of a single block down a mountainside.

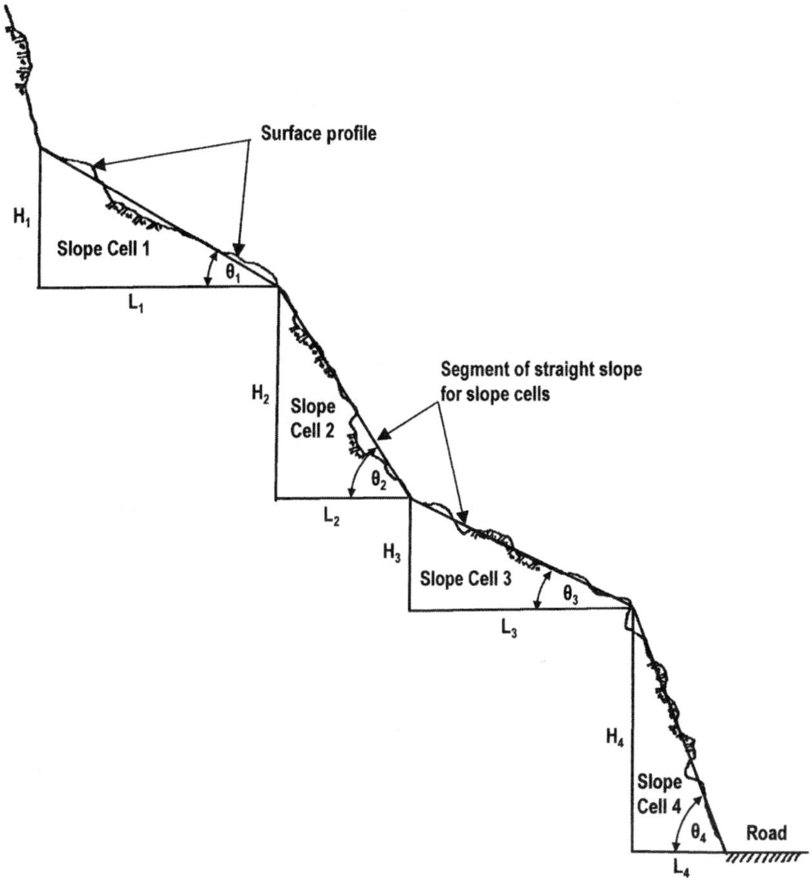

FIGURE 9-31 Typical rock slope definition used by Piteau and Clayton (1977) rockfall model showing subdivision of slope into cells defined by straight-line segments linked together to define geometry of entire slope. Slope roughness conditions are accounted for by appropriate values for coefficients of restitution and friction for each cell (adapted from Piteau and Associates 1980).

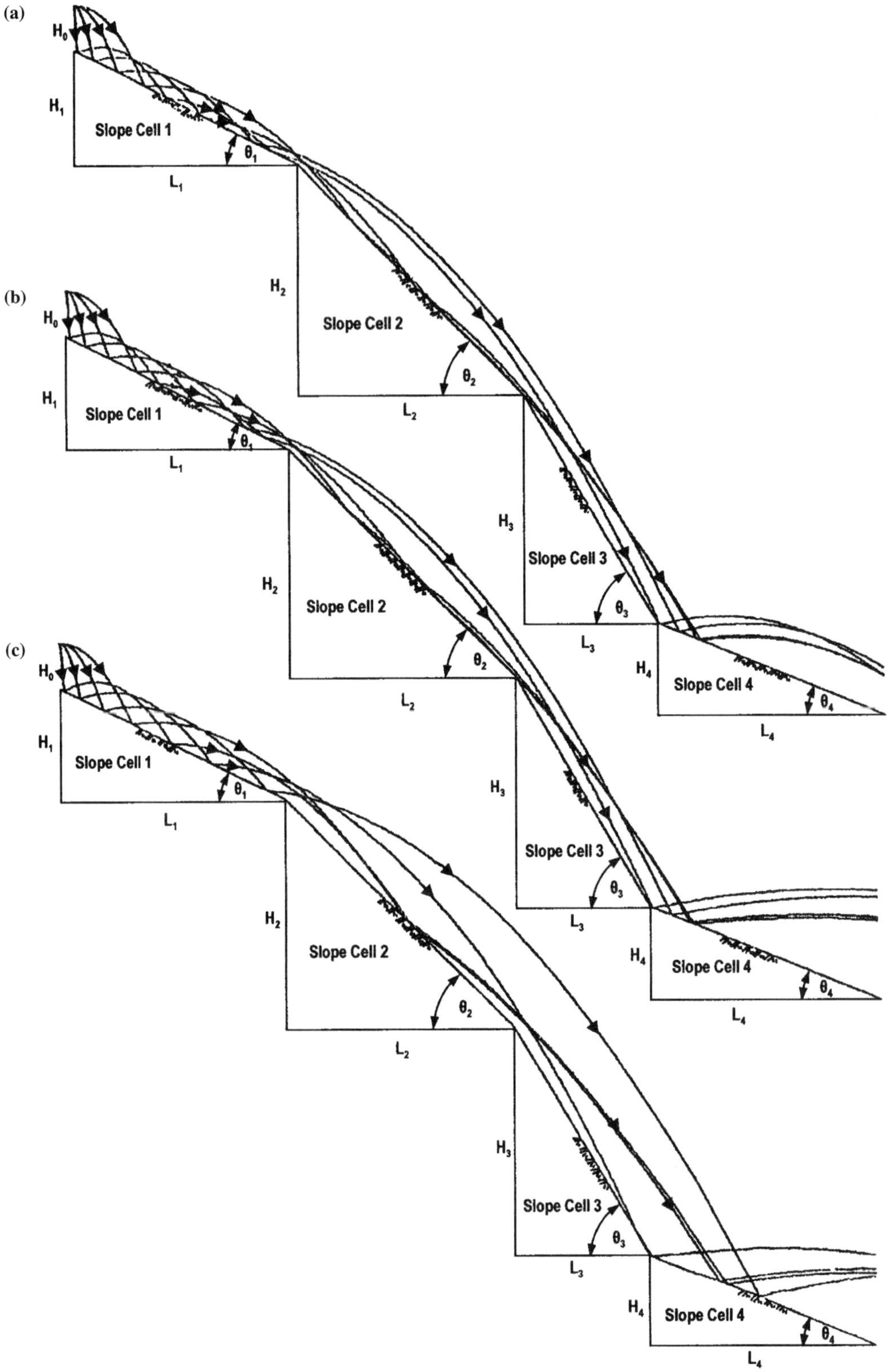

FIGURE 9-32
Calcomp computer plots generated by rockfall program of Piteau and Clayton (1977) illustrating effects of varying coefficient of restitution values in series of slope cells defining complete slope: (*a*) coefficients in slope cells are all less than in Cases *b* and *c;* (*b*) coefficients in Cell 4 are greater than in Case *a;* and (*c*) coefficients in all cells (1 through 4) are greater than in Cases *a* and *b* (adapted from Piteau and Associates 1980).

The falling block is modeled as a rigid ellipsoidal body with axes varying between defined limits. The slope is defined by a series of segments to which data are assigned that define materials and surface conditions. The model accounts for sliding, rolling, impact, and free flight of the individual block, whose motions include both translational and rotational components. To evaluate energy losses during impacts, Bozzolo and Pamini introduced the sophisticated concept of the total energy coefficient of restitution (COR_{TE}), which accounts for changes in both the translational and rotational kinetic energies of the block (see Section 3.3.2 of Chapter 8).

The SASS model provides predictions of the translational and rotational velocities of a block and its bounce height as it travels down the slope. The distance to its stopping point provides an estimate of runout distance. Variations in the model parameters were introduced to evaluate the variability in the predicted trajectories.

Bozzolo and Pamini (1986) used SASS to quantify the behavior of rockfalls observed during a series of detailed rockfall field tests conducted at several locations near Lugano, Switzerland. The tests involved rocks bouncing and rolling on slope surfaces made up of hard rock or covered with fine gravel debris. Because they used an ellipsoidal block shape, Bozzolo and Pamini were able to use SASS to analyze the influence of impact angles on their total energy definition for the coefficient of restitution (COR_{TE}) (see Figure 8-33).

Subsequently, Bozzolo et al. (1988) introduced a somewhat simpler model, named MASSI, that ignored the sliding and rolling motions and evaluated only the impact and free-flight aspects of rockfall events. They compared the results produced by SASS and MASSI by using the same field data as before and concluded that SASS produced better predictions of maximum runout distance, while MASSI provided good estimates of bounce heights for rockfalls over hard-ground conditions.

5.2.4 ROCKSIM Model (1985)

The ROCKSIM model was developed to test the effectiveness of widening a rockfall catchment ditch as part of a relocation project on Interstate 40 in North Carolina (Wu 1985). The algorithms were supported by an experimental program that observed the trajectories of rocks dropped onto an inclined wooden platform. The results of these experiments provided insight into the characteristics of the coefficient of restitution and are reported in Section 4.3.1 of Chapter 8. As a result of these experimental observations, ROCKSIM justified the use of normal and tangential coefficients of restitution and scaled them to account for observed variations in their values as the angle of impact changed.

5.2.5 Modeling by Falcetta (1985)

Direct access to the French-language publication by Falcetta (1985) is limited, but several publications concur that Falcetta used discrete element methods within a sophisticated program to simulate boulder movements down slopes. The program accounted for the shape of the boulder, its dimensions, and its initial translational and rotational velocities. Results of the simulations showed that the shape of the block appreciably affects the motion and trajectory of the block. Several investigators evaluating the development of talus slopes have found the Falcetta model and discussion useful.

5.2.6 Modeling by Bassato et al. (1985)

Bassato et al. (1985) provided a detailed explanation of the algorithms in a program that uses deterministic and lumped-mass kinematic principles to analyze rockfall motions during free fall, rolling, rebounding, or bouncing. A single coefficient of restitution is used to account for velocity losses on impact. A copy of the program code is included in the publication (Bassato et al. 1985).

5.2.7 Modeling by Kobayashi et al. (1990)

Kobayashi et al. (1990) provided a detailed description of the algorithms in an unnamed model used to provide an assessment of hazards posed by rockfalls triggered by future earthquakes. They developed and calibrated their model by using detailed field observations of rockfalls triggered by a 1980 sequence of earthquakes at Mammoth Lakes, California, and by the 1983 earthquake in central Idaho. Their field observations provided data defining topographic profiles and impact locations along several identified rockfall tracks and documenting the size and shape of the individual rockfall blocks (Kobayashi et al. 1990).

The program, as described, uses rigid-body kinematic principles to evaluate the motions and impacts of rolling, bouncing, and free-falling circular 2-D objects. They are taken to represent spherical, discoidal, or cylindrical blocks without any adjustment to their moments of inertia. The ground profile is defined as a series of straight-line segments between surveyed elevation points. The program computations are entirely deterministic. Initial simulations revealed an unrealistic tendency for the boulders to roll along the ground surface, so a stochastic component was added by introducing synthetic surface irregularities. A midpoint with a randomly defined elevation, within limits, was introduced to each segment of the profile. This provided an artificial roughness with an amplitude of 0.2 m to 0.5 m (8 in. to 20 in.), which resulted in simulations with more frequent bouncing motions. However, the model then produced numerous small bounces, so a second modification was introduced that eliminated bounce trajectories shorter than the perimeter of the simulated boulder. The inclusion of these two modifications allowed the model to be calibrated to reproduce the observed rockfall behavior fairly accurately.

5.2.8 ROFMOD Model (1990)

The ROFMOD model was developed to provide 2-D rockfall simulations that accounted for the influence of forests within the rockfall zone (Zinggeler 1990; Zinggeler et al. 1991). The original model has been adopted and enhanced by a Swiss consulting company, GEOTEST AG, and the resulting improved model is sometimes referred to as the *Zinggeler + GEOTEST model*. Forest characteristics used to evaluate impacts of rockfall blocks on trees include mean stem diameter, diameter distribution, tree density, and mean tree-free distance. Rockfalls are specified by a 2-D profile; surface roughness; a coefficient of restitution; the characteristics of the forest stand; and block size and shape, which are defined by the lengths of the block's three major axes. The model assumes that the block always rotates around its shortest axis. The model predicts bounce heights, energy and velocity values, and runout distances. Perret et al. (2004) provided an extensive discussion of the application of the Zinggeler + GEOTEST model and a comparison of its predictions with field observations of a rockfall near Steg in Liechtenstein. These comparisons are discussed in Chapter 8 and are illustrated by Figure 8-47.

5.2.9 CADMA Model (1991)

Azzoni et al. (1991, 1995), with the first English report in 1995, described a new program called CADMA that represented an updating and refinement of the earlier programs SASS and MASSI developed by Bozzolo and Pamini (1982, 1986) and Bozzolo et al. (1988). CADMA shares many of the concepts first introduced by SASS, including the use of the total energy coefficient of restitution (COR_{TE}) to account for changes in both the translational and the rotational kinetic energies of the block at each impact. The use of ellipsoidal block shapes was also retained. CADMA was designed with the experience gained from several field tests. The calculations were based on rigid-body mechanics, and probabilistic methods were used to vary input parameter values to account for model uncertainty.

CADMA was used in evaluating the influence of various topographic resolutions on the accuracy of the predictions as well as in defining appropriate values for the coefficient of restitution and the rolling friction coefficient. The characteristics and potentials of CADMA were evaluated by comparing predictions with observations at two field sites. In all cases, the program produced generally accurate and acceptable predictions in terms of rockfall velocity, energy, bounce height, and runout distance. Azzoni and de Freitas (1995) provide additional information on these aspects.

5.2.10 MOBYROCK Model

Paronuzzi and Artini (1999) developed a specialized 2-D rockfall simulation model named MOBYROCK at the University of Udine in Italy. MOBYROCK focuses on modeling trajectories and impact dynamics by using a lumped-mass representation of the blocks and a single coefficient of restitution COR_V to calculate the decreased translation velocity of the block due to impact processes. Paronuzzi (2009a, 2009b) used MOBYROCK to perform backanalysis of block rebounds measured in the field following a rockfall at Lavone, northern Italy.

5.3 Colorado Rockfall Simulation Program

The initial version of CRSP (CRSP 1.0) was developed in 1988–1989 as a master of engineering research topic at the Colorado School of Mines (Pfeiffer and Higgins 1988; Pfeiffer 1989; Pfeiffer and Bowen 1989). The development of CRSP 1.0 was supported by the Colorado Department of Transportation (DOT) as part of the rockfall hazard assessment efforts along the Glenwood Canyon section of Interstate 70 then under construction. Consequently, CRSP was developed to model rockfall behavior and provide statistical analysis of probable rockfall events at a given site.

CRSP uses a combination of lumped-mass analysis to evaluate free-fall and bounce trajectories and rigid-body analysis to evaluate impacts and rolling motions; thus, it is classified as a hybrid model in Table 9-5. It uses stochastic techniques to modify controlling parameters during each simulation run and thus is classified as a probabilistic model. Experimental verification and calibration of the algorithms used by CRSP 1.0 were provided by data gathered during field tests of rockfall fences conducted at a test site near Rifle, Colorado. CRSP 1.0 subsequently underwent minor modifications, which resulted in the release of CRSP 2.1 (Pfeiffer et al. 1991), CRSP 3.0 (Pfeiffer et al. 1993), and CRSP 3.0a (Pfeiffer et al. 1995). Extensive descriptions of the concepts used by these initial versions of CRSP and associated calibration and verification efforts have been provided by Pfeiffer and Bowen (1989), Pfeiffer and Higgins (1990), and Pfeiffer et al. (1990, 1991, 1993, 1995).

A more extensively revised version, CRSP 4.0, was completed by Jones et al. (2000). This version was reprogrammed by using Visual Basic to be compatible with the Windows 95 and Windows NT operating systems. The Windows environment allowed for the introduction of much more user-friendly interactions, which improved both data entry and data display operations. However, the basic CRSP algorithms remained unchanged. CRSP 4.0 benefited from the availability of additional test results from six rock-rolling field experiments conducted by the California DOT. These data, along with the original Colorado rock-rolling experimental data, allowed for the CRSP algorithms that use the normal and tangential coefficients of restitution to be recalibrated to reflect a greater range of slope-forming materials. Analysis of the revised CRSP 4.0 predictions verified that they tended toward a worst-case scenario when compared with field test observations, but the average predicted values of rockfall velocities, energies, and bounce heights were in close agreement with field test values (Higgins et al. 2003).

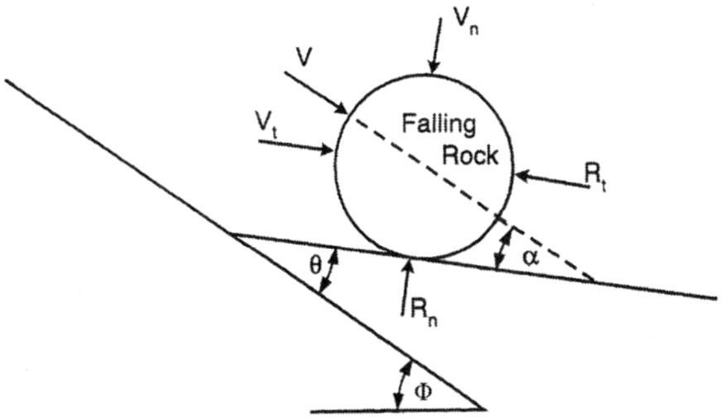

FIGURE 9-33 Geometrical impact relationships used by CRSP. Impact angle (α) is function of block incident trajectory, slope angle (Φ), and slope variation angle (θ). Block velocity (*V*) is converted to normal (V_n) and tangential (V_t) components, which are modified for rebound conditions by normal and tangential coefficients of restitution COR_{VN} and COR_{VT} (Pfeiffer 1989).

All versions of CRSP use a process first proposed by Pfeiffer (1989) to quantify energy dissipation during impacts. At each impact, CRSP uses the incident velocity of the block, its impact angle, and its rotational velocity to calculate the rebound angle, rebound velocity, and a new rate of rotation for the block. Figure 9-33 illustrates the relationships used by CRSP to evaluate impacts. At each impact, the slope angle (Φ) is randomly varied up to a limit corresponding to the maximum probable variation of the slope (θ_{max}), which is determined by field observation of the maximum deviation perpendicular to the slope (S_{max}) within a slope distance equal to the radius (*R*) of the block (Figure 9-34). This random variation in the impact angle deterministically influences the computed rebound conditions and thus is largely responsible for the statistical variation in rockfall trajectories calculated by CRSP.

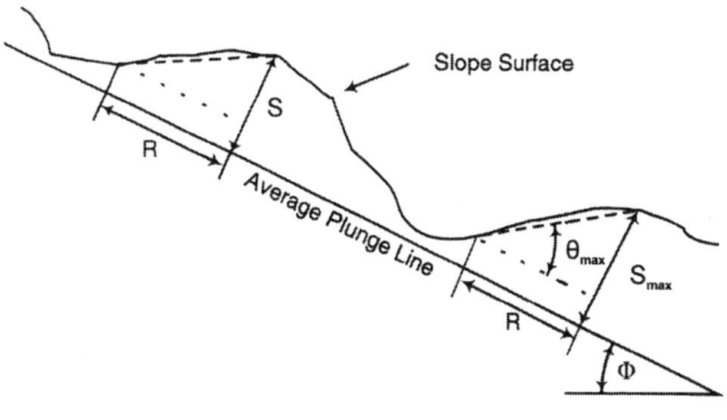

FIGURE 9-34 CRSP computes maximum slope variation angle (θ_{max}), which is determined from field observation of maximum deviation perpendicular to slope (S_{max}) within slope distance equal to radius (*R*) of block (Pfeiffer 1989).

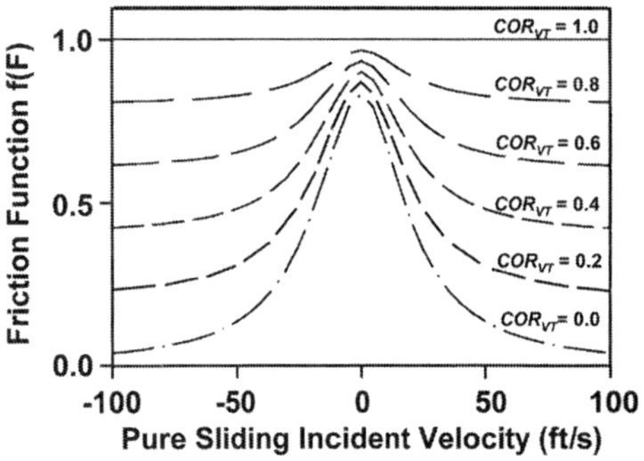

FIGURE 9-35 CRSP friction function *f*(*F*) adjusts tangential coefficient of restitution (COR_{VT}) to account for changes in tangential velocity of block surface relative to ground and anticipated level of frictional resistance (Heidenreich 2004; adapted from Pfeiffer 1989).

CRSP also adjusts the tangential coefficient of restitution (COR_{VT}) to reflect individual impact conditions by using a combination of a friction function *f*(*F*) and a scaling factor SF. The friction function *f*(*F*) adjusts COR_{VT} according to the velocity of the block surface, resulting from both its translational and rotational velocities, relative to the ground surface at the moment of impact, termed the *pure sliding incident velocity*, as shown by Figure 9-35. When this incident velocity is low or when high levels of friction (large COR_{VT}) define an impact, the friction function modifies the COR_{VT} to a small degree. SF further adjusts the effective COR_{VT} value to account for the increased frictional resistance caused by larger normal incident velocities.

CRSP uses the normal coefficient of restitution (COR_{VN}) to determine the rebound normal velocity. However, COR_{VN} is adjusted by a velocity-dependent scaling factor (*B*), shown in Figure 9-36, that reduces the value of COR_{VN} as the normal velocity increases. This accounts for the observed transition from more elastic rebounds at low velocities to more plastic rebounds at higher velocities, which is apparently due to increased tendencies for internal fracturing of the block and cratering of the slope surface at higher impact velocities.

CRSP is typically used to evaluate rockfall hazards at a specific site and to determine optimal locations for rockfall protection barriers or for design of catchment areas. The modeling of rockfalls with CRSP requires the following information:

1. A slope profile, input as a series of straight-line segments referred to as *cells*, designated by the coordinates for the endpoints of each line (Figure 9-37);
2. Slope roughness estimates for each cell on the profile, defined for each rock block size to be evaluated (see Figure 9-34);
3. Normal and tangential coefficients of restitution (COR_{VN} and COR_{VT}) for each cell in the profile; and
4. The number, size, shape (spherical, cylindrical, or discoidal), density, starting location (cell), and initial horizontal and vertical velocities of the simulated rock blocks.

CRSP provides default values for many of the block parameters. A series of simulations allows for the investigation of various block sizes or starting locations. CRSP produces an extensive series of graphical and statistical results that can be saved for further analysis or for inclusion in reports. Among the results are the following:

1. The slope profile (Figure 9-37) showing cell locations and the position of each simulated rock every 0.1 s as it travels downslope;
2. Statistical distributions of velocities, energies, and bounce heights at up to three selected points (analysis points) on the slope (Figure 9-38);
3. Histograms of the distribution of rock velocities and bounce heights at each analysis point (Figure 9-39); and
4. Graphs of the maximum velocities and bounce heights along the slope (Figure 9-40).

In addition, CRSP produces statistical summaries of velocities and bounce heights for each cell in the profile and the numbers of rocks that stopped in each of a series of uniform slope intervals.

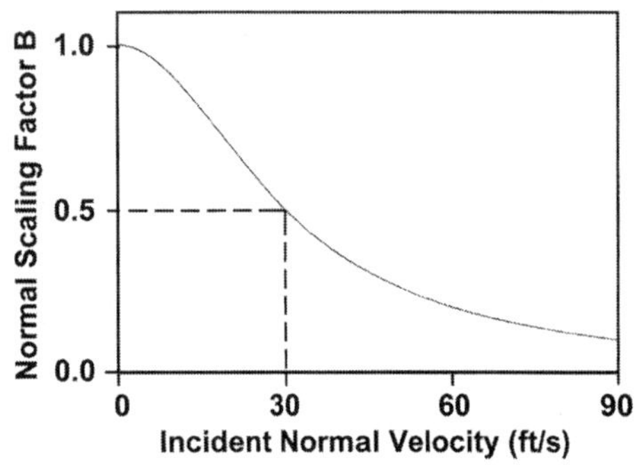

FIGURE 9-36 CRSP normal coefficient scaling factor (*B*) adjusts normal coefficient of restitution (COR_{VN}) to account for observed transition from more elastic rebounds at low velocities to more plastic rebounds at higher velocities (Heidenreich 2004; adapted from Pfeiffer 1989).

5.4 RocFall Program from RocScience, Inc.

The initial version of the RocFall program was developed as a master of science thesis at the University of Toronto (Stevens 1998). It has since been offered as a commercial product by

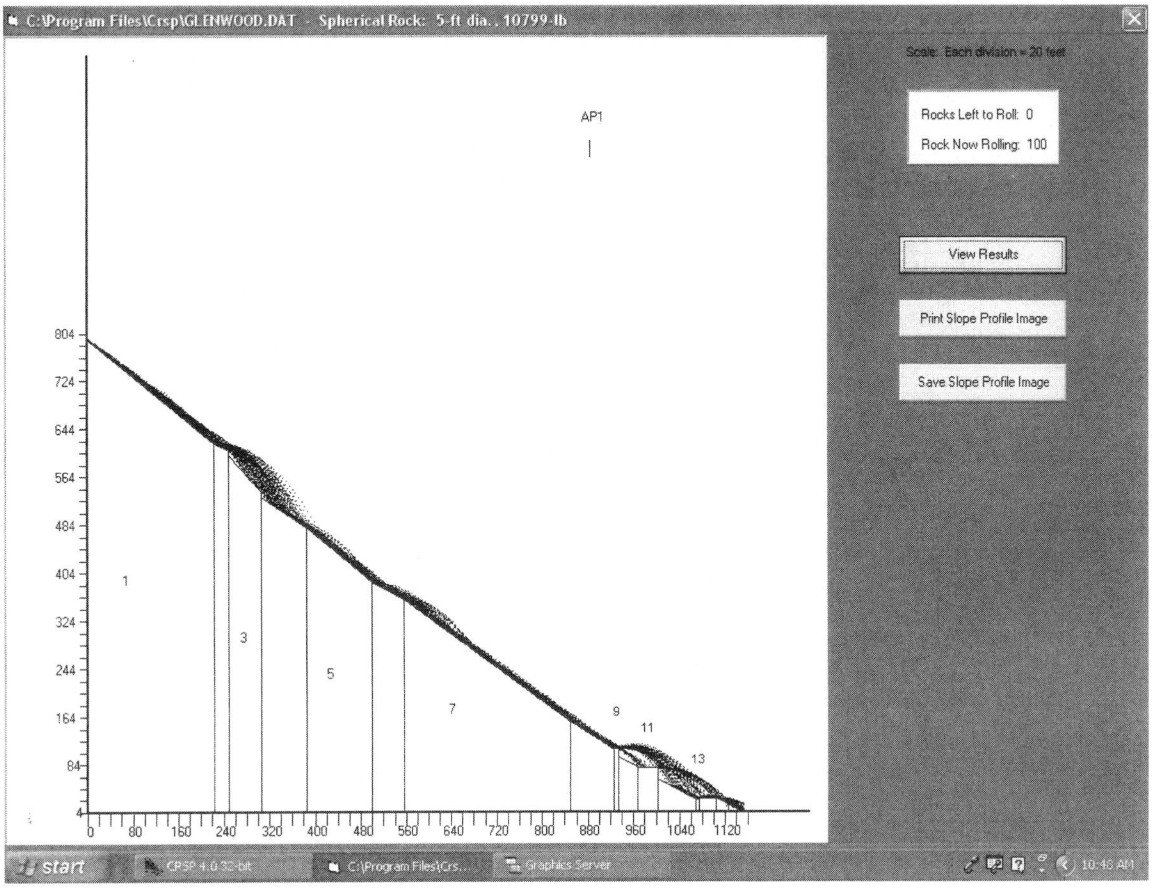

FIGURE 9-37 CRSP 4.0 slope profile window showing completed rockfall simulation using Glenwood Canyon test data set provided with program (Jones et al. 2000).

RocScience, Inc., Toronto, Canada, and has undergone several improvements. The most recent release is RocFall Version 4.0 (RocScience 2011b). RocFall 4.0 provides an attractive and extensive series of user interfaces that facilitate the selection of input parameters, the performance of rockfall simulations, and the export of graphical and statistical results for subsequent analysis or for inclusion in reports.

RocFall simulations rely on probabilistic assignment of values during each simulation step for most of the controlling parameters. Parameters are therefore defined by values describing their mean and standard deviation about the mean. Parameter values are assumed to follow normal distributions. RocFall users are advised to select parameter means and standard deviations with the understanding that, within a normal distribution, approximately 68% of the values will fall within one standard deviation and approximately 95% within two standard deviations of the mean value.

RocFall calculations are based on lumped-mass analysis methods. Thus, block size and shape effects are only approximated and are often only indirectly assessed by adjusting other property parameters. For example, RocFall analyzes blocks sliding across a sloping surface by using a slope friction angle parameter. User instructions recommend

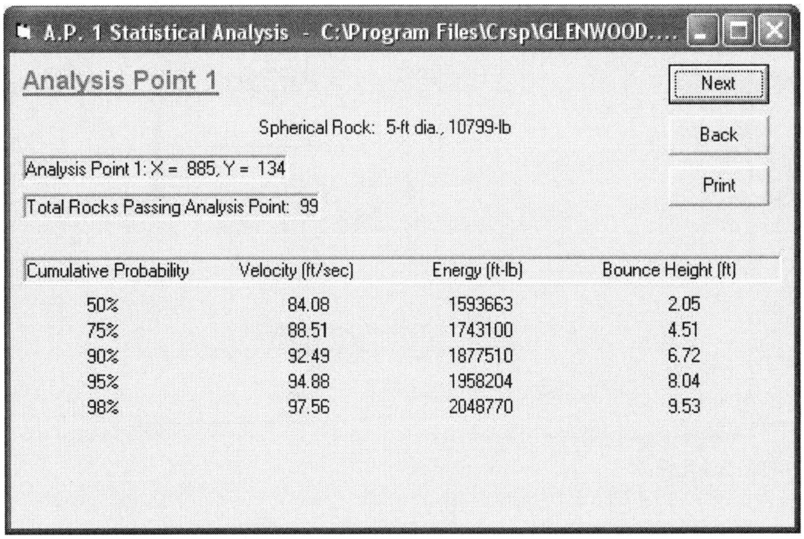

FIGURE 9-38 Example of CRSP 4.0 statistical results for selected analysis point using Glenwood Canyon test data set (Jones et al. 2000).

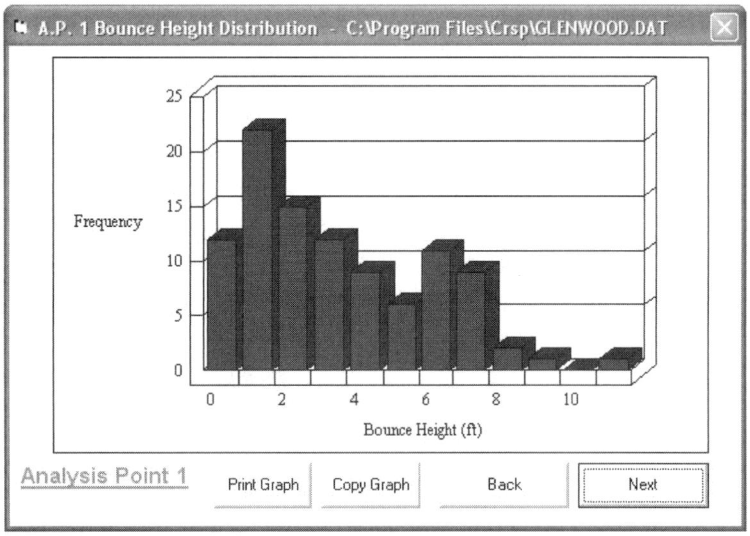

FIGURE 9-39
Example of CRSP 4.0 histogram of bounce heights at selected analysis point for Glenwood Canyon test data set (Jones et al. 2000).

that the user select values of this parameter to reflect not only surface roughness conditions but also the block shape. If highly spherical blocks are to be simulated, their tendency to roll down a slope rather than slide, as an angular block would tend to do, is represented by assigning very low values, even zero, to the slope friction angle parameter.

Before development of RocFall Version 2.0, block rotations were ignored (Stevens 1998; *RocNews* 2003). Angular velocity calculations are enabled by selecting an option to include them; the process includes the optional specification of initial block rotations, which will be selected from a normal distribution on the basis of the provided mean and standard deviation values. The result is randomly assigned to individual blocks. Further details of the angular velocity calculations are not accessible; however, published examples (*RocNews* 2003) have indicated that the inclusion of angular velocities results in blocks rebounding with slightly lower trajectories that extend slightly farther down the slope.

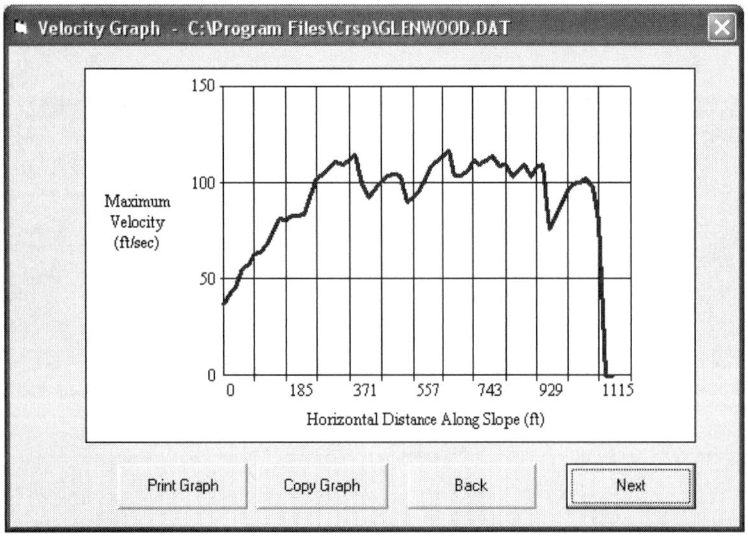

FIGURE 9-40
Example of CRSP 4.0 graph showing maximum velocities along slope profile for Glenwood Canyon test data set (Jones et al. 2000).

RocFall analyzes block trajectories along a 2-D slope profile defined as a series of cells, or segments, in exactly the same manner as CRSP (see Figure 9-37). However, RocFall allows overhanging slope segments, except at the start or end of a profile. RocFall allows for uncertainty in the precise location of each vertex defining the ends of individual slope segments. The uncertainty is defined by a standard deviation in the x- and y-coordinates, and the position of each vertex is then randomly assigned by using a normal distribution. Vertices with positions that are exactly known are given zero standard deviations. It is suggested that this capability allows for uncertainty in slope definition or the simulation of "pseudo-3-D" rockfall events. Slope roughness is introduced in a similar fashion; the average straight-line inclination of each slope segment may be assigned a standard deviation. The maximum recommended standard deviation value for defining slope roughness is 5°, but in some cases higher values are used.

Once the slope profile geometry is defined, the materials underlying each segment are defined, along with the related parameters required to simulate rockfall impacts and motions within the segment. The normal and tangential coefficients of restitution, a friction angle, and the slope roughness parameters are required for each profile segment. A Material Editor (Figure 9-41) assists this process by providing default values for commonly encountered conditions. RocFall 4.0 includes a large table of coefficients of restitution that was compiled after an extensive literature search. RocFall also provides a Material Wizard, which uses statistics to determine the normal coefficient of restitution for a material on the basis of backcalculation of simulated rock trajectories. Recent versions of RocFall have included a version of the CRSP normal coefficient scaling factor (B), shown in Figure 9-36, which adjusts the normal coefficient of restitution to account for the observed transition from more elastic rebounds at low velocities to more plastic rebounds at higher velocities.

RocFall simulations use a true random number generator to sample the parameter distributions, so different analysis results and sequences of rock trajectories are generated by each simulation. This is accomplished by the program internally selecting a different "seed" number to initiate the random number generator. RocFall provides a "pseudo-random" option that allows the same seed number to be used during a sequence of simulations. This causes each simulation to produce exactly the same sequence of trajectories, thereby generating a series of reproducible results for comparative RocFall analysis. Figure 9-42 is an example of a typical RocFall simulation.

RocFall produces graphs (termed *envelopes* in RocFall documentation) of maximum values of velocities, kinetic energies, and bounce heights at a number of evenly spaced horizontal locations along the slope profile (Figure 9-43). The lateral resolution of these graphs is determined by the number of intervals used to subdivide the total profile extent, which is specified by the user before a simulation run. Depending on the file size desired, a large number of analysis points can be chosen. The user has options to request graphs showing translational and rotational velocities and translational, rotational, and total kinetic energies. The graphs provide an overview of block motions and energies along the profile, information that is used to determine the capacity, size, and location of proposed rockfall protection barriers.

RocFall also produces histograms that display the distributions of the locations where simulated rock blocks stopped (Figure 9-44) or the distribution of energies, velocities, or heights of rock blocks at selected analysis points (termed *data collectors* in RocFall documentation), as shown in Figure 9-45. Figures 9-44 and 9-45 also show the initial default presentation of any graph or histogram generated by RocFall. They are presented together with the slope profile shown in the lower half of the screen as a visual reference.

5.5 *Rockfall* Program from Germany

The German consulting firm Dr. Spang Engineering Company for Construction, Geology, and Environmental Technology, Ltd., headquartered in Witten, Germany, continually refines and markets

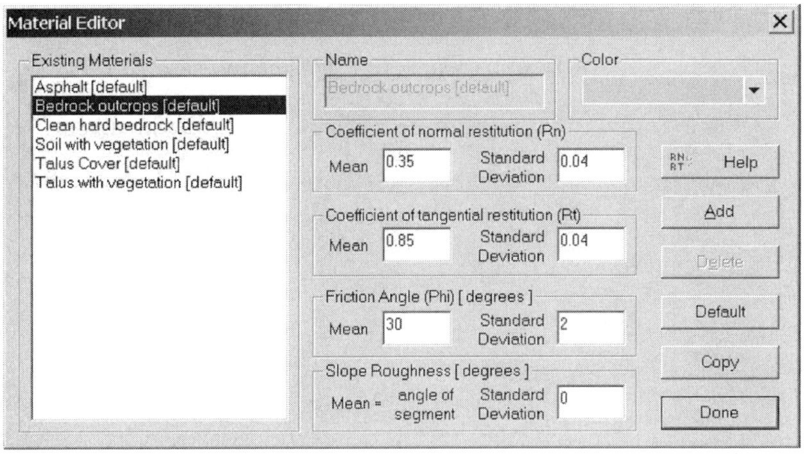

FIGURE 9-41
RocFall 4.0 Material Editor assists with assignment of required parameters to profile segments (RocScience 2011b).

a 2-D rockfall simulation program that is simply called *Rockfall*. An outgrowth of research efforts by Dr. Raymund Spang in the 1970s, the initial versions of the program were described by Spang (1987) and Spang and Rautenstrauch (1988). Subsequently, applications of the Rockfall program were reported by Spang and Sönser (1995), Spang and Krauter (2001), and Spang et al. (2008). The Rockfall program has been used extensively in Europe to support rockfall investigations.

The Rockfall program has undergone a series of refinements and enhancements since it was first released. The latest version is Rockfall 7.1, but the basic computational operations have remained.

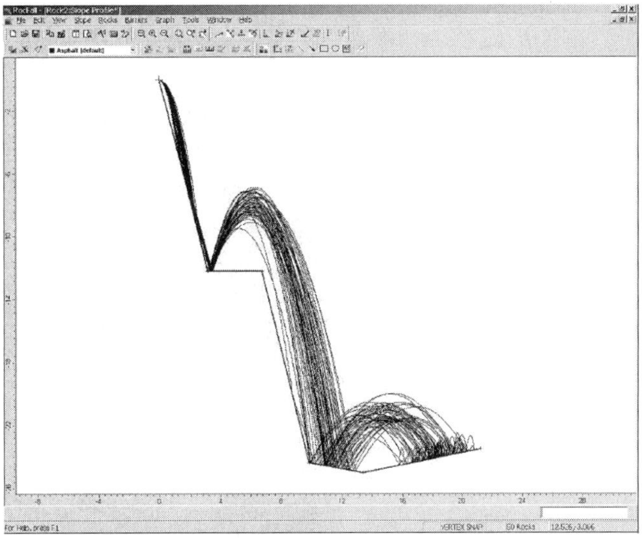

FIGURE 9-42
Results of typical RocFall 4.0 simulation showing trajectories for 50 blocks (RocScience 2011b).

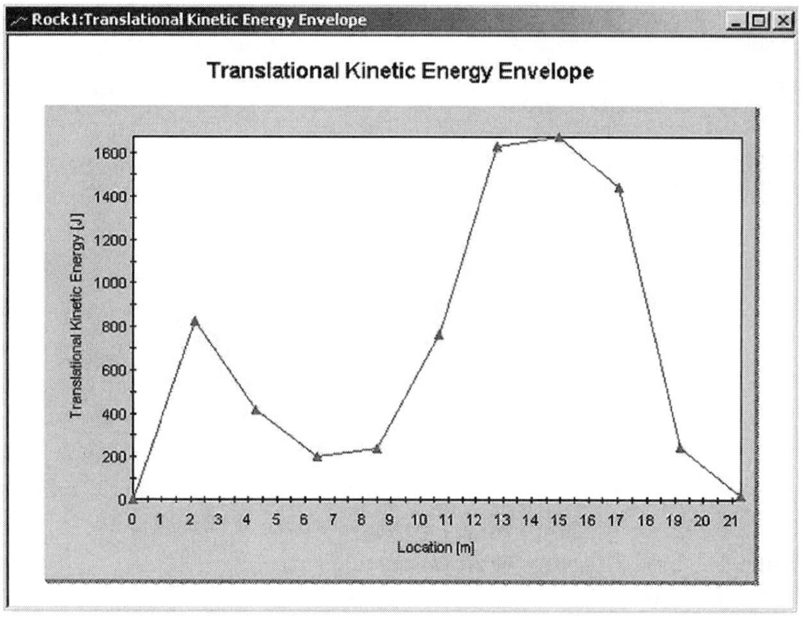

FIGURE 9-43
RocFall 4.0 graphical display of maximum translational kinetic energies along entire slope (RocScience 2011b).

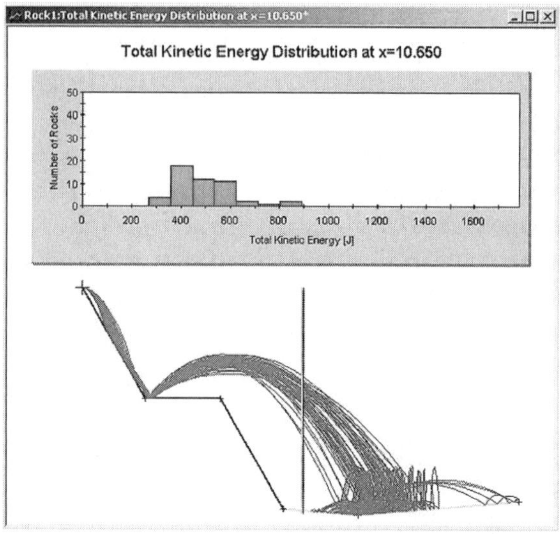

FIGURE 9-45
RocFall 4.0 histogram of distribution of total kinetic energy at data collector location defined by heavy vertical line on profile (RocScience 2011b).

Figure 9-46 was used by Spang et al. (2008) to explain the sequence of algorithms used by Rockfall, but it is essentially identical to an illustration presented by Spang and Rautenstrauch (1988).

The Rockfall program shares many characteristics with CRSP (see Table 9-5). It uses a combination of lumped-mass and rigid-body kinematic principles to evaluate rockfall blocks in free fall or when they are bouncing, rolling, or sliding. Initial versions were restricted to modeling spherical blocks, but current versions allow a variety of rock shapes. Velocity-based definitions of normal and tangential coefficients of restitution are used to analyze impacts. Slope profiles are defined as a series of segments, and material properties are assigned to them. Rockfall results include "envelope curves" displaying the maximum values of velocities or bounce heights along the length of the profile and a series of histograms and summary tabular statistics reporting on energies, velocities, and heights of impact at selected locations where rockfall barriers are being considered.

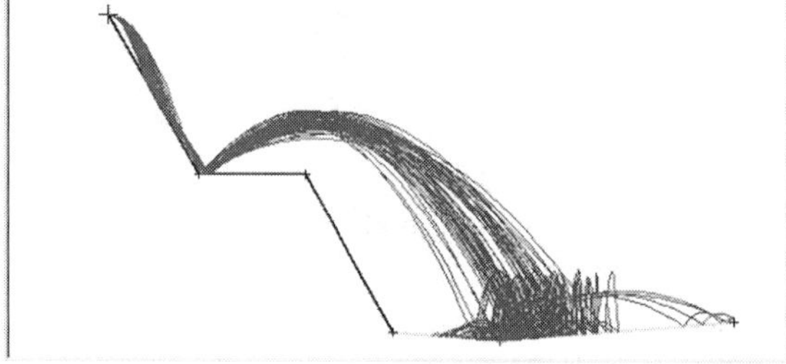

FIGURE 9-44
RocFall 4.0 histogram of distribution of locations where rock blocks stopped (RocScience 2011b).

Later versions of the program provided increasingly sophisticated and extensive graphical and statistical displays and summaries designed to assist in the assessment of proposed rockfall mitigation measures. Figure 9-47 shows a selection of typical displays produced by Rockfall 5.0. They include the slope profile with the simulated rockfall trajectories, envelope curves showing maximum velocity or bounce height values, and histograms of impact energies and bounce heights found at a proposed rockfall barrier.

Spang and Krauter (2001) described the capabilities of a later version, Rockfall 6.0, and illustrated them with an example of a rockfall evaluation at a site on the Koumi Line of Japanese Railways, near Nagano, about 180 km (112 mi) northwest of Tokyo. The railway was threatened by rockfall from a cliff that, along with a talus slope at its base, rose about 130 m (430 ft) above the track. Figure 9-48 shows the simulated rockfall trajectories for 1,000 blocks. A new rockfall barrier was proposed near the track at the foot of the slope (shown near the lower right corner of Figure 9-48). The simulations produced a maximum bounce height of 5.77 m (19.2 ft) at the proposed barrier location, but the distribution of bounce heights indicated that 99.7% of the rocks had a bounce height of less than 4 m (13.3 ft) at this location. The maximum kinetic energy was 2,512 kJ (929 ft-tons), but 99.6% of the rocks had a kinetic energy of less than 1,500 kJ (555 ft-tons). Thus, it was concluded that a barrier 4 m high with a capacity of 1,500 kJ would provide adequate protection without excessive maintenance.

Spang et al. (2008) included examples of displays produced by the current version of Rockfall (Version 7.1). They used Rockfall 7.1 to evaluate rockfall hazards along the Mosel Valley in Germany. Figures 9-49 through 9-51 show three displays for one of these sites. Rockfall 7.1 now includes multiple envelope curves showing the distribution of trajectories at each point along the profile (Figure 9-50). In addition to the original "maximum-value" envelope curve, Rockfall 7.1 automatically displays three additional curves representing the 98th, 95th, and 50th percentiles of the velocity, energy, or bounce height distributions. The user can request curves representing other percentiles. The histogram displays (Figure 9-51) now are accompanied by tabular summaries of the data.

6. ROCKFALL SIMULATIONS THAT INCORPORATE 3-D TERRAIN VARIATIONS

Reliable prediction of rockfalls is complicated by the influence of 3-D variations in slope geometry because of the presence of chutes and channels, convexities, and longitudinal ridges on the trajectories of falling rock blocks and the partition of their kinetic energies into translational and rotational components. According to Bozzolo et al. (1988), Azzoni et al. (1995), and Crosta and Agliardi (2004), the most important effect of 3-D topographic variability on rockfall trajectories is the occurrence of *lateral dispersion*, which is defined as the ratio between the lateral distance separating the two extreme fall paths and the slope length. Lateral dispersion results in the deviation of trajectories from the direction of steepest gradient.

According to Azzoni et al. (1995), lateral dispersion usually amounts to 20% of slope length; it is less for short and steep slopes, but it can be much larger for irregular slopes. The occurrence of lateral

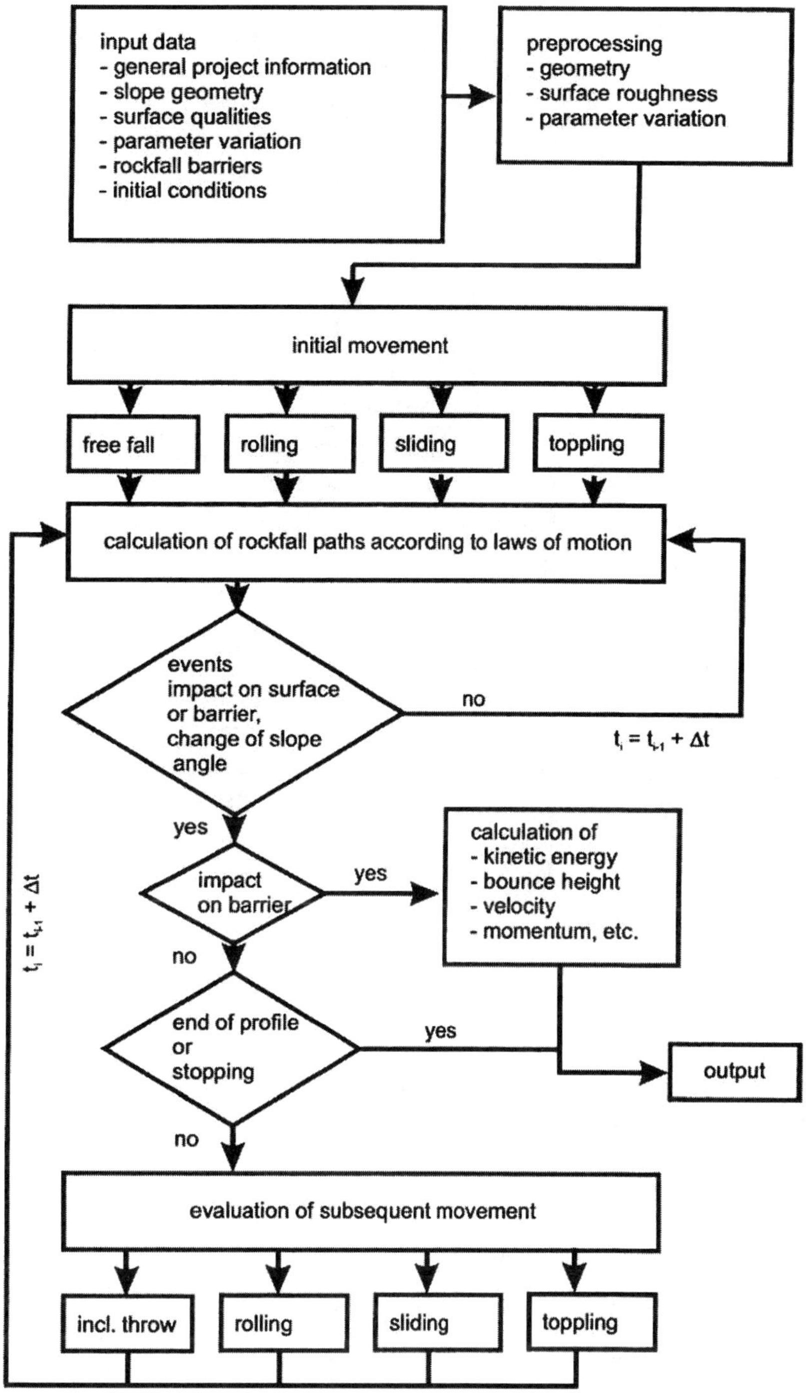

FIGURE 9-46 Flowchart showing sequence of algorithms used by all versions of Rockfall (Spang et al. 2008).

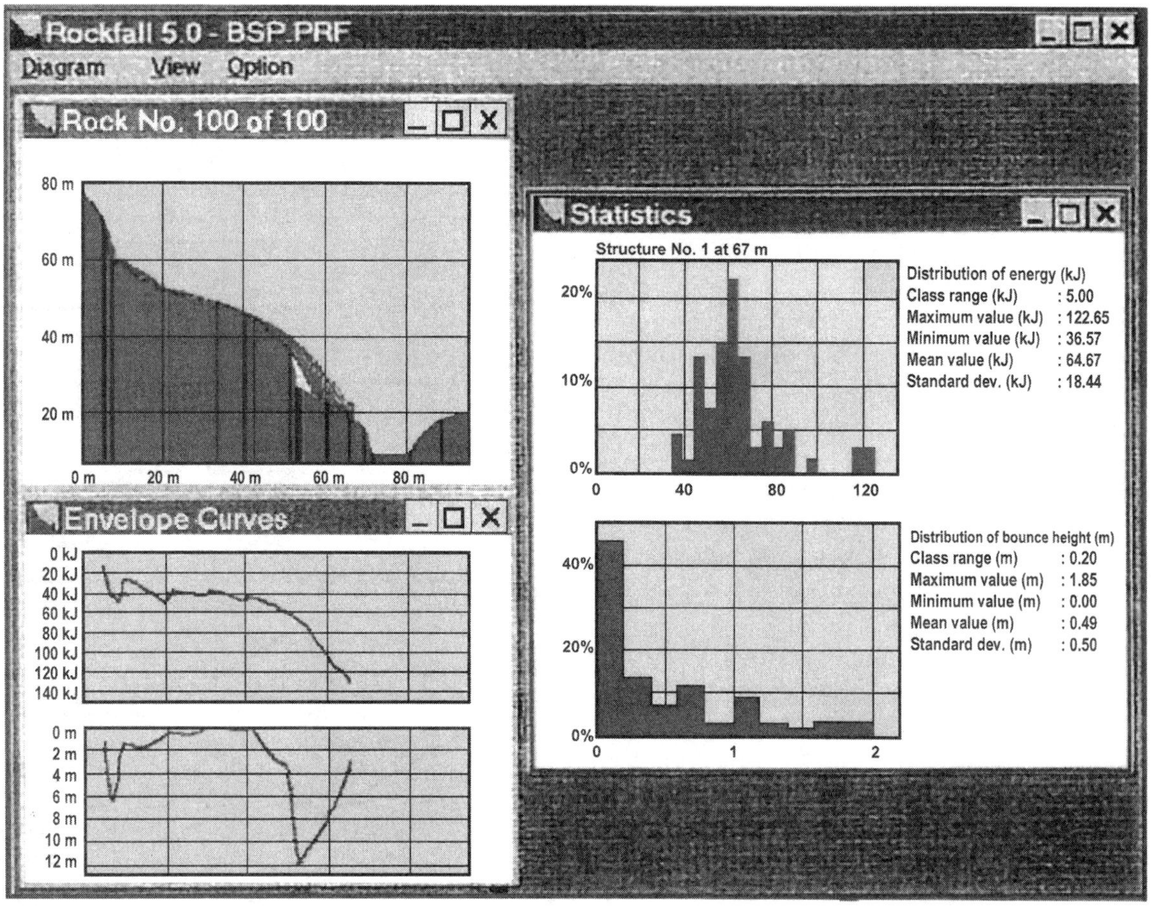

FIGURE 9-47
Example of selected typical displays produced by Rockfall 5.0. They include slope profile with simulated trajectories, envelope curves, and histograms of impact characteristics at proposed rockfall barrier.
(IMAGE COURTESY OF J. D. DUFFY.)

FIGURE 9-48
(below)
Rockfall 6.0 produced this profile display of 1,000 simulated rockfall trajectories at Japanese Railways site. Proposed new rockfall barrier is located at lower right, and only three trajectories overtop it (Spang and Krauter 2001).

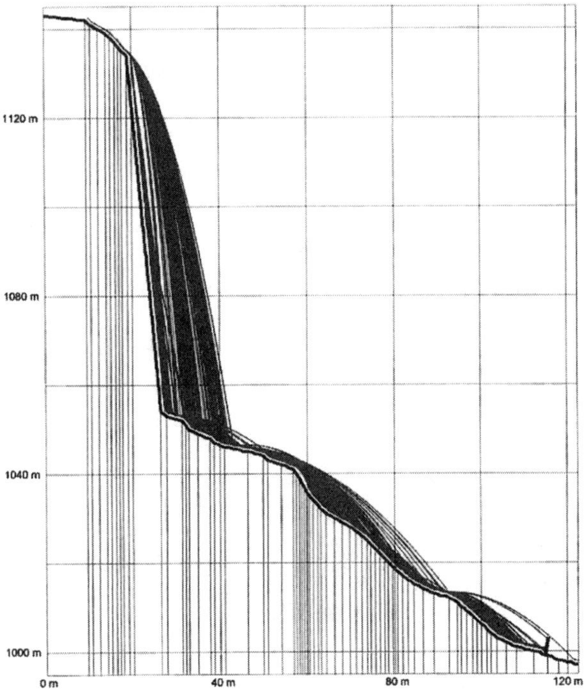

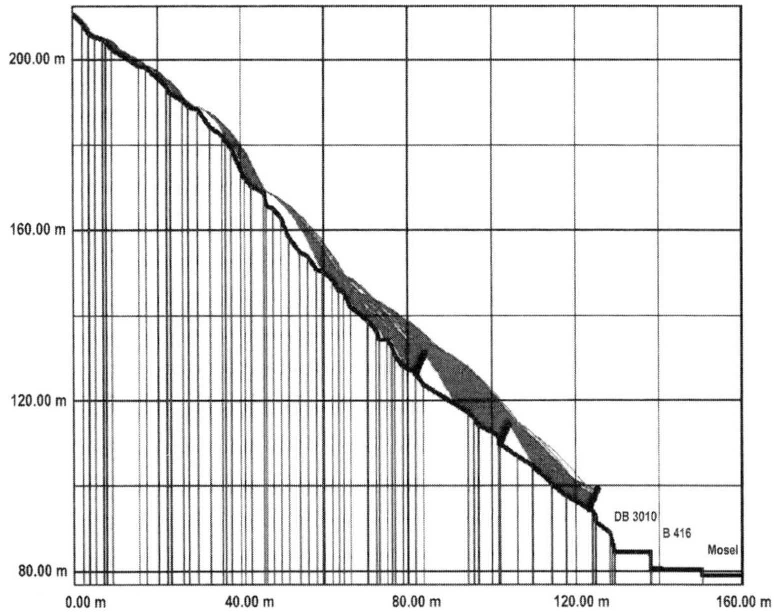

FIGURE 9-49
Profile of Mosel Valley site, Germany, produced by Rockfall 7.1, showing how simulated rockfall trajectories are being stopped by three proposed fences (Spang et al. 2008).

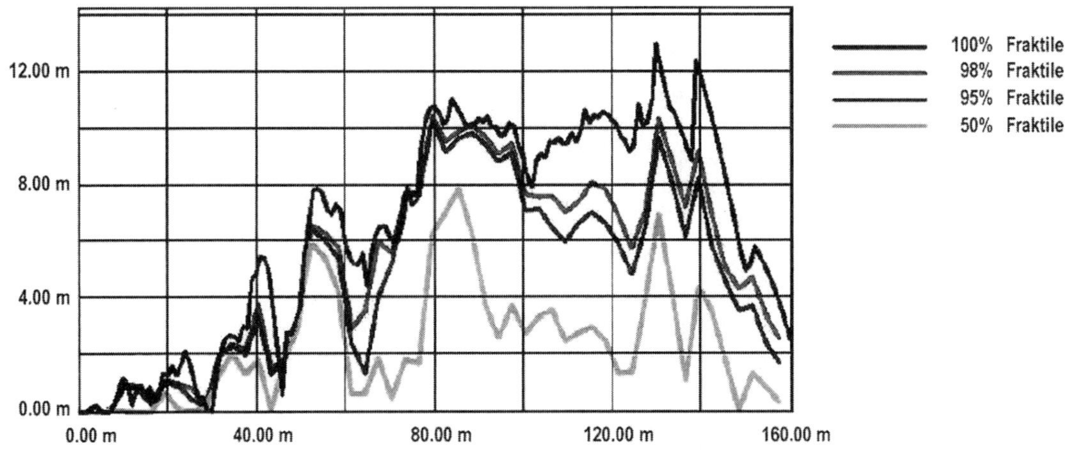

FIGURE 9-50 Envelope curves of bounce heights along entire slope profile at Mosel Valley site. Rockfall 7.1 now automatically displays three curves representing 98th, 95th, and 50th percentiles of bounce height distribution, in addition to maximum values (Spang et al. 2008).

dispersion causes difficulties during the selection of the most appropriate profile for evaluating rockfalls under a 2-D approach. The subjectivity inherent in making such a choice often has been underestimated; much more attention has been paid to the description of block geometry and the implementation of complex dynamic models (Bozzolo and Pamini 1986; Bozzolo et al. 1988). Nevertheless, the 3-D effects of topography are as important as the influence of block geometry on the rigid-body dynamics of falling blocks. Furthermore, the influence of 3-D topographic conditions becomes increasingly relevant as the fall path becomes longer, because the variability of trajectory increases with the number of impacts and changes in conditions encountered by each falling block. A complete description of 3-D topographic variability, combined with a probabilistic approach to rockfall parameter estimation, provides an opportunity for modeling a larger number of block impacts and fall paths, for improving evaluations of spatially varying parameters, and for assessing less frequent trajectories.

These considerations have been apparent to many investigators for many years. Nevertheless, most rockfall simulation programs developed since the late 1970s were limited to 2-D evaluations, as described in Section 5. The limitations of computing power available on desktop personal computers were important constraints that favored 2-D, rather than 3-D, rockfall analyses. However, the lack of

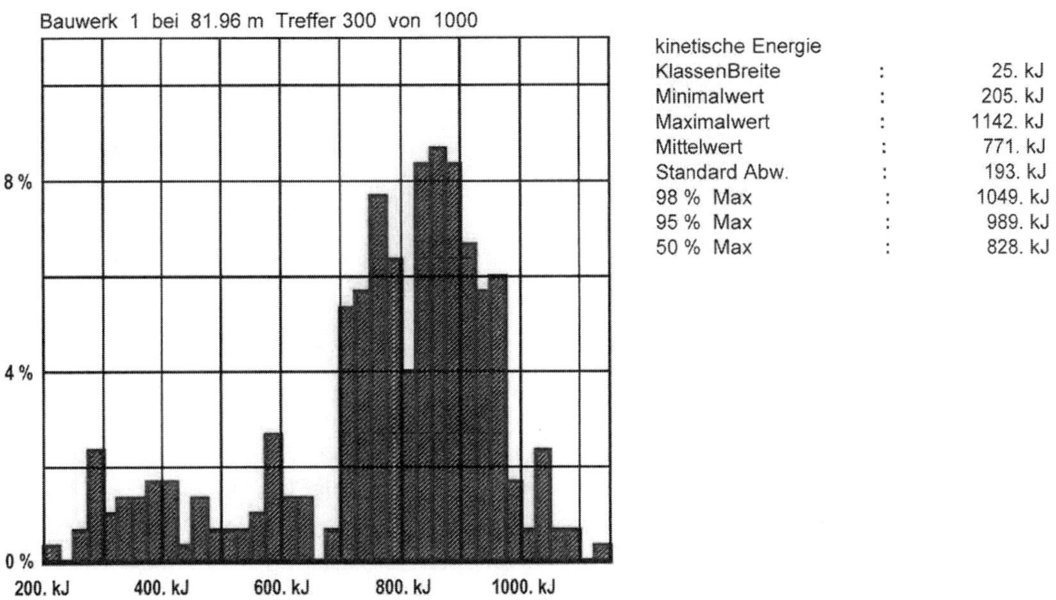

FIGURE 9-51 Rockfall 7.1 histogram and statistical summary of rockfall energies encountered at uppermost fence on slope at Mosel Valley site (Spang et al. 2008).

economical methods for obtaining detailed topographic data for rockfall sites was another factor. In recent years, both constraints have largely disappeared. The computing power of personal computers has grown enormously, so the computational aspects of performing 3-D rockfall simulations are no longer a serious issue. As described in Chapter 7, new site investigation technologies, especially airborne and terrestrial LiDAR instruments, have made detailed representations of topographic conditions at rockfall locations feasible to obtain economically. At the same time, many detailed investigations of rockfall mechanics, as described in Chapter 8, coupled with field observations of both naturally occurring rockfall and field tests at planned rockfall events, as described in Chapter 10, have provided information supporting the development of improved simulation algorithms.

Consequently, several sophisticated 3-D rockfall simulation programs have been developed in the past 25 years (Table 9-6). The earliest program was developed in 1987, a remarkable achievement (Descoeudres and Zimmermann 1987). Since then, more programs have been developed and described in the literature. Several have been applied to multiple rockfall evaluations. The programs are briefly reviewed in this section. A detailed analysis of the operating characteristics and limitations of these programs is beyond the scope of this book. References to more extended instructions, such as user guides, are provided where they have been identified. The development of 3-D rockfall simulation models is a topic of continuing research interest. New programs, or new versions of these programs, are anticipated.

Several of the programs listed in Table 9-6 rely on GIS technologies to describe topographic conditions. The degree to which these 3-D rockfall simulation programs are linked to commercial GIS software varies. For example, RockFall Analyst is closely linked, whereas ROTOMAP provides internal capabilities and thus essentially forms a standalone independent system. Most programs accept topographic data sets that conform to GIS industry standards and produce results that can be further analyzed or displayed by using widely available GIS or graphical display programs. Most programs use the grid-oriented DEM method to describe topographic conditions. As discussed in Section 4, the DEM representation has some limitations; thus, several programs use other methods to represent the terrain.

These 3-D rockfall simulation models represent a compromise between competing rockfall modeling requirements. Although from a theoretical point of view a fully dynamic, rigid-body assessment of block motions that accounts for both rotational and translational kinetic energies is desirable, in practice a computationally simpler lumped-mass approach often provides an acceptable solution. Because of computational issues and the uncertainty affecting the modeling parameters, simplified algorithms are often suitable. This approach makes the performance of 3-D rockfall simulations on widely available computers possible with relatively short calculation times. A probabilistic approach permits investigation of uncertainties in the controlling parameters as well as limitations in describing local site conditions.

6.1 First 3-D Rockfall Simulation Program

Descoeudres and Zimmermann (1987) reported the development of an unnamed 3-D rockfall program to assist in the evaluation of a large rockfall event that had occurred in 1985 at Les Crétaux in the Rhône Valley of south-central Switzerland (Descoeudres 1990; Rouiller 1990). At Les Crétaux, the rockfall hazard zone is partially confined within an eroded zone, probably caused by fracturing of the bedrock (Figure 9-52). In August 1985, approximately 800,000 m^3 (28 million ft^3) failed from this rock cliff, producing a combination of debris slides and rockfall. About 600,000 m^3 (21 million ft^3) reached the valley, and about 200,000 m^3 (7 million ft^3) remained on the upper slope as an actively moving rock mass. This situation resulted in continuing rockfall events, and several blocks between 1 and 10 m^3 (35 and 350 ft^3) reached populated areas. A 2-D rockfall simulation was not believed sufficiently rigorous to evaluate this hazard because the rockfall blocks followed diverging paths on the lower talus slope.

Descoeudres and Zimmermann (1987) used a finite element mesh to describe the topography (Figure 9-53). At locations of each rockfall block impact, four adjacent topographic elevations within the finite element mesh were identified and the surface orientation at each rockfall impact location was computed by bilinear interpolation (Figure 9-54).

Block motions were evaluated by using rigid-body mechanics principles; thus, both translational

Table 9-6
3-D Rockfall Simulation Programs That Incorporate Assessments of Terrain Variations Within the Rockfall Zone

Initial Release[a]	Program Name	Kinematic Principle[b]	Simulation Method[c]	3-D Topography Representation[d]	Source	Comments
1987	No name	Rigid body	Deterministic	Finite element mesh	Descoeudres and Zimmermann 1987	First 3-D rockfall simulation model; entirely deterministic.
1991	ROTOMAP	Lumped mass	Probabilistic	DEM raster supported by ISOMAP	Scioldo 1991; Castelli and Scavia 2008; Geo and Soft 2009; Nocilla et al. 2009; Longo and Oreste 2010	Self-contained modeling approach provides basic GIS tools and probabilistic rockfall analysis when combined with ISOMAP.
2001	EBOULEMENT	Rigid body	Probabilistic	DEM raster based on finite element mesh	Dudt and Heidenreich 2001	Revised and improved version of Descoeudres and Zimmermann model; probabilistic analysis evaluates uncertainty.
2002	STONE	Lumped mass	Probabilistic	DEM raster internally converted to triangulated regular network	Guzzetti et al. 2002; Agliardi and Crosta 2002, 2003	Extensively used 3-D program. Inputs and results designed to interface with commercial GIS programs.
2007	RockFall Analyst	Lumped mass	Probabilistic	DEM raster	Lan et al. 2007, 2010	Programmed as an extension to ArcGIS with two major components: 3-D rockfall trajectory simulation and geostatics-based raster spatial modeling of rockfall hazard.
2008	PiR3D	Lumped mass	Probabilistic	TIN	Cottaz and Faure 2008; Cottaz et al. 2010a, 2010b	Designed for use by engineering consultants. Uses TIN to define topography and links to CAD systems.
2008	HY-STONE	Lumped mass and hybrid	Probabilistic	DEM raster—internally converted to triangulated regular network	Frattini et al. 2008; Agliardi et al. 2009	Improved version of STONE with some enhanced algorithms. Retains links to GIS programs.

(*continued on next page*)

Table 9-6 *(continued)*
3-D Rockfall Simulation Programs That Incorporate Assessments of Terrain Variations Within the Rockfall Zone

Initial Release[a]	Program Name	Kinematic Principle[b]	Simulation Method[c]	3-D Topography Representation[d]	Source	Comments
2010	Rockyfor3D Version 3.0	Lumped mass	Probabilistic	DEM raster	Dorren 2010	10 to 15 raster files define simulation parameters. Models impacts of rocks with trees. Sophisticated methods for evaluating surface roughness, rebound from soft soils, and deviations on impact.
2011	CRSP-3D	Discrete element	Probabilistic	DEM raster	Documentation currently under review by the Technology Development Program of the Central Federal Lands Highway Division of the Federal Highway Administration. On approval, will be at http://www.cflhd.gov/programs/techDevelopment/.	Uses an entirely new algorithm based on discrete element concepts. May be used in either 2-D or 3-D modes.

[a]Initial release dates are estimated on the basis of the earliest published descriptions of the program.
[b]Kinematic principles refer to methods used to analyze the rockfall block motions. *Rigid body* refers to motion analysis that takes into account the size and shape of the body; it allows for analysis of translational and rotational velocities and energies. *Lumped mass* refers to motion analysis that ignores the body shape and size; only translational velocities and energies are evaluated. *Hybrid* refers to a combination of lumped-mass and rigid-body methods; usually lumped-mass methods are used for free-fall portions of the rockfall trajectory while rigid-body methods are used to evaluate bouncing and rolling motions. *Discrete element* refers to use of discrete element methods discussed in Section 2.1 that permit block fragmentation.
[c]Simulation methods are either deterministic or probabilistic. Under deterministic methods, rockfall trajectories are entirely defined by fixed properties and Newtonian mechanical principles. Under probabilistic methods, rockfall trajectories are defined by parameters that vary from simulation to simulation on the basis of predefined ranges of values and probability distributions; this introduces a range of uncertainty in rockfall trajectories.
[d]The following are methods of representing the local topographic conditions within the rockfall zone:
- DEM raster—a series of equally spaced grid cells (usually square) of specified size (see Section 4.1);
- Triangulated regular network—a series of triangles developed from a DEM raster (see Section 4.4.4);
- TIN—a series of triangles of varying dimensions define the terrain, with small triangles in rough (irregular) areas and large triangles in smooth areas; and
- Finite element mesh—an irregularly sized mesh of quadrilaterals defined by irregularly spaced elevation observations (it is similar to a TIN, but the elements may be shapes other than triangles).

FIGURE 9-52
View of Les Crétaux rockfall hazard area, Rhône Valley, Switzerland (Jaboyedoff 2003a).

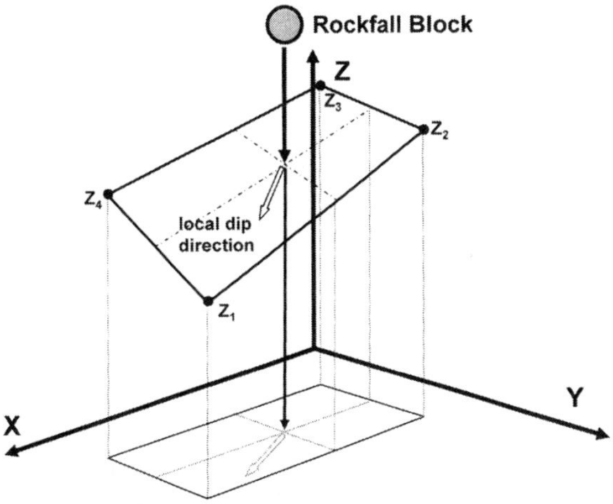

FIGURE 9-54
Determining surface orientation at rockfall impact location (adapted from Descoeudres and Zimmermann 1987).

and rotational velocities were computed at each time step. The following are the principal parameters required by the model:

1. Location of the rockfall starting point,
2. Initial free-fall height above the terrain,
3. Size of block,
4. Shape of block (hexagonal prism or cube),
5. Normal coefficient of restitution,
6. Coefficient of sliding friction, and
7. Plasticity factor.

The plasticity factor accounted for additional energy loss by impacts with less than perfectly elastic ground materials. It allowed for the coefficient of restitution at each impact to vary with values at or below the expected reference value of the coefficient. Blocks stopped once the velocity was below a threshold value and the block exhibited a stable static equilibrium, as illustrated in Figure 9-55.

The program produced several simulations involving variations in starting location, block sizes, and block shapes. A typical analysis involved evaluation of eight to 10 blocks. Additional runs investigated variations in the coefficient of restitution and variations in block velocities and bounce heights. The results were compared with field observations, including film recording

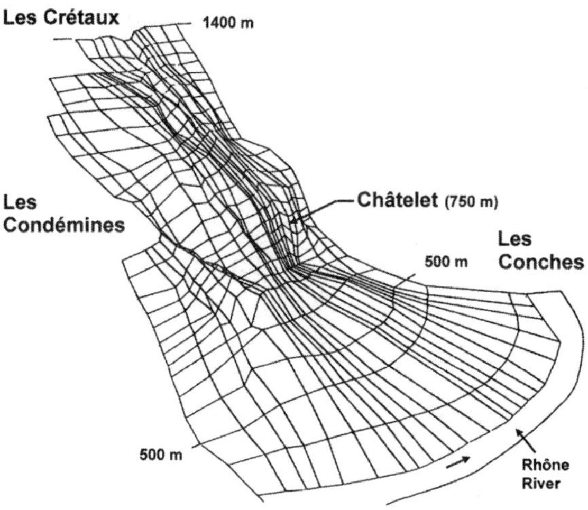

FIGURE 9-53
Perspective view of finite element mesh used to define topographic conditions at Les Crétaux (adapted from Descoeudres and Zimmermann 1987).

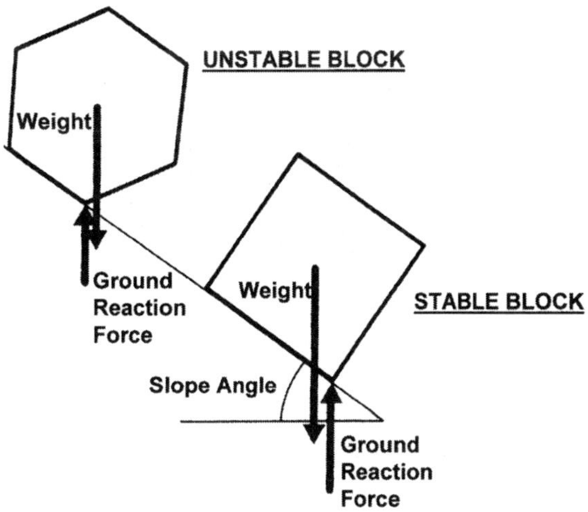

FIGURE 9-55
Evaluation of static equilibrium to establish block stopping point is determined by block shape (adapted from Descoeudres and Zimmermann 1987).

of actual rockfall trials. From the comparisons, Descoeudres and Zimmermann (1987) concluded the following:

1. The initial free-fall height has little influence on the runout distance because apparently much of the free-fall energy is dissipated on the initial impact.
2. Larger blocks have straighter trajectories and tend to stop at shorter distances, apparently a result of the greater influence of the plastic response of the ground materials.
3. Variations in the coefficients of restitution and coefficient of sliding friction caused predictable and expected changes in rockfall trajectories and total runout distances.
4. The model predicted runout distance, block velocities, and bounce heights that were in good agreement with observed rockfall events at this site.

Figure 9-56 shows the trajectories for one simulation run. Because this run used plasticity factor values that represented fairly elastic ground conditions, the trajectories are longer than the observed conditions.

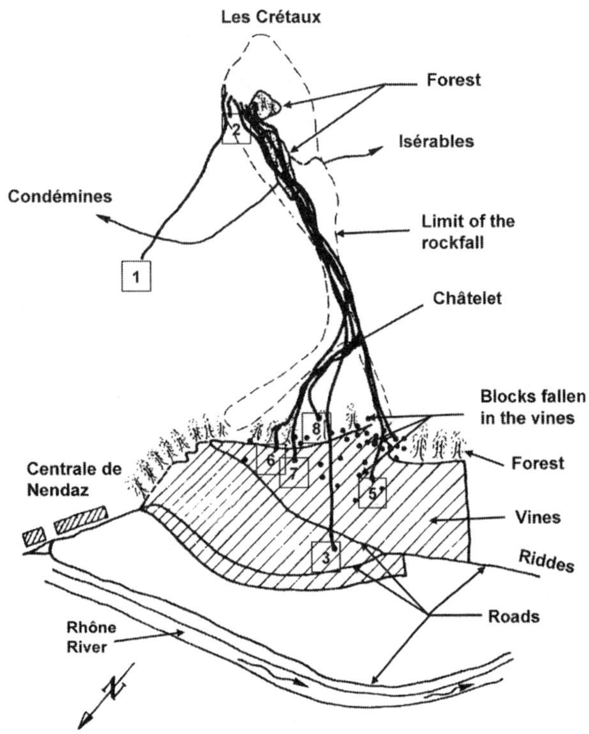

FIGURE 9-56 Simulated trajectories for eight 10-m^3 (350-ft^3) blocks at Les Crétaux. Black circles show locations of actual fallen blocks. Simulation used relatively elastic ground characteristics, so simulated trajectories tend to be longer than observed conditions (adapted from Descoeudres and Zimmermann 1987).

6.2 ROTOMAP Program

The ROTOMAP concept was first presented in an Italian document by Scioldo (1991); its applications have been presented more recently by Castelli and Scavia (2008), Nocilla et al. (2009), and Longo and Oreste (2010). Currently, the ROTOMAP software is available from Geo and Soft International, which is located in Turin, Italy. ROTOMAP performs probabilistic assessment of rockfall when it is combined with ISOMAP, which supports the development and display of DEM topographic data. Thus, as discussed in Section 4.3, the combination of ROTOMAP and ISOMAP represents a GIS-based probabilistic modeling approach to rockfall analysis.

Little precise information is available that describes the algorithms used by ROTOMAP. A user guide for ISOMAP and ROTOMAP (Geo and Soft 2009) describes how to install and operate the programs but does not provide internal computational details. With the limited information available, it appears that ISOMAP and ROTOMAP provide a self-contained approach for basic GIS techniques of DEM construction and evaluation of probabilistic rockfall trajectories. This differentiates ROTOMAP from all the other 3-D rockfall simulation programs described here.

Figure 9-25 (discussed in Section 4.3) shows an example of rockfall trajectory paths developed by ROTOMAP on the basis of ISOMAP DEM data. Recent applications of ROTOMAP (Castelli and Scavia 2008; Nocilla et al. 2009; Longo and Oreste 2010) demonstrated its use in evaluating rockfall hazards at several Italian locations.

6.3 EBOULEMENT Program

Dudt and Heidenreich (2001) discussed the development of the program EBOULEMENT ("landslide" in French), a redevelopment of the earlier Descoeudres and Zimmermann model described in Section 6.1. Their model retained the rigid-body approach of kinematic analysis proposed by Descoeudres and Zimmermann but enhanced the earlier model in several respects. The chief improvement involved the conversion from a purely deterministic assessment to a probabilistic analysis by allowing the model parameters to vary within defined limits. Dudt and Heidenreich

(2001) reported an application of EBOULEMENT to Les Crétaux, which facilitated a close comparison with the earlier model of Descoeudres and Zimmermann.

Dudt and Heidenreich were interested in evaluating the uncertainty of rockfall mechanics, which they noted could be related to the characteristics of the rockfall blocks (including their shape, size, initial location, angular orientation, and velocity), the topography (slope geometry and roughness), and the ground materials (Figure 9-57). Block dimensions, mass, and shape (ellipsoid, cube or parallelepiped, or hexagonal prism) were assigned. Topographic conditions were represented with a DEM. Ground characteristics were evaluated by dividing the rockfall zone into homogeneous geotechnical zones, with each zone being assigned parameter values (and ranges) for factors affecting rebound, sliding, or rolling motions (i.e., normal coefficient of restitution, plasticity factor, coefficient of friction, and rolling resistance).

Parameters were divided into three classes: deterministic, unknown, or random (Dudt and Heidenreich 2001). The deterministic parameters include block shape and the topography. The "unknown" category includes parameters for which a value is not fixed; it can vary within a simulation analysis but remains unchanged throughout a single simulation run. Block size is an example of an "unknown" parameter; a single simulation run involves a rock block of fixed size, but a complete simulation involves multiple rocks of various sizes within predefined limits. In contrast, random parameters can vary within specified limits during a single simulation run. For example, ground characteristics at a series of impact locations are evaluated as random parameters (Figure 9-58).

Figure 9-59 shows 50 possible rockfall trajectories predicted by EBOULEMENT for Les Crétaux. The predictions are more reflective of inherent uncertainties in the rockfall process than the earlier results of Descoeudres and Zimmermann but are in equally good agreement with rockfall observations at this location. EBOULEMENT also produces histograms of predicted bounce heights and kinetic energies at representative locations. They can be used to evaluate the feasibility of mitigation measures at these locations.

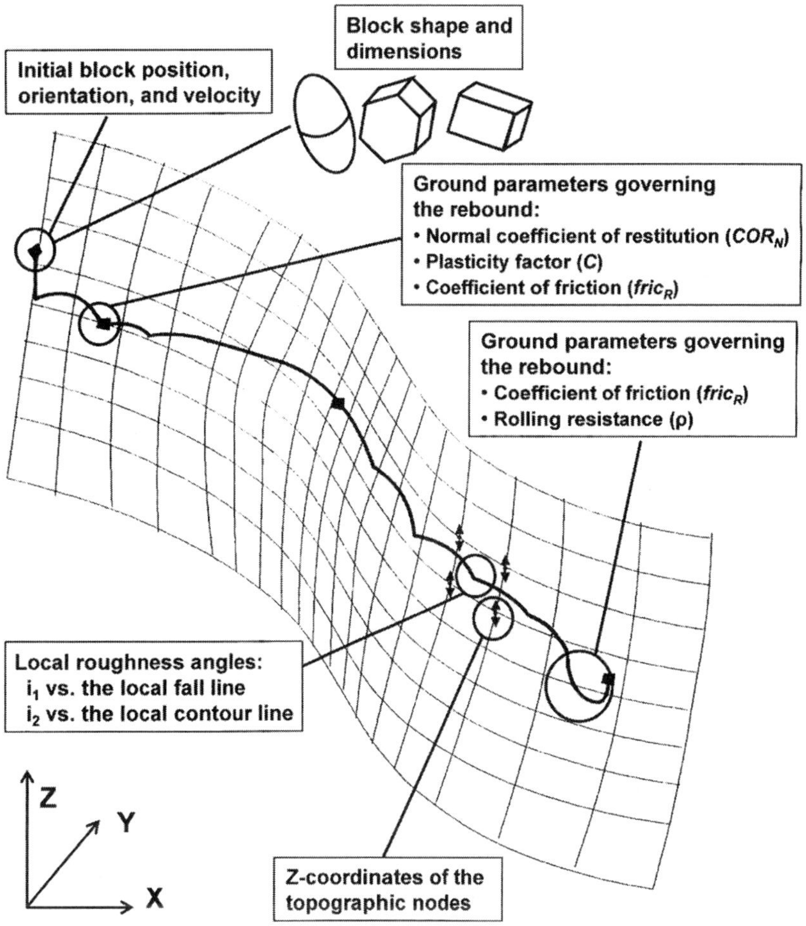

FIGURE 9-57
Rockfall uncertainties evaluated by EBOULEMENT (adapted from Dudt and Heidenreich 2001).

6.4 STONE and HY-STONE Programs

Guzzetti et al. (2002) provided a complete description of STONE, a 3-D rockfall simulation program developed jointly by the Italian National Research Council and the University of Milano Bicocca for research purposes. Additional brief explanations of STONE have been provided by Agliardi and Crosta (2002, 2003), by Crosta and Agliardi (2003), and by Guzzetti et al. (2004). STONE has been used in several rockfall investigations (Agliardi and Crosta 2002, 2003; Guzzetti et al. 2002, 2003, 2004). STONE has also been featured in several investigations comparing 2-D and 3-D rockfall simulation programs (Guzzetti et al. 2002; Tagliavini et al. 2009).

STONE computes 3-D rockfall trajectories and produces maps showing spatial distributions of the

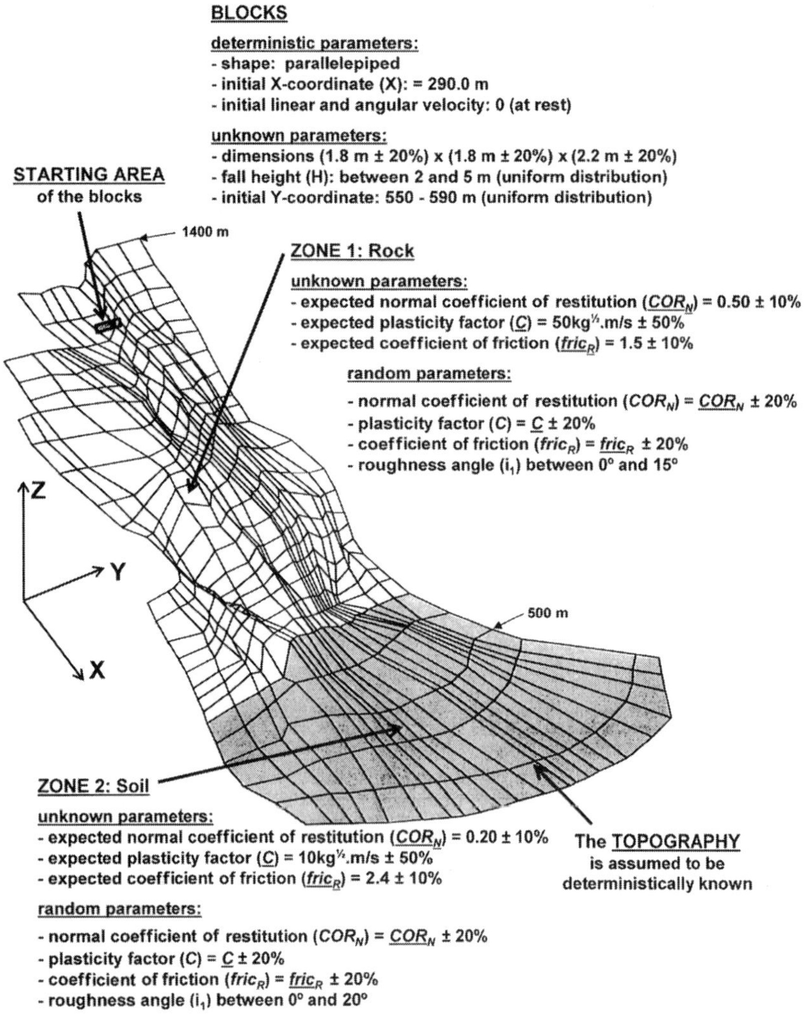

FIGURE 9-58
Perspective view of Les Crétaux, Switzerland, showing parameters used for rockfall simulation by EBOULEMENT (adapted from Dudt and Heidenreich 2001).

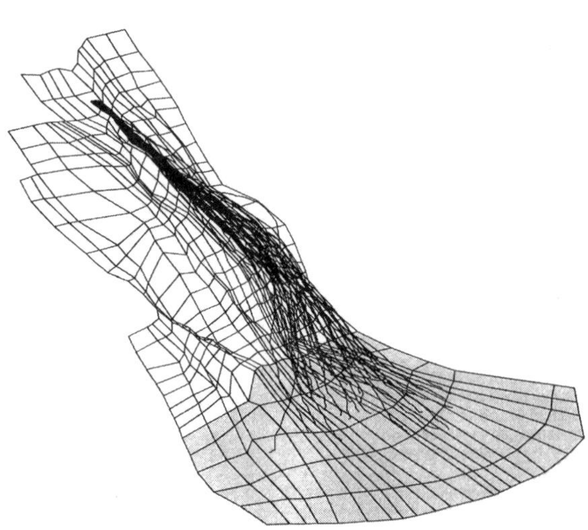

FIGURE 9-59
Rockfall trajectories at Les Crétaux predicted by EBOULEMENT for 50 blocks on basis of parameters shown in Figure 9-58 (Dudt and Heidenreich 2001).

frequency and kinematics of rockfalls. The modeling approach has the following features:

- A lumped-mass algorithm models free-fall, impact, and rolling-block motions in a 3-D framework.
- Topography is defined by DEM data sets of appropriate resolution; the DEM raster is converted to a triangulated regular network to provide a more continuous and sensitive representation of the topography (see Section 4.4.4 of this chapter and Figure 9-30).
- Raster GIS data sets are used to describe surface material and land use conditions.
- Rockfall source areas are defined within a GIS database as points, lines, or polygons and converted to raster representations.
- Variable numbers of blocks are initiated from each source cell, allowing for the simulation of different onset probabilities of rockfalls.
- Stochastic modeling is achieved by a pseudorandom approach during the analysis of a series of individual rockfall block motions.
- Input data and results are produced in an industry-standard raster format recognized by ArcGIS (ESRI 2011). This allows for integration within GIS environments for pre- and postprocessing of input data and model results.

For simplicity, STONE ignores air drag during free fall, and block fracturing is not considered. STONE requires the following input data:

- A raster grid containing elevation data (DEM),
- A raster data set defining the rockfall source areas and the numbers of blocks to be launched from each cell,
- Three raster data sets containing the values for the normal and tangential coefficients of restitution and the rolling friction coefficient (shown as COR_N, COR_T, and $fric_R$ in Figure 9-60), and
- A parameter file that specifies the input file names and the values for the controlling parameters and their variability ranges.

STONE permits representation of the inherent natural variability of rockfall events by launching a variable number of blocks from each source cell. This allows the simulation of both individual zone-by-zone onset probabilities of rockfall and the stochastic nature of the rockfall process. The

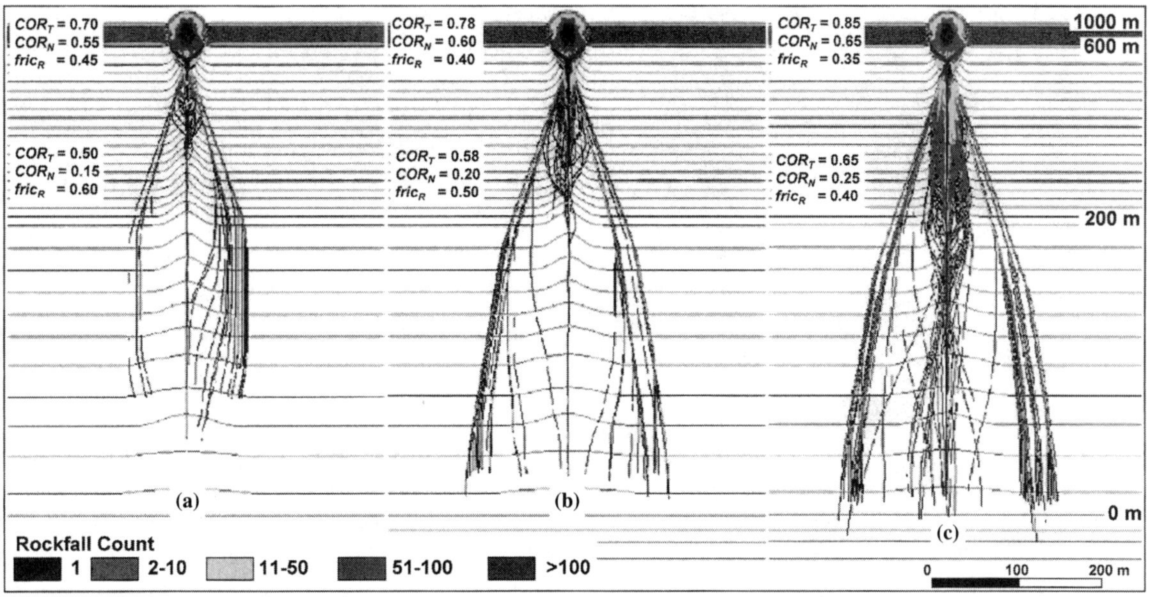

FIGURE 9-60 STONE rockfall trajectory simulations for simple 3-D topographic setting that show effects of changing values for normal and tangential coefficients of restitution (COR_N and COR_T) and coefficient of rolling friction ($fric_R$) (adapted from Guzzetti et al. 2002).

initial projection angle, the dynamic rolling friction coefficient, and the normal and tangential coefficients of restitution are varied randomly within predefined ranges for each of the launched blocks. To allow for the dependence of the coefficients of restitution on block mass and velocity, STONE includes semiempirical relationships originally proposed by Pfeiffer and Bowen (1989) and Stevens (1998). They cause the normal and tangential coefficients of restitution to be scaled according to the block mass or the computed impact velocity. However, STONE does not adjust the normal coefficient of restitution to reflect the influence of the impact angle on its value.

Figure 9-60 illustrates rockfall trajectory simulations produced by STONE for a simple, idealized topographic setting—a cliff forming a source area, a steeply inclined lower slope containing a single rockfall chute or depression on an otherwise planar homogeneous slope, and a more gently inclined basal slope that has a channel crossing it. The intention is to provide an idealized cliff, lower talus slope, and gently sloping valley surface similar to those found in many mountainous areas. Figure 9-60 shows how STONE predicts rockfall trajectories as values of the normal and tangential coefficients of restitution (COR_N and COR_T) and the coefficient of rolling friction ($fric_R$) are changed. The trajectories become longer (Figure 9-60*c* versus Figure 9-60*a*) as the normal and tangential coefficients of restitution are increased and the coefficient of rolling friction is decreased.

STONE relies on commercial GIS displays to produce 2-D and 3-D graphical displays of its results. Typically, both 2-D raster maps and 2-D and 3-D vector maps and views are produced. Two-dimensional raster maps are used to show cell-by-cell cumulative counts of rockfall transits, maximum computed velocities, and maximum bounce heights, while 2-D vector maps are used to show details of the simulated rockfall trajectory paths. Figure 5-21 (in Chapter 5) provides a side-by-side comparison of 2-D raster and vector maps of STONE program simulations. Figure 9-61 provides an example of a 2-D raster map showing maximum predicted rockfall velocities for part of the Camonica Valley in the Lombardy region of northern Italy. Figure 9-62, *a* and *b*, shows, respectively, 2-D and 3-D vector displays of rockfall trajectories in the Camonica Valley. These maps and displays rely on color sequences to illustrate the different velocities or other displayed values. Color versions of these figures are provided on the DVD that accompanies this volume.

Guzzetti et al. (2002) provided considerable details about their use of commercial GIS and CAD systems to produce these displays. They also described statistical and other methods of analyzing or evaluating the trajectories predicted by STONE. The results produced by STONE were compared with predictions made by two 2-D simulation programs—CRSP Version 4 (Jones et al. 2000) and RocFall (Stevens 1998). STONE predicted higher velocities across the gentle lower slopes than did either of the 2-D models.

FIGURE 9-61
Example of 2-D raster map produced by STONE. Map shows maximum predicted rockfall velocities for part of Camonica Valley in Lombardy region of northern Italy. Velocities 50 to 300 km/h (30 to 190 mph) are defined by color scale ranging from green to violet (adapted from Guzzetti et al. 2002). (See accompanying DVD for image in color.)

HY-STONE is an updated version of the original STONE program. It simulates rockfall trajectories in 3-D with a multiscale stochastic approach that incorporates both the 3-D kinematic algorithm originally applied in STONE (Agliardi and Crosta 2003) and a hybrid kinematic-dynamic algorithm based on modifications of techniques used in the 2-D models CRSP Version 1 (Pfeiffer and Bowen 1989) and CADMA (Azzoni et al. 1995). This combination allows the modeling of free-fall, impact, and rolling-block motions, with various damping relationships used to simulate energy losses.

HY-STONE accepts spatially distributed input data as raster matrices to define locations of source cells, distributions of ground materials, and values of damping coefficients accounting for energy loss. Slope topography is described at high resolution by raster DEM data obtained from either airborne or terrestrial LiDAR sources. Other relevant parameters, including the number, shape, and mass of blocks, are defined by random sampling of parameters using appropriate probability density distributions. As with the earlier STONE model, results are provided in raster and vector formats allowing interactions with commercial GIS and CAD programs. The 2-D and 3-D displays describe rockfall frequencies, bounce heights, and rotational and translational velocities and kinetic energies, and they give information about block motions, impact locations, and impact and rebound angles at each computed point along the trajectory paths.

Frattini et al. (2008) used HY-STONE in combination with GIS-based statistical assessments to evaluate the potential for rockfall hazards in the Val di Fassa area in the Dolomite Mountains of Italy (see Section 4.2.2). Agliardi et al. (2009) used the HY-STONE code to evaluate rockfall risk at the village of Fiumelatte, on the eastern shore of Lake Como near Varenna, Italy, where a large

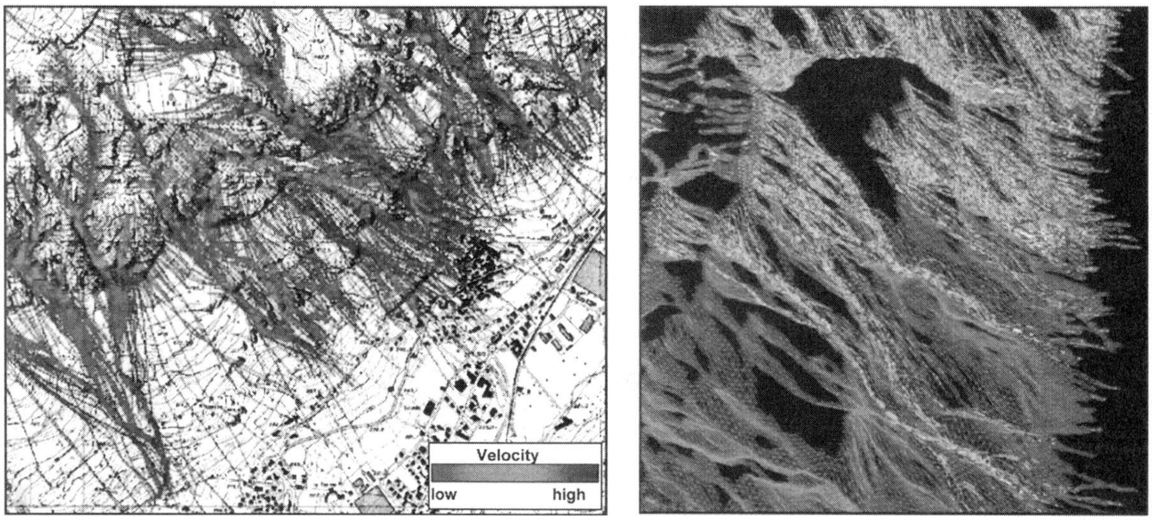

(a) (b)

FIGURE 9-62
Examples of vector map displays by STONE showing point-by-point velocities along rockfall trajectories: (*a*) 2-D map display and (*b*) 3-D view. Both displays are for small central region of Camonica Valley shown in Figure 9-61 (adapted from Guzzetti et al. 2002). (See accompanying DVD for image in color.)

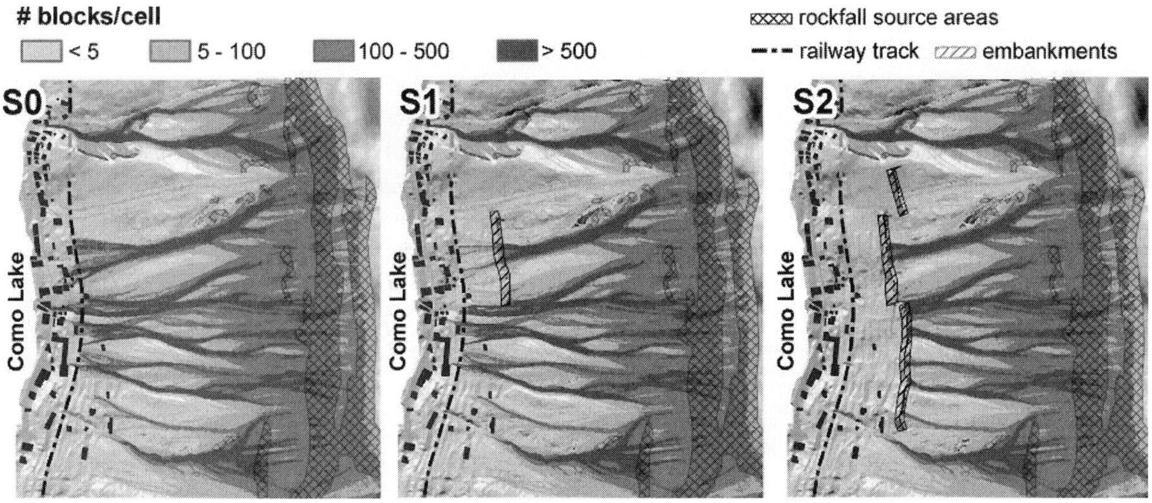

FIGURE 9-63 HY-STONE raster display showing analysis of alternative mitigation measures at village of Fiumelatte, Lake Como, Italy (adapted from Agliardi et al. 2009). (See accompanying DVD for image in color.)

rockfall in November 2004 resulted in two casualties, destruction of several buildings, and damage to transportation corridors. The HY-STONE model evaluated interactions between rockfall blocks and proposed countermeasures or structures by assessing their geometries and energy absorption capacities. An elasto-viscoplastic model (Di Prisco and Vecchiotti 2006) was used to model block impact on soft ground, whereas rockfall impacts against vegetation and fragmentation phenomena were simulated as stochastic processes (Crosta et al. 2006). The HY-STONE simulations were calibrated by backanalysis of the 2004 event, and then predictions were made for the whole area at risk by considering scenarios without protection (S0), with a provisional embankment (S1), and with a series of long-term protection embankments (S2), as shown in Figure 9-63.

6.5 RockFall Analyst

RockFall Analyst differs from the other described 3-D rockfall simulation programs in that it is not a stand-alone program but is an extension of ArcGIS (Lan et al. 2007, 2010). This linkage provides it with powerful analytical capabilities and an attractive user interface but also requires access to the commercial ArcGIS software. RockFall Analyst has two major components: 3-D rockfall trajectory simulation (Component 1) and geostatistics-based raster spatial modeling of rockfall hazard (Component 2). Figure 9-64 summarizes the typical sequence of steps used in assessing rockfall hazards with the two components of RockFall Analyst.

Component 1 of RockFall Analyst performs 3-D rockfall trajectory simulation by using a lumped-mass approach to the evaluation of rock block motions. Thus, block rotations are ignored, and the analysis is concerned with free fall (including airborne rebound trajectories) and combined rolling and sliding motions and with impact dynamics. RockFall

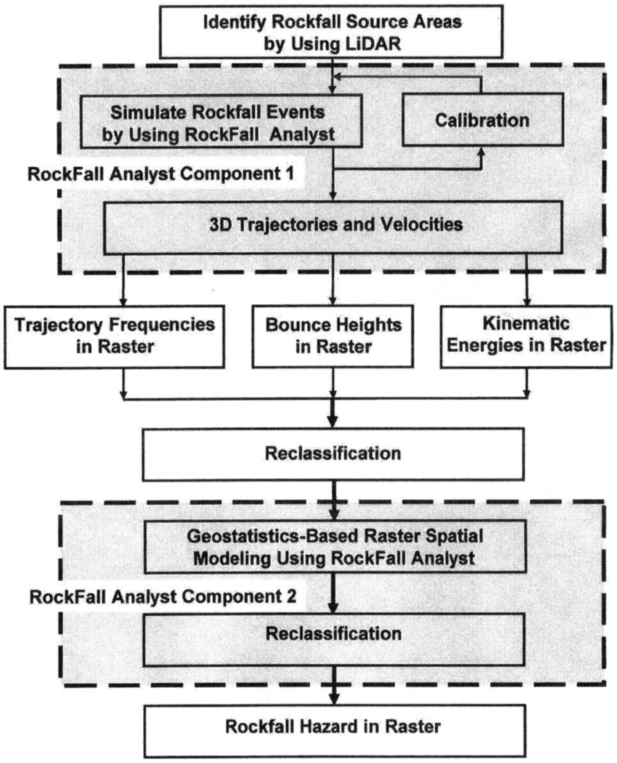

FIGURE 9-64
Rockfall hazard assessment workflow using RockFall Analyst (adapted from Lan et al. 2010).

Analyst relies on ArcGIS tools to compute slope and aspect relationships for each grid cell within a DEM. These GIS functions are described in Section 4.1.3. RockFall Analyst uses the orientation of the DEM cell at the point of impact to compute arrival and rebound trajectory angles. The values of the normal and tangential coefficients of restitution are allowed to vary within predefined limits, which provides a probabilistic character to the RockFall Analyst assessments. Lan et al. (2007) provided a detailed description of the computational algorithms used by RockFall Analyst to determine impact responses and to compute rolling and sliding block motions.

RockFall Analyst uses DEM data to describe the topographic conditions within the rockfall zone. Lan et al. (2010) emphasized the importance of a high-resolution DEM in defining these topographic conditions and recommended the use of LiDAR technology to obtain suitable DEM data. They compared the results produced by RockFall Analyst when DEM data sets with 1- and 5-m resolutions were used. They discovered the accuracy of the trajectory paths to be considerably less when 5-m grid cells were used, because many terrain irregularities were not identified. Consequently, lower-resolution DEM data produce trajectories with much less lateral dispersion.

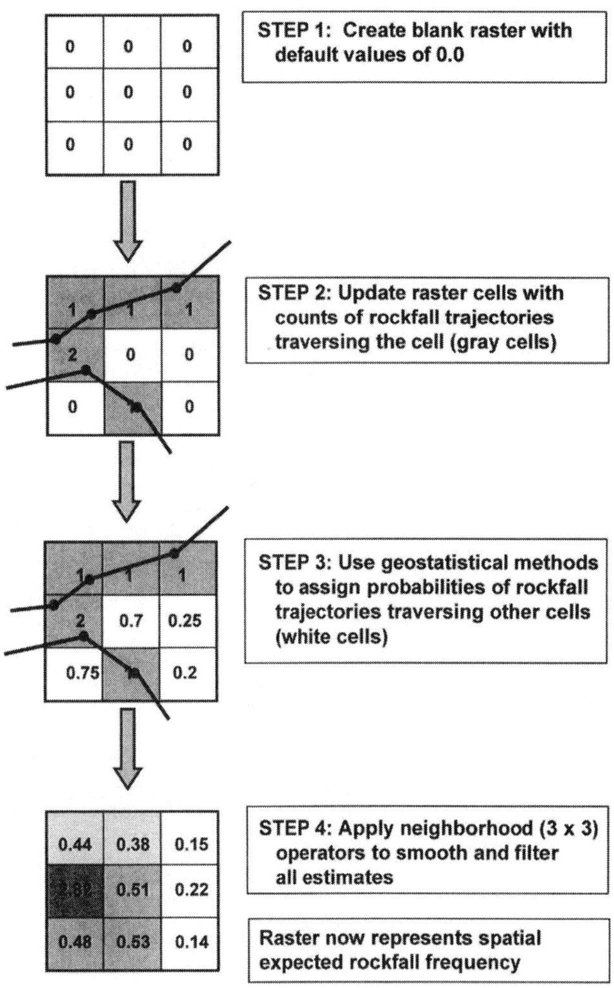

FIGURE 9-65
Four-step process used to compute spatial raster of frequency of rockfall trajectories (adapted from Lan et al. 2007).

As shown in Figure 9-64, RockFall Analyst Component 1 produces information describing a series of 3-D rockfall trajectories and translational velocities of blocks along these trajectories. Since the assumed mass of each block is also known, the velocities are readily converted to kinetic energies. RockFall Analyst evaluates this information by first summarizing the spatial distributions of the predicted rockfall trajectories by using raster GIS functions. Figure 9-65 illustrates the four-step process for converting the Component 1 trajectory information produced in vector form related to each trajectory path to a raster form required by RockFall Analyst Component 2. As shown in Figure 9-65, the process first identifies raster cells traversed by one or more trajectories. However, cells that have not been traversed by any rockfall block but that are close to cells that have been traversed have significant probabilities of being subjected to rockfall. Geostatistical spatial interpolation procedures and neighborhood filtering and smoothing functions, which are both available as ArcGIS tools, allow for the computation of these probabilities. Figure 9-65 illustrates the process for identifying the frequencies of rockfall trajectories. The same process is used to compute the probabilities for bounce heights (related to rockfall potential energies) and kinetic energies. Lan et al. (2007) provided further details of these analyses.

As shown in the middle portion of Figure 9-64, three raster products, representing the spatially distributed probabilities of the frequency of rockfall trajectories, maximum bounce heights, and kinetic energies of rockfall blocks, are used by RockFall Analyst Component 2 to undertake a probabilistic assessment of the spatial distribution of rockfall hazard. The process involves a series of reclassifications of the three input rasters, their spatial combination, and reclassification of the result. Lan et al. (2007) described this process. Figure 9-66

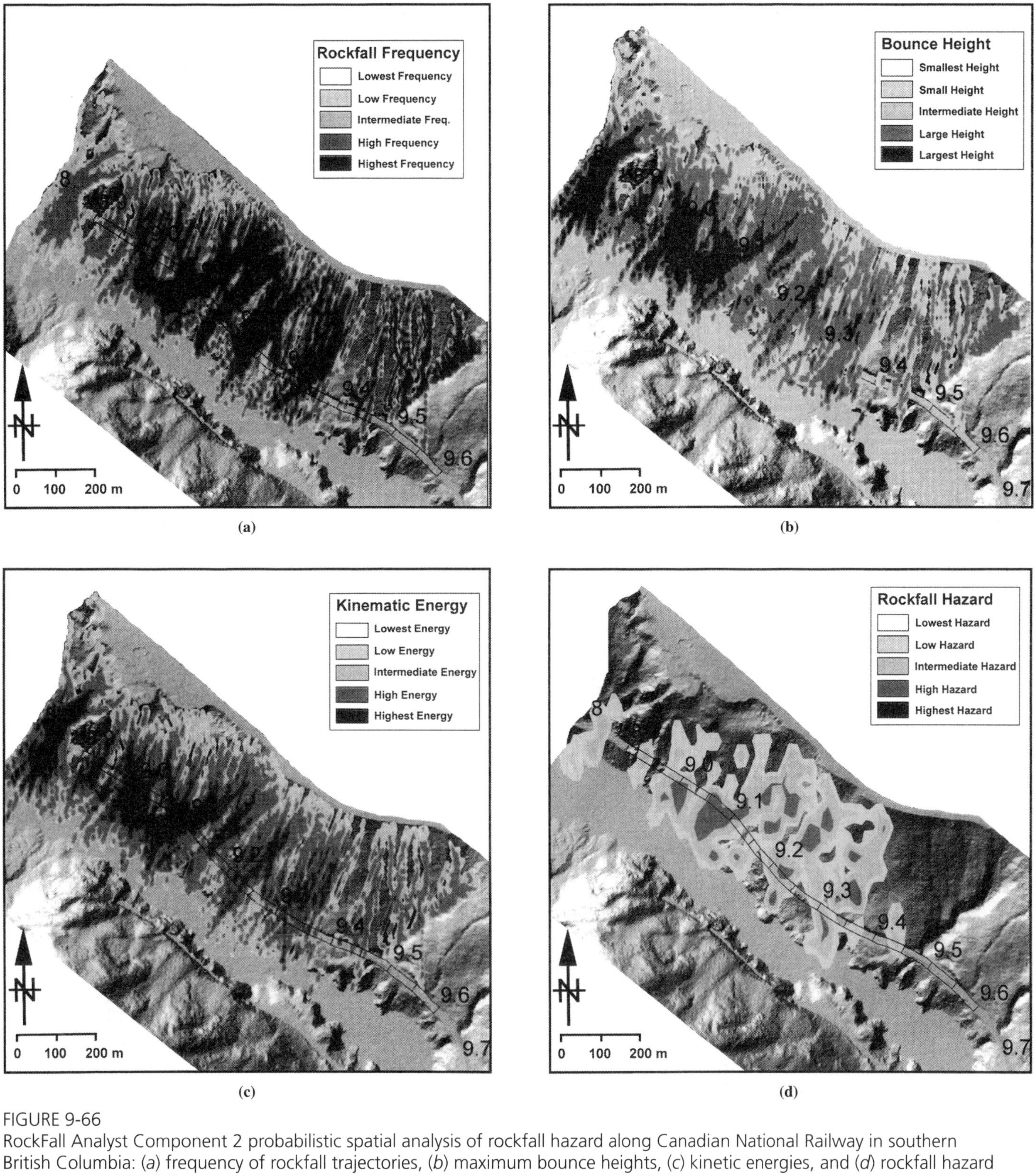

FIGURE 9-66
RockFall Analyst Component 2 probabilistic spatial analysis of rockfall hazard along Canadian National Railway in southern British Columbia: (*a*) frequency of rockfall trajectories, (*b*) maximum bounce heights, (*c*) kinetic energies, and (*d*) rockfall hazard assessment produced by combining Images *a* through *c* (adapted from Lan et al. 2010).

illustrates the RockFall Analyst Component 2 assessment of rockfall hazard for a location along the Canadian National Railway in southern British Columbia, described in detail by Lan et al. (2010).

6.6 PiR3D Program

In 2008, an experimental program was completely redeveloped to produce a new 3-D rockfall simulation program that could be easily used by engineering consultants (Cottaz and Faure 2008). The program, named PiR3D, was developed to provide a fully integrated solution that supported all the numerical evaluation aspects of a rockfall hazard investigation. Its design thus reflects the computational tools and expectations of rockfall practitioners.

Reflecting its connections to the engineering community, PiR3D has links to commercial CAD programs rather than links to commercial GIS programs, which are features of several other 3-D rockfall programs. A major difference is its use of a triangulated irregular network (TIN) to define the topographic conditions within the rockfall zone. The TIN method of representing topographic variations uses varying sizes of triangles to connect irregularly spaced elevation values. Small triangles allow for the accurate representation of highly dissected (rough) areas, while large triangles are used to represent relatively uniform (smooth) areas (Figure 9-67). TIN representations are widely used by CAD software, although most GIS programs can also use them.

PiR3D uses a simple lumped-mass approach for evaluating rockfall block motions. Thus, it only considers free fall, a combination of rolling and sliding, and impact mechanics, since it does not assess block rotations. PiR3D applies a probabilistic approach in evaluating block trajectories. It uses a combination of uniform and Gaussian probability distributions to vary values of parameters within specified limits during sequences of block motions defining a trajectory. Probabilistically selected values are assigned to the normal and tangential coefficients of restitution, a measure of dynamic friction, and perturbations of the horizontal and vertical trajectory orientations after an impact. Greater parameter uncertainties, and thus larger variations in these parameters, are allowed during lower block velocities. At high velocities, the parameters assume nearly constant assigned values. After each impact, PiR3D decides whether the trajectory continues as a free fall (a ballistic bounce) or as a rolling–sliding motion along the ground surface. If the rebound trajectory rises above the local topographic surface at an angle greater than a predefined limiting angle, the block is considered to be in a bouncing mode and the trajectory is computed as a free-fall event.

Results of a PiR3D evaluation include a number of 2-D and 3-D visualizations (Figure 9-68). In addition to a variety of 3-D visualizations of the rockfall trajectories, PiR3D produces several graphical summaries of rockfall impacts at selected locations. For example, if a rockfall fence is proposed, PiR3D can provide a graph showing the energy distribution of rockfall impacts on the fence (Figure 9-68*a*) or a representation of the locations of impacts and their energies along the fence (Figure 9-68*b*). Several other analytical displays may be produced by PiR3D (Cottaz et al. 2010a).

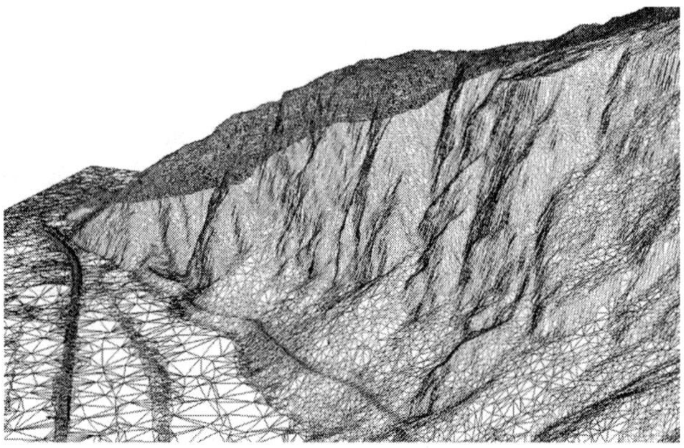

FIGURE 9-67
PiR3D uses TIN to represent topographic conditions within rockfall zone (adapted from Cottaz et al. 2010b).

6.7 Rockyfor3D Version 3.0

Rockyfor3D Version 3.0 combines physically based, deterministic algorithms with stochastic approaches; thus it is appropriately classified as a probabilistic process-based 3-D rockfall trajectory model. Rockyfor3D Version 3.0 has evolved from research on 2-D, and subsequently 3-D, models since 1998. The evolution of Rockyfor3D has been recorded under different names (including Rocky3 and RockyFor) in a series of scientific articles (Dorren and Seijmonsbergen 2003; Dorren and Heuvelink 2004; Dorren et al. 2004, 2006; Stoffel et al. 2006). Dorren (2010) documented Rockyfor3D Version 3.0.

Rockyfor3D simulates the rockfall trajectory as 3-D vector data by calculating sequences of classical parabolic free-fall trajectories and rebounds on the slope surface, as well as impacts against trees. Figure 9-69 summarizes the steps within the Rockyfor3D simulation process. Rolling is represented by a sequence of short rebounds; sliding of rocks is not modeled. The topographic conditions within the rockfall zone are defined by a DEM raster; several additional rasters are used to define the locations of the rockfall source areas, characteristics of blocks within each source area, and characteristics of the slope surface. Rockyfor3D uses rasters in an industry-standard GIS format.

If forest conditions are to be evaluated, Rockyfor3D requires either a file explicitly defining the locations and sizes of individual trees or a series of four rasters defining the tree density (number of trees per unit area), the mean and standard deviation of tree-trunk diameters, and the distribution of tree stands. On the basis of these raster files, Rockyfor3D randomly assigns trees to locations and develops a file containing simulated tree locations and sizes. Rockyfor3D requires an additional raster file providing the percentage of tree cover that consists of coniferous trees. Thus, a Rockyfor3D simulation requires 10 to 15 raster files; Dorren (2010) defined these requirements in detail.

Rockyfor3D uses a sophisticated method of defining surface roughness conditions over the rockfall zone. This procedure is defined in Chapter 8 (Section 4.5 and Figure 8-35). Homogeneous zones of surface characteristics are determined in the field and mapped as polygons within the model study area. Three rasters describe the surface roughness characteristics of three size–probability classes that define the typical height of an obstacle encountered by a falling block during impact in 70%, 20%, and 10% of the cases. During each rebound calculation, the local roughness value is randomly chosen from the three size–probability classes according to their probability of occurrence (thus, the 70% class is selected most frequently, the 20% class occasionally, and the 10% class infrequently). Rockyfor3D also evaluates the influence of soft soils on the rebound dynamics by incorporating information concerning the ground material density and a procedure for estimating a penetration depth that is similar to the methods described in Section 4.6 of Chapter 8.

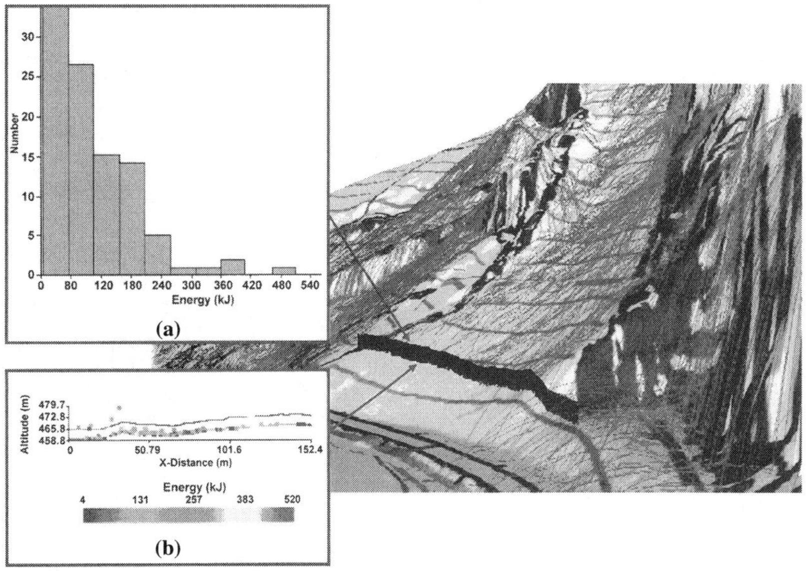

FIGURE 9-68
Results of PiR3D evaluation of rockfall trajectories along road in Jura Mountains of northern Switzerland, including graphical summaries of impacts on proposed rockfall protection fence [(*a*) energy distribution of impacts and (*b*) locations of impacts on fence and their energies] (Cottaz et al. 2010a).

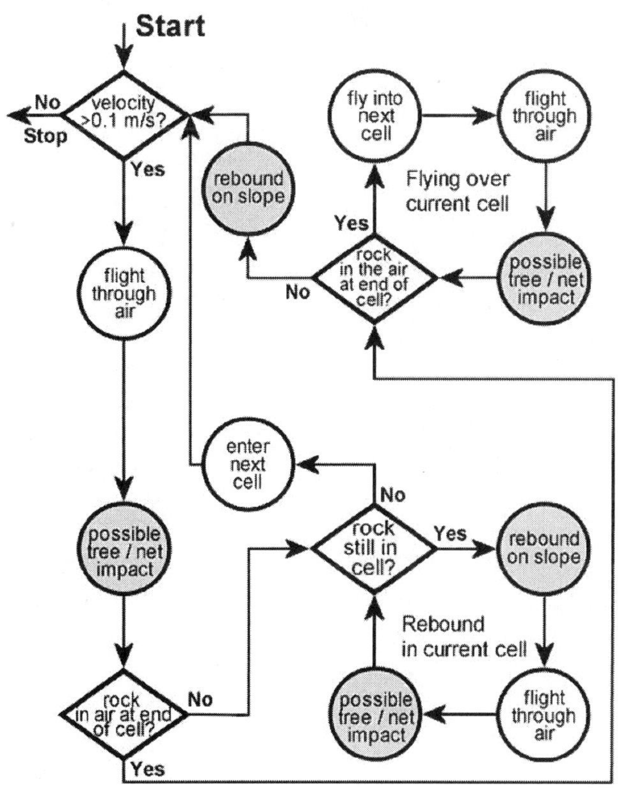

FIGURE 9-69
Flowchart of Rockyfor3D simulation procedure. Gray circles are modeling steps where changes in fall direction of simulated rockfall block may occur (Dorren 2010).

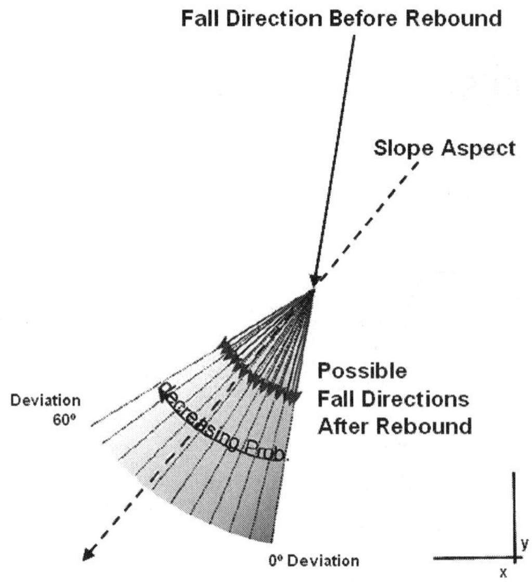

FIGURE 9-70 Principle used in algorithm for calculating deviation of block fall direction after rebound from slope (Dorren 2010).

Rockyfor3D also contains extensive considerations for determining deflections of rockfall trajectories on impact. Figure 9-69 shows five locations within the analysis process where changes in the fall direction may occur on impact. These include rebounds on any slope caused by surface roughness irregularities or impacts with trees on a forested slope. Chapter 8 extensively discusses the interactions between rockfall and forests (Section 4.7 of Chapter 8 and Figures 8-39 through 8-49), so these aspects of Rockyfor3D will not be repeated here. However, Rockyfor3D also includes a method for assessing the tendency of a block to deviate from its direction before impact toward the direction of steepest descent down the slope as it rebounds (Figure 9-70). This process appears unique to Rockyfor3D and is used during impacts in nonforested as well as forested locations. The degree to which the block deviates from its previous direction is determined by a random number that selects a deviation amount from predefined distributions on the basis of field observations. The distributions reflect observed probabilities of occurrence for deviations experienced by three classes of block velocity (Dorren 2010).

6.8 CRSP-3D Program

As discussed in Section 5.3, CRSP was initially developed in 1987 and underwent several updates, which culminated with CRSP Version 4.0 in 2001. In 2008, a study was initiated by the Colorado DOT with the Colorado School of Mines to modernize the CRSP software by using numerical techniques that had become available since CRSP was first developed. A new 2-D version of CRSP, CRSP Version 5.0, was developed. It used a combination of particle flow and discrete element methods to support improved simulation of rockfall kinematics (Bartingale 2009; Bartingale et al. 2009).

The new approach provided a better approximation of the interactions of rock blocks with the slope on impact and during bouncing and rolling trajectories than did previous versions of CRSP. The method of determining surface roughness was revised and simplified. CRSP Version 5.0 defined surface roughness by field measurement of the greatest repeatable perpendicular variation of the slope surface from the average slope angle (within a single cell) over a distance of 2 m (6 ft) (Figure 9-71). However, to model rockfall trajectories correctly, this roughness value must be adjusted to reflect the rock block diameter, with larger values used for small blocks and smaller values for larger blocks. CRSP Version 5.0 automatically adjusted the surface roughness value according to the size of the simulated block, thereby allowing evaluation of various block diameters within a single simulation while using a single roughness value (Bartingale et al. 2009).

CRSP Version 5.0 was calibrated with data from controlled rock-rolling experiments in a fashion similar to the original CRSP programs. This information was supplemented with field observations of rockfall runout distances and bounce heights defined by tree scars. Results of the study were published by Bartingale (2009) and Bartingale et al. (2009), but CRSP Version 5.0 was not released as a functioning program to the public.

During the development and calibration of CRSP Version 5.0, certain limitations in the 2-D approach to modeling spherically shaped blocks by using a combination of particle flow and discrete element methods became apparent. They included an inability to estimate runout distances and bounce heights accurately for some combinations of slope geometry and block shape. Specifically, the estimated runout distances for large-diameter blocks tended to be too short, while bounce heights for small blocks tended to be too large for certain slope configurations (Bartingale 2009). The limitations resulted from the inability of the program to incorporate the surface variations of rock blocks and then evaluate their influence on the behavior of a block as it decelerates and comes to rest.

Subsequently, a new program capable of accurately modeling the motions of variably shaped rock blocks as they traveled across slopes defined by 2-D profiles or as 3-D surfaces, called CRSP-3D, was developed with support from the Technology Development Program of the Central Federal Lands Highway Division of the Federal Highway Administration. The CRSP-3D algorithm is different from the algorithm used by previous versions of CRSP, which were semiempirical in nature. The new algorithm uses a combination of general equations of motion and discrete element methods to simulate rock block velocities and the contact forces between rock blocks and the slope surface. This results in a more detailed and fundamental approach that supports computations of motion and modeling of block trajectories. CRSP-3D uses a linear elastic model with a viscous damping coefficient that represents the viscosity and normal stiffness of both the block and the surface material to evaluate the normal contact forces during impacts. Tangential velocities of rock blocks are estimated by a single input value, the *hardness coefficient*, which models slope material properties. The hardness coefficient varies as a function of the coefficient of restitution, which represents the elasticity of the collision between a falling block and the slope, and a damping coefficient, which models how much tangential resistance the rock is subjected to on sliding on the slope surface. CRSP-3D uses the simplified procedure for evaluating surface roughness first implemented in CRSP Version 5.0 (Figure 9-71).

CRSP-3D evaluates rockfalls in either 2-D or 3-D modes. The 3-D mode requires a terrain model (DEM), which is developed either by manually entering the coordinates for each ground surface point in an *x*-, *y*-, *z*-format or by downloading 3-D terrain data produced with traditional survey methods or with LiDAR surveys. Because of the large file sizes of many DEM data sets, particularly those produced from LiDAR data, a parsing feature was included in CRSP-3D. This feature allowed a reduction in the DEM cell density (and thus the number of elevation points) used in describing the site. This allows the program to run more quickly yet provide sufficient information about the 3-D behavior of the falling blocks. Material properties are delineated on the terrain surface by manually identifying the perimeter of discrete material zones or by entering the coordinates of the zones. Figure 9-72 is an example of a CRSP-3D graphical display. It shows the estimated spatial dispersion of rockfall velocities at a site within Glenwood Canyon in central Colorado. CRSP-3D produces similar displays of spatial distributions of bounce heights and rockfall energies. It also produces a series of rockfall simulation statistics similar to those produced by the earlier 2-D versions of CRSP. In 3-D mode, it accurately models the interactions between a slope with 3-D surface irregularities and 3-D rock blocks of variable shapes and sizes.

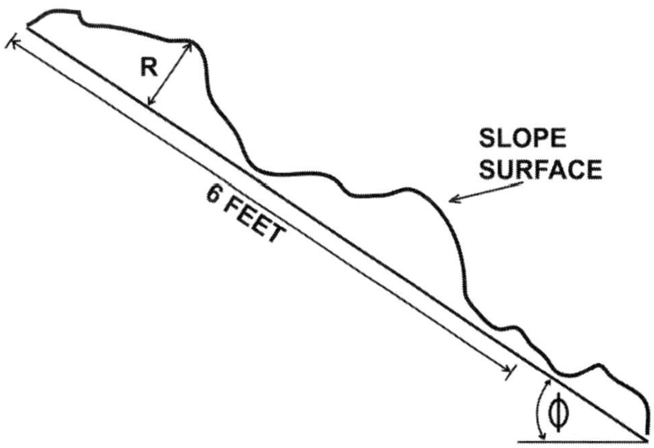

FIGURE 9-71
CRSP 5.0 defines surface roughness (*R*) based on largest repeatable perpendicular variation of slope surface from average slope angle (Φ) over distance of 6 ft (Bartingale et al. 2009).

However, computations in the 3-D mode require a powerful computer, and considerable time is needed for simulations involving numerous blocks. Consequently, CRSP-3D can also be operated in a 2-D mode providing more rapid calculations that can be completed on less powerful computers. Simulations in the 2-D mode do not provide estimates of the 3-D spatial dispersion as accurate as those provided by the 3-D mode. The 2-D mode uses a slope profile, just as in the earlier CRSP versions. The profile is developed by manually entering 2-D (*z* and *y*) coordinates defining the extents of a series of slope segments. Each slope segment represents either a change in slope material or a change in slope angle. CRSP-3D represents each slope segment as a plane rather than a line segment as in earlier versions of CRSP. This allows the program to model the interactions between slope segments and 3-D rock blocks with variable shapes and sizes accurately. However, the lack of lateral variations in the slope profile results in the 2-D mode's providing estimates of lateral

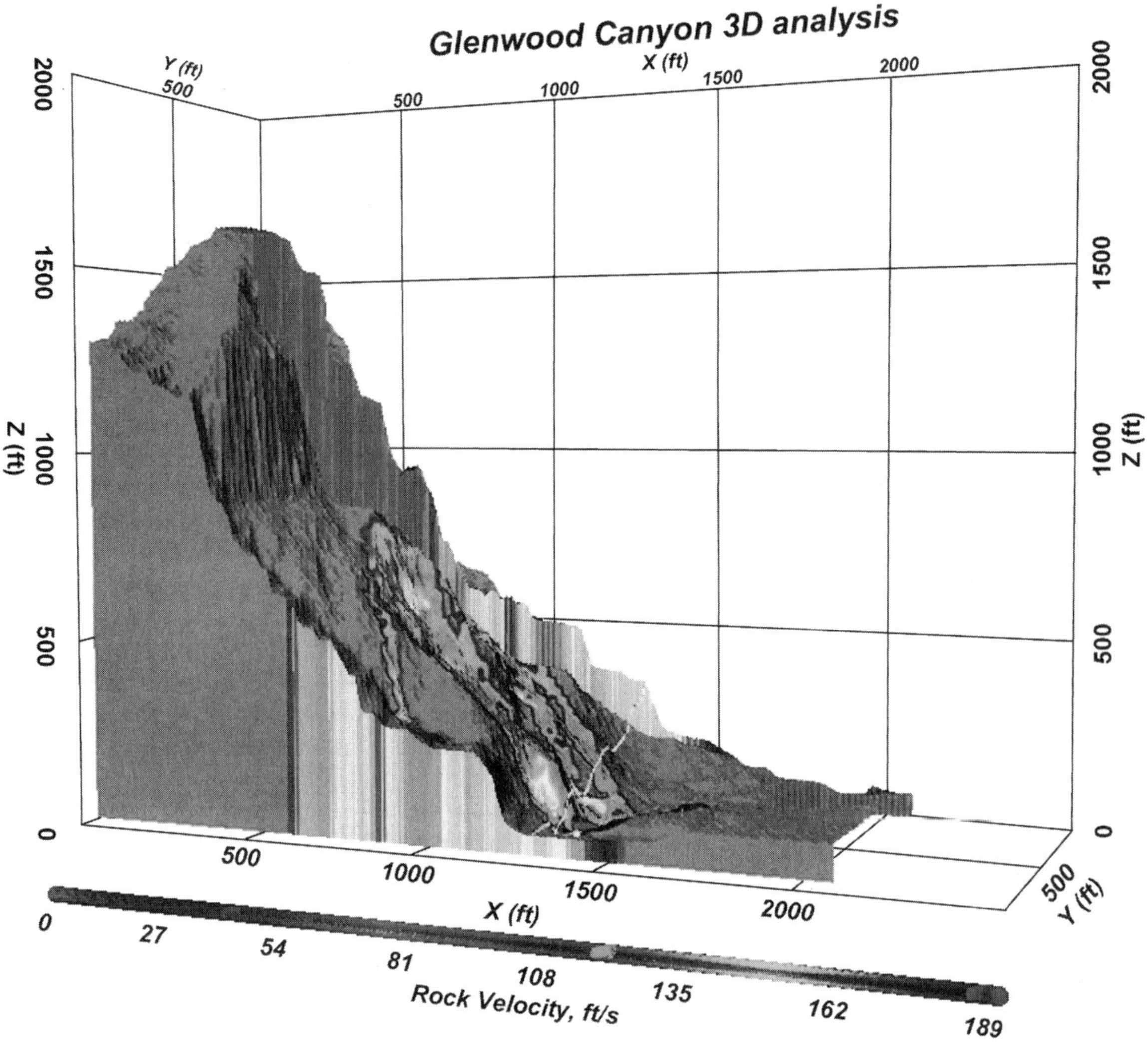

FIGURE 9-72 CRSP-3D display showing spatial distribution of estimated rockfall velocities for site along Interstate 70 in Glenwood Canyon, Colorado. Velocities are displayed by color scale ranging from magenta (zero) to red (high). (See accompanying DVD for image in color.) (IMAGE COURTESY OF R. D. ANDREW.)

spatial dispersions of block trajectories that are less accurate than those produced by the 3-D mode.

To verify the 2-D functionality of CRSP-3D, calibrations were conducted by using the same data and approach as were used for CRSP Version 5.0 and previous versions of CRSP. Calibration of the terrain model and 3-D rockfall behavior modeling was undertaken by using full-scale rockfall testing conducted near Georgetown, Colorado, and through observations and measurements taken at three other sites in Colorado, California, and Utah where natural rockfall events had occurred.

CRSP-3D documentation materials and the program itself are distributed by the Technology Development Program of the Central Federal Lands Highway Division of the Federal Highway Administration. They may be accessed at http://www.cflhd.gov/programs/techDevelopment/.

7. CONSIDERATIONS WITH REGARD TO 2-D AND 3-D MODELING

Reliable prediction of rockfalls is complicated by the influence of 3-D variations in slope geometry caused by the presence of chutes and channels, convexities, and longitudinal ridges on the trajectories of falling rock blocks and the partition of their kinetic energies into translational and rotational components. A complete description of 3-D topographic variability, combined with a probabilistic approach to rockfall parameter estimation, provides an opportunity for modeling a larger number of block impacts and fall paths, for improving evaluations of spatially varying parameters, and for assessing less frequent trajectories. Bozzolo et al. (1988), Azzoni et al. (1995), and Crosta and Agliardi (2004) identified "lateral dispersion," which often amounted to 20% of the slope length on relatively uniform slopes and much more on irregular slopes, as the most important effect of 3-D topographic variability on rockfall trajectories. However, the runout distance achieved by each rock block is also highly dependent on the block shape and its interaction with the 3-D variability of the slope surface.

While these considerations have been apparent for many years, rockfall simulation programs developed since the late 1970s initially focused on 2-D evaluations, primarily because of the limited computing power available on desktop personal computers at that time. However, sophisticated 3-D rockfall simulation programs have been developed in the past two decades. Thus, individuals who wish to undertake rockfall simulation assessments as part of a rockfall investigation are faced with a choice of using 2-D or 3-D models, and this decision is usually made after the characteristics of 2-D and 3-D approaches are considered.

The use of 2-D rockfall simulation programs has become an important and routine component of the current state of practice in rockfall hazard investigation and for the specification of rockfall mitigation methods. Observations during many field and laboratory tests have significantly increased knowledge and understanding of rockfall. If the programs are well calibrated, their simulation results, on evaluation by experienced professionals, produce information that is sufficiently accurate for identifying rockfall-prone areas and can form the basis for the design and specification of rockfall protective structures.

Although 2-D models are thus commonly considered as part of the state of practice for rockfall investigations, they have limitations. They have a limited ability to evaluate the influence of 3-D rock block shapes on the characteristics of rockfall trajectories. The programs simply represent the 3-D block shape as a 2-D object. As a consequence, 2-D programs tend to overpredict bounce heights and block velocities. The subjectivity inherent in selecting the most appropriate profile for evaluating rockfalls with a 2-D approach is a frequently underestimated limitation influencing the accuracy of predictions made by 2-D programs. Because they do not consider lateral variations in the slope profile, 2-D models do not accurately predict the lateral dispersion of rockfall trajectories on most slopes. Despite these limitations, their relatively simple data collection requirements and rapid computation speed make 2-D rockfall simulations economically attractive.

Three 2-D rockfall simulation programs have achieved widespread acceptance, received long-term technical support, or undergone progressive refinement over time: CRSP; RocFall, which was developed in Canada; and Rockfall, which was developed in Germany (see Sections 5.3, 5.4, and 5.5). Most rockfall site investigations use one or more of these three simulation programs to provide a basis for the analysis and specification of rockfall mitigation measures.

Significant advances in computer technologies have occurred in recent years, while the rapid development of both airborne and terrestrial LiDAR instruments has made it economically feasible to obtain detailed representations of topographic conditions at rockfall locations. As a consequence, the development of 3-D rockfall simulation programs that overcome some of the inherent limitations of a 2-D analysis approach has become more feasible. They are appropriately termed "state of the art." They are not widely used in practice because they require significantly greater investments to gather the required site characterization information and because their computations take longer and require enhanced computer resources.

Three-dimensional rockfall simulation programs are capable of modeling the lateral dispersion of rockfall trajectories more accurately. Most 3-D rockfall simulation programs use GIS technologies to define topographic conditions. These models represent a compromise between competing rockfall modeling requirements. Although from a theoretical point of view a fully dynamic, rigid-body assessment of block motions that accounts for both rotational and translational kinetic energies is desirable, in practice a computationally simpler lumped-mass approach often provides an acceptable solution. Because of computational issues and the uncertainty affecting the modeling parameters, simplified algorithms are often suitable. This approach makes the performance of 3-D rockfall simulations on widely available computers possible with relatively short calculation times. A probabilistic approach permits investigation of uncertainties in the controlling parameters as well as limitations in describing local site conditions.

Many detailed investigations of rockfall mechanics, coupled with field observations of both naturally occurring rockfalls and field tests at planned rockfall events, have provided information supporting the development of improved simulation algorithms, including the use of discrete element methods, which were briefly discussed in Section 2.1 of this chapter. Discrete element methods are routinely used by those in the civil and mining professions to analyze rock slopes. Salciarini et al. (2009) used a 2-D discrete element method to evaluate rockfall from a natural rock slope near the town of Postignano in the central Italian Apennines. The model simulated individual falling blocks of various sizes and modeled their fragmentation as they moved down the slope. Comparison with predictions produced by a commonly used algorithm (CRSP) indicated that, because of its ability to model fragmentation of the blocks, the discrete element model predicted shorter and more realistic runout distances. The availability of suitable computer codes, along with appropriate computer capacity, makes further applications of discrete element methods to rockfall simulation modeling increasingly feasible. These applications need not be restricted to 2-D simulations. Although the computations are even more extensive in 3-D, 3-D discrete element methods for rockfall simulation are implemented in the recently released CRSP-3D.

REFERENCES

Abbruzzese, J. M., C. Sauthier, and V. Labiouse. 2009. Considerations on Swiss Methodologies for Rock Fall Hazard Mapping Based on Trajectory Modelling. *Natural Hazards and Earth System Sciences*, Vol. 9, pp. 1095–1109.

Adachi, T., Y. Ohnishi, and K. Arai. 1991. Investigation of Toppling Slope Failure at Route 305 in Japan. *Proc., 7th International Congress on Rock Mechanics* (W. Wittke, ed.), Aachen, Germany, Balkema, Rotterdam, Netherlands, pp. 843–846.

Agliardi, F., and G. B. Crosta. 2002. 3-D Numerical Modelling of Rockfalls in the Lecco Urban Area (Lombardia Region, Italy). *Proc., EUROCK 2002, International Symposium on Rock Engineering for Mountainous Regions* (C. Dinis da Gama and L. Ribeira e Sousa, eds.), Madeira, Portugal, Sociedade Portuguesa de Geotecnia, pp. 79–86.

Agliardi, F., and G. B. Crosta. 2003. High Resolution Three-Dimensional Numerical Modelling of Rock Falls. *International Journal of Rock Mechanics and Mining Sciences*, Vol. 40, No. 4, pp. 455–471.

Agliardi, F., G. B. Crosta, and P. Frattini. 2009. Integrating Rockfall Risk Assessment and Countermeasure Design by 3D Modelling Techniques. *Natural Hazards and Earth System Sciences*, Vol. 9, No. 4, pp. 1059–1073.

Aksoy, H., and M. Ercanoglu. 2006. Determination of the Rockfall Source in an Urban Settlement Area by Using a Rule-Based Fuzzy Evaluation. *Natural Hazards and Earth System Sciences*, Vol. 6, No. 6, pp. 941–954.

Ayala-Carcedo, F. J., S. Cubillo-Nielsen, A. Alvarez, M. J. Domínguez, L. Laín, R. Laín, and G. Ortiz. 2003. Large Scale Rockfall Reach Susceptibility Maps in La Cabrera Sierra (Madrid) Performed with GIS and Dynamic Analysis at 1:5,000. *Natural Hazards*, Vol. 30, pp. 325–340.

Azimi, C., P. Desvarreux, A. Giraud, and J. Martin-Cocher. 1982. Méthode de Calcul de la Dynamique des Chutes de Blocs—Application à l'Étude du Versant de la Montagne de La Pale (Vercors) [Calculation Method for Rockfall Dynamics—Application to the Study of the Side of La Pale Mountain (Vercors)]. *Bulletin de Liaison des Laboratoires des Ponts et Chaussées*, Vol. 122, pp. 93–102 (in French).

Azzoni, A., and M. H. de Freitas. 1995. Experimentally Gained Parameters, Decisive for Rock Fall Analysis. *Rock Mechanics and Rock Engineering*, Vol. 28, No. 2, pp. 111–124.

Azzoni, A., G. La Barbera, and G. Mazzà. 1991. Studio con modello matematico e con sperimentazione in sito del problema di caduta massi. *Bollettino Associazione Mineraria Subalpina*, Vol. 28, No. 4, pp. 547–573 (in Italian).

Azzoni, A., G. La Barbera, and A. Zaninetti. 1995. Analysis and Prediction of Rock Fall Using a Mathematical Model. *International Journal of Rock Mechanics and Mining Sciences and Geomechanical Abstracts*, Vol. 32, No. 7, pp. 709–724.

Baillifard, F., M. Jaboyedoff, J.-D. Rouiller, and D. Tosoni. 2001. Présentation des Méthodes d'Études: Matterock (Presentation of Methods of Analysis: Matterrock). In *Programme Interreg 2c—"Falaises," Prévention des Mouvements de Versants et des Instabilités de Falaises; Confrontation des Méthodes d'Études des Éboulements Rocheux dans l'Arc Alpin ("Cliffs," Prevention of Movements of Slopes and Cliff Instabilities; Evaluation of the Methods of Analysis of Rock Landslides in the Alpine Arc)* (K. Carere, S. Ratto, and F. Zanolini, eds.), pp. 70–79 (in French).

Baillifard, F., M. Jaboyedoff, and M. Sartori. 2003. Rockfall Hazard Mapping Along a Mountainous Road in Switzerland Using a GIS-Based Parameter

Rating Approach. *Natural Hazards and Earth System Sciences*, Vol. 3, No. 5, pp. 435–442.

Bartingale, R.J. 2009. *Calibration of Colorado Rockfall Simulation Program (CRSP) Version 5.0*. MS thesis. Colorado School of Mines, Golden.

Bartingale, R.J., J.D. Higgins, R.D. Andrew, and A. Rock. 2009. Colorado Rockfall Simulation Program Version 5.0. *Proc., 60th Highway Geology Symposium*, Buffalo, N.Y., pp. 189–200.

Bassato, G., S. Cocco, and S. Silvano. 1985. Programma di Simulazione per lo Scoscendimento di Blocchi Rocciosi (Simulation Program for Landslides of Rock Blocks). *Dendronatura*, Vol. 6, No. 2, pp. 34–36 (in Italian).

Bergamin, S., J. Kolberg, and P. Fritz. 2004. Stability Analysis of a Rock Slope on the Bristen Road Using AutoBlock. *Proc., 3rd Asian Rock Mechanics Symposium, International Society for Rock Mechanics*, Mill Press, Rotterdam, Netherlands, pp. 127–132.

Bozzolo, D., and R. Pamini. 1982. *Modello Matematico per lo Studio della Caduta dei Massi* (*Mathematical Model for the Study of Rock Fall*). Laboratorio di Fisica Terrestre ICTS, Dipartimento Pubblica Educazione, Switzerland (in Italian).

Bozzolo, D., and R. Pamini. 1986. Simulation of Rock Falls down a Valley Side. *Acta Mechanica*, Vol. 63, pp. 113–130.

Bozzolo, D., R. Pamini, and K. Hutter. 1988. Rockfall Analysis—A Mathematical Model and Its Test with Field Data. *Proc., 5th International Symposium on Landslides* (C. Bonnard, ed.), July 10–15, Lausanne, Switzerland, Balkema, Rotterdam, Netherlands, Vol. 1, pp. 555–560.

Burrough, P.A., and R.A. McDonnell. 1998. *Principles of Geographical Information Systems*. Oxford University Press, New York.

Carrara, A., M. Cardinali, F. Guzzetti, and P. Reichenbach. 1995. GIS Technology in Mapping Landslides Hazard. In *Geographical Information Systems in Assessing Natural Hazards* (A. Carrara and F. Guzzetti, eds.), Kluwer Academic Publishers, Dordrecht, Netherlands, pp. 135–176.

Castelli, M., and C. Scavia. 2008. A Multidisciplinary Methodology for Hazard and Risk Assessment of Rock Avalanches. *Rock Mechanics and Rock Engineering*, Vol. 41, pp. 3–36.

Chau, K.T., Y.L. Sze, M.K. Fung, W.Y. Wong, E.L. Fong, and L.C.P. Chan. 2004. Landslide Hazard Analysis for Hong Kong Using Landslide Inventory and GIS. *Computers and Geosciences*, Vol. 30, No. 4, pp. 429–443.

Chen, G., and Y. Ohnishi. 1999. Slope Stability Analysis Using Discontinuous Deformation Analysis. In *Rock Mechanics for Industry: Proceedings of the 37th U.S. Rock Mechanics Symposium, Rock Mechanics for Industry* (B. Amadei, R.L. Kranz, and G.A. Scott, eds.), Vail, Colo., Balkema, Rotterdam, Netherlands, pp. 535–541.

Chung, C.F., A.G. Fabbri, and C.J. van Westen. 1995. Multivariate Regression Analysis for Landslide Hazard Zonation. In *Geographical Information Systems in Assessing Natural Hazards* (A. Carrara and F. Guzzetti, eds.), Kluwer Academic Publishers, Dordrecht, Netherlands, pp. 107–133.

Coggan, J.S., and R.J. Pine. 1996. Application of Distinct-Element Modelling to Assess Slope Stability at Delabole Slate Quarry, Cornwall, England. *Transactions of the Institution of Mining and Metallurgy, Section A: Mining Industry*, Vol. 105, pp. 22–30.

Coggan, J.S., D. Stead, and J.M. Eyre. 1998. Evaluation of Techniques for Quarry Slope Stability Assessment. *Transactions of the Institution of Mining and Metallurgy, Section B: Applied Earth Science*, Vol. 107, pp. 139–147.

Coggan, J.S., D. Stead, and J.H. Howe. 2000. Characterisation of a Structurally Controlled Flowslide in a Kaolinised Granite Slope. *Proc., 8th International Symposium on Landslides—Landslides in Research, Theory and Practice* (E. Bromhead, N. Dixon, and M.-L. Ibsen, eds.), Cardiff, Wales, United Kingdom, Vol. 1, pp. 299–304.

Copons, R., and J.M. Vilaplana. 2008. Rockfall Susceptibility Zoning at a Large Scale: From Geomorphological Inventory to Preliminary Land Use Planning. *Engineering Geology*, Vol. 102, pp. 142–151.

Copons, R., J.M. Vilaplana, and R. Linares. 2009. Rockfall Travel Distance Analysis by Using Empirical Models (Solà d'Andorra la Vella, Central Pyrenees). *Natural Hazards and Earth System Sciences*, Vol. 9, No. 6, pp. 2107–2118.

Corominas, J. 1996a. The Angle of Reach as a Mobility Index for Small and Large Landslides. *Canadian Geotechnical Journal*, Vol. 33, pp. 260–271.

Corominas, J. 1996b. The Angle of Reach as a Mobility Index for Small and Large Landslides: Reply. *Canadian Geotechnical Journal*, Vol. 33, pp. 1029–1031.

Cottaz, Y., and R.M. Faure. 2008. PiR3D, An Easy to Use Three Dimensional Block Fall Simulator. In *Landslides and Engineered Slopes: From the Past to the Future: Proceedings of the 10th International Symposium on Landslides and Engineered Slopes* (Z. Chen, J. Zhang, Z. Li, F. Wu, and K. Ho, eds.), Xi'an, China, CRC Press, pp. 319–322.

Cottaz, Y., J. D. Barnichon, N. Badertscher, and F. Gainon. 2010a. PiR3D, An Effective and User-Friendly 3D Rockfall Simulation Software: Formulation and Case-Study Application. Institut National de l'Environnement Industriel et des Risques Rock Slope Stability Symposium, Paris.

Cottaz, Y., J. D. Barnichon, N. Badertscher, and F. Gainon. 2010b. PiR3D, An Effective and User-Friendly 3D Rockfall Simulation Software: Formulation and Case-Study Application. Powerpoint presentation. Institut National de l'Environnement Industriel et des Risques Rock Slope Stability Symposium, Paris. http://www.ineris.fr/rss2010/sites/default/files/RSS2010-Cottaz%20.pdf. Accessed March 2011.

Crosta, G. B., and F. Agliardi. 2003. A Methodology for Physically Based Rockfall Hazard Assessment. *Natural Hazards and Earth System Sciences*, Vol. 3, pp. 407–422.

Crosta, G. B., and F. Agliardi. 2004. Parametric Evaluation of 3D Dispersion of Rockfall Trajectories. *Natural Hazards and Earth System Sciences*, Vol. 4, pp. 583–598.

Crosta, G. B., P. Frattini, S. Imposimato, and F. Agliardi. 2006. Modelling Vegetation and Fragmentation Effects on Rock Falls. *Geophysical Research Abstracts*, Vol. 8, 07694. Sref-ID: 1607-7962/gra/EGU06-A-07694.

Cundall, P. A. 1971. A Computer Model for Simulating Progressive, Large-Scale Movements in Blocky Rock Systems. *Proc., Symposium on Rock Fracture of the International Society of Rock Mechanics*, Nancy, France, Balkema, Rotterdam, Netherlands, Paper II-8.

Cundall, P. A., and O. D. L. Strack. 1979. A Discrete Numerical Model for Granular Assemblies. *Geotechnique*, Vol. 29, pp. 47–65.

Descoeudres, F. 1990. L'Éboulement des Crétaux: Aspects Géotechniques et Calcul Dynamique des Chutes de Blocs. *Publication de la Société Suisse de Mécanique des Sols et des Roches*, Vol. 121, pp. 19–25 (in French).

Descoeudres, F., and T. H. Zimmermann. 1987. Three-Dimensional Dynamics Calculation of Rock Fall. *Proc., 6th International Congress on Rock Mechanics* (G. Herget and S. Vongpaisal, eds.), Montreal, Canada, Balkema, Rotterdam, Netherlands, Vol. 1, pp. 337–342.

Di Prisco, C., and M. Vecchiotti. 2006. A Rheological Model for the Description of Boulder Impacts on Granular Strata. *Geotechnique*, Vol. 56, pp. 469–482.

Dorren, L. K. A. 2003. A Review of Rockfall Mechanics and Modelling Approaches. *Progress in Physical Geography*, Vol. 27, No. 1, pp. 69–87.

Dorren, L. K. A. 2010. *Rockyfor3D Revealed: Description of the Complete 3D Rockfall Model.* ecorisQ, Saint Hilaire du Touvet, France. http://www.ecorisq.org.

Dorren, L. K. A., and G. B. M. Heuvelink. 2004. Effect of Support Size on the Accuracy of a Distributed Rockfall Model. *International Journal of Geographical Information Science*, Vol. 18, pp. 595–609.

Dorren, L. K. A., and A. C. Seijmonsbergen. 2003. Comparison of Three GIS-Based Models for Predicting Rockfall Runout Zones at a Regional Scale. *Geomorphology*, Vol. 56, Nos. 1–2, pp. 49–64.

Dorren, L. K. A., F. Berger, and U. S. Putters. 2006. Real Size Experiments and 3D Simulation of Rockfall on Forested and Nonforested Slopes. *Natural Hazards and Earth System Sciences*, Vol. 6, pp. 145–153.

Dorren, L. K. A., B. Maier, U. S. Putters, and A. C. Seijmonsbergen. 2004. Combining Field and Modelling Techniques to Assess Rockfall Dynamics on a Protection Forest Hillslope in the European Alps. *Geomorphology*, Vol. 57, No. 3, pp. 151–167.

Duarte, R. M., and J. Marquínez. 2002. The Influence of Environmental and Lithologic Factors on Rockfall at a Regional Scale: An Evaluation Using GIS. *Geomorphology*, Vol. 43, pp. 117–136.

Dudt, J.-P., and B. Heidenreich. 2001. Treatment of Uncertainty in a Three-Dimensional Numerical Simulation Model for Rock Falls. In *International Conference on Landslides: Causes, Impacts and Countermeasures* (M. Kühne, H. H. Einstein, E. Krauter, H. Kapperich, and R. Pöttler, eds.), Davos, Switzerland, VGE, Essen, Germany, pp. 507–514.

Eberhardt, E. 2006. From Cause to Effect: Using Numerical Modelling to Understand Rock Slope Instability Mechanisms. In *Landslides from Massive Rock Slope Failure* (S. G. Evans, G. Scarascia Mugnozza, A. Strom, and R. L. Hermanns, eds.), Springer-Verlag, pp. 85–101.

Eberhardt, E., P. K. Kaiser, and D. Stead. 2002. Numerical Analysis of Progressive Failure in Natural Rock Slopes. *Proc., EUROCK 2002, International Symposium on Rock Engineering for Mountainous Regions* (C. Dinis da Gama and L. Ribeira e Sousa, eds.), Madeira, Portugal, Sociedade Portuguesa de Geotecnia, Lisbon, pp. 145–153.

ESRI. 2011. ArcGIS Desktop Help 9.3, Online Help and Documentation. http://webhelp.esri.com/arcgiSDEsktop/9.3/index.cfm?TopicName=welcome. Accessed March 15, 2011.

Evans, S.G., and O. Hungr. 1993. The Assessment of Rock Fall Hazard at the Base of Talus Slopes. *Canadian Geotechnical Journal*, Vol. 30, pp. 620–636.

Fairfield, J., and P. Leymarie. 1991. Drainage Networks from Grid Digital Elevation Models. *Water Resources Research*, Vol. 27, No. 5, pp. 709–717.

Falcetta, J.L. 1985. Un Nouveau Modèle de Calcul de Trajectoires de Blocs Rocheux (A New Calculation Model of Rock Block Trajectories). *Revue Francaise de Géotechnique*, Vol. 30, pp. 11–17 (in French).

Fell, R., J. Corominas, C. Bonnard, L. Cascini, E. Leroi, and W.Z. Savage. 2008a. Commentary: Guidelines for Landslide Susceptibility, Hazard and Risk Zoning for Land-Use Planning. *Engineering Geology*, Vol. 102, pp. 99–111.

Fell, R., J. Corominas, C. Bonnard, L. Cascini, E. Leroi, and W.Z. Savage. 2008b. Guidelines for Landslide Susceptibility, Hazard and Risk Zoning for Land-Use Planning. *Engineering Geology*, Vol. 102, pp. 85–98.

Frattini, P., G.B. Crosta, A. Carrara, and F. Agliardi. 2008. Assessment of Rockfall Susceptibility by Integrating Statistical and Physically-Based Approaches. *Geomorphology*, Vol. 94, Nos. 3–4, pp. 419–437.

Fritz, P., and S. Bergamin. 2004a. Computer Program AutoBlock for Analyzing the Stability of Foundations and Slopes in Rock Based on a Digital Terrain Model. *Proc., 72nd Annual Meeting, International Commission on Large Dams*, Seoul, South Korea. http://www.igt.ethz.ch/www/search/PublicationDisplay.asp?publication=761.

Fritz, P., and S. Bergamin. 2004b. Rock Slope Analysis Based on Digital Terrain Models. *Proc., 3rd Asian Rock Mechanics Symposium*, Kyoto, Japan. http://www.igt.ethz.ch/www/search/PublicationDisplay.asp?publication=765.

Fritz, P., and M. Hauser. 2007. Re-Evaluation of an Anchored Slope Using a Digital Terrain Model. *Proc., 11th Congress of the International Society for Rock Mechanics*, Lisbon, Portugal, Balkema, Rotterdam, Netherlands. http://www.igt.ethz.ch/www/search/PublicationDisplay.asp?publication=955.

Geo and Soft International. 2009. *User Guide: ISOMAP and ROTOMAP 3D Surface Modeling and Rockfall Analysis*. http://www.geoandsoft.com/manuali/english/rotomap.pdf.

Geo and Soft International. 2011. *ROCK3D: Three-Dimensional Rock Block Analysis Based Key Block Theory, User Guide*. http://www.geoandsoft.com/english/download.htm.

Goodman, R.E. 1993. *Engineering Geology: Rock in Engineering Construction*. John Wiley and Sons.

Goodman, R.E., and S.D. Kieffer. 2000. Behavior of Rock in Slopes. *Journal of Geotechnical and Geoenvironmental Engineering*, Vol. 126, No. 8, pp. 675–684.

Greenlee, D.D. 1987. Raster and Vector Processing for Scanned Linework. *Photogrammetric Engineering and Remote Sensing*, Vol. 53, No. 10, pp. 1383–1387.

Guzzetti, F., A. Carrara, M. Cardinali, and P. Reichenbach. 1999. Landslides Hazard Evaluation: A Review of Current Techniques and Their Application in a Multi-Scale Study, Central Italy. *Geomorphology*, Vol. 31, pp. 181–216.

Guzzetti, F., G. Crosta, R. Detti, and F. Agliardi. 2002. STONE: A Computer Program for the Three-Dimensional Simulation of Rockfalls. *Computers and Geosciences*, Vol. 28, pp. 1079–1093.

Guzzetti, F., P. Reichenbach, and S. Ghigi. 2004. Rockfall Hazard and Risk Assessment in the Nera River Valley, Umbria Region, Central Italy. *Environmental Management*, Vol. 34, No. 2, pp. 191–208.

Guzzetti, F., P. Reichenbach, and G.F. Wieczorek. 2003. Rockfall Hazard and Risk Assessment in the Yosemite Valley, California, USA. *Natural Hazards and Earth System Sciences*, Vol. 3, No. 6, pp. 491–503.

Hart, R.D. 1993. An Introduction to Distinct Element Modeling for Rock Engineering. In *Comprehensive Rock Engineering: Principles, Practice and Projects* Vol. 2 (J.A. Hudson, ed.), Pergamon Press, Oxford, United Kingdom, pp. 245–261.

Heidenreich, B. 2004. *Small- and Half-Scale Experimental Studies of Rockfall Impacts on Sandy Slopes*. Doctoral thesis. École Polytechnique Fédérale de Lausanne, Lausanne, Switzerland.

Heim, A. 1932. Bergsturz und Menschenleben. *Bieblatt zur Vierteljahrsschrift der Naturforschenden Gesellschaft in Zurich*, Vol. 77, pp. 1–217 (in German). [*Landslides and Human Lives* (N. Skermer, translator), BiTech Publishers, Vancouver, British Columbia, Canada, 1989.]

Higgins, J.D., R.D. Andrew, and T.J. Pfeiffer. 2003. Computer Simulation for Mitigation of Colorado Rockfall. In *Engineering Geology in Colorado: Contributions, Trends, and Case Histories* (CD-ROM) (D.D. Boyer, P.M. Santi, and W.P. Rogers, eds.), Special Publication 15, Association of Engineering Geologists.

Hoek, E., and J. W. Bray. 1981. *Rock Slope Engineering*. Institution of Mining and Metallurgy, London.

Holm, K., and M. Jakob. 2009. Long Rockfall Runout, Pascua Lama, Chile. *Canadian Geotechnical Journal*, Vol. 46, pp. 225–230.

Horn, B. K. P. 1981. Hill Shading and the Reflectance Map. *Proceedings of the Institute of Electrical and Electronics Engineers*, Vol. 69, No. 1, pp. 14–47.

Hsü, K. J. 1975. Catastrophic Debris Streams (Stürzstroms) Generated by Rockfalls. *Geological Society of America Bulletin*, Vol. 86, pp. 129–140.

Hudson, J. A., and J. P. Harrison 1997. *Engineering Rock Mechanics: An Introduction to the Principles*. Elsevier Science, Oxford, United Kingdom.

Hungr, O., and S. G. Evans. 1988. Engineering Evaluation of Fragmental Rockfall Hazards. *Proc., 5th International Symposium on Landslides* (C. Bonnard, ed.), Vol. 1, Balkema, Rotterdam, Netherlands, pp. 685–690.

Hylland, M. D., W. E. Mulvey, J. P. Johnson, and M. Butler. 2003. *Geologic Hazards of Moab-Spanish Valley, Grand County, Utah*. Special Study 107. Utah Geological Survey.

Itasca International, Inc. 2011a. 3DEC Overview. http://www.itascacg.com/3dec/overview.php.

Itasca International, Inc. 2011b. FLAC Overview. http://www.itascacg.com/flac/overview.html.

Itasca International, Inc. 2011c. FLAC3D Overview. http://www.itascacg.com/flac3d/overview.html.

Itasca International, Inc. 2011d. UDEC Overview. http://www.itascacg.com/udec/overview.php.

Jaboyedoff, M. 2002. *Matterocking 2.0 User's Guide*. http://www.crealp.ch/images/files/logiciels/manual_matterocking.pdf.

Jaboyedoff, M. 2003a. *Brief Field Trip Guidebook About Slope Instabilities in the Rhône Valley and the Vispa Valley Between Lausanne and Zermatt (Switzerland)*. Quanterra Guide Book-01. http://www.quanterra.org/guide/guide1_0.htm. Accessed March 2011.

Jaboyedoff, M. 2003b. *CONEFALL 1.0 User's Guide*. Open Report Soft-01. Quanterra, Lausanne, Switzerland. http://www.quanterra.org/Manual_conefall.pdf.

Jaboyedoff, M., and V. Labiouse. 2003. Preliminary Assessment of Rockfall Hazard Based on GIS Data. *International Society for Rock Mechanics 2003—Technology Roadmap for Rock Mechanics*, South African Institute of Mining and Metallurgy, pp. 575–578.

Jaboyedoff, M., and V. Labiouse. 2011. Technical Note: Preliminary Estimation of Rockfall Runout Zones. *Natural Hazards Earth and System Sciences*, Vol. 11, No. 3, pp. 819–828.

Jaboyedoff, M., F. Baillifard, R. Couture, J. Locat, and P. Locat. 2004. New Insight of Geomorphology and Landslide Prone Area Detection Using DEM. In *Landslides: Evaluation and Stabilization* (W. A. Lacerda, M. Ehrlich, S. A. B. Fontoura, and A. S. F. Sayão, eds.), Balkema, Rotterdam, Netherlands, pp. 199–205.

Jaboyedoff, M., F. Baillifard, C. Marro, F. Philippossian, and J.-D. Rouiller. 1999. Detection of Rock Instabilities: Matterock Methodology. In *Joint Japan–Swiss Scientific Meeting on Impact Load by Rock Falls and Design of Protection Structures*, Kanazawa, Japan, pp. 37–43.

Jaboyedoff, M., F. Baillifard, F. Philippossian, and J.-D. Rouiller. 2004. Assessing Fracture Occurrence Using "Weighted Fracturing Density": A Step Towards Estimating Rock Instability Hazard. *Natural Hazards and Earth System Sciences*, Vol. 4, No. 1, pp. 83–93.

Jaboyedoff, M., J.-P. Dudt, and V. Labiouse. 2005. An Attempt to Refine Rockfall Hazard Zoning Based on the Kinetic Energy, Frequency and Fragmentation Degree. *Natural Hazards and Earth System Sciences*, Vol. 5, No. 5, pp. 621–632.

Jenson, S. K., and J. O. Domingue. 1988. Extracting Topographic Structure from Digital Elevation Data for Geographic Information System Analysis. *Photogrammetric Engineering and Remote Sensing*, Vol. 54, No. 11, pp. 1593–1600.

John, K. W. 1968. Graphical Stability Analyses of Slopes in Jointed Rock. *Proceedings, Soil Mechanics and Foundations Division*, American Society of Civil Engineers, SM2, Paper No. 5865.

Jones, C. L., J. D. Higgins, and R. D. Andrew. 2000. *Colorado Rockfall Simulation Program Version 4.0*. Colorado Department of Transportation, Colorado Geological Survey.

Kalkani, E. C., and D. R. Piteau. 1976. Finite Element Analysis of Toppling Failure at Hell's Gate Bluffs, British Columbia. *Bulletin of the International Association of Engineering Geology*, Vol. 13, No. 4, pp. 315–327.

Keylock, C., and U. Domaas. 1999. Evaluation of Topographic Models of Rockfall Travel Distance for Use in Hazard Applications. *Arctic, Antarctic, and Alpine Research*, Vol. 31, No. 3, pp. 312–320.

Kobayashi, Y., E. L. Harp, and T. Kagawa. 1990. Simulation of Rockfalls Triggered by Earthquakes. *Rock Mechanics and Rock Engineering*, Vol. 23, pp. 1–20.

Kovári, K., and P. Fritz. 1984. Recent Developments in the Analysis and Monitoring of Rock Slopes. *Proc., 4th International Symposium on Landslides*, Toronto, Ontario, Canada.

Krautblatter, M., and M. Moser. 2009. A Nonlinear Model Coupling Rockfall and Rainfall Intensity Based on a Four Year Measurement in a High Alpine Rock Wall (Reintal, German Alps). *Natural Hazards and Earth System Sciences*, Vol. 9, pp. 1425–1432.

Lan, H., C. D. Martin, and C. H. Lim. 2007. RockFall Analyst: A GIS Extension for Three-Dimensional and Spatially Distributed Rockfall Hazard Modeling. *Computers and Geosciences*, Vol. 33, pp. 262–279.

Lan, H., C. D. Martin, C. Zhou, and C. H. Lim. 2010. Rockfall Hazard Analysis Using LiDAR and Spatial Modeling. *Geomorphology*, Vol. 118, Nos. 1–2, pp. 213–223.

Lateltin, O., C. Haemmig, H. Raetzo, and C. Bonnard. 2005. Landslide Risk Management in Switzerland. *Landslides*, Vol. 2, pp. 313–320.

Leroi, E. 1996. Landslides Hazard-Risk Maps at Different Scales: Objectives, Tools and Developments. *Proc., 7th International Symposium on Landslides* (K. Senneset, ed.), Trondheim, Norway, Balkema, Rotterdam, Netherlands, Vol. 1, pp. 35–51.

Lied, K. 1977. Rockfall Problems in Norway. *Proc., Meeting on Rockfall Dynamics and Protective Works Effectiveness*. Publication No. 90, Instituto Sperimentale Modelli e Strutture SpA, Bergamo, Italy, pp. 51–53.

Londe, P., G. Vigier, and R. Vormeringer. 1969. The Stability of Rock Slopes, a Three Dimensional Study. *Journal of the Soil Mechanics and Foundations Division*, American Society of Civil Engineers, Vol. 95, pp. 235–262.

Longo, S., and P. Oreste. 2010. Ceppo Morelli Block-Falls Probability Study to Support the Decision of Excavating a By-Pass Tunnel. *American Journal of Engineering and Applied Sciences*, Vol. 3, No. 4, pp. 723–727.

Loye, A., M. Jaboyedoff, and A. Pedrazzini. 2009. Identification of Potential Rockfall Source Areas at a Regional Scale Using a DEM Based Geomorphometric Analysis. *Natural Hazards and Earth System Sciences*, Vol. 9, No. 5, pp. 1643–1653.

Loye, A., A. Pedrazzini, and M. Jaboyedoff. 2008. Preliminary Regional Rockfall Hazard Mapping Using LiDAR-Based Slope Frequency Distribution and CONEFALL Modelling. *Proc., 4th Canadian Conference on Geohazards* (J. Locat, D. Perret, D. Turmel, D. Demmers, and S. Leroueil, eds.), Quebec City, Quebec, Canada, Laval University Press, pp. 445–452.

Marquínez, J., R. M. Duarte, P. Farias, and M. J. Sánchez. 2003. Predictive GIS-Based Model of Rockfall Activity in Mountain Cliffs. *Natural Hazards*, Vol. 30, pp. 341–360.

Matsuoka, N., and H. Sakai. 1999. Rockfall Activity from an Alpine Cliff During Thawing Periods. *Geomorphology*, Vol. 28, pp. 309–328.

McConnell, R. G., and R. W. Brock. 1904. The Great Landslide at Frank Alberta. *Geological Survey of Canada Annual Report 1902–1903*, Part 8, Appendix.

Meisina, C., A. Piccio, and A. Tocchio. 2001. Some Aspects of the Landslide Susceptibility in the Sorba Valley (Western Alps, Italy). In *International Conference on Landslides: Causes, Impacts and Countermeasures* (M. Kühne, H. H. Einstein, E. Krauter, H. Kapperich, and R. Pöttler, eds.), Davos, Switzerland, VGE, Essen, Germany, pp. 547–556.

Meißl, G. 1998. *Modellierung der Reichweite von Felsstürzen: Fallbeispiele zur GIS-gestützen Gefahrenbeurteilung aus dem Bayerischen und Tiroler Alpenraum*. PhD thesis. Innsbrucker Geographische Studien, Band 28, Universität Innsbruck, Innsbruck, Austria (in German).

Meißl, G. 2001. Modelling the Runout Distances of Rockfalls Using a Geographic Information System. *Zeitschrift für Geomorphologie Supplement Band*, Vol. 125, pp. 129–137.

Nicholson, D. T. 2004. Hazard Assessment for Progressive, Weathering-Related Breakdown of Excavated Rockslopes. *Quarterly Journal of Engineering Geology and Hydrogeology*, Vol. 37, pp. 327–346.

Nocilla, N., A. Evangelista, and A. Scotto di Santolo. 2009. Fragmentation During Rock Falls: Two Italian Case Studies of Hard and Soft Rocks. *Rock Mechanics and Rock Engineering*, Vol. 42, pp. 815–833.

Norrish, N. I., and D. C. Wyllie. 1996. Rock Slope Stability Analysis. In *Special Report 247: Landslides: Investigation and Mitigation* (A. K. Turner and R. L. Schuster, eds.), Transportation Research Board, National Research Council, Washington, D.C., pp. 391–425.

O'Callaghan, J. F., and D. M. Mark. 1984. The Extraction of Drainage Networks from Digital Elevation Models. *Computer Vision, Graphics and Image Processing*, Vol. 28, pp. 323–344.

Onofri, R., and C. Candian. 1979. *Indagine sui Limiti di Massima Invasione di Blocchi Rocciosi Franati durante il Sisma del Friuli del 1976, Reg. Aut. Friuli-Venezia Giulia (Survey on the Maximum Invasion Limits of Collapsed Boulders During the Earthquake of Friuli, 1976, Autonomous Region Friuli-Venezia Giulia)* (in Italian).

Parise, M. 2002. Landslide Hazard Zonation of Slopes Susceptible to Rock Falls and Topples. *Natural Hazards and Earth System Sciences*, Vol. 2, Nos. 1–2, pp. 37–49.

Paronuzzi, P. 2009a. Field Evidence and Kinematical Back-Analysis of Block Rebounds: The Lavone Rockfall, Northern Italy. *Rock Mechanics and Rock Engineering*, Vol. 42, pp. 783–813.

Paronuzzi, P. 2009b. Rockfall-Induced Block Propagation on a Soil Slope, Northern Italy. *Environmental Geology*, Vol. 58, pp. 1451–1466.

Paronuzzi, P., and E. Artini. 1999. Un Nuovo Programma in Ambiente Windows per la Modellazione della Caduta Massi (A New Program in the Windows Environment for Modeling Falling Rocks). *Geologia Tecnica e Ambientale*, Vol. 1/99, pp. 13–24 (in Italian).

Perret, S., F. Dolf, and H. Kienholz. 2004. Rockfall into Forests: Analysis and Simulation of Rockfall Trajectories—Considerations with Respect to Mountainous Forests in Switzerland. *Landslides*, Vol. 2, pp. 123–130.

Petje, U., M. Ribčič, and M. Mikoš. 2005. Computer Simulation of Stone Falls and Rockfalls. *Acta Geographica Slovenica*, Vol. 45, No. 2, pp. 93–120.

Pfeiffer, T.J. 1989. *Rockfall Hazard Analysis Using Computer Simulation of Rockfalls*. Masters thesis. Colorado School of Mines, Golden.

Pfeiffer, T.J., and T. Bowen. 1989. Computer Simulation of Rock Falls. *Association of Engineering Geologists Bulletin*, Vol. 26, No. 1, pp. 135–146.

Pfeiffer, T.J., and J.D. Higgins. 1988. *Colorado Rockfall Simulation Program User's Manual*. Colorado Department of Transportation, Denver.

Pfeiffer, T.J., and J.D. Higgins. 1990. Rockfall Hazard Analysis Using the Colorado Rockfall Simulation Program. In *Transportation Research Record 1288*, Transportation Research Board, National Research Council, Washington, D.C., pp. 117–126.

Pfeiffer, T.J., J.D. Higgins, R.D. Andrew, R.K. Barrett, and R.B. Beck. 1993. *Colorado Rockfall Simulation Program Version 3.0 User's Manual*. Report MI-39. Colorado Transportation Institute, Colorado Department of Transportation, and Colorado Geological Survey, Denver.

Pfeiffer, T.J., J.D. Higgins, R.D. Andrew, R.J. Schultz, and R.B. Beck. 1995. *Colorado Rockfall Simulation Program Version 3.0a User's Manual*. Colorado Department of Transportation, Denver.

Pfeiffer, T.J., J.D. Higgins, R.J. Schultz, and R.D. Andrew. 1991. *Colorado Rockfall Simulation Program User's Manual for Version 2.1*. Report CDOT-DTD-ED3/CSM-89-2B. Colorado Department of Transportation, Denver.

Pfeiffer, T.J., J.D. Higgins, and A.K. Turner. 1990. Computer Aided Rockfall Hazard Analysis. *Proc., 6th International Congress, International Association of Engineering Geology* (D.G. Price, ed.), Symposia Volume, Balkema, Rotterdam, Netherlands, pp. 93–103.

Pine, R.J., J.S. Coggan, Z.N. Flynn, and D. Elmo. 2006. The Development of a New Numerical Modelling Approach for Naturally Fractured Rock Masses. *Rock Mechanics and Rock Engineering*, Vol. 39, No. 5, pp. 395–419.

Piteau and Associates, Limited. 1980. Slope Stability Analysis for Rockfall Problems: The Computer Rockfall Model for Simulating Rockfall Distributions. In *Rock Slope Engineering, Part D*, Federal Highway Administration Reference Manual FHWA-TS-79-208, U.S. Department of Transportation, pp. D62–D80.

Piteau, D.R., and R. Clayton. 1977. Computer Rockfall Model. *Proc., Meeting on Rockfall Dynamics and Protective Works Effectiveness*, Publication No. 90, Instituto Sperimentale Modelli e Strutture SpA, Bergamo, Italy, pp. 123–125.

Priest, S.D. 1993. *Discontinuity Analysis for Rock Engineering*. Chapman and Hall, London.

Quinn, P., K. Beven, P. Chevallier, and O. Planchon. 1991. The Prediction of Hillslope Flow Paths for Distributed Hydrological Modelling Using Digital Terrain Models. *Hydrological Processes*, Vol. 5, No. 1, pp. 59–79.

Raetzo, H., O. Lateltin, D. Bollinger, and J.P. Tripet. 2002. Hazard Assessment in Switzerland: Codes of Practice for Mass Movements. *Bulletin of Engineering Geology and the Environment*, Vol. 61, pp. 263–268.

Rochet, L. 1987. Application des Modèles Numériques de Propagation à l'Étude des Éboulements Rocheux (Application of Numerical Propagation Models to the Study of Rocky Landslides). *Bulletin Liaison des Ponts et Chaussées*, Vols. 150–151, pp. 84–95 (in French).

Rockware. 2011. RockPack III Overview. http://www.rockware.com/product/overview.php?id=119.

RocNews. 2003. How Input Parameters Affect Your RocFall Analysis. Quarterly newsletter, RocScience, Inc., Fall. http://www.rocscience.com/library/rocnews/Fall2003.htm. Accessed March 2011.

RocScience, Inc. 2011a. *Dips Overview*. http://www.rocscience.com/products/1/dips.

RocScience, Inc. 2011b. *RocFall 4.0 Overview*. http://www.rocscience.com/products/12. Accessed March 2011.

Rouiller, J.-D. 1990. L'Éboulement des Crétaux—Isérables—Riddes (Valais). *Publication de la Société Suisse de Mécanique des Sols et des Roches*, Vol. 121, pp. 15–17 (in French).

Rouiller, J-D., and C. Marro. 1997. Application de la Méthodologie "MATTEROCK" à l'Évaluation du Danger lié aux Falaises (Application of the Methodology "MATTEROCK" to the Evaluation of the Danger Linked to Cliffs). *Eclogae Geologicae Helvetiae*, Vol. 90, pp. 393–399 (in French).

Salciarini, D., C. Tamagnini, and P. Conversini. 2009. Numerical Approaches for Rockfall Analysis: A Comparison. In *18th World International Association for Mathematics and Computers in Simulation Congress and MODSIM09 International Congress on Modelling and Simulation* (R.S. Anderssen, R.D. Braddock, and L.T.H. Newham, eds.), Modelling and Simulation Society of Australia and New Zealand and International Association for Mathematics and Computers in Simulation, July, pp. 2706–2712. http://www.mssanz.org.au/modsim09/G3/salciarini.pdf.

Sartori, M., F. Baillifard, M. Jaboyedoff, and J.-D. Rouiller. 2003. Kinematics of the 1991 Randa Rockslides (Valais, Switzerland). *Natural Hazards and Earth System Sciences*, Vol. 3, No. 5, pp. 423–433.

Scioldo, G. 1991. ROTOMAP: Analisi Statistica del Rotolamento dei Massi (Statistical Analysis of Rolling Boulders). *Guida di Informatica Ambientale (Guide to Environmental Information)*, Pàtron, Milan, Italy, pp. 81–84 (in Italian).

Shi, G.H., and R.E. Goodman. 1989. Discontinuous Deformation Analysis—A New Numerical Model for the Statics and Dynamics of Deformable Block Structures. *Proc., 1st U.S. Conference on Discrete Element Methods* (G.G.W. Mustoe, M. Henriksen, and H.P. Huttelmaier, eds.), Colorado School of Mines Press, Golden.

Soeters, R., and C.J. van Westen. 1996. Slope Instability Recognition, Analysis, and Zonation. In *Special Report 247: Landslides: Investigation and Mitigation* (A.K. Turner and R.L. Schuster, eds.), Transportation Research Board, National Research Council, Washington, D.C., pp. 129–177.

Spang, R.M. 1987. Protection Against Rockfall—Stepchild in the Design of Rock Slopes. *Proc., 6th International Congress on Rock Mechanics* (G. Herget and S. Vongpaisal, eds.), Montreal, Quebec, Canada, Balkema, Rotterdam, Netherlands, Vol. 1, pp. 551–557.

Spang, R.M., and E. Krauter. 2001. Rockfall Simulation—A State of the Art Tool for Risk Assessment and Dimensioning of Rockfall Barriers. In *International Conference on Landslides: Causes, Impacts and Countermeasures* (M. Kühne, H.H. Einstein, E. Krauter, H. Kapperich, and R. Pöttler, eds.), Davos, Switzerland, VGE, Essen, Germany, pp. 607–613.

Spang, R.M., and R.W. Rautenstrauch. 1988. Empirical and Mathematical Approaches to Rockfall Protection and Their Practical Applications. *Proc., 5th International Symposium on Landslides* (C. Bonnard, ed.), Vol. 2, Balkema, Rotterdam, Netherlands, pp. 1237–1243.

Spang, R.M., and T. Sönser. 1995. Optimized Rockfall Protection by ROCKFALL. *Proc., 8th International Congress on Rock Mechanics*, Tokyo, Vol. 3, pp. 1233–1242.

Spang, C., B. Romunde, J. Waldecker, and H. Ganser. 2008. *Entwurf und Bemessung von Steinschlagschutzzäunen mit ROCKFALL 7.1, Fels- und Hangsicherung Müden II (Design and Calculation of Falling-Rock Protection Fences with ROCKFALL 7.1, Rock and Cliff Security Identified II)*, Computerorientierte Geologie der ÖGG Universität Salzburg (in German).

Stead, D., E. Eberhardt, and J.S. Coggan. 2006. Developments in the Characterization of Complex Rock Slope Deformation and Failure Using Numerical Modelling Techniques. *Engineering Geology*, Vol. 83, pp. 217–235.

Stevens, W. 1998. *RocFall: A Tool for Probabilistic Analysis, Design of Remedial Measures and Prediction of Rockfalls*. Master's thesis. University of Toronto, Ontario, Canada.

Stoffel, M., A. Wehrli, R. Kühne, L.K.A. Dorren, S. Perret, and H. Kienholz. 2006. Assessing the Protective Effect of Mountain Forests Against Rockfall Using a 3-D Simulation Model. *Forest Ecology and Management*, Vol. 225, pp. 113–122.

Tagliavini, F., P. Reichenbach, D. Maragna, F. Guzzetti, and A. Pasuto. 2009. Comparison of 2-D and 3-D Computer Models for the M. Salta Rock Fall, Vajont Valley, Northern Italy. *Geoinformatica*, Vol. 13, pp. 323–337.

Tarboton, D.G. 1997. A New Method for the Determination of Flow Directions and Upslope Areas in Grid Digital Elevation Models. *Water Resources Research*, Vol. 33, No. 2, pp. 309–319.

Terlien, M.T.J., T.W.J. Van Asch, and C.J. van Westen. 1995. Deterministic Modeling in GIS-Based Landslide Hazard Assessment. In *Geographical Information Systems in Assessing Natural Hazards* (A. Carrara and F. Guzzetti, eds.), Kluwer Academic Publishers, Dordrecht, Netherlands, pp. 57–77.

Toppe, R. 1987. Terrain Models: A Tool for Natural Hazard Mapping. In *Avalanche Formation, Movement and Effects* (B. Salm and H. Gubler, eds.), International Association of Hydrological Sciences, Wallingford, United Kingdom, Vol. 162, pp. 629–638.

Van Dijke, J. J., and C. J. van Westen. 1990. Rockfall Hazard: A Geomorphologic Application of Neighbourhood Analysis with ILWIS. *ITC Journal*, Vol. 1, pp. 40–44.

Van Westen, C. J., E. Castellanos, and S. L. Kuriakose. 2008. Spatial Data for Landslide Susceptibility, Hazard, and Vulnerability Assessment: An Overview. *Engineering Geology*, Vol. 102, pp. 112–131.

Varnes, D. J. 1978. Slope Movement Types and Processes. In *Special Report 176: Landslides: Analysis and Control* (R. L. Schuster and R. J. Krizek, eds.), Transportation Research Board, National Research Council, Washington, D.C., pp. 11–33.

Wieczorek, G. F., M. M. Morrissey, G. Iovine, and J. Godt. 1998. *Rock-Fall Hazards in the Yosemite Valley*. U.S. Geological Survey Open-File Report 98-467. http://pubs.usgs.gov/of/1998/ofr-98-0467/.

Wieczorek, G. F., M. M. Morrissey, G. Iovine, and J. Godt. 1999. *Rockfall Potential in the Yosemite Valley, California*. U.S. Geological Survey Open-File Report 99-578. http://pubs.usgs.gov/of/1999/ofr-99-0578/.

Wieczorek, G. F., G. M. Stock, P. Reichenbach, J. B. Snyder, J. W. Borchers, and J. W. Godt. 2008. Investigation and Hazard Assessment of the 2003 and 2007 Staircase Falls Rock Falls, Yosemite National Park, California, USA. *Natural Hazards and Earth System Sciences*, Vol. 8, pp. 421–432.

Wittke, W. 1990. *Rock Mechanics: Theory and Applications with Case Histories*. Springer-Verlag, Berlin.

Wu, S.-S. 1985. Rockfall Evaluation by Computer Simulation. In *Transportation Research Record 1031*, Transportation Research Board, National Research Council, Washington, D.C., pp. 1–5.

Wyllie, D. C., and C. W. Mah. 2004. *Rock Slope Engineering*, 4th ed. Spon Press, New York.

Zhou, Q., and X. Liu. 2004. Analysis of Errors of Derived Slope and Aspect Related to DEM Data Properties. *Computers and Geosciences*, Vol. 30, pp. 369–378.

Zinggeler, A. 1990. *Steinschlagsimulation in Gebirgswäldern: Modellierung der relevanten Teilprozesse (Rockfall Simulation in Mountain Woods: Modeling of the Relevant Partial Processes)*. Master's thesis. University of Bern, Bern, Switzerland (in German).

Zinggeler, A., B. Krummenacher, and H. Kienholz. 1991. Steinschlagsimulation in Gebirgswäldern (Rockfall Simulation in Mountain Woods). In *Berichte und Forschungen (Reports and Researches)*, Geographisches Institut, Universität Freiburg, Switzerland, Vol. 3, pp. 61–70 (in German).

Chapter 10

JOHN D. DUFFY AND
A. KEITH TURNER

CONDUCTING FIELD-TEST EXPERIMENTS

1. INTRODUCTION

There are two principal methods of analyzing and evaluating rockfall. One method is to perform field tests that involve the initiation of controlled rockfall events and then to observe and document the behavior of the falling rocks. As noted in Chapter 8, although the principles of Newtonian mechanics describe the basic phenomena of falling, striking, bouncing, and rolling rocks, in practice the characteristics of motion after impact are complex and are influenced by slope characteristics and by several parameters related to the block as well as to the kinematics during impact (Labiouse and Heidenreich 2009). A series of field-test experiments at different locations permits the assessment of responses of falling blocks to varying slope characteristics.

The second method of analyzing and evaluating rockfall employs mathematical models to simulate the mechanics of the rockfall processes. As discussed in Chapter 9, a considerable number of computer programs have been developed for this purpose. These programs provide quantitative and probabilistic estimates of rockfall events, including traces of trajectories, velocities, impact energies, bounce heights, and runout distances. However, the value of predictions produced by these programs depends significantly on their calibration to observed responses of falling rocks under defined real-world conditions.

Therefore, the ultimate success of both methods depends greatly on the accuracy of data gathered during actual field rock-rolling trials. Many variables may be evaluated during a field trial, but the significance of several variables depends upon the primary purpose of the evaluation. The field-trial procedures may collect different data if the purpose is to calibrate computer models or if the purpose is to specify proposed rockfall protection measures. In many cases, the main incentive for the evaluation or analysis of rockfalls is to select and design effective and economic protection measures. With this objective, the primary concern is to accurately establish the bounce heights and velocities of the falling blocks and hence to establish the design impact energy for the barrier. In other cases, the field trials are undertaken to determine the degree of hazard associated with rockfall. With this objective, the primary concerns are the trajectories of the falling blocks and the areal limits of the runout zone. Thus, when a field trial to evaluate rockfall behavior is planned, consideration is given to the purpose of the evaluation and the utility of the data being collected. In many cases, field-test experiments are conducted with multiple objectives, as shown in Table 10-1.

Field trials, depending on their location and the objectives of the testing, are often expensive and may pose safety considerations. Trials conducted by rolling representative rocks down the actual slope under consideration for rockfall mitigation

Table 10-1
Summary of Field Rock-Rolling Experiments Performed Since the 1960s

Test Sponsor and Approximate Date	Source Describing Test	Purpose of Test	Number of Rocks Rolled During Test
1960s			
Washington DOT, 1963	Ritchie 1963	Define rockfall trajectories; design fences and ditches	Hundreds
1970s			
City of Lecco, Italy, 1974	Broili 1974	Define rockfall trajectories	10
North Carolina DOT, 1978	Evans 1989	Define rockfall trajectories	146
British Columbia Ministry of Highways and Public Works, 1978	Elston et al. 1978	Define rockfall trajectories	350
1980s			
North Carolina DOT, 1984	Wu 1984; Evans 1989	Define rockfall trajectories	Not defined
Caltrans, 1985	McCauley et al. 1985	Fence, berms, trajectories	223
Golder Associates, British Columbia, Canada, 1987	Wyllie 1991	Barrier designs	60
Caltrans, 1987	Duffy 1987	Define rockfall trajectories	12
University of Arizona, Department of Mining and Geological Engineering, Tucson 1988	Evans 1989	Define rockfall trajectories	50
Colorado DOT, 1989	Barrett and Pfeiffer 1989	Barrier designs	13
Caltrans, 1989	Smith and Duffy 1990	Barrier designs	76
1990s			
Colorado DOT, 1991	Hearn 1991; Hearn et al. 1992	Barrier designs	70
Caltrans, 1991	Duffy 1991	Barrier designs	6
Geobrugg Inc., Switzerland 1991	Duffy 1992	Barrier designs	90
Geobrugg Inc., Switzerland, 1992	Duffy and Haller 1993	Barrier designs	18
University of the Pacific, Stockton, California, 1993	Kane and Duffy 1993	Barrier designs	24
Oregon DOT, 1994	Pierson et al. 1994	Define rockfall trajectories	2,790
ISMES SpA, Bergamo, Italy; and Department of Geology, Imperial College of Science, Technology and Medicine, London, 1992–1995	Azzoni and de Freitas 1995	Define rockfall trajectories	60
Caltrans, 1995	Beck 1995	Define rockfall trajectories	15
Caltrans, 1996	Duffy and Hoon 1996a	Barrier designs	16
Caltrans, 1996	Duffy and Hoon 1996b	Barrier designs	25
Chung Cheng Institute of Technology, Taiwan, 1997	Hwu and Spang 1997	Barrier designs	Not defined
Protec Engineering, Japan, 1998	Hoshida and Nomura 1998	Barrier designs	9
Laboratorio di Fisica Terrestre, Lugano-Trevano, Switzerland, 1998	Bozzolo et al. 1998	Define rockfall trajectories	7
Chama Valley Productions, LLC, Chama, New Mexico, 1998	Andrew et al. 1998	Barrier designs	31
Caltrans, 1998	Duffy et al. 1998	Barrier designs	56

(continued)

Table 10-1 *(continued)*
Summary of Field Rock-Rolling Experiments Performed Since the 1960s

Test Sponsor and Approximate Date	Source Describing Test	Purpose of Test	Number of Rocks Rolled During Test
2000s			
Caltrans, 2000	Duffy and Jones 2000	Barrier	25
Oregon DOT, 2001	Pierson et al. 2001	Define rockfall trajectories	11,250
Geological Survey of Bolzano, Geoproject, Inc., and Bolzano Engineering Office, Bolzano, Italy, 2003	Schweigl et al. 2003	Define rockfall trajectories	19
Colorado DOT, 2004	Arndt et al. 2009	Define rockfall trajectories	10
Department of Civil and Environmental Engineering and Architecture, Università degli Studi di Parma, Parma, Italy, 2003	Giani et al. 2004	Define rockfall trajectories	83
Colorado DOT, 2005	Arndt et al. 2009	Barrier designs	7
Cemagref Grenoble, St. Martin d'Hères, France, 2005	Dorren, Berger, Le Hir, et al. 2006; Dorren, Berger, and Putters 2006	Define rockfall trajectories and evaluate influence of trees	202
Caltrans, 2006	Whitman and Duffy 2006	Define rockfall trajectories	56
IGOR, Inc., Trento, Italy, 2007	Badger et al. 2008	Barrier designs	3
Colorado DOT, 2009	Arndt et al. 2009	Barrier designs	10
Caltrans, 2009	Salisbury et al. 2009	Define rockfall trajectories	70

provide the most accurate information to define rockfall behavior and to support rockfall analyses. When the slope is located adjacent to a transportation corridor, however, traffic diversion or control for brief periods is necessary to allow sufficient time to safely undertake a limited number of rock-rolling events. With such precautions, a sequence of such events often results in a total of 10 to 20 rock-rolling trials (Figure 10-1). This procedure is acceptable if the test area is located and prepared for the rock-rolling experiments so that the rocks do not strike the traveled way.

Rock-rolling trials are frequently performed outside the transportation corridor. This method eliminates traffic impacts, safety concerns, and potential damage to roadway surfaces or existing mitigation measures. In these circumstances, the rock-rolling trials are conducted on a slope with geometry, rock type, and other characteristics similar to those of the location of interest.

2. PREPARING FOR FIELD EXPERIMENTS

Successful field tests require careful preparation. A number of site-preparation tasks are typically undertaken before the rock-rolling experiments are conducted:

- Selection, measurement, and weight determination of all the rocks involved in the rock-rolling experiments;
- Preparation and accurate surveys of the test slope selected for the experiments, including establishing spatial reference lines so that rock velocities and trajectories may be efficiently determined from images collected during the experiments;

FIGURE 10-1 Traffic can sometimes be diverted or controlled for brief periods along transportation corridors, allowing enough time to undertake rock-rolling events. (PHOTOGRAPH COURTESY OF THE CALIFORNIA DEPARTMENT OF TRANSPORTATION.)

FIGURE 10-2 Site-preparation procedures for field test.
(PHOTOGRAPH COURTESY OF THE CALIFORNIA DEPARTMENT OF TRANSPORTATION.)

- Selection of locations for the image-recording equipment and installing and testing this equipment to ensure that it is fully operational;
- Installation and inspection of any mitigation devices, such as rockfall catchment fences or barrier walls, that are to be tested as part of the field experiments and recording information concerning their initial characteristics and geometries; and
- Verification of the performance of any desired instrumentation installed on these mitigation devices.

Safety during the field tests is typically ensured by establishing appropriate safety procedures and fully training all personnel concerning these procedures. Figure 10-2 shows a test site in preparation.

3. SELECTION OF TEST SLOPE

Successful field experiments depend on the selection of an appropriate slope. Ritchie (1963) concluded that the behavior of rockfall is influenced by slope geometry, slope material properties, rock geometry, and rock material properties. Labiouse and Heidenreich (2009) reported that on the basis of their experiences with numerous natural rockfalls and field rockfall trials, the characteristics of motion after impact are affected not only by the slope characteristics but also by several parameters related to the falling blocks as well as to the kinematics during impact.

Thus, field trials typically involve tests performed on the actual slope being evaluated for rockfall hazard or on an alternate slope with similar attributes. Some field trials are undertaken on a series of slopes with varying characteristics for rockfall research purposes. Specifically, slope characteristics that affect rockfall events include

1. Slope angles, heights, and lengths;
2. Cross-sectional shape and contour;
3. Surface roughness or variability;
4. Slope material (soil or rock type or both); and
5. Vegetative cover.

Three practical considerations typically influence the selection of a test slope for rock-rolling experiments: land ownership, accessibility, and traffic control and public safety issues. Agreement by landowners to have rock-rolling experiments conducted on their property is an obvious prerequisite to the selection of any proposed test site. Many reported field experiments have been conducted on government-owned sites with limited or controlled public access or on commercial sites such as quarries (Figures 10-3 and 10-4).

Accessibility to the site affects the cost of performing the experiments, data collection procedures, and the safety of those undertaking the rock-rolling tests. Other important considerations are reasonable access to the top of the test slope and an adequate supply of test rocks. One of the most important considerations is a good field of view for the image-capturing equipment. Recording the rockfall event on film, video, or digital images is paramount for successful velocity and trajectory analyses. In addition to good access to the launching position at the top of the slope, it is important that the slope face be readily accessible so that the spatial reference system is easily prepared and maintained and there is adequate room at the base of the slope for constructing test structures or measuring runout distances.

Safety is critically important because a rapidly rolling large rock is a dangerous object whose path and trajectory are often highly unpredictable. Rock-rolling events are capable of attracting curious onlookers. Therefore, desirable test slopes are sites amenable to restricted public access. Traffic control along existing transportation routes is closely associated with the public safety issue. Where rockfall runout zones exist near, or extend

FIGURE 10-3
Example of government-owned site used for field rock-rolling experiments.
(PHOTOGRAPH COURTESY OF THE CALIFORNIA DEPARTMENT OF TRANSPORTATION.)

FIGURE 10-4
State-owned quarry site used for field rock-rolling experiments.
(PHOTOGRAPH COURTESY OF THE OREGON DEPARTMENT OF TRANSPORTATION.)

onto, busy roads or railroads, conducting field rock-rolling activities at these locations not only raises safety issues but often causes damage to the facilities and poses significant economic costs if unacceptable traffic disruptions result.

When it is possible to conduct rock-rolling experiments at the actual slope where rockfall hazard mitigation concerns exist, a number of uncertainties are obviously removed from the experimental results. Not surprisingly, however, this actual design slope frequently is not an acceptable or feasible location for conducting rock-rolling experiments, and a similar slope that closely resembles the design slope has to be selected. In fact, of the 37 rock-rolling studies cited in Table 10-1, only 12 were undertaken within transportation corridors.

3.1 Test Slope Survey

A detailed survey of the selected slope is undertaken to produce cross sections and slope-profile contour information. Ideally, all such surveys cover the entire test slope and extend from the rockfall source area, or proposed launching point for the experiments, to beyond the anticipated maximum extent of the runout, or impact, zone at the foot of the slope (Figure 10-5).

Typically this information is obtained by detailed topographic or photogrammetric surveys or from a LiDAR-derived digital elevation model (DEM). In some circumstances, surveys use elevation observations spaced as close together as 0.6 m (2 ft) to accurately identify changes in the slope

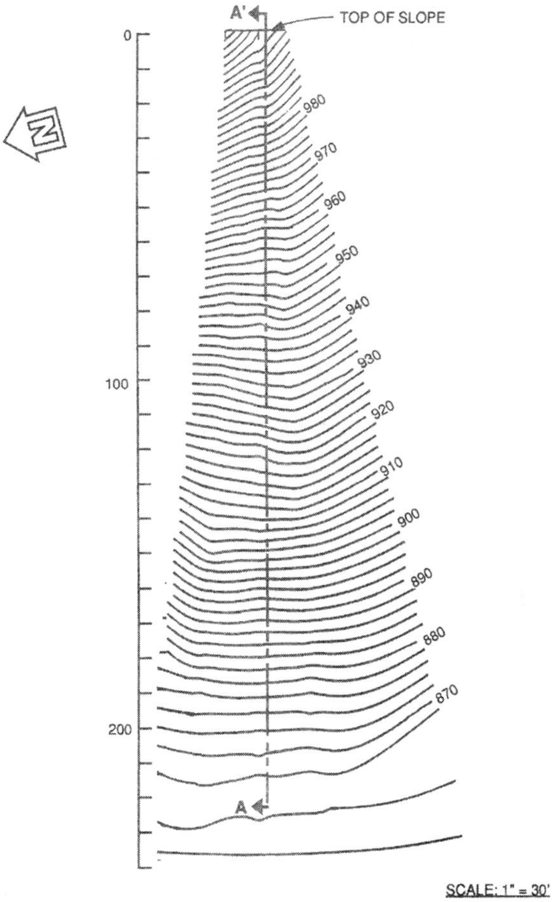

FIGURE 10-5
Detailed contour map of selected slope produced by site survey.
(IMAGE COURTESY OF THE CALIFORNIA DEPARTMENT OF TRANSPORTATION.)

FIGURE 10-6 Ground-based LiDAR surveys offer economical solution to providing detailed topographic data (Kemeny and Turner 2008).

that could affect rockfall trajectories (FHWA 1993). The increasing coverage of many areas by airborne LiDAR surveys along with the increased availability of ground-based LiDAR surveys, as described in Chapter 7, offer economical techniques for producing detailed DEMs of test sites (Figure 10-6).

3.2 Slope Geometry

Jones et al. (2000) specified three categories of slope geometry parameters as primary influences on the behavior of rockfall:

- Lateral variability of the slope surface (the lateral profile),
- Slope profile including the slope inclination and slope length, and
- Surface roughness.

Careful attention to the slope geometry will aid in keeping the test rocks within the field of view of the image-recording equipment and within established limits of the test site.

3.2.1 Lateral Variability of Slope Surface (Lateral Profile)

If the slope extends laterally as a planar slope, rockfall trajectories will disperse to each side to only a moderate extent. In contrast, if the slope has a convex cross profile, such as is found on a talus cone or with a protruding portion of a cliff face, the individual rocks will bounce and roll along highly erratic and unpredictable paths (Crosta and Agliardi 2004). These rocks will be hard to monitor and pose dangers to personnel conducting the experiments. More desirable, from a safety and monitoring viewpoint, is a slope that has a concave cross profile that forms a chute or gully. This shape will tend to confine the lateral spread in the rockfall trajectories and make it easier to successfully monitor the rock-rolling events in comparative safety. Field rock-rolling tests were conducted in an avalanche chute in Forêt Communale de Vaujany, France, and, as shown in Figure 10-7, the upper portion of this chute confined the rockfall trajectories, whereas the lower portion provided no confinement and the rocks began to spread out.

The test site at Rifle, Colorado (Figure 10-8), had a channel excavated about 20 m (60 ft) wide in the lower 33 m (100 ft) of the slope to guide rocks into the area containing the barriers being tested (Parsons De Leuw 1992).

Experience has demonstrated that an occasional rock will take a "lucky bounce" and effectively escape all but the most severe lateral confinement. Severe lateral confinement, however, introduces additional factors that may affect the experimental results, making it more difficult to accurately transfer the results to design locations.

Azzoni and de Freitas (1995) evaluated the lateral dispersion of rockfall trajectories based on an analysis of 60 rock-rolling field tests undertaken by the Istituto Sperimentale Modelli e Strutture (ISMES) laboratory at Bergamo, Italy. As shown in Figure 10-9, they defined the dispersion of rockfall trajectories as the ratio between the distance (D) separating the two most widely diverging rockfall paths as observed when facing the slope and the length of the slope (L). The data came from rockfall tests conducted on two slopes with average slope angles ranging from 40° to 50° with sections ranging from 30° to 65°. As shown in Figure 10-10, longer slopes generally produce a larger maximum separation between observed paths, unless the slope is constrained within a valley or gully profile that provides lateral constraints. The dispersion (D/L) is generally in the range of 10% to 20% of the slope length.

Crosta and Agliardi (2004) provided additional parametric evaluations of three-dimensional (3-D) dispersion of rockfall trajectories. Their studies related the amount of dispersion to the mean slope gradient, surface roughness, and spatial resolution of the ground-surface definition, as well

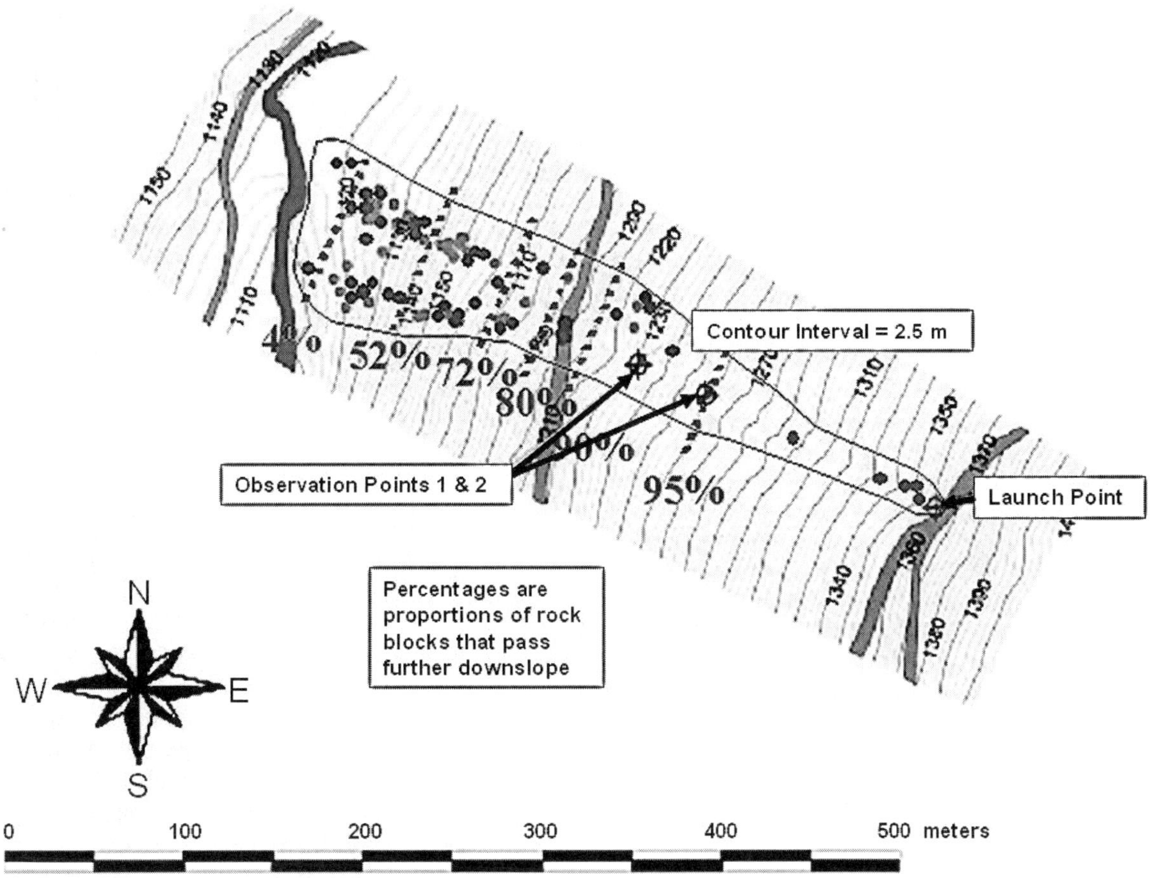

FIGURE 10-7
Map of observed stopping points and outline of dispersion of rockfall trajectories during rock-rolling experiments in avalanche chute in Forêt Communale de Vaujany, France (adapted from Berger and Dorren 2006).

FIGURE 10-8
Test site at Rifle, Colorado, including excavated channel in lower portion of slope to guide rocks into area containing barriers being tested (Parsons De Leuw 1992).

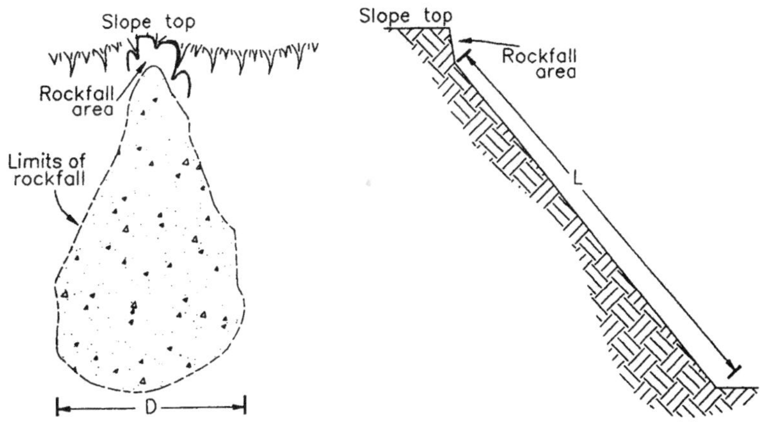

FIGURE 10-9
Dispersion of rockfall trajectories, defined as ratio between distance (*D*) separating two most widely separated rockfall paths, as observed when facing slope, and length of slope (*L*) (Azzoni and de Freitas 1995).

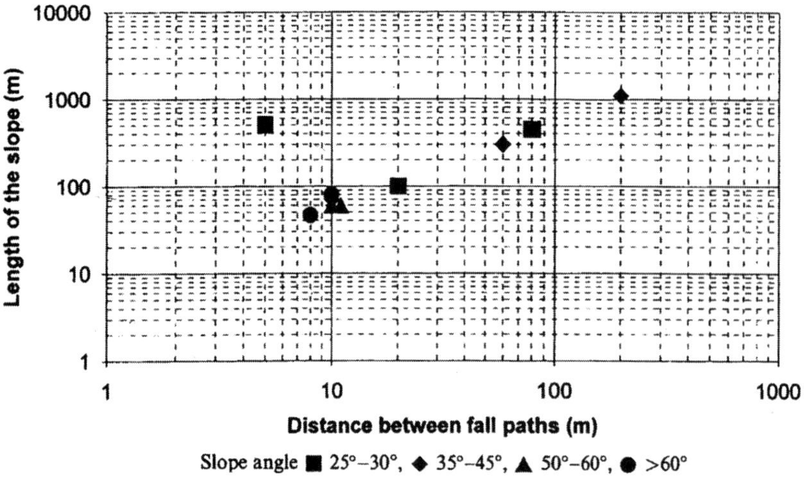

FIGURE 10-10
Rockfall test data collected from tests conducted on two slopes with average slope angles ranging from 40° to 50° and with sections ranging from 30° to 65° (Azzoni and de Freitas 1995).

as to kinematic factors of block motion, including velocity and bounce height.

Although based entirely on computer simulations, their results suggest the following relationships, which are typically observed during field trials:

- Smooth and gently dipping slopes result in only moderate lateral dispersion, especially if rounded and nearly spherical boulders predominate. These move by a combination of making low, short jumps and rolling along the ground surface.
- On very rough slopes, lateral dispersion typically increases with increasing slope inclination up to a threshold value of about 45°, for which *D*/*L* ratios (as defined in Figure 10-9) of about 30% to 35% are observed. On such steep slopes, the blocks continuously accelerate and move by multiple and fairly large bounces.
- Lateral dispersion is low, typically less than 15% for short, very steep slopes where the falling blocks are largely in free fall and experience few impacts with the ground surface.

3.2.2 Slope Profile (Slope Inclination and Length)

The geometry of the *slope profile*—the inclination and length of the various components of the terrain over which the falling rock blocks will pass—obviously influences the velocities, trajectories, and motions of free-falling, bouncing, or rolling rock blocks. Small-scale irregularities of the surface, the composition of the materials forming the slope, and the presence of vegetation also will influence the behavior of the falling blocks. The latter two influences are described in the following sections.

With the exception of excavated artificial rock cuts, rockfall hazards are typically found at locations where exposed rock cliffs provide a source of rock blocks. Structural weaknesses within the rock mass increase the number of rock blocks that will be dislodged from the cliff face. Below these cliffs the accumulating rockfall fragments create talus slopes (Figure 10-11). Because these situations are characteristic of most rockfall hazard locations, many rockfall test sites are chosen to reflect similar conditions: an upper, very steep slope or cliff that serves as the rockfall source location over which the test blocks experience free-fall trajectories, an intermediate transition zone with inclinations typically of 30° to 40° over which the test blocks roll or bounce, and a final gently sloping or level runout zone where the rocks decelerate and come to rest after rolling for a distance.

One practical consideration that is typically evaluated before any rock-rolling tests are undertaken is

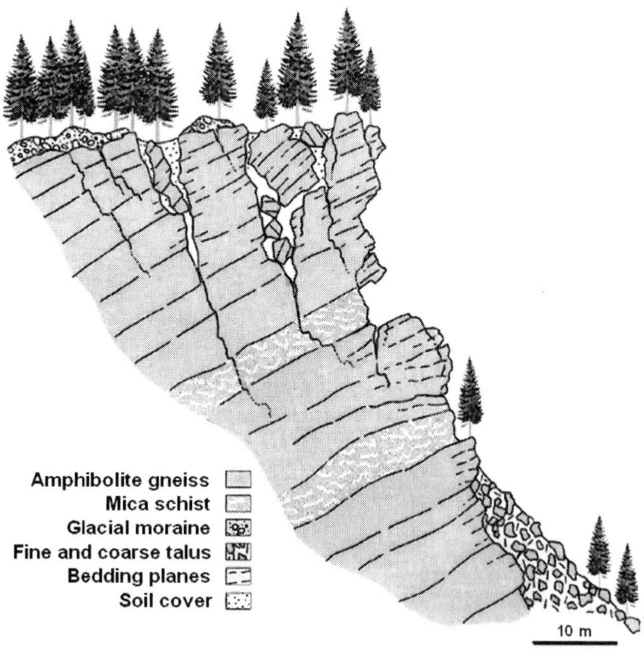

FIGURE 10-11
Typical naturally occurring rockfall hazard zone consisting of unstable rock cliff with talus slope at base located in western Austria; location served as test site for rockfall tests evaluating influence of forests on rockfall hazards (Dorren et al. 2004).

the estimation of the probable maximum distance a bouncing or rolling rock will travel down the slope. This estimate has important safety aspects and often influences the choice of test site. It also defines the extent of topographic surveys and where and how the spatial referencing is established.

Evans and Hungr (1993) investigated 16 locations in southern British Columbia where naturally occurring rockfall encountered talus slopes. They proposed using an empirical relationship to establish the probable maximum extent of a rockfall shadow zone in these situations. Evans and Hungr noted the previous use of the *fahrböschung* concept, initially defined by Heim (1932) in his studies of the very large Elm rockslide in Switzerland. They concluded, however, that a better empirical relationship involved the use of the *minimum shadow angle*, defined as the inclination of a line connecting the most distant rock block to the apex of the talus slope (Figure 10-12). From their observations, they found that a minimum shadow angle of 27.5° provided a useful conservative method for the preliminary estimate of maximum rockfall runout distance. This value was consistent with previously reported observations at several other locations. The empirical rockfall trajectory charts and tables developed by Ritchie (1963) and the Oregon Department of Transportation (DOT) (Pierson et al. 2001) provide additional guidance for designing reference grids, establishing ground monuments, and positioning the image-capturing devices.

Although primarily useful for evaluating and comparing the characteristics of very large and very rapid fragmental rock slides initiated by rockfalls, technically referred to as *stürzstrom*, the fahrböschung concept has some potential analytical value in evaluation of the results of field trials. Hsü (1975) provided an extensive examination of the fahrböschung concept and noted that it is defined as the angle between a horizontal plane and a line from the rockfall source to the farthest stopping point of the rockfall material, measured along the trajectory axis. Hsü evaluated the significance of the fahrböschung and concluded that it defines the rate of energy transfer from the potential energy of the rocks at their starting position to their kinetic energy as they move down the slope. Thus, the slope of the fahrböschung represents the rate of energy dissipation during the rockfall event; a small fahrböschung angle means a low rate of energy loss and thus a great transport distance. Figure 10-13 compares the travel

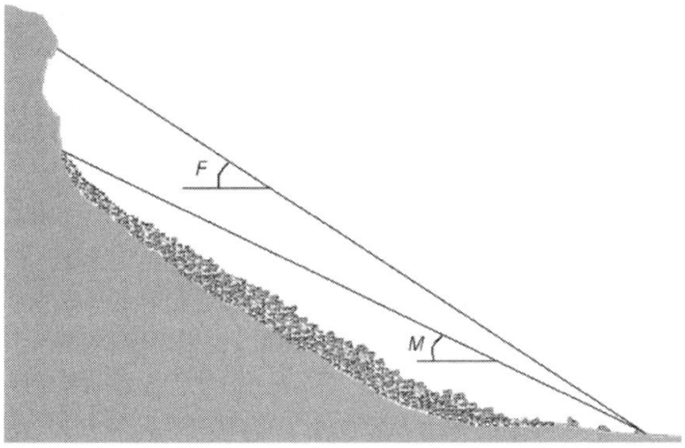

FIGURE 10-12
Fahrböschung (*F*) and minimum shadow angle (*M*) of talus slope (Dorren 2003).

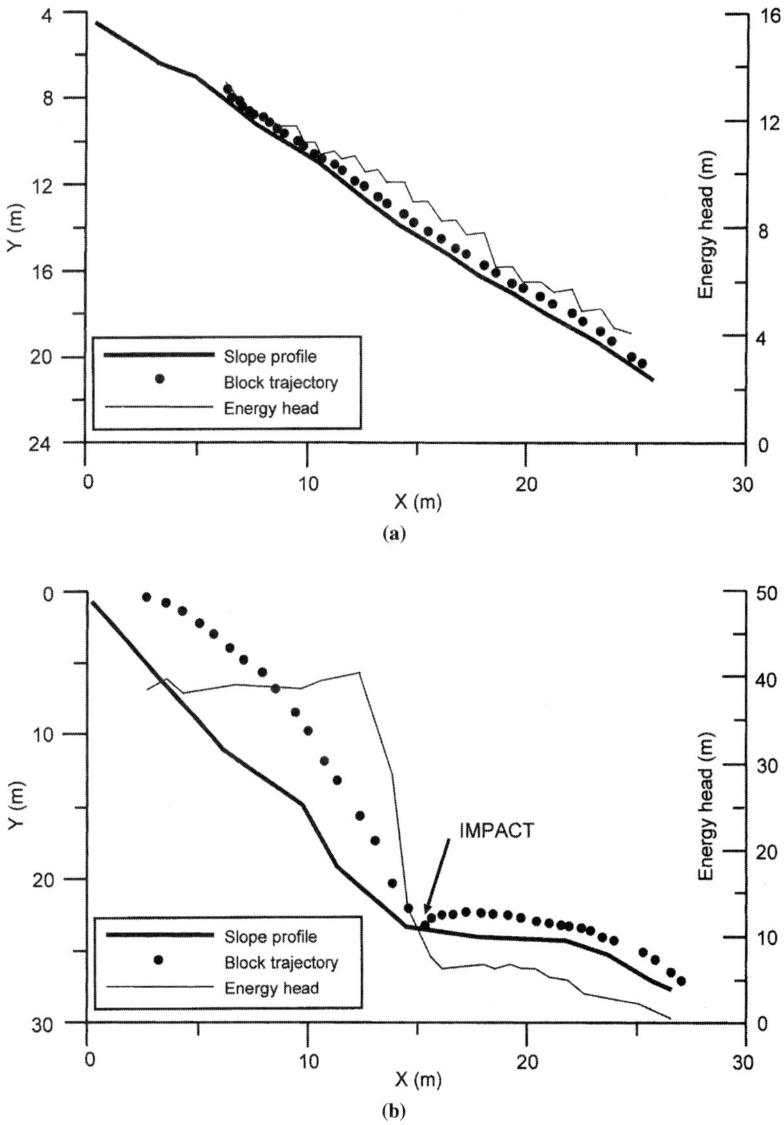

FIGURE 10-13
Travel paths and energy-head diagrams: (*a*) for rolling motion down gentle slope and (*b*) for impact and bouncing motion down irregular slope (Giani et al. 2004).

paths and energy-head diagrams for blocks traveling down a relatively smooth, gentle slope with those for a block striking and bouncing on a more irregular slope.

For detailed and accurate analysis of the rock-rolling events, a two-dimensional (2-D) profile that closely matches the typical trajectory path, such as that shown in Figure 10-14, is constructed from the detailed site topography data, as shown in Figure 10-5. If the slope is relatively uniform from side to side, it is typically characterized with a single 2-D profile. If the slope has numerous lateral irregularities, additional 2-D profiles may be required to represent the different trajectories followed by the various test rocks. The 2-D profile is subdivided into zones that reflect any significant changes in slope geometry or characteristics that could affect rockfall motion. The profile includes both the source area and the entire runout or impact zone. Slope inclination is critical because it defines zones of rockfall acceleration and deceleration and because the lengths of the slope-inclination segments determine the distance over which the rock accelerates or decelerates (Jones et al. 2000). Figure 10-15 is a 2-D profile for a rockfall test site in Switzerland. It shows the multiple initiation points used during the rockfall tests and the location of the barrier being tested.

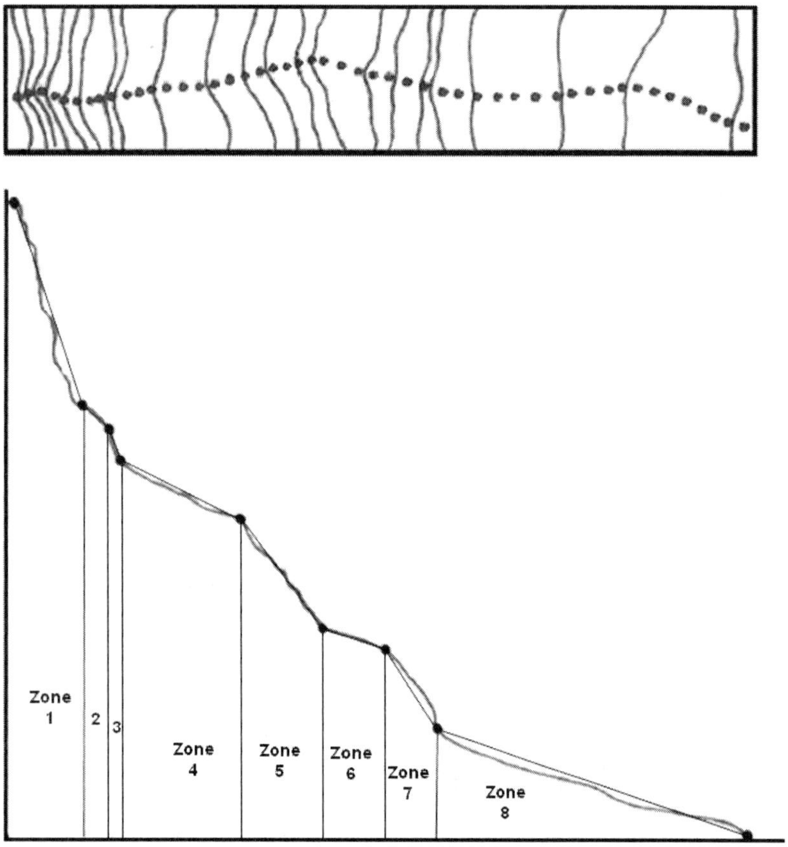

FIGURE 10-14
Two-dimensional slope profile closely following typical rockfall trajectory and subdivided into zones that reflect any significant changes in slope geometry or characteristics that could affect rockfall motion (Dorren 2003).

Predictive computer models are often used in conjunction with rock-rolling tests to provide additional assessments and for comparative analysis. As discussed in Chapter 9, the accuracy of the predictions produced by these computer simulations is influenced by the accuracy of the 2-D profile. The 2-D profile also assists in subsequent interpretations of recorded rockfall events, especially to establish the influence of prominent slope changes on the trajectories and motions of the test rocks.

3.2.3 Surface Roughness

Numerous field observations, laboratory experiments, and computer simulations demonstrate that slope irregularities, commonly referred to as *surface roughness*, materially influence the trajectory of a falling or bouncing rock block after an impact with the ground surface (Azzoni et al. 1992, 1995; Chau et al. 2002; Crosta and Agliardi 2004; Dorren et al. 2004; Dorren, Berger, and Putters 2006).

Field observations on talus (or scree) slopes clearly show that smaller rocks tend to remain near the top of the slope, whereas progressively larger blocks are found downslope (Statham 1976; Broili 1977; Evans and Hungr 1993). Although larger blocks can relatively easily roll or bounce over minor irregularities formed by a series of much smaller blocks, small bouncing or rolling blocks tend to be stopped when they encounter a surface covered with fragments of a size similar to or larger than the moving block (Figure 10-16).

Surface roughness influences the motion of a bouncing or rolling rock in several ways. A rough surface provides hollows and occasional steep adverse surfaces that impede the lateral downslope progress of the block by increasing the frictional resistance of the surface (Giani 1992). A rough surface also provides the potential for an increased impact angle, and observations and laboratory experiments show that larger impact angles result in a larger dissipation of kinetic energy (Bozzolo

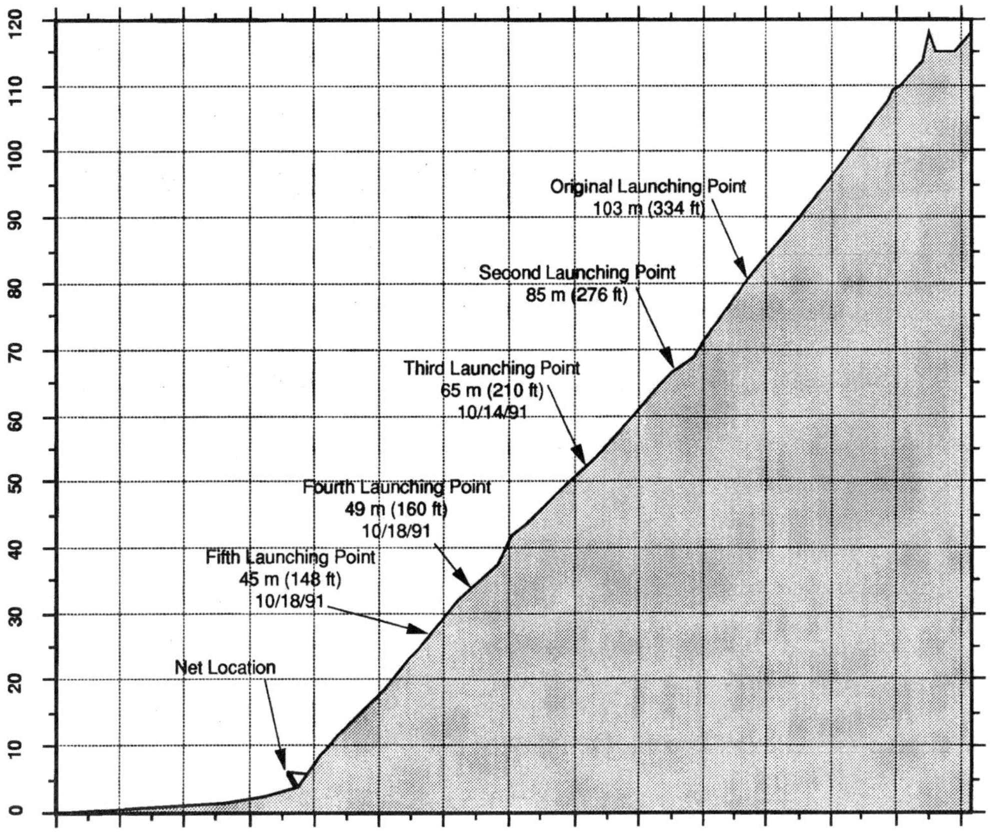

FIGURE 10-15
Rockfall barrier test site cross section, Oberbuchsiten, Switzerland (Duffy and Haller 1993).

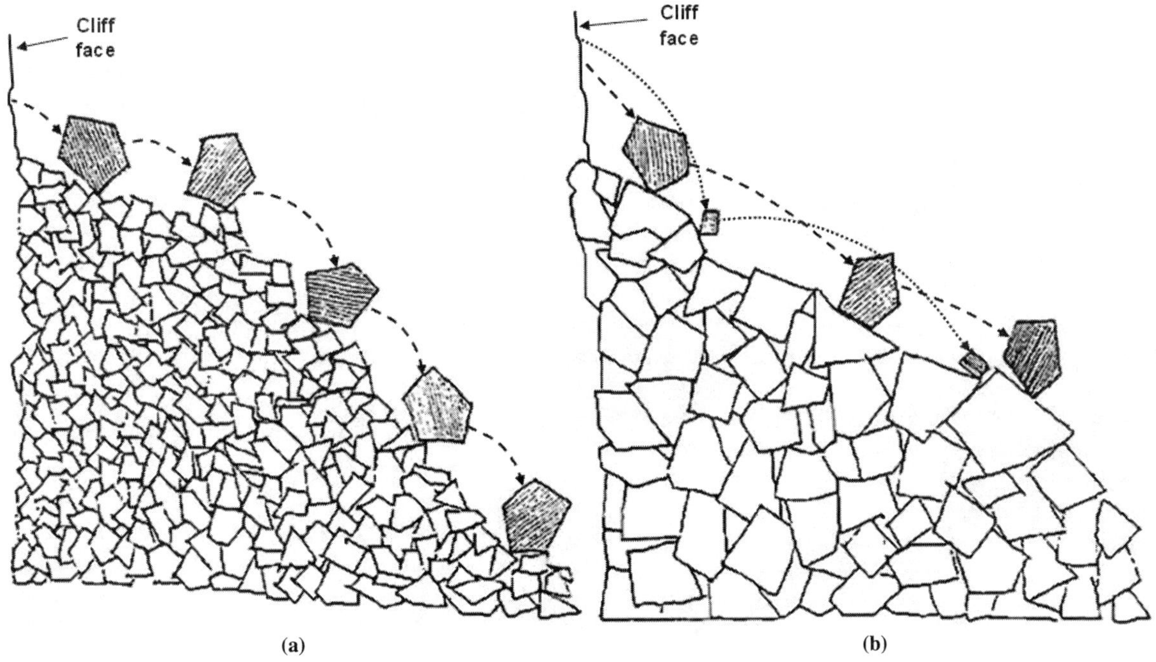

FIGURE 10-16
Influence of size of talus or scree material on distance traveled and motion of falling rock blocks: (*a*) blocks bounce and roll when debris particles are smaller than blocks and (*b*) blocks tend to be stopped when size of debris particles is similar to or larger than blocks (adapted from Giani 1992).

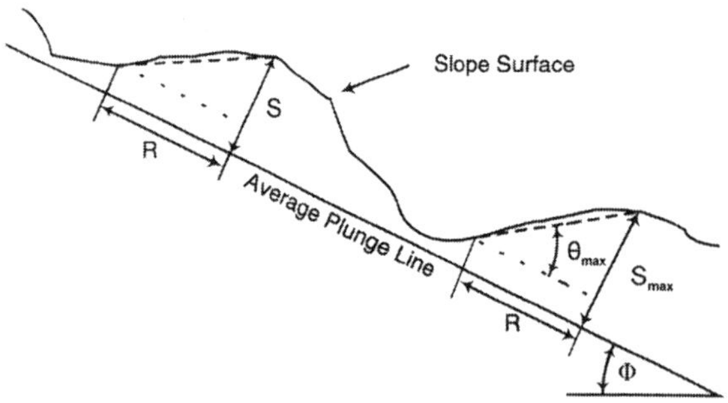

FIGURE 10-17 Surface roughness (S) is perpendicular variation from average plunge line (defined by slope angle Φ) over distance equal to radius of rock (R). Maximum slope variation (θ_{max}) is defined by S_{max} and R (Pfeiffer and Higgins 1990).

and Pamini 1986; Crosta and Agliardi 2004; Bourrier et al. 2009). Thus, a rough slope typically increases the height and number of bounces of a rock block as it travels down the slope but also tends to reduce their velocity and impact energy (FHWA 1993).

Surface roughness is related to the size of the falling rock block and to the dimensions of the irregularities on the ground surface. Pfeiffer and Bowen (1989), Pfeiffer and Higgins (1990), and Jones et al. (2000) defined surface roughness as the perpendicular variation from an average plunge line of the slope surface within a distance equal to the radius of a falling block. As shown in Figure 10-17, the values of surface roughness (S) and the radius of the block (R) are used to calculate the maximum slope variation (θ_{max}), which influences the impact angle and thus the amount of energy dissipated during impact and the subsequent trajectory.

Accurate determinations of surface roughness are obtained by stretching a length of measuring tape equal to the radius of the rock block being tested along the ground surface and then measuring the maximum distance between the tape and the ground surface (Figure 10-17). Once several such measurements have been made, the largest vertical offset distance that is commonly observed defines the value of the surface roughness (S). However, these measurements are typically repeated for each slope zone on a test slope with multiple zones, and if a range in block sizes is being tested, the process is also repeated with different radius lengths according to the different sizes of rock blocks. Higgins et al. (2003) suggested that after a little practice, an estimate of surface roughness is possible without such a series of time-consuming measurements. They also suggested that on very smooth surfaces, such as paved surfaces, the surface roughness is related to the irregularity of the block shape, and in such cases the surface roughness is adequately represented by a value of 25% to 50% of the rock radius.

3.3 Slope Materials

The characteristics of the materials forming the slope surface over which the falling blocks travel influence and change the behavior of a bouncing or rolling rock. Figures 10-18 through 10-20 illustrate the tremendous variety of slope conditions encountered in rockfall hazard areas.

As discussed in Chapter 8, the impacts of falling or bouncing rock blocks on a slope are typically evaluated according to principles of Newtonian mechanics and generally assume constant values of the coefficients of restitution irrespective of the impact energy. Usually the coefficients of restitution used for trajectory calculations are estimated on the basis of a rough description of the slope material (e.g., rock, scree, loose soil), sometimes modified with additional information regarding surface roughness, surficial material properties such as degree of compaction, and the density of vegetation cover (Bozzolo and Pamini 1986; Azzoni et al. 1992; Giani 1992; Evans and Hungr 1993; Azzoni and de Freitas 1995; Bozzolo et al. 1998; Dorren, Berger, Le Hir, et al. 2006; Dorren, Berger, and Putters 2006).

FIGURE 10-18 Typical soil slope. (PHOTOGRAPH COURTESY OF THE CALIFORNIA DEPARTMENT OF TRANSPORTATION.)

For impacts of boulders on hard rock slopes, this approach may be justified (disregarding the fragmentation of the block). But for impacts of blocks onto granular materials found on the surface of most talus slopes below rockfall-prone areas, this assumption does not match reality. Plastic deformation processes develop as the striking boulder penetrates the ground surface. Therefore, the energy losses, and thus the coefficients of restitution, depend on the complex interaction of several factors, including block kinematics.

Although numerous systematic laboratory experimental investigations and some field experiments have been carried out to better understand the bouncing phenomena and to determine the most important parameters influencing block impacts, most of these investigations evaluated impacts on hard ground materials (Fornaro et al. 1990; Azzoni and de Freitas 1995). Surprisingly few investigations have investigated impacts with soft ground materials (Pichler et al. 2005; Kawahara and Muro 2006). For this reason, additional systematic laboratory experimental investigations were carried out at the Rock Mechanics Laboratory of the Swiss Federal Institute of Technology in Lausanne. These investigations used quarter- and half-scale models to evaluate the bouncing phenomena of rock blocks on soft ground materials and to investigate the influence of various impact parameters (Labiouse and Heidenreich 2009).

As a result of these experiments, Labiouse and Heidenreich (2009) made the following general observations:

- Vertical impacts on horizontal ground result in no rotation or lateral translation.
- Impacts of low incident velocity on stable (low inclination) slopes form no distinct crater, but the block gains lateral motion and begins to rotate. The ground behaves elastically and the block rolls along the surface.
- High-velocity impacts on similar surfaces cause the block to penetrate the ground material, creating a distinct crater and causing shearing and failure of the ground material. The crater rim hinders the rolling of the block, which thus tends to rebound normal to the surface.
- On steeper slopes, especially those near their limit of equilibrium, striking blocks lose less energy during the impact and rebound normal to the slope surface to a greater degree. These

FIGURE 10-19
Smooth rock slope.
(PHOTOGRAPH COURTESY OF GEOBRUGG.)

FIGURE 10-20
Rough rock and colluvium slope.
(PHOTOGRAPH COURTESY OF THE CALIFORNIA DEPARTMENT OF TRANSPORTATION.)

trends confirm observations made in the field by other authors (Wu 1985; Chau et al. 2002).

A thorough review of these experimental observations provided by Labiouse and Heidenreich (2009) demonstrates that block motion during and following impact is governed by mechanisms of penetration, sliding, and rotation, which act to a considerable degree in opposition to each other. For a given impact, conditions cause one or another of these mechanisms to predominate. The complexity of the rebound process clearly suggests that two constant coefficients of restitution and lumped-mass models of rock blocks are clearly insufficient for accurately predicting rockfall trajectories of bouncing and rolling rocks.

Vegetation growing on the slope also reduces falling-rock velocities and bounce heights. Forests have been historically employed as rockfall protection measures within the mountainous regions of Austria, France, Italy, and Switzerland (Berger et al. 2002; Brauner et al. 2005; Dorren et al. 2005). Recently, the extent of some of these forests has been reduced by a variety of logging and environmental changes, with a concurrent increase in rockfall hazards. These changes led to an extended series of field tests and modeling studies that have provided considerable quantitative data concerning the effect of forests on rockfall trajectories (Dorren et al. 2004; Perret et al. 2004; Berger and Dorren 2006; Dorren, Berger, Le Hir, et al. 2006; Dorren, Berger, and Putters 2006; Ciabocco et al. 2009). For example, one series of tests performed in France (Figure 10-21) showed that on a 38° slope, collisions with trees stopped 63% of the rocks or significantly altered their trajectory by reducing their velocity by 26% and bounce height by 33% (Dorren, Berger, Le Hir, et al. 2006; Dorren, Berger, and Putters 2006).

3.4 Weather Effects

Most rock-rolling field tests have been conducted in dry weather. Rain or poor visibility conditions have been judged to influence the validity of test results. However, published reports of field tests generally do not indicate antecedent weather conditions. Wet or frozen slope materials will change velocities and trajectories of falling rocks. Snow has been encountered in several naturally occurring rockfall events, but quantitative data are lacking concerning the reactions of falling blocks on snow-covered slopes. Since changing moisture conditions on the slope will influence the motions of the test rocks, consistent weather conditions during the series of field tests remove the effects of unmeasured influences on test results.

4. TEST ROCKS

Many rock-rolling research projects use natural rocks obtained from local stockpiles or quarries or from the rock face being evaluated (Figure 10-22). Others have reported much more consistent results from tests using manufactured concrete blocks with a standard shape (Arndt et al. 2009). A commonly used concrete block shape is a *cuboctahedron*, a polyhedron with eight triangular faces and six square faces (Figure 10-23). These blocks are used to provide a rounded-cube shape with fairly sharp edges and corners. The intent is to provide a

FIGURE 10-21 *(left)* Test slope with trees in Forêt Communale de Vaujany, France (Dorren, Berger, and Putters 2006).

FIGURE 10-22 *(right)* Test rocks obtained from local stockpile. (PHOTOGRAPH COURTESY OF THE CALIFORNIA DEPARTMENT OF TRANSPORTATION.)

FIGURE 10-23
Concrete cuboctahedron artificial rock; highly visible identifying markings are used to identify block orientations by imagery studies of rock-rolling behavior.
(PHOTOGRAPH COURTESY OF B. ARNDT.)

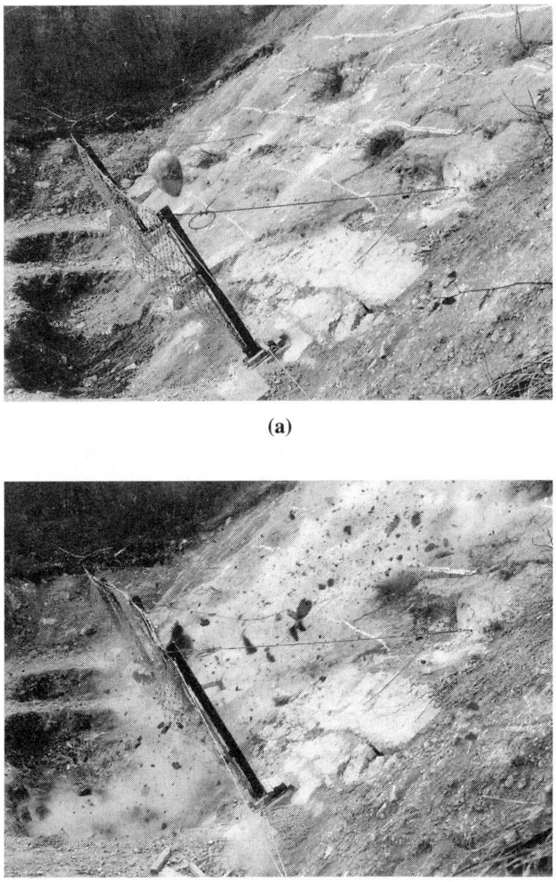

(a)

(b)

FIGURE 10-24
Rock durability:
(*a*) durable rocks produce impacts that are easy to evaluate and
(*b*) multiple impacts from rock fragments are difficult to evaluate.
(PHOTOGRAPHS COURTESY OF GEOBRUGG.)

useful analog to the characteristics of natural rocks but yet to have a more consistent shape and mass (Arndt et al. 2009).

The selection of test rocks is governed by the following considerations:

1. Are the selected rocks durable or are they likely to break up while falling, bouncing, or rolling down the slope?
2. What is the desired size, or range of sizes, of rocks?
3. What shapes of rock are desired?

The purpose of the field rock-rolling tests influences the answers to each of these questions. It is highly desirable to have the test rocks closely mimic the characteristic composition, hardness, shape, and weight of rock blocks so as to duplicate the likely results at the rockfall location being analyzed and mitigated. This feature is especially important when the objective of the field experiments is to test rockfall catchment fences or barriers. The design of these mitigation systems is heavily dependent on an accurate estimation of the kinetic energies, both translational and rotational, that will be imposed by the rock blocks and the bounce heights.

4.1 Durability

More durable rocks are less likely to break during tests and are able to be reused (Figure 10-24). Rocks that remain intact during the test sequence facilitate data reduction from the captured images, and thus analysis results are more reliable. Time is saved during the testing because considerable rock-measuring and -weighing efforts are avoided. When test rocks fragment during a test, some useful information may be obtained, but overall the observations are difficult to quantify.

Block fragmentation also affects how many test rocks are needed. Many experiments use at least some test rocks more than once. The decision to use any test rock more than once depends on the rock durability and the site characteristics of the test facility. In some cases, the equipment costs and time requirements associated with collecting test rocks and transporting them back to the top of the slope for reuse may be too great, and so rocks are only used once. If relatively few test rocks are used multiple times, it is possible to compare the variations in trajectories produced by an individual rock during these tests. Such test results are useful in quantifying the uncertainty and variation predicted by computer simulation models.

There is a real need to improve current understanding of fragmentation mechanisms because predicting the possible size, shape, and number of fragments generated by impact is fundamental to the design of more efficient protection systems. Fornaro et al. (1990) proposed a rockfall model that predicted block breakage at each impact according to a fragmentation energy threshold that was related to the type of rock and the geometry of the block. They proposed that this energy threshold be estimated from earlier experimental results that quantified the energy required to crush blocks with a hammer. When the kinetic energy of the impact reached this level, their model randomly divided the block into several fragments and distributed the remaining kinetic energy and subsequent trajectories among the fragments in proportion to their volumes.

Giacomini et al. (2009) undertook a series of 20 in situ field tests in a quarry located in Crevoladossola, in northwest Italy, to further investigate fragmentation mechanisms. Believing that planar discontinuities strongly influence the tendency for larger blocks to fragment upon impact, they tested two types of orthogneiss, with differing degrees of planar weakness, by dropping blocks from a crane onto the rock floor of the quarry (Figure 10-25). The quarry offered protection from the resulting rock fragments. Of particular interest was the role of the *impact angle*, the angle between the planar discontinuity and the quarry floor (Figure 10-26). Each rock type was subjected to 10 tests with blocks and test procedures selected to provide varying mass, falling height, and striking angle. Block volumes were 0.38 to 1.4 m^3 (15.5 to 57 ft^3) and drop heights were 10 to 40 m (33 to 132 ft), resulting in impact energies of 116 to 900 kJ (35 to 333 ft-tons), with most tests in the range of 200 to 400 kJ (74 to 148 ft-tons) (Giacomini et al. 2009).

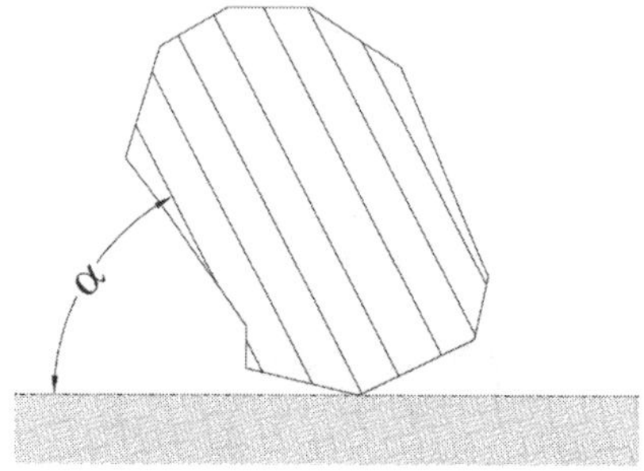

FIGURE 10-26
Impact angle α is defined as angle between planar discontinuity and struck surface (Giacomini et al. 2009).

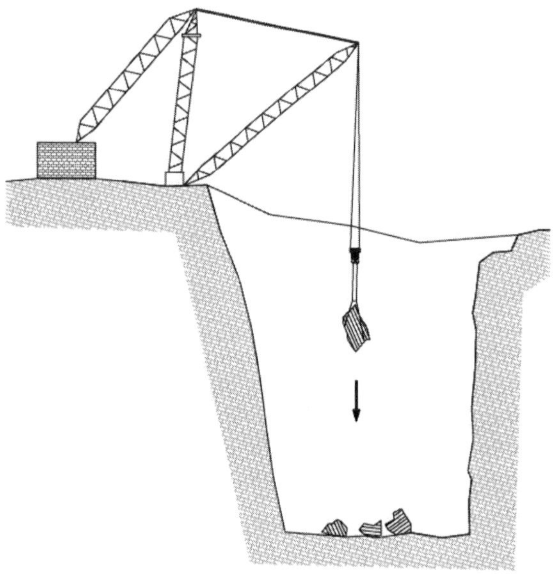

FIGURE 10-25
Schematic representation of test site and experimental procedure for evaluating block fragmentation upon impact (Giacomini et al. 2009).

The fall, impact, and rebound of each block were fully recorded with two digital high-speed and two video cameras positioned in the quarry. These cameras permitted determination of velocities of the block (before impact) and of the fragments (after impact) by backcalculations and also allowed the accurate measurement of the striking angle. Each block was painted before testing to facilitate identification of the fragments produced by the impact.

The rock type with the most pronounced planar discontinuity demonstrated that the striking angle is by far the most important factor controlling the fragmentation of rocks with such anisotropies. Three tests with striking energy of around 400 kJ (148 ft-tons) produced five fragments for a striking angle of 15°, eight fragments for an angle of 30°, and 14 fragments for an angle of 75°. Figure 10-27 shows a block after a 263-kJ (97-ft-ton) impact at an angle of 70°; breakage along the foliation planes is clearly visible. At a field location with a modest slope, this type of fragmentation following the impact of an already spinning block typically generates fragments with a tabular shape that tend to roll a considerable distance down the slope.

The impact energy (within the ranges tested) did not influence the degree of fragmentation for either type of rock used in the tests. Typically the

FIGURE 10-27
View of fragmented block after 263-kJ (97-ft-ton) impact at angle of 70°; breakage along foliation planes is clearly visible (Giacomini et al. 2009).

fragments were of similar size. Giacomini et al. (2009) also discovered that fragmentation events dissipated up to 60% of the intact block energy, with the degree of energy dissipation related to the normal coefficient of restitution.

4.2 Size

Selection of rock size is generally guided by the purpose of the rock-rolling test and practical considerations surrounding the test site itself. The primary goal of most field tests is to use sizes of rocks that are representative of and reproduce the conditions found at the design site that is to be mitigated. Typical logistical questions that affect the testing procedures are

- Will rocks need to be recycled to the top of the slope for reuse?
- Is there space at the top of the slope to stockpile test rocks?
- Is there room at the top of the slope for a large enough piece of equipment to manage and initiate the rockfall events?
- Will the rockfalls be initiated by rock scalers?[1]

Conditions at the base of the slope also impose some constraints on the conduct of field tests. Test sites generally have limited room to collect rolled rocks. Fewer large rocks can be rolled into a narrow runout area without their interfering with each other and affecting the distances that they travel, thus potentially biasing the observed roll-out distances.

Tests conducted to evaluate rockfall mitigation barriers commonly follow the technique developed by the California DOT (Caltrans) in the early 1990s. Small rocks are initially rolled into the barrier and the sizes of rocks are increased gradually until either the barrier system fails or the rocks are of a size that provides the maximum desired design energy (Duffy 1999). The majority of the barrier tests shown in Table 10-1 used rocks with average weights of 70 to 2,720 kg (~150 to 6,000 lb), with a few using larger rocks in the range of 4,090 to 5,900 kg (~9,000 to 13,000 lb) (Barrett and Pfeiffer 1989; Smith and Duffy 1990; Hearn 1991; Hearn et al. 1992; Arndt et al. 2009). One remarkable test in Japan used rocks as large as 18,600 kg (41,000 lb). Figure 10-28 illustrates a series of different-sized test rocks used in various field tests.

Many field tests are performed for the purpose of studying rockfall trajectories, calibrating computer models, and refining tangential and normal coefficients of restitution for various materials. Of the 37 tests in Table 10-1, 21 were performed for these purposes. The majority of these tests used rocks ranging from 70 to 2,720 kg (~150 to 6,000 lb).

Following an extensive series of field tests conducted to evaluate rockfall trajectories, Azzoni and de Freitas (1995) proposed a test methodology that used blocks of different shapes and volumes ranging between 0.1 and 2.0 m^3 (3.5 and 70 ft^3). Rocks of this proposed size weigh approximately 72 to 5,220 kg (160 to 11,500 lb). Giani et al. (2004) reported having rolled a 4.5-m^3 (158-ft^3) rock estimated to weigh 11,793 kg (26,000 lb). Broili (1973) studied the trajectory of a 10-m^3 (353-ft^3) rock with an estimated weight of 26,308 kg (58,000 lb) that was dislodged from a bluff with explosives.

Caltrans undertook field tests at Devils Slide, located along Highway 1 on the Pacific Coast just south of San Francisco, following a large rockfall event, and during the testing they had the opportunity to evaluate several 2.7- to 5.0-m^3 (96- to 180-ft^3) rocks estimated to weigh between 5,450 and 10,900 kg (12,000 and 24,000 lb) (Whitman and Duffy 2006).

Some field tests used rocks within a specific range of sizes. Bourrier et al. (2009) used an

[1]Rock scalers are individuals who, while suspended on ropes near the cliff face, dislodge potentially unstable rock blocks with the use of various tools such as pry bars (see Figure 10-33).

(a) (b) (c) (d) (e) (f)

FIGURE 10-28
Test rocks of varying size and weight used in various field tests: (*a*) rock weighing 231 kg (510 lb); (*b*) rock weighing 695 kg (1,533 lb); (*c*) rock weighing 1,344 kg (2,965 lb); (*d*) rock weighing 2,800 kg (6,173 lb); (*e*) rock weighing 6,170 kg (13,600 lb) [for scale, cable mesh grid spacing is 150 mm (6 in.)]; and (*f*) rock weighing 7,700 kg (16,975 lb).

(PHOTOGRAPHS *a*, *b*, AND *c* COURTESY OF THE CALIFORNIA DEPARTMENT OF TRANSPORTATION; PHOTOGRAPHS *d* AND *e* COURTESY OF GEOBRUGG; PHOTOGRAPH *f* COURTESY OF PROTEC ENGINEERING.)

average rock size of 0.8 m^3 (28 ft^3) for a series of tests that involved a total of 100 rocks. Dorren, Berger, Le Hir, et al. (2006) and Dorren, Berger, and Putters (2006) used rocks within the size range of 0.1 to 1.5 m^3 (3.5 to 53 ft^3) with an average size of 0.49 m^3 (17 ft^3) for a series of tests at two sites that evaluated the trajectories of a total of 202 rocks. The pioneering rockfall field study by Ritchie (1963) used rock scalers to initiate rockfall events that mostly involved hard basaltic blocks of various, but mostly modest, sizes. McCauley et al. (1985) also used rock scalers to dislodge rocks from slopes and thus were restricted to using rocks with longest-axis dimensions of less than 0.9 m (3 ft), but occasional rocks were larger, 1.2 to 1.8 m (4 to 6 ft).

During field tests for research by the Oregon DOT to design rockfall catchment areas (Pierson et al. 2001), each location was tested with a standard suite of 250 rocks. This suite included 100 rocks with an average diameter of 0.3 m (1 ft), 75 rocks with an average diameter of 0.6 m (2 ft), and 75 rocks with an average diameter of 0.9 m (3 ft). These average dimensions were measured along the longest axis, but the rocks tended to be equidimensional. The dimensions for each size category ranged within ±0.15 m (6 in.).

4.3 Shape

Although Ritchie (1963) reported that the size and shape of a rock have little bearing on its falling and rolling characteristics and that large rocks tend to stay close to the face of the cliff, subsequent observations have yielded generally different conclusions. Many rockfall analysis procedures use idealized 3-D geometric shapes to represent, as closely as possible, the shapes of actual rocks. As discussed in Chapter 8, shape imparts an important control to the moment of inertia and hence the rotational kinetic energy of a rotating individual rock block. On steeper slopes, rocks of all shapes generally travel down the slope without stopping. From a practical standpoint, particularly on slopes between 34° and 40°, the most desirable rock shape for field tests is roughly equidimensional because these rocks tend to produce more consistent results.

Azzoni et al. (1995) observed that spheroidal blocks move faster than discoidal or tabular ones and that the velocity achieved by bigger blocks is larger than that of smaller ones on the same slope, because the larger rocks are less affected by slope roughness. Bourrier et al. (2009) selected boulders from a quarry and chose rocks as spherical as possible. Observations during field tests conducted in California revealed that tabular rocks initially rotated around the longest axis for a short distance and then, as their velocity increased, began rotating like wheels around their shortest axis (Smith and Duffy 1990).

Most field tests measure and record the dimensions of the three principal axes of each rock (Figure 10-29). When possible, the measurement axes are marked for reference when they are first measured. The measurements are typically repeated after the rock has been rolled down the slope, and comparisons allow for estimates of volume change or weight loss during the rockfall event. Repeated measurements are especially useful if the rock is reused in later experiments; reuse of rocks has been undertaken during several reported field tests where the availability of suitable rocks was limited and the difficulty of transporting them to the top of the slope was not too great.

The measurement of the three principal axes allows for the field-test team to estimate weight and the moment of inertia of each rock used in the tests (Chapter 8). Azzoni et al. (1995) quantified the shape of rocks by computing the ratio between the volume of the rock and the

FIGURE 10-29 Measuring three principal axes of a rock.
(PHOTOGRAPH COURTESY OF THE CALIFORNIA DEPARTMENT OF TRANSPORTATION.)

minimum sphere included within the block. This ratio was referred to as the shape coefficient. Both the size and shape of blocks were found to influence the maximum and the mean velocity achieved during a rockfall event (Azzoni and de Freitas 1995).

4.4 Weight

Ideally, the weight of each test rock is known before it is subjected to a rock-rolling experiment. Without this information, energy calculations are not possible. Two options are available for obtaining the weight: (*a*) estimation based on volume and specific gravity information or (*b*) direct weighing of each rock.

To estimate rock weight, an approximation of the rock volume is needed. This calculation is accomplished by measuring the three principal axes of the rock (Figure 10-29) and assigning a shape category, such as discoidal, tabular, or spherical. With the appropriate geometric formula for that assigned shape, a close approximation of the rock volume can be estimated (Bourrier et al. 2009). Once the volume has been determined, information on the specific gravity, or density, of the rock material allows for calculation of the weight. Specific gravity can be estimated from published data based on the rock type or can be determined in the laboratory. The weight (or mass) of the rock can then be estimated by multiplying the volume by the specific gravity and by the unit weight of water. This estimating procedure typically produces values within 10% to 15% of the true rock weight (Smith and Duffy 1990). Checking a few actual weights, however, allows for an evaluation of the accuracy of these estimates and may identify the need for refinement of this estimating process.

Where possible, the best practice is to accurately weigh each test rock (Figure 10-30), which can be accomplished with a spring scale or possibly a portable truck weighing unit. More commonly, a load cell is attached to the bucket of a front-end loader and the rock is lifted in a harness attached to the load cell.

(a) (b)

FIGURE 10-30 Methods of weighing test rocks: (*a*) spring scale; (*b*) load cell attached to bucket of heavy equipment offers quick and easy method of weighing individual rocks.
(PHOTOGRAPH *a* COURTESY OF GEOBRUGG; PHOTOGRAPH *b* COURTESY OF THE CALIFORNIA DEPARTMENT OF TRANSPORTATION.)

FIGURE 10-31
Highly visible identifying marks applied to each rock allow for accurate identification of individual rocks during testing process.
(PHOTOGRAPH COURTESY OF THE CALIFORNIA DEPARTMENT OF TRANSPORTATION.)

4.5 Identification

Ideally the surface of each rock has some highly visible identifying marks (Figure 10-31). These marks help identify the rock throughout all phases of the test. They also permit rocks to be stored together in a rock pile and for a particular rock to be located when it is needed for a test. The markings can also quickly identify rocks that have been damaged by impacts and require reweighing.

Markings are also useful when angular velocity is to be determined from captured images (Figure 10-23). Berger, Dorren, and colleagues painted rocks with a biodegradable colored powder to facilitate identification of the rock trajectories on digital film as well as to associate particular impacts on the ground or trees (Berger et al. 2002; Dorren, Berger, Le Hir, et al. 2006; Dorren, Berger, and Putters 2006) (Figure 10-32). In addition, as noted earlier, it is helpful for the locations of the principal axes to be marked on the rock surface when they are first identified. Following each rock-rolling event, these axis markings make remeasurement quicker and reduce interpretative errors.

5. ROCK-ROLLING INITIATION

The method of initiating the rockfall is important because it may produce distinctly different values of initial velocity and thus materially affect the subsequent rockfall process and the calculation of resulting energy values. The initiation method is documented in the record of the field test.

FIGURE 10-32
Prominent surface markings assist in measuring rate of rotation of rock when rotating rocks are observed in images taken during rock-rolling event. Labels t_0 to t_7 identify times that rebounding rock occupied each position and orientation (Berger et al. 2002).

Methods for inducing the rockfall can vary from having a rock scaler pry a rock loose from a slope (Figure 10-33) to constructing a cable system for carrying the rocks up a slope (Figure 10-34) and dropping the rocks from any height as desired. Broili (1973) analyzed rocks dislodged by blasting on a cliff face. Arndt et al. (2009) reported the use of a helicopter to drop rocks onto the upper sections of a very tall rockfall chute above Interstate 70 near Georgetown, Colorado, at a location where more traditional means of access, or the means of safely initiating rockfall with blocks consistent with the size of those characteristic of the site, was all but impossible. More commonly, rocks are pushed or dropped off the top of the test slope by a front-end loader or excavator (Figure 10-35).

To ensure the safety of all personnel involved in the field trial, when the rock scaler or equipment operator is ready to dislodge the rock, he notifies the ground crew. The launch crew and ground

FIGURE 10-33
Rock scalers initiating rockfall by prying rocks off slope by hand.
(PHOTOGRAPH COURTESY OF THE CALIFORNIA DEPARTMENT OF TRANSPORTATION.)

FIGURE 10-34 Transporting rocks upslope via cable lift. (PHOTOGRAPH COURTESY OF GEOBRUGG.)

FIGURE 10-35 Initiating rockfall by dropping rock with excavator; circles outline falling and fallen blocks. (PHOTOGRAPH COURTESY OF THE OREGON DEPARTMENT OF TRANSPORTATION.)

crew are in constant communication so that all actions, including all recording and safety procedures, are in place and set for each rock-rolling event. When all is ready, the filming begins before the crew starts to dislodge the rock.

6. NUMBER OF ROCK ROLLS

Since the 1960s, thousands of rocks have been rolled (Table 10-1). The number for each test naturally varies on the basis of the goals of the study and the number of tests deemed necessary to suitably characterize the variability of the rock-rolling trajectories and energies in question. This decision also affects the duration of the test series and the research costs.

Ritchie (1963) rolled a few hundred rocks on various highway and quarry slopes during his landmark study of rockfall trajectories. The North Carolina DOT undertook some early rockfall field tests at two sites to calibrate some computer modeling efforts, with 40 rocks used at one site and 106 rocks at the other. At about the same time, Elston et al. (1978) rolled 350 rocks at a single location to evaluate bounce heights and rockfall impacts for the Canadian railway. In 1985, Caltrans performed rockfall trajectory and fence testing on various cut slopes throughout the state (McCauley et al. 1985). Twelve sites were used, and a total of 223 rocks were rolled, for an average of 16 rocks tested at each site. Wyllie (1991) rolled about 70 rocks initiated by rock scalers at a slope with difficult access during a barrier test above a Canadian railway. Evans (1989) performed three tests for evaluating bench effectiveness in mines and rockfall modeling accuracy; 20 rocks were rolled during each of the first two tests, and 10 rocks were rolled during the third test.

The first rock-rolling tests into flexible cable net rockfall barriers by Caltrans used a single site and rolled a total of 76 rocks during three tests of three different barriers (Smith and Duffy 1990). In 1991, the Colorado DOT used a total of 70 rocks during tests of a flexible barrier at a test site located near Rifle, Colorado (Hearn 1991; Hearn et al. 1992). In 1994 and 2001, the Oregon DOT performed extensive rockfall catchment studies. During the 1994 testing, 2,790 rocks were rolled, whereas the 2001 tests used a remarkable 11,250 rocks (Pierson et al. 1994, 2001). During the 2001 tests, 250 rocks were rolled at each location (Pierson et al. 2001). More recently, Dorren, Berger, Le Hir, et al. (2006) and Dorren, Berger, and Putters (2006) performed two rock-rolling tests, involving 100 and 102 rocks, respectively, to study the effect of trees on rockfall.

Thus, although there appears to be wide variability in the number of rocks rolled in individual tests, there are some patterns that transcend the test purposes and logistic factors. Many practitioners have found that using 15 to 30 rocks per test sequence provides suitable data. In the early 1990s, Caltrans came to this realization and developed a rock-rolling guideline in which a minimum of 20 rocks are rolled during each test sequence (Duffy 1999). Azzoni and de Freitas (1995), as a result of considerable test experience, developed a methodology for rock-rolling tests and recommended a minimum of 30 rocks per test. The very large number of rocks rolled by Pierson et al. (1994, 2001) was the result of a specific research goal to statistically validate a general design guide for a variety of slopes and conditions.

Bourrier et al. (2009) compared the results of detailed trajectory observations from 100 rock-rolling tests with the predicted trajectories produced by more than 1,000 simulations using a 3-D rockfall model. They noted some differences between the observed and simulated distributions. The maximum values of distance traveled or bounce height were always greater in the simulations. Bourrier et al. suggested that this might be largely the result of the limited number of field observations in which rare extreme events occur. They therefore do not represent the full asymptotic distribution that would have been obtained from a much larger number of experiments. However, with 100 observations, the global shape of the distribution and the most probable values were well defined. Because it is economically infeasible in many cases to undertake extremely large numbers of field experiments, appropriately calibrated computer simulations, with the calibrations reflecting the field trials, are likely to provide better values of extreme events. Thus, results from both field trials and computer simulations are used to provide collaborative support for many rockfall mitigation decisions.

7. ESTABLISHING SPATIAL REFERENCE SYSTEM

A spatial reference system is used to allow analysis of the trajectories and motions of moving rock blocks recorded on film, video, or digital images. These reference systems may use natural features, survey targets, stadia rods, or a grid delineated on the slope. A suitable combination of these features is selected on the basis of conditions found at each site, and they are accurately located and displayed for easy reference. The reference systems typically are designed to provide accurate spatial referencing at areas of critical interest, but most systems provide at least some general spatial reference throughout the test slope. Detailed measurements of rockfall trajectories are possible when the selected reference features are clearly visible on the recorded images.

Motion analysis is used to calculate rockfall translational velocities and bounce heights and distances and to track the trajectory of each block down the slope. The layout and pattern of the reference systems used by various researchers include many variations. Typically, a grid is defined on the slope with high-visibility flagging (red and white or yellow) or athletic field marking chalk. A grid is very useful in measuring the rock movement down and across the slope. To capture bounce heights, stadia rods are sometimes placed on the slope. The implementation of a reference system is facilitated with a detailed plan that specifies the proposed design (Figure 10-36).

Ritchie (1963) used high-visibility flagging tape to provide reference lines on the slope with various spacings. Tape was placed to provide horizontal reference lines with a 3-m (10-ft) spacing at the base of the cliff or slope. An additional vertical reference line was provided by a flagging tape held aloft by a small helium-filled weather balloon. The ground end of the tape was anchored to a rock 6 to 8 m (20 to 25 ft) from the base of the cliff or slope. The balloon end was pulled toward the top of the cliff so that the reference tape was essentially parallel to the face of the cliff. Ritchie also used a movable catchment ditch and placed it below the test slope to measure impact locations as well as ditch effectiveness (Figure 10-37).

To fully use the collected imagery, reference lines are placed perpendicular to the slope axis on the slope and especially in the runout or impact zone. When rockfall barriers are tested, the reference lines are typically spaced at 3-m (10-ft) intervals throughout the anticipated runout or impact zone and for a distance of at least 15 m (50 ft) up the slope above this zone. Grids typically have a larger spacing of 6 to 9 m (20 to 30 ft) near the top of the slope. Additional reference lines are placed at even greater intervals up to the launching point. These spacing dimensions are often modified to reflect the slope geometry of a particular site, however. Reference markers spaced more closely provide greater detail in areas of special concern. In some cases, longitudinal lines are placed running up and down the slope; these lines provide for

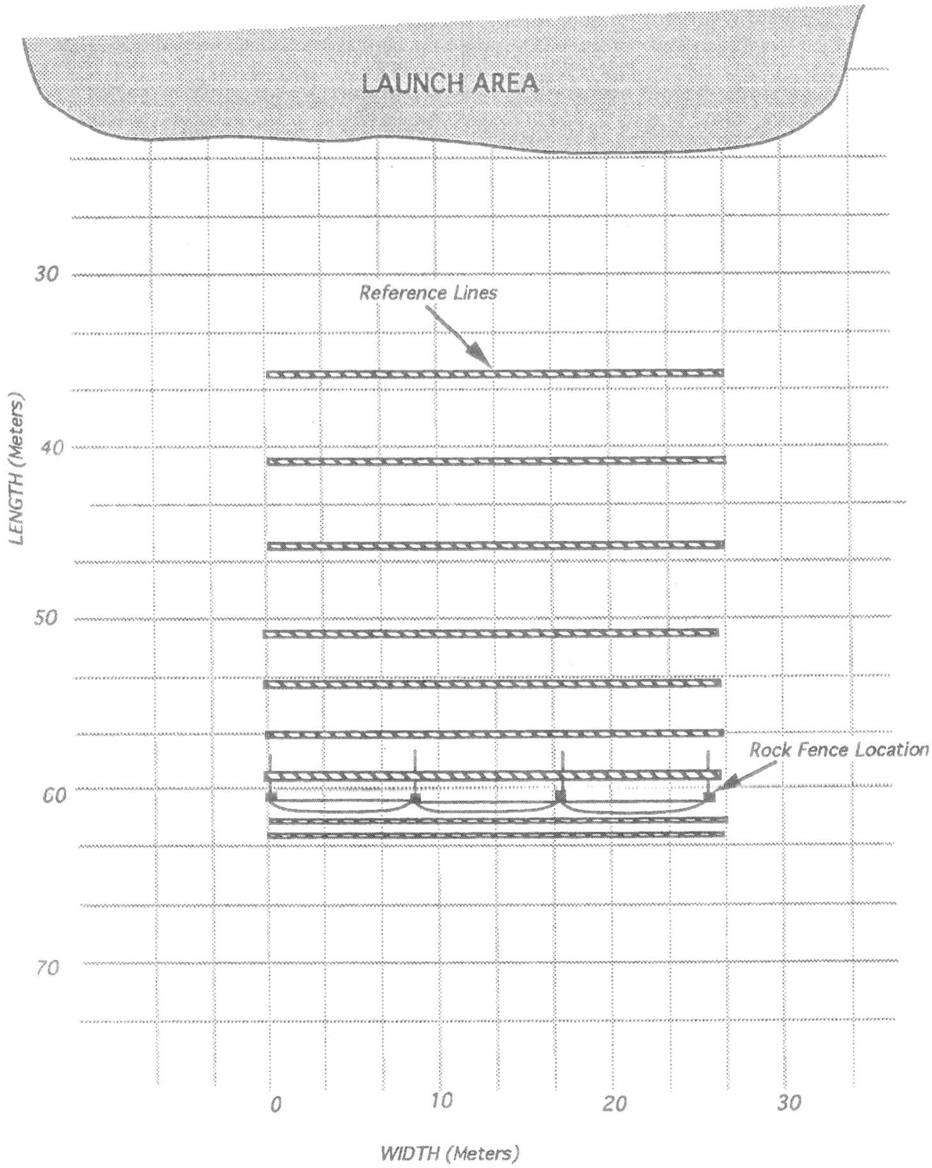

FIGURE 10-36 *(above)*
Plan of Shale Point test site, California, showing proposed placement of reference lines aligned perpendicular to slope axis. To make full use of subsequent collected imagery of tests, reference lines are located on slope and especially in runout–impact zone.
(IMAGE COURTESY OF THE CALIFORNIA DEPARTMENT OF TRANSPORTATION.)

FIGURE 10-37 *(right)*
Movable catchment ditch placed below test slope to measure impact locations as well as ditch effectiveness (Ritchie 1963).

the measurement of lateral dispersion of the block trajectories. Figures 10-38 through 10-41 show examples from several recent field tests.

In addition, stadia rods at least 1 to 2 m (3 to 6 ft) long, depending on the slope, are placed on the slope to aid in bounce height analysis. These stadia rods are often placed randomly because of slope constraints, but an effort is made to place rods at or near the reference lines (Figure 10-42).

Rockfall trajectory studies sometimes use similar reference grids but typically include marks on slope features such as trees and rock outcrops. Elston et al. (1978) set up a grid across the roadway at the base of the slope at 0.6-m (2-ft) intervals. Dorren et al. (2004) painted the test rocks with biodegradable paint and surveyed the impact points on the ground and in trees and on outcrops, now stained with the paint, to map the trajectory after each roll.

Paronuzzi (2009) described field investigations of a rockfall event that involved two blocks that traveled across mostly open fields. The sequences of impact marks caused by the two blocks were measured and analyzed (Figure 10-43). With the aid of subsequent computer simulations, the trajectory of each block was analyzed in detail to provide information on trajectory bounce heights, impact angles, and coefficients of restitution. Similar analyses may be possible at some test sites if the appropriate impact details are measured and recorded. However, this data collection may delay the timely completion of a test series and is not undertaken in most cases.

When terrestrial LiDAR is used to obtain a detailed 3-D model of the slope geometry, sometimes reflective marks are painted onto prominent slope features or reflective targets are attached to the slope (Figure 10-44). These highly reflective locations are readily observed in the LiDAR image and in any subsequent video records. If their locations are accurately determined, they can provide additional spatial control to observations of rockfall tests.

8. FIELD DATA RECORDING

Keeping accurate records during all phases of the rock rolling is extremely important. Those responsible for initiating a rock-rolling event, as well as the ground crew personnel located throughout the test site, are collectively responsible for recording

FIGURE 10-38
Reference system composed of horizontal lines placed across relatively uniform slope where large amounts of lateral dispersion are not expected.
(PHOTOGRAPH COURTESY OF THE CALIFORNIA DEPARTMENT OF TRANSPORTATION.)

FIGURE 10-39
Reference system showing more closely spaced lines at points of special interest, such as near rockfall catchment fence.
(PHOTOGRAPH COURTESY OF THE CALIFORNIA DEPARTMENT OF TRANSPORTATION.)

FIGURE 10-40
Reference system including additional lines running up and down slope to measure amount of lateral dispersion.
(PHOTOGRAPH COURTESY OF THE CALIFORNIA DEPARTMENT OF TRANSPORTATION.)

FIGURE 10-41
Reference lines are typically spaced at 3-m (10-ft) intervals throughout anticipated runout–impact zone and for distance of at least 15 m (50 ft) upslope above this zone.
(PHOTOGRAPH COURTESY OF THE CALIFORNIA DEPARTMENT OF TRANSPORTATION.)

FIGURE 10-42
Stadia rods provide reference measurements for bounce height analysis.
(PHOTOGRAPH COURTESY OF THE CALIFORNIA DEPARTMENT OF TRANSPORTATION.)

FIGURE 10-44
Bright paint marks placed on prominent rock outcrops on slope assist spatial referencing of terrestrial LiDAR images and video records of rockfall tests.
(PHOTOGRAPH COURTESY OF THE CALIFORNIA DEPARTMENT OF TRANSPORTATION.)

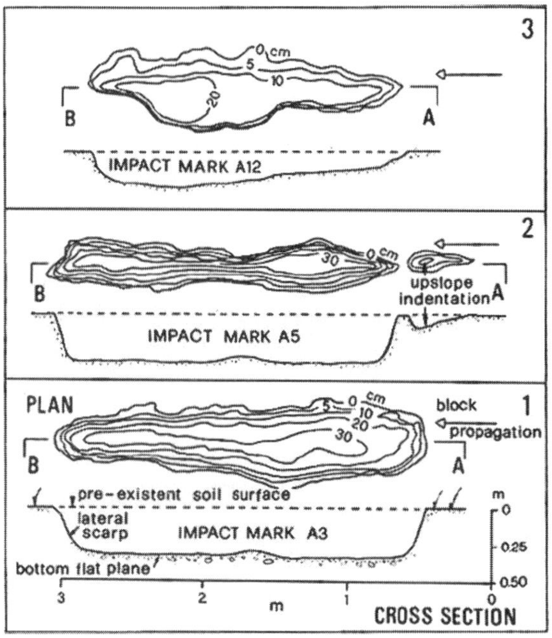

FIGURE 10-43
Planimetric views and longitudinal cross sections of three typical impact marks left by one block as it traveled across grassy field where slope was 15° to 25° (Paronuzzi 2009).

the various aspects of each event. This overlapping responsibility for data recording has obvious advantages in ensuring that important data are not missed, but it also requires careful coordination and a clearly defined delegation of individual responsibilities. At the start of each rock-rolling event, the information describing the rock block is recorded in a data form such as that shown in Figure 10-45. As noted previously, the rock is prepared by applying predetermined high-visibility markings to assist with motion evaluations using the recorded images of the rockfall event. If rock scalers initiate the rockfall, their location is recorded with a range finder and clinometer to determine slope length and slope angle from the point of initiation.

Those operating the cameras recording a rock-rolling event are responsible for identifying the subject of the segment by recording the necessary sequence information before the initiation of each event. This information is frequently written

DATA REQUIRED BEFORE EACH ROCK-ROLLING EVENT

Site ID	Roll ID	Rock ID	Launch Location	Launch Height	Method of Initiation	Initial Values – prior to event					
						Principal Axes			Shape	Weight (Estimated)	Weight (Actual)
						X	Y	Z			

FIGURE 10-45
Data form used to record rock block descriptions before initiation of rock-rolling event.
(FORM COURTESY OF THE CALIFORNIA DEPARTMENT OF TRANSPORTATION.)

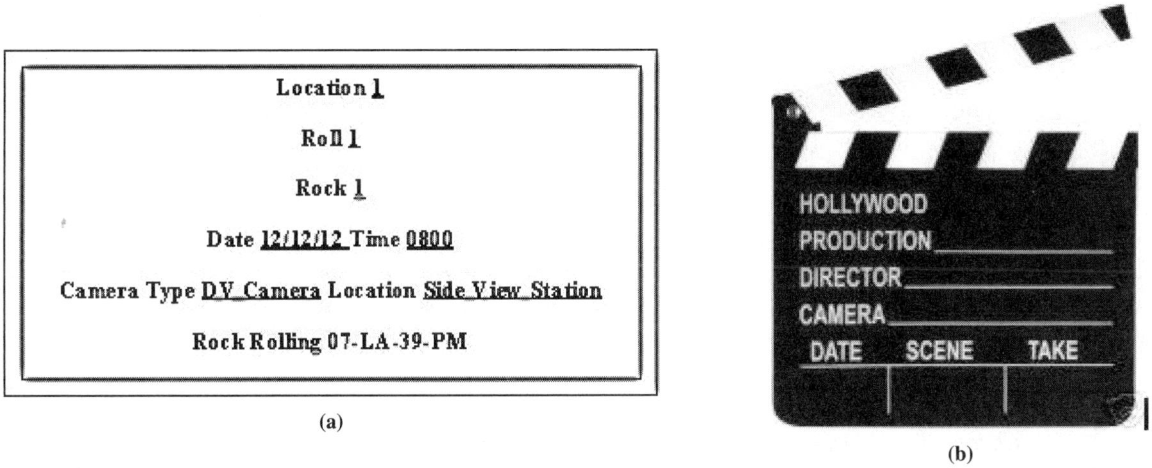

FIGURE 10-46
Examples of film data placards.
(IMAGE COURTESY OF THE CALIFORNIA DEPARTMENT OF TRANSPORTATION.)

on a placard (Figure 10-46) and then the camera records the placard, which identifies, at a minimum, the site location, rock-rolling event number, rock number, date, time, camera, and camera location. Alternatively, a designated person holds up a sign with a number that identifies the rock-rolling event that is visible to the equipment operators or scalers initiating the rockfall and to the film crew. This identification number is recorded on the film of the event (Figure 10-47).

Although it is critically important to document each rock-rolling event on film or by video recording, it is equally important to be able to distinguish between the rocks used in each event. Considering that some tests include hundreds of rolled rocks, keeping good records is imperative.

FIGURE 10-47
Rock-rolling event information recorded on camera image.
(IMAGE COURTESY OF GEOBRUGG.)

All personnel are in radio communication and rock rolling only begins once all parties are confirmed to be ready.

During the actual rock-rolling event, one person is responsible for accurately timing the entire event with a stopwatch. The time is recorded from the initiation of the event through the initial impact in the rollout area and, if possible, to the final stopping point. This record is helpful in estimating velocity during the roll and deceleration rates in the fallout area and as a reference during later image studies. At the end of the rock-rolling event, if any damage is noted, the rock is remeasured to provide an estimate of its postevent weight or is reweighed by using a load cell. The initial impact location and the rollout distance are identified relative to the grid either by surveys or by using a measuring tape. All these postevent data are recorded on a data sheet (Figure 10-48).

Because the data recorded before the initiation of the rock-rolling event are collected at the top of the slope and the postevent data are recorded at the base of the slope, two separate data forms are required (Figures 10-45 and 10-48). Careful coordination and clear delegation of responsibilities are key to the success of field trials. Most practitioners and researchers develop a series of data forms to capture the necessary information in an organized and consistent manner. Figures 10-49 and 10-50 are examples of forms used by Caltrans during rock-rolling experiments to evaluate rockfall trajectories and the effectiveness of chain-link fences and berms (McCauley et al. 1985). Smith and Duffy (1990) developed and used the form shown in Figure 10-51 to aid in data collection and organization of tests geared specifically for testing rockfall barriers.

9. IMAGE DATA COLLECTION FOR ROCK MOTION RECORDING

Captured sequential images of a rock-rolling event are used primarily to determine translational and angular velocities of the falling rock blocks, but other rock and rockfall characteristics may be obtained, such as rock durability and bounce height (Figure 10-52).

Following rock-rolling field trials, information concerning both translational and angular velocities is most readily obtained by using a reduced-speed view of the rockfall events. Traditionally, this procedure was accomplished by using a film camera, which was run at very high speed during the recording. With visible reference lines on the slope and film-editing equipment capable of achieving frame-by-frame control, accurate measurements were obtained of the time required for the rocks to travel between reference lines and the time required for a rock to spin one revolution, as illustrated in Figure 10-32. This procedure was

DATA REQUIRED FOLLOWING EACH ROCK-ROLLING EVENT

Site ID	Roll ID	Rock ID	Time to Impact	Time to Stop	Impact Location	Maximum Roll Out	Final Values – after event					
							Principal Axes			Shape	Weight (Estimated)	Weight (Actual)
							X	Y	Z			

FIGURE 10-48 Data form used to record rock block descriptions and other relevant information following rock-rolling event.
(FORM COURTESY OF THE CALIFORNIA DEPARTMENT OF TRANSPORTATION.)

INDUCED ROCKFALL DATA SHEET

Date	____________
Location	____________
Reviewed by	____________
Weather Conditions	____________
Slope Height	____________ft
Slope Angle	____________degrees
Width of Toe of Slope to Fence	____________ft
Width from Toe of Edge of Traveled Way	____________ft
Width of Paved Shoulder	____________ft
Width of Unpaved Shoulder	____________ft
Reverse Slope at Grade	____________degrees
Fence Height	____________ft
Vertical Clearance Under Fence	____________ft
Fence Reinforcement	_____yes_____no
Overall Effectiveness	____________0–25%
	____________25–50%
	____________50–75%
	____________75–100%
Description of Slope Conditions	____________________
	____________________
	____________________
	____________________
	____________________
Description of Fence	____________________
	____________________
	____________________
	____________________
	____________________

FIGURE 10-49
Data sheet for induced rockfall experiments.
(IMAGE COURTESY OF THE CALIFORNIA DEPARTMENT OF TRANSPORTATION.)

TEST DATA SHEET

Rock Size	______________ft
Rock Shape and Roundness	______________________
	______________________
Mode of Travel	______________Roll
	______________Bounce
	______________Slide
	______________Freefall
Length of Roll in Time t	______________ft
Time	______________sec
Velocity	______________ft/sec
Distance Past Toe of Slope	______________ft
Height of Impact in the Fence	______________ft
Rock Stopped by Fence	______ yes ______ no
Rock Under Fence	______ yes ______ no
Rock Over Fence	______ yes ______ no
Fence Damaged by Rock	______ yes ______ no
Comments:	______________________________
	______________________________
	______________________________
	______________________________
	______________________________

FIGURE 10-50
Test data sheet.
(IMAGE COURTESY OF THE CALIFORNIA DEPARTMENT OF TRANSPORTATION.)

FIGURE 10-51 Data sheet for rockfall restraining net test.
(IMAGE COURTESY OF THE CALIFORNIA DEPARTMENT OF TRANSPORTATION.)

ROCK RESTRAINING NET TEST DATA SHEET

COMPANY: ______________________

TEST OR ROLL #: ______________________

ROCK I.D.: ______________________

ROCKFALL DATA

1. DIMENSIONS:	______METERS (m)	______FEET (ft)
2. VOLUME (V):	______CU. METERS (m^3)	______CU. FEET (ft^3)
3. SHAPE:	______________________	
4. UNIT WEIGHT:	______Kg/m^3	______lb/ft^3
5. WEIGHT (W):	______Kg	______lb
6. MODE OF TRAVEL BEFORE IMPACT:		______ROLL
		______BOUNCE
		______SLIDE
		______FREEFALL
7. TIME OF TRAVEL:		
a) TIME (t_1)		______sec
b) LENGTH OF ROLL (d_1) in TIME t_1	______m	______ft
c) VELOCITY (v_1)	______m/sec	______ft/sec
d) TIME (t_2)		______sec
e) LENGTH OF ROLL (d_2) in TIME t_2	______m	______ft
f) VELOCITY (v_2)	______m/sec	______ft/sec
8. ENERGY:		
a) MASS (m = W/g)	______Kg/m/sec^2	______lb/ft/sec^2
b) VELOCITY (v = d/t)	______m/sec	______ft/sec
c) KINETIC ENERGY (KE =½mv^2)	______Kilojoules	______ft-tons

REMARKS: ______________________

FIGURE 10-52 Reconstruction of block position at different time intervals taken from video camera recording and referred to global (*X, Y*) and local coordinate systems. Labels P_1 to P_8 identify sequential position of falling rock; labels S_1 through S_7 identify sequential sections of trajectory (Giani et al. 2004).

costly, as the film and its processing were expensive. The use of film also affected the field-testing procedures, because loading the camera was time-consuming and the film processing required several days, so velocity estimates were not immediately available.

Recent advances in digital imaging technology have made the use of movie cameras and film increasingly uneconomical and impractical. As described in Chapter 7, modern high-speed video technology makes it possible to record rockfall events with excellent resolution and to immediately view the results on standard video monitors. Digital cameras also offer better exposure control, and digital image enhancement allows images taken under less-than-optimum light conditions to still be reliably analyzed. Digital cameras can be externally triggered, so recording an experimental rockfall event is easier and safer. In addition, video images can be easily reproduced and disseminated as video clips.

It is extremely important to obtain high-quality images of the rock-rolling events. Without this information, the tests cannot be fully analyzed. Poor-quality imagery may obscure important details. The success of the field trial depends on recording the entire rock trajectory. The path followed by a bouncing or rolling rock is highly unpredictable. Although Ritchie (1963) used a single 16-mm camera that viewed the rockfall from the side to observe the rockfall trajectories, current practice generally uses multiple cameras. Typically at least two different camera viewpoints are used, and four or more cameras are preferred (Figure 10-53). Multiple cameras permit the recording of more details from different angles

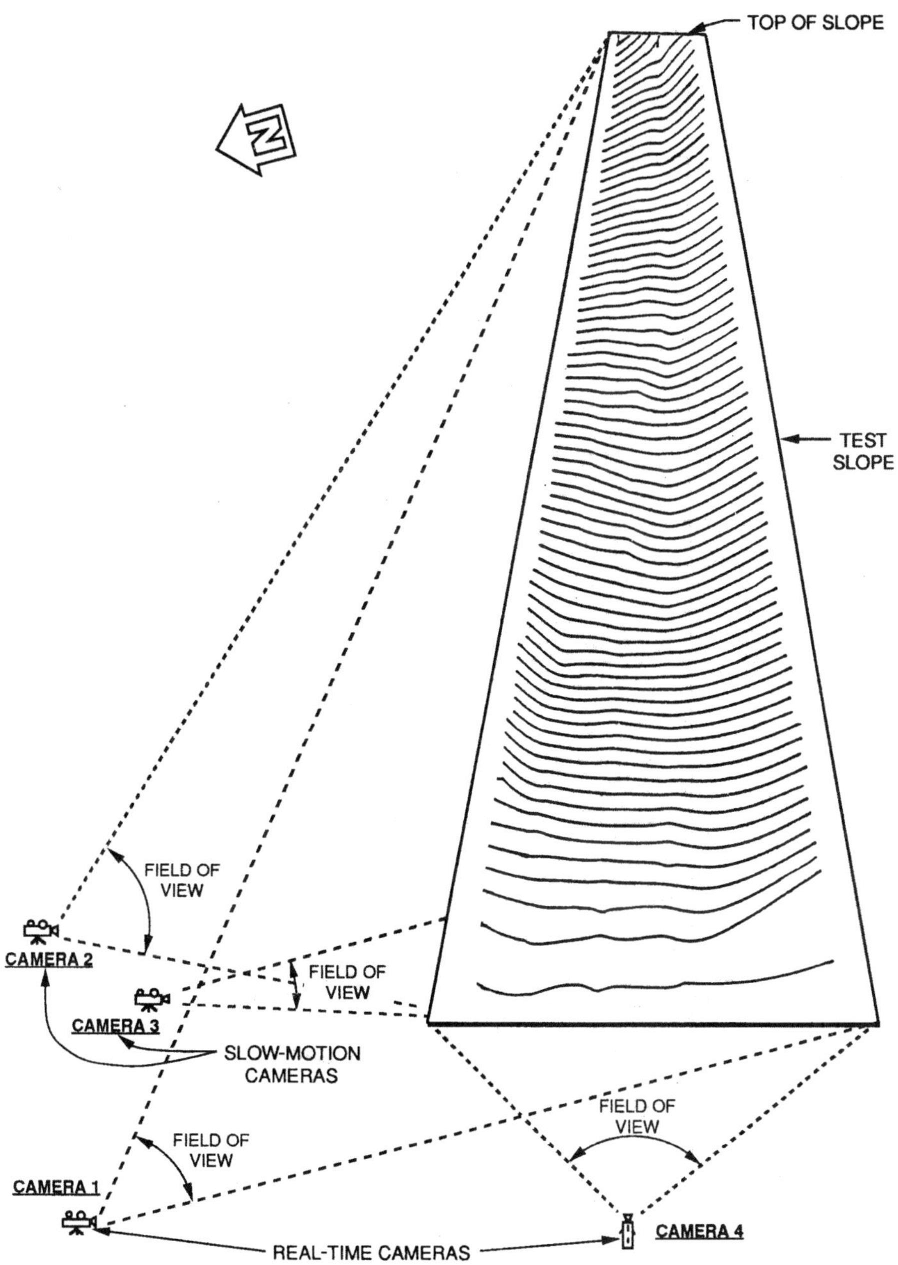

FIGURE 10-53 Caltrans typically uses four cameras during rock-rolling field trials to capture two side views, one oblique view, and one front view. (IMAGE COURTESY OF THE CALIFORNIA DEPARTMENT OF TRANSPORTATION.)

FIGURE 10-54
Planning for establishment of appropriate camera locations.
(PHOTOGRAPH COURTESY OF THE CALIFORNIA DEPARTMENT OF TRANSPORTATION.)

and also provide some redundancy in the event of camera failure or damage. Establishing appropriate camera locations often involves careful planning (Figure 10-54).

During rockfall barrier tests conducted by Caltrans, four cameras were used (Duffy 1990, 1992; Duffy et al. 1990; Smith and Duffy 1990). At least one oblique view of each rockfall event was recorded. One camera was a "sweep camera" that followed the rock down the slope with a minimum field of view and recorded 15 m (50 ft) of slope surrounding the rockfall. The other cameras were stationary, some focusing on side views of the barrier being tested (Figure 10-55). Although Caltrans used one or more cameras to capture images from the base of the slope, other investigators have chosen to gather images looking directly

FIGURE 10-55
Camera located to provide side view along barrier (location highlighted by rectangle) for evaluation of impacts by rock blocks on rockfall fence.
(PHOTOGRAPH COURTESY OF GEOBRUGG.)

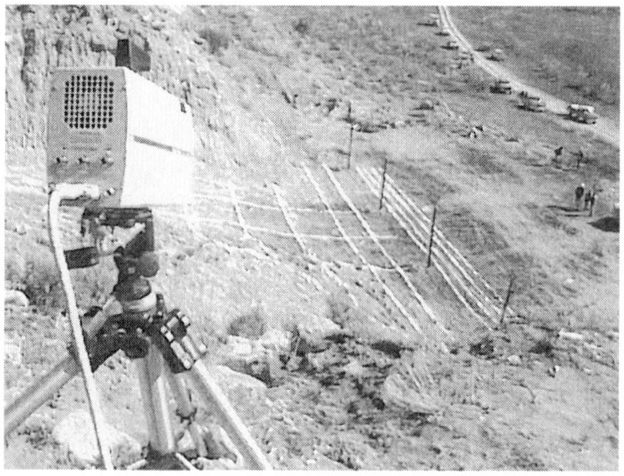

FIGURE 10-56
Camera oriented to produce view directly downslope (Andrew et al. 1998).

down the slope (Figure 10-56). Placement of the camera near the top of the slope provides it with some additional protection from falling rocks.

ACKNOWLEDGMENTS

The authors are grateful for contributions from Richard Andrew, Bruno Haller, Wade Hoon, Howard Ingram, Michael Salisbury, Raymond Spang, and Robert Thommen. The chapter topics significantly benefited from their help. Critical reviews were provided by Larry Pierson, Landslide Technology, and Tim Beck, California Department of Transportation (retired).

REFERENCES

ABBREVIATION

FHWA Federal Highway Administration

Andrew, R. D., D. A. Fry, and R. E. Bookwalter. 1998. *Field Testing and Evaluation of Various Rock Fall Control Systems*. Chama Valley Productions, LLC, Chama, N. Mex.

Arndt, B., T. Ortiz, and A. K. Turner. 2009. *Transportation Research Circular E-C141: Colorado's Full-Scale Testing of Rockfall Attenuator Systems*. Transportation Research Board of the National Academies, Washington, D.C.

Azzoni, A., and M. H. de Freitas. 1995. Experimentally Gained Parameters, Decisive for Rock Fall Analysis. *Rock Mechanics and Rock Engineering*, Vol. 28, No. 2, pp. 111–124.

Azzoni, A., G. L. Barbera, and A. Zaninetti. 1995. Analysis and Prediction of Rockfalls Using a Mathematical Model. *International Journal of Rock Mechanics and Mining Science*, Vol. 32, pp. 709–724.

Azzoni, A., P. P. Rossi, E. Drigo, G. P. Giani, and A. Zaninetti. 1992. In Situ Observation of Rockfall Analysis Parameters. In *Proc., 6th International Symposium on Landslides*, Christchurch, New Zealand (D. H. Bell, ed.), Balkema, Rotterdam, Netherlands, Vol. I, pp. 307–314.

Badger, T. C., J. D. Duffy, F. Sassudelli, P. C. Ingraham, P. Perreault, B. Muhunthan, H. Radhakrishnan, O. S. Bursi, M. E. Molinari, and E. Castelli. 2008. Hybrid Barrier Systems for Rockfall Protection. In *Proc., Interdisciplinary Workshop on Rockfall Protection*, Morschach, Switzerland, June 23–28 (A. Volkwein, V. Labiouse, and K. Schellenberg, eds.), Rock Mechanics Laboratory, Ecole Polytechnique Fédérale de Lausanne, Switzerland, pp. 10–12.

Barrett, R. K., and T. Pfeiffer. 1989. *Rockfall Modeling and Attenuator Testing*, Report CDOH-DTD-ED3/CSM-89-2. Colorado Department of Transportation, Denver.

Beck, T. J. 1995. *Anderson Grade Truck Climbing Lane Rockfall Study*. Internal Memorandum. California Department of Transportation, Sacramento.

Berger, F., and L. Dorren. 2006. Objective Comparison of Rockfall Models Using Real Size Experimental Data. In *Proc., INTERPRAEVENT International Symposium: Disaster Mitigation of Debris Flows, Slope Failures and Landslides*, Niigata, Japan, Sept. 25–29, Universal Academy Press, Inc., Tokyo, Vol. 1, pp. 245–252.

Berger, F., C. Quetel, and L. K. A. Dorren. 2002. Forest: A Natural Protection Means Against Rockfall, but with Which Efficiency? The Objectives and Methodology of the ROCKFOR Project. In *Proc., INTERPRAEVENT International Symposium 2002 in the Pacific Rim*, Matsumoto, Nagano Prefecture, Japan, October 14–18, Congress Publication, Vol. 2, pp. 815–826.

Bourrier, F., L. Dorren, F. Nicot, F. Berger, and F. Darve. 2009. Toward Objective Rockfall Trajectory Simulation Using a Stochastic Impact Model. *Geomorphology*, Vol. 110, pp. 68–79.

Bozzolo, D., and R. Pamini. 1986. Simulation of Rock Falls Down a Valley Side. *Acta Mechanica*, Vol. 63, pp. 113–130.

Bozzolo, D., R. Pamini, and K. Hutter. 1998. Rockfall Analysis—A Mathematical Model and Its Test with Field Data. In *Proc., 5th International Symposium on Landslides*, Lausanne, Switzerland, Balkema, Rotterdam, Netherlands, pp. 550–560.

Brauner, M., W. Weinmeister, A. Agner, S. Vospernik, and B. Hoesle. 2005. Forest Management Decision Support for Evaluating Forest Protection Effects Against Rockfall. *Forest Ecology and Management*, Vol. 207, Nos. 1–2, pp. 75–85.

Broili, L. 1973. In Situ Tests for the Study of Rock Fall. *Geologia Applicata e Idrogeologia*, Vol. 8, No. 1, pp. 105–111 (in Italian).

Broili, L. 1974. Ein Felssturz im Grossversuch. *Rock Mechanics (Suppl. 3)*, pp. 69–78.

Broili, L. 1977. Relations Between Scree Slope Morphometry and Dynamics of Accumulation Processes. In *Proc., Meeting on Rockfall Dynamics and Protective Works Effectiveness*, Publication No. 90, Istituto Sperimentale Modelli e Strutture (ISMES) SpA, Bergamo, Italy, pp. 11–23.

Chau, K. T., R. H. C. Wong, and J. J. Wu. 2002. Coefficient of Restitution and Rotational Motions of Rockfall Impacts. *International Journal of Rock Mechanics and Mining Science*, Vol. 39, pp. 69–77.

Ciabocco, G., L. Boccia, and M. N. Ripa. 2009. Energy Dissipation of Rockfalls by Coppice Structures. *Natural Hazards and Earth System Sciences*, Vol. 9, pp. 993–1001.

Crosta, G. B., and F. Agliardi. 2004. Parametric Evaluation of 3D Dispersion of Rockfall Trajectories. *Natural Hazards and Earth System Sciences*, Vol. 4, pp. 583–598.

Dorren, L. K. A. 2003. A Review of Rockfall Mechanics and Modeling Approaches. *Progress in Physical Geography*, Vol. 27, No. 1, pp. 69–87.

Dorren, L. K. A., F. Berger, C. Le Hir, E. Mermin, and P. Tardif. 2005. Mechanisms, Effects and Management Implications of Rockfall in Forests. *Forest Ecology Management*, Vol. 215, Nos. 1–3, pp. 183–195.

Dorren, L. K. A., F. Berger, C. Le Hir, E. Mermin, and P. Tardif. 2006. Results of Real Size Rockfall Experiments on Forestland and Non-forested Slopes. In *Proc., INTERPRAEVENT International Symposium: Disaster Mitigation of Debris Flows, Slope Failures and Landslides*, Niigata, Japan, Sept. 25–29, Universal Academy Press, Inc., Tokyo, pp. 223–228.

Dorren, L. K. A., F. Berger, and U. S. Putters. 2006. Real-size Experiments and 3-D Simulation of Rockfall on Forested and Non-forested Slopes. *Natural Hazards and Earth System Sciences*, Vol. 6, pp. 145–153.

Dorren, L. K. A., B. Maier, U. S. Putters, and A. C. Seijmonsbergen. 2004. Combining Field and

Modeling Techniques to Assess Rockfall Dynamics on a Protection Forest Hillslope in the European Alps. *Geomorphology*, Vol. 57, pp. 151–167.

Duffy, J. D. 1987. *Rockfall in the Gaviota Pass Area, Highway 101 and in the Red Cut, Highway 154, Santa Barbara County, California, Rockfall Study.* Internal Memorandum. California Department of Transportation Sacramento.

Duffy, J. D. 1990. *Field Testing and Evaluation of Rockfall Restraining Nets.* Transportation Materials and Research Report No. CA/TL-90-05. California Department of Transportation, Sacramento (video recording).

Duffy, J. D. 1991. *Gaviota Rockfall Barrier Test at Site B2.* Internal Memorandum. California Department of Transportation, San Luis Obispo.

Duffy, J. D. 1992. *Field Tests and Evaluation of Flexible Rockfall Barriers.* Brugg Cable Products, Oberbuchsiten, Switzerland.

Duffy, J. D. 1999. *Rockfall Barrier Testing Guidelines.* Internal Memorandum. California Department of Transportation, San Luis Obispo.

Duffy, J. D., and B. Haller. 1993. Field Tests of Flexible Rockfall Barriers. In *Proc., Conference on Transportation Facilities Through Difficult Terrain,* Aspen–Snowmass, Colo., Balkema, Rotterdam, Netherlands, pp. 465–473.

Duffy, J. D., and W. Hoon. 1996a. *Field Tests and Evaluation of Hi-Tech Low Energy Chain Link Rockfall Fence.* Report No. CA/05-96-01. California Department of Transportation, San Luis Obispo.

Duffy, J. D., and W. Hoon. 1996b. *Field Tests and Evaluation of Hi-Tech 50 and 70 Foot-ton Rockfall Fence,* Report No. CA/05-96-02. California Department of Transportation, San Luis Obispo.

Duffy, J. D., and C. Jones. 2000. *Field Tests and Evaluation of Temporary Construction Chain Link Fence.* Report No. CA/05-00-01. California Department of Transportation, San Luis Obispo.

Duffy, J. D., W. Hoon, and D. Serafini. 1998. *Field Tests and Evaluation of Cuesta Grade Chain Link Fence, Gawk Screen, Fence and Jersey Barrier.* Report No. CA/05-98-01. California Department of Transportation, San Luis Obispo.

Duffy, J. D., D. D. Smith, and K. R. White. 1990. Field Tests and Evaluation of Rockfall Restraining Nets. In *Proc., 41st Annual Highway Geology Symposium,* Albuquerque, N. Mex., pp. 125–141.

Elston, M. G., W. A. Richards, and B. W. R. Eastman. 1978. *Slide 5 Trans Canada Highway Slope Stabilization.* Geotechnical and Materials Branch, Highway Engineering Division, British Columbia Ministry of Highways and Public Works, Fort St. John, Canada.

Evans, C. L. 1989. *The Design of Catch Bench Geometry in Surface Mines to Control Rockfall.* Master of Science thesis. Department of Mining and Geological Engineering, University of Arizona, Tucson.

Evans, S. G., and O. Hungr. 1993. The Assessment of Rockfall Hazard at the Base of Talus Slopes. *Canadian Geotechnical Journal,* Vol. 30, pp. 620–636.

FHWA. 1993. *Rockfall Hazard Mitigation Methods.* Publication FHWA SA-93-085. U.S. Department of Transportation.

Fornaro, M., D. Peila, and M. Nebbia. 1990. Block Falls on Rock Slopes: Application of a Numerical Simulation Program to Some Real Cases. In *Proceedings of the 6th International Congress of the International Association of Engineering Geologists,* Amsterdam (D. G. Price, ed.), Balkema, Rotterdam, Netherlands, pp. 2173–2180.

Giacomini, A., O. Buzzi, B. Renard, and G. P. Giani. 2009. Experimental Studies on Fragmentation of Rock Falls on Impact with Rock Surfaces. *International Journal of Rock Mechanics and Mining Sciences,* Vol. 46, pp. 708–715.

Giani, G. P. 1992. *Rock Slope Stability Analysis.* Balkema, Rotterdam, Netherlands.

Giani, G. P., A. Giacomini, M. Migliazza, and A. Segalini. 2004. Experimental and Theoretical Studies to Improve Rock Fall Analysis and Protection Work Design. *Rock Mechanics and Rock Engineering,* Vol. 37, No. 5, pp. 369–389.

Hearn, G. 1991. *CDOT Flex-Post Rockfall Fence.* Report CDOH-R-UCB-91-6. Colorado Department of Transportation, Denver.

Hearn, G., R. K. Barrett, and M. L. McMullen. 1992. CDOT Flexpost Rockfall Fence Development, Testing, and Analysis. In *Transportation Research Record 1343,* Transportation Research Board, National Research Council, Washington, D.C., pp. 23–29.

Heim, A. 1932. Bergsturz und Menschenleben. *Bieblatt zur Vierteljahrsschrift der Naturforschenden Gesellschaft in Zürich,* Vol. 77, pp. 1–217 (translated by N. Skermer as *Landslides and Human Lives,* BiTech Publishers, Vancouver, British Columbia, Canada, 1989).

Higgins, J. D., R. D. Andrew, and T. J. Pfeiffer. 2003. Computer Simulation for Mitigation of Colorado Rockfall. In *Engineering Geology in Colorado: Contributions, Trends, and Case Histories* (D. D. Boyer, P. M. Santi, and W. P. Rogers, eds.), Special Publ. No. 15 and Colorado

Geological Survey Special Publ. 55, Association of Environmental and Engineering Geologists, Denver, Colo.

Hoshida, Y., and T. Nomura. 1998. *Latest Experimental Study for Rockfall Mitigation*. Protec Engineering, Maedakoson Co., Ltd., Japan.

Hsü, K. J. 1975. Catastrophic Debris Streams (Stürzstroms) Generated by Rockfalls. *Geological Society of America Bulletin*, Vol. 86, pp. 129–140.

Hwu, B.-L., and R. Spang. 1997. *Field Tests of a 750 Kilojoule Brugg Rockfall Barrier*. Chung Cheng Institute of Technology for Geobrugg, Romanshorn, Switzerland (video recording).

Jones, C. L., J. D. Higgins, and R. D. Andrew. 2000. *Colorado Rockfall Simulation Program Version 4.0*. Colorado Department of Transportation, Colorado Geological Survey, Denver.

Kane, W. F., and J. D. Duffy. 1993. *Brugg Low Energy Wire Rope Rockfall Net Field Tests*. Technical Research Report 93-01. Department of Civil Engineering, University of the Pacific, Stockton, Calif.

Kawahara, S., and T. Muro. 2006. Effects of Dry Density and Thickness of Sandy Soil on Impact Response due to Rockfall. *Journal of Terramechanics*, Vol. 43, pp. 329–340.

Kemeny, J., and K. Turner. 2008. *Ground Based LiDAR Rock Slope Mapping and Assessment*. FHWA-CFL/TD-08-006. Central Federal Lands, FHWA, U.S. Department of Transportation.

Labiouse, V., and B. Heidenreich. 2009. Half-scale Experimental Study of Rockfall Impacts on Sandy Slopes. *Natural Hazards and Earth System Sciences*, Vol. 9, pp. 1981–1993.

McCauley, M. T., B. W. Works, and S. A. Naramore. 1985. *Rockfall Mitigation*. California Department of Transportation, Sacramento.

Paronuzzi, P. 2009. Rockfall-Induced Block Propagation on a Soil Slope, Northern Italy. *Environmental Geology*, Vol. 58, pp. 1451–1466.

Parsons De Leuw, Inc. 1992. *Full Scale Geotextile Rock Barrier Wall Testing, Analysis and Prediction*. Colorado Department of Transportation, Denver.

Perret, S., F. Dolf, and H. Kienholz. 2004. Rockfall into Forests: Analysis and Simulation of Rockfall Trajectories—Considerations with Respect to Mountainous Forests in Switzerland. *Landslides*, Vol. 1, No. 2, pp. 123–130.

Pfeiffer, T. J., and T. D. Bowen. 1989. Computer Simulation of Rockfalls. *Bulletin of the Association of Engineering Geologists*, Vol. 26, No. 1, pp. 135–146.

Pfeiffer, T. J., and J. D. Higgins. 1990. Rockfall Hazard Analysis Using the Colorado Rockfall Simulation Program. In *Transportation Research Record 1288*, Transportation Research Board, National Research Council, Washington, D.C., pp. 117–126.

Pichler, B., Ch. Hellmich, and H. A. Mang. 2005. Impact of Rocks onto Gravel: Design and Evaluation of Experiments. *International Journal of Impact Engineering*, Vol. 31, pp. 559–578.

Pierson, L. A., S. A. Davis, and T. J. Pfeiffer. 1994. *The Nature of Rockfall as the Basis for a New Fallout Area Design Criteria for 0.25:1 Slopes*. Engineering Geology Group, Oregon Department of Transportation, Salem.

Pierson, L. A., C. F. Gullixson, and R. G. Chassie. 2001. *Rockfall Catchment Area Design Guide*. Final Report SPR-3(032). Research Group, Oregon Department of Transportation, Salem.

Ritchie, A. M. 1963. Evaluation of Rockfall and Its Control. In *Highway Research Record 17*, Highway Research Board, National Research Council, Washington, D.C., pp. 13–28.

Salisbury, M., J. D. Duffy, and P. Huang. 2009. *Highway 39 Rock Rolling Test*. Internal Memorandum. California Department of Transportation, Sacramento.

Schweigl, J., C. Ferretti, and L. Nossing. 2003. Geotechnical Characterization and Rockfall Simulation of a Slope: A Practical Case Study from South Tyrol (Italy). *Engineering Geology*, Vol. 67, pp. 281–296.

Smith, D. D., and J. D. Duffy. 1990. *Field Tests and Evaluation of Rockfall Restraining Nets*. Final Report No. CA/TL-90/05. California Department of Transportation, Sacramento.

Statham, I. 1976. A Scree Slope Rockfall Model. *Earth Surface Processes*, Vol. 1, pp. 43–62.

Whitman, T., and J. D. Duffy. 2006. *Scaling and Rock Rolling Experiments at Devils Slide*. Internal Memorandum. California Department of Transportation, Oakland.

Wu, S.-S. 1985. Rockfall Evaluation by Computer Simulation. In *Transportation Research Record 1031*, Transportation Research Board, National Research Council, Washington, D.C., pp. 1–5.

Wyllie, D. C. 1991. *Evaluation of Performance of a Rockfall Catch Fence Test Section, British Columbia, Canada*. Golder Associates.

PART 3

ROCKFALL MITIGATION

Chapter 11

LAWRENCE A. PIERSON AND
MICHAEL P. VIERLING

MITIGATION SELECTION

1. INTRODUCTION

When corrective measures are selected for a rockfall site, typically the project goals are considered to establish the appropriate level of response and the specific design criteria. The goal of mitigation can range from continued maintenance, to hazard reduction, to hazard elimination. The multiple demands for expenditures from safety budgets often require rockfall mitigation projects to compete with other safety concerns for funding, and ideally project development decisions take this into account.

Clearly, a good understanding of the rockfall problem is necessary. This requires a historical perspective of slope performance and rockfall behavior and completion of detailed studies as described in Chapters 2 ("Rockfall Types and Causes"), 6 ("Site Characterization"), 8 ("Evaluation of Rockfall Mechanics"), 9 ("Modeling and Prediction of Rockfall"), and 10 ("Conducting Field-Test Experiments"). Potential mitigation measures are evaluated on the basis of criteria that include complexity, effectiveness, durability, constructability, environmental and aesthetic effects, cost, and ongoing maintenance requirements, as introduced in Section 3 of this chapter. In the United States, increased emphasis on context-sensitive solutions (CSS) for the design of transportation facilities has caused the Federal Highway Administration to evaluate the application of CSS to rock slope design (Andrew et al. 2011).

2. ROCKFALL MITIGATION MEASURES

Ideally, mitigation decisions balance the cost of mitigation against the cost and associated risks of ongoing maintenance, since the cost of mitigation can be high. A standard benefit–cost analysis can be performed to guide these project development decisions, but unless reductions in the costs of potential injuries, damage to infrastructure, and interruption of traffic are considered along with annual maintenance costs, it is unusual to obtain a ratio greater than 1. More generally, organizations rely on a rockfall management program and policies to guide the expenditure of safety budgets (Pierson and Van Vickle 1993).

Commonly, a combination of maintenance and rockfall mitigation is needed because few mitigation measures result in a maintenance-free end product. For this discussion, rockfall mitigation measures have been grouped into two main categories: engineered and nonengineered. *Engineered measures* are active interventions that diminish the occurrence or effects of rockfalls, while *nonengineered measures* are more passive interventions that include continued or increased maintenance, warning signs, and slope-monitoring systems. A summary of common techniques is included in Tables 11-1 and 11-2.

2.1 Engineered Measures

The purpose of engineered measures is to reduce the potential for or the effects of rockfalls. The

Table 11-1
Summary of Engineered Mitigation Measures

Mitigation Measure	Description or Purpose	Limitations
Stabilization		
Removal		
Scaling	Removal of loose rock from slope by means of hand tools, expansive concrete, or mechanical equipment; commonly used in conjunction with most other design elements.	Temporary measure that usually needs to be repeated every 2–10 years as slope face continues to degrade.
Rock removal, blast scaling	Removal of loose rock or large rock blocks from slope by means of blasting, chemical expanders, or mechanical devices such as hydraulic jacks or air-pillows.	Damage from fly rock and rockfall and possible undermining or loss of support from removal of key block.
Trim blasting	Used to remove overhanging faces or protruding knobs that may act as launch features on slope.	Difficulties with drilling, debris containment, and safety.
Resloping	Cutting rock slope at flatter angle to improve slope stability and rockfall trajectories.	May have right-of-way or environmental issues.
Reinforcement		
Dowels	Untensioned steel bars or bolts installed to increase shear resistance and reinforce a block. Increase normal force friction once block movement occurs.	Passive support requires block movement to develop bolt tension. Slope access difficulties.
Shear pins	Provide shear support at leading edge of dipping rock block or slab by using grouted steel bars.	Need to provide contact (cast-in-place concrete) between bars and leading edge of block. Access difficulties.
Rock bolts	Tensioned steel bars or bolts used to increase normal force friction and shear resistance along potential rock block failure surfaces. Applied in pattern or in specific block.	Less suitable on slopes made up of small blocks. Difficult to access slope.
Shotcrete	Pneumatically applied concrete requiring high velocity and proper application to consolidate. Primarily used to halt ongoing loss of support caused by erosion and spalling. Also helps retain small supporting rock blocks.	Reduces slope drainage. Can be unsightly unless sculpted or tinted. Wire mesh or fiber reinforcement required. Needs minimum 2-in. (5-cm) thickness to resist freeze–thaw. Quality and durability are dependent on nozzleman skills.
Buttresses	Provide support to overhanging rock or lateral support to rock face by using earth materials, cast concrete, or reinforcing steel.	Height limitations. May form roadside hazard and be unsightly.
Cable lashing	Anchored, tensioned cables used to strap rock block in place. May be used in conjunction with cable nets or wire mesh. Also used as temporary support during rock bolt and dowel-drilling activities.	Because of slope or block geometry, typically movement must occur for full cable resistance to develop.
Whalers, lagging	Anchored beams or steel straps used to hold rock blocks in place between bolt locations. Also used as temporary measure to provide support during rock bolt and dowel-drilling activities.	Unsightly as permanent application. Movement must occur for full tensioning and resistance to develop.
Anchored wire mesh, cable nets, high-tensile-strength steel mesh	Free-draining, pinned or anchored-in-place nets or mesh. Used to apply active retention force to retain rocks on slope.	May form pockets of loose rock as rockfall debris accumulates. Can be difficult to clean out.
Drainage		
Weep drains	Reduce water pressures within slope by using horizontal drains or adits. Commonly used in conjunction with other design elements.	Difficult to quantify need and verify improvements achieved.

(continued)

Table 11-1 *(continued)*
Summary of Engineered Mitigation Measures

Mitigation Measure	Description or Purpose	Limitations
Protection		
Mesh or cable nets		
Draped mesh slope protection	Hexagonal wire mesh, cable nets, or high-tensile-strength steel mesh placed on slope face to slow erosion, control descent of falling rocks, and restrict them to catchment area.	Requires debris-collection ditch area. Must consider debris and snow loads on anchors. Depending on mesh type, limited to rocks ≤ 4 ft (1.2 m).
Hybrid drapery attenuator fences, suspended systems	Wire or cable mesh draped by fence posts or suspended across chutes. Fence (impact zone) intercepts and attenuates falling rocks that initiate upslope and directs them beneath mesh and into roadside catchment area.	Require debris-collection ditch areas. Must consider debris and snow loads on anchors. Typically limited to rocks ≤ 4 ft (1.2 m).
Catchment areas or sheds		
Ditches or berms	Shaped catchment area at base of slope used to contain rockfall.	High slopes require wide fallout areas. May have right-of-way or environmental issues.
Rockfall sheds	Covered structure used to intercept and divert rockfalls.	Hazards associated with traffic in confined space. Must consider downslope issues. Expensive.
Barriers		
Rigid barriers (with or without fence extension)	Rigid barrier walls used to intercept falling rocks and restrict them to prescribed fallout area. Examples include Jersey barriers; guardrails; and other concrete, timber, gabion, or mechanically stabilized earth walls.	Rigid systems are more prone to damage by higher-energy events. Complicate debris cleanout and snowplowing.
Flexible barriers	Wire ring, high-strength wire mesh or cable net panels with high-energy absorption capacity supported by steel posts and anchor ropes with braking elements. Typically proprietary systems.	Require room for barriers to deflect during impacts. Must be cleaned out periodically. Complicate snowplowing.
Avoidance		
Tunnels	Avoid slope hazards by moving roadway inside rock mass away from external rockfall sources.	Hazards associated with traffic in confined space. Long tunnels require lighting and special ventilation. Expensive.
Realignment	Full road realignment or facility relocation to move away from rockfall area.	Often old road must be maintained for existing access. Commonly there is limited space for this option. Expensive.
Elevated structures	Used to span anticipated rockfall paths, allowing rockfalls to pass beneath.	Structure must completely span active area to avoid being damaged by rockfalls. Expensive.

Table 11-2
Summary of Nonengineered Mitigation Measures

Mitigation Measure	Description or Purpose	Limitations
Warning signs	Alert users to the potential for rockfall and for fallen rocks to be encountered on the roadway.	Users become accustomed to presence of signs and ignore warnings.
Road patrols	Roadway inspections during periods of higher rockfall activity to find and remove fallen rocks.	May encompass too many miles of road to manage in a timely manner.
Ditch cleaning	Removal of accumulated rockfall debris from ditches to maintain their effectiveness in capturing rocks.	Temporary measure requiring repeated action by maintenance crews and a suitable disposal site.
Monitoring	Use of instruments to detect incipient rockfalls.	Lead time to events can be short.

FIGURE 11-1
Scaling operation.
(PHOTOGRAPH COURTESY OF B. BLACK.)

three main categories of engineered measures are stabilization, protection, and avoidance. The goal of *stabilization* is to reduce the potential for rockfalls to initiate by securing rock blocks in place or by preemptively removing them in a controlled manner. *Protection measures* are intended to intercept or capture falling rocks and restrict them from entering protected areas. *Avoidance measures* relocate facilities or persons from hazardous rockfall areas without necessarily reducing the potential for rockfall. Commonly, an engineered project will include design aspects from all three mitigation categories.

FIGURE 11-2
Rock bolting.
(PHOTOGRAPH COURTESY OF B. BLACK.)

2.1.1 Stabilization

Stabilization measures include changes in the slope to reduce the likelihood of occurrence of a rockfall. This can be accomplished by securing rocks in place, removing loose rocks in a controlled manner, or changing the slope configuration to accommodate the geological structure and reduce the potential for rockfall. Stabilization measures (slope reinforcement, scaling, slope reshaping, etc.) are discussed in Chapter 13. Figure 11-1 shows slope scaling; Figure 11-2 shows drilling for a rock bolt installation.

2.1.2 Protection

Protection techniques are used to control rockfalls once they initiate. The goal of these measures is to change rockfall behavior by dissipating the falling energy or by intercepting rockfalls to prevent them from reaching a road or other facility. Protection options include catchment areas and roadside ditches (Hoek and Bray 1981), draped mesh, hybrid drapery (also referred to as *attenuator fences*), rigid barriers (Figure 11-3), flexible barriers (Figure 11-4), and rock sheds. Protection measures are discussed in Chapters 14, 15, and 16.

FIGURE 11-3
Rigid concrete barrier.
(PHOTOGRAPH COURTESY OF L. PIERSON.)

2.1.3 Avoidance

Avoidance measures rely on relocation or realignment of a road or a facility away from the rockfall source. As discussed in Chapter 12, these techniques often require significant investment and acquisition of additional right-of-way, but they may be the only way to eliminate rockfall risks. Options include new roadway alignments, tunnels, and elevated structures.

2.2 Nonengineered Measures

Nonengineered measures include maintenance activities to keep roadside ditches and roadways clear of fallen rock as well as warning signs to alert users of the potential for rockfalls or for encountering fallen rocks. Maintenance is discussed in greater detail in Chapter 17. Monitoring a slope to identify changes to the rock slope that could indicate an increased potential for rockfall is included as a nonengineered measure and is described in Chapter 17.

2.2.1 Warning Signs

Warning signs, as shown in Figure 11-5, are an inexpensive but important measure. Their presence, or absence, has played a significant role in some rockfall litigation actions. If possible, signs with flashing lights should be used to attract attention and to overcome the tendency of drivers to ignore warning devices.

2.2.2 Monitoring

Slope monitoring uses surveying procedures [traditional as well as terrestrial and aerial light detection and ranging (LiDAR) techniques] or geotechnical instruments such as crack meters and extensometers to detect rock block movements that may indicate incipient failure. Road closures (drop gates) and automated notifications such as those commonly used by railroads can be included with electronic monitoring systems. Figures 11-6 and 11-7 show vibrating wire crack meters used to detect block movements on a rock face.

2.2.3 Maintenance Road Patrols

Road patrols can be intermittent or continuous, depending on the severity of the situation.

FIGURE 11-4 Flexible barrier fence.
(PHOTOGRAPH COURTESY OF L. PIERSON.)

FIGURE 11-5 Rockfall warning sign, Glacier National Park, Montana.
(PHOTOGRAPH COURTESY OF A. K. TURNER.)

FIGURE 11-6 Vibrating wire crack meter.
(PHOTOGRAPH COURTESY OF D. BECKSTRAND.)

FIGURE 11-7 Close-up view of vibrating wire crack meter. (PHOTOGRAPH COURTESY OF D. BECKSTRAND.)

Rockfalls can be removed from the road, and if rockfall conditions worsen, the road can be closed. In addition, maintenance personnel can observe roadside ditches and clean them out as necessary to maintain their effectiveness in capturing rockfalls. These activities are normally part of an overall rockfall maintenance and management program, as discussed in Chapters 17 and 18.

3. SELECTION CRITERIA

The use of either engineered or nonengineered measures depends on rockfall management policies, project design goals and criteria, available budgets, and the effectiveness of the proposed measure with regard to public safety and required levels of future maintenance. The selected measure is also assessed in terms of environmental impacts and aesthetics.

Stabilization, protection, and avoidance measures make up most rockfall mitigation work. Table 11-3 provides a qualitative comparison of each of the mitigation measures given in Tables 11-1 and 11-2. The comparisons are based on routinely considered criteria including complexity, effectiveness, durability, constructability, traffic impacts, environmental and aesthetic effects, cost, and maintenance requirements, as described in the following sections.

3.1 Complexity

Some mitigation measures require a more detailed and extensive investigation as well as greater design experience than do other measures. The complexity of a measure generally relates directly to the duration and cost of the design process (i.e., the level of effort required to advance a project to construction). Although the measures given in Table 11-3 are commonly used, a practitioner's familiarity with each measure will likely vary significantly. In some instances, it may be beneficial for an agency to retain the services of a specialty consultant to assist in the investigation and design of a measure that is most suited to a particular site to better meet the remaining selection criteria.

As indicated in Table 11-3, the complexity of a particular measure can range from low for the use of berms or ditches to very high for rockfall sheds or tunnels. For some mitigation measures, a range of complexity is shown because the complexity is dependent on specific site conditions. For practitioners who are less familiar with the investigation, design, or construction requirements of a particular mitigation measure, it may be advisable to rate complexity higher than indicated in the table.

3.2 Effectiveness

Mitigation measures need to match the defined rockfall conditions. A measure that could be severely damaged or destroyed by the next rockfall event or one that would do little to reduce the maintenance requirements or improve safety would serve little purpose. Quantifying rockfall conditions requires insight into anticipated rockfall size, behavior, and energy, as well as rockfall volume and the recurrence interval of events. A well-documented rockfall history can serve as an excellent indicator of possible future events.

A measure's suitability is also based on the planned use of the facility or roadway as well as the vulnerability of the facility and its users to rockfalls. A higher level of mitigation (greater investment) is generally considered appropriate where vulnerability is higher. However, few measures will eliminate rockfall risks, and for most projects, hazard reduction is the goal. The demands of hazard elimination can be unreasonably high when construction cost, right-of-way needs, and environmental impacts are considered. Certain limitations may not allow construction of the ultimate or preferred option, but in all cases, justification of the construction investment assumes that the end result will be a safer or more easily maintained facility.

In general, stabilization and protection measures reduce the effects of rockfalls by inhibiting their occurrence or controlling their behavior. These

Table 11-3
Qualitative Comparison of Mitigation Measures

	Selection Criteria							
Mitigation Measure	Complexity	Effectiveness	Durability	Constructability, Special Expertise	Road Closure, Traffic Restrictions	Environmental Limitations	Aesthetic Impacts	Cost	Maintenance Requirements
Engineered Measures: Stabilization									
Removal									
Scaling	L–M	L–H	L–M	M	Y	L	L	L–M	L–M
Blast scaling	M–H	L–H	M	M–H	Y	L–M	L	L–M	L–M
Trim blasting	M–H	M–H	M–H	M–H	Y	L–H	L–H	M	L–M
Resloping	L–H	M–H	H	L–M	Y	L–H	L–H	M–H	L
Reinforcement									
Dowels	M	M–H	H	H	P	L	L	M–H	L
Shear pins	M	M	M	H	P	M	M	M	L
Rock bolts	M–H	M–H	H	H	P	L	L	M–H	L
Shotcrete	M–H	M–H	M–H	H	P	M–H	H	M–H	L
Buttresses	M–H	H	H	M	P	L–H	H	M–H	L
Cable lashing	L–H	L–H	M–H	M	P	L–H	L–M	L–M	L–M
Whalers, lagging	M	M	L–M	M	P	L–H	M–H	M	L
Anchored mesh	M	M–H	M	M	P	M	L–H	L–H	L–H
Drainage									
Weep drains	L	L–H	L–M	M–H	P	L–H	L	L	H
Engineered Measures: Protection									
Mesh or cable nets									
Draped mesh	L–M	M–H	M–H	M	Y	L–H	L–H	L	L–M
Attenuator system	L–H	M–H	L–H	M	Y	M–H	M–H	L–M	L–M

(*continued on next page*)

Table 11-3 (*continued*)
Qualitative Comparison of Mitigation Measures

Mitigation Measure	Selection Criteria: Complexity	Effectiveness	Durability	Constructability, Special Expertise	Road Closure, Traffic Restrictions	Environmental Limitations	Aesthetic Impacts	Cost	Maintenance Requirements
Catchment areas or sheds									
Ditches or berms	L	M–H	H	L	P	L–M	L–M	L–H	H
Rockfall sheds	VH	H	M–H	H	P	H	H	H	L–M
Barriers									
Rigid barriers	L	M–H	L–M	L	P	L–H	L–H	M	M–H
Flexible barriers	M	M–H	M–H	M	P	L–M	L–M	M	M–H
Engineered Measures: Avoidance									
Tunnels	VH	VH	H	H	P	L–M	L–H	VH	M–H
Realignment	M–VH	M–VH	H	M	P	M–H	L–H	H–VH	M–H
Elevated structures	M–VH	M–H	L–H	M	P	L–H	L–M	VH	M–H
Nonengineered Measures									
Warning signs	L	L–M	na	na	N	L–M	L	L	L–M
Road patrols	L	L–M	na	na	N	na	na	L–H	H
Ditch cleaning	L	L–H	na	na	Y	L–M	na	L–M	H
Monitoring	M–H	L	L–M	M–H	N	L	L	L–H	L–H

NOTE: L = low, M = medium, H = high, VH = very high, N = no, Y = yes, P = possibly, na = not applicable.

measures are intended to limit the likelihood that a rockfall will reach the facility. Depending on the degree of stabilization or protection, they can vary from inexpensive to costly. Avoidance measures relocate the facility to a safer position with the intent of eliminating the potential for rockfalls, but they tend to require a greater investment.

If rockfalls are made up of relatively small blocks (less than 0.5 to 1 m or 2 to 3 ft in diameter), most stabilization measures other than resloping may be ineffective, and protection measures may be more appropriate to consider. Conversely, if expected rockfalls are made up of larger blocks, stabilization measures may prove to be a more practical and effective means of rockfall mitigation. Because of their typically greater economic, social, and environmental impacts, avoidance measures are usually considered only when user or facility vulnerability is considered too high or where local conditions accommodate this type of mitigation.

3.3 Durability

Durability reflects how well a particular measure will withstand the rigors of exposure and use and continue to serve as intended. The selection of a mitigation measure that does not match the actual rockfall problem can severely lessen service life, especially for measures that must withstand repeated rockfall impacts. Service life can also be shortened by corrosive environments where groundwater or low-pH and low-resistivity soils are present or where acid rain or saltwater spray comes in contact with measures containing steel. In some applications, improper slope drainage or handling of runoff can lead to early failures, as can overly aggressive maintenance practices. These conditions are identified and quantified so that they can be assessed properly during the design phase.

In addition, the quality of construction or the lack of adherence to construction specifications can limit service life. An experienced contractor coupled with thorough construction inspection can reduce the potential for poor construction practices. Inexperience can lead to many problems, including the following:

- Misjudgment of the difficulty in accessing remote slope locations,
- Use of improper equipment or procedures, and
- Increased risk of accidents.

In turn, these problems may result in project delays, an increased likelihood of construction claims, and poor construction quality.

3.4 Constructability and Special Expertise

Constructability is governed by accessibility to the area, the presence of nearby structures or utilities, the level of risk faced by contractor personnel, the availability of specialized materials or rockfall mitigation systems, and the availability of specialty contractors and equipment. The availability of qualified and experienced contractors can be one of the most significant factors in the success of a project (Brawner 1993). The measures detailed by Brawner (1993) are standard and have been constructed worldwide over many years. However, each rockfall site is unique and must be evaluated with constructability in mind.

Worker safety can be affected by slope stability. Temporary stabilization measures may be needed before construction of permanent design elements. In some cases, high-wind areas may preclude the use of cranes, which can limit the ability to install stabilization measures. Adjacent buildings or structures (especially those of historical significance), utilities, or protected streams or rivers may interfere with rock-scaling or blasting operations. Road surfaces and retaining walls may need added temporary protection. Any planned excavation must be wide enough to accommodate standard construction equipment, and narrow (sliver) cuts that cannot be safely worked should be avoided.

3.5 Road Closure and Traffic Restrictions

Many rockfall mitigation measures cannot be constructed unless traffic control is in place (FHWA 1993). Slope scaling and blasting are obvious examples, but whenever workers are on the slope and moving, the risks of a rockfall are heightened. In some cases, outrigger requirements for a crane can take up the entire width of roadway. An adequate detour or the ability to close the road periodically for short durations may be needed. The duration of productive work may be significantly limited when road closures must accommodate both work and road clearing, which, as in the case of slope scaling, has to be undertaken before traffic can move through the area. This restriction will increase project cost and delay the time of completion.

In some cases, the duration or timing of construction work is an important consideration. This can be especially true in high-traffic areas or when environmental requirements limit available construction time, in some instances for entire seasons. For example, unrestricted access may be required in tourist areas or other seasonally open resorts or parks when traffic restrictions could unacceptably reduce the viability of the local economy. Seasonal limitations may also be associated with environmental considerations discussed in the following section or, for safety reasons, with periods of high flow when rivers are involved near the work site.

3.6 Environmental Considerations and Aesthetics

Many potential rockfall projects are located in mountainous areas or steeply incised river valleys. These areas are commonly remote and may include certain protected species of plants, animals, birds, or fish. Environmental regulations may require the project to remain clear of certain areas during nesting periods or when protected fish, plant, or animal species are present. Some areas include historically significant features that cannot be disturbed or modified at all. Such conditions may restrict the types of mitigation measures that will be acceptable to regulatory agencies or that can be built within prescribed environmental or aesthetic limitations.

If necessary, many measures can be modified to be less obtrusive to persons visiting the area or driving by the site, especially when the rock slopes are not the dominant landscape feature. Andrew et al. (2011) evaluated the application of CSS to the design of rock slopes. *Context-sensitive design* is a method of developing transportation facilities that fit within the engineered setting and preserve scenic, aesthetic, historic, and environmental resources while addressing safety and mobility.

Development of CSS starts before the project-scoping stage and incorporates several factors, including community concerns, the effects of new roadway or roadway improvements on the physical character of the surrounding area, and a visual prioritization of design considerations. Before starting construction, the contractor and land management agency should agree on a defined standard of performance and communication protocols to ensure that all project goals are attained.

Modifications to rock slope mitigation measures that reflect context-sensitive design considerations can vary from coloring draped mesh to match the ground to sculpting and tinting shotcrete to match the texture and appearance of the native rock. These requirements generally increase cost, sometimes significantly, but they may be the only way to gain acceptance for the mitigation measures (Wyllie and Norrish 1996). Other environmental and aesthetic factors that may need to be considered on some projects are the accessibility of upper slope areas, disturbance of vegetation (including tree cutting), and the need for acceptable disposal sites and haul routes.

3.7 Cost

Cost studies are a routine part of project development. However, for rockfall mitigation projects, there are numerous levels of mitigation, and thus the costs can vary considerably, depending on the degree of rockfall control. At times, a modest investment can yield a reasonable improvement, such as when minimal reshaping of a fallout area improves the containment of typical rockfalls. Avoidance measures requiring complete realignments, tunnels, or elevated structures tend to be much more costly, but the cost may be justifiable, depending on project goals. A benefit–cost analysis can be performed to compare options and the cost of improvements versus continued maintenance costs, public exposure to rockfall hazards, potential impacts on infrastructure, and costs associated with traffic interruption.

The cost comparisons provided in Table 11-3 are general. Construction costs differ significantly from region to region because of variations in material costs, the availability of specialized equipment that may be required, and the availability and experience of local contractors.

3.8 Maintenance Requirements

Maintenance requirements can vary significantly among measures, but they may not relate directly to the level of improvement. No measure will be maintenance free; however, a reduction in road patrol requirements and the ability to schedule routine maintenance rather than remain in a continuously reactive mode are typically design goals.

4. ROCKFALL MANAGEMENT POLICIES

Organizations responsible for managing rock slopes are required to address rockfall hazards through maintenance efforts and capital expenditures. For transportation agencies, rock slope maintenance deals primarily with removing fallen rock that has accumulated in catchment areas and patrolling roadways during peak rockfall periods to keep travel ways cleared. At some point, these activities may become too demanding on personnel resources or maintenance budgets and may not adequately address safety expectations at certain sites. An organization may elect to invest in additional rockfall mitigation measures to address the rockfall problem (Wyllie and Mah 2004). Ideally, each organization plans for these expenditures and meets its safety responsibilities on the basis of established policies. These efforts may be part of an overall plan that includes a rockfall inventory and hazard rating to prioritize expenditures, as discussed in Chapters 3, 4, and 18, or they may result from reactions to slope failure. Since an established program provides benefits, including the ability to plan expenditures and demonstrate that safety concerns are being addressed, policies that establish a strategic rockfall mitigation program that includes the implementation of a rockfall hazard rating system are desirable (Wyllie and Mah 1998).

5. PROJECT DESIGN GOALS AND CRITERIA

Regardless of the circumstances that bring about the decision to reduce or replace slope maintenance with a capital expenditure, the project development process considers the rockfall mitigation goals and the design criteria needed to meet those goals. Rockfall mitigation, whether it is intended to reduce, control, or avoid rockfalls, can vary from a modest safety improvement to a complete correction of the problem.

The design criteria for selecting rockfall mitigation measures include evaluation of effectiveness and durability to contend with the anticipated rockfall volume, size, and energies; constructability issues; contractor capability and availability; traffic restrictions; and construction time and seasonal requirements. Naturally, environmental limitations and aesthetic goals are also considered, since they may be the overriding considerations for a given project. Once these design criteria are established and mitigation design concepts are developed, the final design and construction costs are evaluated. Typically, the construction cost is compared with long-term maintenance demands and costs.

It is important to quantify and understand the scope of the rockfall problem before performing these evaluations and comparisons. The comparisons do not yield a single correct answer, and more than one mitigation measure, or various combinations of measures, is likely to be suitable. Having experience in rockfall mitigation design is important when making these evaluations.

6. CONCLUSIONS

Chapters 12 through 18 provide a more detailed discussion of these mitigation measures, including their applications and limitations. The selection of an appropriate mitigation measure or approach requires a thorough understanding of the rockfall problem and of the project goals and limitations. Breadth of experience is perhaps most helpful in evaluating and selecting appropriate design options. However, since many practitioners do not have experience with all possible options, a side-by-side comparison relying on the criteria discussed above can be beneficial, especially when the measure requires input from several disciplines.

The opinions of those who will oversee the construction of the rockfall mitigation measure as well as those who will be responsible for ongoing maintenance are included in the decision process. Whenever possible, the project should complement the organization's overall rockfall management program. Ideally, levels of improvement and the corresponding expenditures are compatible with these program goals and other project needs and responsibilities.

Mitigation selection is an early step in the process, and thus it does not necessarily represent the final decision. Often, a better understanding of the rockfall problem and of project limitations will develop over time, especially during the investigation and design phases. Flexibility in considering several appropriate measures and the willingness to change or adjust throughout the project development process are invaluable in achieving success.

REFERENCES

ABBREVIATION

FHWA Federal Highway Administration

Andrew, R. D., R. Bartingale, and H. Hume. 2011. *Context Sensitive Rock Slope Design Solutions*. Publication FHWA-CFL/TD-11-002. Central Federal Lands Highway Division, Lakewood, Colo., Jan. http://www.cflhd.gov/programs/techDevelopment/geotech/css/.

Brawner, C. O. 1993. *Manual of Practice on Rockfall Hazard Mitigation Methods*. National Highway Institute, U.S. Department of Transportation.

FHWA. 1993. *Rockfall Hazard Mitigation Methods*. Publication FHWA SA-93-085. U.S. Department of Transportation.

Hoek, E., and J. Bray. 1981. *Rock Slope Engineering*. Institution of Mining and Metallurgy, London.

Pierson, L. A., and R. Van Vickle. 1993. *Rockfall Hazard Rating System Participant's Manual*. Publication FHWA SA-93-057. National Highway Institute, U.S. Department of Transportation.

Wyllie, D. C., and C. W. Mah. 1998. *Rock Slopes Participant's Manual*. U.S. Department of Transportation.

Wyllie, D. C., and C. W. Mah. 2004. *Rock Slope Engineering: Civil and Mining*, 4th ed. Spon Press, New York.

Wyllie, D. C., and N. I. Norrish. 1996. Stabilization of Rock Slopes. In *Special Report 247: Landslides: Investigation and Mitigation* (A. K. Turner and R. L. Schuster, eds.), Transportation Research Board, National Research Council, Washington, D.C., pp. 474–504.

Chapter 12

THOMAS C. BADGER AND
JOHN D. DUFFY

AVOIDANCE OF ROCKFALL AREAS

1. INTRODUCTION

A*voidance* entails siting a new facility or moving an existing facility away from a hazard—in this case, rockfall. The available alternatives require a shift in either horizontal or vertical position. Realignment of a highway away from the base of a hazardous slope and tunneling through the slope are alternatives involving a *horizontal shift* of the facility. Viaducts, stacked lanes, and separated grades using various retaining wall options rely on a *vertical shift*—elevating facilities to avoid or, often, reduce exposure to a hazard.

If the only design constraint for siting a new facility is mitigating the exposure to a rockfall hazard, avoidance is the prudent and first choice. Such a hypothetical situation is a rarity, because designs are typically constrained by many factors, not the least of which are cost-effectiveness and the minimization of environmental impacts. Primarily because of the latter factors, avoidance alternatives are usually the last considered, after stabilization and protection alternatives. Although the garden-variety rockfall hazards common to many transportation facilities do not warrant the substantial costs and impacts that avoidance typically involves, under certain conditions "avoiding a landslide (rockfall hazard) can be less expensive than remediation" (Cornforth 2005).

Although stabilization and protection alternatives often trump avoidance alternatives for reasons of cost, other factors may drive more serious consideration of avoidance alternatives. Avoidance alternatives have a considerably greater service life than stabilization and protection alternatives, and in many cases they provide permanent and full mitigation of the hazard. The severity of the hazard, as defined by the frequency and size of occurrence, may exclude stabilization or protection, as in the case of the 2006 Ferguson Slide near Yosemite National Park in California, shown in Figure 12-1 (Beck and Duffy 2008; Harp et al. 2008). Avoidance alternatives become increasingly attractive for large slopes with widespread sources of rockfall (Geiger et al. 1991). Keaton and Beckwith (1996) cited the acceptable level of risk associated with a hazard as an important design consideration. Traffic volume, safety, and economic impacts of highway closure largely define the acceptable level of risk associated with exposure to rockfall hazards. Aesthetic value and requirements to meet design standards have also driven the selection of avoidance alternatives over stabilization and protection alternatives. The construction of Interstate 70 through Glenwood Canyon in Colorado used tunnels (Frey et al. 1996) and viaducts over talus deposits, as illustrated by Figures 12-2 and 12-3, to avoid rockfall hazard zones while preserving environmental values (Turner 1996).

Unless an alternative location or route is readily available for relocating a facility, avoidance

FIGURE 12-1
Ferguson Slide, which occurred in 2006 near entrance to Yosemite National Park in California, completely blocked State Highway 140 with rockfall debris, necessitating temporary relocation of highway onto other side of valley.
(PHOTOGRAPH COURTESY OF THE CALIFORNIA DEPARTMENT OF TRANSPORTATION.)

FIGURE 12-3
Hanging Lake Tunnel and Viaduct carry Interstate 70 in Glenwood Canyon, Colorado, past area of high rockfall hazard while maintaining environmental values.
(PHOTOGRAPH COURTESY OF THE COLORADO DEPARTMENT OF TRANSPORTATION.)

solutions typically cannot be implemented rapidly, especially in the case of an emergent situation, principally because of their high costs and significant impacts. Avoidance alternatives are generally costly and require substantial planning and time for project delivery, as demonstrated by the Devil's Slide project, located on the Pacific Coast just south of San Francisco, California.

California State Highway 1 clings to the steep, unstable cliffs along the coast between the towns of Pacifica and Montara (Figure 12-4). The section of this road that crosses the aptly named Devil's Slide has had an extensive history of closures due to rockfall and rockslides since it was constructed in 1937. In 1995, a closure lasted 185 days and cost almost $3 million to repair. In 2006, heavy rains caused extensive rockfall that resulted in a 4-month closure and another $7 million in stabilization costs. Coastal residents and businesses experienced detours of up to 50 km (30 mi) each time the road closed (Caltrans 2006a).

California Department of Transportation (Caltrans) engineers recognized that avoidance of the Devil's Slide offered the best long-term solution to the instability and hazards of that location, and in 1986 Caltrans proposed construction of an inland bypass road across the hilly terrain inland from the coastal cliffs (Caltrans 2006a). However, in 1987,

FIGURE 12-2
Interstate 70 in Glenwood Canyon, Colorado, is carried on viaduct founded on talus slopes to avoid rockfall hazard while retaining environmental values.
(PHOTOGRAPH COURTESY OF THE COLORADO DEPARTMENT OF TRANSPORTATION.)

FIGURE 12-4
California State Highway 1 clings to steep, unstable cliffs along Pacific Ocean coast between towns of Pacifica and Montara, south of San Francisco, California.
(PHOTOGRAPH COURTESY OF THE CALIFORNIA DEPARTMENT OF TRANSPORTATION.)

an injunction resulting from an environmental lawsuit suspended all design work to develop such an inland bypass on the ground surface.

In 1995, public comments on a supplemental environmental impact statement requested further consideration of a tunnel alternative. The subsequent report, completed in 1996, determined a tunnel alternative to be reasonable and feasible.

In November 1996, the Devil's Slide Tunnel Initiative was approved by 74% of the voters in San Mateo County. The initiative amended the county's Local Coastal Plan to allow the tunnel and indicated the tunnel as the only permissible permanent repair alternative for the existing road at Devil's Slide. A second supplemental environmental document identified the tunnel as the preferred alternative, and in 2002 the Federal Highway Administration (FHWA) authorized the state to proceed with the final design of the tunnel alternative (Caltrans 2006a). Thus, almost two decades elapsed between the introduction of an initial relocation proposal and the receipt of authorization to contract for construction of an approved relocation project.

Box 12-1 provides an overview of the approved relocation project, which is under construction with an estimated completion date of 2012. The project is financed by federal emergency relief funds and is estimated to cost $325 million, with the twin tunnels accounting for approximately $272 million of the total (Caltrans 2006a).

Not all avoidance measures require such a lengthy implementation period. The Confusion Hill Realignment Project, which was completed in 6 years, relocated approximately 2.4 km (1.5 mi) of US-101, the major north–south route in northern California. In Mendocino County, between the towns of Leggett and Garberville, US-101 bisects an active rockslide on the side of a steep valley occupied by the South Fork of the Eel River at a location known as Confusion Hill. The slide is approximately 1,000 m (3,000 ft) wide at the roadway level and extends more than 350 m (1,100 ft) above its toe at the river (Figure 12-5).

For many years, US-101 has been regularly closed at this location, and the number and duration of these closures have increased in recent years. Geotechnical studies determined that the stability of the slide was decreasing, suggesting the potential for longer roadway closures and more extensive highway repairs in the near future. Ten closures caused by the Confusion Hill slide during the 2002–2003 winter season generated a high level of concern among residents who needed to use this segment of roadway on a daily basis. A catastrophic slide at Confusion Hill could close US-101 in both directions for 6 months or more. Such a complete closure would require a 402-km (250-mi) detour and cost an estimated $7.1 million per month in travel delays and added vehicle operating costs. During the past 10 years, more than $33 million has been spent on repairs ($14 million during the winter of 2005–2006 alone), and almost $5 million in vehicle-delay costs has been incurred.

Because the frequent closures and high maintenance costs make the existing route undependable and because of the lack of a local detour for highway traffic, in 2003 Caltrans negotiated with FHWA and received approval for $67 million in emergency relief funding to relocate US-101 at Confusion Hill. An environmental impact analysis process ultimately identified six alternative alignments. All involved relocation of the road to the other side of the South Fork of the Eel River, the construction of two bridges, and excavation of a rock cut.

A high level of cooperation between stakeholders and agencies enabled the project to be accelerated by a full year. Construction began in July 2006, the two bridges were completed by the end of 2008, and the entire project was completed in 2009. Box 12-2 provides further details of the Confusion Hill Realignment Project.

2. REALIGNMENT AND RELOCATION

Realignment involves a horizontal shift of a transportation facility away from a hazardous slope (Figure 12-6). Realignment inherently increases the width of the rockfall containment zone. An optimal goal for realignment would be sufficient distance to avoid both the impact and the rollout zones of the rockfall trajectory. At a minimum, the realignment results in the traffic being located beyond the rockfall impact zone. Mitigation of remaining hazards resulting from the rollout of rock blocks can then be addressed by optimizing the effectiveness of the available containment area or by adding a barrier. *Relocation* is defined as a complete resiting of a facility—for example, to the opposite side of or entirely out of a valley bottom. Both

Box 12-1

Devil's Slide Tunnels Project, Near San Francisco, California

The Devil's Slide Tunnels Project (Figure 1) is designed to create a safe, dependable, all-weather highway between Pacifica and Montara by constructing two tunnels beneath San Pedro Mountain. At the northern end, a 304-m (1,000- ft) bridge spans the valley at Shamrock Ranch. South of the tunnels, a realignment of Route 1 provides a safe transition to and from the tunnel portals. An operations and maintenance facility located nearby has an earthen embankment and vegetation-covered roof to help the facility blend in with the natural surroundings.

Throughout the public input and evaluation phases, which lasted several decades, concerns about effects on wetlands, wildlife, plants, noise, and visual impacts received the highest priority. The bypassed section of Route 1, together with 28 ha (70 acres) of state right-of-way, will be available for public access and recreational use after the planned tunnel opening in 2012.

Curved bridges connect the north portal of each tunnel to the existing Highway 1 with a smooth transition (Figure 2). Construction used the cast-in-place cantilever method to avoid building temporary support towers in the environmentally sensitive valley. The bridges were completed and dedicated on September 26, 2008.

Twin tunnels, each about 1,280 m (4,200 ft) long and 9 m (30 ft) wide, are being excavated from south to north through use of the New Austrian Tunneling Method, which relies on inherent rock strength for support. The tunnels are being developed simultaneously, with one tunnel face approximately 40 m (131 ft) ahead of the other to control deformation and stress redistribution. The tunnels will have state-of-the-art ventilation, emissions monitoring, and fire detection and suppression systems. Safety is enhanced by 10 cross-linking passages, closed-circuit television monitoring systems, emergency telephone–call box systems, electronic message signs displaying safety and advisory information, and an over-height-vehicle detection system. A lighting system with brightness transition at each end will assist drivers entering or leaving the tunnels. Each tunnel will be accessible to bicyclists.

FIGURE 1
Map of Devil's Slide relocation.

FIGURE 2
Completed bridges.

Realignment of the existing highway at the south portal has required removal of approximately 65,000 m^3 (85,000 yd^3) of rock from an adjacent rock slope. Two concrete retaining walls are anchored to the cut rock face by 1,200 soil nails. The retaining walls are surfaced with a layer of steel-reinforced shotcrete, which is hand sculpted and colored to resemble the texture of the natural rock (Figure 3).

SOURCE: Caltrans 2006a.

FIGURE 3
Construction site of south tunnel portals with completed retaining wall.

FIGURE 12-5
US-101 crossing Confusion Hill slide area, February 2006.
(PHOTOGRAPH COURTESY OF THE CALIFORNIA DEPARTMENT OF TRANSPORTATION.)

the Devil's Slide (Box 12-1) and the Confusion Hill (Box 12-2) projects involve relocation.

Several other factors influence the feasibility of realignment or relocation alternatives. California Highway 1 traverses the central California Big Sur coast, a region of dramatic topography that attracts millions of visitors, but it is subject to numerous closures due to slope instability (Caltrans 2001; Wills et al. 2001). Maintenance of the highway is expensive, and the disposal of rockslide and rockfall debris is controversial because of potential damage to critical marine habitats in the adjacent Monterey Bay National Marine Sanctuary (Hapke and Green 2005).

The Pitkins Curve slide is the most active landslide on the Big Sur coast (Hapke and Green 2005). Immediately to the south, a rocky promontory named Rain Rocks produces frequent rockfall. As illustrated by Figures 12-7 and 12-8, rockfalls and rockslides in this area have caused numerous closures during the winter months. Between 1998 and 2004, Caltrans spent about $8 million at the Pitkins Curve–Rain Rocks location to keep the highway open, an average annual maintenance cost of $1 million (Caltrans 2005, 2006b, 2006c). In response, Caltrans has gained approval for and is now constructing a bridge spanning the Pitkins Curve rockslide to avoid the instabilities, while a rock shed is being constructed at the adjacent Rain Rocks location (Figures 12-9 and 12-10).

Sufficient space to accommodate the realignment to avoid affecting adjacent facilities or water bodies is desirable. Lack of space to allow for a realignment, combined with concerns for the safety of highway workers and the public, traffic disruption, and environmental impacts, led to the decision to construct a rock shed at the Rain Rocks site (Caltrans 2005). Roadway geometrics are considered; typically, only under more grave situations are deviations from design standards granted. Sometimes, however, improvements in geometrics can be achieved, and these improvements then become an argument in favor of avoidance. Impacts on sensitive resources or changes in access to adjoining areas can also be impediments to realignment or relocation. Environmental considerations were important in the two-decade delay for the Devil's Slide relocation project (Box 12-1).

3. ELEVATED STRUCTURES

Vertical changes in alignment, which usually involve elevating the facility with the use of bridges or retaining walls, are another means of avoiding

Box 12-2

Confusion Hill Realignment Project in Mendocino County, California

The Confusion Hill Realignment Project in northern Mendocino County, California, involved the relocation of approximately 2.4 km (1.5 mi) of US-101. The original highway followed the east side of the South Fork of the Eel River around the outside of an incised meander bend. Erosion by the river caused frequent and extensive slope instabilities that affected the road (Figure 12-5). Rockslides and rockfall debris frequently required partial or complete closure of the road for maintenance and repair work. In the 2002–2003 winter season, the road was closed at Confusion Hill on 10 occasions; during the winter of 2005–2006 continued slide activity required frequent closures. Over a period of 10 years, $33 million was spent on slide repairs, $14 million of this amount during the winter of 2005–2006.

In August 2003, Caltrans negotiated with FHWA and received approval for $67 million in emergency relief funding to relocate US-101 at Confusion Hill. An environmental impact analysis was then conducted to identify project options, constraints, alternatives, and the various impacts and associated costs. The process ultimately identified six alternative alignments and involved many project stakeholders to ensure that various interests were represented. Studies were then conducted of each of these six alternatives and reported in a project environmental impact report and assessment document (Caltrans and FHWA 2005). The relocation alternatives differed slightly in their impact on natural resources, but all the alternatives involved construction of two new bridges across the South Fork of the Eel River and the excavation and removal of more than 300,000 m^3 (392,000 yd^3) of the intervening ridge to complete the relocation of the highway to the other side of the river. The most cost-effective alternative with the least environmental impact was identified and approved in late 2005 (Figure 1).

The project was fast-tracked and construction accelerated through good cooperation between Caltrans and state and federal regulatory agencies. The north bridge was completed in early January 2008 by using conventional falsework methods (Figure 2). The new bridge allowed excavation of the large cut slopes through the ridge between the two bridges. Spoil was trucked out over the north bridge. In contrast, the south bridge used cast-in-place segmental construction. It was completed in late 2008 (Figure 3).

Construction was completed in 2009. Despite the cost of building the two bridges, Caltrans estimates that the project will pay for itself in 10 years by eliminating the maintenance bill for repairs of the existing highway.

FIGURE 1
Map of Confusion Hill Realignment Project.

FIGURE 2
Completed north bridge.

FIGURE 3
Completed south bridge.

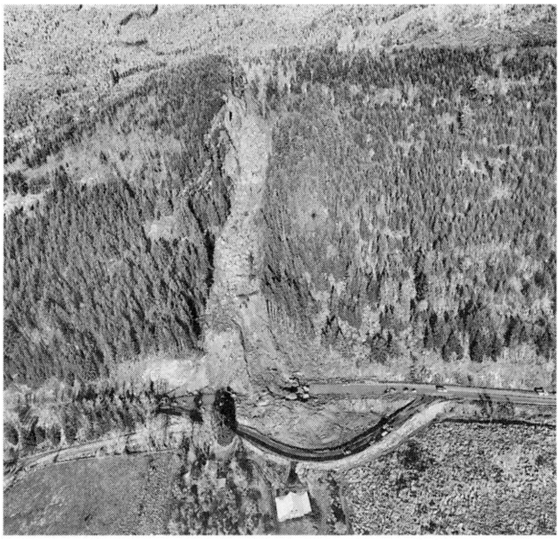

FIGURE 12-6
After two large landslides, and with remaining threat of large rockfalls from depletion zone, US-12 near Randle, Washington, was permanently realigned for several hundred meters. New roadway section consists of large, steep-sided embankment that provides highly effective containment area to address potential rollout of rockfall.
(PHOTOGRAPH COURTESY OF THE WASHINGTON STATE DEPARTMENT OF TRANSPORTATION.)

FIGURE 12-7
Caltrans crews working to reopen Highway 1 after Pitkins Curve slide during winter 2004. One lane of highway is closed (note boulders behind traffic barrier).
(PHOTOGRAPH BY C. J. HAPKE, COURTESY OF THE U.S. GEOLOGICAL SURVEY.)

rockfall hazards. Elevated structures can be used to span narrow chutes or steep-sided ravines or to traverse long, continuous slopes prone to rockfalls and other slope hazards. The use of structures also increases alignment options; a roadway section is less constrained to follow the terrain than is one made up only of cuts and fills.

Single-span bridges may be suitable for avoiding rockfall hazards confined to narrow zones, generally less than 60 m (197 ft) in length. Multispan bridges (viaducts) are used for longer rockfall-prone sections. Figure 12-11 provides examples of viaducts used to avoid rockfall hazards in Washington State and in Saudi Arabia. At Emerald Bay, on the southwest shore of Lake Tahoe, California, viaducts were constructed to realign the highway away from steep slopes generating rockfalls (Figure 12-12). Because of the narrow corridor and the environmental requirements imposed within this scenic area, the realignment had to be augmented with a wall to catch rockfalls (FHWA 2008).

Bridges with stacked decks reduce required widths but may increase heights when multilane facilities are involved. Figure 12-2 illustrates the use of separate alignments within a rockfall hazard zone in Glenwood Canyon, Colorado. Separating opposing lanes by elevating the uphill lane with retaining walls reduces or avoids the rockfall hazard by increasing the distance from the slope and the available catchment area, as illustrated in Figure 12-13 at the location known as Slide Curve along Interstate 90 in the Cascade Mountains of Washington State.

An important advantage of a viaduct over an excavation required for traversing a long slope, such as that shown in Figure 12-11*a*, is the reduced ground disturbance. The alignment of a highway with a sidehill cut not only fails to mitigate its exposure to existing rockfall hazards but also

FIGURE 12-8
Clearing rockfall debris requires road closures at Rain Rocks.
(PHOTOGRAPH COURTESY OF THE CALIFORNIA DEPARTMENT OF TRANSPORTATION.)

FIGURE 12-9
Aerial view of current Highway 1 and its proposed relocation on bridge to avoid Pitkins Curve slide area.
(PHOTOGRAPH COURTESY OF THE CALIFORNIA DEPARTMENT OF TRANSPORTATION.)

potentially generates new rockfall sources from the new cuts. In areas of high scenic value, structures offer opportunities for signature designs that may better conform to the landscape. Figure 12-14 shows an example of the selection of a viaduct structure, the Linn Cove Viaduct on the Blue Ridge Parkway in North Carolina, to preserve the scenic integrity of an area.

In all cases, the deck or roadway surface must be located above the maximum elevation of the expected rockfall trajectory (Figure 12-15). If bridge structures are used, protection of the piers may also

(a)

(b)

FIGURE 12-11
Viaducts for critical, high-volume highways avoid rockfall hazards by elevating alignment and allowing rockfalls to pass beneath structures in (*a*) Cascade Mountains of Washington State and (*b*) Saudi Arabia.
(PHOTOGRAPH *a* COURTESY OF THE WASHINGTON STATE DEPARTMENT OF TRANSPORTATION. PHOTOGRAPH *b* COURTESY OF L. PIERSON.)

FIGURE 12-10
Preferred relocation alternative developed by Caltrans for Pitkins Curve–Rain Rocks site on California Highway 1: bridge across Pitkins Curve slide and rock shed for Rain Rocks rockfall zone to right. Both are under construction.
(PHOTOGRAPH COURTESY OF THE CALIFORNIA DEPARTMENT OF TRANSPORTATION.)

FIGURE 12-12
At Emerald Bay, on southwest shore of Lake Tahoe, viaducts were constructed to realign highway away from steep slopes that generate rockfalls.
(PHOTOGRAPH COURTESY OF THE CALIFORNIA DEPARTMENT OF TRANSPORTATION.)

be necessary (Figure 12-16). Tall bridges sited on sloping ground also require substantial foundations, especially if they are founded in colluvium or talus deposits (Turner 1996). Construction cost and time for implementation are significant, and typically long-term maintenance requirements for bridges are considered.

FIGURE 12-13
Separated grade section will be constructed along Interstate 90 through Cascade Mountains of Washington State to avoid making destabilizing cuts in area known as Slide Curve. Several thousand meters of mechanically stabilized earth walls more than 15 m (50 ft) high will provide width for additional lanes and for rockfall and snow avalanche containment.
(PHOTOGRAPH COURTESY OF THE WASHINGTON STATE DEPARTMENT OF TRANSPORTATION.)

4. TUNNELS

Tunnels largely avoid rockfalls by passing beneath or behind the hazardous slope. Tunnels allow for more direct passage through a land barrier without significantly altering it and often provide more favorable roadway geometrics. Extensions or protective works may be needed at the portals to avoid the hazard of rockfalls completely. Along Interstate 40 in the Pigeon River Gorge in western North Carolina, near the Tennessee border, a major rockfall damaged and closed the eastern portals of a twin tunnel (Figure 12-17*a*). That rockfall resulted in the construction of a large rockfall barrier extending beyond the tunnel portal (Figure 12-17*b*).

Because of their expense, tunnels are rarely used by North American transportation agencies. Tunnels are more frequently encountered along roads in Europe, especially in Alpine countries, and in Asian countries such as South Korea and Japan, where rugged terrain and dense populations coupled with intensive land use make tunnels a more viable solution. In central Switzerland, the famous Axenstrasse was constructed in the 1860s along the steep shores of Lake Uri with numerous tunnels and rockfall protection measures (Figures 12-18 through 12-20).

In contrast to roads, railways have more stringent grade and curvature limits on their alignment. Such limits, coupled with railways' desire to mitigate disruptions to their trains from rockfall, cause them to use tunnels to a much greater extent. Among all transportation modes, the characteristically large expenditures for tunnels are justified when that alternative can address other route deficiencies, such as repeated closures due to snow avalanches and high accident rates. On the downside, especially for highway tunnels, longer tunnels require lighting, fire suppression, ventilation, and drainage systems that add substantially to their already high cost.

FIGURE 12-14
Linn Cove Viaduct on Blue Ridge Parkway skirts rugged and rocky slopes of Grandfather Mountain in western North Carolina. Viaduct avoided construction of large cuts and fills and reduced roadway's impact on scenic value of area.
(PHOTOGRAPH COURTESY OF THE NORTH CAROLINA DEPARTMENT OF TRANSPORTATION.)

FIGURE 12-15
At this location in New Zealand, rockfall hazards are avoided through use of bridge that allows rockfalls to pass beneath the roadway.
(PHOTOGRAPH COURTESY OF J. D. DUFFY.)

FIGURE 12-16
Different view of bridge shown in Figure 12-15. Note diversion structures to guide rockfalls away from bridge bents.
(PHOTOGRAPH COURTESY OF J. D. DUFFY.)

(a)

(b)

FIGURE 12-17
Interstate 40 in Pigeon River Gorge of western North Carolina has twin tunnels: (*a*) major rockfall damaged and closed eastern tunnel portals in 1985; (*b*) large rockfall barrier now protects area.
(PHOTOGRAPHS COURTESY OF THE NORTH CAROLINA DEPARTMENT OF TRANSPORTATION.)

FIGURE 12-18
Axenstrasse, a several-kilometer-long roadway (section identified by arrows), is located midslope along fjordlike Lake Uri in central Switzerland. Constructed in mid-1800s, it includes dramatic series of rock cuts, rockfall sheds, bridges, and tunnels.
(PHOTOGRAPH COURTESY OF T. C. BADGER.)

FIGURE 12-19
Axenstrasse bridge and tunnel near Sisikon, Uri Canton, Switzerland (http://flickr.com/photos/jag9889/2002627401/.)

FIGURE 12-20
Axenstrasse with tunnel in distance in 1904. (Photograph From Wikimedia Commons: http://commons.wikimedia.org/wiki/File:Axenstrasse_with_tunnel_in_distance.jpg. Public domain image reproduced from J. Cook, *Switzerland: Picturesque and Descriptive*, H. T. Coates and Co., Philadelphia, Pa., 1904, p. 123.)

ACKNOWLEDGMENTS

The authors gratefully acknowledge the assistance of Rick Andrew and Larry Pierson, who provided a preliminary draft document for this chapter and reviewed the chapter.

REFERENCES

ABBREVIATIONS

Caltrans	California Department of Transportation
FHWA	Federal Highway Administration

Beck, T. J., and J. D. Duffy. 2008. The Ferguson Rockslide, Mariposa County, California. *Proc., 59th Highway Geology Symposium*, Santa Fe, N.Mex.

Caltrans. 2001. *A History of Road Closures Along Highway 1, Big Sur, Monterey and San Luis Obispo Counties, California*. www.dot.ca.gov/dist05/projects/bigsur/inventory_reports/history_road_closures.pdf. Accessed Jan. 2009.

Caltrans. 2005. Why We Need a Bridge at Pitkins Curve and a Rock Shed at Rain Rocks. www.dot.ca.gov/dist05/projects/pitkins/whybridge_shed.pdf. Accessed Jan. 2009.

Caltrans. 2006a. Devil's Slide Tunnels Project Fact Sheet. www.dot.ca.gov/dist4/dslide/docs/factsheet_042506.pdf. Accessed Jan. 2009.

Caltrans. 2006b. *Final Environmental Impact Report: Highway 1 Improvements at Pitkins Curve and Rain Rocks*. www.dot.ca.gov/dist05/projects/pitkins/feir_web.pdf. Accessed Jan. 2009.

Caltrans. 2006c. Pitkins Curve/Rain Rocks Project Information Sheet. www.dot.ca.gov/dist05/projects/pitkins/factsheet.pdf. Accessed Jan. 2009.

Caltrans and FHWA. 2005. *Final Environmental Impact Report and Environmental Assessment: Confusion Hill Highway Realignment Project*. www.dot.ca.gov/dist1/d1projects/confusionhill/confhill_eir.pdf. Accessed Jan. 2009.

Cornforth, D. H. 2005. *Landslides in Practice: Investigations, Analysis, Remedial/Preventative Options in Soils*. John Wiley and Sons, Inc., Hoboken, N.J.

FHWA. 2008. State Route 89–Emerald Bay, South Lake Tahoe. www.fhwa.dot.gov/environment/flex/cs04.htm. Accessed Jan. 2009.

Frey, C. A., S. L. Danielson, and D. M. Slakey. 1996. Glenwood Canyon Tunnel Aesthetic Design. In *Transportation Research Record 1549*, Transportation Research Board, National Research Council, Washington D.C., pp. 61–68.

Geiger, G., R. W. Humphries, and P. C. Ingraham. 1991. Design of Repairs to Highway Rock Slopes Along the Ohio River in Ohio. *National Symposium on Highway and Railroad Slope Maintenance*, Association of Engineering Geologists, Denver, Colo., pp. 75–90.

Hapke, C. J., and K. R. Green. 2005. Rates of Landsliding and Cliff Retreat Along the Big Sur Coast, California—Measuring a Crucial Baseline. U.S. Geological Survey Fact Sheet 2004-3099. http://pubs.usgs.gov/fs/2004/3099/fs2004-3099.pdf. Accessed Jan. 2009.

Harp, E. L., M. E. Reid, J. W. Godt, J. V. DeGraff, and A. J. Gallegos. 2008. Ferguson Rock Slide Buries California State Highway Near Yosemite National Park. In *Landslides*, Springer-Verlag, Vol. 5, pp. 331–337.

Keaton, J. R., and G. H. Beckwith. 1996. Important Considerations in Slope Design. In *Special Report 247: Landslides: Investigation and Mitigation* (A. K. Turner and R. L. Schuster, eds.), Transportation Research Board, National Research Council, Washington, D.C., pp. 429–438.

Turner, A. K. 1996. Colluvium and Talus. In *Special Report 247: Landslides: Investigation and Mitigation* (A. K. Turner and R. L. Schuster, eds.), Transportation Research Board, National Research Council, Washington, D.C., pp. 525–554.

Wills, C. J., M. W. Manson, K. D. Brown, C. W. Davenport, and C. J. Domrose. 2001. *Landslides in the Highway 1 Corridor: Geology and Slope Stability Along the Big Sur Coast Between Point Lobos and San Carpoforo Creek, Monterey and San Luis Obispo Counties, California*. Department of Conservation, California Geological Survey, Sacramento.

Chapter 13

Richard D. Andrew and
Lawrence A. Pierson

STABILIZATION OF ROCKFALL

1. INTRODUCTION

The goal of stabilization is to reduce the potential for rockfall by preventing or inhibiting material from dislodging at the source area through a reduction in the driving forces or an increase in the resisting forces, or both. Stabilization of rock slopes has been well documented in several publications (Fookes and Sweeney 1976; Wyllie 1980, 1991; McCauley et al. 1985; Schuster 1995; Wyllie and Norrish 1996; Morris and Wood 1999; Wyllie and Mah 2004).

Rock slope stabilization methods can be divided into two categories—*rock reinforcement* and *rock removal*—as shown in Figure 13-1. On a typical rock cut construction project, a combination of methods is usually required, but the degree to which each is used depends on site conditions and accessibility. For example, in cases where the rock quality is good and the joint orientations are generally favorable, scaling and localized rock bolting may be sufficient. However, if the rock quality is low or the rock is susceptible to weathering, a larger investment, perhaps including pattern bolting and shotcreting, may be more appropriate. In many cases, stabilization techniques can provide more highly reliable and cost-effective mitigation than do protection measures such as catchment fences or barrier walls, which have to contend with the uncertainties of rockfall behavior once rocks are in motion, or avoidance measures, which require costly realignments or structures. Where rock quality is poor, rock reinforcement may not be practical. In these cases, the slope may benefit from rock removal or resloping. Alternatively, rockfall protection measures may be considered, as described in Chapter 14. Construction costs of stabilization measures for new rock cuts will be lower if the measures are included in the planning and design phase of the project and are constructed as the rock slopes are developed, while slope access is optimal.

1.1 Rock Reinforcement

Resisting forces are effectively increased with tensioned or untensioned slope reinforcements or by buttressing portions of the slope that are oversteepened or susceptible to differential erosion. Figure 13-2 shows several reinforcement techniques, ranging from tensioned bolts to concrete buttresses. These measures serve to limit further relaxation and loosening of the rock mass that may occur because of excavation, unloading, or weathering along discontinuity surfaces. If the rock is allowed to relax, the small movements that occur reduce the interlocking of adjacent blocks and decrease shear strength along discontinuities. In this regard, rock reinforcement techniques are most effective when they are used before or during excavation.

1.2 Rock Removal

Figure 13-3 shows various rock removal methods that may be used to improve slope stability. Rock removal methods include reshaping zones of unstable rock, trim blasting of overhangs, and, most commonly, slope scaling. Rounding the brow of a rock cut combines reshaping and scaling efforts on a portion of slope that is commonly less stable because of longer exposure to weathering and that is often more heavily damaged by previous construction practices.

Rock excavation normally involves blasting. In most civil engineering projects, blasting is the responsibility of the contractor. The work is specified under a performance-type contract in which the desired results are described, but the methods are left to the contractor.

Production blasting, which is used to break rock and facilitate excavation, can easily damage the rock mass well beyond the design location of the finished cut face unless precautions are taken. Techniques referred to as *controlled blasting* are generally specified in contract documents as a way to form the final cut slope and limit the damage related to production blasting. Controlled blasting measures can also be used during reshaping, trimming, or sculpting of slopes. Sculpting is used in cases where the finished appearance of cut slopes must meet special aesthetic criteria, which is an increasingly common requirement (Cummings 2002).

Slope scaling is a process for removing loose rock from a slope face. It is effective in the short term but less so in the long term because of the ongoing relaxation and weathering of the face. Scaling or other removal methods that do not significantly change the slope configuration will not be adequate by themselves. To maintain the desired improvement, either a recurring scaling program should be planned or a suitable protective measure should be included to address future rockfall.

The discussions in this chapter focus on routine stabilization measures. In accordance with Figure 13-1, rock reinforcement methods are discussed first, followed by rock removal methods. The chapter includes a summary of associated issues, including constructability and considerations and tips with regard to aesthetics. The chapter is presented in the following order:

- Installation of mechanical reinforcement, including tensioned rock bolts and strands or untensioned rock dowels, cable lashing of unstable blocks, and buttresses (Section 2);
- Rock face protection by shotcrete or by installation of anchored mesh (Section 3);
- Stabilization of the rock face by grouting discontinuities (Section 4);

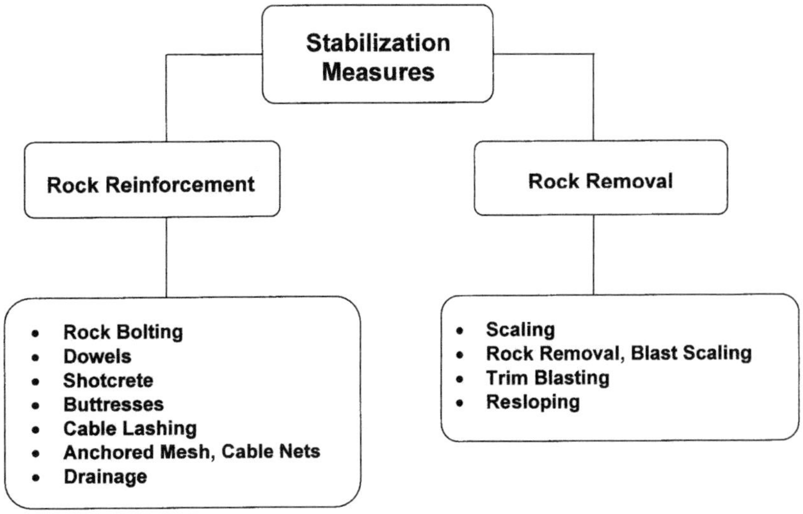

FIGURE 13-1
Rock slope stabilization categories (adapted from Wyllie and Norrish 1996).

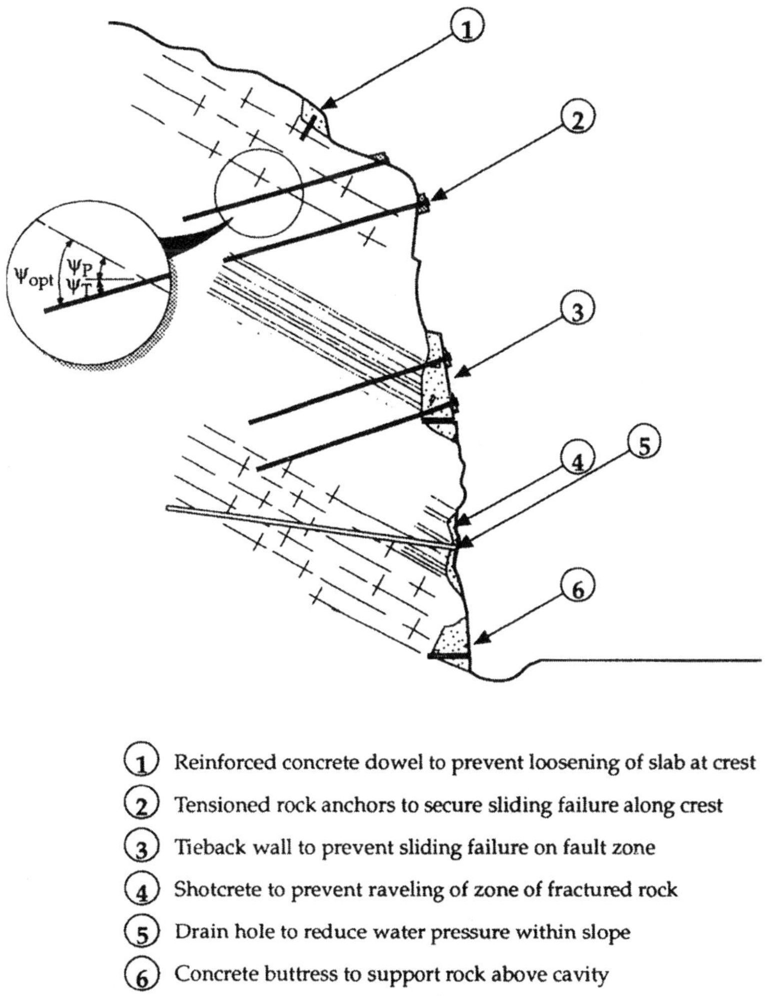

FIGURE 13-2
Rock slope reinforcement methods (Wyllie and Norrish 1996).

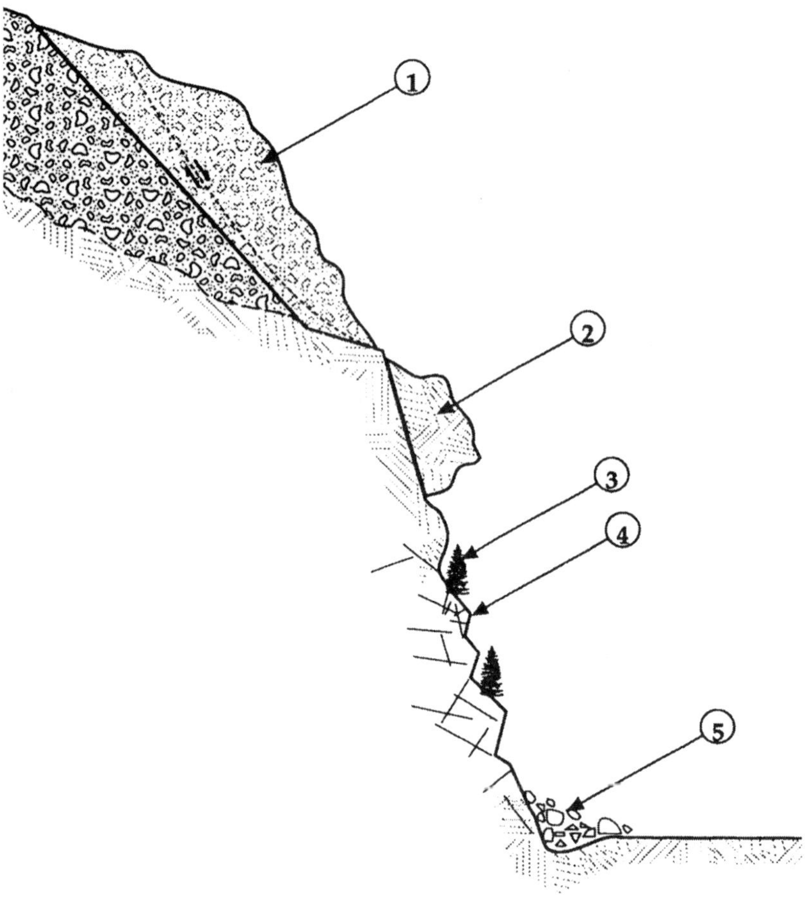

FIGURE 13-3 Rock removal methods (Wyllie and Norrish 1996).

- Improved drainage (Section 5);
- Excavation and reshaping of a rock face by blasting (Section 6);
- Removal of potentially unstable blocks by scaling (Section 7); and
- Constructability and aesthetics considerations (Section 8).

2. INSTALLATION OF MECHANICAL REINFORCEMENT

Reinforcement of a rock slope can be provided by tensioned bars or strands, untensioned bars or shear pins, cable lashing or other forms of lashing, shotcrete, or buttresses. Most of these methods rely on strength improvements within the rock mass, while some act externally on the rock surface.

Tensioned bars (rock bolts) and strands are frequently referred to collectively as *rock anchors*. Rock anchors increase shear strength along discontinuities by increasing the normal force acting along potential failure surfaces. Depending on their orientation, rock anchors can improve resistance to sliding by using the tensile or the shear strength of the steel (Figure 13-4).

Once tensioned, rock anchors are considered active reinforcements. *Untensioned bars (rock dowels) and shear pins* are a passive type of reinforcement. Both provide the shear resistance of the steel bars, but rock dowels do not provide tensional resistance similar to a rock bolt until initial block movement occurs. In certain rock types, the rock mass can relax and rebound after the excavation of material. Rock dowels that are installed as the rock excavation is removed can go into tension as the rock relaxes, effectively loading the dowel. Procedures governing the design and installation of rock anchors are well documented (Littlejohn and Bruce 1975; FHWA 1982; British Standards Institute 1989; Xanthakos 1991; Wyllie and Norrish 1996; Post-Tensioning Institute 2004).

2.1 Rock Bolts and Strands

Rock bolts and cable strands (also referred to as *tendons*) provide tensioned reinforcement. The tensioned elements consist of a steel bar or tendon installed through unstable blocks into a stable base. Such systems typically include an appropriately sized bearing plate and hexagonal nut to hold the applied tension and transfer the load from the bearing plate to the rock. They are generally encased in polyester resin or cement grout to provide corrosion protection and to transfer the bolt tension loads into the adjoining rock mass. A beveled washer may be needed to apply the force evenly if the angle of the rock surface is subperpendicular to the axis of the anchor. Tensioned anchors provide more reliable support in cases where the blocks being stabilized are more than 1 to 2 m (3 to 6 ft) thick or where there has been prior movement.

On slopes where the blocks have already experienced strain along discontinuities, tensioned

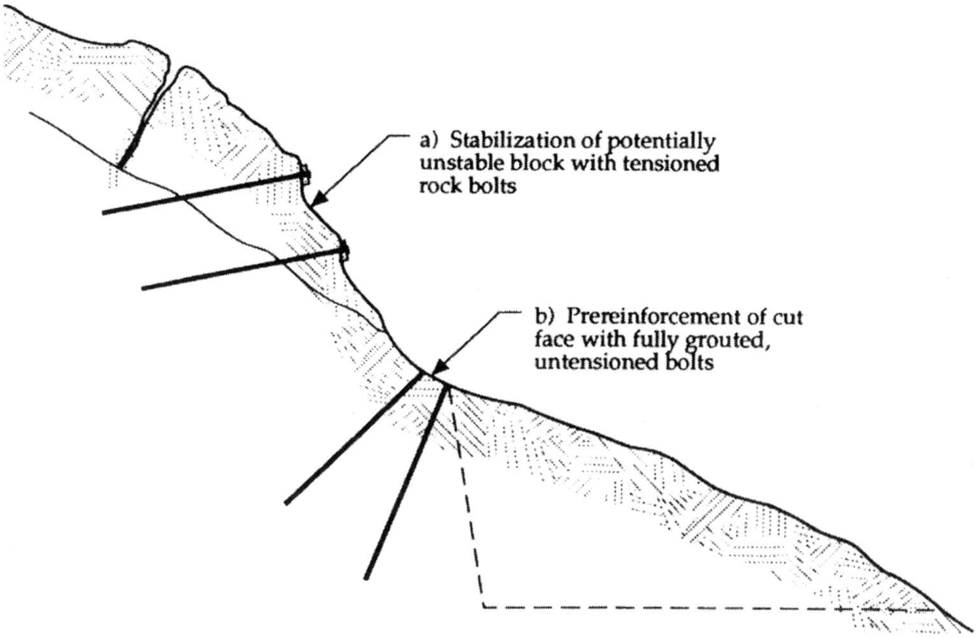

FIGURE 13-4
Reinforcement of rock slope with (*a*) tensioned rock bolts and (*b*) untensioned rock bolts (Wyllie and Norrish 1996).

reinforcements improve the resistance to further sliding or toppling. On new slopes, especially when they are installed as the slope is constructed, tensioned reinforcements maintain slope stability by preventing block movements that reduce shear strength. Factors affecting rock bolt and strand design include desired capacity; spacing; block dimensions; lithology; design life; cost; and constructability, typically related to access (British Standards Institute 1989; Wyllie 1991; Xanthakos 1991; Wyllie and Norrish 1996; Post-Tensioning Institute 2004; Wyllie and Mah 2004).

A component of the applied tension force acts normal to the sliding surface of a block. This increases shear resistance between adjoining block faces in proportion to the applied tension. Block stability is also enhanced by orienting the bolt or tendon so that a portion of the applied tensile load and bolt capacity resists sliding. The length of the bolt or tendon depends on two items: (*a*) the bond strength (adhesion) between the anchoring element and the rock, which is at least partly related to the lithology, and (*b*) the spacing of the discontinuities that form the deepest part of the block being stabilized.

Although it may seem intuitive, establishing the bond zone or anchorage for a rock bolt or tendon beyond the block to be held in place is important. The required length for this portion of the bolt, referred to as the *bond length*, depends on whether the anchorage is established with a mechanical device or with resin or cement grout and on the rock type. If a large number of anchors that rely on resin or cement grout is anticipated for a project, preproduction testing may be valuable in establishing an efficient bond length to reduce material and labor costs.

To estimate the bond length, the following equation may be used:

$$L_b = \frac{Q}{\pi d \tau_a}$$

where

L_b = bond length,
Q = design load at the head of the anchor,
$\pi = 3.1415$,
d = diameter of the drill hole, and
τ_a = working bond stress along the interface between the rock and the grout.

Typical values for the working bond stress (τ_a) and ultimate bond stress (τ_u) between rock and grout are given by the following equations:

$$\tau_a = \frac{\sigma_u}{30}$$

and

$$\tau_u = \frac{\sigma_u}{10}$$

where σ_u is the uniaxial compressive strength of the rock in the bond zone.

These formulas indicate that the factor of safety between the ultimate and working bond stress is 3. The working bond stress (τ_a) for a variety of rock types and rock strengths is shown in Table 13-1 (Littlejohn and Bruce 1975; Post-Tensioning Institute 2004).

Two basic types of resin are available: polyester and epoxy (Wyllie and Norrish 1996; Post-Tensioning Institute 2004; Wyllie and Mah 2004). Epoxy resin is much more fluid and is typically injected through a nozzle applicator. Epoxy resin may be preferred for installations that are oriented nearly vertically downward, but this is rarely the situation. Polyester resin is routinely used for most applications. Most polyester resin is installed by using two-part cartridges referred to colloquially as *sausages*. As the rock bolt is inserted and spun at a specified rate (usually on the order of 60 rpm) to the end of the hole, the cartridges are broken and the two-part material is mixed. Because tendons are flexible and cannot be spun into place, use of two-part cartridges with tendons is not appropriate (Post-Tensioning Institute 2004).

Table 13-1
Approximate Working Bond Stress for Different Rock Types

Rock Type	Working Bond Stress (τ_a) at Rock–Grout Interface: MPa	psi
Strong rock	01.05–1.40	150–200
Medium rock	0.70–1.05	100–150
Weak rock	0.35–0.70	50–100
Granite, basalt	0.55–1.0	80–150
Dolomitic limestone	0.45–0.70	70–100
Soft limestone	0.35–0.50	50–70
Slate, strong shale	0.30–0.45	40–70
Weak shale	0.05–0.30	10–40
Sandstone	0.30–0.60	40–80
Concrete	0.45–0.90	70–130

Source: Wyllie 1999.

If polyester resin is used as a bond for rock bolts, a fast-setting type is used in the bond length and a slow-setting type in the stressing length. The rock bolt is tensioned after the fast-set hardens and before the slow-set gels. Directions from the manufacturer typically include the recommended hole size and bar size for a particular resin cartridge, the required rate of spin and how long to spin the bar to mix the resin completely, set times, and specific properties of the resin such as temperature ranges for acceptable performance. Manufacturer guidance also defines the minimum number of cartridges required to fill the annulus between the rock bolt and the sides of the drill hole.

Cement grouting procedures follow either a two-stage or a single-stage grouting method. The two-stage method uses a bond breaker to separate the stressing and bonding lengths along the bolt and requires the bond-zone grout to cure before the anchor is stressed. Figure 13-5 shows a typical two-stage rock bolt with the first stage of cement grout installed in the bond zone.

If a mechanically anchored rock bolt is used, as shown in Figure 13-6, the anchor head is set by expanding the shell against the sides of the drill hole, the prescribed tension is applied, and then the cement grout is added to fill the hole. This allows for a one-stage grouting process.

For temporary installations, once the anchorage is established, the anchors may be installed without cement grout or resin encapsulation that provides long-term corrosion protection. When appropriate, this can greatly speed up installation and reduce cost. As a precaution, except in cases where the application is clearly for short-term support, corrosion protection should be included. "Temporary" installations commonly become "permanent" and are relied on for much longer periods than originally intended.

In more corrosive environments, cement grout is considered a more appropriate alternative grout material because it creates a higher-pH, reducing environment around the steel. Resin, on the other hand, is inert in this regard and becomes chemically inactive once it sets.

Additional corrosion protection measures for permanent rock bolts include epoxy-coated or galvanized bars with cement grout encapsulation and galvanized bearing plates and anchorage devices. In extreme applications, the use of stainless steel anchors or multiple corrosion protection systems

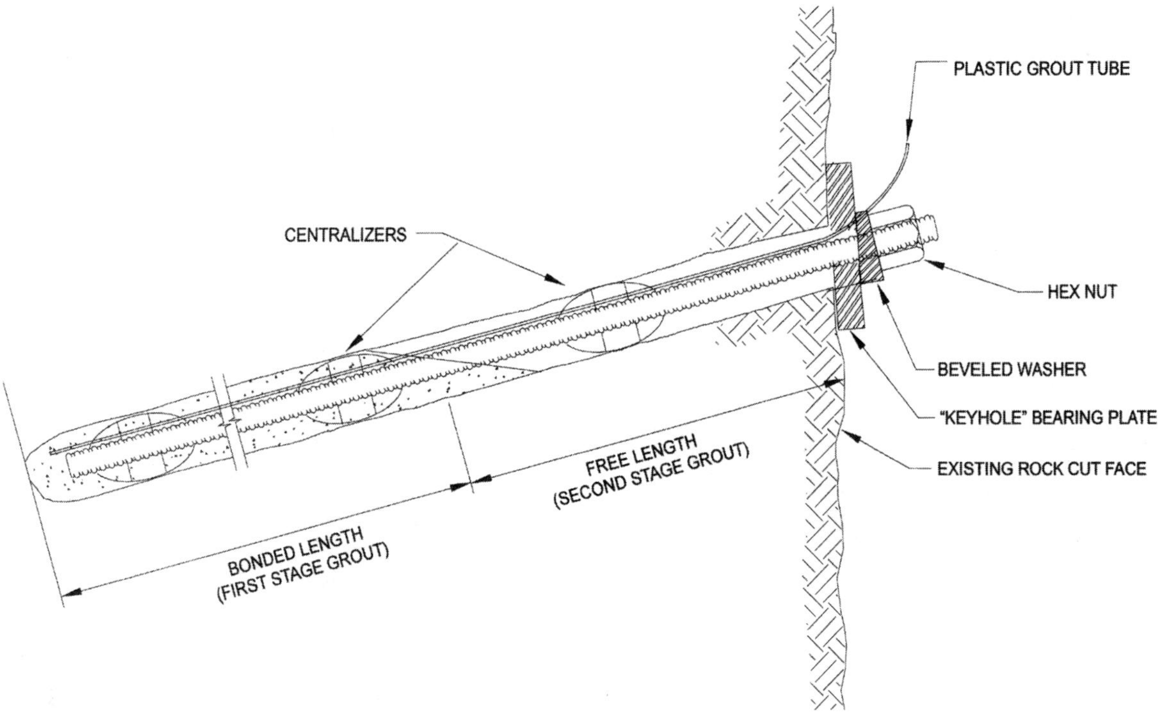

FIGURE 13-5
Typical rock bolt installed with cement grout.
(DIAGRAM BY R. ANDREW.)

may be appropriate. However, because of the substantially higher cost of these applications, their use is typically limited to conditions such as the following:

1. The groundwater or bedrock has low pH and high resistivity;
2. Significant sulfates are present; or
3. Conditions have been identified that produce a corrosive environment, such as those associated with some coal deposits.

Rock bolting applications vary. In cases where the rock mass characteristics produce hazardous

FIGURE 13-6
Mechanical rock bolt anchor with expanding shell.
(PHOTOGRAPH COURTESY OF D. BECKSTRAND.)

blocks of adequate dimension and the relationships between these blocks can be determined, individual or key block bolting may be appropriate. Where repeated sets of discontinuities form numerous smaller blocks or slabs or less competent rock is present, a regularly spaced reinforcement pattern may be more appropriate. Many technical books and manuals are available for designing rock bolt reinforcement for stabilization (Hoek and Bray 1981; FHWA 1982; British Standards Institute 1989; Wyllie and Mah 2004).

Installing rock bolts in locations where access to the slope is limited (Figure 13-7) can be challenging and can dramatically affect the installation cost. Typically, hand-operated equipment can be used to drill small-diameter rock bolt holes up to 6 m (20 ft) where the anchorage is established mechanically or with cement grout. Because of the spinning requirements for mixing the two-part resin cartridges, the length of resin anchorages is generally limited to 5 m (15 ft) unless specialized drilling equipment is used. Some contractors have used a drill mounted on a platform suspended from a crane (Figure 13-8). Others have mounted a drill on the end of a long boom, thereby enabling drilling and installations up to 40 m (130 ft) from the roadway. Rubber-tired drill baskets and drill rigs, commonly referred to as *spider baskets* and *wagon drills*, respectively, are routinely used for

FIGURE 13-7
Rock bolt installation in area of restricted access, Glenwood Canyon, Colorado.
(PHOTOGRAPH COURTESY OF THE COLORADO DEPARTMENT OF TRANSPORTATION.)

FIGURE 13-8
Rock bolting from crane basket.
(PHOTOGRAPH COURTESY OF R. D. ANDREW.)

sites that are difficult to access, and can be cost competitive with crane-supported operations for more accessible slopes. Current practice favors the use of wagon drills (Figure 13-9), which are not limited by the reach of supporting equipment and offer a more economical and efficient means of access to even rugged, steep slopes.

FIGURE 13-9
Wagon drills often permit economical drilling on steep, rugged slopes.
(PHOTOGRAPH COURTESY OF R. D. ANDREW.)

Each site will have unique constructability issues that will directly affect installation costs. Thus, design specifications for reinforcement measures often include experience requirements for installers and controls and directions for drilling, grouting, stressing, and testing, as well as acceptance criteria. Many rock bolts intended to be tensioned have ultimately been installed as untensioned dowels because of poor drilling, improper tensioning procedures, or a general lack of understanding on the part of less-experienced contractors.

2.1.1 Key Block Bolting

Key blocks are specific blocks on the slope face that provide support to adjacent blocks on the basis of their orientation and position (see Figure 2-28). This rock bolting approach is most applicable to slopes made up of larger blocks and stronger rock. The presumption is that securing the key block reduces the ability of surrounding blocks to move past it. Developing an effective key block mitigation strategy requires experienced personnel with the ability to recognize the three-dimensional arrangement of blocks. Since all adjacent blocks interact with each other, it is important that the blocks affecting the greatest area or number of surrounding blocks be correctly identified. Key block bolting can be a more cost-effective method for stabilizing a slope than pattern bolting, which typically uses many more rock bolts to accomplish a similar result.

2.1.2 Pattern Bolting

Pattern bolting is a technique in which the face of a rock excavation is reinforced on a preestablished rectangular spacing designed to increase the slab thickness and strengthen the face. If necessary, the prescribed bolt locations can be adjusted to penetrate particular rock blocks of concern, but the pattern density, rather than the precise rock bolt locations, is relied on to improve slope stability. Weaker or more highly fractured rock typically benefits more from pattern bolting than from the application of a key block approach because it provides overall reinforcement of the face.

2.2 Dowels

Dowels are untensioned steel bars used for passive reinforcement. They are installed as either a rock dowel or a shear pin, depending on the application. Rock dowels are used on steeper slopes in a manner similar to rock bolts. Shear pins are normally installed on flatter slopes, where the rock discontinuities play a major role in forming an overall flatter slope angle and define the potential failure surfaces.

2.2.1 Rock Dowels

Rock dowels are steel bars that are installed before rock movement. They are generally applied to newly constructed rock cuts to prevent ongoing slope relaxation and block movement related to excavation (unloading). Figure 13-10 shows a typical detail of a rock dowel. Rock dowels improve slope stability in two ways:

1. The shear strength of the bar increases the resisting forces along the potential plane of weakness.
2. Once block movement occurs, depending on the dowel orientation, the tensile strength of the bar is engaged, which resists sliding, and the normal force between opposing discontinuity surfaces increases, thereby increasing the shear resistance along the discontinuity.

The block dimensions control the length of the rock dowels. Because dowels initially provide only passive resistance to sliding, they are most effective when there has been less prior slope relaxation or block movement, so that the interlocking

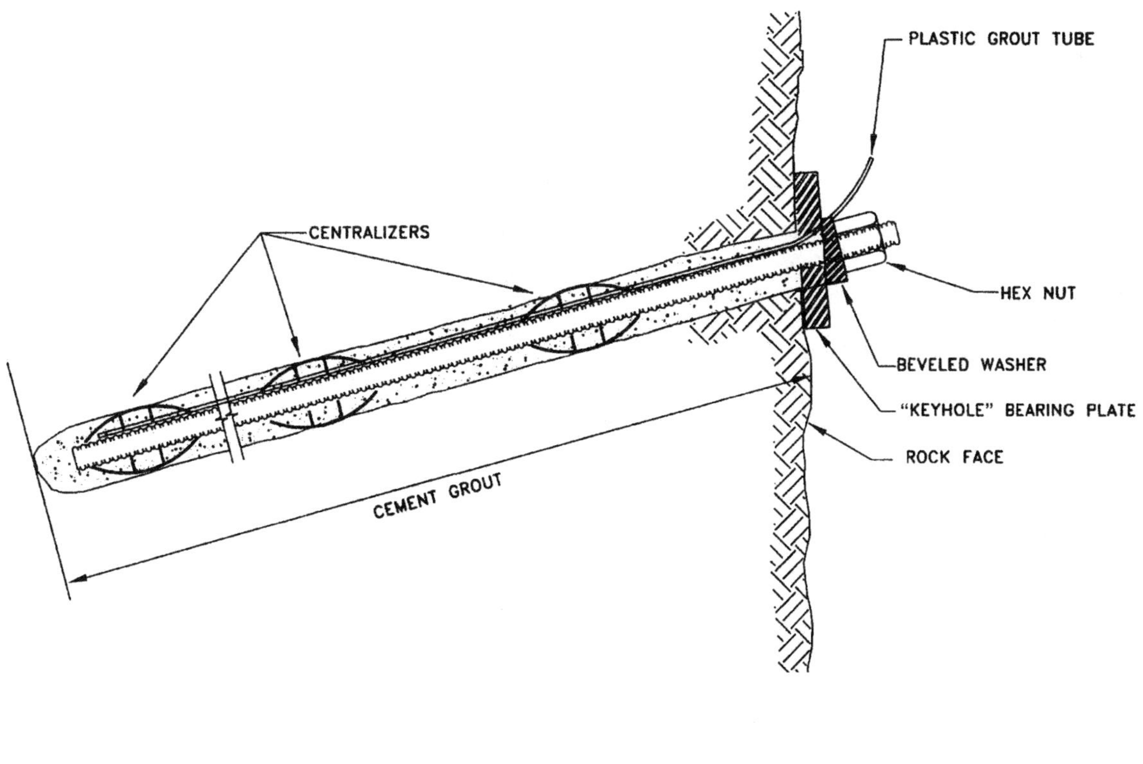

FIGURE 13-10
Typical installation of rock dowel.
(DIAGRAM BY R. ANDREW.)

FIGURE 13-11
Shear pins before encapsulation in concrete.
(PHOTOGRAPH COURTESY OF C. BRAWNER.)

FIGURE 13-12
Installing cable lashing.
(PHOTOGRAPH COURTESY OF R. ANDREW.)

rock bridges between adjacent blocks are retained. Dowels are less expensive than rock anchors because they are faster to install, usually requiring only one setup at each installation site.

2.2.2 Shear Pins

Shear pins are composed of lengths of reinforcing steel, typically less than 1.5 m (5 ft) long, that are grouted into holes so that they extend above the rock surface in front of the leading edge of a rock block or slab resting on an inclined plane. In some applications, longer shear pin dowels are installed roughly perpendicular through the slab into the underlying rock. In both instances, the shear strength of the steel is relied on to provide support. A typical shear pin is usually about 25 mm (1 in.) in diameter. The design goal is to increase shear resistance along the potential failure surface by using the shear strength of the steel.

As shown in Figure 13-11, shear pins are generally installed normal to a bedding plane or other discontinuity that forms a basal sliding surface. Because of the difficulty in drilling and installing shear pins that are in direct contact with the block, concrete is normally cast around the pins and between the block and the pins. The concrete acts as an additional shear block that immediately transfers any block movement to the shear pin, engaging the shear strength of the steel. In other cases, the connection has been established by welding a metal frame between the bar and the rock face.

2.3 Cable Lashing

Potentially unstable blocks that are too difficult or large to remove safely can be effectively stabilized by cable lashing. A basic cable lashing system consists of two anchors and a tensioned cable passed in front of a block or column of rock (Figure 13-12). Generally, the cable is constructed with woven wire rope because it is more flexible and better able to conform to the shape of the rock block.

In some cases, the number and size of the anchors used to secure the cable ends are designed to support the entire estimated weight of the rock, plus a safety factor. In other situations, the anchorage requirements can be reduced, such as when the goal is to prevent the initiation of toppling. On the basis of a simple force diagram with moments to calculate the force required to prevent the rock from tipping, the anchorage requirements can be much lower, making the use of cable lashing more practical and cost-effective.

A key design consideration is the shape of the rock face and the ability to wrap the cable tightly around the front of the rock between the anchor points. Cables stretched across a relatively flat surface, as shown in Figure 13-13, require too much block displacement before the tensile strength of the cable is engaged. In these cases, the blocks may still be able to fail.

Cables can be installed in any orientation from vertical to horizontal, and more than one cable is often used, depending on the block size and shape. The anchorages are typically untensioned rock dowels or cable anchors installed in a stable area on opposite sides of the block. Once installed, the cables are attached and then tensioned. Preferably, the cable will include at least one in-line turnbuckle that can make cable tensioning easier and be accessed later for retightening as the steel cable stretches over time. Cable lashing is particularly

FIGURE 13-13
Improper cable lashing. Note that bottom block has fallen out.
(PHOTOGRAPH COURTESY OF B. BLACK.)

useful in stabilizing a rock on a relatively stable slope where removal of the block by scaling is not practical.

If it is needed, additional support can be provided by using wire mesh panels or cable nets. For the large block shown in Figure 13-14, several cables and a cable net were independently anchored and used to provide permanent support.

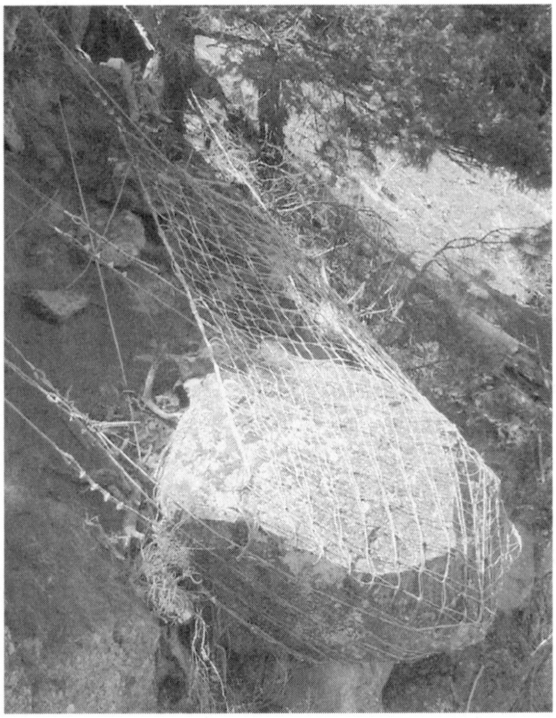

FIGURE 13-14
Cable and cable net lashing system. Note limited base support for block.
(PHOTOGRAPH COURTESY OF B. GEORGE.)

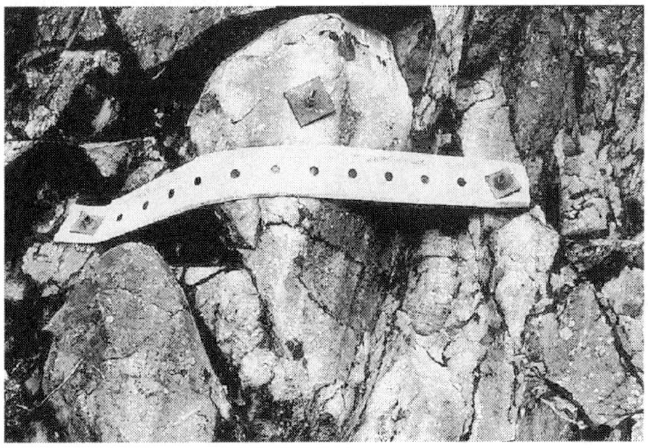

FIGURE 13-15
Temporary mine strap lashing with permanent rock bolt installed.
(PHOTOGRAPH COURTESY OF L. PIERSON.)

As shown in the figure, a "come along" was used to tension the cables. The sharp angle around the face of the block allowed the cables to be tensioned properly at the time of installation.

Cable lashing may also be used as a temporary measure during construction of other permanent measures, such as when support of a block is needed before it is deemed safe to drill and install a permanent rock bolt. A similar technique uses steel mining straps or whalers instead of a cable to provide temporary support (Figure 13-15). Such systems are generally removed once the permanent measure is constructed, since these applications can be noticeable and not acceptable for permanent use.

2.4 Buttresses

When a rock slope face contains a cavity or an overhanging area, typically resulting from a previous rockfall or differential erosion, it may be appropriate to construct a buttress to provide support and prevent further falls. Buttresses typically consist of cast-in-place concrete or strategically placed steel girders and serve two functions: to retain and protect areas of weak rock and to support the overhang.

Buttresses are effective and permanent but tend to be costly to construct because of the design customization and quantity of materials required. They are considered where the area to be supported represents a large portion of the slope or is critical to the stability of other portions of the slope and removal is not advisable. No two buttresses will be the same, because they are specifically tailored to the site geometry and slope conditions. These structures generally are developed at highway level but can be as effective at higher areas of a slope. Examples of buttresses are shown in Figures 13-16 and 13-17.

Whenever possible, buttresses should be designed so that the buttress is placed in compression and

(a)

(b)

FIGURE 13-16 Two views of concrete buttress supporting large rock block, Glenwood Canyon, Colorado. (PHOTOGRAPHS COURTESY OF THE COLORADO DEPARTMENT OF TRANSPORTATION.)

bending moments and overturning forces are minimized. For the buttress to prevent relaxation or creeping of the overhanging block, it should be founded on a sound rock surface and poured so that it is in full contact with the underside of the overhang. If this surface is not at right angles to the direction of thrust, it should be anchored to the base to prevent sliding. It may be necessary to place the last pour through a hole drilled downward into the cavity from the rock face and to use additives in the concrete mix to reduce shrinkage.

FIGURE 13-17 Example of buttress support of rock ledge, Chapmans Peak Drive, Cape Town, South Africa (http://commons.wikimedia.org/wiki/Image:Chapmans_peak_dr.jpg).

If a buttress is needed at a midslope location, construction requires experienced personnel, and helicopters may be needed for delivery of materials (Figures 13-18 and 13-19). The buttress is generally anchored to stable rock by encapsulating rock bolts in a block of concrete (the concept is similar to that for shear pins, which are discussed in Section 2.2.2). The buttress is designed to increase the shear resistance of the unstable block by the shear strength of the steel and the bond between the concrete and the rock surface. This method may require significant quantities of steel. For potential wedge and planar failures, buttressing is generally only considered where other stabilization or protection measures are not constructable or economically feasible.

3. ROCK FACE PROTECTION BY SHOTCRETE OR ANCHORED MESH

Shotcrete and anchored mesh are minor reinforcement systems that provide limited face support to stabilize rock blocks. Both apply a normal force to the slope that helps hold blocks in place, but the improvement in that respect is limited.

3.1 Shotcrete

Shotcrete is a portland cement concrete product containing fine-grained aggregate [typically less than 13 mm (3⁄8 in.) in diameter] that is sprayed onto a slope with compressed air and a nozzle applicator to form layers 75 to 100 mm (3 to 4 in.) in thickness (American Concrete Institute 1995). The quality and strength of the shotcrete depend on the experience and skill of the person applying the shotcrete, since consolidation of the shotcrete occurs as the sprayed material strikes the slope; if the shotcrete slumps for any reason, the strength is lost, and that shotcrete should not be incorporated into the final product. In zones or layers of highly weathered or weaker rock, a shotcrete facing can reduce further degradation and control raveling or erosion of small blocks to maintain support for overlying competent rock. By itself, shotcrete provides only limited tensile strength for reinforcement and will not provide global support or resist larger block movements.

The effectiveness of shotcrete largely depends on the condition of the rock surface during the application. To the extent possible, the surface should be free of loose and broken rock, soil, vegetation, and ice. In addition, the rock surface should be free of standing or running water, but it should be predampened for proper curing and development of good adhesion between the rock and shotcrete. The area covered by the shotcrete should extend at least 0.6 m (2 ft) onto more competent rock to help support the shotcrete layer (FHWA 1993). Rock bolts or dowels can be used to help support or anchor the shotcrete on the rock.

Several admixtures have been used to improve the characteristics of shotcrete, such as increased compressive strength or added viscosity (Wood 1989). The pozzolanic benefits from silica fume can greatly improve shotcrete strength and increase the viscosity, which can allow a greater thickness to be applied to a surface without slumping of the material. Increased viscosity is especially important when shotcrete is applied to locations with wet conditions.

It is important that drain holes be drilled through the shotcrete to prevent buildup of water pressure behind the face; the drain holes are usually about 0.5 m (1.5 ft) deep and are located 1 to 2 m (3 to 6 ft) apart. Otherwise, the elevated water pressure can break the shotcrete bond, and the shotcrete will be damaged or destroyed. Depending on the quality of the underlying strata, they may be lined or unlined. In cases where drainage or seeps are structurally controlled, it is important that the drain holes be drilled before the shotcrete is applied and that they be located to intersect the fractures that carry water; otherwise, the installations may miss these pathways and be ineffective. In these cases, the holes or drainpipes are temporarily plugged with wooden pegs or rags while the shotcrete is applied; the pegs or rags are removed soon after the application is complete and before the shotcrete has set.

For all permanent applications, shotcrete should be reinforced with fibers or welded wire mesh to reduce the propagation of cracks that routinely form as the cement cures. Fibers are made from steel, polyester, nylon, glass, and so forth, with steel being the most commonly used for rock slope applications (Morgan et al. 1989, 1999; Wood 1989). Fibers are now used for shotcrete reinforcement for most applications where welded wire mesh was previously specified (FHWA 1993). Using fibers reduces the cost of shotcrete, since only a single application step is required; however, increased wear on the shotcrete pumping equipment can occur, especially when steel fibers are used.

FIGURE 13-18 Formwork for concrete buttress to support large rock block under construction, Glenwood Canyon, Colorado. (PHOTOGRAPH COURTESY OF THE COLORADO DEPARTMENT OF TRANSPORTATION.)

Welded wire mesh may still be recommended as reinforcement for applications on weathered rock or soil, or across faults where adequate bonding may not occur. Welded wire mesh is fabricated from light-gauge wire woven to form a 100-mm (4-in.) mesh and is attached to the rock face every

FIGURE 13-19 Helicopter delivering concrete for concrete buttress under construction, Glenwood Canyon, Colorado. (PHOTOGRAPH COURTESY OF THE COLORADO DEPARTMENT OF TRANSPORTATION.)

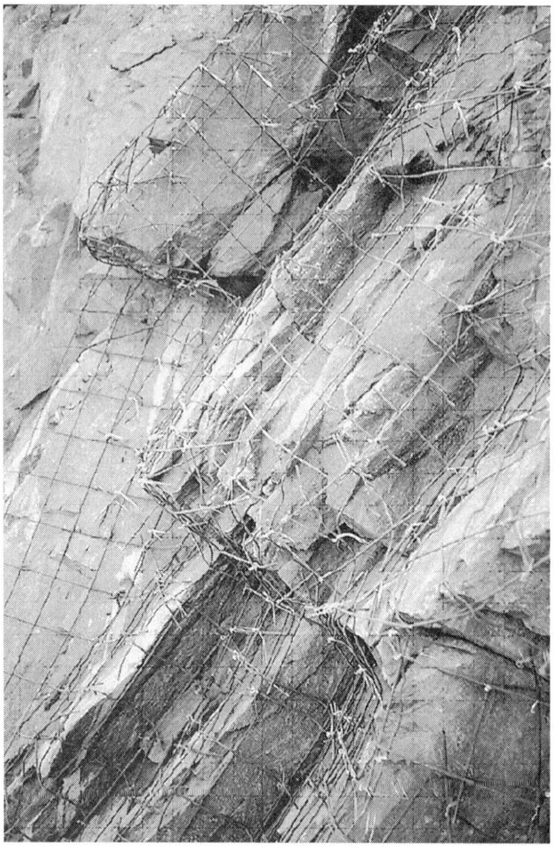

FIGURE 13-20
Welded wire mesh installed before shotcrete application.
(PHOTOGRAPH COURTESY OF C. BRAWNER.)

1 to 2 m (3 to 6 ft) with steel pins, complete with washers and nuts that are grouted into the rock face. The mesh must be attached close to but not in contact with the rock surface for the mesh to be fully encased in shotcrete. Applying welded wire mesh is complicated in locations where the slope surface is irregular (Figure 13-20). The installation of wire mesh in such situations may substantially increase labor costs, and additional problems may be encountered if the mesh is not held tightly in place. If the mesh vibrates or moves during the spraying of shotcrete, the compaction of the shotcrete and its strength and durability will be reduced. In addition, care must be taken to avoid forming voids in the shotcrete behind the mesh.

Shotcrete can be applied as either a wet or a dry mix. With wet-mix shotcrete, the cement ingredients, including the water, are mixed before introduction into the delivery hose. For dry-mix shotcrete, the water is added at the nozzle. Air entrainment is not possible with dry-mix shotcrete, so its ability to resist freeze–thaw cycles is reduced. Shotcrete should be a minimum of 50 mm (2 in.) thick to resist frost damage in colder climates.

During surface preparation, installing gauging pins on the slope before the shotcrete is applied is essential to verify that the minimum shotcrete thickness has been applied. They are installed on roughly 1.8-m (6-ft) centers to protrude from the slope by an amount equal to the specified shotcrete thickness. Once they are covered, it is clear that the proper thickness has been applied. Application problems have occurred when this precaution was not taken. In some cases, later drilling determined that the shotcrete layer was as little as 3 mm (⅛ in.) thick. This error was unintentional and was corrected, but once an area is covered with shotcrete, it is impossible to tell visually how thick the material is without these gauging pins.

If necessary for reasons of appearance, the final shotcrete layer can be sculpted to conform more closely to the native rock; this may reduce the visual impact of the shotcrete. At highly visible locations, if the cost is warranted, the shotcrete can be sculpted and painted or tinted as shown in Figure 13-21, where the three shotcrete applications are practically indistinguishable from the adjacent rock.

FIGURE 13-21
Test sections demonstrating high-quality sculpted and painted or tinted shotcrete.
(PHOTOGRAPH COURTESY OF L. PIERSON.)

3.2 Anchored Mesh

Anchored mesh systems use either woven wire fabric, such as chain link, gabion, or high-tensile-

FIGURE 13-22
Anchored cable nets.
(PHOTOGRAPH COURTESY OF E. ROREM.)

strength steel mesh, or cable nets. Chapter 16 provides additional discussion concerning the design and installation of anchored mesh systems. They are used to cover an entire slope, usually where there is limited or no fallout area available at the base of the slope. If more fallout area is available, the intermediate slope anchors are eliminated to create a drapery system that is self-cleaning and requires only routine maintenance.

Anchored mesh provides some of the same benefits as does cable lashing. An anchored cable net system is shown in Figure 13-22. The drawback to these systems is that if rocks fall, they tend to become trapped behind the mesh close to the source area and form pockets of trapped rock. As more rock dislodges and accumulates, the load applied to the system by the trapped rock may eventually exceed the strength of the nets or anchors, and a system failure can occur with little warning.

System maintenance is difficult. It typically requires periodically disconnecting the mesh from the intermediate anchors to allow trapped debris to drop out for removal, which can be a dangerous procedure. These systems are generally used on slopes composed of smaller blocks [less than 1 m (3 ft) in diameter].

4. GROUTING METHODS TO STABILIZE ROCK FACES

Stabilization of a rock face by grouting techniques, referred to as *rock gluing*, has been practiced since the 1960s. Initially used to stabilize unstable strata units in underground coal mines, this technique has been applied to rock slopes to provide long-term stability in historic or environmentally sensitive areas. The grouting methods fall into two basic categories on the basis of the type of grout: *cementitious* or *chemical*. The viscosity, strength, and setup time can be varied and controlled to a much greater extent with chemical grouts than with cement-based grouts (FHWA 2008).

4.1 Cementitious Grout

Cementitious grout typically consists of portland cement mixed as slurry that can be injected or poured. In some cases, fine aggregate is added to increase strength or consistency. Cement grout has traditionally been used in bonding rock reinforcement (e.g., rock bolts, cables), subgrade improvement, compaction, and mud jacking, to list a few applications. Additives such as deflocculants, accelerators, and expansion and polymeric agents can be used to reduce washout and bleeding of the grout. In addition, fillers such as fly ash, pulverized fuel ash, fine sand, and pea gravel are used to enhance the strength of the grout, particularly where filling of large fissures or cavities is required. For rock-gluing applications, latex- and resin-based products have been added to the cement to improve adhesion qualities. This technique has been used throughout Japan, primarily in areas where rock bolting was not visually acceptable.

4.2 Chemical Grout

More recently, chemical grouts have gained wider acceptance for rock face stabilization than cement-based products. Chemical grouts comprise many systems, including sodium silicate, acrylate, lignin, urethane, and resin grouts. The most commonly used chemical grouts are sodium silicate based to form a polymerized gel capable of binding soil or sediment particles together and filling voids. For rock-gluing applications, the injection of urethane and resin products is preferred because of their relatively high tensile strength properties. These products typically have three to four times the tensile strength of cement- or sodium-silicate-based grouts. In general, cement- and sodium-silicate-based grouts are only used where high-volume, low-cost grouting is required (FHWA 2008).

Polymer injection products are commonly divided into three general classes: polyurethane (PU), polyurethane resin (PUR), and epoxy (EP).

PUR is predominantly used for "gluing" and consolidating weak roof strata in underground coal mine applications. The single-stage PU products are used mostly for water-stop applications, where high shear strength is not required. The PU products generally foam in the presence of water and, as a result, lose strength and density; however, in many cases the product strength is still much greater than that of the surrounding material. The EP products are used primarily for structural foundations that are dry and where only small product quantities are required. EP products are commonly injected with low-pressure pumps that have surface ports attached to a dry surface (FHWA 2008).

PUR grout is applied via an injection rod into the fractures and discontinuities of the rock mass. The product is intended to reinforce the fractured rock to the point where it can support its own weight and the weight of overlying unconsolidated rock by forming a grout-reinforced beam. The beam structure then bridges the weaker or more fractured rock to adjacent abutments with greater supporting strength. Figures 13-23 through 13-25 show the application of rock-gluing techniques on a rock face directly above a tunnel in Poudre Canyon, Colorado.

FIGURE 13-23
Injection of PUR grout via drill hole, Poudre Canyon Tunnel, Colorado.
(PHOTOGRAPH COURTESY OF B. ARNDT.)

FIGURE 13-24
Injection of PUR grout in existing rock fractures, Poudre Canyon Tunnel, Colorado.
(PHOTOGRAPH COURTESY OF B. ARNDT.)

5. DRAINAGE

Controlling surface and groundwater flows improves the stability of rock cuts and reduces potential rockfall hazards. Reshaping upper slopes above unstable areas alleviates surface water ponding and infiltration and directs surface flows away from the slope face. Lining ditches that divert surface water helps limit infiltration.

Subsurface drainage can be established by drilling drain holes at or near the toe of the rock slope or in a series of rows on taller slopes to promote

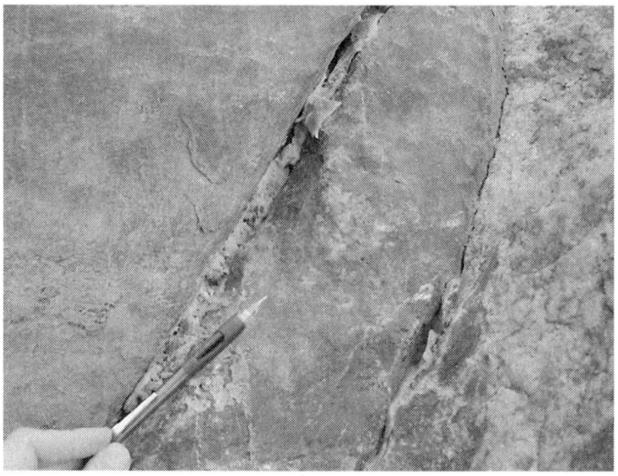

FIGURE 13-25
Rock fracture filled with PUR rock glue, Poudre Canyon Tunnel, Colorado.
(PHOTOGRAPH COURTESY OF B. ARNDT.)

slope drainage. This can lower the groundwater table and reduce hydrostatic pressure on potential failure planes, discontinuities, and tension cracks. What is most important in the design of drain holes for rock slopes is that they intersect fractures that carry the water; little water is contained in the intact rock. Generally, drain holes are inclined upwards by 3° to 5° to maintain a degree of self-cleaning. There are no formulas from which to calculate the required spacing of drill holes because it is more important that the holes be located to suit the geological conditions at the site. Holes are usually drilled on a spacing of 3 to 10 m (10 to 30 ft) to a depth of at least one-third of the slope height. The holes may be unlined in competent rock but they should be lined with a perforated casing through soil zones or in erodible rock, with the perforations sized to minimize infiltration of fines.

Slope drains should be provided for all newly constructed rock cuts, especially whenever shotcrete is applied. Slope drains can also be used to control seepage from the cut face if such seepage is likely to cause differential erosion of underlying weaker layers. The drains may need maintenance and periodic cleaning to continue to perform properly. Icing of outlets can seriously reduce flow in colder climates.

6. EXCAVATION AND RESHAPING OF A ROCK FACE BY BLASTING

Reshaping or reconstructing a slope through modern excavation techniques improves slope performance by lowering the kinematic possibility of rockfalls, as discussed in Chapter 8. To facilitate excavation, blasting is used to fragment the rock. A properly designed blast will partition the rock mass into discrete columns that move outward in response to the gas pressure created by the detonation of the explosives. The displacement fractures the rock according to the blast pattern. Considerable detonation energy is required to displace and break the rock in tension, and this energy can damage the finished cut slope unless precautions are taken (FHWA 1991).

Rock excavation along transportation corridors requires the creation of stable final cuts with slopes that are as steep as possible to minimize excavation and right-of-way requirements (FHWA 1998). These two design criteria can be contradictory, but the stability of cuts is generally enhanced and the safe slope angle steepened if controlled blasting methods are used to reduce damage to the rock behind the final cut face.

Blasting in populated areas requires precautions to control structural damage caused by blast vibrations and to minimize hazards of flyrock, air blast, and noise complaints. Proper blasting techniques and design methodologies are presented in two Federal Highway Administration (FHWA) publications (Konya and Walter 1985; Hoek and Bray 1989) and documented in several textbooks and handbooks (Langefors and Kihlstrom 1973; Hemphill 1981; Atlas Powder Company 1987; Persson et al. 1994; Oriard 2002; International Society of Explosives Engineers 2003; Wyllie and Mah 2004).

6.1 Blasting Principles

Figure 13-26 illustrates several important terms used by blasting practitioners. Definitions are provided below to aid in understanding the following sections. Each of these blast design elements affects the efficiency, safety, and fragmentation achieved by the blasting:

1. *Bench (lift) height*. The vertical limit (depth) of rock to be blasted at one time.
2. *Burden*. The distance from an explosive charge to the nearest free or open face. The effective burden is the distance between the nearest line of blast holes and the free face.

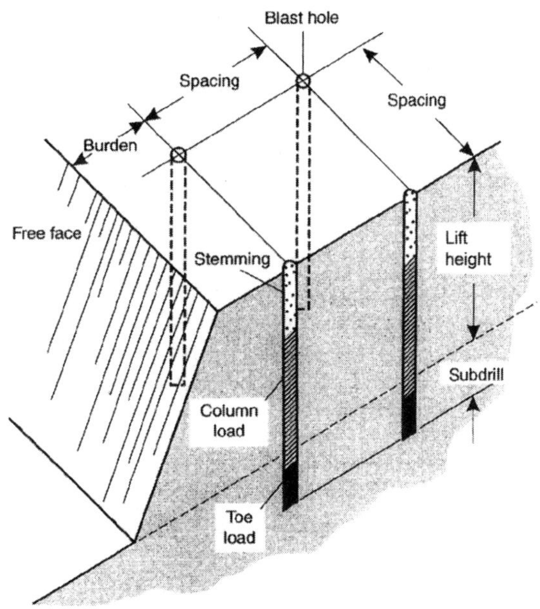

FIGURE 13-26 Definition of bench blasting terms (adapted from Wyllie and Mah 2004).

3. *Column load*. The amount (weight) of explosive placed in a blast hole.
4. *Free face*. The nearest unrestricted surface of the rock mass that will serve as an area of relief for a blasted rock to move toward when a shot is detonated.
5. *Spacing*. The distance between adjacent blast holes within a row.
6. *Powder factor*. The average weight of explosives per volume of shot rock expressed in English units as pounds per cubic yard.
7. *Stemming*. Inert material placed in the collar of a blast hole used to confine the gas pressure from an explosion within the rock mass.
8. *Subdrill depth*. The additional blast hole depth below the design grade needed to break rock to grade throughout the blast area without remnant high points.

Detonation of an explosive charge creates high temperatures and high-pressure gas that expands against the blast hole wall. This in turn creates a compressive strain wave that crushes and pulverizes the rock around the blast hole up to a distance of one or two radii (Figure 13-27*a*). As the compressive wave front expands away from the hole, successive zones of rock are subjected to intense tangential strains, and radial fractures form, extending up to 40 to 50 times the diameter of the blast hole. As the compressive wave continues to expand, concentric shells of rock undergo radial expansion, and tangential relief-of-load fractures develop near each blast hole (Figure 13-27*b*). When the compressive wave reaches a free face, it is reflected as a tensile strain wave. If this wave is sufficiently strong, spalling occurs on the free face, unloading the rock mass and allowing further expansion of the radial cracks (Figure 13-27*b*). These shock waves travel rapidly through the rock, so the process of fracture formation typically occurs within 1 to 2 ms after detonation, while the buildup of explosive gases takes upwards of 10 ms. Thus, the expanding gases enter these fractures, further fragment the rock, and force the rock mass apart (Figure 13-27*c*).

This process is highly dependent on the confinement of the explosive, the burden distance, and the sequencing of the blast. Good results require that the detonation of each hole be adequate for successive free faces to develop. The movement of rock toward successive free faces

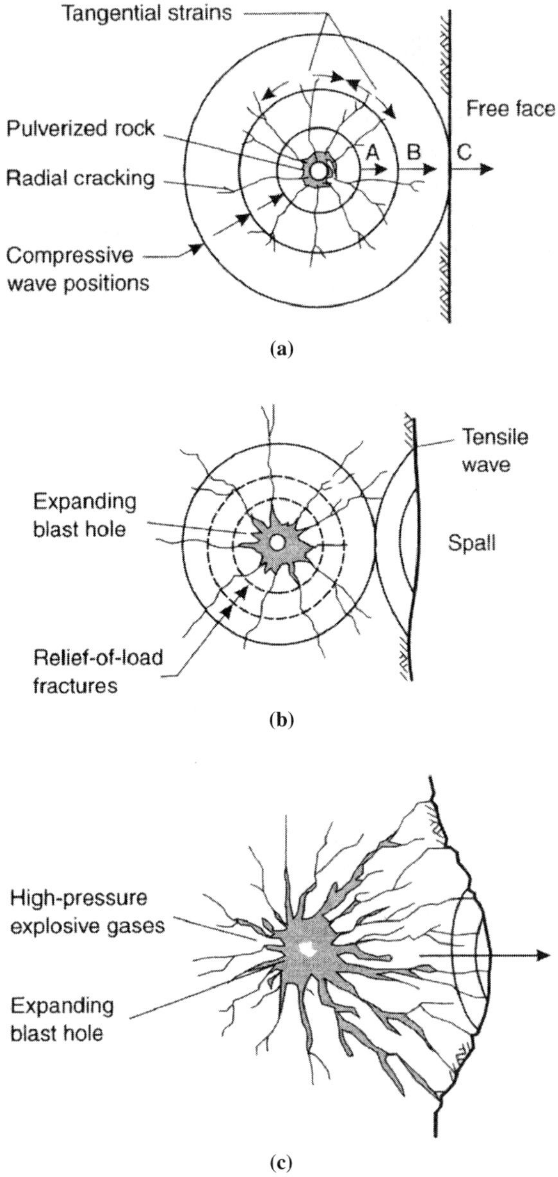

FIGURE 13-27
Rock fragmentation by blasting (Wyllie and Mah 2004).

limits the buildup of gas pressure within the blast area that would otherwise penetrate into surrounding discontinuities beyond the excavation limits and cause overbreak or flyrock, or both.

Bench heights are typically limited to about 9 m (30 ft) by several factors. The burden typically increases with bench height, and the burden near the toe may become excessive if the bench height is very large. In addition, drilling accuracy becomes more difficult as hole depths increase. Precise alignment of holes, particularly along the final face, is important in creating a smooth and undamaged final face.

The relationship between the bench height, H, and the burden, B, can be expressed in terms of the stiffness ratio H/B (Wyllie and Mah 2004). If this ratio is low, so that the burden is about equal to the bench height ($H/B = 1$), the blast will be highly confined, resulting in excessive overbreak, airblast, flyrock from the top of the blast area, and vibration. In contrast, if H/B is greater than about 4, there is little confinement of the free face and the explosive gases will vent along the free face, resulting in airblast and flyrock from the face. A stiffness ratio of 3 to 4 usually produces the best results.

When blasting is required to excavate rock, production blasting is used to fragment the rock that is to be removed between the existing free face and new final face. Controlled blasting techniques are used to form the cut slope and to limit the migration of damage from the high-energy production blasting into the rock behind the final face. The controlled blast holes are drilled on the plane of the final face.

In nearly all cases, controlled blast holes are drilled downward from a working bench or access road, but in some applications or when gaining access, the holes may be drilled roughly horizontally from the ends of the proposed cut. However, when this technique is used, the holes are generally fanned out from a single, equipment-controlled starting point, so their length must be limited to keep the separation between holes from becoming too large to function properly. Controlled blasting techniques may also be appropriate for defining the cut face, even when the rock materials may be rippable, to form a stable cut and limit the potential for overexcavation or undermining that can occur with today's powerful excavation equipment. Controlled blasting includes the following:

1. *Presplitting*—a single line of closely spaced and more lightly loaded holes aligned along the desired excavation limit detonated slightly before the production blasting,
2. *Trim (cushion) blasting*—a single line of closely spaced and more lightly loaded holes aligned along the desired excavation limit detonated after the production blasting, and
3. *Line drilling*—a row of very closely spaced and unloaded holes drilled along the desired excavation limit.

These methods have been used in the transportation construction industry for decades; presplitting is the most common method. It is important to have experienced personnel design and execute blasting operations to limit damage to the new rock face and to protect nearby residences, utilities, or other facilities from ground vibration, air shock, or flyrock damage.

On tall slopes, multiple blasting lifts with lift heights on the order of 9 m (30 ft) are typically used because of drilling limitations and blasting economics. Small horizontal offsets, usually about 0.6 to 1.0 m (2 to 3 ft) wide, are allowed at each lift boundary to allow room for drilling equipment. To maintain the required toe location at the final grade, the cut slopes are either steepened slightly between lifts to make up for the offset, or a setback of the starting drill location at the top of the slope is added for each lift offset.

6.2 Presplitting

Presplitting refers to closely spaced and lightly loaded blast holes, typically spaced 0.75 to 1 m (2 to 3 ft) apart. They are drilled along the neat excavation line and detonated at least 25 ms before the production blast holes are detonated (Hagan and Bulow 2000). The presplitting holes are loaded with continuous, lighter charges of specially manufactured explosives. The amount of stemming (crushed rock placed in the top of the hole to contain the explosive energy) is less than in production blast holes to allow gas pressures to vent more quickly and limit damage.

The goal of presplitting is to break the web of rock between adjacent blast holes and form a continuous open fracture along the planned location of the final cut face. This break in the rock mass serves to vent the high-energy gas pressure produced by production blasting, keeping it from entering into existing discontinuities beyond the cut face. This greatly reduces the likelihood of blast-induced inflation of the rock mass and improves the stability of the new cut face. Without presplitting, production blasting effects and damage can extend more than 15 m (50 ft) into the slope (Hoek and Bray 1989). In good-quality rock, the presplit drill holes will remain visible as "half casts" on the excavated cut face (Figure 13-28).

Presplitting can produce a smoother, less-damaged rock face even if the rock to be removed

FIGURE 13-28 Well-executed presplit rock face showing drill-hole traces (half casts) (http://en.wikipedia.org/wiki/File:P1010759.jpg).

can be excavated by powerful modern equipment. Using excavators to form the final face can also damage or undermine portions of the slope by plucking blocks from the face that extend beyond the limits of the proposed cut. When this happens, the slope may become locally oversteepened or unsupported, leading to increased rockfall potential. In these cases, presplitting allows the use of these excavators, but the design cut face location is preestablished and slope damage is avoided.

Presplitting techniques are typically restricted to slopes steeper than 1V:0.75H, but they have been used successfully on slopes as flat as 1V:1H. As the cut slope flattens, most drilling contractors will have difficulty maintaining adequate hole alignment, and the appearance and performance of the final cut slope may suffer.

Blasting theory assumes that an infinite burden is present in front of the line of presplit holes. In other words, no free face is available to interact with the presplit blasting. This occurs when the burden dimension, the amount of material between the presplit row and the existing rock face, is at least 5 m (15 ft). Where it is less, the burden is considered in the blast design, and trim blasting techniques are used.

6.3 Trim Blasting

Trim blasting, also referred to as *cushion blasting*, is similar to presplitting in that it relies on a single row of closely spaced, lightly loaded blast holes to form the finished cut face. The main difference is that, while presplit holes are detonated before the production blast holes are detonated, the trim blast holes are detonated after the production blasting has taken place. In some cases, they are detonated immediately after the production blast holes are completed and before any material is removed. However, trim blasting is also commonly used after the bulk of the excavation has been completed and the remaining burden dimension is less than 5 m (15 ft). In other words, this technique is used to trim a rock face.

Because the trim blast holes are located close to an existing face, the burden is not infinite and the material will displace when the holes are fired, and the burden dimension must be accounted for in the design of the trim blast hole spacing. Thus, trim blast holes are typically spaced farther apart than presplit holes—on the order of 1 to 1.25 m (3 to 4 ft) apart. Trim blasting is more commonly used with softer or more highly fractured rock, and, because of the poorer rock quality, it is less likely to leave evidence of the drill holes as half casts.

6.4 Line Drilling

In areas where overbreak is a substantial concern, particularly in the corners of building excavations, *line drilling* can be used to control the energy from production blasting. A row of closely spaced holes on the order of 0.3 m (1 ft) apart is drilled along the excavation limit. These holes are left unloaded, and the intent of this technique is to perforate and weaken the rock along the desired excavation limit so that the production blasting energy is concentrated at this location and the rock breaks to form the new cut slope. These holes are unstemmed. As in other cases, the numerous holes and the resulting fracture plane serve to vent production blasting gas pressure, and a less-damaged cut face is produced.

Although this technique is effective, it is much less commonly used because of the cost of the additional drilling. In some cases, the concept can be combined with other controlled blasting methods by adding an unloaded hole between the normally spaced presplit or trim blast holes, such as in the confined corners of excavations. This can improve development of a smoother cut face, especially where the geological structure has a significant influence on the formation of the final face.

6.5 Rock Sculpting

If a dominant structural feature is present within the rock mass, such as bedding or jointing that aligns with the proposed excavation, there may be an advantage in angling the blast holes to conform to these discontinuities. Essentially, this makes use of a preexisting structural element to encourage the rock to break along a discontinuity to form the final face in a manner similar to line drilling. If such a technique is used, these discontinuities will not intersect the new cut, and slope performance will be enhanced. A more naturally appearing cut face that is more visually appealing may also result. The effective application of this technique is contingent on the rock type and the presence of an accommodating rock structure.

Rock staining can also be applied to the final slope face, whether blasted or sculpted, to simulate the natural staining that occurs from exposure to the elements (oxidation) and secondary mineralization. Staining products are used to darken a newly exposed rock surface quickly so that it more closely matches the adjacent rock outcrops that have been exposed for years. The amount of color change and the degree of darkening are controlled by the amount applied to the slope.

Additional rockfall mitigation efforts may be required with sculpted rock cuts. Because naturally appearing variations or benches are introduced in the slope, falling rock blocks may have a tendency to be launched away from the face. Wider catchments at the base of the slope, draped wire mesh, and additional rock bolting can be used to control this increased rockfall potential. Figure 13-29 shows a location on Wolf Creek Pass, Colorado, where rock sculpting techniques were used. Note the draped wire mesh, rock bolts, and wide catchment area at the base of the slope.

6.6 Trimming or Blast Scaling

Trimming uses explosive agents to remove a portion of a slope that may be unstable or that acts as a launch feature that can adversely affect rockfall trajectories. Removing a launch feature can greatly improve the performance of a catchment area at the base of a slope. This technique is typically used on portions of the slope that are too large to hand scale, such as an overhanging rock outcrop, and to round or flatten the slope transition area between a natural slope and a cut slope.

Trimming is similar to trim (cushion) blasting because explosives are confined in drilled holes. After trimming, the affected area is hand scaled to remove any loose rock that remains. Trimming has limitations and advantages similar to those of scaling. However, trimming may produce flyrock, so certain safety precautions must be addressed. As with any blasting operation, trimming can be noisy. In all cases, when scaling or trimming operations are performed, the use of temporary protection systems should be considered to safeguard the public and any sensitive facilities and to control debris in order to promote quicker cleanup.

FIGURE 13-29 Sculpted rock cut, US-160, Wolf Creek Pass, Colorado. (PHOTOGRAPH COURTESY OF R. D. ANDREW.)

7. REMOVAL OF POTENTIALLY UNSTABLE BLOCKS BY SCALING

Scaling is a form of excavation used principally to remove individual loose or unstable rock blocks or a rind of weathered rock from a slope face. It is used on both natural and man-made slopes and is one of the most common rock slope maintenance measures. Because of the ongoing weathering and relaxation of surficial blocks, scaling improvements are temporary and must be repeated periodically. The safe and effective accomplishment of scaling requires experienced personnel. There are two basic approaches: hand scaling and mechanical scaling.

FIGURE 13-30
Routine hand-scaling operation.
(PHOTOGRAPH COURTESY OF T. C. BADGER.)

Hand scaling is done by workers suspended from ropes or working out of a crane or lift basket using handheld pry bars (Figure 13-30). Slopes are always scaled from the top down for safety. The surface of the slope is evaluated, and loose blocks that have the potential to fall are pried off the slope. This is an important distinction, since not all rocks that can be moved with a pry bar have the potential to fall. Hand scaling may be augmented by using expansive concrete products to fracture (reduce in size) blocks that are too large to remove by hand or power-assisted mechanical equipment such as air pillows or splitters that are inserted into open cracks and then expanded.

FIGURE 13-31
Mechanical scaling using hydraulic hammer.
(PHOTOGRAPH COURTESY OF T. C. BADGER.)

Scaling is performed in a controlled manner, and traffic or other access beneath the scaling area is prohibited unless additional barriers or other forms of protection are in place. In recent years, movable temporary barrier fences or suspended drapery systems have been used. Still, wherever possible because of the erratic behavior of rockfalls, scaling is best done under temporary road closure, with traffic queues allowed to clear only while scaling and worker movements on the slope are halted. Regardless of traffic concerns, additional temporary protection may be required, depending on downslope facilities, utilities, and structures.

When scaling is performed with power equipment such as hydraulic hammers or backhoes, the method is termed *mechanical scaling*. Mechanical scaling may be a more efficient method of removing unstable blocks, but final checking and hand scaling of the finished slope are usually still required (FHWA 1993). Figure 13-31 shows mechanical scaling using a hydraulic hammer. Other common mechanical scaling methods include dragging a heavy object, such as a blasting mat or cat track, across the slope to abrade loose rocks and mucking operations using a slusher excavating system. A *slusher system* uses a small excavator bucket that is positioned and dragged over the slope with a system of cables and pulleys to excavate loose surface materials.

Scaling can also be performed by using strategically placed explosives in cracks or drilled holes or by using heavy construction equipment such as a trackhoe. Without confinement, *crack blasting*, the hand loading of explosives into open cracks, can be ineffective and dangerous; it may only produce loud explosions and flyrock.

In all cases, regardless of the technique used, scaling operations should be observed and carefully controlled to prevent concentrated scaling that can lead to unsupported or oversteepened slope areas. This is particularly true when heavy equipment capable of excavating the slope is used.

Scaling is routinely used before constructing other mitigation efforts, such as rock bolts or draped mesh, or on newly excavated rock cuts to remove loose rocks that could pose a risk. Where warranted, scaling may be performed after draped mesh is installed to control the scaled debris near the base of the slope. Any mesh damage is repaired after scaling is completed.

In some situations, because of localized slope disturbances or conditions, rockfall activity may increase to some extent for a short period immediately after scaling. If the scaling was performed adequately, this should diminish quickly. Scaling may only be effective for 2 to 10 years, depending on site conditions, and therefore may not be considered a permanent mitigation measure. Depending on the level and duration of improvement, agencies often rely on routinely repeated scaling operations to maintain certain slopes. In comparison with other measures, scaling is inexpensive and effective in removing potentially unstable blocks. Therefore, it is a widely used rockfall mitigation technique.

8. CONSTRUCTABILITY AND AESTHETICS CONSIDERATIONS

Stabilization projects are measured against economic and constructability criteria. An effective stabilization project will provide benefits—a reduction in costs associated with existing rockfall hazards—exceeding the costs of implementing the stabilization measures. In some circumstances, such as when the rock quality is poor and thus weathering and degradation of the rock face are likely to be rapid, stabilization measures may not offer adequate long-term practical relief, and rockfall protection measures, as described in Chapter 14, may offer more economical alternatives.

Constructability criteria are practical considerations with regard to the undertaking and completion of the stabilization efforts. They include the following:

1. Worksite safety;
2. The availability of workers with appropriate skills and experience;
3. The geological environment, including rock type and structural characteristics;
4. The availability of specialized equipment, including the capacity, size, and reach of some equipment such as cranes; and
5. The characteristics of the site, including accessibility and the location of suitable staging areas for some phases of the stabilization operation.

The urgency and time frame for the proposed stabilization measures, as well as requirements to minimize disruption of surrounding activities, especially traffic on rail or road facilities, are taken into account. Environmental and aesthetic factors and regulations are also considerations. The latter are becoming increasingly important at some locations.

Constructability, economic, environmental, and aesthetic criteria frequently dictate that several stabilization measures be used in tandem to resolve the rockfall hazard at a particular site. Box 13-1 provides an example of the selection of several stabilization measures to mitigate a rockfall hazard at a representative site. Cable anchors provided the best solution and prevented large-scale instability, but additional measures, including trim blasting, scaling, fiber-reinforced shotcrete, and drainage measures, were used to complete the project. This example has been abstracted from a more extensive case study that includes computation of all factors considered in a rock slope stability analysis (Wyllie and Mah 2004).

8.1 Blasting Considerations

Wherever blasting is a possibility, the constructability of the excavation is a consideration during design. The design is developed to accommodate conventional drill and excavation equipment to allow the efficient excavation of new cut slopes. However, conventional equipment typically requires benches at least 5 m (15 ft) wide, with the bench width measured horizontally from the existing rock face to the back of the proposed new cut.

Wider benches increase the amount of excavation, but they allow more effective use of controlled blasting techniques and the use of standard drilling and excavation equipment. Cost savings and reduced duration of construction may be achieved with appropriate bench widths. More important, the ability to use controlled blasting methods generally results in a more stable rock face with reduced future user risks and maintenance costs.

In contrast, narrow benches referred to as *sliver cuts* may require small hand drills or "jack legs," with the operator working from ropes or a manlift. In some cases, larger equipment can be suspended from a crane on a drilling platform. Such designs can add significantly to project cost for the excavation and related items.

Flyrock can occur whenever blasting is undertaken. To minimize and control flyrock, blast holes are appropriately delayed, and the use of blasting mats may be required. Two common drawbacks of blasting mats include the following: (*a*) placing them over a wired shot can disturb the

Box 13-1

Stabilization Case Study

Stabilization was required at a rock cut 38 m (125 ft) high. The cut had been excavated in strong, coarse-grained, fresh, blocky granite containing four prominent joint sets (Figure 1). The slope was surveyed by using reflectorless electronic distance-measuring instruments, and detailed geological observations were made by rappelling down the face. The field observations indicated that blocks were sliding on sheet joints (J1) that dipped out of the slope at 20° to 30°. The blocks were bounded by tension cracks forming along Joint Set J3 and release surfaces at the sides formed along Joint Set J4, which is parallel to the plane of the cross section and thus is not visible in Figure 1. Direct shear tests indicated that a friction angle of 36° should be used to design stabilization measures. Seismic accelerations and transient water pressures in the joints were included in the stability analysis. Wyllie and Mah (2004) provided full details of these calculations.

Two alternative stabilization methods were considered: (*a*) removal of the unstable rock by blasting and then, if necessary, rock bolting the new face, and (*b*) reinforcement of the existing slope by installation of tensioned anchors. The selection considered the need to maintain roadway traffic during construction and the long-term reliability of the stabilized slope.

Blasting would remove the unstable rock blocks and potentially provide a long-term solution, but it presented several uncertainties and required unacceptable road closures. Stabilization using tensioned rock anchors was the alternative selected. The slope reinforcement was provided by steel-strand cables because they were lighter and more flexible than rigid bars and provided a higher tensile strength. Also, they could be installed with a light rig that avoided traffic disruption.

Additional stabilization measures that were included to minimize the risk of subsequent minor rockfalls (Figure 2) included

- Trim blasting to remove the fractured and marginally stable overhang on the face of the upper block (installing anchors through the overhang was deemed to be too dangerous);
- Scaling the fractured rock zones along the sheet joints and then applying steel-fiber–reinforced shotcrete; and
- Installing drain holes, 4 m long on 3-m centers, through the shotcrete to intersect the sheet joints and prevent buildup of water pressure.

SOURCE: Adapted from Wyllie and Mah 2004, pp. 341–347.

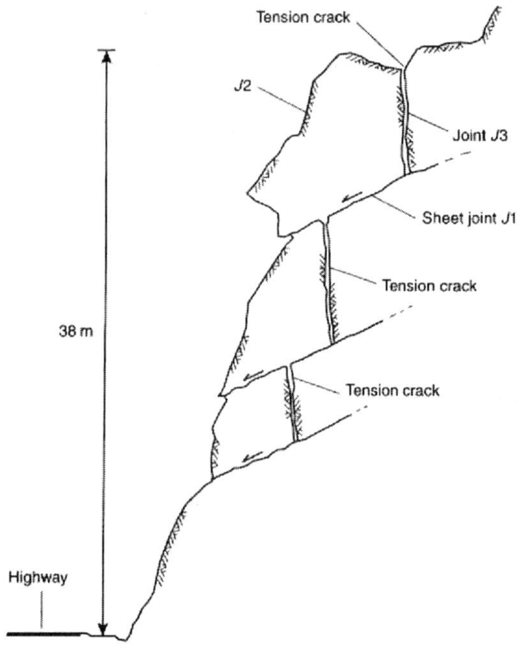

FIGURE 1
Cross section of slope before stabilization.

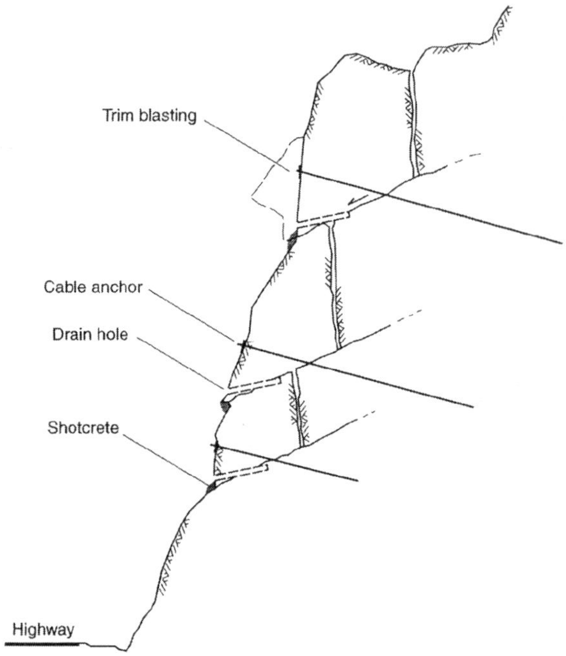

FIGURE 2
Cross section of stabilized slope.

connections, leading to part of the shot not detonating, and (*b*) once a shot is fired, the mats may become entangled in the muck pile, which can interfere with efficient removal of the shot rock.

If the duration of highway closures is an issue or there are environmentally sensitive areas near the site, such as fish-bearing rivers or lakes, the blast detonation sequence is designed to displace the shot rock parallel to the cut slope rather than toward the highway or water body. When highway closures for blasting are specified, sufficient time is allowed to fire the blast, to check the blast area afterwards, and to remove the blasted rock. If cleanup leads to unacceptable closure times, consideration needs to be given to reducing the shot size or creating a sufficient buffer zone at the base of the cut to control and contain shot rock and to provide space for excavating operations. This may require placement of a temporary rockfall barrier near the toe of the slope and possibly the use of part of the travel lanes for shot rock storage.

8.2 Selection of Stabilization Methods to Meet Aesthetic Criteria

Aesthetic requirements are increasingly being associated with highway development or improvement projects in areas with special scenic or recreational characteristics. The requirements are commonly imposed by agencies such as the Bureau of Land Management; the National Park Service; the U.S. Forest Service; and local governmental agencies with an agency goal or mandate to preserve the historical, cultural, or visual resources or recreational opportunities at sensitive sites. In some cases, divisions within state or federal transportation agencies impose similar restrictions or aesthetic requirements (Cummings 2002; Andrew et al. 2011). Increased emphasis on context-sensitive solutions (CSS) for the design of transportation facilities has caused FHWA to evaluate the application of CSS to rock slope design (Andrew et al. 2011).

At times, the construction limitations may not be well defined by these entities, whose approach and focus differ greatly from those of engineers and contractors. This can complicate the design of rock slope stabilization projects. In some cases, the aesthetic requirements may be expressed in nonquantitative terms, such as "reflect the form, line, color, and texture of natural formations," "natural-appearing cut slopes," "roughened cut slopes," or "mimic natural landforms." This can result in extensive debate and confusion, and even with an effective working partnership between an agency construction manager and a contractor, construction claims based on project delays and extra work can result.

At a detailed level, it is relatively easy to avoid excessively large expanses of uniform slopes exhibiting drill-hole half casts. Trim (cushion) blasting techniques, as opposed to presplitting, may reduce this problem and produce more natural-appearing "roughened" slopes. Scars on rock faces produced by excavation equipment can also be managed. The success of such measures rests largely on the skill of the blasting contractor and equipment operators, but it also depends considerably on the rock characteristics and the geological environment.

Sometimes requests are made to create benches or pockets in slopes to serve as vegetative planting areas. However, such irregularities will likely provide surfaces where falling rock fragments can be launched away from the slope, nullifying the effectiveness of roadside fallout areas. In addition, slope plantings, especially trees, can result in root wedging and increase the potential for rockfall. Modern slope design standards discourage the creation of any sort of intermediate slope bench because it is rarely accessible for maintenance and cleaning purposes, and over time such an irregularity tends to form launch features that adversely affect rockfall trajectory. If possible, these benches or planting pockets should be avoided.

Where slope heights are relatively low, compromises are possible. Depending on geological conditions, and with the involvement of a geologist on site, it may be possible to identify and use certain rock zones or strata to create stable and natural-appearing slope breaks that will provide for vegetation. On tall slopes, if such benches are included, the excavated cut slope would likely need to be set back to widen a fallout area at the highway grade.

Rock cuts created in support of a transportation corridor can be blended into the natural terrain if the slope geometry is varied and graded, particularly at the beginnings and ends of cuts. Varying, usually flattening, the cut slope angle at each end can provide a smoother transition to blend with the natural terrain. Related treatments include slope rounding at the crest of the cut to transition gradually into the natural slope above and adjusting the cut slope angle throughout sections requiring lengthy cuts to provide greater visual variability.

In many geological environments, the appearance of a naturally exposed, weathered rock surface may be distinctly different from (usually much darker than) the surface of a new rock cut. In arid regions, chemical weathering produces a dark coating known as *desert varnish*, but the tendency for rock to darken over time is prevalent in all climates. This is easily observed on slopes where light-colored scars represent areas where a rockfall has recently occurred.

Where the color difference produces an unacceptable appearance on a new rock cut, rock staining, using a variety of proprietary products, can darken the newly exposed surface. The stain is typically applied by spraying, but with some skill on the part of the contractor, the stain can be poured over the cut to simulate the natural appearance of drip staining. Care must be taken in applying the products to achieve the desired outcome. Many of the products continue to darken with time. Several light applications with several different colors will produce the most realistic-appearing stains. The success of staining depends on several factors and the applicator's experience with local rock conditions. If shotcrete is used, coloring and sculpting the shotcrete as discussed previously and shown in Figure 13-21 can significantly reduce the visual impact.

It has been shown that many of these project additions may not be especially apparent to a traveler passing by a new rock slope at any reasonably high speed, and the benefits of such aesthetic treatments may be unimportant to most travelers (Cummings 2002). However, a greater aesthetic benefit may result from treatments that affect the long-range perspective as experienced by road users as they approach a newly constructed area.

REFERENCES

ABBREVIATION

FHWA Federal Highway Administration

American Concrete Institute. 1995. *Specifications for Materials Proportioning and Application of Shotcrete*. ACI Report 506.2-95. Farmington Hills, Mich.

Andrew, R. D., R. Bartingale, and H. Hume. 2011. *Context Sensitive Rock Slope Design Solutions*. Publication FHWA-CFL/TD-11-002. Central Federal Lands Highway Division, Lakewood, Colo., Jan. http://www.cflhd.gov/programs/techDevelopment/geotech/css/.

Atlas Powder Company. 1987. *Explosives and Rock Blasting*. Dallas, Tex.

British Standards Institute. 1989. *British Standard Code of Practice for Ground Anchorages*. BS 8081:1989. London.

Cummings, R. 2002. Highway Cut Slopes in Rock—Specialized Excavation and Enhancement Techniques. *Proc., 53rd Annual Highway Geology Symposium*, San Luis Obispo, Calif., pp. 336–348.

FHWA. 1982. *Tiebacks*. Publication FHWA/RD-82/047. U.S. Department of Transportation.

FHWA. 1991. *Rock Blasting and Overbreak Control*. Contract DTFW 61-90-R-0058. U.S. Department of Transportation.

FHWA. 1993. *Rockfall Hazard Mitigation Methods*. Publication FHWA SA-93-085. U.S. Department of Transportation.

FHWA. 1998. *Rock Slopes*. Publication FHWA-HI-99-007. U.S. Department of Transportation.

FHWA. 2008. *Polyurethane Resin (PUR) Injection for Rock Mass Stabilization*. Publication FHWA-CFL/TD-08-004. U.S. Department of Transportation.

Fookes, P. G., and M. Sweeney. 1976. Stabilization and Control of Local Rock Falls and Degrading Rock Slopes. *Quarterly Journal of Engineering Geology*, Vol. 9, pp. 37–55.

Hagan, T. N., and B. Bulow. 2000. Blast Designs to Protect Pit Walls. In *Slope Stability in Surface Mining*, Society for Mining, Metallurgy and Exploration, Denver, Colo., pp. 125–130.

Hemphill, G. B. 1981. *Blasting Operations*. McGraw-Hill, Inc., New York.

Hoek, E., and J. Bray. 1981. *Rock Slope Engineering*, 2nd ed. Institute of Mining and Metallurgy, London.

Hoek, E., and J. Bray. 1989. *Rock Slopes: Design, Excavation, Stabilization*. Publication FHWA-TS-89-045. FHWA, U.S. Department of Transportation.

International Society of Explosives Engineers. 2003. *Blaster's Handbook*, 17th ed. Cleveland, Ohio.

Konya, C. J., and E. J. Walter. 1985. *Rock Blasting: Seminar on Blasting and Overbreak Control*. Publication FHWA-TS-85-225. FHWA, U.S. Department of Transportation.

Langefors, U., and B. Kihlstrom. 1973. *The Modern Technique of Rock Blasting*, 2nd ed. John Wiley and Sons, New York.

Littlejohn, G. S., and D. A. Bruce. 1975. Rock Anchors—State of the Art. Part 1: Design. *Ground Engineering*, Vol. 8, No. 4, pp. 41–48.

McCauley, M. L., B. W. Works, and S. A. Narramore. 1985. *Rockfall Mitigation*. Report FHWA/CA/TL/85/12. FHWA, U.S. Department of Transportation.

Morgan, D. R., R. Heere, N. McAskill, and C. Chan. 1999. Comparative Evaluation of System Ductility of Mesh and Fiber Reinforced Shotcretes. *Proc., 8th International Conference on Shotcrete for Underground Support*, American Society of Civil Engineers, Reston, Va.

Morgan, D. R., N. McAskill, B. W. Richardson, and R. C. Zellers. 1989. A Comparative Evaluation of Plain, Polypropylene Fiber, Steel Fiber, and Wire Mesh Reinforced Shotcretes. In *Transportation Research Record 1226*, Transportation Research Board, National Research Council, Washington, D.C., pp. 78–87.

Morris, A. J., and D. F. Wood. 1999. Rock Slope Engineering and Management Process on the Canadian Pacific Railway. *Proc., 50th Annual Highway Geology Symposium*, Roanoke, Va., pp. 264–275.

Oriard, L. L. 2002. *Explosives Engineering, Construction Vibrations and Geotechnology*. International Society of Explosives Engineers, Cleveland, Ohio.

Persson, P. A., R. Holmburg, and J. Lee. 1994. *Rock Blasting and Explosive Engineering*. CRC Press, Boca Raton, Fla.

Post-Tensioning Institute. 2004. *Recommendations for Prestressed Rock and Soil Anchors*, 4th ed. Phoenix, Ariz.

Schuster, R. L. 1995. Recent Advances in Slope Stabilization. *Proc., 6th International Symposium on Landslides*, Balkema, Rotterdam, Netherlands, Vol. 3, pp. 1715–1745.

Wood, D. F. 1989. Shotcrete–Steel Fiber Reinforced with Silica Fume Additives. *TAC Newsletter*, Tunnelling Association of Canada, April.

Wyllie, D. C. 1980. Toppling Rock Slope Failures, Examples of Analysis and Stabilization. *Rock Mechanics*, Vol. 13, pp. 89–98.

Wyllie, D. C. 1991. Rock Slope Stabilization and Protection Measures. *National Symposium on Highway and Railway Slope Stability*, Association of Engineering Geologists, Denver, Colo., pp. 41–64.

Wyllie, D.C. 1999. *Foundations on Rock*, 2nd ed. E & FN Spon, London.

Wyllie, D. C., and C. W. Mah. 2004. *Rock Slope Engineering: Civil and Mining*, 4th ed. Spon Press, New York.

Wyllie, D. C., and N. I. Norrish. 1996. Stabilization of Rock Slopes. In *Special Report 247: Landslides: Investigation and Mitigation* (A. K. Turner and R. L. Schuster, eds.), Transportation Research Board, National Research Council, Washington, D.C., pp. 474–504.

Xanthakos, P. P. 1991. *Ground Anchorages and Anchored Structures*. John Wiley and Sons, Inc., New York.

Chapter 14

THOMAS C. BADGER, JOHN D. DUFFY,
AND KRISTIAN SCHELLENBERG

PROTECTION

1. INTRODUCTION

The application of protection measures inherently accepts the occurrence of rockfalls and strives to mitigate the associated hazard by stopping, controlling, or deflecting the falling debris. Conditions that typically warrant protection measures include the following:

- Rockfall source areas lie beyond the boundaries of the facility (Figure 14-1).
- The extent or nature of the source area is impractical or excessively costly to stabilize (Figure 14-2).
- Avoiding the hazard by facility relocation is not practical or is excessively costly (Figure 14-3).

1.1 Design Considerations

Unlike stabilization measures that actively reinforce or remove the hazard, protection measures are passive systems. They can provide a barrier, control the trajectory, reduce energy, or provide a catchment. In their most general form, the most common protection measures include the following:

- Catchment areas,
- Rigid barriers,
- Flexible fences, and
- Drapery.

Barriers, fences, and drapery systems provide protection primarily by controlling the trajectory of the rockfall, either by guiding the falling debris into a suitable catchment area or by arresting the trajectory. These measures function as flexible or rigid systems. Draperies and fences represent flexible systems that attenuate energy by internally deforming when hit. Barriers, such as earthen berms and structural walls, function as relatively rigid systems, arresting or deflecting debris by their overall structural stiffness or large mass.

Compared with stabilization and avoidance alternatives, protection measures are often more cost-effective, involve simpler construction, can be constructed within the boundaries of the facility, and result in fewer environmental impacts. For these reasons, protection measures are probably more widely applied for remedial rockfall mitigation than stabilization and avoidance measures. Protection measures, however, are not the panacea for all rockfall hazards. Rockfall severity, slope instability, effectiveness of available measures, maintenance requirements, and environmental constraints may eliminate protection measures as feasible alternatives.

Incumbent with the application of any protection measure and the acceptance of continued rockfalls is maintenance. Maintenance is most critical when protection measures are exposed to repeated, damaging rockfall impacts and debris accumulation. For these problematic conditions,

FIGURE 14-1
Slope in Columbia River gorge in south-central Washington State is about 300 m (1,000 ft) high and frequently produces rockfalls that hit highway and railroad. Although rockfalls most frequently occur from source zones located in oversteepened cuts along highway, rockfalls commonly begin in source zones in basalt cliffs and colluvial deposits far upslope and well outside highway right-of-way. Neither stabilization of basalt cliffs nor avoidance is a practical alternative to addressing rockfall hazards along highway.
(PHOTOGRAPH COURTESY OF THE WASHINGTON STATE DEPARTMENT OF TRANSPORTATION.)

FIGURE 14-2
This 300-m (1,000-ft) section of oversteepened cut slopes in coarse colluvial deposits is a source zone for frequent moderate volumes of rockfall that hit US-12 in Cascade Mountains of Washington State. Steep natural slopes located above preclude development of flatter and more stable cut slope, the most desirable stabilization measure.
(PHOTOGRAPH COURTESY OF THE WASHINGTON STATE DEPARTMENT OF TRANSPORTATION.)

protection measures are best constructed so that they are accessible to maintenance equipment and personnel.

1.2 Selection Criteria

Various alternatives or modifications are generally available for each protection measure (McCauley et al. 1985; Wyllie 1991). Selecting the most appropriate and effective protection alternative typically involves consideration of numerous factors such as the available energy of falling rock blocks and the cost, ease of maintenance, and

FIGURE 14-3
Long section of oversteepened cut slopes in highly fractured metavolcanics is a source zone of frequent rockfalls that hit California State Highway 1 along Big Sur Coast. Steep natural slopes above and in the proximity of the Pacific Ocean on left preclude avoidance and stabilization measures.
(PHOTOGRAPH COURTESY OF THE CALIFORNIA DEPARTMENT OF TRANSPORTATION.)

aesthetics of the protection system. Protection measures are commonly located close to the transportation facility, often because of spatial necessity, such as where a steep rock slope is adjacent to a highway or for ease of construction and maintenance. Protection measures can also be placed near source zones. Limited distance between the base of the slope and the transportation facility may not provide sufficient width for a catchment area or for placement of certain types of barriers. Drapery may be the only alternative in these situations. It is essential that protection systems that deform to dissipate energy, such as flexible fences, be placed in positions with sufficient width to allow them to function properly and to permit maintenance access. Measures that control or reduce the energy of the falling rock blocks, or modify their trajectories, are often used in conjunction with a catchment area at the base of a slope used to contain debris.

Site characterization studies, discussed in Chapter 6, form the basis of all decisions concerning the selection and design of rockfall protection systems. Site characterization provides information on the frequency of rockfalls and on anticipated block sizes, kinetic energies, and trajectories. Rockfall modeling to define anticipated trajectories, as discussed in Chapter 9, and back-calculations based on information gathered during site investigation provide critical design parameters that permit the dimensioning and locating of alternatives such as barriers and fences and the determination of the appropriate size of special catchment areas. Additional design considerations often include an evaluation of the exposure risk at the rockfall location and definition of appropriate design standards for the facility, including cost, constructability, maintenance requirements, and aesthetic and other environmental issues.

Optimizing the available width of potential catchment areas is often the first consideration in the process of selecting the best-suited protection measure for the site conditions and constraints. Minor modifications to the catchment area that often involve little cost or few adverse impacts can greatly enhance the level of protection. Barriers are then considered if bounce heights, rollout distances, or kinetic energies of anticipated rockfall events appear likely to overwhelm the available catchment area. If the catchment area cannot be sufficiently improved, drapery may be the only feasible protection alternative. Rock sheds are considered as an alternative protection measure. They are a high-cost protection measure; they have been used infrequently in North America but extensively in Europe and Japan. They are most effective in locations where the hazard zone is short and well defined and where topographic and road geometric constraints do not allow for the installation of less costly measures. In such locations, rock sheds provide a suitable solution with low maintenance after frequent impacts by rockfall events of low energy.

In many situations, protection measures provide adequate mitigation of rockfall hazards in a cost-effective manner. The much more costly and complex avoidance measures discussed in Chapter 12 are appropriate only if the volumes, frequencies, or kinetic energies of anticipated rockfalls far exceed the capacities of protection alternatives.

2. CATCHMENT AREAS

Rockfall catchment areas are designed to stop moving rocks before they reach the traveled portion of a transportation facility by dissipating their energy on flat or negatively sloped ground. A well-designed catchment area can be a highly effective and reliable mitigation measure. Standard highway construction practice incorporates catchment areas for new rock excavations. Creating or widening catchment areas for existing slopes can be cost prohibitive, or the construction could prove difficult if much excavation is required to gain the necessary width. Also, widening existing rock cuts, required to accommodate enlarged catchment areas, may be infeasible because of right-of-way or environmental constraints. Existing catchment areas can, however, sometimes be modified or protection measures can be added to catchment areas to make them more effective in retaining rockfalls.

2.1 Catchment Area Attributes

The primary attributes required for the design of an effective catchment area are location, width, shape, and substrate (Figure 14-4). Evaluation and selection of each attribute are typically based on the expected trajectory of the characteristic rockfalls at a particular site. This site includes both the zone of impact and the rollout distance.

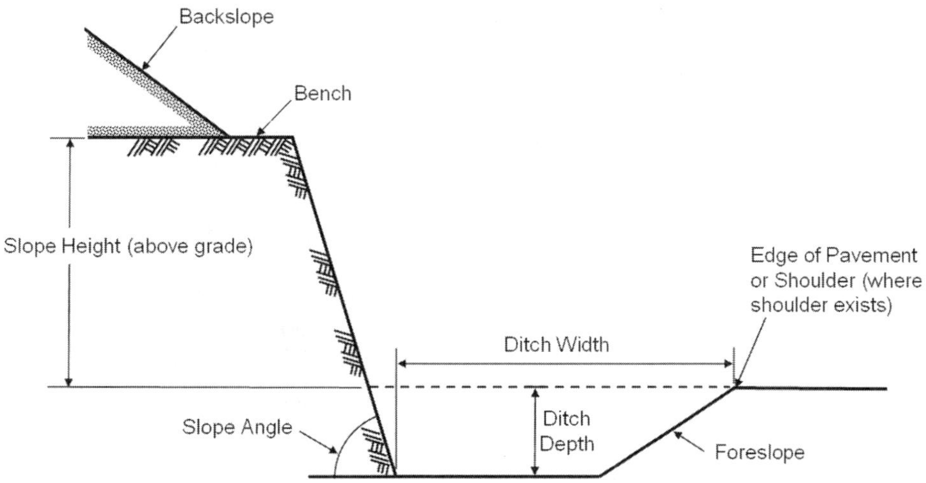

FIGURE 14-4
Basic elements of ditch section, including foreslope, width, depth, and cut slope angle and height (adapted from New York State Department of Transportation 2007).

2.1.1 Position

Catchment areas are most typically located at the base of a slope. Although much less common for highway applications, catchment areas can also be located near an upslope rockfall source area distant from the transportation facility. In the past, midslope benches were often incorporated in designs for larger cut slopes, and they still are used in some geographic regions, particularly in areas where interbedded sedimentary rocks are encountered, and by the mining industry.

2.1.2 Width

Adequate catchment width allows for the initial impact location of the rockfall and the ensuing rollout. For rockfalls initiating from very steep slopes, the rock blocks fall nearly vertically with minimal horizontal deflection. Unless a protrusion is struck, little rollout generally occurs. When slope height increases, protrusions and irregularities typically become more common, and impacts between falling blocks and the slope face become more frequent, causing horizontal velocities to increase along with rollout distances (Pierson et al. 2001). Flatter slopes involve bouncing or rolling trajectories with even larger horizontal components, and thus larger rollout distances are common. Comparison of rollout distances on a flat catchment area for different slope angles and heights illustrates the influence of the slope angle on the width of a catchment area required to contain rockfall debris (Figure 14-5).

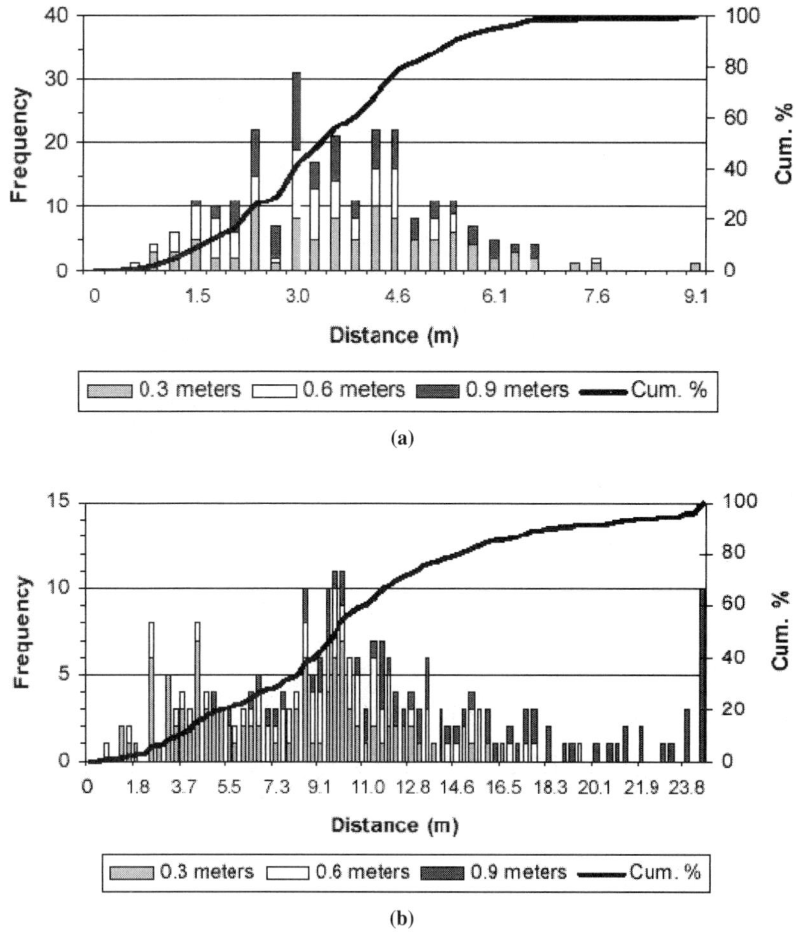

FIGURE 14-5
Rollout distances for (*a*) 24.4-m (80-ft) vertical cut are considerably less than for (*b*) 1V:1H cut slope of same height. Both slopes had flat ditches; block sizes were 0.3, 0.6, and 0.9 m (1, 2, and 3 ft) (Pierson et al. 2001).

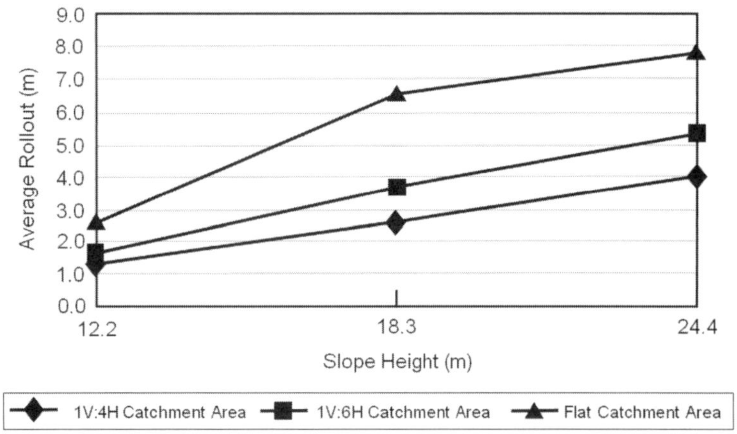

FIGURE 14-6
Average rollout distance for different slope heights (1V:0.25H slope) for flat, 1V:6H–sloped, and 1V:4H–sloped catchment areas. Significant reduction in rollout corresponds to increased foreslope inclination (Pierson et al. 2001).

2.1.3 Shape

The orientation of the foreslope also has a dramatic influence on rollout distances. Rollout distance drops dramatically as the foreslope angle increases (Figure 14-6). A vertical foreslope section effectively stops rolling rocks; however, it poses some safety hazards to passing vehicles, as discussed in Section 2.2.

FIGURE 14-7
Catchment ditch designed to accommodate frequent rockfalls at Waddell Bluffs on California State Route 1 in Santa Cruz County. Constructed in 1930s to catch and contain debris, uniquely designed ditch is 12 m (40 ft) wide and 3 m (10 ft) deep.
(PHOTOGRAPH COURTESY OF THE CALIFORNIA DEPARTMENT OF TRANSPORTATION.)

2.1.4 Substrate

The character of the substrate influences the size of the catchment area. A soft, loose substrate, such as uncompacted sand, has significantly greater ability to attenuate energy than does a hard substrate, such as bedrock or paved surfacing. In addition, rockfalls hitting hard surfaces can shatter and explosively disperse the debris. While the use of uncompacted sand may prohibit easy maintenance access, and errant vehicles that enter a ditch may become stranded, rockfall hazard reduction considerations often promote the avoidance of hard substrates by covering them with a loose material.

2.2 Rockfall Ditches

For transportation routes, the most common catchment area is a rockfall ditch located at the base of the slope immediately adjacent to the traveled way. For highway applications, rockfall ditches ideally incorporate highway design standards for safety, principally clear-zone distance from obstructions and moderate foreslopes (less than 1V:4H) that are recoverable if an errant vehicle enters the ditch. Properly designed and maintained ditches provide highly reliable and effective protection. Ditches typically are simple to construct and maintain, and they are the preferred rockfall mitigation technique if geometric and budget constraints allow.

Properly designed rock-catchment ditches prevent most falling rock from reaching the road (Figure 14-7). There are three general design strategies for catchment ditches: using the *Ritchie ditch* criteria (Ritchie 1963; FHWA 1989), following the specifications of the *Rockfall Catchment Area Design Guide* (Pierson et al. 2001), and using computer modeling to predict rockfall trajectories and velocities and then using these values to specify the dimensions and characteristics of catchment areas and ditches.

The development of formal designs for rockfall ditches began in earnest in 1959 when Arthur Ritchie, then chief geologist with the Washington State Department of Highways and a member of the Highway Research Board Landslide Committee, wrote:

For deep (high) rock cuts it has long been my belief that a deep ditch section is preferable to a wide one, and if the rock mass is sound, a more vertical backslope than a sloping one. A sloping surface induces a horizontal movement to a falling rock, whereas, one in free fall, or nearly so, stops dead in place provided the surface against which it hits is nearly normal to the direction of fall. All of this points up the fact that such ditches should be designed with a slope approaching 2:1 away from the road towards the rock cut. Rather than control rock ravel by expensive netting, let it fall. It can now be caught safely in the ditch section. (Ritchie 1963, quoted by Lowell 1987)

2.2.1 Ritchie Ditch

After recognizing the inadequacy of early designs to control rockfall in deep rock cuts, Ritchie (1963) pioneered the development of rockfall ditches through a rigorous field program of full-scale testing of rockfalls from steep bedrock slopes and moderately inclined talus slopes. Ritchie used a 16-mm slow-motion camera to characterize the trajectories of numerous rockfalls on different slopes, evaluating both impact location and roll-out distance. Important observations were made during these field tests on rock hardness, ditch substrate, limits of behavior (trajectory), and momentum as a function of block size and surface roughness. Through the use of a test trailer with an inclined deck, a steep foreslope (1V:1.25H) was found to improve the ditch effectiveness dramatically. For more severe cases, Ritchie included a lightweight chain-link protection fence to improve ditch effectiveness. Ritchie developed a design table of ditch dimensions for different slope heights and orientations (Figures 14-8 and 14-9); this basic design methodology has become known as the Ritchie ditch.

Ritchie's historic film and video records of other field tests mentioned in this chapter are provided on the DVD included with this book. See Appendix B, "Description of DVD Materials."

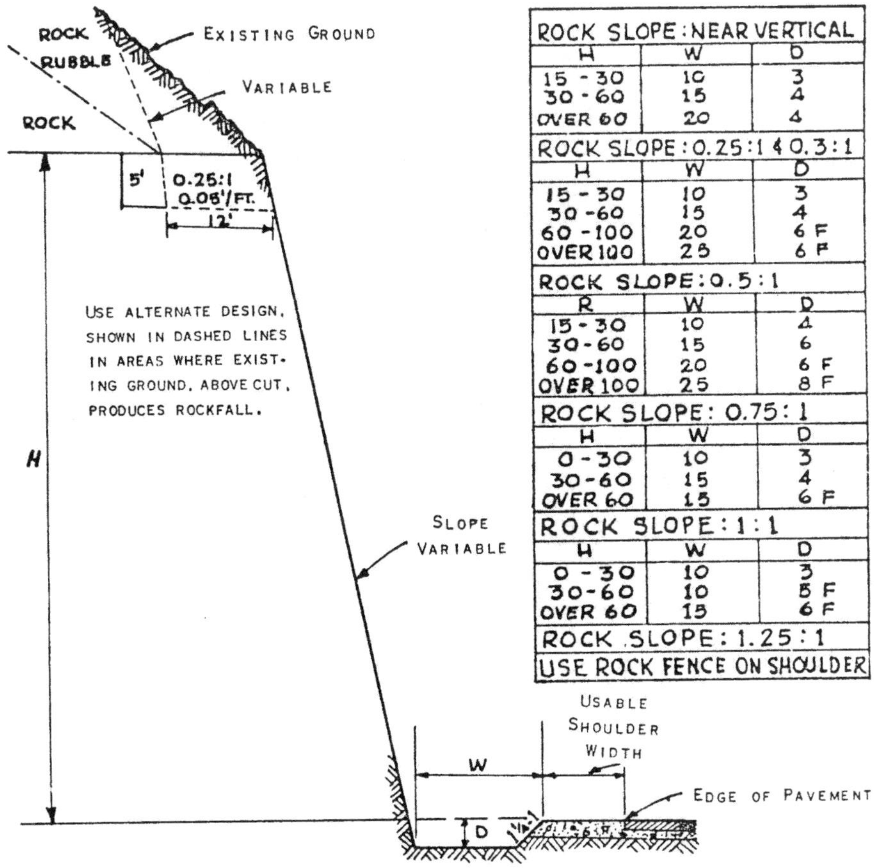

FIGURE 14-8 Ritchie ditch criteria incorporate slope height and inclination to determine dimensions of ditch and, if needed, provide for rock protection fence (Ritchie 1963).

FIGURE 14-9 Ritchie ditch on US-80 near Farad in Nevada County, California.
(PHOTOGRAPH COURTESY OF THE CALIFORNIA DEPARTMENT OF TRANSPORTATION.)

In 1963, the Washington State Department of Highways adopted a standard design for rockfall control based on Ritchie's research. Two basic designs were developed, one for rock cuts (Figure 14-10) and one for talus slopes (Figure 14-11) (Lowell 1987). For rock cuts, Ritchie's design was updated to provide for staged application of ditch modifications, depending on the severity of the rockfall. The Stage 1 ditch incorporated a 1V:6H foreslope with a width that exceeded the Ritchie ditch criteria by 1.2 m (4 ft) but omitted the fence (Figure 14-10). Including a fence (Stage 2) at the shoulder reduced the width by 1.2 m. For more severe rockfall problems, a deep trapezoidal ditch (Stage 3) incorporated Ritchie's 1V:1.25H foreslope with a fence

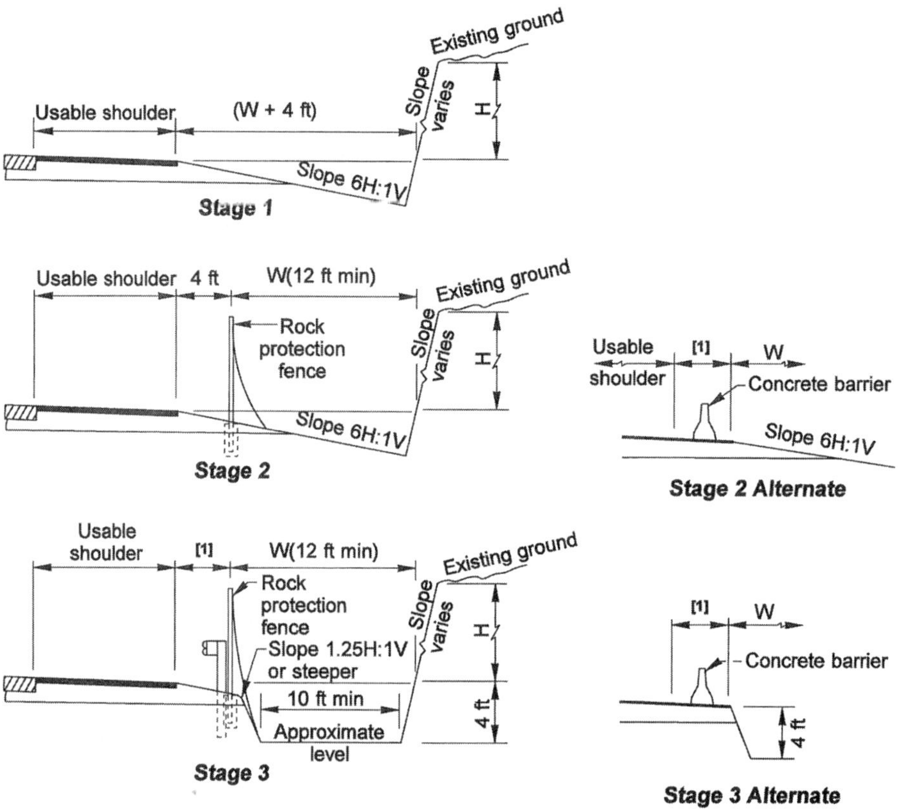

Rock Slope	H (ft)	W (ft)
Near Vertical	20 – 30	12
	30 – 60	15
	> 60	20
0.25H:1V Through 0.50H:1V	20 – 30	12
	30 – 60	15
	60 – 100	20
	>100	25

FIGURE 14-10 Washington State Department of Transportation (DOT) adopted standard ditch designs for highway sections in rock cuts (Design A). Stage 1 ditch, modified from Ritchie's ditch, is standard design used for steep rock cuts. Stage 2 design incorporates fence or barrier in addition to ditch. More severe rockfall problems may warrant Stage 3 design that uses deep trapezoidal ditch with steep foreslope and rock protection fence (Washington State Department of Transportation 2007).

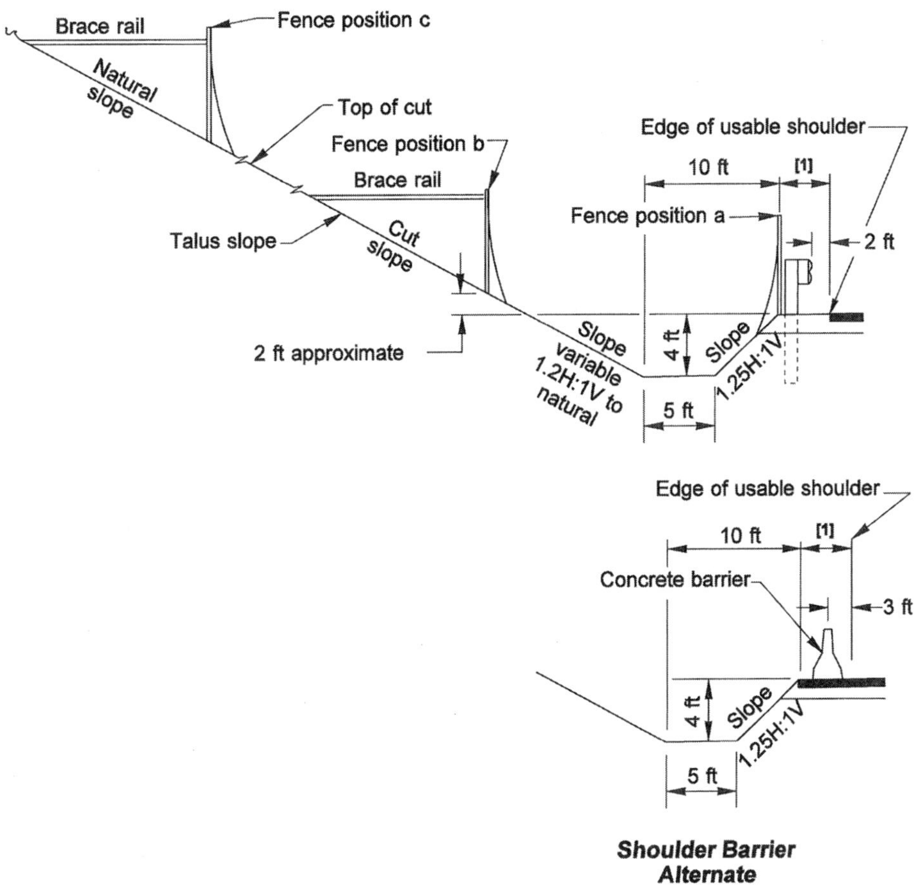

FIGURE 14-11
Washington State DOT uses alternative standard ditch design (Design B) for highway sections with talus slopes. Design B requires ditch with minimum depth of 1.2 m (4 ft), minimum width of 1.5 m (5 ft), and rock protection fence on road shoulder (Fence Position a). If needed, fence can be located just above ditch (Fence Position b) or closer to source area (Fence Position c). Since referenced fence design is long outdated, department now uses tested proprietary fences for these applications. Each fence is designed for site-specific conditions (Washington State Department of Transportation 2007).

(Figure 14-10). Because the capacity of the rock protection fence was never established, the fence was later omitted from the Stage 3 ditch specifications. Figure 14-12 illustrates a Ritchie ditch with a rock protection fence, developed according to these criteria.

FIGURE 14-12
Ritchie ditch with rock protection fence (FHWA 1989).

Ritchie's original design information was adapted into a design chart by the Federal Highway Administration (FHWA 1989), making it easier for designers to interpolate between cut slope heights, cut slope angles, and catchment area (ditch) depths (Figure 14-13).

Ritchie's research produced practical design criteria that can be used to size the width and depth of catchment area on the basis of slope height and slope angle. The basic ditch section is flat-bottomed with a 1V:1.25H foreslope. Unfortunately, the steep foreslope is difficult to construct and maintain by typical construction methods and materials and is deemed a nonrecoverable slope by current highway design standards. Thus, this steep foreslope poses a hazard that is mitigated by using guardrails or barriers. The basic design, including a recoverable 1V:6H ditch, was implemented and still serves as the primary ditch design for numerous state departments of transportation, providing at least 85% effectiveness for rockfall containment (Pierson et al. 2001). Pioneering as it was, the Ritchie study

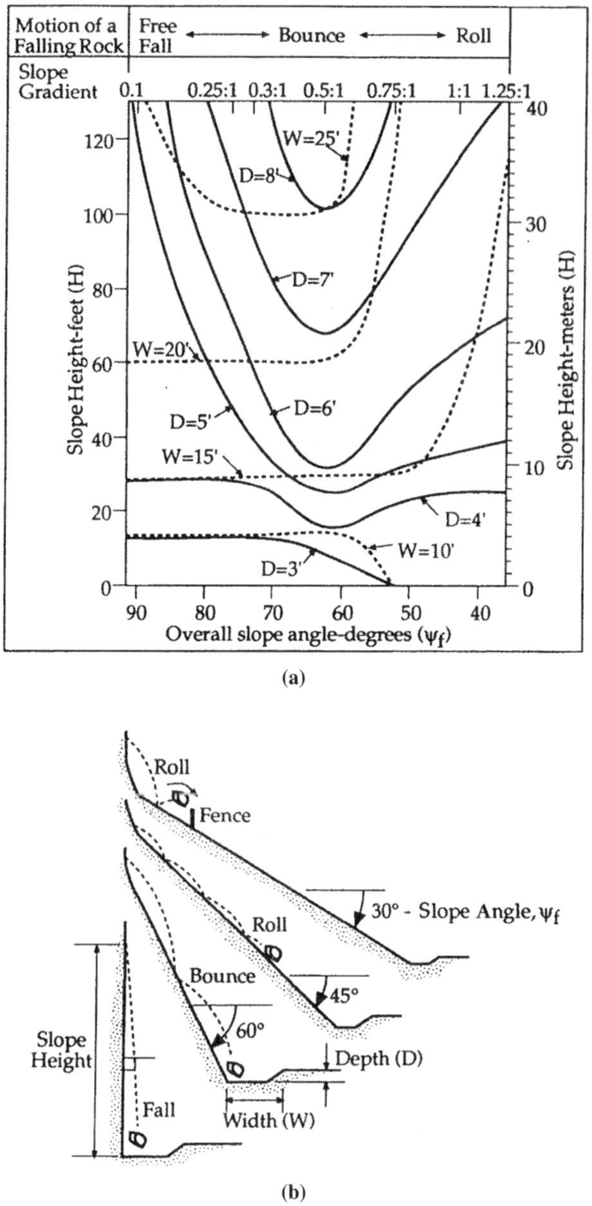

FIGURE 14-13 FHWA design chart provides ditch width and depth dimensions for slopes of different heights and inclinations (adapted from FHWA 1989).

was based on data collected from rolling only a few hundred rocks. Although the 1963 Ritchie rockfall study was a major leap forward, some limitations have since been recognized (Pierson et al. 2001):

- There were no means for varying designs to achieve percent rockfall retention levels.
- There was no standard requiring traversable foreslopes and sufficient clear zones.
- Research was conducted on rough rock cuts that were not constructed with modern controlled blasting methods that result in smoother finished slopes.

2.2.2 Oregon DOT Catchment Design

To modernize rockfall ditch designs and address these limitations, the Oregon DOT completed an exhaustive study by rolling more than 11,000 rocks for varied slope heights, inclinations, and recoverable ditch configurations (Pierson et al. 2001). A suite of design charts was developed to address each slope inclination (vertical through 1V:1H) and slope height [12.2 m (40 ft), 18.3 m (60 ft), and 24.4 m (80 ft)] for a range of percentages of rockfall retained (30% to 99%) (Figure 14-14). From these charts, a designer can select the most appropriate ditch width and inclination for the desired retention objective. Although there is no widely accepted standard for the retention rate, many current designs achieve at least an 85% to 90% retention rate.

2.2.3 Computer Modeling and Nonstandard Ditches

A third method of catchment area design is to use computer modeling of rockfall trajectories to design a nonstandard ditch. This design approach is employed for rockfalls generated from irregular natural slopes and compound cut slopes or those that exceed the heights defined in previous design methods. After the trajectory for a design event is characterized, a nonstandard ditch can be devised that addresses the site conditions. The ditch attributes that are modified to optimize effectiveness include position, width, shape, and substrate character.

Modifications have been made to the basic Ritchie ditch for more severe rockfall hazard areas and higher slopes. These modifications provide for a deepened section and steeper foreslope (Figure 14-15). Rock-filled wire baskets, often referred to as *gabions*, have also been used to create steep foreslopes. Other nonstandard ditch sections have been devised by raising alignment grades to gain catchment depth and steeper foreslopes (Figure 14-16), by laterally shifting the roadway to gain necessary catchment width (Figure 14-17), or by constructing fill walls to provide catchment area above the facility (Figure 14-18).

2.3 Midslope Benches

Midslope benches are flat catchment areas typically constructed at regular elevation intervals within rock cuts. Their primary purpose is to control the degradation of the slope and contain rockfalls. They are commonly used within interbedded strata containing weak, poorly indurated beds, such as shale or coal, and more durable sandstone or limestone layers. For these applications, they are often referred to as *lithologic benches*. Lithologic benches are typically constructed at the base of the durable strata with variable widths, depending on the thickness of the underlying less durable strata. The design attempts to account for degradation of the less-durable strata. For effective, long-term rockfall control, the bench needs to be placed directly beneath a near-vertical (1V:0.25H or steeper) section of slope (M. Woodard, personal communication, 2007).

Like any catchment area, midslope benches require cleanout to maintain their effectiveness. In the 1970s, the widespread practice of incorporating midslope benches for highway cuts experienced a schism in North America. Before the onset of modern rock slope engineering design and the use of controlled blasting for the excavation of highway cuts in the early 1970s, slope designs for many high rock cuts used midslope benches. Because of limited awareness of adverse rock mass conditions and often poor blasting results, many of these slopes deteriorated and these benches filled with debris. Access difficulties, or loss of access resulting from slope failures, as well as insufficient maintenance resources, prevented their necessary cleanout. As a result, these debris-filled benches began to act as launching surfaces for rockfalls, imparting undesirable horizontal components to their trajectories (Figure 14-19). With improved characterization of rock mass conditions and the use of controlled blasting, many designers, particularly in western North America, avoided the use of midslope benches, opting for uniform steep cuts that account for potentially adverse structural control combined with ditches to control rockfall and with localized slope reinforcement, such as anchors and shotcrete. The elimination of midslope benches provided the added benefit of reducing excavation quantities. In central North America, where interbedded limestone–sandstone and shale sequences are common, midslope benches remain standard practice with decades of proven effectiveness in containing rockfall when they are properly maintained (Figure 14-20).

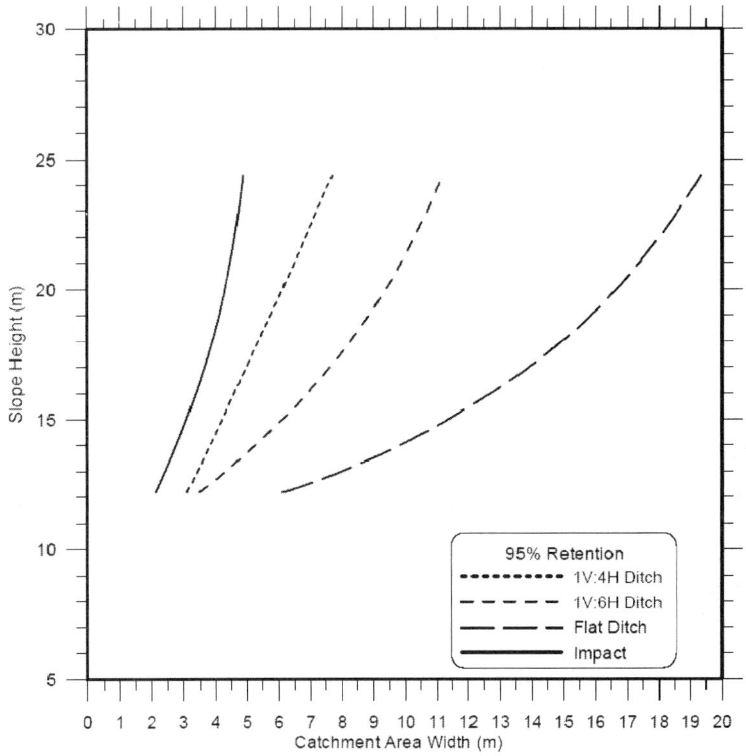

FIGURE 14-14
Design chart from Oregon DOT research project for 4V:1H cut slope relating slope height and catchment area width. Lines depict impact and minimum ditch width to achieve 95% retention from rollout for flat, 1V:6H, and 1V:4H foreslopes (Pierson et al. 2001).

3. BARRIERS

Barriers can deflect or contain rockfall, depending on their location and orientation with respect to the rockfall trajectory. These protection measures can also be used to reduce velocities and thereby dissipate energy; containment is then achieved

FIGURE 14-15
Hybrid ditch along base of steep rock cut with height of 75 m (250 ft) incorporates large concrete blocks within roadway section to provide deep trapezoidal ditch with width of 6 m (20 ft) and near-vertical foreslope with height of 1.2 m (4 ft). Concrete barrier placed along roadway shoulder on top of blocks protects vehicles from entering ditch. (PHOTOGRAPH COURTESY OF THE WASHINGTON STATE DEPARTMENT OF TRANSPORTATION.)

FIGURE 14-16
Elevated roadway creates hybrid ditch along California State Route 2 in San Gabriel Mountains.
(PHOTOGRAPH COURTESY OF THE CALIFORNIA DEPARTMENT OF TRANSPORTATION.)

FIGURE 14-17
(*a*) Viaduct constructed in early 1990s along California State Highway 89 in Eldorado County in Sierra Nevada Mountains near Lake Tahoe to increase catchment width for frequent rockfalls from glacial deposit containing boulders 1 to 3 m (3 to 10 ft) in diameter.
(*b*) Concrete wall backed with timber was included to provide Ritchie ditch profile.
(PHOTOGRAPHS COURTESY OF THE CALIFORNIA DEPARTMENT OF TRANSPORTATION.)

by applying two or more measures in a series. The use of a series of successive barriers is commonly applied in Switzerland and Italy and is especially useful on long or high slopes where large kinetic energies can be generated.

Barriers function either by being sufficiently stiff to withstand the kinetic energy of the rockfall impact or by deforming or flexing to attenuate the impact energy. Stiffness can be gained through material strength, mass, and the interaction of system components. A reinforced concrete barrier is an example of a very stiff, simple element. An earthen berm may be considered stiff through its sizable mass, but its surface also has the ability to deform to dissipate some of the energy. Fences are designed to be highly flexible systems, dissipating energy through the deformation of system components.

These protection measures may be sited close to the transportation facility, close to the rockfall source area, or at intermediate locations. Their footprint and deformation requirements are critical factors in determining their suitability. For instance, earthen berms and gravity walls have large base width requirements, and high-capacity fences typically need several meters or more of downslope clearance to deform properly. Ease of maintenance may also be an important consideration in ensuring optimal performance.

Many materials and systems with different levels of sophistication have been used to serve as barriers; the most common in North America for highway applications include the following:

- Earthen berms,
- Concrete barriers,
- Structural walls, and
- Flexible fences.

When a barrier for a particular site is considered, a comprehensive site characterization process, as discussed in Chapter 6, is an important initial step. It provides the fundamental evaluations of the likely source area, frequency of events, characteristic rockfall trajectories, and realistic estimates of the kinetic energies and bounce heights of rockfall blocks at the proposed barrier location. The last two characteristics are needed in selecting a barrier capable of containing and withstanding the anticipated rockfalls for the requisite service life of the barrier.

3.1 Barrier Testing

The anticipated trajectory of a characteristic rockfall at a site can be modeled or directly observed, and its kinetic energy can be estimated. However, the assurance that the selected barrier can safely contain the design event depends on barrier capacity and durability. Such assurance can be gained only through testing or performance documentation.

The first systematic rockfall barrier tests were performed in 1962 in Brunnen, Switzerland. Rocks weighing 30 kg (66 lb) were dropped from a grain silo at a height of 45 m (148 ft) into wire-rope nets secured by nylon ropes in a horizontal steel frame. The rocks hit with an estimated energy of 23 kJ (8.5 ft-tons) (Spang 2000). Figure 14-21 illustrates a similar Swiss testing facility used in 1976 to test rockfall barrier fences that incorporated friction-dissipating devices. About the same time, the Washington State Department of Highways was also performing field tests aimed primarily at evaluating width criteria of catchment areas, but the impacts of some rocks into low-capacity chain-link fences were also evaluated (Ritchie 1963).

In the following decades, barrier testing continued in earnest, with much of the testing focused on flexible fences. In the 1970s, energy-dissipating devices were developed and incorporated into fence designs (Heierli 1976). In the early 1980s, the California DOT (Caltrans) rolled more than 200 rocks into a variety of catchment ditches and fences. This study was followed by a testing renaissance in California and Colorado in the United States and in Canada, Switzerland, France, Taiwan, and Japan. Today, Switzerland, the European Union, and the United States have fence-testing standards or guidelines (Higgins 2003). As discussed in Chapter 15, proprietary fences now follow testing protocols to confirm their service energy and maximum energy levels.

Contemporaneously, other protection systems were being tested. Extensive testing was performed on catchment ditches, first in the 1960s by the Washington Department of Highways (Ritchie 1963), followed by Caltrans in the 1980s and the Oregon DOT in the 1990s (Pierson et al. 2001). Mechanically stabilized embankments have been tested, although not as extensively as fences, in the United States, Italy, and Japan. Timber

FIGURE 14-18
Metal bin wall constructed at base of large talus slope along Interstate 84 in north-central Oregon increases available catchment area above facility without cutting into talus or moving highway (FHWA 1993).

FIGURE 14-19
Since construction, midslope bench has filled in with rockfall debris, creating potential launching feature for rockfall. Because access was not provided, bench could not be cleaned.
(PHOTOGRAPH COURTESY OF J. D. DUFFY.)

FIGURE 14-20
Midslope benches incorporated into new cut slopes in interbedded limestone–shale provide catchment for rockfall debris generated from rapid deterioration of exposed shale beds.
(PHOTOGRAPH COURTESY OF THE KANSAS DEPARTMENT OF TRANSPORTATION.)

FIGURE 14-21
Drop tests performed in 1976 in Switzerland were among earliest tests of rockfall barrier fences that incorporated friction-dissipating devices.
(PHOTOGRAPH COURTESY OF R. SPANG.)

FIGURE 14-22
Concrete barriers are widely applied to improve ditch effectiveness, but all such systems are best suited for controlling low-energy impacts.
(PHOTOGRAPH COURTESY OF THE WYOMING DEPARTMENT OF TRANSPORTATION.)

walls, concrete barriers, hybrid drapery and fences, and rock sheds have also undergone limited testing in Switzerland, the United States, Italy, and Japan.

Improvement in barrier design continues, primarily advanced through active testing programs. Barrier testing is critical in understanding function and evaluating performance; the assurance of system capabilities under dynamic impacts is best determined through actual tests.

3.2 Concrete Barriers

Concrete barriers, sometimes referred to as *k-rail* or *Jersey barriers*, are widely used along roadways to protect vehicles from roadside hazards, such as nearby obstructions, steep slopes, and rockfalls (Figure 14-22). For rockfalls, they are best suited to control low-impact energies [less than 80 kJ (30 ft-tons)] and rollout from the ditch (Duffy and Hoon 1998). Because of their extreme stiffness, higher impact energies can locally shatter the concrete (sometimes violently) on both sides of the barrier. This effect is sometimes mitigated by placing hay bales, rubber tires, or timber on the concrete to dissipate the impact energy. The relatively low cost, narrow footprints, ready availability, and ease of installation of concrete barriers contribute to their widespread use as rockfall barriers and as improvements to the effectiveness of the catchment-ditch profile.

3.3 Earthen Berms

An earthen berm improves the effectiveness of the available containment area by creating a steepened foreslope. It can provide considerable energy dissipation because of its large mass and loose surface characteristics (Figure 14-23). Factors that influence the containment effectiveness of earthen berms—such as their position, width, shape, and substrate composition—can be evaluated and optimized with rockfall modeling. Earthen berms are generally shaped with steep side slopes and a typical base width–height ratio between 2 and 3. The base width can be reduced by incorporating large rocks or structural elements (Figure 14-24). However, the impact capacity of earthen berms has not been well quantified.

FIGURE 14-23
Earthen berm constructed from on-site materials is low-cost and effective protection measure capable of withstanding moderate- to high-energy impacts. Berm width will be more than twice its height; this considerable width requirement may prevent use of berm at some locations.
(PHOTOGRAPH COURTESY OF THE CALIFORNIA DEPARTMENT OF TRANSPORTATION.)

FIGURE 14-24
Seven-tier gabion facing has been used to maximize available containment width, berm height, and foreslope steepness of earthen berm. WSH-20, located in North Cascade Mountains of Washington State, is threatened by large rockfalls originating more than 500 m (1,500 ft) upslope.
(PHOTOGRAPH COURTESY OF THE WASHINGTON STATE DEPARTMENT OF TRANSPORTATION.)

3.4 Structural Walls

Walls incorporate a variety of structural elements to withstand the often very-high-energy impacts associated with rockfalls while striving to minimize the footprint of the protection structure. By design, walls are typically steep-faced and thus provide a highly effective profile for arresting rolling and bouncing rockfall trajectories. Unlike flexible rockfall fences, structural walls are extremely rigid. Because of their rigidity and inability to deform to dissipate impact energy, thin wall sections, such as masonry walls (Figure 14-25) and soldier pile walls (Figure 14-26), are protection structures with relatively low energy impact capacities.

The most effective structural walls for moderate- to high-energy rockfall impacts are thick-sectioned, gravity-type walls that incorporate structural reinforcing fabrics or elements with earthen materials. The most commonly used structural walls include mechanically stabilized earth (MSE) walls and

(a)

(b)

FIGURE 14-25
(*a*) Masonry walls, either mortared or dry stack, such as this wall in Columbia River gorge in south-central Washington State, were used as early protective barriers for rockfall. (*b*) Typically lacking much structural interlock of individual blocks, masonry walls provide little resilience to rockfall impacts. Note hole in wall and surrounding damage from rockfall impact.
(PHOTOGRAPHS COURTESY OF THE WASHINGTON STATE DEPARTMENT OF TRANSPORTATION.)

(a)

(b)

(c)

FIGURE 14-26
Soldier pile walls can use (*a*) timber, (*b*) concrete, or (*c*) steel lagging and have narrow base widths.
(PHOTOGRAPHS COURTESY OF THE CALIFORNIA DEPARTMENT OF TRANSPORTATION.)

gabion walls with a height–base ratio of about 1; this requisite width, however, limits their potential use at many sites.

3.4.1 MSE Walls and Reinforced Embankments

MSE walls and reinforced embankments consist of horizontal mats of reinforcing elements, such as geogrids, geotextiles, and steel-wire fabrics, placed between layers of coarse earthen material. The reinforcing elements are vertically wrapped at the face for retention of the earthen materials and face stability. MSE walls are characterized by their near-vertical faces and an additional structural treatment of the face (Figure 14-27), while reinforced embankments have more battered faces and lack a structural treatment of the face (Figure 14-28). Overall, MSE walls and reinforced embankments are effective protection structures capable of withstanding high kinetic energies from repeated rockfall impacts.

Research involving full-scale field testing (Barrett and White 1991; Peila et al. 2000) and numerical modeling (Yoshida 1999; Peila et al. 2002) has been

(a)

(b)

FIGURE 14-27
(*a*) MSE wall in New Mexico uses geosynthetic fabric for reinforcement and is faced with modular blocks for aesthetic treatment. (*b*) To attenuate rockfall impact energies, MSE wall is backed with tire bales.
(PHOTOGRAPHS COURTESY OF D. FINDLEY.)

(a)

(b)

FIGURE 14-28
(*a*) Geogrid-reinforced embankment in Italy provides protection structure up to 8 m (27 ft) high to contain large rockfalls. (*b*) Localized cratering around impact from large block.
(PHOTOGRAPH *a* COURTESY OF D. PEILA. PHOTOGRAPH *b* COURTESY OF F. FERRAIOLO.)

conducted on MSE walls and reinforced embankments to quantify their capacity, to understand their performance, and to develop a design methodology. This research shows that kinetic energy from a rockfall impact is dissipated through slip along the reinforcing layers within the zone of impact and by the development of localized cratering of the wall face (Figure 14-28*b*). Peila et al. (2007) demonstrated that a rock block colliding with energy approaching 5,000 kJ (1,850 ft-tons) can be safely stopped by a reinforced embankment with a unit weight per meter of length that is twice the weight of the block without the overall stability of the embankment being compromised (Figure 14-29). A design process proposed by Peila et al. (2007) evaluates the maximum slip along reinforcing layers by using both momentum conservation theory and maximum penetration. The penetration is related to the estimated peak stopping force developed within the block and the density of the earthen material within the wall. Important issues affecting the design of MSE walls for rockfall protection structures include the following:

- Location, available catchment area, and other site-specific constraints;
- Internal and external conditions of static and dynamic stability;

(a)

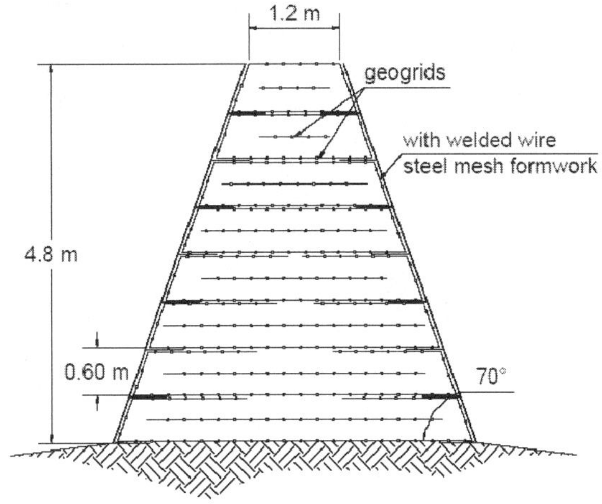

(b)

FIGURE 14-29
Full-scale test of geogrid-reinforced embankment hit by 8,700-kg (19,200-lb) block traveling at approximately 31 m/s (102 ft/s) demonstrates high impact energy capacity (4,180 kJ or 1,540 ft-tons) of this protection measure.
(PHOTOGRAPH AND DIAGRAM COURTESY OF D. PEILA.)

- Minimization of the potential for debris to overtop the barrier;
- Avoidance of excessive, damaging impacts by appropriate design; and
- Facings and other aesthetic issues.

3.4.2 Gabion Barriers

Gabions are rock-filled wire baskets that can be stacked to form rigid barriers for protection from rockfalls (Bertrand et al. 2008). To reduce the potential for overturning and to improve its capacity to withstand impacts, a gabion is generally constructed with a battered face and a height–base ratio of ¾ to 1. Gabions are durable and, consequently, are well suited for construction of protection measures in areas with a high frequency of rockfall. Gabions also have demonstrated an ability to withstand high-energy impacts (Figure 14-30), although their impact capacities have not been published.

3.5 Flexible Fences

Flexible fences are often used to enhance the capability of rockfall catchment areas to retain rockfall debris. A flexible rockfall fence suspends a net in an optimal position to intercept falling rocks (Figure 14-31). The infrastructure typically used to suspend the net panels includes posts anchored to foundations or to competent bedrock; tieback and lateral ropes, typically fitted with energy-dissipation devices; and anchors. Stopping the rock requires deformation of the net panels and use of devices to dissipate the kinetic energy of the block. The flexibility of the fence provides advantages over most rigid barriers through reduced costs, more versatile placement, and simpler design effort, primarily because most current fence installations use well-tested proprietary systems. Through rigorous product development and full-scale testing, a full suite of fence systems is available for energy ratings from 10 to 8,000 kJ (5 to 2,950 ft-tons). The use of flexible fences is discussed more fully in Chapter 15.

(a)

(b)

FIGURE 14-30
(*a*) Colluvium-filled chute in central Cascade Mountains of Washington State is source of frequent rockfalls, sometimes involving large blocks. Gabion barrier 2 m (6 ft) high and 2 m (6 ft) wide at base has been constructed to protect highway. (*b*) In late 1990s, rockfall event that included one block in excess of 5 m (15 ft), locally referred to as *House Rock* and originating from source zone high in chute, hit gabion barrier. Although rockfall event destroyed gabions, barrier contained nearly all debris.
(PHOTOGRAPHS COURTESY OF THE WASHINGTON STATE DEPARTMENT OF TRANSPORTATION.)

4. DRAPERY SYSTEMS

Drapery, sometimes referred to as *slope protection* or *wire mesh,* is a well-proven and widely used protection measure to control rockfalls at or near their source (Muhunthan et al. 2005a, 2005b). Extensive experience and design information are available on draperies, and they are discussed more fully in Chapter 16. In its basic form, drapery consists of a row of ground anchors along the top of the installation that are connected to a horizontal support cable, from which a woven-steel fabric, generically referred to as *mesh,* is suspended down onto the slope (see Figure 16-1). These top-anchored systems are referred to as *unsecured systems*. An unsecured system allows rockfall to occur between the slope and the mesh, controlling its descent into a catchment area at the base of the slope or the mesh terminus. Thus, an unsecured system is designed to avoid external loads caused by accumulating debris behind the mesh.

Conversely, *secured systems* include interior anchors and often a grid of support ropes, because the design goal of secured systems is to retain rockfalls within the mesh system if no catchment or storage area is available at the base of the slope. Chapter 13 on stabilization discusses secured draperies in more detail. Because secured draperies typically provide some active reinforcement and resistance to the initiation of rockfall events, they are considered as providing a degree of stabilization in addition to their role as a protection measure.

Unsecured drapery is most appropriate for rockfalls consisting of a single block or for small-volume failures composed of relatively small block sizes of less than about 1.2 m (4 ft). These rockfall events typically originate from coarse surficial deposits—formed of colluvium, alluvium, or residual soils—or from highly fractured rock masses. Unsecured drapery often covers the entire rockfall source area and thus is typically exposed to rockfall trajectories and impacts that closely conform to the plane of the mesh. In some installations, the upper part of the mesh is raised off the slope with posts or suspended across rockfall chutes to create a section intended to intercept and contain rockfalls originating upslope of the installation (see Figure 16-4). Such systems have been referred to as *hanging nets*, *hybrid drapery* or *barriers*, and *attenuators* and combine the attributes of rockfall fences and unsecured draperies. In this application, the upper portion of a hybrid drapery is exposed to rockfall impacts that are approximately perpendicular to the plane of the mesh.

FIGURE 14-31 Flexible rockfall fence provides effective protection from frequent rock blocks that detach from slope along highway in California. (PHOTOGRAPH COURTESY OF THE CALIFORNIA DEPARTMENT OF TRANSPORTATION.)

5. ROCK SHEDS

The basic configuration of a rock shed, often also called a *rockfall protection gallery*, consists of a reinforced concrete roof slab that is covered with a cushion layer of locally available sand or gravel (Figure 14-32). The roof slab is usually monolithically connected to the back wall and is supported by columns on the valley side. Rock sheds are most often used beneath steep high slopes that are a source of frequent rockfalls.

Because of their high cost compared with other more commonly applied protection measures, rock sheds are mainly used when the hazard zone is short and well defined and where topographic and road geometric constraints do not allow for the installation of less costly measures. Properly designed rock sheds offer several advantages. Compared with other protection measures, rock sheds provide a suitable solution with low maintenance after frequent impacts by rockfall events of low energy. Furthermore, rock sheds limit risk from uncertainties in predicting rockfall trajectory, because they are placed directly above the object to be protected. An added benefit of a rock

FIGURE 14-32 Rock sheds from (*a*) Japan and (*b*) Switzerland (Tanzenbein Gallery, Schoellenen, Gotthard Pass). (PHOTOGRAPHS COURTESY OF N. KISHI AND S. GHADIMI-KHASRAGHY.)

shed is the simultaneous protection from other road hazards, such as snow and snow avalanches, and reduced maintenance costs. Rock sheds also provide protection against high-energy impacts. For instance, typical rock sheds in Switzerland are used for impact energies in the range of 2,000 kJ (740 ft-tons) and can achieve capacities in the range of 1,000 kJ (370 ft-tons) with limited damage (Schellenberg 2009).

The greatest disadvantage of rock sheds is the high cost. Much of the experience with rock sheds has developed in Europe (Jacquemoud 1999), South Africa (Kirsten 1982), and Japan. In Switzerland, more than 3,000 m (10,000 ft) of sheds were constructed between 1960 and 1990 (Schellenberg 2009).

5.1 Types of Sheds

There is a variety of shed types, with the main distinctions being their geometry and the materials used (Figure 14-33). The various types have evolved because of the large number of structural and material combinations and a wide range of design objectives and constraints typically involved in constructing such structures. Local construction preferences and, in particular, required design standards have also played important roles. For instance, in Switzerland many sheds were built during the 1970s, at a time when no design specifications or standards were available.

Sheds can be generally categorized as *slabs* (Figure 14-33*a*) or *shells* (Figure 14-33*b*). The slab type of rock sheds can be subdivided into *horizontal* and *inclined* slabs, with the distinctions between categories being somewhat gradational (Schellenberg 2009). The design inclination for the roof slab is determined by the expected rockfall trajectories and the downslope consequences of not retaining the rockfall and passing it over the shed. Because inclined slabs experience lower impact loads as they deflect rockfalls, they can be built with lighter-weight roofs. The structural viability of a slab-type inclined roof depends on the span length and the reinforcing and bracing needed for stability; thus, slab-type rock sheds are more applicable than other types of sheds for locations where lower impact energies are anticipated. Shell structures have larger capacities because a portion of the impact load is transmitted by compression in the arch. However, their stiffer behavior compared with the slab structures increases the potential for local damage near the impact location. Such local failure and damage can usually be managed by a well-designed cushion layer or by providing anchored shear reinforcement. The type and thickness of the cushion material are largely influenced by the type of structure and impact energy.

The most common type of shed is a reinforced concrete slab. Reinforced cast-in-place concrete allows for a monolithic structure (Figure 14-33*c*) that has a more ductile behavior than precast concrete elements (i.e., girders or frames). However, the use of precast concrete elements can reduce construction costs significantly. Precast concrete elements are widely used in Japanese shed designs. When possible, continuity between the precast elements in the longitudinal direction should also be provided. Without such continuity, a design based on precast elements is structurally less efficient than a shed design that is capable of carrying impact loads in two directions, as provided by cast-in-place concrete slabs.

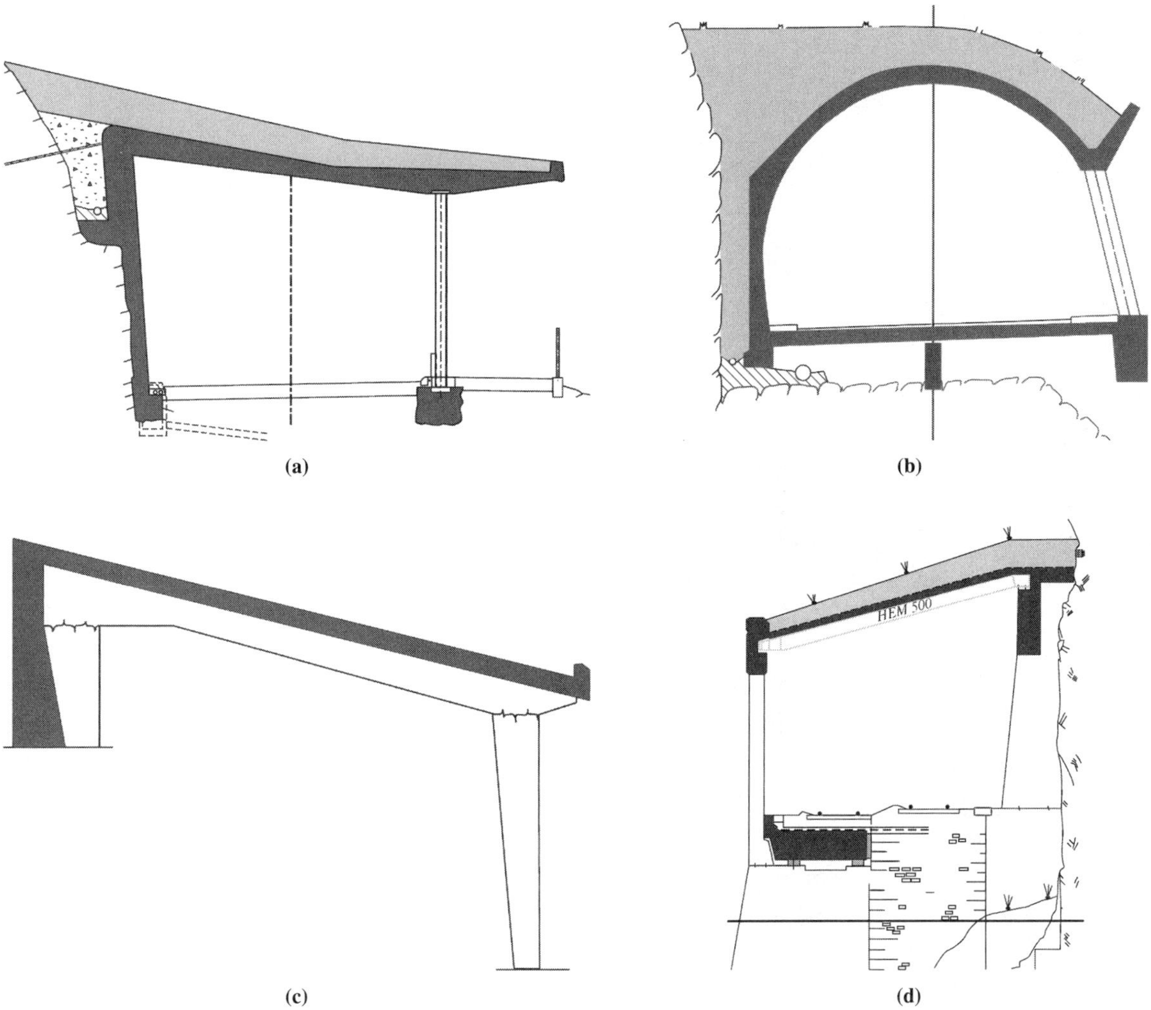

FIGURE 14-33 Different types of rock shed structures: (*a*) reinforced-concrete slab, (*b*) shell type, (*c*) in situ reinforced concrete, and (*d*) steel–concrete composite type (Vogel et al. 2009).

The roof slab is generally not prestressed because of the higher costs. The possible reduction in slab thickness provided by prestressing can be beneficial because the structural stiffness is reduced. However, for the same roof thickness, prestressed members are stiffer than mildly reinforced roof slabs. Stiffer structures increase the impact forces because the deceleration of the rock has to occur within smaller roof deformations. Prestressed concrete members using unbonded tendons, however, can improve structural capacity because larger forces can be accommodated with larger deformations of the prestressed roof and by the increased membrane effect of the continuous roof. Steel–concrete composite structures (Figure 14-33*d*) have the potential to provide considerably more ductility than other types of structures, but they can be much more expensive.

To enhance the impact energy dissipation capacity of flat-roofed structures exposed to direct impacts, a new design has been developed in France (Figure 14-34), referred to as *pare-blocs structurellement dissipant* (PSD) (*structurally dissipating rock shed*). This system incorporates cushioning elements between the roof and the outer support columns and provides nearly 1.8 times greater capacity than conventional rock sheds (IVOR 2001; Berthet-Rambaud 2004; Mougin et al. 2005).

Full-size and model tests, as well as structural analysis, have shown that the structural characteristics of rock sheds, such as their flexibility, significantly influence the magnitude of the force imparted to the structure from a rockfall impact (Masuya et al. 1987; Yokoyama et al. 1993; Schellenberg 2009). The flexibility effect is larger

FIGURE 14-34 Rock shed, or "gallery," with PSD dissipation system, in Val d'Arly, Savoie, France.
(PHOTOGRAPH COURTESY OF H. MASUYA.)

for steel than for concrete structures. The stiffness of steel sheds is relatively low, and the influence of the response of the shed to the impact force is significant. In particular, in the elastic range, reinforced concrete structures have relatively high stiffness, and therefore the influence of the structural response on the magnitude of the impact force can almost be disregarded. However, when the structure approaches failure, the rigidity decreases significantly, and the impact and transmitted forces decrease as well. This decrease in force is accompanied by a corresponding increase in the duration of the impact. Finally, tests show that, for all drop heights, the maximum reaction force decreases as the stiffness of the structure diminishes.

5.2 Cushioning

The cushioning material on the shed roof has three main tasks:

1. Broadly distribute the contact stresses from a localized impact point,
2. Gradually decrease the accelerations in the colliding body, and
3. Increase the impact duration.

These tasks are accomplished through compression and distortion of the material. Ideally, the material should also be inexpensive, have a low density to minimize the dead load, be able to resist the impact forces without damage, and be long lasting.

In Japan, 0.90 m (3 ft) of sand is widely used as the cushioning layer, but Styrofoam, rubber tires, and other multilayer systems are also used extensively (Nakano et al. 1992). In Switzerland, locally available sand and gravel are primarily used. The current Swiss design approach for the cushion layer uses empirical data to limit the penetration depth for the design rock for the given cushion thickness (Montani 1998). Considerable research involving full-scale rockfall tests has attempted to clarify the relationship between the weight impact force and the transmitted impact force for a variety of cushioning materials (Yoshida et al. 1987). The energy-absorbing properties of these materials, on the basis of the tests conducted by Yoshida et al. (1987), are as follows:

- *Sand*. The force–deformation curve of sand is parabolic, with only a small amount of energy absorption in the small deformation range; however, the force rapidly increases as the deformation increases. This behavior is due to sand being loose, so that initial consolidation is necessary before it begins to absorb energy.
- *Rubber tires*. A pile of rubber tires includes considerable air space; thus, they are readily crushed when loaded. Once the tires are fully crushed, the force increases rapidly as the rubber is compressed. As a result of this behavior, the energy-absorbing properties of rubber tires are not adequate for rock shed construction.
- *Styrofoam*. For small strains in the range of about 5%, the force increases rapidly, and then it increases slowly with increasing deformation. Most of the energy absorption occurs between 5% and 70% deformation as the Styrofoam deforms plastically. When the strain reaches 70%, however, the Styrofoam is almost fully compressed, and a sudden increase in the force occurs.

The energy-absorbing properties of various materials have also been studied by Chikatamarla (2006). Large-scale impact tests on a patented lightweight foam glass (Misapor) have also showed promising results (Schellenberg 2009). An appropriate balance between cushion layer and structure has to be found. This balance remains a topic of active research.

5.3 Shed Design

Initial design effort involves characterizing the frequency of rockfalls and their impact energy. Definition of the design impact mass and the corresponding impact velocity allows the impact energy to be quantified, and the design forces imparted to the shed are then evaluated.

5.3.1 Guidelines

Switzerland and Japan have made considerable progress in rockfall protection over the past few decades; in both countries, the responsible authorities have published guidelines for designing rock sheds (galleries) (Japan Road Association 2000; Masuya 2007; ASTRA 2008). The guidelines use static approaches that are briefly introduced here.

In 1998, the first Swiss design guideline for rockfall protection galleries, based on actual impact tests, was published (Figure 14-35). The test series focused on the influence of the cushion layer on the impact force. The application of the guideline is limited to penetrations that are smaller than half the depth of the cushion layer. A static equivalent force, A_d, to consider the impact load, F_k, is used (ASTRA 2008) according to Equation 1:

$$A_d = C \cdot F_k \tag{1}$$

with

$$F_k = 2.8 \cdot e^{-0.5} \cdot r^{0.7} \cdot M_{E,k}^{0.4} \cdot \tan\varphi_k \cdot \left(\frac{m_k \cdot v_k^2}{2}\right)^{0.6} \tag{2}$$

where

- A_d = static equivalent force at the design level (kN),
- C = coefficient to account for ductile (0.4) or brittle failure (1.2),
- F_k = impact load (kN),
- e = thickness of cushion layer (m),
- r = radius of an equivalent sphere of the falling rock (m),
- $M_{E,k}$ = soil modulus of cushion layer (kN/m^2),
- φ_k = internal friction angle of cushion layer,
- m_k = characteristic block mass (tonnes), and
- v_k = impact velocity (m/s).

The most significant document in Japan is a handbook with recommendations for all types of protection measures (Masuya 2007). Here, the impact load, P, on galleries is given by Equation 3, which is based on Hertzian contact theory.

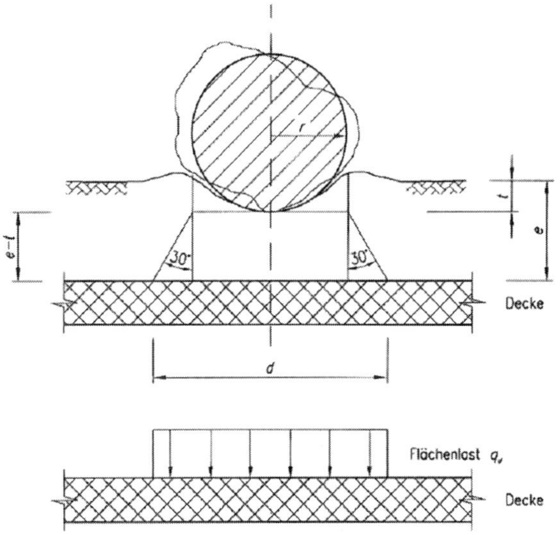

FIGURE 14-35 Load distribution according to Swiss guideline (ASTRA 2008).

$$P = 2.108(m \cdot g)^{\frac{2}{3}} \cdot \lambda^{\frac{2}{5}} \cdot H^{\frac{3}{5}} \cdot \alpha \tag{3}$$

with

$$\alpha = \frac{P_{max}}{P_{max(\lambda=1,000)}} = \left(\frac{T}{D}\right)^{-0.5} \tag{4}$$

where

- m = mass of falling rock (tonnes),
- g = gravitational acceleration (9.81 m/s^2),
- λ = Lame's constant (kN/m^2),
- H = falling height (m),
- α = amplification factor for cushion thickness,
- T = thickness of cushion layer (m), and
- D = diameter of falling rock (m).

Figure 14-36 illustrates these relationships. Because of the relatively high costs of rock sheds, an optimization of design is needed to mobilize the structural reserves of the shed for the given design rockfall event.

5.3.2 Analytical Approach

The design concepts in the guidelines given above are based on application of a static equivalent impact force. This approach is simple to use for practicing engineers. The design method

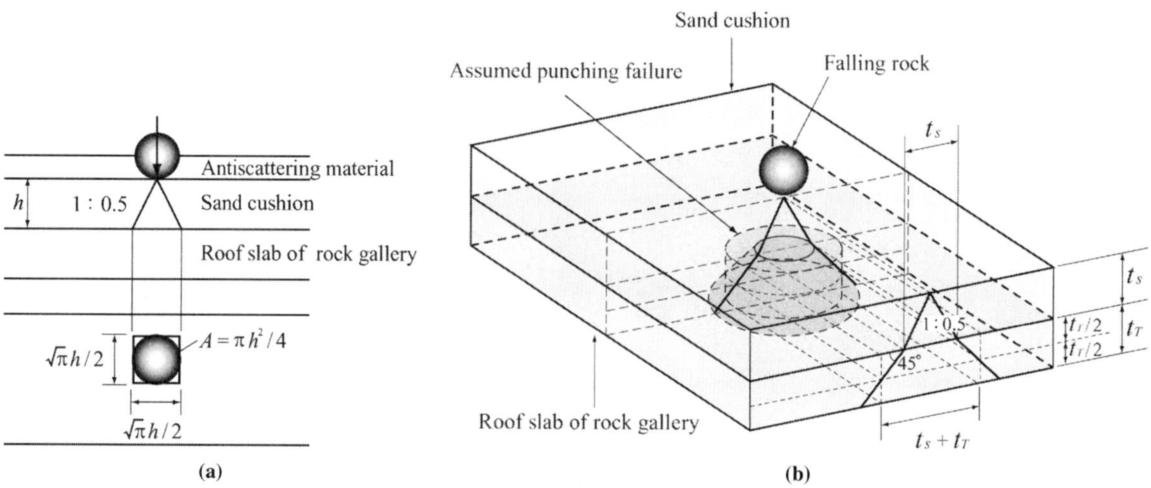

FIGURE 14-36 Load distribution and verification section for punching shear failure according to designs recommended by Japan Road Association in 2000: (*a*) section and elevations view and (*b*) three-dimensional illustration (adapted by N. Kishi from Japan Road Association 2000).

proposed by Yoshida et al. (2007) requires data on two forces: the weight impact force and the transmitted impact force, as shown in Figure 14-37. However, the dynamic behavior of the structure is not adequately taken into consideration.

For modeling the structural behavior during impact, a system of multiple degrees of freedom (SMDF) was proposed by Schellenberg (2009). The SMDF consists of three masses and three nonlinear springs (Figure 14-38). The first mass, M_1, is the mass of the colliding rock. M_2 corresponds to the mass of the assumed punching cone under the loading place, and M_3 is the mass of the surrounding structure. The three nonlinear springs describe the force–displacement relationship of the cushion layer, of the shear failure surface, and of the global structure, respectively. For the cushion layer, a hardening soil behavior is assumed (Figure 14-38c). The spring K_2 describes the shear stiffness for the assumed punching-shear failure of the concrete slab, where the behavior is controlled by three contributors:

- K_{21}: the contribution of concrete,
- K_{22}: the contribution of shear reinforcement, and
- K_{23}: the membrane effect depending on the anchorage of the bending reinforcement.

The third spring, K_3, describes the global stiffness of the concrete slab, taking into account the ratio of the slab spans as well as the boundary conditions. This model considers punching and bending failure of the shed at the same time, where the forces in the springs K_2 and K_3, respectively, are evaluated.

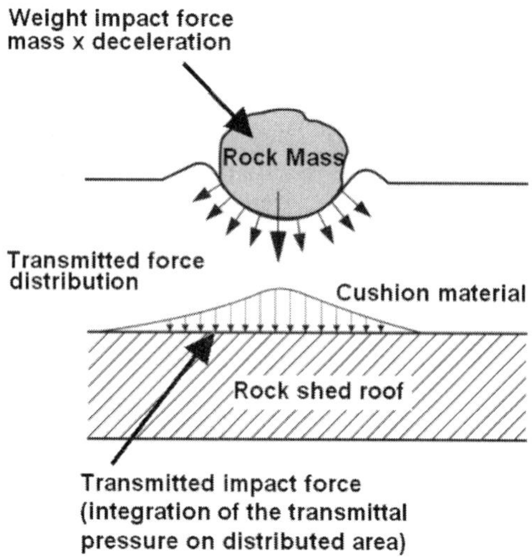

FIGURE 14-37 Rockfall imparts initial weight impact force to cushioning material and transmitted impact force to rock shed roof (Yoshida et al. 2007).

5.3.3 Numerical Simulations

In recent years, numerical simulations have been used for structural design that allow for an even more detailed evaluation of the structure and its response to rockfall impacts (Kishi et al. 2002, 2009). Figure 14-39 provides an example of the use of finite element analysis to evaluate a rock shed hit by rockfall. This approach, however, requires experimental data for calibration and significant resources, limiting its application for routine practice. Such efforts may be worthwhile for developing design guidelines and for evaluating critical sections and parametric influences.

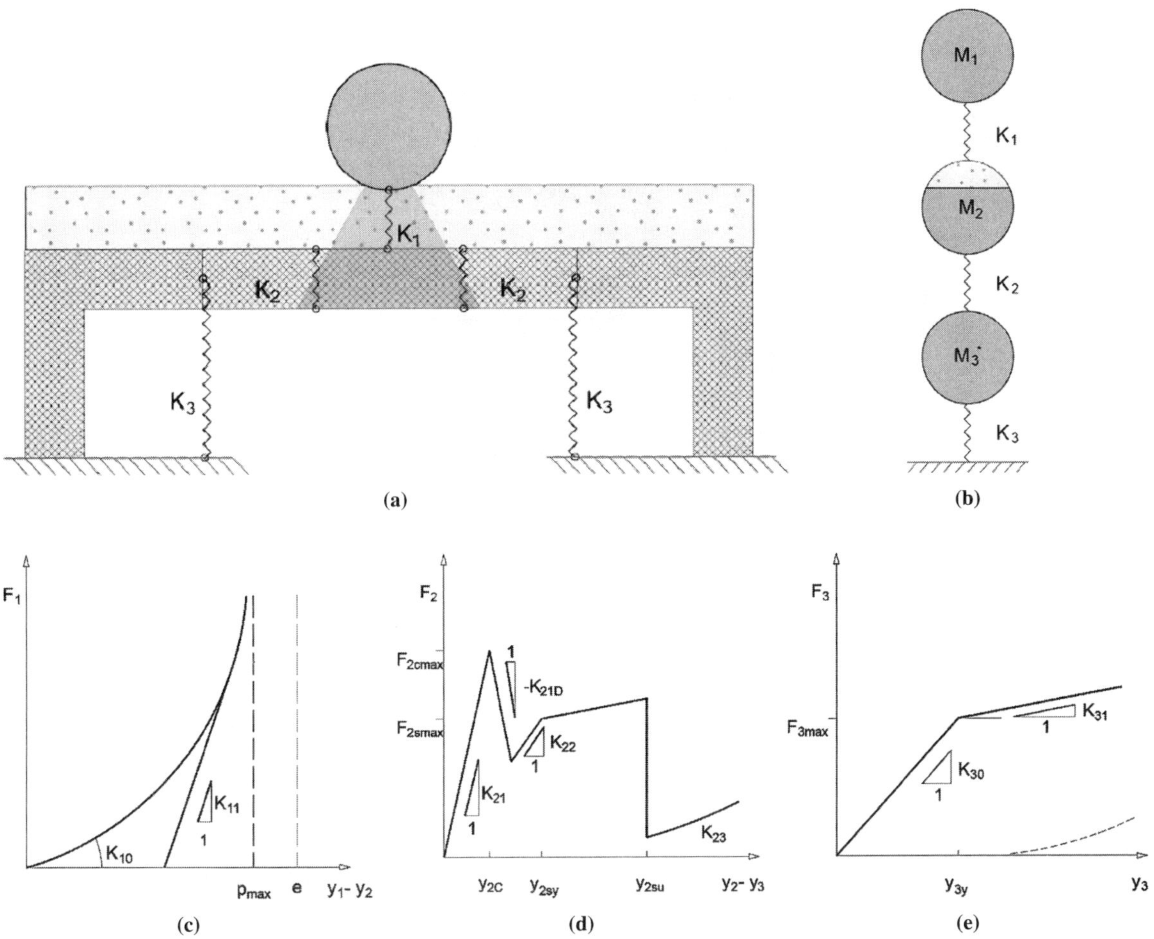

FIGURE 14-38
(*a*), (*b*) SMDF from section of gallery to model definition. Three nonlinear springs describe force–displacement relationship of (*c*) cushion layer, (*d*) shear behavior, and (*e*) global bending stiffness (Schellenberg and Vogel 2009).

Despite the advances in understanding the structural performance of rock sheds, large uncertainties remain on how best to model them, and even larger uncertainties surround the process of selecting a design rock mass and impact velocity. Such uncertainties make probabilistic methods an attractive tool for designing new sheds and for evaluating existing sheds.

6. TEMPORARY PROTECTION MEASURES

Temporary protection measures are often used to support rapid responses to emergency situations immediately after a rockfall event as well as during programmed projects involving earthwork and slope stabilization. Generally, the responsibility for site safety during construction is largely transferred by the owner to the contractor. Contractors commonly use protection measures that are easy and inexpensive to construct and move. However, in some instances, these often improvised and untested measures have proved less than satisfactory (Figure 14-40). While innovation has its benefits, it also carries risks that become shared with a consenting owner or, more often, unknowingly by a facility owner, utility company, or passing motorist. When an emergency slope stabilization project or a work zone cannot be completely secured, contractor-proposed temporary protection measures need careful evaluation with experienced judgment. When possible, proven measures of known capacities and performance are given considerable weight over less-proven techniques.

Many of the previously described protection measures can be used for temporary applications. In some cases, as a temporary measure, the system configuration or the materials used may be modified, because the application usually does not require the durability of a comparable permanent measure. Temporary measures are commonly considered sacrificial and thus often require routine maintenance during their use, especially when they are protecting a construction site.

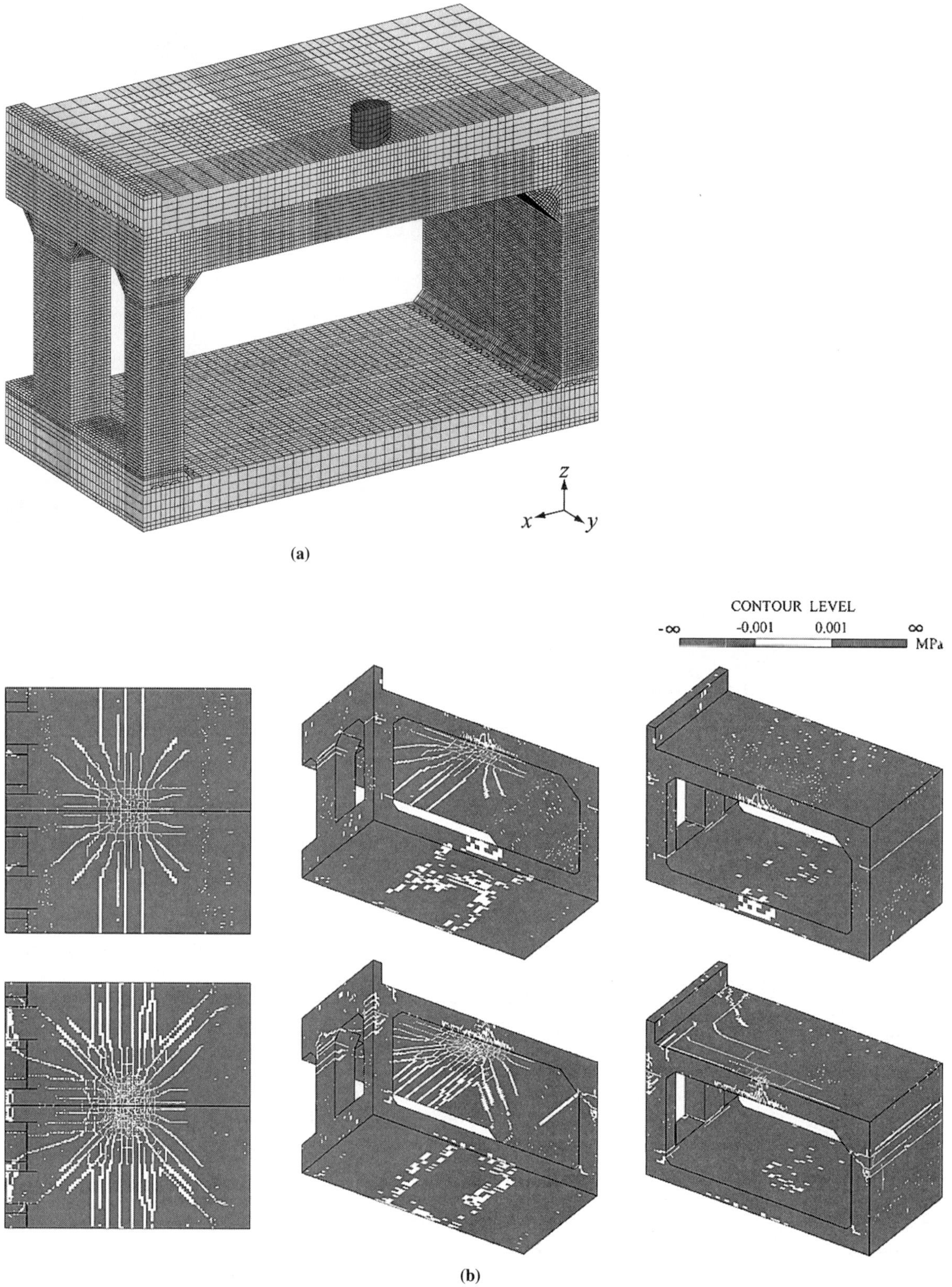

FIGURE 14-39
(*a*) Perspective view of finite element analysis model of rock shed after impact. (*b*) Resulting crack patterns (shown in white) of roof and support columns for two different loading cases (Kishi et al. 2009).

FIGURE 14-40
Testing of frequently used temporary construction barrier consisting of lightweight chain-link fence mounted on concrete barriers revealed that it was inadequate for required protection. Its use was therefore discontinued, and more robust and proven alternatives have been provided.
(PHOTOGRAPH COURTESY OF THE CALIFORNIA DEPARTMENT OF TRANSPORTATION.)

For critical applications, added protection can be realized by combining measures. Desired attributes for temporary applications include proven performance, low cost, ease of deployment, and accessibility for routine maintenance and cleanup of captured rockfall debris. The intent of these temporary measures is to retain debris in a prescribed area to facilitate cleanup and limit the duration and severity of road closures. To provide safety for travelers, traffic through the work zone is halted during active scaling of loose or unstable rocks from rock exposures, except with the most proven of protection measures. Some of the more common and proven measures are presented below.

6.1 Containment Area

Modifications to increase the size of a containment area or to improve its effectiveness are typically an initial consideration for temporary protection measures. This procedure may include lane closures, either for short-duration work periods—as is common during scaling operations—or closures lasting several hours or even 1 or more days, which may be required by more extensive construction activities. For closures of longer duration, temporary detours or realignments are often considered, with the inconvenience and expense offset by significantly reduced construction time and cost.

Cushioning material, such as a layer of loose sand, can be placed to attenuate the kinetic energy of falling rock debris and improve the effectiveness of the available width of an existing containment area. This low-cost measure can also be placed on paved surfaces to reduce impact damage and damage caused by heavy equipment removing rock debris when a portion of the roadway is used as a temporary containment area. On a scaling project involving block sizes in excess of 1.5 m (5 ft) falling from slope heights approaching 100 m (300 ft), the Washington State DOT specified that a layer of loose sand to a depth of 0.5 m (1.5 to 2 ft) be placed on the road surface; there was virtually no damage to the asphalt pavement (B. Romine, personal communication, 2008). Extra cushioning material should be stockpiled to replenish the blanket because debris-removal operations invariably deplete the cushioning material.

Containment areas are also made more effective by inclining the base surface so as to reduce the tendency of debris to roll away from initial impact areas. As demonstrated in the research completed to develop the *Rockfall Catchment Area Design Guide* (Pierson et al. 2001), an inclined base with a rising slope of 1V:4H can dramatically improve the effectiveness of the catchment area, compared with an area with a flat base. In combination with a barrier to restrict errant vehicles from entering the ditch, an even steeper nonrecoverable slope can be used to increase the resistance to rollout and the storage capacity. In all cases, if the containment area is cleaned out regularly, it will function as intended.

6.2 Concrete Barriers

Concrete barriers are commonly used to prevent debris from rolling beyond the available containment area. At low-energy impacts of less than 80 kJ (30 ft-tons) the near-vertical face of the barrier is highly effective at stopping rolling rocks. Higher-energy impacts can cause shattering on both sides of the barrier. An earthen berm or other energy-attenuating devices such as hay bales and timbers can be included on the side of the barrier facing the rockfall source to reduce impact damage by larger rocks; however, some of the benefit provided by the steep-sided barrier will be lost. Concrete barriers are often readily available and can be rapidly placed and removed. They are physically connected to improve their effectiveness.

Concrete blocks (sometimes referred to as *ecology blocks*) have been used as an alternative to standard concrete barriers for temporary containment applications (Figure 14-41). These blocks are typically made from surplus concrete at ready-mix plants and are often available from ready-mix concrete suppliers. They come in various sizes, but blocks of 1 × 1 × 1.5 m (3 × 3 × 5 ft) are relatively common. They have several additional advantages over standard concrete barriers: unlike traffic barriers, they are designed to be stacked, and their characteristically greater thickness means they are capable of withstanding higher impact energies. The energy-absorbing capacity of concrete blocks, however, has not been established.

FIGURE 14-41
Concrete block wall used temporarily to protect hydroelectric facility during scaling operations.
(PHOTOGRAPH COURTESY OF LANDSLIDE TECHNOLOGY.)

6.3 Earthen and Structural Berms

An earthen berm can provide nearly unlimited retention capacity as a temporary retaining structure, as long as adequate mass, height, and storage area are provided. The primary limitations of these berms are the sizable base width required and poor mobility. However, they can be quick and simple to construct if material is readily available. Berms generally require side slopes no steeper than 1V:1H, which creates a base width of at least twice the barrier height.

Geotextile-wrapped earthen berms have proved to be robust effective structures that allow for narrower width, additional height, and improved structural integrity of the berm (Figure 14-42). The construction time and the cost associated with incorporating structural elements typically limit the use of these structural berms to shorter, well-defined work areas.

6.4 Shipping Containers

Shipping containers are regularly used by Caltrans as a temporary barrier to protect facilities during construction operations (Figure 14-43). The containers are made heavier by placing three concrete barriers inside each one. A steel trench plate is then tack-welded to the upslope side of

FIGURE 14-42
Temporary geotextile berm with height and width of 3 m (10 ft) effectively contained several boulders 4 m (14 ft) in diameter. Had containment area been kept clear of debris, large boulder lying on top of berm would likely have been retained behind berm.
(PHOTOGRAPH COURTESY OF THE OREGON DEPARTMENT OF TRANSPORTATION.)

FIGURE 14-43
Shipping containers serve as effective temporary barriers during scaling and grading activities occurring above active traffic flows.
(PHOTOGRAPH COURTESY OF THE CALIFORNIA DEPARTMENT OF TRANSPORTATION.)

the container to reinforce its impact side. This design has been used in landslide removal and cut slope excavation work, where there is a need to maintain traffic flow. In some instances, the containers have been placed on trailers, which are protected from rockfall by standard concrete barriers, enabling the shipping container protection system to be moved daily or to be advanced as the work location progresses (Figure 14-44). The authors have observed that the containers withstand impacts from boulders up to 1.2 m (4 ft) in diameter producing an estimated impact energy of 200 kJ (74 ft-tons).

FIGURE 14-44
Stacked shipping containers serving as temporary rockfall protection are placed on trailer for greater mobility.
(PHOTOGRAPH COURTESY OF THE CALIFORNIA DEPARTMENT OF TRANSPORTATION.)

6.5 Chain-Link Fences

Several chain-link fences that were specifically designed to be used as temporary construction barriers have been tested by Caltrans (Duffy and Jones 2000). All the designs were supported by an infrastructure founded on standard concrete barriers or with posts placed in shallow drill holes backfilled with pea gravel. The tested 4-mm-diameter (6-gauge) mesh was loosely attached to the supporting infrastructure. The testing revealed that maximizing system flexibility was a key design feature, and a design evolved that allowed the mesh to hang on the downslope side of the posts, without additional connections to the posts. Post spacing of up to 6 m (20 ft) is used, and at least 3 m (10 ft) of catchment is required between the base of the slope and the fence (Figure 14-45). These systems are capable of stopping impacts with energies up to 95 kJ (35 ft-tons), although the systems are generally limited to applications with impact energies ranging from about 40 to 70 kJ (15 to 25 ft-tons).

6.6 Movable Flexible Fence and Containment Net

Proprietary flexible rockfall fences have been successfully modified for temporary applications, and some of these modifications permit rapid setup and moving. One example involves a fence 3.5 m (12 ft) high mounted on steel plates. The plates are hinged with simple pin connections, and up to 50 m (150 ft) of barrier can be connected and dragged into place with an excavator or loader

FIGURE 14-45
Temporary chain-link fence developed and tested by Caltrans is used on construction project at Cuesta Grade on California State Route 101 on central California coast.
(PHOTOGRAPH COURTESY OF THE CALIFORNIA DEPARTMENT OF TRANSPORTATION.)

FIGURE 14-46
Proprietary flexible rockfall fence mounted on steel plates allows for rapid deployment. Crane suspends containment net between fence and slope to contain scaled rock debris generated high on slope that otherwise might overtop barrier.
(PHOTOGRAPH COURTESY OF THE WASHINGTON STATE DEPARTMENT OF TRANSPORTATION.)

FIGURE 14-47
Layer of hay bales used temporarily to protect historic wall from rock impacts during scaling activities.
(PHOTOGRAPH COURTESY OF LANDSLIDE TECHNOLOGY.)

(Figure 14-46). The plates are then secured with relatively low-capacity ground anchors to prevent them from sliding or overturning as a result of rockfall impacts. Several state DOTs own and routinely use these movable rockfall fences in conjunction with a crane-suspended cable-net drape to provide additional protection when rock-scaling operations are undertaken adjacent to active traffic flows (Figure 14-46).

6.7 Other Temporary Protection Measures

Stacked standard hay bales and large-diameter hay rolls have proven effective in protecting rigid barriers and other features, such as historic guard walls, hydroelectric facilities, and buildings, from rockfall (Figure 14-47). The hay bales or rolls deflect on impact, which helps attenuate the energies of the moving rockfall blocks. Commonly, plywood is used to spread out impact loads and to protect the twine or wire that binds the bales together.

In other situations, buried utility lines may need to be protected from rockfall impacts during scaling or blasting. Steel plates can provide protection, and they are easy to move as work progresses (Figure 14-48). If the plates are placed properly, traffic can drive across them when queues are cleared.

7. SUMMARY

Rockfall protection measures inherently accept the occurrence of rockfalls and strive to mitigate the associated hazard by stopping, controlling, or deflecting the falling debris. Protection measures can be classified into four broad groups: catchment areas, barriers and fences, drapery, and rock sheds. The process of selecting and designing the most appropriate measure considers the frequency of rockfall events, the energies and trajectories of the

FIGURE 14-48
Layer of steel plates used temporarily to protect buried utility line from rockfall impacts.
(PHOTOGRAPH COURTESY OF LANDSLIDE TECHNOLOGY.)

falling debris, the space available to construct protection measures, construction and maintenance costs, and, possibly, aesthetic constraints.

Optimizing the available catchment area is often the first consideration in the process of selecting the best-suited protection measure for the site conditions and constraints. Minor modifications to the catchment area that often involve little cost and few adverse impacts can greatly enhance the level of protection. Barriers are often considered if bounce heights, rollout distances, or kinetic energies of anticipated rockfall events appear likely to overwhelm the available catchment area. If the catchment area is clearly inadequate, drapery systems may be the only feasible protection alternative. Rock sheds are a potential protection measure if volumes of rockfall or the kinetic energies of the falling blocks are so large that adequate protection cannot be achieved by any of the more commonplace alternatives. However, because of the considerable design complexity and cost of rock sheds, avoidance measures, discussed in Chapter 12, may offer a more appropriate and cost-effective solution. Table 14-1 summarizes the advantages and disadvantages of the commonly used protection methods.

Table 14-1
Summary of Protection Measures

Mitigation	Advantages	Disadvantages	Cost
Catchment area	Highly effective and reliable; low maintenance costs	Sizable width needed; requires regular cleanout	
Rockfall ditch	Standard designs well proven; modifiable for improved catchment	Costly for existing cut slopes if additional width not readily available or easily accessed	Low to high
Midslope bench	Effective for zones and beds of differential erosion; potentially less costly and longer-lived than stabilization measures	Must maintain bench width to provide access for cleanout; increased excavation quantities and cost	Moderate to high
Barrier	Wide range of capacities	Sizable width needed; requires regular cleanout	
Earthen berm	Often can be constructed with on-site materials; moderate to high capacity	Capacity not well proven	Low
Concrete barrier	Highly effective for reducing rollout from ditch; can be rapidly deployed; narrow required width	Very low capacity; very rigid; prone to shattering; can become hazard if shoved into travel lane from impact	Low
Cantilever and soldier pile wall	Narrow required width; wall facings can reduce aesthetic impact	Typically low and mostly unproven capacity; very rigid, prone to damage	Moderate to high
Reinforced embankment and MSE wall	High capacity; durable and low maintenance; wall facings can reduce aesthetic impact	Costly repairs, if needed	Moderate to high
Gabion barrier	Moderate to high capacity; durable and low maintenance	Capacity not well proven	Moderate
Flexible fence	Proprietary designs well proven; wide range of capacities; often requires less width than other barriers	Maintenance requires effort beyond debris cleanout	Moderate to high
Drapery system	Highly effective and reliable; durable and low maintenance; well proven; range of capacities	Requires catchment area; not effective or proven for large rockfalls	Low to moderate
Rock shed	Highly effective; low maintenance; high capacity	Cost prohibitive for most applications	Very high

ACKNOWLEDGMENTS

Duncan Wyllie shared information on rock sheds. Bob Henthorne and Marty Woodard provided input on midslope benches. Daniele Peila, C. Oggeri, and C. Ronco provided much information on reinforced embankments and flexible fences.

REFERENCES

ABBREVIATIONS

FHWA	Federal Highway Administration
IVOR	Innovations Validées sur Ouvrages de Référence

ASTRA. 2008. *Einwirkungen infolge Steinschlags auf Schutzgalerien*. Bundesamt für Strassen, Baudirektion SBB, Eidgenössische Drucksachen und Materialzentrale, Bern, Switzerland (in German).

Barrett, R. K., and J. L. White. 1991. Rockfall Prediction and Control. *National Symposium on Highway and Railway Slope Stability*, Association of Engineering Geologists, Denver, Colo., pp. 23–40.

Berthet-Rambaud, P. 2004. *Structures Rigides Soumises aux Avalanches et Chutes de Blocs: Modélisation du Comportement Mécanique et Caractérisation de l'Interaction "Phénomène-Ouvrage."* Dissertation. Université Grenoble, France (in French).

Bertrand, D., F. Nicot, P. Gotteland, and S. Lambert. 2008. Discrete Element Method (DEM) Numerical Modeling of Double-Twisted Hexagonal Mesh. *Canadian Geotechnical Journal*, Vol. 45, pp. 1104–1117.

Chikatamarla, R. 2006. *Optimisation of Cushion Materials for Rockfall Protection Galleries*. Dissertation. Swiss Federal Institute of Technology, Zurich.

Duffy, J. D., and W. Hoon. 1998. *Field Tests and Evaluation of Chain Link Fence, Gawk Screen Fence, and Jersey Barrier*. Report CA/05-98-01. California Department of Transportation, San Luis Obispo.

Duffy, J. D., and C. Jones. 2000. *Field Tests and Evaluation of Temporary Construction Chain Link Fence*. Report CA/05-00-01. California Department of Transportation, San Luis Obispo.

FHWA. 1989. *Rock Slopes: Design, Excavation, and Stabilization*. Publication FHWA-TS-89-045. Turner-Fairbank Highway Research Center, McLean, Va.

FHWA. 1993. *Rockfall Hazard Mitigation Methods*. Publication FHWA SA-93-085. U.S. Department of Transportation.

Heierli, W. 1976. *Schutz gegen Felssturz und Steinschlag mit Drahtseilnetzen (Protection Against Rock Avalanches and Rock Falls with Wire Rope Nets)*. Berne, Switzerland (in German).

Higgins, J. D. 2003. *Recommended Procedures for the Testing of Rock-Fall Barriers*. Report prepared as part of National Cooperative Highway Research Program Project 20-07, Task 138.

IVOR. 2001. *Couverture Pare-Blocs Structurellement Dissipante*. Innovations Validées sur Ouvrages de Référence. Label IVOR 01.1. Mission Génie Civil, METL/DRAST, Paris. http://www.rgcu.prd.fr/IMG/pdf/Fiche_IVOR_01-1_cle22e8ba.pdf (in French).

Jacquemoud, J. 1999. Swiss Guideline for the Design of Rock Fall Protection Galleries: Background, Safety Concept and Case Histories. *Proc., Joint Japan–Swiss Seminar on Impact Load by Rock Falls and Design of Protection Measures* (P. Masuya and V. Labiouse, eds.), Kanazawa, Japan, pp. 1–61.

Japan Road Association. 2000. *Manual for Anti-impact Structures Against Falling Rocks* (in Japanese). Tokyo.

Kirsten, H. A. D. 1982. Design and Construction of the Konwyn's Pass Rock Fall Shelter. *Civil Engineering, South Africa*, Vol. 24, No. 9, pp. 477–492.

Kishi, N., H. Konno, K. Ikeda, and K. G. Matsuoka. 2002. Prototype Impact Tests on Ultimate Impact Resistance of PC Rock-Sheds. *International Journal of Impact Engineering*, Vol. 97, pp. 969–985.

Kishi, N., S. Okada, and N. Konno. 2009. Numerical Impact Response Analysis of Rockfall Protection Galleries. *Structural Engineering International*, Vol. 19, pp. 313–320.

Lowell, S. M. 1987. Development and Application of Ritchie's Rockfall Catch Ditch Design. In *Federal Highway Administration Rockfall Mitigation Seminar Manual* (R. G. Chassie, ed.), Portland, Ore.

Masuya, H. (ed.). 2007. *Design Method of Structures Under Impact Action by Concept of Performance Based Design* (in Japanese). Japan Society of Civil Engineers, Tokyo.

Masuya, H., K. Maegawa, A. Mizuki, and H. Yoshida. 1987. Impulsive Loads by Rock Falls on Steel Rock-Sheds (in Japanese). *Journal of Structural Engineering*, Vol. 36A, pp. 41–49.

McCauley, M. L., B. W. Works, and S. A. Naramore. 1985. *Rockfall Mitigation*. California Department of Transportation, Office of Transportation Laboratory, Sacramento.

Montani, S. 1998. *Solicitation Dynamique de la Couverture des Galleries de Protection lors de Chutes de Blocs*. Dissertation. École Polytechnique Fédérale de Lausanne, France (in French).

Mougin, J.-P., P. Perrotin, M. Mommessin, J. Tonnelo, and A. Agbossou. 2005. Rock Fall Impact on Reinforced Concrete Slab: An Experimental Approach. *International Journal of Impact Engineering*, Vol. 31, pp. 169–183.

Muhunthan, B., S. Shu, N. Sasiharan, O. A. Hattamleh, T. C. Badger, S. M. Lowell, and J. D. Duffy. 2005a. *Analysis and Design of Wire Mesh/Cable Net Slope Protection*. Report WA-RD 612.1. Washington State Transportation Center, Seattle.

Muhunthan, B., S. Shu, N. Sasiharan, O. A. Hattamleh, T. C. Badger, S. M. Lowell, and J. D. Duffy. 2005b. *Design Guidelines for Wire Mesh/Cable Net Slope Protection*. Report WA-RD 612.2. Washington State Transportation Center, Seattle.

Nakano, O., N. Kishi, Y. Goto, and H. Endo. 1992. Development on 3-Layered Shock-Absorbing System of Rock Sheds. *Journal of Japan Society of Civil Engineering*, Vol. 77, pp. 20–33.

New York State Department of Transportation. 2007. *Geotechnical Design Procedure for Preparing Rock Slope Recommendations*. Geotechnical Design Procedure GDP-13, Revision 1. Geotechnical Engineering Bureau, New York State Department of Transportation, Albany.

Peila, D., C. Castiglia, C. Oggeri, G. Guasti, P. Recalcati, and F. Sassudelli. 2000. Full Scale Tests on Geogrid Reinforced Embankments for Rock Fall Protection. *Proc., 2nd European Geosynthetics Conference*, Patron Editore, Bologna, Italy, pp. 317–322.

Peila, D., C. Oggeri, and C. Castiglia. 2007. Ground Reinforced Embankments for Rockfall Protection: Design and Evaluation of Full Scale Tests. *Landslides*, Vol. 4, pp. 255–265.

Peila, D., C. Oggeri, C. Castiglia, P. Recalcati, and P. Rimoldi. 2002. Testing and Modelling Geogrid Reinforced Soil Embankments to High Energy Rock Impacts. *Proc., 7th International Conference of Geosynthetics* (P. Delmas, J. P. Gourc, and H. Girard, eds.), Taylor and Francis, Balkema, Leiden, Netherlands, pp. 133–136.

Pierson, L. A., C. F. Gullixson, and R. G. Chassie. 2001. *Rockfall Catchment Area Design Guide*. Final Report SPR-3(032). Oregon Department of Transportation, Salem.

Ritchie, A. M. 1963. Evaluation of Rockfall and Its Control. In *Highway Research Record 17*, Highway Research Board, National Research Council, Washington, D.C., pp. 13–28.

Schellenberg, K. 2009. *On the Design of Rockfall Protection Galleries—An Analytical Approach for a Performance Based Design*. SVH Verlag, Saarbrücken, Germany.

Schellenberg, K., and T. Vogel. 2009. A Dynamic Design Method for Rockfall Protection Galleries. *Structural Engineering International*, Vol. 19, pp. 321–326.

Spang, R. M. 2000. Standardisierung von Prüfverfahren für Steinschlagschutzbarrieren: Ziele und aktueller Stand (Standardization of Test Methods for Rockfall Protection Barriers: Objectives and Current Status), *2nd Colloquium on Construction in Soil and Rock*, Jan. 18–19, Esslingen University of Applied Sciences, Esslingen, Germany (in German).

Vogel, T., V. Labiouse, and H. Masuya. 2009. Rockfall Protection as an Integral Task. *Structural Engineering International*, Vol. 19, No. 3, pp. 304–312.

Washington State Department of Transportation. 2007. *Design Manual M 22-01.05*. Olympia. www.wsdot.wa.gov/Publications/Manuals/M22-01.htm.

Wyllie, D. C. 1991. Rock Slope Stabilization and Protection Measures. *National Symposium on Highway and Railway Slope Stability*, Association of Engineering Geologists, Denver, Colo., pp. 41–64.

Yokoyama, K., M. Ohira, and H. Yoshida. 1993. Impulsive Properties due to Rock Fall on a Steel Rock Shed (in Japanese). *Journal of Structural Engineering*, Vol. 39A, pp. 1573–1586.

Yoshida, H. 1999. Recent Experimental Studies on Rockfall Control. *Proc., Joint Japan–Swiss Scientific Seminar on Impact Load by Falling Rocks and Design of Protection Structures* (P. Masuya and V. Labiouse, eds.), Kanazawa, Japan, pp. 69–78.

Yoshida, H., H. Masuya, M. Satou, and T. Ihara. 1987. A Database of Rock Falling Tests and an Estimation of Impulsive Load by Rock Falls (in Japanese). *Journal of Structural Engineering*, Vol. 33A, pp. 571–583.

Yoshida, H., T. Nomura, D. C. Wyllie, and A. J. Morris. 2007. Rock Fall Sheds: Application of Japanese Designs in North America. In *Landslides and Society: Keynote and Invited Presentations at the 1st North American Landslide Conference*, (A. K. Turner and R. L. Schuster, eds.), Special Publication No. 22, Association of Engineering and Environmental Geologists, Denver, Colo., pp. 179–196.

Chapter 15

John D. Duffy and
Thomas C. Badger

FLEXIBLE ROCKFALL FENCES

1. DESCRIPTION AND SITE CRITERIA

Fences, like nearly all barriers, are used to enhance rockfall catchment areas. A fence suspends a net in an optimal position to intercept falling rocks. Stopping the rock requires the fence to deform and dissipate the kinetic energy of the block, and this is done with a suitable safety factor. The fence's flexibility provides advantages over most rigid barriers through reduced costs, more versatile placement, and simpler design effort, primarily through the use of well-tested proprietary systems.

Fences were initially applied to protection from snow avalanches in Switzerland; they involved triangular wire rope nets mounted on timber and, later, steel posts (Figure 15-1) (Spang and Bolliger 2001). The early avalanche fences were exposed to rockfall during snow-free periods and were effective in stopping rocks. This observation prompted Kabelwerke Brugg AG to try a similar system for protection against rockfall. The first known application of a wire rope fence for rockfall was erected in 1958 at Brusio in southern Switzerland for the protection of power transmission lines. The system consisted of individual rectangular nets measuring 3 × 5 m (10 × 16 ft) standing 5 m (16 ft) high, tall even by today's standards. From the 1950s through the 1980s, nearly all rockfall protection fences in the United States were constructed with standard chain-link fabric (Figure 15-2).

During the past 50 years rockfall fence design has changed dramatically, in large part because of increased testing, innovation, and market demand. Flexibility remains the foundation of the design, but materials and infrastructure have evolved. Modern fences are optimized to dissipate energy. Many fence fabrics are available, from lightweight wire mesh to robust ring nets. Supporting infrastructure incorporates energy-absorbing devices. Subtle but important connection features allow potentially damaging energy to bypass fixed points. Through rigorous product development and testing, a full suite of fence systems is currently available for energy ratings from 10 to 8,000 kJ (4 to 2,950 ft-tons).

1.1 Function and Performance

When a fence is struck by a rockfall, energy is distributed from the impact location into adjacent system components to dissipate energy and avoid or minimize damage (Figure 15-3). Within 0.5 s of impact, flexure of the fence panel occurs; any friction energy–dissipation devices, if included as part of the system, may actuate; and the rock block is stopped (Figure 15-4).

Fence performance is evaluated by comparing the relationships between impact loading, maintenance, and efficiency. Impact loading is simply the amount of load transferred to the fence from a rockfall impact. Maintenance refers to necessary

FIGURE 15-1
Early avalanche net using wire rope triangular nets on timber posts. First application was in 1951 on Schafberg at Pontresina in Engadine valley, Switzerland.
(PHOTOGRAPH COURTESY OF R. SPANG.)

FIGURE 15-2
Typical chain-link rockfall fence.
(PHOTOGRAPH COURTESY OF THE CALIFORNIA DEPARTMENT OF TRANSPORTATION.)

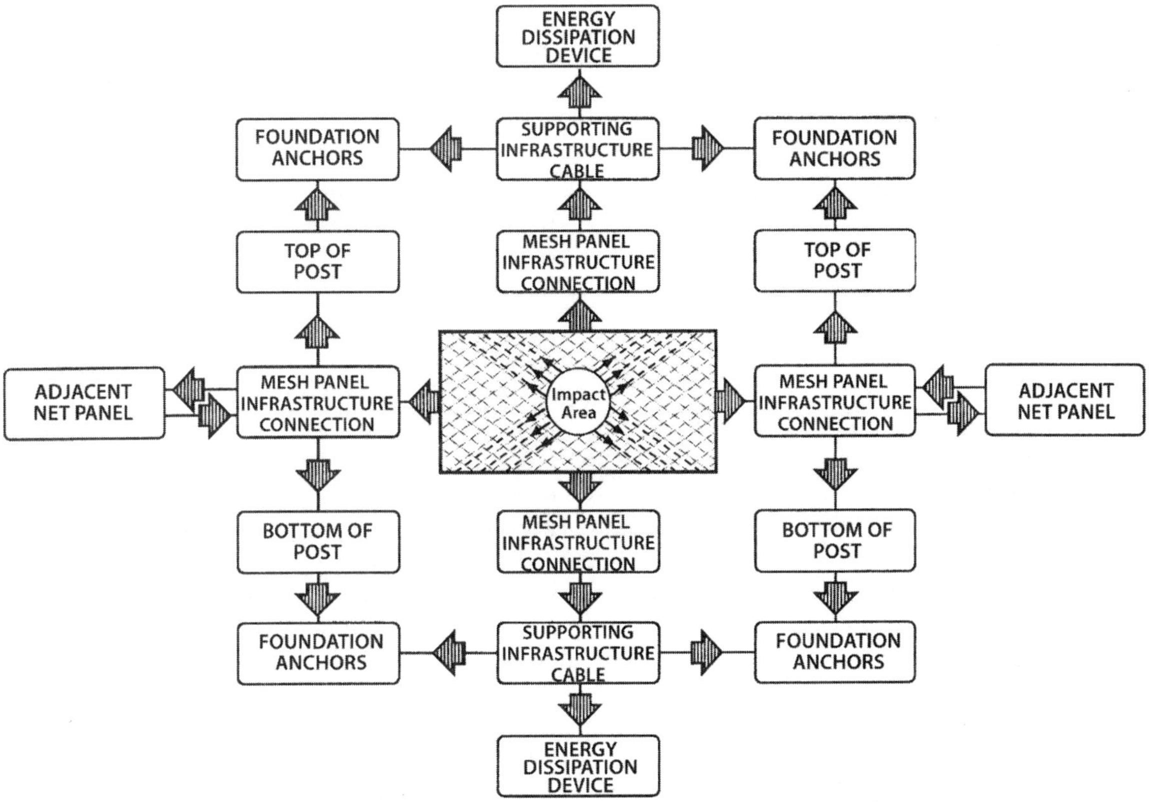

FIGURE 15-3
Energy path diagram of rockfall impact and distribution of loads onto fence components (adapted from Smith and Duffy 1990).

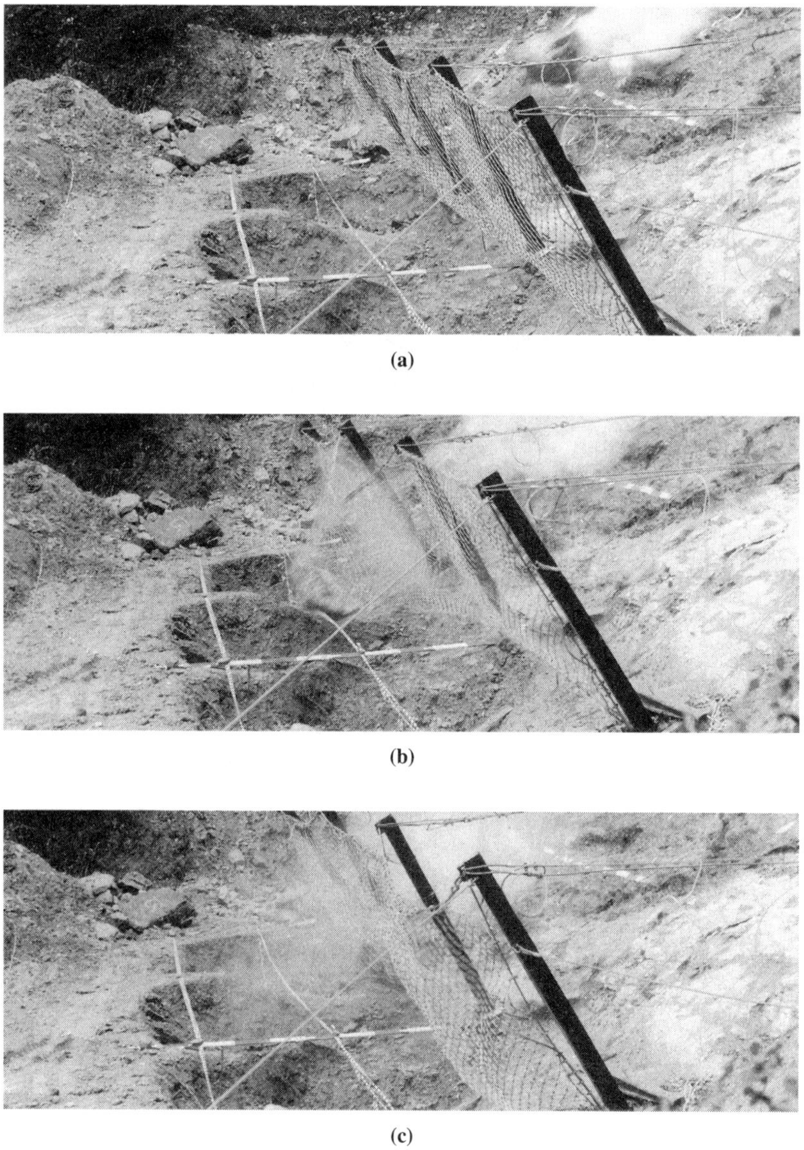

FIGURE 15-4 Fully flexing fence with energy rating of 200 kJ (74 ft-tons) on receiving 200-kJ impact: (*a*) approach of rock to fence, (*b*) maximum deflection of fence, and (*c*) subsequent deformations in adjacent panels and top and tieback ropes.
(PHOTOGRAPHS COURTESY OF J. D. DUFFY.)

repairs from an impact. Efficiency represents the effectiveness of the system in distributing load while minimizing damage; the impact location strongly influences efficiency. The most efficient impact location for a fence is in the center, where fence panels can fully flex [Figures 15-4 (all parts) and 15-5]. As flexibility increases, so does the time provided to dissipate energy and arrest the motion of the rock. Flexibility reduces the loads imparted to fence components and generally increases the overall ability of the system to absorb greater impact energies with less maintenance. Flexibility and energy dissipation decrease as the impact location moves toward the more rigid infrastructure located away from the center. Consequently, the requirement for maintenance of the fence system after the impact event increases.

The capability of the fence system and the demand for maintenance and repairs after a rockfall event are affected by the impact location, especially for large impacts, where the impact loading approaches the fence system design limits. Small rocks traveling at high velocities are an important exception to this general rule. While their energies may be within the design capacity of the fence, small rocks traveling at high speeds often puncture fence panels.

The relationships between impact location and impact load resulting from the combined mass and velocity of the rocks are fundamental to the correct determination of an appropriate fence. The black area in Figure 15-6 represents the design load limit for a fence. Within this range, efficiency is optimal and maintenance is minimal. The impact location generally does not matter. As impact loads extend above this range, the efficiency of the fence system decreases, and consequently the demand for maintenance increases. Impacts within the shaded range will be stopped, but with an increasing likelihood of significant damage. Impacts with energies beyond the "maximum impact load" are to the right of the shaded area in Figure 15-6; rocks exhibiting these energy levels will not be stopped by the fence, but their energy will be attenuated. Significant damage to the fence system has probably occurred. The fence system often will no longer offer any appreciable protection from additional rockfall events, and major repairs will be required to restore the fence system to full functionality.

Common nomenclature used to define performance levels for rockfall fences and similar protective barriers employs two terms: the *service energy level* (SEL), related to the design load impact (as shown in Figure 15-6), and the *maximum energy level* (MEL), related to the maximum impact load (as shown in Figure 15-6). SEL and MEL were developed as descriptors for fence performance and are used in testing guidelines both in Europe and in the United States. More information on fence testing and guidelines for and certification of fence performance capabilities are presented in Sections 5 and 6 of this chapter.

Braking distance, residual height, and forces transferred to the foundations and anchors are also important aspects of fence performance. They can

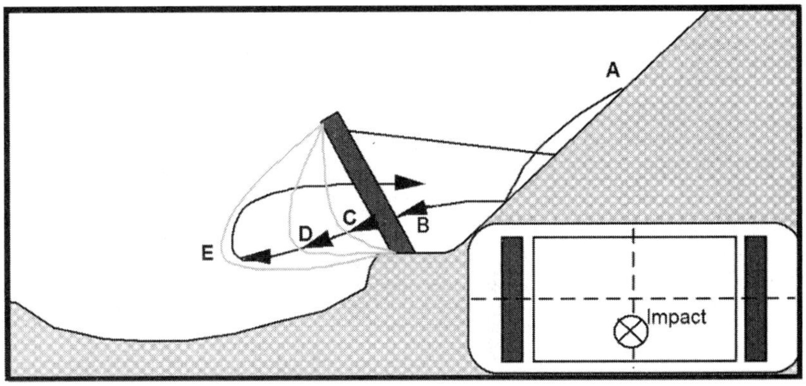

FIGURE 15-5 Fully flexing fence: A, bouncing, rolling rock; B, impact; C and D, fence panel deforms to absorb impact energy; E, maximum fence deflection. Arrow shows possible rebound path of rock after impact (Duffy 1992).

vary widely among fences that have similar impact load capacity but are produced by different manufacturers or use slightly different combinations of fence components. The standards for defining these and other performance characteristics become important when a designer who wishes to achieve specific dimensional limits for some or all of these characteristics has to compare fences from different manufacturers. Current European and American fence-testing guidelines address these aspects to some extent.

The advances in system performance and capacity that have been achieved over the past 50 years would not have been possible without extensive testing of fence systems and components, both in the laboratory and from full-scale impacts. This is illustrated most impressively by the increase in fence capacities from an initial 50 kJ (18 ft-tons) to about 8,000 kJ (2,950 ft-tons) during the period, a nearly 100-fold increase in energy-absorption capacity (Spang and Bolliger 2001).

This chapter describes impact energy levels as low, moderate, high, and very high. For the purposes of the discussion here, these levels are defined as follows: low, less than 100 kJ (37 ft-tons); moderate, 100 to 500 kJ (37 to 184 ft-tons); high, 500 to 2,000 kJ (184 to 738 ft-tons); and very high, greater than 2,000 kJ (738 ft-tons).

1.2 Site Suitability and Characteristics

Fences can be constructed in a variety of locations. Some are located proximate to a transportation facility along the side of the traveled way, while others are placed far upslope above the facility and near the rockfall source area (Figure 15-7). The footprint of a fence is narrow, generally less than 0.5 m (2 ft), not including the space required for tieback anchors. The narrow footprint and lightweight components of most fences facilitate their installation in varied topography. They can skirt obstacles and are readily used in forested terrain, along narrow right-of-way boundaries, and in the midst of urban development. Light construction equipment and helicopters can aid in the installation of many components of a fence. Foundations and ground anchors can, in most instances, be installed with hand equipment. Fences often are constructed with much less environmental impact than many other rockfall protection measures. However, adequate space in front of the fence, on the downhill side, is essential; fence systems undergo large deformations in dissipating energy when rocks strike the fence because of the lightweight infrastructure of the fence system. For example, a high-energy fence can deform horizontally more than 6 m (20 ft) at full flexure.

Fences trap debris. Routine cleanout behind the fence is critical to its optimal performance, so the placement and ultimately the suitability of a fence are related to the frequency of rockfall events and to its accessibility for maintenance operations. While a fence can effectively contain an impact within its energy rating, its ability to withstand dozens of impacts expected at a site with high rockfall frequency can be a critical design issue. Existing guidelines and energy ratings of fence systems provide limited guidance with regard to the loss of barrier efficiency caused by repeated impacts and accumulating debris. While manual cleanout may be feasible, a high frequency of rockfall increases the importance of access by heavy equipment, because the volume of accumulating debris may be too large for manual removal.

Maintenance also generally increases with increasing impact energies. Numerous proprietary

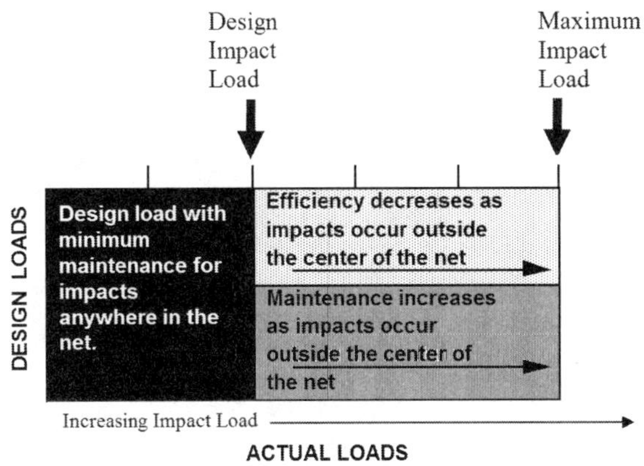

FIGURE 15-6 Design load chart for fence (adapted from Duffy and Haller 1993).

FIGURE 15-7
Fences have been constructed in various locations: (*a*) fence located "at grade," (*b*) fence located high up on steep slopes, and (*c*) fence located in forested terrain.
(PHOTOGRAPHS *a* AND *b* COURTESY OF THE FEDERAL HIGHWAY ADMINISTRATION. PHOTOGRAPH *c* COURTESY OF M. RICHARD, CAN, INC.)

fences are designed to stop rockfalls with energies in excess of 1,000 kJ (370 ft-tons), but impacts on most moderate- to high-capacity fences that exceed a few hundred kilojoules are likely to result in large fence deformations that require maintenance (Figure 15-8). Where rockfall frequency is very high, other barriers that require less maintenance or are more durable, such as earthen or gabion barriers, are often more appropriate. Thus, to minimize maintenance efforts (Spang and Bolliger 2001), it is important to establish the relationship between the energy-absorption capacity of a proposed fence system, the frequency of rockfall events, and the anticipated energies of expected rockfall events.

While rockfall fences were developed on the basis of experience gained from snow avalanche protection, they are not designed to withstand the loads associated with snow creep or sliding. Thus, when fences are located on slopes susceptible to accumulating snow, consideration of these potentially damaging loads is essential in the design process.

2. FENCE DESIGN

Standard practice relies primarily on proprietary rockfall fence systems, so geotechnical design for the installation of fence systems is chiefly concerned with the following:

- Characterization of site conditions, including rockfall source area, frequency, and trajectories; expected impact energies and bounce heights at the fence location; and subsurface conditions for fence foundations and anchors;
- Optimization of the fence location and configuration on the basis of expected braking distance, fence deformations, and maintenance; and
- Evaluation of constructability, aesthetics, and ancillary design issues.

Once site suitability and geotechnical design parameters have been established, the appropriate fence is selected. Several design scenarios are usually considered. For example, consider a fence system for which the anticipated frequent impact energies are within SEL specifications and for which MEL impacts are determined to occur at intervals of several years or decades. Installing this fence system and planning for maintenance when the comparatively rare MEL events occur may be more cost-effective than installing a much more robust fence system capable of responding to the larger rare events without maintenance. A wide variety of tested proprietary fences is available from numerous manufacturers, as are more generic designs that have been developed and tested in the public sector. All modern fence systems have been subjected to some level of engineering and testing to ensure that they can provide for public safety.

2.1 Design Considerations and Component Attributes

Several design considerations influence the selection of fence components and system configurations to optimize fence function. Fence height and inclination to the slope must accommodate the expected rockfall trajectories and bounce heights. The fence length and the terrain influence the location and design of the system components. The response of each fence segment to a rock impact is defined by its post spacing, and post spacing can vary. In general, a wider post spacing results in greater panel deformations than does a narrower spacing. If offset fence sections are required because of terrain obstructions, sections typically overlap by at least one-half segment to ensure containment of the rock blocks (Peila et al. 2006).

Many fence configurations have been developed, but all can be divided into two main geometries: barriers where the net is suspended in the plane defined by the posts, referred to as *planar geometry* (Figure 15-9*a*), and barriers where the net is located outside this plane with a curved geometry, known as *sack geometry* (Figure 15-9*b*) (Peila et al. 2006).

A third configuration, sometimes referred to as a *hybrid* or *attenuator system* (Figure 15-9*c*), incorporates design features at the base of the fence to allow rocks, after impact, to pass beneath the fence under a mesh and into a containment area. The idea is to attenuate the energy of the striking rockfall and control the continued descent of the rock. This type of configuration is also referred to as a *hybrid drapery* and is discussed in detail in Section 4 of Chapter 16.

2.2 System Components

Fence systems are constructed with four principal components (Figure 15-10):

- Fence panels (interception structures),
- Infrastructure (support structure),
- Friction brakes (energy-absorbing devices or connections), and
- Foundations.

FIGURE 15-8 Moderate-capacity fence that effectively arrested multiple rocks from single rockfall event estimated to have had energy impact of several hundred kilojoules and to have been within design rating of fence. Note panel deformation, actuation of several friction brakes, and reduced height, which necessitated system maintenance to restore optimal function.
(PHOTOGRAPH COURTESY OF THE WASHINGTON STATE DEPARTMENT OF TRANSPORTATION.)

The basic fence design consists of posts that are retained in position by a system of cables. The fence panel hangs securely on a system of wire ropes (infrastructure) that is suspended from the posts. Every system has its own combination of components, and the selection of an appropriate combination is controlled largely by the desired energy-management capacity of the system, expected rockfall frequency, and considerations with regard to the level of maintenance (Figures 15-11 and 15-12).

2.2.1 Post Foundations and Ground Anchors

Post foundations provide at least partial moment resistance to the posts during impact and mobilize passive resistance within the substrate to oppose sliding. Some designs allow for movement and rotation of the foundation to dissipate energy.

For soil and mixed ground conditions, posts are most commonly founded on concrete footings (Figure 15-13*a* and 15-13*b*). For rock foundations, posts may be founded on a leveling pad of grout or directly on the rock surface (Figure 15-13*c*). Other designs do not use any concrete footings; they are simply staked to the ground by special spiked footplates. In some designs, the post base is secured to anchors located upslope by means of ground ropes. The ropes provide lateral resistance to sliding by the post and allow reduced dimensions of the post foundation. An important design

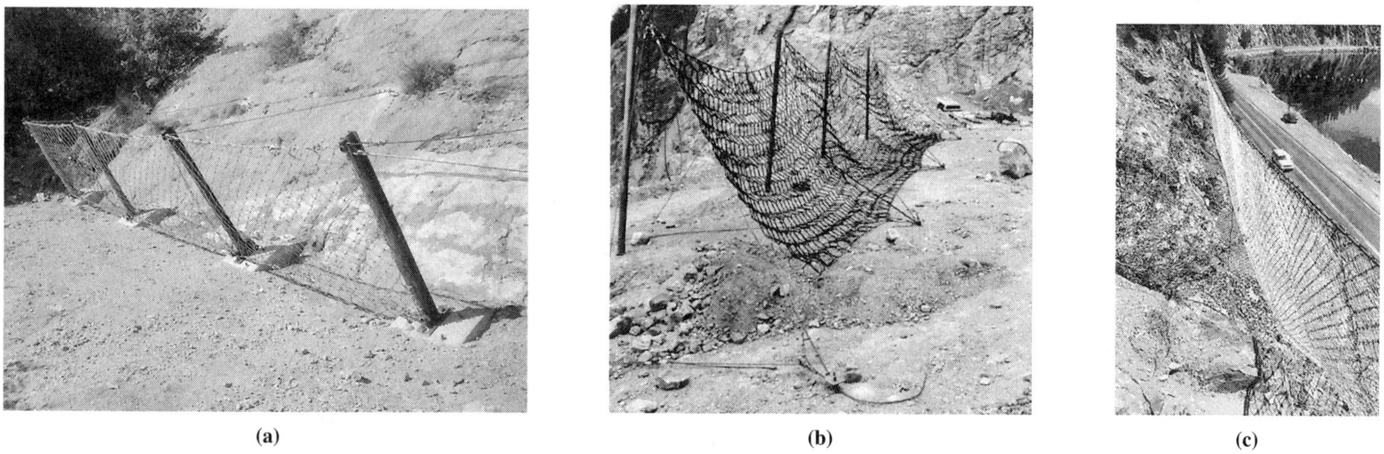

FIGURE 15-9
Examples of various proprietary flexible fence systems: (*a*) planar geometry, (*b*) sack geometry, and (*c*) hybrid or attenuator geometry.
(PHOTOGRAPHS *a* AND *c* COURTESY OF J. D. DUFFY. PHOTOGRAPH *b* COURTESY OF D. PEILA.)

consideration for post foundations is preventing loss of load-bearing strength. When foundations are directly anchored to a rock substrate, anchor elements mobilize a combination of tension, compression, and shear resistance to loads exerted on the posts.

Most post installations are secured by means of heavy-duty threaded dowels incorporated into a concrete foundation on rock (Figure 15-14). The post is then screwed to the foundation over a base plate. Some post bases include breakaway features to accommodate the large forces resulting from direct rock impacts on the post. Therefore, the foundation design only needs to resist the force required to shear the breakaway feature.

The forces derived from impacts on the fence system by rock blocks are ultimately transmitted to the ground through ground anchors. Consequently, the ground anchors receive much of the load, and they should be able to withstand such loads. This demands careful attention to the installation of each anchor. Anchors are usually installed in drilled holes within bedrock and then are grouted and tested. Both rigid (i.e., steel bars)

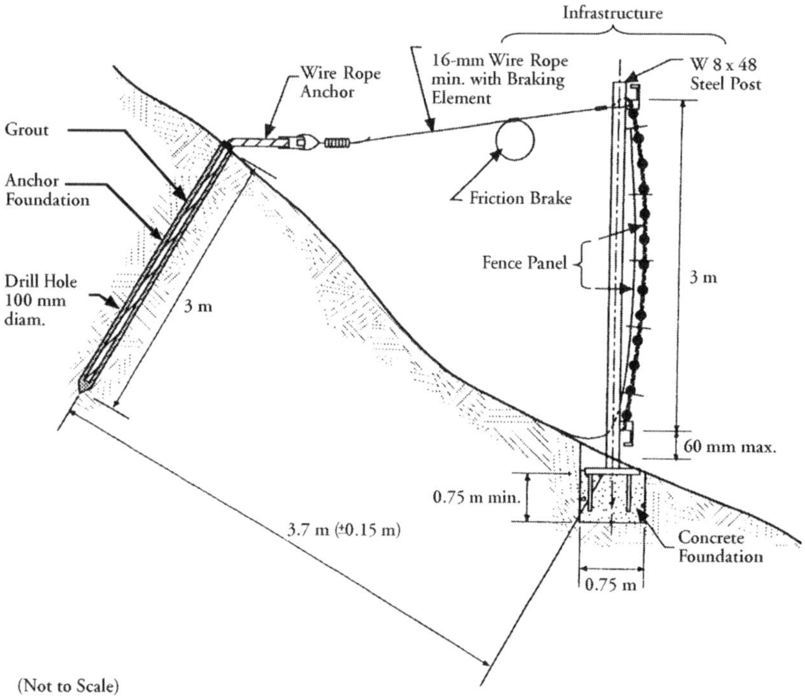

FIGURE 15-10
Side view of typical configuration of wire rope (cable) net fence (adapted from Smith and Duffy 1990).

FIGURE 15-11
Typical fence configuration to enhance capacity of rockfall catchment area. Note rocks trapped behind fence.
(PHOTOGRAPH COURTESY OF J. D. DUFFY.)

FIGURE 15-12
Fence system placed below fairly steep and high exposed rock face protects roadway in conjunction with concrete barrier (Jersey barrier) where available catchment area width is too narrow.
(PHOTOGRAPH COURTESY OF J. D. DUFFY.)

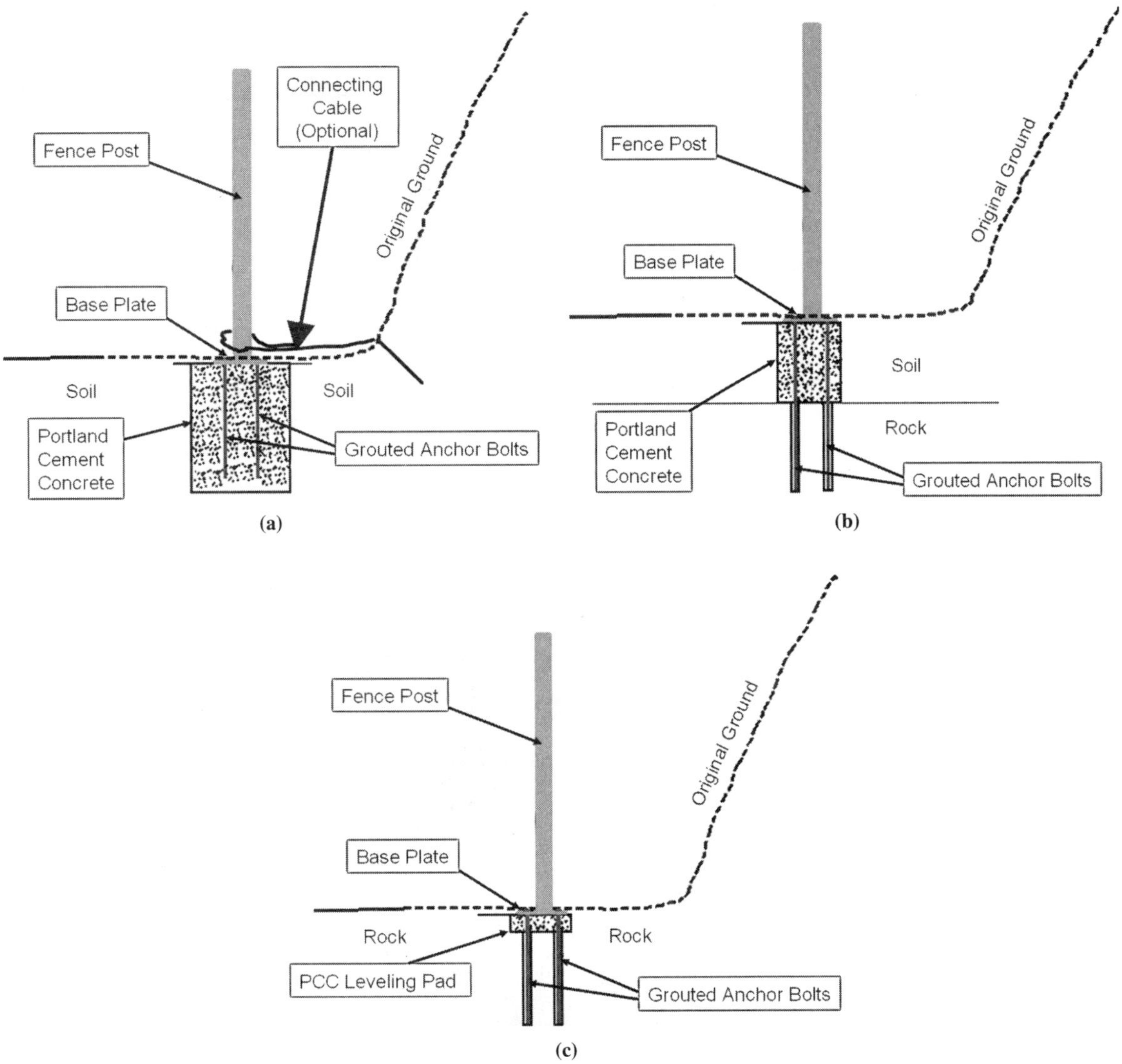

FIGURE 15-13
Typical cross sections of post foundations in (*a*) soil, (*b*) soil and rock, and (*c*) rock (Turner et al. 2009).

(a)

(b)

FIGURE 15-14
Examples of (*a*) concrete post foundation in soil and (*b*) bolted foundation into rock.
(PHOTOGRAPHS COURTESY OF THE CALIFORNIA DEPARTMENT OF TRANSPORTATION.)

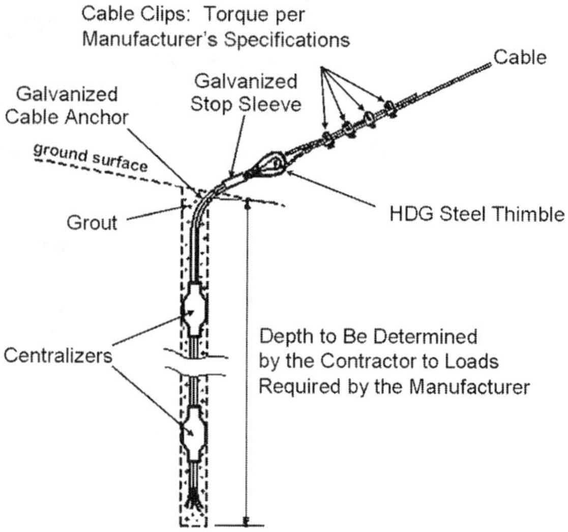

FIGURE 15-15
Cross section showing typical ground anchor system (HDG = hot-dip galvanized).
(DIAGRAM COURTESY OF S. BALABAN, CALIFORNIA DEPARTMENT OF TRANSPORTATION.)

and flexible (i.e., wire ropes) elements are used as anchors. Early systems used steel rods with eyebolts, but they were found to be vulnerable to shearing or excess bending from rockfall impacts. Cable anchors are now generally preferred and most frequently used because of their lack of bending resistance and optimal alignment to the loading direction (Figures 15-15 and 15-16).

Post foundations and ground anchors account for the major portion of the installation costs for any fence system. The fence manufacturer typically specifies the required capacities for ground anchors and post foundations, and the owner and the contractor performing the installation negotiate the responsibility for design details. Excessively large foundations do not necessarily improve system performance. In designing for the required loading capacity, typically every effort is made to determine the nature and properties of the on-site soil and bedrock materials and to provide that information in the contract plans. Verification of required capacities for each foundation component is a critical construction responsibility of the owner, however.

2.2.2 Posts

The principal functions of posts are to suspend the fence panels and to maintain a residual height of the fence so that it continues to provide protection from subsequent rockfalls. Fence posts

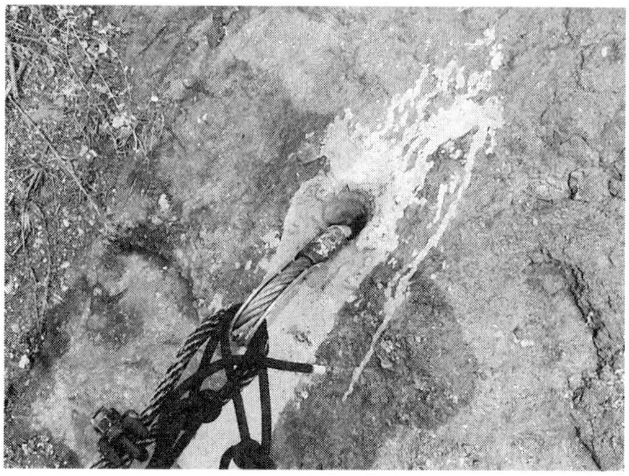

FIGURE 15-16
Detailed view of grouted ground anchor installation.
(PHOTOGRAPH COURTESY OF THE CALIFORNIA DEPARTMENT OF TRANSPORTATION.)

FIGURE 15-17
Post with hinged-design base.
(PHOTOGRAPH COURTESY OF GEOBRUGG.)

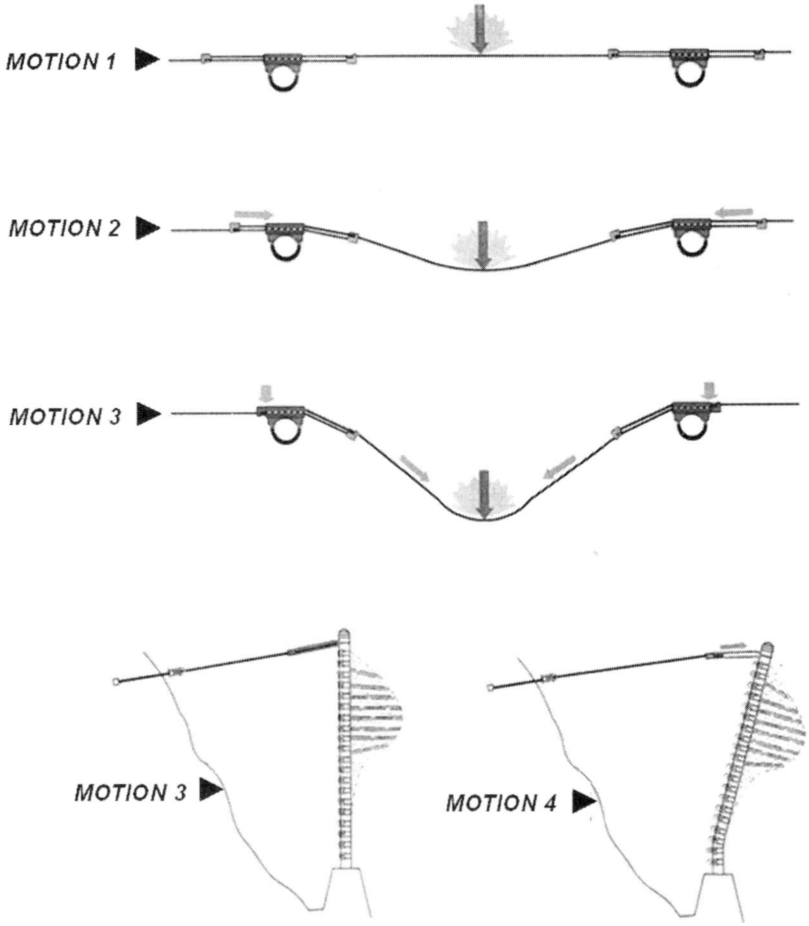

FIGURE 15-19
Deformation and load transfer of rockfall impact from top horizontal rope to post and tieback cable. (Japan Road Association 2000).

are rigid steel elements, usually consisting of H-sections 100 to 300 mm (4 to 12 in.) in size, but tube and cylindrical sections are also used. Some low-energy fence systems have used wooden posts. Posts are secured to their foundations with either a hinged or a fixed design (Figures 15-17 and 15-18, respectively). Both designs typically incorporate breakaway features to reduce damage to the foundations or posts from impacts that exceed the design load. Hinged designs allow for rotation during impact, thus reducing the bending moment on the base element. In all but some low-capacity fences, posts are supported by one or more tieback cables connecting the top of the post to a ground anchor located up the slope from the post (Figure 15-19). For hinged bases where post deflection is anticipated, friction brakes are included on the tieback cables to accommodate the deflection of the post. The resultant force produced by the tieback cables needs to be perpendicular to the plane of the fence system to minimize torsional stresses within the post that are induced by the top longitudinal support rope during impacts. Some posts are designed to collapse on direct impact of a rock, but most post designs balance the strength of the post with anticipated impact energy to maintain the basic functionality of the post and to reduce maintenance requirements.

FIGURE 15-18
Post with rigid-design base. Example shows proprietary fence design with staked post foundation located on US-101 near Gaviota, California.
(PHOTOGRAPH COURTESY OF THE CALIFORNIA DEPARTMENT OF TRANSPORTATION.)

Early fence systems used a close post spacing. A typical spacing for these systems was often only 3 m (10 ft), but by the late 1980s, spacing had increased to 4 to 6 m (13 to 20 ft) (Spang and Bolliger 2001). Today, post spacing is selected according to energy capacity requirements. High-capacity fences often have posts spaced 5 to 10 m (16 to 33 ft) apart, while posts in low-capacity fences are spaced 3 to 5 m (10 to 16 ft) apart (Peila et al. 2006). Post spacing is a design compromise of available deflection distance in front of the system, acceptable level of maintenance, and capacity requirements. A wider post spacing allows for greater fence deflections, thus providing higher energy dissipation and reducing the possibility of rocks striking posts and tiebacks. For these higher-capacity systems, an impact close to the design load may produce deflections as large as 50% to 60% of the distance between the posts (Spang and Bolliger 2001). The larger deflections associated with more widely spaced posts may also result in more frequent maintenance to retain an effective fence height.

2.2.3 Infrastructure

Infrastructure refers to the perimeter (support) wire ropes connected to the posts from which the fence panels are attached, the lateral and tieback wire ropes that provide longitudinal and transverse stability to the fence, and the connections that secure the net panels to the infrastructure (Figure 15-20). The infrastructure is responsible for transmitting the stresses from the rockfall impact on the fence panel to the post foundations and to the ground through the lateral and tieback anchors. Ideally, the perimeter (support) ropes optimize system elasticity and, at the same time, maintain a minimum effective fence height after an impact (Spang and Bolliger 2001). The diameter and number of perimeter (support) ropes are based on the energy capacity that the fence system requires. Most low- to moderate-capacity fence systems use single top-and-bottom perimeter (support) ropes. Higher-capacity fence systems typically have multiple top-and-bottom perimeter (support) ropes. With the exception of some low-capacity fence systems, energy-dissipating devices (friction brakes) are interspersed on these perimeter (support) ropes to provide for their controlled elongation under load. Perimeter (support) ropes are connected and terminated with various clamps or pressed ferrules, which have an ultimate strength equal to or higher than the failure load of the rope. The perimeter (support) ropes are threaded through specially designed fittings at the top and bottom of each post and then tensioned to remove sag.

Fence panels are connected to the perimeter (support) ropes. The connection is typically made with lacing cable, shackles, or cable clips. These connectors allow for the deflection (energy dissipation) of the fence panel, do not restrict its flexibility, and safely transfer impact loads to the infrastructure. There are typically hundreds of connection points, and it is important that each be correctly installed. Impact loads cause damage at any incorrectly installed connections.

Lateral ropes connect the perimeter (support) ropes to the lateral ground anchors, provide longitudinal stability to the fence, and allow for additional tensioning of the support ropes. During an impact, the lateral anchors receive a significant portion of the forces. Each lateral anchor, or the resultant of two or more lateral anchors, is in line with the fence (Figure 15-21). Tieback ropes secure the posts, and, in some cases, perimeter (support) ropes connect tieback anchors to provide transverse resistance and post stability during impact loading. Individual tieback ropes, or the resultant of two or more tieback ropes, must be perpendicular to the fence alignment (Figure 15-21). Some low- to moderate-capacity fences do not use tiebacks; all higher-capacity fences include tiebacks. The energy capacity demands on the fence system define the diameter and the number of lateral and tieback ropes.

FIGURE 15-20 General layout of typical fence infrastructure showing fence panels (net), posts, post foundations, longitudinal perimeter (support) ropes, tieback ropes, and ground anchors (foundations for support cables). (PHOTOGRAPH COURTESY OF D. PEILA.)

2.2.4 Friction Brakes

The terms *energy dissipators* and *friction brakes* refer to devices that are placed on the infrastructure ropes and that are designed to deform plastically when a force is exceeded. They dissipate excess impact energy that cannot be accommodated by the fence panels and infrastructure, and they increase deformation characteristics of the barrier (Figures 15-22 and 15-23). Several designs for these devices exist that involve loops, in-line cables, and springs (Figures 15-24 through 15-27). Friction brakes are located to absorb sudden impact loads; their number and capacity are unique to each system and its energy rating.

The behavior of a friction brake is typically described by its characteristic displacement related to applied force; this is usually represented by a curve of displacement versus applied force. The impact force, and hence the energy, that the friction brake can absorb is represented by the area subtended by the displacement–applied force curve at maximum elongation of the friction brake. The technical characterization of these elements is usually performed by means of pseudostatic elongation tests. The employment of dynamic tests, even if relevant, is not common because of the technical complexity of this type of test (Peila et al. 2006).

It has been suggested that the capacity of a fence might be increased to any desired level by arranging a sufficient number of brake elements in series. This is incorrect. Because of the inertial limitations within the fence system, a single impact cannot fully activate the friction brakes in adjacent sections (Spang and Bolliger 2001).

How much activation a friction brake can undergo before it needs to be replaced is an important maintenance consideration. The answer depends on the remaining elongation in the friction brake element and on the magnitude of anticipated additional rockfalls. Friction brakes must respond to impacts as large as the design capacity of the fence system; the other system components must cope elastically with smaller events.

2.2.5 Fence Panels

Fence panels are the principal interception structure of a system. Lightweight panels are commonly used for low design energies, and heavyweight panels are used for moderate to very high design energies. Panel flexibility is important; stiffer fabrics and panels that are secured tightly to the infrastructure tend to puncture and tear at fixed points, while extremely flexible panels tend to sag excessively. Panel behavior depends on the freedom of the panel to slide on the infrastructure cables, the deformation of the fabric, and the deformability and strength of the wires or cables that

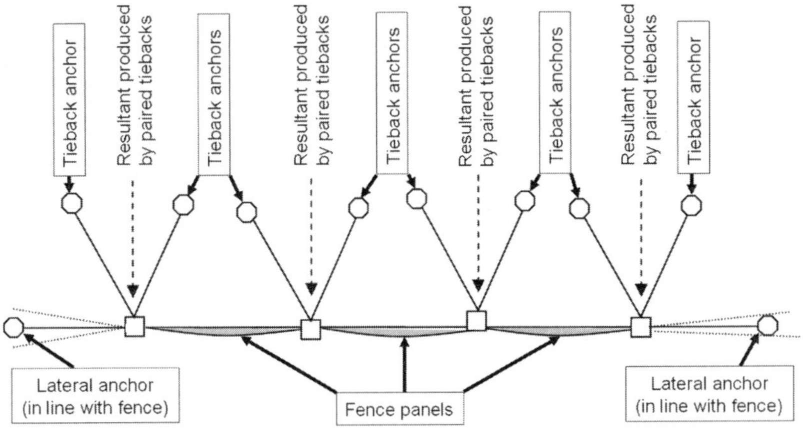

FIGURE 15-21
Typical plan view of fence. View shows how lateral anchors are aligned with orientation of fence and how resultants of each pair of tieback ropes (dashed arrows) are perpendicular to fence alignment (Turner et al. 2009).

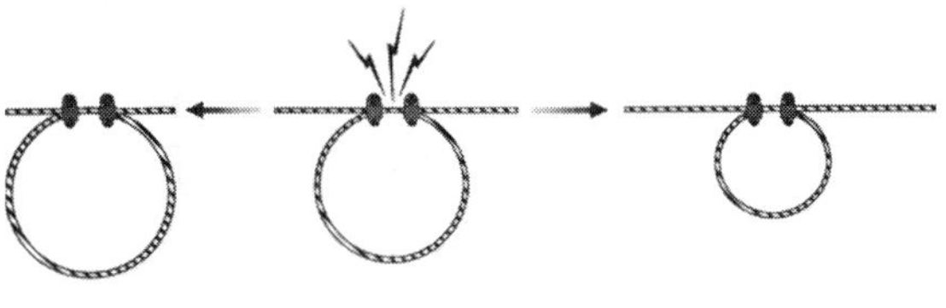

FIGURE 15-22
Example of loop-and-clamp type of friction brake, its activation by tension during impact, and its final deformed state (Duffy 1992).

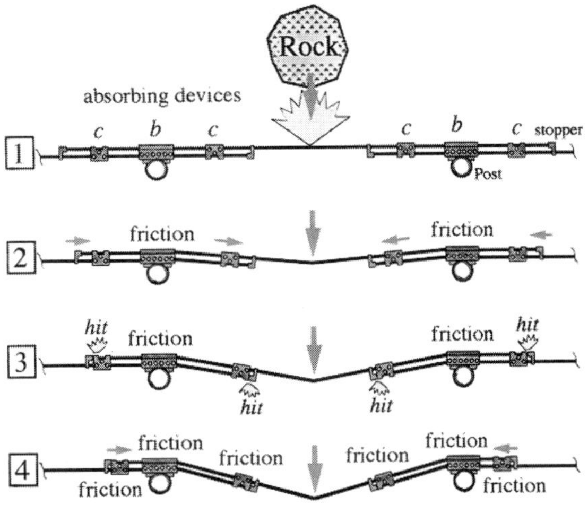

FIGURE 15-23
Illustration in plan view of how energy is absorbed through increased flexibility and plastic deformation and friction (Japan Road Association 2000).

make up the panel fabric. Panel weight is also an important aspect, since its mass (inertia) has to be accelerated on rock block impact—the heavier the mesh, the greater the energy absorption (Spang and Bolliger 2001). Ideally, the fence design ensures that no gaps develop between adjacent fence panels and that the fence panels are not directly

(a)

(b)

FIGURE 15-24
Friction brakes: (*a*) not activated and (*b*) activated. Spring-like device is inserted in line of wire rope.
(PHOTOGRAPHS COURTESY OF T. C. BADGER.)

(a)

(b)

FIGURE 15-26
Friction brakes: (*a*) not activated and (*b*) activated. Wire rope passes through bent pipe of ring shape. Two ends of pipe are compressed together by aluminum sleeve.
(PHOTOGRAPHS COURTESY OF GEOBRUGG.)

connected to a post. These details provide two positive effects: the continued functionality of the fence system is guaranteed, and the undesirable connection of a flexible component (such as a panel) to a rigid component (such as a post) is avoided (Figure 15-28).

The panels are made of a wire or cable "fabric" woven into a pattern of squares, rhombs, rings, or diamonds that are fastened at their intersections by various twists, pressed clips, or wire wraps, or by interweaving of the wire or cable strands. When large-opening, heavyweight fabrics are used, a secondary finer fabric is typically added to intercept small rock fragments. Impact tests of heavy-duty cable nets with large openings, backed on the impact side with lightweight chain-link fabric, have shown that the secondary mesh improved the absorption and distribution of the rotational energy of striking rocks (Smith and Duffy 1990).

Four basic types of fabrics are in use: single-twist wire fabric, double-twisted wire hexagonal mesh,

FIGURE 15-25
Friction brakes: wire rope threaded through holes in steel plate.
(PHOTOGRAPH COURTESY OF PFEIFER ISOFER AG.)

(a)

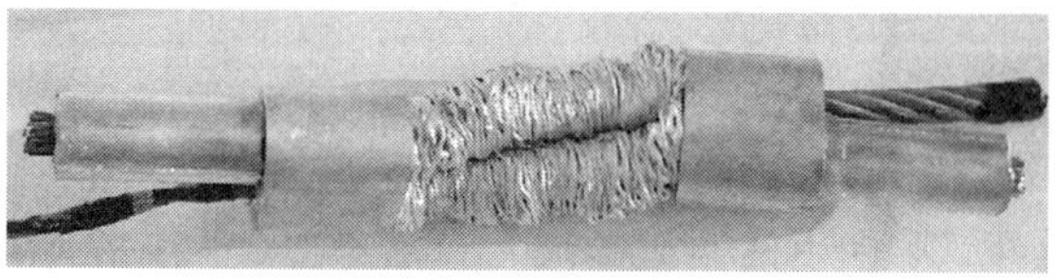

(b)

FIGURE 15-27
Friction brakes: (*a*) not activated and (*b*) activated. Wire rope passes through linear sleeve.
(PHOTOGRAPHS COURTESY OF MACCAFERRI.)

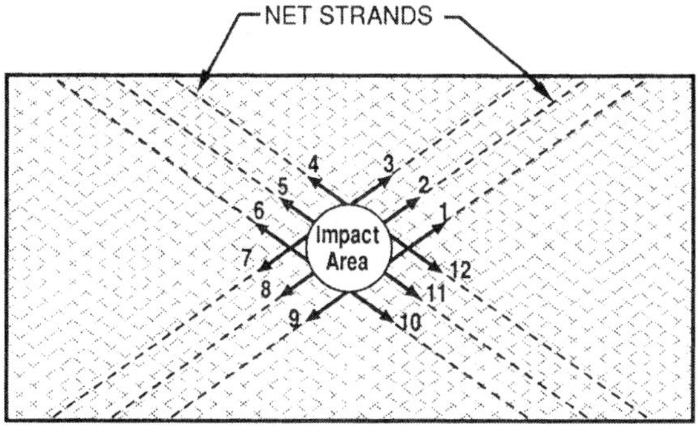

FIGURE 15-28
Transfer of impact energy within fence panel (Smith and Duffy 1990).

wire rope (cable) nets, and ring nets. Each type has variations.

Single-twist fabric is woven by partially bending two intersecting wires in a single twist and comes in a variety of opening sizes, shapes, wire gauges, and types, including some with corrosion protection and colorized coatings. As routinely used in residential property boundary fences, single-twist fabric (commonly referred to as chain-link) typically consists of 50- × 50-mm (2- × 2-in.) diamond openings and is constructed of 3-mm-diameter (9-gauge) mild steel wire (Figure 15-29). Single-twist fabric constructed with either 3-mm-diameter (9-gauge) or 4-mm-diameter (6-gauge) wire is commonly used for rockfall fences. The wire is galvanized for corrosion protection, or it can be coated with plastic [polyvinyl chloride (PVC)] or fused powder. Several manufacturers produce a variety of high-tensile single-twist fabrics for more robust fence and drapery systems (Figure 15-30). With the exception of high-tensile fabrics, single-twist fabrics are used as a primary fabric for very-low- to low-capacity fences and as a secondary fabric for moderate- to high-capacity fences.

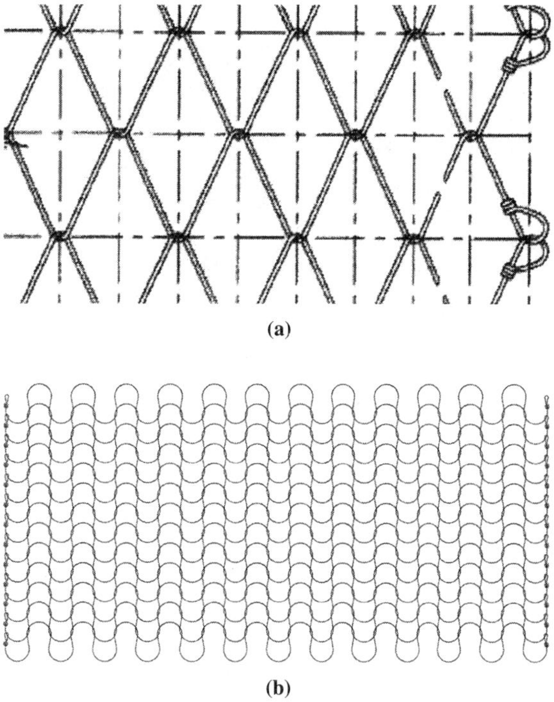

FIGURE 15-30
Examples of high-strength, single-twist fabric include (*a*) TECCO mesh (Geobrugg 2002) and (*b*) OMEGA-Net.
(IMAGE *b* COURTESY OF TRUMER SCHUTZBAUTEN GmbH).

A double-twisted intersection of two wires to form a hexagonal mesh provides another lightweight wire fabric (Figure 15-31). The most common of these products have opening dimensions of 80 × 100 mm (3 × 4 in.) or 100 × 120 mm (4 × 5 in.) and are constructed with 3-mm-diameter (9-gauge) wire. The wire is galvanized for corrosion protection; colorized plastic (PVC) coating or fused-powder coatings can be ordered. Double-twisted wire mesh is more likely to tear when it is used as a secondary mesh because it is one of the stiffer meshes. Double-twisted wire mesh is used in very-low-capacity fence systems.

Cable nets are constructed by weaving a single continuous wire rope, typically with a diameter of 8 to 10 mm (approximately 5⁄16 in.), into a net with

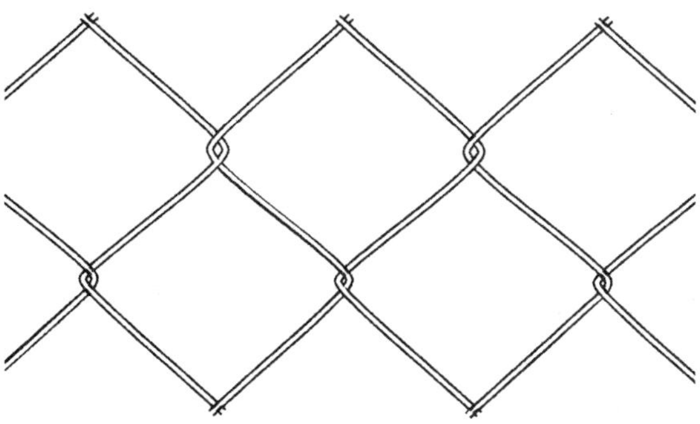

FIGURE 15-29
Standard chain-link fabric (Duffy 1992).

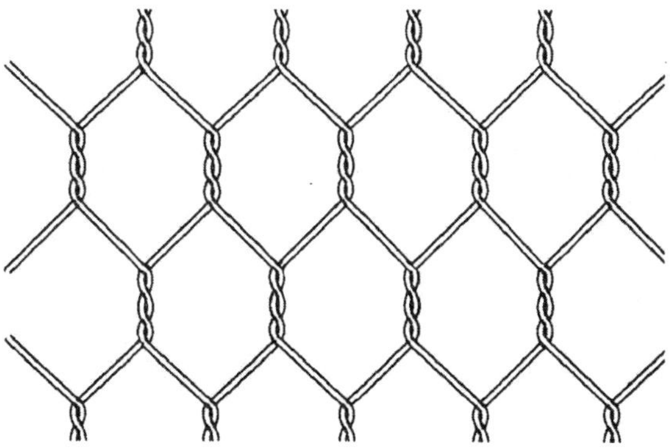

FIGURE 15-31
Double-twisted wire mesh is stronger but stiffer than chain-link fabric for same wire size (Duffy 1992).

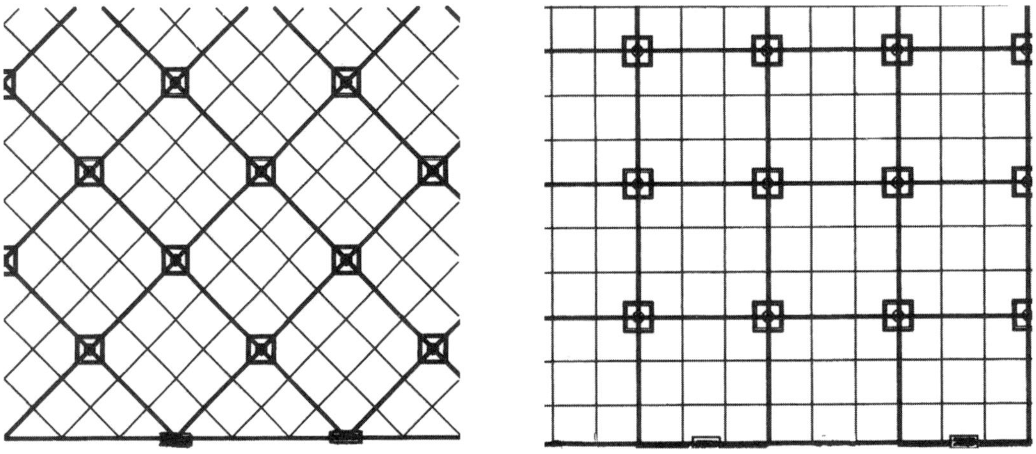
(a) (b)

FIGURE 15-32 Cable nets are woven into either (*a*) diagonal grid or (*b*) square grid.

diagonal or square grid openings (Figure 15-32). Pressed clips, wire twists, or U-bolts join the intersecting wires. A pressed ferrule or C-clip fastens the cable fabric to a border rope having a diameter of 16 to 19 mm (approximately ¾ in.). The wire rope is typically galvanized, although stainless steel is sometimes specified for highly corrosive environments. Wire ropes can be colorized with plastic (PVC) or fused-powder coatings. Net openings vary with design; a typical opening is 150 to 300 mm (6 to 12 in.). Smaller openings result in a heavier fence panel, and greater capacity is achieved because there are more cables per unit area. Diagonal grids, which are superior to square grids for energy transfer to perimeter (support) infrastructure, are used for moderate- to high-energy systems rated up to 2,000 kJ (740 ft-tons) (Spang and Bolliger 2001).

FIGURE 15-33 Large ring net fence installed above Veytaux, at eastern end of Lake Geneva, Switzerland, provides a good example of these high-capacity fences. (PHOTOGRAPH COURTESY OF GEOBRUGG.)

Ring nets were originally developed as antiattack and antitorpedo netting during World Wars I and II. Surplus supplies have since diminished, but because of the superior capacity of ring nets, several manufacturers produce them. The rings are constructed with wire having a diameter of 3 to 4 mm (9 to 6 gauge) by using three to nine turns, thus creating a cable ranging in diameter from 5 to 19 mm (approximately 3⁄16 to ¾ in.). The ring is finished either with a closed splice, where the wire ends are terminated inside the bundle, or with multiple pressed clamps. Ring diameters vary from 250 to 350 mm (10 to 14 in.). As shown in Figure 15-33, ring nets are used for moderate- to very-high-capacity fences, those rated up to 8,000 kJ (2,950 ft-tons). Two ring net designs are available: one with four rings interlinked, the other with six rings interlinked (Figure 15-34). The six-ring interlink has approximately 20% more rings per unit area than does the four-ring interlink (Spang and Bolliger 2001).

2.3 Maintenance

Maintenance and the associated costs are important considerations, especially where rockfall occurs with a high frequency or high impact energies are expected. No fence is maintenance free. Routine activities to maintain optimal performance are often required and include cleanout of accumulated debris, replacement or repair of damaged components, and retensioning of infrastructure elements. Considerable progress toward minimizing maintenance costs was first achieved when the design load dimensioning philosophy was developed, as illustrated in Figure 15-6 (Duffy and Haller 1993). Since this early effort to adapt

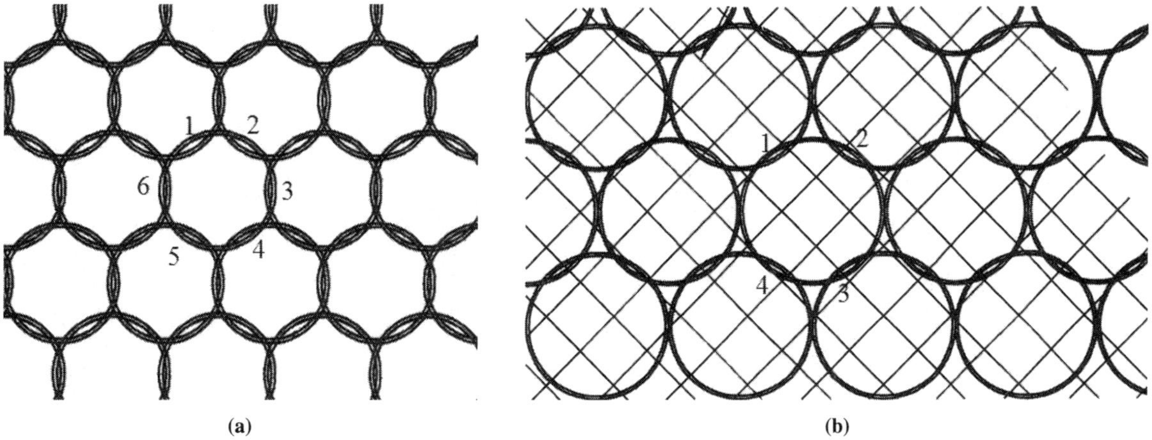

FIGURE 15-34 Ring nets are available with either (*a*) six or (*b*) four interlinked adjacent rings.

fences to meet anticipated impact loads and thus minimize maintenance, much has been learned about fence capabilities and ways to reduce maintenance.

As discussed in Section 1.1 and shown in Figure 15-6, current testing guidelines quantify the SEL, the level of impact loading that a fence can withstand within accepted maintenance requirements, and the MEL, the maximum level of impact load that a fence can withstand. A tested fence should perform well at stopping a rockfall impact corresponding to either of these levels. However, the level of maintenance will vary. It will be negligible for impacts at or below the SEL, while extensive maintenance may be required after an impact approaching the MEL. Thus maintenance demands are related to fence design, capacity, and rockfall frequency. Fences designed to accommodate SEL impacts of no more than 500 kJ (185 ft-tons) effectively stop commonly encountered rock blocks that are from 0.7 to 1 m (2 to 3 ft) in diameter while requiring only routine cleaning and minor maintenance. Above this impact energy level, maintenance requirements increase, although usually not too dramatically, until impact energy levels reach or exceed 1,000 kJ (370 ft-tons). This level is routinely achieved only by blocks that are at least 1.5 m (5 ft) in diameter. These much larger rock blocks are effectively stopped, but maintenance is significant and fence replacement may be necessary.

Routine maintenance, especially cleanout, is a requirement for any fence system; thus, access becomes an important design consideration. Fence heights, quantity and size of rockfall debris, snowpack or accumulation of plowed snow, and heavy equipment requirements can all complicate cleanout efforts (Figure 15-35). Corrosion is also a concern, and most components are galvanized. Other corrosion protection measures include the use of plastic coatings and various proprietary protection coatings (such as Supercoating, the Galfan process, and the zinc–aluminum process).

FIGURE 15-35 Two examples of fence-cleaning operations. (PHOTOGRAPHS COURTESY OF THE CALIFORNIA DEPARTMENT OF TRANSPORTATION.)

2.4 Construction Considerations

Construction of proprietary fence systems has been greatly simplified in recent years. However, the construction of ground anchors and post foundations remains an area of considerable difficulty and complexity, in which contractor specialization continues to be an important consideration.

Fence construction costs are commonly two to three times the cost of materials. In difficult locations, construction costs may reach four to five times the material costs. A large portion of the construction costs is for post foundations and ground anchors. Excessively large foundations do not necessarily improve the system's performance. In most instances, the ability of the post foundation to "wiggle" dissipates energy and reduces overall damage to the fence system. Ground anchors, on the other hand, receive much of the impact load and thus need to be carefully installed. Usually they are installed in holes drilled in bedrock, then grouted and tested. Their proper alignment is also important. Lateral anchors must be aligned in the plane of the fence. Additional lateral anchors that are necessary for alignment changes must also have their resultant in line with the plane of the fence.

Each component of a fence system plays an important role in energy dissipation, and attention to detail is required during construction. Field tests have shown that the forces imparted by a striking rock enter the system and travel to the ground anchors (mostly the lateral anchors). If the fence system is properly constructed, the energy is dissipated within the fence system. If any part of the fence system is improperly constructed, the weak points are likely to fail and cause damage. Improperly torqued or saddled cable clips can slip, or fence panels or infrastructure improperly attached to fixed points will tear.

2.5 Aesthetic Treatments and Environmental Attributes

Fences are generally clearly visible and structurally contrast with their environment. Thus, aesthetics and visual impact are often important factors in gaining public acceptance for the installation of a fence system (Figure 15-36). All of the fence components can be colored to blend with the surroundings. Colorizing can be done with paint, by the application of plasticized coatings during manufacturing, or with a fused powder coating. Plastic coating is more durable than paint, but it can fade where it is exposed to extreme ultraviolet light in perennially sunny locations. Paint or fused powder coating, which is more brittle than plastic coating, can peel when applied to flexible components. Standard galvanizing of cables and wires fades and loses its luster over a period of several years in many environments, which helps subdue their visual impact. Although coloring is somewhat subjective, it can be an important issue in the permitting process.

FIGURE 15-36
Fence blending in with environment.
(PHOTOGRAPH COURTESY OF CAN.)

The size and shape of mesh openings can also influence visual impact. Some meshes have openings as wide as 36 cm (14 in.), large enough for small animals to pass through. This is important when a fence barrier crosses an animal trail or migration path. Vegetation can be planted in, around, and on a fence to blend the fence with its surroundings, as long as the vegetation does not grow to the point of inhibiting system flexibility.

3. PROPRIETARY DESIGNS

Today nearly all fences used are developed, tested, and marketed by industry. Most systems are the product of considerable research and development. Each manufacturer performs a different level of testing, but all systems are tested. Dynamic loading is a fundamental requirement of fence systems; thus, field tests of all system components are essential. Consequently, standard practice favors the use of tested and proven proprietary systems. In some instances, however, fences are designed "in-house" or for a specific project. These systems need to be field-tested. Full-scale testing efforts are sometimes supplemented by computer modeling to aid in optimizing and streamlining the required testing. Computer models can also be used to simulate system modifications under field conditions without requiring expensive and time-consuming testing.

4. FINITE ELEMENT COMPUTER MODELING OF FENCE SYSTEMS

Until recently, the performance of fence systems has been evaluated mainly on the basis of experience, supported by expensive full-scale field tests. Recently, however, explicit finite element computer programs have been developed and applied to studies of the dynamic behavior of fence components subjected to rock impact. The programs take into account large deformations and interactions among various fence system components, including rock contact with the fence, connections among components, sliding, friction, and nonlinear material behavior (Volkwein 2002).

An example of these computer programs is the finite element program called FARO (Figure 15-37). FARO has been developed to model the dynamic behavior and loading of all fence system components while the rock is being stopped. By using computer simulations to study various system configurations, the most favorable designs can be determined. This approach also provides

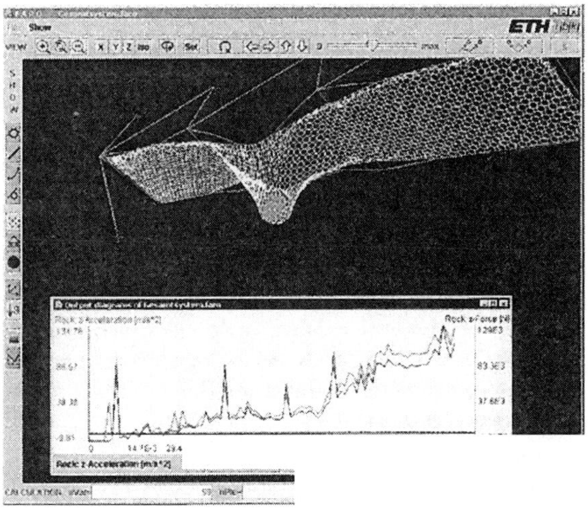

FIGURE 15-37
FARO graphical user interface during rockfall simulation (from finite element program FARO).

an opportunity to reduce the number of expensive full-scale field tests required to support design and development of new fence systems. The information related to the loads experienced by individual system components can be used to develop an optimal design of these components (Volkwein 2002).

Muhunthan and Radhakrishnan (2007) carried out finite element analyses on hybrid cable net draperies to evaluate loads on the support posts and fabric deformation for impact energies up to 200 kJ (74 ft-tons), for a range of slope orientations, slope–mesh interface friction values, and system dimensions. Static benchmark modeling calibrated the model with puncture tests performed on cable net panels with dimensions of 3 × 3 m (10 × 10 ft). The model calculated infrastructure loads, and the modeling results showed comparable envelopment of the rock by the fabric and dramatic attenuation of the horizontal trajectories. The results supported the observations made during the Igor Paramassi field testing described in Section 5.2.

5. FULL-SCALE FIELD TESTING OF FENCES

Full-scale field tests for characterization of fence behavior have been conducted for several decades in both America and Europe. Full-scale tests are the only feasible tool for understanding the behavior of a fence system completely, evaluating the energy that can be dissipated by the fence system, measuring the maximum displacement of the fence during impact, and measuring the force on the ground anchors and post foundations. The knowledge gained from these tests provided the basis for the advances in the capacities of fence systems described in Section 1.1. This experience has culminated in the recent official endorsement by European authorities of appropriate testing methods for certification of rockfall fence systems.

Videos of several of the tests described in the following sections are provided on the DVD included with this book. See Appendix B, "Description of DVD Materials."

5.1 Testing in the United States

In 1989 and 1990, the California Department of Transportation (Caltrans) conducted full-scale field tests to evaluate fences that were expected to absorb impact energies of up to 190 kJ (70 ft-tons) (Duffy 1990; Duffy and Smith 1990; Smith and Duffy 1990). Natural rocks weighing from 136 to 5,897 kg (300 to 13,000 lb) were rolled down a 34° natural slope 40 m (130 ft) high and 76 m (250 ft) long. Surficial materials were reportedly dry and hard and consisted of landslide debris of 2.5- to 46-cm (1- to 18-in.) rock fragments in a clayey silt matrix. More than 80 tests were conducted and recorded on video and high-speed cameras. Results obtained from these field tests led to modifications in fence design that reduced maintenance. Subsequently, another series of tests conducted by the University of the Pacific and Caltrans (Kane and Duffy 1993) demonstrated that wire rope (cable) nets were able to stop impacts of low energy [68 kJ (25 ft-tons)] with minimal maintenance. Additional Caltrans tests using similar methods have been discussed by Duffy (1996) and Duffy and Hoon (1996, 1998).

Between 1987 and 1998, the Colorado Department of Transportation (DOT) conducted several full-scale rock-rolling experiments, primarily for testing a variety of rockfall fence systems. The test site was located near Rifle, Colorado, on a hillside 91 m (300 ft) high. The 37° test slope was covered with colluvial soils weathered from claystone and sandstone bedrock. Full-scale tests conducted in 1987 and 1988 were used as a basis for calibration of the Colorado Rockfall Simulation Program and testing of timber attenuators, tire attenuators, and a combination of tire and timber attenuators (Barrett et al. 1989; Pfeiffer 1989; Pfeiffer and Higgins 1990; Pfeiffer et al. 1990; Andrew 1992). Rocks of basalt composition ranging from 227 to 2,722 kg (500 to 6,000 lb) were rolled down the slope, reaching velocities from 9 to 23 m/s (30 to 75 ft/s). The rocks were launched onto the slope by either pushing or dropping them with a front-end loader. In 1990, the Flexpost fence with wire mesh was tested by using full-scale rock rolling (Hearn 1991); in 1992, timber-faced mechanically stabilized earth barriers were tested (Parsons De Leuw 1992); and in 1992 to 1993, modified Flexpost fence designs using either wire mesh or cable net were tested (Hearn 1992; Hearn 1994). Andrew et al. (1998) reported on full-scale tests of a variety of barrier fence products manufactured by an American company.

A second type of barrier testing was conducted by the Colorado DOT in 1989 with basalt blocks, 0.6 to 0.9 m (2 to 3 ft) in diameter, that were tethered to the boom of a 109,000-kg (120-ton) crane. The rocks were swung on the tether into three barrier designs: Flexpost fence, mechanically stabilized earth barriers, and barriers constructed with gabion baskets (personal communication, R. Andrew, 2009).

FIGURE 15-38 Italian facility established to perform full-scale rockfall tests. (PHOTOGRAPH COURTESY OF FRANCESCO SASSUDELLI.)

5.2 Testing in Other Countries

Additional field tests have been conducted in several European countries, principally Switzerland and Italy, and in Japan. The various individuals and groups conducting the tests have exchanged information and experiences to a considerable extent.

Peila et al. (1998) reported on field tests conducted by Italian researchers. The tests were performed by using a cable aligned from top to bottom of the slope, perpendicular to and extending over the barrier being tested. The test rock was loaded onto a trolley attached to the cable (Figure 15-38). As the trolley slid down the cable, the rock was detached and allowed to fall and strike the barrier directly without bouncing on the ground. The direct impact permitted an accurate calculation of the impact energy. Video cameras were used to monitor and record each test. The stresses or forces on the cables and posts were measured by dynamometers on the cables.

Muraishi and Sano (1999) described the testing procedure used by the Japanese Railway Technical Research Institute. Certification of a fence was determined on the basis of both laboratory-scale static strength tests of individual fence elements (wire rope anchors, brakes, and so forth) and full-scale field testing. The full-scale test was conducted by dropping a "stone" from a crane into a test barrier constructed on a steep slope (Figure 15-39). The fence was installed at 35° from horizontal so that impact of the rock with the fence was at a fixed angle of 55°. The vertical drop test was judged the best method because it provided the most reproducible results. The location, angle, and energy of impact produced by a falling block were essentially identical in each of a series of tests. With such consistency, a much smaller number of tests was required, compared with the number of rock-rolling tests conducted on a hill slope.

Baumann (2002) summarized field testing of fences in Switzerland. More than 350 field tests were conducted in the country between 1988 and 2002, and testing methods and fence designs evolved considerably during that period. An early set of tests (Duffy 1992) was conducted at Oberbuchsiten, Switzerland, with test methods similar to those used in previous Caltrans tests—the rocks were rolled down a slope to strike the fence. The slope was of hard rock and allowed energies at impact to be greater than experienced in the Caltrans tests. After more than 100 rock-rolling experiments, it became obvious that considerable energy was lost from the rocks contacting the ground. The velocities of the rocks reaching the fence varied, ranging from 6 to 30 m/s (20 to 100 ft/s). In addition, the rocks struck the fence panels at unpredictable locations and angles, making the tests difficult to reproduce. Nevertheless, "successful" tests in this series provided impact energies ranging from 14 to 1,000 kJ (5 to 370 ft-tons), and all designs were tested to failure.

A second Swiss test method was developed to reduce the uncertainties of the initial tests. A new test facility was located near Beckenried, Switzerland. The method used an inclined ropeway attached to a crane; the setup was essentially the same as used in the earlier Italian test procedure illustrated in Figure 15-38. The crane transported rocks up the slope, and the ropeway transported the rocks toward the fence and released them in front of it. The rock was allowed to strike the ground and roll into the fence; thus, the impact provided both translational and rotational velocities (energies). The major goal of the test method was to increase the impact rock velocities. However, the velocities achieved were modest, ranging from 10 to 22 m/s (32 to 72 ft/s), and, of even greater concern, the height and angle of impact were not consistent and could not be determined in advance. One improvement was realized: the rocks always hit the central portion of the fence panel.

In 1991, a third Swiss test method, which used an inclined ropeway that launched the rocks directly into the fence, was developed. The method was installed at two sites. Consistent impact velocities were achieved at each site—25 to 27 m/s (82 to 89 ft/s) at the Beckenried site and 29 to 31 m/s (95 to 102 ft/s) at the Giswil site. The more predictable range of velocities allowed accurate calculation of rock energy at impact before the test was conducted.

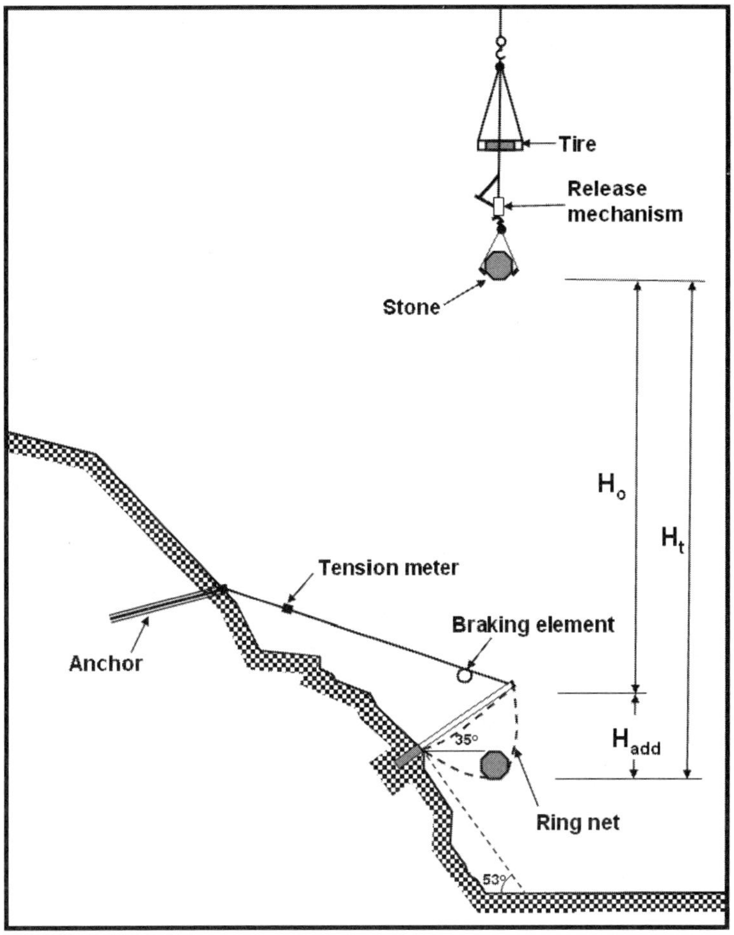

FIGURE 15-39 Vertical drop-test setup used in Japan. H_0 = height of drop to point of initial impact (meters); H_t = height to point of maximum deflection of barrier being tested (meters); and H_{add} = amount of maximum deflection of barrier (meters) (adapted from Muraishi and Sano 1999).

By this time, experience had shown that full-scale tests using the rock-rolling method provided the most realistic testing for fence system behavior under dynamic loading. The tests included angular and rotational energy from the bounding rocks, which was most useful for fence system development (Figure 15-40). However, the tests were not sufficiently reproducible because the impact energy varied considerably from test to test, and the height and angle of impact were not consistent or predictable. In addition, the magnitude of energy achieved by rolling rocks was limited. Subsequent testing methods, such as the second and third Swiss methods, improved somewhat with regard to energy generation, but only the third Swiss method provided reasonably reproducible results.

There remained a perceived need for objective, reproducible, standardized tests of rockfall fences to allow product comparisons and provide relevant information on their capacity and maintenance characteristics (Gerber and Haller 1997). The Swiss Agency for Environment, Forests, and Landscape (SAEFL) undertook development of a certification procedure for rockfall fences (Gerber 2001; Gerber et al. 2001). A permanent testing facility was constructed in an old quarry near Walenstadt, Switzerland. A vertical drop-testing method was designed to overcome deficiencies in the earlier methods. Concrete "rocks" of specified dimension and density are dropped from a crane into fences installed to project out from a vertical rock face (Figure 15-41). This configuration allows excellent control of impact velocity, energy, and location of impact, and those parameters can be adjusted easily. The angle of impact is constant, and the ground does not contribute to energy dissipation. The tests are recorded on high-speed videotape, and several measurements of loading and deformations are made. Although the vertical drop test does not include rock rotation, which can greatly influence fence system behavior and maintenance, the test is reproducible and predictable, and large energies can be produced. These procedures allow various fence systems to be tested under exactly the same conditions, and products from many manufacturers can be compared on the basis of performance at similar loading.

Igor Paramassi, an Italian firm, sponsored a full-scale field-testing program of hybrid attenuator fence systems that involved tests of concrete

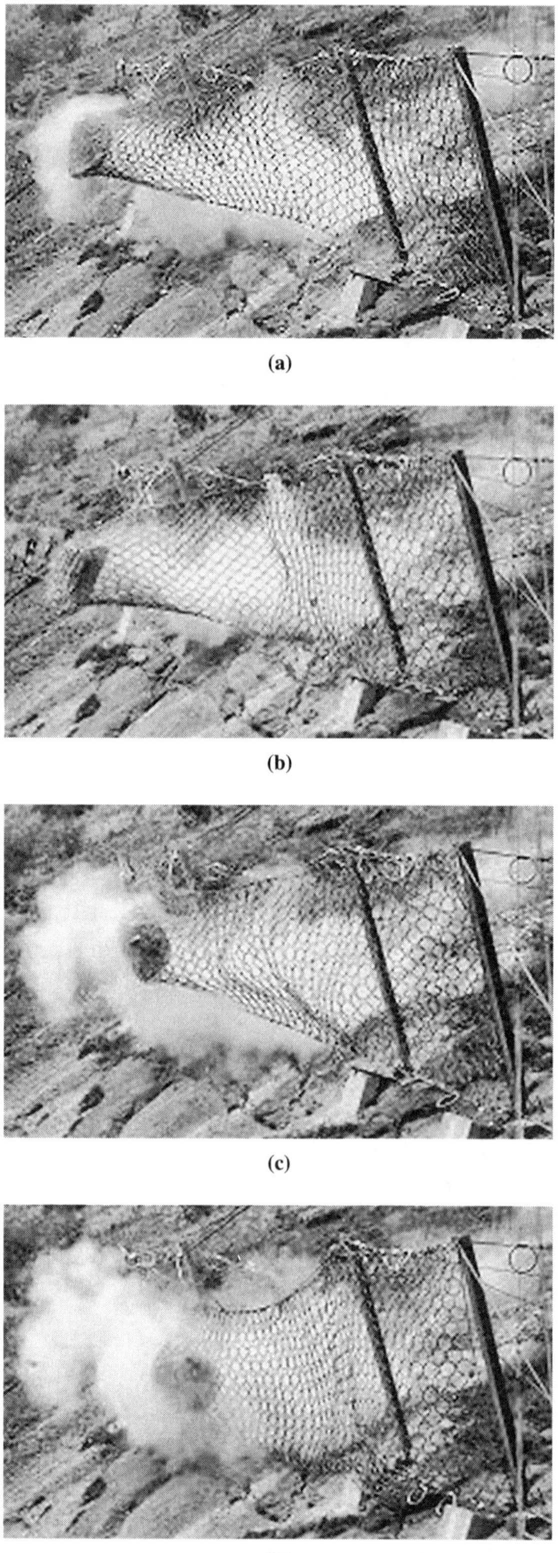

(a)

(b)

(c)

(d)

FIGURE 15-40
Tests of fence systems conducted in quarry at Beckenried, Switzerland, in 1997.
(PHOTOGRAPHS COURTESY OF GEOBRUGG.)

FIGURE 15-41
Rockfall test facility at Walenstadt, Switzerland.
(PHOTOGRAPH COURTESY OF GEOBRUGG.)

"rocks" rolled down a 45° slope (Sassudelli et al. 2007). The rocks ranged in weight from 1 to 5 tonnes (2,200 to 11,000 lb), and the tests produced estimated impact velocities of up to 14.3 m/s (47 ft/s). The impact energies were as high as 500 kJ (185 ft-tons). All rocks were successfully stopped in a flat containment area within 3 m (10 ft) of the base of the slope. Several tests included measurements of infrastructure forces using load cells. Despite several impacts on support ropes and a post, no maintenance of the hybrid attenuator systems was required; they remained completely functional during the succession of tests.

6. PERFORMANCE CERTIFICATION

Rockfall fence designs have rapidly evolved over the past two decades, but no standard field-testing methods to evaluate the designs were accepted before 2001. The lack of standardized testing made comparisons between products from different manufacturers difficult for state transportation departments and design professionals. As a result, there was a worldwide call for standardization of testing to evaluate rockfall fence system designs. In the absence of standardized testing procedures in the United States, some state transportation departments have developed performance criteria, or guidelines, to evaluate various designs.

Official guidelines defining certification procedures and testing standards for fences began in Switzerland. Subsequently, the National Cooperative Highway Research Program (NCHRP) in the United States supported a study to establish a recommended standard testing procedure for American transportation agencies. Concurrently, the European Organisation for Technical Approvals (EOTA) developed a certification standard process for the European Union that was formally endorsed in January 2008 and is now being implemented.

These efforts are described in the following sections.

6.1 American State DOT Guidelines and Performance Standards

Caltrans was the first state DOT to adopt guidelines for rockfall barrier testing. Department personnel use the guidelines to evaluate barrier designs that have been provided by the manufacturer. The guidelines are based on experience gained from the many full-scale rock-rolling experiments conducted by the department. The guidelines require a test that consists of rolling 20 rocks down a slope into the barrier, starting with low impact energies and increasing to the design load. Rocks are allowed to accumulate during the test so that cumulative effects of barrier performance can be observed. The barrier performance is evaluated on the basis of relationships between impact loading, maintenance, and efficiency. *Impact loading* is the amount of energy striking the barrier. Satisfactory performance within the limits of the design load requires the net to stop repeated rockfalls without requiring immediate maintenance. *Efficiency* is the ability of the barrier to stop impacts outside the center portion of a barrier net (Duffy 1999).

Some other state agencies have adopted the Caltrans guidelines or modified them to meet their requirements. Other states have adopted independent criteria or approaches. The Washington State DOT developed an approval process for new products for rockfall barrier net projects (Higgins 2003). Specific tests are not defined, and it is up to the manufacturer to provide test results showing that the products meet Washington State DOT acceptance criteria. The tests used by various manufacturers are likely to be different, however, and thus it remains difficult to compare products from several manufacturers directly.

The Oregon DOT establishes specifications for each rockfall barrier project. As a part of the specifications, a clause requires a "certificate of compliance" from the manufacturer ensuring that the system is capable of absorbing the specified kinetic energies with little or no required maintenance and that the system has been thoroughly tested and has demonstrated satisfactory performance in a similar application and capacity. Such compliance assurances have been difficult to achieve. Only the Swiss testing facility could provide any type of official certificate of compliance, and demonstration of satisfactory performance requires a record of successful use over time, which is impossible to document for new products.

6.2 Establishment of Fence-Testing Guidelines and Certification

Guidelines and certification standards pertaining to the testing of rockfall fence systems began in Switzerland, which in 2001 developed a national certification procedure that was effective. After the Swiss initiative, NCHRP in the United Sates supported a study in 2002 to establish a recommended standard testing procedure. Concurrently, EOTA undertook the development of a certification standard process for the European Union that was formally endorsed in January 2008 and is now being implemented. NCHRP is currently supporting a second study, scheduled for completion in December 2013, to establish new testing procedures that will include procedures outlined in the EOTA certification.

6.2.1 Swiss Certification Program

SAEFL and the Swiss Federal Institute for Forest, Snow, and Landscape Research have developed and presently administer a certification program for testing rockfall barrier systems that became effective in June 2001 (Gerber 2001; Gerber et al. 2001). A standard testing procedure, established as part of the certification program, is administered by a government agency to provide standardization and independent evaluation (Baumann 2002). A copy of the guideline, testing procedures, and the 2006 amendment are available as downloadable PDF files at www.environment-switzerland.ch/publications.

The testing procedure focuses on several important criteria for rockfall fence systems, including the following:

- Design energy (main criterion),
- Maximum displacement of the net after impact with full and half loading,
- Effective height of the net after impact,
- Impermeability to small rocks,
- Anchor and rope forces,
- Simplicity of the installation,
- Maintenance requirements and ease of repair, and
- Ease of construction on rough terrain.

As described in Section 5.2, the official test facility is located near Walenstadt, Switzerland, in an abandoned quarry (Figure 15-41). The full-scale certification test consists of a vertical drop of concrete blocks into the center panel of a three-panel barrier net installed on a vertical rock face.

Fence systems are divided into nine energy categories from 100 to 8,000 kJ (37 to 2,950 ft-tons), and fence heights are specified for each energy category. The testing procedure consists of three individual rock impact tests and observations of fence performance after each test. Concrete blocks are used for the projectile block to control size and density. The block velocity is 25 m/s (82 ft/s), the mass of the block depends on the energy category for the test, and the block must be decelerated and stopped by the structure. The three impact tests include the following:

- Low-energy impact in a boundary section of the barrier net to evaluate the laid-on mesh and to load individual ropes or rings in the net with several small blocks. Deformations of the individual ropes and mesh are permitted and recorded, but no repair work is allowed.
- A 50% energy test (equivalent to the SEL defined in Section 1.1) applied to the middle section of the barrier to evaluate the repair effort required and the ease of maintenance of the fence system.

FIGURE 15-42
High-energy test impact at Walenstadt, Switzerland, test facility. View clearly illustrates residual fence height.
(PHOTOGRAPH COURTESY OF GEOBRUGG.)

- A 100% energy test (equivalent to the MEL defined in Section 1.1) in which the full kinetic energy is absorbed by the fence. On the basis of the energy category, limits on maximum braking distance (fence deformation) and minimum residual fence height after impact must be met (Figure 15-42).

The 50% and 100% tests include measurement of tensile and compressive forces at 12 locations during deceleration of the test block. Videotapes of the impacts and deformations allow observation of motion of the block and individual bearing elements of the structure.

6.2.2 Testing and Certification Guidelines in the United States

To rectify a lack of consistent testing guidelines and certification in the United States, the American Association of State Highway and Transportation Officials requested a study that was conducted as part of NCHRP Project 20-07 (Higgins 2003). The study was guided by a task group composed of rockfall mitigation experts from several state transportation agencies. The purpose of the testing guideline was to define a standardized test sequence, specify limit values of capacity, and classify levels of performance for rockfall barrier systems.

The oversight task group decided that a government-sponsored test site or an agency-run test was unlikely. Therefore, the test procedure and documentation need to be simple enough that manufacturers can afford to conduct tests at temporary test sites to prove their product capacities and performance. In addition, it was considered important that the test procedure be similar to and interchangeable with the Swiss and European Union tests described in Sections 6.2.1 and 6.2.3, to reduce costs for manufacturers and to avoid hindering international business. During the development of this test procedure, the Swiss (SAEFL) certification process, described in Section 6.2.1, was official, but the European Union (EOTA) procedure, described in Section 6.2.3, was only in draft form and evolving. As a result, the NCHRP testing procedure more closely resembles that of SAEFL.

The NCHRP fence-testing guideline focuses on the same criteria for rockfall fence systems and follows the same testing sequence as the SAEFL test, with a few differences as follows:

- The fence manufacturer is responsible for testing.
- The test procedure may use a vertical drop, rolling rock, or an inclined cable system to deliver the load to the fence.
- The test site may be temporary, and the required equipment is readily rented and transported to the test site.
- The rock body may be natural or manufactured.
- The post height tested may be modified by 1.5 times without retesting.

In 2005, Geobrugg North America established a temporary test site in an abandoned portion of an open-pit mine to conduct a test on a low-energy barrier on the basis of the NCHRP guideline (Roth 2006). A vertical drop test was performed by anchoring the fence system post foundations and anchors into a near-vertical face of an open-pit

FIGURE 15-43 Temporary test site configuration for Geobrugg North America test undertaken in 2005 that used NCHRP testing guidelines and vertical-drop delivery method. (PHOTOGRAPH COURTESY OF J. D. HIGGINS.)

mine bench and placing a truck-mounted crane above on a haul road (Figure 15-43). The author of the NCHRP procedure and several authors of this publication were invited to observe and use the experience to evaluate the test procedure. At the conclusion of these tests, minor modifications to the procedure were proposed:

- The number of force and deformation measurements required was reduced,
- The 50% test was established as the SEL test, and
- The 100% test was established as the MEL test.

NCHRP Study 24-35, which is currently under way, with a scheduled completion date of December 2013, will address these observations as well as more closely conform to the protocols and terminology used by the European Union certification procedures discussed in Section 6.2.3.

6.2.3 European Union Certification Procedure

A working group for the European Union has recently developed a certification procedure for rockfall flexible-fence systems (EOTA 2008). The working group includes Austria, Belgium, France, Germany, Italy, Slovenia, and Switzerland. Spang (2002) reported a "vivid and lasting discussion" between working group members concerning the purpose of the intended tests. Some favored testing for comparability of test results only; others favored tests that helped determine limits on energy dissipation and maximum elongation, which can serve as a basis for design decisions and approval of products (Higgins 2003). Spang (2002) and Viktorovitch (2002) described the status of the guideline as it neared its final adoption after its seventh working group session. Moelk et al. (2006) described the finalized testing procedure.

In January 2008, EOTA formally endorsed European Technical Approval Guideline 27 (ETAG 27), *Guideline for European Technical Approval of Falling Rock Protection Kits* (EOTA 2008). The guideline defined the accepted formal process for testing and assessing the fitness for use of rockfall protection fence systems, and it thereby provided the basis for a European certification of these systems.

ETAG 27 focuses on criteria for fence systems similar to those of SAEFL and the NCHRP tests. The certification test may be conducted by an established testing facility or by the manufacturer. Two test methods may be used: an inclined cable that launches blocks in free fall into a fence system or a vertical drop from a crane. Test specifications define the block geometry for the projectile, and three functional panels of the fence system must be erected. Among the tests are two SEL tests and one MEL test. The SEL test includes two successive impacts in the center panel of the fence system with no maintenance allowed between impacts. The test is passed if the following conditions are met:

- After the first SEL impact the rock is stopped by the fence, there are no failures of the connection components, the mesh opening is minimal, and the residual height of the fence is greater than or equal to 70% of its height before the test.
- After the second SEL impact the block is stopped by the fence.

Damage and deformation parameters are measured after the second impact. Maintenance is conducted, if required, and the MEL test (involving three times the energy level of the SEL) is conducted.

One impact is required, and the test is passed if the fence stops the block and the block does not touch the ground until the fence reaches the maximum elongation during the test. Measurements of displacements and high-speed video documentation similar to those used by the SAEFL and NCHRP tests are included in the documentation.

6.2.4 Comparison of the Three Certification Guidelines

Table 15-1 compares primary criteria for the EOTA, SAEFL, and NCHRP testing guidelines. All three guidelines establish a standard test sequence, limit values of capacity, and classify levels of performance such as residual height of the fence after impact. All three require that the projectile be stopped and that the system not be ruptured at the SEL and MEL impact energy levels. However, there are also considerable differences among the guidelines. ETAG 27 and the NCHRP guidelines allow testing to be performed on slopes or in free fall, while the SAEFL guidelines allow only free-fall tests. Energy-level values defining the SEL and MEL energy relationships also differ between the guidelines. ETAG 27 defines the SEL value as one-third the MEL value, while the SAEFL and NCHRP guidelines define the SEL value (or the energy level for the 50% test) as one-half the MEL value.

The guidelines also classify performance. This is done by describing the level of maintenance after a test, evaluating the maximum deflection of a fence during an impact, and evaluating the residual height of the fence after impact. ETAG 27 states "no maintenance" following the first SEL test, while the SAEFL guidelines state "least possible repair." Both ETAG 27 and the SAEFL guidelines require a description of the maintenance that is required after a MEL test, but it is not used as a certifying criterion. The SAEFL guidelines have no residual height requirements for the fence after a SEL test. In contrast, ETAG 27 requires that the fence be at least 70% of its original height. ETAG 27 has no residual height requirements following a MEL test, but the SAEFL guidelines do, although the requirements vary with system energy ratings. The same goes for maximum elongation of the system. ETAG 27 and the SAEFL guidelines require documentation of maximum elongations during SEL tests but do not use these values as certifying criteria. During MEL tests, the SAEFL guidelines specify certifying criteria for elongation; ETAG 27 simply requires a reporting of the elongation. The NCHRP guidelines typically reflect the limits and standards set in the SAEFL guidelines.

6.3 Limitations of Fence-Testing Guidelines and Certification

The certification procedures and testing guidelines serve a specific purpose for rockfall fence system design and selection. However, in view of further development of rockfall fence systems for even higher energies, drop tests without spin and exclusively into the center of the panel are not the end-all of testing (Spang and Bolliger 2001). Rolling tests have provided otherwise unknown insights into fence performance and capabilities. From experience, high angular velocities can lead to a behavior similar to that of a circular saw, particularly in the case of platelike, sharp-edged rockfall, which can cause failures below the design energies. High angular velocities lead to a high loss of usable residual height. Hits in the corners or along the boundaries

Table 15-1
Comparison of Testing Criteria from the European Union, Switzerland, and the United States

Criterion	European Union (EOTA)	Switzerland (SAEFL)	United States (NCHRP)
Test slope angle (degrees)	0–90	90	0–90
Small blocks test	No	Yes	Yes
Pretest (SEL)	2 tests at 33% MEL	1 test at 50% MEL	1 test at 50% MEL
Main test (MEL)	1 test at MEL	1 test at MEL	1 test at MEL
Residual height	Categories A, B, and C	>50%	>60%
Deflection	Declared only	Limited	Limited
Test height	Not defined	Limited	Limited
Maintenance	Not evaluated	Evaluated	Evaluated

cause smaller energy absorptions than in the middle of the panel. Hits on retention ropes can endanger the overall stability of some systems. Therefore, the effects of angular kinetic energy in barrier performance cannot be ignored. Care should be taken in the selection of any system. Standardized tests and modeling are helpful in design but do not, at present, include the effects of a spinning rock.

ACKNOWLEDGMENTS

The chapter benefited significantly from contributions from Daniele Peila and Raymund M. Spang. In addition, the authors are grateful for discussions with and reviews provided by Rick Andrew, Steve Balaban, Tim Beck, Bruno Haller, Wade Hoon, Peter Ingraham, Howard Ingram, Daniel Journeaux, Larry Pierson, Michel Richard, Andrea Roth, and Robert Thommen. Jerry Higgins assisted with Section 5 and contributed a substantial portion of Section 6.

The authors dedicate this chapter to the memory of Raymund M. Spang, whose years of research, study, and practical experience played an important role in the development of rockfall fences.

REFERENCES

ABBREVIATION

EOTA European Organisation for Technical Approvals

Andrew, R. D. 1992. Selection of Rockfall Mitigation Techniques Based on Colorado Rockfall Simulation Program (Abridgement). In *Transportation Research Record 1343*, Transportation Research Board, National Research Council, Washington, D.C., pp. 20–22.

Andrew, R. D., D. A. Fry, and R. E. Bookwalter. 1998. *Field Testing and Evaluation of Various Rockfall Control Systems*. Chama Valley Productions, LLC.

Barrett, R. K., T. D. Bowen, T. J. Pfeiffer, and J. D. Higgins. 1989. *Rockfall Modeling and Attenuator Testing Report*. CDOH-DTD-ED3/CSM-89-2. Colorado Department of Highways, Denver, and Colorado School of Mines, Golden.

Baumann, R. 2002. The Worldwide First Official Approval of Rock-Fall Protection Nets. *Proc., 53rd Annual Highway Geology Symposium*, San Luis Obispo, Calif., pp. 40–51.

Duffy, J. D. 1990. *Field Testing and Evaluation of Rockfall Restraining Nets*. Report CA/TL-9005 (Videos). California Department of Transportation, Transportation Materials and Research, San Francisco.

Duffy, J. D. 1992. *Field Tests of Flexible Rockfall Barriers, Oberbuchsiten, Switzerland*. Brugg Cable Products.

Duffy, J. D. 1996. *Field Tests and Evaluation of Hi-Tech Low Energy Chain Link Rockfall Fence*. Report CA/05-96-01. California Department of Transportation, San Luis Obispo.

Duffy, J. D. 1999. Barrier Testing in North America—A Brief Summary Report. Presented at Seminar on Rockfall Tests and Standardization, Davos, Switzerland, January.

Duffy, J. D., and B. Haller. 1993. Field Tests of Flexible Rockfall Barriers. *Proc., Conference on Transportation Facilities Through Difficult Terrain*, Aspen, Colo., Aug., pp. 465–473.

Duffy, J. D., and W. Hoon. 1996. *Field Tests and Evaluation of Hi-Tech 50 and 70 Foot-Ton Rockfall Fence*. Report CA/05-96-02. California Department of Transportation, San Luis Obispo.

Duffy, J. D., and W. Hoon. 1998. *Field Tests and Evaluation of Chain Link Fence, Gawk Screen, Fence and Jersey Barrier*. Report CA/05-98-01. California Department of Transportation, San Luis Obispo.

Duffy, J. D., and D. D. Smith. 1990. Field Tests and Evaluation of Rockfall Restraining Nets. *Proc., 41st Annual Highway Geology Symposium*, pp. 125–142.

EOTA. 2008. *ETAG 27: Guideline for European Technical Approval of Falling Rock Kits*. Brussels, Belgium.

Geobrugg. 2002. *TECCO Slope Stabilization System Manual*. Fatzer, AG, Romanshorn, Switzerland.

Gerber, W. 2001. *Guideline for the Approval of Rockfall Protection Kits*. Swiss Agency for the Environment, Forests, and Landscapes and the Swiss Federal Research Institute WSL, Bern. www.environment-switzerland.ch/publications.

Gerber, W., and B. Haller. 1997. Safe and Economical Rockfall Protection Barriers. *Proc., Environmental and Safety Concerns in Underground Construction*, Balkema, Rotterdam, Netherlands, pp. 915–920.

Gerber, W., H. Grassl, A. Böll, and W. Ammann. 2001. Flexible Rockfall Barriers: Development, Standardisation and Type-Testing in Switzerland. In *International Conference on Landslides: Causes, Impacts and Countermeasures* (M. Kuehne, H. H. Einstein, E. Krauter, H. Klapperich, and R. Poettler, eds.), Verlag Glueckauf, Essen, Germany, pp. 515–524.

Hearn, G. 1991. *CDOT Flexpost Rockfall Fence*. CDOT Report CDOH-R-UCB-91-6. University of Colorado, Boulder.

Hearn, G. 1992. *High-Capacity Rockfall Fences*. Prepared for Research Branch, Colorado Department of Transportation, Denver.

Hearn, G. 1994. *High-Capacity Rockfall Fences: Addendum on Prototype*. Prepared for Research Branch, Colorado Department of Transportation, Denver.

Higgins, J. D. 2003. *Recommended Procedures for the Testing of Rock-Fall Barriers*. Report prepared as part of National Cooperative Highway Research Program Project 20-07, Task 138.

Japan Road Association. 2000. *Rockfall Protection Measures Manual* (in Japanese), Tokyo.

Kane, W. F., and J. D. Duffy. 1993. *Brugg Low Energy Wire Rope Rockfall Net Field Tests*. Technical Research Report 93-01. University of the Pacific, Stockton, Calif.

Moelk, M., G. Stelzer, and G. Kohlmaier. 2006. Product Certification of Rock Fall Protection Fences in the European Union. *Proc., 59th Canadian Geotechnical Conference*, Vancouver, British Columbia, Canada, pp. 402–408.

Muhunthan, B., and H. Radhakrishnan. 2007. *Finite Element Analysis of Hybrid Barrier for Rock Fall Slope Protection*. Final report. Washington State University, Pullman.

Muraishi, H., and S. Sano. 1999. Full-Scale Rockfall Test of Ring Net Barrier and Components. Presented at Seminar on Rockfall Tests and Standardization, Davos, Switzerland.

Parsons De Leuw, Inc. 1992. *Full Scale Geotextile Rock Barrier Wall Testing, Analysis and Prediction*. Colorado Department of Transportation, Denver.

Peila, D., C. Oggeri, and P. Baratono. 2006. *Barriere Paramassi A Rete: Interventi e Dimensionamento (Rockfall Barriers to Net Interventions and Dimensioning)*. Notebooks of Studies and Documentation 25, Association Geoscience and Environment (in Italian).

Peila, D., S. Pelizza, and F. Sassudelli. 1998. Evaluation of Behavior of Rockfall Restraining Nets by Full Scale Tests. *Rock Mechanics and Rock Engineering*, Vol. 31, No. 1, pp. 1–24.

Pfeiffer, T. J. 1989. *Rockfall Hazard Analysis Using Computer Simulation of Rockfalls*. Master's thesis. Colorado School of Mines, Golden.

Pfeiffer, T. J., and J. D. Higgins. 1990. Rockfall Hazard Analysis Using the Colorado Rockfall Simulation Program. In *Transportation Research Record 1288*, Transportation Research Board, National Research Council, Washington, D.C., pp. 117–126.

Pfeiffer, T. J., J. D. Higgins, and A. K. Turner. 1990. Computer Aided Rockfall Hazard Analysis. In *Proc., 6th Congress of the International Association of Engineering Geology* (D. G. Price, ed.), Balkema, Rotterdam, Netherlands, pp. 93–103.

Roth, A. 2006. *Summary Report on the Approval of the Geobrugg Rockfall Protection Barrier TXI-010 (100 kJ)*. Geobrugg North America, Santa Fe, N. Mex.

Sassudelli, F., P. Perreault, and P. C. Ingraham. 2007. Recent Rockfall Barrier Testing in Italy: Structural Component Tests and Dynamic Testing of Barrier Kits and Hybrid Barrier/Drape Systems in the Framework of European Technical Guidelines. Presented at 50th Annual Meeting of the Association of Environmental and Engineering Geologists, Los Angeles, Calif.

Smith, D. D., and J. D. Duffy. 1990. *Field Tests and Evaluation of Rockfall Restraining Nets*. Final Report CA/TL-90/05. California Department of Transportation, San Luis Obispo.

Spang, R. M. 2002. Certification of Rock-Fall Barriers in Europe. *Proc., 53rd Annual Highway Geology Symposium*, San Luis Obispo, Calif., pp. 52–62.

Spang, R. M., and R. Bolliger. 2001. From the Timber Fence to the High Energy Net: Developments in Rockfall Protection from the Origins to the Present. Presented at Geobrugg Jubilee Conference, Bad Ragaz, Switzerland, June.

Turner, R., J. P. Turner, and J. D. Duffy. 2009. Post Foundations for Flexible Rockfall Fences. *Proc., 60th Annual Highway Geology Symposium*, Buffalo, N.Y., pp. 125–140.

Viktorovitch, M. 2002. The European Technical Approval Guideline for Rock-Fall Protection Net Fences—The Stakes and the General Outlines of the Future Standard. *Proc., 53rd Annual Highway Geology Symposium*, San Luis Obispo, Calif. (addendum CD).

Volkwein, A. 2002. Remarks on the Numerical Simulation of Highly Flexible Rockfall Protection Systems. Presented at International Conference on Numerical Modeling of Rockfall Protection Systems, Walenstadt, Switzerland, Nov. 4–5.

Chapter 16

THOMAS C. BADGER
AND JOHN D. DUFFY

DRAPERY SYSTEMS

1. INTRODUCTION

Since at least the 1950s, woven steel fabrics have been used along North American highways as a cost-effective measure to control rockfall on actively eroding slopes. This protection measure has been termed *drapery*, *slope protection*, or simply *wire mesh*; these systems are here referred to as *drapery*. In its basic form, drapery consists of a row of ground anchors along the top of the installation, typically connected to a horizontal support cable from which a woven steel fabric, generically referred to as *mesh*, is suspended down onto the slope (Figure 16-1). This most basic form of drapery is referred to as *unsecured drapery*, after Muhunthan et al. (2005a). Two variations of drapery have since evolved: secured and hybrid drapery. *Secured drapery*, after Muhunthan et al. (2005a), includes a pattern of ground anchors incorporated within the field of the mesh for the purpose of containing rockfall within the mesh (Figure 16-2). An interior grid of wire ropes secured to the patterned ground anchors may also be included (Figure 16-3). A second variation of drapery involves elevating the upper part of an unsecured drapery by using posts or by suspending the drapery across a chute to create an impact section intended to intercept rockfalls originating upslope of the installation (Figure 16-4). The latter variation is referred to here as *hybrid drapery*. Hybrid drapery is often also referred to as an *attenuator*, a *hybrid barrier*, or *hanging nets*.

2. UNSECURED DRAPERY

Drapery that is only anchored along the top is referred to as unsecured drapery. Unsecured drapery addresses rockfall that originates from the covered slope area and is exposed to rockfall trajectories and impacts that are primarily oriented in the plane of the mesh. Although unsecured drapery can provide some resistance to rockfalls where the mesh is in contact with the slope, the system allows for rockfall between the slope and the mesh, controlling its descent into a catchment area at the base of the slope or the mesh terminus. One important objective of an unsecured system is to avoid external loading caused by accumulating debris behind the mesh. The North American transportation industry relies primarily on unsecured drapery, since its installation costs are significantly lower and it requires less maintenance than secured drapery.

Unsecured drapery has been installed successfully on moderately inclined to overhanging slopes and on very uniform to highly irregular slopes to mitigate rockfall. Some installations exceed 120 m (400 ft) in height. Muhunthan et al. (2005a, 2005b) reviewed the performance of existing unsecured drapery in North America and completed engineering evaluations of their limit conditions for the purpose of developing design guidelines. Study outcomes included a recommended design protocol for evaluating the suitability of this protection measure for the limiting range of site conditions, accounting for external loading conditions to determine anchoring requirements, selecting the appropriate mesh, and detailing the final system (Figure 16-5).

2.1 Site Suitability and Characterization

Unsecured drapery is most appropriate for rockfalls consisting of a single block or small-volume failures composed of relatively small block sizes, such as those originating from coarse surficial deposits (e.g., colluvium, alluvium, residual soils) and closely fractured rock masses. Draperies are well suited for these limiting conditions and have performed effectively with minimal damage for several decades, even when exposed to frequent rockfalls (Muhunthan et al. 2005a).

Many draperies have been installed on slopes that are poorly suited or have been over- or underdesigned for the site and loading conditions (Figure 16-6). Characterization of the rockfall source area and loading conditions is the first and most important step in determining site suitability and in designing an appropriate system for the expected conditions.

2.1.1 *Block and Event Size*

As with any structural system, there are limitations on both the magnitude and repeated sustainable loads for drapery. The size of individual blocks or small-scale instabilities is the most important factor in determining site suitability. Although there are many examples of installations that have sustained apparent extreme debris or impact loads, in practice draperies have well-demonstrated limitations in terms of block size, with a maximum block size capacity on the order of 1.5 m (5 ft). Analyses and case histories presented by Muhunthan et al. (2005a) suggest that mesh secured only at the top cannot sustain dead loads caused by debris much in excess of approximately 10 m^3 (13 yd^3), assuming that full-load transfer of the debris weight occurs. More recent experience of scaling (removal of loose rock) beneath high-strength ring nets suggests that they may be able to support larger block sizes and volumes (D. Journeaux, personal communication, 2008). For general practice with more common meshes, if anticipated modes of slope instability would result in single events larger than approximately 5 to 10 m^3 (6.5 to 13 yd^3) in volume or block sizes characteristically exceeding 1.5 m (5 ft), additional or alternative mitigation measures should be considered (Figure 16-7).

FIGURE 16-1
Typical unsecured drapery, anchored only along top of installation, can provide effective control of rockfalls originating from steep slopes with little available catchment area at base of slope.
(PHOTOGRAPH COURTESY OF THE BRITISH COLUMBIA MINISTRY OF TRANSPORTATION.)

2.1.2 *Slope Conditions*

Drapery placed on a steep slope (>65°) functions somewhat differently from that placed on a flatter slope. Given the orientation and limited slope

FIGURE 16-2
Newly constructed secured drapery, incorporating interior anchors and TECCO mesh with jute matting, installed along visually sensitive highway in Columbia River Gorge to address raveling of oversteepened, bouldery colluvium.
(PHOTOGRAPH COURTESY OF THE WASHINGTON STATE DEPARTMENT OF TRANSPORTATION.)

FIGURE 16-3
Variation in design of secured drapery incorporating interior anchors and wire ropes. Installation along a road in northern Italy with virtually no shoulder width or catchment ditch provides containment of rockfall debris.
(PHOTOGRAPH COURTESY OF P. INGRAHAM.)

contact on near-vertical installations, the mesh imparts little stabilization through its weight, and rocks can generally pass nearly unimpeded between the mesh and the slope. The weight, however, provides energy absorption because of its resistance to movement and plays a key role in controlling the rockfall and its exit from the bottom of the mesh. On moderately inclined slopes (35° to 65°), mesh contact is typically greater, and its weight can impart a significant resistance force on individual blocks. As a result, in many cases rockfall frequency is reduced, and the trajectories of dislodged blocks are slowed considerably.

Slope uniformity can also be an important design consideration (Figure 16-8). Mesh contact is typically greatest on uniform planar slopes and least on concave slopes. Slope uniformity also influences where and how rockfall affects the mesh and the likelihood that debris will accumulate or pass beneath the mesh.

When the extent of slope coverage is determined for drapery, typically the available catchment area at the base of the slope and its effectiveness for containing debris as it clears the mesh are considered in the selection of the bottom elevation. If limited catchment area is available, various treatments can be used along the bottom of the mesh to better control discharge. Adding a layer of mesh 2 m (6 ft) wide along the bottom adds weight and aids in dampening rock velocity. Another method to dampen the velocity of exiting rocks is to partially restrain the bottom of the mesh with an end-anchored cable fitted with a design to allow elongation under loading; friction brakes are often included to allow for some elongation of the cable. An end-anchored cable that does not allow for its elongation is not an appropriate design. The cable should always have a way to extend. In some cases springs have been used to provide for such extension; in other cases, two cables lightly clipped together have been used to allow sliding and extension under load. A single, fixed-length cable without some method of allowing extension is not ideal, because a restrained cable along the bottom can cause rocks to accumulate, which can lead to damage of the system.

Aesthetic concerns or snow accumulation at the base of the installation may also influence the lower terminus. Standard drapery design would typically cover all the observed or anticipated rockfall source areas. Caution is warranted for oversteepened and actively eroding slope brows; the upper terminus should extend beyond the ultimate stable slope configuration. The top of the drapery can also be elevated or suspended to address rockfall source areas upslope of the installation. Although drapery is often a highly economical and effective measure for mitigating rockfall, other containment

FIGURE 16-4
Hybrid drapery intercepts rockfall originating upslope of installation and contains rockfalls initiating from rock slope beneath mesh.
(PHOTOGRAPH COURTESY OF THE WASHINGTON STATE DEPARTMENT OF TRANSPORTATION.)

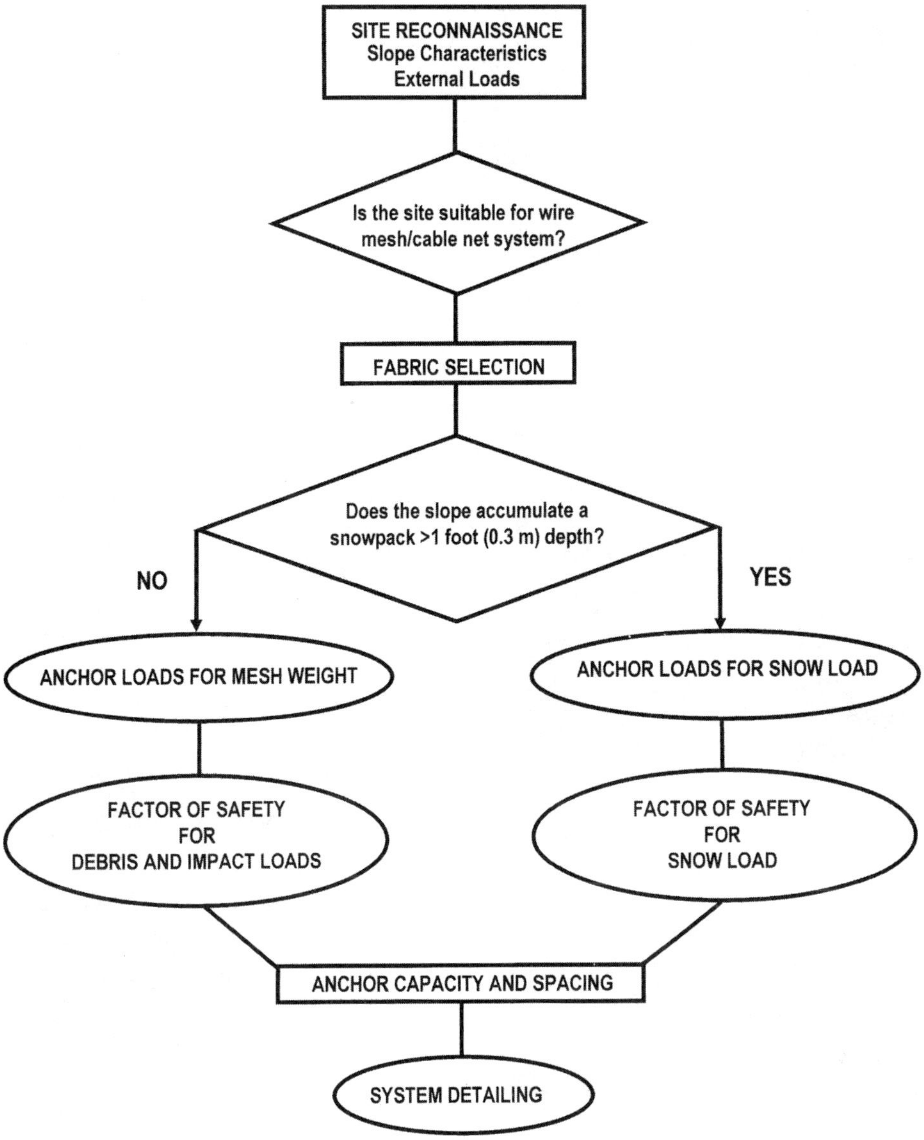

FIGURE 16-5 Design protocol summarizing major design issues and decision points for unsecured drapery (Muhunthan et al. 2005b).

or avoidance alternatives may be more cost-effective if the coverage area becomes very extensive.

Slope characteristics are also evaluated in the assessment of anchoring conditions. Difficult access generally necessitates small portable drills for anchor installation. This characteristic is not usually a problem for installations in bedrock, but loose deposits of cobble or boulders can be challenging for small, hand-operated equipment. More suitable anchoring locations, such as bedrock outcrops, might be located upslope of the drapery. In such cases, cables can be extended down from the anchors to the top of the mesh. These cables are commonly referred to as *tag lines* or *connecting cables*.

2.1.3 Interface Friction

Where the mesh is in contact with the slope, interface friction provides a resistance component to the stability of the system. The interface friction is controlled by the macro- and microscopic roughness of the surface. *Macroscopic roughness* is defined by large-scale irregularities of the slope, and *microscopic roughness* is defined as the texture of the surface. Where the slope is planar and the surface is smooth, minimal interface friction may exist, and the mobilized force on the system is carried largely by the anchors (Figure 16-9*a*). Intermediate and high interface friction may exist

FIGURE 16-6
Large block detached from slope near top of drapery severely damaged low-strength chain-link fabric. Large block sizes in bouldery deposit warrant more robust mesh.
(PHOTOGRAPH COURTESY OF THE WASHINGTON STATE DEPARTMENT OF TRANSPORTATION.)

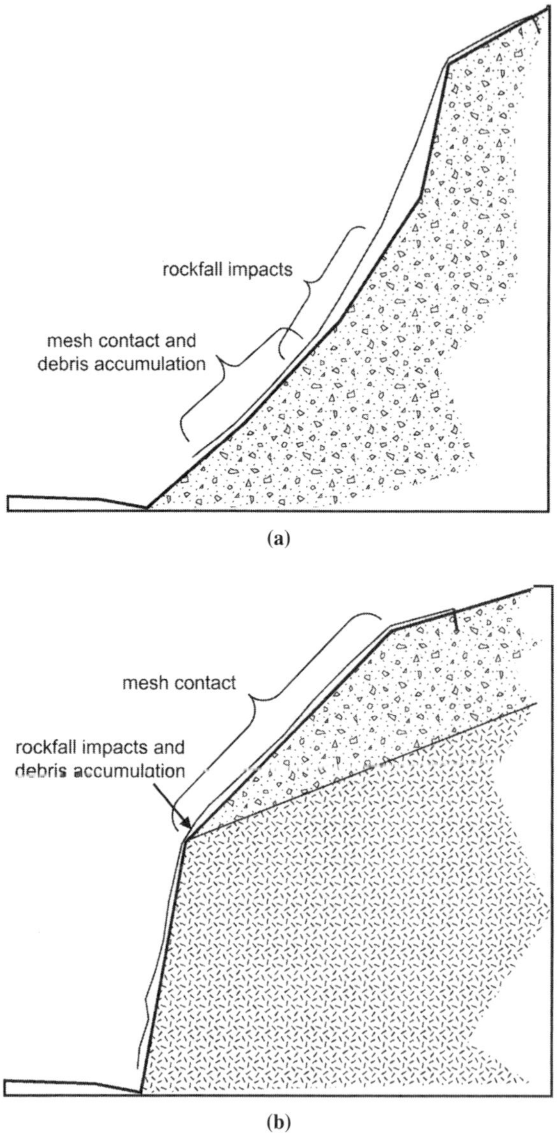

FIGURE 16-8
Slope profiles illustrate typical zones of mesh contact, impacts, and debris accumulation: (*a*) concave profile and (*b*) convex profile (Muhunthan et al. 2005b).

FIGURE 16-7
Slope failure involving up to approximately 10 m^3 (13 yd^3) is contained and controlled under double-twisted wire mesh drapery.
(PHOTOGRAPH COURTESY OF THE CALIFORNIA DEPARTMENT OF TRANSPORTATION.)

on slopes that are irregular and have rough surfaces or abrupt protrusions (Figure 16-9, *b* and *c*); in these cases, little to no mobilized force may be imparted to the anchors.

The effectiveness of unsecured drapery, particularly when installed on moderately inclined slopes, can be strongly influenced by how much contact the mesh has with the slope surface. Generally, maximizing mesh contact provides substantial long-term benefit by inhibiting erosion and promoting

(a)

(b)

(c)

FIGURE 16-9
Examples of interface friction: (*a*) low (typical values 25° to 35°) on planar and highly uniform slope; (*b*) intermediate (typical values 35° to 60°) on slope with some macro- and microscopic roughness; and (*c*) high (typical values > 60°) on uniform slope with considerable microscopic roughness.
(PHOTOGRAPHS COURTESY OF THE WASHINGTON STATE DEPARTMENT OF TRANSPORTATION.)

slope revegetation and by imparting a resistive force to unstable blocks, thereby reducing rockfall frequency and trajectory velocities.

Unfortunately, interface friction is a difficult parameter to quantify, partly because of the near absence of supporting data or documentation. Its inclusion as a resistance force for anchor design is further complicated because the amount of mesh contact is strongly influenced by slope configuration, fabric flexibility, and installation methods. Because of weathering and erosion, interface friction can also be a transient condition. For these reasons, it may be imprudent to include the resistance contribution of interface friction when anchor requirements are determined for mesh weight, debris load, and impact load. Applying a safety factor to the static or dynamic anchor load may be a more appropriate method of addressing a range of potential slope configurations and loading conditions. An important exception exists where snow loads are anticipated; then interface friction should be assessed. In the absence of either backcalculated or field measurements, the interface friction angle can be estimated for the observed slope irregularity and surface roughness (Muhunthan et al. 2005a).

2.1.4 Debris Loads

Debris loading has been a common source of both local and global failure of drapery systems. Commonly used fabrics begin to yield with debris accumulations as small as 5 to 10 m^3 (5 to 10 yd^3). Therefore, it is important to assess the expected type, size, volume, and frequency of slope instabilities. This assessment should also consider how and where debris might accumulate once the mesh has been installed. Common accumulation locations include areas immediately above slope convexities (Figure 16-8*b*), at changes of slope orientation or inclination, along the base of the mesh where the mesh overlaps, and above any restraints or anchors along the perimeter or interior field of the mesh. One often-overlooked restraint is snow and debris covering the base of the mesh, which accumulates either as snow slides off the mesh or from snowplowing. Ice accumulation can also induce load during spring thaw, when large blocks or sheets of ice, sometimes still frozen to the drapery, are released from a slope face. Instability of underlying surficial deposits where rooted vegetation has grown through the mesh has also been observed to impart damaging loads to drapery.

2.1.5 Impact Loads

Rockfall impacts apply transient and short-duration loads on drapery. The actual load imparted is a function of the mass and velocity of the block and the manner and orientation in which the block strikes the system. On near-vertical slopes where the mesh is subparallel to the slope and in limited contact, the rockfall trajectory is generally also subparallel to the slope. Unless the falling rock snags the mesh or is deflected horizontally on striking some asperity, there is little opportunity to transfer a large portion of the kinetic energy to the system.

On moderately inclined slopes, the velocities of rolling and bouncing blocks are significantly reduced by the greater mesh contact. Thus, the kinetic energy should be much less for a rolling or bouncing rock beneath the mesh than that expected on an undraped slope. High impact loads can be imparted to the system when blocks strike subperpendicular to the mesh; such is the case for hybrid drapery. Another common configuration prone to subnormal impact loading occurs on slopes with abrupt convexities, such as a moderately steep slope in surficial deposits overlying a near-vertical cut slope in rock (Figure 16-8*b*), midslope benches, and transitions between excavated lifts. In these cases, rockfall initiating near the top of the installation strikes the mesh just above the slope inflection; this is a frequent location of puncture failures.

The kinetic energy of scenario rockfalls can be estimated by using widely available rockfall-modeling software. How kinetic energy is transferred as a load to the system remains poorly understood, however. Some guidance on resistance capacity for commonly used fabrics in North America has been provided by Muhunthan et al. (2005a, 2005b).

2.1.6 Snow and Ice Loads

Significant loads can be imparted to drapery that accumulates snow and ice. Snow loading, which should be considered separately from that for ice, is only relevant for slopes between 35° and 60°

that are able to accumulate snow. Slopes flatter than 30° to 35° generally do not produce rockfall. Snow loads have been noted to vary temporally and with temperature changes. On an instrumented cable net drapery of moderate inclination, Muhunthan et al. (2005a) noted that anchor loads increased when temperatures rose above freezing, presumably because of decoupling of the snowpack from the slope or accelerated creep. In general, the prediction of actual loading from snow is poorly understood and highly influenced by site conditions and the installation. When snow loads are a design factor, caution is especially warranted on

- All smooth planar slopes (low interface friction),
- Slopes oriented between 45° and 60° (high load transfer to the system),
- Concave slopes where surface contact would be limited, and
- Slopes with an unstable snowpack situated upslope, where forces could be transferred onto the system.

Ice loading can occur on drapery with any slope inclination given requisite temperature and seepage conditions. No information that characterizes this potential loading condition has been published. Despite the potential for extreme loads, the authors are unaware of any drapery failures attributed solely to ice loading. Potential loads might be considered for observed icing conditions through increased anchoring and more robust mesh.

2.2 Design Considerations

After the site conditions are evaluated and the suitability of an unsecured drapery for mitigating rockfall effectively is determined, typically the fabric selection and anchor and support design are considered.

2.2.1 Mesh

North American practice for standard unsecured drapery evolved from lightweight chain-link and welded wire mesh in the 1950s to double-twisted hexagonal mesh in the 1970s and 1980s. Cable nets were added as a more robust alternative beginning in the late 1980s. Widespread application of double-twisted hexagonal mesh for drapery has demonstrated an upper range of effectiveness for block sizes up to about 0.6 m (2 ft), and cable nets have proved effective for block sizes approaching 1.5 m (5 ft). These and other meshes are discussed in more detail in Chapter 15.

Since the late 1990s, alternative drapery meshes have been introduced to the North American market. These include high-tensile single-twist fabrics, ring nets, and composite meshes with both cables and double-twisted hexagonal mesh. Although some of these newer meshes have replaced cable nets for some proprietary flexible fence systems, they generally have not yet seen widespread use in North American drapery.

Mesh performance is influenced by both extrinsic and intrinsic factors. Extrinsic factors include system anchoring, slope conformance, and installation dimensions, which control fabric tautness; tautness in turn affects deformability and energy dissipation. Important intrinsic factors include stiffness (flexibility), weight (inertia), puncture strength, and corrosion resistance. More flexible meshes provide better slope conformance than do stiff meshes and, on moderately inclined slopes, better distribute stabilizing force to the slope (Figure 16-10), reduce erosion and facilitate revegetation, and improve visual subordinance. Tangential and normal forces from rockfall impacts can cause localized tearing or punctures; the selected mesh must have adequate strength to resist these localized and potentially damaging

FIGURE 16-10 Design intent of mesh is to cover slope and control rockfall trajectory by restricting rock movement. Mesh weight provides important stabilizing factor. Double-twisted wire mesh is shown. (PHOTOGRAPH COURTESY OF THE CALIFORNIA DEPARTMENT OF TRANSPORTATION.)

stresses. Corrosion resistance can also be a critically important design consideration in marine environments, in areas where deicing sands and agents are used on roadways, and when mesh is placed in contact with acid-producing rocks.

Seaming of double-twisted hexagonal mesh is generally done with high-tensile-steel hog rings. Lacing wire can also be used, although it is typically more time-consuming to install than hog rings placed with pneumatic guns. When high-tensile-steel fasteners are used, a single cell overlap with one hog ring per cell is all that is required (Muhunthan et al. 2005a, 2005b). This seam detail, however, is less strong than the mesh itself; lacing wire provides seam strength that exceeds the mesh strength (Agostini et al. 1988). A vertical seam that is weaker than the mesh is not necessarily a deficiency. For example, under extreme debris loads it would be better for a seam to fail than for mesh to tear or, worse, for the entire system to fail. Cable nets, ring nets, and other more robust meshes are typically seamed with lacing cable or shackles that match or exceed the strength of the mesh elements. It is important that lacing wire or shackles connect each cell adjacent to the seam.

With large openings [>200 mm (8 in.)] common to more robust meshes, 50-mm (2-in.) chain-link or double-twisted hexagonal mesh is generally included to retain small rocks that might otherwise pass through the larger openings. Current practice includes proponents for placing this lighter-weight backing mesh on the inside of the robust mesh and proponents for placing it on the outside. The favored reason for inside placement is to reduce the potential for debris accumulation by having smaller effective openings against the slope. Favored reasons for placement on the outside are the expectations for less rockfall damage to the backing mesh and the subdued appearance of noncolorized robust mesh with backing mesh, which is typically less expensive to colorize.

There are also arguments both for and against fastening the backing mesh to the robust mesh. Fastening the two meshes together has been observed to reduce the potential for debris to get trapped and accumulate between the meshes. This common practice of attaching the backing mesh has proved effective and efficient for cable nets. Fastening the backing mesh to highly deformable mesh (i.e., ring nets), however, may increase the stiffness of the system and potentially reduce optimal performance for attenuating rockfall impacts. For these reasons, the contrast in stiffness between the nets and the backing mesh should be considered. Attaching the backing mesh on highly deformable ring nets before installation can concentrate stresses in the backing and ultimately cause the fasteners to fail when the ring net panels are lifted and panels are added.

The quantity of mesh required is influenced by

- The area to be covered, considering that slope irregularity and orientation can dramatically increase surface area;
- Incidental overlap between adjacent panels, which can vary because of topography;
- The effort made to optimize mesh contact with the slope; and
- Mesh stiffness and the deformability of mesh during placement.

Greater mesh conformance, and thus more mesh quantity, can be expected for moderately inclined slopes than for steep slopes. Highly irregular slopes require more mesh overlap and greater plan quantities than more uniform slopes. When more deformable meshes are applied to irregular or high slopes, it can be difficult to stretch the nets fully during seaming, which can reduce the expected coverage area of the mesh. A contingency of 10% to 20% may be reasonable for uniform near-vertical slopes, but a contingency of 30% or more may be needed for highly irregular slopes of moderate inclination. If large quantities are involved, consideration might be given to developing a digital terrain model or using stereophotogrammetric methods to estimate quantities more accurately (Figure 16-11).

2.2.2 Anchors

Anchors provide the primary support for drapery. It has been recognized, in part from installations exposed to snow loads, that interface friction can also provide significant support. Observed drapery failures show that anchor failures from debris or impact loads are rare; the mesh yields before the anchors. Snow loads, however, have been responsible for numerous system failures, in all of which anchor yielding occurred either because the strength or passive resistance of the ground

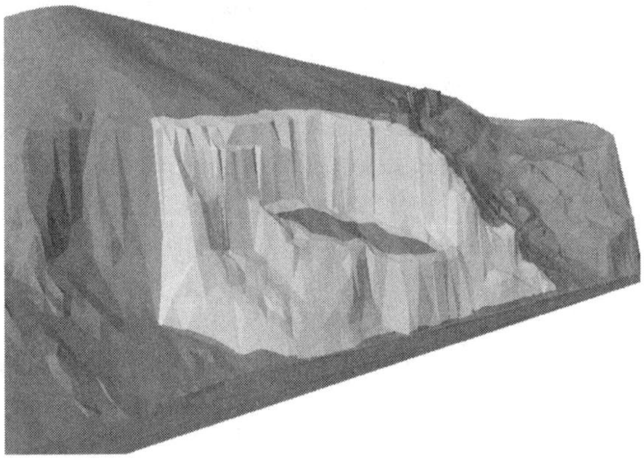

FIGURE 16-11
Quantities for ring net drapery (*light shaded area*) for slope 100 m (300 ft) high were estimated from digital terrain model developed from terrestrial laser scanning. Geographic information system methods were then used to model drapery surface from which surface area could be calculated. Because of expected deformation of ring nets and modeling uncertainties, 20% contingency was included in contract.
(IMAGE COURTESY OF THE WASHINGTON STATE DEPARTMENT OF TRANSPORTATION.)

was exceeded or because the yield strength of the anchor was surpassed. Thus, the design approach considered for debris and impact loading should be different from that for snow loading.

Muhunthan et al. (2005a) developed a series of design charts for determining anchor load and spacing requirements for common meshes (double-twisted wire mesh, TECCO, and cable nets) applied to a range of slope heights, inclinations, and interface friction values (Figure 16-12). To account for potential variability in debris and impact loads for a given site, as well as the current lack of understanding of how impact loads are transferred to the system, they recommended applying a large factor of safety (5 to 10) to the anchor-load requirements for the system weight alone, with no resistance contribution from interface friction (Table 16-1).

Because snowpack develops on moderately inclined slopes and a significant resistance contribution can be mobilized from interface friction on these slopes, the anchor design for snow loading can benefit from consideration of its contribution. The anchor force assuming full load transfer of the snowpack onto the mesh per unit width, F_a, can be estimated as follows:

$$F_a = \rho g H L \sin\theta - \rho g H L \cos\theta \tan\phi$$

where

ρ = overall density of snowpack,
g = gravity constant (for metric units),
H = thickness of snowpack oriented normally to slope,
L = slope length of installation,
θ = slope angle, and
ϕ = interface friction angle.

The design challenge lies in characterizing the interface friction of the entire installation. Muhunthan et al. (2005a) suggested application of a safety factor of 2 to 3 to account for larger-than-anticipated

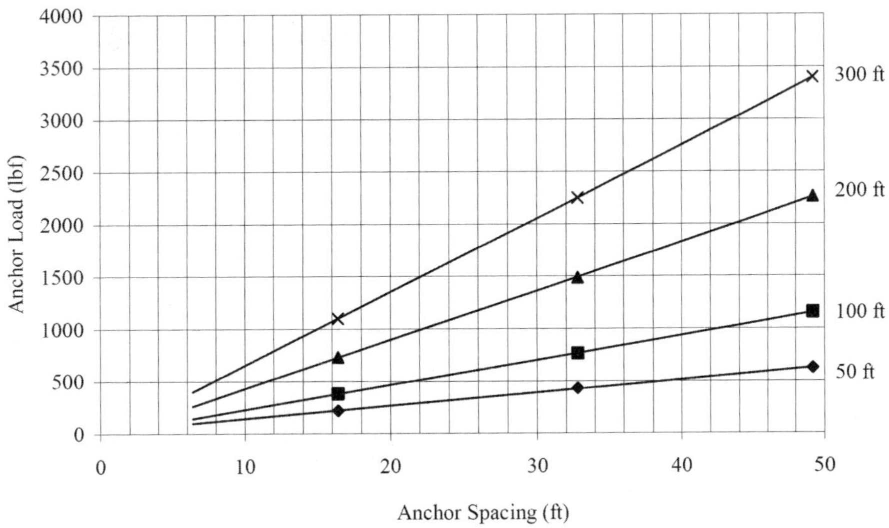

FIGURE 16-12
Example of graph of anchor load versus spacing for double-twisted hexagonal wire mesh for planar (low interface friction), 60° slope with heights of 15, 30, 60, and 90 m (50, 100, 200, and 300 ft). No safety factor for debris, impact, or snow loading was applied to these values (Muhunthan et al. 2005a).

Table 16-1
Recommended Maximum Anchor Spacing as a Function of Slope Height

Slope Height [m (ft)]	Anchor Spacing[a,b] [m (ft)]
≤30 (100)	15 (50)
30–60 (100–200)	10 (35)
60–90 (200–300)	5 (20)

[a]Maximum spacings suitable for double-twisted hexagonal mesh, TECCO G65 mesh, and 300-mm (12-in.) square grid cable nets backed with either double-twisted hexagonal or chain-link mesh.
[b]Anchor spacing is based on a minimum anchor capacity of 90 kN (20,000 lbf).
Source: Muhunthan et al. 2005b.

snowpack and the potential for overestimating interface friction.

Current North American practice generally uses anchor elements that exceed an ultimate tensile strength of 90 kN (20,000 lbf). Common anchors are 25-mm (1-in.) continuously threaded deformed steel bar and 19-mm (¾-in.) wire rope. An obvious advantage of wire rope tendons over steel bars is their ability to bend and optimally orient the load into tension. For most drapery applications, steel bars are generally oriented so that they are loaded in shear, which provides only about 75% of a bar's ultimate tensile strength. Wire rope tendons have increased in favor and use because of their flexibility, greater capacity, and durability.

Passive resistance is an important consideration in the design of anchors founded in loose unconsolidated surficial deposits, particularly for systems that might experience extreme loads. To optimize the passive resistance, anchors are oriented perpendicular to the slope, not vertically (Figure 16-13). Additional passive (and pullout) resistance can also be gained by increasing the diameter of the grout bulb.

A minimum anchor setback of 2 to 5 m (6 to 15 ft) from the active brow is commonly prescribed. However, anchor locations need to consider the potential for ongoing erosion and retrogression of the active brow of the slope over the expected life of the system. For many bedrock slopes, little to no retrogression might be expected, while in some bouldery surficial deposits, rapid erosion might occur (Figure 16-14).

2.2.3 Support Ropes

The mesh is suspended from a top horizontal support rope, which is connected either directly to the anchors or by longitudinal taglines or connecting cables downslope from the anchors. The maximum uninterrupted length of the top support rope is a function of the strength of the rope, the sag provided, and the dead load being suspended from it and typically includes a safety factor to account for additional loads from debris, impact, snow, or all three. Muhunthan et al. (2005b) provided guidance for determining these lengths with appropriate

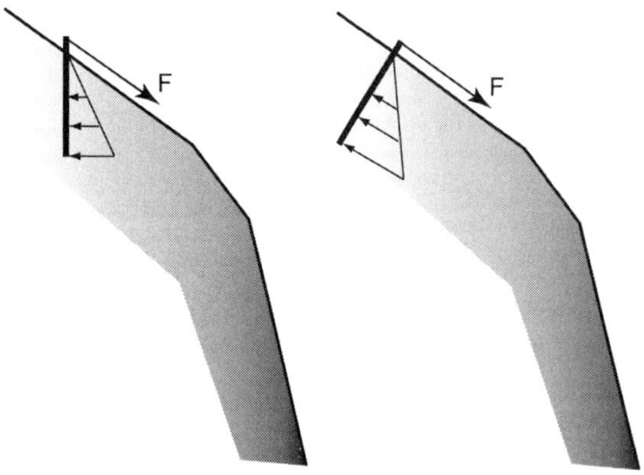

FIGURE 16-13
Additional passive resistance and anchor capacity are achieved by orienting anchor perpendicular to slope.
(Diagram by T. C. Badger.)

FIGURE 16-14
After 20 years, ongoing erosion threatens anchors and entire installation.
(Photograph courtesy of the Washington State Department of Transportation.)

Table 16-2
Recommended Maximum Length for Top Horizontal Support Rope Versus Slope Height for Double-Twisted Hexagonal Mesh

Slope Height [m (ft)]	Maximum Length for 13-mm (1/2-in.) Cable Fabric Weight Only [m (ft)]	Maximum Length for 19-mm (3/4-in.) Cable Fabric Weight Only [m (ft)]
15 (50)	70 (230)	120 (400)
30 (100)	35 (120)	60 (200)
60 (200)	18 (60)	30 (100)
90 (300)	12 (40)	22 (75)

Source: Muhunthan et al. 2005b.

safety factors (Table 16-2). For different loading conditions and materials, a project-specific design could be developed by using a formula for catenary curves.

The only required support rope for an unsecured system is the top horizontal rope. Interior vertical and horizontal or other perimeter ropes provide no mechanical benefit to improve system capacity unless the mesh can be clamped to these ropes at close intervals. Interior horizontal support ropes, particularly when located on the inside of the mesh, trap debris, which has led to many system failures (Figure 16-15). If a horizontal rope is used at the bottom, as is often done to aid in lifting the mesh to clear accumulated debris, it should be located on the outside and the mesh folded outward.

2.2.4 Aesthetic Concerns

Increasingly, the selection of the rockfall mitigation measure is judged not only on its engineering and economic merits but also on aesthetic concerns (Andrew et al. 2011), particularly in areas of considerable scenic or recreational value, where a design objective to visually subordinate engineered facilities may be required or legislated. In some cases, managers and design professionals responsible for aesthetic stewardship have found the use of drapery objectionable. The principal concerns have stemmed from the typically large coverage area, the visual contrast between the mesh and the slope, and the potential for poor mesh contact with the slope. Design efforts to mitigate these aesthetic concerns have included

- Limiting the coverage area,
- Achieving greater mesh contact with the slope, and
- Colorizing system components or promoting vegetation to visually merge the system with the slope.

For some slopes, it may be possible to reduce or limit the coverage area of a drapery and still provide adequate rockfall containment. Rather than covering the entire rockfall source area on a slope, a hybrid drapery might be considered to address upslope source areas. Such an approach is commonly used for chutes and along abrupt convexities; in these cases, impact loading becomes an important design consideration. Alternatively, opportunities may exist to raise the bottom of the mesh to keep it above the immediate view of passing motorists.

FIGURE 16-15 Debris accumulation is common problem where interior horizontal support ropes have been placed between mesh and slope. (PHOTOGRAPH COURTESY OF THE WASHINGTON STATE DEPARTMENT OF TRANSPORTATION.)

FIGURE 16-16 New 1V:0.25H, 60-m (200-ft) high rock cut on highway between Dubois and Moran, Wyoming, incorporates partial drapery of polyvinyl chloride–coated, double-twisted wire mesh to address rockfall initiating in upper colluvium and weathered bedrock. Drapery terminates 5 to 10 m (15 to 30 ft) above ditch (white dashed line) to lessen aesthetic impact for passing motorists. Rockfall ditch included in cut template was designed to contain rock debris exiting drapery and rockfall generated from lower uncovered portion of cut. (PHOTOGRAPH COURTESY OF THE WYOMING DEPARTMENT OF TRANSPORTATION.)

Incorporating design criteria for rockfall catchment ditches or rockfall modeling can aid in determining the height of the bottom terminus of the mesh on the basis of the available catchment area (Figure 16-16). The mesh terminus is located so that debris passing from beneath the mesh is contained in the ditch. Provisions may also be included to increase or improve the available catchment area.

For many moderately inclined slopes (35° to 65°) and some steeply inclined slopes (>65°), efforts to maximize mesh contact with the slope yield both functional and aesthetic benefits. Functional benefits can include greater interface friction and overall system capacity as well as decreased slope erosion and rockfall. Slope contact also reduces or eliminates the gap between the mesh and the slope, which can be discernible when the installation is viewed from the side, as by a passing motorist. On steep slopes, it is often more difficult to achieve slope contact, and because of the reduced component of normal force, it is difficult to maintain contact as the system is loaded.

Reduced erosion promotes revegetation of the underlying slope. Vegetation can significantly reduce visual contrast and aid in visually blending the mesh with the slope as well as increasing the stability of the surface materials (Figure 16-17). More proactive revegetation efforts have been undertaken, such as placing erosion-control fabrics beneath the mesh, hydro-seeding, and installing plantings. Although vegetation most often is beneficial, tree growth through the mesh has to be carefully evaluated because it can cause significant damage and system failures. The damage occurs when trees fall (a common occurrence on steep slopes) or when the substrate in which they have their roots creeps or experiences other shallow instability. Tree growth has also been observed to lift the mesh from the slope, which can lead to downslope accumulation of debris.

Mesh is usually specified to include a galvanized or other corrosion-protection coating. When installed, these coatings have a metallic, often shiny or light-colored appearance that may strongly contrast with the slope. Over time, the shininess of galvanizing typically fades to a dull gray color. Colored mesh, typically with either polyvinyl chloride (PVC) or powder coating, which can greatly reduce the visual contrast between the mesh and the slope, has become commonplace in recent years. Neither coating, however, is without problems. Powder coating reportedly experiences greater damage during handling and is more costly than PVC. High ultraviolet exposure has caused whitening and embrittlement of PVC coatings.

2.2.5 Construction

How a drapery system is installed and inspected during construction determines how well the design objectives of function, aesthetics, and cost-effectiveness are met. Usually, if scaling and rock anchoring are also required, they are performed before mesh installation. Scaling through ring nets before placement of the backing mesh has been done successfully and has reduced traffic-related impacts. Scaling-related damage may require repair before placement of the backing mesh and final acceptance. Damage assessment procedures and the definition of what constitutes damage should be thoroughly considered if this approach is pursued. A significant safety concern with scaling through the mesh is the increased risk that a large block will cause the mesh to fail while scalers are on the meshed slope. Drapery is commonly applied to closely fractured rock masses, which are more hazardous to scalers than slopes with large blocky fractures. Scaling before hanging the mesh typically focuses on the slope preparation necessary for mesh installation and the removal of occasional, potentially damaging oversized blocks or other discrete, more unstable zones. Long-boom mechanical scaling for closely fractured rock masses rather than hand scaling reduces the risk exposure for scaling before mesh placement. As a final point, the slope angle is an obvious limitation for the efficient scaling of debris through mesh, since mesh imparts a significant load to a slope with any orientation less than near-vertical.

Drapery anchor capacities should be verified with load testing; commonly at least 25% are specified for testing. Vertical pullout tests are most often performed. However, since anchors are typically loaded parallel to the slope, anchor testing may be performed laterally to verify not only the anchor pullout resistance but also the passive resistance provided by the ground (Figure 16-18). Lateral load tests might be more appropriate and easier to perform than vertical pullout tests when anchors are located in loose surficial deposits.

A staging area near the slope is generally required for layout and seaming and to stage the lifting and placement of the mesh. For robust fabrics requiring mesh backing, a staging area is essential because it allows the mesh backing to be attached while on the ground. The selection of the staging area is influenced by the method of placement (boom truck, crane, or helicopter) and the size of the installation. For large installations, using a helicopter to lift and place the mesh is usually the fastest and most cost-effective method. However, helicopter placement requires an emergency landing area as part of the staging operations; traffic control and proximity to aerial utility lines are also factors requiring consideration. To mitigate for the potential stretching of the mesh panel when lifted and to improve placement and seaming efficiency, the bottom of the lifted panel is typically placed at the top of the slope and secured to the top horizontal support cable. The mesh panel is then carefully draped down onto the slope.

Efforts to maximize mesh conformance with the slope surface are rewarded with increased resistance to dislodgment and fall of blocks and improved visual blending of the mesh with the slope (Figure 16-19). Walking down the mesh and pushing it into concavities and providing additional mesh for incidental overlapping are the primary means of achieving close slope conformance.

(a)

(b)

FIGURE 16-17
Comparison of vegetation cover of double-twisted wire mesh drapery (*a*) immediately after installation and (*b*) 3 years later (photographs taken from slightly different locations but of same slope).
(PHOTOGRAPHS COURTESY OF THE CALIFORNIA DEPARTMENT OF TRANSPORTATION.)

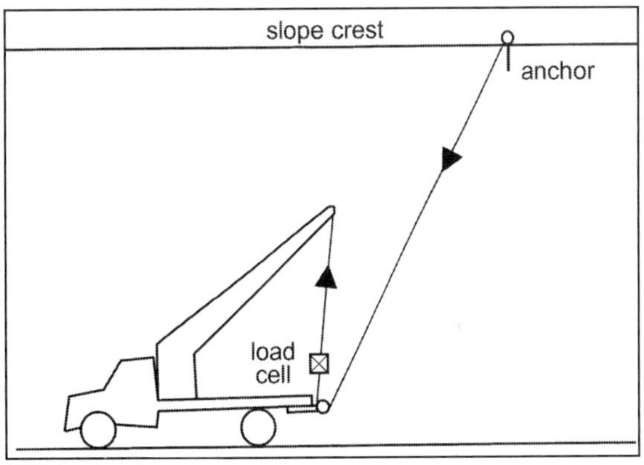

FIGURE 16-18
Lateral load test might better represent actual loading conditions and be simpler to perform than vertical pullout test (Muhunthan et al. 2005a).

FIGURE 16-19 Good construction practices and supervision of colorized, double-twisted hexagonal mesh installation were highly successful in achieving close conformance of mesh to slope. (PHOTOGRAPH COURTESY OF THE CALIFORNIA DEPARTMENT OF TRANSPORTATION.)

2.2.6 Maintenance

For most slope conditions, the design methodology should result in an installation requiring minimal maintenance over the design life of the system. However, maintenance is to be expected under any condition; therefore, all systems should be periodically inspected. Where snow load is anticipated, it is prudent to inspect the anchor and support ropes periodically, particularly after a heavy snow year. Subtle evidence of damage that potentially requires maintenance of the drapery system might be deformation or bulging of the fabric. More direct evidence of potentially required maintenance is large material accumulation, torn seams, and highly loaded anchors and cables.

The dominant maintenance concerns include damage from above-design-event rockfalls and excessive debris accumulation. Often these conditions do not directly load the system but over time can stress the system or cause loss of function. Above-design-event rockfalls can puncture the mesh or slice open intersecting connections of wire or cable. Scree and small-volume rockfalls can accumulate at changes in topography, where the mesh is overlapped, or at cable infrastructure intersections. Small debris slides can also tear the mesh or separate seams.

The standard maintenance strategy is to remove or pass the debris under the mesh to the bottom of the installation. When debris commonly accumulates near the bottom, the mesh can be lifted or pulled away from the slope to remove the material. It is common practice to lift the mesh from several locations so as not to tear it. Some installations include a horizontal bottom cable or cables at other constriction points, laced to the outside of the mesh, to aid in lifting the mesh. When rocks have accumulated farther up the slope, climbers are sometimes needed to shake or pry free the trapped debris. In some cases large rocks can be cut from under the fabric. When the mesh is torn or punctured, effective repair entails patching the fabric with similar materials secured with lacing wire or cables or high-tensile-steel fasteners.

The growth of vegetation on an actively eroding soil slope is generally desirable to reduce erosion and is commonly encouraged in drapery installations. Vegetation growing into the mesh can be and often is a positive attribute. In some cases, revegetation has stabilized the slope surface, resulting in greatly reduced maintenance. However, it is important to consider and monitor for potential conflicts between the engineered system and vegetation. The growth of shrubs and trees through the mesh can result in damage to drapery systems. Toppling of trees that have grown through the mesh can cause localized failures. In addition, creeping of underlying slope materials, common in some surficial deposits, can induce unanticipated load on the anchors if excessive intertwining of shrubby vegetation with the mesh has occurred. In these cases, some management of vegetation growth through the mesh may be required for long-term performance of the system.

3. SECURED DRAPERY

Secured drapery, which incorporates patterned ground anchors within the field of the mesh, is commonly referred to as *anchored mesh, pinned mesh*, or *reinforced netting*. A primary design objective of secured drapery is to retain rockfalls within the mesh system in cases where insufficient catchment or storage area is available at the base of the slope. These systems have been widely used for the past several decades in Europe, where many roads in steep terrain have narrow shoulders and lack catchment ditches. Other desirable attributes of secured drapery are the potential to tension the mesh with closely spaced anchors thereby producing some active slope stabilization, and improved slope conformance of the mesh. The latter attribute can offer benefits over unsecured mesh by reducing slope erosion, promoting revegetation, and improving aesthetic values. The success of system tensioning and slope conformance, however, is highly dependent on the slope and geological conditions, anchor spacing, engineering characteristics of the mesh, and quality of the construction.

3.1 Design Practice

Originally, secured drapery sought only to restrain falling rocks on the slope face and was considered to provide only passive restraint, mobilizing the resistance capacity of the mesh as rockfall debris accumulated. This early design practice primarily developed around the use of lightweight mesh (i.e., chain-link and double-twisted wire mesh). Since the engineering characteristics and loading conditions of the meshes initially were not quantified, design of the interior anchors and cables relied on engineering judgment. In the limited North American installations with which the authors are familiar, anchors and interior ropes were widely spaced (about 10 m or 30 ft). Some of these early installations experienced significant mesh failures when excess debris accumulated behind the mesh, while others have continued to function well and have required minimal or no maintenance.

In the late 1980s, the manufacturer of double-twisted wire mesh, Maccaferri, published a technical manual that provided engineering-based design guidance for secured drapery (Agostini et al. 1988). This work involved characterizing the engineering properties of both chain-link and double-twisted mesh through tensile and puncture tests. Supported by the performance experience from numerous installations, they recommended an anchor for every 15 to 30 m^2 (160 to 320 ft^2) of mesh. Anchors presented in the manual were obviously of low capacity [<50 kN (11,240 lbf)], although no minimum capacity was specified. Strengthening could be achieved by the addition of a diagonal grid of wire ropes to the outside of the mesh, secured to the patterned anchors.

This design approach has been widely used in Italy, where geological conditions produce many rock slopes that generate smaller rock blocks, typically smaller than 0.5 m (18 in.). Many of these installations have been constructed with a 3- by 3-m (10- by 10-ft) diagonal array of wire ropes. The ropes define a series of relatively small tributary areas, or "cells," that are suitable for the type of mesh used and the typical size of the rock fragments. The ropes are most often installed in a diamond pattern, because this offers efficiencies during installation and provides two cables at the bottom of each cell with a decreasing volume toward the cell base. The configuration not only promotes rapid bulking of loose material but also is less likely to develop excessive sag, or "belly," of debris trapped behind the mesh. The restricted base area also limits the dilation of rocks under the mesh and their detachment (Figure 16-20).

In the past decade or so, manufacturers and designers have sought to impart an active or stabilizing resistance to the slope through tensioning of the mesh. This design approach has required research aimed at quantifying the loading conditions and system requirements, which has led to greater anchor capacities and the need for more robust mesh. Around 2000, Geobrugg introduced a high-tensile-steel wire mesh, TECCO, and proprietary design methodology (RUVOLUM) for stabilization of approximately the outer 1.5 m (5 ft) of the slope (Rüegger et al. 2000). Geobrugg introduced another mesh and design methodology, SPIDER, in the past several years as a

FIGURE 16-20 Secured drapery (double-twisted wire mesh) with interior grid of wire ropes installed on slope that produces relatively small rock blocks. Debris accumulation within several lower cells will eventually require cleanout to avoid rupture of mesh. (PHOTOGRAPH COURTESY OF A. BLOOM, MACCAFERRI NORTH AMERICA.)

proprietary alternative to cable lashing (discussed in Chapter 13). Concurrently, full-scale testing was undertaken at the Polytechnic University of Turin in Italy (Bertolo et al. 2009) to evaluate the behavior of a secured drapery using both double-twisted wire mesh and cable nets. A hydraulic ram fastened with a 1.5-m-diameter disk was used to load the mesh, and stresses and deflections were then measured within the anchors, wire ropes, and meshes (Figure 16-21). From these studies, design guidelines for secured drapery are being developed for various meshes (e.g., double-twisted wire mesh, wire rope, and ring nets) that would include serviceability (system deflection) and limit state (failure) criteria.

Bertolo et al. (2009) acknowledged the continuum between the passive (protection) and active (stabilization) capabilities of secured drapery. The configuration of the slope and the stiffness of the mesh largely govern the ability to tension the mesh and impart an active stabilizing force to the slope. A mesh must first be in contact with the slope and be sufficiently stiff or stretched for the design loading condition to impart an active force. If these two conditions are not met, slope movement can occur, and little to no slope stabilization is achieved.

FIGURE 16-21 Field testing of secured cable net performed by Bertolo et al. (2009). (PHOTOGRAPH COURTESY OF T. C. BADGER.)

Where the mesh covers slope concavities, it can be difficult to achieve mesh contact with the slope and even more difficult to tension the mesh. On uniform slopes, anchors are often countersunk in an attempt to tension the mesh. Active force is most easily imparted over slope convexities. Recent inspections of secured drapery installations in California and Washington State found, despite concerted construction efforts, that the mesh commonly was not in contact with the slope or could be easily lifted from the slope (J. Duffy and M. Fish, personal communications, 2011). These findings demonstrated that large portions of the installations were providing only passive restraint for rockfalls and were not actively stabilizing the slope face as intended.

3.2 Construction

As with unsecured drapery, installation and inspection procedures are important factors influencing how well a secured drapery system achieves the design objectives of function, aesthetics, and cost-effectiveness. Obtaining good mesh-to-ground contact is important and is essential if imparting an active force to the slope is a design objective. This is largely accomplished with patterned ground anchors. However, on older slopes with uneven and irregular topography, some slope preparation is helpful in achieving suitable ground contact. This is often accomplished by hand or mechanical scaling and the removal of prominent rock outcrops with trim blasting. Pruning of vegetation on the slope also is done to facilitate good ground contact. The slope preparation work is often paid as a separate work item, commonly defined as clearing or slope scaling. Generally, it is desirable to have anchors and fabric as separate bid items because doing so simplifies payment for field adjustments.

Anchors can be installed before or after the mesh has been placed on the slope. This is generally at the option of the contractor. Designs often include seed-bearing erosion control blankets, which would typically be placed after the anchors are installed and tested and before placement of the mesh (Figure 16-22). For mesh with small openings, some suppliers provide devices that spread the mesh openings to facilitate drilling through the mesh to install anchors. Care must be taken to avoid damaging the mesh during drilling or anchor testing. When anchors need to be countersunk, installation of ground anchors before mesh placement is typically necessary (Figure 16-23).

Anchor capacities should be verified with proof testing by using a vertical pullout test of at least 110% of the design load. This is best done before mesh placement. A typical secured drapery installation may require tens to hundreds of anchors, and anchor testing on the slope face is generally difficult and time-consuming (Figure 16-24). If good inspection can be ensured, 5% to 10% are then typically specified for testing. Verification of additional anchors may then be undertaken by using a torque wrench calibrated from proof-tested anchors.

As discussed in Section 2.2.5 for unsecured drapery systems, a staging area near the slope is generally required when a secured drapery system is installed. The staging area allows for the efficient layout and seaming of mesh panels and for the staging of the lifting and placement of the mesh. A staging area is essential when robust fabrics requiring mesh backing are used because it allows the mesh backing to be attached while the mesh lies on the ground. The selection of the staging area will be influenced by the method of placement (boom truck, crane, or helicopter) and the size of the installation. As discussed in Section 2.2.5, a helicopter is usually the fastest and most cost-effective method for lifting and placing the mesh at sites where large drapery installations are required. However, when a helicopter is used for mesh placement, traffic control and proximity to aerial utility lines must be considered, and an emergency landing area must be provided as part of the staging operations. Excessive stretching of the mesh panel when it is lifted is less likely to occur, and improved drapery placement and seaming efficiency typically result, when the bottom of the lifted panel is initially placed at the top of the slope and secured to the top horizontal support cable before the mesh panel is carefully draped down onto the slope.

As with unsecured drapery, efforts to maximize mesh conformance with the slope surface for secured drapery installations are rewarded with increased resistance to the dislodgment and fall of blocks and improved visual blending of the mesh with the slope (Figure 16-16). Walking down the mesh and pushing it into concavities and providing additional mesh for incidental overlapping are primary ways to achieve close conformance with the slope.

FIGURE 16-22 Installation of ground anchors and erosion control blankets before mesh placement. (PHOTOGRAPH COURTESY OF THE CALIFORNIA DEPARTMENT OF TRANSPORTATION.)

FIGURE 16-23 Countersinking ground anchors after TECCO mesh installation tensions mesh. (PHOTOGRAPH COURTESY OF THE CALIFORNIA DEPARTMENT OF TRANSPORTATION.)

FIGURE 16-24 Proof testing ground anchor for secured drapery installation. (PHOTOGRAPH COURTESY OF THE CALIFORNIA DEPARTMENT OF TRANSPORTATION.)

3.3 Maintenance

Secured drapery systems require periodic inspection and maintenance. A successful design results in an installation requiring minimal maintenance over the design life of the system. Where significant snow loads are anticipated, it is prudent to inspect the anchors and support ropes, particularly after a heavy snow year. Subtle evidence of needed maintenance might be the deformation or bulging of the fabric. More direct evidence of potential maintenance is excessive debris accumulation, torn seams, and heavily loaded anchors and cables.

As with unsecured drapery systems (see Section 2.2.6), the dominant maintenance concerns for secured drapery include damage from large, above-design-event rockfalls and excessive debris accumulation. Often, these conditions do not directly load the system but over time can stress the system or cause loss of function by some components. Above-design-event rockfalls can puncture the mesh or slice open intersecting connections of wire or cable. Scree and small-volume rockfalls can accumulate around the ground anchors, where the mesh is overlapped, or at cable intersections. Small debris slides can also tear the mesh or separate seams.

The standard maintenance strategy is to remove or pass the debris under the mesh to the bottom of the installation. In theory, the nut and washer on the ground anchor are removed and the mesh is lifted from the slope to remove the material. Common practice often finds the anchor head buried, requiring the mesh to be cut to clear the accumulated material and then patched after debris removal. When the mesh is torn, punctured, or cut, effective repair entails patching the fabric with similar materials secured with lacing wire or cables or high-tensile-steel fasteners. Because of the steepness of the slope, climbers often must perform this work.

Secured drapery installations generally benefit from the growth of vegetation on an actively eroding soil slope. Vegetation growing into the mesh helps stabilize the slope surface, resulting in greatly reduced maintenance. It is commonly encouraged by placing seed-impregnated erosion control blankets under the mesh (Figure 16-22). However, monitoring for potential conflicts between the engineered system and vegetation is important. Growth of shrubs and trees through the mesh has resulted in damage to some installations. When trees that have grown through the mesh topple, or if excessively intertwined shrubby vegetation connected with the mesh suffers creep of underlying slope materials, as is common in some surficial deposits, unanticipated loads may be applied to the drapery system anchors, or localized destruction of the mesh may occur. In these cases, management of vegetation growth through the mesh may be required to ensure long-term performance of the system.

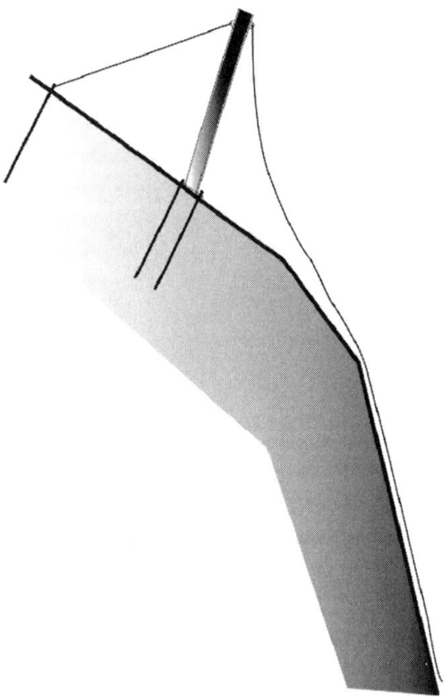

FIGURE 16-25 Hybrid draperies elevate top of mesh with posts or are suspended from end-anchored cable across ravine to intercept rockfalls sourced upslope of as well as beneath lower portion of drapery.
(DIAGRAM BY T. C. BADGER.)

4. HYBRID DRAPERY

Hybrid drapery, also referred to as an *attenuator*, a *hybrid barrier*, or *hanging nets*, provides to a standard unsecured drapery the added benefit of intercepting rockfalls sourced upslope of the installation by elevating the upper part of the drapery off the surface (Figure 16-25). This modification exposes the upper portion of the drapery to impacts orthogonal to the plane of the mesh. Rockfall is then channeled under the drapery and its trajectory confined within the plane of the mesh. Since the mesh is not restrained, it has the ability to deform and attenuate the impact energy and then control the trajectory into a suitable containment area at the bottom of the installation. Hybrid systems have been effectively applied where natural topography channels rockfall trajectories (Figure 16-26) or along abrupt slope convexities.

Although the capacities of these systems have been investigated only recently, hybridization has been incorporated into numerous draperies over the past 20 years. In the past, the application and design basis for hybrid drapery has relied largely on

FIGURE 16-26
Suspended mesh (cable nets) can be highly effective at intercepting rockfalls originating from ravine.
(PHOTOGRAPH COURTESY OF THE CALIFORNIA DEPARTMENT OF TRANSPORTATION.)

(a)

(b)

FIGURE 16-27
Two standard hybrid drapery designs: (*a*) design used by British Columbia Ministry of Transportation to intercept low-energy rockfalls [≤3 kJ (≤1 ft-ton)] sourced from upslope; top of double-twisted hexagonal mesh is raised by extending 44-mm (1.75-in.) diameter ground anchors 1 m (3 ft) above ground, and tieback ropes provide added rigidity; and (*b*) design by California Department of Transportation uses cable nets and friction brakes on tieback ropes to address moderate energy impacts [<500 kJ (<185 ft-tons)].
(PHOTOGRAPH *a* COURTESY OF THE BRITISH COLUMBIA MINISTRY OF TRANSPORTATION AND PHOTOGRAPH *b* COURTESY OF THE CALIFORNIA DEPARTMENT OF TRANSPORTATION.)

empiricism and judgment. Recently, knowledge of hybrid systems has been significantly advanced through performance reviews of existing systems, field testing programs, and numerical modeling (Muhunthan and Radhakrishnan 2007; Badger et al. 2008; Arndt et al. 2009; Glover et al. 2010; Badger et al. 2011; Von Boetticher et al. 2011).

4.1 Field Performance

The performance of existing hybrid drapery installations, many now a decade or more old, has generally been very favorable (Badger 2007; Duffy 2007; Badger et al. 2008). Some transportation organizations even have standard designs for the routine use of these systems (Figure 16-27). In addition to their overall effectiveness in containing rockfalls, the systems typically have required little maintenance. Hybrid drapery systems, however, are not without problems. Debris accumulation and resultant damage have occurred when the mesh was restrained or where its installation over abrupt convexities caused debris entrapment. Puncturing of the mesh has been observed near the top perimeter, where the mesh is least flexible, and rock impacts have also damaged interior support posts. Long narrow installations have also not remained in position when affected by rockfalls.

These experiences have led to refinements in the design, detailing, and installation of hybrid drapery. Effectiveness is generally optimized by minimizing restraint of the fabric, directing impacts into the "sweet spot" of the installation and away from perimeter restraints, limiting the number of vulnerable interior post supports, and minimizing the potential to accumulate debris downslope of the impact zone. More robust mesh (cable nets) has been used in the impact zone to

FIGURE 16-28
Impact of 5,000-kg (11,000-lb) block shows large deflection of ring net fabric around block but over relatively small area of mesh.
(PHOTOGRAPH COURTESY OF IGOR PARAMASSI.)

reduce the potential for puncturing and damage when lighter-weight mesh (double-twisted hexagonal mesh) is used for the tail portion.

4.2 Quantitative Evaluations

Recently, full-scale field tests have been undertaken to quantify the capacities and limitations of hybrid drapery and provide substantive data for its design. The first field tests, sponsored by IGOR Paramassi, an Italian firm, used five concrete "rocks" ranging in weight from 1 to 5 tonnes (2,200 to 11,000 lb) on a 45° slope. They produced estimated impact velocities of up to 14.3 m/s (47 ft/s), resulting in impact energies of up to about 500 kJ (184 ft-tons) (Sassudelli et al. 2007; P. C. Ingraham, personal communication, 2007 and 2008) (Figure 16-28). Several tests included measurements of infrastructure forces using load cells. All rocks were successfully stopped in a flat containment area within 3 m (10 ft) of the base of the slope. Despite several impacts on support ropes and a post, no maintenance was required; the system remained completely functional during the entire series of tests.

A second testing program, conducted by the Colorado Department of Transportation, investigated the effectiveness of different meshes in slowing, but not necessarily fully arresting, rolling rocks (Arndt et al. 2009). The application was referred to as an *attenuator*.

A long, high slope near Georgetown, Colorado, poses significant rockfall hazards to Interstate 70. Access for maintenance of any rockfall mitigation installations is difficult at this location. Thus, a series of attenuators was proposed to control the progressive descent of rockfalls into a suitable containment area adjacent to the highway at the base of the slope. However, information to guide the design and installation of such an application was insufficient. A number of initial installations were rapidly destroyed by rockfalls. Accordingly, a series of full-scale field tests was undertaken at a nearby location (Arndt et al. 2009).

These tests used a crane to drop onto a ramp concrete "rocks" that struck the hybrid barrier with up to 500 kJ (184 ft-tons) of energy (Figure 16-29). Important factors influencing the effectiveness of these systems included the weight of the mesh panel system, the mesh aperture, the length of the mesh extending downslope (tail length), and the resilience of the system to deformation after impact. Heavier tail sections with smaller mesh apertures required shorter tail lengths. It was also noted that, while lighter mesh systems may be effective for applications in which lower-energy impacts [less than 150 kJ (55 ft-tons)] are expected, the longer tail length may make the impact area more rigid and thus more vulnerable to damage during impact and debris and snow accumulation.

FIGURE 16-29
Colorado Department of Transportation testing of hybrid barriers to evaluate different meshes and tail lengths.
(PHOTOGRAPH COURTESY OF THE COLORADO DEPARTMENT OF TRANSPORTATION.)

Verification field tests on the final installation subsequently proved the findings of the previous field tests and the effectiveness of these systems when used in series, even for rockfalls involving multiple blocks. The verification tests (Arndt et al. 2009) further demonstrated the importance of the following:

1. Redundancy produced by a series of these installations when the occasional block bounced over an attenuator,
2. The limiting of the time the tail is lifted off the ground when rockfall events involved multiple blocks, and
3. Optimization of tail lengths to ensure self-cleaning attributes.

Muhunthan and Radhakrishnan (2007) carried out finite element analyses on hybrid cable-net draperies to evaluate loads on the support posts and fabric deformation for impact energies up to 200 kJ (74 ft-tons) and for a range of slope orientations, slope–mesh interface friction values, and system dimensions. Static benchmark modeling calibrated the model with puncture tests performed on 3- by 3-m (10- by 10-ft) panels of cable nets. In addition to calculating infrastructure loads, the modeling results showed comparable envelopment of the rock by the fabric and dramatic attenuation of the horizontal trajectories, as was observed during subsequent field-testing programs.

A third field-testing program undertaken by the Swiss Federal Institute for Forest, Snow, and Landscape Research in collaboration with Geobrugg examined the attenuation capabilities of a hybrid system with a tail section 17.5 m (57 ft) wide by 50 m (164 ft) long (Figure 16-30). High-speed video allowed estimation of impact and exit velocities as well as the trajectory of the rock. Preliminary results obtained from load cells on the supporting infrastructure during low-energy impacts [<150 kJ (55 ft-tons)] demonstrated low infrastructure loading and minimal system damage, similar to results from previous field tests and modeling experience. Subsequent evaluation by Glover et al. (2012) estimated a 50% to 71% reduction in the kinetic energy of rocks exiting the hybrid system as compared with the kinetic energy predicted in the absence of the installation.

FIGURE 16-30
Field test of hybrid drapery (SPIDER nets) completed in 2010 by Swiss Federal Institute for Forest, Snow, and Landscape Research.
(PHOTOGRAPH COURTESY OF THE SWISS FEDERAL INSTITUTE FOR FOREST, SNOW, AND LANDSCAPE RESEARCH.)

Both field performance of existing hybrid draperies and the limited quantitative data available to date demonstrate the potential effectiveness of these systems in withstanding moderate impact energies while sustaining little damage and requiring little maintenance. Having a suitable containment area or structure between the installation and the transportation facility is key to these systems' ability to provide protection from rockfalls.

ACKNOWLEDGMENTS

The authors are grateful for—and the chapter benefited significantly from—discussions with and reviews by Ben Arndt, Matthias Denk, Mike Dowdle, James Glover, Peter Ingraham, Howard Ingram, Daniel Journeaux, Steve Lowell, Norm Norrish, Ty Ortiz, and Larry Pierson.

REFERENCES

Agostini, R., P. Mazzalai, and A. Papetti. 1988. *Hexagonal Wire Mesh for Rock-Fall and Slope Protection*. Officine Maccaferri S.p.a., Bologna, Italy.

Andrew, R. D., R. Bartingale, and H. Hume. 2011. *Context Sensitive Rock Slope Design Solutions*. Publication FHWA-CFL/TD-11-002. Federal Highway Administration, Central Federal Lands

Highway Division, Lakewood, Colo. http://www.cflhd.gov/programs/techDevelopment/geotech/css/.

Arndt, B., T. Ortiz, and A. K. Turner. 2009. *Transportation Research E-Circular E-C141: Colorado's Full-Scale Field Testing of Rockfall Attenuator System*. Transportation Research Board of the National Academies, Washington, D.C. 130 pp. http://onlinepubs.trb.org/onlinepubs/circulars/ec141.pdf

Badger, T. C. 2007. Suspended Draped Rockfall Protection: An Overview of System Performance in Washington. *Program with Abstracts, 50th Annual Meeting of the Association of Environmental and Engineering Geologists*, Los Angeles, Calif., p. 68.

Badger, T. C., B. Dobrev, M. Anderson, H. Radhakrishnan, and B. Muhunthan. 2011. Structural Design of a Hybrid Drapery System. Proc., Interdisciplinary Workshop on Rockfall Protection 2011 (M. Mölk, S. Melzner, T. Sausgruber, and T. Tartarotti, eds.), Innsbruck, Austria, pp. 3–6.

Badger, T. C., J. D. Duffy, F. Sassudelli, P. C. Ingraham, P. Perreault, B. Muhunthan, H. Radhakrishnan, O. S. Bursi, M. Molinari, and E. Castelli. 2008. Hybrid Barrier Systems for Rockfall Protection. *Proc., Interdisciplinary Workshop on Rockfall Protection* (A. Volkwein, V. Labiouse, and K. Schellenberg, eds.), Morschach, Switzerland, pp. 10–12.

Bertolo, P., C. Oggeri, and D. Peila. 2009. Full-Scale Testing of Draped Nets for Rock Fall Protection. *Canadian Geotechnical Journal*, Vol. 46, No. 3, pp. 306–317.

Duffy, J. D. 2007. Hybrid Rockfall Systems. *Program with Abstracts, 50th Annual Meeting of the Association of Environmental and Engineering Geologists*, Los Angeles, Calif., p. 80.

Glover, J., M. Denk, F. Bourrier, A. Volkwein, and W. Gerber. 2012. Measuring the Kinetic Energy Dissipation Effects of Rock Fall Attenuating Systems with Video Analysis. Proc., 12th Congress INTERPRAVENT (G. Koboltschnig and J. Hübl, eds.), Grenoble, France, pp. 151–160.

Glover, J., A. Volkwein, F. Dufour, M. Denk, and A. Roth. 2010 Rockfall Attenuator and Hybrid Drape Systems: Design and Testing Considerations. Third Euro-Mediterranean Symposium on Advances in Geomaterials and Structures, 3rd ed. (F. Darve, I. Doghri, R. El Fatmi, H. Hassis, and H. Zenzri, eds.), Djerba, Tunisia, pp. 379–384.

Muhunthan, B., and H. Radhakrishnan. 2007. *Finite Element Analysis of Hybrid Barrier for Rock Fall Slope Protection*. Final Report. Department of Civil and Environmental Engineering, Washington State University, Pullman.

Muhunthan, B., S. Shu, N. Sasiharan, O. A. Hattamleh, T. C. Badger, S. M. Lowell, and J. D. Duffy. 2005a. *Analysis and Design of Wire Mesh/Cable Net Slope Protection*. TRAC Report WA-RD 612.1. Washington State Transportation Center, Seattle.

Muhunthan, B., S. Shu, N. Sasiharan, O. A. Hattamleh, T. C. Badger, S. M. Lowell, and J. D. Duffy. 2005b. *Design Guidelines for Wire Mesh/Cable Net Slope Protection*. TRAC Report WA-RD 612.2. Washington State Transportation Center, Seattle.

Rüegger, R., D. Flum, and B. Haller. 2000. High-Performance Steel Wire Mesh for Surface Protection in Combination with Nails and Anchors. *Proc., 2nd Colloquium on Construction in Soil and Rock*, Technical Academy of Esslingen, Esslingen, Germany, pp. 10–46.

Sassudelli, F., P. Perreault, and P. C. Ingraham. 2007. Recent Rockfall Barrier Testing in Italy: Structural Component Tests and Dynamic Testing of Barrier Kits and Hybrid Barrier/Drape Systems in the Framework of European Technical Guidelines. *Program with Abstracts, 50th Annual Meeting of the Association of Environmental and Engineering Geologists*, Los Angeles, Calif., p. 122.

Von Boetticher, A., J. Glover, A. Volkwein, and M. Denk. 2011. Modeling Flexible Wire Netting Applied to Rock Fall Attenuating Systems. Proc., Interdisciplinary Workshop on Rockfall Protection 2011 (M. Mölk, S. Melzner, T. Sausgruber, and T. Tartarotti, eds.), Innsbruck, Austria, pp. 71–72.

PART 4

ROCKFALL MAINTENANCE AND MANAGEMENT PROGRAMS

Chapter 17

THOMAS C. BADGER

MAINTENANCE, MONITORING, AND RESPONSE

1. INTRODUCTION

Long-term mitigation of rockfall hazards at a single location or along an entire transportation corridor may not be achievable within a reasonable time frame for financial, environmental, or other reasons. However, a variety of maintenance or monitoring activities may provide a significant reduction of the risk associated with rockfall and improve public safety by reducing the exposure of road users to hazards during critical periods of slope instability. Risk-reduction measures include cautionary signs, observational or instrumentation monitoring, warning systems, regular maintenance patrols, and even preemptive closure of the facility. Interim stabilization or protection measures are also valuable tools for reducing rockfall hazards at a specific location or for a short duration. These risk-reduction measures principally improve safety by increasing awareness of a hazard or by reducing exposure to hazards during critical periods of slope instability.

2. MAINTENANCE PATROLS AND RESPONSE PLANS

Keeping the traveled way clear of hazardous debris is one of the most basic maintenance functions along transportation facilities. At locations where rockfall activity is common or during periods of heightened rockfall activity, regular maintenance patrols can be scheduled to clear fallen debris and routinely inspect active source areas. The patrol frequency can range from once to a few times per regular daytime shift during periods of low rockfall activity to 24 h a day, 7 days a week during periods of high activity. Increased patrol frequency is routinely implemented in transportation corridors through mountainous terrain during wet weather or freeze–thaw periods, in response to severe weather episodes, or following a wildfire event.

Maintenance efforts are part of an ideal clear response plan that

- Defines roles and activities for early responders,
- Requires assessment of site safety and the need for remedial measures by qualified personnel,
- Implements remedial measures as necessary, and
- Determines safe resumption of use following a road closure.

Even in cases in which potential consequences are low, a response plan often implicitly exists within an organization and is well known and followed by first responders. Detailed plans may not be required for guiding routine responses at areas with frequent minor rockfalls; however, the volume and frequency of such rockfall activity should be documented. A well-defined and tested response plan is critical whenever higher hazard consequences or risks to the facility or its users warrant more complex warning systems or rockfalls have greater effects on users, for instance, whenever preemptive closure of the road is necessary.

3. INTERIM REMEDIAL MEASURES

Many of the mitigation measures discussed in preceding chapters of this report may be applied as interim measures to provide some risk reduction of an urgent rockfall hazard. Interim remedial measures may not represent the most desirable, complete, or effective mitigation alternative, and a shorter service life can be expected. However, significant and rapid risk reduction can often be achieved with temporary remedial measures; frequently, these may require little expenditure.

3.1 Scaling

Scaling (removal of loose rock) alone is commonly performed as an interim slope treatment and might focus on a particular unstable feature or may be applied broadly over a large slope area. Because earthen materials making up steep slopes are often prone to rapid degradation through weathering and are exposed to severe climatic conditions and gravity, scaling generally does not achieve long-term hazard mitigation. Depending on geological and climatic conditions, effective improvement of only 5 years or less may be realized. For longer-term objectives, additional, more permanent measures are usually applied, such as wire mesh drapery or rock anchors. A few transportation agencies in North America perform scaling by using specially trained maintenance personnel or execute small scaling contracts, either in a reactive manner or proactively as part of a funded slope-maintenance program.

FIGURE 17-1 Plowed snow on Interstate 90 near Snoqualmie Pass in Cascade Mountains, Washington State, reshaped to improve effectiveness of catchment area to contain frequent rockfalls that occur in spring before melt-out of ditches. (PHOTOGRAPH COURTESY OF THE WASHINGTON STATE DEPARTMENT OF TRANSPORTATION.)

3.2 Catchment Area Improvements

Debris and snow accumulation in the available catchment area can dramatically reduce its effectiveness in retaining rockfalls, and in some cases accumulated debris can worsen rockfall trajectories and increase rollout distances extending from the toe of the slope. Recognition of this phenomenon has shifted current rock slope design practice in some regions away from the routine use of midslope benches because often these benches are not maintained, nor is access provided for cleaning equipment.

To optimize the functionality of available catchment areas, ditch cleaning and reshaping are critically important maintenance functions that should be routinely performed. In areas where snowplowing and snow storage seasonally reduce the extent of catchment areas, reshaping the snow to form berms or steepened faces toward the slope can improve catchment effectiveness (Figure 17-1).

In some cases, widening or deepening available catchment areas may be feasible and may greatly reduce the likelihood of debris reaching the facility. Catchment areas can be widened by realigning or moving the facility away from the hazard or by steepening the base of the slope. Deepening a ditch section, especially when a steep backface can be incorporated, such as with a concrete block or barrier or a series of gabion baskets, may also significantly improve the effectiveness of a catchment area. Photographs illustrating these ditch modifications and the various barrier systems are provided in Chapter 14 of this report.

Beyond the extension of catchment areas, some of the barriers and other mitigation measures may provide appropriate interim rockfall protection. Concrete barriers are probably the most widely employed interim measure for roads and highways. Earthen berms, especially those using available accumulated debris, are inexpensive, rapid to construct, and can be highly effective in containing rockfall. Berms shaped with the steepest possible backslope are the most effective.

3.3 Movable Protection Systems

In recent years, modifications to proprietary rockfall protection fences have made them movable and rapidly deployable (Figure 17-2). Several state departments of transportation (DOTs) own these movable protection systems, which can typically be transported and installed within 24 to 48 h. For example, the Washington State DOT maintains a cache of cable net panels that can be readily suspended from a crane for emergent scaling operations or draped on a slope in the event of an emergent hazard.

4. WARNING SYSTEMS

Warning systems and other preemptive management actions provide the opportunity for risk reduction at locations where there is knowledge of frequent, pending, or existing slope failures. The principal functions of warning systems are any one or all three of the following:

- To increase awareness and urge caution through the hazardous zone,
- To initiate maintenance or emergency response, and
- To restrict access or exposure to the hazard.

The most basic, familiar, and widely used warning systems are rockfall warning signs. For decades railroads have used triggering wires that, if broken, send an electrical signal notifying that a potential impediment exists on the track. With more recent advances in instrumentation, remote sensing, and signal transmission, remote real-time monitoring is being increasingly used to warn of hazardous conditions along roads as well as railways and to protect buildings or other important facilities. However, increasing complexity results in greater equipment and personnel costs to install and maintain the system and often greater occurrence of false alarms. In spite of these potential problems, such instrumentation-based systems can

- Monitor movement or strain within a rock mass or along a slope,
- Incorporate triggering devices indicating recent failure,
- Forecast potentially adverse weather conditions,
- Notify emergency response personnel, and
- Close gates or turn on warning lights.

FIGURE 17-2
Use of movable rockfall barriers to provide protection for traffic when road was reduced from four to two lanes following large rockfall on Interstate 90 near Snoqualmie Pass in Cascade Mountains, Washington State. Barrier consists of proprietary rockfall fence mounted to specially fabricated, interconnected steel plates that are easily secured to pavement and provide full design capacity of fence to absorb rockfall impacts.
(PHOTOGRAPH COURTESY OF THE WASHINGTON STATE DEPARTMENT OF TRANSPORTATION.)

4.1 Warning Signs

Warning signs are widely used to alert road users of rockfall hazards (Figure 17-3). Their widespread use along public roads is largely due to governmentally mandated safety standards. Their simplicity, ease of installation, and lower cost provide obvious benefits over instrumentation-based warning systems. Besides indicating the presence of a hazard, the most useful static signs define the length of the hazardous zone and provide instructions to drivers, such as "No Stopping for the Next 200 m" or "Proceed with Extreme Caution."

Al-Kaisy (2006) evaluated the criteria for the installation and placement frequency of warning signs in North America. Current practice generally relies on engineering judgment consistent with the *Manual on Uniform Traffic Control Devices* (FHWA 2003). Al-Kaisy found that regular users of a facility grow accustomed to the presence of static warning signs, become complacent, or do not perceive the real hazard. Thus, the effectiveness of static warning signs for occasional hazards, such as "Deer Crossing," "Snow and Ice," or "Rockfall," may be limited and, unfortunately, "serious repercussions are likely to result from situations in which risk is present but not perceived" (Bean 1989). Enhanced devices, such as flashing lights and dynamic message boards, are increasingly being used with standard

FIGURE 17-3
Warning signs from areas with rockfall hazards: (*a*) Great Britain (Great Britain Department for Transport 2002), (*b*) Washington State, and (*c*) Bhutan.
(BHUTAN PHOTOGRAPH COURTESY OF P. CONNELLEY; WASHINGTON STATE PHOTOGRAPH COURTESY OF THE AUTHOR.)

signs to provide emphasis and to elicit more caution from drivers.

The principal function of a warning sign is to alert of a potential hazard. Warning signs offer only spatial information about a potential occurrence. They do not indicate when risk of the hazard is elevated nor when the hazard, such as a rockfall, might occur. Generally, only regular road users or experienced practitioners have insights on when the hazard risk may be elevated. For instance, regular travelers on a route may have experienced more frequent rockfall debris on the road during heavy rainfall and so take special precautions during those periods.

In addition to their public safety benefit, warning signs function to reduce the legal liability of the transportation agency that is incumbent with ownership. Although some level of user risk from slope hazards is inherent for many facilities sited in rugged terrain, placement of effective signage demonstrates both recognition of the hazard and proactive notification of the risk to the public. Through this notification, road users assume some of the risk by choosing to use the facility. Conversely, there is a potential for increased liability to the transportation agency if warning signs are not used along public roads when they are expected to be present and are required by current safety standards and design criteria (FHWA 2003).

4.2 Monitoring

Observational or instrumentation monitoring may be used to confirm whether a suspicious feature, such as a portion of a rock slope, is actively deforming; whether the movement is continuous or episodic; and if the process is transient, whether climatic or human-induced conditions influence the movement. Monitoring can also sometimes be used to constrain the timing of a pending failure. The confirmation and quantification of ongoing distress often provide critically important information about the urgency of restricting access to a portion of the facility, such as by lane closure; of requiring preemptive closure of the entire facility; or of performing remedial stabilization work to provide additional protection.

The implementation of a monitoring program requires evidence that a slope failure may be developing, such as identification of features indicative of slope distress, including

- Open tension or shear cracks, particularly those that are steeply oriented or surround a distinct rock mass;
- Recent and especially increased rockfall activity in a localized portion of a slope; and
- Confined zones of seepage or unusual changes (cessation or increase) in discharge flows.

These features are often first identified by those who frequent the site, such as area maintenance personnel. A subsequent inspection by a geotechnical rockfall specialist is usually conducted to evaluate the potential hazard and risk to adjacent facilities and workers and to provide recommendations for further action or remediation. These actions may include the establishment of a monitoring program that includes either direct observations or the installation of instruments or both. The principal objectives of a monitoring program are to

- Identify the extent of ongoing distress within a slope or mass,
- Quantify the rate of change, and
- Constrain the timing and extent of an impending failure.

Available techniques to monitor potential distress include observation, optical surveying, instrumentation, and remote sensing.

4.2.1 Observation

Visual observation may successfully identify a potential rockfall condition. Easily observable evidence of recent distress around open cracks might include fresh exposures of rock or soil or torn vegetation that once spanned an open crack. In most cases, frequent careful inspection is required to identify deteriorating stability. Visual inspection for changes within a slope or mass requires detailed records or photographs of preexisting conditions, and the rate of change is generally not determinable. Nevertheless, frequent observations are often a useful source of information. Access difficulties on steep slopes can inhibit observational data collection of the most critical areas within a distressed mass.

Photographic documentation is a useful and demonstrable data source, especially if scale and date or time can be provided within the photograph. Subsequent photographs can be compared to qualitatively monitor changes within a slope, such as a recent rockfall or opening of cracks. However, quantifying the rate of change and thus possibly predicting or constraining the timing of failure generally cannot be accurately or easily accomplished by photographic or other indirect observational methods.

An inexpensive and easily deployed observational method for detecting movement along or opening of a crack involves the use of survey stakes driven into an open crack (Brawner 1993). If stakes are properly located, changes in their position (either tilting or falling out) would imply movement. The stake should be fastened to a short wire and secured to the slope so that a fallen stake remains near the monitoring point.

Sometimes evidence of pending slope failures cannot be observed from below the slope, especially when natural slopes extend well above cut slopes. If possible, aerial reconnaissance of these areas should be performed to establish a baseline of overall slope conditions and performance. The presence of dense vegetation may limit the usefulness of aerial surveys.

In many cases, visual observation alone is insufficient to quantify slope distress accurately and repeatably for requisite decisions about risk to critical facilities or public safety.

4.2.2 Optical Surveying

High-precision optical survey methods have been widely used as a more quantitative form of observational monitoring. Total-station instruments precisely measure angles and distance from the instrument to a reflective target (prism). In recent years, instruments have been developed that do not require the use of a reflective prism. Although reflectorless instruments are widely used, accuracy, range, and, most important, repeatability are greatly improved when high-quality survey prisms fixed to the rock slope are used, especially in poor weather. A sufficient quantity of critically located points is needed to characterize the extent of movement adequately, provide redundancy, develop a range of monitoring error, and ensure confidence in the data. The case history of the Randle slide, located in Washington State (Box 17-1), provides a good example of the value of optical surveying to monitor and ultimately predict an impending slope failure.

4.2.3 Instrumentation

If properly maintained, instrumentation can monitor deformation and provide significantly more data of higher quality than those obtained with observational methods. As discussed in Chapter 10 of this report, a variety of instrumentation is available to accurately detect and monitor movement, stress, and groundwater. Automated data collection is now routinely deployed with many instrumentation systems. Furthermore, advancements in telecommunications technology allow for greater data-transmission ranges and make remote, near-real-time monitoring much more economical and reliable, and thus feasible for many more situations. A typical recent instrumentation monitoring activity that avoided the use of much more expensive remediation measures while providing a better understanding of the potential risk is described in Box 17-2.

4.2.4 Remote Sensing

Stereographic viewing of paired photographs of rock slopes has been used for some time to compare relative changes in a rock slope. Recent computing advances in three-dimensional modeling of pairs of digital photographs, such as Split-FX and Sirovision, allow for deformational monitoring of rock slopes by comparing successive pairs of photographs. However, quantifying the rate of change and thus possibly predicting or constraining the timing of failure are not best accomplished by photographic or other observational methods.

More recently, high-density surveying using both radar interferometry and light detection and ranging (LiDAR) technology has been used to accurately and repeatably monitor activity and extent and make comparative analyses of slopes prone to rockfall or slopes that are actively failing (Casagli et al. 2004; Rosser et al. 2007). An excellent case history of the use of slope stability radar (SSR) for monitoring the Ferguson slide in central California is presented in Box 17-3.

4.3 Automated Warning Systems

Commonly employed instrumentation-based warning systems generally function by either signal interruption or deformation measurement. A greater level of sophistication and automation is provided when critical thresholds, or triggering events, are defined and incorporated within the instrumentation to develop an automated warning system. Ideally, a system will provide sufficient warning to avoid exposure to an actual rockfall.

The most widely used instrumentation-based warning systems are electrical fences erected by

Box 17-1

Monitoring Rockfall with Optical Surveying Methods

In November 1994, a landslide–debris flow of 100,000 m^3 (130,000 yd^3) occurred on the southwestern side of the Cascade Mountains in Washington State near the town of Randle. A section of U.S. Highway 12 was inundated by the debris. Figure 12-6 illustrates the situation in December 1994 just after a temporary bypass was opened following a 4-week road closure.

The failure extended a considerable distance above the road, and the debris, which included large rock blocks, traveled down a clearly defined chute. In early 1995, continued concerns about additional large-volume rockfalls originating from the source area located near the top of the chute led to the decision to construct a large containment berm upslope of the highway and then shift the highway back to its previous location.

As construction of this containment berm neared completion in October 1995, a fresh, very persistent tension crack that bounded a 200,000-m^3 (260,000-yd^3) wedge failure was discovered high on the left side of the chute. The toe of the wedge was located in a wide zone of clayey fault gouge that formed the center of the chute. An initial plan to remove the wedge proved too unsafe and difficult to accomplish because of accessibility limitations.

A monitoring program was started that involved daily measurements of reflectorless targets on the unstable mass using a total-station survey instrument (a combined electronic distance meter and theodolite) sited upslope of the failing wedge. This setup allowed for monitoring of vertical and horizontal displacements. Figures 1 and 2 show net displacements and rates of movement measured at a typical target location. When displacements began to increase dramatically in late November 1995, the chute was illuminated by high-intensity portable lights. Continuous visual observation for an expected slope failure was initiated along with a 24-h flagging operation at the highway to warn the traveling public. The procedure adopted was to inform the flaggers immediately when any significant movements or rockfalls were observed, and they would stop traffic.

For several months, traffic was continuously maintained on the bypass beneath this hazard. By the first week of January 1996, the trend of the displacements was asymptotic, indicating that a catastrophic failure appeared imminent. At 10:30 p.m. on January 13, while heavy fog and darkness obscured the view, the spotter heard loud noises from the failing mass. The spotter radioed down to the flaggers, who then closed the highway. Loud noises persisted while house-sized blocks could be heard tumbling down the chute but stopping in the large containment area between the base of the slope and the bypass. After nearly 2 h, the noises stopped and the bypass was safely reopened to traffic.

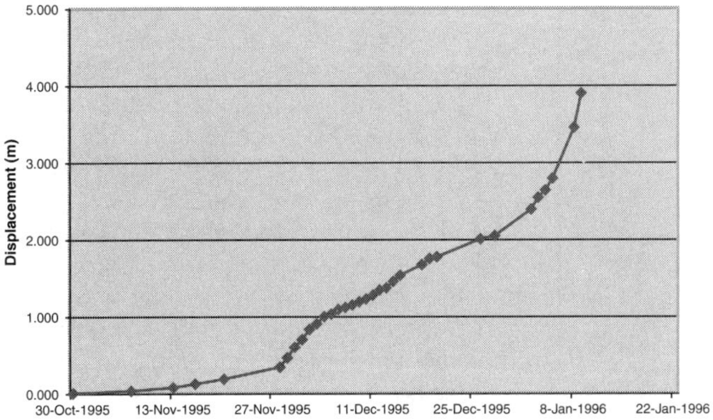

FIGURE 1
Displacement of optical survey target set on failing rock mass between early November 1995 and early January 1996; mass failed January 13, 1996, as the trend of the displacement rate became asymptotic.
(GRAPH COURTESY OF THE WASHINGTON STATE DEPARTMENT OF TRANSPORTATION.)

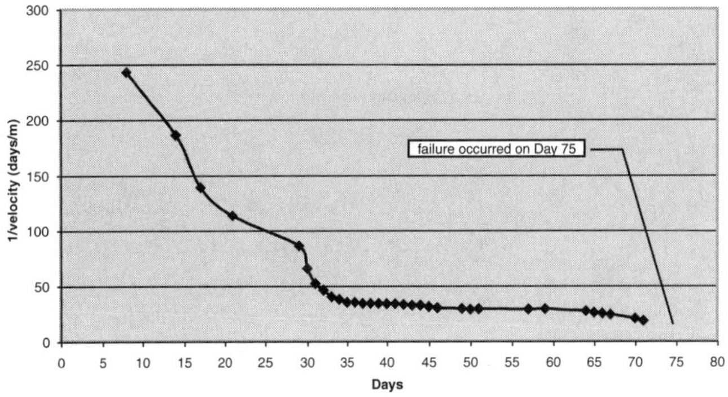

FIGURE 2
Rate of movement displayed as inverse of velocity (days per meter) for 2-month period.
(GRAPH COURTESY OF THE WASHINGTON STATE DEPARTMENT OF TRANSPORTATION.)

Box 17-2

Monitoring Activity That Avoided Use of Expensive Remediation Measures

Over a winter period, Washington State DOT maintenance personnel observed the apparent enlargement of a preexisting, 150-mm (6-in.) wide, continuous tension crack in a near-vertical rock cut adjacent to State Route 14 in the Columbia River Gorge in south-central Washington State.

The tension crack bounds a basalt mass about 25 m (90 ft) high, 3 to 6 m (10 to 20 ft) thick, and at least 5 m (15 ft) wide that contains many moderately to widely spaced fractures. Although there was no observed increase in rockfall frequency, the proximity of the rock face to a busy highway and transcontinental rail lines warranted implementation of a monitoring program.

Three potentiometric extensometers were installed across the tension crack near the top, middle, and bottom of the crack (Figure 1). They were connected to an automated data collection system situated at the base of the slope and read at 6-h intervals. Data stored on the continuously recording loggers are downloaded onto a laptop and analyzed every few months. Since the installation of the extensometers, no movement has been detected. The cost to stabilize the basalt mass with rock anchors was estimated to be about five times the monitoring costs. Monitoring avoided the more expensive remediation alternative for the short term, while the risks to the adjacent facilities were much better quantified and understood.

FIGURE 1
Installation of intermediate extensometer across tension crack; cables extended down to secure box containing data logger at base of slope.
(PHOTOGRAPH COURTESY OF THE WASHINGTON STATE DEPARTMENT OF TRANSPORTATION.)

the railroads (Figure 17-4). When one or more of the closely spaced wires is broken, the electrical circuit is interrupted and a signal is immediately sent to a control device or office. Obviously, these systems only tell of an event immediately after its occurrence, and they can be triggered by events other than rockfalls, for instance, disturbances caused by animals, such as deer, or by strong winds causing vegetation to foul the wires.

Systems that involve deformation measurements are increasingly being used, partly because of their potential predictive value. Both signal interruption and deformational types of automated instrumentation systems are increasingly being routinely coupled with telecommunications or other warning devices for managing risk associated with slope hazards on a real-time basis. A complete instrumentation-based warning system consists of three modules (Gubler 2006):

1. Detection system,
2. Signaling system, and
3. Control system.

The detection system consists of one or several monitoring instruments that detect an impending

Box 17-3

Monitoring Slope Movements with SSR

In late May 2006, rockfall from the reactivated Ferguson slide buried California State Route 140 just west of Yosemite National Park in the Merced River Canyon (Figure 1). The toe of the slide is situated more than 60 m (200 ft) above the roadway. Ongoing rockfall from slide movement made it unsafe to reopen the highway. In response, the California Department of Transportation (Caltrans) constructed a one-lane, signal-controlled detour route across the river.

FIGURE 1
Ferguson slide in June 2006 and burial of California State Highway 140.
(PHOTOGRAPH COURTESY OF THE CALIFORNIA DEPARTMENT OF TRANSPORTATION.)

Slide activity had to be monitored to ensure the continued safety of construction workers and the public after the detour was opened. Monitoring criteria included remote, automated, and continuous operation 24 h a day, 7 days a week. After evaluation of a variety of monitoring alternatives, including reflectorless optical surveys using total stations and terrestrial or airborne laser scanning (LiDAR), Caltrans selected an SSR instrument (Figure 2) previously used by the open-pit mining industry and developed by the Australian company GroundProbe.

The radar unit was set up across the canyon about 180 m (600 ft) from the slide. It scanned the bare rock area at the toe of the slide every 8 min. With proprietary software, an onboard computer processed the scan data and calculated movement of all discretized areas of the rock face with every successive scan. Figure 3 shows generalized monitoring data for the 6 months following the initial failure. Threshold movement rates were selected to warn of critical slide movement. Caltrans connected the radar system to the signals that controlled detour traffic so that the signals would be switched to red and traffic stopped whenever critical slide movement rates were exceeded.

FIGURE 2
SSR equipment used to monitor movement of Ferguson landslide.
(PHOTOGRAPH COURTESY OF THE CALIFORNIA DEPARTMENT OF TRANSPORTATION.)

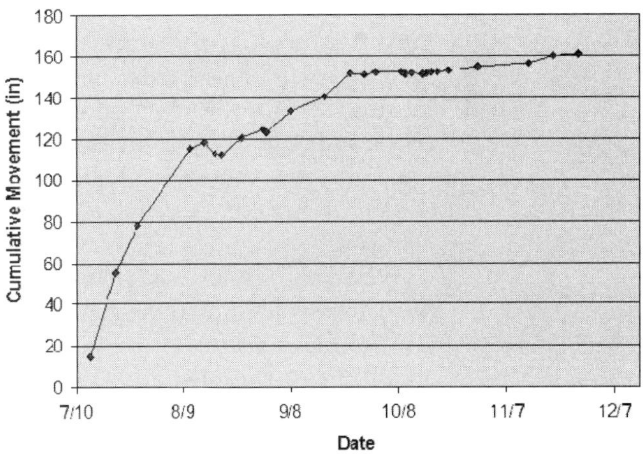

FIGURE 3
Generalized monitoring data measured by radar equipment for 6 months following initial failure in 2006.
(GRAPH COURTESY OF THE CALIFORNIA DEPARTMENT OF TRANSPORTATION.)

FIGURE 17-4 Warning fence in area of frequent rockfall occurrence and risk of impact onto railroad tracks, Glenwood Canyon, Colorado.
(PHOTOGRAPH COURTESY OF J. D. HIGGINS.)

dangerous situation, such as excessive movement of a rock block or mass, or the occurrence of an event. This detection may be done through remote sensors that operate over long ranges, such as ground-based radar monitoring systems recently used for the Ferguson slide in California (Beck and Duffy 2008; Harp et al. 2008); through multiple sensors fixed to the rock mass; or through sensors that respond to the severing of a trip wire.

When the signal is interrupted or a predefined threshold of deformation is exceeded, the sensors making up the detection system send an alarm to the signaling system. Selection of an appropriate threshold value may be difficult and often requires refinement or revision as the block or mass strains toward complete failure. The signaling system then performs one or all of the following steps: activates lights or alarms, closes protective gates across the roadway, and notifies emergency response personnel. The control system, which is commonly integrated within the data collection and signaling devices, allows for interrogation of warning system functions, sets or resets alarms, and manually displays data. More sophisticated control systems may be incorporated into a central maintenance and control office. With increasing system sophistication come greater logistical complexity and cost of operation, response, and maintenance. Such systems would only be considered when the consequence of failure is high, risk tolerance is extremely low, or both.

5. PREEMPTIVE CLOSURE

Although typically unpopular, in some cases the restriction of access to or use of a transportation facility by the public may be the most prudent and safest response to an impending slope failure or extremely hazardous condition. Besides reducing or eliminating public exposure to the hazard, closures may allow for more detailed evaluation and monitoring of the problem, provide a better environment for developing an effective remediation plan, and lessen restrictive conditions to remediate the hazard.

The decision to undertake a preemptive closure ideally reflects a demonstrable safety concern and is based on sound and experienced judgment. A conceptual remedial plan, with desired outcomes, is agreed upon and approved before a closure is implemented. Because conditions often change or become better known, it is common for a plan to evolve during implementation. Site conditions and objectives must be monitored and evaluated constantly to ensure safety and to minimize any impacts on or delays to the public associated with the closure.

5.1 Importance of Public Information

The management of rockfall hazards most typically focuses on mitigation through avoidance, stabilization, or protection measures. Often, however, a crucial and concurrent management activity involves reducing the impacts on the public associated with a significant rockfall event. Thus, communication of timely and accurate public information is extremely important.

Efforts to educate users about the type and extent of any geological (rockfall) hazards to which they may be exposed and the proactive efforts being undertaken to mitigate them demonstrate the values and concerns of the transportation agency. These efforts also provide a basis for a sharing of "ownership" of any potential impacts associated with these hazards. Timely and accurate public information must include a clear justification for actions such as preemptive road closure, an explanation of the remediation plan, alternative routes and travel times, and the potential duration of the closure. A public information campaign undertaken by the Washington State DOT to mitigate travel delays when a major highway was partially blocked by rockfall debris during an important holiday is described in Box 17-4. Because slope conditions are typically difficult to characterize and there are frequent complications with constructing remedial measures on steep slopes, caution should be exercised in predicting when a facility will reopen. Predictions are sometimes misinterpreted as commitments.

5.2 Establishing Conditions for Closure

Site-specific conditions that might warrant a preemptive closure are obviously unique and widely variable. From a regional perspective, increased slope failures and rockfall are commonly associated with extreme events such as storms, forest fires, and earthquakes. Such events usually cannot be predicted, but preemptive closures may be appropriate during or immediately after these events to evaluate new hazards or to allow time for rockfall activity to return to background levels. Although methods to predict rainfall-induced shallow landslides have been improved in recent years, only limited effort has yet been undertaken to predict increased rockfall occurrence (Zvelebil and Moser 2001; Chau et al. 2003; Rosser et al. 2007).

Determining the need for closure depends on experienced judgment about the potential threat. Specific considerations might include the following:

- Are the source area, volume, and extent of potential or ongoing slope failure well defined?
- Is sufficient catchment area available for the potential event and suitable for the expected block sizes?
- Is the trajectory path of falling debris reasonably certain?
- Is the slope failure influenced by weather conditions, and if so, are conditions unfavorable?
- Are interim remedial measures readily available to safeguard the facility?

In some cases, determining when a facility can be safely reopened is more difficult than determining when it should be closed. This situation may arise when an apparent impending failure stalls or fails to occur as expected or when the factors and mechanics controlling potential failure are poorly understood or are not directly observable. In these situations, monitoring might provide crucial predictive information about an impending failure or a basis for determining when a facility could safely reopen.

An example of road management for rockfall hazards after a wildfire is described in Box 17-5.

Box 17-4

Successful Public Information Campaign

Following a rockfall event in early November 2005 along a constricted section of Interstate 90 that crosses the Cascade Mountains in Washington State, traffic was reduced from two lanes to one lane in each direction for a period of 6 weeks while remedial work was undertaken to stabilize the slope.

In an effort to alleviate impacts on the traveling public, live webcams of the construction site, hourly construction and traffic updates, and regular news releases were provided to the public and regional trucking companies by regularly scheduled radio and television news reports and through a website maintained by the Washington State DOT.

The public information campaign proved highly successful. As illustrated by Figure 1, Thanksgiving Day holiday travel, especially during peak periods, was significantly reduced. Comparisons of 2005 total daily volumes with the 2004 daily volumes revealed that peak hourly holiday traffic volumes were reduced by more than half. These reduced peak volumes avoided predicted gridlock conditions and extensive travel delays through the work zone. However, normal travel volumes outside these peak holiday periods were largely unaffected. Thus normal commercial traffic was largely unaffected by the partial closure during the remediation efforts.

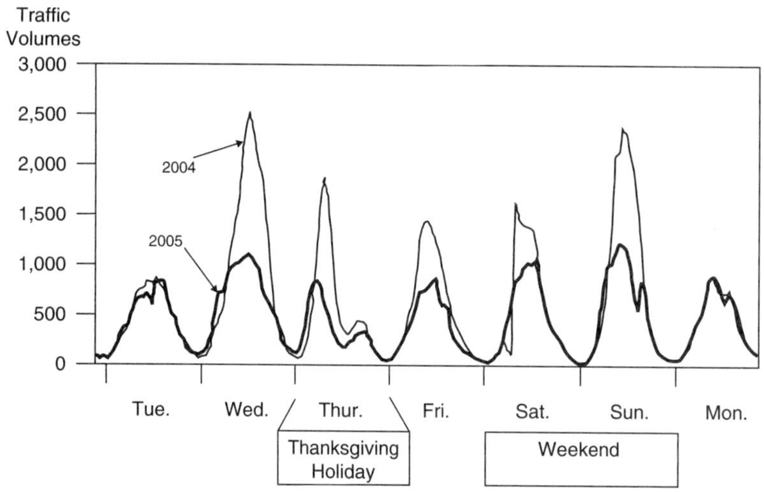

Daily Total Vehicle Volumes

	2004	2005
Tuesday	11,502	11,315
Wednesday	25,326	16,231
Thursday	13,861	8,917
Friday	14,963	9,671
Saturday	13,699	11,390
Sunday	22,143	14,184
Monday	10,281	9,963

FIGURE 1
Hourly eastbound traffic volumes along section of Interstate 90 through rockfall area measured by Washington State DOT over 2004 and 2005 Thanksgiving holidays; total daily volumes are tabulated to right of graph. Public information campaign resulted in a 35% reduction in traffic volumes on the major travel days of Wednesday and Sunday.
(GRAPH COURTESY OF THE WASHINGTON STATE DEPARTMENT OF TRANSPORTATION.)

Box 17-5

Road Management for Rockfall Hazards After Wildfire

In July 2006, the Bear Gulch II wildfire started on the steep slopes above Forest Road 24 along Lake Cushman in the Olympic National Forest in western Washington State (Figure 1). The fire occurred in an area that had previously burned, and the slopes had a history of rockfall and debris slides. During and after the fire, the area experienced an increase in rock and debris fall along a 4-km (2.5-mi) section of road. This rock and debris fall has continued in subsequent years. The road provides access to a number of seasonally occupied private residences and a popular entrance to Olympic National Park.

The road was closed during the fire because of safety concerns resulting from rock and debris fall. The closure was extended until an assessment could be made to determine the risk to the public and what, if any, mitigation or management actions should be implemented. Decisions to close the road have proved relatively easy; developing a rational basis to support decisions to reopen the road was more challenging.

FIGURE 1
Bear Gulch II fire area during early stages of fire; Forest Road 24 traverses base of slope just above reservoir level.
(PHOTOGRAPH COURTESY OF THE OLYMPIC NATIONAL FOREST.)

FIGURE 2
Inventory data included date, rock size, number, and location in roadway; rocks were color marked to distinguish between survey intervals. Locations of rockfall blocks were stratified into four categories from inside to outside edge of road.
(PHOTOGRAPH COURTESY OF THE OLYMPIC NATIONAL FOREST.)

Assessment methods consisted of collecting information concerning geology, burn intensity, rockfall history, and rockfall frequency (Figure 2). In addition, rockfall source areas were examined and interviews were conducted to document historical (anecdotal) and fire-related rockfall accounts. Among the most useful data collected were the

(continued on next page)

Box 17-5 (continued)

Road Management for Rockfall Hazards After Wildfire

direct and anecdotal observations and reports of rockfall, and these were used to create an inventory of rockfall frequency (Figure 3). Travel behavior of rockfall blocks was interpreted from field observations and compared with numerical simulations.

The inventory helped establish relationships of rockfalls with inclement weather, guided decisions by park managers regarding the timing of road closures and openings to ensure safe travel by the public, and formed the basis for scheduling maintenance operations. The continued monitoring of rockfall trends is expected to assist in determining when rockfall activity has returned to background levels and to provide a basis for establishing an acceptable level of risk.

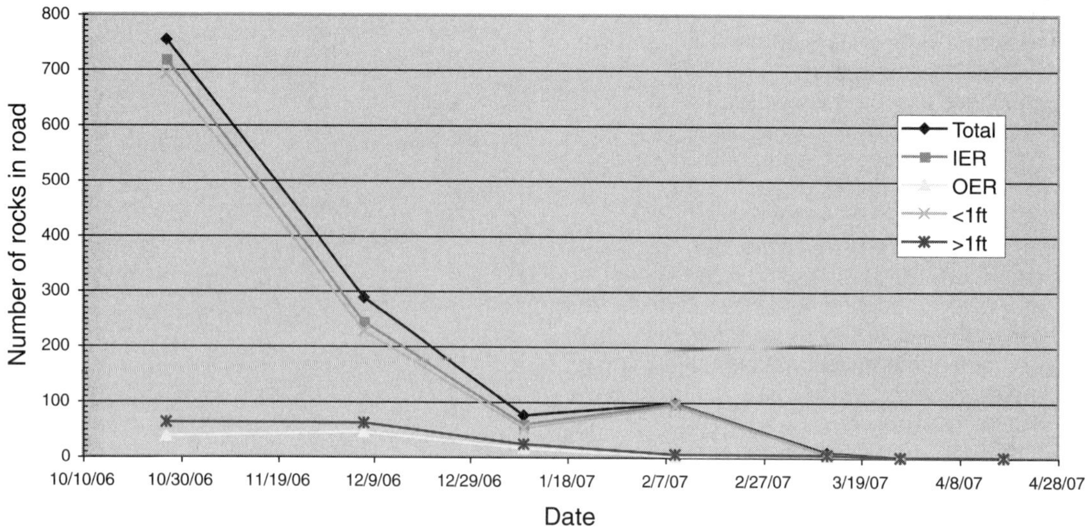

FIGURE 3
Rockfall inventory for high-frequency site over 2006–2007 winter shows blocks smaller and larger than 1 ft (0.3 m) in size and their location in the roadway (IER = inside portion of roadway, OER = outside portion of roadway).
(GRAPH COURTESY OF THE OLYMPIC NATIONAL FOREST.)

6. SUMMARY

Although the maintenance and monitoring measures presented here generally do not provide full or long-term mitigation of the rockfall hazard, some level of risk reduction is achieved through their implementation. Maintenance activities can caution users of the hazard, reduce exposure to rockfall debris through routine patrols and ditch cleaning, or provide some interim protection or stabilization. When large, potentially unstable features are identified, a variety of monitoring efforts can be undertaken to characterize and quantify the extent of the potential instability. These monitoring data can then be defensively used to determine the urgency of repairs or the need for a preemptive closure. Table 17-1 summarizes the advantages and disadvantages of each maintenance and monitoring measure.

ACKNOWLEDGMENTS

Insightful and constructive reviews were provided by Steve Lowell, David Gerraghty, and Larry Pierson. Tim Beck provided the summary of the Ferguson slide and Bill Shelmerdine provided the summary of the wildfire-related rockfalls on the Olympic National Forest.

Table 17-1
Summary of Maintenance and Monitoring Measures

Mitigation	Advantages	Disadvantages	Proactive (P), Reactive (R)
Warning signs	Provide warning of a potential hazard. Are inexpensive. Require little maintenance. Address long or multiple sites.	May not change behavior. Do not reduce rockfall occurrence. Still require slope inspection.	P
Maintenance patrols	Provide regular inspection. Rapidly respond to rockfalls. Cost-effectively cover multiple sites.	Are relatively inexpensive and use available personnel and equipment. Do not reduce rockfall occurrence.	P
Catchment area improvements	Can reduce rockfall that reaches facility. Are generally inexpensive and simple to maintain.	Are limited by available width. May not provide full mitigation.	P
Scaling	Reduces rockfall occurrence. Is relatively inexpensive; may uncover more serious problems.	Generally provides short-term mitigation. Is ineffective for some rock mass conditions. Requires experienced personnel.	P, R
Movable protection systems	Are highly effective in containing design rockfall. Are versatile and quickly deployed.	Are expensive to acquire. Often require regular maintenance. May affect traffic and use. Are limited by available width.	P, R
Observation	Is inexpensive and uncomplicated. Is versatile and quickly deployed.	Requires experienced on-site inspection, with potentially critical data gaps. May not be sufficiently quantitative. May lack consistency.	P
Optical surveying	Provides high-quality, repeatable, quantitative data on a regular basis. Limits exposure of personnel to hazard.	Requires experienced personnel for regular data interpretation. Can be expensive. May not be effective for early detection of rapid failures.	P
Instrumentation	Provides high-quality, repeatable, quantitative data on a regular basis. Can be relatively inexpensive.	Requires experienced personnel for regular data interpretation. Requires some exposure to hazard. Can be prone to damage and malfunction.	P
Remote sensing	Provides high-quality, repeatable, quantitative data on a regular basis. Provides rapid detailed coverage of a large area. Avoids direct exposure to hazard.	Requires experienced personnel for regular data collection and interpretation. Is typically expensive to implement and maintain.	P
Automated warning systems	Provide high-quality, repeatable, quantitative data on a regular basis. Execute preplanned response. Limit exposure of personnel to hazard.	Are typically complex, requiring experienced personnel to implement and maintain. Can generate false alarms. Can be expensive.	P
Preemptive closures	Eliminate exposure to hazard. Provide unfettered access for assessment and mitigation.	Often cause significant traffic disruption. Require rapid, decisive action once implemented.	R

REFERENCES

ABBREVIATION

FHWA Federal Highway Administration

Al-Kaisy, A. 2006. *Static Warning Signs for Occasional Hazards: A Synthesis of Research and Practice*. Final Report, UTC Project 4W0863. Western Transportation Institute, Bozeman, Mont.

Bean, M. E. 1989. Speaking of Risk. *Trends*, Vol. 26, No. 4, pp. 15–19.

Beck, T. J., and J. D. Duffy. 2008. The Ferguson Rockslide, Mariposa County, California. In *Conference Proceedings, 59th Highway Geology Symposium*, Santa Fe, N.Mex.

Brawner, C. O. 1993. *Manual of Practice on Rockfall Hazard Mitigation Methods*. National Highway Institute, U.S. Department of Transportation.

Casagli, N., P. Farina, D. Leva, and D. Tarchi. 2004. Application of Ground-Based Radar Interferometry to Monitor an Active Rockslide and Implications for Emergency Management. In *Landslides from Massive Rock Slope Failure* (S. G. Evans, G. S. Mugnozza, A. Strom, and R. L. Hermanns, eds.), Springer, pp. 157–174.

Chau, K. T., R. H. C. Wong, J. Liu, and C. F. Lee. 2003. Rockfall Hazard Analysis for Hong Kong Based on Rockfall Inventory. *Rock Mechanics and Rock Engineering*, Vol. 36, pp. 383–408.

FHWA. 2003. *Manual on Uniform Traffic Control Devices for Streets and Highways*. U.S. Department of Transportation.

Great Britain Department for Transport. 2002. *The Traffic Signs Regulations and General Directions 2002*. Statutory Instrument 3113. Stationery Office, Norwich, United Kingdom. www.opsi.gov.uk/SI/si2002/20023113.htm.

Gubler, H. 2006. *Alarm Systems to Detect Avalanches, Mudflows and Rockfalls*. AlpuG, Davos, Switzerland. www.alpug.ch/pdf/pdf_06_en.htm. Accessed Sept. 4, 2006.

Harp, E. L., M. E. Reid, J. W. Godt, J. V. DeGraff, and A. J. Gallegos. 2008. Ferguson Rock Slide Buries California State Highway near Yosemite National Park. *Landslides*, Vol. 5, pp. 331–337.

Rosser, N. J., D. N. Petley, S. A. Dunning, M. Lim, and S. Ball. 2007. The Surface Expression of Strain Accumulation in Failing Rock Masses. In *Rock Mechanics: Meeting Society's Challenges and Demands* (E. Eberhardt, D. Stead, and T. Morrison, eds.), Taylor and Francis, London, pp. 113–120.

Zvelebil, J., and M. Moser. 2001. Monitoring Based Time-Prediction of Rock Falls: Three Case-Histories. *Physical, Chemical and Earth Sciences B*, Vol. 26, pp. 159–167.

Chapter 18

STEVE M. LOWELL AND
NORMAN I. NORRISH

ROCKFALL MANAGEMENT PROGRAMS

1. INTRODUCTION

For many decades, rockfalls along many North American transportation corridors have been a chronic problem. Rockfall events can cause traffic disruptions and delays, roadway closures, property damage, and, in rare cases, personal injury. Roadway closures and traffic disruptions and delays equate to direct economic losses to the traveling public and can have indirect economic impacts on adjacent communities if the roadway closures are of considerable duration. In the past, the approach to the rockfall problem was almost universally reactive, with maintenance cleanup and mitigation occurring only after the rockfall event, called the "let-it-fall" policy. This reactive approach has many negative aspects related to public accountability and expectations and the reliability, convenience, and safety of the transportation system.

In North America, modern transportation infrastructure has begun to age, and at the same time large increases in traffic volumes are occurring. The aging of these systems causes the various component assets (including slopes) to require adequate maintenance, preservation, or reconstruction. With regard to rockfall, the existing reactive policies have resulted in a large inventory of deficiencies waiting to be mitigated.

In the early to mid-1990s, a small number of North American transportation agencies began to develop proactive rockfall management programs. In recent years, some transportation agencies in mountainous areas with extensive rockfall problems have begun to develop rockfall inventories as well as internal policies and efforts that together form a comprehensive rockfall management program. The long-term goal of these proactive programs is to mitigate unstable rockfall areas in advance of actual slope failure, thereby reducing the effects of rockfall on the transportation facility. In some transportation agencies, rockfall management programs have advanced to the levels of infrastructure preservation programs that are similar to pavement and bridge preservation programs. In 1999 the Federal Highway Administration (FHWA, p. 21) stated:

> Infrastructure preservation programs represent a departure from traditional approaches to maintenance, in which deficiencies are addressed as they occur. Preservation seeks to reduce the rate of deterioration. The preventive approach is generally less costly and time-consuming than the traditional, more reactive approach.

Ideally, a rockfall management program employs an all-encompassing, systematic, proactive approach that is consistently applied in identifying, planning, programming, priority ranking, designing, and constructing rockfall mitigation projects for an entire transportation infrastructure system or for a specific segment of it.

2. ROCKFALL HAZARD AND RISK

It is useful to distinguish between the terms *hazard* and *risk*, as each plays an important role in the decision-making process of a rockfall management program (Fell et al. 2005):

- *Hazard* can be defined as a natural phenomenon that creates a danger or threat that will occur in a given period of time and can be described by its geometry, failure mechanism, or other characteristics.
- *Risk* can be defined as the consequences that are realized when the rockfall hazard actually fails. Risk is measured in terms of the severity of the event and adverse effects to life, health, property, and so forth.

Being able to distinguish between the levels of hazard and the levels of risk is a fundamentally important concept, as discussed in Box 18-1. Within the context of a rockfall management program, there is a need to address not only the hazard but, more importantly, the risk associated with the hazard. It is logical for transportation agencies to focus their efforts on high-hazard sites that pose moderate to high risk so that the allocated funds for mitigation will have the highest rate of return on investment.

3. ELEMENTS OF A ROCKFALL MANAGEMENT PROGRAM

A rockfall management program consists of a multifaceted "cradle-to-grave" process that takes into consideration (*a*) slope identification, (*b*) hazard and risk evaluation, (*c*) mitigation options and cost estimates, (*d*) project prioritization and project programming, (*e*) final mitigation design, (*f*) construction contract preparation, and (*g*) construction of the mitigation project. Consistency throughout this process is important. Each step in the process should include milestones, decision points, or both. The following sections provide details on the key elements of a rockfall management program.

3.1 Rockfall Inventory

The rockfall inventory is the foundation for any rockfall management program. It provides the definition and scope of the program. Typically, the identification of rockfall sites can easily be made with direct input from the transportation agency's maintenance forces. The maintenance personnel are aware of problems associated with rockfall because they deal with the problems on a day-to-day basis. They are also good sources of information for failure history; failure frequency; accidents related to rockfall; and rockfall cleanup, ditch maintenance, and other costs.

To rank individual slopes accurately by priority within a statewide inventory, a large number of unstable slopes must be rated in a systematic manner with consistent and measurable criteria (Lowell, Moses, and Badger 2005). In North America, this rating is typically accomplished with matrix-based numerical systems, as detailed in Chapter 4 of this report.

Because rockfall inventories can be large, data associated with each site can be input into a simple database for easy retrieval, sorting, querying, reporting, and other purposes. Some transportation agencies with more advanced rockfall management programs have developed web-based slope management systems that are used for slope inventory control and can provide numerical ratings, preliminary design data, cost estimates, and project programming. Users throughout the transportation agency are able to access these management systems from their personal computers via the intranet resources available to agency personnel. Such information is not, however, routinely made available to the general public or to persons outside the agency unless by special request and approval.

3.2 Evaluation of Rockfall Hazard

Once the rockfall inventory has been developed, the transportation agency has data for a large number of slopes, including slope locations, numerical ratings for each slope based on the parameters selected by the agency, and general information relating to rockfall hazards. For a large number of slopes, additional site-specific geotechnical information is required to quantify the rockfall hazard and determine the risk. Because the rockfall inventory may include hundreds if not thousands of rockfall sites, the transportation agency needs to develop methods to prioritize the inventory, as it is

Box 18-1

Distinguishing Between Hazard and Risk

Two large (greater than 1 million m^3, or 1 million yd^3) rock avalanches occurred near Newhalem, Washington, in late October 2003, after a record-breaking rainfall event. The first rock avalanche (Figure 1) occurred in a remote undeveloped valley several miles from State Highway 20. Because of the remote location, it was several weeks before anyone noticed that the rock avalanche had occurred.

The second rock avalanche (Figure 2) occurred in steep, mountainous terrain immediately upslope of State Highway 20, just east of Newhalem. Although the majority of the avalanche was contained in a drainage area adjacent to the highway, large boulders 5 m (15 ft) in diameter fell onto the highway and destroyed a half-bridge, making the highway impassable. Because of the highway damage and the potential for additional large rockfalls, the highway was closed for an extended period of time. The closure resulted in severe social and economic impacts on the community of Diablo, located to the east of the rock avalanche, since the highway provided the sole access to the community.

Both rock avalanches could be considered to be high hazards. The first rock avalanche would be considered a low-risk event, since it occurred in a remote area and the consequences of the failure were minimal. The second rock avalanche represented a high-risk event, since it occurred immediately adjacent to a highway and the consequences of the failure and future potential failures had major impacts on the transportation facility and adjacent communities.

FIGURE 1
Goodell Creek rockfall avalanche.
(PHOTOGRAPH COURTESY OF THE NATIONAL PARK SERVICE.)

FIGURE 2
Rockfall damage on Washington State Highway 20.
(PHOTOGRAPH COURTESY OF THE WASHINGTON STATE DEPARTMENT OF TRANSPORTATION.)

impractical for the entire inventory to be assessed at once.

Fundamentally, in a fully implemented rockfall management program, the slope rating system is the primary tool with which the agency evaluates slopes by hazard or risk. The slope rating system provides guidance for ranking future mitigation projects by priority. Although evaluating slopes with the highest hazard first—commonly called the "worst first" approach—may seem logical, this approach often is not appropriate because the slope with the highest hazard could be located along a transportation facility that has low risk (Lowell and Morin 2000). Mitigating a slope in this situation might not provide the largest benefit for the dollars expended.

To establish priorities, the transportation agency can rank the rockfall inventory in groups by a number of parameters that may be specific to the agency (e.g., facility type, traffic volumes), with the goal of attempting to match high hazards with moderate- to high-risk locations. On the basis of the list of high-hazard and moderate- to high-risk slopes, site reviews are then conducted to gather the following preliminary data: (*a*) site geology, (*b*) detailed slope geometry and cross sections, (*c*) dominant slope failure mechanism or mechanisms based on physical measurement, (*d*) dimensions of potential failures, (*e*) rockfall source areas, and (*f*) other information that may be required. Figure 18-1 is an example of a field sheet used to collect the necessary data. The data gathered during the site review are typically sufficiently detailed to allow for development of preliminary mitigation designs and cost estimates. In particular, the geotechnical information obtained from the site reviews is specifically targeted to allow for a more comprehensive evaluation of the rockfall potential.

3.3 Determination of Preliminary Mitigation Options

A reasonable estimate of the cost of mitigation is required to compare and select candidate sites for mitigation. Accordingly, during the preliminary design stage, multiple mitigation techniques are typically identified for each candidate site. For example, for a slope with chronic rockfall problems, the mitigation options could range from a very expensive excavation of a completely new cut employing state-of-practice drilling and blasting methods, to a more modest option of draping the entire slope face with cable nets, to a relatively inexpensive option of scaling with spot rock bolts. The optimal mitigation option is the one that adequately reduces the risk to the transportation facility at minimal cost for the desired design life. Although quantitative risk-analysis programs, such as those defined by Roberds (2005) and briefly discussed in Chapter 5 of this report, are available for this task, current North American practice tends not to use such approaches and instead relies on the judgment of experienced rock slope specialists.

The selection of mitigation options typically considers factors that go beyond geotechnical conditions. For example, input may be solicited from maintenance personnel, who often are interested in issues such as

- At-grade cleaning of ditches versus on-slope repair of mechanical systems such as rockfall fences and slope protection systems,
- Temporary snow storage in areas that receive heavy snowfall, and
- In-house expertise to maintain mechanical systems.

Opportunities to integrate the mitigation of unstable slopes with collateral benefits, such as the creation of additional lanes to increase capacity, the improvement of sight distance, or the provision of greater snow avalanche protection, are evaluated. Environmental constraints, such as the required protection of endangered species and fish-spawning or bird-nesting habitats, may affect the selection or timing of mitigation options. Finally, aesthetic considerations may constrain the available mitigation options or dictate specific aesthetic measures that are included in the cost estimate. In all cases, the selection of the preferred mitigation options recognizes, in advance, potential external limitations beyond the geotechnical conditions at a given site.

Early cost estimating is an integral part of the rockfall management program. Adequate cost data provide decision makers within the transportation agency with financial information concerning the magnitude of the long-term requirements of the rockfall management program but also provide cost information for preparation of the transportation agency's annual budget. These preliminary estimates typically contain sufficient detail to

USMS - FIELD NOTES
REGION: **NW** SR: **011**
MP: BEG **11.3R** END **11.45R**
Slope#: **2575**

Prepared by: N. Norrish, P.E.
Date: January 13, 2005

Typical Section:

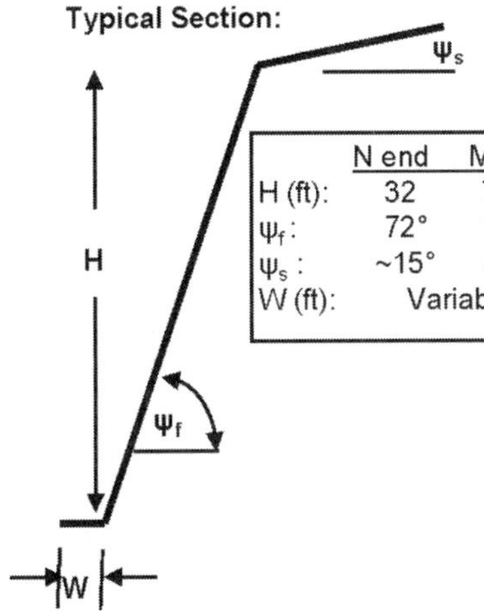

	N end	Mid	S end
H (ft):	32	71	87
ψf :	72°	73°	60°
ψs :	~15°	46°	36°
W (ft):	Variable 1 to 4		

Site Measurements:
Impact of Failure: 640-ft slope length
Sight Distance:
S bound N end: 265 ft
N bound N end: 180 ft
N bound middle: 205 ft
N bound S end: 220 ft
Highway Trend: 342° N end 297° S end

Rock Mass Characterization:
Rock Type: Fresh to slightly weathered, fine-grained, tan–brown, medium-strong sandstone (2- to 10-ft beds) interbedded with slightly weathered, dark gray to black, weak to very weak, locally carbonaceous shale–siltstone (1- to 2-ft beds)

Structures: B 35°/353° P S K 30' + (S end)
J 62°/209° P R K 10'
B 34°/015° P R K 30' + (N end)
J 75°/232° C R K 20' (N end—defines slabs)

Site Photographs: 11/4/04

North end

South end

Types of Instability:
1. Differential weathering.
2. Rockfalls controlled by joints in sandstone where overhangs created by weathering.
3. Planar.

Special Notes:
1. Approximately 90 ft of impact length at S end has been stabilized (rock bolts, drain holes, shotcrete).
2. Masonry wall on outboard side at mid-cut with guardrail on top of wall.

Mitigation Alternatives:
1. Reslope to improve alignment. Sandstone is favorable for controlled blasting. Shale layers will require shotcrete protection to minimize weathering.
2. Remove trees from slope face and crest, scale, spot rock bolts, shotcrete shale layers, slope drape.

FIGURE 18-1
Example of field assessment data sheet (Lowell, Gates, et al. 2005).

reflect accurately the entire cost for any individual rockfall mitigation project, and thus include not only the cost for the geotechnical aspects of the mitigation but also the added costs for items such as preliminary engineering, mobilization, traffic control, construction engineering, and contingencies.

Questions may arise about the accuracy of these preliminary cost estimates and how inaccuracies in the estimates may affect a rockfall management program, the budgets of individual projects, or both. Figure 18-2 provides a comparison of cost estimates for preliminary designs (conceptual design estimates), final designs (engineer's estimates), actual cost (final payment amount), and the low bid for 39 rockfall mitigation projects in Washington State (Lowell, Gates, et al. 2005). These data clearly validate the ability to develop accurate cost estimates for identified mitigation options and their associated costs when current unit cost data and detailed technical input from experienced rockfall mitigation specialists are relied on.

3.4 Program and Project Planning

Because of the potentially large number of unmitigated rockfall sites along transportation routes, the transportation agency needs to develop long-range plans and project programming strategies to rank the mitigation work by priority. These planning and programming efforts focus on high-hazard rockfall sites where substantial reductions in risk can be accomplished at the lowest possible cost. Achieving this goal requires a practical exercise in risk management and prioritization (MacDonald 2006). Consistency is a key factor in a successful planning and programming process. Program planning and project planning are two primary elements in this process.

3.4.1 Program Planning

The main goal of the long-range program plan is to identify high-priority rockfall mitigation projects that will move forward into the construction program of the transportation agency. The plan necessarily spans multiple budget cycles so that the agency can quantify the long-term rockfall mitigation construction program and establish the funding stream required to support that program.

To rank rockfall mitigation projects by priority, each transportation agency needs to develop criteria that are specific to the agency. For example, the Washington State Department of Transportation uses two risk criteria to rank slopes for mitigation. The first criterion is a cutoff on the numerical rating of slopes: numerical ratings must be greater than 350 on high-value facilities, which are defined as Interstates and principal arterials,

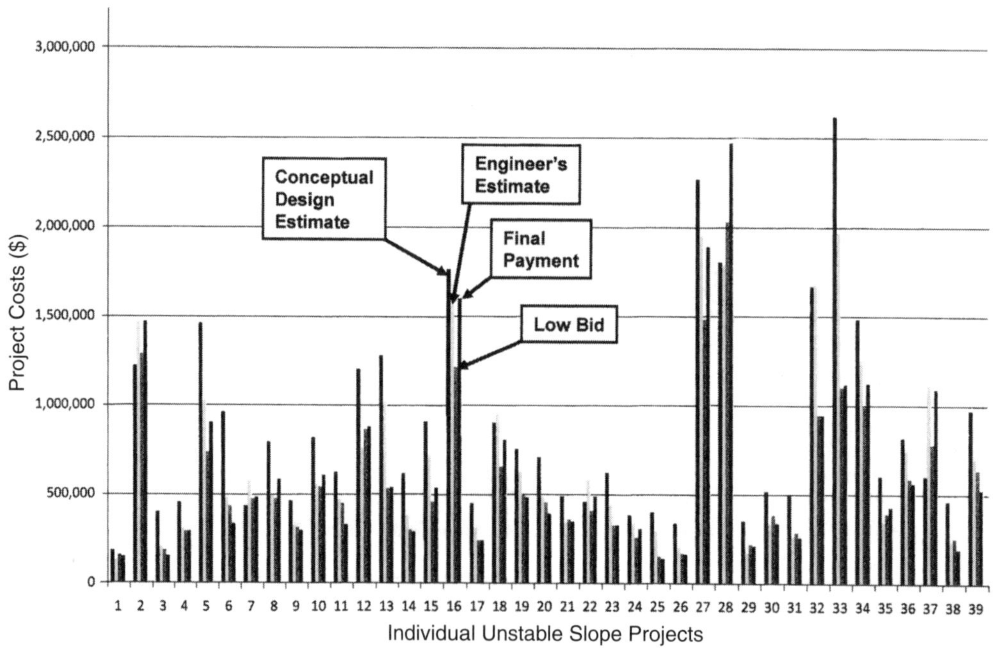

FIGURE 18-2 Comparison of cost estimates for 39 rockfall mitigation projects in Washington State (1999 to 2010). Conceptual design estimates are preliminary designs; engineer's estimates relate to final designs; final payment is actual cost; and low bid is contractor's bid price (Lowell, Gates, et al. 2005).

or on facilities with average daily traffic greater than 5,000 vehicles. The second criterion uses a simplified cost–benefit analysis that compares the cost of mitigating the hazard with the costs associated with a traffic delay combined with the cost of current maintenance practices (Lowell, Moses, and Badger 2005). Application of these two criteria results in a list of priority-ranked rockfall mitigation projects that the department can use for strategizing and programming purposes in current and future budget cycles.

3.4.2 Project Programming

The project programming process allows the transportation agency to make the final selection of projects to be included in the agency's construction program during a given budget cycle. Rockfall mitigation projects are selected from the priority-ranked program plan list. The number of slopes to be programmed will be based on the cost of the individual rockfall mitigation projects and the amount of funding that is available within the budget cycle. Project trade-off decisions are likely made to balance the program with the needs of the transportation agency. For example, an agency may decide it is better to use a given amount of funds to mitigate a number of lower-cost rockfall sites rather than a single high-cost site. The decision to mitigate multiple sites may provide higher potential for risk reduction systemwide than mitigating the single slope. In addition, the transportation agency may consider grouping slopes that are close to each other for mitigation, in the expectation of reducing the cost of the work and the impacts of the work on the traveling public.

3.5 Final Mitigation Design and Construction Contract Preparation

After a site has been selected for mitigation on the basis of project programming criteria specific to the transportation agency, the preliminary design is typically refined to a final design for the development of contract-ready plans, specifications, and cost estimates. Typically this will involve six steps: performance of detailed geological mapping, development of plans and sections, conduct of stability and rockfall analyses, mitigation design, determination of constructability, and determination of contractual issues.

3.5.1 Detailed Geological Mapping

Standard geological mapping techniques are employed to characterize rock types, weathering and alteration, rock strength, overburden depth, and structural discontinuities. Rock mass dilation and discontinuity aperture conditions are observed and noted as possible evidence of slope instability or blast damage. Usually rock face exposures are adequate to provide the information, and geotechnical drilling is not required; however, for cases in which rock slope widening is being designed and the extrapolation of geological conditions at the existing face to those at the proposed face is uncertain, exploratory drilling may be required. Access for rock face mapping can be facilitated by lifts or climbing ropes. Because the rock slope being mapped usually has not been scaled, loose rocks pose a hazard, and great caution must be exercised if climbing ropes are used. Modern digital imagery collection and analysis techniques, especially when combined with digital photogrammetric approaches, are valuable tools for the remote mapping of inaccessible or dangerous rock faces (Haneberg et al. 2006).

3.5.2 Plans and Sections

The site geometry is captured in plan and section for the transportation agency's use in performing stability analyses and estimating quantities and for the contractor's use in selecting types and sizes of equipment. As discussed in Chapter 7, conventional surveying, laser rangefinders, digital photogrammetry, and terrestrial light detection and ranging (LiDAR) techniques provide for accurate development of the required information.

3.5.3 Stability and Rockfall Analyses

Two types of stability analyses are typically considered at each site. Kinematic analyses (sometimes referred to as *Markland analyses*) are performed to determine whether the combination of structural fabric, slope orientation, and slope inclination is conducive to block instability. If the conclusion is in the affirmative, appropriate planar, wedge, or toppling analyses are performed to verify or improve the margin of stability (factor of safety). For weak rock masses, general two-dimensional circular failure analyses are also undertaken. Once larger-scale stability concerns have been satisfied, additional rockfall analyses may be needed to predict the trajectory and energy of individual rock fragments that may become dislodged from the slope. As discussed in Chapter 9, Sections 5 and 6, verified rockfall simulation software supplemented with site calibrations is frequently used for this type of analysis. Burk et al. (2004) described the application of these techniques for the Washington State site discussed in Box 18-1.

3.5.4 Mitigation Design

As shown in Figure 18-3, mitigation strategies include stabilization (with both rock reinforcement and rock removal) and protection measures. The general order of preference is rock removal, slope reinforcement, and, finally, protection measures. As Figure 18-3 indicates, several specific techniques exist within each of these categories. Site conditions dictate the appropriate technique (or combination of techniques) that is to be employed at an individual site. Oblique aerial or terrestrial photographs, or both, are a convenient method for presenting final rockfall mitigation design requirements in the contract documents (Figure 18-4). Chapter 11 provides additional information on the relevant characteristics affecting the selection of rockfall mitigation options. Specific design techniques for each mitigation option are described in detail in Chapters 12 to 17.

3.5.5 Determination of Constructability

Primary issues affecting constructability include access, sequencing, and safety. *Access* includes the selection of equipment for performing the actual on-slope construction activities (e.g., cranes, ropes) as well as the provision of appropriate staging and work areas. Equipment selection is a means-and-methods issue that is the responsibility of the contractor, and the transportation agency typically provides adequate work space to make the construction efficient by, for example, instituting a lane closure at the base of the slope for the duration of the work.

Sequencing is an issue for both the designer and the contractor. Critical stabilization elements that need to be performed in a specific sequence are so stipulated by the designer in the contract plans and specifications. Otherwise, the contractor typically adopts a sequence that minimizes the cost of construction.

Site safety during construction is always the contractor's responsibility. Increasingly, however, transportation agencies want to perform

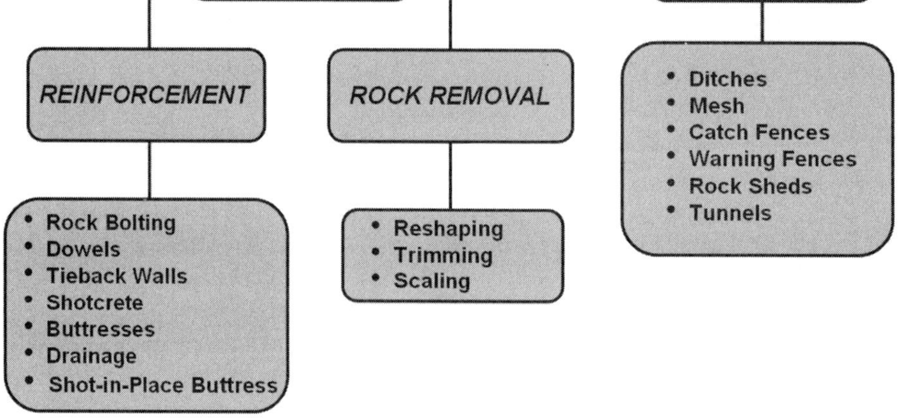

FIGURE 18-3 Rock slope stabilization and protection categories (adapted from Wyllie and Norrish 1996).

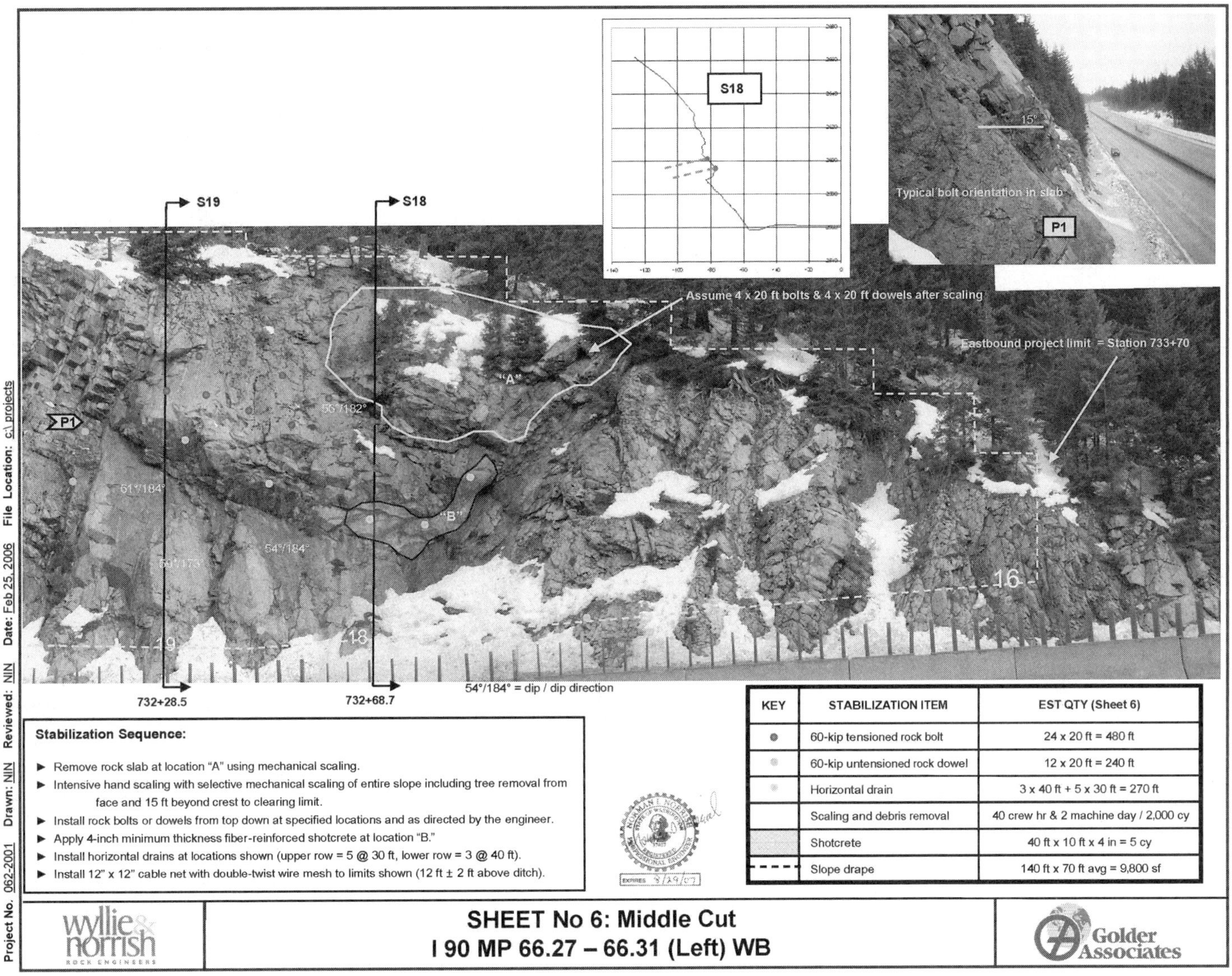

construction work under live traffic conditions. In such cases, the designer specifies the temporary protection measures required. These measures may include protection of the corridor during scaling and rock-bolting activities by, for example, movable rockfall fences along with containment nets suspended by cranes.

3.5.6 Contractual Issues

The procurement process uses competitive bidding procedures that attract specialty contractors. To the extent possible, special provisions, plan details, and materials previously used by the transportation agency are specified to take advantage of bidders' familiarity. This procedure pays dividends to the transportation agency in terms of both overall cost and schedule duration.

The transportation agency includes in the contract documents all appropriate contract plan details and special provisions for all elements of the rockfall mitigation design. Ideally these documents have provisions relating to the qualifications and experience of the rockfall mitigation contractor and submittal requirements on the construction methods that the contractor plans to employ in constructing the rockfall mitigation project as designed.

Quantities for slope stabilization and slope protection work are notoriously difficult to estimate accurately during the design stage. Agencies employ contractual language or bidding practices

FIGURE 18-4 Example of rock slope stabilization specification (Findley and Norrish 2005).

that maximize the agency's ability to vary the quantities without incurring penalties or excessive cost premiums as the result of underruns or overruns in a contract bid item.

3.6 Construction of Mitigation Project

The design philosophy for most slope stabilization and protection projects is to assume that during construction, geological conditions are encountered that require field assessment and potential design modification. Quantity estimates for rockfall mitigation typically accommodate this philosophy. Furthermore, for the duration of the construction project, the transportation agency provides a full-time owner–representative for site inspection in addition to periodic or as-required site review by an experienced rock slope professional. Following completion of the rockfall mitigation project, as-built plans and quantities are prepared and the rockfall inventory is updated to show the date stabilization was completed and a synopsis of the construction that was undertaken.

4. FUNDING

Funding adequate for a comprehensive, proactive rockfall management program is critical to the success of the program. Without adequate and consistent funding, the transportation agency does not have a viable rockfall management program. If a rockfall management program is unfunded or underfunded, the transportation agency continues to face unplanned failures and the mitigation of the rockfall continues in a reactive manner, thereby incurring costs associated with travel delays, potential economic and social losses, elevated expenses resulting from emergency contracting, and potential negative perceptions from the public and the media.

The appropriate level of funding for a rockfall management program is specific to each transportation agency. It may be based on the scope of the rockfall hazard and risk and on the right scale of effort when funding for other programs within the organization is also considered. The level of funding also reflects the explicit or implicit risk-based decision making within the transportation agency and the tolerance of the rockfall problem by those who travel and depend on the affected transportation facility (MacDonald 2006).

Thus, the funding for a successful rockfall management program must consider the following attributes:

- Funding at an appropriate level needs to be a long-term commitment of the transportation agency. Initially, the agency faces a large number of unmitigated rockfall sites and needs to recognize and accept that it may take decades to fully realize the demonstrable benefits of a systemwide rockfall management program.
- The transportation agency typically recognizes that there will be unforeseen and unplanned rockfall events during the life of the rockfall management program. These emergent events need to be part of the overall funding strategy that the agency implements.
- For the program to be successful, the long-term funding for the rockfall management program needs endorsement at the highest decision-making level within the organization. In the case of some state transportation agencies, this endorsement may require legislative action.
- Funding levels need to be stable over time to allow for long-range planning and project programming. Stability is especially important when an agency is attempting to determine the revenue stream for the rockfall management program over multiple budget cycles.
- Establishing a dedicated funding source may be the best approach for a rockfall management program, and this source could be similar to those for other infrastructure preservation programs (e.g., bridge and pavement preservation programs) that already exist within many transportation agencies.
- The funding for a specific budget period is established at levels that can be supported and delivered by the agency. Program delivery and accountability problems will occur if the funding levels are set too high and the program cannot be fully accomplished.
- Ideally, the funding supports the core geotechnical activities of the rockfall management program. These activities include rockfall inventory database management, long-range planning, preliminary design and estimation, project programming support, final design, and construction support.

The benefits from a fully funded proactive rockfall management program will be realized over time as rockfall sites are mitigated. As the hazards are reduced, the risk associated with the transportation facility is also reduced. The focus of the program funding is on the full mitigation of sites. The transportation agency may also want to consider funding a parallel risk-reduction strategy (e.g., slope scaling) for high-risk, high-hazard slopes that are further out in the program schedule.

5. ISSUES IN IMPLEMENTATION OF A ROCKFALL MANAGEMENT PROGRAM

The effectiveness, and ultimately the success, of a rockfall management program is predicated on the finesse with which it is implemented. The primary implementation issues are (*a*) stakeholders and ownership, (*b*) program time frames, (*c*) program technical support, (*d*) comprehensiveness, and (*e*) model programs.

5.1 Stakeholders and Ownership

If a rockfall management program is perceived by a transportation agency to be only a research project, the program will have little chance of success, whereas a program that is viewed as having distributed, agencywide ownership—and therefore collective responsibility for the funding, planning, priority ranking, programming, design, and construction of rockfall mitigation projects—has a better chance of success. It is vitally important that upper management within the agency fully understand the importance and need for a proactive rockfall management program and recognize the benefits derived from such a program in terms of risk reduction, economic impact, and public accountability. The practical reality, especially for mountainous states with significant rockfall problems, is that it is impossible to fully implement an effective long-term rockfall management program without the active participation of all the stakeholders.

5.2 Program Time Frames

Although designing and populating the rockfall inventory database typically takes less than 2 to 3 years, the transportation agency must realize that the time frame for the meaningful reduction in rockfall hazard and associated risk for all slopes within the agency's infrastructure will require a decades-long commitment. For example, the rockfall management program of a national railway operating in British Columbia is realizing benefits from its programmed rock slope maintenance only after three decades of dedicated continuous effort. The actual length of time required for other agencies depends on the size of the annual funding allocations and the number of slopes in the rockfall inventory, but the transportation agency needs to allow decadal time scales and implement rockfall mitigation strategies and associated funding with decade-long time frames in mind.

5.3 Program Technical Support

To successfully implement a rockfall management program, the transportation agency needs to have a core group of geotechnical professionals that specialize in rockfall mitigation. These specialists typically have broad and extensive experience in all aspects of the rockfall mitigation program process and the design and construction of rockfall mitigation projects.

The rockfall management database is critical to the success of the rockfall mitigation program. The practical reality, especially for a mountainous region with a large number of rockfall sites, is that it is impossible to implement an effective, long-term rockfall management program without a comprehensive database. Ideally, the database will capture all the data associated with the rockfall sites and can share these data with others within the transportation agency. It is important that the program stakeholders be able to view exactly the same information when a problem is discussed or reviewed from offices at several distant locations. Web-based communications technology provides an ideal platform to share this distributed information. Data can then be input or viewed from authorized nodes throughout the organization.

5.4 Comprehensiveness

Although rockfalls pose significant hazards and risk, other large mass movements along linear transportation corridors frequently cause significant impacts on transportation facilities for extended lengths of time. These movements can have large economic consequences if a convenient detour is not available. In this context, there is merit in having the rockfall management program designed in such a fashion that it can be expanded to a broader slope management program that considers other geotechnical slope issues, such as landslides, debris flows, erosion, and settlement. The process remains the same, but the slope management program is more comprehensive.

5.5 Model Programs

A transportation agency that is considering the implementation of a rockfall management program needs to resist the temptation to reinvent the wheel. Numerous highly refined programs exist. A better utilization of initial efforts and funding will be to assess these model programs and tailor one, or some, of them to fit the specific needs or conditions of the transportation agency. Every dollar saved on the implementation study can be beneficially spent on the ultimate objective, prevention of rockfalls.

6. BENEFITS OF A ROCKFALL MANAGEMENT PROGRAM

The major benefits of a rockfall management program include (*a*) appropriate and defensible allocation of resources, (*b*) construction efficiency, (*c*) liability reduction, and (*d*) information exchange.

6.1 Allocation of Resources

The most obvious benefit of a rockfall management program is the rational allocation of limited personnel and financial resources to a problem that is almost limitless. The existence of a proactive rockfall management program enables the transportation agency to plan its slope mitigation spending many years in advance and in accordance with the policies and priorities defined and captured by the program. Ideally, the components of the rockfall management program will incorporate both the hazard posed by the rockfall site and site-specific highway conditions and the risk posed by the consequences of a slope failure. The risk can include such issues as economic and environmental impacts and public safety.

6.2 Construction Efficiency

Rockfall management programs provide a further benefit to the transportation agency in that they allow the agency to plan, procure, and perform rockfall mitigation construction work at times determined by the transportation agency, ostensibly under favorable weather, traffic, or environmental conditions. In the absence of such a rockfall management program, rockfall mitigation becomes highly reactive to individual events that often occur at inopportune times, leading to unplanned traffic delays and indirect economic losses. Reactive response usually eliminates competitive bidding, often resulting in expensive emergency mitigation efforts performed under force-account contracts.

6.3 Liability Reduction

The potential liability associated with indirect economic losses, environmental impacts, property loss, and personal injuries as a result of a rockfall is an increasing concern to transportation agencies. As the geotechnical engineering design-and-construction practice for stable rock slopes has improved substantially in the past several decades, the required standard of engineering practice for transportation agencies has also risen. Whereas ignorance of the hazard may have been a defense against rockfall liability in the past, current engineering practice dictates not only knowledge of the hazard (i.e., an inventory system) but also understanding of the mitigation practices employed to reduce the rockfall hazard. Furthermore, it may be possible to develop legal defenses that would limit or exclude liability to transportation agencies for specific rockfall incidents on the basis of discretionary reduction of hazards at other sites. The logic behind these strategies is that all potential rockfall sites cannot be mitigated simultaneously because of limited personnel and available funding. Therefore, the transportation agency should not be held liable if it is reducing overall rockfall risk on the basis of a rational, well-documented, proactive rockfall management program. Some of the elements that the legal system may consider when these rockfall management programs are evaluated are the robustness of the rockfall management process, the degree of consistency within the priority-ranking and decision-making processes, and the level of available funding to actively mitigate rockfall.

6.4 Information Exchange

Rockfall management programs provide a fourth benefit by supporting information exchange within the transportation agency, with outside entities working for the agency, and with the public at large. If properly designed, a rockfall management program database can help integrate and facilitate exchange of information within a transportation agency. Historical information in terms of the location and frequency of rockfall is often important both to regional managers, as they develop and implement their plans for future maintenance allocations, and to the geotechnical personnel who are assigning numerical ratings to slopes. A rockfall database can serve as a convenient storehouse for institutional information that is available to all internal users.

7. LIMITATIONS OF A ROCKFALL MANAGEMENT PROGRAM

Some aspects of the rockfall management program may be perceived as limiting the successful implementation of the program. The primary limitation issues are (*a*) geotechnical constraints, (*b*) inability to predict failure occurrences accurately, (*c*) lack of follow-through, and (*d*) establishment of an appropriate scope for the mitigation program. These limitations need to be addressed and communicated to the stakeholders in the process and efforts made to use them as opportunities to inform and educate the stakeholders on these issues.

7.1 Geotechnical Constraints

Geotechnical engineering, in particular its application to rockfall management programs, is an imperfect science because of site variances and external conditions beyond the control of the rockfall mitigation specialists. Rockfall management programs operate with limited information, and decisions are necessarily based on a combination of observational input, experience, professional judgment, analyses, and inference. The objective is to provide practical and defensible recommendations based on available information and resources.

Slope ratings are designed to be relative so as to enable comparison of all the slopes within a single transportation system. The absolute rating of a slope does not of itself indicate the need to carry out remedial measures or the urgency to do so. Rather, the individual rating in comparison with all other ratings within the inventory is a filtering mechanism through which the highest-priority sites can be identified for follow-up investigation, project programming, design, and mitigation.

7.2 Inability to Predict Failure

Rockfall management programs cannot predict the occurrence or frequency of rockfall at a given site because individual rockfall events result from interactions of a large number of site-specific variables. Thus, it is not possible, when managing a large number of rockfall sites, to predict which slope will fail next, regardless of the assessment rating. In addition, all stakeholders should recognize and accept that a low-rated slope is not a zero-hazard slope.

7.3 Lack of Follow-Through

To be effective, a rockfall management program needs to be comprehensive and sustainable over time. A well-crafted slope inventory system that only documents and numerically rates the rock slopes within a transportation system can, in some instances, become a liability to the transportation agency if it does not lead to a dedicated long-term funding stream that systematically reduces the risk to the users of the transportation system. The existence of a rockfall inventory without follow-up mitigation actions becomes prima facie evidence that the transportation agency was aware of the hazard and the associated risk but consciously chose not to incur the mitigation costs. This shortcoming may raise negative public perception issues and have legal implications.

7.4 Scope of Mitigation Program

The goal of a typical rockfall management program is full mitigation of each site at a reasonable cost and with effectiveness extending over a long design life. In other words, mitigation is all or nothing at a given site. Although this may be a preferred approach, the transportation agency must recognize that there are unmitigated slopes in the inventory and within the program mitigation queue that are still hazardous and pose a risk to the transportation facility. At many locations, a minimal slope treatment—for example, hand scaling or ditch widening—can remediate a large component of the risk. Ideally, methodologies are considered that support the implementation of a risk-reduction program that can complement and balance the effects and demands of a full-treatment rockfall mitigation program.

REFERENCES

ABBREVIATION

FHWA Federal Highway Administration

Burk, R., N. I. Norrish, S. M. Lowell, M. Molinari, R. LaHusen, J. Schick, B. Strickler, B. Dueval, and M. McCabe. 2004. Washington SR 20 Rock Avalanche: Monitoring, Characterization, and Analysis. *Proc., 55th Annual Highway Symposium*, Kansas City, Mo., Sept. 7–10, pp. 129–147.

Fell, R., K. K. S. Ho, S. Lacasse, and E. Leroi. 2005. A Framework for Landslide Risk Assessment and Management. In *Landslide Risk Management: Proceedings of the International Conference on Landslide Management* (O. Hungr, R. Fell, R. Couture, and E. Eberhardt, eds.), Vancouver, Canada, May 31–June 3, Taylor & Francis, London, pp. 3–26.

FHWA. 1999. *Asset Management Primer*. U.S. Department of Transportation.

Findley, D. P., and N. I. Norrish. 2005. *Engineering Report to Washington State Department of Transportation for Rock Slope Stabilization, Milepost 66.00 to 66.58, Interstate 90*. Golder Associates, Inc., and Wyllie and Norrish Rock Engineers, Inc., March.

Haneberg, W. C., N. I. Norrish, and D. P. Findley. 2006. Digital Outcrop Characterization for 3-D Structural Mapping and Rock Slope Design Along Interstate 90 Near Snoqualmie Pass, Washington. *Proc., 57th Annual Highway Geology Symposium*, Breckenridge, Colo., Sept. 27–29, pp. 146–160.

Lowell, S., and P. Morin. 2000. Unstable Slope Management: Washington State. *TR News*, No. 207, March–April, pp. 11–15.

Lowell, S. M., W. Gates, L. Moses, C. Lukkarila, B. Fisher, T. C. Badger, and N. I. Norrish. 2005. Conceptual Design and Cost Estimates: A Critical Step in Managing Unstable Slopes along Washington State Highways. *Proc., 56th Annual Highway Geology Symposium*, Wilmington, N.C., May 4–6, pp. 19–32.

Lowell, S. M., L. J. Moses, and T. C. Badger. 2005. Management of Slope Hazards and Risk Along Washington State Highways. In *Landslide Risk Management: Proceedings of the International Conference on Landslide Management* (O. Hungr, R. Fell, R. Couture, and E. Eberhardt, eds.), Vancouver, Canada, May 31–June 3, Taylor & Francis, London, Supplementary Volume (CD-ROM), CD Submission L100.

MacDonald, D. B. 2006. *Unstable Slopes on I-90 Snoqualmie Pass: Re-assessment and Recommendations*. Washington State Department of Transportation, Olympia.

Roberds, W. 2005. Estimating Temporal and Spatial Variability and Vulnerability. In *Landslide Risk Management: Proceedings of the International Conference on Landslide Management* (O. Hungr, R. Fell, R. Couture, and E. Eberhardt, eds.), Vancouver, Canada, May 31–June 3, Taylor & Francis, London, pp. 129–158.

Wyllie, D. C., and N. I. Norrish. 1996. Stabilization of Rock Slopes. In *Special Report 247: Landslides: Investigation and Mitigation* (A.K. Turner and R.L. Schuster, eds.), Transportation Research Board, National Research Council, Washington, D.C., pp. 474–504.

Appendix A

STEREOGRAPHIC PROJECTIONS FOR STRUCTURAL ANALYSIS

Chapter 2 uses stereographic projections to analyze the orientation of structural discontinuities in rock masses and to identify and evaluate potential rock slope instabilities or failure mechanisms. Figure A-1*a* defines the terms *strike*, *dip*, and *dip direction*, which are used by geologists to define the three-dimensional (3-D) orientations of planar features. Figure A-1*b* illustrates how 3-D orientations are represented on a reference sphere. Figure A-1*c* shows how a stereographic projection, or *stereonet*, can represent these 3-D orientations as a two-dimensional graphic. Descriptions of the use of stereonets for the analysis of rock slope stability are found in Norrish and Wyllie (1996) and Wyllie and Mah (2004).

Additional information on the derivation and use of stereonets may be found elsewhere (Dennison 1968; Ragan 1973). The principles of stereographic projection are discussed in most structural geology textbooks. Two of the most useful stereographic projections, or nets, are reproduced on the following pages:

- The equal-area Schmidt net (or Lambert net) (Figure A-2) and
- The equal-area polar (Billings) net (Figure A-3).

Both of these stereonets have been developed to maintain an equal-area property; this means that areas between lines of latitude and parallels of longitude remain equal over the entire region shown. This property is useful in comparing the frequency distribution of data by point density patterns or contoured results.

The equal-area Schmidt net (Figure A-2) is always used as a graph underlying a transparent or semitransparent sheet attached to the stereonet by a drawing pin placed through the center point. The various orientations of structural features may be shown by tracing great and small circles from the underlying stereonet onto the overlay. Generally it is necessary to rotate the overlay about the stereonet center to develop the desired orientations.

The equal-area polar (Billings) net (Figure A-3) is especially convenient for plotting linear data or poles of planar features in the field. A transparent sheet may be fixed on top of this polar stereonet, and rotations of the sheet are not required. Diagrams prepared with these two stereonets may be interchanged, provided their radii are equal, as shown in Figures A-2 and A-3.

REFERENCES

Dennison, J. M. 1968. *Analysis of Geologic Structures*. W. W. Norton & Company, New York.

Norrish, N. I., and D. C. Wyllie. 1996. Rock Slope Stability Analysis. In *Special Report 247: Landslides: Investigation and Mitigation* (A. K. Turner and R. L. Schuster, eds.), Transportation Research Board, National Research Council, Washington, D.C., pp. 391–428.

Ragan, D. M. 1973. *Structural Geology: An Introduction to Geometrical Techniques*, 2nd ed. John Wiley and Sons, Inc., New York.

Wyllie, D. C., and C. W. Mah. 2004. *Rock Slope Engineering*, 4th ed. Spon Press, New York.

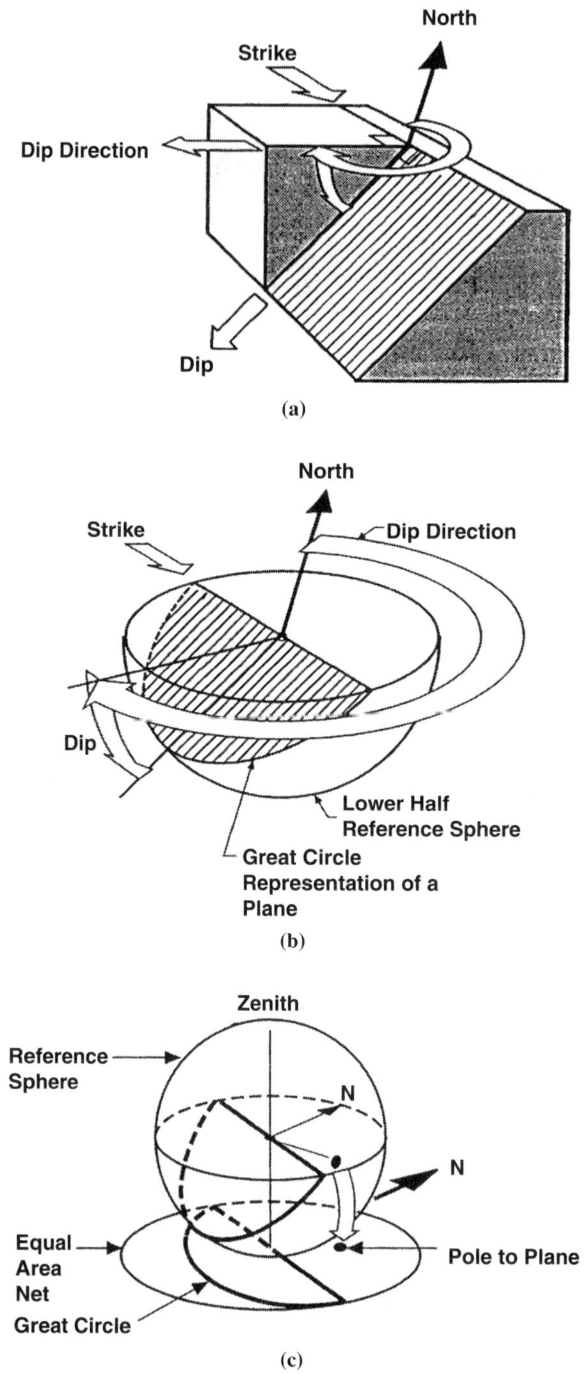

FIGURE A-1
Concepts of stereographic representation of planar features (rock mass discontinuities): (*a*) definition of terms, (*b*) representation on reference sphere, and (*c*) creation of the stereonet projection (adapted from Norrish and Wyllie 1996).

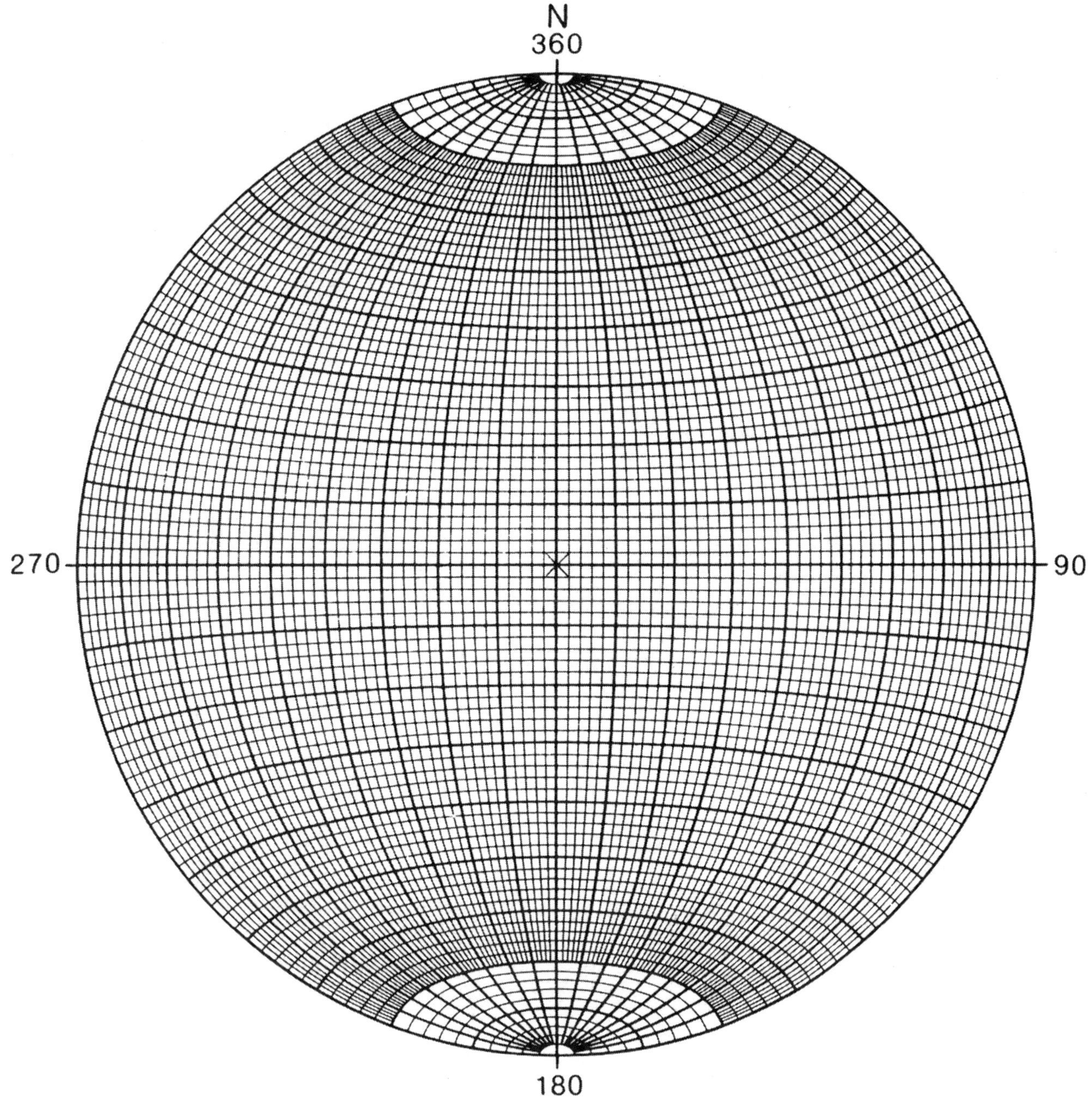

FIGURE A-2
Equal-area Schmidt net.

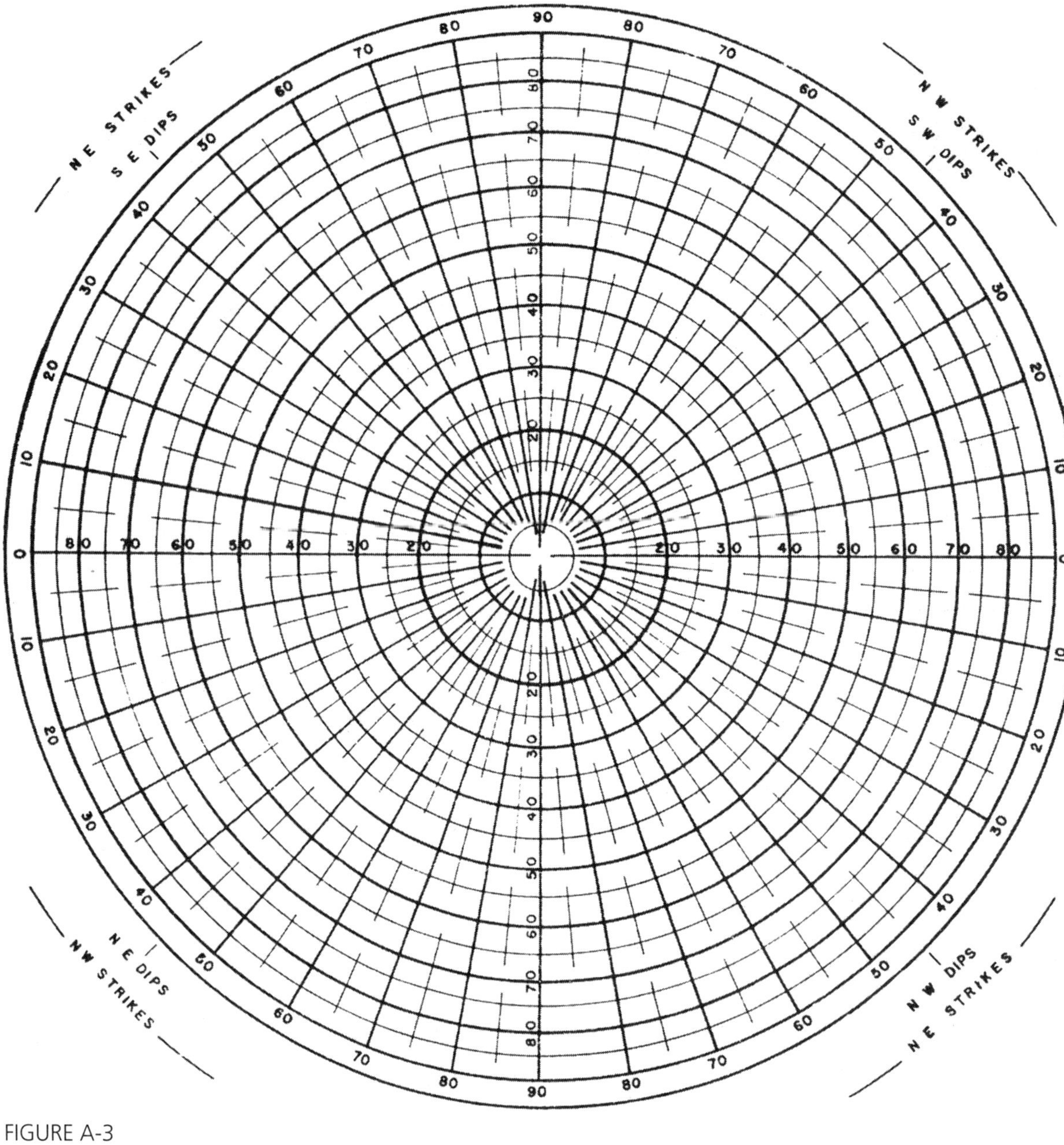

FIGURE A-3
Equal-area polar (Billings) net. The net is graduated for poles of planar elements (lower hemisphere).

JOHN D. DUFFY

DESCRIPTION OF DVD MATERIALS

1. INTRODUCTION

This appendix provides information on the content and use of the supplementary materials provided on the DVD included with this book. This DVD contains the following materials:

- Digital copies of all illustrations in the book, organized by chapter.
- A series of 29 short video clips of rockfall field tests that illustrate different levels of rockfall impact energies and some testing procedures.
- Four longer videos:
  1. The film documenting the historic 1963 rockfall testing procedures conducted in Washington State and reported by Ritchie (1963);
  2. The 1990 field tests and evaluations of rockfall restraining nets conducted by the California Department of Transportation (Caltrans) and reported by Smith and Duffy (1990);
  3. Field tests conducted in conjunction with the 2001 rockfall catchment design guide for the Oregon Department of Transportation (DOT) and reported by Pierson et al. (2001); and
  4. Rock slope scaling techniques recorded by the Colorado DOT.

Figure B-1 illustrates the structure and contents of the DVD, which has three parts. Part A contains digital versions of the book's illustrations in standard digital image file formats that can be viewed with commonly used software (see Section 2 of this appendix). Part B contains 29 short videos of rockfall field tests, which are described in detail in Section 3. Part C contains the four longer videos listed above and described in detail in Section 4.

The videos are provided in the QuickTime Movie (.mov) format, a video compression format developed by Apple Inc. but also available on computers that use the Microsoft Windows operating system. This format was selected for its image quality and broad compatibility with most personal computer systems that use freely accessible software. These files can be viewed on a personal computer, or excerpts can be extracted and used to illustrate particular issues.

Readers should note that extended pauses may occur when any of these videos is accessed directly from the DVD, depending on the capabilities of the computer hardware being used to view the video. Pauses are more prevalent when larger (longer-duration) video files are viewed. Such pauses occur when the computer accesses and loads additional sections of the video from the DVD into memory. Users can reduce or eliminate these pauses by downloading the entire video file from the DVD onto their computer's hard drive and launching the video from the hard drive. Downloading the larger video files typically takes several minutes, and adequate storage space on the hard disk is necessary.

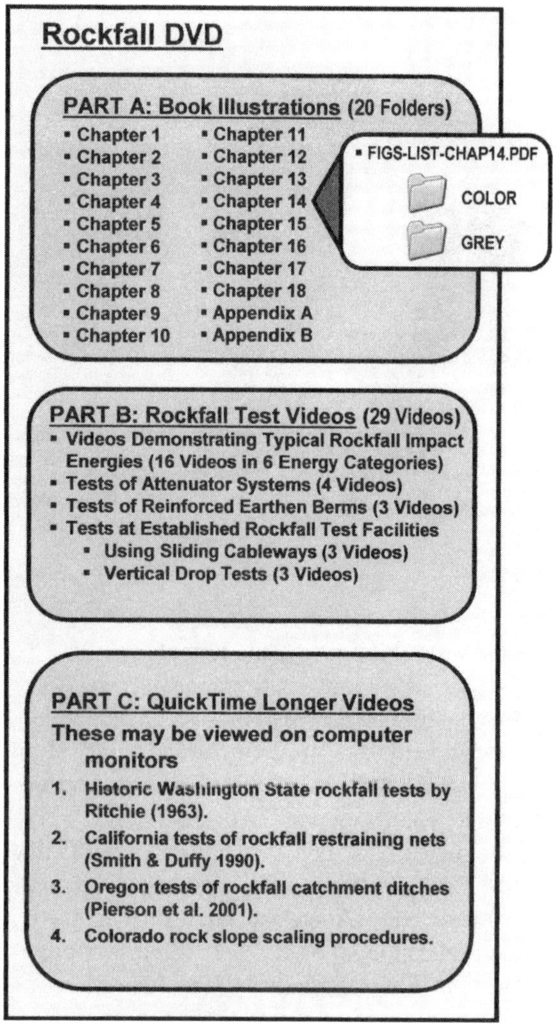

FIGURE B-1 Overview of structure and contents of DVD. Insert to Part A shows contents of folder for Chapter 14. The folder for each chapter and appendix contains equivalent information.

2. DIGITAL COPIES OF BOOK ILLUSTRATIONS

Part A of the DVD contains digital versions of the book's illustrations. These are organized in folders by chapter, and two sets of files are provided for each chapter: (*a*) the complete set of grayscale photographs and black-and-white graphs and line drawings published in the book (subfolder labeled "GREY") and (*b*) a set of color files for those illustrations that were originally in color, which include most of the photographs and some technical drawings (subfolder labeled "COLOR"). Each chapter folder also contains a PDF text file that lists the contents of the COLOR and GREY subfolders. Table B-1 summarizes the number of illustrations and image files by chapter.

Each image is provided as a separate file. These files are in standard digital image file formats that can be viewed with commonly used software. Thus, desired image files can readily be found and copied into PowerPoint lecture presentations or included within other teaching materials or research papers. **Proper source citation should accompany any use of the illustrations in this volume.**[1] **Reproduction for publication of any illustration contained in this volume is subject to copyright and requires permission.**

The image files are named according to the following convention:

- Grayscale illustrations: FIG*cc*-*nnn*-GREY(*xxx*dpi-*ww*in).*ext*;
- Line drawings: FIG*cc*-*nnn*-DWG(*xxx*dpi-*ww*in).*ext*; and
- Color illustrations: FIG*cc*-*nnn*-COLOR(*xxx*dpi-*ww*in).*ext*.

In this naming scheme, *cc* represents the chapter number, *nnn* the figure number (*cc*-*nnn* matches the figure identifier used in the text), *xxx* the figure resolution in dots per inch (dpi) when viewed at 100% size, *ww* the width of the illustration in inches at 100% size, and *ext* the file format. Commonly used file formats include JPG (Joint Photographic Experts Group), GIF (Graphics Interchange Format), PNG (Portable Network Graphics), and TIFF (Tagged Image File Format).

3. VIDEO RECORDS OF ROCKFALL FIELD TESTS

Part B of the DVD contains 29 brief video records that have been selected to illustrate different levels of rockfall impact energy or rockfall field-testing procedures. Each video is identified with a video number (VN). Tables B-2 and B-3 summarize the contents of Part B.

These videos show rockfall events at 11 test sites in the United States, Europe, and Japan. The characteristics of these sites are described in Section 3.1. Section 3.2 discusses 16 videos that illustrate typical rockfall impact energies. These videos are grouped in six energy categories. An additional 13 videos illustrate tests of rockfall

[1]Recommended citation: Turner, A. K., and R. L. Schuster (eds.). 2012. *Rockfall: Characterization and Control*. Transportation Research Board of the National Academies, Washington, D.C.

Table B-1
Digital Files of Illustrations Contained in Part A of DVD, by Chapter or Appendix

		Illustration Files on DVD[a]	
Book Chapter or Appendix	Illustrations in Text[b]	Subfolder GREY[c]	Subfolder COLOR[c]
Chapter 1	23	24	18
Chapter 2	44	45	26
Chapter 3	5	5	3
Chapter 4	28	29	15
Chapter 5	47	49	20
Chapter 6	28	32	19
Chapter 7	82	89	57
Chapter 8	50	51	1
Chapter 9	72	81	40
Chapter 10	56	64	37
Chapter 11	7	7	7
Chapter 12	26	28	26
Chapter 13	33	34	22
Chapter 14	48	67	47
Chapter 15	43	60	44
Chapter 16	30	35	26
Chapter 17	14	17	12
Chapter 18	6	6	5
Appendix A	3	5	0
Appendix B	23	23	12

[a]Number of files may be greater than number of illustrations in text because illustrations in multipart figures may be in separate files.
[b]Number of individually identified and numbered illustrations, including those included in numbered box displays.
[c]Number of distinct illustration files, including files for parts of multipart figures.

attenuator systems (Section 3.3), tests of reinforced earthen berms (Section 3.4), and procedures for conducting tests at established rockfall test facilities (Section 3.5).

3.1 Descriptions of Test Locations

The rockfall field tests documented in the videos contained on the DVD were conducted at 11 test sites in California and Colorado in the United States; Switzerland, Austria, and Italy in Europe; and Japan. Table B-4 lists the general locations and names of these sites and which videos were recorded at each. The following subsections briefly describe each site, including its specific location and dimensions; provide a photograph of the site and, in most cases, a profile; and give additional pertinent information concerning the slope materials, the ownership of the site, and the methods used to initiate a rockfall test.

3.1.1 Shale Point Test Site, California

The Shale Point test site is on Caltrans property located along California State Route 1 at Postmile 8.0 in Monterey County, California (Duffy and Hoon 1996a, 1996b, 1998). The slope site is 20 m (66 ft) high with an actual slope length of 33 m (110 ft) at an angle of 37° (Figures B-2 and B-3). The slope surface consists of soil with occasional rock fragments and very little vegetation. Overall, the slope is smooth with some gullies. Rocks were dropped from a front-end loader. Videos VN-1, VN-2, and VN-8 show tests conducted at this site.

3.1.2 Grey Slip Test Site, California

The Grey Slip test site is on Caltrans property located along the Pacific Ocean coastline on California State Route 1 at Postmile 6.8 in

Table B-2
Video File Contents of Part B of DVD: Typical Rockfall Impact Energies

Video Number	Energy Category (kJ)	Video File	File Size (MB)	Run Time [min (s)]	Test Location (number and name)	Impact Energy [kJ (ft-tons)]	Remarks
VN-1	50	45kJ-ImpactEnergy.mov	16.7	0 (38)	1. Shale Point, California	45 (16)	Single rock impact on fence
VN-2	50	67kJ-ImpactEnergy.mov	44.1	0 (32)	1. Shale Point, California	67 (25)	Single rock impact on concrete barrier
VN-3	100	102kJ-ImpactEnergy.mov	112.8	1 (07)	2. Grey Slip, California	102 (38)	Single rock impact on fence
VN-4	150	144kJ-ImpactEnergy.mov	14.7	0 (28)	6. Oberbuchsiten, Switzerland	144 (53)	Single rock impact on fence
VN-5	150	173kJ-ImpactEnergy.mov	37.8	0 (54)	2. Grey Slip, California	173 (64)	Single rock impact on fence
VN-6	200	180kJ-ImpactEnergy.mov	24.6	0 (38)	2. Grey Slip, California	180 (67)	Single rock impact on post
VN-7[a]	200	198kJ-ImpactEnergy.mov	21.9	0 (36)	6. Oberbuchsiten, Switzerland	198 (73)	Rock 2 impact = 117 kJ:43 ft-tons
VN-8	200	199kJ-ImpactEnergy.mov	25.7	0 (55)	1. Shale Point, California	199 (74)	Single rock impact on fence
VN-9[a]	200	225kJ-ImpactEnergy.mov	20.2	0 (34)	6. Oberbuchsiten, Switzerland	225 (83)	Rock 2 impact = 171 kJ:63 ft-tons
VN-10	500	356kJ-ImpactEnergy.mov	37.4	0 (54)	2. Grey Slip, California	356 (132)	Single rock impact on fence
VN-11	500	404kJ-ImpactEnergy.mov	66.8	1 (02)	2. Grey Slip, California	404 (150)	Rock fragment over barrier
VN-12	500	612kJ-ImpactEnergy.mov	20.3	0 (35)	6. Oberbuchsiten, Switzerland	612 (227)	Single rock impact on fence
VN-13	1,000	862kJ-ImpactEnergy.mov	18.5	0 (30)	6. Oberbuchsiten, Switzerland	862 (319)	Rock overtops cable net
VN-14	1,000	938kJ-ImpactEnergy.mov	18.5	0 (31)	6. Oberbuchsiten, Switzerland	938 (347)	Single rock impact on ring net
VN-15	1,000	1166kJ-ImpactEnergy.mov	46.2	0 (46)	3. Rifle, Colorado	1,166 (430)	Single rock impact on cable net
VN-16	1,000	1260kJ-ImpactEnergy.mov	17.0	0 (30)	6. Oberbuchsiten, Switzerland	1,260 (467)	Single rock impact on ring net

Note: See Section 3.2.
[a]Rock 1.

Table B-3

Video File Contents of Part B of DVD: Tests of Attenuator Systems and Reinforced Earthen Berms and Tests at Established Rockfall Testing Facilities

Video Number	Video File	File Size (MB)	Run Time [min (s)]	Test Location (number and name)	Impact Energy [kJ (ft-tons)]	Remarks
Tests of Rockfall Attenuator Systems[a]						
VN-17	64kJ-Attenuator.mov	43.5	0 (55)	7. Saint Leonard, Switzerland	64 (24)	Single rock impact
VN-18	110kJ-Attenuator.mov	98.6	1 (40)	3. Rifle, Colorado	61–110 (23–41)	Rocks hit railroad tie barrier
VN-19	160kJ-Attenuator.mov	192.2	2 (32)	4. Georgetown, Colorado	118–160 (44–59)	Multiple attenuator tests
VN-20	250kJ-Attenuator.mov	55.4	0 (52)	10. Vigo di Meano, Italy	250 (93)	Single rock hits attenuator
Tests of Reinforced Earthen Berms[b]						
VN-21	1010kJ-Berm.mov	35.5	0 (28)	3. Rifle, Colorado	1,010 (374)	Single rock hits berm
VN-22	1394kJ-Berm.mov	20.2	0 (23)	3. Rifle, Colorado	1,394 (516)	Single rock hits berm
VN-23	2217kJ-Berm.mov	24.1	0 (32)	11. Maeda Kosen Quarry, Fukui, Japan	2,217 (821)	Single rock hits berm
Tests at Established Rockfall Testing Facilities That Employ Sliding Cables[c]						
VN-24	2094kJ-SlidingCable.mov	25.4	0 (33)	9. Styrian Erzberg, Austria	2,094 (775)	Cable launch into fence
VN-25	2507kJ-SlidingCable.mov	21.1	0 (35)	10. Vigo di Meano, Italy	2,507 (928)	Cable launch into earthen berm
VN-26	4474kJ-SlidingCable.mov	17.9	0 (22)	10. Vigo di Meano, Italy	4,474 (1,657)	Cable launch into earthen berm
Tests at Established Rockfall Testing Facilities That Employ Vertical Drop Tests[c]						
VN-27	274kJ-VertDrop.mov	21.5	0 (42)	5. Interstate 70, Georgetown, Colorado	274 (101)	Vertical drop test of fence
VN-28	2045kJ-VertDrop.mov	16.3	0 (44)	8. Walenstadt, Switzerland	2,040 (755)	Vertical drop test of fence
VN-29	8179kJ-VertDrop.mov	17.8	0 (33)	8. Walenstadt, Switzerland	8,000 (2,963)	Vertical drop test of fence

[a]See Section 3.3.
[b]See Section 3.4.
[c]See Section 3.5.

Table B-4
Videos Illustrating Tests at Each Site

LOCATION	NUMBER	TEST SITE	VIDEO NUMBER
California	1	Shale Point	VN-1, VN-2, VN-8
	2	Grey Slip	VN-3, VN-5, VN-6, VN-10, VN-11, VN-31
Colorado	3	Rifle	VN-15, VN-18, VN-21, VN-22
	4	Georgetown	VN-19
	5	Interstate 70, Georgetown	VN-27
Switzerland	6	Oberbuchsiten	VN-4, VN-7, VN-9, VN-12, VN-13, VN-14, VN-16
	7	Saint Leonard	VN-17
	8	Walenstadt	VN-28, VN-29
Austria	9	Styrian Erzberg	VN-24
Italy	10	Vigo di Meano	VN-20, VN-25, VN-26
Japan	11	Maeda Kosen Quarry, Fukui	VN-23

FIGURE B-2
Shale Point test site, Monterey County, California.
(PHOTOGRAPH COURTESY OF THE CALIFORNIA DEPARTMENT OF TRANSPORTATION.)

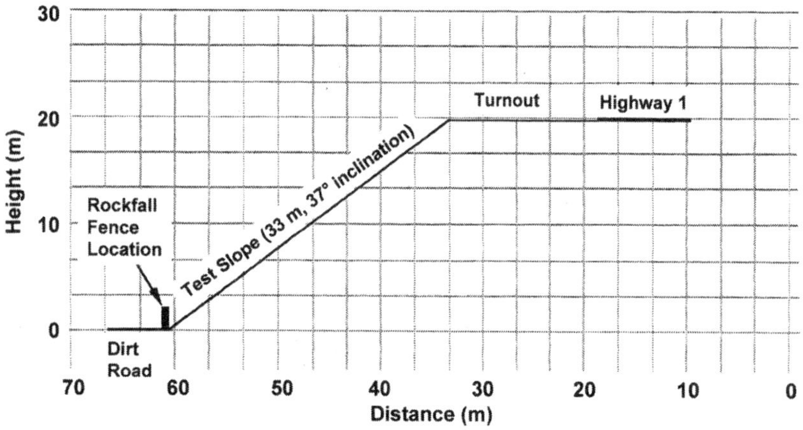

FIGURE B-3
Cross section of Shale Point test site.
(IMAGE COURTESY OF THE CALIFORNIA DEPARTMENT OF TRANSPORTATION.)

Monterey County between Big Sur and San Simeon (Smith and Duffy 1990). The test slope is 39 m (130 ft) high with an overall slope angle of 34° (Figures B-4 and B-5). Vegetation is sparse. The slope material is composed of landslide debris consisting of rock fragments with diameters of 3 to 55 cm (1 to 18 in.) in a matrix of clayey silt. The material was dry and hard during the tests. Rocks were dropped from a front-end loader. Videos VN-3, VN-5, VN-6, VN-10, and VN-11 show tests at this site. One of the longer videos included in Part C of the DVD, Video VN-31, which shows the field procedures used to test new rockfall restraining nets (see Section 4.2), was also recorded at this site.

3.1.3 Rifle Test Site, Colorado

Between 1987 and 1998, the Colorado DOT conducted several full-scale rock-rolling experiments,

FIGURE B-4
Grey Slip test site, Monterey County, California.
(PHOTOGRAPH COURTESY OF THE CALIFORNIA DEPARTMENT OF TRANSPORTATION.)

primarily to test a variety of rockfall fence systems, at a test site located along Interstate 70 a short distance west of Rifle, Colorado. The site has since been reclaimed and is no longer available for rockfall testing. The site consisted of a hillside 91 m (300 ft) high, with a slope angle of 37°. The slope was covered with colluvial soils weathered from claystone and sandstone bedrock. The sparse vegetation apparently did not influence the rock motions.

Test rocks ranged in weight from 227 to 2,722 kg (500 to 6,000 lb). The rocks were launched onto the slope by either pushing or dropping them with a front-end loader. The upper part of the slope (Zone 1 in Figures B-6 and B-7) was convex in plan, and this caused considerable divergence of initial trajectories. The lower portion of the slope (Zone 2 in Figures B-6 and B-7) allowed most rocks to follow more consistent trajectories. As the rocks rolled down the slope, they reached velocities between 9 and 25 m/s (30 and 82 ft/s).

Full-scale tests conducted in 1987 and 1988 were used as a basis for calibration of the Colorado

FIGURE B-6
Rifle test site, Rifle, Colorado (Pfeiffer 1989).

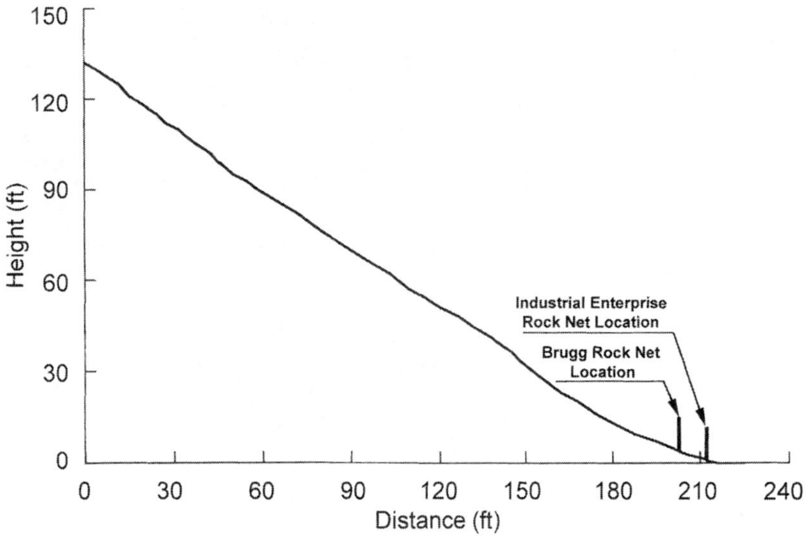

FIGURE B-5
Cross section of Grey Slip test site.
(GRAPH COURTESY OF THE CALIFORNIA DEPARTMENT OF TRANSPORTATION.)

Rockfall Simulation Program, timber attenuators, tire attenuators, and a combination of tire and timber attenuators (Barrett et al. 1989; Pfeiffer 1989; Pfeiffer and Higgins 1990; Pfeiffer et al. 1990; Andrew 1992). In 1992, timber-faced mechanically stabilized earthen barriers were tested (Parsons De Leuw, Inc. 1992). A few of these tests are shown in Videos VN-18, VN-21, and VN-22. In 1998 full-scale tests were performed to test cable net flexible rockfall fences, as seen in Video VN-15 (Andrew et al. 1998).

3.1.4 Georgetown Test Site, Colorado

The 3.5-km (2.2-mi) section of Interstate 70 between the towns of Georgetown and Silver

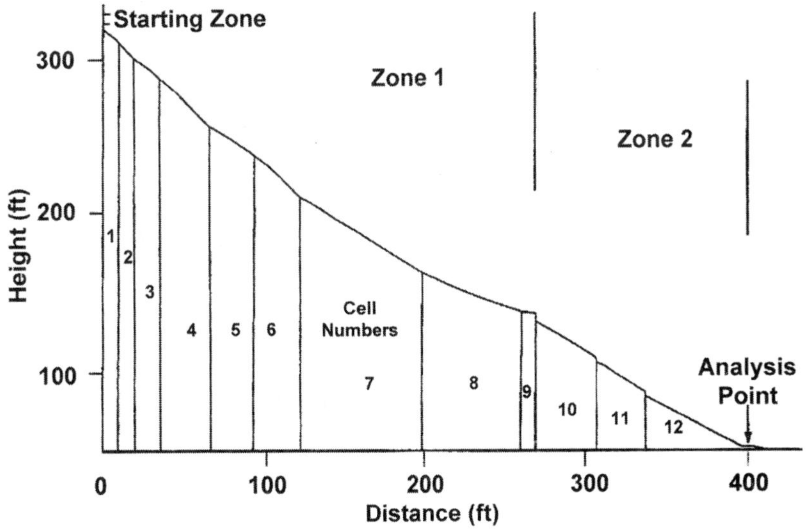

FIGURE B-7
Cross section of Rifle test site (Pfeiffer 1989).

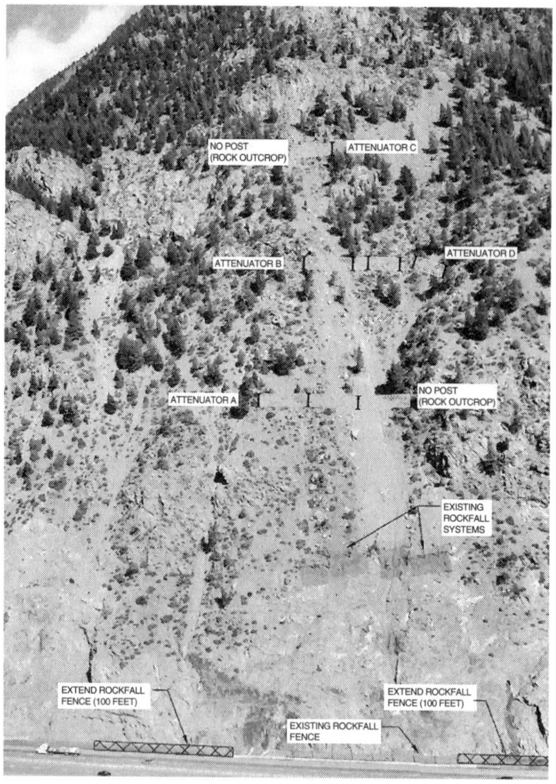

FIGURE B-8 Georgetown test site, Georgetown, Colorado (Arndt et al. 2009).

Plume is locally referred to as the *Georgetown Incline* because the roadway climbs approximately 150 m (500 ft) with a continuous steep grade. The alignment is cut into steep mountainous slopes that exceed 1V:1H in many places. The natural slopes rise more than 518 m (1,700 ft) above the current highway alignment along most of this section.

FIGURE B-9 Cross section of Georgetown test site (Arndt et al. 2009).

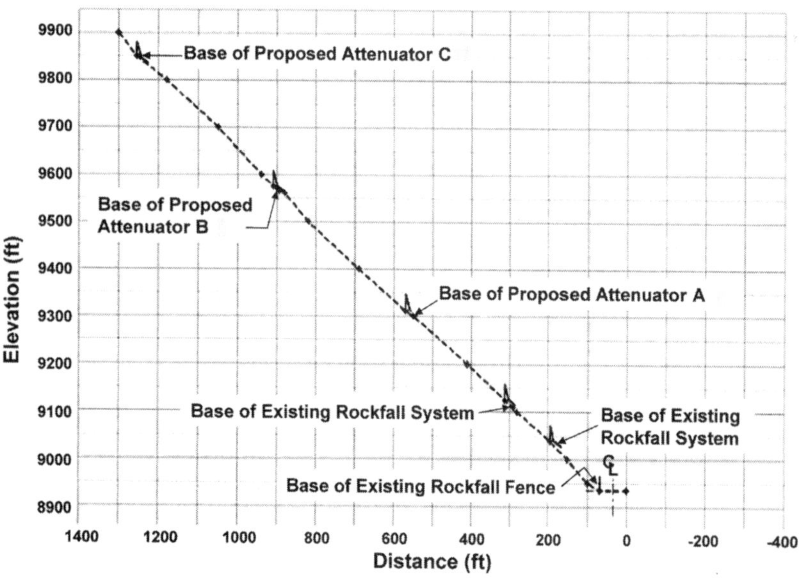

The bedrock consists of gneiss and schist with granite and pegmatite intrusions. Surficial materials covering the steep slopes above the roadway consist of colluvium, talus, and isolated mine tailings; these deposits include materials ranging from silt to boulders. Numerous bedrock outcrops form vertical cliff faces that create potential source areas for rockfall. In addition, the local elevations range between approximately 2,500 m and 3,500 m (8,400 ft and 11,700 ft) and thus experience conditions that are conducive to rockfall, including periods of intense precipitation and extended freeze-thaw events. The combination of climate, steep slopes, fractured bedrock outcrops, and relatively loose surficial materials that include many large fragments creates an extensive rockfall hazard area. Multiple rockfall source areas are located as much as 600 m (2,000 ft) above the roadway.

Just above the highway, rock cuts and disturbed areas that were created to accommodate the multilane roadway generally range from 6 to 45 m (20 to 150 ft) high and have slope angles ranging from vertical to 60°. There was little opportunity to create rockfall catchment ditches of adequate width during construction. Well-defined rockfall chutes channel the rockfall down to the roadway (Figures B-8 and B-9). Video VN-19 shows several rockfall tests at this site before and after the installation of rockfall attenuator systems.

As shown in Figures B-8 and B-9, the potential rockfall hazard at this site has led to the installation of a series of rockfall attenuator systems (Arndt et al. 2009). Initially, two levels of rockfall barrier fences were installed on the lower portions of active rockfall chutes. These were intended to slow rocks down and dampen bounce heights as the rocks travel down the slope. A fence and barrier system was installed along the highway shoulder to serve as a rockfall catchment area. These initial attempts at rockfall attenuation were soon damaged by high-energy rockfall impacts, so additional research was undertaken to produce better system designs (Arndt et al. 2009). In 2009, three additional rockfall attenuator systems incorporating new design criteria were installed at higher locations on the slope. Thus, the Georgetown test site now has five levels of attenuators at increasing elevations above the highway. Video VN-19 provides examples of tests of these systems.

3.1.5 Temporary Vertical Drop Test Facility Along Interstate 70, Georgetown, Colorado

In April 2010, the Colorado DOT established a temporary vertical drop test facility along Interstate 70 at the Georgetown test site described in Section 3.1.4. Tests were conducted to develop an understanding of rockfall net reactions under dynamic loads and minimize panel impact deflections (T. Ortiz, personal communication, January 23, 2011). The Colorado DOT established this temporary test site adjacent to Interstate 70, where a rockfall mitigation project was under construction (Figure B-10). Use of this site eliminated mobilization costs because the contractor for the rockfall mitigation project also performed the testing (T. Ortiz, personal communication, January 23, 2011). Concrete blocks of various sizes were dropped 18.3 m (60 ft) from a crane onto high-strength rockfall cable net panel systems that did not include any braking elements. The blocks were cast as cuboctahedrons—polyhedrons with eight triangular faces and six square faces that provide a rounded cube shape with fairly sharp edges and corners (Arndt et al. 2009). The cable nets were attached to four 4.6-m (15-ft) steel posts installed into the nearly vertical rock face (Figure B-10). The distance between adjacent posts was 9.1 m (30 ft), so the dimensions of the entire cable net installation being tested were 4.6 m × 27.4 m (15 ft × 90 ft). Figure B-11 shows a block just after its initial impact. Further details of these tests are described in Section 3.5.2, and Video VN-27 shows one test conducted at this site. The testing evaluated different panel systems at various impact energies but did not follow any established rockfall barrier testing guidelines (T. Ortiz, personal communication, January 23, 2011).

FIGURE B-10
Temporary vertical drop test facility established along Interstate 70 near Georgetown, Colorado.
(PHOTOGRAPH COURTESY OF THE COLORADO DEPARTMENT OF TRANSPORTATION.)

FIGURE B-11
Vertical drop test under way at the temporary test facility established along Interstate 70 near Georgetown, Colorado. The 1,588-kJ (3,384-lb) concrete block is rebounding from its initial impact with the mesh, which has been stretched by the impact.
(PHOTOGRAPH COURTESY OF THE COLORADO DEPARTMENT OF TRANSPORTATION.)

3.1.6 Oberbuchsiten Test Site, Switzerland

The Oberbuchsiten test site, located outside the town of Oberbuchsiten in the Jura Mountains of Switzerland, is a barren exposure of the Jura Formation, a very hard, competent, dark gray limestone. There is almost no vegetation on the slope surface, which is relatively smooth (Figure B-12). The slope has an angle of 45°, and several possible launch points provide a variety of elevation differences and slope lengths for rockfall trajectories (Figure B-13). The site is located on private property.

FIGURE B-12
Oberbuchsiten test site, Oberbuchsiten, Switzerland.
(PHOTOGRAPH COURTESY OF GEOBRUGG.)

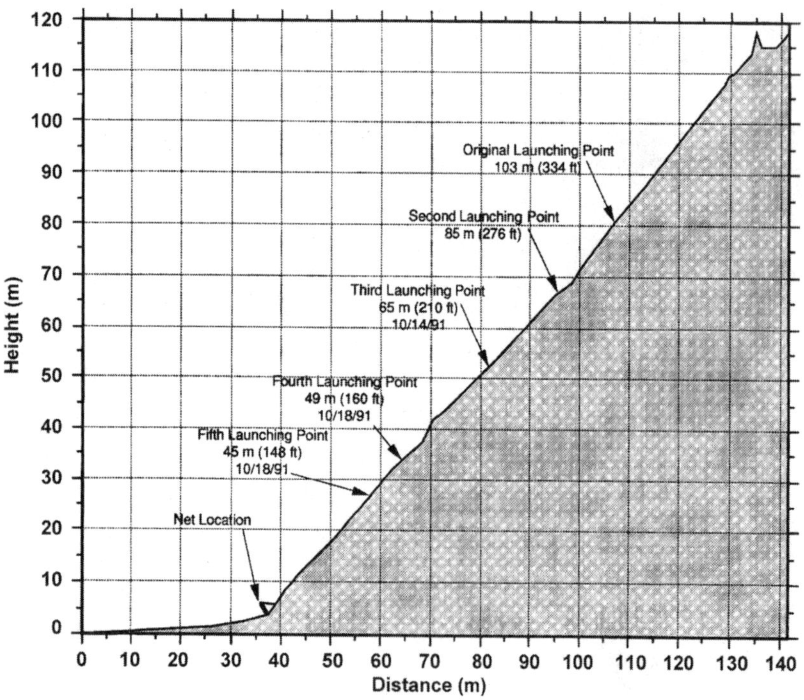

FIGURE B-13
Cross section of Oberbuchsiten test site (Duffy and Haller 1993).

A series of tests was conducted at this site in 1992 (Duffy 1992). The much harder rock surface at this site provided higher rockfall velocities and impact energies than were produced at the previously discussed California and Colorado test sites [Shale Point (Section 3.1.1), Grey Slip (Section 3.1.2), and Rifle (Section 3.1.3)]. At Oberbuchsiten, the rocks were delivered to the selected launch point via a cable lift, temporarily stored on a man-made ledge at the launch point, and then pushed over the lip to begin their trajectory down the slope. Videos VN-4, VN-7, VN-9, VN-12 through VN-14, and VN-16 give examples of a range of energy impacts from tests conducted at this site.

3.1.7 Saint Leonard Test Site, Switzerland

The Saint Leonard test site, developed by the Swiss Federal Institute for Forest, Snow and Landscape (WSL) in cooperation with Geobrugg Protection Systems AG, is located in Saint Leonard, Canton Valais, Switzerland. The site is specifically designed to test rockfall attenuator systems (Glover et al. 2010; WSL 2010). Located within an existing limestone quarry, the test facility consists of an acceleration ramp with a height of 20 m (66 ft) with an automated rock release mechanism. The ramp and release apparatus are at the top of a rock slope with a height of 37 m (121 ft) inclined at 50°, thus providing a slope 50 m (164 ft) long. This arrangement allows the vertical potential energy of test blocks to produce rotation of the blocks and significant horizontal impacts with barrier systems installed at the base of the slope (Figures B-14 and B-15).

The release of both natural rock boulders and artificial test blocks weighing between 500 and 3,000 kg (1,102 and 6,613 lb) has facilitated a wide range of tests. The Saint Leonard facility is designed for testing rockfall attenuator systems, which are described in Chapters 15 and 16. In tests of attenuator systems, mesh is supported at the top of the rock slope by a post-and-cable infrastructure and draped down the slope. A barrier rated for 100 kJ (37 ft-tons) is located at the base of the slope to retain any rocks passing out from under the attenuator mesh (Figure B-15). Video VN-17 shows the testing of a rockfall attenuator system at the Saint Leonard facility. The attenuator mesh was composed of high-tensile steel wire (SPIDER S4-130 mesh). In these tests, a single rock was dropped from a crane onto the ramp located above

FIGURE B-14
Saint Leonard test site, Saint Leonard, Switzerland (Glover et al. 2010).

the slope crest so that the rock was launched into the top of the flexible rockfall attenuator system with a spinning motion.

3.1.8 Walenstadt Test Facility, Switzerland

WSL and the Swiss Agency for the Environment, Forest, and Landscape have developed and currently administer a certification program for testing rockfall barrier systems that became effective in June 2001 (Gerber 2001; Gerber et al. 2001; Baumann 2002). The official test facility is near Walenstadt, Switzerland, in an abandoned quarry about 200 m (656 ft) above the Walensee. The facility was officially opened on May 31, 2001. Access is via a funicular railway with a bottom station about 2 km (1.2 mi) from Walenstadt that is reached by a road along the lake shore. Full-scale certification tests are conducted with artificial blocks, released from a crane, that drop vertically at least 32 m (105 ft) and hit the center panel of a three-panel barrier net with a velocity of up to 30.0 m/s (98.4 ft/s). To prevent ground contact by the deforming net, the barriers are installed on the vertical rock face about 15 m (49 ft) above the quarry floor (Figures B-16 and B-17).

Tests are conducted with blocks of various sizes to obtain impact energies in the range of 100 to 8,000 kJ (37 to 2,963 ft-tons). Each barrier is subjected to a series of tests of increasing impact energy. Tests performed at lower energy levels provide information on the expected service behavior of the barrier system and anticipated required maintenance to the system. The final test, which is carried out at the maximum impact energy classification of the barrier being tested, provides data on the maximum deformation exhibited by the barrier, the operation of rope brakes and other structural components, and any substantial damage to the barrier system components. These results are compared with the Swiss standards for the testing and certification of rockfall barriers (Baumann 2002). Videos VN-28 and VN-29 document tests of impact energies of 2,040 kJ (755 ft-tons) and 8,000 kJ (2,963 ft-tons), respectively, conducted at Walenstadt.

3.1.9 Styrian Erzberg Test Facility, Austria

The Styrian Erzberg test facility, located near Eisenerz in the Erzbach Valley in central Austria,

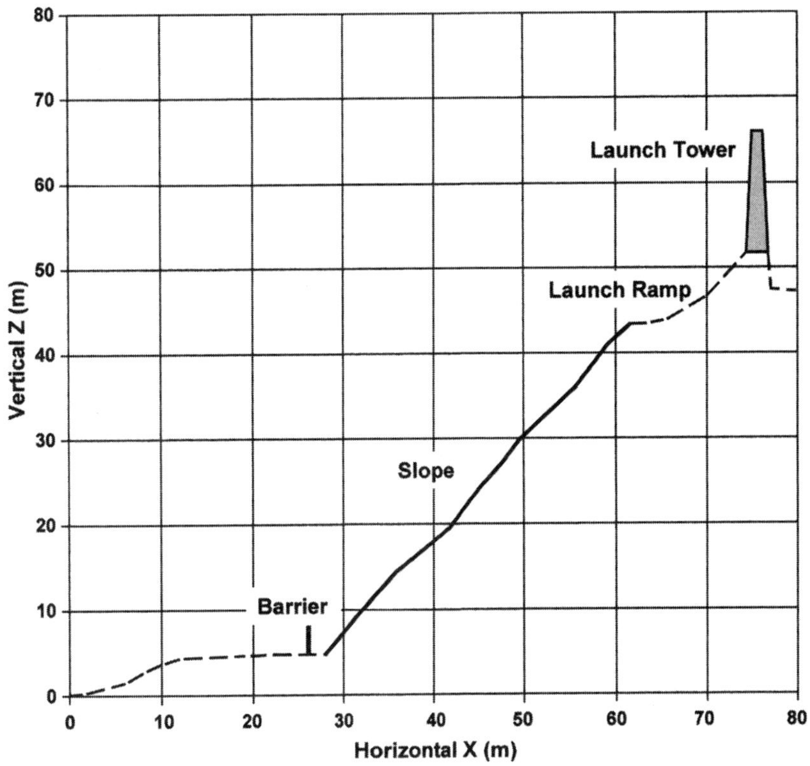

FIGURE B-15
Cross section of the St. Leonard test site (adapted from Glover et al. 2010).

FIGURE B-16
Walenstadt test facility, Walenstadt, Switzerland.
(PHOTOGRAPH COURTESY OF GEOBRUGG.)

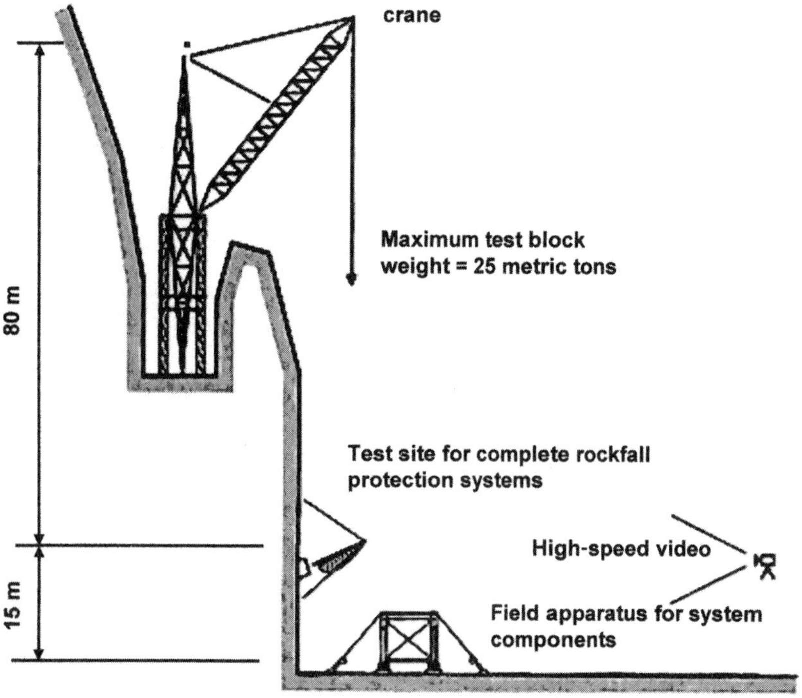

FIGURE B-17
Cross section of Walenstadt test facility (adapted from Baumann 2002).

has been used by various institutes of the University of Leoben. The site is a large operating open-pit mine and is privately owned. At present, field tests can be conducted for energy impacts up to 5,000 kJ (1,852 ft-tons). In 2006, the test site was completely rebuilt and conforms to the new European Union standards (Heiss and Moser 2006). Figure B-18 provides an overview of the facility.

The site uses a sliding cableway capable of handling blocks weighing as much as 14,000 kg (30,864 lb). As shown in Figure B-19, the cableway is designed to launch the test block into the barrier by releasing the rock immediately before impact.

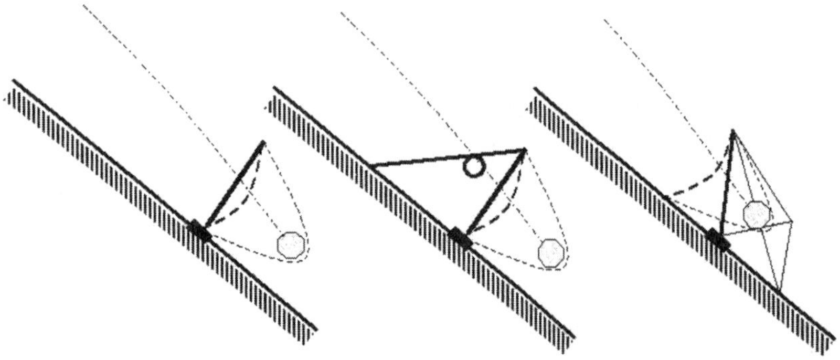

FIGURE B-19
Typical rock block strikes barriers at the Styrian Erzberg test facility (adapted from Peila et al. 2007).

FIGURE B-18
Styrian Erzberg test facility, Eisenerz, Austria.
(PHOTOGRAPH COURTESY OF TRUMER SCHUTZBAUTEN GMBH.)

Test blocks are carried down the cableway and released so that they hit the flexible rockfall fence at angles of 23° to 37° at velocities between 24 and 30 m/s (78 and 98 ft/s). Video VN-24 shows tests undertaken at the Styrian Erzberg facility.

3.1.10 Vigo di Meano Test Facility, Italy

The Vigo di Meano test facility, located near Trento in northeastern Italy, has been used by researchers based at the Polytechnic University of Turin to test the resistance of flexible rockfall fences, rockfall attenuator systems (Badger et al. 2008), and reinforced earthen berms (Peila et al. 2000, 2007) to impacts by large rockfall blocks. Figures B-20 and B-21 provide overviews of the facility.

Attenuator systems were tested by rolling individual rocks from the top of the slope into the rockfall attenuator system (see description of Video VN-20 in Section 3.3). These tests were influenced by the composition of the soil forming the slope, the slope height of 32 m (105 ft), and the steepness of the slope, which was 49°.

FIGURE B-20
Vigo di Meano test facility, Trento, Italy.
(PHOTOGRAPH COURTESY OF FRANCESCO SASSUDELLI.)

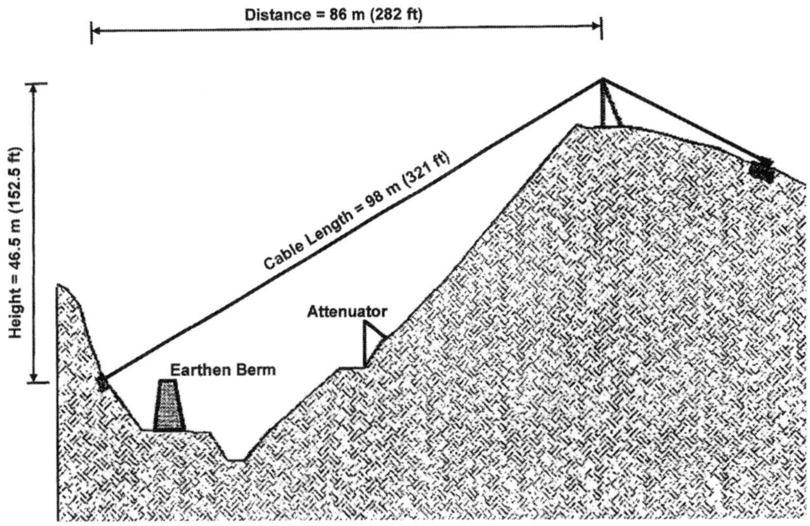

FIGURE B-21
Cross section of Vigo di Meano test facility (adapted from Peila et al. 2007).

slope surface consisted of hard rock surfaces with little to no vegetation covering the slope face. Overall, relative to the size of the test rocks, the slope was smooth with minor irregularities. Test rocks were pushed onto the slope with an excavator. Video VN-23 shows a test conducted at this site.

The site was selected for testing reinforced earthen berms because it allowed for the installation of a sliding cableway capable of handling blocks weighing as much as 10,000 kg (22,046 lb). As shown in Figure B-21, the cableway has a length of 98 m (321 ft) and is inclined at 29° as it extends horizontally 86 m (282 ft) and descends 46.5 m (152.5 ft). Test blocks are carried down the cableway and released so that they hit the reinforced earthen berm at velocities of 31 to 32 m/s (102 to 105 ft/s). Videos VN-25 and VN-26, which are discussed in Section 3.5.1, show two tests of reinforced earthen berms undertaken at the Vigo di Meano test facility.

3.1.11 Maeda Kosen Quarry, Japan

The Maeda Kosen test site is located on private property within an active quarry in Fukui, Japan, in the Chūbu region on the west coast of Honshū island. The tests used a quarry face 40 m (131 ft) high that was inclined at 53°, for a slope length of 50 m (164 ft). Figures B-22 and B-23 illustrate the basic characteristics of the site. The

FIGURE B-22
Maeda Kosen Quarry test site, Fukui, Japan (Yoshida and Nomura 1998).

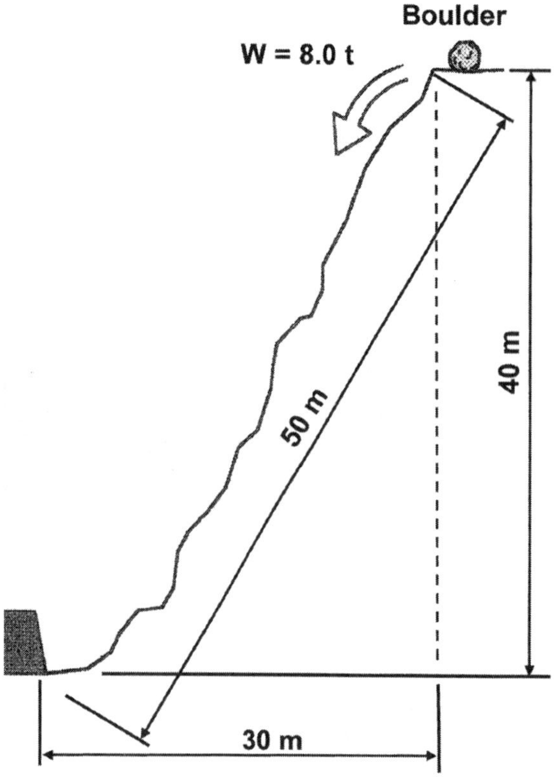

FIGURE B-23 Cross section of Maeda Kosen Quarry test site (adapted from Yoshida and Nomura 1998).

3.2 Videos Demonstrating Typical Rockfall Impact Energies

Of the 29 videos in Part B of the DVD, 16 demonstrate typical impact energies experienced during various rock-rolling tests undertaken between 1989 and 1998. The examples show rock impacts producing energies ranging from 45 kJ (16 ft-tons) to 1,260 kJ (467 ft-tons) and are organized into six energy categories:

1. 50 kJ (18 ft-tons),
2. 100 kJ (37 ft-tons),
3. 150 kJ (55 ft-tons),
4. 200 kJ (74 ft-tons),
5. 500 kJ (185 ft-tons), and
6. 1,000 kJ (370 ft-tons).

The following sections provide brief explanations of what can be observed in these videos.

3.2.1 Impacts in the 50-kJ Category

Two of a series of tests performed by Caltrans at the Shale Point site (Duffy and Hoon 1996b; 1998) are presented. Each test involved a single rock dropped by a front-end loader at the top of the slope. The first test measured only translational kinetic energy, and the second test measured both translational and rotational kinetic energies. The rotational kinetic energy in this particular case was greater than the translational kinetic energy. Table B-5 summarizes the details of these tests.

The barrier being tested in Video VN-1 was a fairly lightweight flexible rockfall fence composed of 6-gauge chain-link mesh supported by a steel post-and-cable infrastructure that incorporated energy-absorbing devices. The rock used in this test weighed 633 kg (1,395 lb) and upon impact had a translational velocity of 12.0 m/s (39.5 ft/s), which produced a translational kinetic energy of 45 kJ (16 ft-tons). The barrier being tested in Video VN-2 was a simple chain-link fence built on top of a concrete barrier. The rock used in this test was 336 kg (740 lb) and hit the concrete barrier with a translational velocity of 11.2 m/s (36.7 ft/s) and a rotational velocity of 80.7 rad/s, which produced a translational kinetic energy of 21 kJ (8 ft-tons) and a rotational kinetic energy of 46 kJ (17 ft-tons), for a total kinetic energy of 67 kJ (25 ft-tons). Pieces of the concrete barrier spalled off upon impact.

Table B-5
Videos Demonstrating Impact Energies in the 50-kJ Category

Video Number	Reference and Location	Test Slope	Rock Weight [kg (lb)]	Translational Velocity [m/s (ft/s)]	Impact Energy [kJ (ft-tons)]	Comment
VN-1	Duffy and Hoon 1996b, Shale Point, California	Soil	633 (1,395)	12.0 (39.5)	45 (16)	Single impact on chain-link fence
VN-2	Duffy and Hoon 1998, Shale Point, California	Soil	336 (740)	11.2 (36.7)	67* (25*)	Single impact on concrete barrier

*Denotes impact energy defined by sum of translational and angular (rotational) kinetic energies.

Table B-6
Videos Demonstrating Impact Energies in the 100-kJ Category

Video Number	Reference and Location	Test Slope	Rock Weight [kg (lb)]	Translational Velocity [m/s (ft/s)]	Impact Energy [kJ (ft-tons)]	Comment
VN-3	Smith and Duffy 1990, Grey Slip, California	Soil	917 (2,021)	13.5 (44.3)	102* (38*)	Single impact on cable mesh fence
VN-7: Rock 2	Duffy 1992, Oberbuchsiten, Switzerland	Rock	320 (705)	27.0 (88.5)	117 (43)	Second rock hits fence

*Denotes impact energy defined by sum of translational and angular (rotational) kinetic energies.

3.2.2 Impacts in the 100-kJ Category

The 100-kJ category is represented by one test performed by Caltrans at the Grey Slip site (Video VN-3) (Smith and Duffy 1990) and by a second rock visible on Video VN-7 during a test at the Oberbuchsiten site (Duffy 1992). Table B-6 summarizes the details of these two tests.

Video VN-3 shows a test with a single rock weighing 917 kg (2,021 lb) that hit a cable mesh fence with a translational velocity of 13.5 m/s (44.3 ft/s), which produced a translational kinetic energy of 83 kJ (31 ft-tons). The rotational kinetic energy was measured at 19 kJ (7 ft-tons), which resulted in a total kinetic energy of 102 kJ (38 ft-tons). The cable mesh was constructed of a cable 8 mm (5⁄16 in.) in diameter diagonally woven into a 203-mm (8-in.) grid that was supported by a steel post-and-cable infrastructure that included energy-absorbing devices.

Video VN-7, taken at the Oberbuchsiten site, shows two rocks successively hitting a fence constructed from a military surplus submarine net (see Section 3.2.4). The smaller (second) rock weighed 320 kg (705 lb) and traveled at 27.0 m/s (88.5 ft/s), which produced a translational kinetic energy of 117 kJ (43 ft-tons).

3.2.3 Impacts in the 150-kJ Category

The 150-kJ category is represented by two tests. Video VN-4 was recorded during tests at the Oberbuchsiten site (Duffy 1992), and Video VN-5 was recorded by Caltrans at the Grey Slip site (Smith and Duffy 1990). Table B-7 summarizes the details of these two tests.

Video VN-4 shows the impact of a rock weighing 320 kg (705 lb) that was dropped from a cable lift onto the slope 65 m (213 ft) above a cable mesh fence. The fence consisted of a cable 8 mm (5⁄16 in.) in diameter diagonally woven into a 203-mm (8-in.) grid supported by a system of steel posts and cables with energy-absorbing devices incorporated into the system. As the rock reached the fence, it was traveling at 30.0 m/s (98.4 ft/s) and produced a translational kinetic energy of 144 kJ (53 ft-tons) (Duffy 1992).

Video VN-5 shows the impact of a rock weighing 1,300 kg (2,866 lb) into a cable mesh fence similar to that tested in Video VN-4. However, this larger rock was traveling at only 14.7 m/s (48.2 ft/s). The translational kinetic energy was measured at 140 kJ (52 ft-tons), and the rotational kinetic energy was measured at 33 kJ (12 ft-tons); thus, the total kinetic energy was 173 kJ (64 ft-tons) (Smith and Duffy 1990).

Table B-7
Videos Demonstrating Impact Energies in the 150-kJ Category

Video Number	Reference and Location	Test Slope	Rock Weight [kg (lb)]	Translational Velocity [m/s (ft/s)]	Impact Energy [kJ (ft-tons)]	Comment
VN-4	Duffy 1992, Oberbuchsiten, Switzerland	Rock	320 (705)	30.0 (98.4)	144 (53)	Single impact on fence
VN-5	Smith and Duffy 1990, Grey Slip, California	Soil	1,300 (2,866)	14.7 (48.2)	173* (64*)	Single impact on fence

*Denotes impact energy defined by sum of translational and angular (rotational) kinetic energies.

3.2.4 Impacts in the 200-kJ Category

The 200-kJ category is represented by four tests undertaken at the Grey Slip and Shale Point sites (Smith and Duffy 1990; Duffy and Hoon 1996a) and at the Oberbuchsiten site (Duffy 1992). Table B-8 summarizes the details of these tests.

Video VN-6 shows a test performed at the Grey Slip site (Smith and Duffy 1990). The test involved a single rock dropped by a front-end loader from the top of the slope. The test was intended to evaluate rock impacts on a fence constructed of a cable 8 mm (5⁄16 in.) in diameter diagonally woven into a 203-mm (8-in.) grid supported by a steel post-and-cable infrastructure with energy-absorbing devices incorporated into the system. However, this test shows the direct impact of the rock on a supporting post. The post was a 203-mm (8-in.) wide-flange steel post weighing 22 kg/m (48 lb/ft). The post foundation was a cast-in-place concrete footing 762 mm × 762 mm (2.5 ft × 2.5 ft) square and 762 mm (2.5 ft) deep that was founded in colluvium. The rock weighed 1,190 kg (2,623 lb) and was traveling at 17.0 m/s (57.0 ft/s). The translational kinetic energy was measured at 172 kJ (64 ft-tons), and the angular kinetic energy was measured at 8 kJ (3 ft-tons), which produced a total kinetic energy of 180 kJ (67 ft-tons).

Video VN-7 shows a test performed at the Oberbuchsiten site (Duffy 1992). Two rocks were dropped from a cable lift onto the slope from a location 65 m (213 ft) above the fence. The fence mesh was constructed of a military surplus submarine net consisting of cables 19 mm (0.75 in.) in diameter woven into a series of rings 406 mm (16 in.) in diameter. Each ring was linked to six adjacent rings to make the mesh panel. The mesh was supported by a steel post-and-cable infrastructure with energy-absorbing devices incorporated into the system. The larger (first) rock weighed 635 kg (1,400 lb) and traveled at 25.0 m/s (82.0 ft/s), which produced a translational kinetic energy of 198 kJ (73 ft-tons). The smaller (second) rock weighed 320 kg (705 lb) and traveled at 27.0 m/s (88.5 ft/s), which produced a translational kinetic energy of 117 kJ (43 ft-tons).

Video VN-8 shows a test performed at the Shale Point site (Duffy and Hoon 1996a). A single rock was dropped from the top of the slope by a front-end loader and hit a fence constructed of 6-gauge chain-link mesh supported by a steel post-and-cable infrastructure that included energy-absorbing devices. The rock weighed 1,770 kg (3,902 lb) and traveled at 15.0 m/s (49.2 ft/s), which produced a translational kinetic energy of 199 kJ (74 ft-tons).

Video VN-9 shows a test performed at Oberbuchsiten (Duffy 1992). Two rocks were dropped from a cable lift onto the slope from a location 65 m (213 ft) above the fence. The fence was constructed of a cable 8 mm (5⁄16 in.) in diameter diagonally woven into a 203-mm (8-in.) grid. It was supported by a steel post-and-cable infrastructure

Table B-8
Videos Demonstrating Impact Energies in the 200-kJ Category

Video Number	Reference and Location	Test Slope	Rock Weight [kg (lb)]	Translational Velocity [m/s (ft/s)]	Impact Energy [kJ (ft-tons)]	Comment
VN-6	Smith and Duffy 1990, Grey Slip, California	Soil	1,190 (2,623)	17.0 (57.0)	180* (67*)	Single rock directly hits post
VN-7: Rock 1	Duffy 1992, Oberbuchsiten, Switzerland	Rock	635 (1,400)	25.0 (82.0)	198 (73)	First rock hits submarine net fence
VN-7: Rock 2	Duffy 1992, Oberbuchsiten, Switzerland	Rock	320 (705)	27.0 (88.5)	117 (43)	Second rock hits submarine net fence
VN-8	Duffy and Hoon 1996a, Shale Point, California	Soil	1,770 (3,902)	15.0 (49.2)	199 (74)	Single impact on chain-link fence
VN-9: Rock 1	Duffy 1992, Oberbuchsiten, Switzerland	Rock	720 (1,587)	25.0 (82.0)	225 (83)	First rock hits cable mesh fence
VN-9: Rock 2	Duffy 1992, Oberbuchsiten, Switzerland	Rock	380 (837)	30.0 (98.4)	171 (63)	Second rock hits cable mesh fence

*Denotes impact energy defined by sum of translational and angular (rotational) kinetic energies.

that included energy-absorbing devices. The larger (first) rock weighed 720 kg (1,587 lb) and traveled at 25.0 m/s (82.0 ft/s), which produced a translational kinetic energy of 225 kJ (83 ft-tons). The smaller (second) rock weighed 380 kg (837 lb) and traveled at 30.0 m/s (98.4 ft/s), which produced a translational kinetic energy of 171 kJ (63 ft-tons).

3.2.5 Impacts in the 500-kJ Category

The 500-kJ category is represented by three tests. Videos VN-10 and VN-11 were recorded by Caltrans at the Grey Slip site (Smith and Duffy 1990), and Video VN-12 was recorded during tests at the Oberbuchsiten site (Duffy 1992). Table B-9 summarizes the details of these tests.

The tests shown in Videos VN-10 and VN-11 involved similarly sized rocks that hit fences of the same design: a cable mesh constructed of a cable 8 mm (5/16 in.) in diameter diagonally woven into a 203-mm (8-in.) grid supported by a steel post-and-cable infrastructure that included energy-absorbing devices. Video VN-10 shows the impact of a rock weighing 2,500 kg (5,511 lb) with a translational velocity of 16.0 m/s (52.5 ft/s). The measured translational kinetic energy was 320 kJ (118 ft-tons), and the measured rotational kinetic energy was 36 kJ (13 ft-tons); thus, the total kinetic energy was 356 kJ (132 ft-tons). Video VN-11 shows the impact of a slightly larger rock weighing 2,930 kg (6,495 lb) traveling with a velocity of 15.2 m/s (49.8 ft/s). The measured translational kinetic energy was 338 kJ (125 ft-tons), and the measured rotational kinetic energy was 66 kJ (24 ft-tons); thus, the total kinetic energy was 404 kJ (150 ft-tons). This impact caused a rock fragment to break off and fly over the fence.

Video VN-12 shows a test performed at the Oberbuchsiten site (Duffy 1992). A rock was dropped from a cable lift onto the slope at a position 65 m (213 ft) above the fence. The fence was constructed of a cable mesh diagonally woven into a 150-mm (6-in.) grid that was supported by a steel post-and-cable infrastructure that included energy-absorbing devices. The rock weighed 1,960 kg (4,321 lb) and traveled at 25.0 m/s (82.0 ft/s), which produced an impact energy of 612 kJ (227 ft-tons).

3.2.6 Impacts in the 1,000-kJ Category

The 1,000-kJ category is represented by three tests undertaken at the Oberbuchsiten site (Duffy 1992) and one test at the Rifle site (Andrew et al. 1998). Table B-10 summarizes the details of these tests.

At Oberbuchsiten, rocks were dropped from a cable lift onto the slope face. The starting position was 65 m (213 ft) above the fence. Video VN-13 shows a test of a fence constructed of a cable 8 mm (5/16 in.) in diameter diagonally woven into a 203-mm (8-in.) grid. Videos VN-14 and VN-16 show tests conducted using fences constructed of military surplus submarine nets that consisted of cables 19 mm (0.75 in.) in diameter woven to form a series of rings 406 mm (16 in.) in diameter. Each ring was linked to six adjacent rings. In each case, the fence mesh was supported by a steel post-and-cable infrastructure that included energy-absorbing devices.

Video VN-15 shows a test at the Rifle site of a fence constructed of a cable 8 mm (5/16 in.) in

Table B-9
Videos Demonstrating Impact Energies in the 500-kJ Category

Video Number	Reference and Location	Test Slope	Rock Weight [kg (lb)]	Translational Velocity [m/s (ft/s)]	Impact Energy [kJ (ft-tons)]	Comment
VN-10	Smith and Duffy 1990, Grey Slip, California	Soil	2,500 (5,511)	16.0 (52.5)	356* (132*)	Single impact on cable mesh fence
VN-11	Smith and Duffy 1990, Grey Slip, California	Soil	2,930 (6,495)	15.2 (49.8)	404* (150*)	Single impact on cable mesh fence; rock fragment flies over fence
VN-12	Duffy 1992, Oberbuchsiten, Switzerland	Rock	1,960 (4,321)	25.0 (82.0)	612 (227)	Single impact on cable mesh fence

*Denotes impact energy defined by sum of translational and angular (rotational) kinetic energies.

Table B-10
Videos Demonstrating Impact Energies in the 1,000-kJ Category

Video Number	Reference and Location	Test Slope	Rock Weight [kg (lb)]	Translational Velocity [m/s (ft/s)]	Impact Energy [kJ (ft-tons)]	Comment
VN-13	Duffy 1992, Oberbuchsiten, Switzerland	Rock	2,760 (6,085)	25.0 (82.0)	862 (319)	Single impact on cable mesh fence; fence down
VN-14	Duffy 1992, Oberbuchsiten, Switzerland	Rock	3,000 (6,613)	25.0 (82.0)	938 (347)	Single impact on submarine net fence; post base sheared
VN-15	Andrew et al. 1998, Rifle, Colorado	Rock and soil	5,312 (11,712)	15.8 (51.7)	1,166[a] (430[a])	Single impact on cable mesh fence
VN-16	Duffy 1992, Oberbuchsiten, Switzerland	Rock	2,800 (6,173)	30.0 (98.4)	1,260 (467)	Single impact on submarine net fence

[a]Denotes impact energy defined by sum of translational and angular (rotational) kinetic energies.

diameter diagonally woven into a 150-mm (6-in.) grid. The test rock was pushed by a front-end loader from a starting position at the top of the slope, 100 m (328 ft) above the fence.

Video VN-13 shows a rock weighing 2,760 kg (6,085 lb) that hit the fence with a velocity of 25.0 m/s (82.0 ft/s) and produced a translational kinetic energy of 862 kJ (319 ft-tons). The impact severely damaged the fence, which allowed the rock to overtop the fence and continue for some distance farther down the slope. Video VN-14 shows the impact of a rock weighing 3,000 kg (6,613 lb) and traveling at 25.0 m/s (82.0 ft/s) that produced a translational kinetic energy of 938 kJ (347 ft-tons). The impact caused the base of a central support post to shear. Video VN-15 shows the impact of a rock weighing 5,312 kg (11,712 lb) and traveling at 15.8 m/sec (51.7 ft/s). This rock produced a translational kinetic energy of 663 kJ (246 ft-tons) and had a rotational kinetic energy of 503 kJ (184 ft-tons), for a total kinetic energy of 1,166 kJ (430 ft-tons). Video VN-16 shows the impact of a rock weighing 2,800 kg (6,173 lb) and traveling at 30.0 m/s (98.4 ft/s) that produced a translational kinetic energy of 1,260 kJ (467 ft-tons).

3.3 Videos of Tests of Attenuator Systems

Tests of rockfall attenuator systems have been conducted in Europe (Badger et al. 2008; Glover et al. 2010) and North America (Barrett et al. 1989; Arndt et al. 2009; T. Ortiz, personal communication, January 23, 2011). Chapters 15 and 16 describe the principles underlying the design and installation of rockfall attenuator systems. Table B-11 summarizes the details of four tests of rockfall attenuator systems.

Video VN-17 shows a test performed at the Saint Leonard test facility (Glover et al. 2010; WSL 2010). This site is described in Section 3.1.7. The attenuator being tested was constructed from SPIDER® S4-130 mesh, a woven-wire cable in a diamond-shaped pattern. The test used a concrete block weighing 1,070 kg (2,360 lb) that entered the attenuator system with a velocity of 11.0 m/s (36.1 ft/s) and produced a translational kinetic energy of 64 kJ (24 ft-tons).

VN-18 shows one in a series of tests performed in 1989 at the Rifle site (Barrett et al. 1989). These tests were undertaken to evaluate alternative designs of potential rockfall attenuator systems that could be employed to protect the section of Interstate 70 passing through Glenwood Canyon that was then under construction. These attenuator systems included attenuator designs that used old timber railroad ties, old rubber tires mounted on wheel rims, and combinations of timber ties and rubber tires suspended from a cable across a rock chute (Barrett et al. 1989). Video VN-18 shows the response of an attenuator constructed with a double row of timber ties to the impact of a test rock that was estimated to weigh between 544 and 680 kg (1,200 and 1,500 lb) and was traveling at a velocity between 15 and 18 m/s (50 and 60 ft/s), which produced a translational kinetic energy of 61 to 110 kJ (23 to 41 ft-tons).

Video VN-19 shows part of a sequence of three field tests undertaken at Georgetown, Colorado,

Table B-11
Videos of Tests of Rockfall Attenuator Systems

Video Number	Reference and Location	Test Slope	Rock or Block Weight [kg (lb)]	Translational Velocity [m/s (ft/s)]	Impact Energy [kJ (ft-tons)]	Comment
VN-17	Glover et al. 2010 and WSL 2010, Saint Leonard, Switzerland	Rock	1,070 (2360)	11.0 (36.1)	64 (24)	Single impact on SPIDER mesh
VN-18	Barrett et al. 1989, Rifle, Colorado	Rock and soil	544–680 (1,200–1,500)	15–18 (50–60)	61–110 (23–41)	Single impact on timber tie barrier
VN-19	Arndt et al. 2009, Georgetown, Colorado	Rock and soil	726 (1,600)	18–21 (60–70)	118–160 (44–59)	Single and multiple impacts
VN-20	Badger et al. 2008, Vigo di Meano, Italy	Soil	5,000 (11,023)	10.0 (32.8)	250 (93)	Single impact on ring net fence

in 2004, before the installation of attenuators, and in 2009, after attenuators were installed. The tests were undertaken to evaluate the efficiency of rockfall attenuator systems installed in 2009 to protect Interstate 70 at Georgetown (Arndt et al. 2009). Traffic on Interstate 70 was stopped while each of these tests was undertaken.

The first portion of Video VN-19 shows one example from a series of 2004 tests that involved rockfall initiated by Colorado DOT personnel, who manually dislodged selected large rock blocks located on upper portions of the Georgetown slope to evaluate the rockfall characteristics of the location before any attenuator systems were installed. The rocks were launched from locations approximately 549 m (1,800 ft) above Interstate 70. The rocks ranged in diameter from 1.2 to 2.1 m (4 to 7 ft). Impact energies were not calculated. On the basis of a comparison of the rock trajectories shown in the video with the height of nearby trees, the bounce heights were estimated to have been in excess of 12 m (40 ft). A portable fence protection system with a design capacity of 80 kJ (30 ft-tons) was located on the shoulder of the road, but, as shown in Video VN-19, many test rocks traveled very rapidly down the slopes, over this fence, and onto, and even across, all the Interstate lanes.

The remainder of Video VN-19 shows two tests performed in 2009 after the attenuator systems were installed. The design of the attenuator system was fully discussed by Arndt et al. (2009) and is summarized in the description of the Georgetown site in Section 3.1.4 and Figures B-8 and B-9. Three new attenuators were installed above two existing attenuators that had been placed near the bottom of the rock slope. Thus, the complete attenuator system consisted of five rows of attenuators spaced along the slope face. The three new attenuators consisted of a cable 8 mm (5⁄16 in.) in diameter diagonally woven into a 203-mm (8-in.) grid supported by a steel post-and-cable infrastructure that did not include any energy-absorbing devices. A rockfall catch fence was installed along the highway shoulder to prevent rocks from rolling out into the traffic lanes.

The second portion of Video VN-19 shows the test of a single 726-kg (1,600-lb) concrete block dropped from a helicopter approximately 304 m (1,000 ft) vertically above the roadway into the top attenuator. The block bounced over the second attenuator at a height estimated to have been in excess of 7 m (23 ft). The block reached the third attenuator with an estimated velocity of 18 to 21 m/s (60 to 70 ft/s); thus, the maximum estimated translational kinetic energy at this location was 118 to 160 kJ (44 to 59 ft-tons). The third attenuator nearly stopped the block, but upon exiting this third attenuator it continued downslope and was eventually deposited in the rockfall catchment area adjacent to Interstate 70.

The third and final portion of Video VN-19 shows the release of five concrete blocks, each weighing 726 kg (1,600 lb). They were released as a group and allowed to roll into the three lowest attenuators shown in Figure B-9 (the lowest new attenuator and the two older attenuators). Four of the five blocks hit the first attenuator and were slowed down. However, the fifth block rolled under the attenuator while the net was lifted into the air by the impacts of the first four blocks and did not slow down. This block was the first to hit the next attenuator downslope and was slowed down at that point. This test demonstrated the value of the redundancy provided by a series of attenuators.

Video VN-20 shows an attenuator test performed at the Vigo di Meano site (Badger et al. 2008). A single concrete block was rolled down a soil slope with a height of 32 m (105 ft) and a slope angle of 49°. The mesh forming the rockfall attenuator system was constructed of ring nets formed of cables 19 mm (0.75 in.) in diameter woven into a series of rings 406 mm (16 in.) in diameter. Each ring was linked to six adjacent rings. The attenuator mesh was supported by a steel post-and-cable infrastructure that incorporated energy-absorbing devices. Video VN-20 shows a concrete test block weighing 5,000 kg (11,023 lb) moving at 10.0 m/s (32.8 ft/s) that delivered an impact of 250 kJ (93 ft-tons) of translational kinetic energy to the attenuator system.

3.4 Videos of Tests of Reinforced Earthen Berms

Three tests of reinforced earthen berms of various designs are shown in Videos VN-21, VN-22, and VN-23 (Table B-12). Berms are capable of absorbing much greater impact energies than most fence systems. Thus, they are often used in areas where very large rocks are likely to produce rockfall hazards. The use of steep-sided barriers permits the establishment of a protection system with a reduced overall footprint as compared with unreinforced embankments. Tests of reinforced earthen berms have been conducted in Colorado (Parsons De Leuw, Inc. 1992), Japan (Yoshida and Nomura 1998), and Europe (Peila et al. 2000, 2007). Videos VN-25 and VN-26, discussed in Section 3.5.1, show tests on reinforced earthen berms performed at the Vigo di Meano site.

Videos VN-21 and VN-22 show tests undertaken in 1992 at the Rifle site (Parsons De Leuw, Inc. 1992). A reinforced earthen berm was constructed with vertical faces covered with timber facing. Video VN-21 shows a test with a rock weighing 5,313 kg (11,713 lb) that hit the berm with a velocity of 19.5 m/s (64.0 ft/s) and produced a translational kinetic energy of about 1,010 kJ (374 ft-tons). Video VN-22 shows a test using a larger rock weighing 8,325 kg (18,354 lb) that hit the berm with a velocity of 18.3 m/s (60.0 ft/s) and produced a translational kinetic energy of 1,394 kJ (516 ft-tons).

Video VN-23 shows a test performed at the Maeda Kosen Quarry in Japan (Yoshida and Nomura 1998). An excavator was used to push a large rock over the quarry wall to test a reinforced earthen berm with steeply sloping sides. The test rock weighed 7,700 kg (16,975 lb) and hit with a velocity of 24.0 m/s (78.7 ft/s), which produced a translational kinetic energy of 2,217 kJ (821 ft-tons).

3.5 Videos of Tests at Established Rockfall Testing Facilities

Table B-13 summarizes the details of five rockfall barrier tests conducted at established rockfall testing facilities in Switzerland, Austria, and Italy and one of a series of tests conducted at a temporary test facility established by the Colorado

Table B-12
Videos of Tests of Reinforced Earthen Berms

Video Number	Reference and Location	Test Slope	Rock Weight [kg (lb)]	Translational Velocity [m/s (ft/s)]	Impact Energy [kJ (ft-tons)]	Comment
VN-21	Parsons De Leuw 1992, Rifle, Colorado	Rock and soil	5,313 (11,713)	19.5 (64.0)	1,010 (374)	Single rock impact on mechanically stabilized earthen berm
VN-22	Parsons De Leuw 1992, Rifle, Colorado	Rock and soil	8,325 (18,354)	18.3 (60.0)	1,394 (516)	Single rock impact on mechanically stabilized earthen berm
VN-23	Yoshida and Nomura 1998, Maeda Kosen Quarry, Fukui, Japan	Rock	7,700 (16,975)	24.0 (78.7)	2,217 (821)	Single rock impact on reinforced berm

Table B-13
Videos of Tests at Established Rockfall Testing Facilities

Video Number	Reference and Location	Rock Weight [kg (lb)]	Translational Velocity [m/s (ft/s)]	Impact Energy [kJ (ft-tons)]	Comment
Facilities That Employ Sliding Cables					
VN-24	Heiss and Moser 2006, Styrian Erzberg, Austria	5,157 (11,369)	28.5 (93.5)	2,094 (775)	Single impact on fence
VN-25	Peila et al. 2000, 2007; Vigo di Meano, Italy	4,989 (10,998)	31.7 (104.0)	2,507 (928)	Single rock impact on reinforced berm
VN-26	Peila et al. 2000, Vigo di Meano, Italy	8,962 (19,757)	31.6 (103.7)	4,474 (1,657)	Single rock impact on reinforced berm
Facilities That Employ Vertical Drop Tests					
VN-27	Ortiz, personal communication, Jan. 23, 2011; Interstate 70, Georgetown, Colorado	1,535 (3,384)	18.9 (62.0)	274 (101)	Impact on cable mesh fence
VN-28	WSL 2009, Walenstadt, Switzerland	4,700 (10,361)	29.5 (96.8)	2,045 (757)	Impact on TECCO mesh fence
VN-29	WSL 2011, Walenstadt, Switzerland	20,000 (44,092)	28.6 (93.8)	8,179 (3,029)	Impact on ROCCO mesh (ring net)

DOT. Several of the European tests were conducted according to testing guidelines for flexible rockfall fence systems established by the Swiss and European testing authorities (Gerber 2001; Gerber et al. 2001; Baumann 2002; Heiss and Moser 2006; ETAG 27 2008).

Permanent rockfall testing facilities were established when it became obvious that simply launching rocks down a natural slope generally produced a wide variation in trajectory paths. This variation made it difficult to produce a series of reasonably consistent impacts on the rockfall barrier systems subjected to testing and evaluation. As shown in Table B-13, two approaches have been used in designing and operating these rockfall testing facilities. Initially, oblique testing procedures were adopted that used an inclined cableway to launch the rocks into the barrier, either directly or after a few impacts and bounces on the ground. Subsequently, it was decided to use a vertical drop test approach in which the test blocks were vertically dropped into a more-or-less horizontally inclined test barrier installed on a vertical rock face. These tests provided highly predictable accelerations, impact energies, and impact locations on the barrier system being tested. In general, neither test procedure caused the test blocks to rotate.

3.5.1 Tests at Facilities That Employ Sliding Cables

Video VN-24 shows a test that followed the European Union rockfall fence-testing guidelines and was performed at the Styrian Erzberg site (Heiss and Moser 2006). A flexible rockfall fence system rated at 2,000 kJ (740 ft-tons) was constructed of Omega mesh supported by a steel post-and-cable infrastructure that incorporated energy-absorbing devices. Video VN-24 shows a 5,157-kg (11,369-lb) test rock with an impact velocity of 28.5 m/s (93.5 ft/s) that produced a total kinetic energy of 2,094 kJ (775 ft-tons), or slightly more than the rated capacity of the system.

Videos VN-25 and VN-26 show tests on reinforced earthen berms performed at the Vigo di Meano site (Peila et al. 2000, 2007). Video VN-25 shows the launching from the cableway of a 4,989-kg (10,998-lb) concrete test block that hit the earthen berm with a velocity of 31.7 m/s (104.0 ft/s) and produced a total kinetic energy of 2,507 kJ (928 ft-tons). Video VN-26 shows a similar test with a much larger test rock. The 8,962-kg (19,757-lb) rock hit the earthen berm with a velocity of 31.6 m/s (103.7 ft/s) and produced a total kinetic energy of 4,474 kJ (1,657 ft-tons).

3.5.2 Tests at Facilities That Employ Vertical Drop Methods

As described in Section 3.1.4, the Colorado DOT established a temporary vertical drop test facility adjacent to Interstate 70 near Georgetown, Colorado (T. Ortiz, personal communication, January 23, 2011). The test facility was located where a rockfall mitigation project was under construction, so mobilization costs were eliminated because the contractor for the rockfall mitigation project also performed the tests. A series of tests evaluated different panel systems at various impact energies, but the test procedures did not follow any established rockfall barrier testing guidelines.

Video VN-27 shows a test of a cable mesh fence constructed from wire rope 13 mm (0.5 in.) in diameter that was diagonally woven into a 203-mm (8-in.) grid. The mesh was supported by a steel post-and-cable infrastructure. No energy-absorbing devices were incorporated in the system. The test block was dropped about 18.3 m (60 ft), weighed 1,535 kg (3,384 lb), and hit the fence at a velocity of 18.9 m/s (62.0 ft/s), which produced a total kinetic energy of 274 kJ (101 ft-tons). The initial impact of the block was near the middle of the cable net configuration, and measurements of the deflection of the net and the loads developed on the bearing ropes were recorded.

Videos VN-28 and VN-29 show typical vertical drop tests performed at the Walenstadt site (WSL 2009, 2011). These tests followed the Swiss rockfall fence-testing guidelines (Gerber 2001; Gerber et al. 2001; Baumann 2002). Two flexible rockfall fence systems are shown. Video VN-28 shows the test of a flexible fence rated at 2,000 kJ (740 ft-tons) that was constructed of TECCO mesh supported by a steel post-and-cable infrastructure that incorporated energy-absorbing devices. The test rock weighed 4,700 kg (10,361 lb) and hit the fence system with a velocity of 29.5 m/s (96.8 ft/s), which produced a total kinetic energy of 2,045 kJ (757 ft-tons). Video VN-29 shows the test of a flexible fence rated at 8,000 kJ (2,963 ft-tons) that was constructed of a ROCCO ring net supported by a cable infrastructure with energy-absorbing devices incorporated into the system. The test rock weighed 20,000 kg (44,092 lb) and hit the fence system with a velocity of 28.6 m/s (93.8 ft/s), which produced a total kinetic energy of 8,179 kJ (3,029 ft-tons). Both of these tests slightly exceeded the maximum rated capacity of the systems being tested, and the videos show an extreme amount of deformation of the fence systems.

4. LONGER VIDEOS

Part C of the DVD contains four longer videos (Table B-14). Because these videos are recorded in the QuickTime movie (MOV) format, they can readily be viewed on a personal computer, and excerpts can be extracted and used to illustrate particular issues. The subjects of these four videos are as follows:

- The historic 1963 rockfall testing procedures conducted in Washington State and reported by Ritchie (1963);
- Field tests and evaluations of rockfall restraining nets conducted in 1990 by Caltrans at the Grey Slip test site and reported by Smith and Duffy (1990);
- Field tests conducted in 2001 during the development of the rockfall catchment design guide for the Oregon DOT and reported by Pierson et al. (2001); and
- Rock slope scaling techniques used by the Colorado DOT.

The following sections provide brief descriptions of the contents of each video.

4.1 Rockfall Characterization and Control, Washington State DOT

Arthur M. Ritchie rolled more than 300 rocks on various highway and quarry slopes during his landmark study of rockfall trajectories (Ritchie 1963). Video VN-30 is a video transcription of the historic 16-mm film made in 1963 by the Washington State Highway Commission (now the Washington State DOT) that provides a dramatic record of these rock-rolling field tests. These tests formed the basis for the "Ritchie Ditch" criteria for containing rockfall debris along highways (Ritchie 1963), criteria that are still employed by many agencies. This 18-min film provides considerable insight into one of the earliest organized rockfall field investigations in North America. It includes several suggestions for rockfall hazard mitigation, including a proposed rockfall attenuator system.

Table B-14
Video File Contents of Part C of DVD

Video Number	Video File Name	File Size (MB)	Run Time [min (s)]	Test Location	Remarks
VN-30	RitchieVideo.mov[a]	2,080	18 (19)	Multiple locations in Washington State	Historic film of field tests undertaken in 1963 by A. M. Ritchie that formed the basis for the "Ritchie Ditch" criteria for containing rockfall along highway rock cuts
VN-31	CaltransFieldTestRestrainingNets.mov[b]	574.4	12 (45)	Grey Slip, California	Field procedures used to test new rockfall restraining nets in California in 1990
VN-32	ODOTRockfallCatchmentDitchTests.mov[c]	137.4	3 (53)	Oregon	Tests conducted to determine pattern of rockfall landing points and rollout distances below vertical rock face
VN-33	CDOTScalingTechniques.mov[d]	275.4	4 (13)	Several locations along Colorado highways	Scaling of rock slopes by hand and by mechanical means, use of inflatable airbags, and blasting

[a]See Section 4.1.
[b]See Section 4.2.
[c]See Section 4.3.
[d]See Section 4.4.

4.2 Field Tests and Evaluation of Rockfall Restraining Nets, Caltrans

The first series of rock-rolling field tests to evaluate flexible cable nets as rockfall barriers was undertaken by Caltrans in 1989. The study was conducted at the Grey Slip site, where a total of 76 rocks were rolled during three tests of three different barriers (Smith and Duffy 1990). Caltrans produced Video VN-31 in 1991 as an educational film that described the procedures used at the Grey Slip site to test rockfall nets developed by Brugg Cable Products. The video runs for almost 12 min and provides background information on the questions that prompted the decision to undertake this test series. The video describes the Grey Slip site, the design of the test fences, and procedures used to install them. It shows how movie cameras were used to capture the rockfall tests and presents a series of rock-rolling events with normal- and high-speed records. The video also shows the methods employed to clean out the accumulated rockfall blocks from the fence and the maintenance procedures needed.

4.3 Rockfall Catchment Ditch Tests, Oregon DOT

In 1994 and 2001, the Oregon DOT performed extensive rockfall catchment studies. During the 1994 study, an impressive 2,790 rocks were rolled, only to be rivaled by the 2001 study, in which a remarkable 11,250 rocks were rolled (Pierson et al. 1994, 2001).

Video VN-32 provides a brief view, lasting less than 4 min, of these tests. Rocks are shown being dropped over a near-vertical quarry face that was 12 m (40 ft) high. The tests were undertaken to determine where the rocks were first likely to hit the ground at the base of such a slope and also how far from the base of the slope they were likely to roll under different ground slope conditions. Video VN-32 also provides information on the conduct of these types of field tests.

4.4 Rock Slope Scaling Techniques, Colorado DOT

In 2008, the Colorado DOT produced a relatively short (slightly more than 4 min) instructional video that illustrates four methods of scaling rock slopes to reduce the hazard posed by loose and unstable rock blocks. Video VN-33 presents the four techniques in sequence, beginning with hand scaling, in which individuals are suspended from ropes or climb upon a rock face and pry loose the unstable blocks. A sequence of views illustrates mechanical scaling techniques, in which a crane drags heavy lengths of old equipment treads across a rock face to dislodge exposed rock blocks. Alternative hand-scaling techniques are shown next. These use inflatable air bags to dislodge or tip over unstable rock blocks. The video concludes with a brief view of trim blasting.

ACKNOWLEDGMENTS

The author acknowledges the efforts of the following people in producing these videos and thanks them for their assistance in making the videos available for inclusion in this book: Eric Rorem, Andrea Roth, and Bruno Haller, Geobrugg, Santa Fe, New Mexico, and Romanshorn, Switzerland; Ahren Bichler and Gernot Stelzer, Trumer Schutzbauten GmbH, Kuchl, Austria; Daniele Peila, Polytechnic University of Turin, Italy; Francesco Sassudelli, IGOR Paramassi, Lavis, Italy; Silvia Beretta, TENAX Spa, Viganò, Italy; Toshimitsu Nomura, Protech Engineering, Niigata City, Japan; Peter Perreault, Mountain Management, Saint-Lazare, Quebec, Canada; Howard Ingram, Chris Ingram, and James Roth, Hi-Tech Rockfall, Forest Grove, Oregon; Rick Andrew, Ben Arndt, and Wade Hoon, Yeh and Associates, Denver, Colorado; Ron Chassie, Federal Highway Administration, U.S. Department of Transportation; Larry Pierson, Landslide Technology, Portland, Oregon; Tim Pfeiffer, Foundation Engineering Company, Beaverton, Oregon; Robert Thommen, Rotec International, LLC, Santa Fe, New Mexico; Robert Barrett, Landslide Solutions, Sacramento, California; Tom Badger, Washington State Department of Transportation, Olympia; Ty Ortiz, Colorado Department of Transportation, Denver; Roy Bibbens, Duane Smith, Marvin McCauley, and Ralph Fitzpatrick, California Department of Transportation, Sacramento. Special thanks also to the many crews, support staff, and supervisors, who are too numerous to mention.

REFERENCES

ABBREVIATIONS

ETAG	European Technical Approval Guideline
WSL	Swiss Federal Institute for Forest, Snow, and Landscape

Andrew, R. D. 1992. Selection of Rockfall Mitigation Techniques Based on Colorado Rockfall Simulation Program. In *Transportation Research Record 1343*, Transportation Research Board, National Research Council, Washington, D.C., pp. 20–22.

Andrew, R. D., D. A. Fry, and R. E. Bookwalter. 1998. Field Testing and Evaluation of Various Flexible Rock Fall Control Systems. Colorado Department of Transportation, Denver.

Arndt, B., T. Ortiz, and A. K. Turner. 2009. *Transportation Research Circular E-C141: Colorado's Full-Scale Testing of Rockfall Attenuator Systems*. Transportation Research Board of the National Academies, Washington, D.C.

Badger, T. C., J. D. Duffy, F. Sassudelli, P. C. Ingraham, P. Perreault, B. Muhunthan, H. Radhakrishnan, O. S. Bursi, M. E. Molinari, and E. Castelli. 2008. Hybrid Barrier Systems for Rockfall Protection. In *Proc., Interdisciplinary Workshop on Rockfall Protection* (A. Volkwein, V. Labiouse, and K. Schellenberg, eds.), June 23–25, 2008, Morschach, Switzerland, pp. 10–12.

Barrett, R. K., T. D. Bowen, J. D. Higgins, and T. J. Pfeiffer. 1989. *Rockfall Modeling and Attenuator Testing Report*. CDOH-DTD-ED3/CSM-89-2. Colorado Department of Highways, Denver, and Colorado School of Mines, Golden.

Baumann, R. 2002. The Worldwide First Official Approval of Rock-Fall Protection Nets, *Proc. 53rd Annual Highway Geology Symposium*. California Department of Transportation, San Luis Obispo, pp. 40–51.

Duffy, J. D. 1992. *Field Tests and Evaluation of Flexible Rockfall Barriers*. Brugg Cable Products, Oberbuchsiten, Switzerland.

Duffy, J. D., and B. Haller. 1993. Field Tests of Flexible Rockfall Barriers. In *Proc., Conference on Transportation Facilities Through Difficult Terrain* (J. T. H. Wu and R. K. Barrett, eds.), Aspen Snowmass, Colo., August 8–12, 1993, pp. 465–473.

Duffy, J. D., and W. Hoon. 1996a. *Field Tests and Evaluation of Hi-Tech 50 and 70 Foot-Ton Rockfall Fence*. Report No. CA/05-96-02. California Department of Transportation, San Luis Obispo.

Duffy, J. D., and W. Hoon. 1996b. *Field Tests and Evaluation of Hi-Tech Low Energy Chain Link Rockfall Fence*. Report No. CA/05-96-01. California Department of Transportation, San Luis Obispo.

Duffy, J. D., and W. Hoon. 1998. *Field Tests and Evaluation of Chain Link Fence, Gawk Screen, Fence and Jersey Barrier*. Report No. CA/05-98-01. California Department of Transportation, San Luis Obispo.

ETAG 27. 2008. *Guideline for European Technical Approval of Falling Rock Protection Kits*. European Organisation for Technical Approvals, Brussels, Belgium.

Gerber, W. 2001. *Guideline for the Approval of Rockfall Protection Kits*. Swiss Agency for the Environment, Forests and Landscapes and Swiss Federal Institute for Forest, Snow, and Landscape, Berne, Switzerland. Original Report and 2006 amendment available at http://www.bafu.admin.ch/publikationen/publikation/00793/index.html?lang=en.

Gerber, W., H. Grassl, A. Böll, and W. Ammann. 2001. Flexible Rockfall Barriers: Development, Standardisation and Type-Testing in Switzerland. In *International Conference on Landslides: Causes, Impacts and Countermeasures,* June 17–21, 2001, Davos, Switzerland (M. Kuehne, H. H. Einstein, E. Krauter, H. Klapperich, and R. Poettler, eds.), Verlag Glückauf Gmbh, Essen, Germany, pp. 515–524.

Glover, J., A. Volkwein, F. Dufour, M. Denk, and A. Roth. 2010. Rockfall Attenuator and Hybrid Drape Systems: Design and Testing Considerations. In *Third Euro-Mediterranean Symposium on Advances in Geomaterials and Structures,* 3rd ed. (F. Darve, I. Doghri, R. El Fatmi, H. Hassis, and H. Zenzri, eds.), Djerba, Tunisia, pp. 379–384.

Heiss, C., and P. Moser. 2006. *Test Report No 206: Testing of the Rock-fall Protection Kits TS-2000-ZD of Trumer Schutzbauten GmbH*. Department of Mineral Resources and Petroleum Engineering, University of Leoben, Austria.

Parsons De Leuw, Inc. 1992. *Full Scale Geotextile Rock Barrier Wall Testing, Analysis and Prediction*. Colorado Department of Transportation, Denver.

Peila, D., C. Castiglia, C. Oggeri, G. Guasti, P. Recalcati, and F. Sassudelli. 2000. Full Scale Tests on Geogrid Reinforced Embankments for Rock Fall Protection. In *Proceedings of the Second European Geosynthetics Conference,* Pàtron Editore, Bologna, pp. 317–322.

Peila, D., C. Oggeri, and C. Castiglia. 2007. Ground Reinforced Embankments for Rockfall Protection: Design and Evaluation of Full Scale Tests. *Landslides*, Vol. 4, No. 3, pp. 255–265.

Pfeiffer, T. J. 1989. *Rockfall Hazard Analysis Using Computer Simulation of Rockfalls*. Master of Engineering Report ER-3614. Colorado School of Mines, Golden.

Pfeiffer, T. J., and J. D. Higgins. 1990. Rockfall Hazard Analysis Using the Colorado Rockfall Simulation Program. In *Transportation Research Record 1288*, Transportation Research Board, National Research Council, Washington, D.C., pp. 117–126.

Pfeiffer, T. J., J. D. Higgins, and A. K. Turner. 1990. Computer Aided Rockfall Hazard Analysis. In *Proceedings of the 6th International Congress, International Association of Engineering Geology* (D. G. Price, ed.), Balkema, Netherlands, pp. 93–103.

Pierson, L. A., S. A. Davis, and T. J. Pfeiffer. 1994. *The Nature of Rockfall as the Basis for a New Fallout Area Design Criteria for 0.25:1 Slopes*. Engineering Geology Group, Oregon Department of Transportation, Salem.

Pierson, L. A., C. F. Gullixson, and R. G. Chassie. 2001. *Rockfall Catchment Area Design Guide*. Final Report SPR-3(032). Research Group, Oregon Department of Transportation, Salem.

Ritchie, A. M. 1963. Evaluation of Rockfall and Its Control. In *Highway Research Record 17*, Highway Research Board, National Research Council, Washington, D.C., pp. 13–28.

Smith, D. D., and J. D. Duffy. 1990. *Field Tests and Evaluation of Rockfall Restraining Nets*. Final Report No. CA/TL-90/05. California Department of Transportation, Sacramento.

WSL. 2009. *Vertical Drop TECCO 2000 kJ, ETAG, Walenstadt, Switzerland*. European Organisation for Technical Approvals, Brussels, Belgium.

WSL. 2010. *Attenuator Testing, ETAG, Saint Leonard, Switzerland*. European Organisation for Technical Approvals, Brussels, Belgium.

WSL. 2011. *Vertical Drop Ring Net 8000 kJ, ETAG, Walenstadt, Switzerland*. European Organisation for Technical Approvals, Brussels, Belgium.

Yoshida, Y., and T. Nomura. 1998. *Latest Experimental Study for Rockfall Mitigation*. Protec Engineering and Maeda Kosen Co., Ltd., Fukui, Japan.

Author Biographical Information

A. Keith Turner, *Editor*, is emeritus professor of geological engineering at the Colorado School of Mines in Golden. Before joining the faculty of the Colorado School of Mines in 1971, he was an assistant professor of civil engineering at the University of Toronto and engineering geologist with the Department of Public Works in Ottawa, Canada. Between 1999 and 2002, Dr. Turner occupied the chair in engineering geology at Delft University of Technology in the Netherlands and pursued new opportunities for international cooperation in teaching and research in geological engineering. He has had a longstanding interest in computer applications to geological and environmental studies and has acted as consultant on computer applications in engineering geology for the United Nations, various national agencies, academic institutions, and private firms in several countries, including the United Kingdom, Poland, the Netherlands, Germany, and Mexico. In 1989, Dr. Turner served as director of an advanced research workshop on three-dimensional geological modeling conducted by the North Atlantic Treaty Organization in Santa Barbara, California. Subsequently he undertook research for the U.S. Geological Survey to develop three-dimensional geographic information system approaches for data management, regional hydrogeological modeling, and scientific visualization of geological data. He is currently a visiting research associate with the British Geological Survey, where he is continuing these research interests. Dr. Turner has served on several committees of the National Academy of Sciences and has been active in the Transportation Research Board (TRB) throughout his career. He chaired several TRB standing committees and task forces; panels for projects of the National Cooperative Highway Research Program (NCHRP) and the Second Strategic Highway Research Program (SHRP 2); and the study committee responsible for the development of Special Report 247, *Landslides: Investigation and Mitigation*, for which he received two national awards. Dr. Turner has been awarded emeritus membership of the TRB Engineering Geology Committee and received the 2010 Roy W. Crum Distinguished Service Award in recognition of his achievements in transportation research. In December 2011, Dr. Turner was recognized as a National Associate of the National Academies for his service to TRB.

Robert L. Schuster, *Editor*, now retired, had a lengthy professional career from 1952 to 2010. He was research geologist, Snow, Ice and Permafrost Research Establishment, from 1952 to 1954; instructor, School of Civil Engineering, Purdue University, from 1956 to 1960; associate professor and then professor, Department of Civil Engineering, University of Colorado, from 1960 to 1967; professor and head, Department of Civil Engineering, University of Idaho, from 1967 to

1974; chief, Engineering Geology Branch, U.S. Geological Survey, from 1974 to 1979; research geologist, U.S. Geological Survey, from 1979 to 1995; scientist emeritus, U.S. Geological Survey, from 1995 to 2009; and international consultant in engineering geology and geotechnical engineering from 1995 to 2010. Dr. Schuster is the author of approximately 300 technical publications in engineering geology and geotechnical engineering, with a specialization in the study of landslides. He is a registered professional engineer in Colorado and Indiana and a registered professional geologist in Idaho and holds a BS in geology from Washington State College, an MS in geology from Ohio State University, an MS and PhD in civil engineering from Purdue University, and a Diploma of the Imperial College in soil mechanics from the University of London.

Richard D. Andrew, a principal at Yeh and Associates, Inc., Denver, Colorado, is a professional geologist with 30 years of experience in the fields of geotechnical engineering and engineering geology. His areas of expertise include geotechnical engineering, geological mapping, rock slope design, rock excavation (blasting), rockfall evaluation and mitigation, landslide evaluation and mitigation, and tunnel stabilization. His experience with the Colorado Department of Transportation and with a civil contractor specializing in geotechnical construction services in sensitive terrain environments gives him a unique perspective on the design and management of transportation improvement projects. Mr. Andrew has managed numerous construction and design teams throughout the United States, primarily in transportation development through difficult terrain and fragile environments. He has participated as an active contributor in value engineering studies and constructability reviews for state departments of transportation and the Federal Highway Administration. Mr. Andrew is the author or coauthor of multiple publications in the fields of rockfall evaluation, rockfall mitigation, and rock slope stabilization.

Ben P. Arndt is an associate at Yeh and Associates, Inc., in Denver, Colorado. He has more than 17 years of experience in the field of geological and geotechnical engineering, including 10 years of working on transportation projects throughout the western United States, 2 years of forensic engineering, and 5 years of underground mining. His areas of expertise include evaluation and mitigation of unstable rock and soil slopes, retaining wall design, and landslide stabilization systems. He currently designs mechanically stabilized earth walls, soil nails, micropiles, ground anchors, lightweight fill, and rockfall mitigation features. He has more than a dozen national publications on topics ranging from rockfall protection systems to micropile foundations and has presented many of these publications at national and international conferences. Representative publications include *Transportation Research E-Circular E-C141: Colorado's Full-Scale Field Testing of Rockfall Attenuator Systems* for the Transportation Research Board and *Polyurethane Resin (PUR) Injection for Rock Mass Stabilization* for the Federal Highway Administration.

Thomas C. Badger is assistant chief engineering geologist for the Washington State Department of Transportation and specializes in the characterization and mitigation of landslides and rockfall, rock slope engineering, and instrumentation and slope monitoring. Since 1984 he has worked on a great variety of transportation-related earthworks and structures as well as many landslide and rockfall remediation projects. Mr. Badger is past chair (2003–2009) of the Transportation Research Board Committee on Engineering Geology.

John D. Duffy is a senior engineering geologist with the California Department of Transportation whose duties include engineering geological studies, landslide investigations and repairs, and rockfall and rock slope engineering and technical specifications. He is a member of several technical committees that address rock slope issues, including those of the National Cooperative Highway Research Program (NCHRP), the Transportation Research Board (TRB), the National Pooled Funds Studies, and the Federal Highway Administration. Mr. Duffy has served as chair of the TRB Committee on Engineering Geology and of the Subcommittee for Rockfall Management. Over the years, he has conducted and participated in several research studies in the United States and Europe that evaluated rockfall and debris flow barriers. Mr. Duffy has authored more than 20 publications on rockfall and landslide topics, including Chapter 6, "Rockfall Analysis," of FHWA's *Rockfall Hazard*

Mitigation Methods. He also contributed material on rockfall barriers to TRB Special Report 247, *Landslides: Investigation and Mitigation*. Mr. Duffy was a licensed California blaster for more than 20 years and continues to instruct Department of Transportation personnel in basic mountaineering and climbing in steep terrain for slope investigations and rock scaling. He received the Karl Moskowitz Award for Exemplary Contribution to the Field of Transportation Engineering, awarded by the California Department of Transportation and the California Transportation Foundation, in 2008.

Jerry D. Higgins has been on the faculty of the Department of Geology and Geological Engineering at the Colorado School of Mines since 1986. Before that appointment, he was a faculty member at Washington State University and held positions in industry and government. He specializes in engineering geology and geotechnics and has taught undergraduate and graduate courses, conducted research, and consulted on rock slope stability issues for more than 30 years. Dr. Higgins is a coauthor of the Colorado Rockfall Simulation Program (CRSP) and the modified Colorado Rockfall Hazard Rating System and authored the National Cooperative Highway Research Program (NCHRP) report on U.S. testing standards for flexible fence rockfall barriers. He was named the 2006 Richard H. Jahns Distinguished Lecturer in Engineering Geology, an award of the Association of Environmental and Engineering Geologists (AEG) and the Geological Society of America, and lectured across North America on slope stability issues. In 2006 he also received the Ivan B. Rahn Education Award from the Society for Mining and Exploration for his distinguished contributions to engineering education. In 2010, AEG presented Dr. Higgins with the Karl and Ruth Terzaghi Outstanding Mentor Award for his achievements as a mentor throughout his career.

G. P. Jayaprakash has served as senior program officer and soils, geology, and foundations engineer in the Technical Activities Division of the Transportation Research Board (TRB) since 1987. Before that he was research geologist with the Kansas Department of Transportation and a research scientist at the University of Kansas. He has a master's degree in applied geology from Karnatak University, India, where he served as lecturer for 2 years, and a master's degree in geology from the University of California, Santa Barbara. His areas of interest are engineering geology and geotechnical engineering. He directed the TRB project that produced Special Report 247, *Landslides: Investigation and Mitigation*, which was published in 1996. Mr. Jayaprakash also directed the TRB project on rockfall characterization and control that produced this book. In 2010, he received the Individual Distinguished Service Award from the National Academies.

Steve M. Lowell was chief engineering geologist for the Washington State Department of Transportation (DOT) in Olympia, Washington. Born and educated in Alaska, he was a graduate of the University of Alaska Fairbanks with a bachelor's degree in geology. Before joining the Washington State DOT in 1985, Mr. Lowell worked 11 years as an engineering geologist and senior reconnaissance engineer for the Alaska Department of Transportation and Public Facilities. He had more than 36 years of experience, and his major areas of professional specialization were rock slope design and construction, rock slope stabilization, rockfall mitigation, landslide stabilization, and unstable slope management. Mr. Lowell participated in numerous technical advisory committees at the national level that provided geotechnical expertise and leadership in rockfall mitigation and control. He was a member of the Transportation Research Board Committees on Engineering Geology and Exploration and Classification of Earth Materials. In 1996, Mr. Lowell received the Dr. L. I. Hewes Award from the Western Association of State Highway and Transportation Officials for his outstanding contributions to highway development in the west. Steve Lowell passed away on June 30, 2012. He will be missed by the rockfall community.

Norman I. Norrish is a founding partner of Wyllie & Norrish Rock Engineers, Seattle, Washington. Born and educated in British Columbia, he is a graduate of the University of British Columbia with a bachelor's degree in geological engineering and a master's degree in mining engineering. Mr. Norrish is a member of the Canadian Institute of Mining, Metallurgy and Petroleum and of the American Association of Civil Engineers and is a registered professional engineer in Washington, Oregon, Wyoming, and British Columbia. He has

40 years of experience in the application of rock mechanics to mining, transportation, and civil construction projects, including senior-level project responsibility for the investigation, design, and construction management of transportation projects in mountainous terrain throughout Western North America. He has worked internationally in Argentina, Peru, Chile, Colombia, Panama, the Philippines, the former Soviet Union, and China. Mr. Norrish contributed significantly to the Transportation Research Board's Special Report 247, *Landslides: Investigation and Mitigation*. Over the past 10 years, he has made 40 presentations of National Highway Institute Training Course 132035, "Rock Slopes," to state departments of transportation throughout the United States.

Lawrence A. Pierson is a senior associate engineering geologist with Landslide Technology, a Division of Cornforth Consultants, Inc., in Portland, Oregon. Previously, he was the chief engineering geologist for the Oregon Department of Transportation. He initiated and completed two national research projects that produced the Rockfall Hazard Rating System (RHRS) and the *Rockfall Catchment Area Design Guide*. His areas of specialization include rockfall mitigation, foundations on rock, rock slope design for transportation and hydroelectric projects, and geotechnical asset management. He has managed many large multidisciplinary projects and provided expert witness services to agencies in the United States and Canada. Mr. Pierson is currently chair of the Transportation Research Board Committee on Engineering Geology, a position he has held since 2009.

Kristian Schellenberg received his diploma in civil engineering from the Swiss Federal Institute of Technology Zürich (ETH Zürich) in 2003. He worked as teaching and research assistant at the Institute of Structural Engineering (IBK) of ETH Zürich, from which he received his PhD in 2008. He then began his practical career as a consulting bridge engineer in the Department of Major Bridges of COWI A/S in Denmark. Since 2011, Mr. Schellenberg has been a consulting engineer in Gruner + Wepf Ingenieure AG in Switzerland and has held a part-time lecturer position at IBK.

Michael P. Vierling is currently an engineering geologist with the New York State Thruway Authority. His duties include work related to engineering geology on both the New York State Thruway and the Erie Canal. Previously, Mr. Vierling worked for the New York State Department of Transportation from 1980 to 2001. He has more than 30 years of experience in the fields of geotechnical engineering and engineering geology, including nearly 30 years as a licensed and certified blaster for New York State. He has also served on several engineering geology committees, national pooled fund studies, and two Transportation Research Board (TRB) committees—the Exploration and Classification of Earth Materials Committee and the Engineering Geology Committee—and as the Chairman of the TRB Subcommittee for Rockfall Management. Mr. Vierling has been a member of the steering committee of the Highway Geology Symposium since 1998 and was the chairman of the committee from 2006 to 2011. In 2009 he was awarded the Medallion Award for his many years of significant contributions to the Highway Geology Symposium.

INDEX

AUTHOR

NOTE: Page numbers in roman type refer to an author's publications cited in this report. Page numbers in bold type refer to an author's chapter in this report.

SUBJECT